T0401974

Lecture Notes in Electrical Engineering

Volume 98

Min Zhu (Ed.)

Electrical Engineering and Control

Selected Papers from the 2011 International Conference on Electric and Electronics (EEIC 2011) in Nanchang, China on June 20–22, 2011, Volume 2

Springer

Prof. Min Zhu
Nanchang University
Xuefu Avenue 999
New Honggutan Zone
Jiangxi
Nanchang, 330047
China
E-mail: yongwang2008@yeah.net

ISBN 978-3-642-21764-7 e-ISBN 978-3-642-21765-4

DOI 10.1007/978-3-642-21765-4

Lecture Notes in Electrical Engineering ISSN 1876-1100

Library of Congress Control Number: 2011929653

© 2011 Springer-Verlag Berlin Heidelberg

This work is subject to copyright. All rights are reserved, whether the whole or part of the material is concerned, specifically the rights of translation, reprinting, reuse of illustrations, recitation, broadcasting, reproduction on microfilm or in any other way, and storage in data banks. Duplication of this publication or parts thereof is permitted only under the provisions of the German Copyright Law of September 9, 1965, in its current version, and permission for use must always be obtained from Springer. Violations are liable to prosecution under the German Copyright Law.

The use of general descriptive names, registered names, trademarks, etc. in this publication does not imply, even in the absence of a specific statement, that such names are exempt from the relevant protective laws and regulations and therefore free for general use.

Typeset & Cover Design: Scientific Publishing Services Pvt. Ltd., Chennai, India.

Printed on acid-free paper

9 8 7 6 5 4 3 2 1

springer.com

EEIC 2011 Preface

The present book includes extended and revised versions of a set of selected papers from the International Conference on Electric and Electronics (EEIC 2011) , held on June 20-22 , 2011, which is jointly organized by Nanchang University, Springer, and IEEE IAS Nanchang Chapter.

The goal of EEIC 2011 is to bring together the researchers from academia and industry as well as practitioners to share ideas, problems and solutions relating to the multifaceted aspects of Electric and Electronics.

Being crucial for the development of Electric and Electronics, our conference encompasses a large number of research topics and applications: from Circuits and Systems to Computers and Information Technology; from Communication Systems to Signal Processing and other related topics are included in the scope of this conference. In order to ensure high-quality of our international conference, we have high-quality reviewing course, our reviewing experts are from home and abroad and low-quality papers have been refused. All accepted papers will be published by Lecture Notes in Electrical Engineering (Springer).

EEIC 2011 is sponsored by Nanchang University, China. Nanchang University is a comprehensive university which characterized by "Penetration of Arts, Science, Engineering and Medicine subjects, Combination of studying, research and production". It is one of the national "211" Project key universities that jointly constructed by the People's Government of Jiangxi Province and the Ministry of Education. It is also an important base of talents cultivation, scientific researching and transferring of the researching accomplishment into practical use for both Jiangxi Province and the country.

Welcome to Nanchang, China. Nanchang is a beautiful city with the Gan River, the mother river of local people, traversing through the whole city. Water is her soul or in other words water carries all her beauty. Lakes and rivers in or around Nanchang bring a special kind of charm to the city. Nanchang is honored as 'a green pearl in the southern part of China' thanks to its clear water, fresh air and great inner city virescence. Long and splendid history endows Nanchang with many cultural relics, among which the Tengwang Pavilion is the most famous. It is no exaggeration to say that Tengwang Pavilion is the pride of all the locals in Nanchang. Many men of letters left their handwritings here which tremendously enhance its classical charm.

Noting can be done without the help of the program chairs, organization staff, and the members of the program committees. Thank you.

EEIC 2011 will be the most comprehensive Conference focused on the various aspects of advances in Electric and Electronics. Our Conference provides a chance for academic and industry professionals to discuss recent progress in the area of Electric and Electronics. We are confident that the conference program will give you detailed insight into the new trends, and we are looking forward to meeting you at this world-class event in Nanchang.

EEIC 2011 Organization

Honor Chairs

Prof. Chin-Chen Chang	Feng Chia University, Taiwan
Prof. Jun Wang	Chinese University of Hong Kong, HongKong

Scholarship Committee Chairs

Chin-Chen Chang	Feng Chia University, Taiwan
Jun Wang	Chinese University of Hong Kong, HongKong

Scholarship Committee Co-chairs

Zhi Hu	IEEE IAS Nanchang Chapter, China
Min Zhu	IEEE IAS Nanchang Chapter, China

Organizing Co-chairs

Jian Lee	Hubei Normal University, China
Wensong Hu	Nanchang University, China

Program Committee Chairs

Honghua Tan	Wuhan Institute of Technology, China

Publication Chairs

Wensong Hu	Nanchang University, China
Zhu Min	Nanchang University, China
Xiaofeng Wan	Nanchang University, China
Ming Ma	NUS ACM Chapter, Singapore

Contents

Targets Detection in SAR Image Used Coherence Analysis Based on S-Transform

Tao Tao, Zhenming Peng, Chaonan Yang, Fang Wei, and Lihong Liu

School of Opto-Electronic Information, University of Electronic Science and Technology of China, Chengdu 610054, China
tt19870419@163.com, zmpeng@uestc.edu.cn

Abstract. A novel method for dim moving target detection of synthetic aperture radar (SAR) image which is based on the S-transform(ST)domain coherent analyzing was proposed in this paper. Firstly, the paper describes the basic principle of ST; and analyzes the mechanism of the second generation of coherent algorithm. On the basis of these algorithms, the coherent formula was obtained which was used in this paper. Making use of the difference between S-Transform domain and background of moving target in SAR image with the same scene, coherent image could be constructed by coherent values which were calculated by the proposed coherent formula. In the coherent image, target can be detected by compare the coherent values. Experiments showed that the proposed method could detect the dim target.

Keywords: S-transform, SAR image, coherence analysis, targets detection.

1 Introduction

The interpretation of SAR images, to which target detection, occupying an important position especially in target identification system, is the key, has been a focus for researchers. The research in this respect attracts considerable attention. Classical target detection is based on CFAR, which is to establish threshold on the foundation of the correctly estimating noise and heterogeneous wave, to detect. After the efforts of numerous scholars, many new methods of target detection have emerged today, such as the improved constant false alarm rate (CFAR), two-parameter CFAR detection, and transform domain-based target detection method [1], etc. Dim target detection has been a hot research, too. As the dim target in the image only occupies a tiny part of pixels, which has no texture, no shape or other information, it's hard to be detected for classical target detection methods, and some special preprocessing must be done to the image in order to better detect the interesting dim target.

SAR image is a kind of complex non-stationary signal, and it can be accurately described using only appropriate method of time-frequency analysis [2]. Traditional Fourier-Transform only does single frequency decomposition to signal, although frequency resolution of that method can achieve the desired level, it loses time resolution, and lacks the function of positioning the signal's time and frequency at the same time, and can't effectively analyze the local properties of the signal, and the

M. Zhu (Ed.): Electrical Engineering and Control, LNEE 98, pp. 1–9.
springerlink.com © Springer-Verlag Berlin Heidelberg 2011

transformed frequency spectrum can only represent the overall effect of signal frequency changing, so it's hard to express the non-stationary changes with time of signal statistical properties. Signal's local property can be accurately described using two-dimensional co-expressed of time domain and frequency domain. S-Transform with a good performance of time-frequency analysis is a new developed method of time-frequency analysis [3-4]. Compared with wavelet-transform, both have good local features and direct relation to Fourier spectrum of signal, therefore it can effectively characterize weak signal's characteristics.

A novel dim moving target detection method of SAR image which is based on the S-Transform domain coherent analyzing is proposed in this paper. Firstly, the paper describes the basic principle of S-Transform, and analyzes the mechanism of the second generation of coherent algorithm. On the basis of these algorithms, the coherent formula is obtained which is used in this paper. Making use of the difference between S-transform domain and background of moving target in SAR image with the same scene, coherent image can be constructed by coherent values which are calculated by the proposed coherent formulas. In the coherent image, target can be detected by comparing coherent values. Experiments show that the proposed method can detect the dim target.

2 Basic Principle of S-Transform

S-transform(ST)first proposed by Stockwell (1996)[5] is a new linear time-frequency representation method, time-frequency resolution of which changes with frequency. It is an extension of the continuous wavelet transform, which takes Morlet wavelet as the basic wavelet. ST has a good performance of time-frequency analysis[6], whose time-frequency window has adjustable nature, with good time resolution properties in the high frequency, while good frequency resolution properties in the low frequency. Here are the basic transform formulas of ST[7].

For a two-dimensional image $h(x,y)$, (τ_1, f_1) is defined as x, (τ_2, f_2) as y, then the transformation formula of two-dimensional ST is as follows:

$$S(\tau_1,\tau_2,f_1,f_2) = \int_{-\infty}^{\infty}\int_{-\infty}^{\infty} h(x,y)\frac{|f_1 f_2|}{2\pi}\bullet\exp\left[\frac{(x-\tau_1)^2 f_1^2+(y-\tau_2)^2 f_2^2}{2}\right]\exp\left[-i2\pi(f_1 x+f_2 y)\right]dxdy \tag{1}$$

where, S is the ST of the two-dimensional image $h(x,y)$, τ_1 and τ_2 are time variables, f_1 and f_2 are the frequency, and (x,y) is the image coordinate.

From Fourier transform and convolution theorem, the implementation formula of two-dimensional ST in frequency domain can be obtained as

$$S(\tau_1,\tau_2,f_1,f_2)= \int_{-\infty}^{\infty}\int_{-\infty}^{\infty} H(f_1+f_a,f_2+f_b)\bullet\exp\left[-\frac{2\pi^2 f_a^2}{f_1^2}-\frac{2\pi^2 f_b^2}{f_2^2}\right]\exp[i2\pi(f_a\tau_1+f_b\tau_2)]\,df_a df_b \tag{2}$$

where, $f_a \neq 0,\ \ f_b \neq 0$.

The direct relation between two-dimensional inverse ST and two-dimensional inverse Fourier transform is established

$$H(f_1,f_2)=\int_{-\infty}^{\infty}\int_{-\infty}^{\infty} S(\tau_1,\tau_2,f_1,f_2)d\tau_1 d\tau_2 \tag{3}$$

Make the two-dimensional S-transform formula (2) discrete, and define $\tau_2 \to kT_2$, $f_1 \to u/MT_1$, $f_2 \to v/NT_2$, $f_a \to m/MT_1$, $f_b \to n/NT_2$, then the implementation formula (4) and (5) of discrete two-dimensional ST and inverse ST can be obtained,

$$\begin{cases} H\left[\dfrac{m}{MT_1},\dfrac{n}{NT_2}\right]=\sum\limits_{j=0}^{M-1}\sum\limits_{k=0}^{N-1} h[jT_1,kT_2]e^{-i\frac{2\pi}{N}nk}e^{-i\frac{2\pi}{M}mj} \\ S\left[jT_1,kT_2,\dfrac{u}{MT_1},\dfrac{v}{NT_2}\right]=\dfrac{1}{MN}\sum\limits_{m=0}^{M-1}\sum\limits_{n=0}^{N-1} H\left[\dfrac{m+u}{MT_1},\dfrac{n+v}{NT_2}\right]e^{\frac{2\pi^2 m^2}{u^2}}e^{i\frac{2\pi}{M}mj}e^{\frac{2\pi^2 n^2}{v^2}}e^{i\frac{2\pi}{N}nk} \end{cases} \tag{4}$$

$$m \neq 0, n \neq 0$$

$$\begin{cases} H\left[\dfrac{u}{MT_1},\dfrac{v}{NT_2}\right]=\sum\limits_{k=0}^{N-1}\left\{\sum\limits_{j=0}^{M-1} S\left[jT_1,kT_2,\dfrac{u}{MT_1},\dfrac{v}{NT_2}\right]\right\} \\ h[jT_1,kT_2]=\dfrac{1}{M}\sum\limits_{u=0}^{M-1}\left\{\dfrac{1}{N}\sum\limits_{v=0}^{N-1} H\left[\dfrac{u}{MT_1},\dfrac{v}{NT_2}\right]e^{i\frac{2\pi}{N}vk}\right\}e^{i\frac{2\pi}{N}uj} \end{cases} \tag{5}$$

From the above transformation formulas, it can be seen that from the time domain to the time-frequency, and then to the frequency domain, and finally back to the time domain, this process is reversible with the rapid non-destructive, without any loss of information in the transformation.

3 Coherent Analyzing in S-Transform Domain

The basic idea of the method in this paper: carry on ST for two SAR images under the same background and the target location with displacement and coherent analysis of the two images in ST domain, make use of energy spectrum feature difference in S domain between the target on image and the background to get coherent values to construct coherent image, and establish threshold to detect targets. Having integrated the advantages of short time-window Fourier transform and wavelet transform, ST provides joint function of time and frequency, describes the signal energy density or signal intensity with variables of time and frequency. ST, with high time-frequency resolution [8], does not have the impact of cross terms. Due to the advantages of ST in time-frequency analysis, when the image is transformed to the S domain, the energy spectrum of the background for the two images in S domain is basically the same. But when the targets of the two images move the target energy spectrum also moves accordingly in S domain and the position corresponding to the target energy spectrum changes.

Coherence technique is very sensitive to mutation of signals [9], which highlights the similarity of signals using mathematical method and then achieves a new technology to detect weak signals and reflect unusual characteristics. The coherent technique is divided into three generations: the first generation algorithm is based on cross-correlation, which has better resolution in the case of high signal-to-noise ratio of data, but bad anti-noise ability relatively; the second generation algorithm is based on the similarity, better for computing coherence compared with the first-generation algorithm and higher resolution, whose coherent processing results, to some extent, are influenced by the data quality still, though; the third-generation algorithm is based on the feature structure.

According to coherence technique in two-dimensional image processing features, the second generation coherent algorithm, proposed by K.J. Marfurt et.al. [10] in 1997. They gave the formula of the second generation algorithm in his article,

$$\sigma(\tau,p,q)=\frac{\sum_{k=-K}^{+K}\{[\sum_{j=1}^{J}u(\tau+\Delta\tau_j,x_j,y_j)]^2+[\sum_{j=1}^{J}u^H(\tau+\Delta\tau_j,x_j,y_j)]^2\}}{J\sum_{k=-K}^{+K}\sum_{j=1}^{J}\{[u(\tau+\Delta\tau_j,x_j,y_j)]^2+[u^H(\tau+\Delta\tau_j,x_j,y_j)]^2\}} \quad (6)$$

where $\Delta\tau_j = k\Delta t - px_j - qy_j$, $u(\tau + k\Delta t - px_j - qy_j, x_j, y_j)$ is the received seismic record when ground coordinate is (x_j, y_j) on the moment of $(\tau + k\Delta t - px_j - qy_j)$, p and q are strata dip parameters. τ is sometime, Δt is the sampling interval, $k\in[-K,\ K]$ is the window sliding factor, J is the number of seismic participating in relevant operations and u^H is the Hilbert transform of seismic u. It can be seen from the formula that u and u^H are made the same processing, so might as well first alone consider the processing done to u, as follows equation,

$$C=\frac{\sum_{k=-K}^{+K}[\sum_{j=1}^{J}u(\tau+k\Delta t-px_j-qy_j,x_j,y_j)]^2}{J\sum_{k=-K}^{+K}\sum_{j=1}^{J}[u(\tau+k\Delta t-px_j-qy_j,x_j,y_j)]^2} \quad (7)$$

The numerator can be expressed as the sum of all elements of the following matrix A which is defined as,

$$A=\sum_{m=k-w}^{k+w}\begin{pmatrix} u_{1m}u_{1m} & u_{1m}u_{2m} & \cdots & u_{1m}u_{Jm} \\ u_{2m}u_{1m} & u_{2m}u_{2m} & \cdots & u_{2m}u_{Jm} \\ \vdots & \vdots & \ddots & \vdots \\ u_{Jm}u_{1m} & u_{Jm}u_{2m} & \cdots & u_{Jm}u_{Jm} \end{pmatrix} \quad (8)$$

where m is sequence number of sampling points, and the denominator part is J times than the sum of diagonal elements of the matrix. Suppose (7) is called as the simplified version of the second generation coherent algorithm, and $\tilde{C}_{2k}$ represents the calculated value, (7) can be written as,

$$\tilde{C}_{2k} = \frac{\sum_{m=k-w}^{k+w}\sum_{i=1}^{J}\sum_{j=1}^{J} u_{im}u_{jm}}{J\sum_{m=k-w}^{k+w}\sum_{i=1}^{J} u_{im}^{2}} \tag{9}$$

What can be proved is $\left|\tilde{C}_{2k}\right| \le 1$. Actually, might as well suppose $|u_{1m}| \le |u_{2m}| \le \cdots \le |u_{Jm}|$ for sampling values $u_{1m}, u_{2m}, .. u_{Jm}$, and then according to the definition of sorting inequality, $\sum_{i=1}^{J} u_{im}^2$ is the positive sequence sum of $|u_{im}|$, $\sum_{i=1}^{J}\sum_{j=1}^{J}|u_{im}u_{jm}|$ is the sum of J-2 groups' positive sequence sum and one group's reverse sequence sum and one group's positive sequence sum, for $|u_{im}|$. According to sorting inequality theory, it can be noted that $\sum_{i=1}^{J} u_{im}^2$ only has one group's sum but $\sum_{i=1}^{J}\sum_{j=1}^{J}|u_{im}u_{jm}|$ has J groups' sum, then there is:

$$J\sum_{i=1}^{J} u_{im}^2 \ge \sum_{i=1}^{J}\sum_{j=1}^{J}|u_{im}u_{jm}| \ge \left|\sum_{i=1}^{J}\sum_{j=1}^{J} u_{im}u_{jm}\right| \tag{10}$$

The conclusion may be deduced from all m between ($k-w$) and ($k+w$), therefore the absolute molecular is always less than or equal to the absolute denominator, then $\left|\tilde{C}_{2k}\right| \le 1$.

There is one characteristic in sorting inequality: the difference between positive sequence sum and reverse sequence sum becomes greater when the difference among various numbers is larger. If various numbers are close to each other, the difference of the sum between positive sequence and reverse sequence is very small. This is the operation mechanism, which makes itself sensitive to incoherency.

It can be supposed from the above analysis that the second generation coherent algorithm can be improved if the incoherency of image data is needed to be emphasized and the algorithm to the sensitivity of incoherency is further improved by using the positive sequence sum divided by the reverse sequence sum instead of all sequences sum divided by J times the positive sequence sum. According to the idea of the second generation coherent algorithm, coherent formula can be defined as:

$$c = \frac{\left(\sum_{m=1}^{N} |(u_m - v_m)|\right)^2}{N \times \sum_{m=1}^{N} (u_m - v_m)^2} \tag{11}$$

where c is coherent coefficient, u_m and v_m are elements in two images, N is the number of elements in the image window, $|\bullet|$ expresses absolute value, and $(u_m - v_m)$ is the difference of pixel values in corresponding position of two images. For two sets of data, the positive sequence sum is greater than or equal to the chaotic sequence sum and the chaotic sequence sum is greater than or equal to the reverse sequence sum, known from the properties of sequencing inequality. The numerator in equation (11) is the chaotic sequence sum of arrays constituted by the absolute value of $(u_m - v_m)$, and the denominator is N times the positive sequence sum, therefore the coherent coefficient c is less than or equal to 1.

When the image is transformed to S domain, the energy spectrum of the position where the target corresponds to is bigger in two energy images of S domain. The target is moving, namely that the target position is different in energy images of S domain, so the corresponding difference $(u_m - v_m)$ is bigger when the computational elements include target elements. Relatively, $(u_m - v_m)$ is smaller when changes in background of two images are smaller. Thereby larger c expresses higher image similarity; lower, contrarily. So, the calculated c is smaller when the computational elements include target elements, which shows a mutation of signal in S domain, where the target exists. According to the above analysis the position of image target can be exposed by calculating coherent coefficients, thereby target detection is achieved by using simple threshold segmentation.

Steps of the algorithm of target detection:

Step 1: implement ST separately for two SAR images $u(x, y)$ and $v(x, y)$ with the same scene to get the energy feature images $S_u(\tau_1, \tau_2, f_1, f_2)$ and $S_v(\tau_1, \tau_2, f_1, f_2)$ in ST domain;

Step 2: use appropriate window size (such as: 3×3 or 5×5) to get the elements in the same position of energy feature images in ST domain in turn, and use equation (11) to calculate coherent coefficients, then construct coherent images by coherent coefficients;

Step 3: set threshold κ for coherent images, and generally κ takes the empirical value $\sqrt{\mu\sigma}$, where μ is a mean value of elements about coherent images and σ is variance, and then carry on target detection.

The flow of the proposed algorithm is shown in figure 1.

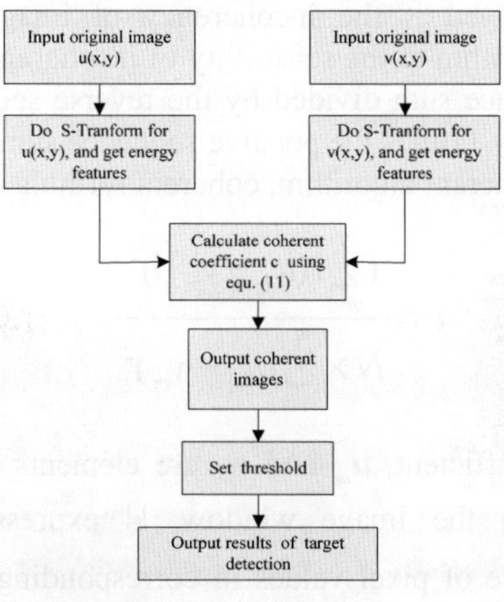

Fig. 1. Algorithm flow about target detection

4 Stimulation of the Algorithm

To test the validity of the method, SAR image of moving target with the same scene is stimulated in the paper. Fig.2 (a) and (b) show two SAR images with the same scene, where there is a moving point target. The two images are detected in accordance with the method proposed in the last section of the paper. Fig.3 (a) shows the coherent image obtained from formula (11) in S domain after ST of the two images. Through coherent image can target location be implemented using relatively simple method, shown as fig.3 (b). Because the position, where the target exists, corresponds to small coherent value, correspondingly the coherent image calculated in S domain corresponds to the low power position, seen from coherent image, the color of the target position is blue which denotes low power.

Known from formula (4), discrete two-dimensional ST has four parameter variables: jT_1, kT_2, u/MT_1, v/NT_2. In order to improve computational efficiency, when calculating discrete two-dimensional ST, fix the value of v to get S domain image. The calculated result is that the resolution is higher in the vertical direction than that in the horizontal direction, as shown in fig.3 (a). Hence the detection result is that the target point was stretched in the horizontal direction, as shown in fig.3 (b). After the segmentation of original image target, calculate the center-of-mass coordinate, then the center-of-mass coordinate of the target point in fig.2 (a) is (78.915, 101.35). The detected result shows that the center-of-mass coordinate of the target point in fig.3 (b) is (79, 101). By comparing the target's center-of-mass coordinate of detection result with that of original image, it can be seen that the method in this paper allows more accurate positioning, as shown in table 1. 100 groups of SAR images with the same scene are simulated using the method proposed in this paper, as a result, there are 8 targets and FAR is 8%. Experiments prove the validity of the method.

Table 1. Performance analysis of algorithm

Target's Real Coordinate	Target's Detection Coordinate	Detection Error	FAR
(78.915, 101.35)	(79, 101)	(0.085, 0.35)	<10%

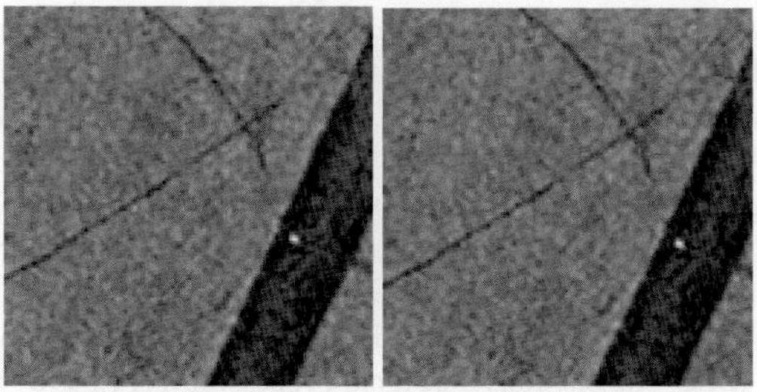

(a) (b)

Fig. 2. Two original SAR images with the same scene

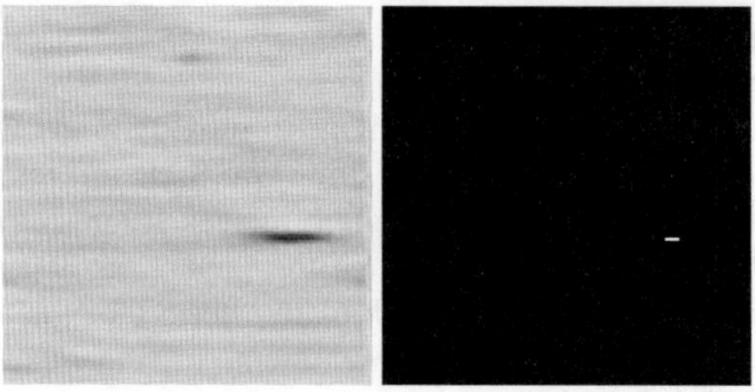

(a) (b)

Fig. 3. Result of target detection: (a) coherent image in S domain, (b) detection result

5 Conclusions

This paper presents SAR image target detection based on ST coherence analysis, using ST time-frequency analysis performance to express the image's energy spectrum with the joint distribution which takes time and frequency as variables. Since the positions of moving target under the same background, in the energy spectrum of ST domain, are different, while the background in the same position has the similar energy, the difference of the target in ST domain can be used to detect targets by coherent techniques. Simulation results verify the validity of the method.

Acknowledgements. This work is supported by National Natural Science Foundation of China (40874066, 40839905), The Key Laboratory Fund of Beam Control, Chinese Academy of Sciences (2010LBC001).

References

1. Peng, Z., Zhang, Q., Wang, J., et al.: Dim Target Detection Based on Nonlinear Multi-feature Fusion by Karhunen-Loeve Transform[J]. Optical Engineering 43(12), 2954–2958 (2004)
2. Peng, Z., Wang, H., Zhang, G., et al.: Spotlight SAR Images Restoration Based on Tomography Model. In: Proc. IEEE of the 2th Asia-Pacific Conference on Synthetic Aperture Radar (APSAR 2009), Xi'an, China, pp. 1060–1063 (2009)
3. Yong, Y., Yang, X.F., Wang, B.X., et al.: A Small Target Detection Method Based on Generalized S-Transform. In: International conference on Apperceiving Computing and Intelligence Analysis, ICACIA 2008, pp. 189–192. IEEE Press, Los Alamitos (2008), doi:10.1109/ICACIA.2008.4770002
4. Peng, Z., Zhang, J., Meng, F., et al.: Time-frequency Analysis of SAR Image Based on Generalized S-transform. In: International Conference on Measuring Technology and Mechatronics Automation (ICMTMA 2009),, Zhangjiajie, Apl, vol. 1, pp. 556–559 (2009)
5. Stockwell, R.G., Mansinha, L., Lowe, R.L.: Localization of the Complex Spectrum: the S-transform [J]. IEEE Transactions on Signal Processing 17(4), 998–1001 (1996)
6. Zhang, J., Peng, Z., Zhang, Q., et al.: Dim Target Detection Based on Image Features Analysis in Generalized S-transform Domain. In: Proc. IEEE International Conference on Intelligent Computation Technology and Automation (ICICTA 2010), Changsha, China, pp. 122–125 (2010)

7. Mansiha, L., Stockwell, R.G., Lowe, R.P.: Pattern Analysis with Two Dimensional Spectral Localization: Application of Two Dimensional S-transform. Physic A 239(3), 286–295 (1997)
8. Schimmel, M., Gallart, J.: The Inverse S-transform in Filter with Time-frequency localization. IEEE Transactions on Signal Processing 53(11), 4417–4422 (2005)
9. Lee, I.W., Dash, P.K.: S-transform-based Intelligent System for Classification of Power Quality Disturbance Signals. IEEE Transactions on Industrial Electronics 50(4), 800–805 (2003)
10. Marfurt, K.J., Sudhakar, V., Gersztenkorn, A., Crawford, K.D., Nissen, S.E.: Coherency Calculations in the Presence of Structural Dip. In: 67th Annual International Meeting Society of Exploration Geophysicists, Expanded Abstracts, pp. 566–569 (1997)

7. Mansinha, L., Stockwell, R.G., Lowe, R.P.: Pattern Analysis with Two Dimensional Spectral Localisation: Applications of Two Dimensional S-transform. Physica A 239(3), 286–295 (1997)
8. Schimmel, M., Gallart, J.: The Inverse S-transform in Filter with Time-frequency localization. IEEE Transactions on Signal Processing 53(11), 4417–4422 (2005)
9. Lee, I.W., Dash, P.K.: S-transform-based Intelligent System for Classification of Power Quality Disturbance Signals. IEEE Transactions on Industrial Electronics 50(4), 800–805 (2003)
10. Marfurt, K.J., Sudhaker, V., Gersztenkorn, A., Crawford, K.D., Nissen, S.E.: Coherency Calculations in the Presence of Structural Dip. In: 67th Annual International Meeting Society of Exploration Geophysicists, Expanded Abstracts, pp. 566–569 (1997)

A Community Detecting Algorithm in Directed Weighted Networks

Hongtao Liu[1,*], Xiao Qin, Hongfeng Yun, and Yu Wu

Institute of Web Intelligence, Chongqing University of Posts and Telecommunications, Chongqing, 400065, China
Liuht@cqupt.edu.cn, qxiaomm@163.com, yunhongfeng@163.com, wuyu@cqupt.edu.cn

Abstract. In this paper, the impact factors of in-degree and out-degree are introduced into community detection, and the directed weighted degree is used to measure the importance of the node. Based on the core nodes, a community detecting algorithm for directed and weighted networks is proposed. Then the community detection on the blog site of Sciencenet is conducted with standard structure entropy as a measure. Experimental results demonstrate that in directed and weighted networks, the proposed algorithm is efficient with shorter execution time. By comparing with the classical algorithm, the detecting results of our algorithm meet the trend of standard entropy better. It means the algorithm proposed is improved to some extent.

Keywords: Directed and Weighted Networks, Community Detection, Standard Structure Entropy.

1 Introduction

Properties of complex networks often excite many researchers' interesting, such as small-world property, scale-free property, rich-club phenomenon and etc. Community structure is one of the most important properties of complex networks, which has become a focus in recent years by Newman and Girvan's research [1].

There are many classical algorithms, such as K-L algorithm[2], Spectral bisection method[3] and GN algorithm[4]. In K-L algorithm and spectral bisection method, the number of communities needs to be pre-determined which is often unrealistic. GN algorithm avoids the defect but the computational complexity is higher relatively. In our previous work, we have studied the evolution of virtual community in BBS by calculating the structure entropy[5]. Since core nodes play an important role in networks, Liping Xiao proposed an algorithm to evaluate importance of the nodes in networks using data field theory[6]. Duanbing Chen[7] proposed a local detecting algorithm in weighted networks based on the core nodes. Since modularity Q was

* Hongtao Liu(1974-), Associate Professor, Ph.D., Research: Emergent Computation, Network Intelligence; Xiao Qin(1985-), Postgraduate, Research: Network Intelligence; Hongfeng Yun(1987-), Postgraduate, Research: Network Intelligence; Yu Wu(1970-), Professor, Ph.D., Research: Emergent Computation, Network Intelligence.

M. Zhu (Ed.): Electrical Engineering and Control, LNEE 98, pp. 11–17.
springerlink.com © Springer-Verlag Berlin Heidelberg 2011

proposed by Newman[1], there was an accepted standard for the results of community detection. Currently, community detection is measured by Q value and computational complexity basically.

From above we find most detecting algorithms view the networks as undirected edges. Thus they lose some useful information, and only conduct a quantitative analysis on the detecting results. In this paper, we consider the direction of edges, and propose a new algorithm with standard structure entropy as a measure.

The rest of this paper is organized as follows. Section 2 gives the previous work. Section 3 describes our algorithm in detail. The algorithm is simulated and analyzed in Section 4. At last, Section 5 concludes our paper and discusses the future work.

2 Previous Work

2.1 Structure Entropy

Entropy is a measure of energy distribution in complex systems. It can reflect the stability and change direction of the system. Entropy has become an important metric to study complex systems, and gains more and more attention.

The network structure entropy is defined as:

$$E = -\sum_{i=1}^{N} I_i \ln I_i \tag{1}$$

where I_i means the important degrees of nodes, which is defined as:

$$I_i = k_i \Big/ \sum_{i=1}^{N} k_i \tag{2}$$

2.2 A Local Detecting Algorithm in Weighted Networks

Duanbing Chen proposed a local detecting algorithm in weighted networks[7]. In this algorithm, the key aspect was selection of node v which had the largest node strength. Through finding all neighbors of node v, an initial community was composed. Then making some adjustment to the initial community, a final community was obtained. Repeated the above steps we could find all communities in the network.

Experimental results demonstrated that the algorithm was rather efficient for detecting communities in weighted networks. However, it ignored the direction of edges and lost some information.

3 A Community Detecting Algorithm in Directed Weighted Networks

3.1 A Novel Model Based on Directed Networks

As we know, if an individual has a lot of direct contacts with others in complex systems, it is more important and has great "power". Under the guidance of this idea, an algorithm(D-W algorithm) for detecting communities in directed and weighted networks is proposed. Duanbing Chen[7] defined node strength as:

$$D_u = \sum_{v \in V} \omega_{uv} , \tag{3}$$

where ω_{uv} denotes the weight of the edge which links node *u* and *v*. In the blog networks, if a user responds to others frequently but obtains few responses, the user can not be a core node. It means direction of the edges have different influences on the importance of nodes. Thus we measure the importance of a node with directed weighted degree. Firstly some reasonable assumptions are presented.

Assumption 1: The in-degree and out-degree of a node make different influence on the importance of the node;

Assumption 2: A node has a broad range of communication in a network if there are more edges which connect with it;

Assumption 3: It means the two nodes have a closer relationship if the edge between them has a greater weight.

Together with the above three assumptions, we give the definition of in-weighted degree D_{ip} and out-weighted degree D_{op} of node *p*.

$$D_{ip} = \sum_{q \in V} \omega(v_q, v_p) \quad , D_{op} = \sum_{q \in V} \omega(v_p, v_q) \quad , \tag{4}$$

where $\omega(v_q,v_p)$ denotes the weight of the directed edge and $< v_q, v_p > \in E$ denotes the directed edge from *q* to *p*.

Therefore, directed weighted degree of node *P* is defined as:

$$D_p = \sum_{q \in V} (\alpha D_{ipq} + \beta D_{opq}) \ , \tag{5}$$

Where α, βare the impact factors of in-degree and out-degree, $0 \le \alpha \le 1$, $0 \le \beta \le 1$, and $\alpha + \beta = 1$.

For a community C_i and a node *p*, the belonging degree $B(p, C_i)$ is defined as:

$$B(p, C_i) = \frac{\sum_{q \in C_i} (\alpha D_{ipq} + \beta D_{opq})}{D_p} . \tag{6}$$

3.2 Algorithm Description

Our algorithm consists of three main components: (1) detecting the initial community, (2) adjusting the initial community and (3) expanding the initial community.

Initially, all nodes in the network are marked with label "*F*".

Input: ***G*** **(*V, E*)**, *α*, *β*

Output: ***C*** **(*C₁*, *C₂*, *C₃*, … , *C*ₘ)**

Step1: Detecting the initial community

A. Calculate the directed weighted degree D_p for each node *p* with label "*F*";

B. Select a node u with the largest directed weighted degree and find its neighbors marked by label "F". These nodes compose an initial community C_i;

Step2: Adjusting the initial community

A. For each node v in community C_i, if the belonging degree $B(v, C_i)$ is less than 0.5(we don't consider node v to be tight enough with community C_i if the belonging degree is less than 0.5), remove node v from community C_i;

B. Repeat A until $\forall v \in C_i$, $B(v, C_i)$ is not less than 0.5, and obtain the initial community C_i;

Step3: Initial community extended

A. Find all neighbors N_c of community C_i. For every node v in N_c, calculate the belonging degree $B(v, C_i)$;

B. For each node v in N_c, if the belonging degree $B(v, C_i)$ is more than 0.5, add node v into community C_i;

C. Repeat B until $\forall v \in N_c$, $B(v, C_i) <0.5$, then obtain a final community and also be denoted by community C_i. Marking all nodes in C_i with label "T";

D. Return to step1 to mine the next community.

4 Algorithm Simulation and Analysis

4.1 Selection of Experimental Object and Parameters

(1) Experimental Object

The experimental data is collected from the blog site of Sciencenet (http://www.sciencenet.cn/blog/). We obtain 4702 active users from Jan 2007 to Mar 2010 and about 135852 comments with our crawler program. The user information includes user ID, name, blog links, reply time and etc.

(2) Algorithm Parameters

Based on the data collected, a directed and weighted network is built in which nodes stand for active blog users and directed edges stand for the comments' relationship. The value of the edge weight is the number of comments.

Table 1. Matching ratio based on different parameters

Number	A	β	Matching Number	Matching Ratio
	0.2	0.8	12	40%
	0.1	0.9	12	40%
	0	1	12	40%
	0.3	0.7	14	46.70%
	0.4	0.6	15	50%
30	0.5	0.5	16	53.30%
	0.6	0.4	16	53.30%
	0.7	0.3	16	53.30%
	0.8	0.2	18	60%
	1	0	18	60%
	0.9	0.1	19	63.30%

The experiment is implemented on Matlab 7.1. We use formula(5) to obtain our Top 30 Bloggers. To find out the optimal parameters, we obtain the matching ratio under different parameters by comparing them with the official top 100 ranking of the Sciencenet , as shown in Table 1.

As seen in Table 1, the matching ratio reaches 63.30%, the highest value, when α,βare set to 0.9 and 0.1. To validate our result, several more experiments are done with greater amount of data e.g. Top 40, Top 50. The results are similar to Top 30.

4.2 Experimental Results and Analysis

(1) Experimental Implementation

The algorithm proposed in this paper is implemented by C++ programming language running on a PC with a 2.9GHz processor and a 2.0GB memory.

The directed and weighted network built in 4.1(2) is used to evaluate the proposed algorithm. As a result, eighteen main communities are detected without consideration of several small communities.

(2)Comparative Experiments

Two sets of experiments are conducted in order to evaluate the algorithm proposed in this paper.

Experiment 1: Taking standard structure entropy as a measure, detecting results based on the proposed algorithm and Duanbing Chen's[7] are compared.

In order to exclude the impact of the number of nodes, the network structure entropy of Sciencenet is normalized, as shown in Fig. 1a. Based on the analysis in 4.1(2), the parameters α, βare set to 0.9 and 0.1, respectively, in our algorithm. Since users in the network interact infrequently in the first four months, community detection is conducted from the 5th month in this paper. By removing several communities that have less significant effect, the main communities are obtained. And the results based on two different algorithms are shown in Fig. 1b.

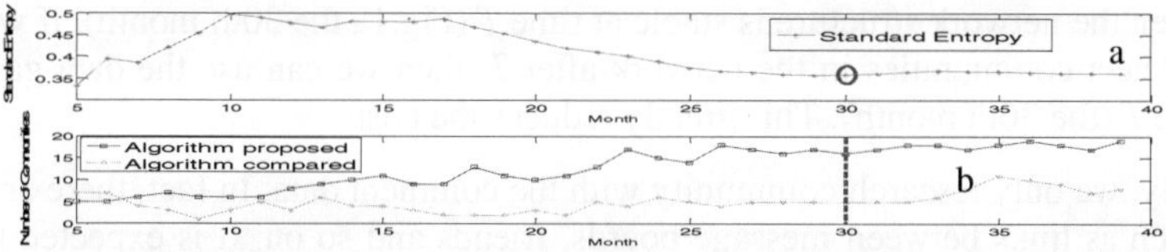

Fig. 1. Results based on different algorithms

As seen in Fig. 1a, the network structure tends to be stable after the 30th month. The reason is that as the standard structure entropy has little fluctuation, the community structure becomes stable. As shown in Fig. 1b, the community structure is more stable while adopting the proposed algorithm, compared to Duanbing Chen's. Thus the proposed algorithm is more efficient.

Experiment 2: Detecting results with different parameters are compared in order to illustrate the rationality of parameters selected in 4.1(2). Here, three circumstances are considered: 1) αis comparatively bigger than β; 2) αis equal to β; 3) αis comparatively smaller than β. Since the second situation has been discussed in Experiment 1, only two conditions need to be further explored. Hence, the parameters are set to α=0.8, β=0.2 and α=0.1, β=0.9, respectively. And the detecting results are shown in Fig. 2.

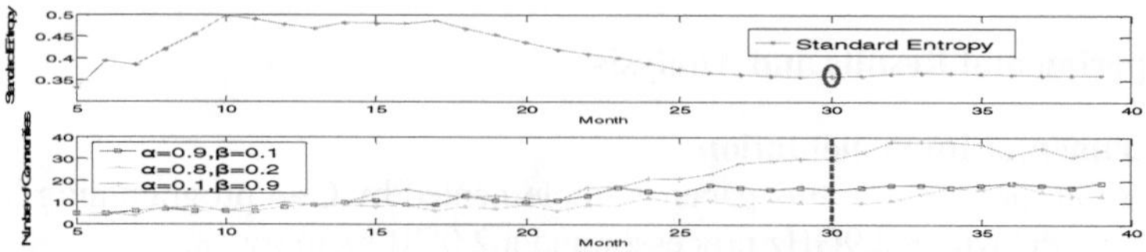

Fig. 2. Results based on different parameters

As seen in Fig. 2, there are large fluctuations in community structure after the 30th month while choosing the above two sets of parameters. This is not consistent with the standard structure entropy. Meanwhile, experiments with other sets of parameters are also conducted, and the results are similar as above. Therefore, the parameters selected in 4.1(2) are reasonable.

5 Conclusion

Taking standard structure entropy as a measure, we propose a community detecting algorithm in directed and weighted networks. It has the following advantages:

(1) It is applicable for detecting communities in weighted and directed networks for considering the information of the directed edges.
(2) The detecting results are reasonable and meet standard structure entropy well, as described in Experiment 1.
(3) When the network structure is stable at time *T* (Fig.1a the 30th month), if we need to detect communities in the network after *T*, then we can use the data gained at time *T* (the 30th month). This greatly reduces the costs.

Currently, we only research community with the comment data. In fact, there are more links such as links between message boards, friends and so on. It is expected to gain more data to research community in the next step.

Acknowledgements

This paper is supported by National Natural Science Foundation of China (60873079, 61040044), (Key) Natural Science Foundation of Chongqing (2008BB2241, 2009BA2089), Program for New Century Excellent Talents in University (NCET).

References

1. Newman, M.E.J., Girvan, M.: Finding and Evaluating Community Structure in Networks. Physical Review E 69(2), 26113 (2004)
2. Kernighan, B.W., Lin, S.: A Efficient Heuristic Procedure for Partitioning Graphs. Bell System Technical Journal 49(2), 291–307 (1970)
3. Barnes, E.R.: An Algorithm for Partitioning the Nodes of a Graph. SIAM J Alg Discr Meth 4(3), 541–550 (1982)
4. Girvan, M., Newman, M.E.J.: Community Structure in Social and Biological Networks. Proc. Natl. Acad. Sci. 99, 7821–7826 (2001)
5. Wu, Y., Xiao, K., Liu, H., Tang, H.: Evolution of BBS Virtual Community and Its Simulation. Systems Engineering Theory & Practice 30(10), 1883–1890 (2010)
6. Xiao, L., Meng, H., Li, D.: Evaluate Nodes Importance in the Network Using Data Field Theory. Journal of Wuhan university 33(4), 379–383 (2008)
7. Chen, D., Shang, M., Lv, Z., Fu, Y.: Detecting Overlapping Communities of Weighted Networks via a Local Algorithm. Physica A 389, 4177–4187 (2010)

Autonomous Rule-Generated Fuzzy Systems Designs through Bacterial Foraging Particle Swarm Optimization Algorithm

Hsuan-Ming Feng

Department of Computer Science and Information Engineering, National Quenoy University,
No. 1 University, Rd., Kin-Ning Vallage,
Kinmen, 892, Taiwan, R.O.C.
hmfenghmfeng@gmail.com

Abstract. An innovative bacterial-foraging-based swarm intelligent algorithm called bacterial foraging particle swarm optimization (BFPSO) is applied for the design of fuzzy systems to balance the car-pole platform. The BFPSO is an efficient evolutionary learning algorithm to deal with complex and global optimization problems. The BFPSO combines the inspired behaviors of bacterial foraging mode and the PSO learning stage to approximate the benefits of fast convergence ability and lower computational load. This paper illustrates the perfect BFPSO algorithm in detail with the simulation to automatically select appropriate parameters of fuzzy systems. Computer simulation results on the nonlinear control problems are derived to demonstrate the efficiency of BFPSO.

Keywords: Fuzzy rule-based systems, bacterial foraging particle swarm optimization, evolutionary learning algorithm.

1 Introduction

Fuzzy systems with the linguistic rules have been successfully known to put on many complicated fields, such as high-dimensional functional approximation [1-3] and nonlinear control [4] problems. In some case studies, there still have some difficulties to generate appropriate fuzzy systems. One of the main problems in approaching the better fuzzy systems is to acquire the suitable fuzzy rules and regulate the membership functions shapes. In traditional search way, the fuzzy rules are determined by the experience oriented way and membership functions are selected by the trial-and-error procedure. The work in obtaining the above-mentioned terms is time-consuming. There are two major approaches to develop the suitable parameters of fuzzy systems. One approach implies that fuzzy rules are tuned by human experts. However, these available fuzzy rules are too rough for complex and ill-defined system. The other training procedure is that the desired fuzzy rules are often extracted from input-output training-data pairs. Traditional trial-and-error and gradient-type learning strategies are difficult for designers when solving nonlinear and complicated problems. Thawonmas and Abe [5] developed a learning method to determine fuzzy

 © Springer-Verlag Berlin Heidelberg 2011

rules from training data pairs, where the drawback is needed to resolve the overlapping problem for accordance reason. Lin, et al. [6] developed a fuzzy partitioning concept to configure the organization of fuzzy rule, but the difficulty is to choose the location of cuts. Wong and Chen [7] developed a clustering algorithm to set the initial architecture of fuzzy rules from input-output data. However, system convergence is becoming a very slow cycle when the training data set is too large. Therefore, these learning methods are not great for user to resolve the complex and high-dimensional problems.

Particle swarm optimization (PSO) is first introduced by Eberhart and Kennedy in 1995 [8]. PSO presents an evolutionary computation and swarm intelligent technique, which is inspired by social behavior of bird flocking or fish schooling. PSO learned from the scenario and used it to solve the optimization problems. The computation of this swarm learning is dependent on only two pieces of important information: every particle's best solution and the swarm's best experience. Due to the simple learning machine, PSO has a great probability to suddenly get the local optimal trap. Bacterial foraging optimization (BFO) is first introduced by Passino in 2002 [9], it is a probabilistic searching procedure by the natural behavior of the Escherichia coli. BFO eliminate animals with great foraging strategies to successfully gain the desired target. BFO mimics the biological motion of the E. coli bacteria, there are chemo taxis, swarming, tumbling, reproduction, elimination and dispersal actions. But the complexity of BFO forced researchers for its simplification and for faster convergence. Therefore, BFPSO learning algorithm is integrated the benefits of the BFO's global search ability and the PSO's fast convergence learning machine. The BFPSO can be considered into a population-based learning cycle to solve the ill-defined, nonlinear and complicated high dimensional optimization problems [10].In the article, we propose a bacterial foraging particle swarm optimization (BFPSO) algorithm to avoid the trial-and-error type trying way. It is an efficient fuzzy rule tuning algorithm to self-generate the appropriate parameters of the fuzzy system. The selected fuzzy systems with desired linguistic rules present a robust ability to achieve a great control performance.

2 Architecture of Fuzzy System Designs

In the design of the organized fuzzy system, an n-inputs, single-output system is proposed with the illustration of fuzzy rules. It can be displayed as follows:

$$R^i: \text{IF } \boldsymbol{X} \text{ is } ME_i \text{ THEN } Y \text{ is } y_i,\ i = 1, 2, \ldots, M, \tag{1}$$

where $\mathbf{X} = (x_1, x_2 \ldots, x_n)$ in the form of vector is to denoted as the input variable, and M is the total number of fuzzy rules. The ME_i denotes as a fuzzy set for the input vector (X) in the premise part. Y is an output fuzzy number and y_i means the real values for the related i-th fuzzy rules in the consequent part. In this paper, the definition of the fuzzy set is described by the membership function formulas:

$$ME_i(\mathbf{X}) = \exp\left(-\left(\frac{(x_1 - c_{i1})^2}{d_{i1}^2} + \ldots + \frac{(x_n - c_{in})^2}{d_{in}^2}\right)\right). \tag{2}$$

Where the set of $(c_{i1}, c_{i2}, ..., c_{in})$ is the centered position value of the hyper-elliptic function and (d_{ij}) denotes as the length of the j-th principal axis for the hyper-ellipsoid. This can be considered that several hyper-elliptic membership functions are distributed in an n-dimensional space. The consequent part (y_i) is delivered as the real value type to simplify the fuzzy system. The appropriate parameters set (a_{ij}, b_{ij} and y_i) is required to be achieved by the novel BFPSO algorithm in this article.

In this study, an efficient weighted average defuzzifier is determined to convert the fuzzy domain into a real output. While the firing value of the premise part in the respective i-th rule is deserved, the actual fuzzy system output ($\bar{y}$) can be calculated by

$$\bar{y} = \frac{\sum_{i=1}^{m} ME_i(\mathbf{X}) \bullet y_i}{\sum_{i=1}^{m} ME_i(\mathbf{X})} . \tag{3}$$

According to the above description, the contour of the membership function $ME_i(\mathbf{x})$ which is regulated by the combination of fuzzy parameters $\{a_{i1}, b_{i1}, a_{i2}, b_{i2}, \dots, a_{in}, b_{in}\}$ and the real value of consequent parameter y_i determine a fuzzy system. Thus, different parameters set $\{c_{i1}, c_{i2}, \dots, c_{in}, d_{i1}, d_{i2}, \dots d_{in}, y_i, 1 \le i \le m\}$ decide different fuzzy system with different performance setting. If there are m fuzzy rules to construct, the m*(2n+l) parameters in this parameter set $\{a_{i1}, a_{i2}, \dots, a_{in}, b_{i1}, b_{i2}, \dots b_{in}, y_i, 1 \le i \le m\}$ need to be determined for designing such fuzzy systems. This searching problem is that the m fuzzy rules with n inputs are proposed in an n dimensional space to choice the proper m hyper-ellipsoids functions.

In this paper, the parameters selection in approaching the appropriate fuzzy system is formulated as a space search problem. The BFPSO-based learning algorithm is applied to determine the proper parameter set R in searching space. It is discussed in the following section.

3 Bacterial Foraging Particle Swarm Optimization Parameters Learning Algorithm

Bacterial foraging particle swarm optimization leaning algorithm combined the natural creature's behavior of bacterial foraging optimization (BFO) and particle swarm optimization (PSO) strategies. One of the main streams of PSO is to observe how natural creatures act as a swarm and simulate the swarm intelligent behavior in a computer computation. The PSO algorithm performs the heuristic exchange of their own and other particle's experiences which have been better so far to discover superior offprint. In the learning cycle, each particle's position value X and velocity value Y are regulated by two best values Pbest_x and Gbest_x. Pbest_x is denoted as the individual particle's best solution (highest fitness) it has achieved so far. Gbest_x is obtained by choosing the overall best value from all particles in populations. In this basic PSO iteration learning step, the velocity of the particle is learned according to the relative Pbest_x and Gbest_x values. The new velocity for each particle is updated by the following equation [8]:

$$v_{p,d}(k+1)=v_{p,d}(k)+\alpha_1(k+1)(Pbest\_x_{p,d}(k+1)-x_{p,d}(k)) \\ +\alpha_2(k+1)(Gbest\_x_{p,d}(k+1)-x_{p,d}(k)) \tag{4}$$

where $v_{p,d}$ is the velocity of the *p*th particle in the *d*th dimension and its related *p*th particle's position in the *d*th dimension is $x_{p,d}$. the p employs the particles number; k denotes as the current state; k+1 represents the next time step, $\alpha_1(k+1)$ and $\alpha_2(k+1)$ are random numbers in the interval between 0 and 1.

While the velocity of the particle is obtained, the particle's location will be modified at the next time step.

$$x_{p,d}(k+1)=x_{p,d}(k)+v_{p,d}(k+1) \tag{5}$$

The BFPSO is a learning algorithm based on the concepts of PSO and BFO to solve the ill-defined and complicated problems. The BFO is a new type bionic algorithm; it mimics the biological movement of the E. coli bacteria to advance the search ability. BFO contains four sequence stages taxis, dispersal, reproduction and elimination to extract the required optimal solution. Detail BFO algorithm is discussed in the following descriptions:

3.1 Bacterium Taxis

Taxis behavior is a nature response, while the E. coli bacterium is stimulated by the surroundings. E. coli bacterium can move in different two ways. One direction is called the swimming state and the other is the tumbling cycle. In the swimming procedure, the flagellum rotates counterclockwise so that it gives an opposite force to the bacterium. This action will push the bacterium cell causing swimming. The otherwise, the bacterium is no movement while it is in the state of tumbling cycle. It is notes that the direction can be changed by the random selection. The movement of the i-th bacterium after one step is represented by

$$P^s(f+1,g,u)=P^s(f,g,u)+C(s)*V(f). \tag{6}$$

Where $P^s(f, g, u)$ denotes the location of s-th bacterium at f-th chemotactic step, g-th reproductive and u-th elimination and dispersal step. C(s) is the length of unit walk. In here, it is a const. V(f) means the direction angle of the f-th chemotactic step. It is between in the range of $[0, 2\pi]$.

3.2 Swarm Dispersal

The Objective of the E. coli bacterium is to find the best solution. The cell-to-cell signaling attraction is that bacterium can dispersal its own message to the others. Each bacterium also releases a repellent function to signal others to be at a minimum distance from it. Based on this behavior of swarm dispersal, the bacterium's position can be approximated into the best one which contains the highest nutrient. The formulas in the presentation of cell-to-cell signaling is present by

$$J_{\alpha}^{s}=(\mathrm{P}, \mathrm{P}^{s}(\mathrm{f,g,u}))=-d_{attract}\exp[-\varpi_{attract}\sum_{x=1}^{q}(p_{opi}-p_{x}^{s})^{2}]$$
$$+h_{repellent}\exp[-\varpi_{repellent}\sum_{x=1}^{q}(p_{opi}-p_{x}^{s})^{2}] \tag{7}$$

Where q is the number of the bacterium. $d_{attract}$ and $\varpi_{attract}$ are the attract coefficients. Otherwise, there are the selected $d_{repellent}$ and $\varpi_{repellent}$ repellent coefficients. p_{opi} is the known best solutions in this learning cycle.

3.3 Reproduction and Elimination

Bacteria have the natural tendency to gather to the nutrient-rich areas by an activity called chemotaxis. After chemotactic and dispersal steps, Reproduction and elimination are taken to improve the total performance in the same swarm size. The 50% of the bacterium population size will be reproduced by means of the evaluated fitness values. So that, the higher half of the bacteria are live and the others are died. In conclusion, the health bacteria will be refilled into the swarm populations. Therefore, the live bacteria may have the higher probability to stick around the initial or local optima positions. The swarm dispersion operation takes place after a certain number of reproduction and elimination. This process will prevent the local minima trapping events. The proposed BFPSO learning algorithm is described in the following steps:

BFPSO 1) Set the initial parameters. Number of bacteria, Number of input variable, number of parameters, swinging unit length, number of chemotactic loop, number of reproduction loop, random generation swarm position. Select the number of the fuzzy rule and the PSO learning rate (c1, c2). Select the attract and repellent coefficients.

BFPSO2) Define the fitness function

$$F_{p} = \exp(-RMSE) \tag{8}$$

Where the root mean square error (RMSE) is calculated by

$$RMSE = \sqrt{\left(\frac{1}{M}\sum_{k=1}^{M}\left(y^{d}(k)-\frac{\sum_{i=1}^{m} HE_{i}(x(k))\cdot w_{i}}{\sum_{i=1}^{m} HE_{i}(x(k))}\right)^{2}\right)} \tag{9}$$

Where $y^{d}(k)$ means the k-th desired value which is selected from the input-output data pairs. HE_{i} is the i-th membership function. The desired objective is to approach the minimal RMSE value. Therefore, the selected parameters are to achieve maximal fitness value, which is defined by the following formula:

$$MAX\left(F_p\right) \tag{10}$$

BFPSO3) Run the bacterium taxis learning loop based on the (6).
BFPSO4) Run the swarm dispersal cycle based on the (7).
BFPSO5) Select the personal best solution (*pBest*) by the following formula.

$$pBest_p^{t+1} = \begin{cases} Y_p^{t+1} & \text{if } F\left(Y_p^{t+1}\right) \geq F\left(pBest_p^t\right) \\ pBest_p^t & \text{if } F\left(Y_p^{t+1}\right) < F\left(pBest_p^t\right) \end{cases} \tag{11}$$

BFPSO6) Select the global best solution (*gBest*) by the following formula.

$$gBest^{t+1} = \begin{cases} pBest_p^{t+1} & \text{if } F\left(pBest_p^{t+1}\right) \geq F\left(gBest^t\right) \\ gBest^t & \text{if } F\left(pBest_p^{t+1}\right) < F\left(gBest^t\right) \end{cases} \tag{12}$$

BFPSO7) Regulate the particle position by formulas (4) and (5)
BFPSO8) Repeat step2 to step7 until g=G.
BFPSO9) Select the best solution to generate the desired fuzzy system.

4 Illustrated Case Studies - Inverted Pendulum Balance Problem

The proposed BFPSO self-generation algorithm is proposed to balance the Inverted Pendulum problems. The control objective is to produce an appropriate actuator force, F, to control the motion of the cart such that the pole can be balanced in the vertical position (θ =0). Let $x_1(t)=\theta$ (angle of the pole with respect to the vertical axis) and $x_2(t)=\dot{\theta}$ (angular velocity of the pole), then the inverted pendulum system state equation can be described by [4]

$$\dot{x}_1 = x_2 \tag{13}$$

$$\dot{x}_2 = H(x_1, x_2, F) = \frac{g \bullet \sin(x_1) + \cos(x_1)\left(\frac{-F - m \bullet l \bullet x_2^2 \sin(x_1)}{m + M}\right)}{l \bullet \left(\frac{4}{3} - \frac{m \bullet \cos^2(x_1)}{m + M}\right)}, \tag{14}$$

where g (acceleration due to the gravity) is 9.8 meter/sec^2, m_c. (mass of cart) is 1.0 kg, m (mass of pole) is 0.1 kg, l (half length of pole) is 0.5 meter, and F is the applied force in Newtons. In this problem, there are two input variables $x_1=\theta$ and $x_2=\dot{\theta}$, so the fuzzy rules can be represented by

$$R^{(i)}: \text{IF } \mathbf{x}=(\theta, \dot{\theta}) \text{ is } ME_i \text{ THEN y is } y_i,\ i=1,2,\ldots,m, \tag{15}$$

and the hyper-ellipsoid type membership function in the designed fuzzy system can be represented by

$$HE_i(\theta, \dot{\theta}) = \exp(-(\frac{(\theta - a_{i1})^2}{b_{i1}^2}) + \frac{(\dot{\theta} - a_{i2})^2}{b_{i2}^2})) \tag{16}$$

The designed fuzzy control system is given 5 fuzzy rules at the start, therefore, 25 parameters $\{a_{i1}, a_{i2}; b_{i1}, b_{i2}; y_i, 1 \le i \le 5\}$ are required to be efficiently chosen by the BFPSO learning algorithm in the search space. It is assumed that no prior knowledge from a human operator's experiment is required to make a good fuzzy rules in balancing this Inverted Pendulum problems. The objective of this BFPSO learning algorithm is to regulate the fuzzy system for controlling the Inverted Pendulum. Therefore, the difference between the desired position and the actual target can be achieved into the zero state. The goal of the BFPSO learning algorithm is to maximize the fitness function value, i.e. minimize the RMSE. The initial conditions are set $\theta(t)=20$ and $\dot{\theta}(t)=0$. Computer simulation results for PSO and PFPSO are illustrated in Figure 1 and Figure 2, respectively. These simulations show the time response for pole angle, angle velocity and input force from time =0 to time=10 sec. Performance comparison of this best result is illustrated in Table 1. Computer simulations demonstrate that the BFPSO learning method has a shorter rise time, smaller RMSE values and almost the similar maximal overshoot than the PSO method.

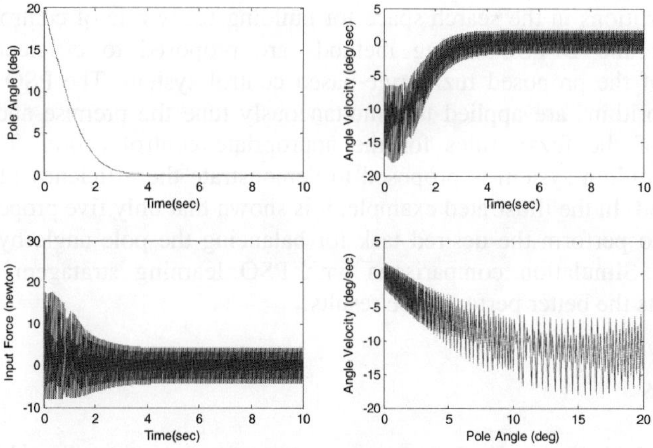

Fig. 1. Simulation results of the inverted pendulum balance problem by the PSO.

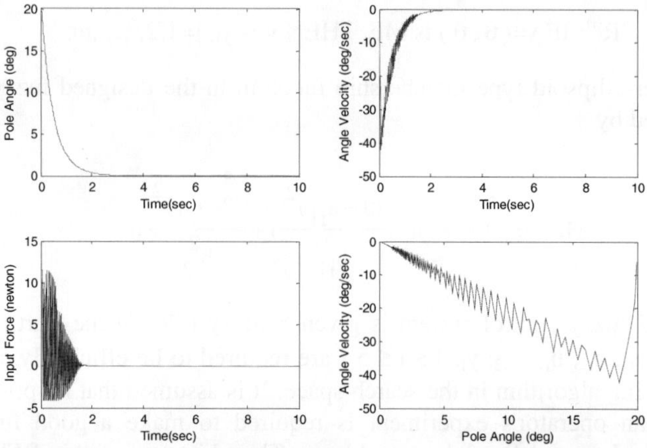

Fig. 2. Simulation results of the inverted pendulum balance problem by the BFPSO.

Table 1. Performance comparison in this illustrated example.

	Rise Time	Maximal Overshoot	RMSE
PSO	1.63	0.0056	0.0454
BFPSO	0.110	0.015	0.0127

5 Conclusions

In this article, a set of hyper-ellipsoids membership functions are developed to define the fuzzy partitions in the search space for building fuzzy rule of control systems. A typical PSO and BFPSO tuning methods are proposed to efficiently tune the parameters of the proposed fuzzy rule-based control system. The PSO and BFPSO learning algorithms are applied to simultaneously tune the premise and consequent parameters of the fuzzy rules for the appropriate control action. The nonlinear Inverted Pendulum system is proposed to demonstrate the efficiency of the BFPSO tuning method. In the illustrated example, it is shown that only five proper fuzzy rules are enough to perform the desired task for balancing the pole angle by the BFPSO tuning type. Simulation comparisons with PSO learning stratagem, the BFPSO algorithm gets the better performance results.

References

1. Dickson, J.A., Kosko, B.: Fuzzy function approximation with ellipsoidal rules. IEEE Trans. on Systems, Man and Cybernetics 26(4), 542–560 (1996)
2. Wong, C.-C., Chen, C.-C.: A GA-based method for constructing fuzzy systems directly from numerical data. IEEE Trans. Systems, Man and Cybernetics 30(6), 905–911 (2000)

3. Fang, H.-M.: Hybrid Stages Particle Swarm Optimization Learning Fuzzy Modeling Systems Design. The Tamkang J. of Scie. And Engin. 9(2), 167–176 (2006)
4. Feng, H.-M., Wong, C.-C.: Fewer Hyper-ellipsoids fuzzy Rules Generation Using Evolutional Learning Scheme. Cybernetics and Systems: An International Journal 39(1), 19–44 (2008)
5. Thawonmas, R., Abe, S.: Function approximation based on fuzzy rules extracted from partitioned numerical data. IEEE Trans. on Systems, Man and Cybernetics 29(4), 525–534 (1999)
6. Lin, Y., Cunningham III, G.A., Coggeshall, S.V.: Using fuzzy partitions to create fuzzy systems from input–output data and set the initial weights in a fuzzy neural network. IEEE Trans. on Fuzzy Systems 5, 614–621 (1997)
7. Wong, C.-C., Chen, C.-C.: A hybrid clustering and gradient descent approach for fuzzy modeling. IEEE Trans. on Systems, Man and Cybernetics 29, 686–693 (1999)
8. Kennedy, J., Eberhart, R.C.: Particle swarm optimization. In: Proc. IEEE Int. Conf. Neural Networks, Perth, Australia, pp. 1942–1948 (1995)
9. Passino, K.M.: Biomimicry of bacterial foraging for distributed optimization and control. IEEE Control Systems Magazine, 52–67 (2002)
10. Gollapudi, S.V.R.S., Pattnaik, S.S., Bajpai, O.P., Devi, S., Bakwad, K.M.: Velocity Modulated Bacterial Foraging Optimization Technique (VMBFO). Applied Soft Computing 11(1), 154–165 (2011)

Evolutionary Learning Mobile Robot Fuzzy Systems Design

Hua-Ching Chen[1], Hsuan-Ming Feng[2,*], and Dong-hui Guo[1]

[1] Department of Electronic Engineering; Xiamen University
Xiamen Fujian 361005; China
galaxy.km@gmail.com, dhguo@xmu.edu.cn

[2] Department of Computer Science and Information Engineering, National Quemoy University
No. 1 University, Rd., Kin-Ning Vallage Kinmen, 892, Taiwan, ROC
hmfeng@nqu.edu.tw

Abstract. The evolutionary particle swarm optimization (PSO) learning algorithm with the image processing technology is proposed to efficiently generate the fuzzy systems for achieving the control adaptability of the embedded mobile robot. The omni-directional image model of the mobile robot system is established to represent the entire tracking environment. The fuzzy control rules are automatically extracted by the defined flexible fitness function for multiple objectives in avoiding obstacles, selecting suitable fuzzy rules and approaching toward the desired targets at the same time. The illustrated examples with various initial positions and different blocks sizes are demonstrated that the selected fuzzy rules can overcome the obstacles and achieve the targets as soon as possible.

Keywords: Particle Swarm Optimization; Fuzzy Control Systems; Embedded Mobile Robots.

1 Introduction

In the last few years, the improvement of computer engineering and image processing devices were rapidly developed in several applications. The embedded system, containing more comfortable operations, lower power consumption, smaller size and higher portable ability, etc., are widely applied in human life applications Especially, the embedded image mobile robot systems are usually developed by the embedded-computer, vision-sensors, mechanism and other electrical elements. In real practical engineering, the developed vision-based mobile robot systems with the identified interesting objects in an unknown environment are successfully applied in several application fields. The challenge in the design of the vision-based software is how to recognize the anomalous behavior, extract inspected features, analysis the treated patterns and start an appropriate control stratagem [5]. In the image processing technology, there are two type image sensors to capture the view of the objects in various environments. The omni-directional image sensor can snatch the pantoscopic scene from the real environment. The advantage of the omni-directional mobile robots

M. Zhu (Ed.): Electrical Engineering and Control, LNEE 98, pp. 29–36.
springerlink.com © Springer-Verlag Berlin Heidelberg 2011

has the higher capabilities to move toward arbitrary way without turning the direction of motor wheels. In the other consideration, it can nimbly attain any desired orientation and position in the traveling line. Based on different dynamics training models, there are more and more research to successfully applied in vision-based robot systems applications [2, 4, 7].

Fuzzy systems first introduced by the Zadeh in 1965 [11] was known linguistic rules and knowledge based machine. It is highly desirable to represent the human thinking to utilize the knowledge in developing autonomous strategies of controlling the mobile robots. Fuzzy system is applied successfully in many different fields, especially in the application of complex modeling, tracking or control problems, while it is very hard to present the mathematic model of robot system. This feature demonstrates their navigating ability to efficiently approach the desired targets due to the adaptable self-organized ability. Fuzzy logic is given a adaptable ability to solve the mobile robot model map problem in the unknown environment [9, 10].

Even the practical mobile robot applications are efficiently developed by the fuzzy systems [2-3, 8], there still have many problems in choosing perfect fuzzy rule base to control the mobile robot. One of the main objectives in design the fuzzy systems are acquiring the favorable parameters of the fuzzy rules. In the traditional generation of fuzzy rules, it derived from expert's experience. Otherwise, the parameters tuning is gathered by the skilled operator by the trial-and-error operation. The above-mentioned terms in obtaining the parameter value causes a time-consuming task. It can be expected that fuzzy rules extraction form the high-dimensional search space is a complicated but crucial procedure.

Eberhart and Kennedy initially introduced the PSO in 1995 [6], which is inspired by the social behavior of bird flocking or fish schooling. This PSO learning algorithm simulates natural creatures behave as a swarm and the individual particles are attracted stochastically toward the positions of evaluated best performance. The concept of the PSO learning stratagem is learning from the scenario of social behavior to approach these global solutions for ill-defined, complicated and nonlinear problems. The PSO has been well-known to solve many mobile robot control problems [1-3]. To improve the training accuracy, the simple but efficient evolutionary PSO is proposed in this article.

2 Mobile Robot Image Model Design

The evolutionary learning fuzzy mobile robot system is illustrated in Figure 1. Robot platform includes with the motor driver and the mobile robot machine. Motor drive can directly regulate the moving speed and rotating angle while receiving the control signal for the fuzzy system. The omni-directional (OD) sensor image captures the desired targets and obstacles scenes from the surrounding environment of mobile robot. Based on the image transformation formulas, the identified patterns of targets and obstacles are mapped into the x-y plane to development the image mathematical kinematics model of mobile robot. The evolutional PSO algorithm with the defined fitness function is evaluated and the desired fuzzy rules are achieved to control the robot. The objective of the fitness function is to minimize the trace between the initial

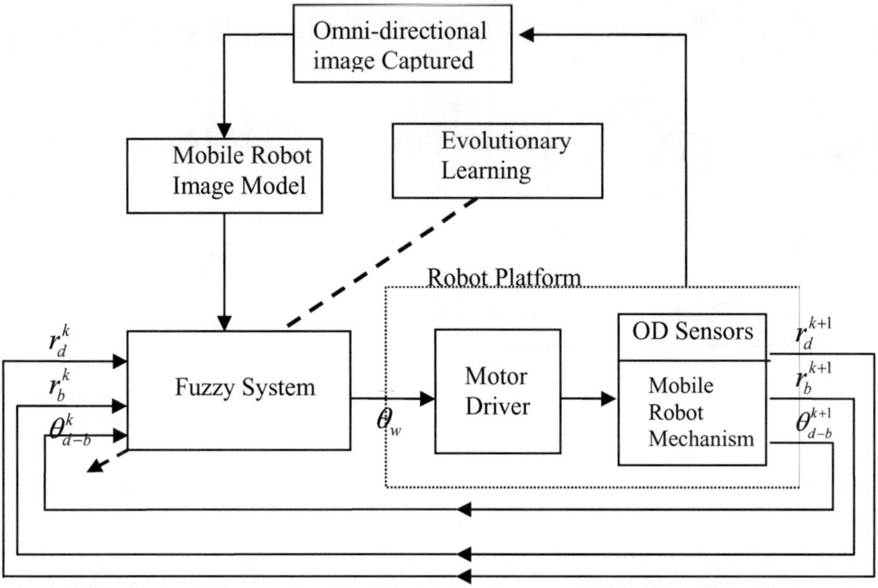

Fig. 1. Evolutionary Learning Mobile Robot Fuzzy Systems structure.

and targets robot positions, respectively. The trained fuzzy system by the evolutionary learning scheme is highly performed to control the mobile robots.

In the image processing stage, the panoramic image is determined by the OD sensor. In general, the captured image usually consists of red, green and blue (RGB) color channels. The original R, G and B channel-values are the main color element to represent the true color space. The HSV space, denoted as the hue, the hue and bright values, is more perfected than that in RGB area in the other experiment results. In HSV color space, H and S are determined to get the angle and length between these interesting objects. The transformed HSV values are determined by the discussed image objects. Their own colors for the destination and block tracking objects will be identified by their (H, S) located position. In order to correctly rebuild the contour and improve the image quality, the image morphology method is applied in this article. Two fundamental operations, openings and closings, are delivered to eliminate the small pellet and reconstruct the completed region in the discussed objects. The erosion and dilation operation are used to reduce the affect of image flat zones. This image procedure is determined to extract the narrow connection and clean small outlier. The basic function of the closing operation is that it can join narrow broken parts and collect thin gap area to mend the small holes in the bad image zone. In addition, the small gap of the shape will be refilled out in this operation. The destination and block size is available to determine after the one-opening-one-closing image procedure is completed. It will be used to detect the required tracking location with the proposed geometry pattern reorganization method.

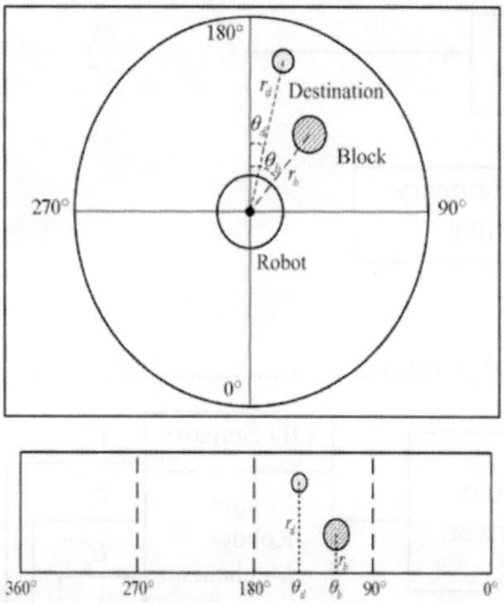

Fig. 2. Mobile robot image coordinate system.

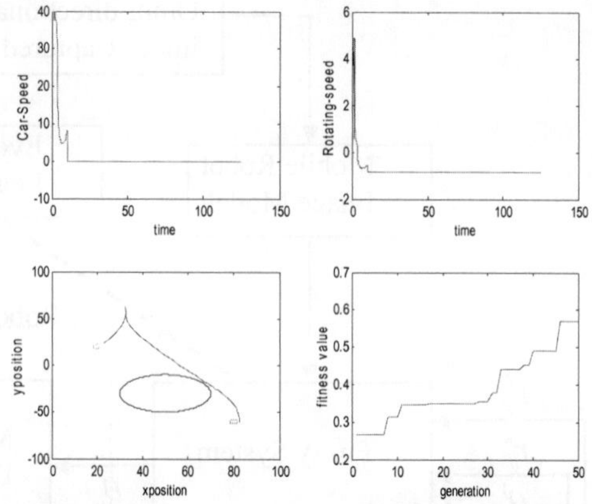

Fig. 3. Simulation results of the proposed fuzzy robot system.

Based on the developed image processing stage as mentioned above, the transformed locations information of mobile robot, destination and block in its workspace are illustrated in Figure. 2. The location (x_d , y_d), (x_b , y_b) and (x_r , y_r) are co-coordinates for the destination, block and robot, respectively. The robot mathematic formulas is proposed by [3]

$$r_d = \sqrt{(x_d - x_r)^2 + (y_d - y_r)^2} \tag{1}$$

$$r_b = \sqrt{(x_b - x_r)^2 + (y_b - y_r)^2} \tag{2}$$

$$\theta_d = \cos\left(\frac{\vec{X}_{dr} \cdot \vec{Y}_{dr}}{\left|\vec{X}_{dr}\right| \cdot \left|\vec{Y}_{dr}\right|}\right)^{-1} \tag{3}$$

$$\theta_b = \cos\left(\frac{\vec{X}_{br} \cdot \vec{Y}_{br}}{\left|\vec{X}_{br}\right| \cdot \left|\vec{Y}_{br}\right|}\right)^{-1} \tag{4}$$

$$\theta_{d-b} = \theta_d - \theta_b \tag{5}$$

Were θ_d and θ_b are denoted as the corresponding angle of the robot for the destination and block in y-axis, respectively. r_d and r_b are the distances between robot and destination; robot and block, respectively. $\vec{X}_{dr}$ and $\vec{Y}_{dr}$ are denoted as vectors which is determined the distance from the robot to destinations with respective to their x-axis and y-axis locations values, respectively. $\vec{X}_{br}$ and $\vec{Y}_{br}$ are also the vectors between the robot to blocks for the x-axis and y-axis, respectively . Symbol $| \, |$ means the vector length.

The close-loop contains three input variables (r_d , r_b , θ_{d-b}) and two output variables (θ_w, v_w) are applied to construct the fuzzy mobile robot system. Where θ_w and v_w are the turn-rate in angle unit to the motor wheels driver and robot's traveling line-speed, respectively.

The mobile robot speed in the current time step k is denoted as v_w^t and its turn-rate angle is setting as θ_w^t. While the mobile robot is derived at a very small increasing time interval (Δk), the mobile robot will move by the (6) and (7) at next time step (k+1).

$$x_r^{k+1} = x_r^k + v_w^k \bullet \Delta k \bullet \cos(\theta_w^k) \tag{6}$$

$$y_r^{k+1} = y_r^k + v_w^k \bullet \Delta k \bullet \sin(\theta_w^k) \tag{7}$$

Based on the equations (1)-(7), the whole mobile robot kinematics in the coordinate x-y space is finished. The novel fuzzy system design by the evolutionary PSO learning algorithm will be discussed in the next section.

3 Evolutionary Learning Fuzzy System Design

If variables (r_d , r_b , θ_{d-b}) considered as X= (x_1, x_2, … , x_n) is the input vector with a whole n-dimensional pattern, which is regarded as the premise part of the fuzzy inference system. The constructed fuzzy rules can be illustrated as follows:

$$R^{(i)} : \text{IF } \boldsymbol{X} \text{ is } HE_i \text{ THEN } \boldsymbol{Y} \text{ is } y_i, \; i = 1, 2, \ldots, m, \tag{8}$$

Where m is the total number of fuzzy rules and HE_i is denoted as the symbol of fuzzy membership function with respect to the input vector (***X***). The y_i is denoted as the real value in the consequent part. In this article, the definition of fuzzy membership function is described by the following formula:

$$HE_i(X) = \exp\left(-\left(\frac{(x_1 - a_{i1})^2}{b_{i1}^2} + \ldots + \frac{(x_n - a_{in})^2}{b_{in}^2}\right)\right) \tag{9}$$

This presentation shows that there are several hyper-ellipsoid type functions combination search in the n-dimensional space. Where parameter set ($a_{i1}, a_{i2}, \ldots, a_{in}$) is the center value and (b_{ij}) is the length of the j-th principal axis of the hyper-ellipsoid, respectively. The consequent part (y_i) is simplified as a single real number.

In this study, the simple defuzzifier with the weighted average calculation is used to convert the fuzzy area domain into a real value. While the firing strength from the premise part of the i-th rule is deserved, the output value of fuzzy system (y^o) can be calculated by (10)

$$y^o = \frac{\sum_{i=1}^{m} HE_i(\mathbf{x}) \bullet y_i}{\sum_{i=1}^{m} HE_i(\mathbf{x})} \tag{10}$$

According to the above description, the contour of the membership function $HE_i(\boldsymbol{X})$ is to approach the better behavior of mobile robot. These (a_{ij}, b_{ij} and y_i) parameters are selected by the efficient PSO evolutionary algorithm from the amount of parameters combination set to improve the performance of the mobile robot system.

The evolutionary PSO learning algorithm simulates the swarm behavior inside a computer computation to yield the best of the characters among the comprehensive old population. At each iteration step, Pbest is every particle's best solution, which has been achieved so far and Gbest is obtained by choosing the overall best value from all particles of populations. The new velocity for each particle is updated by the following formulas:

$$\begin{aligned} v_{np}(k+1) &= \tau \cdot v_{np}(k) + \beta_1 * rand(\ ) * \left(p_{bestnp}(k) - Y_{ni}(k)\right) \\ &+ \beta_2 * rand(\ ) * \left(G_{bestnp}(k) - Y_{np}(k)\right) \end{aligned} \tag{11}$$

Therefore, the new particle position will be regulated by

$$Y_{ni}(k+1) = Y_{ni}(k) + v_{ni}(k+1) \tag{12}$$

Here, n and p is denoted as the dimensional number and particle, respectively. k employs current state, k+1 descript the next time step, β_1 and β_2 are constant learning rate by the designer.

The fitness function FIT(.) is defined by (13) to find the near optimal solutions in the selection of the highest fitness value. When R_i presents the parameter set of fuzzy system, FIT(R_i) is denoted as the evaluated fitness value with the input parameter R_i.

$$FIT(R_i) = \exp(-\frac{RMSE}{5}) * OB \tag{13}$$

Here RMSE means the mean square errors of the distance between the desired destinations and the robot. The OB presents the robot state which is selected as 1 when the robot not collides with the block. Otherwise, OB is 0 when the robot successfully goes through the obstacle from the tracking path. The goal of fitness function of PSO is derived to maximize the fitness function vale, i.e. minimize the root mean square error (RMSE) and successfully reach the target. The evolutionary PSO learning algorithm is represented to generate the fuzzy system. It is determined as following steps:

Step 1) Give the maximal iteration number (G), and set initial generation g=0. Set the fuzzy rules number and the PSO learning rate (β_1 , β_2).Randomly construct the initial system parameters.

Step 2) Based on (11)-(12) to determine the appropriate parameters of the fuzzy system by the evolutionary PSO learning formulas.

Step 3) Gets the fitness value for each individual particle according to the evaluated fitness function by (13).

Step 4) Compare each individual's evaluation fitness value with the personal best value (Pbest) to replace the new Pbest. The best evaluation value among the Pbest is setting to Gbest.

Step 5) g=g+1.

Step 6) If g=G, then go to exit, otherwise go to step 2.

Step 7) Chooses final Gbest value to develop the desired fuzzy mobile robot system.

4 Illustrated Example and Conclusions

In this illustrated example, the developed fuzzy system is considered as an identifier to describe the image feature of the mobile robot. The near-optimal parameters of fuzzy system are automatically built by the evolutional PSO learning algorithm. The PSO swarm sizes =30; generation=50 and $\beta_1 = \beta_2 = 0.75$. In this cased, the fuzzy system is given 5 rules in the initial condition, then 30 particles are randomly generated with the computer simulations.

In this example, the mobile robot is starting at (20, 20), the target location is to approach toward (80, -60) and the block position is located at (50, -30) for the x-axis and y-axis, respectively. It is note that the diameter of block is 20. Based on the defined fitness function, the selected fuzzy system can drive the robot from the initial position to the target in avoiding the obstacle. The near-optimal parameters in solving this robot travel tracking problem is deserved in this training cycle. Simulation results for this example are illustrated in Figure 3. In computer simulation results, Fig. 3(a) and Fig. 3(b) are displayed in the line speed and the rotating angle of the mobile robot, respectively. The best fitness value with respective to the generation number is displayed in Fig. 3(c). This trace shows that the extracted robot fuzzy system is containing the near best solutions with respective to the highest fitness value. In the Fig. 3(d), it shows the simulation results of the mobile robot in dynamic environment. From this simulation, mobile robot is gradually and smoothly moving into the target. The selected fuzzy rules are both applied to achieve the target within a shorter time and to avoid the obstacle.

Acknowledgments. This research was partly supported by the National Science Council of the Republic of China under contract NSC 96-2221-E-507-004.

References

1. Chatterjee, A., Matsuno, F.: A Geese PSO tuned fuzzy supervisor for EKF based solutions of simultaneous localization and mapping (SLAM) problems in mobile robots. Expert Systems with Applications 37(8), 5542–5548 (2010)
2. Chen, C.-Y., Feng, H.-M.: Hybrid Intelligent Vision-Based Car-Like Vehicle Backing Systems Design. Expert Systems with Applications 36(4), 7500–7509 (2009)
3. Feng, H.-M., Chen, C.-Y., Horng, Wei, J.-H.: Intelligent Omni-Directional Vision-Based Mobile Robot Fuzzy Systems Design and Implementation. Expert Systems with Applications 37(5), 4009–4119 (2010)
4. Basu, F.M., Anup: Robot navigation using panoramic tracking. Pattern Recognition 37(11), 2195–2215 (2004)
5. Freda, L., Oriolo, G.: Vision-based interception of a moving target with a nonholonomic mobile robot. Robotics and Autonomous Systems 56(6), 419–432 (2007)
6. Kennedy, J., Eberhart, R.C.: Particle swarm optimization. In: Proc. IEEE Int. Conf. Neural Networks, Perth, Australia, pp. 1942–1948 (1995)
7. Labrosse, F.: Short and long-range visual navigation using warped panoramic images. Robotics and Autonomous Systems 55(9), 675–684 (2007)
8. Mbede, J.B., Ele, P., Mveh-Abia, C.-M., Toure, Y., Graefe, V., Ma, S.: Intelligent mobile manipulator navigation using adaptive neuro-fuzzy systems. Information Sciences 171(4), 447–474 (2005)
9. Samsudin, K., Ahmad, F.A., Mashohor, S.: A highly interpretable fuzzy rule base using ordinal structure for obstacle avoidance of mobile robot. Applied Soft Computing 11(2), 1631–1637 (2011)
10. Wang, M., Liu, J.N.K.: Fuzzy logic-based real-time robot navigation in unknown environment with dead ends. Robotics and Autonomous Systems 56(7), 625–643 (2008)
11. Zadeh, L.A.: Fuzzy sets. Inform. Control 8, 338–353 (1965)

Radar Waveform Design for Low Power Monostatic Backscattering Ionosonde

Ming Yao[1], Zhengyu Zhao[2], Bo Bai[1], Xiaohua Deng[1], Gang Chen[2], Shipeng Li[2], and Fanfan Su[2]

[1] Institute of Space Science and Technology Nanchang University Nanchang, Jiangxi, China
Ym008@126.comg

[2] School of Electronic Information Wuhan University Wuhan, Hubei, China
chen@whu.edu.cn

Abstract. This paper introduces radar waveform design for a low power monostatic backscattering ionosonde. Bi-phase interpulse coded pulse train is analyzed. The waveform can achieve high pulse compression gain, long unambiguous range, highrange resolution and high Doppler resolution, and are specially suitable for low power monostatic backscattering ionosonde.

1 Introduction

Using coding technique to realize extremely low-power HF vertical incidence Ionosonde was reported by Barry [4], Reinisch et al. [5] [6], and Grubb et al. [7]. However, because of the much larger group range and loss of propagation, the amplitude of oblique backscattering echo is about 60 dB weaker than the amplitude of vertical incidence echo with the same transmitted power. Therefore, using lowpower HF backscattering radar to obtain the observations with long ranges (several kilometers) has rarely been reported. Wuhan Ionosphere Comprehensive Sounding System (WICSS), a monostatic ionosonde, was successfully developed at Wuhan University [8]. The most prominent characteristic of this ionosonde is its low power. It can carry out backscattering sounding within 200-W transmitting power. WICSS utilizes different bi-phase inter-pulse coded pulse trains in backscattering sounding. These waveforms are specially suitable for low Doppler applications, such as ionosphere sounding. By choosing the appropriate coding sequences, these kinds of pulse compression radar waveforms can achieve high pulse compression gain with low peak sidelobe level (PSL); by adopting the suitable timing, they can get long unambiguous range and high-range resolution; and by using long coherent processing interval (CPI), they can accomplish high Doppler resolution.

However, the timing of these waveforms require the coding sequences to have "good" periodic autocorrelation function (PACF) performance, namely low PSL. Maximum length sequences (m-sequences) [1] and Wolfmann-Goutelard sequences (WG sequences) [2] [3] possess such "good" PACF properties. In order to fulfill their good PACF performance, the waveforms coded with these sequences should be

M. Zhu (Ed.): Electrical Engineering and Control, LNEE 98, pp. 37–43.
springerlink.com © Springer-Verlag Berlin Heidelberg 2011

transmitted periodically and continuously. As a consequence, many oblique and backscattering ionosondes adopt bistatic CW scheme to achieve the good PACF property. But, in normal monostatic radar systems, CW waveform is not suitable because the receiver is saturated by the nearby strong transmitted signals and cannot receive echoes. WICSS successfully solves such problem by inter-pulse coded pulse train waveforms [9].

Section 2 introduces the timing of the waveform. Section 3 analyzes the waveform with Ambiguity Function (AF) [10]. In Section 4, backscattering ionograms obtained from WICSS using the waveform are illustrated. Conclusions are given in Section 5.

2 The Timing of the Waveforms

Fig. 1 demonstrates the timing. U_i (i=0,...L-1, L equals the length of the sequence) represents the coding chips within one pulse. R_i(i=0,N-1, N equals the number of the range bins) denotes received echoes. T_p denotes the pulse width (PW). T_{r1} represent pulse repetition period (PRP), T_{r2} denotes the sequence repetition period and equals LT_{r1}. The coded pulses are transmitted with equal pulse interval, and the intervals between two adjacent pulses are used for receiving. In real system, the first L-1 echoes (S_1 to S_{L-1}) of the first period should be abandoned, because they contain incomplete range bin information. Useful echoes can be received from the second period [9].

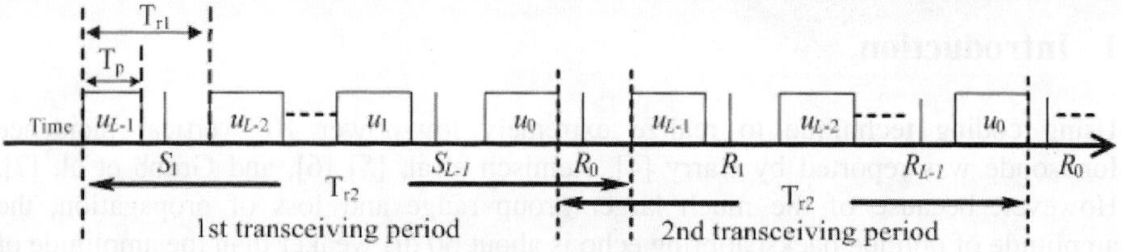

Fig. 1. Timing of bi-phase inter-pulse coded pulse train.

3 Waveform Analysis

The Ambiguity Function (AF) was introduced by Woodward (1953) [10] and is the main tool in radar signal analysis. It represents the time response of a filter matched to a given finite energy signal when the signal is received with a delay τ and a Doppler shift υ relative to the nominal values (zeros) expected by the filter. The AF in this study is defined by (1):

$$|\chi(\tau, \upsilon)| = \left| \int_{-\infty}^{\infty} u(t)u^*(t+\tau)e^{j2\pi\upsilon t}dt \right| \quad (1)$$

where u(t) is the complex envelope of the signal.

To analyze the delay τ and Dopplers υ separately, onedimensional cuts can be obtained from its two-dimensional AF. Taking the zero-Doppler cut along the delay axis, we get the autocorrelation function (ACF) of u(t). And taking the zero-delay cut along the Doppler axis, we can get the Fourier transform of the magnitude squared of u(t).

Bi-phase inter-pulse coded 15 bits m sequence ($T_{r1}=3T_p, T_{r2}=45T_p$) is analyzed in this section. In the simulation, this waveform is transmitted 2 times and 30 times separately. We use Wave-A and Wave-B here for convenience (See Table. I).

Table 1. Waveform Parameters

Waveform	Wave-A	Wave-B
T_p	1	1
T_{r1}	3	3
T_{r2}	45	45
Coding Sequence	15 bits m sequence	15 bits m sequence
Transmitted periods	2	30

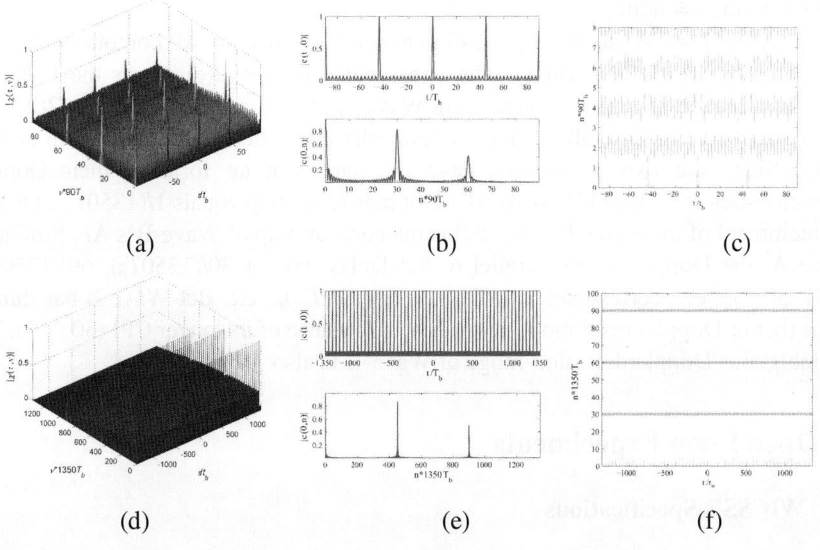

Fig. 2. (a) AF (Positive Doppler part), (b) AF cuts (Positive Doppler part), and (c) AF contour map (central Positive Doppler part) of of Wave-A (with 2 periods); (d) AF (Positive Doppler part), (e) AF cuts (Positive Doppler part), and (f) AF contour map (central Positive Doppler part) of of Wave-B (with 30 periods)

Fig. 2(a) ~ Fig. 2(c) show AF, one-dimensional AF cuts and AF contour of Wave-A (the waveform with 2 periods). The upper panel of Fig. 2(b) shows the zero Doppler AF cut (i.e. PACF) of Wave-A. The height of the main peak, i.e. L, determines pulse compression gain, which equals 15 in the simulation. For 511 bits m sequence, it can reach 27 dB (10 log_{10}(511) dB), which is high enough for ionospheric backscattering sounding with low transmitting power. The evenly distributed small pinnacles between the peaks correspond to the PSL. The half-power width of the main peak, i.e. T_p, equals the delay (range) resolution. Set T_p to 25.6 s, the delay (range) resolution equals 3.84 Km, which is precise enough for ionospheric backscattering sounding. The distance between the main peak located at zero delay and peaks at 45 T_p corresponds to the unambiguous delay-detection-range (UDLDR), which equals 45 T_p, i.e. T_p. Under the condition of L = 511, T_p = 25.6 s, and T_{r1} = 10 T_p, T_{r2} equals 130.8 ms (511×10×25.6 s), corresponds to a UDLDR of 19622 Km, which is long enough for the application. The lower panel of Fig. 2(b) shows the Doppler characteristics of the waveform at zero delay. The half-power width of the main Doppler peak, 1/(90 T_p), i.e. 1/(2 T_{r2}) equals the Doppler resolution. It corresponds to the reciprocal of the CPI. If T_{r2} equals 130.8 ms and transmits the waveform 128 times, the Doppler resolution can reach 0.06 Hz (1/(128×130.8 ms)), which is precise enough. In Fig. 2(c), the Doppler strips parallel to the Delay axis at 2/(90 T_p), 4/(90 T_p), 6/(90 T_p)... etc, correspond to 1/(T_{r2}), 2/(T_{r2}), 3/(T_{r2})... etc. The distance between Delay axis and the first Doppler strip denotes unambiguous Dopplerdetection-range (UDDR) of the waveform, i.e. 1/ T_{r2}. For T_{r2} equals 130.8 ms, it gets to 7.6 Hz (1/(130.8 ms)), which is also suitable for ionospheric backscattering sounding.

Fig. 2(d) ~ Fig. 2(f) show AF, one-dimensional AF cuts and AF contour of Wave-B (the waveform with 30 periods). The upper panel of Fig. 2(e) shows the PACF of waveform, which is 15 times longer than Wave-A in the upper panel of Fig. 2(b). The lower panel of Fig. 2(e) shows the Doppler characteristics of the waveform at zero delay. Since the two waveforms have the same timing formats, their Doppler characteristics are alike [11]. Wave-B's Doppler resolution equals 1/(1350T_p), i.e. also the reciprocal of the total CPI. Fig. 2(f) is the contour map of Wave-B's AF. Similar to Wave-A, the Doppler strips parallel to the Delay axis at 30/(1350T_p), 60/(1350T_p), 90/(1350T_p)... etc, correspond to 1/(T_{r2}), 2/(T_{r2}), 3/(T_{r2})... etc. But Wave-B has thinner strips (better Doppler resolution) than Wave-A,because of its longer CPI (30 vs 2). The unambiguous Dopplerdetection-range of Wave-B is also 1/ T_{r2}.

4 Open Loop Experiments

4.1 WICSS's Specifications

The experimental platform is WICSS, which utilizes a pair of log-periodic antennas in oblique backscattering sounding. WICSS's major system parameters are listed below:

Power of transmission
200 W (typical)
Operating frequency range
2-30 MHz

Chip width	*25.6 s (minimum)*
Receiver bandwidth	*39.0625 kHz*
Receiver sensitivity	*-153 dBm/Hz*
Sounding range	*3.84~3000 Km*
Radial resolution	*3.84 km (minimum)*
Waveform	*bi-phase coded inter-pulse and intra-pulse*
Sounding mode	*Fixed-frequency/ swept-frequency*

WICSS is located at Wuhan University, Wuhan, Hubei Province, China (30.35 °N, 114.33 °E).

4.2 Backscattering Ionograms

In the experiment, major waveform parameters are: L = 511, T_p = 25.6 s, and T_{r1} = 10 T_p , and T_{r2} = 130.8 ms. The corresponding waveform abilities are: 27 dB pulse compression gain, 3.84 Km range resolution, 19622 Km UDLDR, 0.06 Hz Doppler resolution, and 7.6 Hz UDDR. 64 pulse trains are sent continuously with 200 W peak power.

Fig. 3(a) ~ Fig. 3(e) are original backscattering ionograms. Fig. 3(a) is a fixed frequency sounding ionogram at 10.4 MHz. Vertical-Incidence echo at about 500 km and backscattering echoes from 600 km to 1700 km can clearly be seen. Echoes obtained from different group ranges have different Doppler frequencies.

In Fig. 3(b), a fixed frequency sounding ionogram at 15.8 MHz is shown. Backscattering echoes from 1300 km to 2000 km are clear. Echoes obtained from short group ranges (1300 km to 1400 km) have larger Doppler frequencies than echoes obtained from long group ranges (1500 km to 2000 km).

Fig. 3(c) is a fixed frequency sounding ionogram at 18.4 MHz. Backscattering echoes from 1100 km to 1900 km are clear. Echoes obtained from short group ranges (1100 km to 1300 km) have larger Doppler frequencies than echoes obtained from long group ranges (1400 km to 1900 km).

In Fig. 3(d), a fixed frequency sounding ionogram at 20.4 MHz is depicted. Backscattering echoes from 1300 km to 2400 km are clear. Echoes obtained from all the group ranges have similar Doppler frequency (-0.2 Hz to -0.1 Hz).

Fig. 3(e) is a swept frequency sounding ionogram of 6~30 MHz frequency range and 200KHz frequency step (Fig. 3(a) ~ Fig. 3(d) are taken out form Fig. 3(e)). The ionogram clearly depicts the distribution of backscattered power as a function of group range and frequency in the range of 2500 km. The evenly distributed blank strips are dead zones [9] under suchwaveform. The growing leading edge of the ionogram with frequency is typical of F-layer propagation.

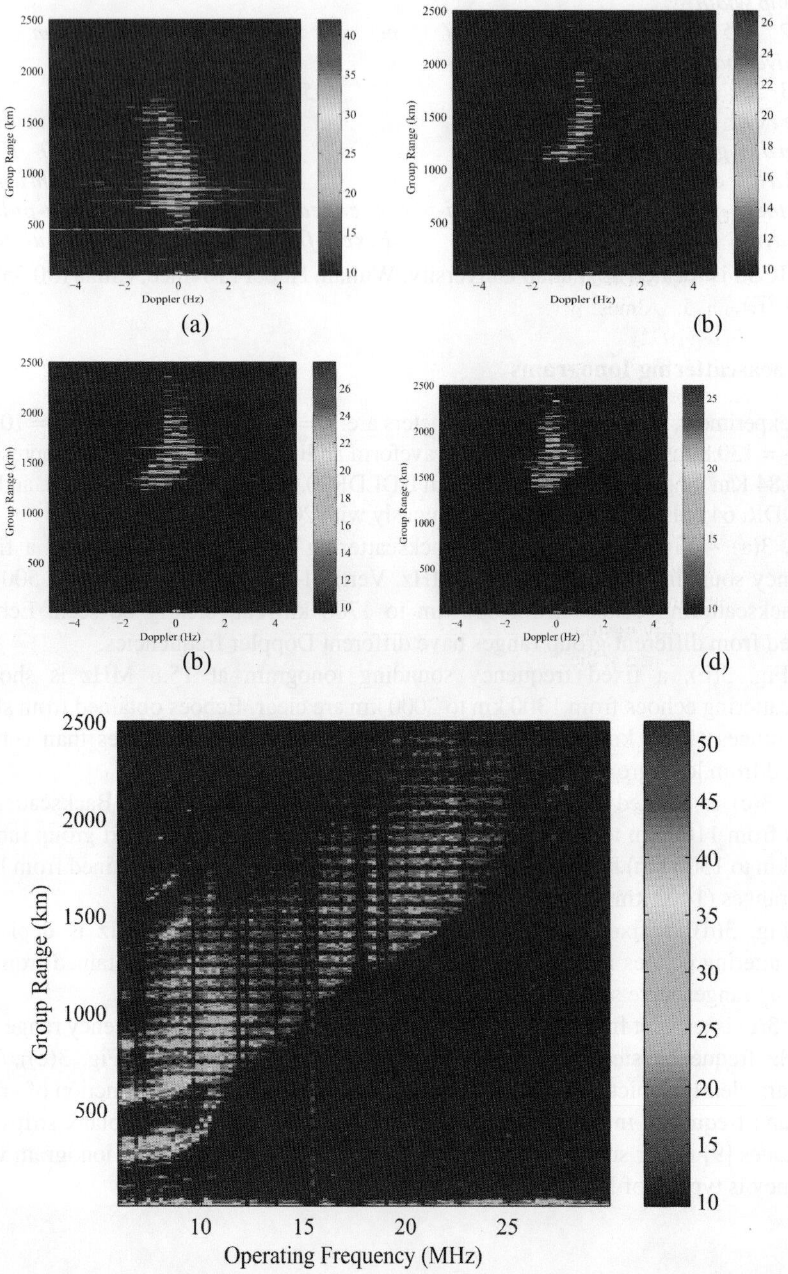

Fig. 3. Oblique backscatter ionogram of WICSS. (a) Fixed frequency sounding at 10.4 MHz. (b) Fixed frequency sounding at 15.8 MHz. (c) Fixed frequency sounding at 18.4 MHz. (d) Fixed frequency sounding at 20.4 MHz. (e) Swept frequency sounding, 6 30 MHz, 200KHz step, started at 14:21 L.T., ended at 14:36 L.T., 27 December 2009. The colorbar represents SNR (dB)

5 Conclusion

Bi-phase inter-pulse coded pulse trains can achieve high pulse compression gain, long unambiguous range, high-range resolution and high Doppler resolution, and are specially suitable for low power monostatic backscattering ionosonde.

References

1. Golomb, S.W.: Shifted register sequences. rev. edn. Aegean Park, Laguna Hills (1982)
2. Wolfmann, J.: Almost perfect autocorrelation sequences. IEEE Trans. inf. Theory 38(4), 1412–1418 (1992)
3. Goutelard, C.: Coding and sounding: extremely low power sounder. In: Proc. IEE Int. Conf. HF radio systems and techniques, pp. 110–114 (1997)
4. Barry, G.H.: A low-power vertical-incidence ionosonde. IEEE Trans.Geosci. Remote Sens. GRS-9(2), 86–89 (1971)
5. Reinisch, B.W., Haines, D.M., Bibl, K., Galkin, I., Huang, X., Kitrosser, D.F., Sales, G.S., Scali, J.L.: Ionospheric sounding insupport of over-the-horizon radar. Radio Sci. 32(4), 1681–1694 (1997)
6. Reinisch, B.W., Galkin, I.A., Khmyrov, G.: Advancing digisondetechnology: The DPS-4D. Radio Sounding Plasma Phys. 974, 127–143 (2008)
7. Grubb, R.N., Livingston, R., Bullett, T.W.: A new general purpose high performance HF radar. In: Presented at the XXIX URSI General Assembly, Chicago, IL (August 2008) Paper GH.4
8. Shi, S.-z., Zhao, Z.-y., Su, F.-F., Gang, C.: A Low-Power and Small-Size HF Backscatter Radar for Ionospheric Sensing. IEEE Geosci. Remote Sens. Lett. 6(3), 504–508 (2009)
9. Yao, Y.: Study on Pulse Compression Coded Radar System of Ionospheric Sounding. Ph.D. Dissertation, Department of Electrical Engineering, Wuhan University (2002)
10. Woodward, P.M.: Probability and Information Theory with Applications to Radar. Pergamon Press, New York (1953)
11. Levanon, N., Mozeson, E.: Radar Signals. John Wiley, Hoboken (2004)

Space –Time Wireless Channel Characteristic Simulations Based on 3GPP SCM for Smart Antenna Systems

Junpeng Chen[1,2], Xiaorong Jing[1,2], Qiang Li[1], Zufan Zhang[1], Yongjie Zhang[2]

[1] Key lab of Mobile Communicaiton Technology, Chongqing University of Posts and Telecommunications, Chongqing 400065, China
[2] Science and Technology on Information Transmission and Dissemination in Coummunication Networks Laboratory, Shijiazhuang 050081, China
chenjp05020118@163.com

Abstract. The paper firstly analyzes the channel characteristic of the smart antennas systems based on spatial channel model(SCM), which includes the time-correlated feature, frequency-correlated feature, spatial-correlated feature and its corresponding spectral features for Suburban Macro, Urban Macro and Urban Micro. Furthermore we study the fading depth of it. At last, considerable simulations are utilized to verify and compare their features in different communication scenarios.

Keywords: smart antenna, channel characteristic, fading depth.

1 Introduction

In order to improve system capacity and the link quality, smart antenna technology was introduced into the Base Station (BS). Due to the complex scattering environment around BS, the elements of the antenna array are in different spatial position and experience the different fading, which led to space selective fading, the wireless communication channel model must contain spatial factor except for time and frequency information. Up to now, many channel models are proposed about smart antenna system, such as Lee Model [1], Discrete Uniform Distribution Model [2], Time Varying Vector Channel Model [3] and so on, but these channel models are base on certain assumption. Lee Model assumed that scatters were uniform distribution in circumference. Discrete Uniform Distribution Model assumed the angle of arrival was uniform distribution. Time Varying Vector Channel Model assumed the signal energy was Rayleigh fading. Therefore, these channel models can not reflect the real channel environment.

Based on a large number of measured data, 3GPP proposed a channel model [4] named spatial channel model (SCM). SCM utilize the temporal and spatial parameters obtained. In this paper, we analyze channel characteristics of the SCM, such as spatial characteristics, time domain characteristics, frequency domain characteristics and fading depth) for smart antenna system in different scenarios [4], and also we use numerical simulation tools to verify them.

M. Zhu (Ed.): Electrical Engineering and Control, LNEE 98, pp. 45–52.
springerlink.com © Springer-Verlag Berlin Heidelberg 2011

2 Smart Antenna System Based on 3GPP SCM

Figure 1 shows BS part of the smart antenna system, which is assumed as an S-element uniform linear antenna array. From the figure, antenna *0* is a reference antenna. The distance between neighboring antenna elements is d and direction of arrival of the signal is θ. The Mobile Station (MS) has a single antenna element, and the transmitted signal from it is represented by $s(t)$. In a typical wireless scenario, the transmitted signal generally experiences the multipath fading. If let the vector channel response $\mathbf{h}(\tau,t)=\sum_{i=1}^{N}\mathbf{a}(\theta_i)a_i(t)\delta(\tau-\tau_i)$,and $\mathbf{x}(t)$ denote the received signal, we have the following [5]:

$$\begin{aligned}\mathbf{x}(t)&=s(t)*\mathbf{h}(\tau,t)\\&=\sum_{i=1}^{N}\mathbf{a}(\theta_i)a_i(t)s(t-\tau_i)+\mathbf{n}(t)\end{aligned}, \tag{1}$$

where L, $\mathbf{a}(\theta_i)$, θ_i, τ_i , $a_i(t)$ and $\mathbf{n}(t)$ denote the number of the path, steering vector , angle of arrival is, the delay of *i*th path, complex amplitude of the channel and the noise vector at receiver. The $\mathbf{a}(\theta_i)$ and $\mathbf{n}(t)$ can be further written as

$$\mathbf{a}(\theta_i)=\left[1,e^{j2\pi d\cos\theta_i/\lambda},\ldots,e^{j2\pi(S-1)d\cos\theta_i/\lambda}\right]^T,\ \mathbf{n}(t)=\left[n_1(t),n_2(t),\ldots,n_{N-1}(t)\right]^T.$$

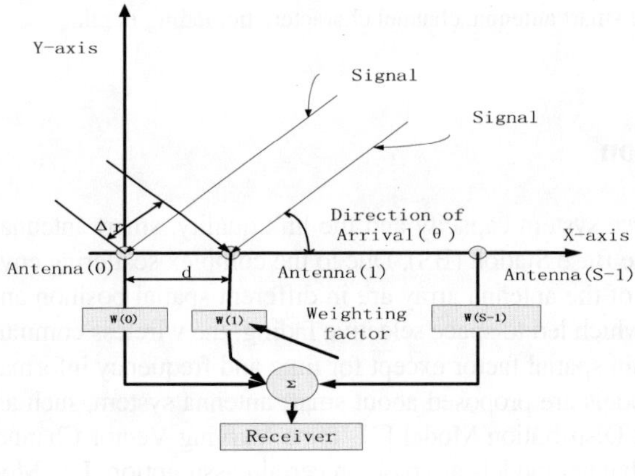

Fig. 1. Smart Antenna System

3 Theoretical Analysis about Channel Characteristics of Smart Antenna System

The superimposition of different multipath delay signals will bring about frequency selective fading, the movement of MS will bring about time selective fading and the different fading of elements at different spatial position will bring about spatial selective fading. The depth of channel fading will decide the launch approach of smart antenna.

According to the description of 3GPP TR 25.996, the *n*th multipath channel coefficient corresponding to antenna s can be given by

$$h_{s,n}(t)=\sqrt{\frac{P_n\sigma_{SF}}{M}}\sum_{m=1}^{M}\left\{\begin{array}{l}\sqrt{G_{BS}(\theta_{n,m,AoA})}\exp\left(j\left[kd_s\sin(\theta_{n,m,AoA})+\varphi_{n,m}\right]\right)\times\\ \sqrt{G_{MS}(\theta_{n,m,AoD})}\times\\ \exp\left(jk\|v\|\cos(\theta_{n,m,AoA}-\theta_v)t\right)\end{array}\right\} \tag{2}$$

where P_n, σ_{SF}, M, d_s, $\theta_{n,m,AoD}$, $\theta_{n,m,AoA}$, $G_{BS}(\theta_{n,m,AoD})$, $G_{MS}(\theta_{n,m,AoA})$, $\|v\|$ and θ_v respectively denote the power of the nth path, the lognormal shadow fading, the number of sub-paths for per-path, the distance in meters from BS antenna element s from the reference (s = 0) antenna, the angle of departure (AOD)for the mth sub-path of the nth path, the AOA for the mth sub-path of the nth path, BS antenna gain, MS antenna gain, the magnitude of the MS velocity vector and the angle of the MS velocity vector.

In order to analysis the channel space-time autocorrelation function easily, generally assume $G_{BS}(\theta_{n,m,AoD})$ = $G_{MS}(\theta_{n,m,AoA})$ =1, so the space-time autocorrelation function is given by:

$$\rho(\Delta d_s,\tau)=E\left\{\frac{h_{s_1,n}(t)h^*_{s_2,n}(t+\tau)}{\sigma_{h_{s_1,n}}\sigma_{h^*_{s_2,n}}}\right\} \tag{3}$$

where $E\{*\}$ is statistical average, $\sigma_{h_{s_1,n}}=\sigma_{h^*_{s_2,n}}=\sqrt{P_n}$ is standard deviation of $h_{s_1,n}(t)$ and $h_{s_2,n}(t)$. Substituting $h_{s,n}(t)$ into the equation (3), we have

$$\rho(\Delta d_s,\tau)=\frac{1}{M}\sum_{m=1}^{M}\left\{\begin{array}{l}\exp\left(j\left[k\Delta d_s\sin(\theta_{n,m,AoA})\right]\right)\times\\ \exp\left(-jk\|v\|\cos(\theta_{n,m,AoA}-\theta_v)\tau\right)\end{array}\right\} \tag{4}$$

where $\Delta d_s=\left|d_{s_1}-d_{s_2}\right|$ is the distance between BS antenna elements.

Through the above formulas we know that channel space correlation and time correlation are mutual correlated, for simplicity, we analyze channel space correlation without considering the influence of time correlation.

(1) Spatial Characteristics

The correlation distance is an important parameter to describe channel spatial characteristics. It can be obtained from spatial correlation function (SCF) and limit the interval of antenna array. SCF is related to the wave number spectrum through a Fourier transform [6]. Assuming τ=0, SCF is given by:

$$\rho(\Delta d_s) = \frac{1}{M}\sum_{m=1}^{M} \exp\left(j\left[k\Delta d_s \sin\left(\theta_{n,m,AoA}\right)\right]\right) \tag{5}$$

Channel correlation distance D_c is Δd_s when $\rho(\Delta d_s)$ =0.5. Channel wave number spectrum can be given by following,

$$S(k) = \int_{-\infty}^{\infty} \rho(\Delta d_s)\exp(-jk\Delta d_s)\Delta d_s \tag{6}$$

(2) Time Characteristics

Correlation time describes how fast channel varies. Within the correlation time, channel can be assumed as constant. Time correlation function (TCF) is related to the Doppler power spectral through a Fourier transform [6]. Assuming Δd_s=0, TCF is given by:

$$\rho(\tau) = \frac{1}{M}\sum_{i=1}^{M} \exp\left(-jk\|v\|\cos\left(\theta_{n,m,AoA} - \theta_v\right)\tau\right) \tag{7}$$

Channel correlation time T_c is τ ,when $\rho(\tau)$=0.5. Doppler power spectral of the channel is given by:

$$S(\omega) = \int_{-\infty}^{\infty} \rho(\tau)\exp(-j\omega\Delta t)\,d\Delta t \tag{8}$$

(3) Frequency Characteristics

Correlation bandwidth will limit the bandwidth of transmit signal. Frequency correlation function (FCF) is related to the power delay profile through a Fourier transform [6]. Since the model is using only specula components, FCF is given by the sum equation [7]:

$$\rho(\Delta f) = \sum_n P_n \exp(-j2\pi\tau_n\Delta f) / \sum_n P_n \tag{9}$$

where P_n is the power and τ_n is the delay of the nth path, Δf is the frequency different. Channel correlation bandwidth F_c is Δf ,when $\rho(\Delta f)$ =0.5. Channel power delay profile is given by:

$$S(\tau) = \int_{-\infty}^{\infty} \rho(\Delta f)\exp(-jk\Delta f)\Delta f \tag{10}$$

(4) Fading Depth

The amplitude or power of signal is rapid change, in order to study transient change of signal; we introduced the fading depth, which can be given by [8].

$$\sigma_{fading} = \varepsilon / \mu \tag{11}$$

where μ is the mean of channel coefficients $h_{s,n}(t)$ of nth path ε is the standard deviation of channel coefficients $h_{s,n}(t)$ of nth path.

4 Computer Simulation and Analysis

In this section, the channel characteristic is verified by simulations for the uplink of the antenna system. In the simulation, a uniform linear antenna (ULA) with 8-element is considered with carrier frequency 2GHz. For the purpose of comparison, we use the carrier wavelength λ as the normalized distance.

(1) Simulation of Spatial Correlation

Space correlation under three different communication scenarios of smart antenna system is shown in Figure 2. The result show that channel space correlation reduces with the distance increasing of antenna elements. Furthermore, we find that the channel space correlation of Suburban Macro and Urban Macro is stronger than the Urban Micro. The correlation distance of Suburban Macro and Urban Macro are 59 λ and the Urban Micro's is 25 λ. For Urban Macro, we simulate space correlation corresponding to different AOA, as shown in Figure 3. The result show that if the AOA is closer to the normal direction (AOA=90 degree) of the antenna, the channel spatial correlation is stronger. Figure 4 show the wave number spectrum of Urban Macro.

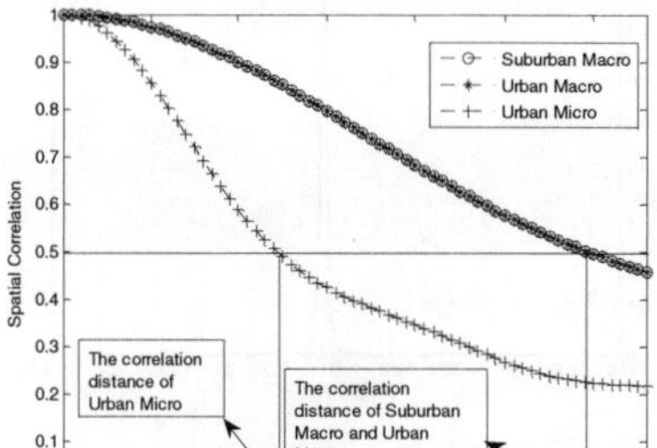

Fig. 2. Spatial Correlation Corresponding to Different Scenarios.

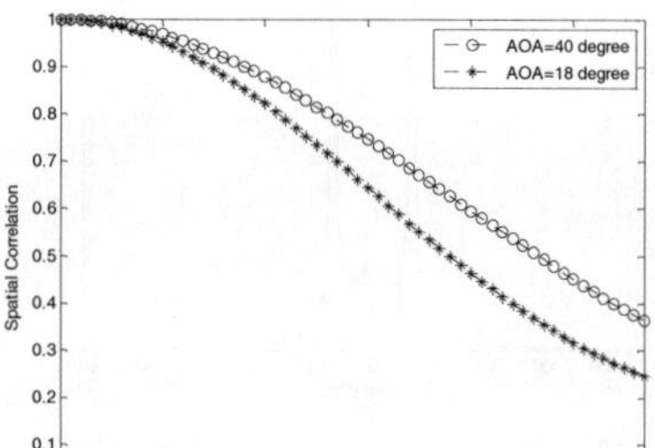

Fig. 3. Spatial Correlation Corresponding to Different AOA.

(2) Simulation of Time Correlation

For simulation of time correlation, we only consider the Urban Macro. We show that the characteristic of time correlation of the channel will reduce with normalize time increasing. In addition, the characteristic of time correlation will reduce correspondingly with mobile speed increasing. From the Fig 5, correlation time is 0.25T, 0.55T and 1.1T when velocity is 20 m/s, 10m/s and 5m/s, where the T denotes the carrier cycle.

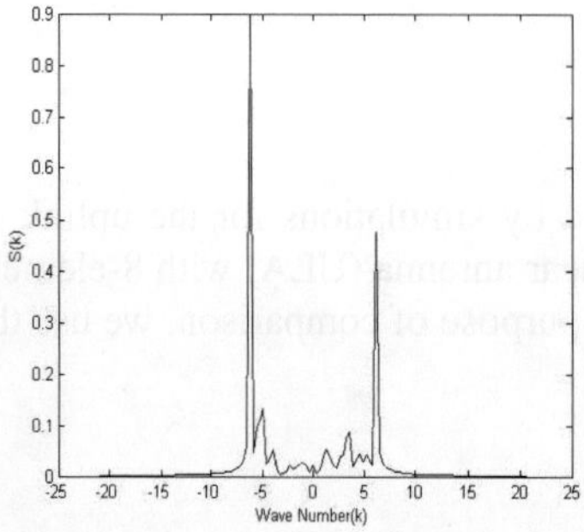

Fig. 4. Wave Number Spectrum of Urban Macro.

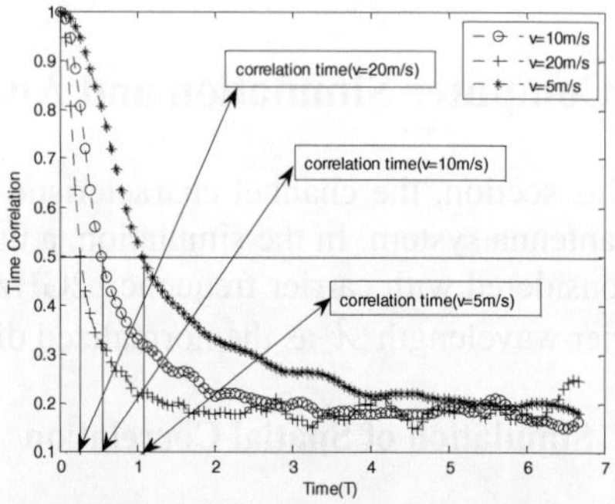

Fig. 5. Time Correlation Corresponding to Different Velocity.

For Urban Macro, we simulate the Doppler Power Spectral corresponding to different Angle Spread as shown in Fig 6 and Fig 7. From them, the Doppler Power Spectra is close to Classical Power Spectrum [9] for angle spread (AS) is 35 Degree, but when AS is 2 Degree, it is close to Gaussian one.

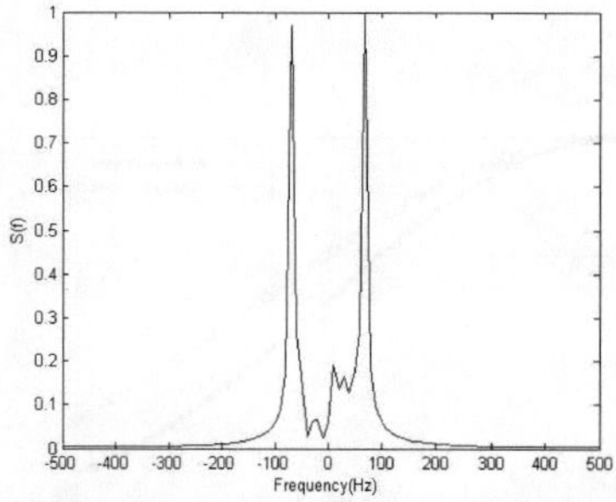

Fig. 6. Doppler Power Spectral (AS=35°)

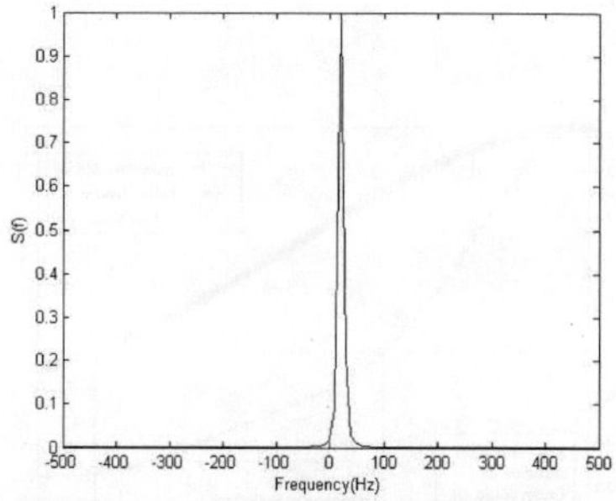

Fig. 7. Doppler Power Spectral(AS=2°)

(3) Simulation of Frequency Correlation

Fig 8 simulates the frequency correlation in Suburban Macro, Urban Macro and Urban Micro scenarios. Correlation bandwidths of Suburban Macro, Urban Macro and Urban Micro are 2.1MHz, 0.3MHz and 0.75MHz. Fig 9 simulate channel

Frequency Correlation of Suburban Macro scenario corresponding to different average delay. Figure 10 show Power Delay Profile of Suburban Macro scenario, generally, it obey exponential distribution.

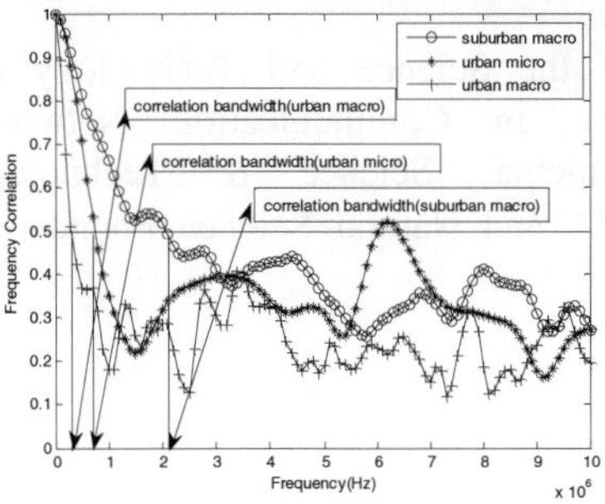

Fig. 8. Frequency Correlation Corresponding to Different Scenarios.

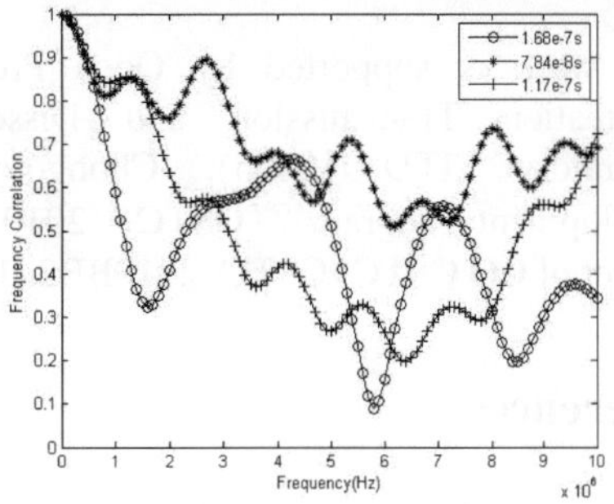

Fig. 9. Frequency Correlation Corresponding to Different Average Delay.

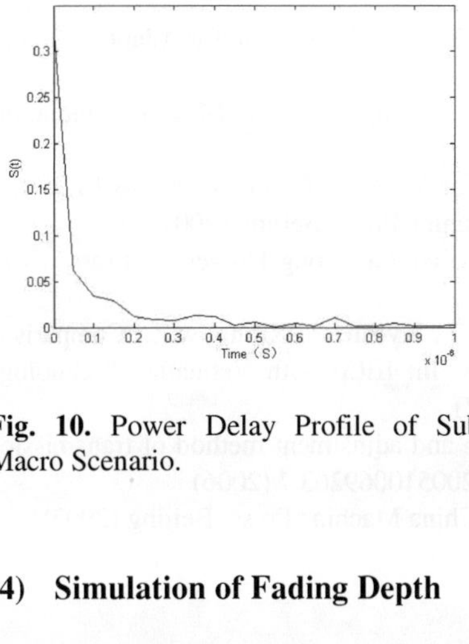

Fig. 10. Power Delay Profile of Suburban Macro Scenario.

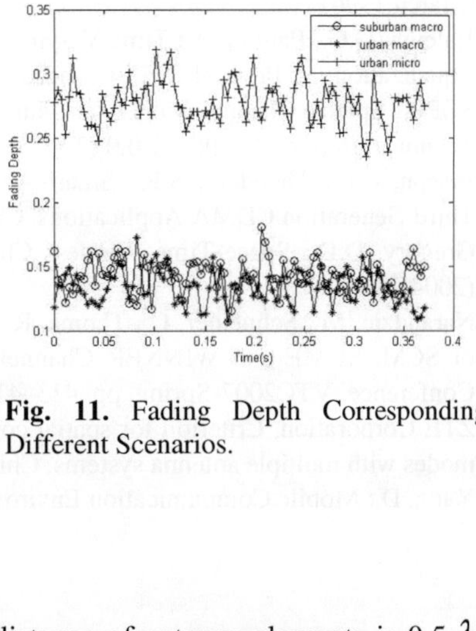

Fig. 11. Fading Depth Corresponding to Different Scenarios.

(4) Simulation of Fading Depth

Assuming the MS velocity is 10 m/s, the distance of antenna elements is 0.5 λ .We simulate the fading depth corresponding to different communication scenarios. The result show that the fading depth is random variation and the value of fading depth of Urban Micro is greater than Suburban Macro and Urban Macro.

5 Conclusion

In the paper we analyze the spatial correlation, time correlation, frequency correlation and fading depth of wireless communication channel which is described in 3GPP TR 25.996. By computer simulation, we verify its time-correlated feature, frequency-correlated feature, spatial-correlated feature for Suburban Macro, Urban Macro and

Urban Micro and obtain its correlation time, correlation bandwidth and correlation distance. Also we get the random characteristic of the fading depth.

Acknowledgment

This work is supported by Open Project of the Science and Technology on Information Transmission and Dissemination in Communication Networks Laboratory (ITD-U10005), Chongqing Municipal Science & Technology Development Program （CSTC, 2010AC2143） and Natural Science Foundation Project of CQ CSTC (CSTC, 2010BB2417).

References

1. Lee, W.C.Y.: Mobile Communications Engineering. McGraw Hill Publications, New York (1982)
2. Aszetly, D.: On Antenna Arrays in Mobile Communication Systems: Fast Fading and GSM Base Station Receiver Algorithms.:Ph.D. Dissertation, Royal Institute of Technology (March 1996)
3. Raleigh, G.G., Paulraj, A.: Time Varying Vector Channnel Estimation for Adaptive Spatial Equalization. In: Proc. of IEEE Globecom, pp. 218–224 (1995)
4. 3GPP.: Spatial Channel Model for Multiple Input Multiple Output (MIMO) Simulation. Techincal Report, 25.996 V7.0.0 (2007)
5. Joseph, C.L., Theodore, S.R.: Smart Antennas for Wireless Communications:IS-95 and Third Generation CDMA Applications. China Machine Press, Beijing (2002)
6. Gregory, D.D.: Space-Time Wireless Channel. Xi’an Jiao Tong University Press, Xi’an (2004)
7. Narandzic, M., Schneider, C., Thoma, R., Jamsa, T., Kyosti, P., Xiong, w.Z.: Comparison of SCM, SCME and WINNER Channel Models. In: IEEE 65th Vehicular Technology Conference, VTC2007-Spring, pp. 413–417 (2007)
8. ZTE Corporation, Criterion for spatial correlation and adjustment method of transmission modes with multiple antenna systems. China: CN200510069203.7 (2006)
9. Yang, D.: Mobile Communication Environment. China Machine Press, Beijing (2003)

Knowledge Reduction of Numerical Value Information System Based on Neighborhood Granulation

Yu Hua

Department of Information,
Business college of Shanxi University,
Taiyuan 030031 China

Abstract. Numerical value information systems have a wide application in our living. This paper, based on neighborhood granulation, presented distributing reduction, assigned reduction and rules reduction in numerical value information systems, and gave the concepts of approximate classified precision in neighborhood relation. Finally, obtaining methods of decision-making regularity were described by interzone form, and this is Pawlak's popularization.

Keywords: neighborhood; numerical value information systems; knowledge reduction.

1 Introduction

Early in 1980s, professor Z.Pawlak, a Poland scientist, presented the theory of rough set, and granulated field theory to equivalent class by equivalent relation, as a kind of basic information particle. The concept of fluctuation approximation was given. This was data analysis tool of dealing fuzzy and uncertain knowledge, and was applied in several ways such as decision-making and analysis, mode identification, machine study and knowledge discovery.

In reference [1], T.Y.Lin presented that neighborhood was the expression of granulated computation, which was an extension of the theory of rough set. In reference [2],Y.Y.Yao mainly discussed the basic question of granular computing and the construction of granular computing. Granular computing was a generalized model, mainly applied in data mining, knowledge discovery, computing with words and theory of quotient space.

As an computation of numerical value information granulation, this paper used the method of neighborhood granulation to construct neighborhood of point in the space, and divided numerical value information systems by regions. In addition, distributing reduction and regularity reduction were presented, and the formalized expressions of decision-making regular were given. Finally, the feasibility of this method was analyzed and experimented by simple example[3-7].

M. Zhu (Ed.): Electrical Engineering and Control, LNEE 98, pp. 53–60.
springerlink.com © Springer-Verlag Berlin Heidelberg 2011

2 Basic Concept of Neighborhood Granulation

2.1 Neighborhood Granulation

Definition 1. $S=(U,A,V,f)$ is supposed as numerical value information system. U is non-empty finite set of object. A is a finite set of attribute, which is made of conditional attribute set C and decision-making attribute set D. C and D are satisfied $C\bigcup D=A$, $C\bigcap D=\varnothing$. For each $a_k\in A$, $f:U\to V_k$, there into V_k ($V_k=[0,1]$) is valued region of f in a_k. For $\forall x_i\in U$, neighborhood information particle of x_i in A is defined as:

$\delta_A(x_i)=\{x_j \mid x_j\in U,\Delta(x_i,x_j)\le\delta\}$, there into, $\Delta(x_i,x_j)=\sqrt{\sum_{l=1}^{m}|f(x_i,a_l)-f(x_j,a_l)|^2}$, $a_l\in A$ is Euclidean distance formula.

Definition 2. $S=(U,A,V,f)$ is supposed as numerical value information system. U is non-empty finite set of object. For $\Delta:U\times U\to U$, the character below is satisfied as:

(1) $\forall x_i,x_j\in U$, $\Delta(x_i,x_j)\ge 0$, $\Delta(x_i,x_j)=0$ only when $\forall a_l\in A$, $f(x_i,a_l)=f(x_j,a_l)$;

(2) $\Delta(x_i,x_i)=\Delta(x_i,x_i)$;

(3) $\forall x_i,x_j\in U$, $\Delta(x_i,x_j)=\Delta(x_j,x_i)$;

(4) $\forall x_i,x_j,x_k\in U$, $\Delta(x_i,x_k)\le\Delta(x_i,x_j)+\Delta(x_j,x_k)$

The character below is obtained by definition 1 and definition 2:

(1) $x_i\in\delta(x_i)$;

(2) $x_i\in\delta(x_j)\Leftrightarrow x_j\in\delta(x_i)$;

(3) $U=\bigcup_{i=1}^{n}\delta(x_i)$。

That is, neighborhood information particles constructed the coverage of U, not the partition.

2.2 Neighborhood Numerical Value Information Systems

Definition 3. $S=(U,A,V,f)$ is supposed as numerical value information system. U is non-empty finite set of object. A is a finite set of attribute, which is made of conditional attribute set C and decision-making attribute set D. C and D are satisfied $C \bigcup D = A$, $C \bigcap D = \varnothing$, For each $a_k \in A$, $f: U \to V_k$, there into V_k ($V_k = [0,1]$) is valued region of f in a_k. The relationship is defined: $R_A = \{(x, y) \in U \times U \mid \Delta(x, y) \le \delta\}$, So $\delta_A(x) = \{x \mid x \in U, (x, y) \in R_A\}$ there into, $B \subseteq A$ generates R_B, the equivalent relationship R_D is obtained from D, $U / R_D = \{D_1, D_2, \cdots, D_r\}$, note:

$D(D_j / \delta_B(x)) = \frac{|D_j \cap \delta_B(x)|}{|\delta_B(x)|}$, $\mu_B(x) = \{D(D_j / \delta_B(x)) \mid j \le r\}$,

$\lambda_B(x) = \{D_j \mid D_j \bigcap \delta_B(x) \ne \varnothing, j \le r\}$,

$\gamma_B(x) = \{D_{j0} \mid D(D_{j0} / \delta_B(x)) = \max D(D_j / \delta_B(x)), j \le r\}$,

$m_B(x) = \max_{j \le r} D(D_j / \delta_B(x)) = D(D_{j0} / \delta_B(x))$,

$C_B(D) = \{x \mid x \in U, D \subseteq U / R_D, \gamma_B(x) = D\}$,

$q_B(D_j) = \min\{m_B(x) \mid x \in C_B(D_j)\}$

Theorem 1 $S = (U, A, V, f)$ is supposed as numerical value information system, let $C = \{C_B(D) \mid C_B(D) \ne \varnothing, D \subseteq U / R_D\}$, so C composes one partition of information systems U.

Prove:

(1) First, we prove $C_B(D) \bigcap C_B(D') = \varnothing$

Reduction to absurdity: we suppose there are different $D, D' \subseteq U / R_D$, and $C_B(D) \bigcap C_B(D') \ne \varnothing$, then $\exists x_0 \in U$ can get $x_0 \in C_B(D)$ and $x_0 \in C_B(D')$. So $\gamma_B(x_0) = D = D'$, but it is contradiction. Thus, $C_B(D) \bigcap C_B(D') = \varnothing$.

(2) Second, we prove $\bigcup\{C_B(D)\} = U$. In fact, $\bigcup\{C_B(D)\} \subseteq U$ is obviously right. In addition, for $\forall x \in U$, we suppose $\gamma_B(x) = D$, then $x \in C_B(D)$, that is $U \subseteq \bigcup\{C_B(D)\}$ can be obtained. Thus, $\bigcup\{C_B(D)\} = U$.

From theorem 1, we can divide U to $C = \{C_1, C_2, \cdots, C_t\}$, $t = |C_B(D)|$, $D \subseteq U / R_D$。

Therefore, we can get the decision-making regular $C_j \Rightarrow D_j$, there into, $C_j = C_B(D_j) \ne \varnothing$, $j \le t$. We note $q_B(D_j) = \min\{m_B(x) \mid x \in C_B(D_j)\}$ is the reliability of decision-making regular.

Definition 4. $S = (U, A, V, f)$ is supposed as numerical value information system, then lower approximation and upper approximation of neighborhood fuzzy set are defined as:

$$\underline{R}X = \{x_i \mid x_i \in U, \delta(x_i) \subseteq X\}, \overline{R}X = \{x_i \mid x_i \in U, \delta(x_i) \bigcap X \ne \varnothing\}$$

The set $bn(X)=\overline{R}X-\underline{R}X$ is entitled approximate boundary of X ; $pos(X)=\underline{R}X$ is entitled positive region of X ; $neg(X)=U-\overline{R}X$ is entitled negative region of X . Obviously, $\overline{R}X=bn(X)\bigcup pos(X)$.
$\underline{R}X$ is the biggest merging set of neighborhood information particles which surely belong to X , $\overline{R}X$ is the smallest merging set of neighborhood information particles which possibly belong to X , $bn(X)$ is the set of neighborhood information particles which neither surely belong to X nor surely belong to $\sim X$, $neg(X)$ is the set of neighborhood information particles which don't belong to X affirmatively.

Definition 5. $S=(U,A,V,f)$ is supposed as numerical value information system. U is non-empty finite set of object. A is a finite set of attribute, which is made of conditional attribute set C and decision-making attribute set D . C and D are satisfied $C\bigcup D=A$, $C\bigcap D=\varnothing$. The decision-making attribute set D divides U to n equivalence classes $D_1,D_2,\cdots,D_r$ 。The lower approximation and upper approximation of decision-making attribute set D about $B\subseteq A$ are defined as:

$\underline{R_B}(D)=\bigcup_{i=1}^{r}\underline{R_B}(D_i)$, $\overline{R_B}(D)=\bigcup_{i=1}^{r}\overline{R_B}(D_i)$, there into,

$\underline{R_B}(D_i)=\{x_i\mid\delta_B(x_i)\subseteq D_i,x_i\in U\}$, $\overline{R_B}(D_i)=\{x_i\mid\delta_B(x_i)\bigcap D_i\neq\varnothing,x_i\in U\}$.

Therefore, the approximate precision of decision-making classification is defined as:

$$\alpha=\frac{|\underline{R_B}(D)|}{|\overline{R_B}(D)|}.$$

And the approximate classified quality of decision-making classification is defined as:

$$\beta=\frac{|\underline{R_B}(D)|}{|U|}.$$

The approximate classified precision describes the percentage of correct decision-making in possible decision-making, and the approximate classified quality indicates the percentage of dividing into decision-making class correctly.

3 The Rules Reduction of Numerical Value Information System

Definition 6. $S=(U,A,V,f)$ is supposed as numerical value information system. The threshold $\delta\geq 0$ is given, for $B\subseteq A$:

(1) If for any $x\in U$, $\mu_B(x)=\mu_A(x)$ can be right, then it is noted that B is δ distributing harmonic set of S . If B is the biggest δ distributing harmonic set

of S , and $\mu_{B'}(x) \neq \mu_A(x)$ can be right for any real subset $B' \subseteq B$, then B is δ distributing reduction of S .

(2) If for any $x \in U$, $\gamma_B(x) = \gamma_A(x)$ can be right, then it is noted that B is the biggest δ distributing harmonic set of S . If B is the biggest δ distributing harmonic set of S , and $\gamma_{B'}(x) \neq \gamma_A(x)$ can be right for any real subset $B' \subseteq B$, then B is the biggest δ distributing reduction of S .

(3) If for any $x \in U$, $\lambda_B(x) = \lambda_A(x)$ can be right, then it is noted that B is δ assigned harmonic set of S . If B is the biggest δ distributing harmonic set of S , and $\lambda_{B'}(x) \neq \lambda_A(x)$ can be right for any real subset $B' \subseteq B$, then B is δ assigned reduction of S .

The distributing harmonic set is attribute set of keeping subjection degree of object neighborhood particles unchangeable in every decision classes However, the biggest distributing harmonic set keep the biggest distributing decision classes of every neighborhood particle unchangeable.

Definition 7. $S = (U, A, V, f)$ is supposed as numerical value information system. The threshold $\delta \geq 0$ is given, for $B \subseteq A$:

(1) If $C_B = C_A$, then it is noted that B is δ rules harmonic set of S . If B is rules harmonic set of S , and $C_{B'} \neq C_A$ can be right for any real subset $B' \subseteq B$, then B is rules reduction of S .

(2) If B is rules harmonic set of S , and $q_B = q_A$, then it is noted that B is precise rules harmonic set of S . If B is precise rules harmonic set of S , and $q_{B'} \neq q_A$ can be right for any real subset $B' \subseteq B$, then B is precise rules reduction of S .

The rules harmonic set is that the regular of attribute set B is the same with the proposition regular of A . The precise rules harmonic set is that B and A have the same proposition regular, and the same reliability.

Theorem 2 $S = (U, A, V, f)$ is supposed as numerical value information system;

(1) If B is δ distributing harmonic set of S , then B is surely rules precise harmonic set of S .

(2) If B is the biggest δ distributing harmonic set of S , then B is surely rules harmonic set of S .

Prove:

(1)If B is δ distributing harmonic set of S , then $\mu_B(x) = \mu_A(x)$ for $\forall x \in U$. That is $D(D_j / \delta_B(x)) = D(D_j / \delta_A(x))$ for $\forall j \leq r$, so $\gamma_B(x) = \gamma_A(x)$. Moreover, for $\forall D_j \subseteq U / R_D$,

$C_B(D_j)=\{x \mid x\in U, \gamma_B(x)=D_j\} =\{x \mid x\in U, \gamma_A(x)=D_j\} =C_A(D_j)$。

Therefore, $C_B = C_A$.

In addition, $q_B(D_j)=\min\{m_B(x) \mid x\in C_B(D_j)\}$

$= \min_{x\in C_B(D_j)} \max_{j\le r} D(D_j / \delta_B(x)) = \min_{x\in C_A(D_j)} \max_{j\le r} D(D_j / \delta_A(x))$

$= \min\{m_A(x) \mid x\in C_A(D_j)\} = q_A(D_j)$,

That is $q_B = q_A$, so B is rules precise harmonic set of S.

(2)If B is the biggest δ distributing harmonic set of S, then $\gamma_B(x)=\gamma_A(x)$ for $\forall x\in U$. Therefore for $\forall D_j \subseteq U / R_D$,

$C_B(D_j)=\{x \mid x\in U, \gamma_B(x)=D_j\}=\{x \mid x\in U, \gamma_A(x)=D_j\}=C_A(D_j)$, $C_B = C_A$。

So B is rules harmonic set of S.

Definition 6 and Definition 7 put forward some methods of getting reduction. From theorem 1, $C_j \Rightarrow D_j$ ($q_B(D_j)$). Usually we get condition attribute value as antecedent of regular, and get decision-making attribute value as consequent of regular, and note the average value of all attribute as the criterion, $Z_B(D_j) = (a_1^j, a_2^j, \cdots, a_m^j)$. There into, $a_i^j = \frac{\sum_{x\in C_B(D_j)} f_i(x)}{|C_B(D_j)|}$.

In this way, the regular $C_j \Rightarrow D_j$ ($q_B(D_j)$) converts to $Z_B(D_j) \Rightarrow \vee_{D_k\in D_j} (d,k)$ ($q_B(D_j)$).

If $a_i^j = \max a_p^j$, $(p\le m)$ for any i, j, then let $h_i^j = a_i^j$, otherwise let

$h_i^j = \frac{a_i^j + \min\{a_i^k \mid a_i^k > a_i^j\}}{2}$;

If $a_i^j = \min a_p^j$, $(p\le m)$, for any i, j, then let $l_i^j = a_i^j$, otherwise let

$l_i^j = \frac{a_i^j + \max\{a_i^k \mid a_i^k < a_i^j\}}{2}$.

In this way, the interzone decision-making rules can be obtained:

$(l_1^j, h_1^j), (l_2^j, h_2^j), \cdots, (l_m^j, h_m^j) \Rightarrow \vee_{D_k\in D_j} (d,k)$ ($q_B(D_j)$)。

4 Analysis of the Example

A numerical value information system is given.

Table 1. A numerical value information system

U	a1	a2	d
X1	0	0	1
X2	0.2	0	1
X3	0.4	1	2
X4	0.6	1	2
X5	0.6	1	3
X6	1	0	3

d divides U to $U/R_d=\{D_1,D_2,D_3\}$, there into $D_1=\{x_1,x_2\}$, $D_2=\{x_3,x_4\}$, $D_3=\{x_5,x_6\}$, let $\delta=0.3$, then $\delta(x_1)=\{x_1,x_2\}$, $\delta(x_2)=\{x_1,x_2\}$, $\delta(x_3)=\{x_3,x_4,x_5\}$, $\delta(x_4)=\{x_3,x_4,x_5\}$, $\delta(x_5)=\{x_3,x_4,x_5\}$, $\delta(x_6)=\{x_6\}$. It's easy to calculate that $\mu_A(x_1)=(1,0,0)$, $\mu_A(x_2)=(1,0,0)$, $\mu_A(x_3)=(0,2/3,1/3)$, $\mu_A(x_4)=(0,2/3,1/3)$, $\mu_A(x_5)=(0,2/3,1/3)$, $\mu_A(x_6)=(0,0,1)$

$\lambda_A(x_1)=\{D_1\}$, $\lambda_A(x_2)=\{D_1\}$, $\lambda_A(x_3)=\{D_2,D_3\}$,

$\lambda_A(x_4)=\{D_2,D_3\}$, $\lambda_A(x_5)=\{D_2,D_3\}$, $\lambda_A(x_6)=\{D_3\}$

$\gamma_A(x_1)=\{D_1\}$, $\gamma_A(x_2)=\{D_1\}$, $\gamma_A(x_3)=\{D_2\}$

$\gamma_A(x_4)=\{D_2\}$, $\gamma_A(x_5)=\{D_2\}$, $\gamma_A(x_6)=\{D_3\}$

Therefore,

$D_1=\{D_1\}$, $C_1=C_A(D_1)=\{x_1,x_2\}$, $q_A(D_1)=1$

$D_2=\{D_2\}$, $C_2=C_A(D_2)=\{x_3,x_4,x_5\}$, $q_A(D_2)=2/3$

$D_3=\{D_3\}$, $C_3=C_A(D_3)=\{x_6\}$, $q_A(D_3)=1$

For other $D\subseteq U/R_d$, $C_A(D)=\varnothing$ is existing. As the same theorem, if let $B=\{a_1\}$,

then $\mu_B(x_1)=(1,0,0)$, $\mu_B(x_2)=(2/3,1/3,0)$, $\mu_B(x_3)=(1/4,2/4,1/4)$, $\mu_B(x_4)=(0,2/3,1/3)$, $\mu_B(x_5)=(0,2/3,1/3)$, $\mu_B(x_6)=(0,0,1)$

$\lambda_B(x_1)=\{D_1\}$, $\lambda_B(x_2)=\{D_1,D_2\}$, $\lambda_B(x_3)=\{D_1,D_2,D_3\}$,

$\lambda_B(x_4)=\{D_2,D_3\}$, $\lambda_B(x_5)=\{D_2,D_3\}$, $\lambda_B(x_6)=\{D_3\}$

$\gamma_B(x_1)=\{D_1\}$, $\gamma_B(x_2)=\{D_1\}$, $\gamma_B(x_3)=\{D_2\}$

$\gamma_B(x_4)=\{D_2\}$, $\gamma_B(x_5)=\{D_2\}$, $\gamma_B(x_6)=\{D_3\}$

Moreover,

$D_1=\{D_1\}$, $C_1=C_B(D_1)=\{x_1,x_2\}$, $q_A(D_1)=2/3$

$D_2 = \{D_2\}$, $C_2 = C_B(D_2) = \{x_3, x_4, x_5\}$, $q_A(D_2) = 1/2$

$D_3 = \{D_3\}$, $C_3 = C_B(D_3) = \{x_6\}$, $q_A(D_3) = 1$

For other $D \subseteq U / R_d$, $C_B(D) = \varnothing$ is existing too, but $C_A = C_B$, $q_A \neq q_B$, so B is rules harmonic set of A. Consequently $Z_B(D_1) = (0.1)$, $Z_B(D_2) = (0.5)$, $Z_B(D_3) = (1)$.

Therefore, the interzone decision-making rules are

a1 d

$(0.05, 0.15) \Rightarrow (d, 1)$

$(0.45, 0.55) \Rightarrow (d, 2)$

$(0.6, 1) \Rightarrow (d, 3)$

Here, the approximate classified precision is reflected by the neighborhood relationship of A, $\alpha_A = |\underline{R_A}(D)| / |\overline{R_A}(D)| = 1/2$, and the approximate classified precision is reflected by the neighborhood relationship of B, $\alpha_B = |\underline{R_B}(D)| / |\overline{R_B}(D)| = 1/3$.

5 Conclusions

The fuzzy set theorem given by Z.Pawlak granulated field theory using equivalent relationship. He also put forward upper approximation and lower approximation generating regular to simulate the process of human study. This paper granulated field theory by the method of neighborhood granulation, and put forward the methods of harmonic reduction and getting regular about neighborhood relationship. Moreover, the validity of the methods is testified by the example.

References

1. Lin, T., Granular, Y.: Computing on binary relations I: Data mining and neighborhood systems. In: Skoworn, A., Polkowshi, L. (eds.) Proc. of the Rough Sets in Knowledge Discovery, pp. 107–121. Physica-Verlag, Heidelberg (1998)
2. Yao, Y.: Relational interpretation of neighborhood operators and rough set approximation operators. Information Sciences 111(198), 239–259 (1998)
3. Qing-Hua, H.U., Da-Ren, Y.U., Zong-xia, X.I.E.: Numerical Attribute Reduction Based on Neighborhood Granulation and Rough Approximation. Journal of Software 19(3), 640–649 (2008)
4. Shao, M.W., Zhang, H.Y.: Dominance relation and rules in ordered information system. J. Chinese Journal of Engineering Mathematics 22(4), 697–702 (2005)
5. Yang, X.B., Yang, J.Y., Wu, C., et al.: Dominance based rough set approach and knowledge reduction in incomplete ordered information system. J. Information Science 178, 1219–1234 (2008)
6. Qian, Y.-h., Liang, J.-y., Dang, C.-y.: Interval ordered information systems. Computers and Mathematics with Applications 56, 1994–2009 (2005)
7. Xu, W.-h., Zhang, X.-y., Zhang, W.-x.: Knowledge granulation, knowledge entropy and knowledge uncertainty measure in ordered information systemsss. Applied Soft Computing 9, 1244–1251 (2005)

Effects of HV Conductor Aging Surface Elements, on Corona Characteristics

Nick. A. Tsoligkas

Department of Electrical Engineering,
Technological and Educational Institution of Chalkida
Psahna, Evia, Greece, GR-34 400
Tsoligas@teihal.gr

Abstract. For power lines with operating voltages in excess of 400 kV, the key environmental factors of the design are generally imposed measures taken to limit radio and audible noise generated by corona. In this study a chemical analysis of the long term formatted black coating on the surface of differently treated single stranded aluminium conductors, was carried out. The effect of the constituent elements on the corona characteristics, particularly inception voltage and radio ultrasonic noises were studied, for the purpose of developing a corona free conductor for ac overhead transmission lines. It was found that aged samples with a higher sulphur or carbon content in their surface, presented a lower inception voltage and a considerable reduction in radio and ultrasonic interference.

Keywords: HV conductor aging, corona inception voltage, radio noise, ultrasound noise level.

1 Introduction

Electromagnetic interference from transmission lines is primarily caused by partial discharges of the air (corona) in the immediate vicinity of the conductor, when the electric field intensity at the conductor surface exceeds the breakdown strength of the air. The produced corona induces impulse currents on the line. These currents, in turn, cause wide band radio noise frequencies that fill the entire frequency spectrum from below 100 MHz [1], [2]. Power line noise can impact radio and television reception including cable TV, mobile and Internet service, amateur radio and critical communications, such as police, fire and military. It also defines the boundary conditions for communications via overhead power lines [3], [4] and [5]. Concern has been expressed that power line corona could degrade the performance of DGPS receivers in the 283.5-325 kHz band [6].

Due to the increasing number of new wireless communication sources and radio receptors the study of radio noise emitted from transmission lines, has assumed much greater importance in recent years by researchers. Investigations have shown that when a transmission line has been in operation for some time there is a formation of a

M. Zhu (Ed.): Electrical Engineering and Control, LNEE 98, pp. 61–68.
springerlink.com © Springer-Verlag Berlin Heidelberg 2011

natural black deposit on the surface of the conductor. This is known as conducting aging. This surface blackening appears to be desirable for the following reasons:

- lines with a relatively high electric stress in the air at the surface of the conductors, produce lower levels of audio [7] and radio noise when surface blackening has taken place [8]; the conductors become less visually obstructive when the reflection of light is reduced.

With audio noise being a very important design criterion for overhead line conductors, the process of the conductor aging must be accurately known. The aim of the present work was to analyze chemically the black deposit formed on the surface of the conductor and determine the chemical composition of the surface deposit. Then to investigate possible relation of any of the deposit substances with the reduction of radio and ultrasonic noise in an effort to develop technical solutions for a corona free HV conductor.

The chemical analysis of the surfaces was done by Auger Electron Spectroscopy (AES) technique. This technique is based on the Auger process according to which when a core level of a surface atom is ionized by an impinging electron beam, the atom may decay to a lower energy state through an electron arrangement which leaves the atom in a double ionized state. The energy difference between these two states is given to the ejected Auger electrons and will have a kinetic energy characteristic of the parent atom. These electrons are ejected from the surface and give rise to peaks in the secondary electron distribution function. The energy and shape of these Auger features can be used to unambiguously identify the composition of the solid surface [9].

The AES system consists of an ultrahigh vacuum system, an electron gun for specimen excitation, and an energy analyzer for detection of Auger electron peaks in the total secondary electron energy distribution. The Auger peaks are detected by differentiating the energy distribution function N(E). Thus the conventional Auger spectrum is the function:

$$\frac{dN(E)}{dE}. \tag{1}$$

The peak-to-peak magnitude of the Auger peak in a differentiated spectrum is directly related to the surface concentration of the element, which produces the Auger electrons. In the calculations the atomic concentration is expressed as:

$$C_X = \frac{\dfrac{I_X}{S_X\,\mathbf{dx}}}{\sum_a \dfrac{I_a}{\mathbf{Sada}}} \tag{2}$$

where I_X is the peak-to-peak Auger amplitude from the sample and S_X is the relative sensitivity between any element X and silver, and dx is the scale factor.

2 Experimental Procedure

2.1 Experimental Set-Up

The configuration used in the experiments, whether performed inside the HV laboratory (University of Manchester) or outside, included the test conductor above a metal earthed plate, a HV transformer, a coupling capacitor for radio noise measurements and the surrounding metal security frame.

All test samples were 3 m long single strand aluminium conductors of ⅛ inch diameter used to form the stranded aluminium conductors used in the grid system in Great Britain. The voltage gradient at the surface of the conductor was controlled by the metal ground plate, supported by an adjustable stand.

For radio noise measurements we used a radio noise meter designed to measure radio frequency noise voltage and the fields generated by electrical equipment in the frequency ranges of 150-400 kHz and 0.55-30 MHz. The meter was connected to the test circuit through a decoupled unit, which consisted of a matching resistor, a radio noise inductor and a spark gap. The radio noise characteristics were obtained at frequency of 1 MHz.

For recording ultrasonic noise, a fiber optic acoustic waveguide, with a 40 kHz ultrasonic transducer was implemented [10], placed very close to the test conductor so that the attenuation of the sound wave and its contamination from the ambient noise were highly reduced.

2.2 Test Conductors

Test conductors can be divided into two groups:

1. Those aged in the HV laboratory where the atmospheric pressure, ambient temperature and relative humidity remained approximately C0 20, 755 mm Hg and 55% respectively;
2. Those aged outdoors as in an industrial area.

All test samples were carefully handled to avoid mechanical stress and abrasion defects. As a result the conductor surfaces were completely untouched and in the same condition at the time of testing as they were while aging.

Conductor samples with following surface conditions were tested:

1. Samples with a clean surface;
2. Samples coated with oil of the type used by the manufacturer for drawing the conductors. This oil contained a high percentage of sulphur.

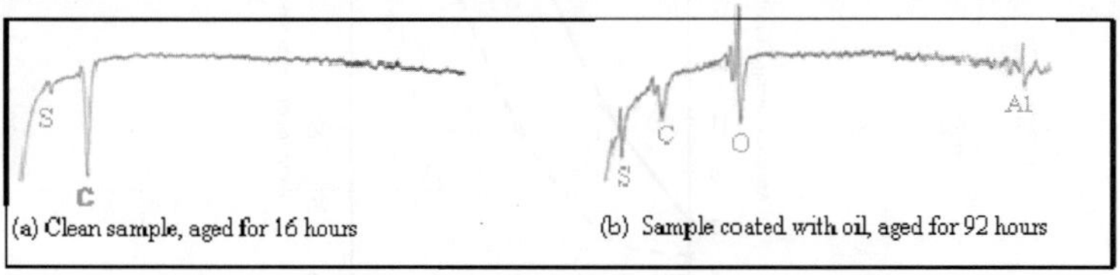

Fig. 1. Auger spectrums for two different samples

2.3 Chemical Analysis of Surface Coating

All samples brought to the laboratory for the analysis were cut from the middle of the tested conductor and were protected against accidental damage. The Auger spectrums received from two samples with different surface treatments and aging periods are shown in Fig 1.

The main elements identified in the survey scans, as can be seen in the above figures were sulphur (S), carbon (C) and oxygen (O).

2.4 Corona Inception Voltage Tests

Tests were performed to investigate the influence of concentration of the two main elements of surface coating, sulphur and carbon, on the corona inception voltage of the aged conductors.

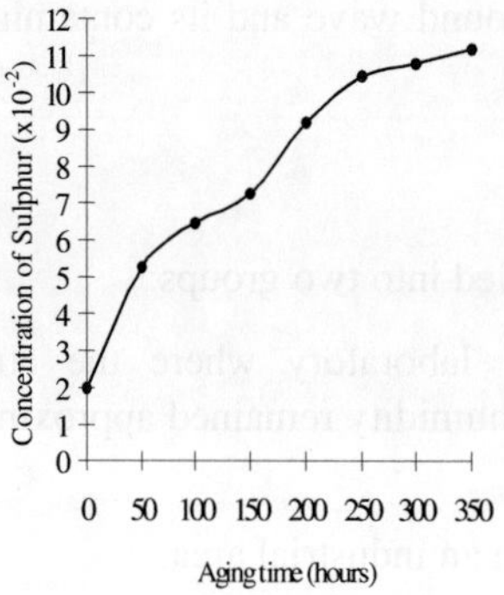

Fig. 2. Concentration of sulphur versus aging time characteristics, for clean samples aged indoors at 37.6 kV/cm

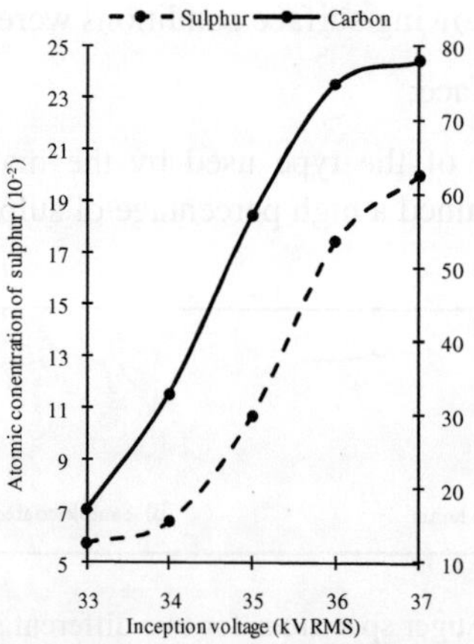

Fig. 3. Atomic concentration of sulphur and carbon versus corona inception voltage characteristics, for various samples aged outdoors.

A preliminary test performed to verify that concentration of sulphur, in the surface blackening, is increased as aging time progresses. The results are shown in Fig.2

The relation of the corona inception voltage with the concentrations of sulphur and carbon is shown in Fig. 3. It can easily be seen that both concentrations follow the same pattern: as they increase, the inception voltage decreases.

Corona inception voltage versus aging time characteristics, of two conductors being tested under the same conditions but with different surface treatment, are shown in Fig. 4.

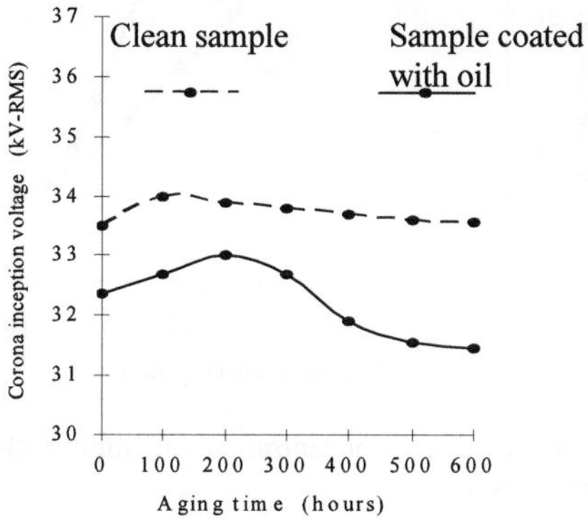

Fig. 4. Corona inception voltage versus time, with surface treatment as a parameter. Samples aged indoors for 600 hours at 37.6 kV/cm

It can be seen that corona inception voltage produced by the sample coated with oil which contained a high percentage of sulphur, is lower than that produced by the clean sample. The difference is becoming slightly higher as the aging period is emerged.

2.5 Radio and Ultrasound Noise Measurements

Tests for measuring the influence of the concentrations of sulphur, carbon and oxygen on the radio and ultrasonic noise levels of samples aged under different surface conditions and different voltage stresses were performed.

The influence of the concentration of sulphur on ultrasound level for various samples aged outdoors or indoors is shown in Fig. 5.

As can be seen as the concentration is increased the ultrasonic noise level becomes lower. The influence of the concentrations of sulphur, carbon and oxygen on ultrasonic noise level for various samples aged outdoors is shown in Fig. 6.

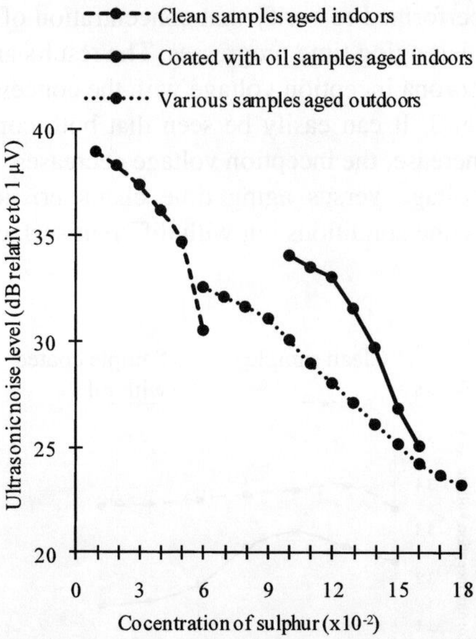

Fig. 5. Ultrasound noise level versus concentration of sulphur characteristics, for various samples aged indoors or outdoors.

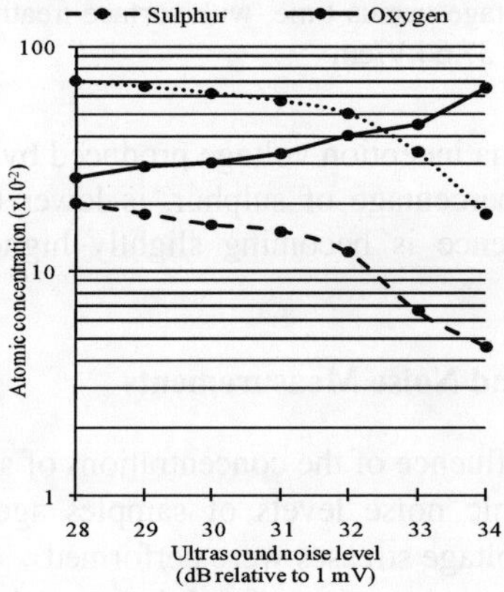

Fig. 6. Ultrasound noise level versus concentrations of sulphur, oxygen and carbon characteristic for various samples aged outdoors. All readings taken for an applied voltage of 50 kV (RMS)

It is clearly shown that samples with higher atomic concentration of sulphur or carbon exhibit a lower radio noise level. It can easily be seen that as the concentration of sulphur or carbon are quadruplicated, the emitted ultrasound noise is reduced by an average of 6 dB and 4 dB respectively. Oxygen follows a different patent. As its concentration is tripled, increment of the emitted ultrasound noise reaches 3 dB.

The influence of the concentrations of the three above mentioned elements on the emitted radio noise is depicted in Fig.7 from where it can easily be concluded that the correlation between the concentrations and radio noise is more or less the same with that of the ultrasound level depicted in Fig. 6.

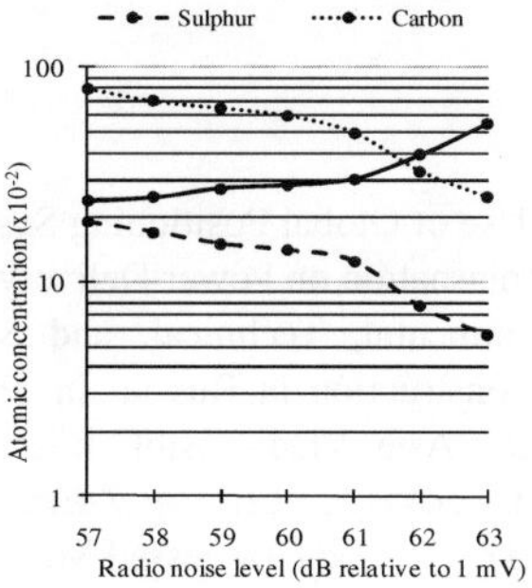

Fig. 7. Radio noise level versus concentration of sulphur oxygen and carbon characteristics, for various samples aged outdoors. Readings taken for an applied voltage of 50 kV(RMS)

It is clear that the ultrasonic noise levels are directly proportional to the generated radio noise levels [11].

2.6 Corona Oscillographic Observations

Many oscillograms of individual corona pulses were obtained for various samples at various test voltages. Some of the oscillograms related to radio noise emitted by two samples with different surface treatment.

All of the corona oscillograms, referred to radio noise extended from 55 dB to 65 dB, indicated negative corona pulses. It found that for oscillograms near corona inception voltage, the number of pulses were small. For higher voltages the number of the pulses increases. Radio noise in this case exceeded 65 dB.

3 Conclusions

Chemical analysis of the surface of conductors aged in different surroundings show that the main substances identified in the blackish coat formed after a period of operation, were sulphur, carbon and oxygen. Investigations carried out on those aged conductors show that:

- conductors with a greater concentration of sulphur or carbon, exhibited a lower inception voltage
- aged conductors exhibited lower inception voltage as time passes
- there was a direct proportionality between radio noise and ultrasonic noise levels emitted by tested conductors

- conductors with a greater concentration of sulphur or carbon exhibited lower radio and ultrasound noise
- emitted radio noise in the range of 55 to 65 dB, seemed to be related with negative corona
- emitted radio noise exceeded 67 dB seemed to be related with positive corona.

References

1. Silva, J.M., Olsen, R.G.: Use of Global Positioning System (GPS) Receivers Under Power-Line Conductors. IEEE Transaction on Power Delivery 4, 938–944 (2002)
2. Gerasimov, A.S.: Environmental, Technical, and Safety Codes, Laws and Practices Related to Power Line Construction in Russia. In: Proc. 3rd Workshop on Power Grid Interconnection in Northeast Asia, Vladi-vostok, Russia, September 30 - October 3 (2003), http://www.nautilus.org/archives/energy/grid/2003Workshop/Gerasimov%20paper_final1.pdf (accessed May 25, 2009)
3. Suljanovic, N., et al.: Computation of High-Frequency and Time Characteristics of Corona Noise on HV Power Line. IEEE Trans. on Power Delivery 20(1), 71–79 (2005)
4. Pighi, R., Raheli, R.: Linear Predictive Detection for Power Line Communications Impaired by Colored Noise Power. In: Proc. IEEE International Symposium on Line Communications and Its Applications Orlando, Florida, USA, March 26-29, pp. 337–342 (2006)
5. Mujicic, A., et al.: Corona noise on a 400 kV overhead power line: Measurements and computer modeling. Electrical Engineering 96(3), 61–67 (2004)
6. Silva, J.M.: Evaluation of the potential for power line noise to degrade real time differential GPS messages broadcast at 283.5-325 kHz. IEEE Transactions on Power Delivery 17(2), 326–333 (2002)
7. Kiyotomi, M., Kazuo, T.: 'Evaluation of audible noise from surface processing conductors for AC overhead transmission line. Electrical Engineering in Japan 159(3), 19–25 (2007)
8. Larsson, C., Hallberg, B., Israelsson, S.: 'Long term audible noise performance from the operating 400-kV transmission line. IEEE Trans. on Power Delivery 3(4), 1842–1846 (1988)
9. Smith, G.C.: Surface analysis by electron spectroscopy: measurement and interpretation, p. 11. Springer, Chester (1994)
10. Halkiadis, I.S., Theofanous, N., Greaves, D.: A high-voltage low cost wide band fiber optic transmission system with improved linearity. Electric Power System Research 37, 121–128 (1996)
11. Halkiadis, I.S.: An Electromagnetic and Ultrasonic Interference on Conductor Aging. WSEAS Transactions on Systems 8(4), 1198–1205 (2005)

Energy Efficiency Methods in Electrified Railways Based on Recovery of Regenerated Power

M. Shafighy[1,2], S.Y. Khoo[1], and A.Z. Kouzani[1]

[1] School of Engineering, Deakin University, Geelong, Victoria 3217, Australia
[2] Rail Group, Transport, AECOM, Brisbane, Australia
{mehran.shafighy,sui.khoo,kouzani}@deakin.edu.au

Abstract. Various attempts have been made to minimise energy consumption of rail vehicles by means of regenerative power from electric braking of traction motors. This paper describes energy efficiency methods in electrified railways based on recovery of energy. Direct recovery methods that return regenerative power to electrified networks, and recovery methods based on energy storage systems are elaborated. The benefits of developing recovery methods and advantages of energy storage systems are discussed.

Keywords: Energy efficiency, energy recovery, regeneration, storage systems, electrified railways, rolling stock, propulsion system, traction power supply.

1 Introduction

Improvement of the energy efficiency of traction systems has been an important subject for industry and research community for decades. There are two main areas of energy consumption in rail systems: traction use and non-traction use. For traction use, some effective methods have been developed. However, the recovery of regenerated power is the most effective solution among the existing approaches due to the high rate of the regenerated energy. Regenerating mode of the traction motors is the most prominent aspect of rail vehicles that could allow recycling of kinetic energy in an electrified network. It means that there is a chance to use the power, regenerated from a decelerating train, for other accelerating trains or return it to a traction feeder through the contact lines. In addition, following the recent development of energy storage devices, storing of regenerated power has become more practical. The devices can be utilised on-board of the rail vehicles and/or in track-side.

2 Electric Propulsion and Regeneration

The function of the power conversion stages within a traction package is to provide suitable smooth power flow to the traction motor. There are four basic configurations for the system that are in common use:

1. DC supply to a suitably modulated voltage for DC motors.
2. DC supply to AC motors via VVVF inverter.

M. Zhu (Ed.): Electrical Engineering and Control, LNEE 98, pp. 69–77.
springerlink.com © Springer-Verlag Berlin Heidelberg 2011

3. AC supply, through an AC-DC converter and a modulated voltage for DC motors.
4. AC supply to AC motors via AC-DC converter, DC link, and VVVF inverter.

Fig. 1 shows a block diagram of the advanced electric traction propulsion. The supply conditioner can be a filter, voltage stepper in DC traction feeders or transformer with converter in AC traction feeders. Any input filters act primarily to decouple the equipment from the supply by establishing sufficient impedance for harmony and train loading against supply. The output of the conditioner is a DC link which must have stable voltage within the defined margins to secure the perfect performance of the drive of the traction motors. Traction choppers and inverters should work in four quadrants: forward and reverse motoring for driving maneuver, and forward and reverse generating for braking maneuver.

Railcars with DC traction motors are equipped with DC-fed choppers in a wide variety of circuit arrangements. A basic arrangement of a DC-fed chopper as a 4Q drive (see Fig. 2) usually has at least two high speed power switching devices: GTO or IGBT. A self commutation semiconductor switch operating at between 200 and 500 Hz, together with flywheel diodes form the basis of the chopping part.

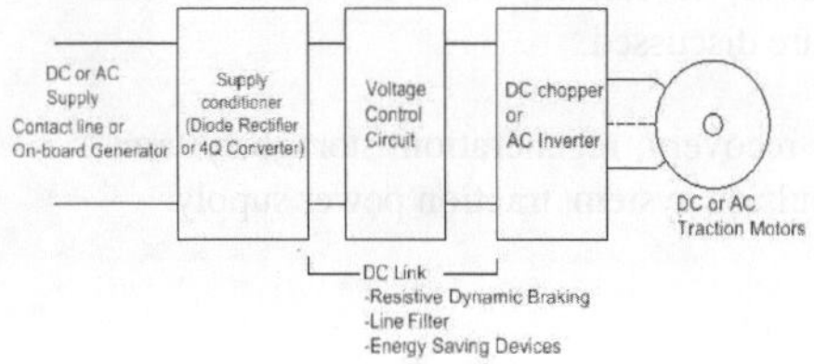

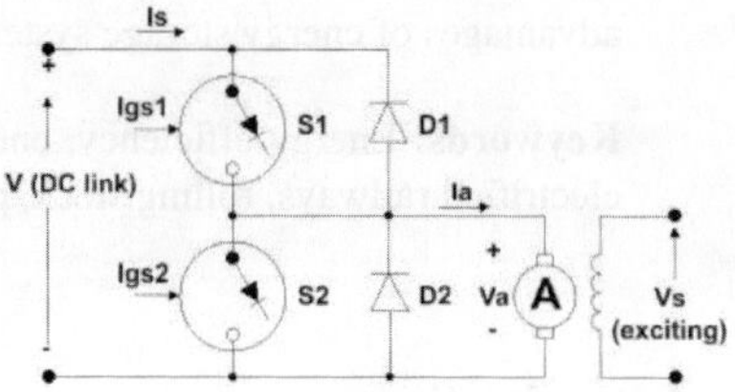

Fig. 1. Typical propulsion blocks of a train [1]. **Fig. 2.** 4Q chopper as DC drive [2].

AC traction motors are induction motors because they are electrically made based on the same design and structure, so that they have the same simplicity. The generating mode of AC traction motor could commence in speeds higher than synchronous speed while the motor is connected to the receptive supply (see Fig. 3). In AC traction vehicles, the three phase voltage source inverter (VSI) with six fast power semiconductor switches such as IGBT and GTO, controlled by space vector direct torque topology, feed the AC traction motors. These inverters (see Fig. 4) are more complex than the DC chopper; however, the operator can get the benefit of full regenerating capability. Regenerating occurs when the load torque becomes greater than the motor's electromechanical torque so the torque summation is negative and puts the motor in the regenerating mode at second quadrant. In Fig. 5, I_s and V are the effective values of the phase current and voltage and Φ is phase angle between them. The Φ>90 results the negative I_s and energy flows from motor to the DC link.

3 Direct Recovery and Regenerative Networks

Electric traction supplies are categorised in two systems: AC and DC. The type of supply is a key factor for the feasibility and operability of recovery of the regenerated power. It is less costly to develop regenerative equipment for DC tractions, yet their ability to accept regenerated power can be limited. Regenerative equipment related to AC traction supplies is more complicated but more receptive to regenerated energy.

Most DC traction power supplies use a three-phase twelve-pulse diode rectifier to produce DC line voltage to feed DC trains with low harmonic and good stability. Returning the regenerated power from electrically braking trains to the traction network raises the DC line voltage, so the feeding system becomes unstable. That leads to the reduction of braking performance and wasting of regenerated power. In these lines, the regenerated power would be efficiently recycled among all vehicles that are operating at the same time within the same network at the effective distance. However, about 40% of the excess energy generated by a braking vehicle is wasted when there are no vehicles in operation in an effective distance to absorb it. To solve this problem, a regenerative inverter and storage devices are used.

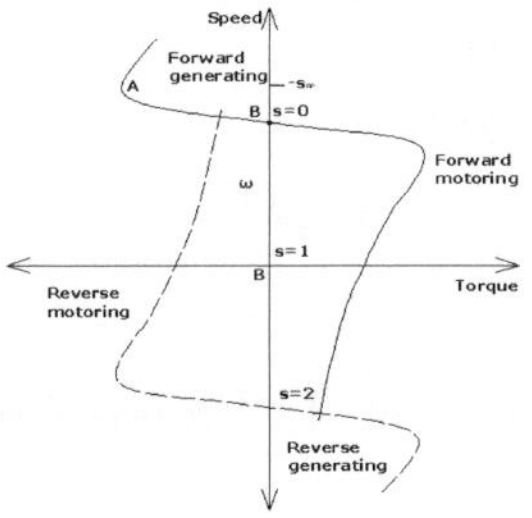

Fig. 3. Generating area of AC induction motors.

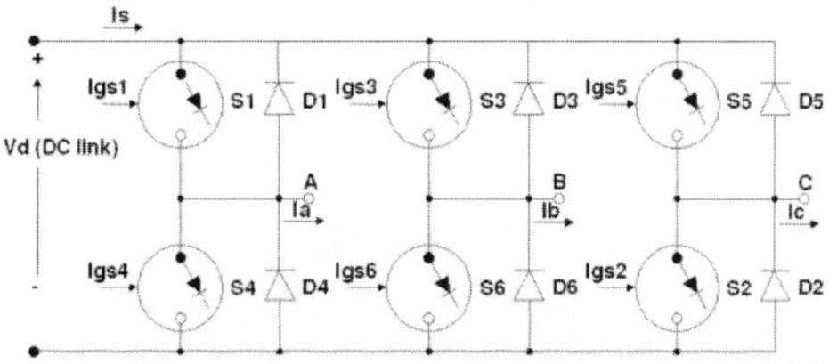

Fig. 4. 4Q AC traction inverter.

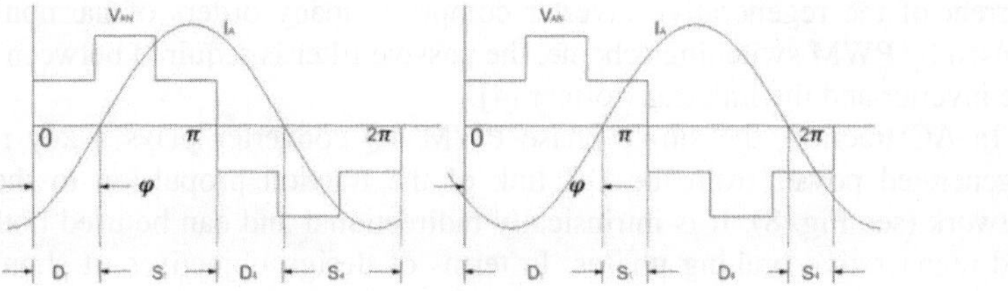

(a) (b)

Fig. 5. Traction motor curves, (a) motoring, (b) generating: *(P=3VIsCosΦ, Is<0 and Φ>90).*

One solution in DC traction is to absorb the surplus regenerated power by installing an inverter in DC substations to deliver this power to the primary side of the transformer as shown in Fig. 6. This inverter is made of IGBT switching devices and works based on the regenerative control algorithm of the three-phase PWM AC/DC converter. It is connected to the high voltage side of the AC supply at substation via link transformer to carry regenerative power from DC feeder lines to AC HV supply. The operation of the DC substation with regenerative inverter is classified in three modes in terms of energy flow (see Fig. 7).

1. Rectification: Voltage of the DC line is lower than no load voltage owing to train load diode rectifier operates.
2. Circulation Current: Circulating current flows between the rectifier and inverter in parallel operation. This mode is to provide fast response and good stability when changing the inversion mode into rectification mode.
3. Inversion: As the DC line voltage is higher than the operation voltage of the inverter, it begins operating.

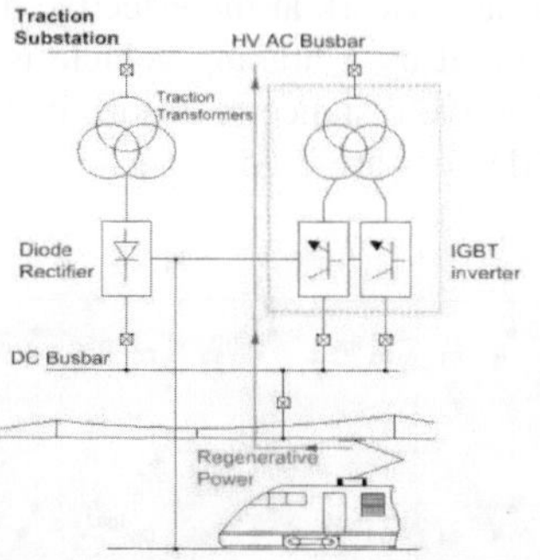

Fig. 6. Regenerative inverter in DC traction.

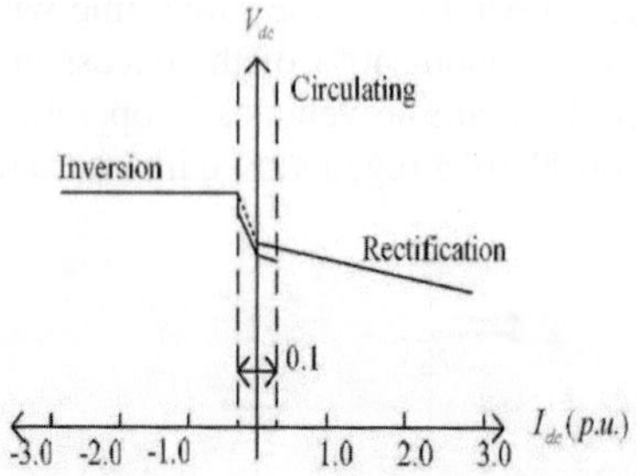

Fig. 7. Modes of DC regenerative substation [3].

The control system of the regenerative inverter detects the rise of the line voltage due to electric braking, and starts to convert regenerated power. In inversion, the output current and the line voltage have a 180° phase difference. Since the output current of the regenerative inverter comprises many orders of harmonic distortions caused by PWM switching scheme, the passive filter is required between the output of the inverter and the link transformer [4].

In AC traction, the single phase PWM 4Q converter plays a key role to return regenerated power from the DC link of the traction propulsion to the AC supply network (see Fig. 8). It is intrinsically bidirectional and can be used both for traction and regenerative braking phases. In terms of design objectives, it should be able to transfer the bidirectional power in motoring and generating modes, correct the power factor and limit the harmonic contents. Its main benefit is to give a nearly sinusoidal line current in both directions of energy flow and mitigate the reactive power drawn from the line.

On the other hand, harmonic is the major drawback of 4Q converters [5, 6]. This would be a massive source of pollution for utility grid energising a traction network. Moreover, one of the most serious problems is the resonance phenomena produced by the interaction of many traction propulsions with 4Q converter operating in the same contact line supply. Interlacing, applying dedicated control logic and using line filter are the solutions to minimise the harmonic of 4Q converter and its impacts. Considering the above advantages and disadvantages, some operators decide not to return regenerated power passing from 4Q converter to supply grid. Fig. 9 presents a typical configuration for interlacing.

4 Storage Devices in Traction Systems

Another contribution to higher energy efficiency of railways can be achieved by intermediate storing of the braking energy in stationary (track-side) or mobile (on-board) energy storage systems [7]. These systems make it possible to have internal interactions of energy within the traction network without requiring it to be returned to the supply grid. Storing of regenerated energy has become important through the development of new technology for storage devices. Different types of storage devices are used in construction of the mobile storage systems in the rail vehicles: batteries, flywheels and double-layer capacitors (Ultracaps).

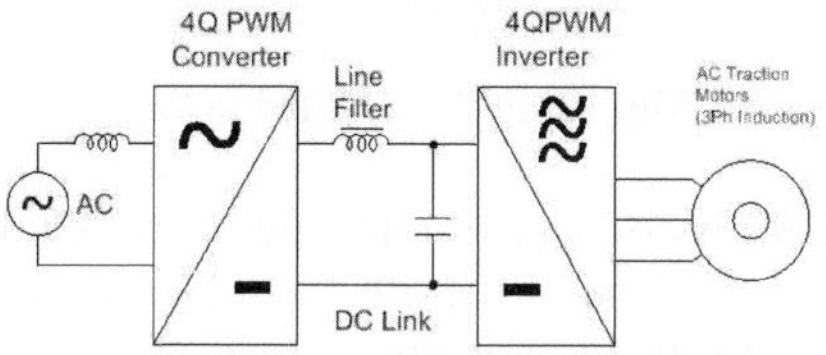

Fig. 8. Traction propulsion for AC supplies.

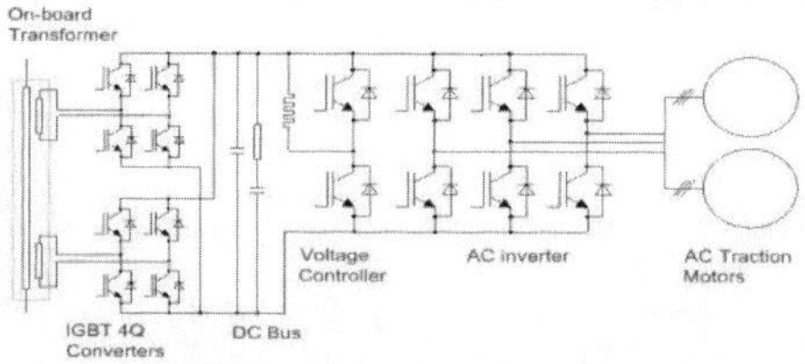

Fig. 9. Interlaced converter.

4.1 On-Board Installations

There are different circuit configurations to install a storage device such as EDLC on-board for various types of traction propulsions and power supplies. A good parallel connection for an ESS is across the DC link before the traction inverter [8, 9]. Conjunction is typically done by a DC/DC chopper matching the wide voltage variations. Fig. 10 shows this configuration for a DC-fed railcar. Basically, the ESS is charged during braking by the regenerated power through inverter when the voltage of the DC link tends to rise up. Afterwards, it will be discharged by the energy demand in the accelerating mode when the voltage of the DC link begins dropping. By providing impedance, ESS assists the voltage of the DC link to reduce jumping up.

Other options have also been introduced by manufacturers for DC supply (see Fig. 11 & Fig. 12) [10].: modular ESS connected outside the traction converter at the input voltage point for better flexibility and a series connection with the main propulsion system to boost its input voltage as well as recycling the energy.

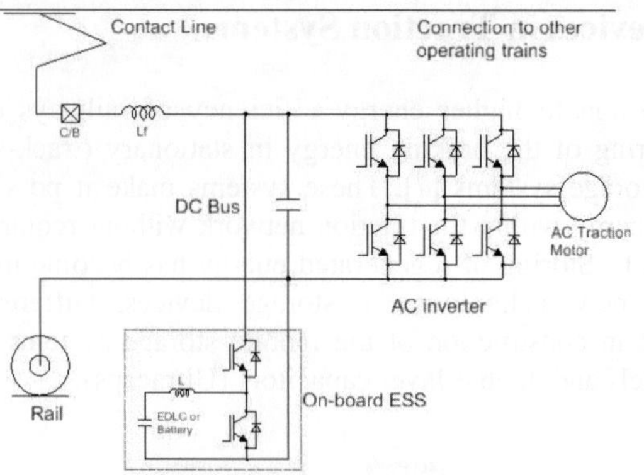

Fig. 10. ESS in DC-fed railcar.

In AC traction, the ESS is connected in DC link between the 4Q converter and the traction inverter (see Fig.13). One of the great achievements obtained by using an on board ESS is the correction of line voltage fluctuations. In Fig. 14, the sums of line vehicle currents are compared with and without on-board ESS with the upper graph. A 50% line current reduction besides a 50% reduction of voltage drop over the line resistance results from the application of on-board ESS as shown in the graph.

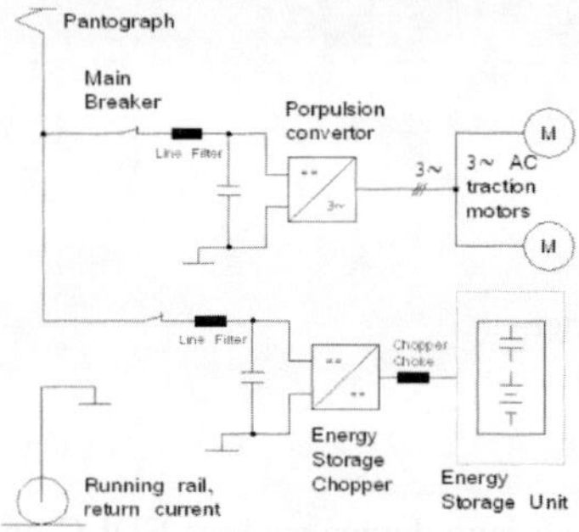

Fig. 11. Independent connection.

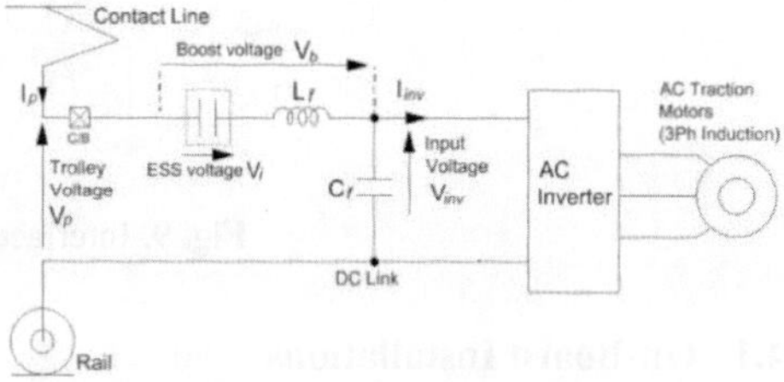

Fig. 12. Series connection, boosting input voltage.

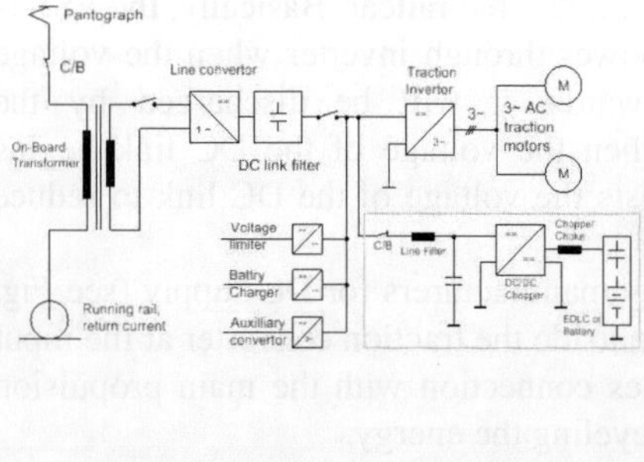

Fig. 13. ESS in AC-fed traction.

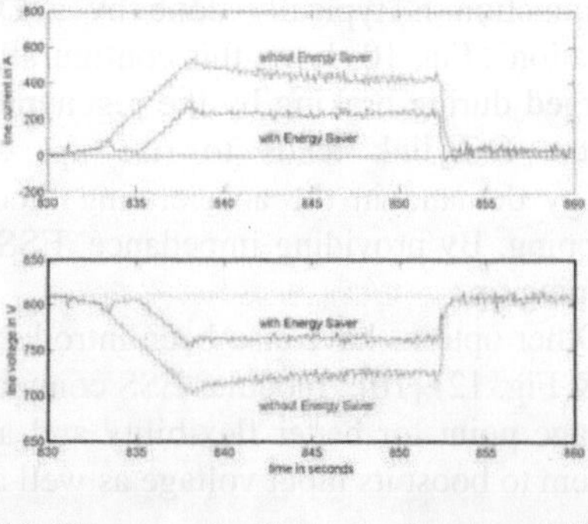

Fig. 14. Current and voltage graph, adapted [8].

4.2 Track Side Applications

Another reliable and more efficient solution to reduce energy wasting in DC traction system is to install ESS in trackside at stationary locations such as stations and/or substations (see Fig.15) [11]. The storage device is connected to the supply network through a power converter and a common DC link. The power converter is a classical bidirectional DC/DC boost converter, similar to the on-board system. Besides the energy saving, this stationary storage system is able to counteract the catenaries voltage variation during operation conditions.

5 Discussions

The methods of recovery of the regenerated energy in electrified railways were discussed. Advantages and disadvantages of the method were presented. Now, it is useful to compare two approaches; returning regenerated energy directly to the power supply and using storage devices. On one hand, recovery of the regeneration in an AC system is simple and operational. No extra cost or equipment in fixed installation and railcars is needed, but harmonic distortion is a serious drawback. In both AC and DC traction systems, 4Q converter is responsible to produce harmonics into the primary AC supply. Filtering is costly in terms of construction and maintenance. On the other hand, the application of ESS in railways has some excellent merits which can make it a desirable approach. Such as:

1. Saving energy and reducing costs by lowering the average peak power.
2. Reduction of voltage fluctuation due to low voltage sags in contact lines.
3. Reduction CO2 emission: less power consumption means less CO2.
4. Possible saving on infrastructure by increasing distance between substations.
5. Possible improving of the operations via more railcars and reducing travel time.
6. Ability to run with downed pantograph in on-board ESS for short distances.

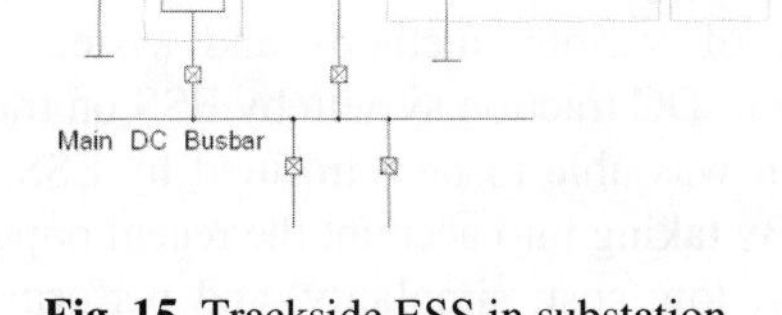

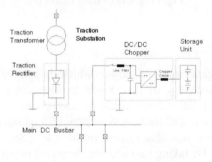

Fig. 15. Trackside ESS in substation.

Besides, using ESS integrated in fixed installation on DC substations or train stations offers significant energy saving achievements (see Fig. 16). For example, Fig. 17 is a bar chart that shows energy saved in the Seoul (South Korea) mass transit system by ESS installed in DC traction substations [12]. Despite these great benefits of ESS, considerable costs of infrastructure, complexity of the system and fragility against short circuit can be some main drawbacks of this method. Table 1 gives a comparison of the described methods, ranking each parameter in quality terms of poor, medium and good.

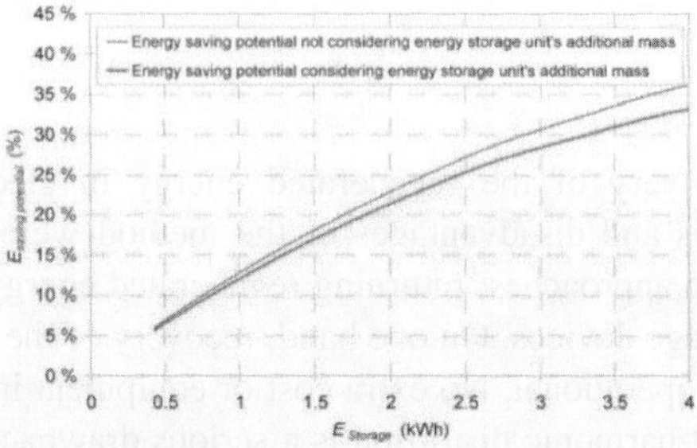

Fig. 16. Energy saving potential [11].

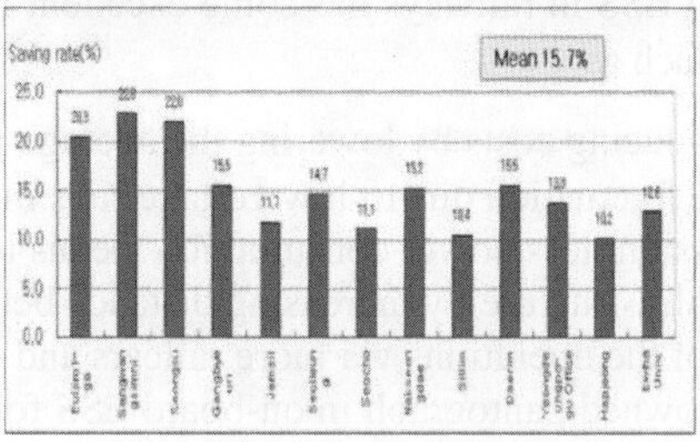

Fig. 17. Energy saving chart [11].

6 Conclusions

We discussed the regeneration of power which is one of the basic aspects of electrified railways. Recovery of this power to save energy and improve system performance by means of various methods and systems was investigated. The recovery of regeneration in DC traction system by ESS on trackside can offer the best result. If the AC traction was able to be retrofitted by ESS, it could be expected to provide similar results. By taking into account the recent popularity of the AC traction because of its reliability, low cost, simplicity, and performance, it can become the most desirable option in terms of energy as well recovery.

Table 1. Comparison of recovery by regeneration methods.

Comparison Parameters	Regeneration Recovery				
	Direct		ESS		
	ACT	DCT	ACT	DCT-OB	DCT-TS
Infrastructure Cost Savings	good	poor	good	good	poor
Train Cost Saving	good	good	poor	poor	good
Train performance	good	good	med	med	good
System performance	poor	poor	med	med	good
Power quality	poor	poor	med	good	good
Energy saving rate	good	med	good	good	good

ACT: AC Traction Supply; DCT: DC Traction Supply;
DCT-OB: DC Traction Supply with On-Board ESS; DCT-TS: DC traction supply with trackside ESS.

References

1. Gunselmann, W.: Technologies for increased energy efficiency in railway systems. In: European Conference on Power Electronics and Applications, p. 10 (2005)
2. Fletcher, R.G.: Regenerative equipment for railway rolling stock. Power Engineering Journal 5(3), 105–114 (1991)
3. Bae, C.H., Han, M.S.: Simulation Study of Regenerative Inverter for DC Traction Substation. In: The Eighth International Conference on Electrical Machines and Systems, ICEMS, vol. 2, p. 1452 (2005)
4. Nomura, J., Kataoka, A., Inagaki, K.: Development of a Hybrid Inverter and a Hybrid Converter for an electric railway. In: Power Conversion Conference, p. 1167. IEEE, Los Alamitos (2007)
5. Bhim, S., Bhuvaneswari, G.: Improved power quality AC-DC converter for electric multiple units in electric traction. In: Power India Conference, p. 6. IEEE, Los Alamitos (2006)
6. Cheok, D., Kawamoto, S.: High power AC/DC converter and DC/AC inverter for high speed train applications. In: Proceedings of TENCON 2000, vol. 1, pp. 423–428 (2000)
7. Wang, X., Yang, Z.: A study of electric double layer capacitors improving electric network voltage fluctuation for urban railway transit. In: Vehicle Power and Propulsion Conference, VPPC 2008, p. 1. IEEE, Los Alamitos (2008)
8. Hase, S., Konishi, T.: Fundamental study on energy storage system for DC electric railway system. In: Power Conversion Conference,, PCC Osaka, vol. 3, p. 1456 (2002)
9. Meinert, M.: New mobile energy storage system for rolling stock. In: 13th European Conference on Power Electronics and Applications, EPE 2009, p. 1 (2009)
10. Takahara, E., Wakasa, T., Yamada, J.: A Study Electric Double Layer Capacitor (EDLC) Application to Railway Traction Energy Saving including Change over between Series and Parallel mode. In: Power Convertion Conference IEEE, PCC Osaka, p. 855 (2002)
11. Romo, L., Turner, D.: Cutting traction power costs with wayside energy storage systems in rail transit systems. In: Rail Conference ASME/IEEE Joint, p. 187 (2005)
12. Gilgong Kim, H.L.: A study on the application of ESS on Seoul Metro Line 2. In: International Conference on Information and technology IEEE, p. 38 (2009)

Prediction of Transitive Co-expressed Genes Function by Shortest-Path Algorithm

Huang JiFeng

College of Information Mechanical and Electrical Engineering
Shanghai Normal University Shanghai 200234, China
jfhuang@shnu.edu.cn

Abstract. The present paper predicted the function of unknow genes by analyzing the co-expression data of *Arabidopsis thaliana* from biological pathway based on the shortest-path algorithm. This paper proposed that transitive co-expression among genes can be used as an important attribute to link genes of the same biological pathway. The genes from the same biological pathway with similar functions are strongly correlated in expression. Moreover, the function of unknown genes can be predicted by the known genes where they are strongly correlated in expression lying on the same shortest-path from the biological pathway. Analyzing the *Arabidopsis thaliana* from the biological pathway, this study showed that this method can reliably reveal function of the unknown *Arabidopsis thaliana* genes and the approach of predicting gene function by transitiving co-expression in shortest-path is feasible and effective.

Keywords: Arabidopsis thaliana, Biological pathway, Shortest-path.

1 Introduction

As more and more genome sequencing projects completed, researchers can obtain a mass of sequence information via the internet. Therefore, it becomes more and more important that effective availing of these sequences predict the functions of unknown genes to guide further experiments. At present, the common method of predicting the function of new gene in the international arena is to make sequences alignment between the gene with unknown function and the genes with known functions in sequence database and find the sequences having high similarity with the gene with unknown function. Finally, the function of unknown gene is predicted through gene with known function that is highly similar with the unknown gene in sequence. However, this approach has the disadvantages of relaying manual operation and low predicting accuracy. The advent of Gene Ontology (GO)[1] reduced the disadvantages of the above approach, which provides a set of standards for gene functional annotation. Gene Ontology offers a semantic framework agreement for the storage, retrieval and analysis of biological data, thus setting a fundament for the interactive operations and mutually understanding the contents among the different database systems. At present, there are only 40% of the genes annotated in *Arabidopsis*

M. Zhu (Ed.): Electrical Engineering and Control, LNEE 98, pp. 79–87.
springerlink.com © Springer-Verlag Berlin Heidelberg 2011

Gene Ontology database. As a result, it is very important to perfect *Arabidopsis* gene ontology database and make annotation for more unknown gene.

Gene annotation is a research focus in the post-genome era . Currently, the common approach on gene annotation is gene clustering based on the similarity of the expression model. The most popular clustering methods include hierarchical clustering[2], K means clustering[3] and self-organizing map[4]. These approaches, using expression data of gene chips to analyze gene function, usually assume that the genes have similar function when they have similar expression structure. In addition, the functions of unknown genes can be deduced by the known genes having the similar expression structure. However, in fact, the genes with similar expression structure might not always have similar function. Firstly, the genes with similar function may not exhibit their similarity on gene. Secondly, the measurement of gene expression similarity is not always accurate. For example, when the Pearson correlation coefficient and Euclidean distance are used to calculate the similarity of gene expression, the relationship between two genes with similar gene expression structure may be not accurately calculated likely because of some reasons of delay[5].

In order to determine the function relationships between genes, the authors proposed an approach different from clustering methods. In recent years, Tapan Mehta et al. found that the genes belonging to the same metabolic pathway have higher co-expression than that do not belong to the same metabolic pathway, namely, gene co-expression in the same metabolic pathway[6].

The methods described by them were designed for the genes in the same metabolic pathway. However, the expression correlation between the genes in the same metabolic pathway may not be very high, because in the same metabolic pathway, high expression correlation does not always exist between the gene pairs. It was proposed in the paper that transitive co-expression among genes can be used as an important attribute to link genes of the same biological pathway.The genes in the same metabolic pathway are constructed into an undirected weighted graph, and then find the shortest path in the graph. Based on statistical analysis of genes with known functions, it was found that the genes with similar functions had high expression correlation. Therefore, for the genes in the same shortest, the function of unknown genes can be predicted through the known genes if high expression correlation exists among them. This approach has been verified in yeast gene[7]. By analysis of *Arabidopsis* anther gene in the same metabolic pathway and referencing *Arabidopsis* existing Go annotation, the authors predicted the functions of part of anther unknown genes and again verified the feasibility of predicting gene function by the method of calculating the shortest path in the metabolic pathway.

2 Data and Methods

2.1 Data

Gene Chip Data. The *Arabidopsis* Information Resource (TAIR) provides resources for all aspects of *Arabidopsis* thaliana[8]. All the *Arabidopsis* metabolic pathway data were downloaded from the Internet `ftp://ftp.arabidopsis.org/home/tair/pathways/`. Most of the *Arabidopsis* gene chip data were collected in ATTED-II (http://atted.jp/), where these chips data were integrated for analysis to obtain related data of gene co- expression among *Arabidopsis* genes[9]. All the *Arabidopsis* genes co-expression data were downloaded for study in the paper.

GO Database. GO, the abbreviation of Gene Ontology, is a collaborative project developed by Gene Ontology Consortium. The goals of GO are to design a structured, precisely defined universal thesaurus, which can be used to describe genes and the actions of gene products in any organisms. So far, GO database built three mutually independent ontologies of biological processes, molecular functions and cellular components. Now GO has become an extremely important methods and tools in the field of bioinformatics for gene function annotation, revealing and integrating biological data and databases, and the establishment of biological association among data, and so on. Usually, if two gene products have similar function, the annotation terms in GO will be similar. Therefore, this article referred the functions annotation in *Arabidopsis* GO database to speculate the functions of unknown genes. TAIR website `ftp://ftp.arabidopsis.org/home/tair/ontologies/` offers existing *Arabidopsis* genes GO annotation[10]. All the *Arabidopsis* genes GO annotations were downloaded for analysis in the study.

2.2 Methods

Transitive Co-Expression. Transitive co-expression means two genes with low expression correlation can transitively co-express if both of the genes have a high expression correlation with another gene. For example, b gene has a high expression correlation with a gene and c gene respectively. However, the expression correlation between a gene and c gene is not high. Under this circumstance, it is described as that a gene and c gene can transitively co-express through b gene, and b gene is called as transitive gene. The gene pair with strong expression correlation is linked by an edge, which is called as similar biological path. The relationship between genes increases as the decrease of edge length. Therefore, the shortest similarity biological path can accurately expressed function relationship between genes. If two known genes connect each other through a similar biological path, the function of transitive gene can be predicted through the two known genes.

Graph Constructing. The genes in the same metabolism pathway are constructed into an undirected and weighted graph. The vertexes of the graph represent the genes and the edges represent gene expression correlation. The genes having high expression correlation are connected by an edge. Through statistical analysis of gene expression correlation coefficient in the *Arabidopsis thaliana* metabolic pathways, it is found that gene expression correlation is quite high when the expression correlation coefficient between genes ca, $b > 0.6$. Therefore, the genes with ca, $b > 0.6$ are connected with an edge, which could remove some of the data and retain large number of genes pairs. Edge length of da,b is calculated as the following formula.

$$d_{a,b} = f(c_{a,b}) = (1 - c_{a,b})^k \quad (1)$$

In the formula, parameter of k is the power factor, which is used to strengthen the difference between the high and low of genes correlation. Because the length of a shortest path is sum of each edge, the shortest path can cover transitive genes as much as possible by increasing the difference in edge length. Experiments has verified that when the number of transitive genes is relatively stable (For detailed results, visit the website of `www.biostat.harvard.edu/complab/sp/`. As a result, the parameter of was selected in the study. Through analysis of genes in the shortest path, it is found that when the length of the shortest path is greater than 0.008, functional similarity between genes begin to decrease. Therefore, in order to ensure the accuracy of the shortest path, it is believed when the total length of the shortest path is greater than 0.008, the genes do not transfer function among them. Consequently, in order to ensure the quality of the shortest path method, only the paths with the length shorter of 0.008 are considered in the study.

Algorithm of Shortest Path. For the objective is to predict unknown gene function, the shortest path between any nodes can be calculated. Floyd algorithm was adopted it the study to calculate the shortest path [11]. Floyd algorithm, adopting the idea of dynamic programming, solves any of the shortest paths between two points through breaking down the problem into sub-problems. Setting $G = (V, E, W)$ is a weighted graph, its edges are $\nu = \{\nu_1, \nu_2, \ldots, \nu_n\}$. For $k < n$, the subset of node V, $V_k = \{\nu_1, \nu_2, \ldots, \nu_n\}$, is taken into account. For any two nodes of ν_i and ν_j in V, considering the interval nodes between ν_i and ν_j are all the paths in ν_k, $p_{i,j}^{(k)}$ is assumed the shortest among them and the path length of $p_{i,j}^{(k)}$ is assumed as $d_{i,j}^{(k)}$ in the study. If the node ν_k is not the shortest path from ν_i to ν_j, then $p_{i,j}^{(k)} = p_{i,j}^{(k-1)}$. Otherwise $p_{i,j}^{(k)}$ can be divided into two sections, a section from ν_i to ν_k, another paragraph from ν_k to ν_j. By this, the formula of $p_{i,j}^{(k)} = p_{i,k}^{(k-1)} + p_{k,j}^{(k-1)}$ can be acquired. The above discussions can be summarized as the following recursive formula.

$$d_{i,j}^{(k)} = \begin{cases} \omega(\nu_i, \nu_j) & \text{if} k = 0 \\ min\{d_{i,j}^{(k-1)}, d_{i,k}^{(k-1)}, d_{k,j}^{(k-1)}\} & \text{if} k \geq 1. \end{cases} \quad (2)$$

The original problems is transformed to calculate $d_{i,j}^{(n)}$ for each i and j, or calculate the matrix of $D^{(n)} = (d_{i,j}^{(n)})$.

3 Experiment and Results

3.1 Experiment Steps

Experimental flow chart is as Fig.1.

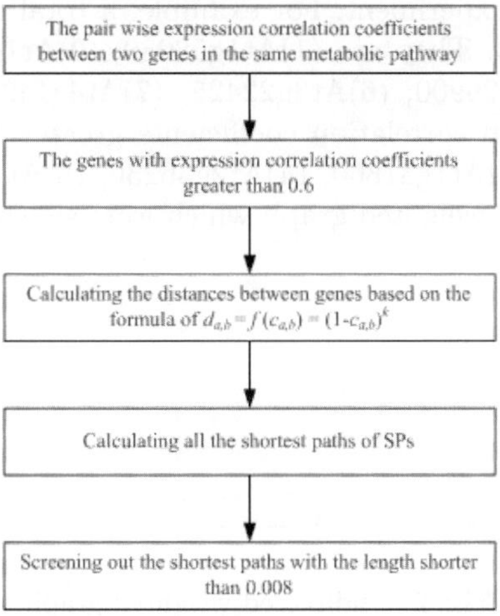

Fig. 1. Flow of the experiment

For the data in every *Arabidopsis* metabolic pathway downloaded from internet, the pair wise expression correlation coefficients between genes are calculated by writing Perl programs.

According to graph constructing methods, the genes with correlation coefficient of $c_{a,b} > 0.6$ are further screened out. The perl programs are written to filter out genes meeting our experimental conditions.

According to the formula of $d_{a,b} = f(c_{a,b}) = (1 - c_{a,b})^k$, the distances of $d_{a,b}$ between genes screened in the previous step are calculated by Perl programs.

The shortest paths in every metabolic pathway are calculated with Floyd algorithm by writing c++ programs. The shortest path that is shorter than 0.008 in length is screened out by the Perl program.

3.2 Experimental Results Validation

First, a total of 175 metabolic pathway data were downloaded from TAIR Website `ftp://ftp.arabidopsis.org/home/tair/pathways/plantcyc_dump`. In which, 8 metabolic pathway where genes of Atmg00280,Atcg00490, Atcg00420, Atmg00285, Atmg01320, Atcg01250, Atcg00890, Atcg00430, Atmg01280,Atmg00 220,Atcg00500,Atcg00460 located were removed. In the remaining 166 pathways,

103 metabolic pathways with the expression correlation coefficients greater than 0.6 between genes were filtered out. The study focused on the shortest paths in the 103 metabolic pathways.

The experimental method is based on transitive co-expression in the shortest path to predict the function of unknown genes in the same metabolic pathway. This method has been used for yeast gene function prediction [7].Here this approach we have validated was feasible for *Arabidopsis* anther gene function prediction in the experiment. For example, a total of 8 genes are in histidine synthesis pathway. They are (1)At1g58080, (2)At1g09795, (3)At1g31860, (4)At2g36230, (5)At4g26900, (6)At3g22425, (7)At4g14910 and (8)At5g63890. 4 genes with expression correlation coefficients greater than 0.6 were selected among the 8 genes, (3)At1g31860, (4)At2g36230, (5)At4g26900,(8)At5g63890, to build the undirected weighted graph which were shown in Fig.2.

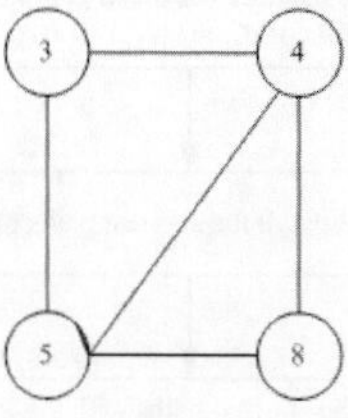

Fig. 2. Undirected weighted graph

Where the shortest path from gene (3)At1g31860 to (4)At2g36230 is (3)At1g 31860→(5)At4g26900→(4)At2g36230. The Arabidopsis GO annotations were downloaded from TAIR Website `ftp://ftp.arabidopsis.org/home/tair/`. which showed the function of (3)At1g31860, (4)At2g36230, (5)At4g26900 genes were exactly the same and from the chloroplast. This result proved the assumption was correct that in the same metabolic pathway the genes with high correlation had similar function. The high expression map of glucosinolate biosynthesis in phenylalanine metabolism pathway was shown in Fig.3.

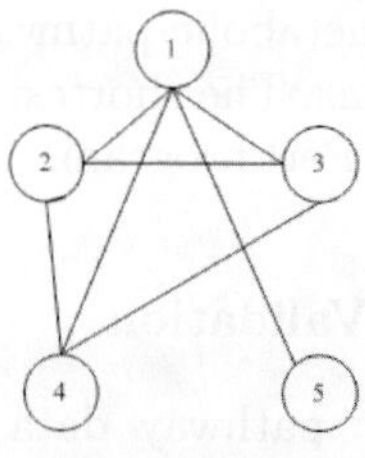

Fig. 3. High-expression of phenylalanine metabolic pathways

This metabolic pathway includes the genes of (1)At2g20610, (2)At1g74100, (3)At1g24100, (4)At4g31500, (5)At4g13770 and (6)At5g05260. Except for gene (6)At5g05260, the other five gene has a high expression correlation. The shortest paths were calculated based on the graph constructed by the five high expression genes. Of which, the shortest path from (4)At4g31500 to (5)At4g13770 was (4)At4g31500→(1)At2g20610→(5)At4g13770. Actually, the functions of gene (4) At4g31500 and gene (1)At2g20610 in the GO annotation are all membrane. The function of gene (5)At4g13770 is the endometrium. The result indicated genes' functions are very similar in the shortest path.

Experimental results validated genes' functions in the shortest path are the same or highly similar and it is feasible to predict gene function in *Arabidopsis* metabolic pathway by this method. Consequently, the function of unknown genes can be predicted through the function of known gene in the shortest path.

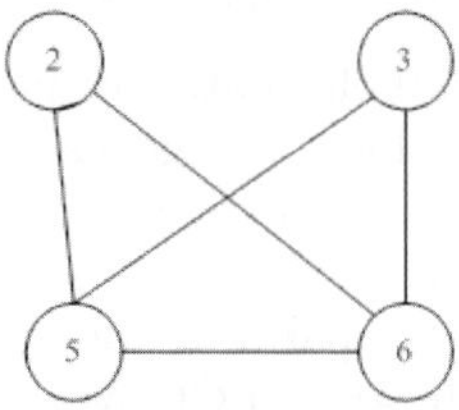

Fig. 4. The related genes for the high-expression of citric acid under the metabolic pathways of the synthesis of acetyl coenzyme A

The genes with high expression in the citric acid Coenzyme A synthesis of metabolic pathway was shown in Fig.4. There are 6 genes in the citric acid coenzyme A synthesis metabolic pathway. They are (1)At1g09430, (2)At3g06650, (3)At1g60810, (4)At2g44350, (5)At5g49460 and (6)At1g10670 respectively. As it was shown in Fig.4, the expressions of (2)At3g06650, (3)At1g60810, (5)At5g49460 and (6)At1g10670 gene are highly related. In the calculated shortest path, the shortest path from At3g06650 gene to (6)At1g10670 gene is (2)At3g06650→(5)At 5g49460→ (6)At1g10670. In the GO annotation, gene (2)At3g06650 and gene (6)At1g10670 are all citric acid lyase, and gene (5)At5g49460 is unannotated gene. According to the assumption that the genes with similar expression in the shortest path have similar function, it can be predicted that (5)At5g49460, (2)At3g06650 and (6)At1g10670 have similar function.

4 Experimental Results and Analysis

4.1 Parameter Settings

In the formula of calculating distances between the nodes in the undirected weighted graph, the methods for the setting of parameter k and selecting the shortest path the length with the length shorter than 0.008 were taken according to the reference of [7]. For the gene with $c_{a,b} > 0.6$, it was believed that had a high expression correlation. This criteria was set just based on the statistical analysis of gene expression correlation coefficient in the metabolic pathway. Among the downloaded 177 metabolic pathways, except the genes in photosynthesis metabolism pathway which average expression correlation coefficient was higher than the 0.8, the average expression correlation coefficient for genes in other *Arabidopsis* metabolism pathway were under 0.6, which indicated that the gene expressions were highly related.

4.2 Biological Significance

According to Tapan Mehta's research, the genes belonging to the same metabolic pathway have a higher capacity of co-expression than the genes that does not belong to the same metabolic pathway. While the experiment is based on the assumption that the genes with high co-expression have similar function, so selecting the genes in metabolic pathway for analysis makes experimental data to be certain degree of reliability.

The genes in the same metabolic pathway participate in the same biological process. In the view of biology, the genes participating in the same biological process likely have the same or similar function, which makes the experiment might hypothesize have biological significance. For example, there are 27 genes in the metabolic pathway of photosynthesis. In which, the genes of At1g08380, At1g30380, At1g31330, At4g12800, At5g64040, At4g02770, At3g21055, At4g28660, At1g44575, At5g66570, and At1g79040 have no GO annotations. The genes of AT4G05180, AT4G21280, and AT1G06680 were annotated as oxygen complex. The GO annotation of AT2G06520, AT2G30570, and AT1G67740 was chloroplast thylakoid membrane. The remaining 10 genes OF At2g20260, At1g03130, At1g52230, At1g55670, At3g16140, At4g28750, At2g46820, At1g14150, At3g01440, and At3g50820 were annotated as chloroplast by GO. Based on the GO annotations of these genes, it can be concluded that the genes in the same metabolic pathway have the same or highly similar functions. As a result, the prediction results in the experiment possessed a certain degree of reliability. The information acquired in the study can provide valuable references for further improving the *Arabidopsis* GO annotation. Forty percent of the genes in biochemical pathways downloaded from TAIR website have no gene annotation, so this method has some practical significance.

5 Discussions

High co-expression of genes in the metabolic pathway of *Arabidopsis* were chose to construct undirected weighted graph, which was used to calculate the shortest paths between genes. The function of unknown gene could be predicted by known gene with high transitive co-expressing in the shortest path. The study proved that this approach has a certain degree of reliability. In addition, compared with the traditional approach of clustering analysis of genes by assuming that the genes with similar expressing structure have similar function, the new approach has certain of feasibility. However, there are some shortcomings for this approach. Firstly, only the genes with highly relative expression were selected for analysis, which makes the function of some genes can not be predicted. Secondly, since the data in the *Arabidopsis* GO database is limited at present, it makes the approach is absent of reference in particle application. The accuracy of function prediction for unknown genes in *Arabidopsis* metabolic pathway will be improved when combining with other prediction methods.

References

1. The Gene Ontology Consortium.Gene Ontology: Tool for the unification of biology. Nature Genet. 25, 25–29 (2000)
2. Eisen, M.B., Spellman, P.T., Brown, P.O., et al.: Cluster analysis and display of genome-wide expression patterns. Proc. Nat. Acad. Sci. USA 95, 14863–14868 (1998)
3. Tavazoie, S., Hughes, L.D., Campbell, M.J., et al.: Systematic determination of enetic network architure. Nature Genet. 22, 281–285 (1999)
4. Golub, T.R., Slonim, D.K., Tamayo, P., et al.: Molecular classification of cancer: Class discovery and class prediction by gene expression monitoring. Science 286, 531–537 (1999)
5. Qian, J., Dolled-Filhart, M., Lin, J., et al.: Beyond synexpression realtionships:local clustering of time-shifted and inverted gene expression profiles identified new biologically relevant interactions. Journal of Molecular Biology 314, 1053–1066 (2001)
6. Srinivassainagendra, V., Page, G.P., Mehta, T., et al.: Cress express: a tool for large-scale mining of expression data from *Arabidopsis*. Plant Physiology 147(7), 1004–1016 (2008)
7. Zhou, X.H., Kao, M.C.J., Wong, W.H.: Transitive function annotation by shortest-path analysis of gene expression data. PNAS 90(20), 12783–12785 (2002)
8. Swarbreck, D., Wilks, C., Lamesch, P., et al.: The *Arabidopsis* Information Resource(TAIR): Gene structure and function annotation. Nucleic. Acids Res. 36, 1–6 (2007)
9. Obayashi, T., Hayashi, S., Saeki, M., et al.: ATTED-II provides coexpressed gene networks for *Arabidopsis*. Nucleic. Acide Res. 37, 987–991 (2009)
10. Berardini, T.Z., Mundodi, S., Reiser, R., et al.: Functional annotation of the Arbidopsis genome using controlled vocabularies. Plant Physiol. 135(2), 1–11 (2004)
11. Floyd, R.W.: Algorithm 97:Shortest Path. Communication of the ACM 5(6), 345 (1997)

approach is absent of reference in particle application. The accuracy of function prediction for unknown genes in *Arabidopsis* metabolic pathway will be improved when combining with other prediction methods.

References

1. The Gene Ontology Consortium: Gene Ontology: Tool for the unification of biology. Nature Genet. 25, 25–29 (2000)
2. Eisen, M.B., Spellman, P.T., Brown, P.O., et al.: Cluster analysis and display of genome-wide expression patterns. Proc. Natl. Acad. Sci. USA 95, 14863–14868 (1998)
3. Tavazoie, S., Hughes, J.D., Campbell, M.J., et al.: Systematic determination of genetic network architecture. Nature Genet. 22, 281–285 (1999)
4. Golub, T.R., Slonim, D.K., Tamayo, P., et al.: Molecular classification of cancer: Class discovery and class prediction by gene expression monitoring. Science 286, 531–537 (1999)
5. Qian, J., Dolled-Filhart, M., Lin, J., et al.: Beyond synexpression relationships: local clustering of time-shifted and inverted gene expression profiles identified new, biologically relevant interactions. Journal of Molecular Biology 314, 1053–1066 (2001)
6. Srinivasasainagendra, V., Page, G.P., Mehta, T., et al.: Cress express: A tool for large-scale mining of expression data from Arabidopsis. Plant Physiology 147(3), 1004–1016 (2008)
7. Zhou, X.J., Kao, M.C.J., Wong, W.H.: Transitive function annotation by shortest-path analysis of gene expression data. PNAS 99(20), 12783–12788 (2002)
8. Swarbreck, D., Wilks, C., Lamesch, P., et al.: The *Arabidopsis* Information Resource (TAIR): Gene structure and function annotation. Nucleic Acids Res. 36, 1–6 (2007)
9. Obayashi, T., Hayashi, S., Saeki, M., et al.: ATTED-II provides coexpressed gene networks for *Arabidopsis*. Nucleic Acids Res. 37, 987–991 (2009)
10. Berardini, T.Z., Mundodi, S., Reiser, R., et al.: Functional annotation of the Arabidopsis genome using controlled vocabularies. Plant Physiol. 135(2), 1–11 (2004)
11. Floyd, R.W.: Algorithm 97: Shortest Path. Communication of the ACM 5(6), 345 (1962)

Fast Implementation of Rainbow Signatures via Efficient Arithmetic over a Finite Field

Haibo Yi, Shaohua Tang*, Huan Chen, and Guomin Chen

School of Computer Science & Engineering, South China University of Technology, Guangzhou, China
haibo.yi87@gmail.com, shtang@IEEE.org, huangege@qq.com, sarlmolapple@gmail.com

Abstract. An efficient hardware implementation of Rainbow signature scheme is presented in this paper. It introduces an effective way to accelerate the generation of multivariate signatures by using optimized arithmetics including multiplication, multiplicative inverse and Gaussian elimination over finite fields. Not only the speed but also the area are considered in the design. 27 parallel multipliers are adopted and the design has been fully implemented on a low-cost Field Programmable Gate Array. Compared with other public key implementations, the proposed implementation with 15490 gate equivalents and 2570 clock cycles has better performance. The cycle-area product of this implementation shows that it is suitable for fast multivariate signature generation in the resource-limited environments, e.g.smart cards.

Keywords: Multivariate Public Key Cryptosystems (MPKCs), Field Programmable Gate Array (FPGA), digital signature, Rainbow, finite field.

1 Introduction

Since many cryptographic schemes used to protect electronic information are based on the near impossibility of factoring large numbers, building a working quantum computer would be not only a mathematical coup but a cryptographic one as well, potentially putting much of the world's most secret information in jeopardy. Besides, an algorithm that Peter Shor discovered can be used to conquer RSA, DSA and ECDSA in polynomial time. Multivariate Public Key Cryptosystems (MPKCs) [1] are the cryptosystems which can resist the attacks by quantum computation. They are based on the difficulty of the problem of solving multivariate quadratic equations over a finite field, which is considered to be NP-hard. Oil-Vinegar signature schemes are a group of MPKCs. They can be grouped into three families: balanced Oil-Vinegar, unbalanced Oil-Vinegar and Rainbow [1], a multilayer construction using unbalanced Oil-Vinegar at each layer. The efficiency of signature generation is one of the focuses of many

* Corresponding Author.

M. Zhu (Ed.): Electrical Engineering and Control, LNEE 98, pp. 89–96.
springerlink.com © Springer-Verlag Berlin Heidelberg 2011

researchers. At present, there are some methods to optimize multivariate signature schemes, e.g.TTS [2], [3], some instances of MPKC [4], and Rainbow scheme [5].

In this paper, we show how to implement Rainbow signatures via efficient arithmetic over a finite field. Not only its speed but also its implemented area are our focuses. The efficiency of Rainbow signature generation relies heavily on the arithmetics over finite fields. The main contribution of this paper is to propose an efficient hardware implementation for Rainbow signature generation via using optimized arithmetics. The design is fully implemented on a low-cost FPGA. Experimental results show that our design is suitable for resource-limited environments.

The structure of this paper is as follows: In Section 2, an overview of Rainbow signature scheme is presented. In Section 3, three basic operations in Rainbow and how to optimize these operations are presented. The proposed design is fully implemented on a low-cost FPGA and compared with other public key signature schemes in Section 4. Conclusions and discussions are summarized in Section 5.

2 Overview of Rainbow Signature Scheme

Rainbow scheme belongs to the class of mixed schemes under MQ constructions. The scheme consists of a quadratic system of equations involving Oil variables and Vinegar variables that is solved iteratively. Such an Oil-Vinegar polynomial over these variables can be represented by the form

$$\sum_{i\in O_l, j\in S_l} \alpha_{ij}x_ix_j + \sum_{i,j\in S_l} \beta_{ij}x_ix_j + \sum_{i\in S_{l+1}} \gamma_i x_i + \eta. \quad (1)$$

O_i is a set of Oil variables in the the ith layer and S_i is a set of Vinegar variables in the the ith layer. The number of Vinegar variables used in the ith layer of signature construction is denoted by v_i. The number of Oil variables used in the ith layer of signature construction is denoted by $o_i = v_{i+1} - v_i$. The hardware implementation of Rainbow signature generation is the focus in this paper. Therefore, a brief introduction of signature generation is presented in the following section.

2.1 Signature Generation

The message is defined by $Y = (y_1, ..., y_{n-v_1}) \in k^{n-v_1}$. The message size is fixed. Thus we use a hash function to produce a fixed-size one-way hash value of the document. After that, it is crucial to find a solution of the equation $\bar{F} = L_1 \circ F \circ L_2 = Y$. In fact, ${L_1}^{-1}$ and ${L_2}^{-1}$ are used during the signature generation. Firstly, we should compute $\bar{Y'} = {L_1}^{-1}(Y)$. Next the equation $Y' = F$ is needed to be solved. In the first layer, the v_1 Vinegar variables in the Oil-Vinegar polynomials are randomly chosen. Therefore, these polynomials are substituted into a set of linear equations of Oil variables. If these equations have no solution, a new set of Vinegar variables should be chosen. The Vinegar variables in the second layer of Rainbow are all the variables in the first layer. The equations in the first layer are solved to find the values of all the variables. Thus the values of Vinegar variables are known now. This procedure for each successive layer is repeated until the solutions of all the equations are found. If these equations have no solution, we have to return to the first layer and choose a new set of Vinegar variables. In this step, the values $\bar{X} = (\bar{x_1}, ..., \bar{x_n})$ of all the variables are found. Finally, compute $X' = {L_2}^{-1}(\bar{X}) = ({x_1}', ..., {x_n}')$. Then X' is the signature of Y.

2.2 Design Parameters of Rainbow

The parameters of Rainbow we choose is shown in Table 1. Basic operations in Rainbow scheme include multiplication, multiplicative inverse and Gaussian elimination over the finite field. Besides, it is also important to choose the primitive polynomial, because primitive polynomial determines the structure of the finite field. Also, choosing different primitive polynomial can affect the efficiency of signature generation.

Table 1. A Rainbow structure.

Parameter	Rainbow
Ground field size	$GF(2^8)$
Message size	27 bytes
Signature size	37 bytes
Number of layers	4
Set of Vinegar variables	(10, 20, 24, 27, 37)

3 Acceleration of Fundamental Arithmetic of Rainbow

3.1 Choice of the Primitive Polynomial

Each $GF(q)$ has more than one primitive element, and each element except zero over $GF(q)$ can be expressed by the power of the primitive element. It is important that every binary number which is longer than 8 should be computed modulo the primitive polynomial, therefore how to choose primitive polynomial is related to the efficiency of implementation.

Primitive polynomial over $GF(2^8)$ can be expressed as the form: $x_8 + x_k + ... + 1 (0 < k < 8)$. It has been proofed that the fewer the number if coefficients valued one, the more efficient the operations over the finite field will be. There are 16 polynomials can be chosen as a primitive polynomial. Among these polynomials, the number of polynomials which have five terms is 12 and the number of polynomials which have seven terms is 4. Therefore, the optimizations of multiplication, multiplicative inverse and Gaussian elimination are based on these 16 polynomials.

3.2 Optimization of Multiplication over the Finite Field

Actually multiplication over finite fields is modular multiplication. Multiplication is the most time consuming operation in implementing Rainbow. Because it is used a lot in the matrix multiplications and Gaussian elimination. It is crucial to improve the efficiency of multiplication over $GF(2^8)$ to accelerate multivariate signatures [6].

Table 2. Efficiency of the multiplication.

Polynomial	*AND*	*XOR*	Logic units	Delay (ns)
100011101	64	76	55	19.542
101110001	64	77	53	18.426
100101011	64	81	51	18.852
110101001	64	79	55	19.610
100101101	64	71	49	17.647
101101001	64	85	52	20.287
101100011	64	79	47	19.473
110001101	64	79	50	19.797
111001111	64	82	48	21.431
111100111	64	79	50	21.975
101001101	64	81	50	18.712
101100101	64	80	51	20.451
101011111	64	78	50	24.251
111110101	64	79	53	23.768
110000111	64	81	51	18.481
111000011	64	81	51	17.566

Suppose $a(x)$ and $b(x)$ are both elements in $GF(2^8)$ and $c(x) = a(x) \otimes b(x) \bmod f(x)$. And $f(x)$ is the irreducible polynomial. Then we have

$$s_0 = a_0b_0, s_1 = a_1b_0 \oplus a_0b_1, s_2 = a_2b_0 \oplus a_1b_1 \oplus a_0b_2, \tag{2}$$

$$s_3 = a_3b_0 \oplus a_2b_1 \oplus a_1b_2 \oplus a_0b_3, s_4 = a_4b_0 \oplus a_3b_1 \oplus a_2b_2 \oplus a_1b_3 \oplus a_0b_4, \tag{3}$$

$$s_5 = a_5b_0 \oplus a_4b_1 \oplus a_3b_2 \oplus a_2b_3 \oplus a_1b_4 \oplus a_0b_5, \tag{4}$$

$$s_6 = a_6b_0 \oplus a_5b_1 \oplus a_4b_2 \oplus a_3b_3 \oplus a_2b_4 \oplus a_1b_5 \oplus a_0b_6, \tag{5}$$

$$s_7 = a_7b_0 \oplus a_6b_1 \oplus a_5b_2 \oplus a_4b_3 \oplus a_3b_4 \oplus a_2b_5 \oplus a_1b_6 \oplus a_0b_7, \tag{6}$$

$$s_8 = a_7b_1 \oplus a_6b_2 \oplus a_5b_3 \oplus a_4b_4 \oplus a_3b_5 \oplus a_2b_6 \oplus a_1b_7, \tag{7}$$

$$s_9 = a_7b_2 \oplus a_6b_3 \oplus a_5b_4 \oplus a_4b_5 \oplus a_3b_6 \oplus a_2b_7, \tag{8}$$

$$s_{10} = a_7b_3 \oplus a_6b_4 \oplus a_5b_5 \oplus a_4b_6 \oplus a_3b_7, s_{11} = a_7b_4 \oplus a_6b_5 \oplus a_5b_6 \oplus a_4b_7, \tag{9}$$

$$s_{12} = a_7b_5 \oplus a_6b_6 \oplus a_5b_7, s_{13} = a_7b_6 \oplus a_6b_7, s_{14} = a_7b_7. \tag{10}$$

The first step is to compute the coefficients. Then the next step, the product should be computed modulo the primitive polynomial. In the case of a primitive polynomial 101001101, we have

$$c_7 = s_7 \oplus s_9 \oplus s_{11} \oplus s_{12}, c_6 = s_6 \oplus s_8 \oplus s_{10} \oplus s_{11}, \tag{11}$$

$$c_5 = s_5 \oplus s_{10} \oplus s_{11} \oplus s_{12} \oplus s_{14}, c_4 = s_4 \oplus s_9 \oplus s_{10} \oplus s_{11} \oplus s_{14} \oplus s_{14}, \tag{12}$$

$$c_3 = s_3 \oplus s_8 \oplus s_9 \oplus s_{10} \oplus s_{12} \oplus s_{13} \oplus s_{14}, c_2 = s_2 \oplus s_8 \oplus s_{13} \oplus s_{14}, \tag{13}$$

$$c_1 = s_1 \oplus s_9 \oplus s_{11} \oplus s_{13} \oplus s_{14}, c_0 = s_0 \oplus s_8 \oplus s_{10} \oplus s_{12} \oplus s_{13}. \tag{14}$$

For the purpose of functional simulation and verification, we use the Quartus II Version 8.0. For the FPGA implementation, we choose the ALTERA Cyclone FPGA family. The data in Table 2 is tested on the device of EP1C12Q240C8 in Quartus II.

3.3 Optimization of Multiplicative Inverse over the Finite Field

Multiplicative inverse is used in Gaussian elimination to find the inverse element of a pivot element. According to our design, we use a parallel multiplicative inverse based on Fermat's theorem [7]. Suppose β is an element in $GF(2^8)$. According to Fermat's theorem, we have

$$\beta^{2^8} = \beta, \beta^{-1} = \beta^{2^8-2} = \beta^{254}, \tag{15}$$

$$2^8 - 2 = 2 + 2^2 + 2^4 + ... + 2^{8-1}, \beta^{-1} = \beta^2 * \beta^4 * \beta^8 ... \beta^{2^{8-1}}. \tag{16}$$

Computing $\beta^2, \beta^4, \beta^{16}, \beta^{32}, \beta^{64}$ and β^{128} are parallel. The data in Table 3 is tested on the device of EP1C12Q240C8. The efficiency of multiplicative inverse mainly depends on multiplication. Since multiplicative inverse is not used a lot in Rainbow and parallel multiplication must be used so much, we can multiplex multiplication in multiplicative inverse operations so that we can reduce the space of logic gates.

3.4 Optimization of Gaussian Elimination over the Finite Field

Gaussian elimination is a bottleneck in Rainbow. The efficiency of Gaussian elimination determines the efficiency of Rainbow. Multiplication and multiplicative inverse over the finite field are used a lot in Gaussian elimination and have been designed in this paper. Multiplication and multiplicative inverse can be used as components. They could be instantiated when they are used. We adopt an efficient method, multiplexing components, to reduce the resources, because Gaussian elimination consumes a lot of resources.

Generally, Gaussian elimination over the finite field consists of 5 steps. The first step is to choose a column and find a pivot element. Each nonzero element in this column can be chosen as the pivot element. The second step is to exchange the place of the current row and the row in which the pivot element is located.

Table 3. Efficiency of the multiplicative inverse.

Polynomial	*AND*	*XOR*	Logic units	Delay(ns)
100011101	384	581	446	35.104
101110001	384	603	474	35.074
100101011	384	634	478	36.163
110101001	384	598	433	34.881
100101101	384	567	357	34.491
101101001	384	624	337	36.589
101100011	384	607	353	38.547
110001101	384	617	372	36.122
111001111	384	610	367	41.096
111100111	384	613	347	39.929
101001101	384	624	424	35.731
101100101	384	614	470	36.472
101011111	384	589	470	35.421
111110101	384	617	469	35.292
110000111	384	623	466	35.531
111000011	384	616	463	34.385

And the next step is to normalize the current row. The fourth step, eliminate each element in each column, and then each element in current column is zero except the pivot element. Then we choose the next column to repeat the above four steps to do all jobs until the final column is chosen.

To consider both time complexity and space complexity, there is no need to use too many multipliers. 11 multipliers are used in Gaussian elimination. Therefore, the space complexity equals $O(n)$. Since normalizing operation could be executed within one clock cycle and eliminating operation could be executed within n clock cycles, the time complexity equals $O(n)$ too. The data in Table 4 is tested on the device of EP2S180F1020C3. Four linear equations are needed to be solved. It has data dependencies during Gaussian elimination in Rainbow, thus it can not do other things. Parallel multiplications are free so that they can be used by Gaussian elimination. These multiplication are active in their lifetime so that it improves Rainbow a lot.

Table 4. Efficiency of Gaussian elimination.

Round	Matrix	Multiplication	Multiplicative inverse	Delay (ns)
1	10*11	28	10	3.86
2	4*5	10	4	1.43
3	3*4	7	3	1.03
4	10*11	28	10	3.86
Total		73	30	10.18

4 Implementation and Comparison

To consider both speed and area, an effective way of implementing Rainbow scheme is adopted. This architecture is programmed in VHDL. Basically, fundamental arithmetics including multiplication and multiplicative inverse are encapsulated into components. On the whole, the signature generation is controlled by finite state machines. 27 parallel multipliers as well as one inverter are used in Rainbow signature generation. The number of clock cycles required for the core of signature generation is 2570. The area of our hardware implementation is measured in terms of gate equivalents. Table 5 lists the gate equivalents of our implementation.

Table 5. Gate equivalents for proposed implementation.

Components	Cost gate equivalents
Coefficient matrix	6160
Multipliers	6959.25
Inverters	1884
Adders	486
Total	15489.25

We now compare some public key signature methods with our implementation of Rainbow in terms of performance. The comparison results are given in Table 6. The cycle-area products are normalized to the proposed Rainbow. Our design has a good performance in cycle-area products.

Table 6. Comparison of Rainbow with other signature schemes.

Scheme	Area	Clock cycles	Cycle-area products
RSA-PSS [8]	250000	348672	2189.64
EN-TTS [2]	21000	60000	31.65
Rainbow [5]	63593	804	1.28
Proposed	15490	2570	1

5 Conclusion and Discussion

An efficient hardware implementation for Rainbow signature generation is proposed by using optimized operations including multiplication, multiplicative inverse and Gaussian elimination over a finite field. Not only the speed but also the area are considered in our proposed design. This design is fully implemented on a low-cost FPGA. Compared with other public key hardware implementations, our implementation has a better performance in cycle-area product with 15490 gate equivalents and 2570 clock cycles. Since it is sensitive to the area rather than the running time in limited resource environments, this method is more suitable for implementing Rainbow scheme in these areas, e.g. smart cards.

References

1. Ding, J., Schmidt, D.: Multivariate public key cryptosystems. In: Advances in Information Security. Citeseer (2006)
2. Yang, B.Y., Cheng, C.M., Chen, B.R., Chen, J.M.: Implementing minimized multivariate PKC on low-resource embedded systems. Security in Pervasive Computing, pp. 73–88 (2006)
3. Yang, B.Y., Chen, J.M., Chen, Y.H.: TTS: High-speed signatures on a low-cost smart card. Cryptographic Hardware and Embedded Systems, 318–348 (2004)
4. Chen, A., Chen, C.H., Chen, M.S., Cheng, C.M., Yang, B.Y.: Practical-sized instances of multivariate PKCs. Post-Quantum Cryptography, 95–108 (2008)
5. Balasubramanian, S., Carter, H.W., Bogdanov, A., Rupp, A., Ding, J.: Fast multivariate signature generation in hardware: The case of Rainbow. In: International Conference on Application-Specific Systems, Architectures and Processors, pp. 25–30. IEEE, Los Alamitos (2008)
6. Wang, C.C., Troung, T.K., Shao, H.M., Deutsch, L.J., Omura, J.K., Reed, I.S.: VLSI architectures for computing multiplications and inverses in $GF(2^m)$. IEEE Transactions on Computers, 709–717 (1985)
7. Schroeder, M.R., Schroeder, M.R.: Number theory in science and communication. Springer, Heidelberg (1986)
8. Großschädl, J.: High-Speed RSA Hardware Based on Barret's Modular Reduction Method. In: Paar, C., Koç, Ç.K. (eds.) CHES 2000. LNCS, vol. 1965, pp. 95–136. Springer, Heidelberg (2000)

Novel Resonant-Type Composite Right/Left Handed Transmission Line Based on Cascaded Complementary Single Split Ring Resonator

He-Xiu Xu[1], Guang-Ming Wang[1], and Qing Peng[2]

[1] Missile Institute, Air Force Engineering University, sanyuan, 713800, China
[2] School of Foreign Language, Nanchang University, Nanchang, 330031, China
hxxu20008@yahoo.cn, Wgming01@sina.com, ellen07622@yahoo.cn

Abstract. In this paper, a novel electrically-small composite right/left handed transmission line cell (CRLH TL) is presented based on cascaded complementary single split ring resonator(CCSSRR). For deep insight, a circuit model is also presented and validated by the simulated full wave S-parameters. The nagative refractive index and backward wave propogation is demomonstrated by the constitutive effective electromagnetic parameters. This type of CRLH TL cell features an additional transmission zero above the edge of right handed (RH) band which in turn enhances the selectivity and harmonic suppression to a great extent. The resultant left handed (LH) and RH bands can be merged, which results in a smooth continuous transition for broadband design. A set of CRLH TL cells by cascading different numbers of basic CCSSRR are studied and compared. The proposed CRLH TL cell should be a good candidate for small-size broadband devices design.

Keywords: Composite right/left handed transmission line, cascaded complementary single split ring resonator, constitutive electromagnetic parameters, electrically-small, selectivity.

1 Introduction

The concept of resonant-type composite right/left handed transmission line (CRLH TL) has not been formed and become a subject of intensive research until the seminal work done by F. Falcone et al. [1]. In this seminal work, the complementary split ring resonators (CSRRs) etched in the ground plane of microstrip line is firstly proposed and demonstrated with a negative effective permittivity in the vicinity of resonant frequency. At short notice, the possibility to fabricate planar CRLH TL by incorporating the CSRRs and series gap is also proved [2]. Soon after that, the applicability of this type of CRLH TL in the design of many microwave compact devices with competitive performances [3], [4] has been widely validated.

In recent years, resonant-type CRLH TLs also have gone through a great development, for instance, the hybrid approach, which incorporates the series capacitive gap, CSRRs, and additional ground inductors, enables a further degree of

M. Zhu (Ed.): Electrical Engineering and Control, LNEE 98, pp. 97–104.
springerlink.com © Springer-Verlag Berlin Heidelberg 2011

flexibility in CRLH TL synthesis [5], followed by CSRRs etched on the conductor strip which were exploited as a good strategy to those systems where the ground plane cannot be etched [6], using complementary spiral resonators (CSRs) for further miniaturization and open complementary split ring resonator (OCSRR) for the high selectivity [7], introducing the fractal geometry in CSRRs for a significant lower resonant frequency and improved selectivity [8], [9], and also complementary single split ring resonator(CSSRR)[10], etc.

In view of them, the body of this work is aimed to propose a novel CRLH TL based on cascaded CSSRR (CCSSRR). This paper is well organized as follows. In section 2, the configuration as well as corresponding equivalent circuit model of CCSSRR-loaded cells based on different basic CSSRR is proposed, and analysis of the structure through Bloch theory is also systematically carried out. In section 3, electromagnetic (EM) characteristic (S-parameters) is provided and the constitutive EM parameters are extracted for a deep insight into the working mechanism of CCSSRR. Finally, a major conclusion is summarized in section 4.

2 CCSSRR-Loaded CRLH TL: Topology, Circuit Model and Theory

In this section, we will show a type of planar CRLH TL cells in microstrip technology. Note that these cells also can be carried out in coplanar waveguide (CPW) technology. The proposed typical unit cells and the correlative circuit model are illustrated in Fig.1. As can be observed, Novel CRLH TL cells consists of a CCSSRR element (depicted in white) etched in the ground plane (depicted in light grey), a stepped-impedance conductor line (depicted in dark grey), and a series capacitive gap etched in the center of the top side above the CCSSRR which is constructed by cascading several identical CSSRR elements and symmetrically locating splits on each CSSRR. Note that every adjacent two CSSRR own a community vertical slot. It is worth to mention that the number of cascaded basic CSSRR elements are even for symmetry and can be infinite. This configuration is engineered to benefit the proper electric excitation from the gap to CCSSRR. As a consequence, the resultant CCSSRR is also associated with the negative permittivity and will be demonstrated later this section.

In the circuit model (see Fig.1(d)), where L_s models the line inductance, C_g models the gap capacitance, C_1 represents the electric coupling between the stepped-impedance conductor line and the biggest CSSRR which is described by means of a parallel resonant tank L_{p1}, C_{p1}. In like manner, C_2 models the electric coupling between the line and the fundamental CSSRR which is characterized by means of a resonant tank L_{p2}, C_{p2}. It is useful to mention that circuit model of conventional CSRRs-loaded CRLH TL is a special case of proposed circuit model by ruling out one shunt branch or set values of them to be zero. As a consequence, novel LH particle with additional lumped-element parameters in the circuit model enables enhanced design flexibility compared with its conventional counterpart.

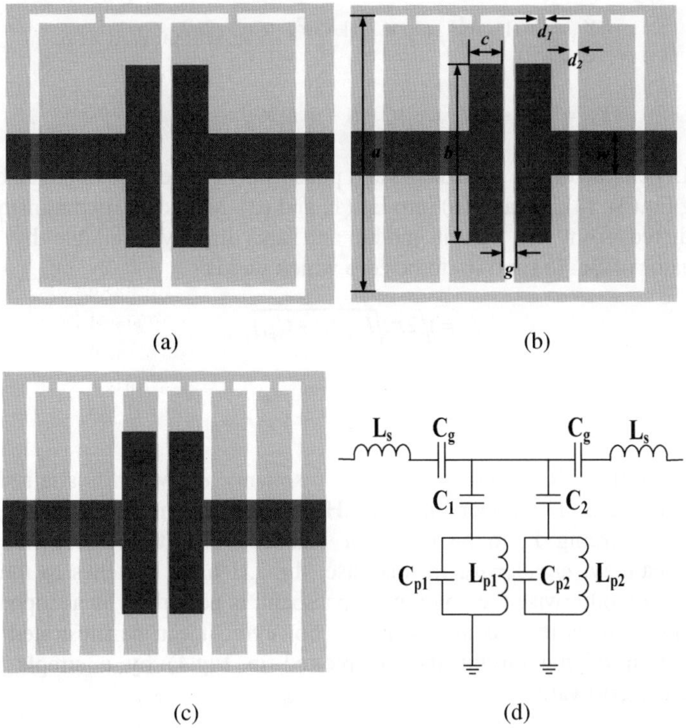

Fig. 1. Topology of the CCSSRR-loaded CRLH TL unit cells (a) based on two, (b) four and (c) six fundamental CSSRR elements, and (d) equivalent T-type circuit model. Note that these CCSSRR are in square shape for simplicity in this paper and are with identical physical parameters including side length of square slot *a*, split width d_1, slot width d_2, gap separation *g*, height *b* and width *c* of stepped-impedance line, and width of microstrip line *w*. Detailed physical parameters are d_1=d_2=0.2mm, *g*=0.3mm, *a*=6mm, *b*=4mm, *c*=0.85mm, and *w*=1mm.

Let us now study the structure through Bloch theory based on the derived circuit model. The phase shift per cell Φ and characteristic impedance Z_β, which are required as real numbers for EM wave propagation, are given by

$$\phi = \beta l = 1 + Z_s(jw)/Z_p(jw) \tag{1}$$

$$Z_\beta = \sqrt{Z_s(jw)[Z_s(jw) + 2Z_p(jw)]} \tag{2}$$

where $Z_s(jw)$ and $Z_p(jw)$ are series and shunt impedance, respectively, which reads

$$Z_s(jw)] = (1 - \omega^2 L_s C_g)/j\omega C_g \tag{3}$$

$$Z_p(jw)] = 1/Y_p = 1/(Y_{p1} + Y_{p2}) \tag{4}$$

Y_{p1} and Y_{p2} are the admittances of the two shunt branches formed by C_1 L_{p1}, C_{p1} and C_2 L_{p2}, C_{p2}, respectively. They are formulated as follows after some simple manipulations.

$$Y_{p1} = jw(1 - w^2 L_{p1} C_{p1}) C_1 / (1 - w^2 L_{p1} C_{p1} - w^2 L_{p1} C) \tag{5}$$

$$Y_{p2} = jw(1 - w^2 L_{p2} C_{p2}) C_2 / (1 - w^2 L_{p2} C_{p2} - w^2 L_{p2} C_2) \tag{6}$$

The CRLH characteristic impedance and phase shift per cell are obtained explicitly by inserting Eq.(3), (4), (5) and (6) into Eq.(1) and (2). Moreover, two transmission zeros are predicted from the circuit model and are directly available by forcing the denominator of Eq. (5) and (6) to be zero which yields

$$f_{z1} = 1 / 2\pi \sqrt{L_{p1}(C_1 + C_{p1})} \tag{7a}$$

$$f_{z2} = 1 / 2\pi \sqrt{L_{p2}(C_2 + C_{p2})} \tag{7b}$$

The lower limit of RH band is the series resonance ω_s which is attained by forcing Eq.(3) to be zero. The upper limit of LH band is the shunt resonance ω_p which is acquired by forcing Y_p to be null. For balanced condition, these frequencies are identical, namely, $\omega_s = \omega_p = \omega_0$. In this case, the LH band switches to the RH region without a gap, otherwise the continuous passband is perturbed by a stopband. Explicit expression of ω_p is tedious and complex, however, it can be illustrated through the representation of the circuit model (plotted in Fig.4) by a simple program in mathematical software.

3 S-Parameters and Constitutive Effective EM Parameters

For characterization, novel CRLH TL cells depicted in Fig.1 are built on the F4B-2 substrate with a thickness of 0.8 mm and a dielectric constant of 2.65. They are analyzed by means of planar full-wave EM simulation through Ansoft Designer as well as electrical simulation through circuit software Ansoft Serenade. During the electrical parameters extraction process (electrical simulation), S-parameters of the circuit model are obtained and driven to match the EM simulated ones.

Fig.2 shows the simulated full-wave S-parameters of these CRLH TL cells. From this figure, three most important aspects should be emphasized. First, two obvious transmission zeros (attenuation poles) are located below the lower edge of LH band and above the upper edge of RH band, respectively, which have enhanced the selectivity and out-of-band suppression. Note that the exhibited LH and RH band will be demonstrated sooner. Second, the CRLH TL cells in three cases operate in balanced condition, thus the LH band changes continuously to the RH band without a stopband. Third, as the number of CSSRR increases, the fundamental transmission zero and LH band are with a very small variation which can be negligible, however, the upper transmission zero obviously shifts toward higher band, which significantly increases the passband bandwidth characterized by 10dB return loss.

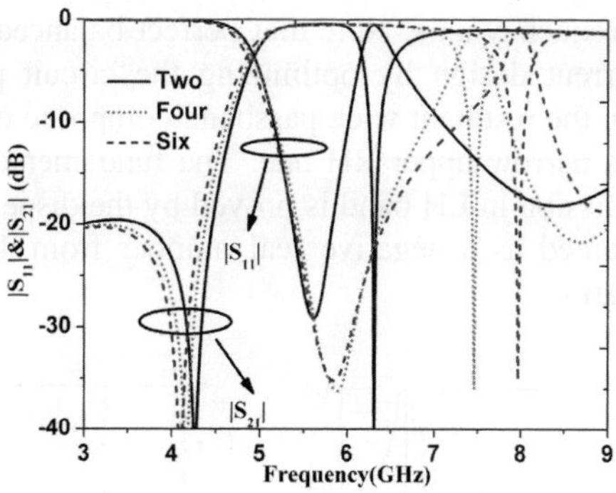

Fig. 2. Comparison of simulated full-wave S-parameters between different CCSSRR-loaded CRLH TL cells plotted in Fig.1.

The S-parameters of CCSSRR-loaded CRLH TL cell constructed by six CSSRR elements obtained from EM and electrical simulation are compared in Fig.3. It is obvious that these results are in reasonable agreement which has fully validated the rationality of the circuit model. The slight discrepancies are due to the wide frequency range that observed, nevertheless, they are in normal level. EM simulated S-parameters reveal that all return loss is better than 10dB from 5.1 to 7.3GHz. Moreover, two transmission zeros occur at 4.1GHz and 8GHz, respectively.

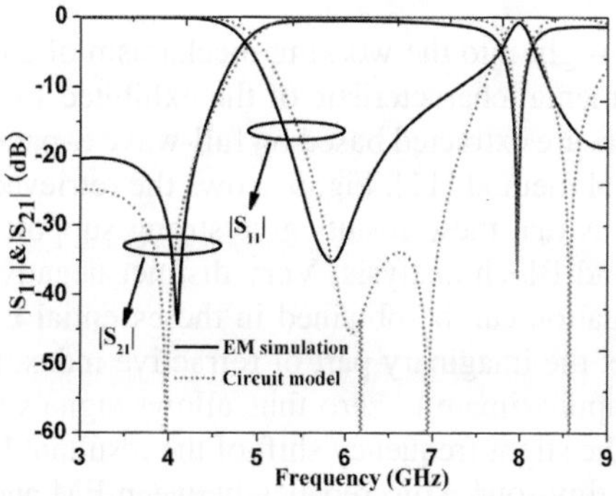

Fig. 3. Comparison of S-parameters between EM and electrical simulation of CCSSRR-loaded CRLH TL in the case of six CSSRR elements plotted in Fig.1(c). Extracted lumped elements are: L_s=2.1nH, C_g=0.306pF, C_1=1.53pF, C_2=0.658pF, C_{p1}=0.246pF, L_{p1}=0.89nH, C_{p2}=19.99pF, and L_{p2}=0.019nH.

Fig. 4 depicts the representation of CCSSRR-loaded CRLH TL including series impedance, shunt admittance, characteristic impedance and dispersion relation based on the lumped-element circuit parameters derived above. As expected, two transmission zeros, i.e., two resonances from the blue curve of Y_p, are located in the vicinity of lower and upper edge of passband where ϕ and Z_β are real numbers. It also reveals that the CRLH TL cell is working in balanced condition. As envisaged, the balanced point occurs at the intersection (6.2GHz) of shunt admittance curve and series impedance curve on frequency axis. Note that perfect balanced condition is rigorous and is fulfilled in current design by optimizing the circuit parameters. The most important aspect lies in the resultant wide passband composed of not only a lower LH contribution but also a narrow upper RH one. The fundamental characteristic of the backward wave transmission in LH band is proved by the dispersion curve. As can be observed, Φ is maintained as a negative real number from 4.94 to 6.2GHz which indicates a LH propagation.

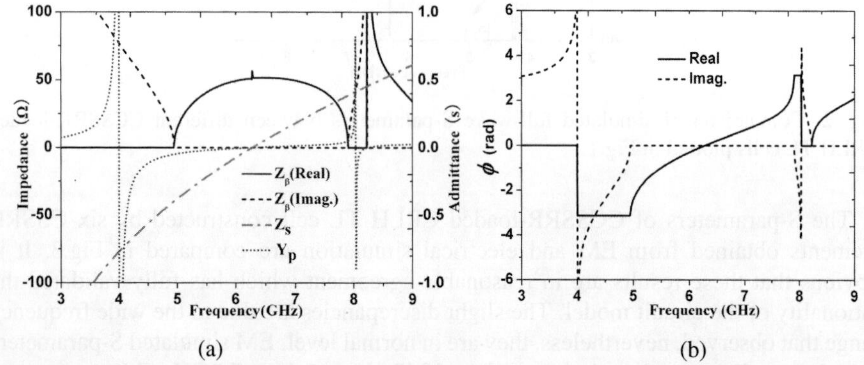

(a) (b)

Fig. 4. Representation of (a) the series impedance, shunt admittance, characteristic impedance and (b) dispersion curve for the CCSSRR-loaded CRLH TL cell in Fig.1(c).

To provide a deep insight into the working mechanism of the CCSSRR and further demonstrate the fundamental characteristic of the exhibited LH band, the constitutive effective EM parameters are extracted based on full-wave S-parameters plotted in Fig.3 by an improved retrieval method [11]. Fig. 5 shows the retrieved effective constitutive parameters. It is obvious that these results give strong support to all results obtained from EM simulation and Bloch analysis. Very distinct negative refractive index and backward wave propagation can be obtained in the essential LH band within 4.78 to 5.75GHz. In this range, the imaginary part of refractive index, accounting for electric and magnetic loss, is approximate to zero thus allows signals to transmit freely. It is worth to mention that the slight frequency shift of the resultant LH band between Fig.4 and F.g.5 is due to the tiny source discrepancy between EM and electrical simulation. Nevertheless, the tendency of these curves is in excellent agreement.

Consulting Fig. 5(b), we conclude that novel CCSSRR is responsible to the exhibited negative effective permittivity occurred from 4.67 to 5.75GHz. Below 5.75GHz, the effective permeability is negative, thus the simultaneous negative LH band solely depends on negative permittivity band. Moreover, the upper transmission zero can be successfully interpreted through the exhibited single negative permeability around 7.9GHz. It is worth to mention that basic CSSRR (responsible to the magnetic resonance) is externally driven through the component of the magnetic field contained in the plane of the particle, while the biggest CSSRR (responsible to electric response) is still excited electrically.

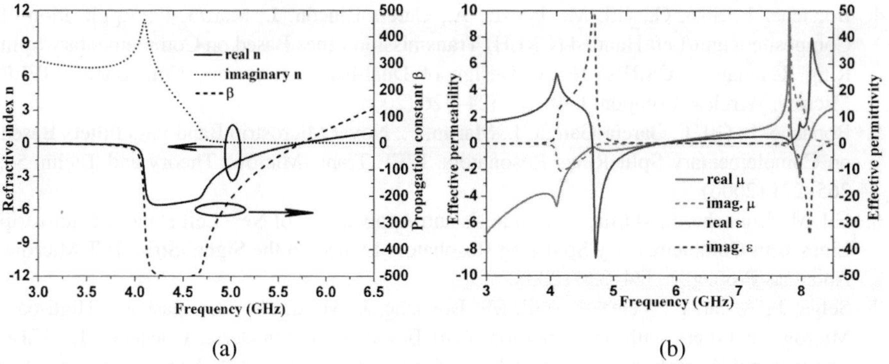

Fig. 5. Constitutive EM parameters of proposed CRLH cell based on full-wave S-parameters plotted in Fig.3 (a) refractive index and propagation constant (b) effective permeability and permittivity.

4 Conclusion

In conclusion, it has been demonstrated that a novel electrically small cell based on CCSSRR and series gap exhibits a composite right/left handed behavior. The validated circuit model is very useful for the synthesis of compact size and high-performance CRLH TL cell. The proposed particle shows many merits over previous CSRRs-loaded ones, e.g., enhanced flexibility, additional transmission zero, and controllable bandwidth which can be tuned by a reduction or an increase of the basic CSSRR element. Consequently, we will exploit and demonstrate possible applications of this particle in novel microwave devices design in the next step.

Acknowledgments

This work is supported by the National Natural Science Foundation of China under Grant Nos. 60971118.

References

1. Falcone, F., Lopetegi, T., Baena, J.D., Marques, R., Martin, F., Sorolla, M.: Effective Negative-ε Stop-band Microstrip Lines Based on Complementary Split Ring Resonators. IEEE Microw. Wirel. Compon. Lett. 14, 280–282 (2004)
2. Falcone, F., Lopetegi, T., Laso, M.A.G., Baena, J.D., Bonache, J., Beruete, M., Marque´s, R., Martin, F., Sorolla, M.: Babinet Principle Applied to the Design of Metasurfaces and Metamaterials. Phys. Rev. Lett. 93, 197401 (2004)
3. Wu, H.-W., Su, Y.-K., Weng, M.-H., Hung, C.-Y.: A Compact Narrowband Microstrip Band Pass Filter with a Complementary Split Ring Resonator. Microw. Opt. Technol. Lett. 48, 2103–2106 (2006)

4. Bonache, J., Sisó, G., Gil, M., Iniesta, Á., García-Rincón, J., Martín, F.: Application of Composite Right/Left Handed (CRLH) Transmission Lines Based on Complementary Split Ring Resonators (CSRRs) to the Design of Dual-band Microwave Components. IEEE Microw. Wireless Compon. Lett. 18, 524–526 (2008)
5. Bonache, J., Gil, I., García-García, J., Martín, F.: Novel Microstrip Band Pass Filters Based on Complementary Split Rings Resonators. IEEE Trans. Microw. Theory and Techn. 54, 265–271 (2006)
6. Gil, M., Bonache, J., Martin, F.: Synthesis and Applications of New Left Handed Microstrip Lines with Complementary Split-ring Resonators Etched on the Signal Strip. IET Microw. Antennas Propag. 2, 324–330 (2008)
7. Selga, J., Aznar, F., Vélez, A., Gil, M., Bonache, J., Martín, F.: Low-pass and High-pass Microwave Filters with Transmission Zero Based on Metamaterial Concepts. In: IEEE International Workshop on Antenna Technology, Santa Monica, CA, United states, pp. 1–4 (2009)
8. Crnojevic-Bengin, V., Radonic, V., Jokanovic, B.: Fractal Geometries of Complementary Split-ring Resonators. IEEE Trans. Microw.Theory Tech. 56, 2312–2321 (2008)
9. Xu, H.-x., Wang, G.-m., Zhang, C.-x., Hu, Y.: Microstrip Approach Benefits Quad Splitter. Microwaves&rf 49, 92–96 (2010)
10. Xu, H.-x., Wang, G.-m., Zhang, C.-x., Lu, K.: Novel Design of Composite Right/Left Handed Transmission Line Based on Fractal Shaped Geometry of Complementary Split Ring Resonators. Journal of Engineering Design 18, 71–76 (2011)
11. Xu, H.-x., Wang, G.-m., Liang, J.-G.: Novel CRLH TL Metamaterial and Compact Microstrip Branch-line Coupler Application. Progress In Electromagnetics Research C 20, 173–186 (2011)

Design of Network Gateway Based on the ITS

Zhu Jinfang and Fan Yiming

Zhejiang Industry Polytechnic College,
312000 Shaoxing, Zhejiang, China
Zjipc_ec@Hotmail.com, fyim@21cn.com

Abstract. This article first introduces lightweight XMPP protocols and embedded operating system, Contiki, the establishment of standardized, open network server architecture. Its architectural style is based on the most basic of IPV6 network protocol has able to send short messages, radio-subscribe to the mechanism and the authentication mechanism and the protection of data security and other functions, is also a lightweight, scalable and good, ópen-ended. Then the test platform, successfully transplanted to Contiki within the XMPP protocol.

Keywords: Internet of Things, Network Gateway, XMPP protocols.

1 Introduction

Currently, the United States, European Union, China attaches great importance to things and so the development of their countries have to a strategic height to shaping the economy, promoting economic and social development [1].

Internet of Things is a concept, refers to all kinds of information sensing devices, such as radio frequency identification devices, infrared sensors, global positioning systems, laser scanners and various other devices and the Internet combine to form a huge network, which aims to All items connected with the network to facilitate the identification, location, tracking, monitoring and management.

In this paper, Contiki operating system and protocol stack XMPP made in detail, to explore a specific protocol stack to the operating system Contiki XMPP transplant method and had connectivity test [2].

2 Method

Contiki embedded operating system to support multi-tasking, open source operating system, first developed by the Adams Dunkels designed specifically for embedded systems and wireless sensor networks design. After the effective cut, only use 40KB ROM and 4K RAM [3]. The ARM7-based LPC2220 development board and no built-in reserve too much space to the ROM, which requires external expansion memory.

M. Zhu (Ed.): Electrical Engineering and Control, LNEE 98, pp. 105–108.
springerlink.com © Springer-Verlag Berlin Heidelberg 2011

2.1 Contiki Embedded Operating System

Contiki embedded operating system supports multi-tasking, open source operating system, first developed by the Adams Dunkels design, it is designed for embedded systems and wireless sensor networks design. Today, Contiki has been widely used in embedded operating systems, the United States, General Motors, NASA are using it [4].

Is based on event-driven Contiki embedded operating system, which is also based on the program execution flow of events, event, or request by the underlying hardware request to trigger preemptive multi-threading to the optional libraries to achieve. Contiki hardware resources in the form of an abstract realization of the system libraries, and any other procedures connected. In order to achieve the implementation of the principle of priority, Contiki does not prohibit the interruption.

In terms of Internet connectivity, Contiki has been based on the basic IPV6 protocol stack UIP. UIP is a lightweight network protocols, special use and embedded systems, its code is very small, with very low memory usage, UIP handle protocol also uses TCP / IP RFC standards.

2.2 XMPP Protocols

XMPP is an open XML protocol, designed for near real-time messaging and presence information, and request-response service. The basic syntax and semantics of the first mainly by Jabbe: the open source development community in 1999. In 2002, XMPP Working Group was authorized to take over the development and adaptation of the Jabber protocol to meet the IETF's messaging and presence technology [5].

XMPP is client-server architecture, also known as the C / S mode. From client to server and server to server access mode TCP connection, the server is also a communication between TCP. The XMPP abstract diagram is shown in Fig.1.

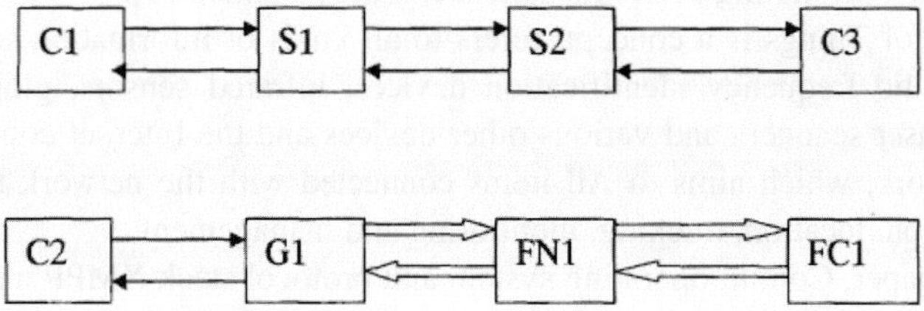

Fig. 1. XMPP abstract diagram

XMPP server using a client mode, can be combined, it can be dispersed. This allows the user to easily control server according to their wishes, so that the server provides the functions they need.

2.3 Contiki Embedded Operating System Based on the XMPP Protocol Stack

XMPP protocol stack is based on the Contiki operating system tailored for the XMPP protocol effectively, so that it can meet the low-power embedded systems. XMPP protocol stack supports XML stream, and only one of the three basic elements: <message/>, <presence/>, <iq/>.

Event handler function to handle from the perception layer (wireless sensor networks), and UIP that the remote server time, that XMPP protocol stack has been connected with the perception of the role layer and the Internet gateway.

This model established the first to do the following work: 1, to establish TCP traffic using the UIP; 2, the information on the <message/> and <presence/> pushed to the network layer; 3, using the <<iq/> mode to send And receive information.

UIP network protocol based on the work process is as follows:

1, open XMPP <stream/> label, XMPP protocol stack start <<stream/>, remote server or client sends a <<stream/> to respond.
2, XMPP protocol stack and the remote server or client authenticate each other.
3, XMPP protocol stack to establish a publication - subscribe mechanism.
4, XMPP protocol stack to establish a push mechanism.
5, The remote server or on client sends a message / query request, XMPP protocol stack to send one on information / queries answered.
6, Close the XMPP protocol stack <stream/> label.
7, Off a remote server or client <<stream/> label.

The XMPP protocol stack and the perception layer, That the wireless sensor network data exchange is through the application layer of the API interface implementation. XMPP protocol stack to achieve the following function body wireless sensor network and data exchange.

3 Gateway to the Internet Communication Module

Network communication can be divided into C / S and B / S two structures. C / S (Client / Server, the client and server architecture. commitment to the client and server are different tasks. Client will be submitted to the needs of users Server, then Server returns the results to be provided in the form to the user. Client Server task is to receive service requests made to carry out the appropriate treatment, and the results returned to the Client. Server process normally in "sleep" state until the server sends the client connection request, to "wake up". The client running the browser, the browser to the form of hypertext access the database to the Web server made the request. Web server accept the client request, the request to the SQL usage, and database access, and then return the results to the Web server, Web server, then the result is converted to HTML documents, returned to the client browser to web pages Form displayed. B/S structure, Web browser, the most important is the client software, the core part of the system function to focus on the server. C / S structure of the rational allocation of tasks to the client and the server, reducing communication overhead in the system. Many tasks on the client side processing, then submitted to the server, the server running the load lighter, the client response is fast. However, this structure requires the client to install special client software. B / S The biggest advantage is no need to install any special software, just install the browser client.

4 Conclusion

Using this gateway to the Internet further improve the performance and universal gateway. Further in-depth study of things related to knowledge, and emerging wireless networks, embedded operating system, the gateway protocol stack, able to compare their advantages and disadvantages, on this basis, further reducing energy consumption and the gateway protocol stack Network bandwidth utilization.

References

1. Logsdon, T.: The Navastar global positioning system. Van Nostrand Reinhold, New York (1992)
2. Zhang, Z.: Handbook of wireless networks and mobile computing. John Wiley&Sons, New York (2002)
3. Rohmer, K.: The lighthouse location system for smart Dust. In: Proc. lst ACM/LJSENIX Conf. Mobile Syst. Applications, Services, pp. 15–30. ACM Press, New York (2003)
4. Addlesee, M., et al.: Implementing a sentient computing system. IEEE Computer 34(8), 50–56 (2001)
5. Drane, C., Macnaughtan, M., Scott, C.: Positioning GSM telephones. IEEE Commun.Mag. 36(4), 46–54 (1998)
6. Lance, S., Michael, A.M., Sebastien, G.: an XMPP-based framework for many task computing applications. In: Proc. 2nd Workshop on Many-Task Computing on Grids and Supercomputers, ACM Press, New York (2009)

The Research of Improved Grey-Markov Algorithm

Geng Yu-shui and Du Xin-wu

School of Information Science and Technology,
Shandong Institute of Light Industry, Jinan, China
gys@sdili.edu.cn,
dxwary@163.com

Abstract. When using Grey-Markov model for prediction, the forecast curve of GM(1,1) model must pass through the point (1,x(0)(1)) is usually assumed as the known prerequisite condition before moving on to the next step. However, this kind of means is groundless, and the result we get is not necessarily the optimal solution. This paper introduces a different Grey-Markov model which is based on a kind of improved GM (1,1) algorithm and applied it to the sales forecast. The result shows that this model is working well in sales forecast. At the same time, the comparison and analysis of the prediction result of this model also helps to prove that there is no direct connection between GM(1,1) model and the first data element of the raw time series.

Keywords: Grey-Markov, Sales Forecast, Data Mining.

1 Introduction

Sales Forecast refers to the estimate of all or some specific kinds of products' sales in future time[1]. In full consideration of the various factors affecting the future and based on the combination of the enterprise's performance, It is a feasible sales target proposed through a certain analysis method. For factors that affecting the product sales are complicated, such as factors come from internal and external of enterprise, substitution, and the change of customer's preferences, the sales of products is always changing. Therefore, sales forecast itself is a complicated multi-factor and multi-layer system. This system has known information as well as unknown information, so it is extremely difficult to construct an accurate sales forecast model. According to the grey system theory, we can use the time sequence already known to find out useful information, using its dynamic memory characteristic to establish grey model and to find and reveals the message inside the system, so as to avoid the research of complex systems and the relationship between internal factors.

For grey prediction is a GM(1,1) model based forecast, the solution of GM(1,1) model is an exponential curve and its predicted geometrical graph is a relatively smooth curve, this makes it less useful when the time series fluctuate a lot. Markov probability matrix is a random dynamic system oriented forecast model. It is based on the probability of transfer between state to predict the future development of the system. The probability of transfer between state is a reflection of the influence from

M. Zhu (Ed.): Electrical Engineering and Control, LNEE 98, pp. 109–116.
springerlink.com © Springer-Verlag Berlin Heidelberg 2011

random factors, which reviews the internal connections between different states. Therefore, Markov probability matrix forecast model can be used to deal with prediction of random time series with high volatility. Through the above analysis, Grey-Markov forecast model, which is a combination of grey GM(1,1) and Markov probability matrix forecast model, can make full use of historical data and greatly improve the predict precision of random time series with high volatility, as well as improve the prediction results[2].

2 Model Construction

Details about model construction are described down below.

2.1 Constructing GM(1,1) Model

1) Data accumulation

Assuming the original data sequence is:

$$x^{(0)} = \{x^{(0)}(1), x^{(0)}(2), \cdots, x^{(0)}(n)\}$$

After one time accumulation, we get a new data sequence:

$$x^{(1)} = \{x^{(1)}(1), x^{(1)}(2), \cdots, x^{(1)}(n)\}$$

Each number of the sequence is calculated according to the following formula:

$$x^{(1)}(k) = \sum_{i=1}^{k} x^{(0)}(i), k = 2,3,\cdots,n$$

2) Model construction

After obtaining the accumulated data sequence $x^{(1)}$, the next step is to tectonic the background values sequence :

$$z^{(1)} = \{z^{(1)}(2), z^{(1)}(3), \cdots, z^{(1)}(n)\}$$

Each element of the sequence can be calculated according to the following formula :

$$z^{(1)}(k) = [x^{(1)}(k-1) + x^{(1)}(k)]/2, k = 2,3,\cdots,n$$

The albinism differential equation of GM(1,1) model is:

$$\frac{dx^{(1)}}{dt} + ax^{(1)} = u \tag{1}$$

After discretization, we can get its discrete predictive formula:

$$x^{(0)}(k) + az^{(1)}(k) = u, k = 2,3,\cdots,n \tag{2}$$

3) Parameter estimation

There are two unascertained parameters in the formula (2), "a" and "u". "-a" is called the developing coefficients, its value determines the increase rate of $x^{(0)}$. Investigation shows that the GM(1,1) model is very precisive when the original series change slowly and evolution parameter is tiny. Parameter u here is called the grey value[3]. Because there is no linear relation between source data sequence, so least square method is used for estimation. The parameter estimation formula is:

$$[a \quad u]^T = (B^T B)^{-1} B^T Y \tag{3}$$

Y and B here in the formula are both an indication of a matrix:

$$Y = [X^{(0)}(2), X^{(0)}(3), \cdots, X^{(0)}(n)]^T,\ B = \begin{bmatrix} -z^{(1)}(2) & 1 \\ -z^{(1)}(3) & 1 \\ \vdots & \vdots \\ -z^{(1)}(n) & 1 \end{bmatrix}$$

4) Establish forecasting formula and error calculation

Usually, according to the parameter "a" and "u", we can establish the forecasting formulas as follows:

$$x^{(1)}(k+1) = (x^{(0)}(1) - u/a)e^{-ak} + u/a \tag{4}$$

From the formula above, we can find that the following equation $\hat{x}^{(1)}(1) = x^{(1)}(1) = x^{(0)}(1)$ is first assumed as the default known condition. Therefore the corresponding fitting curve $\hat{x}^{(1)}$ must pass through the point $(1, x^{(1)}(1))$ of the coordinate plane $(k, \hat{x}^{(1)}(k))$. However, according to the theory of least square method, the fitted curve does not necessarily through the first data points . There is no theoretical basis for taking $\hat{x}^{(1)}(1) = x^{(1)}(1) = x^{(0)}(1)$ as the known conditions. Further more, considering $x^{(1)}(1)$ is the oldest data of the sequence, it does not have close relation with future and is not get from the accumulation process, so it would be more wise to abandon the traditional way of making $\hat{x}^{(1)}(1) = x^{(1)}(1)$ as the prerequisite of problem solving. Other data can be used as known conditions[4], such as :

$$\hat{x}^{(1)}(m) = x^{(1)}(m) \ (m = 2, 3, \cdots, n)$$

then we can get a new prediction formula accordingly.

$$\hat{x}^{(1)}(k+1) = [x^{(1)}(m) - u/a]e^{-a(k-m+1)} + u/a,\ k = 1, 2, \cdots, n \tag{5}$$

This formula can be viewed as a generalized form of (4). This is because when "m" here equals to 1, the two formula are equivalent[3]. According to the definition of sequence $x^{(1)}$, we know that $\hat{x}^{(0)}(k+1) = \hat{x}^{(1)}(k+1) - \hat{x}^{(1)}(k)$, combined with formula (5) , we can get the ultimate form of prediction formula for $\hat{x}^{(0)}(k)$.

$$\hat{x}^{(0)}(k)=[x^{(1)}(m)-u/a](e^{-a}-1)e^{(k-m)} \quad (6)$$

5) Parameter estimation

In order to get the best forecasting formula, we must first find a "m" which has the minimum predicted bias. To begin with, according to the value of m $(m=1,2,\cdots,n)$ we can get a corresponding prediction sequence:

$$\hat{x}_m^{(1)}=\{\hat{x}_m^{(1)}(1),\hat{x}_m^{(1)}(2),\cdots,\hat{x}_m^{(1)}(n)\}\ k=2,3,\cdots,n$$

Then, according to the following formula:

$$\overline{\varepsilon}_m=\sum_{i=1}^{n}|(\hat{x}^{(0)}(i)-x^{(0)}(i))/n|$$

We can get a mean prediction error for parameter "m". Finally, we have a mean prediction error sequence $\varepsilon=(\overline{\varepsilon}_1,\overline{\varepsilon}_2\cdots,\overline{\varepsilon}_n)$. Before establishing the best forecasting formula, we must select a best m from 1 to n whose corresponding prediction error is the least among the sequence. Assuming that m=j is our final selection, then we can build the best forecast formula like this:

$$\hat{x}^{(1)}(k+1)=[x^{(1)}(j)-u/a]e^{-a(k-j+1)}+u/a \quad k=1,2,\cdots,n$$

Accordingly, the prediction formula for $\hat{x}^{(0)}$ is:

$$\hat{x}^{(0)}(k)=[x^{(1)}(j)-u/a](e^{-a}-1)e^{(k-j)} \quad k=2,3,\cdots,n$$

2.2 Constructing Markov Model

Markov Chain is a time-discrete and state-discrete stochastic process. It has the property of non-after effect, that is to say the state of time t is only related to previous state, and it has nothing to do with states of other time. A n-rank Markov Chain is determined by a n-size state collections and a group of state transition probability. This random process can only be in one state in a time. If it was in state Si at time t, it will transfer to state Sj at time t+1 in probability p. According to the transferring probility between different states, the Markov Chain can help to predict the developments and changes of the system's future.

1) State division

Taking curve $\hat{x}^{(0)}(t)$ as the base, the whole coordinate plane can be divided into a number of parallel strip area, each strip is an indication of a state. The curve itself also reviews the changing trend of the original data sequence. For Markov Chain that has clear state boundaries, traditional state division method is selecting several constants as boundary. However, for those states that keep changing as time goes on, the boundary is also changing[5]. When original sequence is an unstable random one, we can divide them into k groups, remembered as $E_1, E_2, \cdots, E_k$. Each state is constructed according to standards like this:

$$E_i \in [E_{iL}, E_{iR}] \ (E_{iL} = \hat{x}^{(0)}(t) + e_{i1}\overline{x}, \ E_{iR} = \hat{x}^{(0)}(t) + e_{i2}\overline{x} \quad i = 1,2,\cdots n)$$

$\overline{x}$ here is the mean value of the original data sequence. There is no clear limitation on the assignment of e_{i1} and e_{i2}, usually, information we know about the object that is going to be predicted and some personal experience can help to find a better solution.

2) Constructing state transition probability matrix

The formula for state transfer probability calculation is:

$$P_{ij}(m) = M_{ij}(m) / M_i \quad (i, j = 1,2,\cdots,k)$$

$M_{ij}(m)$ is the number of original data that transfer to state E_j from E_i after a m-step state transferring, M_i is the number of data that belongs to the group of E_i, $P_{ij}(m)$ shows the probability of changing to state E_j from E_i after a m-step state transferring. With the help of the above definition, we can construct a state transferring probability matrix like this:

$$P(m) = \begin{bmatrix} p_{11}(m) & p_{12}(m) & \cdots & p_{1k}(m) \\ p_{21}(m) & p_{22}(m) & \cdots & p_{2k}(m) \\ \vdots & \vdots & \ddots & \vdots \\ p_{k1}(m) & p_{k2}(m) & \cdots & p_{kk}(m) \end{bmatrix}$$

This matrix is a reflection of the law of transfer between states. With the help of transferring probability matrix, if we already know the current state, then the future state of the system can be predicted. The transferring probability matrix is constructed according to historical data, some further modification is needed to cope with accidental changes in future better.

3) Getting the predicted result

If the future state of the predicted object after state transfer is determined, so it the changing interval of predicted value. Assuming E_i will be the next state, then we can know that its changing interval is $[E_{iL}, E_{iR}]$. So the middle point of this changing interval can be selected as the final predicted value: $\hat{x}^{(0)}(i) = (E_{iL} + E_{iR})/2$.

3 Model Application

For a better understanding of this improved Grey-Markov algorithm, we take a glass firber company's sales data as an example to illustrate the application of the algorithm. The following table shows the company's sales performance from July 2007 to August 2008.

Table 1. Sales Performance From July 2007 To August 2008

1 (Jul)	2 (Aug)	3 (Sep)	4 (Oct)	5 (Nov)
3526.2221	4791.2718	4883.866	4850.5931	4813.0319
6 (Dec)	**7 (Jan)**	**8 (Feb)**	**9 (Mar)**	**10 (Apr)**
4932.1256	4703.7828	5436.3688	5357.2408	5299.4797
11 (May)	**12 (Jun)**	**13 (Jul)**		
5081.8351	5036.52	4917.0305		

3.1 Constructing Optimal GM(1,1) Model

To construct optimal GM(1,1), we have to determine the best "m" value. In this example, the value of "m" can change from 1 to 13, according to formula (5) and the definition of mean prediction error, we can get the following mean prediction error table.

Table 2. Mean Prediction Error(%)

1 (Jul)	2 (Aug)	3 (Sep)	4 (Oct)	5 (Nov)
4.6795	4.6842	4.6927	4.6862	4.6644
6 (Dec)	**7 (Jan)**	**8 (Feb)**	**9 (Mar)**	**10 (Apr)**
4.6507	4.5948	4.6342	4.6519	4.6514
11 (May)	**12 (Jun)**	**13 (Jul)**		
4.6118	4.5575	4.4794		

From table 2 we find that when m=1, the corresponding mean prediction error is not the optimal one. So, assuming $\hat{x}^{(1)}(1) = x^{(1)}(1)$ as known condition, the predicted result of GM(1,1) model is not necessarily what we want it to be. However, when m is equal to 13, the predicted value we get from GM(1,1) model is more closer to true value, and the mean prediction error is only 4.4794%. Obviously, 13 is the best choice we can make. Therefore, we can make the conclusion that the GM(1,1) model corresponding to m=13 is the optimal one. This optimal GM(1,1) is:

$$\hat{x}^{(1)}(k+1) = [x^{(1)}(13) - u/a]e^{-a(k-12)} + u/a \tag{7}$$

3.2 GM(1,1) Prediction

In order to be more intuitive, some reduction is needed for formula (7). The final result obtained is shown below.

$$\hat{x}^{(0)}(k) = [x^{(1)}(13) - u/a](e^{-a} - 1)e^{(k-13)} \tag{8}$$

According to formula (8), the predicted values we can get from it and some other related data are listed in table 3.

Table 3. Optimal GM(1,1) Model Prediction Result

	Oct	Nov	Dec	Jan	Feb
True Value	4850.5931	4813.0319	4932.1256	4703.7828	5436.3688
Predicted Value	4915.6529	4903.4103	4843.9218	4921.8508	4791.5540
Prediction Error	1.3413	1.8778	-1.7884	4.6360	-11.8611
	Mar	**Apr**	**May**	**Jun**	**Jul**
True Value	5357.2408	5299.4797	5081.8351	5036.52	4917.0305
Predicted Value	5152.4804	5307.4495	5378.3228	5336.6196	5287.0021
Prediction Error	-3.8221	0.1504	5.8343	5.9585	7.5243

3.3 State Division

The average monthly sales of this company from October 2007 to July 2008 is $\overline{x} = 5042.80083$. According to the actual situation, here the original data sequence can be divided into four different state.

$$E_1 : E_{1L} = \hat{x}^{(0)}(t) - 0.08\overline{x}\,;\, E_{1R} = \hat{x}^{(0)}(t) - 0.04\overline{x}$$
$$E_2 : E_{2L} = \hat{x}^{(0)}(t) - 0.04\overline{x}\,;\, E_{2R} = \hat{x}^{(0)}(t)$$
$$E_3 : E_{3L} = \hat{x}^{(0)}(t) \qquad ;\, E_{3R} = \hat{x}^{(0)}(t) + 0.05\overline{x}$$
$$E_4 : E_{4L} = \hat{x}^{(0)}(t) + 0.05\overline{x}\,;\, E_{4R} = \hat{x}^{(0)}(t) + 0.13\overline{x}$$

3.4 Constructing State Transition Probability Matrix

According to the above state division standard, the company's monthly sales distribution from October 2007 to July 2008 is: $E_1 = 3, E_2 = 3, E_3 = 2, E_4 = 1$, and the sales of July 2008 is selected as the prediction target. The number of data that transfer from state E_1 to E_1, E_2, E_3 and E_4 after one step transition are: $M_{11} = 2$, $M_{12} = 0$, $M_{13} = 0$, $M_{14} = 1$. With the help of state transition probability formula $P_{ij}(m) = M_{ij}(m)/M_i$, we can know that the probability value of transferring from state E_1 to E_1, E_2, E_3 and E_4 after one-step transition are $P_{11} = 1$, $P_{12} = 0$, $P_{13} = 0$, $P_{14} = 0$. Similarly, we can get other state transition probability value. Finally, we can construct a one-step transition probability matrix using those probability values.

$$P(1) = \begin{bmatrix} 2/3 & 0 & 0 & 1/3 \\ 0 & 1/3 & 1/3 & 1/3 \\ 1/3 & 1/3 & 1/3 & 0 \\ 0 & 0 & 1 & 0 \end{bmatrix}$$

3.5 Get the Predicted Value

The state of June 2008 is E_1, according to the above transition probability matrix, we can infer that the most probable state of July 2008 is E_1. Therefore, we can predict that the sales volume for July 2008 is approx $(E_{1L} + E_{1R})/2 = 5034.0516$. Comparing to the real sales volume of July 2008, the prediction error is just 2.38%. Similarly, with the help of transition probability matrix, the predicted sales volume for August 2008 is approx 5008.0516.

4 Summary

Grey-Markov prediction model, which both have inheritaged the advantages of grey prediction and Markov model, can make full use of the information based on historical data for sales forecast. Using the forecast curve of Grey GM (1,1) model to reflect the general development trend of products sales, and then find the next most probable state with the help of Markov model, analysis shows that the predicted value of Grey-Markov model is very precise and accurate when predicting with random time series with high volatility. At the same time, this article is also a proof that there is no positive connection between the first element of original time sequence and the GM(1,1) model.

References

1. Sales Forecast, `http://wiki.mbalib.com/wiki`
2. Chen, L.-J.: System Security Index Grey- Markov Forecast Model And Its Application. Industrial Safety and Dust Control 9, 18–21 (1995)
3. Zhang, D.-H., Jiang, S.-F., Shi, K.-Q.: Theorietical Defect Of Grey Prediction Formula and Its Improvement. Systems Engineering-Theory & Practice, 140–142 (2008)
4. Hai, H.: Prediction Formula's Defect Of GM(1,1) And Its Improvement. Journal Of Wu Han University Of Technology 26(7), 81–83 (2004)
5. Yu, H.- r., Mo, J., Li, J.: Market Demand Prediction Based on GreyMarkov Model. Commercial Research 17(276), 43–45 (2009)

Vehicle Mass Estimation for Four In-Wheel-Motor Drive Vehicle

Zhuoping Yu, Yuan Feng, Lu Xiong*, and Xiaopeng Wu

Automotive College, Tongji University, CaoAn Road. 4800,
201804 Shanghai, China
yuzhuoping@fcv-sh.com, fengyuan0015@gmail.com,
xionglu.gm@gmail.com, wwwxpen@163.com

Abstract. Vehicle mass is an important parameter which is utilized in the vehicle dynamics control system. This paper proposes a new mass estimation algorithm for four in-wheel-motor drive vehicle, based on the accuracy and availability of driving torque and wheel speed signals. The algorithm estimates the vehicle mass and the driving resistance simultaneously using the recursive least square method with multiple forgetting. The results with experimental data show that the estimated values converge rapidly and the mass can be estimated with good accuracy.

Keywords: mass estimation, in-wheel-motor, electric vehicle.

1 Introduction

The importance of vehicle mass estimation has lead to considerable research in this area in the past few years [1]. Different approaches have been proposed, and most of them could be classified into two groups: sensor-based and model-based estimation.

Sensor-based methods, mainly concentrating on suspension dynamics (such as LVDT), provide an excellent opportunity for mass estimation [2]. As convenient as these ways are, the equipment of these special sensors may lead to higher price of an individual vehicle.

Model-based methods use a vehicle dynamics model, together with data (e.g., vehicle speed, engine torque) from the vehicle CAN-bus, in order to estimate unknown system parameters. In longitudinal dynamics model, the mass, longitudinal forces and acceleration are linked by an algebraic relationship. Ardalan Vahidi [3][4] proposed a recursive least squares with multiple forgetting estimation algorithm to reflect a time-varying road grade and a constant mass. Michael L. McIntyre [1] proposed a two-stage approach for the estimation. In the first stage, a least-squares estimator is developed to estimate the vehicle mass and a constant road grade. And secondly, a nonlinear estimator is developed to provide a more accurate estimation of the road grade. Vincent Winstead [5] developed an active on-line estimation scheme

* Corresponding author.

M. Zhu (Ed.): Electrical Engineering and Control, LNEE 98, pp. 117–125.
springerlink.com © Springer-Verlag Berlin Heidelberg 2011

for road grade and vehicle mass. The scheme combines an Extended Kalman Filter to generate on-line parameter estimates and a Model Predictive Controller to enhance parameter identifiability. Limitation in such method is to estimate accurately with variable road slope. Hosam K. Fathy [6] has designed a mass estimator based on the proposition that when the motion is predominantly longitudinal, the resistances can be ignored in the high-frequency component. But the effectiveness of obtaining the high-frequency component is needed to be discussed. However, the variability of the driving resistance under different road conditions hasn't been taken into consideration.

This paper proposes a new algorithm for a four in-wheel-motor drive vehicle. Containing the road gradient information in the measurement of the longitudinal acceleration sensor, the algorithm takes both vehicle mass and driving resistance as unknown parameters. And a test with several experimental data is carried out to prove the effectiveness of the algorithm.

2 Mass Estimation Algorithm

Once the vehicle starts, the vehicle mass can be regarded as a constant, so after the estimation ends during a standing start process, the result of the mass estimation will be kept until the vehicle stops. The flow chart of the algorithm is shown in Fig.1.

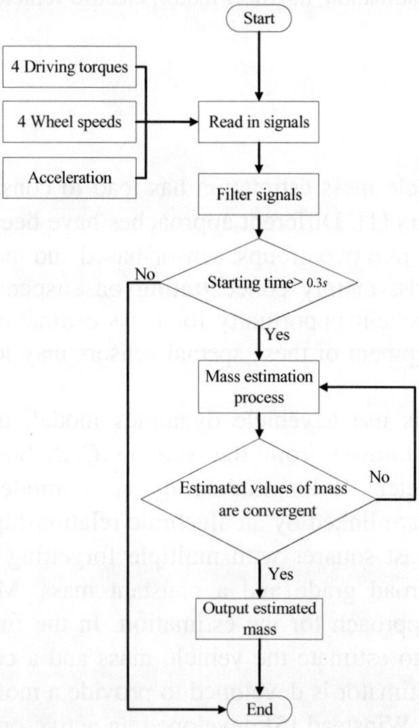

Fig. 1. The flow chart of the mass estimation algorithm

2.1 Vehicle Dynamics

The vehicle longitudinal dynamics model is shown in Fig.2. The wheel rotational dynamics model can be presented in the following form:

$$J_w\dot{\omega}_i = T_i - F_{xi}R - F_{ri}R \tag{1}$$

sum up $i = 1,2,3,4$, and $F_x = \sum F_{xi}$, $F_r = \sum F_{ri} = \mu_r mg\cos\beta$, then:

$$F_x = \sum_{i=1}^{4}[\frac{T_i - J_w\dot{\omega}_i}{R}] - \mu_r Mg\cos\beta \tag{2}$$

where $i = 1,2,3,4$ represents different wheels of the vehicle, J_w is the rotary inertia of each wheel, ω_i is the rotational speed of each wheel, T_i is the driving torque, F_{xi} is the longitudinal force, F_{Ri} is the rolling resistance, R is the radius of the wheel and μ_r is the rolling resistance coefficient.

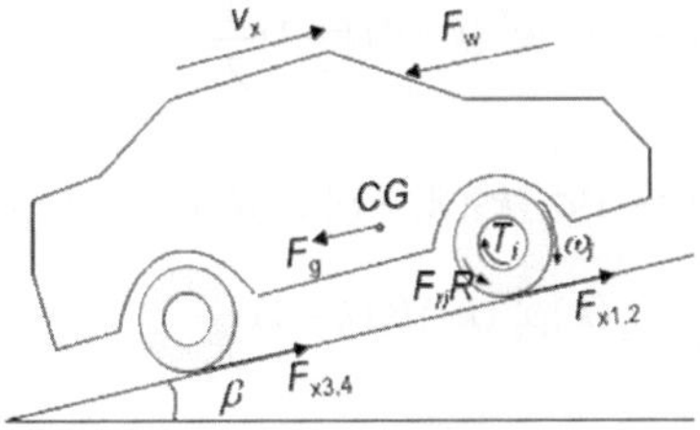

Fig. 2. The vehicle longitudinal dynamics.

The vehicle longitudinal dynamics model can be presented in the following form:

$$M\dot{v}_x = F_x - F_g - F_w = \sum_{i=1}^{4}[\frac{T_i - J_w\dot{\omega}_i}{R}] - \mu_r Mg\cos\beta - Mg\sin\beta - \frac{1}{1.63}C_d A v_x^2 \tag{3}$$

where F_g is the gradient resistance, F_w is the aerodynamic resistance, β is the road gradient, C_d is the drag coefficient and A is the frontal area.

The acceleration sensor can be used to calculate the tilted angle [7], therefore the measured acceleration comprises both the acceleration of the movement and the acceleration of gravity along the road slope:

$$a_{sensor,x} = \dot{v}_x + g\sin\beta \tag{4}$$

Using equation(4), equation(3) can be transformed into :

$$\sum_{i=1}^{4}[\frac{T_i - J_w\dot{\omega}_i}{R}] = Ma_{sensor,x} + \mu_r Mg\cos\beta + \frac{1}{1.63}C_d A v_x^2 \tag{5}$$

if $F_R = \mu_r Mg\cos\beta + \frac{1}{1.63}C_d A v_x^2$, equation(5) can be rewritten as:

$$\sum_{i=1}^{4}[\frac{T_i - J_w \dot{\omega}_i}{R}] = Ma_{sensor,x} + F_R \tag{6}$$

2.2 Recursive Least Square Method with Multiple Forgetting

The equation(6) can be written in the following linear parametric form:

$$y = \phi^T \theta \tag{7}$$

where $\phi = [\phi_1, \phi_2]$, $\theta = [\theta_1, \theta_2]^T$, $\theta_1 = M$, $\theta_2 = F_R$ are the estimated parameters, $y = \sum_{i=1}^{4}[\frac{T_i - J_w \dot{\omega}_i}{R}]$, $\phi_1 = a_{sensor,x}$, $\phi_2 = 1$. Using the classical form of the least square method, the solution of the recursive least square method with multiple forgetting [3] can be deduced:

$$\begin{cases} L_1(k) = P_1(k-1)\phi_1(k)(\lambda_1 + \phi_1^T(k)P_1(k-1)\phi_1(k))^{-1} \\ P_1(k) = (I - L_1(k)\phi_1^T(k))P_1(k-1)\frac{1}{\lambda_1} \end{cases} \tag{8}$$

$$\begin{cases} L_2(k) = P_2(k-1)\phi_2(k)(\lambda_2 + \phi_2^T(k)P_2(k-1)\phi_2(k))^{-1} \\ P_2(k) = (I - L_2(k)\phi_2^T(k))P_2(k-1)\frac{1}{\lambda_2} \end{cases} \tag{9}$$

$$\begin{bmatrix} \hat{\theta}_1(k) \\ \hat{\theta}_2(k) \end{bmatrix} = \begin{bmatrix} 1 & L_1(k)\phi_2(k) \\ L_2(k)\phi_1(k) & 1 \end{bmatrix}^{-1} \begin{bmatrix} \hat{\theta}_1(k-1) + L_1(k)(y(k) - \phi_1(k)\hat{\theta}_1(k-1)) \\ \hat{\theta}_2(k-1) + L_2(k)(y(k) - \phi_2(k)\hat{\theta}_2(k-1)) \end{bmatrix} \tag{10}$$

λ_1 and λ_2 are forgetting factors. The initial value of P_1 and P_2 should be large enough (1×10^6), the initial value of $\hat{\theta}_1$ should be close to the common mass, the one of $\hat{\theta}_2$ should be close to the common driving resistance.

2.3 Convergence Determination of the Algorithm

Once the estimation starts, the values of estimated mass are sampled every t_s second, we store only the last n values, and calculate the variance as follows:

$$\varepsilon = \sum_{i=1}^{n}(\frac{\hat{m}_i - \overline{m}}{\overline{m}})^2 \tag{11}$$

where $\hat{m}_i$ is one of the values estimated mass(sample time= t_s seconds), and $\overline{m}$ is the average of these last n values. When ε is smaller than ε_0, it is believed that the values of estimated mass are convergent, and the estimation stops. The values of t_s, n, ε_0 will be adjusted in the test.

Fig. 3. The test vehicle with four in-wheel-motors

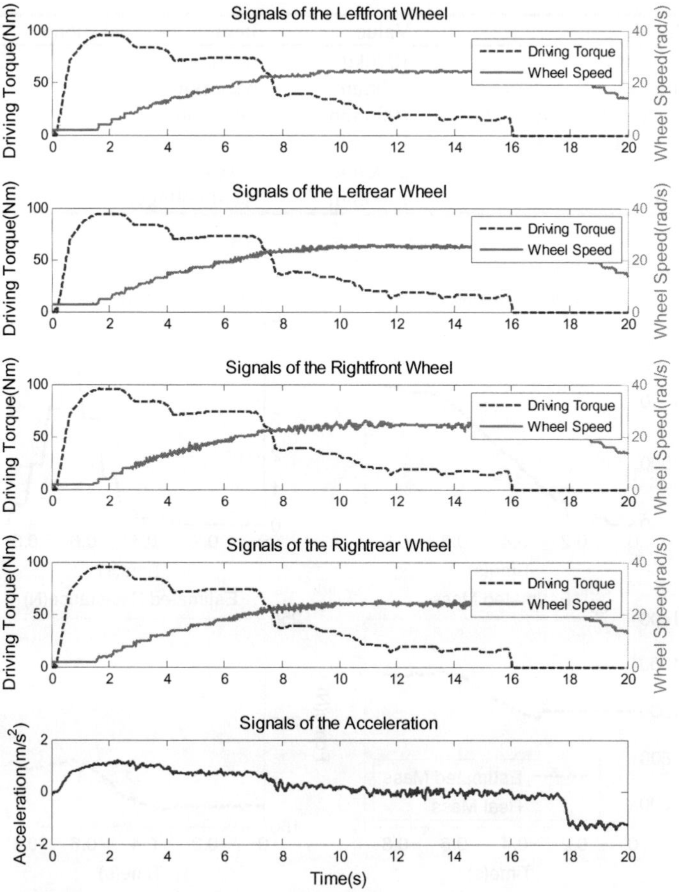

Fig. 4. The experimental data includes torque signals and rotational speed signals of four wheels, longitudinal acceleration signal of the vehicle. The vehicle starts from standing, and runs for about 100m on a flat road. The sample time is 0.02 second.

3 Results with Experimental Data

The data we used is obtained from the preceding test of a four in-wheel-motor drive vehicle (Fig. 3), and the specification of the vehicle is shown in Table 1. The experimental data is shown in Fig. 4.

The reasonable choose of λ_1 and λ_2 will be beneficial to the accurate estimation. After lots of simulations, we choose $\lambda_1 = 0.98$, $\lambda_1 = 0.95$. The estimation process of test 1, 2, 3 is shown in Fig.5, 6, 7. The estimated results are shown in Table 2.

Table 1. Specification of the test vehicle and the in-wheel-motor.

Item of vehicle	Value	Item of in-wheel-motor	Value
total vehicle mass	1070kg	rated power	2.5kW
height of the CG	380mm	peak power	7.5kw
distance between CG and front axle	1080mm	rated torque	55.7Nm
distance between CG and rear axle	1220mm	peak torque	167Nm
wheel base	2300mm	max speed	1250rpm
vehicle tread	1200mm	rated voltage	120V DC

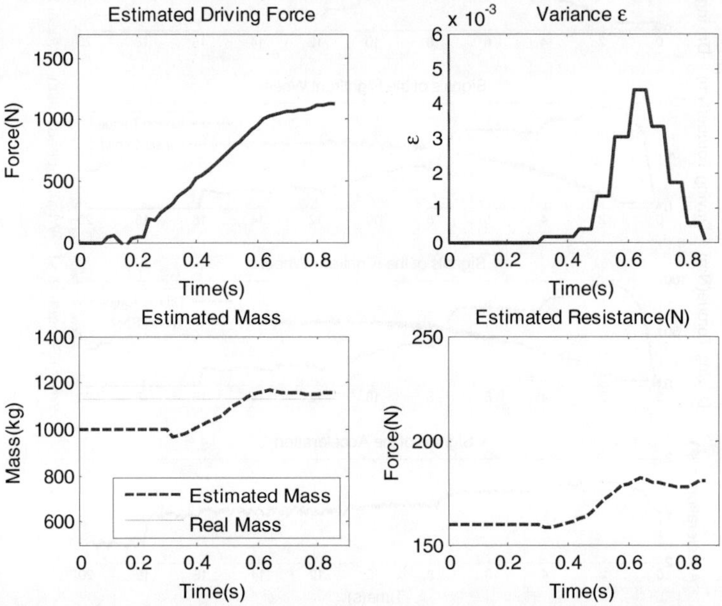

Fig. 5. The estimation process with a vehicle mass of 1130kg: $t_s = 0.06s$, $n = 6$. The estimation lasts from 0.3s to 0.86s, the time of the estimation is 0.56s. The final estimated mass is 1156kg with the estimated error of 2.30%.

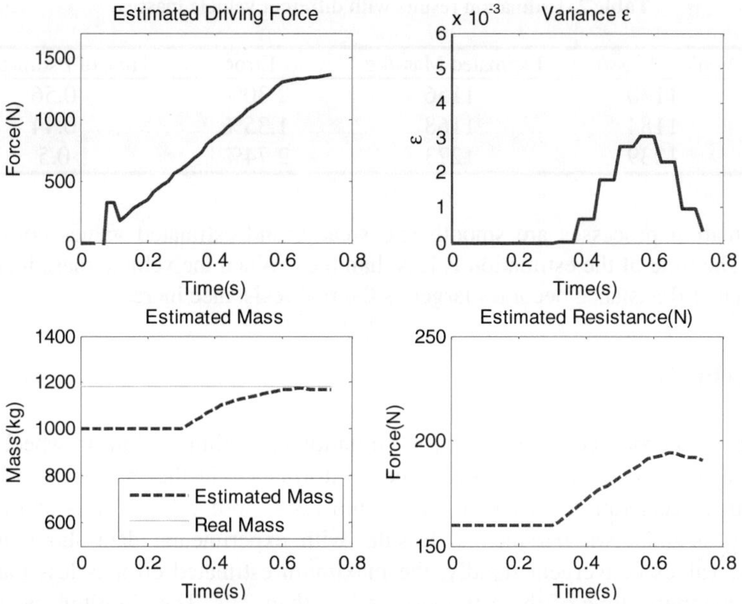

Fig. 6. The estimation process with a vehicle mass of 1184kg: $t_s = 0.06s$, $n = 6$. The estimation lasts from 0.3s to 0.74s, the time of the estimation is 0.44s. The final estimated mass is 1168kg with a error of 1.35%.

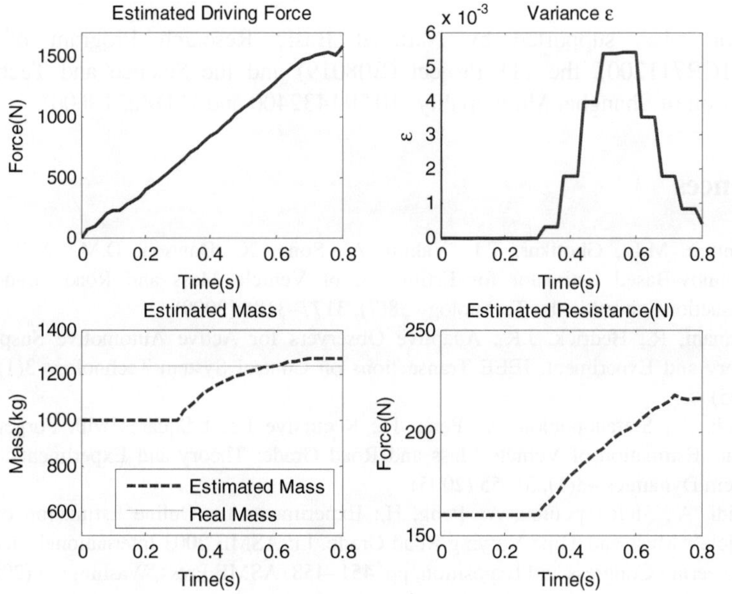

Fig. 7. The estimation process with a vehicle mass of 1239kg: $t_s = 0.06s$, $n = 6$. The estimation lasts from 0.3s to 0.8s, the time of the estimation is 0.5s. The final estimated mass is 1273kg with the estimated error of 2.74%.

Table 2. Estimation results with different vehicle masses.

	Vehicle Mass/kg	Estimated Mass/kg	Error	Time of Estimation/s
NO.1	1130	1156	2.30%	0.56
NO.2	1184	1168	1.35%	0.44
NO.3	1239	1273	2.74%	0.5

The estimation processes are smooth and steady, and estimated values convergent rapidly. The time of the estimation is less than 0.6s. When the vehicle mass increases, the estimated Resistance becomes larger as the real resistance increases.

4 Conclusion

This paper has proposed a vehicle mass estimation algorithm for four in-wheel-motor drive vehicle. Containing the road gradient information in the measurement of the longitudinal acceleration sensor, the algorithm takes both vehicle mass and driving resistance as unknown parameters. Results with experimental data show that the estimated values convergent rapidly, the maximum estimated error is less than 3%, and the maximum time of the estimation is less than 0.6s. The algorithm performed effectively.

Acknowledgments

This work was supported by National Basic Research Program of China (No.2011CB711200), the 111 Project (B08019) and the Science and Technology Commission of Shanghai Municipality (10ZR1432400 and 10DZ2210600).

References

1. McIntyre, M.L., Ghotikar, T.J., Vahidi, A., Song, X., Dawson, D.M.: A Two-Stage Lyapunov-Based Estimator for Estimation of Vehicle Mass and Road Grade. IEEE Transactions on vehicular Technology 58(7), 3177–3185 (2009)
2. Rajamani, R., Hedrick, J.K.: Adaptive Observers for Active Automotive Suspensions: Theory and Experiment. IEEE Transactions on Control System Technology 3(1), 86–93 (1995)
3. Vahidi, A., Stefanopoulou, A., Peng, H.: Recursive Least Squares with Forgetting for Online Estimation of Vehicle Mass and Road Grade: Theory and Experiments. Vehicle System Dynamics 43(1), 31–55 (2005)
4. Vahidi, A., Stefanopoulou, A., Peng, H.: Experiments for Online Estimation of Heavy Vehicle's Mass and Time-Varying Road Grade. In: ASME 2003 International Mechanical Engineering Congress and Exposition, pp. 451–458. ASME Press, Washington (2003)

5. Winstead, V., Kolmanovsky, I.V.: Estimation of Road Grade and Vehicle Mass via Model Predictive Control. In: Proceedings of the 2005 IEEE Conference on Control Applications, pp. 1588–1593. IEEE Press, New York (2005)
6. Fathy, H.K., Kang, D., Stein, J.L.: Online Vehicle Mass Estimation Using Recursive Least Squares and Supervisory Data Extraction. In: 2008 American Control Conference, pp. 1842–1848. IEEE Press, New York (2008)
7. Tuck, K.: Tilt Sensing Using Linear Accelerometers. Freescale Semiconductor Application Note, `http://www.freescale.com/files/sensors/doc/app_note/AN3461.pdf`

5. Winstead, V., Kolmanovsky, I.V.: Estimation of Road Grade and Vehicle Mass via Model Predictive Control. In: Proceedings of the 2005 IEEE Conference on Control Applications, pp. 1588–1593. IEEE Press, New York (2005)
6. Fathy, H.K., Kang, D., Stein, J.L.: Online Vehicle Mass Estimation Using Recursive Least Squares and Supervisory Data Extraction. In: 2008 American Control Conference, pp. 1842–1848. IEEE Press, New York (2008)
7. Tuck, K.: Tilt Sensing Using Linear Accelerometers, Freescale Semiconductor Application Note, `http://www.freescale.com/files/sensors/doc/app_note/AN3461.pdf`

A Method of State Diagnosis for Rolling Bearing Using Support Vector Machine and BP Neural Network

Jin Guan, Guofu Li, and Guangqing Liu

Faculty of Mechanical Engineering and Mechanics
Ningbo University, Ningbo, China
litswee@163.com

Abstract. By utilizing the SVM and neural BP network, a method of state diagnosis for rolling bearing is presented. The SVM is used to establish a classifier for the normal and fault state, then two kinds of samples caused by the distinct states are trained to judge whether the rolling bearing is normal or false. If the rolling bearing is in the fault state, all the fault samples were trained by the classifier composed of BP neural network to recognize which fault state it is in, otherwise, the state diagnosis is finished. The final experiment results show that the proposed method can diagnose the fault type more quickly and effectively in the small sample circumstances compared with the one using the BP neural networks solely.

Keywords: Support vector machine; BP neural network; State diagnosis; Rolling bearing.

1 Introduction

The rolling bearing is the most widely used standardized components in the mechanical equipment and one of the most easily damaged machine parts [1, 2], so its running status is directory related to the equipment performance of the whole machine. In the continuous production enterprise, rolling bearings is used extensively in the vital parts of the rotating equipment. According to the statistics, about 30% of the mechanical failure is due to the result of the damage of the rolling bearings in the rotating machinery using the rolling bearings. Therefore, there is extremely vital significance of the rolling bearing condition monitoring and fault diagnosis.

The most common method of the rolling bearing fault diagnosis is to make a correlation analysis of the bearing vibration signal, such as resonant demodulation method, the inverse spectrum analysis technology, the wavelet analysis, etc. In recent years, artificial intelligence is used more and more widely in the bearing fault diagnosis, such as using the highly nonlinear characteristics of the neural network and distributed storage, self-organizing and fault-tolerance properties of the information to establish pattern classifier for fault state classification, the expert system based on neural network for fault diagnosis, the dynamic prediction model for fault forecast. The RBF is used to establish the mapping relationship between the characteristic

M. Zhu (Ed.): Electrical Engineering and Control, LNEE 98, pp. 127–134.
springerlink.com © Springer-Verlag Berlin Heidelberg 2011

vector and fault mode to diagnose the fault of the rolling bear [3]· In addition, the combination of the neural network and the wavelet transformation, genetic algorithm and so on is also often used in the mechanical fault diagnosis, but because neural network belongs to the large sample study method, it possesses the disadvantages of large amount of calculation, causing excessive learning extremely easily when the training sample size is small, easy to fall into the local extreme value pole, not easy to find the best weights, lacking the theoretical basis to determinate the number of the neurons in the hidden layer and the choosing of it is subjective, so sometimes the diagnosis effect of using the fault diagnosis system set up on the neural network is not good.

The Support Vector Machine (SVM) is a learning algorithm for limited samples designed based on the statistical learning theory, structural risk minimization principle and VC Dimension (Vapnik Chervonenks Dimension) theory. It solved the problem of little sample, nonlinear, local minimum etc and provides a new way for the intelligent diagnosis which has been restricted because the lack of large fault samples. Yang Zhengyou etc found that, in limited fault sample conditions, SVM classifier had better classification performance than BP neural network classifier [4]. So there will be obvious advantages if it is applied to the rolling bearing fault diagnosis. The diagnosis model established on the combination of SVM and the BP neural network under the small fault samples was used in rolling bearing state diagnosis in this paper and achieved good diagnosis effect.

2 Support Vector Machine Classifier Principle

Support vector machine (SVM) is to construct the optimal hyperplane in the sample feature space to make the distance between different samples which nearest to the hyperplane largest and then reach the biggest generalization ability [5]. In linear can divide case if given the training sample points: (x_1, y_1) ,..., (x_l, y_l) , $x_i \in R^n$, $y_i \in \{-1,+1\}$, l stands for the sample number and n input dimension. There will be a separating hyperplane which makes these two types of samples separate completely. A splitting plane is constructed in the feature place and it is described as:

$$w \cdot x + b = 0 \tag{1}$$

This makes

$$\begin{cases} (w \cdot x_i) + b \geq 0 & y_i = +1 \\ (w \cdot x_i) + b < 0 & y_i = -1 \end{cases} \Leftrightarrow y_i[(w \cdot x_i) + b] \geq 1 \tag{2}$$

where $i = 1, 2, \ldots, l$.

According to the related theory, this problem can be transformed into solving quadratic programming which made $\min \phi(w) = \frac{1}{2}\|w\|^2$ the smallest. Quadratic optimization function $\phi(w)$ and the linear constraints (2) constituted a quadratic programming problem, and this could be solved by Lagrange multiplier method. When we introduce the Lagrange multiplier: $a_i \geq 0, i = 1,2,\cdots,l$, there is:

$$L(w,b,a) = \frac{1}{2}\|w\|^2 - \sum_{i=1}^{l} a_i \left[y_i(x_i w + b) - 1\right] \tag{3}$$

The solution of the constraint optimal problem is determined by the saddle points of the Lagrange function [6]. For this function, w and b must be minimized and a must be maximized first, and then through the original problem under the variable linear divisible condition the problem became a dual problem, finally, we can calculate the optimal weight vector and optimal partial values through solving the maximum of the following formula.

$$\max Q(a) = \sum_{i=1}^{l} a_i - \frac{1}{2}\sum_{j=1}^{l}\sum_{i=1}^{l} a_i a_j y_i y_j x_i x_j \tag{4}$$

The satisfying constraint condition of the Lagrange multipliers is as follows:

$$\sum_{i=1}^{l} a_i y_i = 0, a_i \geq 0 \tag{5}$$

In linear inseparable cases, the linear inseparable problem turned into linear divisible problem through using kernel function $K(x_i, x_j)$ in the optimal classifier plane. And then the sample x is mapped to High-dimensional Hilbert space where the linear classification after the nonlinear transform can be realized though using the linear classifier but the complexity did not increase, so the objective function is as follows:

$$Q(a) = \sum_{i=1}^{l} a_i - \frac{1}{2}\sum_{j=1}^{l}\sum_{i=1}^{l} a_i a_j y_i y_j K(x_i, x_j) \tag{6}$$

The satisfying constraint condition of the Lagrange multipliers is as follows:

$$\begin{cases} \sum_{i=1}^{l} a_i y_i = 0 \\ 0 \leq a_i \leq C \end{cases} \tag{7}$$

The C here is a penalty factor to control the punishment extent to the misclassification sample. Then the corresponding classification turned into follow

$$f(x) = \text{sgn}\sum_{i=1}^{l} y_i a_i^* K(x_i, x_j) + b^* \tag{8}$$

The a_i^* here is the optimal solution. The polynomial kernel function, RBF kernel function and S form (Sigmoid) kernel function are the commonly used kernel functions.

3 The Extraction of Rolling Bearing State Feature

The vibration signal is the research object of the fault diagnosis in this paper. The normal state, inner ring fault state, outer ring fault state, roller fault state and retainer fault state of the rolling bearing operation state are selected. Some characteristic parameters which can reflect the bearing state are extracted and analyzed. They are used as the characteristic quantities of the bearing's working state to train and learn and at last to achieve the purpose of state recognition. The characteristic parameters chosen here are: Amplitude domain statistical characteristic parameters mean square root value(rms) , Kurtosis coefficient(Kv) ,Peak coefficient (Cf) , Demodulation amplitude (Er) and the characteristic value of the inner ring ,outer ring, roller and retainer in the frequency domain feature parameters of demodulation spectrum(Spo , Spi , Spb , Sph). The experimental data is an actual data which from a railway vehicle bearing, the bearing number is 197726. Details please refer to reference [7]. In the literature, the fault diagnosis of the rolling bearing has been realized and a relatively good result also has been achieved through BP neural network method. It is conducted when the total training sample are 54, including normal for 43 and fault for 11. In order to improve the efficiency of the diagnosis and reduce the number of the training sample, the combination of the SVM and the BP network method is tried in the diagnosis of the rolling bearing state. The data in table 1 is part of the data in the

Table 1. Bearing characteristic parameters of the training sample in different states

Fault type	rms	Kv	Cf	Er	Spo	Spi	Spb	Sph
Sort A	0.1454	0.2051	0.5452	0.1205	0.0469	0.0898	0.0860	0.1204
	0.1975	0.2595	0.5838	0.5004	0.0496	0.2281	0.1130	1.0000
	0.1209	0.4380	0.8463	0.3734	0.0425	0.0904	0.4287	0.0883
Sort B	0.2086	1.0000	1.0000	0.2558	0.5796	0.1153	0.1479	0.0830
	0.0482	0.7666	0.9471	1.0000	1.0000	0.1555	0.1067	0.0661
	0.2123	0.6208	0.9176	0.1193	0.4045	0.1453	0.0860	0.1169
Sort C	0.1191	0.1528	0.4867	0.9727	0.0568	1.0000	0.0944	0.0887
	0.4725	0.2287	0.5725	0.1631	0.0339	0.6586	0.1187	0.0627
	0.2598	0.4771	0.8405	0.0918	0.0652	0.7672	0.1217	0.0332
Sort D	0.3899	0.1825	0.4941	0.6669	0.0634	0.0705	1.0000	0.0408
	0.1786	0.5067	0.8811	0.0630	0.0252	0.1125	0.6110	0.2238
	1.0000	0.1756	0.5594	0.8733	0.0732	0.1104	0.4449	0.0586
Sort E	0.5796	0.2825	0.6874	0.2271	0.0299	0.3143	0.1502	0.0693
Sort F	0.3766	0.2621	0.6588	0.7661	0.1287	0.1090	0.1239	0.5003

literature where when training completely using the BP neural network method. All these data are normalized, so they are representative and can reflect the various states of the bearings. The data of sort A in Table 1 represents the bearing working in normal condition, sort B the inner ring fault state, sort C the outer ring fault state, sort D the roller fault state, sort E the comprehensive faults state and sort F the retainer fault state.

4 State Diagnosis Model of Rolling Bearing

SVM is designed to solve two clustering problem. It can effectively solve the small sample learning problems, but the classification problems it often encountered are the Multi-cluster problems. There are four main methods to solve SVM multi-classification problems now. They are "One against One" clustering method, directed acyclic graph SVM method, "one against all" clustering approach and "half against half" clustering method.

In which, "One against One" clustering method constructed C_k^2 decision functions between fault types among the k fault types. The SVM decision function is trained by the two corresponding fault samples. Then the trained $k(k-1)/2$ two types of classifiers are able to be tested by data input. The commonly used method is Voting method. If a group of data need to be tested, they should to be input to the $k(k-1)/2$ classifiers successively, 1 will be added to the classifier which the test result shows the type it belongs to, the type of the testing data is determined by the type of classifier which has the highest vote number in the end. The number of the classifiers increases dramatically as the number of categories increases, it will result the Gate-level increased and the training and test speed reduced. If the two types get the same votes, it is hard to determine which kind they belonged to. Generally speaking, when the fault type is not so much, or the number of constructed classifiers is not so much, this kind of method is of superiority performance [8].

The directed acyclic graph SVM method generated the $k(k-1)/2$ classifiers through one-on-one arbitrary combinations method of the k fault types in the training.

All of the classifiers make up directed acyclic graph when classified. It consists of C_k^2 nodes and k leafs. Each node is a classifier. Classification can be completed only by k-1 steps [9].

"One Against all" clustering approach needs k two kinds of classifier to solve the classification problem of k fault types. If the training target of i kind classifier is defined as positive class, target outputs of the other sample data will be the negative classes. Calculate the Decision-making function value of each classifier separately in the training, so the category with the biggest function values is the type of the tested sample. So the category with the biggest function values is the type of the tested sample. The advantage of this method is that the number of the classifiers is equal to the number of categories, the number is relatively less and the classification speed is faster. However, its shortcoming is that all the k samples have to participate in the operation to construct each category classifier. Based on the principle that all the classifiers must have a maximum output, there will be erroneous judgment when testing sample does not belong to any kind of training sample [10].

Since the bearing is running in normal condition most of the time, if the normal state can not be told from fault state correctly, or made classification mistakes, it may cause a great lose. So in order to improve the efficiency of fault identification, a state diagnosis model for the rolling bearing using the method of the combination of SVM and BP neural network was proposed in this paper. In this mode, "One Against all" clustering approach was chosen. Firstly, as the SVM has the advantage of solving the small sample question better, the SVM is used to establish a classifier for the normal state and all the fault state on the basic of they are taken as two kinds, and the training is made to the two kinds of samples. During the training, the output of normal state is defined as 1, and the output of fault state is defined as-1. The second classifier consisted of BP networks and it is set up by all fault samples to train and study. Thus the rolling bearing fault diagnosis model is constructed by the combination of SVM and BP neural network. Testing began when the two classifiers had been set up and the testing data are showed in table 2. The fault type represented by the fault category in table 2 is as the same as table 1, in which, because there is no test data, sort E and sort F are revised based on the sample data. At first, the data in chart 2 are put into SVM classifier, then the target output of testing sample is （1 1 1 -1 -1 -1 -1 -1 -1 -1 -1 -1 -1 -1）. It is perceived that the method is able to distinguish the fault state from free-fault state correctly. If the target output is 1, the bearing is normal and the diagnosis ended. If the target output is -1, a BP neural network classifier began to work to check the kind of fault diagnosis.

BP network state classifier is designed by 3 layers in BP network. There are eight corresponding characteristics, so there are 8 nerve cells in the input layer. Test result show that adopting 31 inter layers neurons is able to get the best effect. The output layer has 5 neurons. The transfer function of hidden layer neurons is S-type tangent function and the transfer function of output layer neurons is S-type Logarithmic function. Finally, the corresponding target output of each state should be defined. （0, 0, 0, 0, 1） is used to represent inner ring fault or sort B, （0, 0, 0, 1, 0） outer ring fault or sort C, （0, 0, 1, 0, 0） roller fault or sort D, （0, 1, 0, 0, 0） resultant fault or sort E and （1, 0, 0, 0, 0） cage fault or sort F. So the network is trained by the few samples in table 1. Then the test output is produced by the defective test sample in table 2. The output shows in table 3. Results indicate that the proposed method is able to classify the various faults. If the sample is tested only by BP networks directly, it is tested not to be able to classify the states correctly and the test time is long relatively.

Table 2. Characteristic of the testing bearing sample in different conditions

Fault types	*rms*	*Kv*	*Cf*	*Er*	*Spo*	*Spi*	*Spb*	*Sph*
Sort A	0.2686	0.1566	0.5070	0.2394	0.0637	0.0986	0.2829	0.0696
	0.1389	0.1739	0.5407	0.0649	0.0505	0.1025	0.0664	0.0820
	0.2220	0.2048	0.5700	0.1296	0.0445	0.0790	0.0789	0.0849
Sort B	0.4070	0.1626	0.5175	0.0903	0.3865	0.0734	0.0800	0.0409
	0.1920	0.5444	0.8499	0.5326	0.3365	0.1064	0.2075	0.0735
Sort C	0.1440	0.4256	0.8303	0.2298	0.0391	0.8399	0.0808	0.0435
	0.1966	0.5643	0.8401	0.1584	0.0341	0.8956	0.0849	0.0862
Sort D	0.4532	0.2113	0.6005	0.1904	0.0471	0.0826	0.6234	0.1447
	0.5934	0.1636	0.5133	0.2138	0.0366	0.0930	0.3976	0.0946
Sort E	0.5786	0.2824	0.6974	0.2271	0.0279	0.3153	0.1582	0.0693
Sort F	0.3746	0.2661	0.6578	0.7761	0.1287	0.1098	0.1249	0.5003

Table 3. Output of the BP network test

Fault types	Desired output	Actual output
Sort B	0 0 0 0 1	0.0000 0.0485 0.0000 0.0000 0.9997
	0 0 0 0 1	0.0005 0.0000 0.0000 0.0000 0.9696
Sort C	0 0 0 1 0	0.0000 0.0000 0.0000 1.0000 0.0000
	0 0 0 1 0	0.0000 0.0000 0.0000 1.0000 0.0000
Sort D	0 0 1 0 0	0.0003 0.0001 1.0000 0.0000 0.0000
	0 0 1 0 0	0.0000 0.0577 0.9955 0.0000 0.0000
Sort E	0 1 0 0 0	0.0000 1.0000 0.0000 0.0000 0.0000
Sort F	1 0 0 0 0	1.0000 0.0000 0.0000 0.0000 0.0000

5 Conclusions

The running status of the rolling bearing is directory related to the equipment performance of the whole machine. Because rolling bearings are running in the normal state at most of the time, so in order to improve the efficiency of fault identification, under the condition of small fault samples, SVM is introduced to combine with the BP neural network for the diagnosis of the rolling bearing. Compared with the fault diagnosis method of the BP neural network, the presented state diagnosis method lowers the training samples apparently, applies easier algorithm and provide higher accuracy. The proposed method provides a new research technique to solve the fault diagnosis problems under the circumstance of limited samples.

Acknowledgements

This paper is supported by the Natural Science Foundation of Zhejiang Province, China (Y1080429), the National Basic Research Program of China (2009CB326204), the Natural Science Foundation of China (70871062), the Key Subject Program of Zhejiang Province, China (szxl1055).

References

1. Chang, J., Li, T., Luo, Q.: Fault dignosis of rolling bearing based on time domain parameters. In: Chinese Control and Decision Conference (CCDC),5498857, Xuzhou, China, May 26-28, pp. 2215–2218 (2010)
2. Han, Q., Wang, H.: Based on rolling bearing failure diagnosis in wavelet analysis. In: International Conference on Computer, Mechatronics, Control and Electronic Engineering (CMCE), Changchun, China, August 24-26, vol. 6, pp. 59–63 (2010)
3. Liu, L., Wei, L., Song, X., et al.: Fault diagnosis method of rolling bearing based on RBF neural network. Journal of Agricultural Machinery 37(3), 163–165 (2006)
4. Yang, Z., Peng, T.: Fault diagnosis method of rolling bearing based on Vibration signal analysis and SVM. Journal of Hunan University Of Technology 23(1), 96–99 (2009)
5. Shi, Z.: Neural networks. Higher Education Press, Beijing (2009)
6. Vapnik, V.N.: The nature of statistical learning theory[M]. Springer, NewYork (1999)
7. Ding, F., Shao, J., Zhang, Y., et al.: The application of neural network in fault diagnosis method of rolling bearing. Journal of Vibration Engineering 17, 425–428 (2004)
8. Sun, L., Yang, S.: Fault diagnosis method of rolling bearing based on SVM "one against one" cluster structure. Journal of Hefei University of Technology 32(1), 4–8 (2009)
9. Jaehe, Y., Azer, B., Ibrahim, M.: Adaptive reliable multicast. In: IEEE International Conference on Communications, New Orleans, US, July 18-22, vol. 3, pp. 1542–1546 (2000)
10. Zhang, Z., Li, L., He, Z.: Fault classifier and application based on SVM. Mechanical Science and Technology 23(5), 536–538 (2004)

RLM-AVF: Towards Visual Features Based Ranking Learning Model for Image Search

Xia Li[1], Jianjun Yu[2], and Jing Li[3]

[1] College of Sciences, Beijing Forestry University, 100083
leexia66@sina.com

[2] Computer Network Information Center, Chinese Academy of Sciences, 100190
yujj@cnic.ac.cn

[3] College of Computer Science and Technology, Huaqiao University, Xiamen, Fujian, 361021
lijing@nlsde.buaa.edu.cn

Abstract. With the development of Internet, large scale of images are published and searched in the Web. How to find those related images has been the research focus in the field of image processing. And one of the important problems is how to efficiently rank the image search results. In this paper, we present a learning to rank model named as RLM-AVF (Visual Features based Ranking Learning Model), which is based on the large margin method, under the framework of structural SVM(Support Vector Machine). Firstly we incorporate the visual features into the ranking model together with related textual features. Then an optimal problem of learning the ranking parameters based on the large margin method is schemed. Finally the cutting plane algorithm is introduced to efficiently solve the optimal problem. This paper compared the performance of RLM-AVF with the textual ranking methods using the well-known learning to rank algorithms and reranking methods. The experimental result shows that RLM-AVF performs considerable search efficiency than the other algorithms.

Keywords: Image Search, Learning to Rank, Visual Reranking, Visual Feature.

1 Introduction

A large scale of images are publishing at WWW, and how to find those related images has been the research focus. One of the most important problems is how to rank those image search results efficiently. In this paper, we would choose the learning to rank model to rank the search results, and present the most similar images (i.e. the top ranking results) to the users. Learning to rank is a type of supervised or semi-supervised machine learning problem, which aims to automatically construct a ranking model from training data. This training data consists of lists of items with some partial order specified between items in each list. Most of the research work on image search are based on textual learning to rank algorithms, like RankSVM[1], ListNet[2]. Or based on visual reranking algorithm, such as Visual Rank[3], Bayesian Reranking[4]. Textual

M. Zhu (Ed.): Electrical Engineering and Control, LNEE 98, pp. 135–140.
springerlink.com © Springer-Verlag Berlin Heidelberg 2011

learning to rank algorithms would apply the related textual information tagged on the images, such as the title, label, and the context web information on the image, to rank the image search results. Whereas these textual information includes the noise, uncertain, and even wrong tagged information, which would certainly reduce the accuracy of the ranking results. Visual reranking algorithm adds a separate reranking step based on the preliminary results searched with the textual learning to rank algorithm, which aims to promote the accuracy with the visual features of the images. Whereas the reranking step would enhance the errors if the preliminary results are overfitting and offset imported from the training sets.

In our option, the visual features would not help to efficiently promote the performance for the preliminary ranking model. In this paper, we analyze the shortcomings of these two approaches, and combine the advantages of the textual features and the visual features based learning to rank algorithms. We then scheme a novel learning to rank model named as RLM-AVF (Visual Features based Ranking Learning Model) to solve the ranking problems of image search results.

2 RLM-AVF: Learning to Rank Model for Image Search

2.1 The Construction of the RLM-AVF Model

RLM-AVF is a model that applies image visual features for the learning to rank algorithm based on textual features of image search results.

Assume we would learn to rank with the given search set Q. For each query $q^i = \{x^i, v^i, y^i\} \in Q, y^i = [y_1^i, ..., y_{N^i}^i] \in \Upsilon$, we define it as the ranking scores of manual label for each query q^i on the all of the images. N^i is expressed as the corresponding number of the images of query q^i. x is represented as the textual features space, v as the image visual features space. and Υ as the ranking score space.

The traditional ranking models only consider the textual features. That means for each query with textual feature x^i, the ranking function would be defined as follows:

$$\widetilde{y} = f(x^i) = arg\max\nolimits_{y \in \Upsilon} F(x^i, y; w) \tag{1}$$

In the above function, $x^i = [x_1^i, ..., x_{N^i}^i] \in \chi$ is the textual features of the query/image pairs, w is the ranking parameter trained from the given training set. The function $F = (x^i, y; w)$ can be defined as $F = (x^i, y; w) = w^T \psi(x^i, y)$, where $\psi(x^i, y)$ maps the textual features x^i and ranking prediction results y into the real number. For example, $\psi(x^i, y)$ can be mapped as $\sum_{j=1}^{N^i} x_j^i y_j$.

Based on the above model, we add the visual features into the learning to rank model based on the textual features. Intuitively for each query, similar images have the characteristics of the same vision. That means the similar images would have approximately same ranking results.

With this assumption, we expect that our ranking results should have the same homogeneity of the visual features in addition to the similar textual features. So we provide a visual features based learning to rank model combining the visual feature as follows:

$$\begin{aligned}\widetilde{y} &= arg\max_{y\in\Upsilon} F(x^i, v^i, y; w) \\ &= arg\max_{y\in\Upsilon} w^T\psi(x^i, y) - \gamma \textstyle\sum_{m,n=1}^{N^i} G^i_{mn}(y_m - y_n)^2 \end{aligned} \quad (2)$$

Where $v^i = [v^i_1, ..., v^i_{N^i}] \in \nu$ represents the visual features of the images.

$\gamma > 0$ is a parameter to balance the value between the score prediction item $w^T\psi(x^i, y)$ and visual homogeneity item $\sum_{m,n=1}^{N^i} G^i_{mn}(y_m - y_n)^2$.

G^i is a matrix that aims to measure the similarity between each pair of images, which can be defined as the follows:

$$\begin{aligned}&G^i_{mn} = Sim(v^i_m, v^i_n) \text{ if } v^i_n \text{ is the KNN (k-nearest neighbors) of } v^i_m \\ &\text{else } G^i_{mn} = 0 \text{ if } v^i_n \text{ is not the KNN (k-nearest neighbors) of } v^i_m.\end{aligned} \quad (3)$$

Where $Sim(v^i_m, v^i_n)$ is the similarity between v^i_n and v^i_m, which can be measured by the vector space distance of these two images.

According to the above function, G^i is a sparse matrix applied with KNN policy. The minimum value of visual homogeneity item would make those similar images labeled with the similar prediction results. This is because those similar image pairs would have the bigger value of G^i_{mn}. To minimize the value of visual homogeneity item, we should make the value $(y_m - y_n)^2$ as minimal as possible. That means we should make the value y_m and y_n as same as possible.

With the function (3), we can compute the values of different images with visual features and textual features. The consequent question is to rank the search results and give a considerable computing model.

2.2 The Optimization of w in RLM-AVF Model

Given a labeled training set of Q, we try to find a weighted vector w, which can predict the ranking result of each query in Q with the learning to rank model.

We borrowed the idea of structural support vector machines and the large margin method [5] to solve the optimization problem of the learning to rank parameter w. They generalize large margin methods to the broader problem of learning structured responses, which aims to learn mappings involving complex structures in polynomial time despite an exponential (or infinite) number of possible output values.

Thus we define the optimization problems of w as follows:

$$\begin{aligned}&min_w \tfrac{1}{2} \parallel w \parallel^2 \\ &s.t. \forall q^i \in Q, y \neq y^i, \parallel y \parallel = 1, F(x^i, v^i, y^i; w) - F(x^i, v^i, y; w) \geq 1\end{aligned} \quad (4)$$

To reduce the noise in the training dataset, we import a slack variable to reduce the strong constrain of the function. Thus we can transform the above optimization problem as the following function:

$$min_{w,\xi} \tfrac{1}{2} \parallel w \parallel^2 + C\Sigma\xi^i$$

$$s.t. \forall q^i \in Q, \xi^i \geq 0, y \neq y^i, \| y \| = 1,$$
$$F(x^i, v^i, y^i; w) - F(x^i, v^i, y; w) \geq \Delta(y^i, y) - \xi^i \quad (5)$$

Where $C > 0$ is a parameter to balance the values between the model complex $\|w\|^2$ and the upper bound of prediction loss $\sum \xi^i$.

$\Delta(y^i, y)$ aims to measure the loss of prediction ranking results y comparing to the base ranking results y^i.

3 Model Solutions

3.1 The Solutions of Ranking Parameter w

It's difficult to optimize the function (5) directly for the constrained number is almost unlimited. That means it's impossible to compute the parameter w in a limited time.

In this paper, we apply the cutting plane method to solve the optimization problem. In mathematical optimization, the cutting plane method iteratively refines a feasible set or objective function by means of linear inequalities, termed cuts. Such procedures aim to find integer solutions to mixed integer linear programming problems, as well as to solve general, not necessarily differentiable convex optimization problems. Referring to the cutting plane method, The optimization algorithm to compute the parameter w is introduced as the following steps:

1) Executed from an empty constrained set. And then executed several rounds iteratively.

2) During each round, tries to find the most possible value $\hat{y}$ that would violate the predict constraint under current w.

3) If the correspondent constraint for the $\hat{y}$ is violated with a predefined tolerance ε, $\hat{y}$ would be added into the work set W^i for the query q^i.

4) Seek the best solution for the w under the current all constraints $W = \bigcup_i W^i$.

Theoretically the optimization algorithm would assure the convergence under monomial iterative steps constrained by the parameter ε. Detailed algorithm analysis is described in the reference [5].

3.2 Ranking Prediction

Given a test query q^t combining the visual features with the textual features x^t, v^t, the correspondent ranking results of the images can be predicted by the learning to rank parameter $\tilde{w}$. That means we would make $w = \tilde{w}$ in the function (2), and then compute the result of function (2) as follows:

$$\tilde{y} = arg\max_{y \in \Upsilon} \tilde{w}^T \psi(x^t, y) - \gamma \sum_{m,n=1}^{N^t} G^t_{mn} (y_m - y_n)^2 \quad (6)$$

Thus the optimization problem can be computed by the dual Lagrange method and one-dimensional linear search efficiently.

4 Experiment Evaluation

4.1 Experiment Setup

In this paper, we would choose 150 frequently used web query keywords, and use these keyword to acquire the correspondent images from the business image search engine. The search results of images would be used as the experiment dataset.

In this paper, for each query q, we would choose the top 10 ranking images from the search engines, i.e. we would choose 1,500 images as the experiment dataset. The query keywords and the correspondent search result of images would be represented as the two tuple $(query, images)$. For each $(query, images)$ pair, we extract the textual features related with the query and 5 types of visual features that irrelated with the query, includes Attention Guided Color Signature, Wavelet, SIFT, Multi-Layer Rotation Invariant EOH, Histogram of Gradient. Each $(query, images)$ pair would be labeled as $'relevant'$, $'notrelevant'$ or $'highlyrelevant'$ manually.

We compared our RLM-AVF model with those famous model, like the textual based learning to rank algorithm RankSVM, ListNet, and visual reranking algorithm Visual Rank and Bayesian Reranking on the ranking performance. Because the visual reranking algorithm needs a preliminary ranking result based on the textual learning to rank algorithm, we apply RankSVM and ListNet separately as the ranking base of textual feature for each visual reranking algorithm.

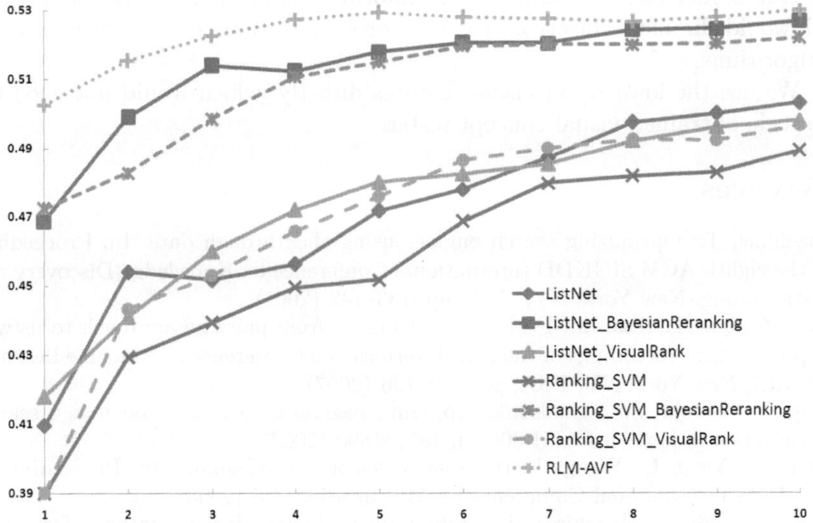

Fig. 1. The Comparison Between RLM-AVF and Other Algorithms

We got four visual reranking algorithms: RankSVM-VisualRank, RankSVM-BayesianReranking,ListNet-VisualRank, ListNet-BayesianReranking. These algorithms would be evaluated with the NDCG(Normalized Discounted Cumulative Gain) approach.

4.2 Experiment Results

We give an experiment that compares the performance between RLM-AVF algorithm and other six algorithms as shown in Figure 1.

We can conclude that the performance of RLM-AVF is better than other algorithms. This experiment proves that our RLM-AVF processes the noise in the textual features of sample images efficiently during the construction of combining the visual homogeneity into learning to rank model, which helps to promote the performance of the learning to rank model.

5 Conclusions

In this paper, we put forward a novel learning to rank model named as RLM-AVF, which is proved to provide the better performance than other learning to rank model for image search. RLM-AVF combines the visual features with the textual features of images, and imports the lower level visual features to promote the ranking results. So RLM-AVF has three advantages comparing to other learning to rank algorithms:

1) RLM-AVF inherites the advantages of the visual features based learning to rank model, that means we consider the visual features of images to rank the results.

2) Our model can be learned in a uniform framework with the single step, which would eliminate the error broadcast problems existing in the visual reranking algorithms.

3) We use the lower level visual features directly, which would not need the large scale of trained visual concept tester.

References

1. Joachims, T.: Optimizing search engines using clickthrough data. In: Proceedings of the eighth ACM SIGKDD International Conference on Knowledge Discovery and Data Mining, New York, NY, USA, pp. 133–142 (2002)
2. Cao, Z., Qin, T., Tsai, M.-F.: Learning to rank: from pairwise approach to listwise approach. In: Proceedings of the 24th International Conference on Machine Learning (ICML), New York, NY, USA, pp. 129–136 (2007)
3. Jing, Y., Baluja, S.: Visualrank: Applying pagerank to large-scale image search. IEEE Transactions on PAMI 30(11), 1877–1890 (2008)
4. Tian, X., Yang, L., Wang, J.: Bayesian video search reranking. In: Proceedings of the ACM International Conference on Multimedia, MM (2008)
5. Tsochantaridis, I., Joachims, T., Hofmann, T.: Large Margin Methods for Structured and Interdependent Output Variables. Journal of Machine Learning Research (JMLR) 6, 1453–1484 (2005)

Design and Implement of Intelligent Water Meter Remote Monitor System Based on WCF

Lilong Jiang, Jianjiang Cui, Junjie Li, and Zhijie Lu

Room 118 at Complex Building, Software College,
Northeastern University, Shenyang, Liaoning Province, China
jianglilong0225@gmail.com

Abstract. According to the needs of water meter remote monitor system for interoperability and security, based on WCF, a loose-coupling, service-oriented and distributed intelligent water meter remote monitor system was designed. With the system, the users can sell water, carry out data analysis and monitor the water meters and so on. This system can serve to the informatization in Water Supplies Department. The relevant techniques can be referenced by other meters.

Keywords: WCF, Intelligent Water Meter, Asynchronous Socket, ADAM, FusionCharts.

1 Introduction

With the development of the informatization in Water Supplies Department, the remote monitor for water meters is become more and more important. At present, the water meter remote monitor system in developed countries is relatively mature, while in China water meter remote monitor system is at the early-stage. Considering the differences in the informatization in different districts, such as the adopted web server and operating system, the design of the system must consider the cross-regional and cross-platform, namely interoperability; at the same time, because of the secrecy of the information relevant to water supplies, security should be taken into consideration. WCF (Windows Communication Foundation) is the extension of .NET Framework. WCF provides the uniform framework for constructing secure, reliable and transaction services. The developer can develop the service-oriented application based on WCF. The system is intelligent water meter remote monitor system based on WCF. The water meter can establish the bi-directional connection with the central server. The end user can take advantage of the system to sell water, carry out data analysis and monitor water meters and so on. The system deploys the operations as WCF services, so a loose-coupling, service-oriented and distributed network platform was constructed.

M. Zhu (Ed.): Electrical Engineering and Control, LNEE 98, pp. 141–147.
springerlink.com © Springer-Verlag Berlin Heidelberg 2011

2 Introduction of WCF

WCF, a group of application developing interfaces for data communication, was developed by Microsoft. The communication ways between both sides was defined by the contract. The communication methods obeyed by both sides were enacted by the protocol binding. The security during the communication was implemented by security layer. The contract was embodied by the interface while the actual service must derive from the contract. That a WCF client uses a proxy to access service is showed in Fig.1.

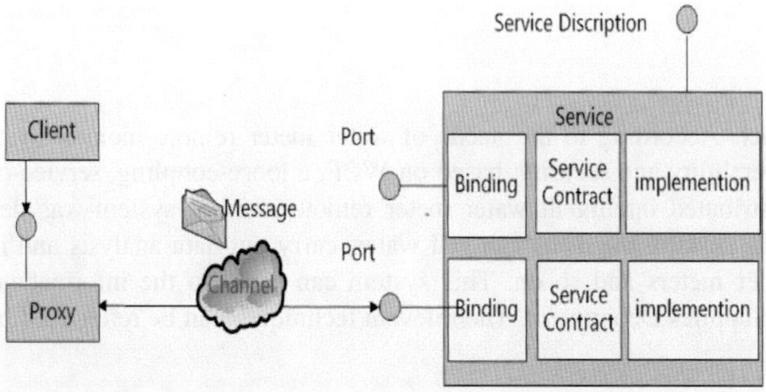

Fig. 1. A WCF client uses a proxy to access service.

3 Architecture of the Intelligent Water Meter Remote Monitor System Based on WCF

The overall architecture of the intelligent water meter remote monitor system based on WCF is illustrates in Fig.2. WCF services were deployed in WebHost, while website was deployed in WebApp. WebApp invokes the relative service by the proxy

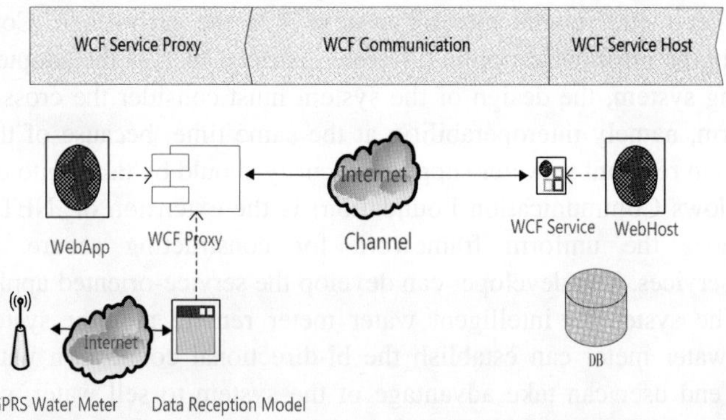

Fig. 2. This shows a figure of the architecture of the intelligent water meter remote monitor system based on WCF.

WCFProxy. For the communication with the water meters, the data reception model was deployed for listening and receives the message through GPRS from water meters, and then data reception model accesses WCF service to accomplish the data exchange between the server and water meters.

4 Design and Implement of the Intelligent Water Meter Remote Monitor System

4.1 Design and Implement of Communication with Water Meters

In the system, the communication between the server and the water meters is accomplished by asynchronous socket. Without adopting the multithreading technology, synchronous socket may block after listening the connection, not suitable for high concurrent accesses. If adopting multithreading technology, when many water meters established connections with the server simultaneously, the server needs to create many threads to handle the connections, as a result, much resource in server would be consumed and even to cause the loss of package. Considering all above questions, asynchronous socket was adopted to handle the communication.

In the system, Class SocketAsyncEventArgs in .NET was used to implement the asynchronous socket. The class was especially for the design of application in need of high-perform socket server. The process of communication is showed in Fig.3.

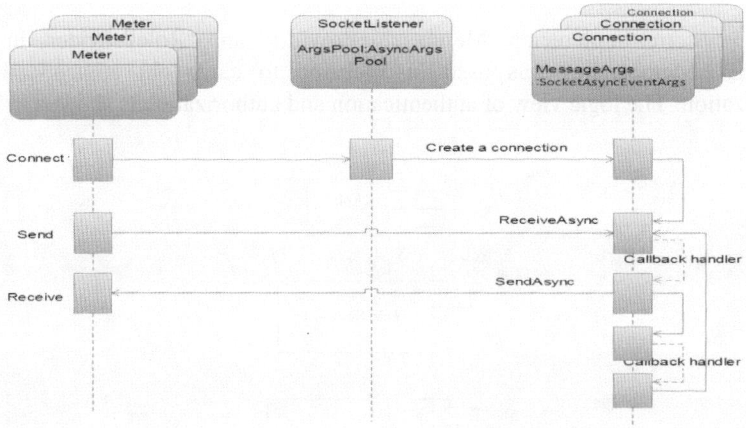

Fig. 3. This figure describes the communication with SocketAsyncEventArgs.

4.2 Design and Implement of Authentication and Authorization

The authentication and authorization of the system was implemented by adopting the ADAM and RBAC. ADAM (Active Directory Application Model) is a new kind of Active Directory, especially supplying LDAP directory service for application. The system stores the authentication and authorization information in directory by using

ADAM. The tenant's archive information and users' information was stored in Identity Database. The information in ADAM and Identity Database accomplish the management of users jointly. The system encapsulates AuthenticationSerivce and AuthorizationService as WCF services. The class diagram of authentication and authorization is showed in Fig.4. Class AuthenticationService mainly accomplishes operations about the users and the operations relative to roles are mainly carried out by Class AuthorizationSerivce.

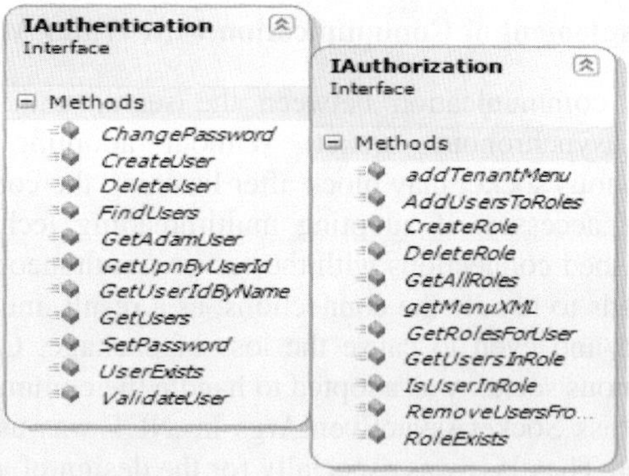

Fig. 4. This figure describes the class diagram of AuthenticationService and AuthorizationService.

The system implements MembershipProvider and RoleProvider in .NET. ASP.NET Membership was used in WebApp to carry out authentication and authorization. The logic view of authentication and authorization is showed in Fig.5.

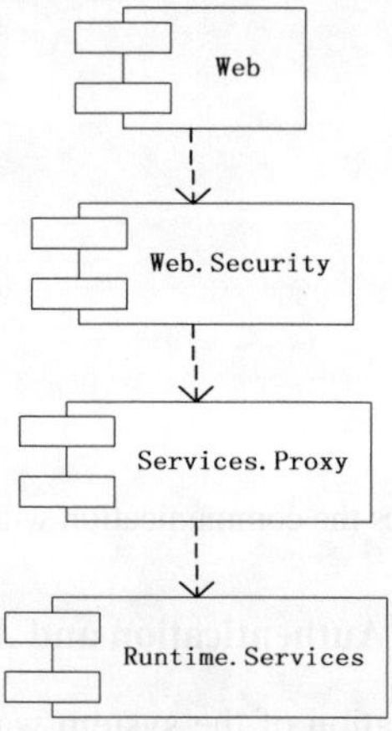

Fig. 5. This figure describes the logic view of authentication and authorization.

4.3 Design and Implement of Sale of Water

Sale of water accomplished the process of users' paying water charges. The system deploys functions relative to sale of water as WCF services. Clients in WebApp invoke the services by proxies. The sequential chart for sale of water is showed in Fig.6.

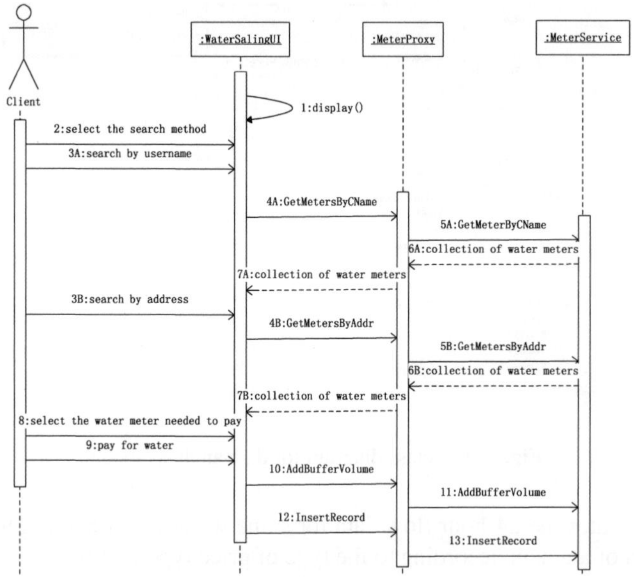

Fig. 6. The sequential chart for sale of water.

At the same time, making use of the transaction fundamental structure provided by Class System.Transactions in .NET, the operation of selling water was regarded by a transaction.

4.4 Design and Implement of Data Analysis

FusionCharts Free is a across-platform solution for creating animated and interactive charts. Here, FusionCharts Free is used to display statistics generated in this monitor system. The statistics includes both the 24-hour flow data of water meters and water consumption of a tenant according to the type of price type during a specific time.

In WebApp, FuChartsProxy is deployed as a proxy for charts service, FuChartsService is deployed as charts service in WebHost. In WebApp, application obtains serialized objects such as DateFlow and WaterConsum from FuChartsService by invoking FuChartsProxy. At last the application sends both data and flash to the client browser, thus charts is generated on the client side.

The class diagram of data analysis model is shown in Fig.7.In this model, the service contract is IFuChats and data contract is DateFlow and WaterConsum. DateFlow stores the flow data information ,WaterConsum works as a container which is filled with different water consumption during a specific time.

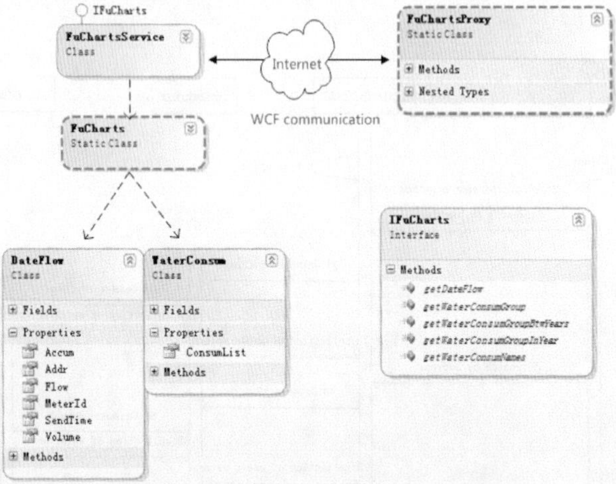

Fig. 7. The class diagram for data analysis model.

Fig.8 illustrates the 24-hour flow data for some water meter and Fig.9 shows water consumption of a tenant according to the type of price type in 2011.

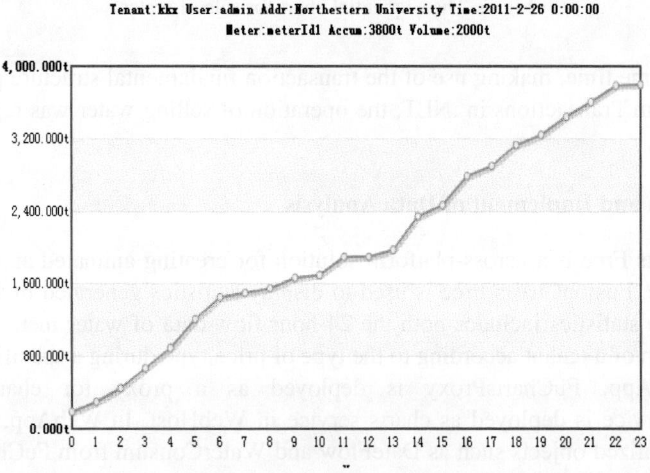

Fig. 8. 24-hour flow data for some water meter.

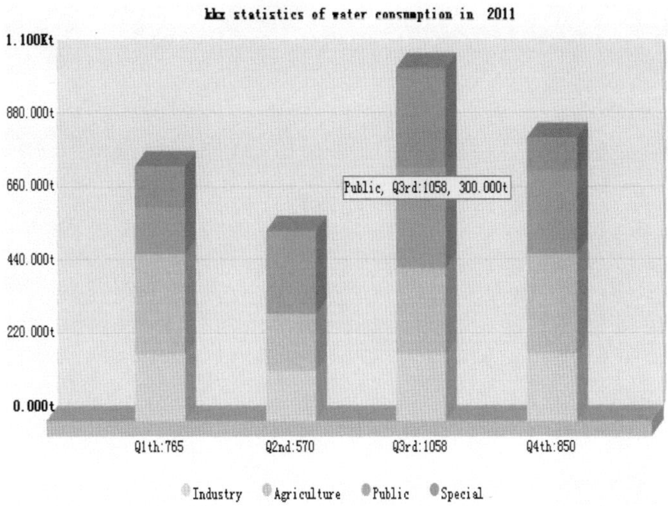

Fig. 9. Water consumption of a tenant according to the type of price type in 2011.

5 Conclusions

Applying the technology of WCF into the intelligent water meter remote monitor system enhanced the degree of loose-coupling, embodied the idea of SOA and provided a solution for wireless devices accessing the server simultaneously. The intelligent water meter remote monitor system implements the water reading, monitor of water meters, selling water and data analysis of water-consuming information, especially serves to the informationization of the Water Supplies Department. At the same time, the system provides reference to the distributed monitor for wireless devices and the service-oriented system architecture.

References

1. Molina, F.J., Barbancho, J., Luque, J.: Automated meter reading and SCADA application for wireless sensor network. J.Comput.Sci. 2865, 223–234 (2003)
2. Brasek, C.: Urban utilities warm up to the idea of wireless automatic meter reading. J. Comput. Control Eng. 15, 10–14 (2005)
3. Evjen, B., Hanselman, S., Rader, D.: Professional ASP.NET 3.5 SP1 in C# and VB. Tsinghua University, Beijing (2010) (in Chinese)
4. Klein, S.: Professional WCF Programming. Tsinghua University, Beijing (2008) (in Chinese)
5. Nagel, C., Evjen, B., Glynn, J.: Professional C# 2008. Tshinghua University, Beijing (2008) (in Chinese)
6. MSDN, http://msdn.microsoft.com/en-gb/default
7. Stunning charts for Web and Enterprise, http://www.fusioncharts.com

Protection of Solar Electric Car DC Motor with PIC Controller

Ahmed M.A. Haidar[1], Ramdan Razali[1], Ahmed Abdalla[1], Hazizulden Abdul Aziz[1], Khaled M.G. Noman[2], and Rashad A. Al-Jawfi[3]

[1] University Malaysia Pahang, Malaysia
[2] Sana'a University
[3] Ibb University, Yemen

Abstract. The electric car may represent new opportunities for any country and its electric utilities. Widespread use of electric cars can reduce the consumption of both imported and domestic oil, substitute abundant fuels such as coal and nuclear power. Since the charging of electric cars DC motor can be accomplished to a large extent during utility off-peak hours, electric cars can contribute to improve the utility load factors, as a result, reducing the average cost of generation. The problem arising when the DC motor does not stop automatically due to the abnormal condition and cause the loss of energy and the damages to the motor itself. This paper is mainly about controlling ON and OFF of Electric Car DC motor when the load is sharply varied. In this study, the DC motor is connected to the current sensor interfaced with the PIC controller and a booster IC is used to boost 12V to 24V from a rechargeable battery which is supplied by a solar panel. The PIC is utilized for automatically stopping the DC motor in order to save the Electric Car and reduce the lost of energies. The Microcode Studio software has been used for PIC coding incorporating with the Protues 7.5 simulation. Since the DC motor is being used extensively in machineries and vehicles, the proposed controlling system in this paper could reduce the cost in the industries and improve the quality of Electric Cars.

Keywords: Electric Car, DC motor, PIC controller, Temperature sensor.

1 Introduction

Nowadays, in many countries, the solar cell energy has been used to replace the fuel. Solar cell energy does not create the pollution, but still has to be developed into effective use. Many car manufacturers introduce a hybrid electric car in their gamut. The required motor power has been determined from typical traffic load diagrams. For a top speed of 80 km/h on slopes up to 20 % and a targeted vehicle weight of 650 kg and the motor power needs to be 17.5 kW nominal (30 kW peak) [1]. Of the several technologies available, the ones adopted by the automotive industries classified broadly into hybrid electric vehicles (HEV's) and electric vehicles (EV's). The concept of hybrid electric vehicles initially came up as a stop gap approach to facilitate a smoother transition from conventional to electric vehicles as and when

M. Zhu (Ed.): Electrical Engineering and Control, LNEE 98, pp. 149–161.
springerlink.com © Springer-Verlag Berlin Heidelberg 2011

EV's became viable. Compressed air based technologies, plug-in HEV's; solar powered automobiles are some of the several alternative propulsion technologies being worked on right now [2].

For a hybrid car, low fuel consumption is one of the most important goals because the additional cost of the hybrid components must be amortized by the cost reduction of fuel consumption. A long life cycle for the hybrid component is also an important goal, because an amortization can only be achieved if all car components reach their estimated age. With the respect to the life time, a battery is the most sensitive component in electric car [3]. The current designs have relatively poor performance when discharged at high rates and their use in automobiles is likely to be restricted to low-power or reserve duties where their high specific energy is the main attractive factor [4]. Currently, an electric car incentive is the possibility of using renewable, liquid hydrocarbon fuels, such as methanol, in efficient reformer systems to replace pure hydrogen storage and improve the fuel storage capacity [5].

Due to the technical demands and the progress in the field of car networks, enforce continuing development of the existing electrical car sensors [6]. In the automotive industry, sensors are mainly used to give information to the driver. In some cases sensors are connected to a computer that performs some guiding actions, attempting to minimize injuries and to prevent collisions. The application of sensors in the intelligent transportation system is to provide assistance to some control elements of the vehicle, like the throttle pedal and consequently, the speed-control assistance [7]. A cruise control system is a common application of these techniques. It consists of maintaining the vehicle speed at a user (driver) pre-set speed and computer control speeds their response times to road hazards so that the cars can travel more safely [8]. The method of electrical array reconfiguration that is suitable to a photovoltaic powered car is differentiating from that of the water pump which needs only the maximum power. But photovoltaic powered car needs both torque and speed which is changed by the appropriate situation. The switching of electrical reconfiguration is controlled for automatic reconfiguration by fuzzy controller [3].

The DC motors that are used in the electric cars do not have an intelligent controller to stop automatically during over load situation. This case will force the DC motor to work more than its capacity and could cause a malfunction of the DC motor or any electrical short at the wiring system. Therefore, the current protection of electric cars should be modified in order to avoid this problem. This paper proposes an effective protection based on temperature sensors to detect the changes of temperature. A thermocouple converts temperature to an output voltage and the output signal will be sent into the peripheral interface controller (PIC) which is programmed to send commends to the switching systems for taking a further action to stop the motor instantly via IRF 540 [9]. The effective protection controlled by PIC is interfaced with the analog temperature sensor to switch ON or OFF the DC motor. The designed source codes of PIC are considering the required conditions of the system protections and the behavior of electric cars DC motor.

2 System Description

Since the contribution of this work is mainly focusing on the DC electric car protection, more details describing the PIC and temperature sensor are given in this section. Whereas the other parts involved directly to the electric system are briefly outlined.

2.1 Solar Panel

PV panels convert sunlight to electrical energy used to supply power directly to the electrical car. Generally, PV is considered as an expensive method of producing electricity but it offers a cost-effective alternative to expensive grid. The development of new PV technologies for applications of PV in public electricity has grown rapidly. Solar system panel were the main supply for the system which supplies a maximum of 12 V to the battery.

2.2 Solar Charge Controller

A solar charge controller is needed in virtually all solar power systems that utilize batteries. The function of the solar charge controller is to regulate the power going from the solar panels to the batteries. Overcharging batteries will significantly reduce battery life. The most basic charge controller simply monitors the battery voltage and to open the circuit (stopping the charging) when the battery voltage rises to a certain level. Older charge controllers used a mechanical relay to open or close the circuit.

2.3 PIC Microcontroller

PIC microcontrollers have attractive features and they are suitable for wide rang of applications. PIC is a family of a microcontrollers and use Harvard architecture. The instruction is set to be non-overlapping or mutually independent. The Harvard architecture makes use of separate program and data memories. The separation of data and address buses is allowing increased data flow to and from the CPU and making the different widths between these busses. PIC microcontroller can be used to model the behavior of the machine as a logic program. The PIC 16F877 used in this work is shown in Figure 1, for the protection system, the OSC pin is connected with the crystal 8Mhz, RAD/ AND connected to the temperature sensor. The pin kl is connected to the LED, Pin P1 is connected to the motor and the Pin Rbs are connected to the LCD display. All the features of the PIC are illustrated in Table 1 and given as follows:

- Full Speed USB 2.0 (12Mbit/s) interface
- 1K byte Dual Port RAM + 1K byte GP RAM
- Full Speed Transceiver
- 16 Endpoints (IN/OUT)
- Streaming Port
- Internal Pull Up resistors (D+/D-)
- 48 MHz performance (12 MIPS)

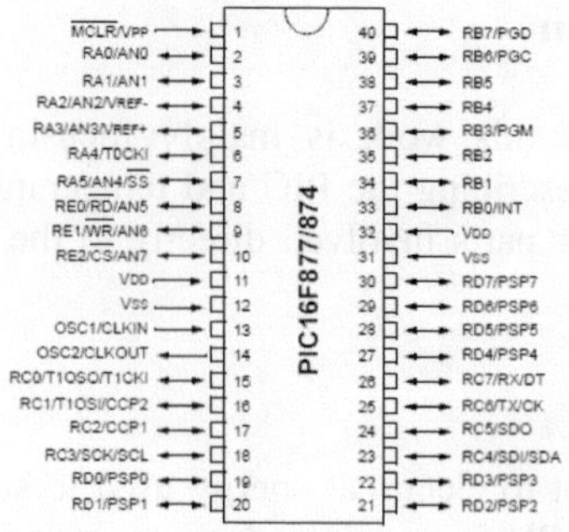

Fig. 1. PIC 16F877.

Table 1. Features of PIC16F877.

Specification	Value
Program Memory Type	Flash
Program Memory Type (KB) CPU	32
Speed (MIP)	12
RAM Bytes	2,048
Data EEPROM (bytes)	256
DigitalCommunication Peripherals	1-A/E/USART,1-MSSP (SPI/I2C)
Capture/Compare/PWM Peripherals	1 CCP, 1 ECCP
Timers 1 x 8-bit,	3 x 16-bit
ADC 13 ch,	10-bit
Comparators	2
USB (ch, speed, compliance)	1, Full Speed, USB 2.0
Temperature Range (C)	-40 to 85
Operating Voltage Range (V)	2 to 5.5
Pin Count	40
Packages	40 PDIP, 44 TQFP, 44 QFN
I/O pins	34

2.4 Temperature Sensor (LM35)

The LM35 series are precision integrated-circuit temperature sensors, and its output voltage is linearly proportional to the Celsius temperature. It does not require any external calibration or trimming to provide typical accuracies of ±¼°C at room temperature and ±¾°C over a full -55 to +150°C temperature range. The LM35's low output impedance, linear output, and precise inherent calibration make interfacing to readout or control easily. In LM35 temperature sensor, the output voltage is 10mV per degree centigrade. If output is 300mV then the temperature is 30 degrees. It can be used with single or more power supplies and has very low self-heating that is less than 0.1°C because it only draws a 60 μA from its supply. Basically, there are 3 type of sensors; thermistor, thermocouple and resistance temperature detectors (RTD). A thermistor is a type of resistor whose resistance varies significantly with temperature. Thermistors are widely used as inrush current limiters, temperature sensors, self-resetting over current protectors, and self-regulating heating elements. A thermocouple temperature sensor which is the junction between two different metals that produces a voltage related to a temperature difference. An RTD is mainly a temperature sensitive resistor. It is a positive temperature coefficient device, which means that the resistance increases with temperature [10]. The Characteristics of LM35 are:

- Calibrated directly in ° Celsius (Centigrade)
- Linear + 10.0 mV/°C scale factor
- 0.5°C accuracy guaranteable (at +25°C)
- Rated for full −55° to +150°C range
- Suitable for remote applications
- Low cost due to wafer-level trimming
- Operates from 4 to 30 volts
- Less than 60 μA current drain
- Low self-heating, 0.08°C in still air
- Nonlinearity only ±1⁄4°C typical
- Low impedance output, 0.1 W for 1 mA load

For the above mentioned characteristics, the electrical specification, storage temperature and lead temperature of the LM35 packages are given in Table 2, 3 and 4 respectably.

Table 2. Distributors for availability and specifications of LM35

Input/Output Sources	Voltage/Current
Supply voltage	+35V to −0.2V
Output voltage	+6V to −1.0V
Output current	10 mA

Table 3. Storage Temperature for LM35 packages

Package	Temperature range
TO-46	−60°C to +180°C
TO-92	−60°C to +150°C
SO-92	−65°C to +150°C
TO-220	−65°C to +150°C

Table 4. Lead Temperature for LM35 Packages Storage.

Package	Lead Temperature
TO-46	(Soldering, 10 seconds) 300°C
TO-92	(Soldering, 10 seconds) 260°C
TO-220	(Soldering, 10 seconds) 260°C
SO	Vapor Phase (60 seconds) 215°C
	Infrared (15 seconds) 220°C
	ESD Susceptibility 2500V
	Specified Operating Temperature Range: TMIN to T MAX

2.5 MOSFET

The Metal oxide semiconductor field-effect transistor (MOSFET) is a device used for amplifying or switching electronic signals. In MOSFETs, a voltage on the oxide-insulated gate electrode can induce a conducting channel between the two other contacts called source and drain [11]. The channel can be of n-type or p-type (see article on semiconductor devices), and is accordingly called an nMOSFET or a pMOSFET (also commonly nMOS, pMOS). It is a common transistor in both digital and analog circuits [9].

2.6 Boost Converter

A boost converter (step-up converter) is a power converter with an output DC voltage greater than its input DC voltage. It is a class of switching-mode power supply (SMPS) containing at least two semiconductor switches (a diode and a transistor) and at least one energy storage element [12]. Filters made of capacitors (sometimes in combination with inductors) are normally added to the output of the converter to reduce output voltage ripple [13]. In this project, a 12 to 24 V boost converter are needed to supply voltage to 24 V DC motor [14]. A replacement had been made to the boost circuit with TDA 2004 fixed among components of application circuit as given in the data sheet. TDA 2004 is an amplifying IC which has many advantages compared to electronic switch [12].

2.7 Battery

A 12 V lead acid battery is chosen to ensure maximum battery performance, this battery can be both charged and desolated at the same time. Lead acid batteries are designed by considering all the important aspects to avoid the main problems and failures due to the sulfation build-up on the battery plates. It is made using six identical two volt cells. All the cells contain lead plates of different types of sitting in dilute sulphuric acid.

2.8 DC Motor

DC motors were the first practical device to convert electrical energy into mechanical energy. Although AC motors and vector-control drives now offer alternatives to DC, there are many applications where DC drives offer advantages in operator friendliness, reliability, cost, effectiveness, and performance. The DC motor is extensively used as a positioning device because of its speed as well as torque that can be controlled precisely over a wide range. The wide Applications of DC motor in automobiles, robots, movie camera, electric vehicles, in steel and aluminum rolling mills, electric trains, overhead cranes, control devices, etc [15].

3 The Approach for Electric Car Protection

This work is mainly about the DC motor protection of electric car which is controlled automatically using PIC when there is an increase in load indicated by the temperature. The circuit protection of DC Motor should be designed first. Then a simulation must be carried out to evaluate the performance of the system before assembling the hardware protection system. For this purpose software Proteus 7 is used to check the accuracy of PIC coding.

3.1 Construction of Electric Car Protection

The constructed block diagram of the system is shown in Figure 2. The PIC reads the data from the temperature sensor which is connected to the DC motor. The normal operating temperature for the 24V DC motor is 35 to 36 C° (degree celcius). When there is an increase in the temperature above 36 C°, the motor will be stopped by PIC

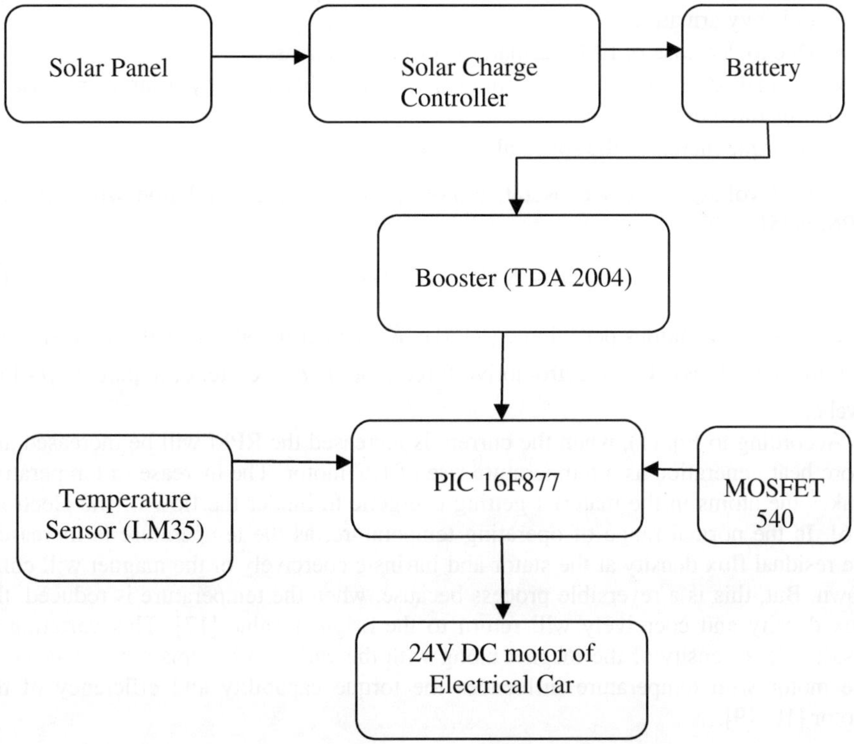

Fig. 2. Block diagram of the system

using PWM signal. The TDA 2004 IC is used in the system to boost up voltage from 12 V to 24 V. The solar panel is the main power source to charge the 12 V lead acid battery which is connected into charge controller and the 12 V output is boosted to the 24 V using the TDA 2004. The output is connected to the motor through the PIC 16F877 for interfacing. A Temperature sensor is linked to the PIC and the MOSFET is directly connected to the motor. For the Hardware implementation, the components were selected and tested according to the data of tables as given in section 2. Then, all components are collected based on the created design. The maximum operating temperature of the selected DC motor was measured to check whether the data of the datasheet are correct or not before inserting the coding values. The written codes were compiled into the HEX file before burning into the PIC. A 5V DC was supplied into the PIC and LCD display. The motor connected into 24V DC and the heavy load was applied to the mechanical output of the DC motor.

3.2 Temperature Effect on the DC Motor

Due to the load increase, the flux in the stator will be increased as well as the torque. If the torque is decreased by decreasing the field current, the following sequences are found [16]:

- Back EMF drops instantly, the speed remaining constant because of the inertia of heavy armature
- Due to decrease of EMF, armature current is increased
- A small decrease of flux is more than counterbalanced by a large increase of armature current
- If torque increases the speed also increases

If applied voltage is kept constant, motor speed has inverse relation with flux and RPM is [5],

$$N = K(V - IR)/\phi \tag{1}$$

where, N - revolutions per minute (RPM); K - proportional constant; R - resistance of armature (ohms); V - electromotive force (volts), I - current (amperes); ϕ - flux (webers).

According to Eq. (1), when the current is increased the RPM will be increased and more heat generation as a temperature rise of DC motor. The increase in temperature makes the atoms in the material getting energetic to hinder the flow of the electrons [16]. In the normal range of operating temperature, as the temperature is increased, the residual flux density at the stator and intrinsic coercively of the magnet will come down. But, this is a reversible process because, when the temperature is reduced, the flux density and coercively will return to the original value [17]. This variation in residual flux density of the magnet along with the variation in armature resistance of the motor with temperature influences the torque capability and efficiency of the motor [18, 19].

3.3 PIC Coding of Electric Car Protection

Microcode studio was used to compile the PIC coding. The constructed cods have been compiled and simulated using Protues software to insure the accuracy of the coding commands. For the hardware implementation, the PIC was loaded by the following cods:

```
Define crystal being used
'
DEFINE OSC 8 'Using 8Mhz external crystal
'
```

```
'Define LCD
'
DEFINE LCD_DREG PORTB 'Set LCD data port to PORTB
DEFINE LCD_DBIT 0 'Set data starting bit to 0
DEFINE LCD_RSREG PORTB 'Set RS register port to PORTB
DEFINE LCD_RSBIT 4 ' Set RS register bit to 4
DEFINE LCD_EREG PORTB 'Set E register port to PORTB
DEFINE LCD_EBIT 5 'Set E register bit to 5
DEFINE LCD_BITS 4 ' Set 4 bit operation
DEFINE LCD_LINES 2 ' Set number of LCD rows
'
'Define A/D converter parameters
'
DEFINE ADC_BITS 8 'A/D number of bits
DEFINE ADC_CLOCK 3 'Use A/D internal RC clock
DEFINE ADC_SAMPLEUS 50 'Set sampling time in us
'
'Definition for HPWM
'
DEFINE CCP1_REG PORTC 'Channel 1 port
DEFINE CCP1_BIT 2 'Channel 1 bit
DEFINE CCP2_REG PORTC 'Channel 2 port
DEFINE CCP2_BIT 1 'Channel 2 bit
'
'Variables
'
Res Var Word 'A/D converter result
Temp1 Var Byte 'Temperature in degrees C
TRISA = 1' RA0 (AN0) is input
TRISB = 0 'PORTB is output
PAUSE 500 ' Wait 0.5sec for LCD to initialize
D VAR BYTE 'HPWM calculated duty cycle
LED_D1 VAR PORTC.0  'PORTC.0 as LED_D1
OUTPUT LED_d1   'PORTC.0 as output
'
' Initialize the A/D converter
'
ADCON1 = 0 'Make AN0 to AN4 as analog inputs,
'make reference voltage = VDD
ADCON0 = %11000001'A/D clock is internal RC, select AN0
' Turn on A/D converter
LCDOUT $FE, 1 'Clear LCD
'
'Main Programs
'
MAIN:
    '
    'Start A/D conversion
```

```
'
ADCIN 0, Res ' Read Channel 0 data
Temp1 = 2 * Res ' Convert to degrees C
LCDOUT $FE,2,"TEMP = ",DEC2 Temp1, "C" ' Display
decimal part
PAUSE 500
'
'HPWM programs
'
d = 8*(Temp1 - 25) + 127
IF Temp1  >= 36 THEN
HPWM 1, 0, 2000
    PAUSE 500
    ENDIF
IF Temp1 < 36 THEN
    hpwm 1, 255, 2000
    PAUSE 500

ELSE
    hpwm 1, D, 2000
    PAUSE 500
ENDIF
'
'LED programs
'
IF  Temp1 < 36 THEN
    HIGH LED_D1
    PAUSE 100
    LOW LED_D1
    PAUSE 100
    ENDIF
 IF Temp1 >= 36 THEN
    HIGH LED_D1
ELSE
    LOW LED_D1
ENDIF

GOTO MAIN 'Repeat Main
END
```

4 Case Study

A simulation has been carried out on the selected components of the circuit given in Figure 3 and the HEX file inserted into the PIC for real testing of control commands interfaced with Proteus 7.0. The output is obtained by setting the temperature above and below 36 C° as seen from Figure 4. The LCD LM 016L was used to display the temperature based of the commands sent by PIC coding. The protection system was

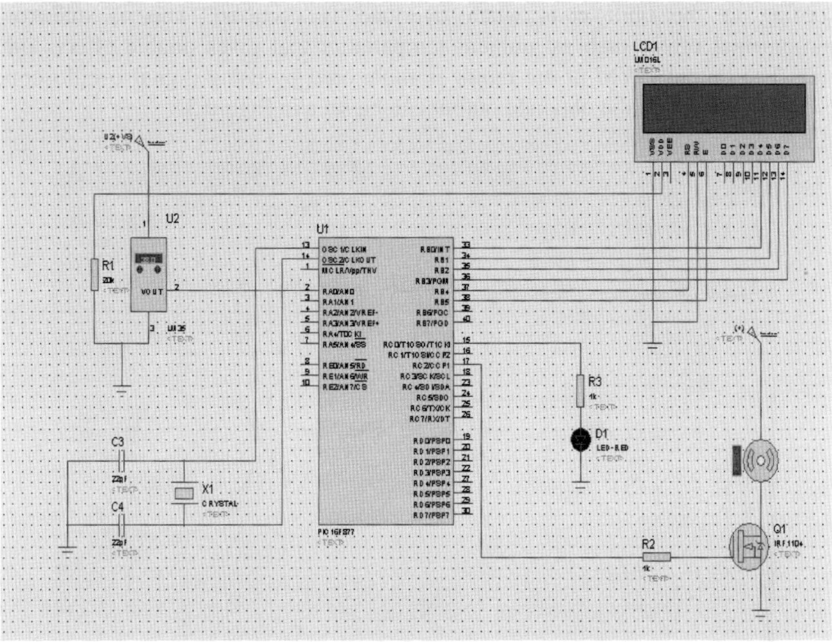

Fig. 3. Simulation of Electric Car Protection

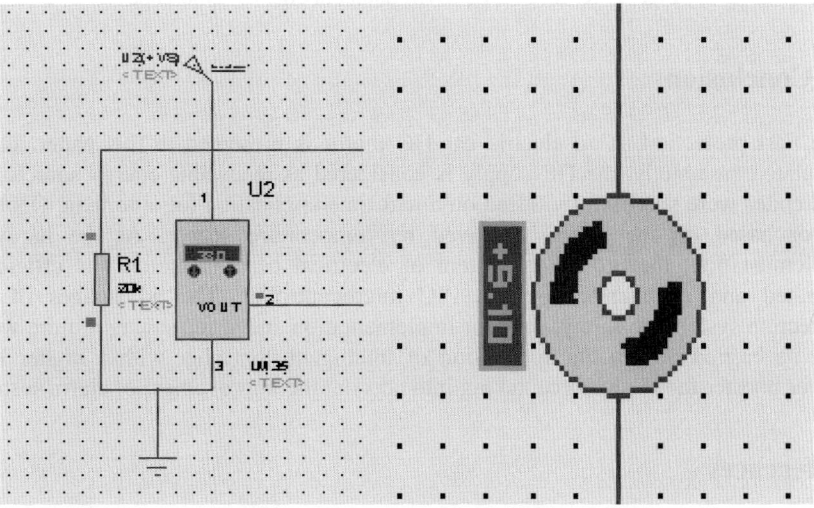

Fig. 4. Temperature is below 36 C° motor is running

tested on real time application to validate the results of simulated coding. The hardware components of protection system have been connected and then turned on by supplying a 5V to the PIC and LCD display with a 24 V applied to the DC motor. It was found that in some cases of real time testing, the motor is still functioning according to the built cods when overload is occurred. This is due to the environmental effect such as ambient temperature that gives double effects on the motor. The output PWM at the motor is monitored by using digital oscilloscope at 50% Duty Cycle, and the obtained output can be seen in Figure 5.

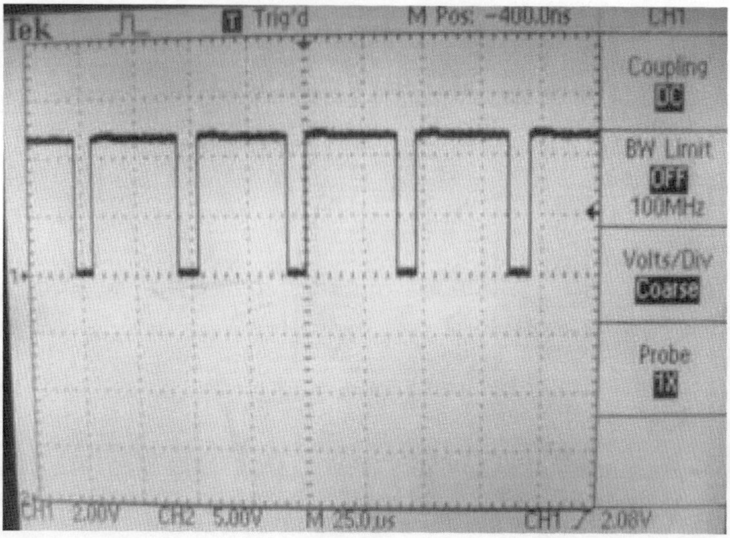

Fig. 5. PIC 16F877A's PWM output in digital oscilloscope

5 Conclusion

The Fast protection of an electric car DC motor is presented in this paper. In the simulated case study, the DC supply is considered as renewable energy source. The PIC codes were successfully tested on a real time application for automatic ON/OFF motor based on the signals received by temperature sensor. At the hardware implementation, the protection system of electrical car DC motor was efficiently operated and commended by the PIC microcontroller. The advantages of this protection system are safety, easy for implementation and space saving. Future works will be carried out on the application of intelligent controller with a higher level power booster for solar source, taking into account the improvement of alarm system.

References

1. Nobels, T., Gheysen, T., Vanhove, M., Stevens, S.: Design considerations for a plug-in hybrid car electrical motor. In: Proceeding of International IEEE Conference on Clean Electrical Power, pp. 755–759 (2009)
2. Mangu, R., Prayaga, K., Nadimpally, B., Nicaise, S.: Design, Development and Optimization of Highly Efficient Solar Cars: Gato del Sol I-IV. In: Proceeding of the IEEE Conference on Green Technologies Conference, pp. 1–6 (2010)

3. Stiegeler, M., Bitzer, B., Kabza, H.: Improvement of Operating Strategy for Parallel Hybrid Cars by Adaptation to Actual Route and Electrical On-Board Load. In: Proceeding of the 13th International IEEE Annual Conference on Intelligent Transportation Systems, pp. 19–22 (2010)
4. Peters, K.: Design Options for Automotive Batteries in Advanced Car Electrical Systems. Journal of Power Sources 88, 83–91 (2000)
5. Andreasen, S.J., Ashworth, L., Remón, I.N.M., Kær, S.K.: Directly Connected Series Coupled HTPEM Fuel Cell Stacks to a Li-ion Battery DC Bus for a Fuel Cell Electrical Vehicle. International Journal of Hydrogen Energy 33, 7137–7145 (2008)
6. Świder, J., Wszołek, G., Baier, A., Ciupka, G., Winter, O.: Testing Device for Electrical Car Networks. Journal of Materials Processing Technology, 1452–1458 (2005)
7. Naranjo, J.E., González, C., Reviejo, J., García, R., de Pedro, T.: Adaptive Fuzzy Control for Inter-Vehicle Gap Keeping. IEEE Transactions On Intelligent Transportation Systems 4(3), 132–142 (2004)
8. Kim, H.M., Dickerson, J., Kosko, B.: Fuzzy Throttle and Brake Control for Platoons of Smart Cars. Fuzzy Sets and Systems 84, 209–234 (1996)
9. Hwu, K.I., Liaw, C.M.: DC-link Voltage Boosting and Switching Control for Switched Reluctance Motor Drives. In: Proceeding of International IEEE Conference on IEE Electric Power Application, vol. 147(5), pp. 337–344 (2004)
10. Mallik, A., Gupta, S.D.: Modelling of MEMS Based Temperature Sensor and Temperature Control in a Petrochemical Industry Lab VIEW. In: Proceeding of IEEE International Conference on Computer and Automation Engineering, on Digital Object Identifier, pp. 287–292
11. El-Kholy, E.E.: AC/DC Flyback Converter with a Single Switch Controlled DC Motor Driver. In: IEEE PEDS Proceeding of IEEE international conference on Power Electronics and Drives Systems, vol. 1, pp. 395–400 (2005)
12. Song, P., Lin, J.: A Hybrid AC-DC-AC Matrix Converter with a Boost Circuit. In: Proceeding of 9 International IEEE Conference on Electronic Measurement & Instruments, pp. 2-416 – 2-421. (2009)
13. Hwu, K.I., Liaw, C.M.: DC-link Voltage Boosting and Switching Control for Switched Reluctance Motor Drives. In: Proceeding of IEEE Conference on Electric Power Application, vol. 147(5), pp. 337–344 (2000)
14. Muhammad, H.R.: Power Electronics. Pearson Education, California (2004)
15. Beaty, H.W.: Electric Motor Handbook. McGraw-Hill, California (1998)
16. Kim, C.G., Lee, C.G.J.H., Kim, J.H.H.W., Youn, H.W.M.J., Study, M.J.: on Maximum Torque Generation for Sensorless Controlled Brushless DC Motor with Trapezoidal Back EMF. In: Proceeding of IEEE Conference on Electric Power Application, vol. 152(2), pp. 277–291 (2005)
17. Peters, K.: Design Options for Automotive Batteries in Advanced Car Electrical Systems. Journal of Power Sources 88, 83–91 (2000)
18. Boussak, M., Jarray, K.: A High-Performance Sensorless Indirect Stator Flux Orientation Control of Induction Motor Drive. IEEE Transaction on Industrial Electronics 53(1), 41–49 (2006)
19. Khalifa, F.A., Ismail, S.S., Basem Elhady, M.M.: Effect of Temperature Rise on the Performance of Induction Motors. In: Proceeding of IEEE International Conference on Computer Engineering & Systems, pp. 549–552 (2006)
20. Sebastian, T.: Temperature Effects on Torque Production and Efficiency of PM Motors Using NdFeB Magnets. IEEE Transactions on Industry Applications 31(2), 78–83 (1993)

Parameter Design and FEM Analysis on a Bearingless Synchronous Reluctance Motor

Yuanfei Li, Xiaodong Sun, and Huangqiu Zhu

School of Electrical and Information Engineering,
Jiangsu University, Zhenjiang 212013, China
530851639@163.com

Abstract. Based on the analysis of generation principle of radial force and the generation characteristic of radial force and torque in the same air-gap field, the design procedure of a bearingless synchronous reluctance motor is introduced. With air-gap magnetic field analytic approach and the torque coefficient iteration, optimal design of radial force and torque is complemented. The calculated results by finite element method show that magnetic circuit and magnetic field distribution in air-gap are reasonable. The design methodology for the bearingless synchronous reluctance motor is verified.

Keywords: bearingless synchronous reluctance motor, radial force, air-gap magnetic field analytic approach, finite element analysis.

1 Introduction

The requirements for motor drive systems become higher and higher in the modern automation manufacture and advanced manufacturing equipments, especially in high speed numerically-controlled machine tool, turbine molecular pump, centrifugal machine and flywheel power storage. At the same time, life science, pure sterile space and some special transmission areas also need high quality transmissions of no lubrication, no pollution. Bearingless motors have advantages of non-friction and non-wear, high speed, high precision and long life. Compared with other types of bearingless motors, the bearingless synchronous reluctance motor has some special advantages such as simple structure, rotor without windings and simple control. Therefore, the research of the bearingless synchronous reluctance motor has obvious application value in projects.

A large amount of theoretical and experimental researches for bearingless motors has been carried on in the world. The effects of rotor eccentricity, magnetic saturation on radial force of the bearingless synchronous reluctance motor have been studied [1], [2], [3]. Magnetic couplings and a decoupling control method have been studied [4], [5]. Through the analysis of characteristic of the rated values of radial force and torque, a basic approach for design of the bearingless motor is presented [6], [7]. The radial force production capabilities in different type of bearingless motors are studied [8]. The rotor structure with multi-flux barriers has been reported with good torque

M. Zhu (Ed.): Electrical Engineering and Control, LNEE 98, pp. 163–171.
springerlink.com © Springer-Verlag Berlin Heidelberg 2011

and radial force characteristics [9]. The study of bearingless synchronous reluctance motors still stay in theory and experimental stage. The design hasn't formed a mature theory and methodology. In this paper, a general research for the design scheme in the bearingless synchronous reluctance motor is presented and verified.

2 Principle of Radial Force Generation

To generate torque and radial force simultaneously, a bearingless motor inserts two winding sets with different pole pairs in stator slots. Fig. 1 shows the principle of radial force generation with 4-pole torque windings N_a and 2-pole radial force windings N_y [3]. When the two winding carry currents, 4-pole magnetic fluxes Ψ_a and 2-pole magnetic fluxes Ψ_y are produced. By the superposition of two magnetic fields, the flux density in the air-gap 1 increases, the flux density in the air-gap 3 decreases oppositely. As a result, the radial force is produced in y-axis positive direction. In addition to N_y, a bearingless motor has another set of 4-pole windings N_a in order to produced revolving magnetic field for torque production. The bearingless motor also has another 2-pole windings N_x, which is perpendicular to N_a windings, in order to produce radial force in any desired directions.

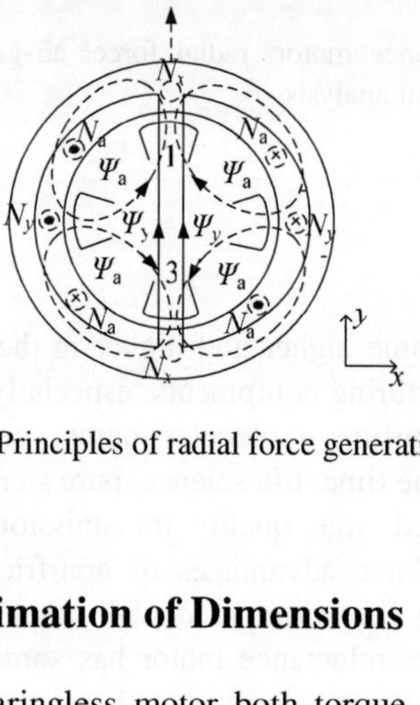

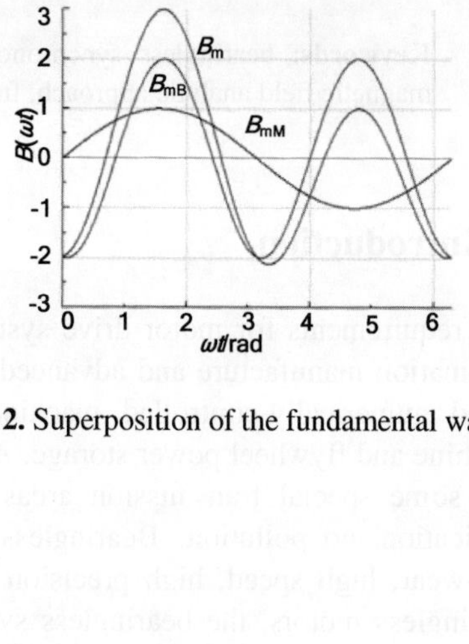

Fig. 1. Principles of radial force generation. **Fig. 2.** Superposition of the fundamental waves.

3 Estimation of Dimensions

In a bearingless motor both torque and radical force are generated in one air-gap magnetic field. The magnetic saturation and heating in a real motor influence the production of torque and radial force. For simply analysis, ignore the effect of stator teeth, the terminal, copper losses and so on. The air-gap magnetic induction fundamental waves are mainly torque windings magnetic induction fundamental waves and radial force windings magnetic induction fundamental waves. The superposition of the air-gap fundamental waves can be written as

$$B(\omega)=B_{\mathrm{B}}\cos(\omega t+\mu)+B_{\mathrm{M}}\cos(2\omega t+\gamma)\ \cdot \qquad (1)$$

Where B_B and B_M are amplitudes of radial force windings and torque windings magnetic field fundamental waves, μ and γ are the initial phase angles. ω is the angular frequency of the windings currents.

Fig. 2 shows the worst case, which means that the fundamental waves of radial force windings magnetic field and torque windings magnetic field have the same angular position. The superposition magnetic induction reaches the maximum. The magnetic circuit is easy to come into magnetic saturation state. By the increase of the superposition magnetic induction, the strong influence of magnetic saturation becomes clearer.

In order to output radial force and torque in a real bearingless motor, the bearingless motor system must be divided into the motor subsystem and the bearing subsystem. The superposition magnetic induction B_m must be divided into radial force windings magnetic induction B_{mB} and torque windings magnetic induction B_{mM}. The superposition magnetic induction B_m can be written as

$$B_m = B_{mB} + B_{mM} \ . \tag{2}$$

Where B_m, B_{mB} and B_{mM} are the magnetic induction of the air-gap, radial force windings and torque windings magnetic field.

When the rotor is positioned to the stator center, the distribution of the air-gap is uniform on the stator armature, $\Lambda(\theta) = \mu_0/\delta_0$. The magnetic induction B_{mB} and B_{mM} can be written as

$$\begin{cases} B_{mB} = \dfrac{\mu_0}{\delta_0}\left(\dfrac{3}{2}\cdot\dfrac{4}{\pi}\cdot\dfrac{I_B}{2}\cdot\dfrac{N_1 k_{N1}}{P_B}\right) \\ B_{mM} = \dfrac{\mu_0}{\delta_0}\left(\dfrac{3}{2}\cdot\dfrac{4}{\pi}\cdot\dfrac{I_M}{2}\cdot\dfrac{N_2 k_{N2}}{P_M}\right) \end{cases} . \tag{3}$$

Where μ_0 is the permeability of vacuum. δ_0 is the uniform length of the air-gap. I_B and I_M are the currents of the radial force windings and the torque windings, respectively. N_1 and N_2 are the number of series windings of the radial force windings and torque windings, respectively. k_{N1} and k_{N2} are the winding coefficient.

The torque of the bearingless motor is produced by the revolving torque windings magnetic induction. So the torque windings induction magnetic factor must be considered. The factor C_{red} can be written as

$$C_{red} = \frac{B_{mM}}{B_m} = \frac{N_2 I_M k_{N2} P_B}{N_1 I_B k_{N1} P_M + N_2 I_M k_{N2} P_B} \ . \tag{4}$$

Where C_{red} is the torque windings magnetic induction factor.

It is important to choose a proper value of C_{red}. When C_{red} =1, B_{mB} = 0 and B_{mM} = B_m are obtained with (4) and (2). There is no radial force magnetic field, only torque system is existed. In this case it is equivalent to the normal AC motor; When C_{red} =0, B_{mM} = 0 and B_{mB} = B_m are obtained with (4) and (2). There is no torque magnetic induction field, no torque can be generated, only radial force magnetic induction field and radial force system is existed.

Because both torque and radial force should be generated, the value of the torque windings magnetic induction factor C_{red} must be between 0 and 1. Meanwhile, the torque reduction is caused by synchronous radial force generation. So the torque coefficient K_{red} is defined by

$$K_{red} = \frac{T}{T_{max}} . \quad (5)$$

Where K_{red} is the torque coefficient. T is the output torque. T_{max} is the maximum output torque when the motor is only used as the motor subsystem.

According to principle of the synchronous reluctance motor, the torque of a bearingless synchronous reluctance motor can be written as

$$\begin{cases} i_M = \sqrt{i^2_{Md} + i^2_{Mq}} \\ i_{Md} = i_{Mq} \\ T = \frac{3}{2} P_M \left(L_{Md} - L_{Mq} \right) i_{Md} i_{Mq} \end{cases} . \quad (6)$$

Where i_M is the torque windings current. i_{Md} and i_{Mq} are the torque windings d-axis and q-axis current. P_M is the number of pole pairs of the torque windings. L_{Md} and L_{Mq} are the d-axis and q-axis inductance of the torque windings.

The estimation of main dimensions is done according to the wellknown electrical motor design. The electromagnetic load of AC-machines is determined by the inner apparent power [7]

$$S_N = \frac{\pi^2}{\sqrt{2}} \xi_1 AB\, D^2 l_i n_N = CD^2 l_i n_N . \quad (7)$$

Where S_N is the rated output power. ξ_1 is the winding factor. A is the electrical loading. B is the amplitude of the air-gap induction. C is the utilization factor. D is the inner diameter of the stator. l_i is the stack length, n_N is the rated speed.

Fig. 3 shows the sequence of the design in principle. Initial values are the rated power S_N, the rated speed n_N and required maximum radial force F_m. For the windings, it is advised to choose the pole pair numbers according to $P_M = P_B \pm 1$. The equation of relative length λ is

$$\lambda = \frac{l_i}{\tau} = \frac{2P_M l_i}{\pi D} . \quad (8)$$

Where P_M is torque magnetic field pole pairs. τ is polar distance.

The utilization factor C is given as a function of the output power or power per pole for the different AC-motors. The utilization factor C contains electric loading and air gap magnetic induction. It can be obtained from the empirical value of realized motors. The main dimension D can be calculated with (7) and (8).

$$D = \sqrt[3]{\frac{2P_{\mathrm{M}}S_{\mathrm{N}}}{\pi\lambda n_{\mathrm{N}}CK_{\mathrm{red}}}} \ . \tag{9}$$

The estimation of maximum radial force to the rotor surface can be calculated from (10).

$$F_{\mathrm{m}} = F_{\mathrm{s}} \cdot \pi D l_{\mathrm{i}} \ . \tag{10}$$

Where F_s is the specific radial force. It is not only related to C , but also related to the motor type and size. In generally, it can be obtained from the empirical value of realized motors from the similar power motor [7].

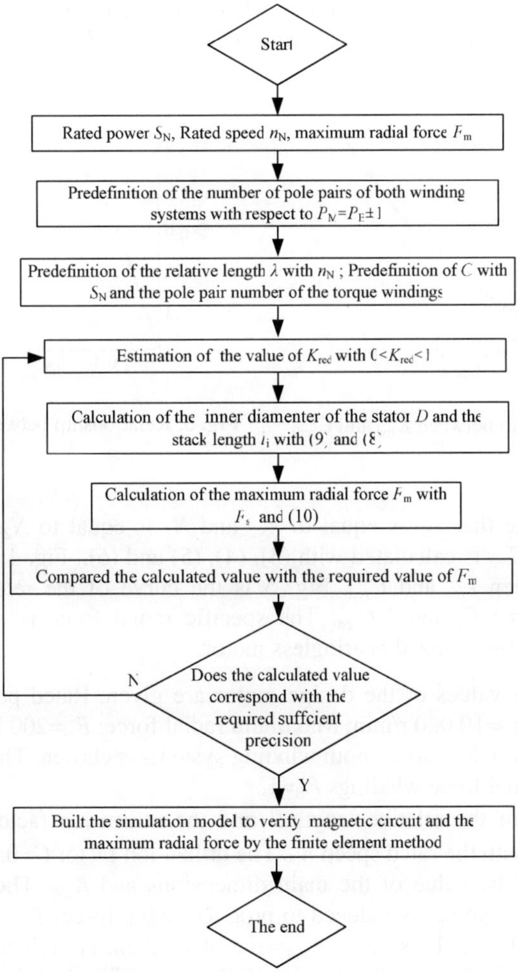

Fig. 3. Design diagram of the bearingless synchronous reluctance motor.

The maximum radial force is estimated with (10). Compared the calculated value with the required value F_m, if the calculated value doesn't correspond with the required sufficient precision, K_{red} will be corrected upwards or downwards and then iteratively until the calculated value meet the precision requirement. After reaching the design requirement, the magnetic circuit and windings are designed. The FEM-model is built. The effective magnetic circuit and the maximum radial force are verified by the finite element method.

4 Parameter Design and FEM Analysis

4.1 Parameter Design

In this section, it is assumed that the whole slot space can be used for torque windings. An electrical loading of 15kA/m can be gained [7].

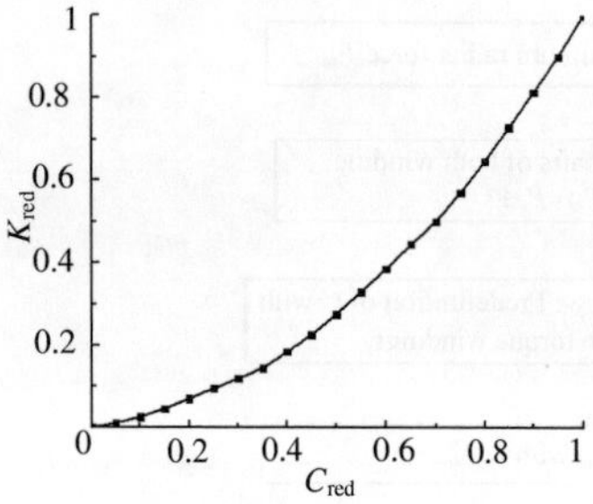

Fig. 4. Relationship between K_{red} and C_{red}.

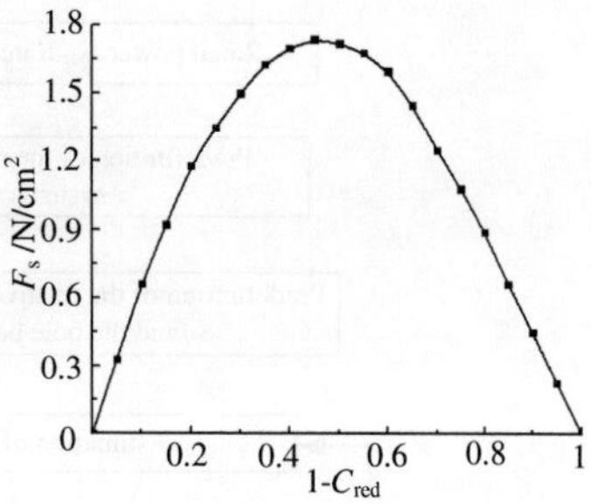

Fig. 5. Relationship between F_s and 1-C_{red}.

First we assume that k_{N1} is equal to k_{N2} and N_1 is equal to N_2. The relationship between K_{red} and C_{red} is calculated with (3), (4), (5) and (6). Fig. 4 is the curve of the relationship between k_{red} and C_{red}. Fig. 5 is the curve of the relationship between specific radial force F_s and 1-C_{red}. The specific radial force is obtained from the empirical value of the realized bearingless motor.

(1) The parameter values of the design motor are given. Rated power: S_N=1.0 kW; Rated speed: n_N=10 000 r/min; Maximum radial force: F_m=200 N.
(2) The number of pole pairs of both winding systems is chosen. The torque windings P_M=2. The radial force windings P_B=1.
(3) Predefinition of the relative length λ and the utilization factor C. The relative length λ=1.4 with the rated speed n_N. The utilization factor C=0.42 kW·min/m^3.
(4) Estimation of the value of the main dimensions and K_{red}. The question is how much the motor power is reduced to produce radial force. The torque coefficient K_{red} range is 0 to 1. Lets taken K_{red} =0.5 at random. Now inner diameter of the stator can be calculated with (9), it is D=0.067 m. The stack length l_i also can be calculated with (8), it is l_i=0.074 m.

(5) Calculated the maximum radial force. First, with K_{red} =0.5 at random, C_{red}=0.7 and F_s=1.5 N/cm^2 are obtained in Fig. 4 and Fig. 5, respectively. The radial force can be calculated with the above dates and (10). It is F=233 N. The required maximum radial force is 200 N, so the torque windings magnetic induction factor C_{red} has to be adjusted. Through incessant adjustment of the torque coefficient K_{red}, K_{red}=0.7 is suitable. The radial force F=204N meet the precision requirement. The main dimensions D=0.067 m and l_i=0.074 m are calculated with (9) and (8). Design the magnetic circuit and the stator windings. It means that the main dimensions are found. Table 1 lists the main dimensions of the bearingless synchronous reluctance motor.

Table 1. Main size parameters of the prototype motor.

The main dimensions of stator		
Torque windings	Number of pole pairs	2
	Inner diameter	80 mm
	External diameter	130 mm
	Slot number	24
	Number of turns of each slot	30
Radial force windings	Number of pole pairs	1
	Number of turns of each slot	30
The main dimensions of rotor		
Rotor	Diameter on salient poles	79.5 mm
	Diameter between salient poles	40 mm
	Axial length	90 mm
	Pole arc	30°
	Air -gap length	0.25 mm

4.2 Verification of FEM

The rated currents of both windings can be calculated with the above dates. When K_{red} is 0.7, the value of radial force windings is 1 A, and the rated current of torque windings is 4 A. ANSYS software is used to calculate the above designed.

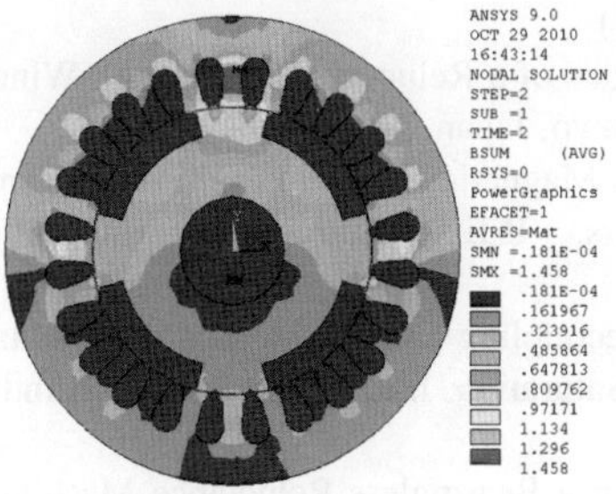

Fig. 6. Magnetic flux density distribution.

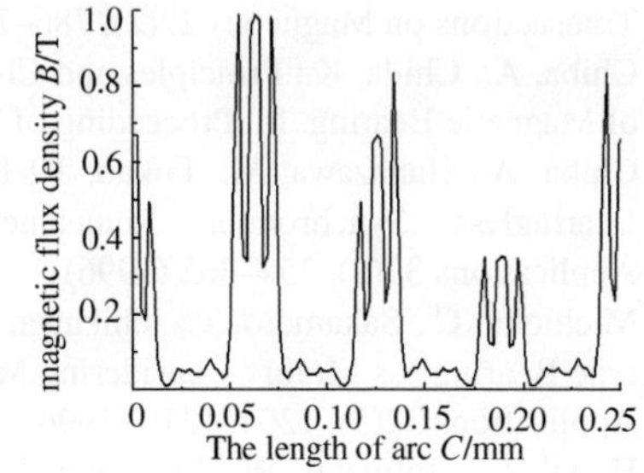

Fig. 7. Magnetic flux density distribution in air.

Fig. 6 shows the magnetic flux density distribution with the torque windings current 4 A and the radial force windings current 1 A. The magnetic flux density of the upper section is increased, while the opposite is decreased. Radial force is generated along the *y*-axis positive direction. The magnetic flux density of the left section and the right section in the motor is symmetrical about *y*-axis. No radial force is generated along the *x*-axis. Fig. 7 shows the distribution of magnetic flux density in the air-gap. The magnetic flux density over the pole width is about 0.45T. This value is according to the requirement of motor design. As stated previously, the designed magnetic circuit and the air-gap magnetic flux density is corrected.

Lastly, the maximum radial force can be calculated by FEM. The value of the prototype motor's radial force is 210 N. It is hit with the requirement value F_m, the design approach of the bearingless synchronous reluctance motor is verified.

5 Conclusion

In this paper, a basic approach for the design of the bearingless synchronous reluctance motor is presented. By the present method of AC motors design and the torque coefficient iteration, the main parameters of the bearingless synchronous reluctance motor is calculated. The magnetic circuit and the magnetic field are analyzed by the FEM. The maximum radial force is also verified. The proposed method has provided a reference for the design of similar bearingless motors.

Acknowledgements

This work is sponsored by the National Natural Science Foundation of China (60974053), the Natural Science Foundation of Jiangsu Province (BK2009204), and the Research Fund for the Doctoral Program of Higher Education of China (20093227110002).

References

1. Chiba, A., Rahman, A.M.: Radial Force in a Bearingless Reluctance Motor. IEEE Transactions on Magnetics 27(2), 786–790 (1991)
2. Chiba, A., Chida, K.: Principles and Characteristics of a Reluctance Motor with Windings of Magnetic Bearing. In: Proceeding of IPEC, Tokyo, Japan, pp. 919–926 (1990)
3. Chiba, A., Hanazawa, M., Fukao, T.: Effects of Magnetic Saturation on Radial Force of Bearingless Synchronous Reluctance Motors. IEEE Transactions on Industry Applications 32(2), 354–362 (1996)
4. Michioka, C., Sakamoto, T., Ichikawa, O.: A Decoupling Control Method of Reluctance-type Bearingless Motors Considering Magnetic Saturation. IEEE Transactions on Industry Applications 32(5), 1204–1210 (1996)
5. Hertel, L., Hofmann, W.: Magnetic Couplings in a Bearingless Reluctance Machine. In: Proceeding of ICEM, vol. 5(3), pp. 1776–1780. Helsinki University of Technology, Helsinki (2000)

6. Hertel, L., Hofmann, W.: Design and Test Results of a High Speed Bearingless Reluctance Motor. In: Proceeding of EPE, pp. 1–7. European Power Electronics and Drives Association, Lausanne (1999)
7. Hertel, L., Hofmann, W.: Basic Approach for the Design of Bearingless Motors. In: 7th International Symposium on Magnetic Bearings, ETH, Zurich, pp. 341–346 (2000)
8. Castorena, A.M., Soong, W.L., Ertugrul, N.: Analysis and Design of the Stator Windings of a Bearingless Motor for Comparisons of Radial Force Capabilities with Different Rotors. IEEE Transactions on Power Electronics and Drives Systems 2, 1161–1166 (2005)
9. Takemoto, M., Yoshida, K., Tanaka, Y., et al.: Synchronous Reluctance Type Bearingless Motors with Multi-flux Barriers. In: Proceeding of Power Conversion Conference, pp. 1559–1564. Nagoya University, Nagoya (2007)

A Hybrid and Hierarchy Modeling Approach to Model-Based Diagnosis

Dong Wang, Wenquan Feng, and Jingwen Li

School of Electronics and Information Engineering,
Beijing University of Aeronautics and Astronautics,
Beijing 100191, China
wangdong1106@gmail.com, buaafwq@buaa.edu.cn,
lijingwen@buaa.edu.cn

Abstract. The attributes of spacecrafts diagnosis are analyzed. To make up for the deficiencies of existing modeling methods, a hybrid and hierarchy approach to model-based diagnosis is presented. The model description capacity is improved through a hybrid describing way. The advantage of easy abstraction of qualitative knowledge is reserved and the state-space explosion due to discrete abstraction is handled. Based on the analysis of hierarchy modeling process, a model describing language is proposed. To validate the feasibility, a distribution circuit is modeled, which is typical in the electrical power system of spacecrafts and the modeling process is simplified through module reuse.

Keywords: model-based diagnosis, model-building, hybrid system, hierarchical system.

1 Introduction

Fault diagnosis is an important way to insure the reliability of space-based missions. Ruled-based expert system is the most commonly-used technology in applications for ages. Due to the difficulty of capturing all the rules in the expert system, the model-based way has been studied for spacecraft diagnosis in recent years. It is also a considerable and applicable method of realizing the autonomous management for deep space exploration.

Model-based diagnosis uses system's structural, behavioral or functional relationships to reason out a solution to explain the differences between observations and expectations. The accuracy of diagnosis depends on the model, which is an abstract of the real world. But no knowledge expressing method can describe all the details. Thus modeling technologies receives more and more attention. Quantitative approach [1, 2] is a classical method of describing dynamic system accurately. But the complexity and the model's errors in quantitative approach make the diagnostic result hard to control. Qualitative approach [3, 4, 5] which is a more abstract way has been widely used in model-based diagnosis [6]. The main obstacle to applications is the explosion of state spacc as a result of domain abstraction. Most diagnostic approaches are centralized on the use of the global model [7]. But it is not realistic to model large

M. Zhu (Ed.): Electrical Engineering and Control, LNEE 98, pp. 173–180.
springerlink.com © Springer-Verlag Berlin Heidelberg 2011

scale systems globally. Thus hierarchy [8, 9] is a solution to the modeling difficulties, which makes hierarchical diagnosis possible.

Based on the characteristics and requests of spacecraft diagnosis, this paper is concerning the modeling process in model-based diagnosis. A hierarchy modeling approach and a description language are proposed, and the modeling process is simplified through module reuse. Finally the typical distribution circuit of the electrical power system is modeled as an example.

2 Modeling Process

The model of system should be able to describe the expecting behaviors accurately under specific applications and situations. Three parts are included in the modeling process [10]: ontological choices, representational choices and behavioral choices. The ontological choices describe the modeling purpose. The domain knowledge is reflected through representational choices. Finally, proper variables and methods are chosen to describe the expected system behaviors according to relevant behavioral choices.

The purpose of diagnosis is to localize faults through the conflicts between expectations and observations. Considering the specialty of spacecrafts, the following aspects should be taken into account:

1) To increase the reliability of spacecrafts, redundant design is adopted in hardware equipment and control routes. Therefore, module reuse will reduce the complexity of the modeling process.

2) Rule-based expert system is widely used in diagnosis at present. But it is difficult to use the rules to describe complicated faults related with several subsystems. So the accurate description of relationships is important in modeling.

3) The resources of satellite-ground links are limited and sensors can not be placed at will. Thus, observables are not sufficient in spacecraft systems. The model should be able to reflect the actual measurability which is the base of diagnosis.

3 Hierarchy Modeling Approach

A hierarchy modeling approach is introduced in [9]. A system is composed by several components, which are connected to each other through the interface. For a component, input signals and output signals are defined. But in a system, there are inner signals as well. An event means that a signal is received or emitted through the interface, which is the abstracted notion of port, message and link.

3.1 Model of Component

A component c_i receives exogenous events from the environment $\sum_{exo}^{i}$, and communication events from other components $\sum_{com\_rcv}^{i}$. It also emits observable events $\sum_{obs}^{i}$ that can be observed, and communication events to other components $\sum_{com\_emit}^{i}$.

Definition 1. (Model of component). The model of the component c_i is described by a finite state machine:

$\Gamma_i = (\sum_{rcv}^{i}, \sum_{emit}^{i}, Q_i, E_i)$

1) $\sum_{rcv}^{i}$ is the set of received events $\sum_{rcv}^{i} = \sum_{exo}^{i} \cup \sum_{com\_rcv}^{i}$;

2) $\sum_{emit}^{i}$ is the set of emitted events $\sum_{emit}^{i} = \sum_{obs}^{i} \cup \sum_{com\_emit}^{i}$;

3) $\sum_{rcv}^{i} \cap \sum_{emit}^{i} = \varnothing$;

4) Q_i is the set of component states;

5) E_i is the set of transitions. A transition t is noted as $q \xrightarrow{rcv(t)/emit(t)} q'$, $rcv(t)$ is the event which triggers t, $emit(t)$ is the set of events emitted by t;

3.2 Model of System

$\sum_{rcv}^{\gamma}$ is the set of received events of the subsystem γ : $\sum_{rcv}^{\gamma} \triangleq (\bigcup_{j\in\{1,\cdots k\}} \sum_{rcv}^{i_j}) \setminus (\bigcup_{j\in\{1,\cdots k\}} \sum_{emit}^{i_j})$. There are two types of received events:

1) $\sum_{exo}^{\gamma}$ is the set of exogenous events $\sum_{exo}^{\gamma} \triangleq \sum_{rcv}^{\gamma} \cap \sum_{exo}$;

2) $\sum_{com\_rcv}^{\gamma}$ is the set of communication events received by γ whose source is a component which is not in γ;

$\sum_{emit}^{\gamma}$ is the set of emitted events of γ : $\sum_{emit}^{\gamma} \triangleq (\bigcup_{j\in\{1,\cdots k\}} \sum_{emit}^{i_j}) \setminus (\bigcup_{j\in\{1,\cdots k\}} \sum_{rcv}^{i_j})$. There are two types of emitted events:

1) $\sum_{obs}^{\gamma}$ is the set of observable events $\sum_{obs}^{\gamma} \triangleq \sum_{emit}^{\gamma} \cap \sum_{obs}$;

2) $\sum_{com\_emit}^{\gamma}$ is the set of communication events emitted by γ whose destination is a component which is not in γ;

$\sum_{int}^{\gamma}$ is the set of internal events of the subsystem γ : $\sum_{int}^{\gamma} \triangleq (\bigcup_{j\in\{1,\cdots,k\}} \sum_{emit}^{i_j}) \cap (\bigcup_{j\in\{1,\cdots,k\}} \sum_{rcv}^{i_j})$. An internal event is a communication event between two components in γ. Therefore, the events occurring in γ are the set $\{\sum_{rcv}^{\gamma}, \sum_{emit}^{\gamma}, \sum_{int}^{\gamma}\}$.

Definition 2. (Model of subsystem). The model of the subsystem $\gamma = \{c_{i1}, \cdots, c_{ik}\}$ is a set of finite state machines $\{\Gamma_{ia}, \cdots \Gamma_{ik}\}$.

Definition 2. (Model of global system). The model of the global system $\Gamma = \{c_1, \cdots, c_n\}$ is a set of subsystems' models $\{\Gamma_1, \cdots, \Gamma_n\}$. It only receives exogenous events and emits observable events.

4 Behavior Description Methods

Whether a model can describe the expected system behaviors is decided not only by the domain knowledge, but also by the behavior description method. According to variable types, three description methods are included: the quantitative, the qualitative and the hybrid.

The quantitative method is a traditional way which uses differential/integral functions to express the accurate behavior and map the feature space to decision space with threshold logic. Because of the complexity, the high dimension, the non-linear and the data deficiency, it is often difficult to build the quantitative model. The qualitative modeling is a discrete abstraction of the value of variables. Low precision makes it more suitable to describe the uncertain and incomplete knowledge and the model is easier to understand. There are two ways of qualitative abstraction: signs and orders of magnitude [11]. The value domain of continuous variables is transferred into a finite number of ordered symbols: the boolean or the enumeration.

The discrete abstraction of qualitative modeling may lead to state space explosion which slows down the diagnosis. Thus, the combination of the qualitative and the quantitative methods will give consideration to description capability and scale of state space.

5 Hybrid and Hierarchy Modeling Language

A bottom-top and object-oriented way is adopted to describe the structure and the connection of a system.

1) Each tree node stands for a system and the edge linking the parent and the child refers to the affiliation of the system.

2) The root node is the top system without any ancestors.

3) A leaf node is the bottom system without any descendants, which is corresponding to an undivided component in system. And all statuses could be described only by one mode variable. Mode variables only exist in leaf nodes and the combination of modes is used in upper levels.

4) There could be multiple instances for one type of system, but an instance belongs only to one instance of upper level.

5.1 Grammar

The structure of a model is as follows.

```
system name(port definitions){
Inner variable definitions;
Mode variable definitions;
Subsystem definitions;
```

```
Constrain definitions;
Transition definitions;}
```

1) Port definitions

```
type name1, type name2 …
```

2) Inner variable definitions

```
type name1, type name2 … or type name=value
```

3) Mode variable definitions

```
mode type name
```

The mode variable describes the working state. The value unknown stands for all the unexpected states to make the value domain complete.

4) Subsystem definitions

```
system type name1, name2 …or system type name1(ports),
name2(ports)…
```

This is the definition of the subsystem instance. The order of ports-parameters should be consistent with ports list in the system definition.

5) Constrain definitions

```
expressions or If (expressions) then {expressions} else
{expressions}
```

The expressions are in the forms of arithmetic, inequality or Boolean calculations. And the if-then constrain can be nested for complicated relations.

6) Transition definitions

```
Transition (init_mode) {conditions} (probability 1, …,
probability N)
```

A transition triggered by the conditions expresses that the mode variable transfer from the initial mode to others with certain probabilities. The number N equals to the scale of value domain of the mode variable.

5.2 Modeling Application

The electrical power system is one of the most important systems in spacecraft. If the power supplement is abnormal, other systems may not stay in order and the mission will even fail in the worst case. There are many components, like relays and breakers, which are discrete event systems and suitable for the model-based diagnosis approach. But the global modeling and global diagnosis are not realistic due to the exponential relation between complexity and system scale. Figure 1 is a typical power distribution in the electrical power system. The power is transmitted from the battery to loads through the distribution circuit. Two redundant batteries are placed for high reliability. When one battery is broken-down, one of the other two will be chosen to be the power generator through the exogenous command. The voltage converter CONV includes two parallel DC/DC units controlled by relays, as figure 2 shows.

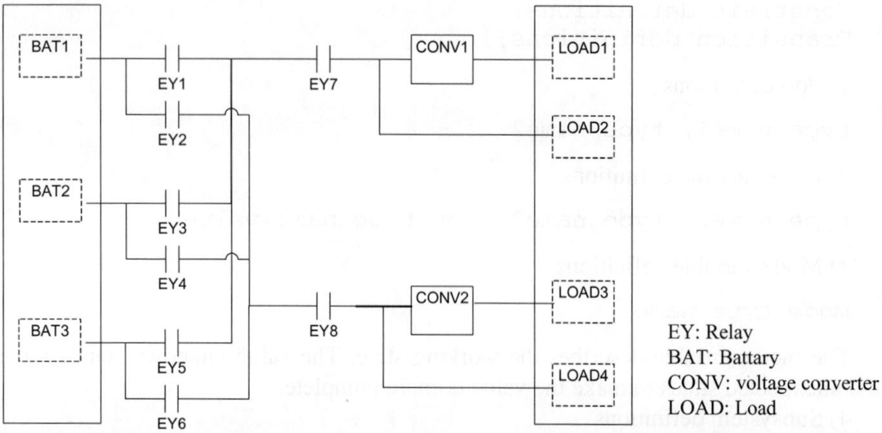

Fig. 1. The typical distribution circuit in the electrical power system.

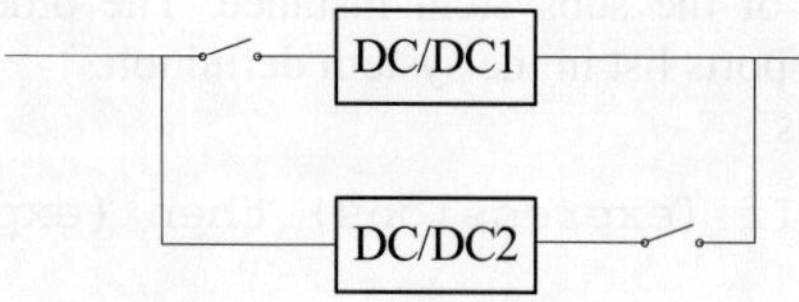

Fig. 2. The redundant design in the voltage converter unit.

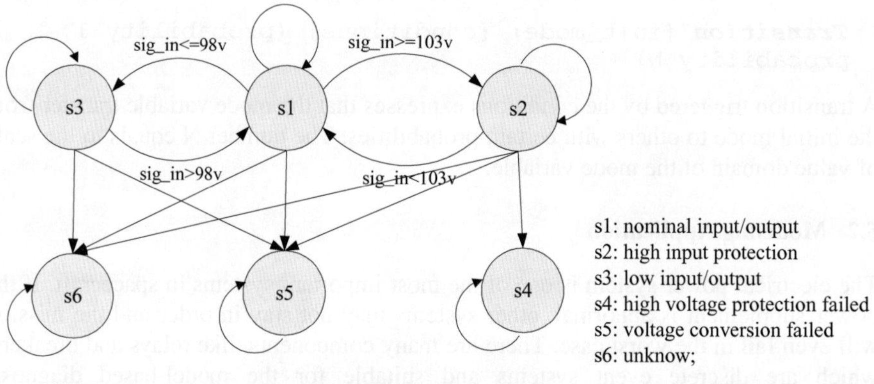

Fig. 3. The state graph of DC/DC unit

In the hierarchy model, only the behaviors of components in bottom level, like relays and DC/DC units, are described and all the other levels are a composition of reusable modules. Therefore, the modeling process is simplified through module reuse.

Take the DC/DC unit as an example. There are six modes: three nominal modes, high voltage protection failed, voltage conversion failed and unknow. The normal interval of the input voltage is [98v, 103v] and the output is [27v, 30v]. Under the nominal mode, if the input voltage rises above 103v, the converter will shut down to protect itself. If the input drops to 98v or less, the output will be less than 27v. Under the high voltage protection failed mode, the output rises above 30v. Under the voltage conversion failed mode, the output is 0v. The state graph is shown in figure 3. And finally the model file of DC/DC unit is given.

The model file of DC/DC unit is as follow:

```
typedef modetype=enum {nominal, overvoltagefailed,
failed, unknow};
system DCDC (float sig_in, float sig_out) {
  mode modetype state_var(0.7999, 0.1, 0.1, 0.001);
  if (state_var=nominal) {
  if (sig_in>=98.0) {
         sig_out>=27.0;
         sig_out<=30.0; }
  else {
     sig_out<27.0; }}
  else if (state_var=overvoltagefailed) {
     sig_out>30.0; }
  else if(state_var= failed) {
     sig_out=0;} }
```

6 Conclusion

Based on the characteristics and requests of spacecraft diagnosis, a hybrid and hierarchy approach to model-based diagnosis and a modeling language are presented. To validate the feasibility, a distribution circuit is modeled, which is typical in the electrical power system of spacecrafts. The hybrid description avoids the state space explosion of discrete abstraction and the module reuse simplifies the modeling process, which is suitable for the complicated and redundantly designed system. Moreover, the hierarchy style is of benefits to hierarchical diagnosis which is the trend of large-scale system diagnosis.

References

1. Korn, G.A., Wait, J.V.: Digital continuous-system simulation. Prentice-Hall, Englewood Cliffs (1989)
2. Venkatasubramanian, V., Rengaswamy, R., Yin, K., et al.: A review of process fault detection and diagnosis: Part I: Quantitative model-based methods. Computers & Chemical Engineering 27(3), 293–311 (2003)
3. De Kleer, J., Brown, J.S.: A qualitative physics based on confluences. Artificial Intelligence 24(1-3), 7–83 (1984)
4. Forbus, K.D.: Qualitative process theory. Artificial Intelligence 24(1-3), 85–168 (1984)
5. Kuipers, B.: Qualitative simulation. Artificial Intelligence 29(3), 289–338 (1986)
6. Venkatasubramanian, V., Rengaswamy, R., Kavuri, S.N.: A review of process fault detection and diagnosis: Part II: Qualitative models and search strategies. Computers & Chemical Engineering 27(3), 313–326 (2003)

7. Sampath, M., Lafortune, S., Teneketzis, D.: Active diagnosis of discrete-event systems. IEEE Trans. Automat. Contr. 43(7), 908–929 (1998)
8. Falkenhainer, B., Forbus, K.D.: Compositional modeling: finding the right model for the job. Artificial Intelligence 51(1-3), 95–143 (1991)
9. Pencol, Y., Cordier, M.: A formal framework for the decentralised diagnosis of large scale discrete event systems and its application to telecommunication networks. Artificial Intelligence 164(1-2), 121–170 (2005)
10. Leitch, R.R., Shen, Q., Coghill, G.M., et al.: Choosing the right model. IEE Proc. Control Theory Appl. 146(5), 435–449 (1999)
11. Trave-Massuyes, L., Ironi, L., Dague, P.: Mathematical foundations of qualitative reasoning. AI Magazine 24(4), 91–106 (2003)

A New Codebook Design Algorithm of Vector Quantization Based on Hadamard Transform

Shanxue Chen*, Jiaguo Wang, and Fangwei Li

Chongqing University of Posts and Telecommunications,
Mobile Communication Security Technology Laboratory
Chongqing 400065, China
287366171@qq.com

Abstract. This paper presents a new codebook design algorithm improving the existing design algorithm. The algorithm combines the excellence of conventional LBG algorithm, and uses the exclusion algorithm which greatly reduce the computational complexity. The proposed algorithm uses category average of statistical characteristics and also uses Hadamard-transform to speed up the search algorithm. The experimental results show that the proposed algorithm reduce the computation and improve the codebook performance, especially in case of larger codebook size and high-detail images.

Keywords: vector quantization, Hadamard transform, codebook design, exclusion algorithm.

1 Introduction

With the rapid development of multimedia communication technology, it becomes more and more important to find effective means of digital and data processing. Vector Quantization (VQ) [1,2] based on muti-dimension because of its high compression ratio and fast coding and encoding are widely used in speech and image compression and pattern re-cognition system, and it is also highly efficient method for loss data compression[3,4]. The amount of medial images is huge and increasing rapidly every year. Thus, the image compression is needed to reduce the data volume of these radiological images.

In 1980, Linde, Buzo, and Gray proposed a VQ design algorithm based on a training sequence. At present, the most widely used technique to design VQ codebooks is Generalized Lloyd Algorithm (GLA) [5]. Usually we call it LBG algorithm. Recently VQ is considered to be a most popular technique for image Compression [6]. In a conventional image VQ compression method, an input image is firstly divided into a series of non-overlapping smaller image blocks. Then, VQ encoding is implemented block by block in a raster scanning order. A k-dimension,

* The Natural Science and Technology Specific Program of China(No.2009ZX03001-004), the National Science Foundation of China(No.61071116), and the Project of Key Laboratory of Signal and Information Processing of Chongqing(No.CSTC2009CA2003), China.

M. Zhu (Ed.): Electrical Engineering and Control, LNEE 98, pp. 181–187.
springerlink.com © Springer-Verlag Berlin Heidelberg 2011

N-level vector quantizer is defined as a mapping from a *k*-dimensional vector space R^k into a certain finite subset $Y=\{Y_j, j=1,...,N\}$. The subset *Y* is called a codebook and its elements are called code words. Kekre proposed a 2-level Vector Quantization method for codebook design using Kekre's Median codebook generation algorithm [7]. The algorithm cannot use the exclusion algorithm to eliminate unnecessary code words.

Now VQ which has been successfully used in various application involving VQ-based encoding and VQ-based recognition, due to its excellent rate-distortion performance. [8,9] The codebook design algorithm of the fundamental purpose is to find an effective way to seek to the global optimum as possible or at least near global optimal codebook to improve the performance of the codebook. [10] The quality of codebook design is the key factor of the performance of vector quantizer.

2 LBG Codebook Design Algorithm

2.1 The Drawbacks of LBG Algorithm

The traditional LBG algorithm is a classic algorithm of codebook design, and LBG lay a foundation for the development of vector quantization technology, but there exits three main drawbacks as follows:

1) Very sensitive to the initial codebook.
2) That codebook is not adaptive enough.
3) Large computation.

2.2 LBG Algorithm Is Summarized as Follows

1) *step 1*: An initial codebook $Y_N^{(0)} = \{Y_j; j=1,2,...N\}$ with size N is given, set average distortion $D^{(-1)} \to \infty$ and iteration *n=0*, the threshold $\varepsilon(0<\varepsilon<1)$.
If you have more than one surname, please make sure that the Volume Editor knows how you are to be listed in the author index.

2) *step 2*: The training aggregation $X = \{Y_j; j=1,2,...M\}$ include the vector x_m and divide into *N* different sub-interval $R_i(n)(i=1,2,..N)$.

3) *step 3*: The average distortion $D_n = \frac{1}{M}\sum_{j=0}^{M-1} \min_{y\in y_N^{(n)}} d(x_j, y)$ if the nearest two average distortion meet the inequality $(D_{n-1} - D_n)/D_n \le \varepsilon$ or meet the given iterations, then stop the algorithm, and get the last codebook $Y_N^{(n)}$ .Otherwise go to *Step 4*.

4) *step 4*: Calculate the centroid $Y_j(n) = \frac{1}{|R_i(n)|}\sum_{x_m\in R_i(n)} x_m$, update every sub-interval codebook, set $n = n+1$, go to *Step 2*.

3 Hadamard Transform

The definition and nature of the Hadamard-transform:

Set H_n be $2^n \times 2^n$ Hadamard square matrix with elements with in the set {1 -1}, assume that all of the following vectors for k-dimensional vector, set $k=2^n$ (n>0), then the following basic definitions and properties can be obtained:

1) $H_1 = \begin{bmatrix} 1 & 1 \\ 1 & -1 \end{bmatrix}$, $H_{n+1} = \begin{bmatrix} H_n & H_n \\ H_n & -H_n \end{bmatrix}$

2)Set vector x through the hadamard transform be vector X, we can define that:
$X = H_n x$

3) $X_1 = S_x$

Where X_1 is vector X's 1st – dimensional component, S_x is the sum of the input vector X.

4) $D(X, Y_j) = kd(x, y_j)$

Where the codeword y_j through Hadamard-transform to be Y_j;

Based on $X = H_n x$, we have:

$$\|X\|^2 = X^T X = (H_n x)^T H_n x = x^T H_n^T H_n x = kx^T x = k\|x\|_2^2$$

That is to say, the norm of transform vector is k times the norm of the corresponding spatial vector. So we search the nearest neighbor codeword in the Hadamard domain is equivalent to in the airspace.

4 Proposed Algorithm

This paper provides a new codebook design algorithm, reduce the computational, and more importantly improve the codebook performance greatly.

4.1 Previous Work

For a given codebook $Y=\{Y_j; j=1,\ldots,N\}=\{(y_{j1}, y_{j2},\ldots,y_{jk}); j=1,\ldots,N\}$ of size N and a query vector $X=(x_1, x_2, \ldots, x_k)$ in the k-dimension vector space, we denote the sum, the variance and the L_2 norm of X and Y_j as (S_x, V_x, L_x) and (S_j, V_j, L_j), respectively. They are defined as:

$$S_x = \sum_{i=1}^{k} x_i, \quad S_j = \sum_{i=1}^{k} y_{ji}. \tag{1}$$

$$V_x = \sqrt{\sum_{i=1}^{k} (x_i - S_x/k)^2}, \quad V_j = \sqrt{\sum_{i=1}^{k} (y_{ji} - S_j/k)^2}. \tag{2}$$

$$L_x = \|X\| = \sqrt{\sum_{i=1}^{k} x_i^2}, \quad L_j = \|Y_j\| = \sqrt{\sum_{i=1}^{k} y_{ji}^2}. \tag{3}$$

4.2 Three-Step Exclusion Algorithm

Let $H(k)$ be $k\times k$ Hadamard matrix. The Hadamard-transfromed vectors h_x and h_{yj} of the vector X and codeword Y_j are defined as $h_x=XH(k)$ and $h_{yj}= Y_jH(k)$, respectively. The Hadamard-transformed variance and norm of vector X and codebook Y_j are defined as:

$$hv_x = \sqrt{\sum_{i=2}^{k} hx_i^2}\ , \quad \|hx\| = \sqrt{\sum_{i=1}^{k} hx_i^2}\ . \tag{4}$$

$$hv_j = \sqrt{\sum_{i=2}^{k} hy_{ji}^2}\ , \quad \|hy_j\| = \sqrt{\sum_{i=1}^{k} hy_{ji}^2}\ . \tag{5}$$

Let D_{min} be the current minimum transform domain distortion. Let hx_1 and hy_{j1} be the first element of hx and hy_j, respectively. The proposed algorithm presented a fast codeword exclusion algorithm, its three-step test flow method can be expressed as:

$$\left(hx_1 - hy_{j1}\right)^2 \geq D_{\min}\ . \tag{6}$$

$$\left(hv_x - hv_j\right)^2 \geq D_{\min}\ . \tag{7}$$

$$\left(hx_1 - hy_{j1}\right)^2 + \left(hv_x - hv_j\right)^2 \geq D_{\min}\ . \tag{8}$$

4.3 Off-Line Calculation

Before encoding, calculate Hadamard-transfromed vectors hy_j of codeword Y_j, then calculate hy_{j1}, the first element of the L_2 norm of codebook Y_j, sort in ascending order. In the encoding process, calculate hx of the vector X and V_x Hadamard-transformed variance. Choose the codeword nearest hx_1 Hadamard-transform of inputting vector X to be the initial match codeword.

4.4 On-Line Calculation

During the on-line stage, the encoding process for each input vector X can be illustrated as follows: perform the Hadamard-transform on the input vector X to obtain hx_1 and compute hx_1 and hy_j, then initialize the current closest codeword of hx_1, there after a three-step exclusion test flow will be introduced to reject unlikely codewords. It can be expressed as:

$$\left(hx_1 - hy_{j1}\right)^2 \geq D_{\min}\ . \tag{9}$$

$$\left(hv_x - hv_j\right)^2 \geq D_{\min}\ . \tag{10}$$

$$\left(hx_1 - hy_{j1}\right)^2 + \left(hv_x - hv_j\right)^2 \geq D_{\min} . \tag{11}$$

The proposed algorithm combines Hadamard transform and K-means theory, uses the classification method of mean of the statistical characteristics to generate the initial codebook. Moreover, improves the frequency of centroid. Whenever a training vector is classified into cell lumen, calculate the centroid of the corresponding cell and replace the original codeword.

Compared with LBG algorithm, the adjusted code words represents the characteristics of the cell lumen, accelerates the speed of convergence of codebook, improves the performance of codebook.

5 Experiment Results

The standard images are used in the experiment to evaluate the efficiency of the proposed method. Computer simulation using real images is performed on a PC to evaluate the proposed algorithm in comparison with some fast algorithms. For the experiment, we use training and test images of size *512×512* with *256* gray levels. Lena and Peppers are used as training images to design the codebook, and the resulting codebook are used to compare with the LBG algorithm.

From Table 1, we find that the proposed algorithm is improved compared the coding results of the other algorithm, especially when the codebook size is large, the advantage is quite obvious, the Peak Signal to Noise Ratio (PSNR) is improved *0.1-0.9 db* compared the LBG algorithm.

Table 1. Comparison of the PSNR of one training and twenty trainings of Lean and Peppers.

Image	Methods	Iterations	Codebook size			
			128	256	512	1024
Lena	LBG	1	29. 5014	30. 4018	31. 2355	32. 2211
	New	1	30. 2946	31. 2376	32. 1692	33. 1973
	LBG	20	30. 6698	31. 6475	32. 5198	33. 4066
	New	20	30. 6042	31. 7155	32. 7534	33. 9017
Peppers	LBG	1	29. 0239	29. 9973	30. 8187	31. 7110
	New	1	29. 8354	30. 7536	31. 5887	32. 6122
	LBG	20	30. 1531	31. 1408	31. 9311	32. 7231
	New	20	30. 1119	31. 1661	32. 3233	33. 3903

Fig. 1. shows the Peak Signal to Noise Ratio (PSNR) of the two algorithms under the different iterations and the codebooks of size *512*. From Fig. 2., we can find that the new algorithm is more efficient when the iteration is equal, moreover, the new algorithm can also exceed the conventional LBG algorithms when the iteration of the new is less than the other iteration.

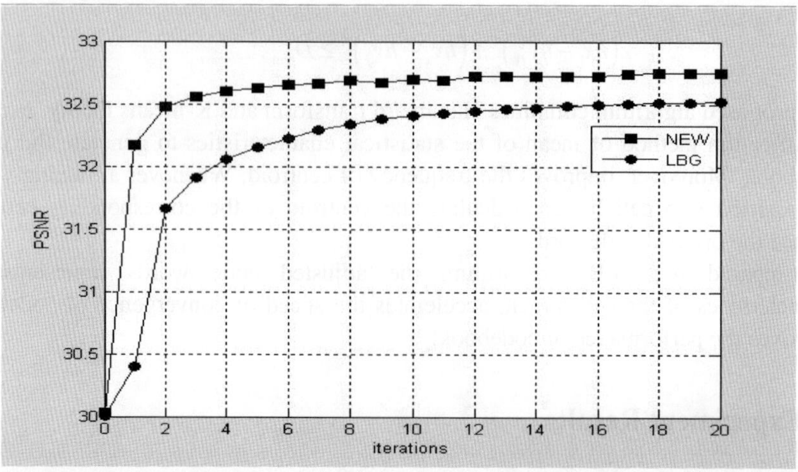

Fig. 1. The PSNR of the two algorithms under different iterations and the codebook size 512

6 Conclusion

This article presents a new codebook design algorithm for VQ technology based on Hadamard-transfrom. The proposed algorithm combines the excellence of the conventional LBG algorithm. First, using the Hadamard-transfrom and K-means theory, the vector can be quantized with a considerably lower complexity than direct vector quantization of the original vector. Then, using the fast exclusion algorithm, it use three-inequality which can preclude some unnecessary codeword fast. The simulation results show that the PSNR of the proposed algorithm is improved significantly. It is also found that the proposed algorithm outperforms the existing codebook design algorithm when the iteration of the new algorithm is less than the others.

References

1. Gersho, A., Gray, R.M.: Vector Quantization and Signal Compression. Kluwer Academic Publishers, Boston (1992)
2. Linde, Y., Buzo, A., Gray, R.M.: An algorithm for vector quantize design. IEEE Trans. Commun. 28(1), 84–95 (1980)
3. Shanxue, C., Fangwei, L.: Fast encoding algorithm for vector quantization based on subvector L2-norm. Journal of Systems Engineering and Electronics 19(3), 611–617 (2008)
4. Torres, L., Huguet, J.: An improvement on codebook search for vector quantization. IEEE Trans. on Com. 42(2/3/4), 208–210 (1994)
5. Hwang, W.J., Jeng, S.S., Leou, M.R.: Fast codeword search technique for the encoding of variablerate vector quantisers. IEE Proc. of Vis. Image Signal Process. 145(2), 103–108 (1998)

6. Chen, S.-x.: Initial codebook algorithm of vector quantization. Computer Engineering and Applications 46(11) (2010)
7. Kekre, H.B., Sarode Tanuja, K.: 2-level Vector Quantization Method for Codebook Design using Kekre's Median Codebook Generation Algorithm, vol. 2(2) (2009)
8. Yea, S., Pearlman, W.A.: A Wavelet-based Two-Stage Near-Lossless Coder. IEEE Trans. on Image Processing 15(11), 3488–3500 (2006)
9. Cohen, A., Daubechies, I., Feauveau, J.C.: Biorthogonal bases of compactly supported wavelets. Comm. Pure and Appl. Math. 45(5), 485–560 (1992)
10. Thangavel, K., Kumar, D.A.: Optimization of codebook in vector quantization. Ann. Oper. Res. 143, 317–325 (2006)
11. Sun, H.W.: Efficient vector quantization using genetic algorithm. Neural Comput. & Applic. 14, 203–211 (2005)

Studies on Channel Encoding and Decoding of TETRA2 Digital Trunked System

Xiaohui Zeng[1], Huanglin Zeng[2], and Shunling Chen[2]

[1] Dept. of Comm. Eng., Chengdu University of Information Technology,
Chengdu, 610225, P.R. China
zxhui@cuit.edu.cn

[2] College of Auto. & Inf. Eng., Sichuan University of Science and Technology,
Sichuan, 643000, P.R. China
zhl@suse.edu.cn, csl_1220@163.com

Abstract. This paper presents a design and analysis of some efficient channel encoding and decoding schemes for enhanced data services provisioning within next generation personal mobile radio systems, specifically Terrestrial Trunked Radio 2 system. Two different turbo codes schemes are compared by way of the parallel convolution codes and the serial convolution codes. An efficient channel coding scheme applied in the TETRA Release 2 system is also proposed.

Keywords: Terrestrial Trunked Radio 2, Channel Encoding, Channel Decoding, Turbo Code parallel convolution codes (PCCC) and the serial convolution codes (SCCC).

1 Introduction

With the growth of wireless communication, interference and noise, multipath fading and bit error rate also is increased. It needs to adopt appropriate channel coding to improve the data wireless communication quality. TETRA (Terrestrial Trunked Radio) [1] is recommend as a kind of the 2nd generation (2G) mobile communication standard for digital trunking system in China. TETR digital trunking system has been fast development in personal mobile communication network since its stringent architecture and an open standard.

The European Telecommunications Standards Institute ETSI has developed TETRA—TETRA2 in the 3rd generation(3G) mobile communication standard for a new generation of wireless communication. A packet-switched technique is used in the 3G mobile communication network to increase the data transmission rate and meet with different service of the QOS required and effective management of user who ask the flexibility physical application.This technique is not only applied to such as GSM, UMTS GRPS public mobile telecommunications system, just as for personal mobile communication network. Turbo code is being used as a standard channel encoding and decoding in the 3rd generation mobile communication standards [2-5].

In this paper we will do comparison on parallel Convolution codes PCCC and serial Convolution codes SCCC, and put forward a new kind of encoding and decoding of Turbo codes for TETRA2.

M. Zhu (Ed.): Electrical Engineering and Control, LNEE 98, pp. 189–194.
springerlink.com © Springer-Verlag Berlin Heidelberg 2011

2 Optimization of Terrestrial Trunked Radio of 3G

Turbo code being used as a standard channel encoding and decoding in the 3rd generation mobile communication standard will face with some problems such as error rate increasing since changes of the transport channel and communication conditions of time-varying characteristics [6-8]. In order to meet the needs of the high-speed multimedia applications and reduce the error rate, increase information throughput, improve communication capability, and enables communication systems become reliable systems without delay time-varying characteristics, TETRA2 will be of the following characteristics:

(1) Selecting best channel encoding scheme to enhance communication quality so that TETRA2 can realize an interworking with other 3G mobile system.

(2) Improving public mobile communication network such as GSM, GPRS, etc. in the Internet and roaming capabilities.

(3) Increasing the coverage of network and reducing scheduling overhead.

(4) Providing a high speed data service for multimedia communications.

(5) Increasing network interface performance to optimize the utilization of spectrum efficiency and improve communication power consumption.

Based on the above-mentioned requirements, a new interface standard known as TEDS (TETRA Enhanced Data system) is proposed as following:

The main performance for TEDS:

(1)TEDS adopts 8、16、32 or 48 channel transmission where channel bandwidth is 25kHz、50kHz、100 kHz or150 kHz respectively.

(2) Modulation mode is 4-QAM, 16-QAM, 64-QAM.

(3) Channel coding rate of Turbo codes is 1/2 or 2/3.

Figure 1 an optimization of communication channel coding system implemented on Turbo codes is shown as following.

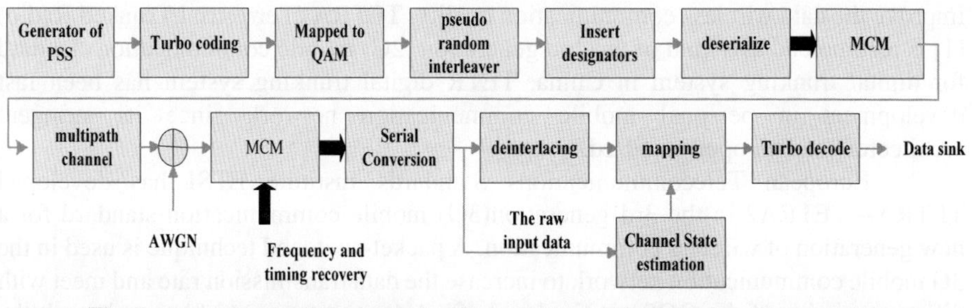

Fig. 1. A communication channel coding system implemented on Turbo codes

3 A New Encoding and Decoding of TETRA2 Digital Trunked System

Turbo code is of format of the parallel convolution codes (PCCC) and the serial convolution codes (SCCC). The parallel convolution codes (PCCC) and the serial convolution codes (SCCC) are Turbo codes based on binary Convolution encoder (BCE). A cascading encoding structure diagram of BCE base unit is shown as Figure 2.

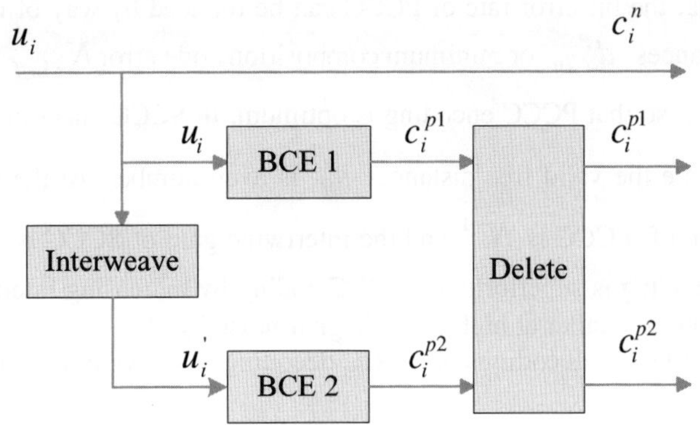

(a) PCCC cascading encoding structure diagram（n，k）

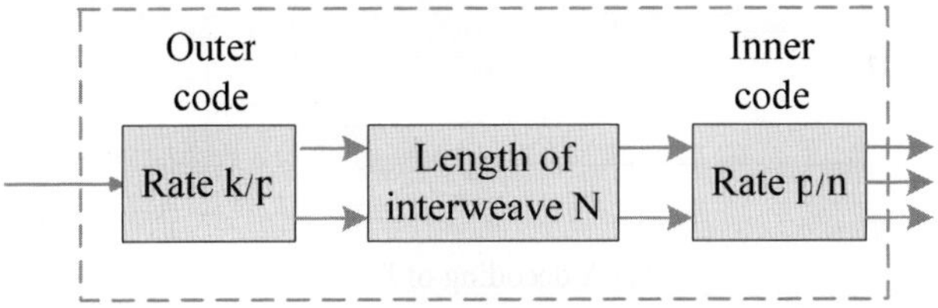

(b) SCCC cascading encoding structure diagram（n,k,N）

Fig. 2. A cascading encoding structure diagram of BCE base unit

The code rate is 1/3 in the PCCC which can be increased1/2 or 2/3 by deleting method. The code rate in the SCCC can be expressed by

$$R = R_c^o R_c^i = k / n \tag{1}$$

Turbo code is of small bit-error rate (BER), low noise ratio in wireless channel coding standards in 3G since Turbo code based on Shannon's theory.The bit-error rate of PCCC and SCCC can be expressed as below [9-10].

$$\text{PCCC:} \quad P_b(e) \approx K N_{f1,eff} N_{f2,eff} N^{-1} Q\sqrt{\frac{2 d_{f,eff}^P R_c E_b}{N_O}} \tag{2}$$

$$\text{SCCC:} \quad P_b(e) \approx k_{even} N^{-\frac{d_f^O}{2}} Q\sqrt{\frac{d_f^O d_{f,eff}^i R_c E_b}{N_0}} \quad d_f^O \text{ even number} \tag{3}$$

$$k_{odd} N^{-\frac{d_f^O+1}{2}} Q\sqrt{\frac{(d_f^O - 3)d^i_{f,eff}}{2} + h_m^{(3)}\frac{2R_c E_b}{N_0}} \quad d_f^O \text{ odd number}$$

It is shown that the bit-error rate of PCCC can be reduced by way of maximize code valid free distances $d^P_{f,eff}$ or minimum composition code error $N_{f1,eff}$ and code class number $N_{f2,eff}$ so that PCCC encoding is optimum. In SCCC encoding, bit-error rate is smallest while the valid free distances d_f^O is even number. At the same time, the intertwine gain of PCCC is N^{-1} and the intertwine gain of SCCC is N^{-2}、N^{-3}、N^{-4}, SCCC coding is superiority on PCCC coding by increasing intertwine gain and the bit-error rate is smaller at high rate of signal-noise [6-8].

A scheme of PCCC decoding and SCCC decoding is shown in Figure 3.

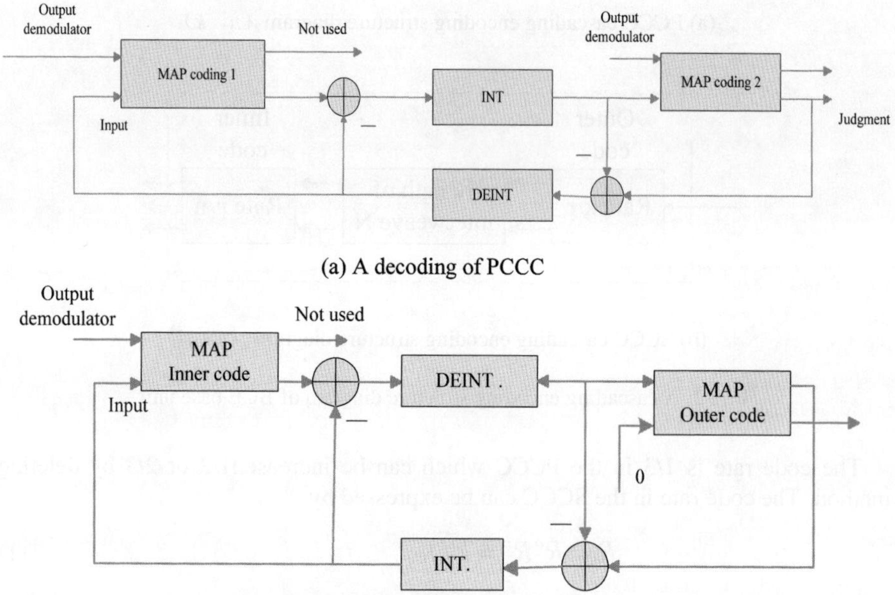

(a) A decoding of PCCC

(b) A decoding of SCCC

Fig. 3. A scheme of PCCC decoding and SCCC decoding

In Figure 3, BCJR decoding algorithm in the Turbo decoding is used in a Maximum Posteriori probability (MAP) decoding models for decoding single input and single output (SISO) of PCCC and SCCC. It is noted that the SCCC decoder is a typical serial cascading decoder in which external encoder corresponding input is 0 due to external encoding does not contain information bits of information, and Turbo codes circuit is a kind of feedback implementation.

The polynomial generated of PCCC and SCCC coding is expressed as following:

The polynomial generated of PCCC at 16 state coding:

$$g_0 = (31)_8 \text{ and } g_1 = (33)_8 \tag{4}$$

The polynomial generated of SCCC at 16 external state coding:

$$g_1 = (46)_8 \text{ and } g_2 = (72)_8 \tag{5}$$

The polynomial generated of SCCC at 8 internal state coding:

$$g_0 = (13)_8 \text{ and } g_1 = (15)_8 \tag{6}$$

Here the encoding parameter is analyzed with additive white Gaussian noise. It is seen that the states of SCCC coding is reduced by 25% than that of PCCC coding. SCCC coding is of better performance such as the error rate, data throughput, and so on than that of PCCC coding.

4 Summaries

This article discusses an enhanced data services can be made by way of use of an appropriate channel coding to implement a TEDS standard. SCCC coding is an appropriate channel coding since it can provides better performance such as the universality, effectiveness, bit-error rate, time-delay and data throughput in channel encoding compared with the PCCC which are applied to the 3G standard of Turbo codes. These advantages of SCCC coding are good for video phones, electronic monitoring, and streaming media, and so on high-speed multimedia applications of TETAR2.

Acknowledgement. This work was supported by a grant from Foundation of major special project of key laboratory of liquor and biotechnology of Sichuan province under No. NJ2010-01, and a Foundation of major special project of applied science of Sichuan Province Education Department under No. 09ZX002.

References

1. ETSI. EN 300 395-2 V1.3.1 Terrestrial Trunked Radio (TETRA); Speech codec for full-rate traffic channel; Part 2: TETRA codec . ETSI (2005)
2. Liu, F.: An Improved RBF Network for Predicting Location in Mobile Network. In: Proceeding of Fifth International Conference on Natural Computation, vol. 3, pp. 345–348 (2009)
3. Dwivedi, A.D.D., Chakrabarti, P.: Sensitivity analysis of an HgCdTe based photovoltaic receiver for long-wavelength free space optical communication systems. Optoelectronics Letters 01 (2009)
4. Li, W., Mei, J.-y., Han, Q.-s.: Influence of the nonlinear phase and ASE noise on DPSK balanced optical receiver in optical fiber communication system. Optoelectronics Letters 01 (2009)

5. Juanjuan, F., Shanzhi, C.: An Autonomic Interface Selection Method for Multi-interfaces Mobile Terminal in Heterogeneous Environment. In: Proceeding of WRI World Congress on Computer Science an Information Engineering, Los Angeles,USA, pp. 107–111 (2009)
6. Cui, J., Shi, H., Chen, J.: The transmission link of CAPS navigation and communication system Science in China. Series G: Physics,Mechanics & Astronomy, vol. 03 (2009)
7. Wei, L., Xu, Y., Cai, Y.: Robust And Fast Frequency Offset Estimation For Ofdm Based Satellite Communication. Journal of Electronics (China) 01 (2009)
8. Wang, W.Y., Wang, C., Zhang, G.Y.: Arbitrarily long distance quantum communication using inspection and power insertion. Chinese Science Bulletin 01 (2009)
9. Wang, S., Chen, C.: A Framework for Wireless Sensor Network Based Mobile Mashup Applications. In: Proceeding of WRI World Congress on Computer Science an Information Engineering, Los Angeles,USA, pp. 683–687 (2009)
10. Zeng, H., Huang, Y., Zeng, X.: A New Approach of Attribute Reduction Based on Ant Colony Optimization. In: Proceeding of Fifth International Conference on Natural Computation, vol. 3, pp. 3–7 (2009)

Selective Block Size Decision Algorithm for Intra Prediction in Video Coding and Learning Website

Wei-Chun Hsu[1], Yuan-Chen Liu[2], Dong-Syuan Jiang[2], Tsung-Han Tsai[1], and Wan-Chun Lee[3]

[1] National Central Unervisity, Department of Electrical Engineering, Zhongda Rd. 300, Zhongli City, Taoyuan County 320, Taiwan (R.O.C.)
[2] National Taipei Unervisity of Education, Department of Computer Science, Heping E. Rd. 134, Da'an Dist., Taipei City 106, Taiwan (R.O.C.)
[3] Taipei Municipal Ren-Ai Elementary School
Liu@tea.ntue.edu.tw

Abstract. In this paper, an algorithm of Selective Block Size Decision for Intra Prediction in Video Coding and Learning Website will be presented. The main purpose is to reduce the high complexity of video when predict in intra prediction to select the mode.

Video coding decision is divided into two stages: the selective block size and reduce the prediction mode. When selective block size predict by the block smooth, it can divided to three stages. The smooth block use the16x16 intra prediction mode and the complexity use the 4x4 intra prediction mode. When reduce the prediction mode, we use the Elyousfi' algorithm.

On experimental results, we decide the variance by 500 and use the smooth block to control the selective block size then we can get the lower complexity.

Keywords: intra prediction, H.264, variable.

1 Introduction

Because of the completeness with the third generation mobile communication technology, there are more and more portable products which can watch multimedia, such as MP3, I-Pod, I-Phone, Mobile Phone and so on. However, these portable products, not enough storage space for multimedia storage, the video compression technology is a topic of growing importance. In 2003, the latest video encoding is proposed, an increase of applied on the network, but also improves the video coding rate of the original, making the image on the network can be used more widely.

In this paper, we focus on the internal prediction based on the video coding techniques and modify the original prediction of the high complexity [1], to make further improvements.

M. Zhu (Ed.): Electrical Engineering and Control, LNEE 98, pp. 195–202.
springerlink.com © Springer-Verlag Berlin Heidelberg 2011

2 Related Work and Researches

2.1 JM Intra Predict Mode

JM (Joint Model) is a kind of basic prediction method, because it is using Full Search internal prediction model, in particular, is also using Lagrangian multiplier JM (λ_{MODE}) which is the best model selection method, through the multiplier to consider the overall picture distortion (Distortion) and bit rate (Bitrate) [3], so the use of a rate - distortion optimization (Rate-Distortion Optimization: RDO) [4] method of calculation and formula is as follows:

$$J = Distortion + \lambda_{MODE} \times Rate \tag{1}$$

Distortion of which represents the prediction model to predict the block distortion with the original block, but Rate is the representative of the aforementioned code generated by the encoding bit rate (Bitrate). The Lagrangian multiplier is expressed as:

$$\lambda_{MODE} = 0.85 \times 2^{(QP-12)/3} \tag{2}$$

Therefore, the size of the Lagrangian is determined by the size of quantization parameter (QP), as part of distortion is used by the calculation of sum of squared difference (SSD). SSD is calculated as follows:

$$SSD = \sum_{i,j} (block(i,j) - prediction(i,j))^2 \tag{3}$$

Block represents the original block of pixels, prediction is a prediction pixel value, the best of ratio-distortion is looking out for each block of the predicted residual value of the minimum, and representatives for the block predicted the value of the minimal distortion. RDO complete formula such as (4).

$$J = SSD + \lambda_{MODE} \times BitRate \tag{4}$$

These are the predictions for the 4x4 block of the used rate-distortion optimization, predictive models for the 16x16, JM modifies the distortion assessment of SSD in favor of a new assessment method which called sum of absolute transformed difference (SATD), and the residual Hadamaed converted value as the new residual value for the absolute value and then add up the pixel value divided by 2, such as (5).

$$SATD = \left(\sum_{i,j} \left| (newdiffer(i,j)) \right| \right) / 2 \tag{5}$$

The RDO formula for 16x16 just changes the SSD of (4) to SATD, and the determination method for the RDcost which is calculated by J follow the following steps:

(1) Select 4x4 block prediction mode, calculating nine models RDcost.
(2) Select the smallest RDcost, and record the minimum cost of the prediction model.
(3) Prediction of 16 4x4 blocks minimum RDcost sum, assuming the sum of A.
(4) Select the 16x16 block prediction model to calculate four models RDcost.
(5) Select the smallest RDcost, and record the minimum cost of the prediction models and assumptions for the minimum Rdcost B.
(6) A and B, compare the cost, if B is small then this block uses a 16x16 prediction mode; the other hand, use the 4x4 prediction mode, and record the minimum cost prediction model, it is predicted by the block the best prediction model.

By the above steps, for JM, all blocks will be used nine 4x4 prediction modes and four 16x16 prediction mode, so a higher complexity.

2.2 Elyousfi's Fast Intra Mode Decision

Pan's fast intra mode decision was made in 2003 [6], the way he used through the pixel values within each block first to the block for this prediction, as shown in Figure 1(a) to (f).He observed the difference between pixel values to determine a possible direction for this block. And in 2007 Elyousfi proposed his fast intra mode decision [6] is based on Pan's algorithm to do the amendment.

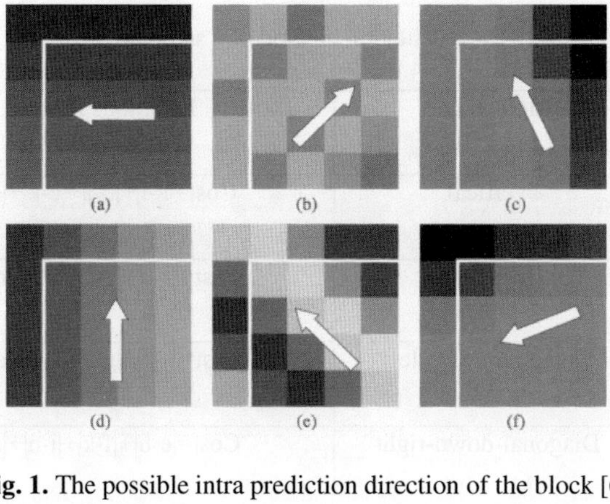

Fig. 1. The possible intra prediction direction of the block [6].

In Pan's algorithm, based on horizontal and vertical pixel values of the difference between 4x4 and 16x16 blocks as a predictive model to determine a possible model, as shown in Figure 1 (a) and (d), and pursuant to Table 1 shows that the most common mode of 0,1,2of 4x4 will be selected to the three models.

Table 1. The three most commonly used model [7].

Name	Sum of 3 proportion
Akiyo	73.6%
Coastguard	68.7%
Foreman	48.1%
Mother and daughter	60.9%
Salesman	58.5%
News	62.3%
Trevor	69.5%
Carphone	71.6%
Container	66.2%

The Elyousfi's algorithm in addition to horizontal and vertical addition, also adds the judge of the mode 3 and 4. The ways show in Table 2.

Table 2. The calculation method of Candidate model [6].

Num	Mode name	Equations for cost
0	Vertical	Cost=|a-i|+|b-j|+|c-k|+|d-l|
1	Horizontal	Cost=|a-c|+|e-f|+|i-k|+|m-o|
3	Diagonal-down-left	Cost=|b-e|+|g-j|+|l-o|+|d-m|
4	Diagonal-down-right	Cost=|c-h|+|f-k|+|i-n|+|a-p|

3 Selective Block Size Decision Algorithm for Intra Prediction

3.1 Prediction Process of the Original

In the above two related work, JM algorithm in predicting the completion of all 4x4 blocks, the prediction is again all 16x16 blocks, JM, respectively, with a screen for the last two predictions. To assume that if a scan can be divided into 4x4 blocks or 16x16 blocks for the best predictors of the best prediction block, and can reduce the use of predictive models, so we can reduce the complexity. The Elyousfi's algorithm only simplifies in the mode of the prediction, so the prediction for the 16x16 block is 4x4 blocks, etc. must be completed first prediction, and predict more than a 8x8 block to calculate the minimum RDcost complexity, then do 16x16 prediction block. For the above two problems, this paper decided to use two separate blocks, can be seen from Figure 2, the part inside the red box, the block pixel values are very close to the nearby, so in this block which will choose 16x16 prediction mode. From here we can see that when the screen block is smooth, we select the large block prediction mode as the prediction block of prediction models.

Fig. 2. Foreman.

So for the judge to find the block at the time, it was suggested that the threshold value [2] of the judge, the judge in setting the threshold is very important. If the threshold set too high, represent a part of the predictive coding would be predicted error, although the complexity will lead to decrease, the error code could lead to a waste of coding. So for the theory of the threshold value, we propose a different way to judge, and we propose the usage of variance (Standard Deviation) to analysis. We turn the selected block size is 16x16, then the 4x4 block is divided into 16 blocks, 16 blocks from this variance for analysis, first of all, such as the variance formula (6):

$$\sigma = \frac{\sum_{k=1}^{N} (x_k - \mu)^2}{N} \tag{6}$$

x_k represents each pixel value inside the 4x4, and μ is expressed as the average. It would seem that the 16x16 block size will have 16 variable value, then we can see from Figure 3,4, when the value of each 4x4 pixel block differences are smaller, the smaller the variance, whereas the larger. This result can be that, assuming the 16 variable values are small, then we can determine these 16 blocks are all relatively smooth block.

60	55	60	63
58	63	59	57
59	62	61	58
59	62	60	60

Fig. 3. Block with smaller pixel value difference.

67	43	182	123
27	101	254	234
33	190	246	210
110	244	175	86

Fig. 4. Block with higher pixel value difference.

But how to judge the value of variance value is belong to smooth picture? We assume that the variance is less than a fixed value, this block is relatively smooth block, and then for 16 variable values to determine the results of the classification of blocks judged.

4 Experimental Results

4.1 Test Environment and Sequences

Tool: MATLAB 7.6.0 (R2008a)
Hardware: Pentium® Centrino Duo with 1.00GB memories
System: Windows® XP SP2
Sequences: Akiyo、Bridge-close、Foreman、Hall、Mobile
Size: QCIF (176X144 gray)

4.2 Experimental Results

Table 3 shows that the PSNR value of the proposed method has higher value than Elyousfi's algorithm and better than JM algorithm. As the hall in the rear of the set of images will not change, so for the more stringent threshold for judging in this paper, it will have better prediction results, but more stringent limits because of block prediction method, so relatively speaking, will sacrifice time. In the 4x4 and 16x16 if prediction block is more then this would also increase the overall complexity, so if we go to get rid of the middle of the choice of 4x4 and 16x16 blocks of uncertainty, as a 4x4 forced 4x4 blocks to predict the pattern of blocks, although we can get better PSNR (31.9208 dB), relatively speaking, as mentioned in the previous section, the result of coding errors, it will result in low performance.

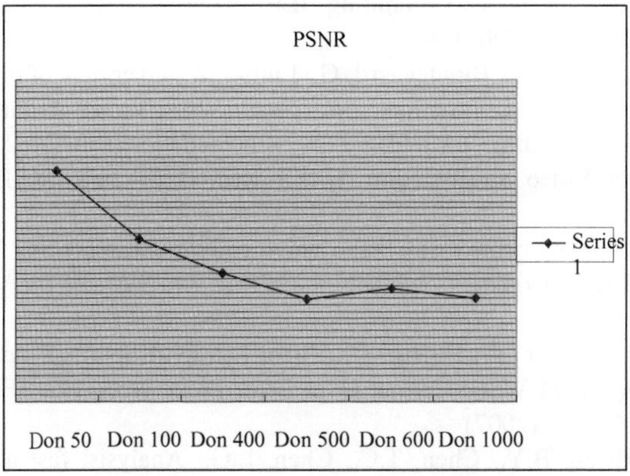

Fig. 5. The PSNR value of Akiyo.

Table 3. The predicted block PSNR values in the three algorithms.

sequence	JM		Elyousfi's		Proposed	
	PSNR	TIME	PSNR	TIME	PSNR	TIME
Akiyo	31.45927	257.3	31.48312	250.0781	31.59936	253.7781
Bridge-close	30.37561	240.2031	30.30629	245.0969	30.65883	247.8906
Foreman	29.91378	250.8094	29.92498	247.8906	30.50395	248.0688
Hall	30.46308	244.1344	30.36744	250.4281	31.55669	251.8063
Mobile	28.48973	247.1438	28.44292	249.4344	29.32731	250.05

5 Conclusion

This paper set the value as 500, and we use the variance, as the threshold for image prediction process, while in the fourth chapter of the experimental results also show that when the value is set to 500, showing a better performance, but poor performance in time, the future may be able to make improvements for this point, in addition to the gray area in the middle for the judge, and perhaps the future is a part can be studied. And in determining the value we may be able to choose images for different habits to develop more of the fixed value, or judgments for the color image.

References

1. Topcuoglu, H., Hariri, S., Wu, M.Y.: Performance-Effective and Low-Complexity Task Scheduling for Heterogeneous Computing. IEEE Transactions on Parallel and Distributed Systems 13(3), 260–274 (2002)
2. Wiegand, T., Sullivan, G.J., Bjøntegaard, G., Luthra, A.: Overview of the H.264/AVC video coding standard. IEEE Transactions on Circuits Syst.Video Technol. 13(7), 560–576 (2003); Joint Video Team, Draft ITU-T Recommendation and Final Draft International Standard of Joint Video Specification, ITU-T Rec. H.264 and ISO/IEC 14496-10 AVC (May 2003)
3. Ding, W., Liu, B.: Rate control of MPEG video coding and recording by rate-quantization modeling. IEEE Transactions on Publication Circuits and Systems for Video Technology 6, 12–20 (1996)
4. Zhang, Y., Gao, W., Lu, Y., Huang, Q., Zhao, D.: Joint Source-Channel Rate-Distortion Optimization for H.264 Video Coding Over Error-Prone Networks. IEEE Transactions on Multimedia 9, 445–454 (2007)
5. Huang, Y.W., Hsieh, B.Y., Chen, T.C., Chen, L.G.: Analysis, fast algorithm, and VLSI architecture design for H.264/AVC intra-framecoder. IEEE Trans. Circuits Syst. Video Technol. 15(3), 378–401 (2005)
6. Liu, C.H., Chen, O.T.-C.: Data hiding in inter and intra prediction modes of H.264/AVC. In: IEEE International Symposium on, Circuits and Systems, ISCAS 2008, May 18-21, pp. 3025–3028 (2008)
7. ITU-T Rec, H.264 Advanced Video Coding for Generic Audiovisual services, H.264 Standard (2005)

Layer Simulation on Welding Robot Model

Guo-hong Ma and Cong Wang

School of Mechanical & Electrial Engineering, Nanchang University,
Nanchang 330031, China
ghma2006@gmail.com

Abstract. This paper designed a full information simulation way to replace inhibic arc simulation with software according to complicate information of welding robot system. In order to simplify simulation model, a layer simulation way is adopted and developed. Simulation experiments showed that simulation model had stable structure. Full information simulation way solved questions of information pursuing one by one. With initial information increases, simulation time would decrease with same simulation steps, which indicated that running efficiency of robot system can be increased.

Keywords: PETRI net, Welding robot, Model; Simulation.

1 Introduction

Due to the procedure of flexible system is more complex, it refers to control, technology and motion, etc. If make a simulation before the operation of system, not only can avoid the potential malfunction, but also can reduce the cost. On the macro level, it provides a more comprehensive understanding of system performance. The research is of great significance towards system's plan and scheduling. And on the micro side, established models of different research systems have different characteristics, the structure and performance of these models are also different. And simulation can conduct a further validation about the Feasibility and correctness of the established model.

Current researches based on the simulation of PETRI net are restricted by the level of computer technology, so few studies about the simulation of welding robot system have desirable results.

According to complexity of information flow movement and its distribution, which PETRI net model of welding robot system has, this paper designs some relevant simulation models and simulation experiments, and develops correlative simulation research.

2 Simulation Model

The theory of PETRI net is a graphical and digital means which can be used in multiple systems. The simulation which adopted the theory can applicable to research system that has parallelism, asynchronism, distributed, randomness and some other characteristics. Through the simulation research, the whole process of complex system can be shown in

M. Zhu (Ed.): Electrical Engineering and Control, LNEE 98, pp. 203–207.
springerlink.com © Springer-Verlag Berlin Heidelberg 2011

the form of graph, and can make researcher clear. As for the deepen system research, the simulation can further provide a theoretical guarantee towards the system control.

According to the PETRI net model which based on the Robot welding system, this paper carries out a simulation model design.

The design of simulation model is shown in these charts. Figure 1 is user layer simulation model, Figure 2 is interface layer simulation model, and Figure 3 is execution layer.

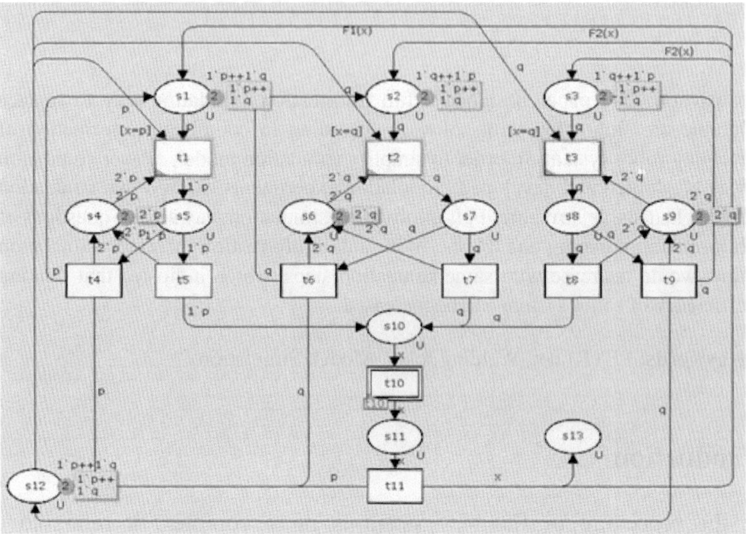

Fig. 1. Simulation chart of user layer model

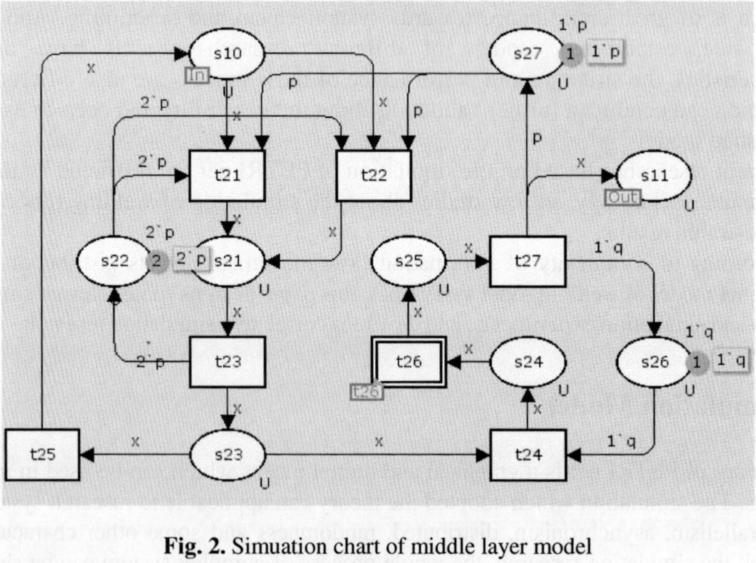

Fig. 2. Simuation chart of middle layer model

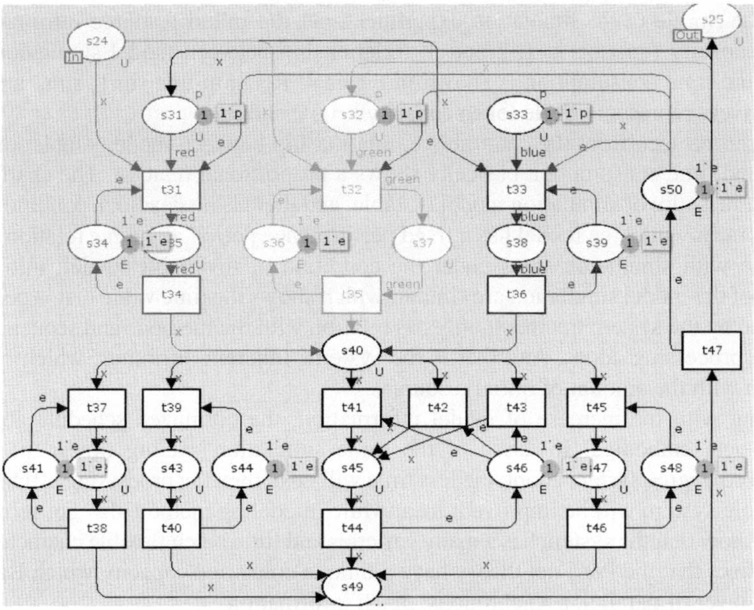

Fig. 3. Simulation chart of execute layer

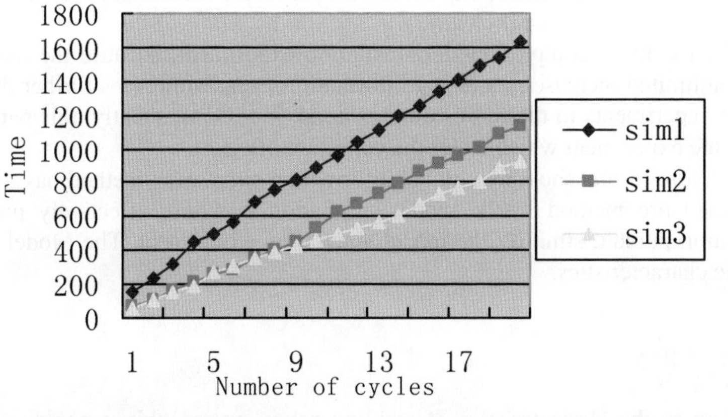

Fig. 4. Relation chart between Simulation cycles and time with different initial information

3 Simulation Experiments

In order to analyze the characteristics of concurrent model of information, this paper utilizes time as the evaluation parameters to analyze the model performance. In this model, presume that there only exists transitional consuming time, and token resources 0 ms. Under this circumstances, Simulation experiment mode and cycle simulation approach, which includes time concrete parameters, carries out the simulation experiment of the whole model, which contains simulation way and multiple cycle simulation.

In the Single cycle simulation experiment, set the initial resource information as token capacity function k=1, 2 and 3. Relationship between model simulation cycle index and time consumption is shown in Figure 4. Relation line sim1, sim2 and sim3 respectively represents while token capacity k=1, 2 and 3.

As can be seen from the chart above, relationship between model simulation cycle index and time consumption basically shows a proportional relation. The chart shows that the structure of simulation model is stable, and there is no deadlock condition in the model; meanwhile, the model has a fixed performance period, the three relation line all increase with simulation steps under the condition of invariable model; initial time values of the model simulation are similar, which shows the time value that is necessary required by the system treatment only has relation with the methods and sequences that system processing adopt. And that is the systems inherent attributes, which have no relation with the amount of initial resources.

Along with the increase of model information, the optimized schedule ability of model gets embodied. In terms of the same simulation cycling times, the higher optimized degree system adopt, the less time will cost, then the processing efficiency of the whole system will be improved; meanwhile, modeling method also got proved, all of this show that the system has a many varieties and little batch flexible characteristics, also reflect that the Petri net theory have a unique concerted capacity which Based on logical layer of parallel, asynchronous, much information.

The reason that system optimization phenomenon occurs due to the increase in the system resource allocation. Although the number of machining processes increases, the chance that system simultaneously deals with processes increases. And thereby reduce all running time, improve efficiency.

As for the more complex model simulation experiment, because the model itself cannot unlimited increase processing information, meanwhile, it' s further difficult to emulate experiments in the more complex condition. Consequently, this paper hasn't discuss the experiment which under the condition of $k \geq 4$.

Simulation results show that global information simulation method has replace the constrained arc method in the model simulation program, effectively prevent the information pursuit, simplify the model simulation experiment. The Model has good dynamic characteristics.

4 Summary

According to the characteristics of welding robot system, this paper discusses and designs a simulation model system. Meanwhile, on the basis of the system model designed in, carries through single cycle model and multiple cycle simulation experiment by means of automatic simulation. The structure of experiment proves that there exists concurrent synchronous information, shows the necessity of the optimization scheduling in the complex system, and through the time parameter value reflects the structure of the system validity and stability.

These simulation experiments provide a theory security to the system control experiment which based on the model, and laid a foundation for future system optimization.

Acknowledgements

This research is supported by NSFC Fund (No.50705041).

References

1. Zhang, W.J., Li, Q., Bi, Z.M.: A generic Petri net model for Flexible manufacturing systems and its use for FMS control software testing. J. Int. J. Prod. Res. 38(5), 1109–1131 (2000)
2. Martens, C., Graser, A.: Design and Implementation of a Discrete Event Controller for High-level Command Control of Rehabilitation Robotic System. In: Proceedings of the 2002IEEE/RSJ Intl. Conference on Intelligent Robots and Systems, pp. 1457–1462. EPFL, Lausanne (2002)
3. Ma, G.H., Chen, S.B., Qiu, T., et al.: Model on Robots of Flexible Manufacturing System with Petri Net in Welding. In: The Eighth International Conference on Control, Automation, Robotic and Vision, pp. 1885–1888 (2004)
4. Guohong, M., Tao, L., Shanben, C.: Controlling of complex welding robot system. J. Journal of Shanghai Jiaotong University 40(6), 881–883 (2006)
5. Ma, G.H., Chen, S.B.: Modeling and controlling the FMS of a welding robot. J. Int.J. Manufa Technol. 34, 1214–1223 (2007)
6. Ma, G.H., Chen, S.B.: Model on simplified condition of welding robot system. J. Int. J. Manufa Technol. 33, 1056–1064 (2007)

Welding Seam Information Acquisition and Transmission Method of Welding Robot

Guo-hong Ma, Bao-zhou Du, and Cong Wang

School of Mechanical & Electrial Engineering, Nanchang University, Nanchang 330031, China
ghma2006@gmail.com

Abstract. To be aimed at the transmission of welding seam video information for binocular vision welding robot, this paper uses these hardware devices, such as DaHeng image acquisition card DH-VT140, two CCD cameras and PC,in order to acquire welding seam image. And with DirectShow technology, in Visual C++6.0 environment, the paper realizes transmission and displaying of welding seam video through capturing video from the image acquisition card and function of MPEG-4 coding and decoding in DivX. Experiment results show that this method can obtain a very good welding seam video, which will provide a reference for the welding seam video information in LAN transmission and monitor in the future.

Keywords: DirectShow; MPEG-4; DivX; Welding robot; Welding seam video.

1 Introduction

Welding process automation, robotic nanomanipulation and intelligence have become the development trend of welding technology. Using the relevant technologies to acquire the feature information of welding process can effectively ensure welding quality. However, the acquisition of welding seam characteristic information is inseparable from the welding video image extracted. At present, image acquisition equipment can be divided into two kinds [1] by technology through the core of the CCD: one is made up of CCD camera, image acquisition card and PC. The principle is that the image acquisition card changes the video signal originating in CCD into digital image signal. Another is the CCD camera itself with digital equipment, which can directly transfer digital image information into the computer through computer port. Among them, the former is dominated as a classic image acquisition system in video collection applications.

Video capture methods play an important part in realizing welding image collection. In Windows environment, in order to support collection compression, decompression and playback of multimedia information, Microsoft provides two multimedia development frameworks [2]: the one is VFW (Video for Windows), the other is DirectShow. Because DirectShow supports for multiple audio and video CODEC, manipulating with the flexibly and conveniently, it is used very extensively.

M. Zhu (Ed.): Electrical Engineering and Control, LNEE 98, pp. 209–215.
springerlink.com © Springer-Verlag Berlin Heidelberg 2011

This paper makes related research into the welding seam image transmission and capture on welding robot.

2 Technical Background

2.1 Video Code and Decode Standards

The image encoding technology has been a rapid development and wide application of customs and maturity, which was marked by a few on the image coding development of international standards, the International Organization for Standardization ISO and the International Electrotechnical Commission IEC on the still image coding standards JPEG, the International Telecommunication Union ITU-T on the TV phone /video conference video coding standard H261, H.263, H.264 and ISO/IEC on the activities of the image encoding standard MPEG-1, MPEG-2 and MPEG-4 and so on.

JPEG is a static image compression standard. MPEG-2 standard is in the MPEG-1 standard, based on further expansion and improvement, mainly for digital video broadcasting, high-definition television and digital video discs, etc. MPEG-2 and MPEG-1 target the same, still is to improve the compression ratio, improved audio, video quality, using the core technology or a sub-block DCT, and interframe motion compensation prediction. MPEG-4 is no longer a simple video and audio codec standards, it will content and interactivity as the core, so as to provide a more extensive multi-media platform.

H.264/MPEG-4 Part 10 or AVC (Advanced Video Coding) is a standard for video compression, and is currently one of the most commonly used formats for the recording, compression, and distribution of high definition video. H.264 is perhaps best known as being one of the codec standards for Blu-ray Discs. It is also widely used by streaming internet sources, such as videos from Vimeo, the iTunes Store.

This paper uses the DivX, which is based on Microsoft's version of MPEG-4 encoding technology, called as Windows Media Video V3. Besides, the audio of DivX uses MP3 technology to compress digital multimedia.

2.2 Video Acquisition and Display Technology

According to the different drivers, at present, there are two video capture cards in the market: VFW (Video for Windows) card and WDM (Windows Model) card. The former is a tend to be abandoned drive model, but the latter is master stream driver. WDM also supports more new features, such as video conference, television receive, and 1394 interface equipment, etc. Video capture card interface can be based on way of PCI or AGP inserting PC chassis, or USB interface.

Microsoft introduced a media streaming layer on top of DirectX that was meant to handle pretty much any type of media you could throw at it, called DirectShow. It's included in DirectX Media, which also includes DirectAnimation (mostly web page stuff), DirectX Transform (more web page stuff), and the old Direct3D Retained Mode that used to be part of the standard DirectX collection.

As with DirectX, the DirectShow API is accessed through COM interfaces. DirectShow is set up with the ideas of a number "filters" joined together to create a "graph". Each box represents a filter. Arrows connecting boxes represent the output of one filter being passed to the input of another filter. Arrows also show the flow of data in the graph. GraphEdit is nice for those just getting started with DirectShow, as it gives a nice visual equivalent to what to do in software. GraphEdit also lets the drag filters around, connect them to other filters, and to run the final complete graph.

3 System Hardware Components

3.1 Binocular Vision Welding Robot

Binocular vision welding robot is mainly made up of the cross slipper structure with control system, a wheeled car, two CCD visual cameras, a welding torch, ect(shown as Figure 1). Wheeled mobile car chassis has a powerful magnet device, which has the robot steadily move on a magnetic plane and oblique plane.

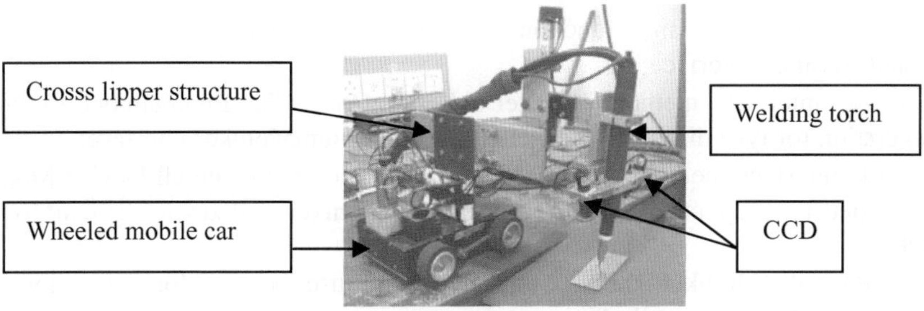

Fig. 1. Binocular vision welding robot

3.2 Video Acquisition Equipment

This paper makes use of DH-VT140 image capture card as a video acpuire equipment. The card is based on computer PCI bus, WDM driver, and has four thoroughfare image videos. Besides, each thoroughfare image video has two groups of video input sources. Two CCD cameras are made by SHENZHEN SHINAIAN ELECTRONICS TECHNOLOGY Co., LTD, which belongs to SNA-M325 series with micro-using pinhole cameras, SONY 1/3 "CCD image sensor. The power is 12V, DC.

4 System Software Components

4.1 Capture and Creation of Welding Seam Video

The capture of welding seam video makes use of “filter” in the DirectShow. And “filter” makes video capture devices registered under video capture sources directory. The types of video capture devices registered by the “filter” and the corresponding relations between them are shown as table 1.

Table 1. Directories of video capture devices registered by the "filter"

Type directory	CLSID Category	MERIT Category
Audio Capture Sources	CLSID_AudioInputDeviceCategory	MERIT_DO_NOT_USE
Video Capture Sources	CLSID_VideoInputDeviceCategory	MERIT_DO_NOT_USE
WDM Streaming Capture Devices	AM_KSCATEGORY_CAPTURE	MERIT_DO_NOT_USE

Before video capture, we need to acquire the amount of capture device. Fortunately, DirectShow provides a special system device named enumeration components (CLSID_SystemDeviceEnum), which can be used to get the video acquire device that supports DirectShow. How to get the creation and capture of welding seam video in application by using DirectShow from Visual C++ 6.0 as the follows [3]:

① use CoCreateInslance function to create system enumeration object and gain ICreateDevEnum interface;

② use interface methods(ICreateDevEnum::CreateClassEnumemtor) to create enumeration for type directory designated,and get IEnumMoniker interface;

③ use interface methods(IEnumMoniker::Next) to enumerate all Device Moniker in specified type directory. Such, each Device Moniker realizes its IEnumMoniker interface;

④ after call IMoniker::BindToStorage function, use Device Moniker attributes to get a Display Name, Friendly Name, and so on;

⑤ make Device Moniker generate DirectShow Filter by means of call IMoniker::BindToObject function, after that, IFilterGraph::AddFiher function in Filter Graph would work.

If hope DirectShow work well, you must make video capture equipment be in the state of the "Filter". After that, each "Filter" can collaboratively work. In fact, the Filter of acquisition device created is also enumeration process. After select Daheng DX Video Capture, get this device with Display Name and Friendly Name, afterwards, you can use the name for parameters to create "Filter". Such as, this paper has realized the Capture and Creation of Welding Seam Video.

4.2 Display and Storage of Welding Seam Video

After equipment of welding seam video was successfully created, if we create a complete the Filter Graph again, we can get the output video. The method is as follows [4]: fistly, using Capture Graph Builder component technology provided by DirectShow; secondly, having attribute of IGraphBuilder be setted to SetFilterGraph; last, calling Capture Graph Builder. In this way, we can successfully construct capture a Filter Graph. The image acquisition card DH-VT140 is based on PCI interface with WDM driver. Here is one Preview pin for video image preview, another Capture pin for video data capture(shown as Figure 2).

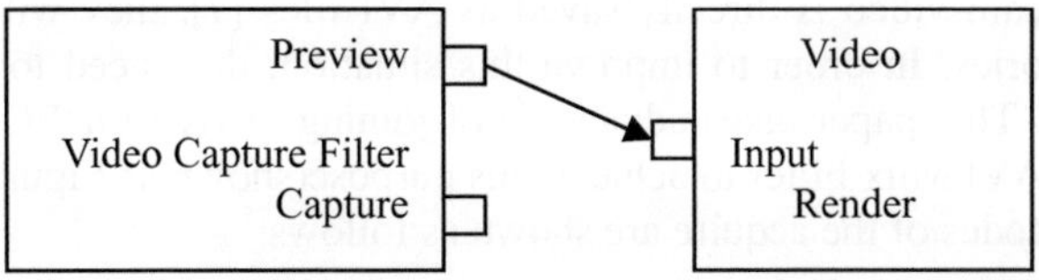

Fig. 2. Principle of video previewed with the Filter Graph

The way that filters are connected in a graph is through their "pins" [5-6]. Every filter, no matter what type, must have at least one pin to connect to other filters. When attempting to connect two filters, the pins on both the filters pass information back and forth to determine if the downstream filter (the one accepting data) can handle the data passed in by the upstream filter (the one sending data). If the pins successfully negotiate a data type they both know, a successful connection has been made between the two filters. For example, the MPEG-4 Stream Splitter filter needs to send the audio and video portions of MPEG-4 data to separate decoder filters.

After acquiring welding seam video image, in order to go a step further to comprehend the quality of welding seam, we need to store welding seam video. The Principle is shown as Figure 3.

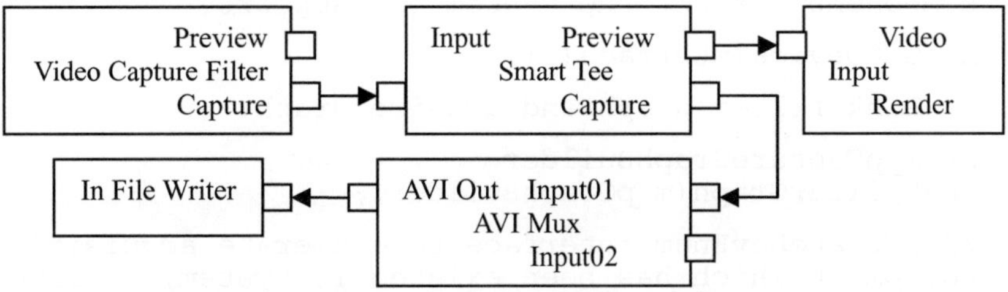

Fig. 3. Principle of video stored with the Filter Graph

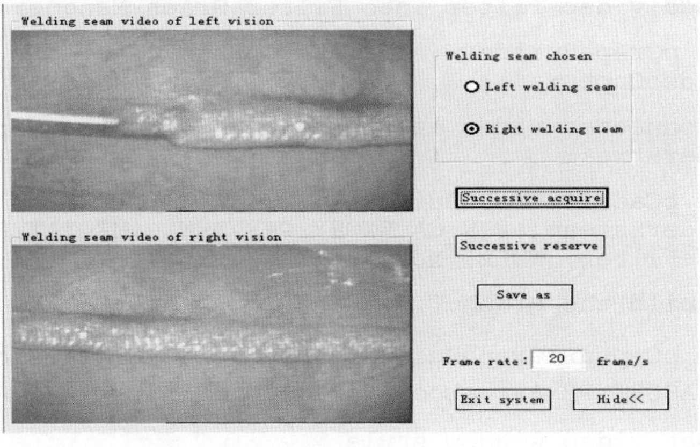

Fig. 4. Video operation software

If welding seam video is directly saved as AVI files [7], they will take up many computer memories. In order to improve this situation, they need to be compressed and then saved. This paper takes advantage of joining DivX CODEC between video output Pin and AVI Mux Filter to achieve this purpose(shown as Figure 3).

The core codes of the acquire are shown as follows:

```
HRESULT CaptureYideo( )
{
HRESULT              hr;
IBaseFilter         *pBaseFilter=NULL;
ImediaControl       *m_pMediaControl;
IGraphBuilder       *m_pGraphBuilder;
ICaptureBuilder2    *m_pCaptureGraphBuilder;
IVideoWindow        *m_pViewWindow;
// Initialize the COM library.
CoInitialize(NULL);
// Use QueryInterface method to inquire about the
component whether the system can support.
hr=GetDeviceInterface( );
// Link filter graph and acquire graph.
Hr=m_pCaptureGraphBuilder-
>SetFilterGraph(m_pGraphBuilder);
// ICreateDevEnum interface to enumerate acquisition
equipment which has been existed in system, to find
available equipment.
hr=FindCaptureDevice(&pBaseFilter);
// Add video filter into Filter Graph Manager.
hr=m_pGraphBuilder->AddFilter(pBaseFilter,
L"VideoCapture");
// Connected video manager and preview manager with
RenderStream.
hr=m_pCattureGraphBuilder-
>RenderStream(&PIN_CATEGORY_PREVIEW, &MEDIATYPE_Video,
pBaseFilter, NULL, NULL);
// Build the graph.
Hr=m_pCaptureGraphBuilder->SetOutputFileName(&
MEDIASUBTYPE_Avi, L"c: \example.avi", &ppf, NULL);
// Set video window style
hr=SetupVideoWindow( );
```

```
// Run the graph and start to preview video acquired
hr=m_pMediaControl->Run( );
return S_OK;
}
```

5 Conclusion

After this paper studies the basic principle of DirectShow, using CCD camera, image acquisition card and PC as the welding seam video acquire equipments, in Visual C++ 6.0 environment, have compiled a software with video acquire. The experiment shows the software(shown as Figure 4) is stable and reliable, and welding seam video image would transmitted clear. This method provides a reference for transmission and monitor of welding video information on the local area network(LAN) in the future.

Acknowledgments

Thanks to anonymous peer reviewers and editors and fund of National Science Foundation of China under Grant 50705041.

References

1. Wei, L.-c., Zhu, G.-l.: Design and Implementation of Video Capture System Based on DirectShow. J. Computer Engineering 31(14), 187–189 (2005)
2. Lu, Q.-m.: DirectShow pragmatic chosen. Science&Technology Press, Beijing (2004)
3. Shao, M.-m., Hu, B.-j., Zhao, L.: Implementation of video capture based on DirectShow. J. Electronic Instrumentation Customer 16(4), 84–85 (2009)
4. Yan, S.-c., Quan, H., Li, Q.: Research on image acquisition techniques on Windows platform. J. Control & Automation 22(4), 252–254 (2006)
5. Pan, P.: Video Surveillance and Capture System Based on DirectShow. J. Modern Computer 8, 156–158 (2010)
6. Weng, C.-r., Zheng-jing.: The technology of remote video monitor system based on DirectShow. J. Wfujian Computer 24(11), 163–164 (2008)
7. Li, Q.-c., Wang, C.-m., Duan, H.-l.: Several video collection method design and implementation based on Windows. J. Silicon Valley 12, 67–68 (2008)

Search Pattern Based on Multi-Direction Motion Vector Prediction Algorithm in H.264 and Learning Website

Tsung-Han Tsai[1], Wei-Chun Hsu[1], Yuan-Chen Liu[2], and Wan-Chun Lee[3]

[1] National Central Unervisity, Department of Electrical Engineering, Zhongda Rd. 300, Zhongli City, Taoyuan County 320, Taiwan (R.O.C.)
[2] National Taipei Unervisity of Education, Department of Computer Science, Heping E. Rd. 134, Da'an Dist., Taipei City 106, Taiwan (R.O.C.)
[3] Taipei Municipal Ren-Ai Elementary School
Liu@tea.ntue.edu.tw

Abstract. Motion estimation is the most important part that needs about 80% of computation in the procedure of image compression coding. So there are many fast algorithms for motion vector search proposed to reduce time consumed and keep good qualities.

In this paper some traditional algorithms and algorithms using prediction values are investigated, including DS (diamond search), HEXBS (hexagon-based search) and UMHEXagonS. The mechanism of search pattern based on multi-direction motion vector prediction algorithm is proposed to develop a fast algorithm based on motion estimation and combining search patterns and directions.

At the moment, we just speed up the encoding time and perform better high PSNR values in the proposed algorithm. But that is very similar with MDS or SPS, and the factors of software and hardware like the selection of algorithms or frame number and the phase of early termination or the difference of hardware are not discussed.

Keywords: motion estimation, motion vector, search pattern, multi-direction.

1 Introduction

Recently, the applications of multi-media have become more popular in the people's life. In the data of multi-media, video data has larger than that of text data. So there is a problem that video data consumes a lot of the bandwidth, storage and computing resources in the typical personal computer.

For this reason, video compression has become an important issue of research areas. In recent years, many video compression standards have been proposed, such as MPEG-1/2/4 and H.264/AVC. H.264/AVC is ratified by both the ISO/IEC and ITU-T. It has better coding efficient than other standards, but it has too higher computational complexity in implementation. The main computation is about block matching motion estimation (ME), variable block size motion compensation, multiple reference frames motion compensation and Rate-Distortion Optimization(RDO).

M. Zhu (Ed.): Electrical Engineering and Control, LNEE 98, pp. 217–224.
springerlink.com © Springer-Verlag Berlin Heidelberg 2011

In this paper, we proposed a new algorithm in H.264 for motion estimation (ME) with the direction of search range and search patterns to advance the efficiency of compression.

2 Related Work and Researches

2.1 The Design of H.264/AVC Standard

H.264/AVC Encoding Standard is formed by both the ISO/IEC and ITU-T in March 2003[1][2][3]. It focuses on the efficiency of compression and the high reliability of transmission. The diagram of H.264/AVC Structure is shown in Fig. 1.

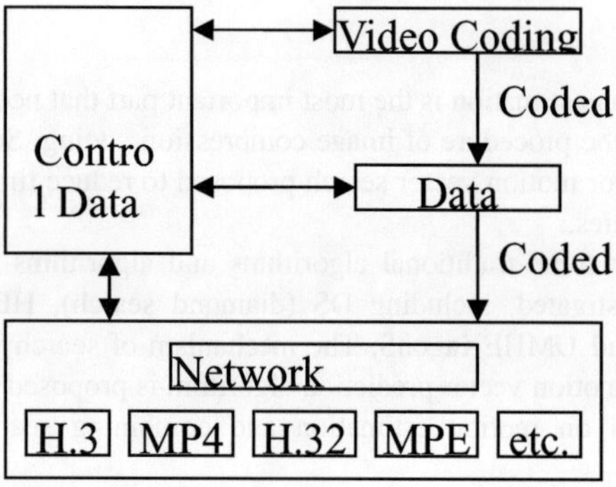

Fig. 1. The diagram of H.264/AVC Structure.

For inter prediction coding, we use current frame and previous frames to do motion estimation and get the motion vector for motion compensation. For the block of MC, we can get good results if ME has good performance.

2.2 Motion Estimation Algorithms

In H.264, we often use block matching algorithm (BMA) to compress the video for temporal redundancy [4][5][6][7][8]. Block Matching Algorithm is that divided the video frames into the blocks of static size and not overlapping, and every block searches the similar block in the search window of reference frames. Motion Estimation is to decide which macro-block of reference frame is similar with the macro-block of current frame. The first step is to define the search range in search window. Second step is to find the all points in the blocks with minimum block distortion. If we find the smallest one, we can confirm that block is correct.

Besides BMA, we use Full Search, Diamond Search (DS)[10], HEXBS[9], BBGDS[11], UMHexagonS[12],SPS[13]and MDS[14]to compare the efficiency of motion estimation.

3 Search Pattern Based on Multi-Direction Motion Vector Prediction Algorithm

3.1 Prediction Vector Selection and Computation

Because there are high correlations and similarity between the frames of video, and we call that temporal redundancy. In this paper, we use the relation to select one prediction candidates set. And we will explain the definition of prediction vectors and the reasons of selection.

We use the high correlation and similarity between the frames of video in temporal and spatial domain to select one prediction candidate set. Like Fig.2, t is the current frame, and t-1 is previous first reference frame. The formula of motion vector is as follows:

$$\overrightarrow{mv(P,I,J)} = (X, Y) \tag{1}$$

If it is reference frame, shown as follows:

$$\overrightarrow{mv(P,I,J)} = (F, X, Y) \tag{2}$$

The definition of above parameters is shown as follows:

P: the frame number of current frame.
I: the value of X position of macro-block in current frame
J: the value of Y position of macro-block in current frame
F: the absolutely value of reference frame compare with current frame
X: the value of X position of motion vector
Y: the value of Y position of motion vector

And we select $\overrightarrow{MV_{A0}}$, $\overrightarrow{MV_{B0}}$, $\overrightarrow{MV_{C0}}$, $\overrightarrow{MV_{zero}}$, $\overrightarrow{MV_{Median}}$, $\overrightarrow{MV_{T0}}$ to use in this algorithm. $\overrightarrow{MV_{A0}}$, $\overrightarrow{MV_{B0}}$, $\overrightarrow{MV_{C0}}$, $\overrightarrow{MV_{zero}}$ is used in UMHexagonS.

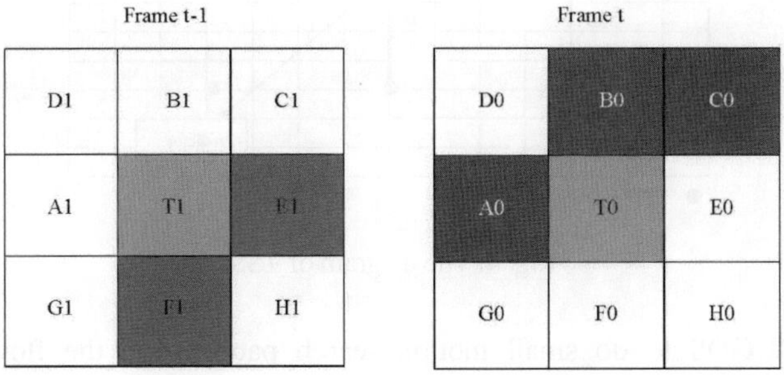

Fig. 2. The diagram of relative blocks with spatial and temporal domain.

3.2 Multi-Direction Search

We divide the search range into some regions, like Fig. 3

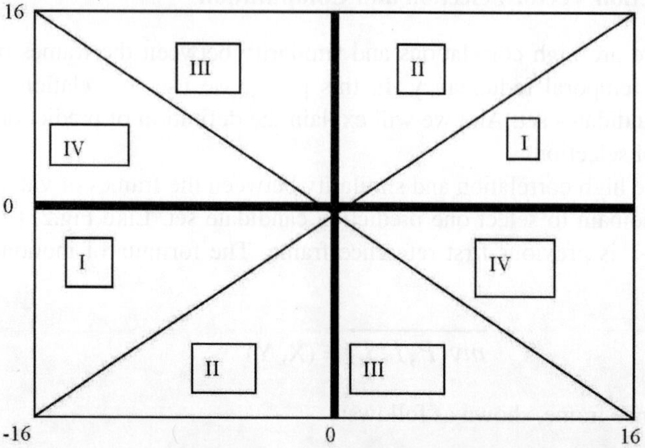

Fig. 3. The diagram of search ranges.

We first perform vertical-direction and horizontal-direction search to find the best points. If we don't find the points, then perform diagonal-direction search and find all best points.

We use TSS to expand the search range in order to avoid the situation of local minima, like Fig. 4.

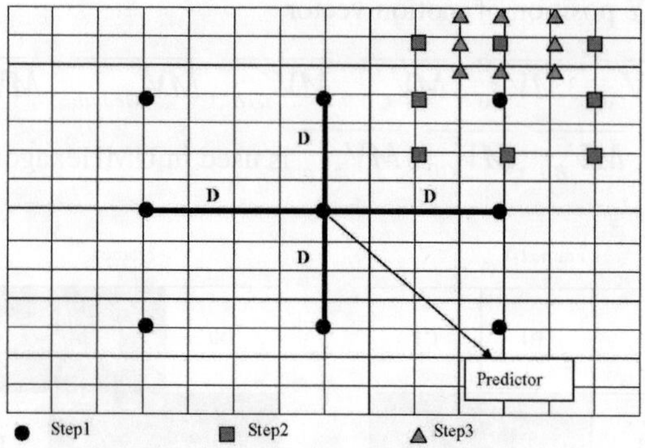

Fig. 4. The diagram of TSS.

We use BBGDS to do small motion search pattern and the flowchart of the Proposed Algorithms is as follows:

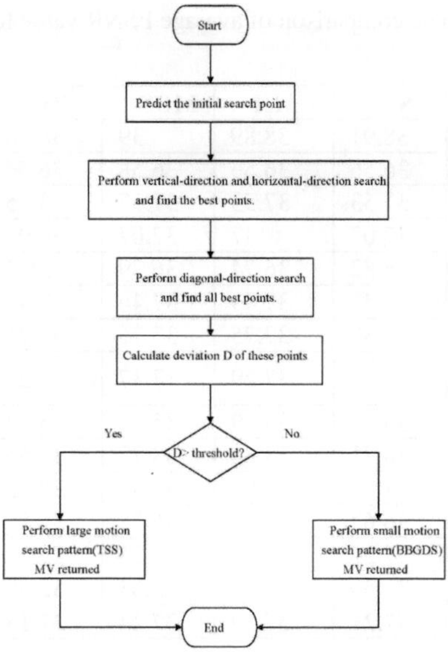

Fig. 5. The flowchart of SPBMD.

4 Experiment and Discussion

4.1 Test Environment and Sequences

Tool: JM 15.1
Hardware: Pentium® Centrino Duo with 1.00GB memories
System: Windows® XP SP2
Sequences: Akiyo(300 Frames),Bus(300 Frames), Coastguard(300 Frames),Container(300 Frames), Flower(250Frames),Football(125 Frames), Foreman(300 Frames), Hall Monitor(330 Frames), Mobile(300 Frames),Mother & Daughter(300 Frames),News(300 Frames),Silent(300 Frames),Stefan(300 Frames),Tempete(260 Frames)
Size: CIF(352×288)

4.2 The Comparison of Performance

For some sequences like Akiyo and Mother, the value is higher than other sequences, and the values of the difference between the maximum and the minimum are higher than 0.3db.

The value of the difference between the maximum and the minimum with most sequences are less than 0.2db.

We can observe that the average PSNR value with our proposed algorithm is higher than other algorithm between 0.05db and 0.15db.

Table 1. The comparison of average PSNR value for QP=28.

Sequence	FS	UMH	MDS	SPS	Proposed
Akiyo	38.91	38.89	39	38.76	39.13
Bus	36.56	36.56	36.58	36.56	36.58
Coastguard	37.53	37.53	37.57	37.5	37.61
Container	37.07	37	37.07	36.95	37.14
Flower	36.82	36.74	36.79	36.7	36.79
Football	37.43	37.44	37.46	37.45	37.48
Foreman	37.83	37.75	37.75	37.71	37.78
Hall Monitor	37.37	37.29	37.37	37.24	37.45
Mobile	35.37	35.39	35.39	35.37	35.41
Mother	38.71	38.68	38.77	38.55	38.94
News	37.84	37.84	37.88	37.75	37.92
Silent	36.67	36.68	36.78	36.55	36.85
Stefan	37.26	37.25	37.25	37.25	37.27
Tempete	35.59	35.61	35.63	35.58	35.65
Average	**37.21**	**37.19**	**37.24**	**37.14**	**37.29**

Table 2. The comparison of average encoding time for QP=28.

Sequence	FS	UMH	MDS	SPS	Proposed
Akiyo	2231.974	369.706	369.545	369.635	369.5
Bus	2520.559	551.141	550.653	550.561	550.344
Coastguard	2415.466	435.464	434.711	435.01	434.323
Container	2186.88	382.604	381.425	381.625	381.226
Flower	2482.373	506.798	506.42	506.7	506.682
Football	2373.264	525.289	524.19	525.159	525.128
Foreman	2309.872	437.357	437.252	437.334	437.147
Hall Monitor	2258.216	386.817	386.649	387.451	386.685
Mobile	2616.169	615.604	615.538	615.478	615.16
Mother	2092.179	357.104	357.084	357.119	356.883
News	2265.806	402.525	402.521	402.505	402.509
Silent	2254.769	400.33	400.305	400.306	400.269
Stefan	2469.178	526.893	526.433	526.542	526.317
Tempete	2442.829	516.214	516.191	516.109	516.116
Average	**2351.395**	**458.132**	**457.78**	**457.967**	**457.735**

We use SPBMD to compare with Full Search, UMHexagonS, MDS and SPS. We can see the performance of the proposed algorithm average above the other algorithms, better than UMHEXagonS in encoding time and has higher average PSNR values than other algorithms. For some sequences like Mother, News and Silent, the performance of PSNR with SPBMD is the best one in all algorithms. Besides, the search points of the proposed algorithm are similar with SPS, expect some sequences like Coastguard and News. The reason may be that some frames of the sequences keep errors in judge motion vector. But our algorithm is more suitable in various kinds of images for the results.

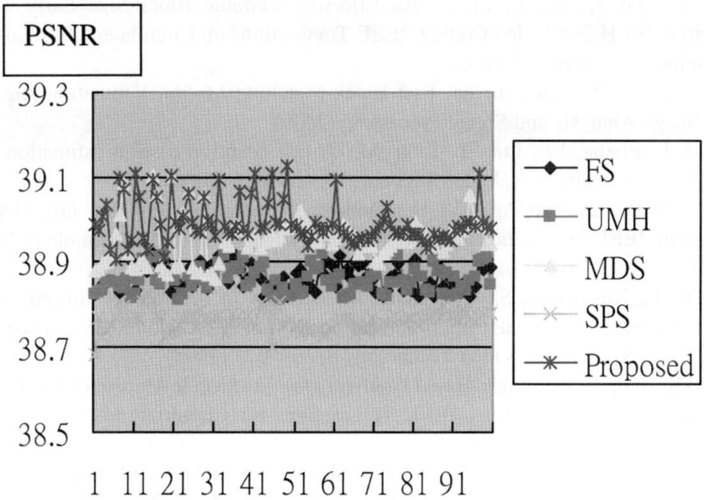

Fig. 6. The PSNR value for Akiyo.

5 Conclusion

In this paper, we combine the characters of search patterns and the direction of search range and use adaptive prediction method to reduce the computational complexity. According to our experiments, the correct ratio of prediction vector is very important in the searching quality and searching time. Our method-SPBMD is more suitable than traditional fast search algorithm in real-time application and high resolution and high motion videos. The experiments show that our method similar with MDS or SPS and faster than UMHexagonS. It has high PSNR and the encoding time is less than other algorithms.

References

1. Kuo, T.Y., Lu, H.J.: Efficient Reference Frame Selector for H.264. IEEE Transactions on Circuits and Systems for Video Technology 18(3) (March 2008)
2. Joint Video Team, Draft ITU-T Recommendation and Final Draft International Standard of Joint Video Specification, ITU-T Rec. H.264 and ISO/IEC 14496-10 AVC (May 2003)
3. Huang, Y.W., Hsieh, B.Y., Chien, S.Y., Ma, S.Y., Chen, L.G.: Analysis and Complexity Reduction of Multiple Reference Frames Motion Estimation in H.264/AVC. IEEE Transactions on Circuits and Systems for Video Technology 16(4) (April 2006)
4. Nie, Y., Ma, K.K.: IEEE Transactions on Adaptive rood pattern search for fast block-matching motion estimation 11(12) (December 2002)
5. Chen, C.Y., Chien, S.Y., Huang, Y.W., Chen, T.C., Wang, T.C., Chen, L.G.: Analysis and architecture design of variable block-size motion estimation for H.264/AVC. IEEE Transactions on, Circuits and Systems I: Regular Papers 53, 578–593 (2006)

6. Yang, L., Yu, K., Li, J., Li, S.: An Effective Variable Block-Size Early Termination Algorithm for H.264 Video Coding. IEEE Transactions on Circuits and Systems for Video Technology 15(6) (June 2005)
7. He, S., Zhang, X.: An Efficient Fast Block-Matching Motion Estimation Algorithm. In: Proc. Image Analysis and Signal Processing (2009)
8. Ahmad, I., Zheng, W., Luo, J., Liou, M.: A fast adaptive motion estimation algorithm. IEEE Trans. Circuits Syst. Video Technol. 16, 420–438 (2006)
9. Ce, Z., Xiao, L., Lap-Pui, C.: Hexagon-based search pattern for fast block motion estimation. IEEE Transactions on Circuits and Systems for Video Technology 12, 349–355 (2002)
10. Jo Yew, T., Ranganath, S., Ranganath, M., Kassim, A.A.: A novel unrestricted center-biased diamond search algorithm for block motion estimation. IEEE Transactions on Circuits and Systems for Video Technology 8, 369–377 (1998)
11. Liu, L.-K., Feig, E.: A Block-Based Gradient Descent Search Algorithm for Block Motion Estimation in Video Coding. IEEE Transactions on Circuits And Systems For Video Technology 6(4) (August 1996)
12. Chen, Z.B., Zhou, P., He, Y.: Fast integer pel and fractional pel motion estimation for JVT, JVT-F017. In: 6th Meeting, Awaji, Japan, December 5-13 (2002)
13. Huang, S.Y., Cho, C.Y., Wang, J.S.: Adaptive fast block-matching algorithm by switching search patterns for sequences with wide-range motion content. IEEE Trans. Circuits Syst. Video Technol. 15(11), 1373–1384 (2005)
14. Lin, C.C., Lin, Y., Hsieh, H.J.: Multi-direction search algorithm for block motion estimation in H.264/AVC. IET Image Process. 3(2), 88–99 (2009)

The Formation and Evolution Tendency of Management Philosophy

Wu Xiaojun and Si Hui

Wuhan Textile University, Hubei, China
xiaojun9669@163.com, 45266135@qq.com

Abstract. This paper clarified the forming of management philosophy was inevitable to management science's development. First of all, this paper clearly explained the content of management philosophy from the forming reason. Secondly, this paper analyzed the double importance of management philosophy amid theory and practice of management science research, thus proposed the necessity of managing philosophy research. Finally, this paper put up with prediction to the evolution tendency of managing philosophy's development through above analysis and the professional knowledge in management science.

Keywords: management science, management philosophy, innovation, knowledge economics.

1 Introduction

Management philosophy's rising, has verified the historical transformation of management values, the management thinking mode, is containing profoundly the epoch-making surmounting to the modern management, It will lead management to a new time from "the science" to "the philosophy", will thus manifest and is attributing the modern management theory field of vision and the value direction unceasingly.

2 The Forming of Management Philosophy: From Management Science to Management Philosophy

Management science's born, is the management jump to theory significant. But management science's appearance, does not mean that the philosophy is expelled in the management domain, in the opposite, was precisely the philosophy promoted the management concept sublimation, has boosted the management science growth. As scientific management founder's Taylor, in its management science the philosophy implication is very remarkable. Its representative work "Scientific management Principles" is not long, but actually has 10 to concern the philosophy in the management function. Daniel.A.Wren's book of "The Evolution of Management concept" simply called him as "the philosopher who everywhere lectures", and believed that "before

M. Zhu (Ed.): Electrical Engineering and Control, LNEE 98, pp. 225–230.
springerlink.com © Springer-Verlag Berlin Heidelberg 2011

Taylor, there are no portraits can develop management question to one system approach to this degree, and simultaneously combined the management with the philosophy category."[1]

From philosophy stratification plane ponder management question, is not only theorist's responsibility and the theory idea, moreover also manages practitioner's indispensable value pursue. The management practice needs to draw support from the philosophy thought to the theory sublimation, but is precisely this kind of theory and the practice interaction, causes the management philosophy from to move toward the onstage secretly gradually.[2] Management philosophy discipline system's construction, not only need carries on the philosophy think to the existing management practice and the management theory, More importantly, through to manages the practice and the management theory and reciprocity inventorying and the inspection, tries hard to carry on the Trans-Culture, the cross domain omni-directional assurance in the top level to the modern management. But this kind of assurance causes the management philosophy not to be impossible to become the edge of management study domain, its discipline traits decided that it must move toward the center inevitably from the edge.

Management philosophy's entering the stage, symbolizes to "the denial" and surmounting to management science time. Although the management science is indispensable as one technological means or the technical tool at the modern age, but it is the modern management essential condition merely but non-sufficient condition, Because depends upon the management science merely is unable to see clearly the management the intrinsic value and the spiritual essence, is unable to give dual attention to the management understanding from multi-dimensions and the multi-angles of view, is unable to dispel the humanities crisis and ecological crisis that the management science bring[3]. Along with management practice's deepening, as well as to management practice and the management theory's reconsideration, the flaw to the management science will come forth day after day and to enlarge unceasingly. Thus, to confirm again modern management connotation, and, therefore realizes the management thought transformation and the management value hypothesis, is the significance to precisely manages the philosophy surmounting management science.

Here, what we need to further stress: The management philosophy's born, first is carries on the profound reconsideration to the management science from the scientific theory and the reality operation dimension is the premise that surveys it's reasonable and the legitimate scope. At the same time, the rising of manages the philosophy intrinsic request modern management must implant two fundamental factors of the culture and the science, achieves to has the new determination to the culture, must have the new determination to the science, thus realizes the synthesis of humanities management and the scientific management. And management philosophy inquires about the management's new balance in person's perception and the rationality, between the individual, the organization and the society. Thus it can be seen, management philosophy to management science intrinsic surmounting, is the necessity to modern management theory and the practice evolution. Although the management philosophy's production walks with difficulty, but transforms from the management science to the management philosophy's in-depth actually has the realistic agent which the intrinsic logical inevitability and is unable to resist.

3 The Preliminary Study of Management Philosophy's Evolution Tendency

When people's discussion in the 90s US-Japan economical rise and fall inversion, more and more scholars attribute the US success to the knowledge economy rising, in-depth conclusion is the American culture victory. American culture tradition long-standing to innovation's esteem has poured into the charm in the new times for the economic growth. In the 90s the emerging of the knowledge management theory, although in also occupies forms, but the management center of gravity tends to by might and main to the innovation ability value and cultivates oneself to be obvious. r the innovation philosophy becomes, the modern management's deep and essential request.

3.1 The Organizational Theory of Seeking Active and Adaptation

Following the strength of the enterprise size and the complication of external environment day by day, to the focus on every g managerial technique more and more yields to the organization system's arrangement which constant has an effect. Since the modern management stage, in many manages scientist there, the organization and the management use nearly in the identical significance. Because organizational structure's adjustment means that also promotes the mode of administration to have the deep transformation inevitably.

What the innovation needs is the spirit of keen, nimble, is accommodating the atmosphere and the risk ask change, is the organizational structure elasticity and the dynamic[4]. Drucker as early as published "New Organization's Appearance" in 1988, made the concrete tentative plan to the future enterprise: "20 year later model big enterprise, its management level will be inferior to today's half, the administrative personnel also to be inferior to today 1/3. In the organizational structure, the management object and in the control area, these enterprises will rise with the 50s later, were presented today still by the textbook for the classics big manufacture company not slightly similarity, but possibly approaches in these the personnel and manages not surely by present's manager the organization which the scientist neglects. In mine mind, future typical enterprise should be called the information organization. It take the knowledge as the foundation, is composed of various experts. These experts act according to come from the colleague, the object and higher authority's massive information, the independent decision-making, the self-control, its organization type is likely the hospital or the symphony orchestra."[5]

If Drucker has provided the structural design for era of knowledge economy's organization, then Peter.Senge's learning organization has provided the organization cultural guiding[6][7]. In "Fifth Practice - Study Organization's Art and Practice", he made the popular explanation to learning organization, explained the learning organization: "in here, the people expand unceasingly their ability, Creates their true expecting result; in here, the people may release their depressed own long fervor; in here, how can the people learn to study together unceasingly[6]. In order to obtain the above result, the Senge proposed to carries on five big practice of the system thought that self-surmounting, the improvement mental pattern, the establishment common prospect and the team study and so on. The era of knowledge economy, more and more enterprises indicated by own success practice: Many enterprises benefit in personally set an example "five practice", thus has made a connection with the network shape organization "the channels and collaterals", specially formed one batch of high quality association to the organization to play the enormous role, but this kind of high quality association, also for study organization "place primarily" the management has created the intermediate perspective foundation. The network is faculty system condition of the organization has the adaptive, but the intrinsic creativity of organization mainly also to depend on the study, so long as establishes the study organization effectively, adaptive can enhance.

3.2 Take Knowledge as Centre of Theory

Enter knowledge-based economy, innovation become "God's hand" that dominates the business'existence or death" fate. In existing economic environment, "uncertainty" is the only factor that can recognize, anti-innovation will break up, slow innovation necessarily fall behind, numerous tides of businesses rise the tide fall, life and death the rise and fall, deduce again and again this tape to contain law phenomenon. However, the innovation is an empty slogan in no way, an abstract idea, it is one kind of ability essentially, but this kind of ability nourishes own springhead running water by the knowledge achievement[8]. Furthermore, innovation ability is decided by an enterprise to the knowledge and the accumulation, the development, sharing and the use, namely knowledge management level. Because the knowledge is the soil which freely innovation ability grows, but must transform the knowledge as reality innovation ability, to a great extent is decided by to the knowledge management.

Obviously, what Drucker respected is the knowledge transforms into the innovation "we thought the knowledge is the knowledge demonstration in the motion. We said now the knowledge is in the motion the effective information, emphatically in effect information. Effect is outside human body, in society and economy, or during knowledge's enhancement."[9] Why as for the knowledge from the non-practical evolution for the practical reason, after France famous, modernism philosopher Lyotard has the profound elaboration. He thought that to sell is produced, for multiplies in the new production expended commercialized the knowledge, because with the production unified in together, the cause knowledge became the value one form, Everything has the close direct relation knowledge with the production is legitimate, only then possibly receives takes seriously, otherwise then cannot obtain the company and enterprise's subsidization, will lose the value, will lose the existence the validity.

The knowledge has taken the human resources unprecedented importance in economic growth's overwhelming function. The knowledge emphasis and talented person's value, certainly begins in the knowledge economy, "the knowledge is the strength", "the talented person results in the world" in the ancient and moderns in China and abroad is the common truth. But the agricultural society land's mother of wealth, the land function crushes the agriculture technology absolutely; The industrial society work disassimilation for the capital slave, the capital becomes the economic growth the fountainhead, was still essentially the physical resources plays the control role in the industrialization early intermediate stage. Enters the industry person later period, after specially the knowledge economy appears, the knowledge becomes the economic growth the predominant factor, carries negative knowledge the human resources obtain the impetus economic growth from this "the first productive forces" status. Entering the industry person later period, specially after the knowledge economy appears, the knowledge becomes the economic growth the predominant factor, carries negative knowledge the human resources obtain the impetus economic growth from this "the first productive forces" status. "human capital's and so on person's knowledge, ability, health enhancements to economic growth's contribution material, labor force quantity increase far are much more important." The profound understanding human resources' importance, brings the mode of administration inevitably the profound transformation. Since the 90s, management practices present the noticeable new change, the most obvious characteristic is the based on- human melts are getting more and more thick. Is corresponding with it, in the management theory research "person's sun" to raise, the most prominent achievement mainly concentrates in the human resources theory and in the innovation management progress.

A western people- oriented of realization's course makes clear us step by step: The solution society history progresses depends on the mode of administration the transformation, only the management is only incorrect, but if thought that so long as established the advanced social system certainly to be able to carry out humanist, similarly was not good. Human's full scale development relies on the advanced social system and the science mode of administration unification.

4 Conclusion

There is no doubt, total trend of development of the management science, the management science is experienced by "the substance-oriented " to "the people-oriented" 's transformation, management culture and the management ethics of understanding and analyzing has become the modern management science research the hot spot and the front.

The management philosophy, take the new value principle, the thinking mode and the theory logic assurance "the management", the utilization "the management", the advancement "the management", has brought the management idea revolutionary transformation as us. Because of it, the management science "the subversion" with to set at variance the quilt, the modern management will realize the historical big spanning under the management philosophy's guidance, the management new times - person and the human nature the time which highlights truly in the management and makes widely known arrives.

References

1. Wren, D.A.: Management concept's evolution. Social Sciences in China Publishing house, Beijing (2000)
2. Sun, Y.: Western management science of famous work abstract. Jiangxi People's Publishing Agency, Nanchang (1995) (in Chinese)
3. Cheng, Z.: C theory: Chinese management philosophy. Studies the forest publishing house, Shanghai (1999) (in Chinese)
4. Pudy, J.M., et al.: The management science abstract: Asian. Mechanical industry publishing house, Beijing (1999) (in Chinese)

5. Weiping, M., et al.: Management philosophy. CPC Central Party School Publishing house, Beijing (2003) (in Chinese)
6. Drukker: Knowledge management. Renmin University of China Publishing house (1999)
7. Senge, P.: The Fifth Discipline. Shanghai San Lien Book Store Shop (1998) (in Chinese)
8. Peters, T.: Wins superiority. Business management publishing house (1989) (in Chinese)
9. Drucker: Management: The duty, the responsibility and practice. Social Sciences in China Publishing house (1987)

Mobility and Handover Analysis of Existing Network and Advanced Testbed Network for 3G/B3G Systems

Li Chen[1], Xiaohang Chen[1], Bin Wang[1], Xin Zhang[1], Lijun Zhao[2], Peng Dong[2], Yingnan Liu[2], and Jia Kong[2]

[1] Wireless Theories and Technologies (WT&T), Beijing University of Posts and Telecommunications, Beijing, 100876, China
alibupt@gmail.com
[2] China Mobile Research Institute (CMRI), Beijing, 100053, China
zhaolijun@chinamobile.com

Abstract. The wireless testbed network is generally used to evaluate the equipment performance and the network functions. This paper focuses on the mobility and handover analysis in a novel advanced testbed network, T-Ring, for 3G/B3G system which is a miniature environment compared with the real commercial network, aiming at obtaining the similar handover performance. Theoretical analysis, system-level simulation, and outfield testing are adopted for the handover analysis. The simulation and the outfield test are implemented to analyze the fast fading of wireless channel as the size of network changes into miniature scale. The conclusion indicates by appropriately adjust the handover parameters in the testbed network, the mobility is able to stay the same and the simulation proves that the fast fading characteristic of simulation in the testbed network is resemble to that in the actual network.

Keywords: 3G/B3G Systems, Advanced Testbed Network, Outfield Test, System-level Simulation, T-Ring, Mobility, Handover.

1 Introduction

The outfield testing is one of the most effective methods in the wireless network deployment and equipment evaluation because of the real propagation condition in the open air. Usually, some special locations will be chosen or some dedicated environment will be built to execute the outfield test. The existing test method has the drawback of long test period, low test efficiency, and non-repeatability, so it is difficult to evaluate the performance of actual 3G/B3G networks rapidly and accurately. The advanced T-Ring network testing system developing project of China Mobile Communication Corporation (CMCC) Research Institute (CMRI) and Wireless Theories and Technologies (WT&T) lab established an advanced testbed network to emulate the actual network for 3G/B3G systems.

M. Zhu (Ed.): Electrical Engineering and Control, LNEE 98, pp. 231–238.
springerlink.com © Springer-Verlag Berlin Heidelberg 2011

This advanced testbed network is a miniaturized telecommunication environment. Due to the miniature network scale and the shortened cell radius compared with the real network, the parameter configurations for the testbed will be changed, in order to maintain the similar performance to the actual network. The development on the testbed network for mobile communication system has obtained a few achievements, including both the indoor and the outfield scenarios [1-4]. But these testbed networks seldom considered the similarity between the testing environment and the real network. So a conception of fitting degree is proposed by T-Ring to measure the similarity and reliability of the testbed. In this paper, based on the study in the physical layer, the mobility and handover performance in the testbed will be further studied.

In this paper, we will focus on the handover management. Factors that make the mobility of the testbed network be different to the actual network will be analyzed. Firstly, speed adjustment will be discussed in order to adapt the time-stamp to the scaling network. Secondly, we will study the handover process in detail and discuss the adjustment of the handover parameters. Besides, the performance of the testbed network after adjusting the handover parameters will also be evaluated. The fast fading characteristic of the wireless channel will be the metric for the performance evaluation. We will take the outfield test and the system-level simulation at the same time to carry out the number results analysis.

The rest of this paper is organized as follows. Sector 2 introduces the system model and mobility management. Sector 3 presents the theoretical analysis for the mobility management adjustment in the testbed network. Sector 4 analyzes the handover parameters adjustment through outfield test and system-level simulation. Finally, conclusions are presented in Section 5.

2 System Model and Analysis Method

In this section, we will introduce the system model and the structure of the advanced testbed network. The method for the theoretical analysis, outfield test, and simulation are also discussed here.

2.1 System Model and Testbed Network Structure

In this paper, we consider the actual network and the advanced testbed network for 3G/B3G systems. The advanced testbed network is established by CMRI to emulate the actual network for 3G/B3G systems. TD-SCDMA and TD-LTE is considered in the following simulation and testing. Outfield test is implemented in the actual network for TD-SCDMA system deployed in Beijing. We choose several typical dense urban areas of TD-SCDMA system to evaluate the performance.

The advanced testbed network, as T-Ring network, will be used in the system-level simulation. Fig. 1 presents the structure of the T-Ring network briefly. The system contains base stations, a lot of mobile stations (MSs), and railways. Several base stations are settled around the railway with the site-to-site distance ranging from 200 meters to 300 meters. The terminals for test are placed aboard a specific train on the railway while testing. Inside and outside the railway area, there are some base stations and a certain amount of mobile phones acting like the source of interference to the system. MSs are uniformly distributed in the network like that in the practical network.

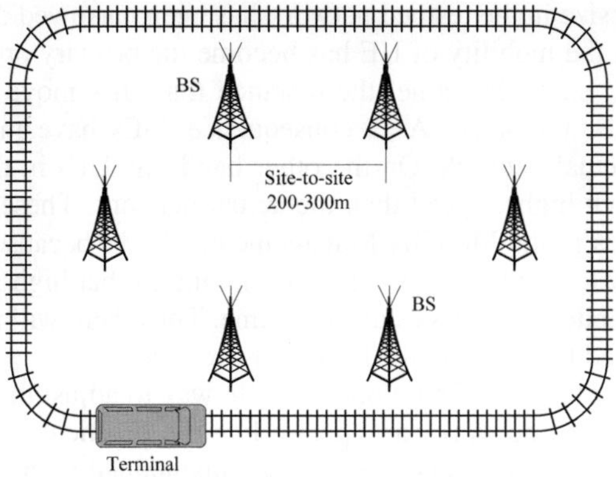

Fig. 1. Illustration of the structure of T-Ring system

2.2 Mobility Management and Handover

Mobility Management is used to provide sustaining network service for the traveling UEs within the coverage of wireless network. It contains Location Management and Handover Management. Location Management is used to track and locate UE in the wireless network, while Handover Management is used to preserve the connection between the network and the UE.

The major work of Location Management is Paging and Location Update. Paging is the process in which the network searches for UE and RNC sends. Hard Handoff is widely used in the cellular communication network. When the UE begins to handover from one sector to another sector, the communication link between UE and original BS is interrupted before the new one to be set up. However during the handover process, some information may be lost because of the transient disconnection.

Baton Handover is a specific handoff mode of TD-SCDMA system. The most important characteristic is the uplink pre-synchronization technology, which is able to provide UE the information of uplink transmit time and power in advance. With the pre-synchronization technology, UE in advance acquires the synchronous parameters, and keep synchronous as the target cell with the open loop mode. Once the network decides to handover, UE can rapidly handover from the source cell DCH state to the target cell DCH state. During the handover process, the transmission of the traffic data can be held, which can reduce the handover time, improve the success rate of handover and lower the DCR of handover. The Baton Handover process is similar with the hard handover. The difference is that the value of Synchronization Parameters of ul_timing advance in the physical channel re-configuration information is empty. In addition, the open-loop power control and the open-loop synchronization method of UE are also different. And there are not the UE uplink processes of transmitting UpPCH and receiving FPACH from NodeB.

3 Theoretical Analysis for Mobility and Handover

In this section, we derive the theoretical analysis for mobility, handover and fast fading performance in the testbed network and actual network.

Because the cell size in the testbed network has been narrowed down, when the UE is traveling around, the mobility of UE has become the primary problem. If the speed of the traveling UE stays the same, the distance that UEs move in the testbed cell network will also be the same. As a consequence, UEs have moved much longer distance than the actual network. On the other hand, the UEs in the testbed network are traveling in much higher speed than the actual network. These will certainly lead to greater difficulties in the Mobility Management. That is because the handover will be a lot more likely to happen when UEs are moving in that high speed. Besides, the time used for the handover process stays the same. Thus there will not be enough time for UEs in high speed to complete the handover process.

In our paper, we attempt to figure out a simple way to adjust the traveling speed of UEs, so that the testbed network can operate like the practical network. We suggest that these processes mentioned above should be maintained with the time dimension. In a period of time, the distance UE has traveled in the testbed network is in proportion to that in the actual network. The proportion of that is equal to the distance scaling factor alpha. Therefore, we can derive the expression of the speed of UE in the testbed network as followed.

$$v = \frac{dS}{dt} \cdot \tag{1}$$

$$S_1 = \alpha S, 0 < \alpha < 1 \cdot \tag{2}$$

$$v_1 = \frac{dS_1}{dt} = \frac{d(\alpha S)}{dt} = \alpha \frac{dS}{dt} \cdot \tag{3}$$

Where v is the original speed of UE traveling in the actual network, S is the distance, alpha is the scaling factor, v_1 is the speed of UE in the testbed network and S_1 is the distance.

It is the simplest way to adjust the traveling speed to adapt to the minimized environment. Making the UE traveling speed in proportion to the distance scaling factor maintain the time-stamp as the actual network. In this way, the mobility management, including location management and handover management, are able to stay the same regarding the spending time and the system overhead. Therefore, in the testbed network, we will not need to change the configuration of the network to keep the network operating well. Besides that the speed adjustment has to be correspondence to the actual network in the time dimension, the change of speed is associated with other factors such as cell coverage and test methods.

Firstly, we analyze the handover process in detail, including the handover parameters and handover performance. For TD-SCDMA system, the primary parameters that are involved with the handover process are Handoff Hysteresis, Trigger Time, Cell Individual Offset and RSCP Threshold. [5-6] Then we study the influences on these parameters caused by the distance scaling. What changes does the testbed network bring into the mobility management? The factors that affect the performance of handover will be analyzed. What reasons that make the handover process be different to the practice will be analyzed.

Since the scale of distance decreases, the geography environment of the testbed network becomes different. Although the large scale fading is proportion to the distance between BS and UE, the shadow fading is quite different due to the change of the distance. Therefore, the UEs on the edge may have greater probability to request for handover. Accordingly the conditions for handover need to be lowered.

The travel of UE causes the Doppler Effect and in the testbed network, the speed change may result in greater Doppler Effect on the frequency offset. Besides, the speed variety also changes the size of the area where the UE is carrying out the handover process. The Cell Individual Offset is only needed in a complex wireless environment. When the wireless channel of the local environment is quite bad for communication, the mobile network operator adds a positive Cell Individual Offset to complete the handover process.

On our previous work [1], we find that the transmit power the distance decreasing will lead to the drop of the transmit power and the UE traveling speed. Therefore, it is necessary to appropriately adjust the parameters related to the handover process for the demand of reflecting the actual network.

Since the cell distance metric is decreased, the handoff area on the cell edge is relatively reduced in proportion to the scaling factor. On the other hand, in terms of the UE moving speed, the traveling speed should be lowered in order to simulate the high speed situation of the actual environment. If the configuration parameters of handover stay the same, the handover processes of UE will be slower. The purposes of the adjustment are that the handover process in the testbed network should be completed in the handover area and the handover performance is close in on the actual network.

The mobility and traveling speed in the testbed network are both different to those in the actual network. In terms of the network performance, the fast fading caused by the UE mobility is what we concern about in the testbed network. Because we lack of the methods to evaluate the handover performance in the testbed network, we consider the fast fading as the main aspect to evaluate the mobility both in the testbed network and the actual network. We assume that fast fading characteristic of wireless channel is used to evaluate the performance effects caused by the mobility and traveling.

We compare the measurement values and the simulation values of the UE received power. From the analysis of the power comparison, we can know the fast fading characteristic both in the actual environment and the simulation environment is similar。The fitness between the simulation results and the measurement results is very good. It shows that the fast fading characteristic of the simulation testbed environment is very similar to the actual network environment.

4 Outfield Test and System-Level Simulation

In the following, we apply the outfield test and system-level simulation to the analysis of the mobility management performance in the testbed network. The simulation results are compared with the measured results to study the difference of fast fading characteristic. Moreover, the theoretical results are also presented to compare with the simulation and test results in order to verify the theoretical derivation.

4.1 Outfield Test and System-Level Simulation

We have carried out the outfield test on the TD-SCDMA network operated by CMCC in [1]. The voice traffic in Release 4 and the data traffic in Release 5 are selected to collect the output performance metrics. We drove the testing car with the testing terminals and GPS device along the pre-defined testing circuit to collect the data of the actual operating network. The output metrics of RSCP and C/I for the actual network are collected.

Moreover, we resort to the system-level simulation to evaluate the performance of the actual network and the testbed network. The parameters in the simulation of actual environment are equal to the actual network. The configuration parameters for the testbed network are studied in [1].

The simulation platform is established based on [7-8] by Visual Studio C++. The COST-231 Hata model [9] is adopted to calculate the path loss. In the simulation, the combination of time driven and snap shot is used. All UEs are randomly dropped in a layout of 1-tier 19 hexagonal cells with 3 identical sectors in each cell. Wraparound model is employed to simulate interference from neighboring cells. Some other main simulation parameters for both actual network and testbed network are provided in Table 1. In the following, the simulation results are analyzed below by comparing with the test results.

Table 1. Parameter Configurations for Simulation and Testing

Parameter		Value
Frequency		2 GHz
Bandwidth		1.6 MHz
BS-to-BS Distance	Actual Network	528 m
BTS Tx Power	Actual Network	43 dBm
Propagation Model		COST 231 Hata Model
BTS Antenna Height	Actual Network	35 m
	Testbed Network	30 m
BTS Antenna Gain		15 dBi
UE Antenna Height		1.5 m
UE Antenna Gain		0 dBi
Voice During / Voice Interval		60 s / 15 s

4.2 Results Analysis

In this section, we take the outfield test and the system-level simulation to analyze the handover process and the handover parameters adjustment performance. During the outfield test we record the measured values before the handover and after the handover between the sectors or the cells of the TD-SCDMA system. According to the handover proceeding, the control channel information such as PCCPCH is used to describe the handover process. As the UE is traveling in constant speed, handover happens when the UE cannot receive strong enough PCCPCH signal power from the original BS. We note that the cell ID and the PCCPCH RSCP received by the UE from every handover process happens during the outfield test.

Table 2. Measurement of handover and handover time in the Outfield test

PCCPCH Measured Values(dB)	RSCP Mean	C/I Mean
Before Handover	-69.93	8.73
After Handover	-65.57	11.75
Handover Time(Baton Handover)(ms)	595	
Handover Time(Hard Handover)(ms)	224	

Table 2 shows that PCCPCH measured values and the handover time-spending in the outfield test. The means of PCCPCH RSCP and the PCCPCH C/I of all handover measured values are calculated. From the results, we can know that in what PCCPCH signal power degree will the UE request the BS for a handover process or will the BS start a handover process for the UE. Besides, the UE gets nearly 5 dB higher RSCP after successfully completing handover to the target BS. The same phenomenon can be seen in the PCCPCH C/I values. In addition, the mobility of UE, which will cause some difficulties on the handover management, is measured by the handover spending time. From Table II, we know that how much time will be spent in one Hard Handover process and one Baton Handover. These results give us the references to adjust the handover parameters in the testbed network.

The system-level simulation is carried out when the handover parameters are all appropriately adjusted based on the measured values in the outfield test. Besides, the mobility of UE in the simulation of the testbed network is also changed to simulate the practical mobility of the real environment. The simulation results are the statistical data of PCCPCH RSCP received by UE.

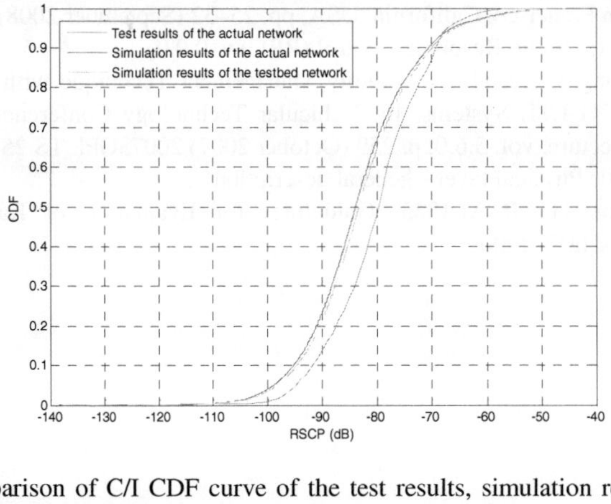

Fig. 2. The comparison of C/I CDF curve of the test results, simulation resluts of the actual network and the simulation results of the testbed network.

The comparison of RSCP CDF curves between the test and the final simulation of the actual network is shown in Fig. 2. The other output metrics have similar performance, which are omitted here. We can see that the simulation results are very similar to the test results in the statistic dimension. Although there is about 4 dB difference between the test results and the simulation results, the RSCP of the test results has almost the same statistics trend as the simulation, which indicates that the wireless channel characteristics including the fast fading are both similar. The change of the mobility management caused by the various mobility of UE in the testbed network has scarcely effects on the wireless channel fast fading characteristic.

5 Conclusion

In this paper, we discuss the Mobility Management in the TD-SCDMA system. We focus on the handover management in the testbed network. The various factors caused by the miniature scale of the network are analyzed. The numerical simulation and outfield test are taken to analyze the wireless channel fast fading characteristic. The results show that the mobility changes in the testbed network caused by the miniature scale hardly affect the mobility management, which can maintain the same configurations as the actual network.

References

1. Huang, X., Zhao, L., Yang, G., et al.: The Construction of T-Ring- A Novel Integrated Radio Testing Ring. to be published in China Commun.
2. Raychaudhuri, D., Seskar, I., et al.: Overview of the ORBIT Radio Grid Testbed for Evaluation of Next-Generation Wireless Network Protocols. In: Proc. of WCNC 2005, vol. 3, pp. 1664–1669 (March 2005)
3. De, P., Raniwala, A., Sharma, S., Chiueh, T.: MiNT: A Miniaturized Network Testbed for Mobile Wireless Research. In: Proc. of INFCOM, vol. 4, pp. 2731–2742 (March 2005)
4. Su, Y., Gross, T.: Validation of a Miniaturized Wireless Network Testbed. In: Proc. of WiNTECH, San Francisco, California, USA, pp. 25–32 (September 2008)
5. 3GPP TS 25.331: "Radio Resource Control (RRC) v10.0.0
6. Kim, T.-H., Yang, Q., et al.: A Mobility Management Technique with Simple Handover Prediction for 3G LTE Systems. In: Vehicular Technology Conference, Radio Interface Protocol Architecture, vol. 5.6.0, p. 259 (October 2007) 20073GPP TS 25.301
7. 3GPP TS 25.201: Physical layer - general description
8. Recommendation ITU-R M.1225. Guidelines for Evaluation of Radio Transmission Technologies for IMT-2000

High-Layer Traffic Analysis of Existing Network and Advanced Testbed Network for 3G/B3G Systems

Li Chen[1], Bin Wang[1], Xiaohang Chen[1], Xin Zhang[1], Lijun Zhao[2],
Peng Dong[2], Yingnan Liu[2], and Jia Kong[2]

[1] Wireless Theories and Technologies (WT&T),
Beijing University of Posts and Telecommunications,
Beijing, 100876, China
alibupt@gmail.com
[2] China Mobile Research Institute (CMRI),
Beijing, 100053, China
zhaolijun@chinamobile.com

Abstract. On the basis of the previous establishment of the testbed network structure for the simulation test, this paper focus on how the high-layer traffic acts like in the existing network and the advanced testbed network for 3G/B3G systems. The advanced testbed network structure is established previously by China Mobile Communication Corporation (CMCC) for the simulated testing. Theoretical analysis and numerical simulation are adopted for the high-layer traffic analysis. The conclusion indicates the studied advanced testbed network could simulate the actual system from the aspect of high-layer traffic service.

Keywords: 3G/B3G Systems, Advanced Testbed Network, High-Layer Traffic, T-Ring,Queueing Theory.

1 Introduction

The outfield testing in 3G/B3G commercial network aims to optimize and evaluate the performance and facilities. The existing test method has the drawback of long test period, low test efficiency and non-repeatability, so it is difficult to evaluate the performance of actual 3G/B3G networks rapidly and accurately. The advanced T-Ring network testing system [1] developing project of China Mobile Research Institute (CMRI) established an advanced testbed network to emulate the actual network for 3G/B3G systems.This advanced testbed network is called a novel integrated radio testing ring, which is a miniaturized telecommunication environment. The configuration parameters for the testbed network are studied by Wireless Theories & Technologies Lab and CMRI in [1].

The development on the testbed network for mobile communication system has obtained a few achievements, including both the indoor and the outfield scenarios [2-4]. In the previous work, we built an integrated framework of fitness evaluating and analysis method [1]. We mainly focus on the fitness performance of physical-layer indices, such as the analysis of RSCP or C/I in aspects of statistic, time and

M. Zhu (Ed.): Electrical Engineering and Control, LNEE 98, pp. 239–246.
springerlink.com © Springer-Verlag Berlin Heidelberg 2011

space dimensions [5]. Meanwhile, some physical-layer related performance, such as the transmit power and cell coverage, etc., in the testbed network are also analyzed. How does the high-layer traffic act like in the testbed network?

The rest of this paper is organized as follows. Sector 2 introduces the system model and analysis method for the high-layer traffic. Sector 3 presents the theoretical analysis for this problem. Sector 4 evaluates the high-layer traffic performance in the testbed network and actual network through numerical simulation. Finally, conclusions are presented in Section 5.

2 System Model and Analysis Method

In this section, we will introduce the system model and the structure of the advanced testbed network. The method for the theoretical analysis and simulation are also discussed here.

2.1 System Model and Testbed Network Structure

In this paper, we consider the actual network and the advanced testbed network for 3G/B3G systems. The advanced testbed network, as T-Ring network, is established by CMRI to emulate the actual network for 3G/B3G systems. The structure of the T-Ring network is presented briefly in Fig. 1. The system contains BSs, a lot of mobile stations (MSs), and railways. Several BSs are settled around the railway with the site-to-site distance ranging from 200 meters to 300 meters. The terminals for test are placed aboard a specific train on the railway while testing. Inside and outside the railway area, there are some BSs and a certain amount of mobile phones acting like the source of interference to the system. MSs are uniformly distributed in the network like that in the actual network.

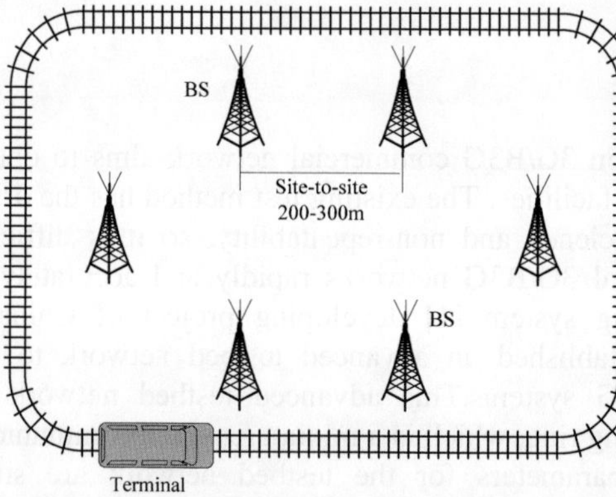

Fig. 1. The structure of T-Ring network for 3G/B3G systems.

Since the deployment of the testbed network is very different from the actual network, we are motivated to study how the high-layer traffic acts like in the testbed network. To approach the actual situation, we assume the transmit power of all cells in the testbed network is the same, which means that they have the same service capability. Transmit power should vary with the cell radius in the network, since the received power mainly determined by the distance from the terminal to the source. As a result, the service capacity of each cell in the testbed network is different from that in actual network. In the following, we will study the high-layer traffic in the testbed network and actual network for 3G/B3G systems.

2.2 Analysis Method

Typical queuing theory is used here to analyze the high-layer traffic service performance in the testbed network and actual network. Queuing theory model that can also be regarded as random serving system is well studied in [6-7]. It mainly resolves the application problem related to the random arrival and queue service. The queuing theory is first proposed by Er lang [8], the originator of the queue theory, to solve the capacity design problem of the telephone switch. It is suitable for all kinds of service system, including communication system, computer system etc. It is generally believed that all systems which have the congestion phenomenon are the random service systems.

There are three variables in queuing theory, m, λ, and μ, which are regarded as the three key elements in queuing model. m is the number of service window, which represents the system resource. It indicates how many servers can be provided to customers at the same time. λ is the arrival rate. Customers usually arrive at the queuing system randomly. μ is the system serving rate, which is a variable represents the system serving capacity. The serving time is a random variable. The performance of queuing system mainly depends on the distribution of arrival interval t and serving time τ, and the queuing rule of the customers. They can determine the statistical state of queuing system.

The queuing theory was applied in wireless communication in [9-11], which analyzed the related system performance. The parameters in communication systems can be mapped into the variable in the queuing theory. Average user number in unit time is mapped to the arrival rate in queuing theory. Average occupation time of each call is almost equal to average service time. In wireless communication system, service time is related to band width and SINR.

3 Theoretical Analysis for High-Layer Traffic

In this section, two typical models in queuing theory, single-server model (M/M/1) and multi-server model (M/M/C), are used to analyze different representation of the high-layer traffic in the testbed network and actual network. The service performance is studied and compared between the testbed network and actual network.

In order to analyze the performance of application layer in mobile communication networks, capacity-limit M/M/1 queuing system (M/M/1/N/∞/FCFS) and capacity-limit M/M/c queuing system with multiple parallel servers (M/M/c/N/∞/FCFS) are adopted. Without losing generation, we assume the interval of customer arrival time and customer serving time follow negative exponential distribution. The maximal system capacity is N, the maximal number of customers in queue is N-1. If there are already N customers in the system, a new customer will be refused to be served. First-come-first-served mechanism is adopted. Servers are independent to each other in multi-server system.

3.1 M/M/1 Model

In this model, the system has a single server queuing. In the steady state, the state transition diagram is shown in Fig. 2.

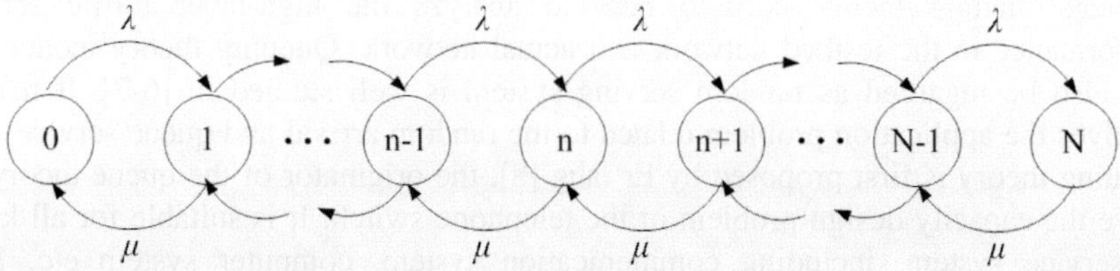

Fig. 2. The state transition diagram for M/M/1 model.

The steady state of the state probability is as below,

$$\begin{cases} \mu.P_1 = \lambda.P_0 \\ \mu.P_{n+1} + \lambda.P_{n-1} = (\lambda + \mu).P_n \,, n \le N-1 \,. \\ \mu.P_N = \lambda.P_{N-1} \end{cases} \tag{1}$$

Solve this difference equation, since $P_0 + P_1 + ... + P_N = 1$, we assume $\rho = \lambda / \mu$, then

$$\begin{cases} P_0 = \dfrac{1-\rho}{1-\rho^{N+1}} & \rho \ne 1 \\ P_n = \dfrac{1-\rho}{1-\rho^{N+1}}.\rho^n & n \le N \end{cases} . \tag{2}$$

In the situation that the system capacity is infinite, $\rho \le 1$, this is requirement of the practical problem and the necessary condition for infinite series to convergence. If the system capacity is finite N, there is no need to request $\rho \le 1$. In the following, we analyze the performance indices of the system. From [6], the different indices can be expressed as follows.

1) system queue length (mean value)

$$L_s = E(n) = \sum_{n=0}^{N} nP_n = \frac{\rho}{1-\rho} - \frac{(N+1)\rho^{N+1}}{1-\rho}, \quad \rho \ne 1 \,. \tag{3}$$

2) queue length (mean value)

$$L_q = E(n-1) = \sum_{n=0}^{N} (n-1)P_n = L_s - (1-P_0) \,. \tag{4}$$

3) customer waiting time (mean value)

$$W_q = W_s - 1/\mu \ . \quad (5)$$

When the cell size is miniaturized, the user number in the cell decreases, the load of the cell decreases, and the user average arrival rate λ decreases, but the system average serving rate μ does not change. Thus, the corresponding queue length, residence time and waiting time will change therewith.

3.2 M/M/C Model

In this model, the customers arrive independently. The number of the arriving customer in a given period of time follows the Poisson distribution, the arriving process is steady. The average serving rate of each server is the same, $\mu_1 = \mu_2 = \cdots = \mu_c = \mu$. The serving rate of the system is $c\mu$ when $n \geq c$, $n\mu$ when $n<c$.

The state transition diagram of the steady state for this model is shown in the following figure.

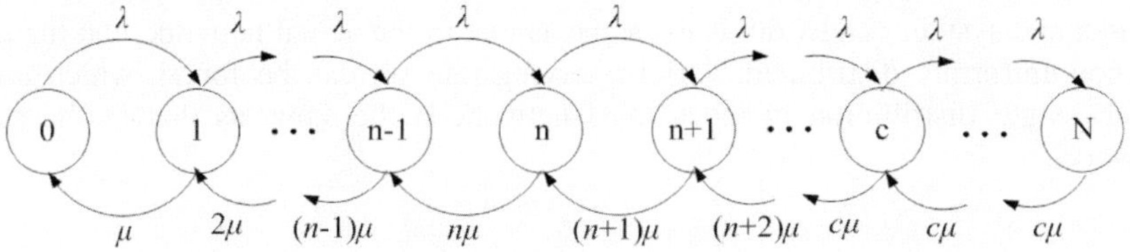

Fig. 3. The state transition diagram of the steady state for M/M/C model.

Similarly, we can get the different performance indices can be expressed as:

$$\begin{cases} L_q = \dfrac{P_0 \rho (c\rho)^2}{c!(1-\rho)^2}[1-\rho^{N-c} - (N-c)\rho^{N-c}(1-\rho)] \\ L_s = L_q + c\rho(1-P_N) \\ W_q = \dfrac{L_q}{\lambda(1-P_N)} \end{cases} \quad (6)$$

4 Numerical Simulation

In the following, we adapt numerical simulation to analyze the high-layer traffic service performance in the testbed network and actual network. Besides, the special cases that the users are not uniformly distributed in the cell are also evaluated through simulation.

We consider the application-layer performance comparison in two scenarios. The first one is that the user number and system bandwidth, the user arrival rate is the same as that in actual network. However, the users in the testbed network are non-uniformly distributed. It can be indicated that several servers' serving rate is different.

In the simulation, the main simulation parameters are provided in Table 1, the user arrival rate is 8,. When users are uniformly distributed, two servers' serving rate is 5. Through the simulation, the mean value of queue length is 8.8129. When scale is miniaturized, two sets of serving rate are 4, 6 and 3.5, 6.5, respectively. The system queue lengths are 8.8048 and 9.0745, shown in Table I. The distributions of the queue length in three cases are depicted in Fig. 4.

Table 1. Parameter Configurations and Simulation Results

Case	λ	μ_1	μ_2	L_s
I	8	5	5	8.8129
II	8	4	6	8.8048
III	8	3.5	6.5	9.0745

From the results, we can see that when the network is miniaturized, if the user number and system bandwidth is the same as that in the actual network, and the users are non-uniformly distributed, various serving rate set can be found, which makes queue length distribution in the testbed network is the same as that in the actual network.

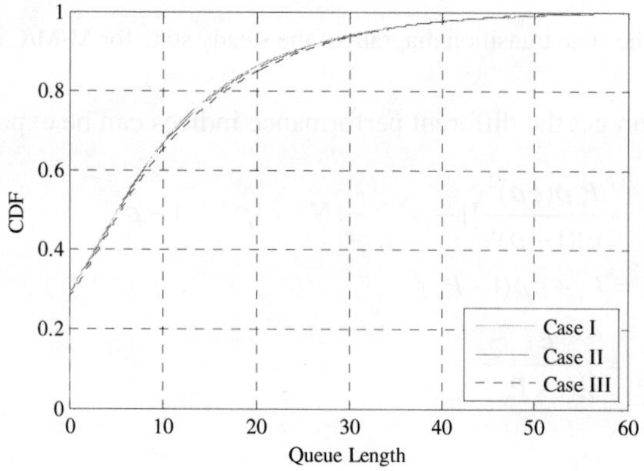

Fig. 4. The distribution of queue lengthes under different serving rates.

The second scenario is that the user number and system bandwidth changes, and the users in the testbed network is non-uniformly distributed. Two servers model is simulated here. The user arrival rate in actual network is configured as 8, if the users are uniformly distributed, the two servers' serving rate is the same as 5. When the cell scale is miniaturized and the user arrival rate is 6, if the users are non-uniformly distributed in the cell, the two servers' serving rate is different, when one server's serving rate is determined as 4, the other server's serving rate can be found to be 3.25, making the mean value of queue length is the same in the testbed network and actual network. Similarly, when the user arrival rate is 4, the two servers' serving rate is 2 and 2.65, respectively. The simulation configurations and results are shown in Table 2.

Table 2. Parameter Configurations and Simulation Results

Case	λ	μ_1	μ_2	L_s
I	8	5	5	8.8129
II	6	4	3.25	8.6721
III	4	2	2.65	8.8514

The queue length distributions in two cases are depicted in Fig. 5. From the simulation results, wWe can see that when the network is miniaturized, for different user arrival rates, when a server's serving rate is determined, the other server's serving rate can be found, making the mean value of queue length is comparable with that in the actual network. Besides, the distribution of queue length is not greatly different between the testbed network and actual network. However, the system serving rate is related to the system bandwidth, the difference between two servers' serving rates is related to the user distribution in the cell. Considering the physical-layer and application-layer fitness performance, the users need to be distributed as uniformly as they can in the testbed network.

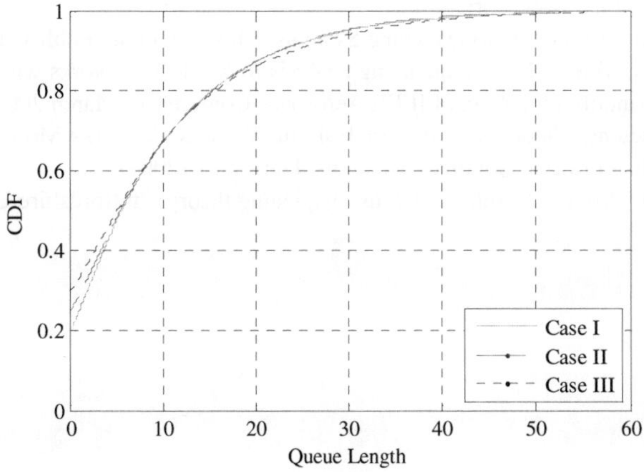

Fig. 5. The distribution of queue lengthes under different bandwidths and serving rates.

5 Conclusion

In this paper, we analyze the high-layer traffic in the testbed network and actual network for 3G/B3G systems. The typical queuing theory is used to study the traffic service performance through M/M/1 model and M/M/C model. The performance under the scenario that several test terminals are bound together is also discussed.

Finally, numerical simulation is adopted to evaluate the performance high-layer traffic service in the testbed network and actual network. From the results, we can conclude that the proposed testbed network can emulate the actual network in aspect of high-layer traffic service. The gap between the actual network and the testbed network can be controlled under a reasonable range if they have similar physical-layer performance.

References

1. Huang, X., Zhao, L., Yang, G., et al.: he Construction of T-Ring- A Novel Integrated Radio Testing Ring. to be published in China Commun
2. Raychaudhuri, D., Seskar, I., et al.: Overview of the ORBIT Radio Grid Testbed for Evaluation of Next-Generation Wireless Network Protocols. In: Proc. of WCNC 2005, vol. 3, pp. 1664–1669 (2005)
3. De, P., Raniwala, A., Sharma, S., Chiueh, T.: MiNT: A Miniaturized Network Testbed for Mobile Wireless Research. In: Proc. of INFCOM, vol. 4, pp. 2731–2742 (March 2005)
4. Su, Y., Gross, T.: Validation of a Miniaturized Wireless Network Testbed. In: Proc. of WiNTECH, San Francisco, California, USA, pp. 25–32 (September 2008)
5. Chen, L., Wang, B., Chen, X., et al.: Performance Evaluation of Existing Network and Advanced Testbed Network for 3G/B3G Systems (to be published)
6. Leonid, B.K., Sinai, Y.G.: Theory of Probability and Random Processes. Springer, Heidelberg (2007)
7. Gross, D., Shortle, J.F., Thompson, J.M., Harris, C.M.: Fundamentals of Queuing Theory. Wiley & Sons, Inc., New York (1985)
8. Bose, S.J.: An Introduction to Queueing Systems. Kluwer/Plenum Publishers (2002)
9. Chen, X., Lyu, M.R.: Message queueing analysis in wireless networks with mobile station failures and handoffs. In: Proc. of IEEE Aerospace Conference (March 2004)
10. Sylla, B.: Queuing Theory Systems Analysis In Wireless Networks Mobile Stations with Non-Preemptive Priority. Southern Methodist University (2004)
11. Carrington, A., Yu, H.: Wireless NCS using queuing theory. Staffordshire University,

Analysis of Cased Hole Resistivity Logging Signal Frequency Effect on Detection

Yinchuan Wu, Jiatian Zhang, and Zhengguo Yan

The Key Laboratory of photoelectricity Gas & oil Logging and Detecting Ministry of Education, Xi'an Shiyou University, Xi'an 710065, China

Abstract. The skin effect of metal casing is analyzed, a ultra-low frequency signal source is put forward in cased hole resistivity logging technology. In this technology, the direct detection signal order is microvolt, the useful signal order is nanovolt. Based on the characteristics of this technology, the phase sensitive technique is put forward in this paper for improving the accuracy of data acquisition. The logging signal frequency effect on detection is analyzed, and the specific solution is given. The results have been applied in design of signal source and weak signal acquisition system.

Keywords: Cased Hole Resistivity, Skin Effect, Phase Sensitive Detection.

1 Introduction

In oil production, the cased hole resistivity logging technology is one of the advanced new technologies being studied in our country.[1] It is widely used in types of engineering such as confirming the saturation of remaining oil, identifying dead oil and gas formation, evaluating flooded oil layers, monitoring the saturation of fluids, locating the surface of dividing water & oil. In the cased well, by injecting high current into the cased hole, the voltage drop of weak signal on the hole is measured at fixed distance so that currents is that leaked on the target layers are estimated, and by inversion algorithm calculating formation resistivity and quantitatively confirming the oil situation. In this technology, the direct detection signal order is microvolt, the useful signal order is nanovolt. The key problem to be solved in the project is the accurate detection of nanovolt signal. In this paper, the analysis was on options of signal resources frequency, it is also put forward that digital phase sensitive detection technique should be applied to realize the weak signal detection. And the logging signal frequency effect on detection is analyzed, then the specific solution is given.

2 Signal Resource Frequency Option

2.1 Skin Effect Phenomenon

In the metal hole, as the exciting signal frequency increasing, injected power tends to flow towards the surface of the hole, which makes the ground outside the hole free of

M. Zhu (Ed.): Electrical Engineering and Control, LNEE 98, pp. 247–252.
springerlink.com © Springer-Verlag Berlin Heidelberg 2011

leaked currents, as a result, cased hole resistivity unable to be detected. It is learned from electromagnetic field that the better the quality of the conductor is, the faster the current decreases along in depth by the index, the more apparent the skin effect is, which is called skin effect phenomenon. electromagnetic wave weakens rapidly in good conductor usually decreasing almost to zero at the distance of micron orders so high frequent magnetic field only exists in a thin layer of conductor surface. Therefore exciting currents tend to flow towards the inner surface of metal holes. When vibration range of magnetic field decreases to 1/e of the surface, is 36.8% in depth, it is called skin depth (or penetration depth) δ , where is

$$\delta = \sqrt{\frac{2}{\omega\mu\sigma}} \tag{1}$$

Which w is signal angle frequency, μ is permeability of conductor, σ is conductivity. It can be seen from formula (1) that skin depth, signal frequency, permeability of conductor and conductivity are related with one another.

2.2 The Relationship between Exciting Signal Frequency and Skin Depth

Suppose the skin depth of hole is δ_c, put $\omega = 2\pi f$ 、 $\mu = \mu_0\mu_r$ into the formula (1), and

$$\delta_c = \sqrt{\frac{2}{\omega\mu\sigma}} = \sqrt{\frac{2}{2\pi f\mu\sigma}} = \sqrt{\frac{1}{\pi f\mu_r\mu_0\sigma}} \tag{2}$$

$\mu_0 = 4\pi\times10^{-7}$ (H/m) , $\rho = \frac{1}{\sigma}$ is known, and put it into the above formula and result is

$$\delta_c = \frac{10^3}{2\pi}\sqrt{\frac{10}{\mu_r f\sigma}} = 5.03\times10^2\sqrt{\frac{\rho}{\mu_r f}} \tag{3}$$

Generally speaking, permeability of conductor of the petroleum industry steel casing μ_r is 40~100, conductivity ρ is $(2\sim3)\times10^{-7}\,\Omega m$, the thickness of the hole is among 7.52~11.51. This, we let μ_r equal 50. When the resistivity ρ is $2\times10^{-7}\,\Omega m$, the skin depth δ_c is

$$\delta_c = 3.18\times10^{-2}\sqrt{\frac{1}{f}} \tag{4}$$

The relationship between exciting signal frequency and skin depth is shown in figure 1, as the frequency is increasing, the skin depth is reduced.

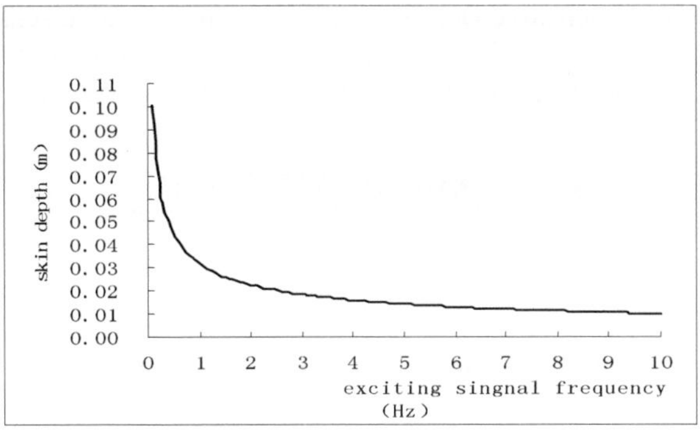

Fig. 1. The relationship between exciting signal frequency and skin depth

2.3 Exciting Signal Frequency Option

Cased hole resistivity logging techniques consist of three modes: logging model (leaking current), scale model (hole resistivity model), reference potential model (full resistivity model), under kinds of modes, exciting signal frequency options directly influences the speed, efficiency of instruments. Therefore, when certain accuracy is achieved, exciting signal frequency should be boosted increased as possible as it can be to boost the equipment efficiency. Tabarovsky etc. in 1994 analyzed frequency effects & concluded: (1) under the logging mode , only if skin effect depth of hole is equal to or more than two , the measurement accuracy will be less than 10%. For the sake of analysis, in the following discussion, the hole thickness is supposed as 10mm. According to this theory, it is requested that the skin depth should be less than 20mm. hence, following formula (4), it can be achieved that exciting signal frequency is less than 2.5hz; (2) under the scale mode, only if the cased hole skin depth or cased hole thickness is more than 1, the measurement accuracy will be less than 10%. According to this theory, it is requested that the skin depth should be more than 10mm , and based on formula (4), it can be realized that the exciting signal frequency is less than 10hz; (3) under the reference potential mode, between the cables with injected currents and the cables with potential measurement have the problem of coupling, the longer the cable is, the more serious the coupling is. When exciting signal is 1hz, the coupling phenomenon will be very serious. Therefore, under this mode, the signal frequency is much lower, e.g 0.1hz.

3 Signal Detection

3.1 Phase-Sensitive Detection Technology

According to the above analysis, the detection signal achieved through cased hole resistivity logging technique is the known low frequency sine weak signal. The logging mainly measures the amplitude width of weak signals. In the real logging practice, detecting system noise makes signal noise ration as low as -30db~-60db. The most effective way of detecting known frequency weak sin signal width is correlation detection, is the phase-sensitive Detection. The principle of phase-sensitive Detection is shown in figure 2, where x(t) is the detecting signal, r(t) is the known reference signal having the same frequency as x(t), k is addition of multiplying device.

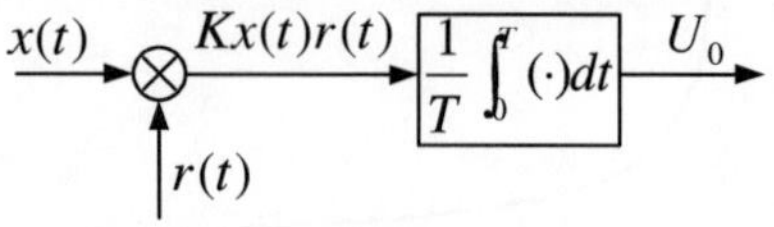

Fig. 2. The principle of phase-sensitive detection

Suppose the detected signal is

$$x(t) = U_s \cos\omega_0 t + n(t) \tag{5}$$

Then, in the above formula, U_s is the width of the detected sin signal, ω_0 is the angle frequency of the detected sine signal, n(t) is Zero_mean Stationary Gaussian White Noise.

Suppose the reference signal is

$$r(t) = U_r \cos(\omega_0 t + \varphi) \tag{6}$$

According to figure 2, its output is

$$U_0 = \lim_{T\to\infty}\frac{1}{T}\int_0^T K[U_s\cos\omega_0 t + n(t)]U_r\cos(\omega_0 t + \varphi)dt \tag{7}$$

Since the sine signal is irrelevant to the noise, Integral average is zero and

$$U_0 = \frac{KU_sU_r}{2}\cos\varphi \tag{8}$$

From formula (8), U_s can be worked out, when $\varphi = 0$, is the reference signal and the detected signal have the same phase,

$$U_s = \frac{2U_0}{KU_r} \tag{9}$$

Then U_s can be measured. But the above result can be achieved only if the endless time is given. However, it is impossible in the practical engineering project. Also, the cased hole logging resistivity direct measuring signal is nanovolt and the useful signal is nanovolt order, under which the noise is very huge. Therefore, the U_0 achieved based on the above principle is very unstable, which greatly influences the measurement accuracy of U_s. So the digital phase-sensitive Detection approach has to be applied in order to increase the measurement accuracy of weak signals, when k is one, the digitalized U_0 is as follows:

$$U_0 = \frac{1}{N}\sum_{i=1}^{N} x(iT_S)r(iT_S) \tag{10}$$

Where N is the length of sampling data, T_S is sampling period.

3.2 Phase-Sensitive Detection of the Amplitude-Frequency Characteristic

In the above analysis, it is supposed that the reference signal is the same as the frequency detection signal, but when they are different in value, there will be errors in measurement results. The following part is about the study of the relations between the detecting wave output result and frequency difference. Based on (8), it can be inferred that the digital Phase sensitive detection calculating width features should be:

$$H(j\Delta f) = \frac{1}{N}\cdot\left|\frac{1-e^{-jN\Delta f T_S}}{1-e^{-j\Delta f T_S}}\right| = \frac{1}{N}\left|\frac{\sin(N\Delta f T_S/2)}{\sin(\Delta f T_S/2)}\right| \tag{11}$$

When Δf is the deduction value of the reference signal frequency & the detection signal frequency .Here, it is supposed that the sample data length N is 500, $f_S = 1/T_S = 500$ hz, then the digital Phase sensitive detection of the amplitude-frequency characteristic are shown as in figure 3. From figure 3, it can be seen that the digital Phase sensitive of the amplitude-frequency characteristic have many peak values among which the main peak value width is

$$W = 2f_S/N = 2\text{Hz} \tag{12}$$

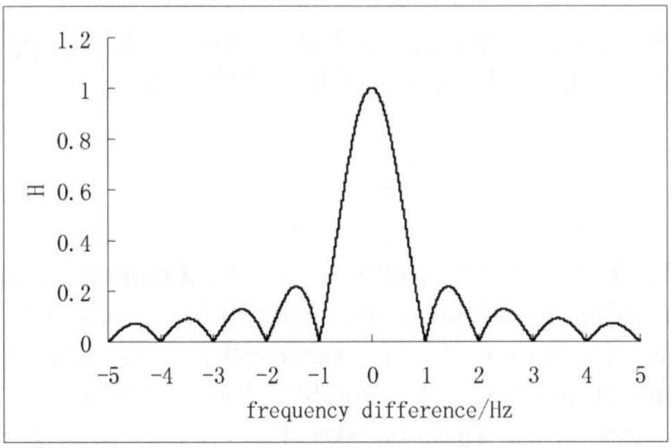

Fig. 3. Phase sensitive detection of the amplitude-frequency characteristic

The digital Phase sensitive detection is actually a Narrowband filter. The narrower the main peak width is, the better the quality of Phase sensitive detection is, the stronger the restraining noise ability is. In the practical cased hole resistivity detection, frequency deduction value range is among the main peak width value. The narrower the main peak is, the better the wave detecting quality is, however, the stronger the effect of frequency deduction is.

3.3 The Influence of Exciting Signal Frequency Stability on Measurement

When the cased hole logging resistivity is measured, the exciting signal source is on the ground, which hence transmitted to the well through thousands of long kilometer-cables. According to the above analysis, phase sensitive detecting calculating methods require that the frequency of the reference signal should be the same as that of the exciting signal. The reference signal can be achieved directly from the exciting signal as well as from computer calculations. In the practical logging, as long as the exciting signal frequency stability meets the conditions, the reference signal can be achieved from computer calculations. The reference signal frequency achieved from computer is stable, so the frequency difference of phase sensitive detecting calculating methods is mainly influenced by the exciting signal. In order to increase the measurement accuracy, the exciting signal frequency stability has to be improved.

4 Conclusions

Because the skin depth phenomenon appeared in cased hole, is accurate detection signal, cased hole resistivity logging require the frequency of the exciting signal less than 10Hz. The frequency stability of exciting signal directly influence the accuracy of the signal detection, when confirming the length of the sample data and the frequency of sample, according above analysis, we can quantitative calculate the range of permissible frequency deviation of exciting signal source, and determine the frequency stability of exciting signal source. This conclusion is used to instruct the exciting signal of the cased hole and the design of weak signal detection system.

Acknowledgements

The authors thank Miss Qin for the active help. This work is supported by the Natural Science Foundation of Shaanxi Province, China (2007D01).

References

1. Wu, Y., Zhang, J., Yan, Z.: An overview of The Logging Technology of Formation Resistivity through casing. Petroleum Instruments 20(5), 1–5 (2006)
2. Singer, B.S., Fanini, O., Strack, K.-M., Tabarovsky, L.A., Zhang, X.: Measurement of Formation Resistivity Through Steel Casing. SPE 30628 (1995)
3. Tabarovsky, L.A., Cram, M.E., Tamarchenko, T.V., et al.: Through-casing resitivity(TCR): physics, resoulution and 3-D effects. In: SPWLA 35th Annual Logging Symposium (1994)
4. Dai, Y.: The Performance Analysis of Digital Phase Sensitive Detection Algorithm used to Measure Voltage at Low Signal Noise Ratio. Acta Metrologica Sinica 18(2), 126–132 (1997)
5. Liu, Y., Liu, F., Dai, Y.: Study on the DPSD Algorithm Automatically Adjusting the Frequency of Reference Signal. Acta Metrologica Sinica 19(4), 312–316 (1998)
6. Yan, Z., Zhang, J.: Error Analysis of Weak Signal Detection in Cased Hole Formation Resistivity Logging. Well Logging Technology 31(4), 486–488 (2007)

Communication Software Reliability Design of Satelliteborne

Shuang Dai[1] and Huai Wang[2]

Changchun Institute of Optics, Fine Mechanics and Physics,
Chinese Academy of Sciences Dong Nanhu Road 3888, 130033,
Changchun, Jilin, China
[1]dai-dai123@163.com, [2]playsnail@sina.com

Abstract. Firstly, the failure model of Satelliteborne communication software is given for space environment, especially SEU(Single Event Upset), unlike common communication software, then the design flow for satelliteborne software is set , and some measures are taken such as register renewed periodically, memory voting and redundancy design. Finally fault injection is applied to validate. Practical results show that the design is necessary and effective.

Keywords: satelliteborne communication software; SEU; fault tree; reliability.

1 Introduction

Satelliteborne communication software is real time communication embedded software, which received control data from satelliteborne system and download state parameter of operating system. it is a bridge of satelliteborne system and ground control system. To make sure finishing the task of satelliteborne system, we should take accuracy and reliability of satelliteborne communication software into accout. The most important characteristics of satelliteborne communication software is inflected by radiation effect of space environment. As well as we known, among the influence of radiation effect, SEU (SEU - Single Event Upset) is directly influent satelliteborne communication software.

SEU can make logic error of device or circuit, for example: data rolling over in the memory make confusion of logic faction or make program run away, even make disaster result. As far as communication software concerned, it will make memory or register of communication interface device roll over, and make confusion of system communication logic, then whole system can't communicate with external system, and can't carry out mission.

This paper takes measures to SEU, which is as software method, to make sure reliability of satelliteborne communication software.

M. Zhu (Ed.): Electrical Engineering and Control, LNEE 98, pp. 253–260.
springerlink.com © Springer-Verlag Berlin Heidelberg 2011

2 Failure Mode Analysis of Satelliteborne Communication Software

Like common communication software, satelliteborne communication software has communication initialization, receiving function, sending function and interruptin function and so on. Because of influence of space environment, it will arise error of dataflow or control flow. When data changed between 0 and 1 of memory, it will make frame error or important variable error, and pass error during the reading or writing, then make failure of sending or receiving.

On the other side, when SEU occurred, register and memory initialization area is modified, then interruption can not be respond, which result in reading and writing data failure. As discussed above, we make failure mode analysis which is different from common communication software. We set fault tree of satelliteborne communication software shown as figure 1.

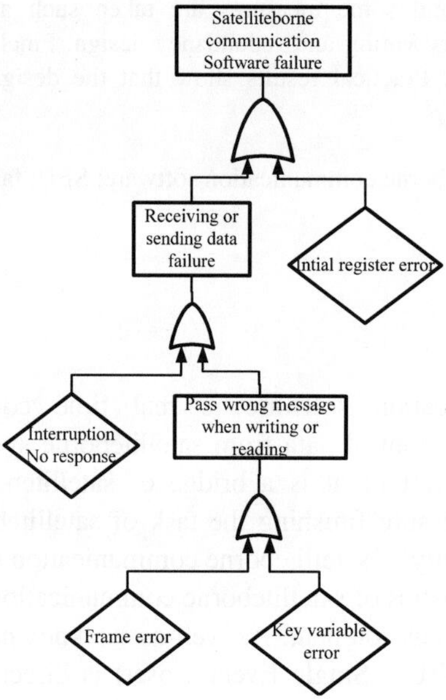

Fig. 1. Satelliteborne communication software fault tree

In figure 1, we can conclude middle event which conclude failure of communication software is sending or receiving data failure and reading or writing process pass error information. Bottom event includes initial register error, interruption no response, frame error, important variable error and so on. The failure mode we discussed, will be taken measure of reliability in the below.

3 Constitution of Satelliteborne Communication Software

3.1 Satelliteborne Communication Software Topology

As reference of web seven layers mode, as same as common communication software, we divide satelliteborne communication software into 4 layers: physical layer, data link layer, driver layer and application layer. Which shown as figure 2.

Physical layer transmitted by the difference Manchester code. Data link layer transmitted by data format defined by the 1553B bus special protocol. Nowadays, Driver layer is supported by special protocol chip such as BU-65170, application layer which we should emphasize on and realized is divided into two parts: MBI bottom driver program and MBI communication program. MBI bottom driver program mainly realize initialization, interruption, address set of RAM and register, self-test and so on. MBI communication program realized, depending on MBI bottom driver program and interface communication data (ICD), chip initialization, management 1553 data sending and receiving and so on.

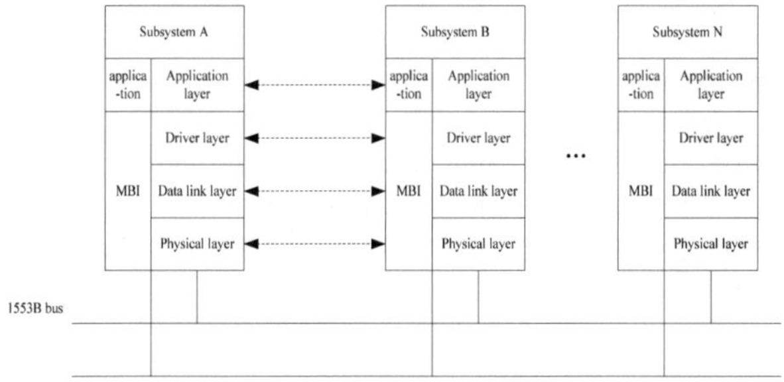

Fig. 2. Communication software topological structure

3.2 Satelliteborne Communication Software Design Flow

In the software, the software engineer technology is the base of reliability, which put forward the essential principle and requirement. For example, there is a rule that different people take charge of software design, program coding and testing; keeping design document , diagraph and testing record, which make the procedure of software design and usage be seen. A main factor of reliability is software size and complexity, and use the way of from top to bottom to design which can make the reasonable module and realize low coupling and high cohesion, furthermore lower complexity of system, which is good for realibility[1].

In the satelliteborne communication software, except for the common principle we talked above, we put forward design flow as figure 3. This design flow includes design of ICD table, MBI bottom layer development and MBI communication program design. Some problem should be concerned. Firstly the cycle and data size should be defined by ICD. In ICD table, there is definition of upload and download parameters (including unit, variability range and so on), allocation of parameter

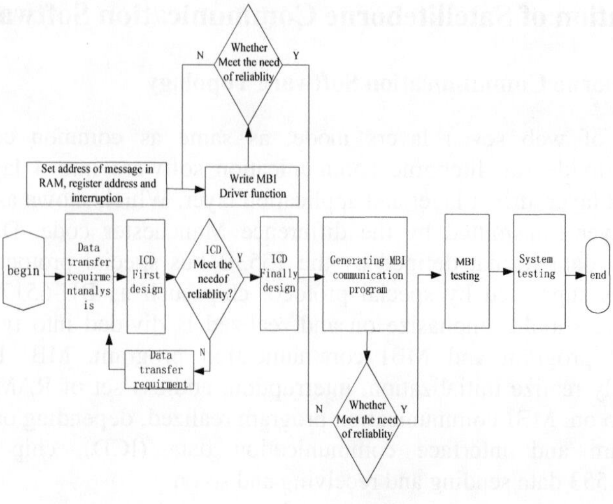

Fig. 3. Satelliteborne communication software design flow

address, cycle and so on. For some important variable, there should be reliability design for space environment. Secondly, base of MBI driver program, MBI communication program use ICD table to allocate size and time of communication, which divide the communication task into cycle and no cycle task, make reasonable scheduling to make system design requirement. Lastly, we make MBI testing and system testing to certification.

4 Satelliteborne Communication Software Realize

Based on design flow and characteristics of space environment, we set 1553B RT communication software as an example, in the aspects of composing ICD, design of MBI bottom driver and MBI communication program to the design of reliability.

4.1 1553B Communication Software MBI Realization

As RT software, for BU-65170 chip, it realized the function introduced as below:

- Configuration register initialization

It realizes the setting of main work mode, choice of memory style and interruption style and so on.

- Message storage style and memory address arrangement

There are three kind of storage style: single buffer, double buffer and circular buffer. We choose special style depending on the size of message block. If data block is no more than 32 Bytes, we choose single buffer, if data block is bigger than 32 Byte. As for receiving message, we usually choose double buffer, but when the interval of sending cycle is very short (less than 100ms) or length of message is long(more than 32 Bytes), we choose the style of loop buffer

- Setting illegal map for illegal message

To make sure the reliability of 1553B communication, we set illegal map, which set these message that didn't satisfy the requirement we set as illegal message, and these message will eliminate by chip automatically.

- Interruption management

Message interrupt realize the movement of data. For receiving message, it will move the message from the buffer of interface chip to user processing area. For sending message, it will move the message to the buffer of interface chip, then set service requirement bit.

We set an example as below: Setting subaddress 3 as receiver that receives 32 words and broadcasting style message. subaddress 5 as receiver that receives 5 words and broadcasting style message, subaddress 1 as sender that sends 21 words message. The code of intialization 1553B communication as RT is written as below:

```
/*initial register:*/
AceRegWrite(CONFIG_3_REG,0x8000);
//set RT as forbidden running mode
AceRegWrite(INTERRUPT_MASK_REG,0x0010);
/*enable end of message interruption: */
 AceMemWrite((int *)StackPtrA,0);
// set stack pointer as 0
  AceRegWrite(CONFIG_1_REG,0x0cf80);
  AceRegWrite(CONFIG_3_REG,0x8019);
  AceRegWrite(CONFIG_2_REG,0x981a);
  AceRegWrite(CONFIG_4_REG,0x0000);
  AceRegWrite(CONFIG_5_REG,0x0700);
/*receiving and sending buffer address initialization*/
  AceMemWrite((int *)0x0143,(int)AddrR3);
// set address of R3memory
  AceMemWrite((int *)0x0145,(int)AddrR4);
// set address of R5memory
AceMemWrite((int *)0x0161,(int)AddrT1);
// set address of T1memory
/*set momory mode*/
AceMemWrite((int *)0x01a1,0x4000);
//T1 as single buffer;
AceMemWrite((int *)0x01a3,0x8200);
```

```
// R 3 as double buffer
AceMemWrite((int *)0x01a4,0x8200);
// R 5 as double buffer
/* set illegal map*/
AceMemWrite((int *)0x03c3,0x0ffdf);
//T1 21word
AceMemWrite((int *)0x0306,0x0fffe);
//R3 32word broadcasting
AceMemWrite((int *)0x0308,0x0ffdf);
//R4 5word  broadcasting
Other set as 0xffff ;
```

After initial chip, communication interrupt program can be written to communition, we didn't discuss.

4.2 Satelliteborne Communication Software Reliability Design

- Design of ICD table

Definition of important variability should not be by one bit; especially using 0 and 1 define different station. More than one bit can be used, for example system power on state using 16 bits define when it=0x5a5a,it means power off, and it =0x6161 means power on.

- MBI reliability design

For reliability, one of method is to judge variable values; another method is decision making system in memory to avoid SUE. Refresh register periodically to avoid interrupting no response or error of initialization of register.

1. Refresh register and memory periodically

For some important registers, such as controller register of interface chip, if SEU occurred in these registers, it would make whole communication interface chip malfunction. If interruption mask register some bits are turned over, it would make interruption request no response. So these registers should be refreshed periodically to avoid value of register turning over.

During the time of registers refreshing, the communication will stop, so we broadcast message for refreshing in whole system. It helps all subsystem refreshing in the same time.

2. Decision making system in data memory

Receiving data are stored in three or more areas; the principle of "two in three" is taken to use data.

3. Data redundancy design

Key variables mainly include loop control variable, finite state machine controller variable, important global pointer, important global marking and so on. Those variables have a big deal with whole process of program running. It has bigger chance to occur SEU than other private variable, so the result is more serious. The way of redundancy can reduce the probability of SEU.

We set global variable iSystemState as an example to show as below:

When no redundancy:

```
If (iSystemState==0x33)
{
   Running signal process;
   iSystemState=0x55;
}
```

When has redundancy :

```
iSystemState = voter(is1,is2,is3);
if(iSystemState==0x33)
{
  run signal process;
      iSystemState=0x55;
  is1= iSystemState;
  is2= iSystemState;
  is3= iSystemState;
}
```

When redundancy is taken, we copy variable iSystemState, and use is1 and is2 and is3 to make decision (two in three), it enhance the degree of belief.

- Others

To make sure the reliability of 1553B communication, we set illegal map, which set these message that didn't satisfy the requirement we set as illegal message, and these message will eliminate by chip automatically.

5 Conclusion

In the experiment, the approach of failure injection proofs the reliability [3]. It injects simulated SEU failure in fixed location of memory. When application reads this special unit, monitor will monitor the state of program running to judge the ability of application to recovery processing.

In the aspect of software approach, depending on the characteristics of space environment, not only a common design flow is put forward to satelliteborne communication software, but also useful design to prevent SEU. The experiment shows this approach enhance the ability of satelliteborne communication software.

References

1. Duan, X.H., Jianwen, D.Z.: Research on Designing-Methods for Reliability of On-Board Software. Computer Engineering 35, 73–75 (2009)
2. Liu, H.F.: Fault-tolerance Design and Verification of Satelliteborne Software. Journal of China Academy Electronics and Information Technology 4, 313–316 (2009)
3. Chunru, Y., Taian, L.: Embedded software reliability estimation based on testing-coverage. Computer Engineering and Design 30, 2198–2200 (2009)

A Moving Objects Detection Method Based on a Combination of Improved Local Binary Pattern Texture and Hue

Guo-wu Yuan, Yun Gao, Dan Xu, and Mu-rong Jiang

Department of Computer Science and Engineering,
Yunnan University, Kunming 650091, China
yuanguowu@sina.com

Abstract. Moving objects detection is one of the key techniques in intelligent video surveillance. A new moving objects detection algorithm is proposed in this paper, which combines improved local binary pattern(LBP) texture and hue information to describe background and adopts the idea of Gaussian mixture model that uses multiple modes to describe background model. In order to reduce matching complexity and reach real-time, many kinds of LBP are cut down. Experiments show that the proposed algorithm can remove effectively the effect of shadow, reaches real-time in common resolution videos and has better performance than other ones.

Keywords: background modeling, background subtraction, Gaussian mixture model, local binary pattern, hue.

1 Introduction

Moving objects detection is often one of the first tasks in computer vision, which obtains moving objects' region from a sequence of video. It is used widely in visual surveillance, intelligent transport systems, industrial vision etc.

Many methods for detecting moving objects have been proposed. Background subtraction is the most usual method in the present. The key technology is how to build a background model to accurately describe background information. A robust background model is able to adapt to various changes of environment, for example, illumination changes, object shadows, swaying vegetation, rippling water and so on.

There are 2 categories to describe a background: (1) using color, brightness or other pixel information; (2) using edge, texture or other structural information. The first category includes Gaussian mixture model (GMM) [1,2], filtering prediction model[3,4], codebook model, kernel density model[5] and so on, in which the best famous one is GMM proposed by Stauffer and Grimson[1]. But GMM is sensitive to shadows. Because of illumination, weak or strong shadows are existed widely, and it causes that the moving area detected includes shadows' region, so it is necessary to overcome the influence.

M. Zhu (Ed.): Electrical Engineering and Control, LNEE 98, pp. 261–268.
springerlink.com © Springer-Verlag Berlin Heidelberg 2011

At the same time, background subtraction based on edge, texture or other structural information is discussed. In [6], the background model was constructed by dividing a background image into same size blocks and by calculating an edge histogram for each block, and this method has no good performance in the area without plentiful edges. The algorithm presented in [7] is based on the texture features consisted of local binary pattern (LBP). It can decrease the obstruction of shadow, but it is incorrect to detect moving object using the LBP texture in the regions of few textures.

If two methods are combined to describe the background, it is greatly possible to improve the accuracy. In [8], it is proposed to construct a background model combined GMM with gradient image division. The algorithm has a good performance, but it is very complex and is out of real-time. The method proposed in [9] is to establish a background model combined improved D-LBP textures with color information. It has a good performance, but it only achieves a processing speed of 13 fps for a video with the resolution 160×120 pixels, and it has no real-time for a common video with the resolution 320×240 pixels.

This paper presents a background modeling method based on a combination of improved local binary pattern texture and hue. The method divides each frame image into equal size blocks, in each block the local texture is denoted by the LBP uniform pattern histogram, and the local hue information is denoted by the normalized hue value. We adopt the idea of GMM to use more than one pattern to model each block. It is more accurate because of combining textures with color, and it has a high speed in comparing the vector features between the current frame and the background models because of employing the LBP uniform pattern and the normalized hue value. Our experiments show that the proposed algorithm can reach real-time in common video resolution of 320×240 pixels and has better performance than other ones.

2 Local Binary Pattern (LBP)

2.1 Basic LBP

LBP is a powerful means of texture description. The operator labels the pixels of an image region by thresholding the neighborhood of each pixel with the center value and by arraying the result to binary codes [10, 11]. The basic LBP is defined as follows:

$$LBP_{P,R}(x_c, y_c) = \sum_{i=0}^{P-1} s(g_i - g_c)2^i, s(u) = \begin{cases} 1, u \geq 0 \\ 0, u < 0 \end{cases} \tag{1}$$

where g_c is the gray-scale value of the center pixel (x_c, y_c) and g_i is the values of P equally spaced pixels on a circle of radius R. If a value of neighbor does not fall exactly on a pixel, it is estimated by bilinear interpolation. Fig. 1 demonstrates the distribution of neighbor sampling points. Eq. (1) indicates that various textures are described by P-bit binary codes, resulting in 2^P distinct values for the LBP. In Fig. 2, the sequence of P points is prescribed, and the positions of number 0~7 correspond to the 0th~7th binary code of LBP value. Fig. 3 shows an example, the center pixel's LBP value is 191.

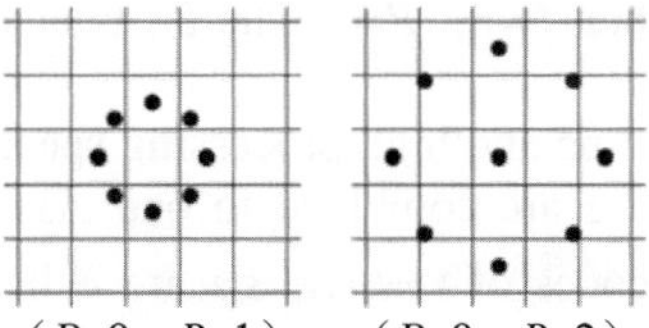

Fig. 1. Circularly symmetric neighbor sets for different values of *P* and *R*

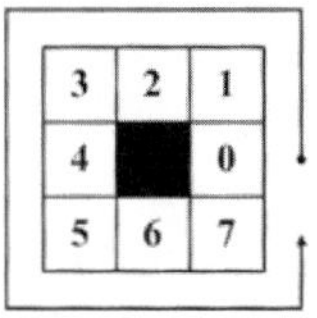

Fig. 2. Calculating sequence

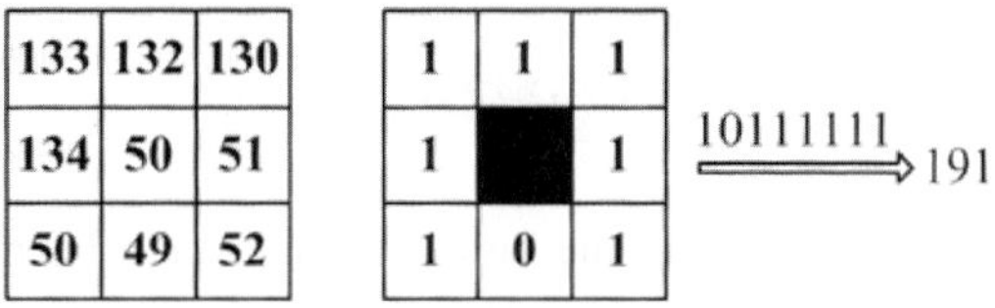

Fig. 3. Example for calculating the original LBP code

In fact, there are 4 pixels with gray-scale values close to the center in the Fig.3, so the 4 pixels should be classified with the center one. In order to make the LBP operator more robust against these negligible change in gray values, we redefine the thresholding function *s(u)* and let *s(u)* =1 as $u \geq T$ and *s(u)* =*0* otherwise. A relatively small value for *T* should be used, for example, $2 \leq T \leq 5$. In our experiments, *T* was given a value of 4. The basic LBP is redefined as follows:

$$LBP_{P,R}(x_c, y_c) = \sum_{i=0}^{P-1} s(g_p - g_c)2^i, s(u) = \begin{cases} 1, u \geq T \\ 0, u < T \end{cases} \tag{2}$$

2.2 LBP Uniform Pattern

The LBP operator produces *P*-bit binary codes. If placing the binary codes end-to-end and counting the times of 0 jumping to 1 and 1 to 0, the jumping times represent the frequency of texture change. The jumping times *Variance* is defined as follows:

$$Variance_{P,R}(x_c, y_c) = \sum_{i=0}^{P-2} (s(g_i - g_c) \oplus s(g_{i+1} - g_c)) + s(g_{P-1} - g_c) \oplus s(g_0 - g_c), s(u) = \begin{cases} 1, u \geq T \\ 0, u < T \end{cases} \tag{3}$$

In Eq. (3), ⊕ is an XOR logical operator in bit arithmetic operations. For example, patterns 0000 0000 and 1111 1111 have *Variance* of 0, 0001 1000 has 2, and 0011 0001 has 4. It can clearly be seen that *Variance* has a value of even numbers between 0 and *P*.

Because a frame in a video has powerful correlations between neighborhood pixels and has little possibility in the sudden change of neighborhood pixels' value, the binary codes from the LBP have few diversifications between the neighborhood bits. We count the standard visual surveillance testing videos from http://cvrr.ucsd.edu/aton/shadow/, and it is obtained that the pixels of $Variance_{8,1} > 2$ have less than 4% portions in all pixels. However, there are 198 binary codes which meet $Variance_{8,1} > 2$ and 58 ones which have $Variance_{8,1} \leq 2$ in the 256 binary codes produced from the LBP when P=8, R=1. The 58 binary codes represent more than 96% pixels.

In order to achieve real-time and high processing speed in the following matching, the 198 codes of $Variance_{8,1} > 2$ are combined to one class, which are called by non-uniform patterns. The 58 codes of $Variance_{8,1} \le 2$ are called by uniform patterns. The uniform patterns, which have the same number of 1, are combined to one class in order to reduce complexity again. Let us denote a LBP uniform pattern by $LBP_{P,R}^{riu}$ as follows:

$$LBP_{P,R}^{riu}(x_c, y_c) = \begin{cases} \sum_{i=0}^{P-1} s(g_i - g_c), & if \quad Variance_{P,R}(x_c, y_c) \le 2 \\ P+1 \quad , & otherwise \end{cases} . \tag{4}$$

It is seen from the above formula that the model of P points has P+1 uniform patterns. According to the number of 1, we assign a unique serial number (0~P) to the patterns, which means that the unique number of a uniform pattern is the quantity of 1 in its binary codes. The non-uniform patterns are assigned to a unique number of P+1. So the 2^P LBP textures are reduced to P+2.

The transformation from $LBP_{P,R}$ to $LBP_{P,R}^{riu}$ could use a calculated code table. The table is calculated only once in the whole procedure so the transformation is very fast.

3 Moving Object Detection Based on Improved LBP Texture and Hue Information

Gaussian mixture model (GMM) proposed by Stauffer and Grimson [1] is one of the most famous background modeling methods. We adopt the idea of GMM to present a background modeling method based on a combination of improved local binary pattern texture and hue information.

3.1 Local LBP Texture and Local Hue Information

Our method divides each video frame into equally sized blocks by using partially overlapping grids, and then the LBP uniform pattern histogram and local hue are counted in each block. We denote the LBP uniform pattern histogram at time instant t by X_t. Because the LBP uniform patterns have only P+2 kinds, the histogram has only P+2 bars. This paper uses the histogram intersection as the proximity measure of $LBP_{P,R}^{riu}$ in the experiments:

$$D_{LBP}(X_1, X_2) = \sum_{i=0}^{P+1} \min(X_1(i), X_2(i)) . \tag{5}$$

where X_1 and X_2 are the histograms and P+2 is the num of histogram bars. The proximity measure neglects the features which only occur in one of histograms, and the measure method has very low complexity.

We denote the local hue vector at time instant t by H_t:

$$H_t = \{h_0, h_1, h_2, \cdots, h_{blockW*blockH-1}\} . \quad (6)$$

where *blockW* and *blockH* are the divided blocks' width and height respectively. The hue values from h_0 to $h_{blockW*blockH-1}$ are obtained according to the sequence from row to column and they are normalized to 0.0~1.0.

The paper uses the following method as the proximity measure of local hue vectors in the experiments:

$$D_{Hue}(H_1, H_2) = 1 - \frac{2*\sum_i \min(|H_1(i) - H_2(i)|, 1 - |H_1(i) - H_2(i)|)}{blockW*blockH} . \quad (7)$$

where H_1 and H_2 are the local hue vectors, and *blockW* and *blockH* are the divided blocks' width and height respectively. The more similar the vectors of H_1 and H_2 are, the bigger the value of $D_{Hue}(H_1, H_2)$ is.

3.2 Background Modeling, Updating and Foreground Extracting

At time instant *t*, a block's background model is represented by *K* $LBP_{P,R}^{riu}$ histograms $\{X_t^1, \cdots, X_t^K\}$ and local hue vectors $\{H_t^1, \cdots, H_t^K\}$ with weights $\{w_t^1, \cdots, w_t^K\}$ respectively, where *K* is a constant integer (usually $3 \le K \le 5$) and $0 \le w_t^i \le 1$. The model is denoted by $g_t^i = \{X_t^i, H_t^i, w_t^i\}$, where $i = 1, \cdots, K$, and each g_t^i is a model. In the following, we explain the background model's updating procedure for one block, and the procedure is same for each block.

At time instant *t*, when capturing a new frame, we calculate the block's X_t and H_t, and then compare the two vectors with *K* background models of corresponding block. Their proximity measure is defined as following:

$$D(g_t, g_t^i) = \lambda D_{LBP}(X_t, X_t^i) + (1-\lambda) D_{Hue}(H_t, H_t^i) . \quad (8)$$

where $i = 1, \cdots, K$. λ, which is a constant of [0, 1], represents the mixture parameter of judging results.

It is supposed that the variety of $D(g_t, g_t^i)$ appears a Gaussian distribution, and that is as following:

$$D(g_t, g_t^i) \sim N(0, \sigma_{t,i}^2) . \quad (9)$$

Therefore, each block's background is described by *K* Gaussian distributions:

$$p(g_t) = \sum_{i=1}^{K} w_t^i * \eta(D(g_t, g_t^i), 0, \sigma_{t,i}^2) . \quad (10)$$

where η is a Gaussian probability distribution function with an average value of 0:

$$\eta(g_t, 0, \sigma_{t,i}^2) = \frac{1}{\sqrt{2\pi}\sigma_{t,i}} e^{-\frac{g_t^2}{2\sigma_{t,i}^2}} . \quad (11)$$

A block is judged to background if $D(g_t, g_t^i) < 2.5\sigma_{t,i}$ is achieved when i exists and $1 \le i \le K$, and is foreground otherwise.

Afterward, the background model is updated. If we suppose that *matchIndex* is the matched model' index in a background block, the background is updated as following:

For (i=1;i++;i≤*K*)
  If (i == *matchIndex*)
    $w_t^i = (1-\alpha)w_{t-1}^i + \alpha$
    $\rho = \alpha\eta(D(g_t, g_t^i), 0, \sigma_{t,i}^2)$
    $X_t^i = (1-\rho)X_{t-1}^i + \rho X_t$
    $H_t^i = (1-\rho)H_{t-1}^i + \rho H_t$
    $\sigma_{t,i}^2 = (1-\alpha)\sigma_{t-1,i}^2 + \alpha(D(g_t, g_t^i))^2$
  else
    $w_t^i = (1-\alpha)w_{t-1}^i$
  endIf
endFor

If a block is judged to foreground, we replace the model of minimum weight with the new model from the new frame, reduce other models' weights as following, and make other parameters invariable.

$$w_t^i = (1-\alpha)w_{t-1}^i \ . \tag{12}$$

After updating the background models, because w_t^i is changed, w_t^i should be united in order to let $\sum_{i=1}^{K} w_t^i = 1$.

4 Experiments and Analysis

We used a PC with an Intel i3-350 CPU processor and a 2 GB memory in our experiments, in which Visual C++6.0 and OpenCV1.0 are installed. The performance of our method was evaluated using the standard testing videos come from http://cvrr.ucsd.edu/aton/shadow/. The videos include indoor, outdoor and shadow examples. We contrast our method to GMM [1] and general LBP method [7], and the results are shown in Fig. 4.

Fig . 4 (a) are original test frames come from the standard testing videos. The results in Fig. 4 (b) are gotten by GMM. It is seen that the results include numerous shadows, which obstruct the following process. Fig. 4 (c) is gotten by general LBP method. It is obvious that shadows are eliminated greatly, but the contours of moving objects are not precise, and some noises exist. Fig. 4 (d) is our method's result, in which shadows are eliminated completely and noises are mostly removed.

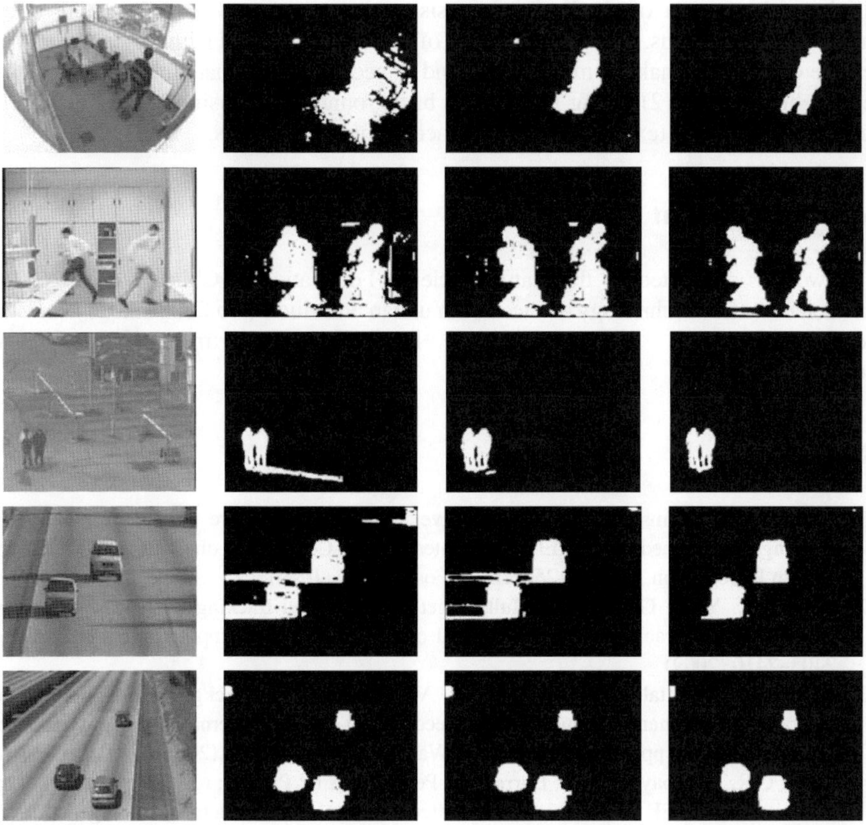

(a) Test frames (b) GMM (c) General LBP (d) Our method

Fig. 4. Comparison results of our method and other methods on standard testing videos

In our experiments, we use R=1 and P=8 to calculate $LBP_{P,R}^{riu}$. Let *blockW*=4 and *blockH*=4 when calculating the local LBP texture histogram and local hue information. Let K=3, when constructing Gaussian background model. Our experiments get good results when the fusion parameter λ is in [0.6,0.8], and $\lambda = 0.65$ in Fig. 4. The update rate parameter α of background model controls background's updating speed, and α is generally in [0.001, 0.05]. On a resolution of 320 × 240 video, our algorithm can achieve a processing speed of 14 fps and can reach real-time basically in the aforementioned software and hardware platforms.

5 Conclusions

This paper presents a background modeling method based on a combination of local texture and hue to detect moving objects. Our method was tested by the standard testing videos, and the experimental results show that the algorithm can effectively eliminate the impact of shadows, can resist some noises, and achieves more effect than other algorithms. The contribution of this paper are: (1) improving the LBP texture operator to make it more robust and to accelerate the matching speed of local texture histograms; (2) trying to describe background model using a combination of improved LBP local texture and hue, and achieving good effects.

Acknowledgment

This work is supported by the Natural Science Foundation of China (No.11026225), the Science and Technology Project of Yunnan Province (No.2009CA021) and the Science Research Foundation of Education Department of Yunnan Province(09Y0044).

References

1. Stauffer, C., Grimson, W.E.L.: Adaptive background mixture models for real-time tracking. In: Proceedings of IEEE Computer Society Conference on Computer Vision and Pattern Recognition, pp. 246–252. IEEE, Fort Collins (1999)
2. Huang, C., Yuan, G., Xu, D.: Multi-target detection and tracking by Gaussian mixture model and blob tracking analysis. Journal of Information & Computational Science 6(6), 2403–2410 (2009)
3. Monnet, A., Mittal, A., Paragios, N., Visvanathan, R.: Background modeling and subtraction of dynamic scenes. In: Proceedings of the 9th International Conference on Computer Vision, pp. 1305–1312. IEEE, Washington D. C., USA (2003)
4. Wren, C.R., Azarbayejani, A., Darrell, T., Pentland, A.P.: Pfinder: real-time tracking of the human body. IEEE Transactions on Pattern Analysis and Machine Intelligence 19(7), 780–785 (1997)
5. Elgammal, A., Duraiswami, R., Harwood, D., Davis, L.S.: Background and foreground modeling using nonparametric kernel density estimation for visual surveillance. Proceedings of IEEE 90(7), 1151–1163 (2002)
6. Mason, M., Duric, Z.: Using histograms to detect and track objects in color video. In: Proceedings of the 30th Applied Imagery Pattern Recognition Workshop, pp. 154–159. IEEE, Washington D. C., USA (2001)
7. Heikkila, M., Pietikainen, M.: A texture-based method for modeling the background and detecting moving objects. IEEE Transactions on Pattern Analysis and Machine Intelligence 28(4), 657–662 (2006)
8. Hu, S.S., Fu, L.C., Hsiao, P.Y.: Region-level motion-based background modeling and subtraction usnig MRFs. IEEE Transactions on Image Processing 16(5), 1446–1456 (2007)
9. Jian, X., Ding, X.-q., Wang, S.-j., Wu, Y.-s.: Background subtraction based on a combination of local texture and color. Acta Automatica Sinica 35(9), 1145–1150 (2009)
10. Ojala, T., Pietikainen, M., Maenpaa, T.: Multiresolution gray-scale and rotation invariant texture classification with local binary patterns. IEEE Transactions on Pattern Analysis and Machine Intelligence 24(7), 971–987 (2002)
11. Ahonen, T., Matas, J., He, C., Pietikäinen, M.: Rotation Invariant Image Description with Local Binary Pattern Histogram Fourier Features. In: Salberg, A.-B., Hardeberg, J.Y., Jenssen, R. (eds.) SCIA 2009. LNCS, vol. 5575, pp. 61–70. Springer, Heidelberg (2009)

Stationary Properties of the Stochastic System Driven by the Cross-Correlation Between a White Noise and a Colored Noise

Yun Gao[1], Shi-Bo Chen[2], and Hai Yang[3,*]

[1] Editorial Department, Journal of Yunnan Normal University, Kunming 650092, China
[2] Physics Department, Kunming College, Kunming 650214, China
[3] Physics Department, Yunnan Normal University, Kunming 650092, China
kmyangh@263.net

Abstract. The stationary properties of the stochastic system driven by cross-correlated additive white noise and multiplicative colored noise were investigated. The stationary probability distribution (SPD) of the stochastic system was derived. The SPD of the different correlation time between white and colored noises (τ_1) with the correlation time between the colored noises (τ_2) and the coefficient of correlation (r) fixed, shows the onset of trimodal-unimodal transition. Those two correlation time have different effects on the stationary properties of the stochastic system.

Keywords: Stochastic analysis methods; Fluctuation phenomena; Stochastic processes.

1 Introduction

The stochastic systems driven by cross-correlated white noises have attracted extensive studies, since Fulinski and Telejko pointed out which noises in some stochastic processes may have a common origin and may be cross-correlated [1]. Noise-induced transition has got wide application in the field of physics, chemistry, and biology [2-10]. The fluctuating environment is often regarded as a source of external noise with statistical characteristics. The reason to do so is that white noise provides an idealization of real noise with a small correlation time for a large variety of applications. Noise is either additive or multiplicative [7, 11-13], and either white or colored. The main theme of these investigations is to study the steady state properties and the state transitions induced by two multiplicative white noises of the systems in most of the previous works, however, realistic models of physical systems require considering a colored noise source.

In this paper, the stochastic system driven by a white noise and a colored noise are considered. Although a stochastic system with two Ornstein-Uhlenbeck (O-U) noises [13] and a bistable system with a white noise and a colored noise [14] were discussed,

---

* Corresponding author.

M. Zhu (Ed.): Electrical Engineering and Control, LNEE 98, pp. 269–279.
springerlink.com © Springer-Verlag Berlin Heidelberg 2011

the authors neglected some special situations that the stochastic systems with a linear restoring force and specific functions of noises. In Ref [15], the stochastic system driven by two white noises with a linear restoring force and specific functions of noises was investigated, the authors only considered white noises and some special parameter values. However, the system driven by a white noise and a colored noise has not been studied.

The paper is organized as follows. In Section 2, the stationary solutions of the Fokker-Planck equation are given which correspond to the Langevin equation with dimensionless state. In Section 3, both the nonequilibrium phase transition and the stationary state properties of the nonlinear stochastic systems driven by the cross-correlation between an additive white noise and a multiplicative colored noise are obtained. The conclusions are summarized in Section 4.

2 Theories

The nonlinear stochastic system driven by two multiplicative noises is described as:

$$\frac{dx(t)}{dt} = f(x(t)) + G_1(x(t))\gamma_1(t) + G_2(x(t))\gamma_2(t) \tag{1}$$

$f(x(t))$ 、 $G_1(x(t))$ and $G_2(x(t))$ are deterministic functions that can depend explicitly on t as well. $\gamma_1(t)$ and $\gamma_2(t)$ are intensities of colored noise and white noise. To obtain the broadest variety of application, we assume that each Gaussian noise $\gamma_i(t)$ is characterized by its own parameter λ_i ($0 \le \lambda_i \le 1$). It is well known that the equation (1) is meaningless until an appropriate interpretation for the integral of the noise term has been adopted. λ_1 and λ_2 are specified in order to determines the points of time at which $G_1(x(t))$ and $G_2(x(t))$ are evaluated in the corresponding integral sum. $\gamma_1(t)$ and $\gamma_2(t)$ are Gaussian noise with the following statistical properties:

$$<\gamma_1(t)\gamma_1(t')> = \frac{\Delta_{11}}{\tau_1}\exp[-\frac{|t-t'|}{\tau_1}]$$

$$\to 2\Delta_1\delta(t-t') \text{ as } \tau_1 \to 0, \tag{2}$$

$$<\gamma_2(t)\gamma_2(t')> = 2\Delta_{22}\delta(t-t') = 2\Delta_2\delta(t-t'), \tag{3}$$

And we assume

$$<\gamma_1(t)\gamma_2(t')> = <\gamma_2(t)\gamma_1(t')> = \frac{r\sqrt{\Delta_1\Delta_2}}{\tau_2}\exp[-\frac{|t-t'|}{\tau_2}]$$

$$\rightarrow 2r\sqrt{\Delta_1\Delta_2}\delta(t-t') \text{ as } \tau_2 \rightarrow 0. \tag{4}$$

$\Delta_1(\geq 0)$ and $\Delta_2(\geq 0)$ are the intensities of the noises $\gamma_1(t)$ and $\gamma_2(t)$, respectively, $\Delta_{12} = \Delta_{21} \equiv r\sqrt{\Delta_1\Delta_2}$, r is the coefficient of correlation between $\gamma_1(t)$ and $\gamma_2(t)$, and $\delta(t)$ is the Dirac δ function. The correlation time τ_2 between $\gamma_1(t)$ and $\gamma_2(t)$ is different from the correlation time τ_1 between the colored noises.

The Fokker-Planck equation corresponding to Eq. (1) for the system driven by two Gaussian noises was given by Densiov, Vitrenko and Horsthemke [16]. The Fokker-Planck is:

$$\frac{\partial}{\partial t}p(x,t) = -\frac{\partial}{\partial x}[f(x)+h(x)] + \frac{\partial^2}{\partial x^2}d(x)p(x,t) \tag{5}$$

Where

$$h(x) = 2\sum_{i=1}^{2}\sum_{j=1}^{2}\lambda_i\Delta_{ij}G_i'(x)G_j(x), \tag{6}$$

$$d(x) = \sum_{i=1}^{2}\sum_{j=1}^{2}\Delta_{ij}G_i(x)G_j(x). \tag{7}$$

In addition, $h(x) = 2\lambda\Delta G'(x)G(x)$ and $2d(x) = 2\Delta G^2(x)$ are the noise-induced drift and the diffusion coefficient of the Fokker-Planck equation. We can capture the conclusion that $h(x) \propto d'(x)$ if we assume $d(x) > 0$. The stationary solution of the Fokker-Planck equation (7) is obtained as follows [16]:

$$p_{st}(x) = \frac{N}{d(x)}\exp\left[\int^x \frac{f(x')+h(x')}{d(x')}dx'\right] \tag{8}$$

N is normalization constant. The extreme of $p_{st}(x)$ obey a general equation:

$$f(x) + h(x) - d'(x) = 0 \tag{9}$$

The transition induced by the cross-correlation of the noises is seen from equation (9). In other words, equation (9) has the same number of roots as equation $f(x) = 0$ for $r = 0$ and that number changes for $r = r_{cr}$.

Using the stationary probability distribution function we can obtain the moments of thc state variable x:

$$< x^n >= \int_{-\infty}^{+\infty} x^n P_{st}(x)dx. \quad (10)$$

The mean and variance of the state variable are given by the numerical integrations of equation (10).

The mean of the state variable is:

$$< x >= \int_{-\infty}^{+\infty} x\, p_{st}(x)dx. \quad (11)$$

The variance of the state variable is:

$$\delta x^2 =< x^2 > - < x >^2. \quad (12)$$

In this paper, simply, we take $G_2(x) = 1$, then the stationary solutions of equation (8) are obtained and we can study the possibility of nonequilibrium transitions induced by the cross-correlation between a multiplicative colored noise and an additive white noise.

3 The Possibility of Nonequilibrium Transitions

We consider the system with a linear restoring force $f(x) = -kx \ (k > 0)$, and

$$g_1(x) = l\frac{x^4}{1+x^4} \quad (l > 0). \quad (13)$$

The stationary probability distribution function is given by

$$p_{st}(x) = N\frac{\exp\left[-m\int_0^x \frac{dz}{\left(\frac{z^2}{1+z^2}\right)^2 + 2rv\frac{z^2}{1+z^2} + v^2}\right]}{\left[\left(\frac{x^4}{1+x^4}\right)^2 + 2rv\frac{x^2}{1+x^4} + v^2\right]^{1-\lambda 1}}, \quad (14)$$

$$\text{Where} \quad m = \frac{2(1-\lambda_1)}{\eta},\ v = \frac{1}{l(1+k\tau_2)}\sqrt{\frac{\Delta_2}{\Delta_1}},\ \eta = \frac{4(1-\lambda_1)\Delta_1 l^2}{k(1+k\tau_1)}. \quad (15)$$

To simplify numeral computation, we take $k = 1$ and $l = 1$, then we obtain the specific expression of the probability distribution functions:

(1) For $\tau_2 < \sqrt{\frac{\Delta_2}{\Delta_1}} - 1$,

$$p_{st}(x)=\frac{Z_1}{[d(x)]^{1-\lambda_1}}\exp\{-m[AU_1(x)+\frac{2B}{\sqrt{v-1}}U_2(x)+\frac{C}{v}U_3(x)+\frac{2C\sqrt{v}}{(v-1)^{\frac{3}{2}}}U_4(x)]\}$$
$$(r=-1). \tag{16}$$

Where

$$d(x)=(\frac{x^2}{1+x^2}-v)^2, \tag{17}$$

$$U_1(x)=x^2\{1+tg[\frac{\pi}{4}+\frac{1}{2}arctg\, x^2]tg[\frac{\pi}{4}-\frac{1}{2}arctg\, x^2]\}, \tag{18}$$

$$U_2(x)=arctg\frac{1+\sqrt{v}tg[\frac{1}{2}arctg\, x^2]}{\sqrt{v-1}}+arctg\frac{-1+\sqrt{v}tg[\frac{1}{2}arctg\, x^2]}{\sqrt{v-1}}, \tag{19}$$

$$U_3(x)=-\frac{\sin[2arctg\, x^2]}{(v-1)\{1-\frac{\sin^2[arctg\, x^2]}{v}\}}, \tag{20}$$

$$U_4(x)=arctg\frac{1+\sqrt{v}tg[\frac{1}{2}arct(x^2]}{\sqrt{v-1}}+arctg\frac{-1+\sqrt{v}tg[\frac{1}{2}arctg\, x^2]}{\sqrt{v-1}}, \tag{21}$$

$$A=\frac{1}{2(v-1)^2},\ B=\frac{1-3v}{4v^{3/2}(v-1)^2},\ C=\frac{1-v}{4v(1-v)^2}. \tag{22}$$

(2) For $\tau_2=\sqrt{\frac{\Delta_2}{\Delta_1}}-1$,

$$p_{st}(x)=\frac{Z_2}{[d(x)]^{1-\lambda_1}}\exp[-m(x^2+\frac{2}{3}x^6+\frac{1}{5}x^{10})]\quad (r=-1). \tag{23}$$

(3) For $\tau_2>\sqrt{\frac{\Delta_2}{\Delta_1}}-1$,

$$p_{st}(x)=\frac{Z_3}{[d(x)]^{1-\lambda_1}}\exp\{-m[AU_1(x)+[\frac{B}{\sqrt{1-v}}+\frac{C}{(1-v)^{3/2}}]U_2(x)+$$
$$[\frac{B}{\sqrt{1-v}}-\frac{C}{(1-v)^{3/2}}]U_3(x)+C\sqrt{v}U_4(x)]\}.$$
$$(r=-1) \tag{24}$$

where

$$U_1(x) = tg(\frac{\pi}{4} + \frac{1}{2} arctg\, x^2) - tg(\frac{\pi}{4} - \frac{1}{2} arctg\, x^2), \tag{25}$$

$$U_2(x) = \ln \left| \frac{\sqrt{v tg(\frac{1}{2} arctg\, x^2) + 1} - \sqrt{1-v}}{\sqrt{v tg(\frac{1}{2} arctg\, x^2) + 1} + \sqrt{1-v}} \right|, \tag{26}$$

$$U_3(x) = \ln \left| \frac{\sqrt{v tg(\frac{1}{2} arctg\, x^2) - 1} - \sqrt{1-v}}{\sqrt{v tg(\frac{1}{2} arctg\, x^2) - 1} + \sqrt{1-v}} \right|. \tag{27}$$

When we change the value of the correlation coefficient for $r = 1.0$, the stationary solution of equation(8) is obtained:

$$p_{st}(x) = \frac{Z}{[d(x)']^{1-\lambda_1}} \exp\{-m[A' U_1(x)' + B' U_2(x)' + C' U_3(x)']\} \ (r = 1.0). \tag{28}$$

Where

$$d(x)' = (\frac{x^2}{1+x^2} + v)^2, \tag{29}$$

$$U_1(x)' = tg(\frac{\pi}{4} + \frac{1}{2} arctg\, x^2) - tg(\frac{\pi}{4} - \frac{1}{2} arctg\, x^2), \tag{30}$$

$$U_2(x)' = \frac{\sin(2 arctg\, x^2)}{(v + \sin^2(arctg\, x^2))}, \tag{31}$$

$$U_3(x)' = arctg(\sqrt{1 + \frac{1}{v}}\, x^2), \tag{32}$$

$$A' = \frac{1}{2(1+v)^2},\ B' = \frac{1}{4v(1+v)^2},\ C' = \frac{1}{\sqrt{v(1+v)^5}} + \frac{1+2v}{2\sqrt{v^3(1+v)^5}} \quad (r = 1.0). \tag{33}$$

The phase diagrams of the system for different values of correlation time of colored noises τ_1 were plotted in Fig. 1. It is shown that there is a phase transition of the state function (the probability distribution density function) with the change of the values of

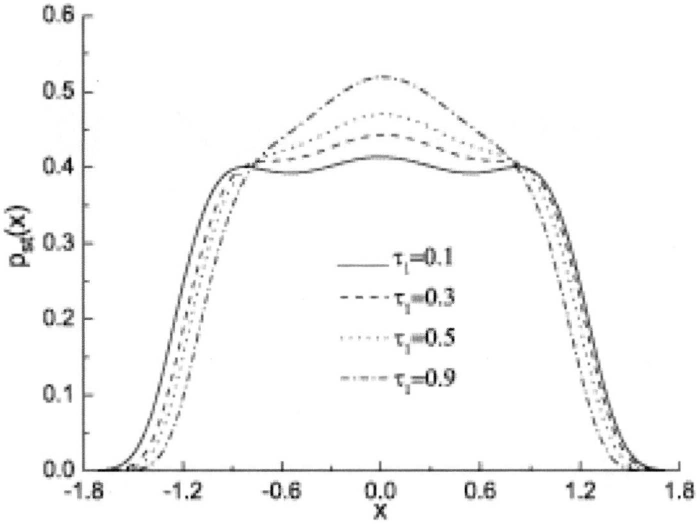

Fig. 1. Plot of $p_{st}(x)$ (probability density) vs x (state variable) for different values of τ_1, $r=-1$, $\lambda_1=0.5$, $\Delta_1=1.0$, $\Delta_2=2.0$, $\tau_2=0.2$, respectively.

τ_1. The curves are trimodal with small values of τ_1, and are unimodal with large values of τ_1. The probability distribution density increases with increasing τ_1 in the little absolute values of x, but decreases with increasing τ_1 in the large absolute values of x. The parameter τ_1 plays an important role on the state variable probability distribution functions. The parameter λ_1 has a converse effect on the state variable probability distribution function.

The curves of probability distribution density functions are given from equation (23) in Fig. 2. The probability distribution densities are changed in the different values of λ_1. The values $\lambda_1 = 0 \text{and} \lambda_1 = 0.5$ are corresponding to the Ito [16] and Stratonovich [17] interpretation of equation (1) respectively. Equation (1), with different λ_1, represents a valuable tool for modeling a great variety of phenomena and processes, including stochastic resonance [18], noise-induced transitions [19], resonant activation [20], and directed transport [21].

The curves of probability distribution function with different τ_2 are shown in Fig. 3. The bigger the correlation time between colored and white noises (τ_2), the higher the peaks near $x=0$. The overlaps of the probability distribution occur at $x=\pm1.4$. During the region $8\rangle|x|\rangle1.4$, The bigger the correlation time between colored and white noises (τ_2), the smaller the probability density.

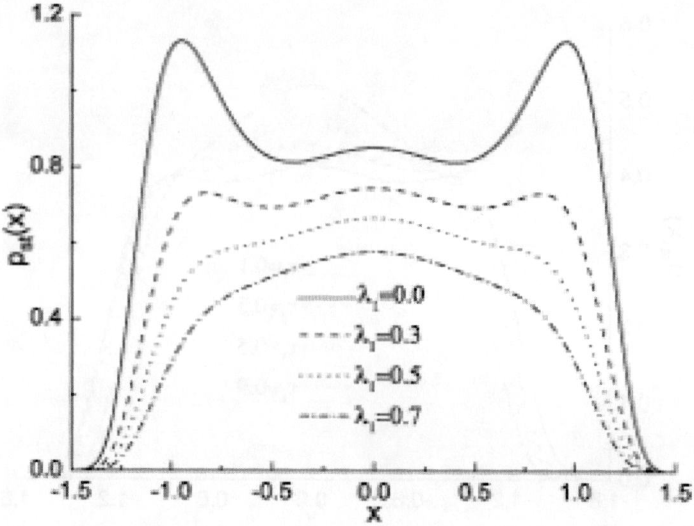

Fig. 2. Plot of $p_{st}(x)$ (probability density) vs x (state variable) for different τ_2, $r=-1$, $\Delta_1=1.0$, $\Delta_2=0.64$, $\lambda_1=0.5$, $\tau_1=0.3$, respectively.

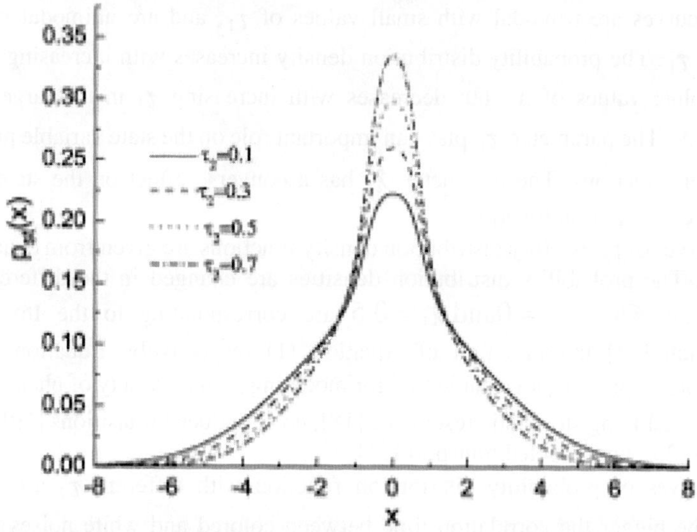

Fig. 3. Plot of $p_{st}(x)$ (probability density) vs x (state variable) for different τ_2, $r=1$, $\Delta_1=1.0$, $\Delta_2=4.0$, $\lambda_1=0.5$, $\tau_1=0.2$, respectively.

The effects of the correlation time τ_2 and τ_1 on the variance of the system are explicitly shown in Fig.4 and Fig. 5. The curves are almost parallel to the others in

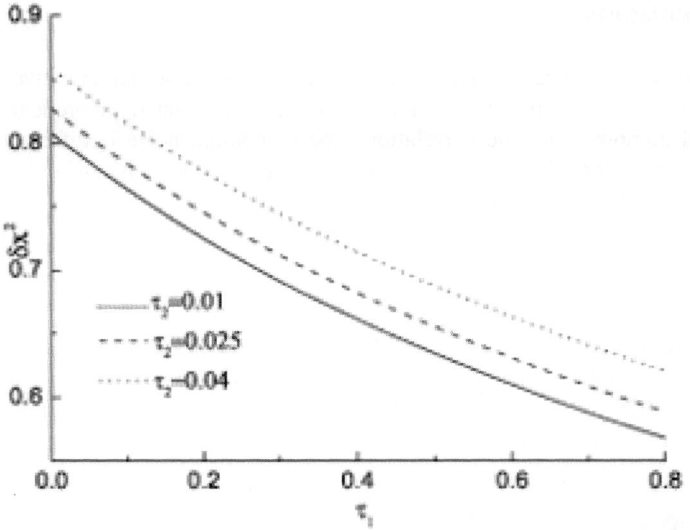

Fig. 4. The variance of the state variable δx^2 as a function of τ_1 for different τ_2, $r = -1$, $\Delta_1 = 1.0$, $\Delta_2 = 2.0$, $\lambda_1 = 0.5$, respectively.

Fig. 4. With the τ_2 increasing, variances increase in Fig 4, However the variances decrease with increasing τ_1 in Fig 5. When τ_2 larger than 1.25, the variances overlap.

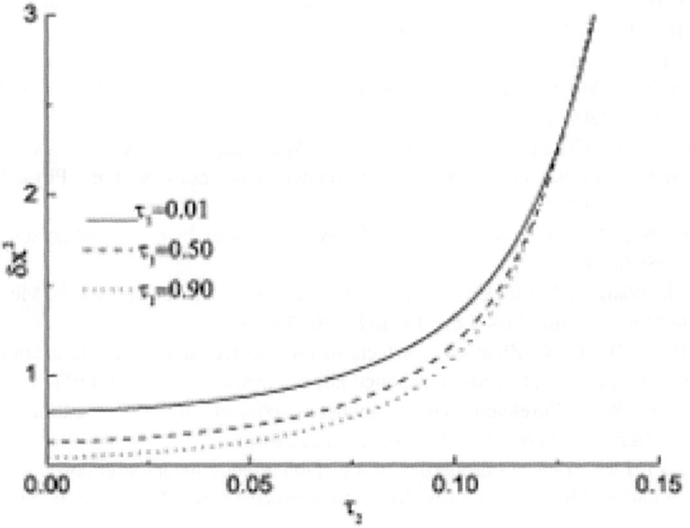

Fig. 5. The variance of the state variable δx^2 as a function of τ_2 for different τ_1, $r = -1$, $\Delta_1 = 1.0$, $\Delta_2 = 2.0$, $\lambda_1 = 0.5$, respectively.

4 Conclusions

To sum up, the nonlinear stochastic process with colored correlation between additive white noise and multiplicative colored noise is investigated theoretically through numerical computation. The correlation time of colored noise is different from the correlation between white noise and colored noise, and they have different effects on the stationary properties of the stochastic system.

Acknowledgement

This research was supported by the National Natural Science Foundation of China (Grant No. 10864009), the Natural Science Foundation of Yunnan Province, China (Grant No. 2008CD109) and by the State Key Program of National Natural Science of China(Grant No.50734007).

References

1. Fulinski, A., Telejko, T.: On the effect of interference of additive and multiplicative noises. Phys. Lett. A. 152, 11–14 (1991)
2. Xie, W.X., Xu, W., Cai, L., Jin, Y.F.: Upper bound for the time derivative of entropy for a dynamical system driven by coloured cross-correlated white noises. Chin. Phys. B 14, 1766–1777 (2005)
3. Bidhan, B.C., Chin-Kun, H.: Escape through an unstable limit cycle: Resonant activation. Phys. Rev. E. 73, 61107 (2006)
4. De Queiroz, S.L.A.: Wavelet transforms in a critical interface model for Barkhausen noise. Phys. Rev. E. 77, 21131–21138 (2008)
5. Galán, R.F.: Analytical calculation of the frequency shift in phase oscillators driven by colored noise: Implications for electrical engineering and neuroscience. Phys. Rev. E. 80, 36113–36117 (2009)
6. Denisov, S.I., Vitrenko, A.N., Horsthemke, W., Hänggi, P.: Anomalous diffusion for overdamped particles driven by cross-correlated white noise sources. Phys. Rev. E. 73, 36120–36127 (2006)
7. Cao, L., Wu, D.J.: Stochastic dynamics for systems driven by correlated noises. Phys. Lett. A. 185, 59–63 (1994)
8. Ai, B.Q., Wang, X.J., Liu, L.G., Nakano, M., Matsuura, H.: A Microscopic Mechanism for Muscle Motion. Chin. Phys. Lett. 19, 137–140 (2002)
9. Mei, D.C., Xie, C.W., Zhang, L.: Effects of cross correlation on the relaxation time of a bistable system driven by cross-correlated noise. Phys. Rev. E. 68, 051102–051110 (2003)
10. Banik S K.: Correlated noise induced control of prey extinction (2001), `http://arxiv.org/abs/physics/0110088`
11. Jia, Y., Li, J.R.: Transient properties of a bistable kinetic model with correlations between additive and multiplicative noises: Mean first-passage time. Phys. Rev. E. 53, 5764–5769 (1996)
12. Jia, Y., Li, J.R.: Reentrance Phenomena in a Bistable Kinetic Model Driven by Correlated Noise. Phys. Rev. Lett. 78, 994–998 (1997)
13. Wei, X.Q., Cao, L., Wu, D.J.: Stochastic dynamics for systems driven by correlated colored noise. Phys. Lett., A. 207, 338–342 (1995)

14. Jia, Y., Li, J.R.: Stochastic system with colored correlation between white noise and colored noise. Phys. A. 252, 417–423 (1998)
15. Denisov, S.I., Vitrenko, A.N., Horsthemke, W.: Nonequilibrium transitions induced by the cross-correlation of white noises. Phys. Rev. E. 68, 46132–46140 (2003)
16. Ito, K.: Stochastic differential equations in a differentiable manifold. Nagoya Math. J. 1, 35–47 (1950)
17. Stratonovich, R.L.: A new representation for stochastic integrals and equations. SIAM J. Control 4, 362–371 (1966)
18. Benzi, R., Sutera, A., Vulpiani, A.: The mechanism of stochastic resonance. J. Phys. A 14, L453–L458 (1981)
19. Horsthemke, W., Lefever, R.: Noise-Induced Transitions. Springer, Berlin (1984)
20. Doering, C.R., Gadousa, J.C.: Resonant activation over a fluctuating barrier. Phys. Rev. Lett. 69, 2318–2321 (1992)
21. Magnasco, M.O.: Forced thermal ratchets. Phys. Rev. Lett. 71, 1477–1481 (1993)

Improved Power Cell Designed for Medium Voltage Cascade Converter

Liang Zhang*, Guodong Chen, and Xu Cai

Wind Power Research Center & State Key Laboratory of Ocean Engineering, Shanghai Jiao Tong University, Minhang District, Shanghai 200240, China

Abstract. This paper presents a novel power cell with an active rectifier for the medium voltage cascade converter. The technology innovations make a low line current distortion and a small DC voltage ripple for the converter. Meanwhile, high power factor and no extra expensive multi pulse transformer are other advantages for the improvement. The complete control strategy is also presented for it. A 150 kW prototype with improved power cell base on IGBT is designed, and then experiments are carried out to verify the innovations. Both the steady and transient results show a good performance.

Keywords: Cascade converter, Power Cell, AC drive, Active rectifier.

1 Introduction

Medium voltage multilevel converters [1]-[2] have been widely employed in AC drive. The available medium voltage multi level converter is broadly classified into two topologies [3]. One is cascade H-bridge multi level converter, and the other is diode clamped multi level converter. Of them, the cascade multi level converter is more popular thanks to a separate DC voltage in each power cell and modularization by low voltage semiconductor. However, an uncontrolled input rectifier is used as the front end part in its power cells, which makes the medium voltage cascade converters have some drawbacks as follows:

i) An expensive multi pulse transformer is necessary to solve the input current harmonic distortion.
ii) A large size of electrolytic filter capacitor is required at DC voltage side to reduce the DC voltage ripple against the heavy load.
iii) A special chopper circuit is equipped at the DC link to consume the braking energy of a regenerative operation.

To avoid these drawbacks, an improved power cell with active input rectifiers [4] has been developed for the medium voltage multilevel converter. So, in this paper, a grid voltage vector oriented control scheme is presented for the improved power cell. Meanwhile, a voltage feed-ward method is implemented to decouple the d and q components. And some experiments are carried out to validate the change.

* This work is supported in part by shanghai technique innovative plan 08DZ1200504, in part by the fund 09DZ1201303.

M. Zhu (Ed.): Electrical Engineering and Control, LNEE 98, pp. 281–288.
springerlink.com © Springer-Verlag Berlin Heidelberg 2011

This paper is organized as follows. The improved configuration for the medium voltage cascade converter with the active front end rectifier is described in Section 2. The control scheme for the improved power cell is introduced in Section 3. The digital control unit for it is designed in Section 4. Experiments and results are shown in Section 5. Some conclusions are given in Section 6.

2 Improved Configuration with Active Front End

Fig. 1 gives a diagram of three phase medium voltage multi level converter for AC drive based on cascade H-bridge power cells.

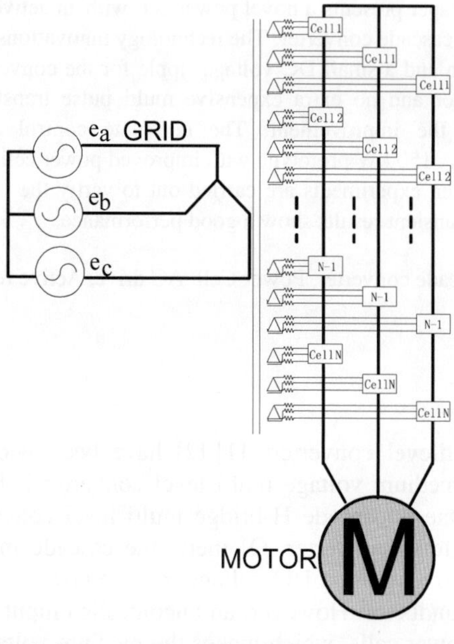

Fig. 1. System configuration for cascade converter.

As seen, power cells in one phase are connected in series to constitute a bridge leg. In each power cell, there are a rectifier and an inverter. In traditional topology, the rectifier is an uncontrolled bridge rectifier. The improved topology replaces it with an active rectifier. Fig.2 shows a power cell contrast between passive front end and active front end.

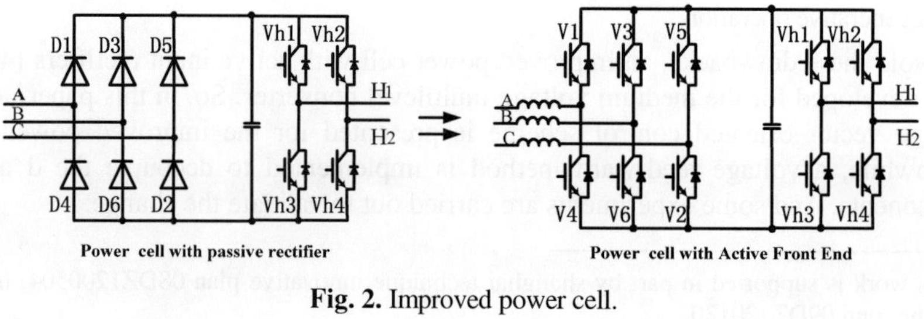

Fig. 2. Improved power cell.

3 Control Scheme Design

In order to obtain the good anticipations, a standalone control scheme should be designed for the replaced circuit. Define the grid voltage of any power cell as the following.

$$\begin{cases} e_{ak} = U_m \sin(\theta) \\ e_{bk} = U_m \sin(\theta - \frac{2\pi}{3}) \\ e_{ck} = U_m \sin(\theta + \frac{2\pi}{3}) \end{cases} \tag{1}$$

Where, e_{ak} , e_{bk} and e_{ck} are the phase voltage of cell k. U_m is the peak value of the phase voltage and θ is the angle of the grid.

Describe the phase current of any power cell as:

$$\begin{cases} i_{ak} = I_m \sin(\theta) \\ i_{bk} = I_m \sin(\theta - \frac{2\pi}{3}) \\ i_{ck} = I_m \sin(\theta + \frac{2\pi}{3}) \end{cases} \tag{2}$$

Where, i_{ak} , i_{bk} and i_{ck} are the phase current, I_m is the peak value of them.

Aligning the d axis of the park reference with the grid voltage vector, we have the d axis symbolized for active power, and the q axis for reactive power. Then, in d-q frame, the input voltage and current have a relation [5]:

$$\begin{bmatrix} v_{dk} \\ v_{qk} \end{bmatrix} = -\begin{bmatrix} Lp+R & -\omega L \\ \omega L & Lp+R \end{bmatrix} \begin{bmatrix} i_{dk} \\ i_{qk} \end{bmatrix} + \begin{bmatrix} e_{dk} \\ e_{qk} \end{bmatrix} \tag{3}$$

Where, v_{dk} , v_{qk} are the components of the AC side voltage of active rectifier, e_{dk} , e_{qk} are the components of the grid, i_{dk} i_{qk} are the input current components, and R is the grid resistance, L is the inductance.

Considering the basic function for input rectifiers is to support a steady isolated DC voltage for the inverter side and regulate the power flow for the converter, the control strategy is a classic two loops. The outer loop is a voltage control for DC link. The inner loop is a current control to regulate the active power and reactive power for the system. And a feed forward method is applied for the inner loop to eliminate the coupling between d-component and q-component as shown in equation (3). The inverter operates as a variable frequency voltage source. It outputs different sinusoidal voltage according to the different motor speed demands. Fig.3 shows the structure of the control scheme for the improved power cell. All the power cell have the same topology and functions, the strategy is similar for them.

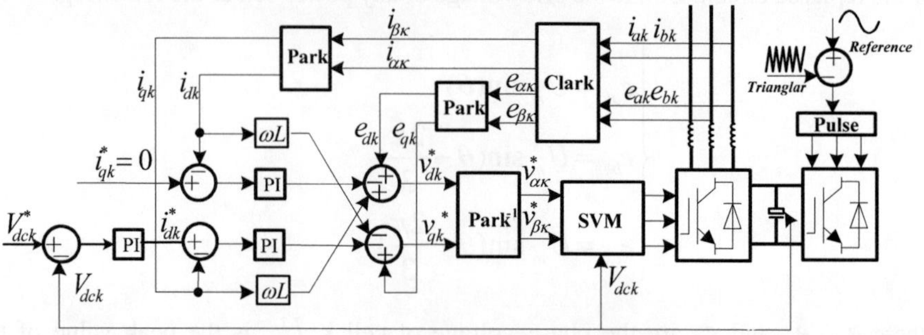

Fig. 3. Control structure for improved cell.

4 Totype and Digital Control Unit Design

Which parameters are proper for the power cell? The number of power cells a phase in traditional cascade multi level converters is 12 for a 10 kV AC drive. However, with the improved power cell and setting the cell line voltage 690V, the number could be 8. In this section, a power cell is designed as an example for a 10 kV 3 MW cascade multilevel converter. The operation conditions for each power cell are listed in Table 1.

Table 1. Parameters of each power cell

Rated power	150 Kw
Peak value of phase voltage	563 V
Peak value of the input current	300 A
DC voltage	1100V
Angular frequency of the input voltage	100π rad/s
Power factor	1.0
Maximum DC voltage utilization ratio	1.15

IGBT FF300R17MES is chosen as the power switching device. The power modules are arranged on a heat sink, and connected with the driver. Low inductance bus bar is designed to minimize the stray inductor of the power circuit Passive components and digital control unit are arranged on a printed board to minimize the volume of the converter.

The bottom digital control unit has the following tasks:

i) Generate the PWM driver pulses for the active rectifier and the inverter.
ii) Regulating the DC voltage to a constant value, immunity to grid voltage variability and load current fluctuation.
iii) Control the start stage of the converter when connecting to the grid.
v) Different fault protection and data exchange with the main control unit.

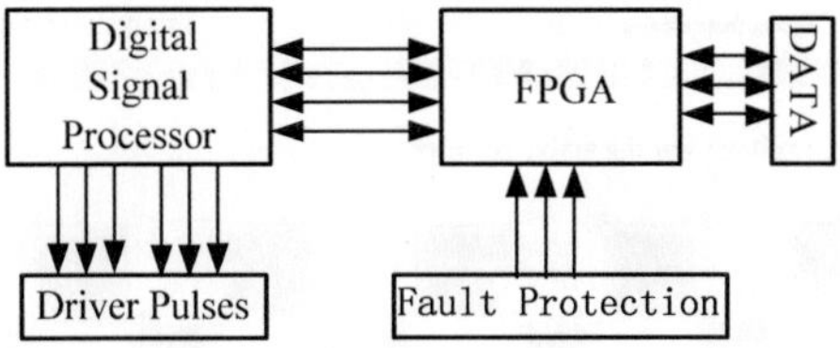

Fig. 4. Block diagram of bottom control unit.

The Fig.4 shows the main function of the bottom digital control unit for the power cell.

5 Experimental Results

In this section, experiments are carried out to test the improved power cell. Firstly, a power cell prototype with active input rectifier is established as shown in Fig.5.

Fig. 5. Power cell prototype with active rectifier.

The improved DC voltage ripple for the load fluctuation and input current harmonics are verified for the front end rectifier alone with a resistance load at the DC side. The switching frequency is 3 kHz and the value of the resistance is 16.5 Ohm.

5.1 Steady Operation of the Active Rectifier

Fig.6 shows a waveform of the input current and the voltage at the rectifier side. With no extra passive filter, the input current is almost sinusoidal and the power factor is nearly unity.

Fig.7 gives the THD measurement result of the input current. The total current harmonic distortion of phase A for example is 3.102%, which extremely meets the international standard IEEE 519-1992.

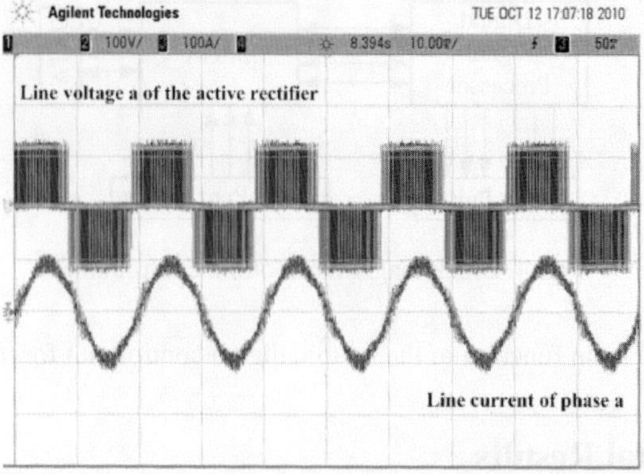

Fig. 6. Steady operation of the front end rectifier.

THD

PLL	I2
Freq	50.004 Hz
U1	670.505 V
I1	65.210 A
P1	37.4270kW
S1	43.7038kVA
Q1	-22.5665kvar
λ1	0.85638
φ1	328.912 °
Uthd1	1.027 %
Ithd1	3.102 %
Pthd1	0.004 %
Uthf1	1.147 %
Ithf1	1.226 %
Utif1	53.705
Itif1	48.998

Or.	I1 [A]	hdf[%]	I2 [A]	hdf[%]
Tot.	65.210		64.687	
dc	0.139	0.213	-0.025	-0.039
1	65.178	99.952	64.657	99.954
2	0.747	1.145	0.185	0.286
3	0.210	0.322	0.520	0.804
4	0.335	0.514	0.305	0.472
5	0.804	1.233	0.706	1.091
6	0.071	0.109	0.249	0.384
7	1.331	2.040	1.362	2.106
8	0.559	0.858	0.575	0.889
9	0.169	0.259	0.132	0.204
10	0.276	0.424	0.206	0.318
11	0.375	0.575	0.284	0.439
12	0.061	0.093	0.060	0.093
13	0.417	0.640	0.506	0.782
14	0.234	0.359	0.252	0.389
15	0.036	0.055	0.076	0.118
16	0.029	0.045	0.044	0.067
17	0.193	0.296	0.247	0.383
18	0.022	0.033	0.052	0.081
19	0.071	0.109	0.102	0.158
20	0.117	0.180	0.153	0.236

Fig. 7. THD measurement result for input current.

5.2 Transient Operation of the Front End Rectifier

The front end rectifier has a quick response to the fluctuation of the load. The DC voltage shows a good robust for the disturbance.

Fig.8 gives the input current response when the rectifier is loaded by a resistance. At the same time, Fig.9 gives the variation of DC voltage.

Combined with the active input rectifier and the inverter, another experiment is also done to test the DC voltage ripple output voltage quality of the power cell, when a high power load is set at the output side. The switching frequency for the active rectifier is 3 kHz, and 3.3kHz for the inverter.

5.3 Output of the Power Cell

Fig.10 gives the output voltage and DC voltage waveforms when the power cell is heavily loaded. The DC ripple is less than 40V, so the output voltage quality is satisfactory.

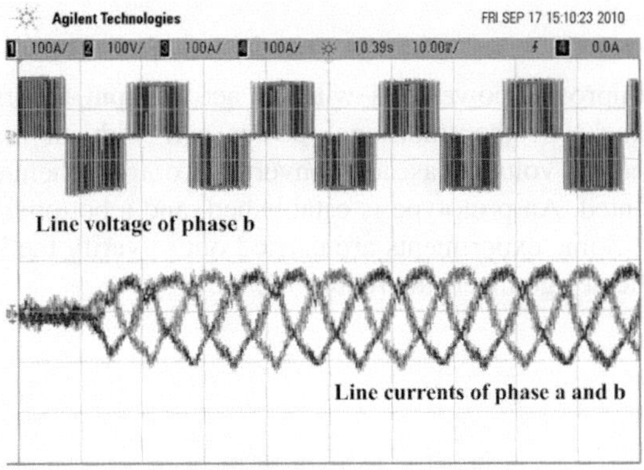

Fig. 8. Transient result.

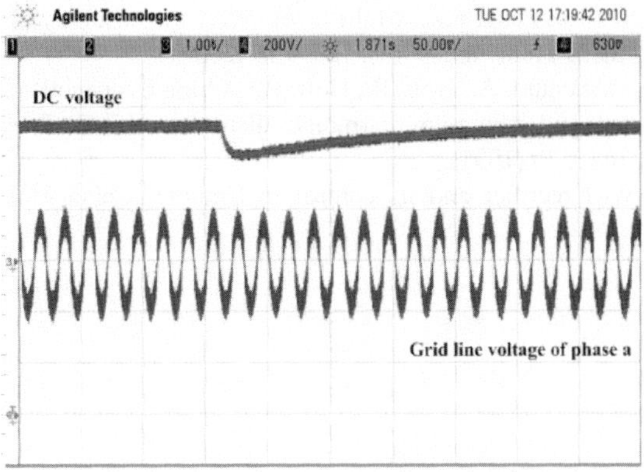

Fig. 9. DC voltage fluctuation

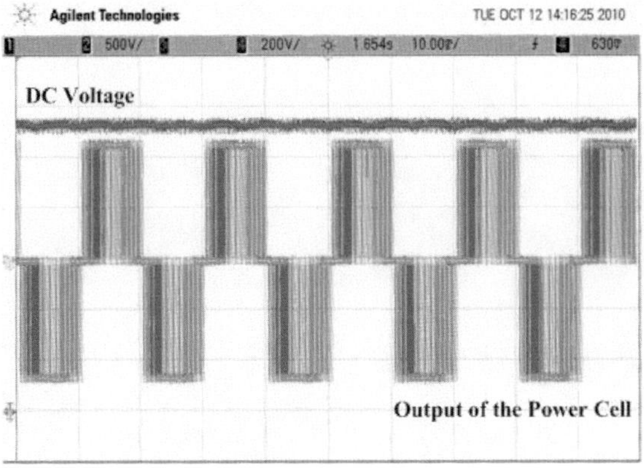

Fig. 10. Output of the power cell.

6 Conclusion

In this work, an improved power cell with an active input rectifier is designed to remove the drawbacks of the traditional power cell with an uncontrolled bridge rectifier for the medium voltage cascade converter. Control scheme for the improved power cell is presented. An prototype is established, and a bottom digital control unit is developed for it. Some experiments are carried out to verify the improvement. The results show that the design meets the requirements.

References

1. Tolbert, L.M., Peng, F.Z., abetler, T.G.: Multilevel converters for large electric drives. IEEE Transactions on industry Application 35(1), 36–44 (1999)
2. Lai, J.S., Peng, F.Z.: Multilevel converter-A new breed of power converters. In: Proc. IEEE-IAS Annual Meeting 1995, pp. 2348–2356 (1995)
3. Rodriguez, J., Lai, J.-S., Peng, F.Z.: Multi level inverters: "a survey of topologies, control, and applications. IEEE Trans. Ind. 49(4), 724–738 (2002)
4. Wernekinck, E., Kawamura, A., Hoft, R., Hoft, R.: A high frequency AC/DC converter with unity power factor and minimum harmonic distortion. IEEE Transactions on Power Electronics 6(3), 364–370 (1991)
5. Chong, W.Z.: PWM rectifier and its control in Chinese. China Machine Press, Beijing (2003)

GIS in the Cloud: Implementing a Web Coverage Service on Amazon Cloud Computing Platform

Yuanzheng Shao[1,2], Liping Di[2], Jianya Gong[1], Yuqi bai[2], and Peisheng Zhao[2]

[1] State Key Laboratory of Information Engineering in Surveying, Mapping and Remote Sensing (LIESMARS), Wuhan University, 129 Luoyu Road, Wuhan, China, 430079
yshao3@gmu.edu, geojy@163.com
[2] Center for Spatial Inforamtion Science and Systems (CSISS), George Masion University, 10519 Braddock Road Suite 2900, Fairfax, VA 22032
ldi@gmu.edu, ybai1@gmu.edu, pzhao@gmu.edu

Abstract. With the continuously increment of the available amount of geospatial data, a huge and scalable data warehouse is required to store those data, and a web-based service is also highly needed to retrieve geospatial information. The emergence of Cloud Computing technology brings a new computing information technology infrastructure to general users, which enable the users to requisition compute capacity, storage, database and other service. The Web Coverage Service supports retrieval of geospatial data as digital geospatial information representing space and time varying phenomena. This paper explores the feasibility of utilizing general-purpose cloud computing platform to fulfill WCS specification through a case study of implementing a WCS for raster image on the Amazon Web Service. Challenges in enabling WCS in the Cloud environment are discussed, which is followed by proposed solutions. The resulting system demonstrates the feasibility and advantage of realizing WCS in Amazon Cloud Computing platform.

Keywords: Cloud Computing, Web Coverage Service, Geospatial data, Amazon Web Service.

1 Introduction

The OGC Web Service (OWS) are crucial to serve geospatial data and discover underlying and useful geospatial information. OGC WCS provide an interface allowing requests for geographical coverage across the web using standard and implementation-neutral web interfaces. A series of operation, such as reformat, re-projection and subset, could be executed on the dataset based on WCS specification. Because the volume, the spatial and temporal coverage, and the update frequency of geospatial data are increasing, the available data are characterized as massive, heterogeneous and distributed features, which bring the pressure on both the storage size and processing performance in GIS system. Achieving high level of reliability and performance in GIS when deploying OWS services is challenging: such services are often processor- intensive and disk-intensive and many of the small data providers

M. Zhu (Ed.): Electrical Engineering and Control, LNEE 98, pp. 289–295.
springerlink.com © Springer-Verlag Berlin Heidelberg 2011

do not have the capability to run highly-available server infrastructures of this nature. A powerful, dependable and flexible information infrastructure is required to store those data and process those data into information.

Cloud computing is rapidly emerging as a technology almost every industry that provides or consumes software, hardware, and infrastructure can leverage. The technology and architecture that cloud service and deployment models offer are a key area of research and development for GIS technology. From a provider perspective, the key aspect of the cloud is the ability to dynamically scale and provide computational resource in a cost efficient way via the internet. From a client perspective, the ability to access the cloud facilities on-demand without managing the underlying infrastructure and dealing with the related investments and maintenance costs is the key [1]. Some Cloud Computing platforms are already available, including Amazon Web Service (AWS), Google App Engine, and Microsoft Azure. Powered by Cloud Computing platform, the users can requisition compute power, storage, database, and other services–gaining access to a suite of elastic IT infrastructure services as demands.

In this study, we explore the feasibility of utilizing general-purpose cloud computing platform to fulfill WCS specification through a case study of implementing a WCS for heterogeneous geospatial data on the Amazon Web Service. Section 2 of this paper introduces the related work and background. Section 3 addresses the challenges in enabling WCS in the Cloud environment and the proposed approaches. The implementation details are describes in Section 4. Section 5 presents the conclusion and discussed planned future work.

2 Background

2.1 Web Coverage Service

The OGC WCS enables interoperable access to geospatial coverages across the web. The term "coverage" refers to any data representation that has multiple values for each attribute type, where each direct position within the geometric representation of the feature has a single value for each attribute type [2]. Examples of coverage include satellite images, digital elevation data, and other phenomena represented by values at each measurement point. Coverages are most commonly associated with the raster data model.

OGC released WCS 1.0 and 1.1 specifications, and WCS 2.0 is available but its Earth observation Application Profiles (EO-AP) are not completely finished so far. Taking WCS version 1.1.0 as example in this paper, it provides three mandatory operations: GetCapabilities, DescribeCoverage and GetCoverage. The GetCapabilities operation allows WCS clients to retrieve service metadata from a WCS server. The response to a GetCapabilities request shall be an XML document containing service metadata about the server, usually including summary information about the data collections from which coverages may be requested. The user could issue a DescribeCoverage request to obtain a full description of one or more coverages available. The GetCoverage operation allows retrieval of subsets of coverages.

2.2 Cloud Computing

The recent emergence of cloud computing brings new patterns about service deployment. Services and applications can be deployed in cloud environments that can be made to scale up or down as required.

Several large Information Technology companies such as Google and Amazon have already created their own cloud products. Since Amazon AWS platform is selected in the implement of this paper, a brief introduction about Amazon AWS will be given in this part. Amazon Web Services is more than a collection of infrastructure services. With pay as you go pricing, the user can save time by incorporating compute, database, storage, messaging, payment, and other services. All AWS services can be used independently or deployed together to create a complete computing platform in the cloud [3]. Amazon Elastic Compute Cloud (EC2) is a web service that provides resizable compute capacity in the cloud. Amazon Simple Storage Service (S3) can be used to store and retrieve large amounts of data, at any time, from anywhere on the web.

2.3 Related Work

Several efforts on using Cloud Computing in geospatial application have been reported. Baranski et al. (2009) creates a cloud enabled spatial buffer analysis service in the Google App Engine and conducts a stress test for scalability evaluation. Wang et al. (2009) proposes the use of OGC as Well Known Binary (WKB) and Well Known Text (WKT) for data exchange in the Cloud Computing environment, and demonstrates how spatial indexes can be created in the Google App Engine. Gong et al. (2010) presents an implementation of geoprocessing service that integrates geoprocessing functions and Microsoft Cloud Computing technologies to provide geoprocessing capabilities in a distributed environment [5]. Blower (2010) implements a Web Map Service for raster imagery within the Google App Engine environment.

3 Challenges and Solutions

To deploy WCS on the Cloud Computing platform and achieve high level of reliability and performance, there two challenges need to be addressed.

The first challenge is to build the proper running environment in the Cloud Computing environment for WCS. WCS instances rely on the specified libraries and proper configuration on standalone environment. When it comes to Cloud Computing platform, those libraries need to be pre-built to support WCS running environment. For example, the WCS running in Windows operating system may needs some dynamic link libraries, but as to the Cloud Computing platform which provides Linux operating system as instance running environment and use shared objects other than dynamic link libraries, the server side has to compile and rebuild the dependent libraries and configures the environment for WCS instance in Cloud environment.

The second challenge is with regard to the data storage in cloud. Particularly, how to store and access the data in the Cloud Computing platform? The storage limit in standalone environment is a bottleneck when the serving data becoming massive. The server side has to upgrade the hardware and make corresponding modification to existing WCS service to deal with new storage space. Cloud computing could address the storage issue since it provides unlimited storage space, however, migrate the data to the cloud and manage the data in the Cloud Computing is a challenge since the server side need to deal with complicated Cloud environment to access and manage the data with Cloud-specified protocols and Application Program Interfaces (APIs).

To overcome the aforementioned challenges, we propose the following solution to enable WCS on Amazon AWS platform. Fig. 1 illustrates the proposed architecture of the integrated system.

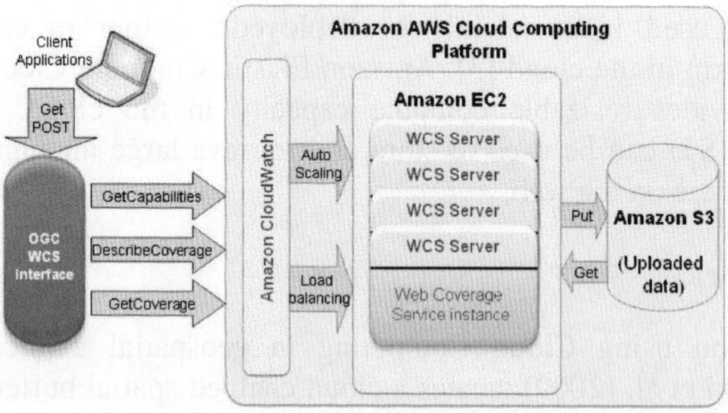

Fig. 1. A high level frameworks for WCS in Amazon AWS Platform.

As to the first challenge, Amazon AWS provides EC2 to give the resizable compute capacity in the cloud, which presents a true virtual computing environment, allowing the user to use web service interfaces to launch instances with a variety of operating systems, load them with the custom application environment, manage network's access permissions. To build a self-contained environment for geoprocessing functions in Cloud, Amazon Machine Image (AMI) is created to start EC2 instance. An Amazon Machine Image (AMI) is a special type of pre-configured operating system and virtual application software which is used to create a virtual machine within the EC2. It serves as the basic unit of deployment for services delivered using EC2.

As to the second challenge, Amazon S3 is selected to manage the application data and store the output data. Amazon S3 provides a highly durable storage infrastructure designed for mission-critical and primary data storage. Objects are redundantly stored on multiple devices across multiple facilities in an Amazon S3 Region. Moreover, Amazon S3 has the higher latency throughput. Amazon AWS provide both console and command line tools to use S3 service.

4 Implementation

Amazon AWS is selected to implement WCS. The AWS services could be managed through both Amazon Management Console and command line tools. The implementation includes the following steps.

4.1 Uploading Data to Amazon S3

Amazon S3 stores data as objects within buckets. An object is comprised of a file and optionally any metadata that describes that file. In order to store an object in Amazon S3, the user could upload the file to a bucket, and set the permission on the object as well as any metadata. The user could manage the upload process through Amazon Management Console, which provide a point-and-click web-based interface for accessing and managing all of the user's Amazon S3 resources, as Fig.2 shows.

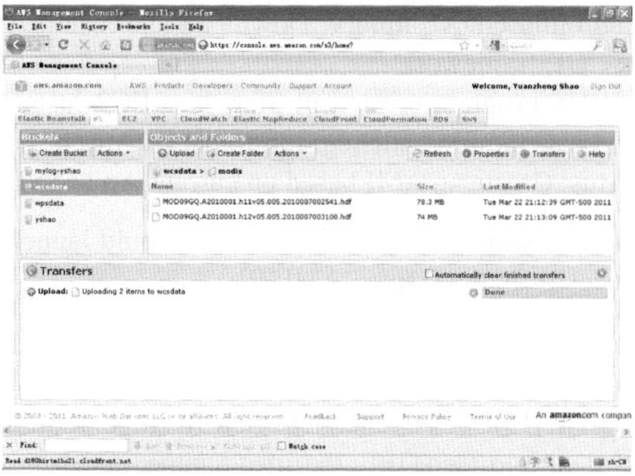

Fig. 2. Amazon S3 upload interface through AWS console.

To get started with Amazon S3, the following steps should be implemented:

(1) Create a Bucket to store your data. The user can choose a Region where the bucket and object(s) reside to optimize latency, minimize costs, or address regulatory requirements.

(2) Upload Objects to the Bucket. The data is durably stored and backed by the Amazon S3 Service Level Agreement.

(3) Optionally, set access controls. The user can grants others access to the data from anywhere in the world.

Once the data are uploaded to Amazon S3, the other Amazon Web Service and applications could access the data through its URL.

4.2 Creating Amazon EC2 for WCS

To deploy the WCS into Amazon Cloud environment, the Amazon EC2 instances, which provide the running environment for WCS service, should be created.

Geospatial Data Abstraction Library (GDAL) is selected as the dependent library for WCS. GDAL is a translator library for raster geospatial data formats that is released under an X/MIT style Open Source license by the Open Source Geospatial Foundation. As a library, it presents a single abstract data model to the calling application for all supported formats [7]. It also comes with a variety of useful command line utilities for data translation and processing.

The following steps are used to describe how to set up WCS under Amazon EC2 instance (Suppose the user has created an account in Amazon AWS).

(1) Create a WCS program on standalone environment based on OGC WCS version 1.1.0 specification, with GDAL supported in the back end.

(2) Select a basic 32-bit Amazon Linux AMI; and launch the Amazon EC2 instance; Keep the public DNS name for the instance.

(3) Access the created instance using any SSH client based on the public DNS name and the private key file.

(4) Download and build GDAL library; Download and install Apache as Common Gateway Interface (CGI) framework.

(5) Upload the WCS program created in Step 1 to EC2 instance, and rebuilt the project. Put the WCS execute file under cgi-bin folder of Apache.

(6) Test WCS status. Create an AMI based on current Amazon EC2 instance.

Once the AMI is created successfully, the server side could launch the multiple same Amazon EC2 instances from the created AMI [9].

4.3 Validation

The following is a URL of KVP encoded GetCoverage request for reformatting HDF-EOS MODIS data into GeoTIFF, the image generated from WCS is shown in Fig.3.

http://ec2-67-202-54-157.compute-1.amazonaws.com/cgi-bin/wcs/service=wcs&version=1.1.0&request=getcoverage&identifier=HDFEOS:https://s3.amazonaws.com/wcsdata/modis/MOD09GQ.A2010001.h12v05.005.2010007003100.hdf:sur_refl_b01_1&format=image/geotiff

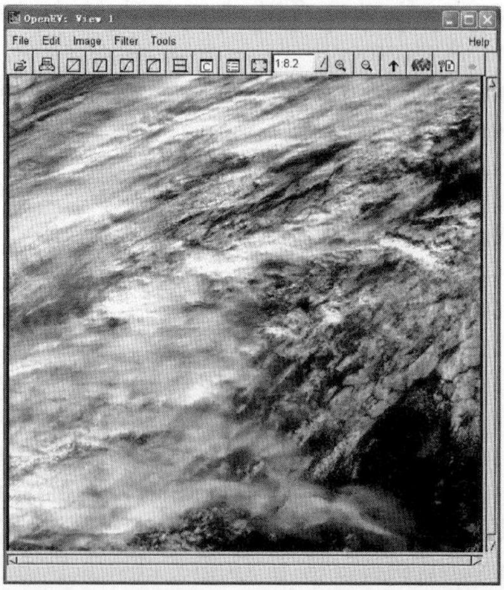

Fig. 3. The result image generated from WCS deployed on Cloud.

5 Conclusion and Future Work

This paper presents an implementation of deploying OGC WCS on Cloud Computing platform. Approaches on how various services in Amazon platform can be utilized to meet the storage and computing requirements of WCS are described.

Deploying WCS on Cloud Computing platform require considerable work, since the infrastructure, deployment requirements and APIs of different Cloud Computing platform are also different. The platform-depend APIs complicate the development of geoprocessing functions in different Cloud Computing platform. Based on the unlimited computing capacity provided by Cloud Computing platform, the users could concentrate on domain business logic without worrying about the hardware and software limitation.

Since the most current commercial Cloud Computing platforms do not have free offers, the user should take the economic cost into account when deploying application in Cloud environment. The implement in this paper is to demonstrate the feasibility and flexibility of deploying WCS on Cloud Computing platform for academic purposes. Comparison for performance price ratio between different Cloud Computing platforms will be made in the future work.

References

1. 52n North WPS geoprocessing Community, `http://52north.org/communities/geoprocessing`
2. Whiteside, A., Evans, J.D. (eds.): Web Coverage Service (WCS) implementation standard, Version 1.1.0, OGC06083-r8, Open Geospatial Consortium (2008)
3. Amazon Web Service, `http://aws.amazon.com`
4. Granell, C., Dıaz, L., Gould, M.: Service-oriented applications for environmental models: Reusable geospatial services. Environmental Modelling and Software, 182–198 (2010)
5. Jianya, G., Peng, Y.: Geoprocessing in the Microsoft Cloud Computing platform – Azure. In: A special joint symposium of ISPRS Technical Commission IV and AutoCarto in conjunction with ASPRS/CaGIS 2010 Fall Specialty Conference, Florida (2010)
6. Yang, C., Raskin, R., Goodchild, M., Gahegan, M.: Geospatial cyberinfrastructure: past, present and future. Computers Environment and Urban Systems 34(4), 264–277 (2009)
7. GDAL - Geospatial Data Abstraction Library, `http://gdal.org`
8. Blower, J.D.: GIS in the cloud: implementing a web map service on Google App Engine. In: Proceedings of the 1st International Conference and Exhibition on Computing for Geospatial Research and Application (2010)
9. George, R.: Cloud Application Architectures – Building Applications and Infrastructure in the Cloud. O'Reilly Media Inc., Sebastopol (2009)

The Influence of Humidity on Setting of DC Bus in Converter Station

Guo Zhihong[1], Xu Mingming[2,4], Li Kejun[2], and Niu Lin[3]

[1] Shandong Electric Power Research Institute, Jinan, Shandong, China
guozh05@126.com
[2] School of Electrical Engineering, Shandong University, Jinan, Shandong, China
[3] State Grid of China Technology College, Jinan, Shandong, China
[4] hawkeagle57@163.com

Abstract. The corona characteristic of HVDC tube bus is the main factor taken into account for choosing tube bus of HVDC converter substation. The intension of corona could affect the electromagnetic environment of converter station and generate audible noise. For the HVDC transmission line has long distance, it is inevitable for converter station to built in high humidity area. It is necessary for studying the corona characteristics of converter station bus and choosing the best type of the bus to consider the humidity. By establishing the calculation models of electric field strength along the tube bus surface, the paper computes the influence of humidity on corona inception electric field, high and electrode spacing of tube bus. And gives the high and electrode spacing, at which the tube bus in ±660kV HVDC converter substation will not generate corona.

Keywords: Corona; Humidity; DC; Converter Station; Tube Bus.

1 Introduction

The corona characteristic of DC bus in converter station is the major reason for choosing bus. The intension of bus corona could affect the electromagnetic environment of converter station, for instance: radio interference and audible noise. In extreme cases, corona current could cause the severe electromagnetic interference (EMI) that interfere the control and secondary protection of converter station. It also could cause the serious EMI that causes corona loss and disturbs the communication equipments [1, 2, 3]. Therefore it is very important for design and the economic operation of the converter station that studying the corona characteristics of converter station bus and choosing the best type of the bus [4, 5, 6].

Electric field at the surface of conductor is the most significant factor of corona inception. There are several main factors influence the corona inception electric field at the surface of tube bus, for example, the radius of tube bus, material, surface roughness and ambient humidity, temperature. In decades, international use the Peek's calculation formula to calculate the corona inception voltage gradient on wire surface [7]:

M. Zhu (Ed.): Electrical Engineering and Control, LNEE 98, pp. 297–304.
springerlink.com © Springer-Verlag Berlin Heidelberg 2011

$$E_c = 30\delta m\left(1+\frac{0.308}{\sqrt{\delta r}}\right) \tag{1}$$

where r is the conductor radius in cm; m is roughness coefficient of conductor surface ($m \leq 1$); δ is air relative density

$$\delta = \frac{P}{101.3\times10^3}\cdot\frac{293}{273+t} \tag{2}$$

where p is atmospheric pressure in Pa and t is Celsius temperature in ℃. For standard atmospheric pressure and temperature is 20℃, δ=1. Peek believes that there is no influence of air humidity on the corona, so the Peek's calculation formula does not consider the humidity.

For the HVDC transmission line has long distance, it is inevitable for converter station to built in high humidity area, such as ±660 converter station in Qingdao. Therefore it is necessary for studying the corona characteristics of converter station bus and choosing the best type of the bus to consider the humidity. The paper computes the influence of humidity on corona inception electric field, high and electrode spacing of tube bus. And give the high and electrode spacing for different radius tube buses, at which the tube buses in ±660kV HVDC converter substation will not generate corona.

2 Theory of Corona Phenomena

The corona discharge is defined by the self-sustained discharge near the conductor, and occurs when the electric field strength on the surface of conductor reaches a critical value. When the electric field strength in the vicinity of conductor reaches the threshold value for ionization of gas molecules by electron collision, an electron avalanche (primary avalanche) starts to develop. With the growth of avalanche, more electrons are emitted in all directions and more positive ions are left in the avalanche wake. When there are enough space electrons (or space charge) in the primary avalanche, a successor avalanche (second avalanche) starts to develop and the discharge will transit to streamer discharge. This is the result of the strength of local electric field, caused by space charge, and the generation of space photoionization. It is the starting point of corona. At this time, the maximum value of electric field strength on the surface of conductor is the corona inception electric field. And the voltage is the corona inception voltage. With the increasing of air humidity, the effective ionization coefficients will increases. It expands the ionization zone and enhances the ability of impact ionization. They are the main reasons that the corona inception electric field on the surface of conductor decreases with the increasing of air humidity.

3 Corona Inception Electric Field

3.1 Corona Inception Electric Field Criterion

According to the mechanism of corona discharges, when the number of electron in the primary avalanche arrives to one point, the primary avalanche will transit to streamer discharge. The number of electron in the primary can be calculated by

$$n = n_0 \exp\left(\int_{r_0}^{r_e} (\alpha(r) - \eta(r))dr\right) \tag{3}$$

where n is the number of free electron in electron avalanche at the position r_e, n_0 is the initial number of the free electron in the space, α is the Townsend's first ionization coefficient, η is attachment coefficient, and r_e is the radius of ionization zone, r_0 is the radius of tube bus.

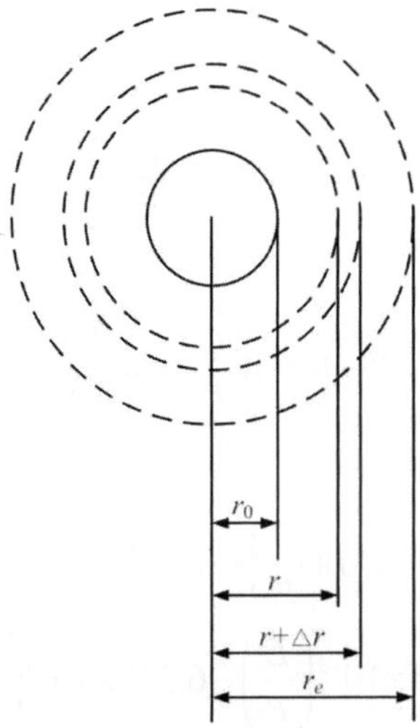

Fig. 1. Ionization zone around a conductor

In the ionization zone, the value of α is greater than η. The number of free electrons is increased until the avalanche reaches r_e. Maximum value of free electrons exists at the boundary of ionization zone ($\alpha=\eta$). The number of free electrons at this boundary can be used as a criterion to calculate corona inception voltage. Since the exact number of initial free electrons at the surface of conductor is unknown, the ratio n/n_0 will be used. This critical value is adopted about 3500 for air gaseous insulation by experimental measurements [8].

Both α and η are functions of the electric field and gas pressure. In order to calculate the corona inception voltages for humid air, considered as a mixture of air molecules and water molecules, it is necessary to have the effective coefficients for the mixture. In the following, subscripts d, w refer to dry air and water vapour. The mixture coefficients were obtained using the linear expression proposed by [9]:

$$\alpha = \frac{P_w}{P}\alpha_w + \frac{P_d}{P}\alpha_d \tag{4}$$

$$\eta = \frac{P_w}{P}\eta_w + \frac{P_d}{P}\eta_d \tag{5}$$

where P is the normal atmospheric pressure. The expressions for α and η in dry air and water vapour is:

$$\frac{\alpha_d}{P} = \begin{cases} 4.7786\exp\left(-0.221\cdot\frac{P}{E}\right) & 25 \le \frac{E}{P} \le 60 \\ 9.682\exp\left(-0.2642\cdot\frac{P}{E}\right) & 60 \le \frac{E}{P} \le 240 \end{cases} \tag{6}$$

$$\frac{\alpha_s}{P} = 5.6565\exp\left(-125.47\cdot\frac{P}{E}\right) \tag{7}$$

$$\frac{\eta_d}{P} = 0.01298 - 0.541\left(\frac{E}{P}\right) + 8.7\left(\frac{E}{P}\right)^2 \tag{8}$$

$$\frac{\eta_s}{P} = \begin{cases} -0.026137 - 0.0261\left(\frac{E}{P}\right) - 3.817\left(\frac{E}{P}\right)^2 & 1.9 \le \frac{E}{P} \le 45.6 \\ 0.1385 - 3.575\times10^{-3}\left(\frac{E}{P}\right) - 6.313\times10^{-5}\left(\frac{E}{P}\right)^2 & 45.6 \le \frac{E}{P} \le 182.4 \end{cases} \tag{9}$$

where E is the electric field in V/cm.

3.2 Calculate Corona Inception Electric Field

Calculating of corona inception electric field is based on a simple algorithm which has shown in Fig. 2. In the beginning, the electric field around the surface of conductor is determined using an initial value of voltage. Then the ratio of the numbers of free electrons at the ionization zone boundary to the numbers of free electrons at the surface of conductor is calculated. If this value is less than the specified threshold then the value of voltage is increased a specified step and the above process goes on. If this value is more than the specified threshold then the value of voltage is considered as corona voltage inception.

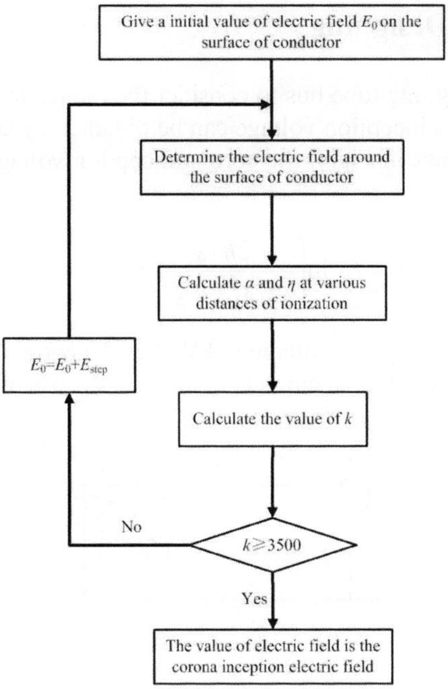

Fig. 2. Process of calculation of corona inception voltage

To calculate the corona inception electric field of different humidity for different radius tube buses(10cm、12.5cm、15cm) by the algorithm showed in Fig.2. The influence of humidity and radius of bus on the corona inception electric field shows in Fig.3.

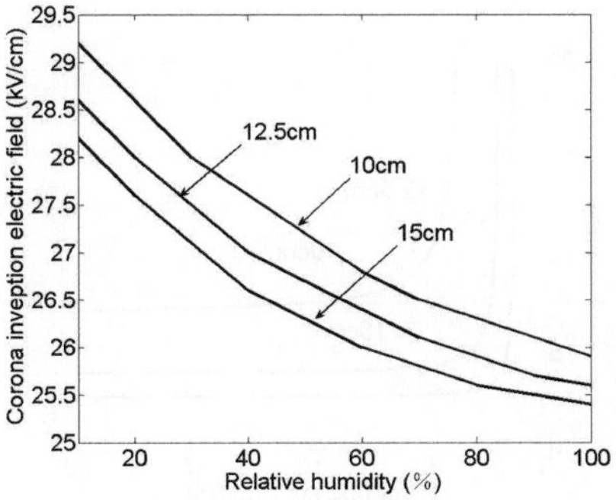

Fig. 3. Influence of humidity on corona inception electric field

The Fig.3 shows that the corona inception electric field on the surface of tube bus with the increasing of air humidity and the corona inception electric field decreases with the increasing of radius of tube bus.

4 The Tube Bus Designing

It is important for designing tube bus to consider the radius, the electrode spacing and high of bus. The corona inception voltage can be obtained by corona electric field. By the image method, shows in Fig.4, the corona inception voltage can be calculated by formula (10).

$$U = Er\ln\left(\frac{2h\cdot s}{r\sqrt{4h^2+s^2}}\right) \quad (10)$$

where, U is the corona inception voltage in kV, E is the corona inception electric field in kV/cm, r is the radius of bus in cm, h is high in m, s is electrode spacing in m.

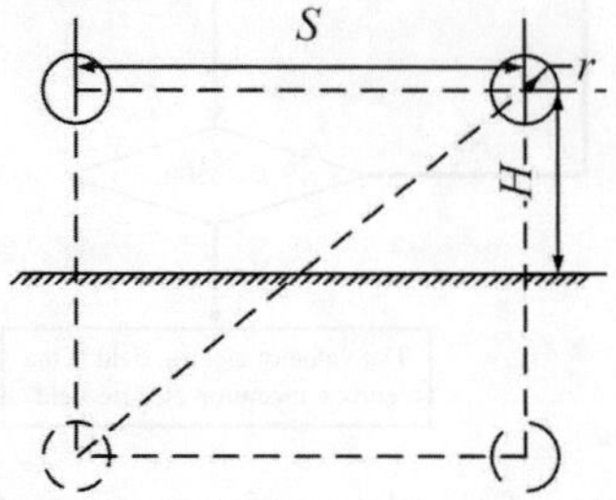

Fig. 4. Mirror image of tube bus

Without consider the air humidity, the h-s curve for different tube buses in ± 660 DC converter station can be obtained by Eq.10, shows in Fig. 5. The tube bus will not generate corona in good weather, when it is installed at the high and electrode spacing. The E is calculated by Eq.1.

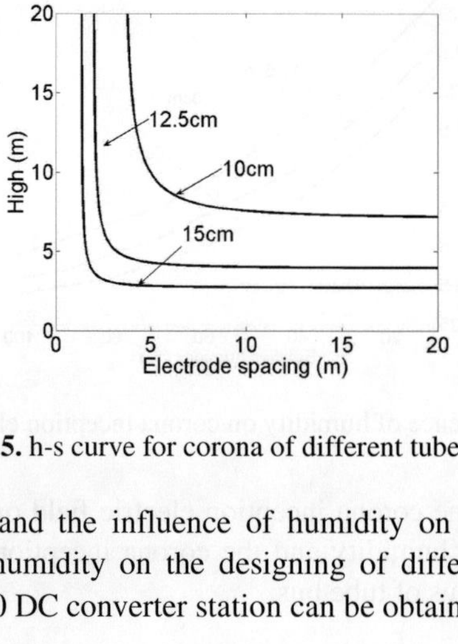

Fig. 5. h-s curve for corona of different tube buses

According to Eq.10 and the influence of humidity on corona inception electric field, the influence of humidity on the designing of different tube buses (10cm、12.5cm、15cm) in ± 660 DC converter station can be obtained, shows in Fig.6.

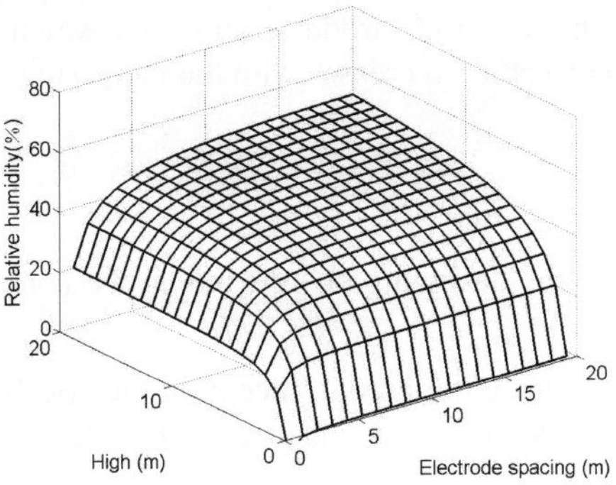

(a) 10cm

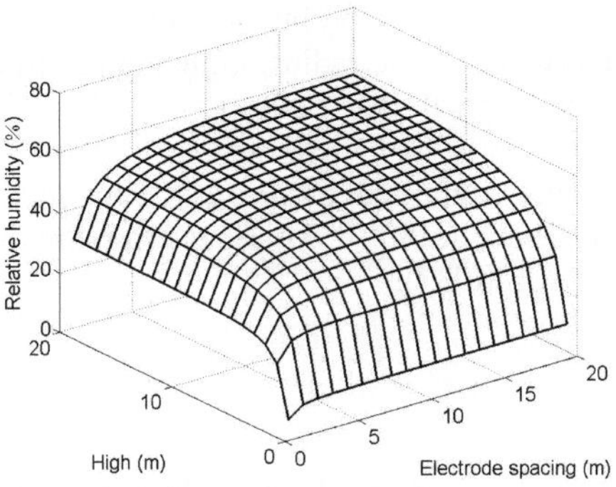

(b) 12.5cm

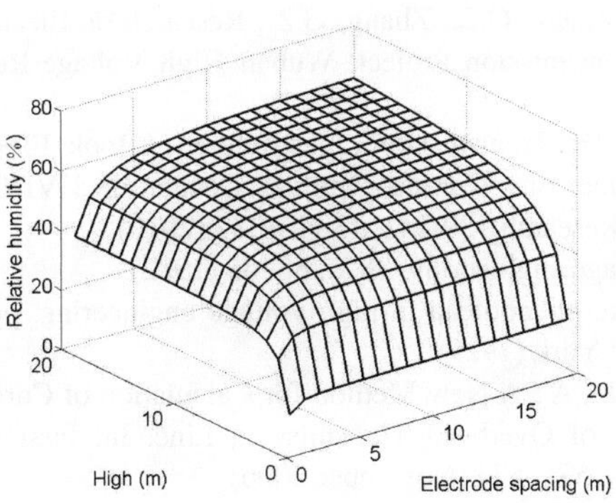

(c) 15cm

Fig. 6. Influence of humidity on tube buses

The pictures show that the air humidity affects the designing of tube bus in converter station. The high and electrode spacing, on which the tube bus will not generate corona in good weather, increases with the increasing of air humidity.

5 Summary

The air humidity affects the designing of tube bus in the converter station. It affects main factors as:

1) The air humidity affects the corona inception electric field of tube bus in the converter station. The corona inception electric field decreases with the increasing of air humidity.

2) The air humidity affects the high and electrode spacing of tube bus in converter station. The high and electrode spacing, at which the tube bus will not generate corona in good weather, increases with the increasing of air humidity.

So, it is necessary for designing the tube bus in converter station to consider the air humidity. The air humidity affects the radius, high and electrode spacing of tube bus by affecting the corona inception electric field.

Acknowledgments. Project supported by Project Supported by National Natural Science Foundation of China (50977053); Shandong Provincial Natural Science Foundation, China (ZR2010EM033).

References

1. Wu, X., Wan, B.Q., Zhang, G.Z.: Research on Electromagnetic Environmental Parameters Control Indexes of ±800 kV UHVC Transmission Lines. Wuhan High Voltage Research Institute, Wuhan (2005)
2. Harvey, J.R., Zaffanelia, L.E.: DC conductor development. EPRI, New York (1982)
3. Wu, X., Wan, B.Q., Zhang, G.Z., Zhang, G.Z.: Research on Electromagnetic Environment of 750 kV Power Transmission Project. Wuhan High Voltage Research Institute, Wuhan (2005)
4. EPRI report ±600 HVDC Transmission Lines Reference Book. EPRI, California (1993)
5. Xu, M., Li, K.: Tube Bus Corona Optimal Design of HVDC Converter Substation. Advanced Materials Research 139-141, 1384–1387 (2010)
6. Zhu, T.C.: High Voltage engineering. 8(3), 58–65 (1982)
7. Peek, F.W.: Dielectric phenomena in high voltage engineering, pp. 9–108. Mc-Graw-Hill Book Company, New York (1929)
8. Zangeneh, A., Gholam, A.: A New Method for Calculation of Corona Inception Voltage in Stranded Conductors of Overhead Transmission Line. In: First International Power and Energy Coference, pp. 571–574 (November 2006)
9. Sarma, M.P., Janischewskyj, W.D.C.: Corona on smooth conductors in air steady-state analysis of the ionization layer. Proc. of IEEE 116(3), 161–166 (1969)

Design and Study of Professional PWM Core Based on FPGA

Guihong Tian, Zhongning Guo, Yuanbo Li, and Wei Liu

Guangdong University of Technology, Guangdong 510006, China

Abstract. In view of the requests of Micro Electric Discharge Machining Pulse Generator, a professional core supporting PWM output based on FPGA has been proposed in this paper. A high precision PWM with frequency and duty adjustable and an interrupt signal which is useful for MCU are produced. The PWM pulse and interrupt output are simulated and analyzed in Quartus II. The PWM pulse duration is up to 10ns. In this paper a 40ns PWM pulse width for MOSFET is realized. It has provided a possible technological assurance for nanofabrication micro-EDM.

Keywords: FPGA; Micro-EDM; MCU; PWM output.

1 Introduction

In the field of micro electrical discharge machining (EDM), one of the factors which determine the surface performance is the discharge energy per pulse. The energy suit for micro-EDM is about at the level of 10-6~10-7J. In order to obtain micro discharge energy, micro-EDM Pulse Generator is the key point [1]. It was proved that to reduce PWM pulse duration when keeping peak current big enough was a more effective way [2]. And to reduce the pulse width is the essential method reducing discharge energy. Mainly as the micro energy is needed, it's very important to keep the pulse wide at the level of nanosecond or even picosecond and with fast turn-on rise time and as short turn-off fall time as possible. It put forward the very high request in the aspects of power switching device output waveform performance. On the other hand the traditional MCU (Multipoint Control Unit) and DSP (Digital Signal Processor) control have far short of demand [3]. With the rapid development of the electron & information technology, PLD (Programmable Logic Device) has been applied to wider fields. More and more researchers in China have realized the advantages of CPLD (Complex Programmable Logic Device) /FPGA (Field Programmable Gata Array). As except for realizing high frequency and high-precision counting, CPLD/FPGA Can greatly reduce system complexity, simplify circuit, stable circuit characteristics and achieve the hardware flexible design [4]. As MCU is used so widely because of its easily grasp, flexible application, rich peripheral device, etc., and CPLD/FPGA is so suitable for high-frequency and high-precision pulse output that it's perfect to combine MCU and CPLD/FPGA to realize the high-precision pulse generator with ultra short pulse width. In the design of micro EDM power supply, MCU is mainly used as function control, such as pulse's parameter settings, power's

M. Zhu (Ed.): Electrical Engineering and Control, LNEE 98, pp. 305–311.
springerlink.com © Springer-Verlag Berlin Heidelberg 2011

mode settings, Keyboard and LED display, and transmit data to FPGA module. And FPGA generate PWM (Pulse Width Modulation) pulse according to the settings of MCU to drive the power switching device such as MOSFET in order to realize the pulse machining [5]. In micro-energy pulse generator, PWM is the main method to control and realize the digital power supply and it's the key factor to make power supply system stable and reliable. Nowadays, it's more and more combining MCU with FPGA to optimum circuit structure in motor control, automatic detection, micro-EDM power supply, etc [6]. So it's of great practical value to design a professional PWM core based on FPGA. This professional PWM core can not only support PWM pulse output and also can provide an interrupt output signal to MCU. In this design, FPGA is mainly used to interface to MCU, PWM pulse generator and an interrupt output signal for MCU control. It can realize the circuit highly integrated, high speed and stabilization, what's more important it can shorten power design cycle as soon as possible.

2 Professional PWM Core Overall Structure Design

In this design, the core mainly contains four function modules: the interface to MCU system, the internal register and data buffer, internal combinational logic and temporal logic circuit module and independent external I/O ports. The overall structure chart is shown in Figure1.

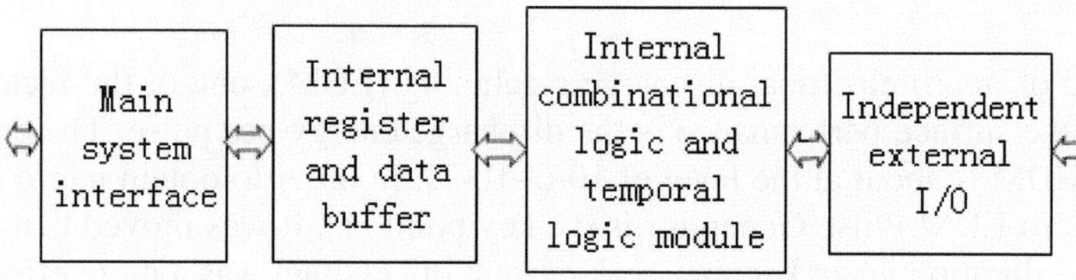

Fig. 1. Professional PWM core overall structure chart

2.1 Main System Interface

The main system interface is mainly for the communication between this core and main system such as MCU. It's convenient for calling the core. The main system transmits data and sends function requests through the main system interface. The interface not only realizes real-time communication and interoperability between the core and the main system, but also has high reliability and high portability. And the core mainly outputs the PWM pulse or starts the timer and counter to output an interrupt signal. In this paper, the interface adopts SoC (System-on-Chip) circuit interconnect interface specification based on WISHBONE standard [7]. This interface standard defines two types of interfaces: the host interface from the master port to the interconnect interface and the slave interface from the slave port to the interconnect interface. The design uses the main system interface is the slave interface. The core independently processes the data transmitted from the main system and then generates the appropriate PWM pulse or an interrupt output signal, but not returns the processed data to the main system.

2.2 Internal Register and Data Buffer

In the FPGA, the register is to achieve the data buffer when transferring data between the circuit entity area of the core and the main system interface. In this case, the main system transmits data to the core through the interface. And at first, the data has to be buffered by a register to make the data be stable, reliable and effective at the following processing. In this design, four types of registers are mainly used, including the main count register, the two butter registers which is used for data storage of PWM pulse width and pulse interval and the control registers which make the data transferred orderly. All this registers work together to determine the core mode and the type of operation. Among them, the main count register is a 32-bit wide register, its counter/timer clock cycle can either use the system clock or be input via an external clock port named gate_clk_pad_i; the two butter registers used for buffer the PWM pulse on-time and off-time data are also 32-bit wide registers; the control register contains 9 control bits, and it's mainly used to control other registers to work timely and effectively and realize the precise communication between the designed core and the main system such as MCU.

2.3 Internal Circuit Module

This internal circuit module mainly contains three modules, including the module of PWM data register and RS flip-flop output control, the timer/counter comparing module and interrupt signal generator module. Using these modules, the designed core provides two types of work modes, that is professional PWM pulse outputs and a timer/counter interrupt output, and according to the internal controller-register's data, it's very flexible and easy to select the core's work mode.

On the one hand, when the main system communicates with the core and the PWM pulse output is let out by the controller register, the registers named rptc_hrc and rptc_lrc respectively store the pulse on-time and off-time data, and then the controller register makes the timer/counter work to trigger the RS flip-flop circuit and eventually get the flexible duty cycle of the output pulse according to the designers settings.

On the other hand, when the main system communicates with the core and the control register enables the core work in the timer/counter mode, the timer/counter can use either the internal system clock or an external reference clock; it can not only used for a single loop control, but also a continuous loop control. When the timer/counter uses external reference clock and at the same time the corresponding control bit is enabled, the clock would be accept by the core through the specified port called gate_clk_pad_i and the timer / counter would be counted by an external clock cycle. And then the MCU send a full-decoding address to the core and the register rptc_lrc is initialized. This initial data is just the timer/counter's needed data. The register named rptc_cntr plus one by one according to the work frequency the timer/counter selected. When the value of the register rptc_cntr is up to the value of register rptc_lrc, the core would use an interrupt generator to output an interrupt signal to MCU. But the interrupt signal's final output must be enabled by the relevant control bits of the control register.

3 Function Realization

According to the functional description design above and adopting a very common and easily grasped hardware description language called Verilog HDL to program this design [8] and adopting Ateral company's FPGA chip called EP3C4 to simulate it, a 10ns PWM pulse width output has been achieved, which is suitable so much for the micro-EDM. This 10ns pulse width PWM wave output is of adjustable pulse duty cycle. And the realized core's function chart with I/O port is shown in Figure 2.

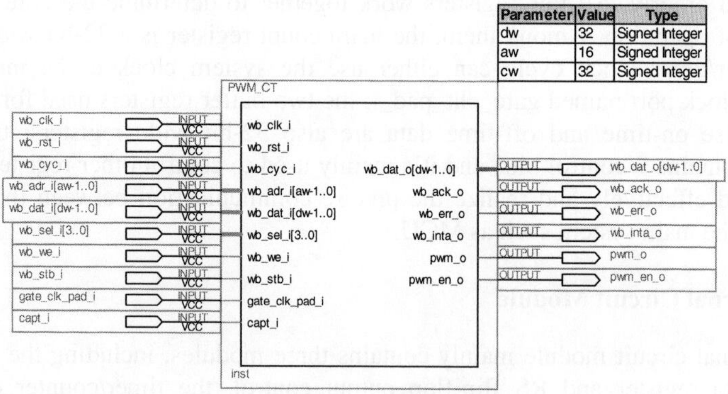

Fig. 2. The core's function chart with I/O ports

In this chart, the main ports defined as following:
wb_clk_i: the input clock of main system interface circuit interface;
wb_rst_i: the reset signal of main system circuit interface;
wb_adr_i: 16bits, the input address; wb_we_i: written enable signal;
wb_data_i: 32bits, input data; gate_clk_pad_i: external clock input;
wb_data_o: 32bits, output data; wb_ack_o: output acknowledgments;
wb_err_o: error verification signal; wb_inta_o: interrupt signal output ;
pwm_en_o: pwm pulse output enabled signal ; pwm_o: pwm pulse output.

4 Simulation and Analysis

In the paper, the designed PWM pulse timing for the power switching device and the interrupt signal output were logic synthesis and function simulated in Quartus II . The simulation results were shown in Fig.3 and Fig 4. It can be seen from Figure 3, as long as the timer / counter value to the rptc_lrc, an interrupt request signal would be output at the next rising edge of system clock. The first signal interruption was wrongly produced as the timer/counter was equal to the rptc_lrc register when they two were not initialized. As the MCU can judge this interrupt request signal according to the internal status in the core settings, this design was also to meet the requirements. And in Fig. 4, the PWM waveform pulse duration and interval were set by register rptc_hrc and rptc_lrc, and the duty was adjustable. Its minimum output pulse duration 10ns was realized. This ultra short pulse generator was very suitable for micro-EDM.

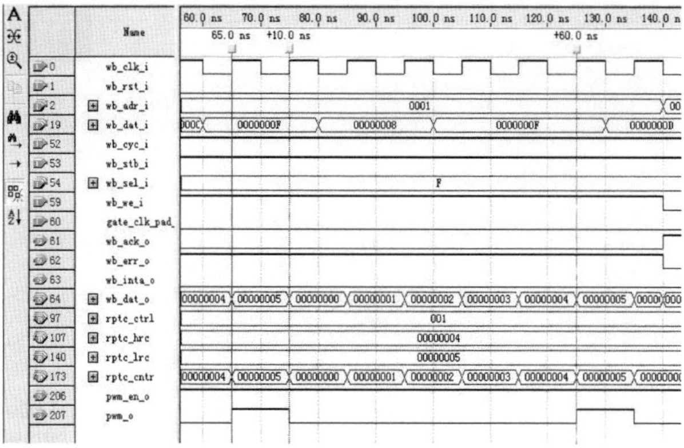

Fig. 3. Simulation of interrupt output

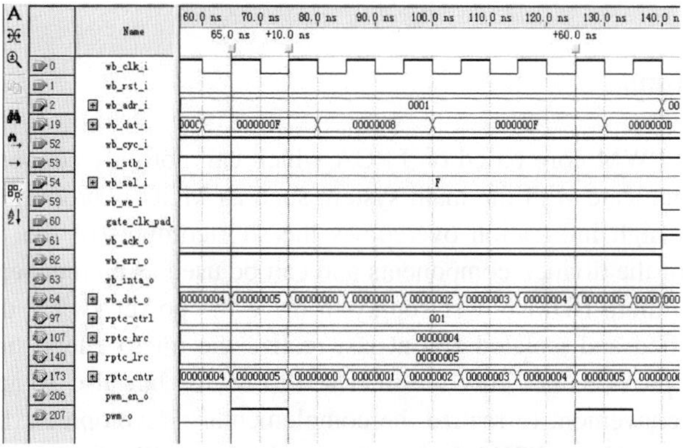

Fig. 4. Simulation of PWM pulse output

5 Results

In order to facilitate results of actual output observed by oscilloscope, the pulse duration was set as 40ns. Connect the USB-blaster and personal computer, switch on the power and download the program files and other specific files to the EP3C40 chip, and then the PWM pulse output was observed by oscilloscope. The duty of 1:5 pulse output was shown in fig.5. It can be seen that the generated pulse was steep rising and falling. The beginning of the pulse has some overshoot which was much use for accelerating the switch's opening process and at the switch-off transient, the pulse overshoot can provide sufficient negative voltage so that the switch's drain current can decrease rapidly, accelerating the shutdown process.

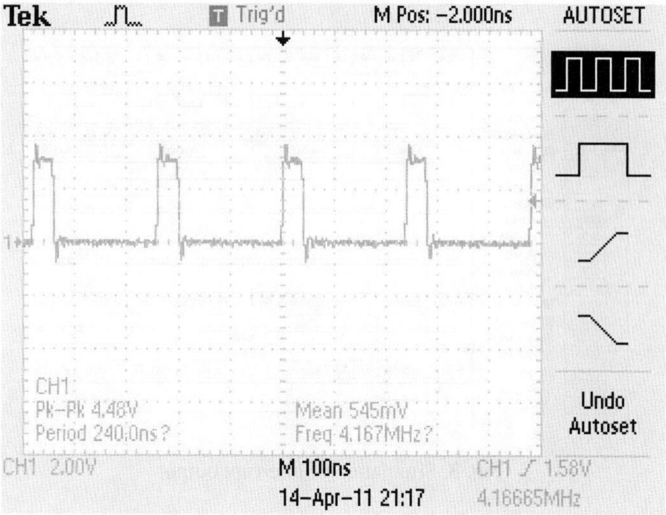

Fig. 5. The PWM pulse output with duty of 1:5

6 Conclusions

A professional PWM core based on FPGA which can not only support PWM output but also communicate with the main system such as MCU is proposed in this paper. The core with high integration overcomes the shortcomings of the long delay and poor stability of the divided components and can be used as an independent hardware module in the micro-EDM power supply to drive the power switch device after the pulse is amplified and isolated and at last realize the micro energy machining. The designed core provides not only the interface to MCU, but also some useful output ports and it's convenient to realize the complementary development between FPGA and MCU system. Using FPGA to design a professional PWM output core in a MCU system can easily realize the ultra short pulse duration power supply for micro-EDM, improve the power's integrated level and reliability of system and decrease system's volume. It's of extensive adaptability and practical value.

References

1. Wansheng, Z.: Advanced electrical discharge machining technology. National Defence Industry Press, Beijing (2003)
2. Huiyu, X., Di, Z.: Development of high aspect ratio microstructures technology from abroad. Journal of Transducer Technology 23(12), 4–6 (2004)
3. Xutao, C., Rijie, Y., You, H.: Development of signal system based on DSP and FPGA. Chinese Journal of Scientific Instrument 28(5), 918–922 (2007)
4. Lijiang, D.: FPGA application based on EDA technology. Tianjin Polytechnic University Master Thesis (2004)

5. Zhifu, Z.: Research and Design of Switch Power PWM Comparator. Southwest Jiaotong University Master Thesis (2008)
6. Mingyu, L.: Develop of Ultra Short Pulse Generator for Micro-EDM. Harbin Institute of Technology Master Thesis (2007)
7. Haili, W., Jinian, B.: Research on Interface Synthesis Based on WISHBONE Bus Protocol in SoC Design. In: The 8th Computer Engineering and Technology National Conference, pp. 361–365 (2004)
8. Coffman, K., Shuqun, S.: Practical FPGA Design based on Verilog HDL. Science Press, Beijing (2004)

5. Zhiru, Z.: Research and Design of Switch Power PWM Comparator. Southwest Jiaotong University Master Thesis (2008)
6. Mingyu, L.: Develop of Ultra Short Pulse Generator for Micro-EDM. Harbin Institute of Technology Master Thesis (2007)
7. Haili, W., Jinian, B.: Research on Interface Synthesis Based on WISHBONE Bus Protocol in SoC Design. In: The 8th Computer Engineering and Technology National Conference, pp. 361–365 (2004)
8. Chuman, K., Shoqun, S.: Practical FPGA Design based on Verilog HDL. Science Press, Beijing (2004)

Ancient Ceramics Classification Based on Case-Based Reasoning

Yu Wenzhi[1] and Yan Lingjun[2]

[1] Department of Information and Technology,
Guilin University of Electronic Technology, Guilin 541004, China
[2] Department of Historical Culture and Tourism,
Guangxi Normal University, Guilin 541004, China
yuwenzhi2005@qq.com, yanlingjun@foxmail.com

Abstract. For studying the classification relationship of Ru Guan porcelain and Jun Guan porcelain, the method based on case-based reasoning was presented, with ancient ceramics examples as base cases, and the study samples as target cases. By means of calculating the similarity between base cases and target cases, the similar series were determined, and the most similar base case to the ancient ceramics target case was found out. Finally, the classification of target cases was evaluated. The results show the correct recognition rates as 94.1% and 100% to ancient ceramics body and glaze respectively. Case-based reasoning is a very valuable new method for ancient ceramics classification.

Keywords: case-based reasoning, ancient ceramics, classification, proton induced x-ray emission.

1 Introduction

The Ru Guan kiln site is located in Qingliangsi, Baofeng County, and the Jun Guan kiln site is located in Juntai, Yuzhou City, Henan Province, China. They are two famous kilns in the ancient China. As the two kilns are located close to each other, the Ru and Jun porcelains look quite similar, and therefore, it is often difficult to distinguish one from the other [1]. For studying the sources of raw materials, ingredients and their classification relationships, some hybrid models combining chemical elements measure technology [2-4] and classification method appeared. Common classification methods are: fuzzy cluster [1], grey theory [5], neural network [6], support vector machine algorithm [7], rough set theory [8], et al.

The case-base reasoning (CBR) is a new artificial intelligent technology, presented by Schank in 1982. Kolodner realized it by computer in 1983 [9]. CBR uses a human-inspired philosophy: it tries to solve new cases by using old and previously solved ones [10, 11]. The process of solving new cases also updates the system, by providing new information and new knowledge to the system. This new knowledge can be used for solving other feature. CBR has achieved success in planning, design, medicine, fault diagnosis and forecasting, since its presentation [12]. Ancient ceramics classification is very complex and its knowledge is difficult to obtain, so CBR may be

M. Zhu (Ed.): Electrical Engineering and Control, LNEE 98, pp. 313–318.
springerlink.com © Springer-Verlag Berlin Heidelberg 2011

a valuable new method for its classification. The classification of some ancient ceramics was determined by other researchers, which are taken as base cases in CBR classification system in this paper.

2 Method of Case-Based Reasoning of Similar Porcelains

2.1 Base Case Library

This paper collects the ancient ceramics, whose classification was determined by other researchers, and takes them as base cases in its CBR classification system. Each record is a case, and each field is a feature.

2.2 Weights [13]

The importance of each feature is evaluated by weight. The more important the feature, the bigger the weight. Therefore, this paper uses variable weights to reflect environmental sensitivity of features. The weight of the feature is computed as:

$$w_h = \sum_{i=1}^{n} \left[\frac{N_h(T,C_i)}{\sum_{j=1}^{n} N_h(T,C_j)} \right]^2 \tag{1}$$

$$N_h(T,C_j) = Num\left(r\left\{r \in C_j \; and \left(v_r(h) = r_T(h) \; or \; v_T(h) \in [L,U]\right)\right\}\right) \tag{2}$$

In this equation, w_h means the weight of h th feature, n means the total number of the ancient classifications, T means the target case, C_i means the i th classification of the base cases, $N_h(T,C_j)$ means the number of relation with T in C_j classification, $v_r(h)$ means the value of h th feature of r th base case, $v_T(h)$ means the value of h th feature of T th target case, and $[L,U]$ means the domain of h th feature of T th target case, which is defined as follows:

$$\begin{cases} L = \inf\left[\bigcap_{k=1}^{n} dom_k(h)\right], v_T(h) \in dom_k(h) \\ U = \sup\left[\bigcap_{k=1}^{n} dom_k(h)\right], v_T(h) \in dom_k(h) \end{cases} \tag{3}$$

In this equation, $dom_k(h)$ means the domain of h th feature in C_i th classification, and L,U means the supremum and infimum of $dom_k(h)(k = 1,2,...,n)$ intersection.

In formula (1), $N_h(T,C_i)$ measures the frequency of target case T which belongs to classification C_i by feature h th. $\sum N_h(T,C_j)$ means the total summation of target case T which belongs to each classification by feature h th. $N_h(T,C_i)/\sum N_h(T,C_j)$ means the probability of target case T which belongs to classification C_i by feature h th.

2.3 Similarity

The notion of similarity between two cases is computed using different similarity measures, such as euclidean, manhattan and infinite mould distance [11]. In this paper we choose the euclidean distance.

$$d_{iT} = \sqrt{\sum_{i=1}^{n} w_h \left(v_i(h) - v_T(h)\right)^2} \quad (4)$$

Here d_{iT} is the euclidean distance between base case T and i th target case, $v_i(h)$ means the value of the h th feature of i th base case, and $v_T(h)$ means the value of the h th feature of target case T.

The similarity S_{iT} between T and i is defined as follows:

$$S_{iT} = 1 - d_{iT} \quad (5)$$

2.4 Classifying

The target case and the base case, which is most similar with target case, are classified by the most similarity principle.

3 Experiment Results

In this paper, we choose eighty-four Ru Guan and Jun Guan porcelains from reference [2] as research objects, whose classification was determined. The classification and the major compositions of Ru Guan and Jun Guan porcelain body samples are listed in Table1, partly; choose seventeen samples (seven of Ru Guan porcelains and ten of Jun Guan porcelains) as target cases from ancient ceramics samples, and the others as base cases. Then, CBR method is used to determine the classification.

From table 1, we know the ancient ceramics are divided into two classifications: Ru and Jun porcelains. Each porcelain has seven features: Al_2O_3、SiO_2、K_2O、CaO、TiO_2、MnO、Fe_2O_3. To eliminate ill effects by different dimension, the experimental data is reduced.

$$v_i'(h) = \frac{v_i(h) - v_{\min}(h)}{v_{\max}(h) - v_{\min}(h)} \quad (6)$$

In this equation, $v_{\max}(h)$ and $v_{\min}(h)$ mean respectively the maximum and minimum value of h th feature of the base case.

Table 1. Major composition of Ru Guan porcelain body and Jun Guan porcelain body [wt%]

No	Al2O3	SiO2	K2O	CaO	TiO2	MnO	Fe2O3	Classification
1	30.56	62.21	1.77	0.86	1.13	0.01	1.94	Ru
2	26.77	66.26	1.82	0.44	1.27	0	1.94	Ru
3	26.66	66.31	1.86	0.6	1.23	0.01	1.84	Ru
4	27.38	64.91	2.07	1.01	1.09	0.02	2.03	Ru
5	27.63	65.39	1.77	0.42	1.33	0.02	1.95	Ru
6	31.16	61.72	1.51	1.07	1.07	0.01	1.96	Ru
7	31.75	60.94	1.77	0.84	1.1	0	2.1	Ru
					……			
27	31.17	62.29	1.55	0.45	1.1	0.01	1.94	Jun
28	26.5	65.3	2.46	0.52	0.97	0.02	2.55	Jun
29	27.5	64.3	2.55	0.54	0.92	0	2.51	Jun
30	26.6	65.2	2.29	0.65	0.92	0.02	2.67	Jun
31	25.4	66.4	2.37	0.65	0.98	0.02	2.61	Jun
32	27	63.9	2.71	0.62	0.99	0.02	2.46	Jun
33	26.2	65.4	2.51	0.66	0.9	0.02	2.88	Jun
					……			
68	27.91	64.98	1.68	1.06	1.08	0.02	1.77	Ru
69	27.56	65.74	1.77	0.52	1.15	0.02	1.73	Ru
70	26.21	66.06	2.18	0.73	1.08	0.02	2.24	Ru
					……			
75	25.8	66.3	2.2	0.88	1.02	0.02	2.58	Jun
76	25.8	65.5	2.66	0.61	0.93	0.04	2.85	Jun
77	25.7	66.4	2.29	0.47	1.06	0	2.45	Jun
					……			

The weight of T_{68} is $w(68) = [0.522, 0.516, 0.528, 0.52, 0.68, 0.51, 0.51]$, computed by Eq.1. Table 2 shows the similarity between ancient ceramics target cases and base cases. The accurate recognition rate of porcelain body and glaze are 94.1% and 100%, respectively.

Table 2. The similarity between base cases and target cases of Ru Guan porcelain body/glaze and Jun Guan porcelain body/glaze

Target case	Target case classification	Code No. of the most similar base case	Classification of the most similar base case	The maximum value of the similarity
R39	Ru	R288/ R384	Ru/Ru	0.7792/0.8519
R40	Ru	R379/ R271	Ru/Ru	0.8509/0.8636
R41	Ru	J106/ R291	Jun/Ru	0.7966/0.7977
R42	Ru	R291/ R387	Ru/Ru	0.8744/0.8847
R43	Ru	R385/ R288	Ru/Ru	0.6211/0.8947
R44	Ru	R356/ R291	Ru/Ru	0.8431/0.8229
R47	Ru	R386/ R334	Ru/Ru	0.8016/0.7841
J123	Jun	J89 / J116	Jun/Jun	0.8552/0.8266
J124	Jun	J95 / J87	Jun/Jun	0.8008/0.4000
J125	Jun	J118/ J91	Jun/Jun	0.7362/0.8786
J126	Jun	J98 / J105	Jun/Jun	0.8816/0.7646
J127	Jun	J86 / J106	Jun/Jun	0.9171/0.8961
J128	Jun	J90 / J87	Jun/Jun	0.9736/0.8896
J129	Jun	J103/ J117	Jun/Jun	0.7897/0.8943
J130	Jun	J116/ J89	Jun/Jun	0.9053/0.8930
J131	Jun	J118/ J114	Jun/Jun	0.9250/0.7577
J132	Jun	J98 / J106	Jun/Jun	0.6574/0.8423

4 Conclusions

(1) CBR considers the influence of many uncertainties, gets knowledge from cases and finds out the similar base cases, and therefore can apply to ancient ceramics classification.
(2) Attribute weight method is presented according to the concept of variable weights. It makes connection between the result and the current classification environment, and reflects the sensitivity of weights to environmental.

References

1. Li, R., Li, G., Zhao, W., et al.: A Provenance Relation Study of Qingliangsi Ru Guan Kiln and Juntai Guan Kiln by Fuzzy Cluster Analysis. Journal of Beijing Normal University (Natural Science) 39, 628–631 (2003)
2. Li, G., Zhao, W., Li, R., et al.: Discrimination of Major Compositions of Ru Guan Porcelain and Guan Porcelain. J. Chinese Ceramic Society 35, 998–1006 (2007)

3. Wang, S., Li, G., Sun, H., et al.: The Nondestructive Identification of the Kindred Glazing Color o f Ancient and Ru Porcelain by Using PIXE. Journal of Henan Normal University (Natural Science) 35, 71–74 (2007)
4. Wang, J., Chen, T.: Study on the Pre-Qin Pottery and Proto-Porcelain from Boluo County, Guangdong Province with INAA. J. Nuclear Techniques 26, 454–462 (2003)
5. Liu, T., Jian, Z., Lin, W., et al.: Application of Grey Theory in Period Determine of Ancient Ceramic. J. Computer Engineering 36, 259–261 (2010)
6. Guo, J., Chen, N.: Pattern Recognition-Artificial Neural Network Method Applied to The Classification of Ancient Ceramics. J. Chinese Ceramic Society 25, 614–617 (1997)
7. Fu, L., Zhou, S., Peng, B., et al.: Classification of Ancient Ceramic Pieces form Unearthed Official Ware an Hangzhou Based On Least Square Support Vector Machine Algorithm. J. Chinese Ceramic Society 36, 1183–1186 (2008)
8. Xiong, Y., He, W., Wang, B.: Introduction on Rough Set Theory and Its Application to Ancient Ceramics Classification. J. Sciences of Conservation and Archaeology 14, 298–308 (2002)
9. Kolodner, J.: An introduction to case-based reasoning. J. Artificial Intelligence Review 6, 3–34 (1992)
10. Riesbeck, C., Schank, R.: Inside Case-Based Reasoning. Lawrence Erlbaum Associates, Hillsdale (1989)
11. Kolodner, J.: Case-Based Reasoning. Morgan Kaufmann Publishers, Los Altos (1993)
12. Shi, Z.: Advanced Artificial Intelligence. Xian Jiaotong University, Xian (1998)
13. Liu, M.: Intelligent approach to slope stability evaluation based on case-based reasoning. Wuhan University of Technology, Wuhan (2001)

Distributed Power Control Algorithms for Wireless Sensor Networks

Yourong Chen[1], Yunwei Lu[2], Juhua Cheng[1], Banteng Liu[1], and Yaolin Liu[1]

[1] College of Information Science and Technology, Zhejiang Shuren University, 8 Shuren Street, Hangzhou, Zhejiang, China
[2] Department of Humanities & Information, Zhejiang College of Construction, Higher Education Zone, Xiaoshan, Hangzhou, China
jack_chenyr@163.com, lyw1103@163.com, sjhcjh@126.com, 173299467@qq.com, liuyaolin_610@qq.com

Abstract. To prolong network lifetime and reduce average energy consumption when nodes are unable to measure the distance to neighbor nodes or have not distance measuring hardwire, distributed power control algorithms for wireless sensor networks are proposed. SPCA and PPCA are used to calculate the transmission power according to local information (residual energy). In the SPCA, the interval of residual energy is uniformly divided into α sections. Each section corresponds to one transmission power from maximum power to minimum power. In the PPCA, the transmission power is polynomial function of residual energy. When network starts, all nodes transmit data to sink node with DLOR routing and calculated transmission power. Simulation results show that by selecting the appropriate parameters, SPCA and PPCA can prolong network lifetime and reduce average energy consumption. Under certain conditions, it is fit for wireless sensor networks in special situation.

Keywords: Wireless Sensor Networks, Distributed, Power Control, Lifetime.

1 Introduction

Wireless sensor networks were originated in the military field. Now, low-cost wireless sensor networks have been used in target tracking, environmental monitoring, flood warning, farm management, smart home and other fields. They get more and more attention from academia and industry. In the wireless sensor networks, transmission power control is one direction of topology control research. It primarily regulates the transmission power of each node. If the premise of network connectivity with power control algorithm is meted, the number of single-hop neighbor nodes is balanced, unnecessary communication links are removed, and an optimal forwarding network topology comes out [1].

At present, scholars proposed many new power control algorithms such as VRTPC[2], PCAP[3], LMA and LMN [4], CBTC [5] and etc. Reference [6] proposed DLOR (distributed lifetime optimized routing algorithm) for wireless sensor networks. In the DLOR, every node maintains a neighbor information table. According to the information, the link weight is calculated with new function and

M. Zhu (Ed.): Electrical Engineering and Control, LNEE 98, pp. 319–326.
springerlink.com © Springer-Verlag Berlin Heidelberg 2011

distributed asynchronous Bellman-Ford algorithm is used to construct the shortest routing tree. Then data are gathered along the shortest routing tree to sink node. But DLOR needs a method to measure the distances to neighbor nodes. In the real environment, the RSSI value is greatly influenced by the surrounding environment, especially harsh environment. It is difficult to accurately measure the distances. Other methods need additional hardware and energy costs. In order to prolong network lifetime when the nodes can't measure the distances or have distance measuring hardwire, this paper develops the research accomplishment mentioned in [6], and proposes two distributed power control algorithms for wireless sensor networks. In the algorithms, transmission power is only influenced by the residual energy. The algorithms can prolong the network lifetime and reduce network energy consumption.

2 System Assumptions and Link Energy Consumption Model

In the research, assume that:

- Sink node and other nodes are static with fixed location. Moreover, sink node doesn't know the topology information of whole network;
- Ordinary nodes have the same performance (such as initial energy, energy consumption parameters, communication radius). But every node has different transmission power and can changes;
- Ordinary nodes have the same energy model;
- Sink node gathers data regularly, and ordinary nodes transmit data to sink node directly or in multi-hop way;
- Each ordinary node's energy is limited and sink node's energy is not limited;
- Ordinary nodes are in the harsh environment, or have not distance measuring hardwire, so nodes are unable to know its own location and the distance to neighbor nodes.

Typical node energy consumption is mainly generated by wireless data transceiver. Transmission energy consumption E_{Tx} contains the electronic energy consumption of transmission circuit $E_{\text{Tx-elec}}$ and energy consumption of signal amplifier $E_{\text{Tx-amp}}$. $E_{\text{Tx-elec}}$ is fixed at kE_{elec}, k represents the amount of transmitting data. $E_{\text{Tx-amp}}$ relates to transmission power P_{t}. Receiver only considers the electronic energy consumption kE_{elec}. The formulas can be described as follows:

$$E_{\text{Tx}}(k,d) = E_{\text{Tx-elec}}(k) + E_{\text{Tx-amp}}(k,d) = \begin{cases} kE_{\text{elec}} + kE_{\text{amp}}P_t^{\gamma}, & d < d_{\max} \\ 0, & d \ge d_{\max} \end{cases}. \quad (1)$$

$$E_{\text{Rx}}(k) = E_{\text{Rx-elec}}(k) = kE_{\text{elec}}. \quad (2)$$

Therefore, according to (1) and (2), energy consumption for transmitting k bits data from node i to node j is:

$$C_{i,j}(k, d_{i,j}) = 2kE_{\text{elec}} + kE_{\text{amp}}P_t^{\gamma}. \quad (3)$$

Energy consumption for transmitting k bits data from node i to sink node is:

$$C_{i,s}(k,d_{i,s}) = kE_{\text{elec}} + kE_{\text{amp}}P_t^{\gamma}. \tag{4}$$

3 Algorithm Description

3.1 Basic Principle

If nodes can't measure the distances to neighbor nodes, in DLOR they transmit data with default maximum power. The network lifetime is great decline. But in this situation, nodes can get the residual energy of itself and neighbor nodes. When its residual energy reduces, the node can adjust the transmission power to save energy. Therefore, SPCA (stepwise power control algorithm) and PPCA (polynomial power control algorithm) are proposed. Their transmission power relates to the residual energy [7].

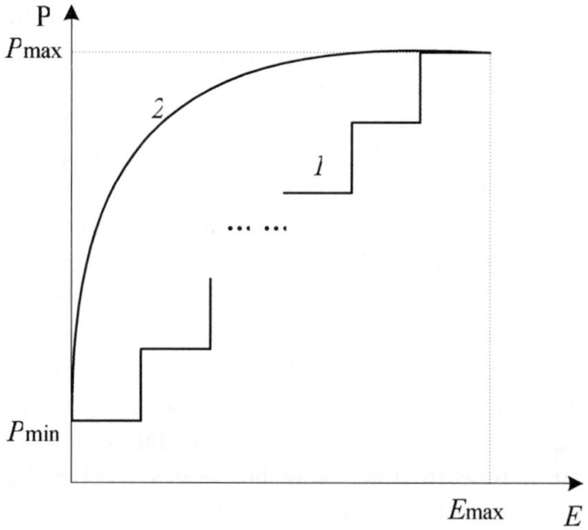

Fig. 1. The curves of SPCA and PPCA

As shown in fig.1, the curve 1 represents SPCA. The interval of residual energy is uniformly divided into α sections. Each section corresponds to one transmission power. The transmission power is uniformly distributed in the interval between maximum and minimum. The formula is as follow.

$$P_t(i) = \begin{cases} P_{\max} & E(i) = E_{\max}\ or\ \alpha = 1 \\ \dfrac{(P_{\max} - P_{\min}(i))}{\alpha - 1}(\text{floor}(\dfrac{\alpha * E(i)}{E_{\max}})) + P_{\min}(i) & 0 < E(i) < E_{\max}, \alpha \geq 2, \alpha \in \mathbb{N}^+ \end{cases} \tag{5}$$

Where, $P_t(i)$ represents the transmission power of node i; P_{max} represents default maximum transmission power; P_{min} represents minimum transmission power which node i can use to communicate with at least one neighbor node; $E(i)$ represents the residual energy of node i; E_{max} represents initial energy of nodes; α represents the number of sections which the energy interval is divided into; Function floor(x) rounds the element x to the nearest integers less than or equal to x.

As shown in fig.1, the curve 2 represents PPCA. In the PPCA, the transmission power is polynomial function of residual energy. Its curve passes through the points (E_{max}, P_{max}) and (0, P_{min}). The formula is as follow.

$$P_t(i) = -\frac{P_{max} - P_{min}(i)}{E_{max}^n}(E(i) - E_{max})^n + P_{max}, 0 \le E(i) \le E_{max}, n \ge 2, n \in \mathbb{N}^+ \quad (6)$$

Where, n represents the polynomial order.

3.2 Realization of SPCA and PPCA

If power control algorithm works, it needs a routing algorithm. So the routing algorithm DLOR is used to verify the algorithm's effectiveness. The realization of SPCA and PPCA with DLOR is as follow.

When routing starts, nodes have zero transmission power and need to get the minimum transmission power. They increase transmission power gradually to broadcast querying packets. Neighbor nodes receive the packet and relay an informing packet. Then, the nodes receive the packet and record minimum transmission power. After all nodes have the minimum transmission power, the DLOR algorithm works to construct the shortest routing tree. In the DLOR, each node has a neighbor information table which records residual energy of itself and neighbor nodes. Then according to information table, nodes execute (5) or (6) to calculate the transmission power P_t. The nodes transmit data with P_t along the shortest path.

In routing maintenance, nodes transmit the latest forecast packets to all neighbor nodes from time to time. Neighbor nodes receive the forecast packets, update the neighbor information table, and execute (5) or (6) to update transmission power.

4 Simulation Result and Analysis

The simulation doesn't consider the energy consumption of routing establishment, routing maintenance, routing failure, timeout retransmission, data calculation and some others, but only consider the energy consumption of wireless communication. Network lifetime is defined as the period of time until the first node runs out of energy or isolated node exists. DGC (data gathering cycle) represents network lifetime. DGC is the time that all nodes successfully transmit data to sink node. Average energy consumption = the total number of energy all nodes consume / (DGC*number of nodes).

In the simulation, $500*500m^2$ network simulation area is chosen. 40, 50, 60, 70, 80, 90 and 100 numbers of nodes and one sink node are generated. According to given numbers of nodes, 20 different network topologies are randomly generated. Then the network lifetime and average energy consumption of algorithms are compared. The other needed parameters are as follows [8]. The initial energy of nodes E_{max} is 1000J. The maximum transmission power P_{max} is 0.37W. Electronic energy consumption for transmitting and receiving data E_{elec} is 50nJ/bit. Energy consumption for amplifying signal E_{amp} is $10.8*10^{-6}$. Amplification factor γ is 1. Data size that node transmits to sink node every time k is 1Kb. Route update interval of network (times of node data transmission) is 1000.

4.1 Research on Algorithm's Parameters

Selection of α in the SPCA. When $\alpha \in \mathbb{N}^+$, $\alpha = 1,2,\ldots,50$ are chosen in the network with 100 nodes. As shown in fig.2, it is concluded that when $\alpha<10$ and α increases, SPCA has longer network lifetime. It is reason that when α increases, the interval of energy is divided into more sections, and nodes have more transmission power changing. Nodes which have lower residual energy transmit data with lower power. Therefore, it affects the network topology, reduces the energy consumption and prolongs network lifetime. But when $\alpha>10$, the interval is divided too much sections and the small change of transmission power don't affect network topology. Therefore, the network lifetime fluctuates around the optimal value. In summary, choosing the appropriate α ($\alpha=10$), DLOR with SPCA has longer network lifetime.

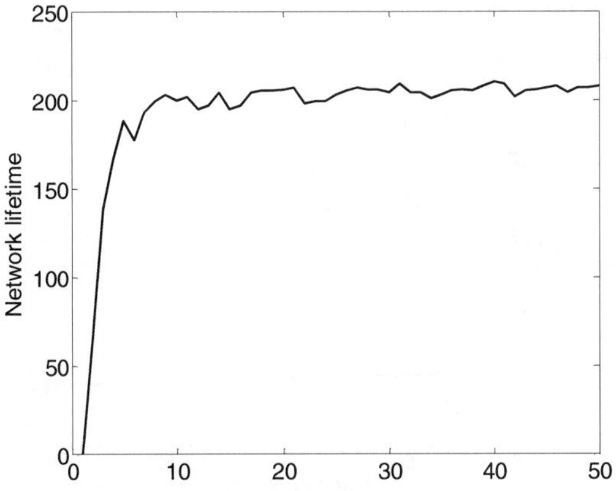

Fig. 2. The network lifetime when α changes

Selection of *n* in the PPCA. When $n \in \mathbb{N}^+$, $n = 2,\ldots,30$ are chosen in the network with 100 nodes. As shown in fig.3, it is concluded that when n increases, network lifetime will shorten. It is reason that when n increases, all nodes tend to transmit data with larger power. Thus, the network lifetime shortens. In summary, choosing the appropriate n (n=2), DLOR with PPCA has longer network lifetime.

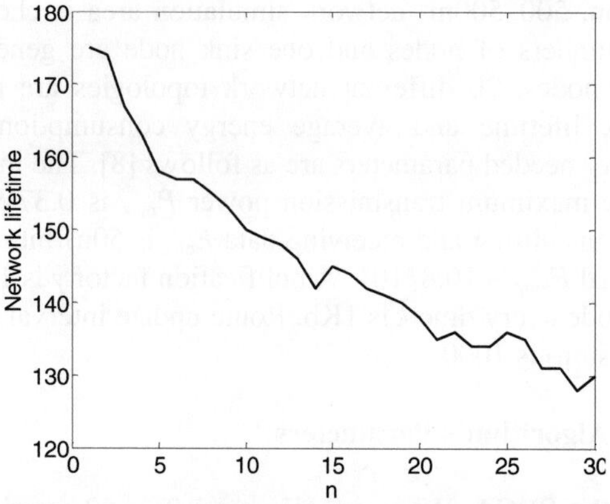

Fig. 3. The network lifetime when n changes

4.2 Simulation Results

According to the parameter research, α=10 and n=2 are chosen. Then the performances of the DLOR with maximum transmission power, SPCA and PPCA are compared and analyzed.

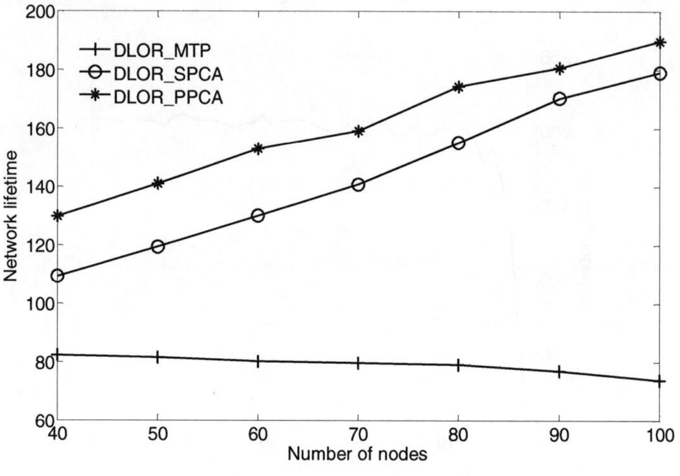

Fig. 4. Network lifetime comparison

As shown in Fig.4, when network has the same routing algorithm DLOR, PPCA has the optimal network lifetime. SPCA has a little lower. But they have far longer network lifetime than maximum transmission power. The reason is that SPCA and PPCA lower transmission power to save energy when the residual energy of nodes reduces. In the PPCA, the change of transmission power is second-order. The decline rate is faster and it fit for the topology. It has longer lifetime than SPCA. Therefore, SPCA and PPCA can prolong the network lifetime.

As shown in Fig.5, SPCA has the lowest average energy consumption. PPCA has a little higher. But they have far lower average energy consumption than maximum transmission power. It is reason that SPCA and PPCA balance the node energy consumption by adjusting the network topology. The PPCA has longer lifetime. Relatively nodes transmit more data and the network energy consumption is higher. Therefore, SPCA and PPCA can reduce average energy consumption.

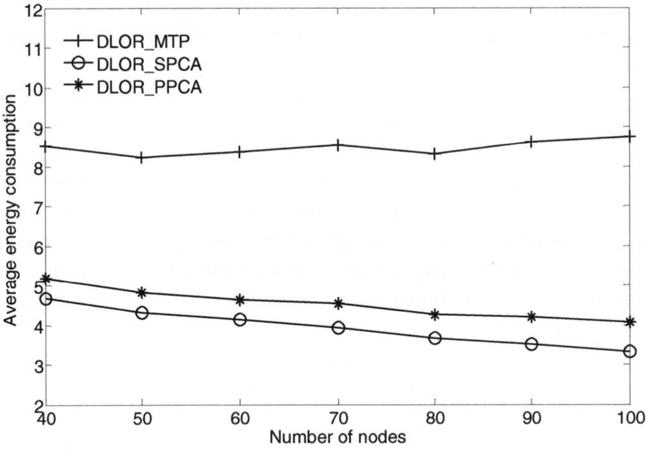

Fig. 5. Average energy consumption comparison

5 Conclusion

If nodes are unable to measure the distance to neighbor nodes or have not distance measuring hardwire, distributed power control algorithms are used to reduce the transmission power with residual energy for wireless sensor networks. First SPCA and PPCA are introduced. In the SPCA, the interval of residual energy is uniformly divided into α sections. Each section corresponds to one transmission power from maximum power to minimum power. In the PPCA, the transmission power is polynomial function of residual energy. Then the realization of SPCA and PPCA are described. All nodes use the local information to reduce energy consumption. Finally, network lifetime and average energy consumption of algorithms are compared and analyzed. SPCA and PPCA can prolong network lifetime and reduce average energy consumption. In the hardware environment, SPCA is easy to realize and PPCA needs certain resources to realize mathematical operation. The future work is that SPCA and PPCA are applied to the actual nodes with operating system TinyOS.

Acknowledgments. This project is supported by a general project, Zhejiang provincial education department of China under grant Y201018705.

References

1. Akyildiz, I.F., Su, W.l.: A survey on sensor networks. IEEE Communications Magazine 40, 2-116 (2002)
2. Gomez, J., Campbell, A.T.: Variable-range transmission power control in wireless Ad hoc networks. IEEE Transactions on Mobile Computing 6, 87–99 (2007)
3. Wen, K., Guo, W., Huang, G.J.: A power control algorithm based on position of node in the wireless Ad hoc networks. Journal of Electronics and Information Technology 31, 201–205 (2009)
4. Kubisch, M., Karl, H., Wolisz, A., et al.: Distributed algorithms for transmission power control in wireless sensor networks. In: Proceedings of IEEE Wireless Communications and Networking Conference, pp. 132–137. IEEE, New Orleans (2003)
5. Li, L., Halpern, J.Y., Bahl, P., et al.: A cone-based distributed topology control algorithm for wireless multi-hop networks. IEEE/ACM Transactions on Networking 13, 147–159 (2005)
6. Chen, Y.R., Lu, L., Dong, Q.F., Hong, Z.: Distributed lifetime optimized routing algorithm for wireless sensor networks. Applied Mechanics and Materials 40-41, 448–452 (2011)
7. Minhas, R.M., Gopalakrishnan, S., Leung, V.C.M., Leung, V.C.M.: An online multipath routing algorithm for maximizing lifetime in wireless sensor networks. In: 2009 Sixth International Conference on Information Technology, New Generations, pp. 581–586 (2009)
8. Wendi, B.H.: Application-specific protocol architectures for wireless networks, pp. 56–109. Massachusetts Institute of Technology, Boston (2000)

System Framework and Standards of Ground Control Station of Unmanned Aircraft System

Jia Zeng

China Aero Polytechnology Establishment, 1665 mailbox, Beijing, 100028, China
zengjia1980@sina.com

Abstract. The basic functions and the system framework of Ground Control Station(GCS) of Unmanned Aircraft System(UAS) are introduced firstly. The actuality of and the gap between the standards relative to the GCS of UAS at home and abroad are analyzed secondly. The suggestions on research and development of the GCS standardization are proposed finally, in which the function, the framework and the domestic usage requirements of GCS and UAS are considered.

Keywords: Unmanned Aircraft System(UAS), Ground Control Station (GCS), Standards, System Framework.

1 Introduction

Ground Control Station (GCS) is one of the key components of Unmanned Aircraft System(UAS). It is a data processing/distribution, integrated display and control unit consisting of computers and other equipments. GCS implements the measurement and control of the payloads on the UAV by the operators.

With the development of the information technology, the application of UAS becomes more and more versatile. GCS must adapt to the complicated application requirements, adopt modular design/implementation approach and formulates modules based on GCS function but independent of each other. Then GCS can be assembled by the basic functional modules flexibly according to application requirements, which is its openness.

With the traction of the joint usage mode of UAS, the idea of "One GCS, Multiple UAVs" gradually replace that of "One GCS, One UAV". Thus GCS must adopt general interface, to improve its universality, and make it possible to form communication link with arbitrary UAVs and payloads. This will enhance the interoperability between UASs, and reduce the influence on UAS when the GCS fails.

The openness and interoperability of GCS is the direction of development and its essential features. And research on standardization is an effective approach to improve such characteristic of its system framework. Thus, to conclude the function and system framework of GCS, and analyse the current situation of related standards and the gap between home and abroad has significant importance for the standardization for CGS of our own.

M. Zhu (Ed.): Electrical Engineering and Control, LNEE 98, pp. 327–334.
springerlink.com © Springer-Verlag Berlin Heidelberg 2011

2 System Framework of GCS

2.1 Fundamental Functions of GCS

Basic functions of CGS are approximately the same, i.e. to implement the effective control of UAS, mentioned in[1] to[3]. Detailed descriptions are as follows.

(a) Mission planning(including monitoring of UAV positions and display of trajectories) --- Considering to the environment and mission information, GCS schedule the mission plans and trajectories both on time domain and space domain. UAV then complete the mission according to the planned trajectories.

(b) Attitude control of UAV(including each flight phase) --- According to flight status data from on-board sensors, GCS calculates control input by the control law and sends it to flight control computer. Flight control computer then processes these information and sends out the control command, implementing the attitude control of UAV.

(c) Payload control(including display of payload parameters) --- Payloads are those devices mounted on UAV for a certain mission. They are the mission execution units of UAV, and the UAV is the platform and carrier for these devices. GCS controls the payload according to the mission requirements, and monitor the mission via display of the payload parameters.

(d) Navigation and positioning --- The navigation of GCS functions when unexpected situation occurs while UAV is on the planned trajectory. It controls and navigates the UAV, and make it fly within safe trajectory. The positioning function of GCS implements the positioning of both the target and UAV, for the convenience of recovery and mission accomplishment of the UAV.

(e) Communication and data processing --- The data link of GCS is used to command/control and distribute the information UAV collects, implementing timely and effective data and command transmission. GCS has data fusion and analysis functions for the acquired information from sensors.

(f) Integrated display and monitoring --- The integrated display provides effective display of all UAS parameters. While the monitoring mainly measure and analysis important data during the UAV mission.

2.2 Description of System Framework

The fundamental functions of CGS, together with its direction of development determine the system framework. Based on the above description of the functions, the system frameworkof CGS can be illustrated in Fig. 1, concluding of [4] and [5].

In Fig.1. Flight operation and management module includes function of launch and recovery control, aircraft attitude control, flight condition monitoring, navigation and positioning, etc. Mission planning and management module includes functions of mission planning, map and trajectory display, etc. Communication and data processing module includes functions of record and replay, link communications, data processing, etc. Thus, GCS facilitates with equipments and system corresponding to those functions. Besides, GCS interacts with UAV operator through integrated display and I/O devices, and interconnects with other C^4I system in cooperative network through data-link communications.

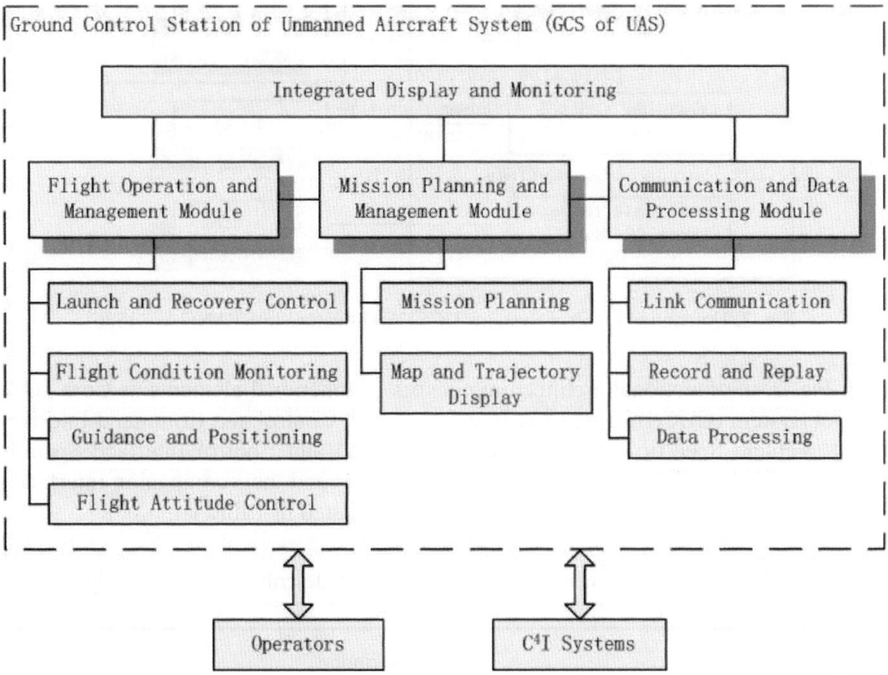

Fig. 1. System Frameworkof CGS

Not only does the framework of GCS characterise the internal functional modules and external relationships between each equipment which highlight the thought of modular design, it also describes the interconnection with other systems. For the internal data communication and external interconnection, general interface standards is essential for GCS. Currently, research and establishment of related standards on GCS has started both at home and abroad.

3 Related Standards on GCS of UAS Worldwide

3.1 Overseas Related Standards

NATO developed the generic standard interface for the interoperability of GCS, named STANGAG 4586 "NATO Unmanned Control System (UCS) Architecture", see [6], which established the functional architecture and interface requirements for UAV control system. This architecture is illustrated in Fig.2.

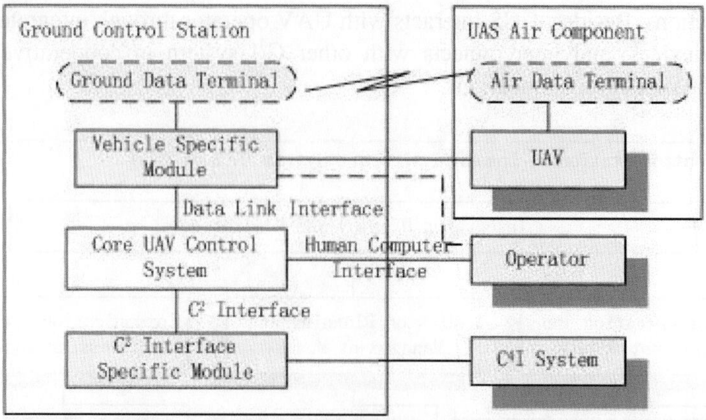

Fig. 2. Unmanned Control System (UCS) Architecture in STANGAG 4586

In Fig. 2, the functional architecture consists of functional elements as Core UAV Control System (CUCS), Data Link Interface (DLI), Control and Command Interface (CCI), Vehicle Specific Module (VSM), Command and Control Interface Specific Module (CCISM), Human Computer Interface (HCI) and their companion interfaces. [reference]Detailed description of the functions are listed in Table 1.

Table 1. Function components and Interfaces of UCS described in STANAG 4586.

Component & Interface	Description
CUCS	Provides a user interface, including graphical user interface. It shall support the requirements of the DLI, CCI, and HCI. Besides, it provides the UAS operator with the functionality to conduct all phases of a UAS mission and enables a qualified UAS operator the ability to control different types of UAVs and payloads.
VSM	Provides specific communication protocols, interface synchronization and data formats needed by certain aircraft. It shall support the conversion for DLI protocols and message formats according to specific requirements of aircraft.
CCISM	Provides CCI data conversion and packaging for the compatibility of physical communication link between UCS and C^4I system.
DLI	Regulates DLI between CUCS and VSM, which enable the UCS the ability to create and understand the prescribed message. It shall apply to control and condition monitoring on UAVs and payloads.
CCI	Regulates protocols on the layer of the message contents and formats between C^4I system and CUCS. It means the data requirements should be satisfied in a general and standard interface.
HCI	Regulates the requirements of operator display and input supported by CUCS. It includes mission planning, aircraft management, operator and payloads monitoring, communication management, etc.

Besides, American Society for Testing and Materials (ASTM) Committee F-38 Sub-committee F38.03 Work Group specializes in the research and establishment of standards on the UAV system operator training, Certification requirements and other critical fields. Until the end of 2010, two standards had been released which are all suitable for the training of operators for GCS of UAV.

3.2 Domestic Standards

Domestic research and establishment of standards on GCS can be concluded as three parts as follows.

(a) General standards for UAS, including general requirements for GCS.

The standard "General Requirements for Unmanned Aerial Vehicle System" established the general specifications for the GCS equipments, cabin and wiring and requirements for flight operation/management and mission planning equipment. The standard "General Requirements for UAV Measurement and Control System" established the definition of GCS and data link, and the design of GCS. The standard "General specifications on Communication Jamming UAV " established the performance of GCS.

(b) A batch of measurement and control information system interface and data standards of GCS are established.

Research has been carried out on information transmission structure standards of measurement/ control and information transmission systems. Several standards as "General data link protocol for UAV measurement and control information transmission system" are established.

(c) General requirements on some GCS subsystems are established based on GCS structure.

Some subsystem level standards are established by considering the composition of the subsystems and integrating the requirements on research and development of UAS. These standards include general requirements on the subsystem design, functionality and performance, environmental adaptability, Man-Machine Engineering, packaging and transportation.

3.3 Current Status of and Gap between the Standards at Home and Broad

From the above, institutes at home and abroad have carried out research on standards of GCS and established some of them. The current status can be concluded as follows.

(a) GCS standards at home and abroad both focus on the promotion of interoperability.

Current standards on GCS at home and abroad both focus on the internal and external information interfaces of GCS, and have special attention on the interoperability of UAS. This improves the UAS self-integration and the integration with other joint platforms, thus promoting the practice of "One station, multiple UAVs" operation mode. Currently, standards system on the information interface need to be further consummated.

(b) Deficiency of GCS subsystem level standards.

At the moment, no research has been found on subsystem level standards of GCS. Besides, the number of subsystem level standards at home is also not large. Research on the GCS subsystem level standards helps to promote the modularization of GCS subsystems and improve the openness of GCS. Thus further research and establishment on related standards should be carried out.

(c) Deficiency of standards on GCS operator and Human-Machine Engineering design requirements.

Currently, standards on GCS operator has been released in United States ASTM F-38 standard. While the number and coverage for related standards on GCS operator is insufficient at home, and standards on Human-Machine Engineering design is blank. Thus further research and establishment on those standards should be carried out.

4 Suggestions on Research and Establishment of Standards on GCS of UAS

Based on the above analysis, GCS should include the following items for the implementation of its basic functions. (a) Equipments or platforms carrying fundamental modules. i.e. the hardware. (b) Computer coding implementing fundamental functionalities like information interaction, on the equipment and platform. i.e. the software. (c) Related staff for the operation of UAV through GCS, i.e. the operator.

Thus, to improve the modularization and generality of hardware/software and standardize the operator of GCS is an effective approach to improve the openness and interoperability of GCS.

Based on the current status of GCS at home and abroad and with the consideration of the basic functions and requirements, suggestions on the research and establishment of GCS standards at home are proposed from the aspects of hardware, software and operator.

4.1 Suggestions on Research and Establishment on GCS Hardware Standards

The hardware device and platform implementing fundamental functionalities of GCS have certain degree of independence. Based on these functionalities, to establish corresponding standards on hardware devices and platforms could help improve the modular design of GCS and both its openness and interoperability.

Consequently, research on GCS subsystem level standards should be enhanced, including the general requirements, computer in each module and signal processing hardware configurations standards, data link ground terminal and communication equipment standards, display equipment standards, image acquisition and remote measurement facility standards, environment control and survivability protective equipment standards, etc.

4.2 Suggestions on Research and Establishment of GCS Software Standards

(a) Standards on information interface

Standards on information interface forms the basis for the generalizability of GCS, which determines the efficiency of information transmission and interaction between UAVs. According to the fundamental functionalities of GCS, the information data and remote control/measurement commands required by UAV during mission roughly includes the following items. (1)GCS uploads the mission planning result to UAV. (2) Information of flight status acquired by the on-board sensors are transmitted to GCS. (3) GCS uploads remote control/measurement commands to on-board Flight Control Computer. (d)flight data acquired by the control unit in GCS are transmitted to the visual module and supervision module. (4) Data transmission between the control module and navigation module of GCS. The content of the information includes flight parameters, image data,etc. Thus standards on information interface should cover various contents and formats of the data and commands as described above.

Besides, in future applications of UAS, various new payloads will be introduced, such as sensors for space environment sensing. Those new payloads will reveal the limitations of current standards on information interface. Thus, research and establishment of standards on information interface should adopt the existing standards and also carry out exploratory research on the new payloads standards.

(b) Standards on computer software

In GCS, although the computer software implementing fundamental functionalities are only contained in the modules themselves. But coherent standards on the software could improve its readability and portability and thus the generalizability of the modules. So, establishment of standards on GCS software should also be carried out.

4.3 Suggestions on Research and Establishment of GCS Operators Standards

(a) Standards on operator training ,certification and qualification

Establish standards on operator training, certification and qualification based on related standards on GCS operator in ASTM F-38 and regularize the operation of GCS.

(b) Standards on human computer interaction

GCS demands large number of human computer interactions. The most direct and important device in interactive environment is the monitor. And the sensation of the input information by the commander is also accomplished through the monitor. Display interface is the most prominent mode in HCI interface of GCS, including the visual interface, navigation interface, measure and control interface,etc. High-quality design of standards on HCI interface could effectively improve the publicity and friendliness of GCS.

5 Conclusions

With the development of UAS technology, demands on the openness and interoperability of GCS becomes more and more prominent. In the meantime, requirements on the research and establishment of related standards of GCS gets more and more urgent. For the current status and requirements of UAS at home, to investigate, analyse, adopt standards from abroad, complete the standards system of GCS of UAS at home and start the establishment of essential standards is a imperative task, which will promote the openness and interoperability of GCS and give basis for the design, research, production and maintenance of the information transmission products of UAS.

References

1. Natarajan, G.: Ground Control Stations for Unmanned Air Vehicles. J. Aeronautical Development Establishment, Bangalore 560075, 5–6 (2001)
2. Zhang, Z.S.: System Design and Development of UAV GCS. D. North-west polytechnic university, Master Thesis (2007)
3. Zhou, Y.: A Review of UAV GCS Development. J. Avionics Tec. 4(1), 1–6 (2001)
4. Zhang, Y.G.: Development of man-machine interface design and simulation system on UCAV GCS. D.North-west polytechnic university, Master Thesis (2006)
5. General requirements for unmanned aerial vehicle system .S (2005)
6. STANAG 4586.Unmanned Control System (UCS) Architecture.S (1999)

Stabilization of Positive Continuous-Time Interval Systems

Tsung-Ting Li[1], Shen-Lung Tung[2], and Yau-Tarng Juang[1]

[1] Department of Electrical Engineering, National Central University, Taoyuan, Taiwan 320, ROC
985201011@cc.ncu.edu.tw,
ytjuang@ee.ncu.edu.tw
[2] Telecommunication Laboratories Chunghwa Telecom Co., Ltd., Taoyuan, Taiwan 326, ROC
tung168@cht.com.tw

Abstract. This paper is concerned with the stability and stabilization of continuous-time interval systems. For stability analysis and stabilization of positive continuous-time interval systems, new necessary and sufficient conditions are derived. In particular, the proposed conditions can be easily implemented by using linear programming method. It is utilized to stabilize the system being positive and asymptotically stable via dynamic state feedback control. Finally, we provide an example to demonstrate the effectiveness and applicability of the theoretical results.

Keywords: Positive systems, stabilization, linear programming.

1 Introduction

Positive systems mean that the state variables are nonnegative at all times whenever the initial conditions are nonnegative. In the literature [1]-[4], many physical systems and applications are positive in the real world. For example, some applications include population numbers of animals, absolute temperature, chemical reaction, heat exchangers. Since positive systems have numerous applications in various areas, the stability analysis and synthesis problems for positive systems are important and interesting. In the recent years, therefore many results of positive systems have been presented [5]-[19].

Kaczorek [14] use Gersgorin's theorem and quadratic programming to establish a sufficient condition. Recently, some results of state-feedback controller have been obtained by linear matrix inequality (LMI) and linear programming (LP) in [15] and [16], respectively. The necessary and sufficient conditions by using a vertex algorithmic approach are obtained in [17]. Shu [18] fully investigates the observers and dynamic output-feedback controller problems of the positive interval linear systems, with time delay is presented in [19].

M. Zhu (Ed.): Electrical Engineering and Control, LNEE 98, pp. 335–342.
springerlink.com © Springer-Verlag Berlin Heidelberg 2011

In this paper, a necessary and sufficient condition is proposed to solve the positivity and stability problem of interval systems. Combining the proposed condition with linear programming to establish the state-feedback controller, then the closed-loop system is not only asymptotically stable, but also positive.

The paper is organized as follows. Section 2 gives the notations and characterizations of positive linear systems. In Section 3, we derive necessary and sufficient conditions for stability and stabilization problems. Example is given in Section 4 to illustrative the proposed method. Finally, we conclude the paper in Section 5.

2 Notations and Preliminaries

Let $R^n(R_+^n)$ be the *n*-dimensional real (positive) vector space; $R^{n\times m}$ is the set of $n\times m$-dimensional matrix. For a matrix $A\in R^{n\times m}$, $[a_{ij}]$ denotes the element located at *i*th row and *j*th column. A matrix A is said to be Metzler, if all off-diagonal entries are nonnegative. The notation $A\in[\underline{A},\overline{A}]$ means that $[\underline{a}_{ij}]\le[a_{ij}]\le[\overline{a}_{ij}]$. A^T represents the transpose of matrix A.

Consider a continuous-time linear system:

$$\dot{x}(t)=Ax(t) \tag{1}$$

where $x(t)\in R^n$ is the system state and $A\in R^{n\times n}$.

Definition 1 [2]. The system (1) is said to be positive if and only if for any initial conditions $x(0)=x_0\ge 0$, the corresponding trajectory $x(t)\ge 0$ for all $t\ge 0$.

Next, we recall some characterizations of continuous-time positive linear systems.

Lemma 1 [2]. The system (1) is positive if and only if the matrix A is Metzler matrix.

Lemma 2 [16]. Continuous-time positive linear system described by (1) is stable if and only if there exist a $d\in R_+^n$ with $Ad<0$.

In next section, we will use above lemmas to derive necessary and sufficient conditions.

3 Main Results

3.1 Stability of Continuous-Time Interval Systems under the Positivity Constraint

Consider the following continuous-time interval linear system described by the following state equation:

$$\dot{x}(t)=Ax(t),\ x_0\in R_+^n \tag{2}$$

where $A \in [\underline{A}, \overline{A}]$. Then, a necessary and sufficient condition of system (2) is provided as follow.

Theorem 1. Continuous-time interval linear system is positive and asymptotically stable if and only if $\underline{A}$ is Metzler matrix and $\overline{A}d < 0$ with $d \in R_+^n$.

Proof:
Necessity: According to Lemma 1 and 2, the system (2) is positive and asymptotically stable, then $\underline{A}$ is Metzler matrix and $\overline{A}d < 0$.

Sufficiency: Owing to the matrix $\underline{A}$ is Metzler, we know

$$[\underline{a}_{ij}] \geq 0 \tag{3}$$

for $1 \leq i \neq j \leq n$. For any $A \in [\underline{A}, \overline{A}]$, it obvious that

$$[a_{ij}] \geq [\underline{a}_{ij}] \geq 0 \tag{4}$$

for $1 \leq i \neq j \leq n$. Eq. (4) shows that the off-diagonal entries of the matrices A are nonnegative. Therefore, the system (2) is positive. Because the transpose matrices of Metzler A are also Metzler, we know that the system

$$\dot{x}(t) = A^T x(t), \; x_0 \in R_+^n. \tag{5}$$

Next the positive system (5) with $x(t) \geq 0$ that there exists Lyapunov functions of the form

$$v(x) = x^T d, \; d \in R_+^n. \tag{6}$$

Because $x(t) \geq 0$ and $d \in R_+^n$, we obtain

$$v(x) > 0. \tag{7}$$

Then by taking the derivative of Eq. (6), we have

$$\dot{v}(x) = x^T Ad. \tag{8}$$

By $A \in [\underline{A}, \overline{A}]$ and $d \in R_+^n$, we obtain

$$\begin{aligned} \dot{v}(x) &= x^T Ad \\ &\leq x^T \overline{A} d \\ &< 0. \end{aligned} \tag{9}$$

From Eqs. (7) and (9), we know the positive system (5) is asymptotically stable. The positive system (5) is asymptotically stable if and only if the positive system (2) is asymptotically stable. Therefore, the system as describe in (2) is positive and asymptotically stable if the matrix $\underline{A}$ is Metzler and $\overline{A}d < 0$.
The proof is complete. Q.E.D.

3.2 Stabilization of Continuous-Time Interval Systems under the Positivity Constraint

Consider the following continuous-time interval linear system:

$$\dot{x}(t) = Ax(t) + Bu(t)\,,\; x_0 \in R_+^n \tag{10}$$

with the dynamic state-feedback control $u = Kx$, the matrix $K \in R^{m\times n}$ is defined as follow:

$$K = ZD^{-1} \tag{11}$$

where $Z = [z_{ij}] \in R^{m\times n}$ and $D = diag\{d_1, d_2, ..., d_n\}$. Here, $[a_{ij}] \in [\underline{a}_{ij}, \overline{a}_{ij}] \in R^{n\times n}$ and $[b_{ij}] \in [\underline{b}_{ij}, \overline{b}_{ij}] \in R^{n\times m}$. Then the closed-loop system is

$$\dot{x}(t) = (A + BK)x(t)\,. \tag{12}$$

A necessary and sufficient condition of positive and stabilization for the closed-loop system (12) is given as follows.

Theorem 2. The closed-loop system (12) is positive and asymptotically stable if and only if there exist $[d_j]$ and $[z_{ij}]$ such that

$$\underline{a}_{ij} d_j + \sum_{e=1}^{m} \min\left\langle \underline{b}_{ie} z_{ej}, \overline{b}_{ie} z_{ej} \right\rangle \geq 0\,,\; 1 \leq i \neq j \leq n \tag{13}$$

$$\sum_{j=1}^{n}\left(\overline{a}_{ij} d_j + \sum_{e=1}^{m} \max\left\langle \underline{b}_{ie} z_{ej}, \overline{b}_{ie} z_{ej} \right\rangle \right) < 0\,,\; 1 \leq i \leq n \tag{14}$$

Proof:
Necessity: According to Theorem 1, the closed-loop system (12) is positive and asymptotically stable, then we have $\underline{A+BK}$ is a Metzler matrix and $\left(\overline{A+BK}\right)d < 0$. Owing the off-diagonal entries of $\underline{A+BK}$ be nonnegative and $d \in R_+^n$, the components of $\left(\underline{A+BK}\right)d$ include

$$\left(\underline{a}_{ij} + \sum_{q=1}^{m} \min\left\langle \underline{b}_{iq} k_{qj}, \overline{b}_{iq} k_{qj} \right\rangle \right) d_j = \underline{a}_{ij} d_j + \sum_{q=1}^{m} \min\left\langle \underline{b}_{iq} z_{qj}, \overline{b}_{iq} z_{qj} \right\rangle \geq 0 \tag{15}$$

for $1 \leq i \neq j \leq n$. Moreover, $\left(\overline{A+BK}\right)d < 0$ is equivalent to

$$\sum_{j=1}^{n}\left(\overline{a}_{ij} + \sum_{q=1}^{m} \max\left\langle \underline{b}_{iq} k_{qj}, \overline{b}_{iq} k_{qj} \right\rangle \right) d_j = \sum_{j=1}^{n}\left(\overline{a}_{ij} d_j + \sum_{q=1}^{m} \max\left\langle \underline{b}_{iq} z_{qj}, \overline{b}_{iq} z_{qj} \right\rangle \right) < 0 \tag{16}$$

for $1 \leq i \leq n$.

Sufficiency: We will prove that the conditions (13) and (14) hold, then the closed-loop system (12) is positive and asymptotically stable. The remaining proof is divided into two parts: positivity and stability.

(a) Positivity

Due to $\underline{a}_{ij}d_j + \sum_{q=1}^{m}\min\left\langle \underline{b}_{iq}z_{qj}, \overline{b}_{iq}z_{qj}\right\rangle \geq 0$, $b_{iq}z_{qj} \geq \min\left\langle \underline{b}_{iq}z_{qj}, \overline{b}_{iq}z_{qj}\right\rangle$, $\overline{a}_{ij} \geq a_{ij} \geq \underline{a}_{ij}$, and $d_j > 0$, for $1 \leq i \neq j \leq n$, $q = 1,\ldots,m$, then we have

$$\begin{aligned} a_{ij}d_j + \sum_{q=1}^{m} b_{iq}z_{qj} &\geq \underline{a}_{ij}d_j + \sum_{q=1}^{m} b_{iq}z_{qj} \\ &\geq \underline{a}_{ij}d_j + \sum_{q=1}^{m}\min\left\langle \underline{b}_{iq}z_{qj}, \overline{b}_{iq}z_{qj}\right\rangle \\ &\geq 0 \end{aligned} \tag{17}$$

for $1 \leq i \neq j \leq n$. Owing to $k_{ij} = z_{ij}d_j^{-1}$, we obtain

$$a_{ij} + \sum_{q=1}^{m} b_{iq}k_{qi} \geq 0. \tag{18}$$

for $1 \leq i \neq j \leq n$. According to Eq. (18), we can know the off-diagonal entries of the matrices $A + BK$ are nonnegative. By Lemma 1, the closed-loop system (12) is a positive system.

(b) Stability

Due to $\sum_{j=1}^{n}\left(\overline{a}_{ij}d_j + \sum_{q=1}^{m}\max\left\langle \underline{b}_{iq}z_{qj}, \overline{b}_{iq}z_{qj}\right\rangle\right) < 0$, $b_{iq}z_{qj} \leq \max\left\langle \underline{b}_{iq}z_{qj}, \overline{b}_{iq}z_{qj}\right\rangle$, $\overline{a}_{ij} \geq a_{ij} \geq \underline{a}_{ij}$, and $d_j > 0$, for $1 \leq i, j \leq n$, $q = 1,\ldots,m$, then we have

$$\begin{aligned} \sum_{j=1}^{n}\left(a_{ij}d_j + \sum_{q=1}^{m} b_{iq}z_{qj}\right) &\leq \left(\sum_{j=1}^{n}\overline{a}_{ij}d_j + \sum_{q=1}^{m} b_{iq}z_{qj}\right) \\ &\leq \sum_{j=1}^{n}\left(\overline{a}_{ij}d_j + \sum_{q=1}^{m}\left\langle \max \underline{b}_{iq}z_{qj}, \overline{b}_{iq}z_{qj}\right\rangle\right) \\ &< 0 \end{aligned} \tag{19}$$

for $1 \leq i \leq n$. Owing to $k_{ij} = z_{ij}d_j^{-1}$, we obtain

$$\sum_{j=1}^{n}\left(a_{ij} + \sum_{q=1}^{m} b_{iq}k_{qj}\right)d_j < 0. \tag{20}$$

for $1 \leq i \leq n$. According to Eq. (20), we know

$$(A + BK)d < 0. \tag{21}$$

From Eq. (21) and Lemma 2, we know the closed-loop system (12) is asymptotically stable.

The proof is complete. Q.E.D.

Remark 1. It should be pointed out that we do not impose any restriction on the matrices A and B. When the system is not positive, the stabilization problem can be interpreted as enforcing the system to be positive and asymptotically stable. Theorem 2 presents a necessary and sufficient condition for the existence of desired controller. Conditions (13) and (14) are all linear problems. Thus, the variables $[d_j]$ and $[z_{ij}]$ can be solved by using linear programming optimal toolbox [20].

Corollary 1. The system in (12) is positive and asymptotically stable if the following LP problems in variables $[d_j]$ and $[z_{ij}]$ are feasible:

$$\underline{a}_{ij}d_j+\sum_{e=1}^{m}\min\left\langle \underline{b}_{ie}z_{ej},\overline{b}_{ie}z_{ej}\right\rangle \geq 0,\ 1\leq i\neq j\leq n \tag{22}$$

$$\sum_{j=1}^{n}\left(\overline{a}_{ij}d_j+\sum_{e=1}^{m}\max\left\langle \underline{b}_{ie}z_{ej},\overline{b}_{ie}z_{ej}\right\rangle\right)<0,\ 1\leq i\leq n \tag{23}$$

$$d_j>0,\ 1\leq j\leq n \tag{24}$$

4 Illustrative Example

Example 1. Consider a continuous-time interval linear system in (10) described as

$$\overline{A}=\begin{bmatrix}-0.23 & 0.44 & 0.26\\ 0.43 & -0.3 & 0.17\\ -0.05 & -0.06 & -1.49\end{bmatrix},\ \overline{B}=\begin{bmatrix}0.35 & 0.27\\ 0.23 & -0.4\\ -0.3 & -0.4\end{bmatrix}, \tag{25}$$

$$\underline{A}=\begin{bmatrix}-0.25 & 0.41 & 0.21\\ 0.39 & -0.32 & 0.15\\ -0.07 & -0.14 & -1.52\end{bmatrix},\ \underline{B}=\begin{bmatrix}0.24 & 0.25\\ 0.22 & -0.5\\ -0.4 & -0.4\end{bmatrix}. \tag{26}$$

By Corollary 1, we use MATLAB LP Toolbox and obtain

$$Z=\begin{bmatrix}-38.4495 & -9.9963 & 23.8601\\ 20.3118 & 5.7343 & 12.6046\end{bmatrix},\ d=\begin{bmatrix}22.1724\\ 2.2925\\ 3.1949\end{bmatrix}. \tag{27}$$

Then the upper and lower matrices of the closed-loop system (12) are obtained respectively as

$$\overline{A+BK}=\begin{bmatrix}-0.3152 & 0.2483 & 1.9349\\ 0.0817 & -1.2119 & 0.2336\\ 0.1779 & 0.4769 & -2.6890\end{bmatrix}, \tag{28}$$

$$\underline{A+BK} = \begin{bmatrix} -0.4137 & 0.0228 & 1.4744 \\ 0.0079 & -1.3259 & 0 \\ 0.1251 & 0.3415 & -3.9372 \end{bmatrix}. \tag{29}$$

Fig. 1 shows that trajectories of uncompensated systems and compensated systems with initial condition $x(0) = [0.5 \quad 1 \quad 1.5]^T$. The results show that the proposed method can efficiently drive the system to be stable and positive.

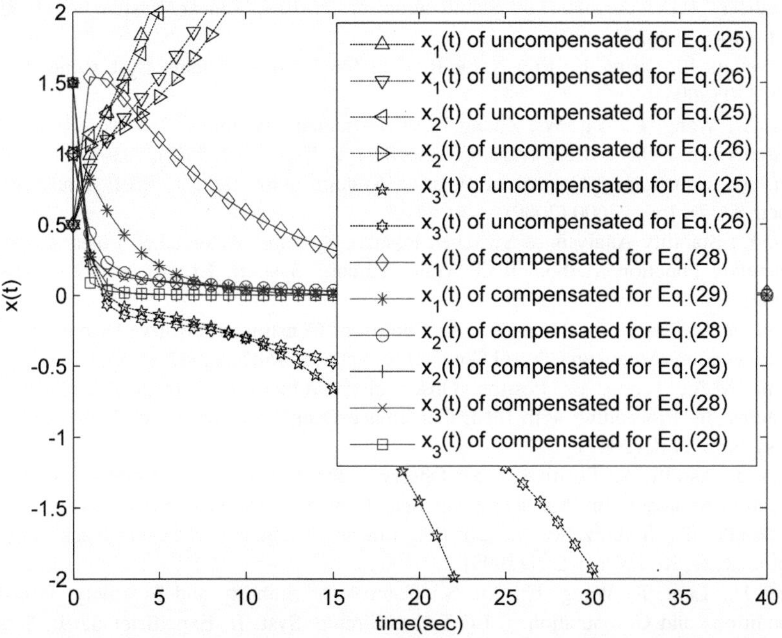

Fig. 1. Trajectories of systems

5 Conclusion

In this paper, some necessary and sufficient conditions are proposed for continuous-time interval linear systems under the positivity constraint. Both stability and stabilization conditions are presented. The proposed conditions are solvable in terms of simple linear programming problems. The illustrative example shows that the obtained conditions are efficient and easy to use.

References

1. Berman, A., Neumann, M., Stern, R.J.: Nonnegative Matrices in Dynamic Systems. Wiley, New York (1989)
2. Farina, L., Rinaldi, S.: Positive Linear System: Theory and Application. Wiley, New York (2000)
3. Kaczorek, T.: 1-D and 2-D Systems. Springer, Berlin (2002)
4. Luenberger, D.G.: Introduction to Dynamic Systems. Wiley, New York (1979)
5. Caccetta, L., Foulds, L.R., Rumchev, V.G.: A Positive Linear Discrete-Time Model of Capacity Planning and Its Controllability Properties. Math. Comput. Model 40(1-2), 217–226 (2004)
6. Leenheer, P.D., Aeyels, D.: Stabilization of Positive Linear System. Syst. Contr. Lett. 44(4), 259–271 (2001)
7. Benvenuti, L., Farina, L.: Eigenvalue Regions for Positive Systems. Syst. Contr. Lett. 51, 325–330 (2004)
8. Liu, X., Wang, L., Yu, W., Zhong, S.: Constrained Control of Positive Discrete-Time Systems with Delays. IEEE Trans. Circuits Syst. II, Exp. Brief. 55(2), 193–197 (2008)
9. Liu, X.: Constrained Control of Positive Systems with Delays. IEEE Trans. Autom. Contr. 54(7), 1596–1600 (2009)
10. Liu, X.: Stability Analysis of Switched Positive Systems: A Switched Linear Copositive Lyapunov Function Method. IEEE Trans. Circuits Syst. II, Exp. Brief. 56(5), 414–418 (2009)
11. Rami, M.A., Tadeo, F., Benzaouia, A.: Control of Constrained Positive Discrete Systems. In: Proceeding American Control Conference, pp. 5851–5856 (2007)
12. Rami, M.A., Tadeo, F.: Positive Observation Problem for Linear Discrete Positive Systems. In: Proceeding 45th IEEE Conference Decision Control, pp. 4729–4733. IEEE Press, New York (2007)
13. Back, J., Astolfi, A.: Positive Linear Observers for Positive Linear Systems: A Sylvester Equation Approach. In: Proceeding American Control Conference, pp. 4037–4042 (2006)
14. Kaczorek, T.: Stabilization of Positive Linear System by State-Feedback. Pomiary, Automatyka, Kontrola 3, 2–5 (1999)
15. Gao, H., Lam, J., Wang, C., Xu, S.: Control for Stability and Positivity: Equivalent Conditions and Computation. IEEE Trans. Circuits Syst. II, Exp. Brief. 52(9), 540–544 (2005)
16. Rami, M.A., Tadeo, F.: Controller Synthesis for Positive Linear Systems with Bounded Controls. IEEE Trans. Circuits Syst. II, Exp. Brief. 54(2), 151–155 (2007)
17. Roszak, B., Davison, E.J.: Necessary and Sufficient Conditions for Stabilizability of Positive LTI Systems. Syst. Contr. Lett. 58, 474–481 (2009)
18. Shu, Z., Lam, J., Gao, H., Du, B., Wu, L.: Positive Observers and Dynamic Output-Feedback Controllers for Interval Positive Linear Systems. IEEE Trans. Circuits Syst. I, Regul. Paper 55(10), 3209–3222 (2008)
19. Ling, P., Lam, J., Shu, Z.: Positive Observers for Positive Interval Linear Discrete-Time Delay Systems. In: Proceeding 48th IEEE Conference Decision Control, pp. 6107–6112. IEEE Press, New York (2009)
20. Coleman, T., Branch, M., Grace, A.: Optimization Toolbox for Use with Matlab. The Math Works Inc., Natick (1999)

Influencing Factor Analysis and Simulation of Resonance Mechanism Low-Frequency Oscillation

Yang Xue-tao[1], Song Dun-wen[2], Ding Qiao-lin[1], MA Shi-ying[2], Li Bai-qing[2], and Zhao Xiao-tong[2]

[1] Department of Electrical Engineering, North China Electric Power University, Baoding 071003, Hebei Province, China
[2] China Electric Power Research Institute, Haidian District, Beijing 100192, China
xuetaoyoung@163.com, yn8800@163.com

Abstract. The enhancement of structure and the access of various kind of new energy in power grid do not reduce the risk of the occurrence of low frequency oscillation. It's very important to analyze the influencing factors of low frequency oscillation theoretically in view of low frequency monitoring and analysis. In order to analyze the all kinds of factors of resonance frequency oscillation that caused by continuous cycle small disturbance, this article deduces response characteristics of forced power oscillation from mathematics. Its influencing factors is analyzed from the frequency, amplitude, phase three key elements and electrical disturbances propagation, meanwhile this paper points out the corresponding relationship between the various factors and actual disturbance. Finally the correctness of the analytical results is verified using 3-machine 9-bus system.

Keywords: Low-frequency oscillations, resonance mechanism, influencing factors, BPA simulation.

1 Introduction

Forced oscillation (resonance mechanism) in power system can well explain on many occasions which are non-negative damping power oscillation in recent years. Such oscillation starts fast, maintains oscillation amplitude after start-up and decays rapidly after the disappearance of vibration sources. Literature [1] proposed the concept of resonance mechanism, which focused on the analysis of the main factors of oscillation amplitude. From different engineering disturbance scenario, paper [2]-[6] pointed out the periodic pertubation in turbine, excitation system, generator, load and grid side would cause power oscillation in varying degrees. For low-frequency oscillation of resonance theory can better guide prevention and control in engineering applications, this article references the common features of mechanical resonance and electrical resonance caused by disturbance, analyzes and simulates the influencing factors of resonant oscillation systematically from the three essential elements of response signal and propagation of electromechanical disturbance.

M. Zhu (Ed.): Electrical Engineering and Control, LNEE 98, pp. 343–349.
springerlink.com © Springer-Verlag Berlin Heidelberg 2011

2 Response Theory Analysis of Resonant Low-Frequency Oscillation

Using classical second-order model in generator, the motion equation of single-machine infinite system in figure [1] can be shown as follow:

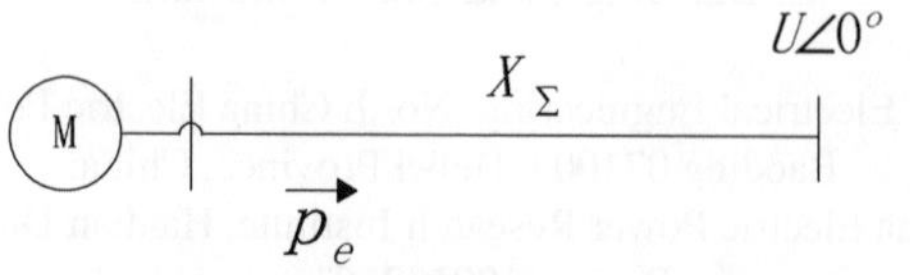

Fig. 1. Single machine infinite bus system diagram

$$\ddot{x}+2\zeta\omega_n\dot{x}+\omega_n^2 x = h\sin\omega t \tag{1}$$

Where ζ is the damping ratio, ω_n is the undamped natural angular frequency,

$$x=\Delta\delta\,,\quad 2\zeta\omega_n=\frac{D}{M}\,,\quad \omega_n^2=\frac{K}{M}\,,\quad h=\frac{F_0}{M}\,.$$

General solution $x_1(t)$ is corresponding to the homogeneous equation of damped free vibration. In the case of weak damping, $0<\zeta<1$, at this time,

$$x_1(t)=Ae^{-\zeta\omega_n t}\sin(\sqrt{\omega_n^2-\zeta^2\omega_n^2}\,t+\phi) \tag{2}$$

Equation (2) is called the transient vibration.

Particular solution can be obtained as follows:

$$x_2(t)=\frac{h}{\sqrt{(\omega_n^2-\omega^2)^2+4\zeta^2\omega_n^2\omega^2}}\sin(\omega t-\psi) \tag{3}$$

Combined with Eq (2) and (3), the full expression of response characteristics of forced oscillation can be obtained.

3 Factors Analysis

3.1 Frequency

From the derivation in the first section we can see that in the harmonic funtion the system response in the initial stage:

$$x=x_1+x_2=Ae^{-\zeta\omega_n t}\sin\left(p't+\phi\right)+B\sin\left(\omega t-\psi\right) \tag{4}$$

From the equation (4), we can see that The influencing factors of oscillation frequency in this stage involve 1) damping ratio, 2) inertia time constant, 3) synchronous moment coefficient and 4) the frequncy of disturbance signal.

3.2 Amplitude

Suppose $B_0 = \dfrac{h}{\omega_n^2}$, $\lambda = \dfrac{\omega}{\omega_n}$ and introduce the concept of amplitude magnification factor β, we can get:

$$\beta = \frac{B}{B_0} = \frac{1}{\sqrt{\left(1-\lambda^2\right)^2 + \left(2\zeta\lambda\right)^2}} \tag{5}$$

Curves as follows can be obtained:

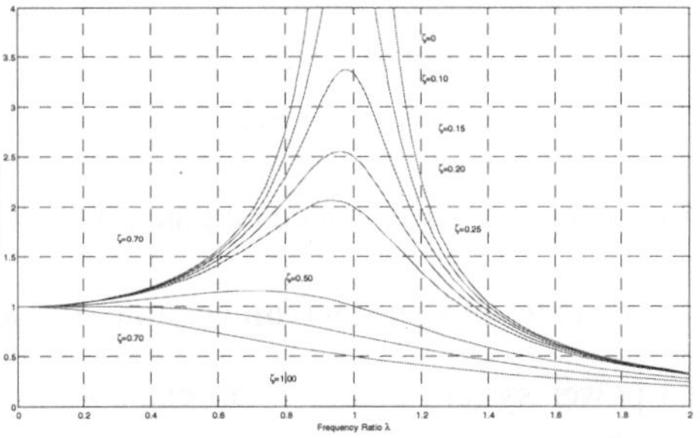

Fig. 2. The impact of frequency ratio and damping coefficient for the amplification

From $B = \dfrac{F_0/K}{2\zeta}$, $K = \dfrac{E'U}{X_\Sigma}\cos\delta_0$, and $\zeta = \dfrac{D}{2M\,\omega_n} = \dfrac{D}{2\sqrt{KM}}$, we can get:

$$B = \frac{F_0\sqrt{MX_\Sigma}}{D\sqrt{E'U\cos\delta_0}} = \frac{F_0}{|D|\sqrt{K/M}} \tag{6}$$

Factors of oscillation amplitude include several aspects as follows,

[1] Formula [6] shows that, when the damping ratio and frequency ratio is fixed, the amplitude of disturbance and power is proportional.

[2] From figure [2] we can see that the amplitude is the largest when the disturbance frequency ratio is approximately equal to 1.

[3] Figure shows that when frequency ratio is fixed, the greater the damping ratio is, the smaller the amplitude is.

[4] When operating parameters are fixed, the greater the reactance is, the larger the oscillation amplitude is.

[5] When the operating parameters and disturbance signal are fixed, the larger the generator inertia is, the greater the oscillation amplitude is.

[6] The greater the initial power angle of generator is, the greater the amplitude caused by resonance is.

[7] The lower the system voltage is, the greater the amplitude of the oscillation

[8] The amplitude of the resonant oscillation will be effected by the load change weakly.

[9] The greater of the participation factor indicates that the activity of the state variables of the related generator is relatively large. At this point, the amplitude caused by the disturbance is larger.

3.3 Phase

The phase difference between the power angle of the generator and the disturbance can be calculated by (7):

$$tg\psi = \frac{2\zeta\omega_n\omega}{\omega_n^2 - \omega^2} = \frac{2\zeta\lambda}{1-\lambda^2} \tag{7}$$

The phase relationship between the disturbance and the power angle depends on λ.

3.4 The Delay Characteristics of the Electromechanical Disturbance

Taking the distributed power system model as an example, paper [9] has given the speed formula of propagation:

$$v^2 = \left(\omega V^2 \sin\theta\right) / 2h|z| \tag{8}$$

4 Simulation

Taking typical 3 machines 9 nodes system as an example, using BPA simulation, the influencing factors such as amplitude, frequency ratio, resonant sites and the propagation are verified.

4.1 The Impact of Amplitude of the Disturbance Signal

Adding the disturbance $2.5\%S_N\sin(12.46t)$ and $5.0\%S_N\sin(12.46t)$ on the side of Generator-3 respectively and setting the disturbance time 0-8 seconds:

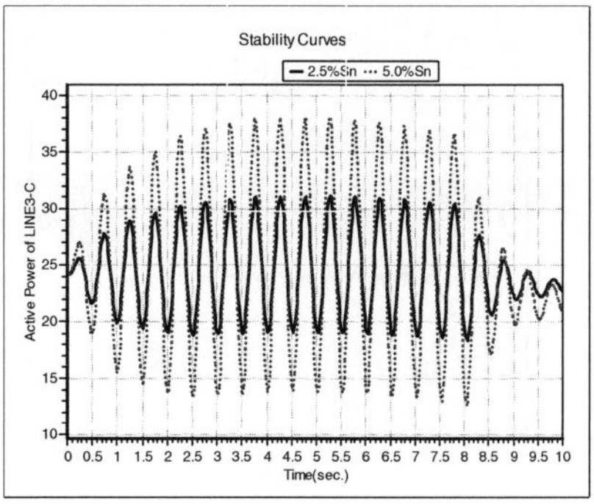

Fig. 3. The impact of disturbance amplitude for the resonant amplitude

4.2 The Influence of Frequency of the Disturbance

Similarly adding the disturbance $5.0\%S_N \sin(12.46t)$ and $5.0\%S_N \sin(37.38t)$ on the side of Generator-3 respectively and setting the disturbance time 0-8 seconds:

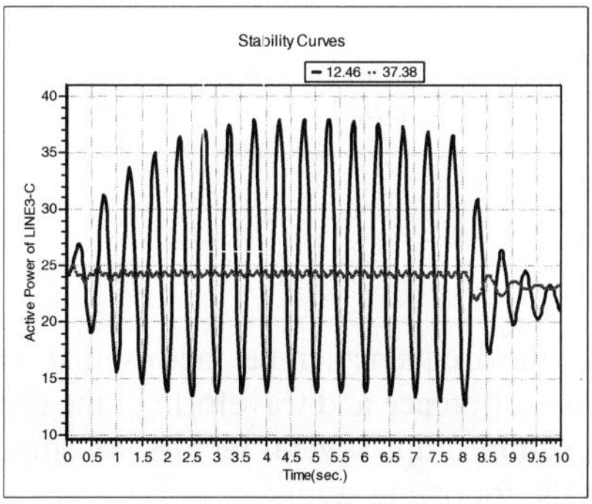

Fig. 4. The impact of frequency ratio for the oscillation amplitude

4.3 The Effects of the Disturbance Location

Adding the same disturbance $5.0\%S_N \sin(12.46t)$ on the side of Generator-1 and Generator-2 respectively, simulation results obtained are as follows:

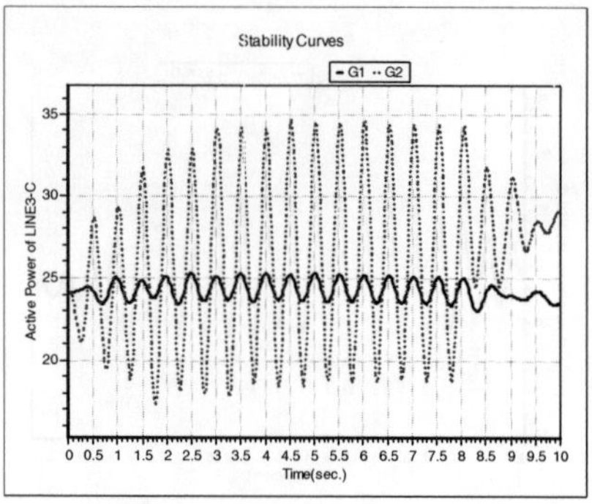

Fig. 5. The impact of disturbance location for the oscillation

4.4 The Delay Characteristic Simulation of the Electromechanical Disturbance

Adding the same small disturbance on the generator G1, G2 and G3 respectively and using curve contrast functions, datas in table 1 are obtained as follows:

Table 1. The first time to achieve extreme of active power of each branch

Disturbance point	LINE2-C	LINE3-C	LINE2-A	LINE1-A	LINE1-B	LINE3-B
G1	0.40	0.40	0.30	0.30	0.28	0.28
G2	0.26	0.26	0.30	0.30	0.36	0.36
G3	0.24	0.24	0.32	0.32	0.26	0.26

5 Conclusion

Four factors that impact the frequency, nine factors that affect the amplitude, the relationship between phase difference and waveform of the oscillation and propagation delay characteristics are summed up. Several factors are simulated using BPA, which lays theoretical basis for the following study:

1) According to the PMU information in the key branches and nodes, through the analysis of the performance characteristics of the current operating parameters, we should research the disturbance source location method which can be applied to practice.
2) Study the propagation law of the oscillation in the power grid and search the related localization method using delay characteristics of electromechanical disturbance.

References

1. Tang, Y.: Fundamental Theory of Forced Power Oscillation in Power System. Power System Technology,2006,30 30(10), 29–33 (2006) (in Chinese)
2. Wang, T., He, R., Wang, W., et al.: The mechanism study of low frequency oscillation in power system[J]. In: Proceedings of the CSEE 2002, vol. 22(2), pp. 21–25 (2002) (in Chinese)

3. Han, Z., He, R., Xu, Y.: Study on resonance mechanism of power system low frequency oscillation induced by turbo-pressure pulsation. In: Proceedings of the CSEE2008, vol. 28(1), pp. 47–51 (2008) (in Chinese)
4. Xu, Y., He, R., Han, Z.: The Cause analysis of turbine power disturbance inducing power system low frequency oscillation of resonance mechanism. In: Proceedings of the CSEE 20007, vol. 27(17), pp. 83–87 (2007) (in Chinese)
5. Han, Z., He, R., Ma, J., et al.: Comparative analysis of disturbance source inducing power system forced power oscillation. Automation of Electric Power Systems 33(3), 16–19 (2009)
6. Zhu, W., Zhou, Y., Tan, X., et al.: Mechanism analysis of resonance-type low-frequency oscillation caused by networks side disturbance. In: Proceedings of the CSEE 2009, vol. 29(25), pp. 37–42 (2009) (in Chinese)
7. Ni, Y., Chen, S., Sun, B.: Dynamic power system theory and analysis, pp. 260–262. China Electric Power Press, Beijing (2002)
8. Faulk, D., Murphy, R.J.: Comanche peak unit no. 2 100% load rejection test-underfrequency and system phasors measured acorss TU electric system. In: Proc. Annu. Conf. Protective Relay Engineers, College Station, TX (March 1994)
9. Thorp, J.S., Seyler, C.E., Phadke, A.G.: Electromechanical wave propagation in large electric power systems. IEEE Trans. On CAS-I 45(6), 614–622 (1998)

Real-Time Control Techniques Research of Low-Frequency Oscillation in Large-Scale Power System Based on WAMS and EMS

Song Dun-wen[1], Ma Shi-ying[1], Li Bai-qing[1], Zhao Xiao-tong[1], Yang Xue-tao[2], Hu Yang-yu[3], Wang Ying-tao[1], and Du San-en[1]

[1] China Electric Power Research Institute, Haidian District, Beijing 100192
[2] Department of Electrical Engineering, North China Electric Power Univercity, Baoding 071003, Hebei Province, China
[3] Henan Electric Power Dispatching Centre, Zhengzhou 450052, Henan Province, China
Yn8800@163.com, xuetaoyoung@163.com

Abstract. With real-time data from WAMS and EMS, the paper proposes a kind of comprehensive dynamic stability prevention and control method including low-frequency vibration monitoring, oscillation source search and dynamic stability control strategy. This method on one hand observes multiple oscillation modes and damping characteristics that may be existing in the power grid on line from EMS and through real-time small signal stability analysis, combining the sensitivity analysis, dynamic stability control aided decision for every kind of mode is generated. On the other hand, data from WAMS and online Prony calculating method are used to fast detect the occurrence of low frequency oscillation. Once it occurs, oscillation propagation way will be tracked by the algorithm based on transient energy function and dynamic hybrid simulation to identify possible initial oscillation point. Oscillation pattern matching calculations start to choose the feasible real-time auxiliary control strategy from the matching small disturbance mode. In ideal condition, initial point position and oscillation inhibit quell measures will be showed which provides online automatic monitoring method for preventing dynamic instability of power grid. The method gets practical application in Henan interconnected power grid.

Keywords: Low frequency oscillation; Real-time monitoring; Weak damping; Forced oscillation; Wide area measurement; Parallel computing.

1 Introduction

The interconnection of the large regional power grid improves the economic of the operation of the system, while the whole dynamic process of the interconnected systems becomes more complex, which may lead the stability margin to become smaller and induce low frequency oscillation.

M. Zhu (Ed.): Electrical Engineering and Control, LNEE 98, pp. 351–357.
springerlink.com © Springer-Verlag Berlin Heidelberg 2011

Selecting the appropriate data sources, the three basic tasks of prevention and control of low frequency, especially in real-time prevention and control are to identify the occurance of low frequency accurately, to location the disturbance source and to produce quell and control measures.

The emergence of the WAMS (Wide Area Measurement System) provides a powerful tool to inhibit the interregional LFO (Low Frequency Oscillation). Regional generator relative rotor angle, rotor angular velocity and other global information can be gotten through WAMS, which can be used as feedback signal of the damping controller to constitute closed-loop control. At the same time, effective WAMS data can be also used as the reliable sources to search the disturbance source.

Perturbation calculation is the effective method in the off-line dynamic stability analysis of large power grid. QR algorithm and implicitly restarted algorithm are often used in practical. Applying perturbation calculation to online system and generating the dynamic stability auxiliary control decision go through many technical aspects including oscillation recognition, vibration unit clustering, factor analysis, control performance assessment and quantitiative analysis.

Initial range research and oscillation source orientation are important and difficult in the prevention of low frequency oscillation. On one hand, the incentives of the oscillation are very complex and the characteristics of the phenomenon are uncertain. On the other hand, the workload of the analysis and calculation of the localization is very large and the available results are difficult to be obtained in real time.

How to combine the real-time data from EMS (Energy Management System) with the observations from the limited PMU (Phasor Measurement Unit) and use off-line perturbation analysis techniques to extract low frequency information effectively and comprehensively and provide online and practical scheduling control strategy is an urgent problem needed to be searched and solved.

2 Integrated Control Technology for LFO

The analysis of the forms and impact factor of the LFO shows that it is very difficult to prevent and control various low frequency oscillation in a large power grid.

For guiding engineering practice, considering the various types of advanced algorithms, this paper presents a kind of comprehensive prevention and control method including making a joint application of WAMS and EMS, using prallel computing techniques, comprehensive analysis of various types of calculations, online discriminant of low frequency oscillation, proposing online control measures and seeking strating positions of the oscillation.

2.1 The Overall Design of Integrated Prevention and Control

For using the real-time data from EMS and WAMS, layering design method is used and data processing, calculation analysis and results judgment are relegated to different levels to be realized. The whole prevention and control system of low frequency oscillation is shown in figure 1:

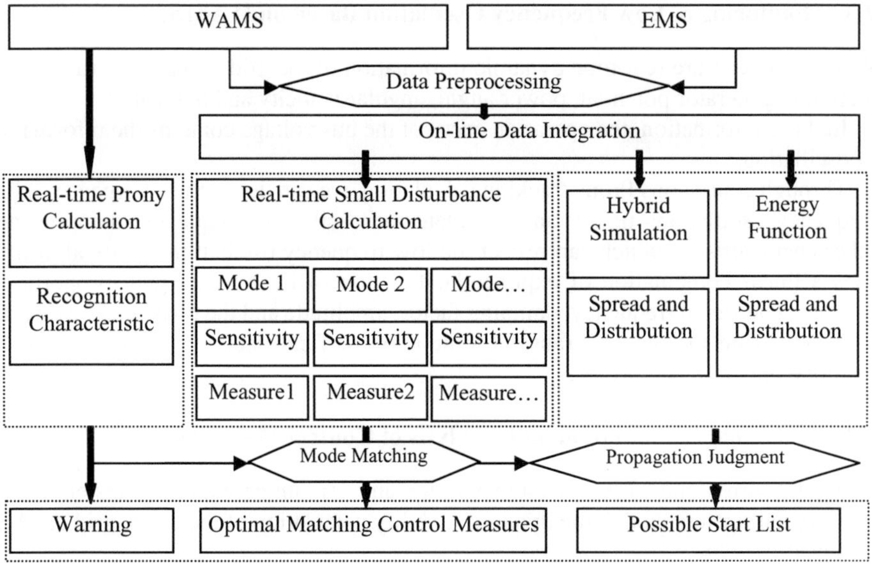

Fig. 1. Schematic diagram of overall design of integrated prevention and control

Online Prony analysis based on data from WAMS is used to detect the occurance of low frequency oscillation rapidly. At the same time, the modes and characteristics can be calculated.

Using online integrated data, on the one hand, the dynamic stability calculation is used to search the weak or negative damping mode in the system and sensitivity method is used to look for the inhibitory control measures under the different weak damping modes. On the other hand, hybrid simulation analysis and transient energy function analysis are paralleled used to explore the law of the spread and distribution.

Once low frequency occurs monitored by WAMS, LFO warning will be given. At the same time, oscillation mode matching calculation is opened immediately and the possible real-time auxiliary control strategies are selected from the matched small perturbation modes. Meanwhile, spread path of the oscillation is selected and determined and the possible start-up modes are listed at last.

2.2 Data Integration

In order to improve the quality of the EMS data for the calculation of oscillation control, limit filtering, redundancy checking, qualitative judgments and other methods are used to integrate the data from EMS and WAMS effectively to improve the availability of real-time data.

In the proposed method about the prevention and control, for low frequency oscillation warning, data in each monitoring point is directly used to be calculated and analyzed to provide the warning information. The online integrated data is used to be calculated and analyzed to provide auxiliary control measures and vibration source searching.

2.3 Monitoring of Low Frequency Oscillation Based on WAMS

WAMS can capture real-time dynamic information of the grid at the same time frame including generator potential, power angle, angular velocity and bus voltage.

In these information, the operation curve of the bus voltage contains the information of oscillation.

Through continuous Prony tracking calculation, we can detect the emergence of low frequency oscillations timely and calculate the oscillation frequency, damping and other characteristics which can reflect the low frequency oscillation. Prony algorithm uses a linear combination of exponential function to fit the data with proportional spacing. The signal frequency, damping factor, amplitude and the phase can be directly estimated from the data of transient simulation or field datas. Prony is an actually mature engineering algorithm.

2.4 Real-Time Computation and Analysis of Control Measures

Using the current model from data integration and combining with the transient stability parameters model, the dynamic stability calculation of the entire grid can be achieved.

Use implicitly restarted Arnoldi method and then research the system eigenvalue solution in the related frequency range. The left and right characteristic matrix generated in the solution process contain the information including oscillation mode, the oscillation frequency, damping ratio, electromechanical circuit correlation ratio and participation factor. Homology fleet information related low frequency oscillation is implied among them. In order to identify homology fleet leading generator, on one hand, the attribute of oscillation mode should be distinguished. On the other hand, the units participating in computing should be group-dividing. Homology fleet corresponding to the same frequency are usually divided into two groups, in which there is only a small part of unit playing a leading role. The dominant generator should be selected from each fleet. Implicit Ritz Arnoldi method does not give the influcing factors under each oscillation mode. Therefore, sensitivity analysis is adopted after the small disturbance calculation to determine the magnification of the PSS, the unit output change caused by one mode of the oscillation and other influencing factors. Thus control measures corresponding to different weak damping modes can be found and dynamic stability control strategy talble can be formed.

2.5 The Source Research of the Forced Oscillation

In order to reflect the practical of the method, the pattern of the oscillation is scaned by two different methods. On one hand, the scope of the reason analysis is narrowed. On the other hand, the engineering breakthrough of the localization of the source is to be found.

Energy Function [1]

Forced oscillation in multi-machine power system has the disturbance source clearly. Exogenous disturbances continue to work and inject the generated energy into system through the unit where the disturbance source is located. The energy dissipated by the damping is the result of the damping of all of the units and network in the system. The energy changing of units where the disturbance source is located and other units is certainly different.

Restricted by the number of PMU, it is important to determine the general location of the disturbance source according to the key branch and node. The energy injected by exogenous disturbance spreads in the network in the form of non-periodic components. Therefore, there is obvious relationship between the changes of potential of each branch and the location of disturbance.

The approximate direction can be determined by the inflow and outflow of the potential in the key nodes.

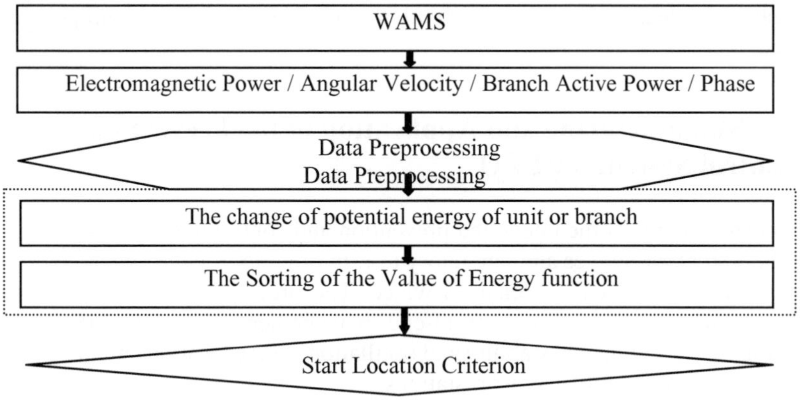

Fig. 2. Transient energy function method

Hybrid Simulation Method [2]-[3]

Synchronous collection data is used to replace one of the areas. When the reserved area does not contain the disturbance, no matter the curves of the lines and buses in the border or in the region can overlap, so preliminarily we can conclude that the disturbance is in external region. In the defective area, no matter the curves of the lines and buses in the border or in the region can not overlap, that is, in the case of without adding any disturbance, the disturbance scenes can not be reproduced, so we can determine that the disturbance is in this region.

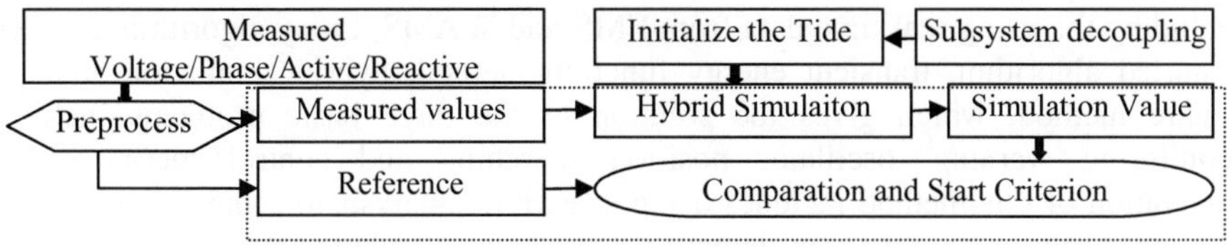

Fig. 3. Schematic diagram of Hybrid dynamic simulation

2.6 The Warning and Decision-Making Control

After the integration completion of the online data, three basic computing functions including oscillation monitoring, the search of the control measures and oscillation source localization proceed at the same time. In this process, the three functions are independent. When low frequency oscillation is detected by WAMS, the integrated output is exported through the three functions.

The monitoring of the low frequency oscillation will be scheduled to provide warning information. At the same time, the mode matching calculation is started. Using least squares, the modes of oscillation from warning are matched to the model table calculated through small disturbance calculation. The optimal or suboptimal auxiliary control measures are recommended to the dispatchers. If the characteristics of the propagation is clear in the warning vibration mode, the possible start-up position will be listed.

3 The Manufacturing and Application of the Prevention and Control System for LFO

The technical points of the integrated prevention and control method in section 3 have been checked by principles and laboratory test. In order to fully test its practical value, the method will be used in Henan Province. The prevention and control system in Henan contains a communication front-end, management server, WAMS monitoring computing servers, dynamic stability and oscillation source location parallel computing server and a number of display workstations.

The communication front-end has been achieved to be connected with WAMS and EMS. The output of data integration, task scheduling and auxiliary control decision are achieved by the management server. Real-time non-parametric Prony analysis based on bus voltage of the monitoring points is realized by WAMS vibration monitoring computing server and provides the results of judgment of the low frequency oscillation. Parallel computering server achieves the analysis of small disturbance, transient energy and hybrid simulation calculation. Oscillation warning results, oscillation control measures and the determination of initial position are displayed to scheduling analyst through display workstations.

4 Conclusion

This paper proposes a set of low-frequency oscillation prevention and control method including the using real-time data from EMS and WAMS, Prony algorithm, implicity restarted algorithm, transient energy function method, hybrid simulation and least square method, which gives the solution to the three basic problems including monitoring warning, oscillator position searching and control measures. The innovation of this method includes using sensitivity analysis to achieve the search of the control measures for weak damping oscillation, calculating the eigenvalue of the mode using the data from WAMS and screening the control measures and the comparison of oscillation law searching. Currently this method is limited to disturbance searching for forced oscillation and its control measures are also limited to active power regulation and the adjustment of PSS amplification factor. The start monitoring method is also relatively single. These question should to be further strengthened and deepened in the following studies.

References

1. Yu, Y., Min, Y., Chen, L., Zhang, Y.: Disturbance Source Location of Forced Power Oscillation Using Energy Function. Automation of Electric Power Systems 34(5), 1–6 (2010) (in Chinese)
2. Bobo: The Research on Locating the Disturbance of Forced Oscillation. North China Electric Power University, Beijing (2008)
3. Wu, S., Wu, W., Hang, B., Zhang, Y.: A Hybrid Dynamic Simulation Validation Strategy by Setting V-θ Buses with PMU Data. Automation of Electric Power Systems~ 34(17), 12-16 (2010) (in Chinese)

Two Methods of Processing System Time Offset in Multi-constellation Integrated System*

Longxia Xu[1,2,3], Xiaohui Li[1,2], and Yanrong Xue[1,2,3]

[1] National Time Service Center, the Chinese Academy of Sciences, China
[2] Key Laboratory of Precision Navigation and Timing Techinology, National Time Service Center, the Chinese Academy of Sciences, China
[3] The Graduate University of the Chinese Academy of Sciences, China
xulongxia@ntsc.ac.cn

Abstract. Two methods on how to deal with the time offset in multi-constellation navigation systems are discussed in this paper. One approach is disposing time offset in receiver end (DRE) and another disposing in system level (DSL). Comparisons and analysis in different aspects of these two methods are involved. Simulation results indicate that the positioning accuracy of multi-constellation navigation systems is affected more obviously by the method of DSL than DRE. When the error of broadcasted time offset is 5ns, it is more advisable to process time offset use the method of DRE.

Keywords: Multi-constellation Integrated Navigation System, System Time Offset, Position Accuracy.

1 Introduction

Multi-constellation navigation technology refers to combine two or more navigation systems together appropriately to achieve better performance than that of any single navigation system. Until the beginning of this century, promoted by the development of European navigation system Galileo and COMPASS of China, combined navigation becomes a research hotspot again.

The combined navigation systems can be the combination of GPS、GLONASS or Galileo navigation systems. However, at the stage of design neither the GPS system nor the GLONASS system considered the interoperability problem [1]. Galileo is designed to be compatible with the other navigation systems and the concept of global navigation satellite system (GNSS) that consisting of all the satellite navigation systems is proposed. With the establishment and implementation of Galileo and the modernization of GPS, combined navigation systems becomes inevitable.

After the introduction of user positioning model in Section 2, two methods are proposed to deal with the time offsets in Section 3. In Section 4 comparisons between the two alternatives is shown in three different aspects by simulations.

* Funded by the project of National Key Natural Science Foundation of China (11033004).

M. Zhu (Ed.): Electrical Engineering and Control, LNEE 98, pp. 359–365.
springerlink.com © Springer-Verlag Berlin Heidelberg 2011

2 The Positioning Equation

It is known that each satellite navigation system has its own system time scale. However, due to errors, the difference among each navigation system is different. The value of the difference is about few tens of nanoseconds called system time offset, time offset for short [2].

In the solo system of GPS, four unknown parameters (three components of user position and one clock offset) are needed to be determined. The standard positioning model is given below:

$$\mathbf{I} = \mathbf{Hu} + \mathbf{v} . \tag{1}$$

Wherein $\mathbf{I}$ is the vector of observations, $\mathbf{H}$ is the geometry matrix corresponding to the direction from user to satellites in view of user, $\mathbf{u}$ is the vector need to be determined consisted of a priori user position and clock offset, and $\mathbf{v}$ is the vector of measurement noise. Vector $\mathbf{u}$ includes three corrections to user 3D position and one to user clock offset. The Least Squares (LS) solution $\hat{\mathbf{u}}$ is given by:

$$\hat{\mathbf{u}} = (\mathbf{H}^T\mathbf{H})^{-1}\mathbf{H}^T\mathbf{I} . \tag{2}$$

When satellites from at least two navigation systems are available, for instances in a combined GPS/GLONASS navigation system user has to cope with an additional unknown parameter (the clock offset between user and GLONASS time scale or the time offset between GPS and GLONASS time scales).

3 Two Methods to Dispose Time Offset

Theoretically speaking, taking the unknown parameter as clock offset or system time offset are both feasible. As we known, the time scale of a navigation system is the result of high-precision atomic clocks group and the time offset behaves sufficiently stable. Therefore, practically the time offset designated with symbol $\mathbf{o}$ is added into the standard positioning model of Eq. (1) and thus get the following positioning model [3]:

$$\mathbf{I} = \mathbf{Hu} + \mathbf{v} + \mathbf{o} . \tag{3}$$

Just as the clock bias, the time offset $\mathbf{o}$ will also leads to a bias in user positioning solution, so it is necessary to determine its value.

There are two methods to estimate the value of time offset $\mathbf{o}$. One method called Disposing in the Receiver End (DRE) introduces the time offset as an unknown parameter as in Eq. (3) into the position model. In order to obtain the LS solution, the total number of satellites available of GPS and GLONASS systems must not be less than five. When this condition can not be met, the recent values of time offset are used to predict the value of time offset of current time. Changing slowly, time offset can be predicted with a linear model. Therefore, when only four satellites in view of user, the predicted value of time offset is used in Eq. (3) to estimate the user position.

Another method referred as Disposing in level of Systems (DSL) obtains the system time offset between two navigation systems from the GPS or GLONASS navigation messages. The value of time offset is broadcasted in the navigation messages and thus user only need to determine the four unknown parameters in Eq. (1). According to reference [1], the accuracy of broadcasted time offset is about five nanoseconds.

In the following section we will simulate and contrast the number of satellites visible, values of DOP and impact of time offset on estimated user positions in GPS stand alone system, combined GPS/GLONASS system and combined GPS/GLOANSS/Galileo system using the two methods discussed above.

4 Simulation

4.1 Simulation of Satellites in View

Take GPS, GLONASS and Galileo navigation systems for example the orbit parameters given in Table 1 are used to simulate the position of satellites in each navigation system. The length of simulation time is one day with one minute being time interval and Lintong station as the user position.

Our simulations are based on the satellite position simulated of the three navigation systems and the number of satellites visible at elevations of 10, 35 and 60 degrees in GPS, GPS/GLONASS and GPS/GLONASS/Galileo navigation systems are simulated in Lintong. Table 2 lists the time percentage of satellites available from less than four to more than six in the GPS only, combined GPS/GLONASS, and combined GPS/GLONASS/Galileo navigation systems at different elevations. Table 3 gives the results of similar simulation scenario in Urumqi [4].

Table 1. Parameters of Orbit

	GPS	GLONASS	Galileo
Satellites	24	24	30
Planes	6	3	3
Altitude	20183km	19100km	23616km
Elevation	55°	65°	56°
Cycle	11h58’	11h15’	14h4’

Table 2. Time percentages of satellites available in different combined systems in Lintong

Time percentages		<4	=4	=5	=6	>=7
GPS	10°	0 %	0 %	4 %	38%	58%
	35°	32%	29%	22%	15%	2 %
	60°	96%	4 %	0 %	0 %	0 %
GPS/ GLONASS	10°	0 %	0 %	0 %	0 %	100%
	35°	3 %	2 %	7 %	11%	77%
	60°	80%	13%	5 %	2 %	0 %
GPS/ GLONASS /Galileo	10°	0 %	0 %	0 %	0 %	100%
	35°	0 %	0 %	0 %	0 %	100%
	60°	47%	20%	10%	12%	11%

Table 3. Time percentages of satellites available in different combined systems in Urumqi

Time percentages		<4	=4	=5	=6	>=7
GPS	10°	0 %	3 %	9 %	23%	65%
	35°	35%	27%	23%	15%	0 %
	60°	97%	3 %	0 %	0 %	0 %
GPS/ GLONASS	10°	0 %	0 %	0 %	0 %	100%
	35°	0 %	2 %	9 %	11%	78%
	60°	70%	14%	10%	6 %	0 %
GPS/ GLONASS /Galileo	10°	0 %	0 %	0 %	0 %	100%
	35°	0 %	0 %	0 %	0 %	100%
	60°	40%	16%	11%	17%	16%

Results in Table 2 and Table 3 give the following conclusions that the larger the elevation, the fewer the satellites in view of user in the same navigation model. When elevations are the same, the more navigation systems in use, the more satellites will be visible, and the greater the proportion of time available for positioning and integrity monitoring [5].

4.2 Value of DOP Using DRE and DSL Methods

When using the method of DSL, the first three columns of geometry **H** is the vector from satellites to user and the fourth are all ones. While in the method of DRE the geometry **H** has the same first three columns and additionally two (corresponds to GPS/GLONASS system) or three (corresponds to GPS/GLONASS/Galileo system) columns consist of zeros and ones. The differences of geometry matrix **H** in the two methods may lead to the variation of their DOP [6, 7]. The mean values of simulated DOP under these two methods are presented in Table 4 and Table 5.

Table 4. Mean values of DOP of two different methods in Urumqi

Navigation Model	Method	10°	35°	60°
GPS/ Galileo	DSL	1.55	4.08	81.33
	DRE	1.91	5.82	92.48
GPS/ GLONASS/ Galileo	DSL	1.25	3.16	65.21
	DRE	1.70	4.69	87.11

The results from Table 4 and Table 5 demonstrate that with the same elevation the more navigation systems are combined used, the smaller the mean values of DOP. In the same navigation model, the values of DOP are of difference under the two presented methods. It can be concluded from the results of Table 4 and Table 5 that the mean values of DOP derived from the method of DSL are better than those from the DRE method and DOP deteriorates rapidly with the increase of elevation in the DRE method.

Table 5. Mean values of DOP of two different methods in Shanghai

Navigation Model	Method	10°	35°	60°
GPS/ Galileo	DSL	1.57	4.18	89.10
	DRE	1.93	8.98	94.50
GPS/ GLONASS/ Galileo	DSL	1.31	3.32	73.64
	DRE	1.79	8.26	93.82

Although the geometry configuration of visible satellites keeps unchanged, the differences of DOP are caused by the two different processing methods of time offset. Compared with the DSL method, the essential reason leads to the increase of DOP is the increased dimensions of geometry matrix $\mathbf{H}$.

4.3 Simulation of Effects on Position Using Methods of DRE and DSL

What kinds of effects do the methods of DRE and DSL impose on the estimated user position? This will be illustrated by the following simulations at five stations Urumqi, Lintong, Changchun, Shanghai and Kunming in China. Simulation parameters are given in terms of the following specification.

A) The clock bias between receiver clock and GPS system time is 100 nanoseconds. The uncertainties of broadcasted system time offset of GLONASS, Galileo and GPS are respectively 0ns, 5ns and 10 ns.

B) The time offset between GLONASS and GPS systems is expressed with a linear function and the coefficient of one degree term is 5 ns per day and the constant term is 10 ns. Similarly, the time offset between Galileo and GPS is also described by a linear model with the coefficient of one degree term being 8ns per day and constant term being 8ns.

C) The error of pseudo-range measurements from GPS system is 1.3 meters and for those from Galileo and GLONASS systems are respectively 1.05 meters and 1.6 meters.

D) Two scenarios are selected in which elevation are 10 and 30 degrees.

In order to contrast the influences of these two methods on estimated user positions, simulations at five stations (denoted with Uq, Lt, Chc, Shh and Km in Table 6 and Table 7) in China are performed using the methods of DRE and DSL in combined GPS/GLONASS system and combined GPS/GLONASS/Galileo system. Table 6 and Table 7 give the simulation results which are measured by the value of RMS of positioning error.

Table 6. Values of RMS under elevation of 10 degrees

Navigation Model	Method		Uq	Lt	Chc	Shh	Km
GPS/ GLONASS	DSL	0ns	2.3	2.4	2.3	2.5	2.5
		5ns	2.6	2.9	2.8	3.0	3.0
		10ns	3.5	4.0	3.9	4.2	4.2
	DRE		2.5	2.7	2.6	2.8	2.7
GPS/ GLONASS /Galileo	DSL	0ns	1.5	1.5	1.5	1.51	1.5
		5ns	1.8	1.9	1.8	1.9	1.9
		10ns	2.4	2.7	2.7	2.7	2.7
	DRE		1.6	1.6	1.6	1.63	1.6

Table 7. Values of RMS under elevation of 30 degrees

Navigation Model	Method		Uq	Lt	Chc	Shh	Km
GPS/ GLONASS	DSL	0ns	9.2	9.2	6.7	10.2	12.0
		5ns	10.3	10.1	7.6	11.5	13.5
		10ns	12.9	12.5	9.8	12.3	15.5
	DRE		15.0	14.3	13.6	11.3	15.3
GPS/ GLONASS /Galileo	DSL	0ns	3.2	3.5	3.3	3.5	3.9
		5ns	3.8	4.1	3.7	4.2	4.4
		10ns	5.1	5.6	4.9	5.8	6.0
	DRE		3.7	4.1	3.6	4.0	4.4

According to the results of Table 6 and Table 7, we have the following three conclusions.

A) The first conclusion is that the lower the accuracy of the broadcasted system time offset, the greater the RMS of positioning error.

B) If there is no error in the broadcasting of system error, the accuracy of positioning result obtained by the method of DSL is superior to that of DRE. If the accuracy of broadcasted system error is 5 ns [2], it is recommended to use the method of DRE instead of DSL. Since the broadcasted accuracy of navigation message of Galileo system is 5 ns, this conclusion provides a reference for multi-mode navigation system positioning.

C) When the elevation is 30 degrees, the positioning accuracy in GPS/GLONASS system with the method of DRE is lower than that of the DSL with the accuracy of broadcasted time offset being 10 ns. In GPS/GLONASS/Galileo system, the positioning accuracy derived from the method of DRE is close to that of DSL when the accuracy of broadcasting is 5 ns. Therefore, it is sensible to use the method of DSL when the environment is severely obstructed.

5 Conclusions

The multi-constellation navigation system has become a significant developing direction in future navigation field. One key factor that impacted on the positioning accuracy of multi-mode navigation is the uncertainty of system time offset. This paper analyses two different methods on how to dispose time offset. Simulation results indicate that when the uncertainty of broadcasting is 5ns, it is better to user the method of DRE in GPS/GLONASS navigation system in open sky. And when the accuracy of broadcasted system time offset is 10 ns, it is more sensible to adopt the method of DSL in combined GPS/GLONASS/Galileo system in badly obstructed environment.

References

1. Hahn, J., Powers, E.: GPS and Galileo Timing Interoperability. In: Proceedings of GNSS 2004, CD-ROM (2004)
2. Moudrak, A.: GPS Galileo Time Offset: How It Affects Positioning Accuracy and How to Cope with It. In: Proceedings of ION GNSS 17th International Technical Meeting of the Satellite Division, pp. 666–669 (2004)
3. Liu, J.: The principle and method of GPS navigation/positioning. Science Press, Beijing (1999)
4. Xiaokui, Y., Yanlei, Y.: Simulation and analysis of Performance for Galileo system. Journal of system simulation 19, 5491–5494 (2007)
5. Sturza, M.A.: Navigation System Integrity Monitoring using Redundant Measurements. Journal of the Institute of Navigation 35(-89), 69–87 (1988)
6. Chunmei, Z., Jikun, O., Yuanlan, W.: Simulation and analysis of system performance of GALILEO and integrated GPS-GALILEO. Journal of System Simulation 17, 1008–1011 (2005)
7. Zemin, W., Yang, M., Yue, W., Shujun, L.: DOP of GPS, Galileo and Combination Navigation System. Geomatics And Informantion Science of Wuhan University 31, 9–11 (2006)

Quick Response System for Logistics Pallets Pooling Service Supply Chain Based on XML Data Sharing

Nan Zhao

College of Management Science and Engineering,
Donghua University
Room 1001C, Buiding 12, No.1882, West Yan'an Road,
Shanghai, 200051
zzxxffdh@126.com

Abstract. XML is a state-of-the art technology which is utilized for integrating the information along the Pallets Pooling Service Supply Chain. This paper presents an XML-based quick response (QR) framework for the Pallets Pooling Service Supply Chain. XML data sharing, data linking and the bidirectional mapping rule of XML and database are studied and XML model of quick respond system for the Pallets Pooling Service Supply Chain is investigated. By means of XML trans-platform information integration ability, the model improves the trans-platform mutual communication mechanism and links all the data concerning material supply and pallets manufacturing in the upstream and pallets operating and using in the downstream so that pallets associated companies can commonly share the real-time data information on the supply chain, thus shortening the lead-time and make a quick respond to the changing market.

Keywords: pallet; pallets pooling service supply chain; quick response; XML; date sharing.

1 Introduction

In modern logistics operation process, pallets can be kept to facilitate loading-and-unloading, transport and storing, greatly enhancing the efficiency of logistics operations [1]. Many developed countries have adopted the Pallets Pooling System to ensure recycling of pallets in the supply chain. Pallets Pooling System is a third-party independent of pallets producers and pallets users, which is a professional service system responsible for pallets rental, recycling, maintenance and updating, and improve pallets utilization, promoting pallets rapid circulation in the supply chain. In order to meet rental demand of different sizes and varying numbers of pallets from different enterprises, a reasonable requirement planning must be developed timely to make manufacturers and raw materials suppliers can be able to provide the corresponding types and numbers of pallets.

XML technology can integrate information based on network environment due to its function of integration mechanism.[2] Based on lots of scholars' literatures about the

M. Zhu (Ed.): Electrical Engineering and Control, LNEE 98, pp. 367–374.
springerlink.com © Springer-Verlag Berlin Heidelberg 2011

Pallets Pooling System[3]-[7], this paper presents an XML-based quick response (QR) framework for the Pallets Pooling Service Supply Chain. The system can improve the pallets supply chain platform for interaction mechanisms and enhance degree of entire Pallets Pooling Service Supply Chain, thereby reducing response time and order lead times.

2 Rapid Response Mechanism of Pallets Pooling Service Supply Chain

Pallets Pooling Service Supply Chain includes pallets raw materials suppliers, pallets manufacturers, pallets carriers, pallets enterprises users (Fig.1 shows the supply chain structure). The pallet carrier plays a significant role as an organizer, an executor and a coordinator in this supply chain; pallets raw materials suppliers, pallets manufacturers and pallets users as strategic partners guarantee rationality of pallets production, convenience and effectiveness of pallets tenancy to achieve pallets recycling among various types of enterprises.

This system can promote consistency, standardization and mechanization of the pallets transit, and reduce logistics costs, improve logistics efficiency and resource utilization and achieve win-win of logistics Pallets Pooling System. Because this supply chain structure involves various information systems of a number of departments and different enterprises including raw materials suppliers, carriers, manufactures, and users, information content is not only large but also from different data sources. Traditional data integration methods require explicit database format, while XML can realize data integration, add XML package body on heterogeneous data sources, and package relevant information via XML definition and description of data sources and creation of heterogeneous data sources on the situation of unchanging dada sources. All the package information of data sources on nodes of supply chain as XML Schema or DTD format are stored in the XML virtual server, and XML data exchange between each other via XML package body.

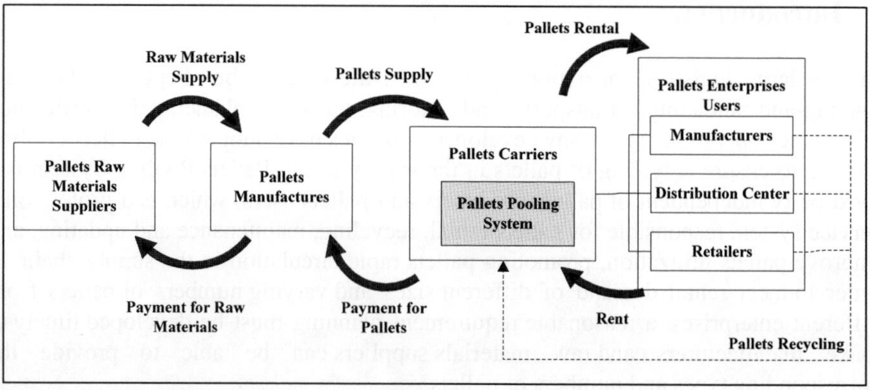

Fig. 1. Structure of the Pallets Pooling Service Supply Chain

XML has some features including separating data model, data content and data display, scalability, self-descriptive and cross-platform. Due to these features, enterprise business rules can be separated from the data sources, and enterprises in the supply chain can transform their information into XML documents through XML-based interface, then they can send information to other members using HTTP protocol by the internet, after receiving the documents other members mapped them into the XML information that can be identified. Supply chain members, on the premise of meeting the authority, can take the initiative to check the information of others via the XML view, and also can publish their information on XML-based view web through XML technology. For example, the information of pallets enterprises users are mapped into XML view, then transformed into the XML Schema or DTD in accordance with schema mapping template using mode processor, after transformed the XML-base view information can be mapped in the manufacture's information system, in order to achieve cross-platform interaction of different data sources and ensure the supply chain information in real time response.

3 The Design of Pallets Pooling Service Supply Chain System

XML-based Pallets Pooling Service Supply Chain System is functioned by MXL documents from pallets enterprises users, pallets carriers, pallets manufacturers and raw materials suppliers. The documents should include but not limit the attributes of pallets rental information, inventory data, pallets specifications and pallets materials, which exit in their own Web server. These servers connect together through internet, implementing communications and junctions via XML technologies: the XML documents of pallets enterprises users stock information of pallets usage periodically and testing report of pallets usage requirements prediction; the XML documents of pallets carriers include pallets rental information, pallets recycling information, pallets inventory information, rental contract management and testing report of pallets requirements prediction; the XML documents of pallets manufactures include pallets production schedule, raw materials orders, pallets specifications; the XML documents of raw materials suppliers include raw material inventories, raw materials production schedule tables.

When enterprises located in the supply chain node access to gather the related information via XML, the real-time information could be informed promptly and all level inventories could be adjusted according to the customers' requirements, which could assure the scheduled delivery time. Fig.2 shows the framework map of the Pallets Pooling Service Supply Chain System based on XML. The different data source of the supply chain node information in the graph are transformed into XML data from Wrapper to stock in the virtual memory, users could obtain XML format or submit the searching command via DOM/SAX interface, then the system searches through recognizable local data converted by wrapper and interact via XML package technology. For example, after carriers attain the pallets usage report, which includes the rental indications of the pallets enterprises users, from the system, the system sends the pallets requirement data to pallets manufacturers and pallets users; Manufacturers could predict the demand of the raw materials and feedback the inventory information to carriers according to the data of production situation; Raw materials suppliers draw up the further production plan and feedback the plan to manufacturers and carriers according to this raw materials requirement report. In this way, not only the manufacturers and raw materials suppliers could interchange the raw materials supply data immediately, but also carriers could be informed the raw materials supply information accordingly, through which all members in the supply chain could monitor the parallels stock conditions or modifications rapidly. Other example, carriers could recycle, update inventory in terms of pallets rental information provided by pallets enterprises users to manage inventory accurately and quickly. Therefore, all members in the supply chain could share the information and attain the latest development of the supply chain via XML technology in rental, order, production and recycling. In addition, the enterprise in the supply chain node could also carry out cash flow transactions via XML.

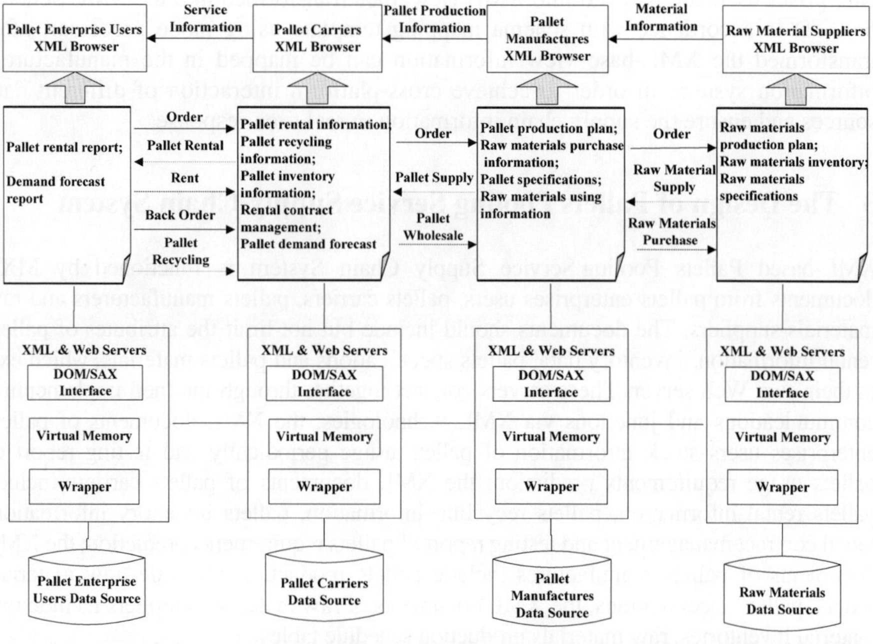

Fig. 2. Frame of the Pallets Pooling Service Supply Chain System

4 The Realization of Pallets Pooling Service System Based on XML Technology

4.1 Data Sharing Mechanisms

Pallets Pooling Service Supply Chain involves in all kinds of information systems and databases of pallets rental, pallets recycling, pallets manufacturing and raw materials purchase which are all belong to different enterprises. Because the XML model is based on web development, all the information and its change can be timely shared by other members on the node in the supply chain, even if indirect information of non-adjacent members can also be linked through the XML data model. For example, the pallets specifications data are hidden information for raw materials suppliers, while through XML technology this information can be conversed by package and associated with the raw information, so that suppliers and manufactures can share the pallets specification data. Other example pallets carriers can always check the pallets condition of service, track dynamic pallets state, and recycle pallets timely.

Specifically, all the files related to pallets production plan including raw materials inventory reports, pallets using reports and so on, these data are attribute data of pallets planning XML documents, such as pallets size, texture, number and other attributes. Different pallets, such as steel pallets, composite pallets, press-wood pallets and plastic-wood pallets, can be associated by pallets attributes. Thus, members in supply chain will be able to know all kinds of information from rental requirements access to purchase orders by clicking on the relevant documents of the supply chain.

The XML data exchange document shown in Fig.3 explains the correlation among the supply chain as followings. When pallets rental markets emerge varying demands, pallets enterprises users put out rental requirements via their own Web server, which means the order data are stored in the database server 1 by order Servlet implementation in SQL query language, then the pallets carrier makes a request in the database server 1 and receive ordering data in XML format. In the case of the pallets out of stock, the out of order data are stored in the database server 2, these data as DSO object data source are returned, and then DSO matches suitable raw materials data among the pallets specification, model, texture from manufactures and suppliers. Manufactures receive the carrier's orders associated with various types of material information in HTML format presented to the producers via browsers. Accordingly, when receiving the carrier's purchase orders, manufactures have also obtained raw materials suppliers' information, thus ensuring the various stages of the supply chain information in real time.

4.2 The Establishment of On-Line Mechanism

With the application of XML technology, pallets carriers can realize real-time pallets demand information tracking, and receive pallets recycling and defects consumption information timely. If pallets out of stock, carriers can even track the information from raw materials suppliers and develop demand plans by checking the inventory, and subsequently manufacturers develop production plans. In this way, XML technology establishes a connection among pallets users, pallets carriers, pallets manufacturers and raw materials suppliers, so that every member in supply chain can obtain all necessary information and data and thus make ordering, rental decision, replenishment decision and other decision-making. This connection mechanism helps carriers predict pallets usage requirements in advance, in order to adjust inventory levels of existing pallets, reducing out of stock rate. Similarly, the connecting mechanism can also help enterprises users find out the pallets inventory, and thus they will be able to make accurate and reasonable orders timely, thereby reducing waste and the possibility of late delivery. In addition, manufacturers and raw material suppliers can track the pallets needs information timely.

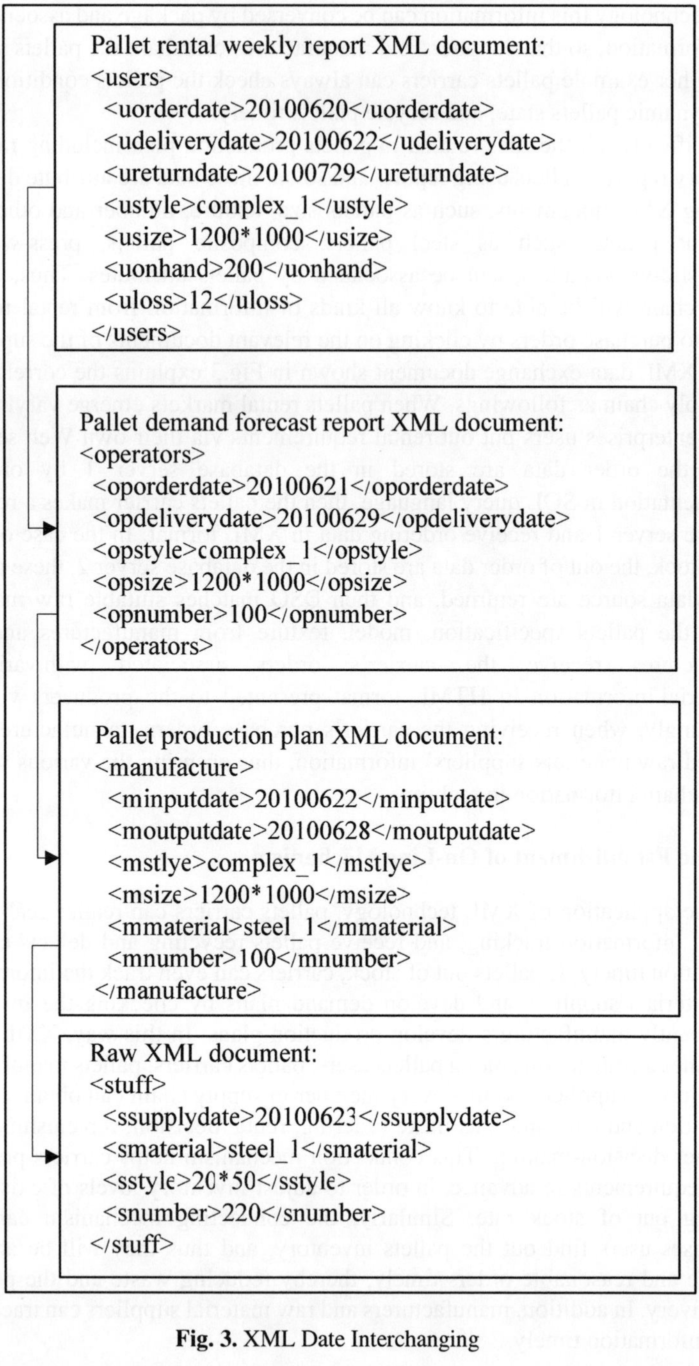

Fig. 3. XML Date Interchanging

4.3 The Design of Model Operating Environment

The operating environment of XML model includes Java Servlets, SQL Server 2005, XML documents, DTD or XML Schema, Java Script and the data source object (DSO) and other intergraded application environment. Accordingly, all members in the supply chain can maintain their own databases and Web servers. Each Web server, containing SQL Server 2005, Java Servlets and 4 components of DSO, provides an interactive information platform for every member in the supply chain. Each database stores the corresponding technical documentation, such as purchase orders, the pallets rental reports, the manufacturers' production plans, raw materials usage, pallets carriers' inventory. When the XML model is running, the Java Servlets implements SQL query language access to the relevant data from other members. For example, when the manufacturer and the raw material supplier send a request to the carrier, Java Servlets responds to the request and execute the SQL query language to find out pallets usage data from pallets rental weekly reports. Java Servlets mainly through the Web server responds to requests and establish a connection with the database to obtain corresponding data inserted into the relevant XML document as XML file output. Following, XML file checked the validity by XML Schema or DTD as a data source is sent to the DSO via Java Servlets, then DSO convert the XML output in HTML format and displayed in web browser.

5 Conclusion

Pallets Pooling Service Supply Chain involves raw materials suppliers, pallets manufacturers, pallets carriers and pallets enterprises users, information sharing of these enterprises are very low before integrated by XML technology, which has the cross-platform information integration trait. This paper narrate the rapid response system of Pallets Pooling Service Supply Chain based on XML technology, which combine the pallet operation situation via XML model and other members in the supply chain. In this system, pallets carriers are in the core position of the supply chain, who could attain the requirement information of pallets enterprises users rapidly and this information could be shared with manufacturers and raw materials suppliers at the same time. Therefore, we could provide raw materials and pallets specifications according to the requirement in the least time to manufacture at the same time, greatly enhancing the operation efficiency of supply chain.

References

1. Xu, Q.: Logistics Pallets Pooling Service Supply Chain System and Optimum Management. J. Modern Logistics (3), 22–25 (2010)
2. Dongbo, W., Runxiao, W., Yijun, S.: Supply Chain Information Integration Technology Based on XML. J. Computer Engineering and Application (10), 22–26 (2004)
3. Qingyi, W.: Again on the Establishment of Pallets Pooling System in China. J. Logistics Technology and Application 1, 1–5 (2004)
4. Taiping, L.: Research on Problems and Countermeasures of Establishment of Logistics Pallets Pooling System in China. J. East China Economic Management (5), 85–90 (2006)

5. Qingyi, W.: Research on the Establishment of Pallets Pooling System in China. J. Logistics Technology and Application 2, 1–4 (2003)
6. Guoqiang, M., Ke, S.: Recommendations on the Application and Development of Pallets. J. China Logistics and Purchasing 12, 12–16 (2004)
7. Xueyan, Z., Jianwei, R.: Research on Concept Model of Pallets Supply System. J. Logistics Procurement Research (45), 24–26 (2009)

Research on the Aplication of HACCP System to Gas Control in Coal Mine Production

Wang Yanrong[1,2], Wang Han[3], and Liu Yu[2]

[1] Institute of Electrical and Information Engineering,
China University of Mining and Technology (Beijing), Beijing, China 100083
wyr223 @126.com
[2] College of Management and Economics, North China
University of Water Resources and Electric Power, Zhengzhou, China 450011
vivian_ly_good@163.com
[3] College of Environmental Science and Engineering, Hohai University,
Nanjing, China, 211100
wanghan1230 @126.com

Abstract. HACCP system has been applied in the fields of drinking water safety, feed safety, animal source food security , risk management of securities companies as well as dangerous goods logistics security .Based on one of coal mines of our country as an example, the paper carries out an explorative study on the HACCP system for its control of gas concentration .Application of fault tree analysis techniques to identify the coal production which may lead to the potential hazards of excessive gas concentration, applied decision tree analysis techniques to determine the critical control points, the establishment of critical limits, monitoring systems and corrective measures, and analysis of the verification process and file and record keeping systems.

Keywords: HACCP, Mine production;Gas, Concentration, Control.

1 Introduction

HACCP, abbreviation of the Hazard Analysis and Critical Control point, is a set of food surveillance system researched and developed jointly by American Pillsbury Company, NASA and the American Army Natick Institute in the 1960s, which was based on science to guarantee food safety through the systematic methods to identify specific harms and control measures.The applications of HACCP system include 12 steps, of which the first five steps are the preparation stages for the HACCP system, and the following seven steps are corresponding to the seven basic principles of HACCP system respectively. Applications mainly include: take hazard analysis to identify the critical control points (CCPs); formulate key limits, establish monitoring system to monitor each key point in the control condition, establish the deviation-correcting measures which should be taken when critical points out of control; establish validation procedures to confirm the effective operation of the HACCP

M. Zhu (Ed.): Electrical Engineering and Control, LNEE 98, pp. 375–383.
springerlink.com © Springer-Verlag Berlin Heidelberg 2011

system; establish relevant principles and necessary procedures of the applications as well as the records of documents and files. At present, HACCP system has been used extensively by many international organizations and nations all over the world, and has been applied in drinking water safety [1-2], feed safety [3], risk management of securities companies [4] and logistics safety of dangerous goods [5-7]. While in the field of the HACCP system applied in coal production to control the production safety, there is still be a blank in the literature neither of theoretical study nor of practical research.The author chose a coal mine enterprise in central China as the research object, with gas control as the goal, identified the main harms of gas exceeding standard in mine production and the control measures in accordance with HACCP application system, ascertained the CCPs, and set up the key limits, monitoring system, deviation-correcting measures and validation procedures of CCPs, etc.

2 Gas Control Procedure of Coal Mine Based on HACCP

2.1 Composition HACCP Group

Composition HACCP group is the premise of HACCP system application, and team personnel structure decides scientificalness, rationality and validity of the HACCP plan. Specifically speaking, when applying HACCP system into gas control in the coal production, the HACCP group should includes experts who are proficient in HACCP system, the gas processing technology, mine designing, chemistry, coal gas monitoring, and familiar to gas processing equipment and automatic control, the mine production operation and management.

2.2 Expected Using Purpose

"Production safety" is the purpose of the process establishing HACCP system In coal mine production, that is to say, try to avoid accidents and making any adverse impact to human and environment in the production processes or links, which include alley of tunneling, extraction, ventilation, drainage, power supply, transportation and production drawing. And more often than not, it should be measured whether the production processes meets the requirements by relevant national laws and regulations. for instance, *Safety In Production Law, Coal Law, Safety In Mines Law, Coal Mine Safety Supervision Regulations, Safety Production Permit Regulations, The Mine Safety Standards, Coal Mine Safety Production Basic Conditions Provisions, Coal Mine Enterprises Implementing Measures of Safety Production License*, etc. These procedures carried on the detailed stipulation on the operation procedures of mining and digging work, appraisal of the gas grade, system of independent ventilation, daily examination signature system of gas, system of safety monitoring, system of power supply, systems of dust proofing and water supply, systems of waterproof and drainage, measures and facilities of fire prevention, system of communication and organization of emergency rescue, This study just took gas control as the research object, set the gas concentration standard as the basis of gas control in mine production, and tried to identify the potential hazards of higher gas concentration in production process.

2.3 Drawing and Site Verifying of the Flow Chart

Choose a well as an example and study on its security control procedures in coal production process. Main procedures include: down into the well, digging, mining, transportation, ascension, etc. HACCP working group should check the flowchart according to the processing in all stages of operation on the spot. And the safety production control flow is shown as in the figure 1.

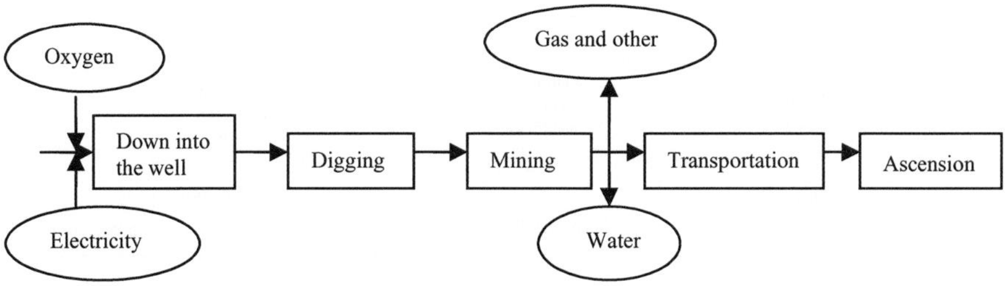

Fig. 1. Coal production flow chart of a mine

2.4 Analysis of all Potential Harms and Control Measures

Using the Fault Tree Analysis (FTA) technique, take the actual operation situation and the possibility of future adverse situation into consideration, set the gas concentration exceeding standard as the final event, analyze out the incidents of precursors from top to bottom until we find the initial origin events, just as figure2 shows.

Except for conditional event X_1, every potential harm(PH) in the figure 2 were presented in table 1, which including their occurrence probability(OP), the influence degree(ID), risk coefficient(RC) and control measures(CM). Among them, occurrence probabilities are corresponding to the numbers from 1 to 5 just like that follows: very low (<once/a), low (once/a), medium (once/month), high(once/week), very high(once/day); and the influence degrees go that way all the same as follows: very low (has no significant influence on concentration), low (slightly elevated concentrations but not exceeding standard), medium (concentration rise but basically not exceed the allowed figure), high(concentration exceeding standard), very high(concentration exceeding standard seriously); risk coefficient is the product of occurrence probability and influence degree, in other words, it could be any integer between 1 and 25.

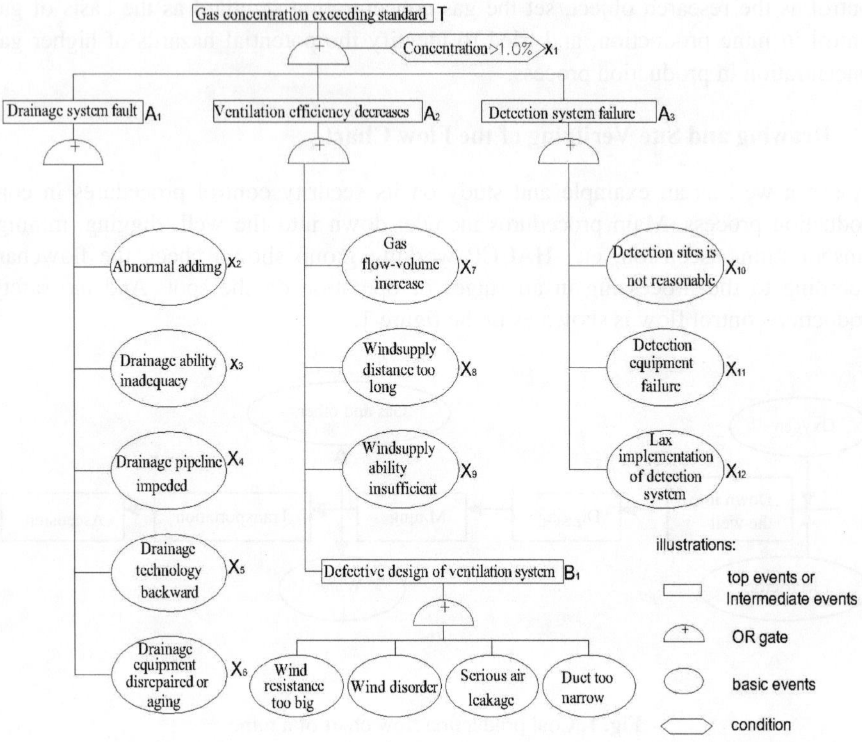

Fig. 2. FTA on gas concentration exceeding standard of the coal production

2.5 Confirmation of the CCPs

Use Decision Tree Analysis (DTA) technique to confirm CCPs, and the figure3 is the flow chart. For those potential harms,shown in table1, whose risk factors are not less than 6, all should be required to answer four questions in turn as the procedures shown in the figure 3. Only in this way can we determine whether it could be the CCP. Take X_2 for example, its preventive control measures are the gas dynamic monitoring and gas drainage, therefore the answer of the first question is "yes";However, the mine itself lack the ability to overcome the problems of gas concentration rising which is led by the abnormal addimg, then the answer of the second question is "no"; But the abnormal addimg may be the reason of gas concentration rising inside the mine, so the answer of the third question is "yes"; The mine does not have a perfect procedures for gas dynamic monitoring and gas drainage, which leads the answer of the fourth question to be "no". Thus we could infer that X_2 is the CCP.

Table 1. CCPs recognition of gas concentration of the coal production

PH	OP	ID	RC	CM	Q1	Q2	Q3	Q4	CCP
X_2	2	5	10	Perfect gas dynamic monitoring and drainage system	Yes	Yes	Yes	No	Yes
X_3	2	4	8	Increase drainage machines or increase its power	Yes	Yes	Yes	Yes	No
X_4	1	3	3	Dredge special drainage pipeline	Yes	Yes	Yes	Yes	No
X_5	2	3	6	Adopt advanced drainage and drilling drainage	Yes	Yes	Yes	Yes	No
X_6	1	4	4	Timely check, repair and update drainage equipment	Yes	Yes	Yes	Yes	No
X_7	2	5	10	Establish gas dynamic monitoring system and increase ventilation and wind speeds	Yes	Yes	Yes	No	Yes
X_8	1	4	4	Add vents	Yes	Yes	Yes	Yes	No
X_9	2	3	6	Timely increase ventilators or increase its power	Yes	Yes	Yes	No	Yes
X_{10}	2	4	8	Adjust stationing of monitoring facilities according to the accumulation of gas	Yes	Yes	Yes	Yes	No
X_{11}	2	4	8	Check and repair monitoring equipment at least once a month and update equipments	Yes	Yes	Yes	Yes	No
X_{12}	2	4	8	Monitoring each shift at least 2 times and establish systems of rewards and punishment	Yes	Yes	Yes	No	Yes
X_{13}	1	4	4	Shorten distance of the wind, dredge channel and working face, and adjust the wind road	Yes	Yes	Yes	Yes	No
X_{14}	2	4	8	Strictly carry on zoning ventilating, timely jam mined-out, adjust auxiliary fans and dredge the duct	Yes	Yes	Yes	No	Yes
X_{15}	3	3	9	Determine the position of ventilation structures, and timely jam to reduce air leakage	Yes	Yes	Yes	No	Yes
X_{16}	1	4	4	Enlarge duct cross-sectional to improve the ventilation ability	Yes	Yes	Yes	Yes	No

After analyzing the potential threats one by one, we could find these CCPs when controlling the gas concentration in the process of this mine's production, which include the expectation of gas emission(X_2), disrepair or ageing of drainage equipment(X_6), increase of gas flow-volume(X_7), insufficient of wind supply ability(X_9), laxation of the monitoring system execution(X_{12}), disorder of wind direction (X_{14}) and the serious air leakage(X_{15}).

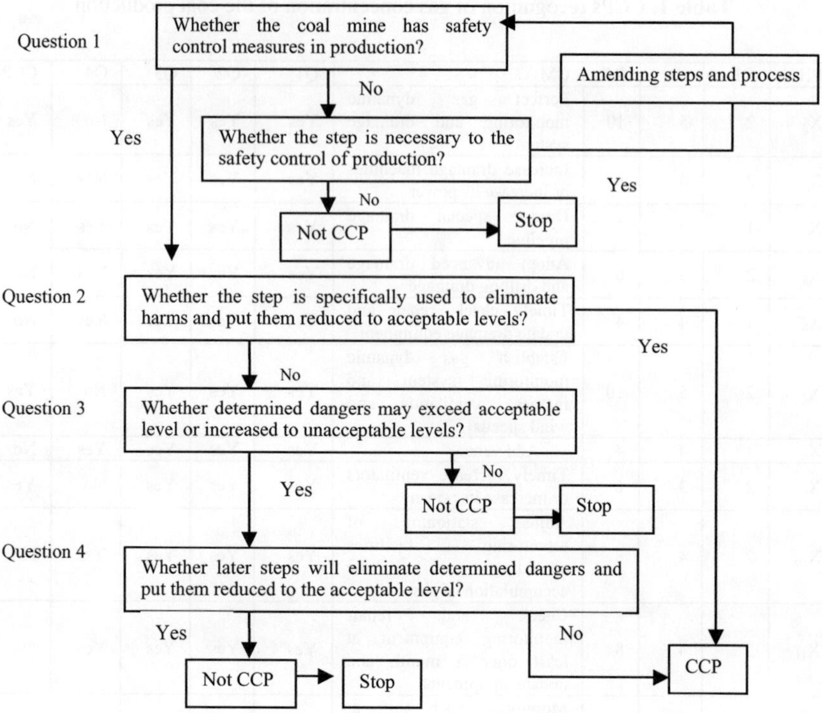

Fig. 3. Decision tree of CCP of safety production in coal mine

2.6 Establishment of Key Limits , Monitoring System and Deviation-Correcting Measures of CCPs

In order to check effectively whether the CCPs are working, we should select control indexes for each CCP, and establish the key limits. Therefore, the selected control indexes of CCPs should work quickly and accurately, thus we could take correcting measures timely when there is something wrong with the running system.

Table2 has listed 7 CCPs which had been identified with their control indexes(CI), key limits(KL), monitoring systems(MS) and rectification measures(RM), and the critical limits could be fatherly divided into expectations(EX) and action value(AV). When the control indexes are in the range of expectation, it shows that the system is running well; While when the control indexes reach the action value level, timely action should be taken according to the rectification measures, and make the system performance back to the expectation level as soon as possible. Meanwhile, the selected control indexes are basically can achieve continuous monitoring through online system(making MO stand for monitoring online), which is helpful to improve the ability of identifying deviations in time of the system to improve the security of the system.

Table 2 only briefly listed the monitoring systems and the rectification measures of each CCP, while the content of HACCP, as a complete system, should be more specific.

In monitoring system, what should be made clear includes the position, frequency, equipment and procedures of sampling, as well as the record, expression, preservation and reporting types of data, etc. In deviation-correcting measures , some contents should be made clear, which include duties of principals and their contact information, specific plan of the deviation-correcting measures, standard operating procedures, address of backup device and some other relevant technical information, etc.

Table 2. Key limits , monitoring system and deviation-correcting measures of CCPs

CCP	CI		KL		MS	RM
			EX	AV		
CCP_1	Peak of gas flow-volume (m3/min)		<30	>30	MO	Survey causes of gas's abnormal addimg, take measures to reduce emission, increase drainage equipments and improve the drainage ability
	Gas drainage rate(%)		100	<100	MO	
CCP_2	Intact rate of drainage equipment(%)		100	<100	MO	Repair and update drainage equipment
CCP_3	The rate of change of airflow needed		<Max	>Max	MO	According to the needs of every part, increase or decrease windsupply to improve wind speed
	Wind speed (m/s)	Wellbore	<8	>8	MO	
		Wind bridge	<10	>10	MO	
		Main return airway	<8	>8	MO	
		Locomotive roadways with lines	1.0-8	<1.0, >8	MO	
		Transporters lanes and return airway of mining zone	1.0-6	<1.0, >6	MO	
		Coal face and coal roadways in tunneling	0.25-4	<0.25, >4	MO	
		Other ventilation pedestrian roadways	>0.15	<0.15	MO	
	Ratio of supply and demand of air volume		1.1	1.0	MO	
CCP_4	Intact rate of windsupply equipment (%)		100	<100	MO	Repair and update windsupply equipment
	Demand rate of safety input (%)		100	<100	MO	Increase investment
CCP_5	Execution rate of responsibility for security (%)		100	<100	MO	Estamblish strict security disciplines and responsibility system
	Attendance rate of security personnel (%)		100	<100	MO	
CCP_6	Intact rate of auxiliary fans (%)		100	<100	MO	Repair and update auxiliary fans
CCP_7	Washroom leakage rate (%)		100	<100	MO	Investigate air leakage reasons, and timely do leak-stopping

2.7 Establishment of Validation Procedures and Archive System of Documents and Records

HACCP system test is refers to using of methods, procedures, test or evaluation besides monitoring, and determine whether the HACCP system can satisfy requirements of gas concentration control. In the gas concentration control of a mine's production, the effectiveness of HACCP system cold be verified by gas safety evaluation system. The archive system of documents and records of HACCP system should contain: the technical documentation, illustritive documentation, standard operating procedure, some orther related methods and procedures of the HACCP system and records on quality monitoring, equipment maintenance, systematic deviation, training of personnel in the implementation process of the HACCP system.

3 Discussion and Analysis

The key of whether the application of HACCP system into gas control in coal mine production safety field could be success lies on three links as follows: the determination of CCPs, establishment of key limits and validators for the HACCP system. At present, in the application of HACCP system, it still gives priority to half quantitative method when determining the CCPs, which depends on the subjective judgment of researchers to some degrees.In this study, CCPs were screened out from those potential harms whose risk factors are not less than 6 in table1. If the limits were lowered down, the number of CCPs would possiblly increase. Therefore, the screening methods and standards of CCPs needs further research. On the other hand, the occurrence probability and risk factor of all sorts of hazards are not fixed, and those indistinctive hazards originally could transform into deadly hazards in a given condition. For example, with the process of the equipment approaching to its service life, the failure frequency would be increasing correspondingly which would be new CCPs . HACCP should be evaluated and updated periodically.

In this case study, monitoring data of the mine at hand is far not enough to identify the statistical characteristics of CCPs' various control indexes when the gas concentration is out of limits, which brings difficulties when establishing the key limits for CCPs. Considering the particularity of safety in production of coal mine, most of the key limits are based on national security regulations. Therefore, it is much strict on gas control, from the perspective of production safety control. As to the need air volume rate, only the logical expression of their key limits were listed in table2, and further study is needed to determine the specific numerical value.

4 Conclusion

In this paper, the author did explorative study on application of HACCP system in gas concentration control of the coal mine, used FTA technique to identify the potential dangers which may lead to gas exceeding standard in the mine production, and put DTA technique applied to determine the of CCPs gas concentration control. Then, we established the key limits of CCPs, the monitoring system and deviation-correcting measures, analyzed the requirements for the verification procedures and documents and records system.However, the establishment and implementation of HACCP system is a cycle that improves gradually and updates constantly. HACCP system of gas concentration control set up in this paper is just preliminary research achievements, which need to be modified and perfected gradually in practice according to massive historical data, operation experiences and productive experiments of mine.

References

1. Jagals.: Application of HACCP Principles As a Management Tool for Monitoring and Controlling Microbiological Hazards in Water Treatment Facilities. Water Sci Technol (1), 69–76 (2004)
2. Fu, S., Jining, C., Yu, Z.S.: Application of HACCP System to Turbidity Control in Conventional Waterworks. China Water & Wastewater (13), 1–6 (2007)
3. Wang, Y.h., Wang, G.y.: Application of HACCP in Feed Enterprise. Feed Research (7), 25–27 (2004)
4. Zhao, Y.h.: Analisis on HACCP System Applied in Risk Management of Security Companies. Journal of Fujian Institute of Financial Administrators (1), 19–23 (2004)
5. Yuan, J.f., Wang, H.y., He, F.: Application of HACCP to Transportation Management of Dangerous Goods. Security and Safety Technology Magazine (3), 23–25 (2006)
6. Su, Y.l., Zhong, L.g.: Application of HACCP in Logistics Operation Management. Logistics Technology (1), 22–23 (2004)
7. Luo, Y.x.: Discussion on Security Management of Dangerous Goods Logistics in China. Engineering Science (2), 29–30 (2006)

Application of Adaptive Annealing Genetic Algorithm for Wavelet Denoising

Huang Yijun and Zeng Xianlin

First Aeronautical College of Air Force, Xinyang, 464000 China
Phyj96@163.com

Abstract. It's very difficult to select the best wavelet denoising threshold. A novel adaptive annealing genetic algorithm is presented to solve this problem. A new adaptive annealing method is given to calculate select probability for improving the convergence of this algorithm. Cross probability and variance probability are selected adaptively for enhancing this algorithm stability and convergence. The simulation shows that the best wavelet denoising threshold parameter can be found effectively by this algorithm.

Keywords: Wavelet analysis, Denoise, Threshold, Genetic algorithm.

1 Introduction

The measurement signals are interfered by noise in the process of obtaining and transmission. The main characters of original signal are obtained by denoising.

The wavelet denoising is a simple and effective method. This method was applied in many domains. The wavelet denoising restructures the signal with larger wavelet transforming coefficients. Those coefficients less than the threshold are give up. So it is very important to select this threshold. Donoho found the better common threshold and proved it in theory[1], but the application effect is not satisfaction. There are faults in Donoho's hard-threshold and soft-threshold denoising. The hard-threshold function is not continuous at the selected threshold. There are oscillations in restructure signal using those thresholds obtained by hard-threshold denoising. The soft-threshold function is continuous but there are constant deviation between the actual signal wavelet transform coefficients and the soft-threshold coefficients. So the restructure signal using soft-threshold coefficients is distortion in a certain level. D. L. Donoho and I . M. Johnstone[2] of Stanford University study these methods in theory unceasingly to improve these methods. Many scholars, such as R. R. Coifman [3] of Yale University, Xiao-Ping Zhang [4] of Texas University, Pan Q[5] of Northwestern Polytechnical University, Li C N[6] of Shanghai Jiaotong University, Mei W B[7] of Beijing Institute of Technology University, give some methods to improve these denoising methods.

A new threshold was given by Pan Quan[8] using static wavelet transform. The threshold is

$$t(m) = c^{\sigma_m}$$

m is scale; σ_m is noise variance at m scale.

M. Zhu (Ed.): Electrical Engineering and Control, LNEE 98, pp. 385–390.
springerlink.com © Springer-Verlag Berlin Heidelberg 2011

This method has a better denoise effect. But the value of c is variety with different signal and noise. It is difficult to select a best value of c. A MSE (mean square deviation) approximate function was presented by Zhang Lei[9] to restructure the signal. This function can be used to select a better value of c.

A novel adaptive annealing genetic algorithm is presented to select the denoising threshold parameter c in this paper. The criterion of threshold is the mini value of the MSE approximate function given by paper [9]. The simulation shows that the best wavelet denoising threshold parameter can be found effectively by this algorithm.

2 Wavelet Denoising and Threshold Selecting

2.1 Wavelet Denoising

Suppose a measurement signal f(t) composed by s(t) and n(t).

$$f(t)=s(t)+n(t)$$

s(t) is an original signal. *n(t)* is a Gaussian white noise, its variance distribution is $N(0, \sigma^2)$. The measurement signal *f(t)* can approximate the *s(t)* as possible by wavelet transform eliminating n(t).

The discrete signal *f(n)*, *n*=0,1,2, …… ,*N-1* can be obtained by discrete sampling. Its wavelet transform is

$$Wf(j,k) = 2^{-1/2}\sum_{n=0}^{N-1} f(n)\psi(2^{-j}n-k) \tag{1}$$

Wf(j,k) is wavelet coefficient.

The calculation of function (1) is very difficult and there isn't generally explicit expression of wavelet function $\psi(x)$. So the recursion expression of wavelet transform can be obtained by dilation equation.

$$Sf(j+1,k) = Sf(j,k) * h(j,k) \tag{2}$$

$$Wf(j+1,k) = Sf(j,k) * g(j,k) \tag{3}$$

The h is low band filter of scale function and g is high band filter of wavelet function. *S f(0 , k)* is original signal *f(k)*, *Sf(j, k)* are scale coefficients. *Wf(j , k)* are wavelet coefficients. The reconstruction function is

$$Sf(j-1,k) = Sf(j,k) * \hat{h}(j,k) + Wf(j,k) * \hat{g}(j,k) \tag{4}$$

Wf(j,k) are discrete wavelet coefficients of *f(k)* = *s(k)+n(k)* (using $w_{j,k}$ denotes *Wf(j,k)*.) The wavelet transform is linear transform. The wavelet coefficients $w_{j,k}$ composed by *Ws (j,k)* (denote $u_{j,k}$) and *Wn(j,k)* (denote $v_{j,k}$) . *Ws (j,k)* are discrete wavelet coefficients of *s(k)*. *Wn(j,k)* are discrete wavelet coefficients of *n(k)*.

The wavelet threshold denoising method:

1) Obtain wavelet coefficients $w_{j,k}$ of measurement signal $f(k)$ by wavelet transform.
2) Obtain $\hat{w}_{j,k}$ estimative wavelet coefficients by threshold processing $w_{j,k}$, causing $\left\|\hat{w}_{j,k} - u_{j,k}\right\|$ small as possible.
3) Compute reconstruction signal $\hat{f}(k)$

2.2 Threshold Parameter

Donoho's threshold parameter is $\lambda = \sigma\sqrt{2\log(N)}$. σ is standard deviation of noise, N is samples number. This threshold is too large while N is large number and too small while N is small number. So the factor $\sqrt{2\log(N)}$ usually substituted by constant c. $\lambda = c\sigma$

The best value of c is variable following the change of signal and noise. The value of c selecting randomly is not best value of c usually. A MSE approximate function was presented by Zhang Lei[9] to restructure the signal. This function can be used to select a better value of c by ergodic calculation.

$$F(c) = 2\left|\sigma^2 - \frac{\sigma}{\frac{8N}{\sqrt{2\pi}}\int_0^c e^{-t^2/2}dt}\int_{-c^\sigma}^{c^\sigma} t^2 \frac{e^{-t^2/(2\sigma^2)}}{\sqrt{2\pi\sigma}}dt\right| - E(\hat{w}_{j,k})^2 \tag{5}$$

Obviously, it will spend a long time to calculate c by function (5). A adaptive annealing genetic algorithm is presented to find the better value of c in this paper.

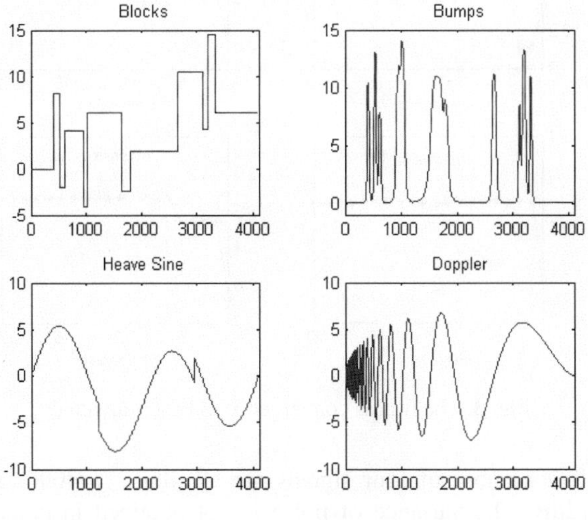

Fig. 1. Four typical signals

3 Experimental Results

Four typical signals shown in figure 1, Blocks、Bumps、HeaviSine and Doppler used in paper [9], are processed in this paper. These signals are polluted by Gauss white noise artificially. The pollution signals are shown in figure 2.

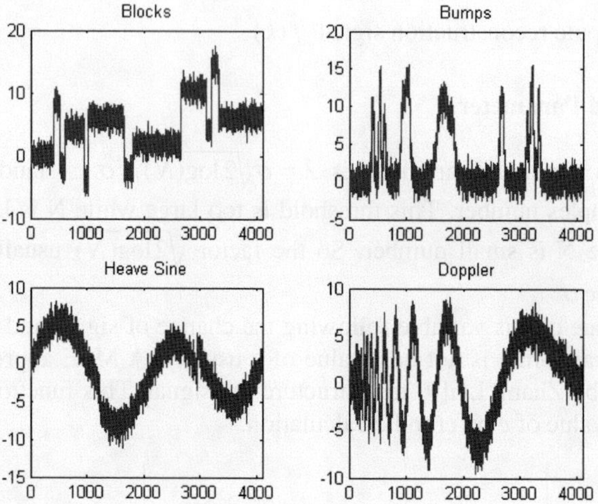

Fig. 2. Pollution signals

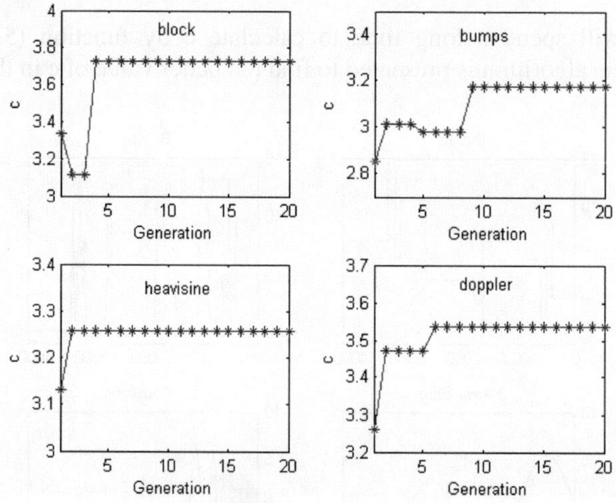

Fig. 3. The fitness convergence of best individuals

The sampling number of four signals are all 4096, N=4096. The signal noise ratio(SNR) is 12dB. The variance of noise is not obtained in practical application. Donoho proved the variance of noise can be given by equation 6.

$$\sigma = median(|w_{j,k}|)/0.6754 \qquad (6)$$

The value of c making $F(c)$ reaching minimum can be obtained by using adaptive genetic algorithm. The best value of c can be obtained in ten generations generally. So the maximum evolving generation is 20 in this paper. The simulation results are shown in figure 3.

Suppose the computing time of $F(c)$ is T, then the practical computing time are 20 (size of population) $*20*T=400T$. The computing time is $600T$ in paper [9]. The numerical results of c obtained by adaptive genetic algorithm are shown in table 1.

Table 1. Numerical results of c.

Signal	Blocks	Bumps	heavisine	Bumps
F(c)	-6.0538	-0.5257	-9.4062	-0.0855
C	3.7265	3.1727	3.2583	3.5379

The results in table 1 are same with them in paper [9]. So the value of c can be selected in interval [3,4] directly.

4 Conclusion

It's very difficult to select the best wavelet denoising threshold. A novel adaptive annealing genetic algorithm is presented to improve convergence and stability of standard genetic algorithm. A new adaptive annealing method is given to calculate select probability for improving the convergence of this algorithm. Cross probability and variance probability are selected adaptively for enhancing this algorithm stability and convergence. The convergence of this algorithm can be ensured by competition in male parent1. There are many merits such as convergence rapidly, avoiding local extremum and global optimization ability in this algorithm. The simulation shows that the best wavelet denoising threshold parameter can be found effectively by this algorithm.

References

1. Donoho, D.L.: De-noising by Soft-thresholding. IEEE Trans. on IT 41, 613–627 (1995)
2. Donoho, D.L., Johnstone, I.M.: Adapting to unknown smoothness via wavelet shrinkage. J. Amor. Star. Assoc. 90, 1200–1224 (1995)
3. Coifman, R.R., Donoho, D.L.: Translation- Invariant de-Noising Tech. Rep., Dept. Statistics, Stanford Univ, to be published in Wavelets and Statistics, A. Antoniadis ed. Springer Verlag, Berlin Germany (1995)
4. Zhang, X.P., Desai, M.D.: Adaptive denoising based on SURE risk. IEEE Signal Processing Letters 5, 265–267 (1998)
5. Pan, Q., Dai, G.Z., Zhang, H.C., Zhang, L.: A Threshold Selection Method for Hard-Threshold Filter Algorithm. Acta Electronica Sinica 26(1), 115–117 (1998) (in Chinese)

6. Li, C.N., Hu, G.R.: A Modified Wavelet Domain Speech Enhancement Method. Journal Of China Institute Of Communications 20(4), 88–91 (1999) (in Chinese)
7. Mei, W.B., Shark, L.-K.: Optimum Wavelet Threshold Technique for Fractal Signal Estimation. Acta Electronica Sinica 26(4), 15–18 (1998) (in Chinese)
8. Quan, P., Lei, Z.: Two Denoising Methods by wavelet transform. IEEE Tran. on SP 47, 3401–3406 (1999)
9. Zhang, L., Pan, Q., Zhang, H.C., Dai, G.Z.: On the Determination of Threshold in Threshold based De-noising by Wavelet Transform. Acta Electronica Sinica 29(3), 400–402 (2001) (in Chinese)
10. Zhang, J.S., Xu, Z.B., Liang, Y.: Integrated anneal genetic algorithm and its sufficient convergence precondition. Sciene in China (Series B) 27(2), 154–164 (1997)

Secure Group Key Exchange Protocol

Lihong He

Computer Center, Anshan Normal University, Anshan Liaoning 114007, China
anshanhlh@yahoo.com.cn

Abstract. For communication security, the article illustrated the security proofs on the nPAKE. The nPAKE protocol use password authentication protocol to replace the name of symmetric encryption password , which can effectively prevent the attack or the server key compromise camouflage leakage attacks. Game sequences in this paper under the random oracle model proved the protocol's security.

Keywords: information security, password-authentication, group key exchange.

1 Introduction

Group key exchange protocol [1-5] is used to establish a shared session key for team members in the open network, and then use that key to ensure the confidentiality and integrity of the multicast message. Group key exchange protocol plays an important role in the environment with the security requirements for applications such as video or telephone conference, but there are large number of participants in the group key exchange protocols, while there are large amount of exchanged information too, which makes the design and analysis protocol more complex.

In 2002, Bresson and the others [6] first proposed key exchange protocol formal model based on group Diffie-Hellman which can resist a dictionary attack. In this model, group members shared authentication password (shared password-authentication, SPWA), that means all members of the group have the same password, trusted server password identifies the group members by the shared password and help group members to establish close Key. However, sharing password authentication model is not applicable in practical applications, first of all, if there are members of the group members to leave or a compromised, then the sharing password must be updated, but the update of the group password is expensive; Second, for any capture of any member of the capture may lead to the collapse of the entire system [3, 7].

In the view of shared insecurity password authentication model in 2005, Byun and the others[3] proposed a model key exchange protocol EKE-M and EKE-U which based on different set of password authentication. The model and the credibility of the group members have different passwords between the server, the server can be certified through the password as a member of the group and help group members establish a secure communication channel. However, there are server key compromise disguised attack against leakage [8] on the password-based authentication protocols, to solve this problem, we present the value of N based on password authentication key exchange protocol side nPAKE, and gives the proof of the protocol Formal .

M. Zhu (Ed.): Electrical Engineering and Control, LNEE 98, pp. 391–399.
springerlink.com © Springer-Verlag Berlin Heidelberg 2011

2 Group Key Exchange Protocol nPAKE

In nPAKE protocol, every clients $C_i (1 \le i \le n-1)$ respectively share their verifiers v_i with server S. The verifiers $v_i = g^{H_1(S,C_i,x_i)}$, where $H_1 : \{0,1\}^* \to \{0,1\}^{\ell_1}$ is a hash function and x_i is the private key of client C_i. And the protocol uses three types of operators :

$$\phi(\{\alpha_1,...,\alpha_{i-1},\alpha_i\},x) = \{\alpha_1^x,...,\alpha_{i-1}^x,\alpha_i\} \in \overline{G}^i \tag{1}$$

$$\xi(\{\alpha_1,...,\alpha_{i-1},\alpha_i\},x) = \{\alpha_1^x,...,\alpha_{i-1}^x,\alpha_i^x\} \in \overline{G}^i \tag{2}$$

$$\pi(\{\alpha_1,...,\alpha_{i-1},\alpha_i\},m,n) = \{\alpha_m,\alpha_{m+1},...,\alpha_{m+(n-1)}\} \in \overline{G}^n \tag{3}$$

The parameters $\{\alpha_1,...,\alpha_{i-1},\alpha_i\} \in \overline{G}^i$ is a set, $x \in Z_q^*$ is private exponent, $m \in [1,i]$, and $n \in [1, i-m+1]$.

The nPAKE protocol proceeds as follows:

(1) Client C_1 chooses three random values $(a_1, b_1, k_1) \xleftarrow{R} [1, q-1]$, computes $X_1 = \{(v_1)^{b_1}\}$ and $EM_1 = E_{v_1}(C_1, X_1, k_1)$, and then send EM_1 to server S.

(2) S uses D_{v_1} decrypt EM_1 to obtain X_1 and k_1, then S encrypts and computes $DM_1 = E_{v_2}(X_1, k_1)$ with v_2, and sends DM_1 to C_2.

(3) In the following steps, $C_i (2 \le i \le n-2)$ can obtain X_{i-1} and k_{i-1} from DM_{i-1}. C_i also chooses random values $(a_i, b_i, k_i) \xleftarrow{R} [1, q-1]$, computes $X_i = \phi(\{X_{i-1}, (v_i)^{b_i}\}, a_i)$ and $EM_i = E_{v_i}(C_i, X_i, k_i)$, then sends EM_i to S. S decrypts EM_i, encrypts and sends $DM_i = E_{v_{i+1}}(X_i, k_i)$ to C_{i+1}.

(4) When the last client C_{n-1} receives DM_{n-2} from S, she can obtain X_{n-2} and k_{n-2}. Then C_{n-1} chooses two random values $(a_{n-1}, b_{n-1}) \xleftarrow{R} [1, q-1]$, computes $X_{n-1} = \phi(\{X_{n-2}, (v_{n-1})^{b_{n-1}}\}, a_{n-1})$ and $EM_{n-1} = E_{v_{n-1}}(C_{n-1}, X_{n-1})$, and sends EM_{n-1} to S.

(5) S obtains X_{n-1}, chooses a random value $r \xleftarrow{R} [1, q-1]$, computes $Y_0 = \xi(\{X_{n-1}\}, r)$ and $EK_0 = E_{k_1}(Y_0)$, then S sends EK_0 to C_1.

(6) $C_i (1 \le i \le n-1)$ receives EK_{i-1} and obtain Y_{i-1} by using the temporary key k_i, computes $Y_{i-1}' = \xi(\{Y_{i-1}\}, a_i)$ and extracts $X_i' = \pi(\{Y_{i-1}'\}, 1, 1)$ by using operator π. Then $C_i (1 \le i \le n-2)$ extracts $Y_i = \pi(\{Y_{i-1}'\}, 2, n-i-1)$, encrypts and sends $EK_i = E_{k_i}(Y_i)$ to C_{i+1}.

According to X_i', every clients can compute the session key $sk_i = H_2(\{C_1, ..., C_{n-1}, S\} \| g^{a_1 \ldots a_{n-1} r})$, where $g^{a_1 \ldots a_{n-1} r} = (X_i')^{b_i^{-1} \cdot (H_1(S, C_i, x_i))^{-1}}$.

3 Proof of Security

3.1 Security Model

Players. Players U include clients C and a server S, Where $C = \{C_1, C_2 ... C_{n-1}\}$. We define U^i as the i-th instance of player U.

Long-term Key. Each client $C_i (1 \le i \le n-1)$ has a secret verifier v_i shared with the server S, so S can authenticate the identity of the client. nPAKE protocol is based on symmetric mechanism. The verifier v_i is called long-term key.

Adversary Capability. The adversary A interacts with the players by making various queries. The queries describe the capabilities of adversary. The queries include:

- $Send(U^i, m)$: This query models A sending a message m to an instance U^i, the player U send back the real message according to the protocol.
- $Execute(U^i)$: This query models passive attacks, where the adversary gets access to honest executions of P by eavesdropping. Therefore, A can obtain the messages exchanged between players.
- $Corrupt(U)$: The adversary A is given the long-lived key.
- $Reveal(U^i)$: This query models the reveal of the session key. The query is only available to A if U^i holds a session key.
- $Test(U^i)$: This query models the semantic security of the session key. The query can be asked at most once by the adversary A and is only available to A if U^i is Fresh. A guesses a random coin b as b', if $b' = b$, A can obtain the real session key in query Reveal, else he can only obtain a random number with equal length.

3.2 Security Notions

Freshness. We define the instance U^i is fress if the following conditions are met: Players involved in the collection $\{U^i, pid(U^i)\}$ have never been asked for a Corrupt query or a Reveal query. $pid(U^i)$ is used to express the identity of U^i.

AKE Security. The security objective of nPAKE protocol is n-1 clients with the help of the server S generates a shared session key for communication, and to authenticate the clients. In an execution of the nPAKE protocol P, we say an adversary A wins if he asks a single Test query to a Fresh player U and correctly guesses the bit b used in the game $Game^{nPAKE}(A, P)$. We denote the AKE advantage as:

$$Adv_{nPAKE}^{ake}(\mathtt{A}) = 2\Pr[b = b'] - 1 \tag{4}$$

Provided that passwords are drawn from dictionary D, we define the advantage of $\mathtt{A}$ in violating the semantic security of the protocol P as $Adv_{nPAKE}^{ake}(\mathtt{A})$. If :

$$Adv_{nPAKE}^{dictionary-att}(\mathtt{A}) \leq \frac{n}{|\mathbf{D}|} + \varepsilon(k) \tag{5}$$

we say nPAKE is secure password authentication protocol. $\varepsilon(k)$ is a function can be ignored based on the secure parameter k.

3.3 Assumptions

We will prove security under the group computational Diffie-Hellman assumption(GCDH) and the group decisional Diffie-Hellman assumption(GDDH). Our definitional approach of Group Diffie-Hellman distribution is from [6], and nPAKE is based on the triangular structure, we illustrate for n = 4 on Figure 1:

g_0					$= X_0$
g_1	$g_1^{x_1}$				$= X_1 = \Phi(X_0, x_1, v_1)$
$g_2^{x_2}$	$g_2^{x_1}$	$g_2^{x_1x_2}$			$= X_2 = \Phi(X_1, x_2, v_2)$
$g_3^{x_2x_3}$	$g_3^{x_1x_3}$	$g_3^{x_1x_2}$	$g_3^{x_1x_2x_3}$		$= X_3 = \Phi(X_2, x_3, v_3)$
$g_4^{x_2x_3x_4}$	$g_4^{x_1x_3x_4}$	$g_4^{x_1x_2x_4}$	$g_4^{x_1x_2x_3}$	$g_4^{x_1x_2x_3x_4}$	$= X_4 = \Phi(X_3, x_4, v_4)$

Fig. 1. Triangular structure defined when *n*=4

GCDH Assumption. The adversary $\mathtt{A}(T, \varepsilon)$ is a probabilistic Tuing machine running in time T in Group G. We assume that the advantage $Adv_{nPAKE}^{gcdh}(\mathtt{A})$ is negligible that $\mathtt{A}$ can compute Group Diffie-Hellman distribution in the time T as:

$$Adv_{nPAKE}^{gcdh}(\mathtt{A}) = \Pr[\mathtt{A}(GDH_{\Gamma}(x_1, x_2, ..., x_n)) = g^{x_1 \cdots x_n}] \leq \varepsilon(n) \tag{6}$$

GDDH Assumption. We define the two following distributions:

$$GDH_n = (g^{x_1}, ..., g^{x_n}, g^{rx_1}, ..., g^{rx_n} \mid x_1, ..., x_n, r \in Z_q) \tag{7}$$

$$Rand_n = (g^{x_1}, ..., g^{x_n}, g^{y_1}, ..., g^{y_n} \mid x_1, ..., x_n, y_n, ..., y_n \in Z_q) \tag{8}$$

The distinguisher $\mathtt{A}(T, \varepsilon)$ is a probabilistic Turing machine running in time T that is able to distinguish the two distributions with advantage $Adv_{nPAKE}^{gddh}(\mathtt{A})$ in Group G. We assume the advantage can be ignored.

Theorem 1. Let Password be a finite dictionary of size N. Let A be an adversary against the AKE security of P within a time bound T, after q_s interactions with the players, q_h hash-queries, and q_e encryption/decryption queries. Then we have:

$$Adv_{nPAKE}^{ake}(A) \le \frac{q_h^2}{2^l} + \frac{(q_e+q_s)^2}{(q-1)^2} + \frac{2q_s}{N} + 2q_s Adv_{nPAKE}^{gddh}(A) + 2q_h Adv_{nPAKE}^{gcdh}(A) + 2(n-1) Adv_{nPAKE}^{cca-se}(A) \quad (9)$$

3.4 Proof of Security

In this section we use Game-playing[9] method to prove the security. We define a sequence of games starting at the real game Game0 and ending up at Game6. We define b and b', and refer to S_i as the event $b = b'$ in game Gamei.

Lemma 1. Let E, E' and F be some events defined on a probability space. Let us assume that $\Pr[E \wedge \neg F] = \Pr[E' \wedge \neg F]$, then $|\Pr[E] - \Pr[E']| \le \Pr[F]$.

Lemma 2. Let G be a finite cyclic group and g be the generator. Given $L = (g^{a_1}, ..., g^{a_i})$, then the collection $\{\Psi(L, \theta, v)\}$ and $\{g^{r_0}, ..., g^{r_i}\}$ is indistinguishable.

Game0: This is the real attack to the protocol nPAKE in which several oracles are available to the adversary: hash oracles, the encryption/decryption oracles, and all the instances of players. All oracles are answered according to the real protocol excution, so the probability of a successful attack for the adversary A is equal to the probability that A guesses b in Test query as:

$$Adv_{nPAKE}^{ake}(A) = |2\Pr[b = b'] - 1| = |2\Pr[S_0] - 1| \quad (10)$$

Game1: We simulate the hash and the encryption/decryption oracles as in Game0 by maintaining lists: a hash list(Λ_H), encryption lists (Λ_E) and decryption lists (Λ_D).

- Hash-query: For a query q such that a record (q,r) appears in Λ_H , the answer is r. Otherwise r is chosen at random from $\{0,1\}^l$ and the record (q,r) is added to Λ_H . We have $H(q) = r$.
- Encryption-query: For an encryption query (k,X) such that a record (k,X,Y) appears in Λ_E , the answer is Y. Otherwise Y is a random ciphertext of length $|X|$. The record (k,X,Y) is then added to encryption list Λ_E .

– Decryption-query: For a decryption query (k,Y) such that a record (k,Y,X) appears in Λ_D, the answer is X. Otherwise X is a random tuple $X = \{g^{r_i}\}_{1\le i\le |Y|}$, where $r_i \in Z_q$. The record (k,Y,X) is then added to decryption list Λ_D, and the record (k,X,Y) is then added to encryption list Λ_E.

From the above simulation we easily see that the Game1 and Game0 are perfectly indistinguishable :

$$\Pr[S_1] = \Pr[S_0] \tag{11}$$

Game2: For an easier analysis in the following, we cancel games in which some collisions appear. We define the collision event in Hash simulation as Collh, and $\Pr[Coll_h] \le q_h^2 / 2^{l+1}$. We also define the collision event in encryption/decryption scripts as Collh. According to birthday paradox, the probability of collision $\Pr[Coll_d] \le (q_e + q_s)^2 / 2(q-1)^2$. By Lemma 1, we have:

$$|\Pr[S_2] - \Pr[S_1]| \le \Pr[Coll] = \frac{q_h^2}{2^{l+1}} + \frac{(q_e + q_s)^2}{2(q-1)^2} \tag{12}$$

Game3: We abort the executions wherein the adversary may have guessed the password and verifier. We define Encrypt as the events that when the adversary execute $Send(U_i, Y)$ query, the record $(v_i, *, Y)$ is added to encryption list Λ_E. If Encrypt events happen, the adversary can verify his password guessing. If Encrypt events do not happen, Game2 and Game3 are perfectly indistinguishable:

$$|\Pr[S_3] - \Pr[S_2]| \le \Pr[Encrypt] = \frac{q_s}{N} \tag{13}$$

Game4: We introduce a simulator ST to attack on the symmetric encryption algorithm under the chosen ciphertext. First, ST costructs two messages: $m_0 = (g^{a_1}, ..., g^{a_n})$ and $m_1 = (R, ..., g^{a_n})$, where $R \in Z_P^*$. Then, ST obtains $c_d = E_{v_i}(m_d)$ by using encryption oracles, where $d \in \{0,1\}$.

During the simulation, the adversary queries all kinds of oracles, and the oracles answer under the controlling of ST. When the adversary executes Send query to instance U_i in the second round, he modifies the answer as c_1. Then in Test query, if b=1, ST answers the real session key, else answers a random number. In the last step, the adversary outputs the guessing bit b'. If $b' = b$, the adversary attacks in symmetric encryption successfully.

If the adversary can not be successful, Game4 and Game3 are indistinguishable:

$$\Pr[S_4]-\Pr[S_3]=(n-1)Adv_{nPAKE}^{cca}(\mathbb{A}) \quad (14)$$

Game5: In this game we consider an adversary against GCDH problem. For n players, we construct an algorithm $A(x_1,x_2,...,x_n)=g^{x_1\cdots x_n}$ and the triangular structure illustrated in figure 1 by using blinding exponents $\{v_1,...,v_n\}$. In figure 1, the lines of the triangular structure will be used in this game to answer to the Send queries. The lines are denoted and contructed as:

$$L_i=\begin{cases}\{g\} & i=0\\ \Phi(L_{i-1},x_i,v_i) & 1\le i\le n\end{cases} \quad (15)$$

We maintain a list Λ_Ψ that keeps track of the exponents θ used to blind a line $L_i:L=\Psi(L_i,\theta,v)$. The list Λ_Ψ contains records of the form (i,θ,v,L), where $\theta=\{\theta_1,\theta_2,...,\theta_n\}$, $L=\{g_1,g_2,...,g_{i+1}\}$ and $v\in Z_q^*$. For L_i, computes $L=\Psi(L_i,\theta,v)$ as

$$\Psi(L,\theta,v)=\{g_1^{v\theta_2\theta_3...\theta_i},g_2^{v\theta_1\theta_3...\theta_i},...,g_i^{v\theta_1\theta_2...\theta_{i-1}},g_{i+1}^{v\theta_1\theta_2...\theta_i}\} \quad (16)$$

- When the adversary executes $Send(U_i,m)$ query, we chooses $\mu_i,\rho_i\in Z_q^*$ randomly. If the record $(i-1,\theta,v,L)\in\Lambda_\Psi$, then we compute $L'=\Psi(L_i,\theta/\rho_i,\mu_i v)$ and update the record as $(i-1,\theta,v,L')$. Otherwise, we use $(\mu_i x_i,\rho_i v_i)$ instead of (x_i,v_i), and add the result into Λ_Ψ. In each query, μ_i and ρ_i are choosed randomly. Even if the adversary excutes the same query several times, the answers are different.
- When the adversary excutes $Send(U_i,E)$ query, we compute $(X'_i)^{\mu_i\rho_i}$ for instance.

By induction, the client can choose secret exponents $(\mu_i x_i,\rho_i v_i)$ to compute K_i and $sk_i=H_2(U\,\|\,K_i)$. By analysis, the operation Φ between L_{i-1} and L_i can be preserved by Ψ. And even if K_i is computed by using the player's own secret numbers $(\mu_i x_i,\rho_i v_i)$, that Game5 and Game4 are indistinguishable:

$$\Pr[S_5]=\Pr[S_4] \quad (17)$$

Game6: In this game, we update the answer returned from $Send(U_i,m)$ query. When the adversary excutes $Send(U_i,m)$ query, he randomly chooses $v\in Z_q^*$ and $\theta=\{\theta_1,\theta_2,...,\theta_n\}\in(Z_q^*)^n$, then he computes $L'=\Psi(L_i,\theta,v)$ and updates the list Λ_Ψ.

If the input is $L=(a_1,...,a_{i-1},a_i)$, the adversary can get $L'=(a_1^{vx},...,a_{i-1}^{vx},a_i^{v},a_i^{vx})$ as it is defined in Game3. But in this game, the adversary gets $L''=(g^{r_0},...,g^{r_i})$. By Lemma 2, we know L' and L'' are indistinguishable as:

$$\Pr[S_6]-\Pr[S_5]\le q_s Adv_{nPAKE}^{gddh}(\mathbb{A}) \tag{18}$$

Game7: In this game, we only give an instance $A(x_1,x_2,...,x_n)$ of GCDH without its output $g^{x_1\cdots x_n}$. If the adversary can obtain some K_i by excuting oracle queries, he have solved the GCDH problem.

However, the adversary can't compute K_i and therefore the answers to Reveal and Test queries are all random numbers. We denote by aH the event that the adverary asks a Hash query of the form $(U \parallel K_i)$. The event shows that the adverary can solve the GCDH problem as $\Pr[S_7 \mid \neg aH]=\Pr[S_6 \mid \neg aH]$. According Lemma 1, we get:

$$\Pr[S_7]\text{-}\Pr[S_6]\le q_h Adv_{nPAKE}^{gcdh}(\mathbb{A}) \tag{19}$$

The answers to Reveal and Test queries are purely random, so:

$$\Pr[S_7]=1/2 \tag{20}$$

Put all these together, the Theorem 1 is proved.

4 Conclusion

Based on N side authentication key exchange protocol - nPAKE agreement, this paper gives the definition of security model and security under the random oracle model .We proved the security of the assumption on GDDH under the random oracle model by building game series and using Hybird.

References

1. Byun, J.W., Lee, D.H., Lim, J.: Password-based group key exchange secure against insider guessing attacks. In: CIS 2005, Xi'an, China, pp. 143–148 (2005)
2. Lee, S.M., Hwang, J.Y., Lee, D.-H.: Efficient password-based group key exchange. In: Katsikas, S.K., López, J., Pernul, G. (eds.) TrustBus 2004. LNCS, vol. 3184, pp. 191–199. Springer, Heidelberg (2004)
3. Byun, J.W., Lee, D.-H.: N-party encrypted diffie-hellman key exchange using different passwords. In: Ioannidis, J., Keromytis, A.D., Yung, M. (eds.) ACNS 2005. LNCS, vol. 3531, pp. 75–90. Springer, Heidelberg (2005)
4. Kwon, J.O., Jeong, I.R., Lee, D.-H.: Provably-secure two-round password-authenticated group key exchange in the standard model. In: Yoshiura, H., Sakurai, K., Rannenberg, K., Murayama, Y., Kawamura, S.-i. (eds.) IWSEC 2006. LNCS, vol. 4266, pp. 322–336. Springer, Heidelberg (2006)

5. Sayed, R.M., Ibrahim, M.H., Nossair, Z.B.: Group key exchange protocol for users with individual passwords. Journal of Engineering and Applied Science 55(8), 327–342 (2008)
6. Bresson, E., Chevassut, O., Pointcheval, D.: Group diffie-hellman key exchange secure against dictionary attacks. In: Zheng, Y. (ed.) ASIACRYPT 2002. LNCS, vol. 2501, pp. 497–514. Springer, Heidelberg (2002)
7. Wan, Z., Deng, R.H., Bao, F., Preneel, B.: nPAKE A hierarchical group password-authenticated key exchange protocol using different passwords. In: Qing, S., Imai, H., Wang, G. (eds.) ICICS 2007. LNCS, vol. 4861, pp. 31–43. Springer, Heidelberg (2007)
8. Sun, H.-M., Chen, B.-C., Hwang, T.: Secure key agreement protocols for three-party against guessing attacks. The Journal of Systems and Software 75(1-2), 63–68 (2003)
9. Shoup, V.: Sequences of games: a tool for taming complexity in security proofs. In: Yung, M., Dodis, Y., Kiayias, A., Malkin, T. (eds.) PKC 2006. LNCS, vol. 3958, pp. 329–362. Springer, Heidelberg (2006)

Application of Wavelet Packet Analysis in the Fault Diagnosis for Flight Control Systems

Jiang xiaosong

First Aeronautical College of Air Force, Henan Xinyang, 464000, China
Wuhan University of Technology, Hubei Wuhan 430000 China
jxs2000624@sohu.com

Abstract. The decomposition and restructuring algorithm of wavelet packet analysis (WPA) is introduced. The initial data collected by sensor group are denoised by WPA. The viewpoint about data information identification based on 'energy-information' is shown. An algorithm of using WPA extracting data feature is presented. A new method of flight control system fault diagnosis is given. Simulation results shows that three kinds of common faults of flight control system can be distinguished by using this method.

Keywords: wavelet packet analysis, denoising, feature extraction, fault diagnosis.

1 Introduction

The method of failure testing and isolation of flight control system of a type of fighter is using residuals getting from equivalence space [1]. This method was effective in failure testing, but it was weak in failure isolation.

In many aspects wavelets analysis is a synthesis of older ideas with new elegant mathematical results and efficient computation algorithms. Wavelet analysis is in some cases complementary to existing analysis techniques (e.g. correlation and spectral analysis) and in order cases capable of solving problems for which little progress had made prior to the introduction of wavelet analysis. The wavelet analysis is internationally recognized up to the minute tools for analyzing time-frequency. In recent years, the wavelet analysis was researched widely in fault detect domain. Wavelet analysis has localization feature in time domain and frequency domain. Because the wavelet function grows and decays in a limited time period, the wavelet transforms can focus any details of a high frequency signal. The wavelet analysis has become a more effective tool in fault detect domain.

After introducing wavelet packet analysis (WPA) in signal translation, the failure diagnosis system designed by the author of paper [1] was improved. The failures of actuator were isolated effectively by this system.

2 Wavelet Packet Analysis

The signal is decomposed into low-frequency band a1 and high-frequency band d1 by wavelet analysis. The lost information in a1can be obtained in d1. In next level signal

M. Zhu (Ed.): Electrical Engineering and Control, LNEE 98, pp. 401–405.
springerlink.com © Springer-Verlag Berlin Heidelberg 2011

analysis, the a1 is decomposed into low-frequency band a2 and high-frequency band d2. The lost information in a2 can be obtained in d2. By analogy with this method, the signal can be decomposed into higher level. This is the multiresolution analysis (MRA). The wavelet packed system was proposed by Ronald Coifman to allow a finer and adjustable resolution of frequencies at high frequencies [2]. It also gives a rich structure that allows adaptation to particular signals or signal classes. In order to generate a basis system that would allow higher resolution decomposition at high frequencies, the highpass wavelet branch of the Mallat algorithm will be iterated (split and down-sample) as well as the lowpass scaling function branch.

In order to split wavelet subspace W_j (the closure of wavelet function$\psi(t)$), the scale space V_j and wavelet subspace were expressed by a new subspace U_j^n. U_j^n is the closure space of function $u_n(t)$ and U_j^{2n} is the closure space of function $u_{2n}(t)$. Here the $u_n(t)$ fits the two-scale function:

$$\begin{cases} u_{2n}(t) = \sqrt{2}\sum_{k\in Z} h_k u_n(2t-k) \\ u_{2n+1}(t) = \sqrt{2}\sum_{k\in Z} g_k u_n(2t-k) \end{cases}$$

The h_k and g_k are filter coefficients and $g_k = (-1)^k h_{1-k}$

The decompose algorithm of WPA is [3]:

$$\begin{cases} d_m^{j+1,2n} = \sum_l d_l^{j,n} \overline{h}_{l-2m} \\ d_m^{j+1,2n+1} = \sum_l d_l^{j,n} \overline{g}_{l-2m} \end{cases} \quad m \in Z$$

The recover algorithm of WPA is:

$$d_l^{j,n} = \sum_m [d_m^{j+1,2n} h_{k-2m} + d_m^{j+1,2n+1} g_{k-2m}]$$

3 Signal Denoising by Wavelet Packet Analysis

There are many noises in signal as the influences of environment and circuit in general. So the signal denoising must be done before using the signal. The wavelet analysis has better local character for analyzing time-frequency; it can get better effect in signal denoising.

Many wavelet packet basic functions can be chosen when the signal was decomposed using wavelet packet analysis. The best wavelet packet basic function would be got according to the entropy norm of signal [4]. The general stage of signal denoising and data compression using WPA is shown as

(1) choosing a wavelet function and the layer N of wavelet decompose, decomposing the signal with wavelet transform.

(2) Calculating the best tree according to the entropy norm, thus getting the best wavelet packet basic function.
(3) Choosing a properly threshold for quantized the WPA coefficients.
(4) Recovering the signal using the quantized the WPA coefficients.

The keys of these stages are choosing the threshold and quantized the WPA coefficients. The result of signal denoising would be affected by the keys. In general, the threshold would be chosen with experiment.

4 Simulation Results

The actuator of a type of fighter aircraft is composed of steering engine and strengthener, its input signal is provided by control program and its output signal is obtained by liner movement pick-up at the strengthener terminal. The actuator system is a high-order non-linearity segmental apparatus, its mathematic model can be simplified second-order segmental apparatus and its transfer function is:

$$G(s) = \frac{k \cdot \omega_n^2}{s^2 + 2 \cdot \zeta \cdot \omega_n \cdot s + \omega_n^2} = \frac{\delta_L(s)}{U(s)}$$

This second-order segmental apparatus can be simulated by computer. The simulation system is shown in figure 1.

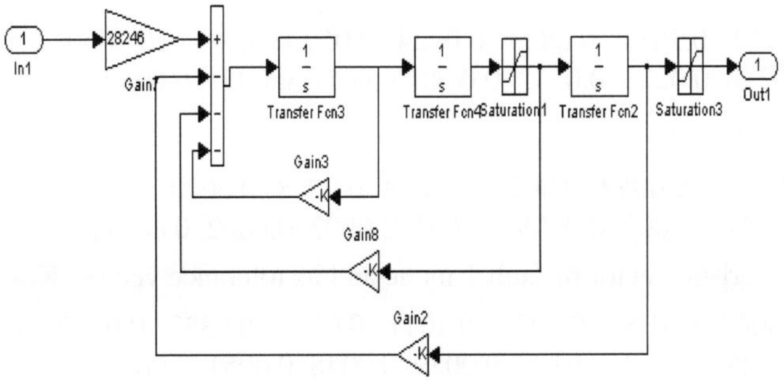

Fig. 1. Simulation block graph of actuator

The saturation constraints of rudder deflection angle and rudder deflection velocity were considered in the simulation. Three kinds common faults of this actuator can be shown by output signals.

The sample data of three kinds common faults of this actuator can be obtained by actuator fault simulation. The characteristic vectors of these faults were created. The length of these characteristic vectors n is 15 as the experiment data can be got repeatability. Because the experiment numeric values is small, these experiment numeric value needn't transform the normalized coefficients. The test data of normal mode and fault 1 mode are list in table 1 and table 2.

Table 1. Data of normal mode.

	E_{40}	E_{41}	E_{42}	……	$E_{4,15}$
1	0.8751	0.1881	0.0833		0.0042
2	0.8750	0.1881	0.0832		0.0042
.	.	.	.	.	.
.	.	.	.	.	.
.	.	.	.	.	.
15	0.8751	0.1882	0.0832		0.0041

Table 2. Data of fault 1 mode.

	E_{40}	E_{41}	E_{42}	……	$E_{4,15}$
1	1.8047	0.1058	0.1022		0.0051
2	1.8045	0.1058	0.1021		0.0050
.	.	.	.	.	.
.	.	.	.	.	.
.	.	.	.	.	.
15	1.8048	0.1057	0.1021		0.0050

The characteristic vector of normal mode can be got after calculating.

T 正常=[0.8751 0.1881 0.0833 0.0924 0.0570 0.0617 0.0640 0.0660 0.0192 0.0218 0.0223 0.0243 0.0045 0.0051 0.0051 0.0042]

Tolerance vector is (K=4)

ΔC 正常=[0.0277 0.0059 0.0026 0.0029 0.0018 0.0020 0.0020 0.0021 0.0006 0.0007 0.0007 0.0008 0.0001 0.0002 0.0002 0.0001]

The characteristic vector of fault 1 mode and its tolerance vector (K=4) can be got.

T 故障$_1$=[1.8047 0.1058 0.1022 0.1011 0.0262 0.0332 0.0317 0.0327 0.0087 0.0125 0.0113 0.0127 0.0039 0.0048 0.0051 0.0042]

ΔC 故障$_1$=[0.0261 0.0033 0.0032 0.0032 0.0008 0.0011 0.0010 0.0010 0.0003 0.0004 0.0004 0.0004 0.0001 0.0002 0.0002 0.0001]

The relation of fault signature and characteristic vector was found and it was verified by new experiment data. The verified result show that the prerealcomplete uniformity ratio of fault 1 (actuator lock) is 99%, and the prerealcomplete uniformity ratios of fault 2 (actuator deviation) and fault 3 (gain fault) are 97% and 98.5%. These results show that this assorting method using characteristic vector is effective.

5 Conclusion

The wavelet packet analysis is finer than wavelet analysis. As a result, wavelet packet analysis methodology has had a significant impact in areas as signal processing, fault diagnosis and image processing. The viewpoint about data information identification based on 'energy-information' is shown. A fault diagnosis method for flight control system was presented. The simulation results shown that three kinds of common faults of flight control system can be distinguished exactly by using this method.

References

1. Luo, C.: Software procedure and test for faulure testing and isolation of self-repair flight control system. Northwestern Polytechnical University, Xi'an (2000)
2. Percival, D.B., Walden, A.T.: Wavelet methods for time series analysis, vol. 47. Cambridge University Press, Cambridge (2000)
3. Li, S., Wu, J.: Fractal and wavelet, vol. 280. Science Press, Beijing (2002)
4. Aminian, F., Aminian, M.: Fault Diagnosis of Nonlinear Analog Circuits Using Neural Networks with Wavelet and Fourier Transforms as Preprocessors. Journal of Electronic Testing: Theory and Applications 17, 471–481 (2001)

Design of Intelligent Carbon Monoxide Concentration Monitoring and Controller System

Zou Tao, Xu Hengcheng, and Zeng Xianlin

The First Aeronautical Institute of the Air Force Xinyang, 464000 China
hnxyzqy1@sina.com

Abstract. Under the situation of gas and silo operating, accidents caused by carbon monoxide occurred frequently, for example the gas poisoning, explosion and fire hazard. For this reason, it is very important to the safety of people's life and property to measure the concentration of the gas accurately at any time. A kind of intelligence type carbon monoxide supervision system is introduced in the paper. The system is mainly composed of three parts: detector system, signal amplifying and processing system and display system. The functions, i.e. alarm setting, sound and light alarm, dynamic display of the concentration, error checking and auto adjustment is provided in this system. The principle, the structure and the main specification are introduced. The key technology and relevant solution are discussed in the paper.

Keywords: carbon monoxide monitoring and controlling, infrared spectrum absorption, concentration measurement, gas sensor, sound and light alarm.

1 Introduction

In recent years, with the development of the industry, the pollution caused by carbon monoxide from boiler is gradually serious. In the situation of gas and silo operating, accidents caused by carbon monoxide occurred frequently. The carbon monoxide would make people dizzy and vomited, and even cause poisoning, explosion and fire hazard etc. For this reason, it is very important to the safety of people's life and property to measure the concentration of the gas accurately and achieve auto alarm at any time, so that we can control it in time. [1, 2]

Recent years, various carbon monoxide concentration measuring, controlling and alarming instruments adopted the ionic conductibility solid electrolyte type detector. However, those equipments generally are of the low sensitivity, bad stability and easily influenced by environment, as a result, they can not work for a long period under the bad condition of the industrial environment. So, we developed the carbon monoxide supervision system based on the SCM, which consumedly raised the safety, function of the system and degree of autoimmunization. The main function of the system is as follows: realizing the monitoring and controlling for density, following the manifestation of the carbon monoxide density, the function of sound and light alarm. According to manifestation of the carbon monoxide density, the system will close the production facilities and open releaser or evacuate field personnel automatically. [3]

M. Zhu (Ed.): Electrical Engineering and Control, LNEE 98, pp. 407–413.
springerlink.com © Springer-Verlag Berlin Heidelberg 2011

2 Principle of Detecting the Carbon Monoxide Density

The experiment and theoretical calculation all point out that there is a strong vibrate absorption peak in 4.7μm of the carbon monoxide and the absorbing relation submits to Lambert-beer laws. So when the certain strength narrow band infrared light of 4.7μm wave-length passes hybrid air of implying the carbon monoxide, the relationship between the incident light intensity and emergent light intensity is as follows:

$$I_2^{'} = I_2 \exp(-ucl) \quad (1)$$

In the formula, μ for the absorption coefficient of the carbon monoxide; C for density to be measured; l for the length of the transmitted light .The principle structure sketch of the detector is as the figure 1. If light source 1 and 2 give out light alternately, then:

Only when light source 1 gives out light, can we easily draw a conclusion from figure 1:

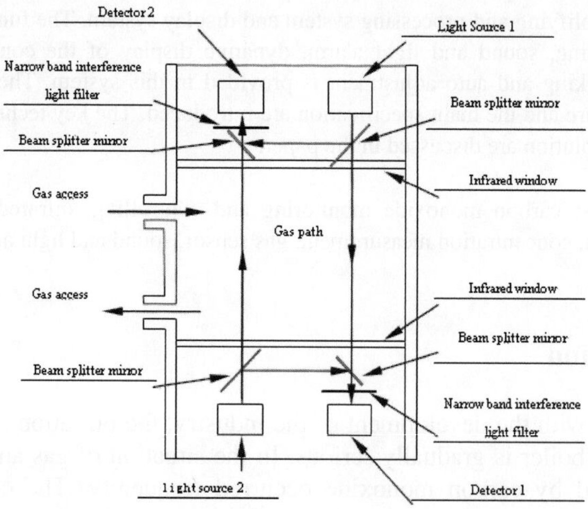

Fig. 1. Principle structure figure of the detecting system

$$I_{21} = \alpha_1 \alpha_4 I_1 \quad (2)$$

$$I_{11} = (1-\alpha_1)\beta_2 I_1^{'} \quad (3)$$

In the formula, I_{21}、 I_{11} are infrared energy received from detector 1 and detector 2 respectively; α_1、 α_4 are the mirror ratio of the beam splitter 1 and beam splitter 4 when there is only incidence; β_2 for mirror ratio of the beam splitter 2 when there is only I_1 incidence. I_1 for luminescent intensity of light source 1; $I_1^{'}$ for emergent light intensity of light source 1 when it passes gas circuit. From formula (1), $I_1^{'}$ is:

$$I_1^{'} = I_1 \exp(-ucl) \tag{4}$$

Put (4) into (3):

$$I_{11} = (1-\alpha_1)\beta_2 I_1 \exp(-ucl) \tag{5}$$

Only when light source 2 gives out light, can we draw a conclusion from figure 1:

$$I_{22} = (1-\alpha_3)\beta_4 I_2^{'} \tag{6}$$

$$I_{12} = \alpha_3 \alpha_2 I_2 \tag{7}$$

In the formula, I_{22}、I_{12} are infrared energy received from detector 1 and detector 2 respectively; α_2、α_3 are the mirror ratio of the beam splitter 2 and beam splitter 3 when there is only I_2 incidence. β_4 for mirror ratio of the beam splitter 4 when there is only I_2 incidence. I_2 for the luminescent intensity of light source 2; $I_2^{'}$ for emergent light intensity of light source 2 when it passes gas circuits. From formula (1), $I_2^{'}$ is:

$$I_2^{'} = I_2 \exp(-ucl) \tag{8}$$

Substituted (8) into (6):

$$I_{22} = (1-\alpha_3)\beta_4 I_2 \exp(-ucl) \tag{9}$$

From (2) × (7) / (5) × (9) , we can draw the conclusion:

$$\frac{I_{12} \times I_{21}}{I_{11} \times I_{22}} = \frac{\alpha_1\alpha_2\alpha_3\alpha_4}{(1-\alpha_1)(1-\alpha_3)\beta_2\beta_3} \exp(-2ucl) \tag{10}$$

If ream

$$\alpha = \frac{\alpha_1\alpha_2\alpha_3\alpha_4}{(1-\alpha_1)(1-\alpha_3)\beta_2\beta_4} \tag{11}$$

Then (10) become:

$$c = \frac{1}{ul} lu(\alpha \frac{I_{12} \times I_{21}}{I_{11} \times I_{22}}) \tag{12}$$

Obviously, α is a constant and light source 1 and light source 2 just gives out light alternately once. We can get density of the carbon monoxide from formula (12).

3 Description of Measuring and Controlling System

3.1 Gross Structure

This intelligent measuring and controlling system is mainly composed of detecting system, voltage following amplification system, sampling and retaining circuits, multiplex analog switch, A/D circuit and SCM -AT89C51 etc. The principle structure is as figure 2.

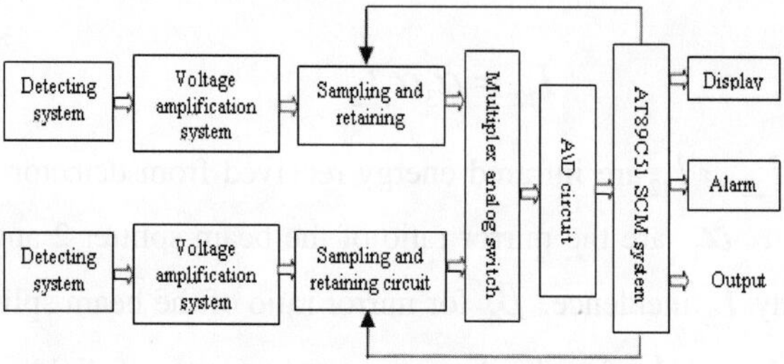

Fig. 2. Intelligent measuring and controlling system

3.2 Working Process

When the instrument works, the light source send out of impulse radiation, which is divided into two parts by beam splitter of higher reflectance ratio ,one (stronger reflectance) is reflected again and comes into the photo-translating system on the same side through the narrow band interferometer filter (central wavelength 4.7μm、bond width 0.1μm), which is converted into electric signal; the other (weaker reflectance)passes gas circuits (length 20cm)and comes into the photo-translating system on the different side through the narrow band interferometer filter (central wavelength 4.7μm、bond width 0.1μm), which is converted into electric signal. Two-way signal is first differently magnified by voltage following amplification circuits and then converted by A/D circuits, which is calculated in the SCM at last, in order to get the density of the gas.

3.3 Monitoring and Controlling System

The monitoring principle of system is that the carbon monoxide density is converted into analog voltage signal by gas sensor. The signal is sent to A/D converter after magnified by amplified and sent to CPU after converted into digital signal, and then the CPU carries on the data analysis. When the carbon monoxide density reaches the set value, CPU will output numeral quantity, which can drive the control installation to close the equipment and open the exhaust gear and output digital signal to drive photoelectric alarm, in order to make supervision personnel deal. After removing accidence, we can manipulate key-press in control panel to open production facility or recover work.

Monitoring and controlling system is an effective system on the base of configurationally software package to display the carbon monoxide density on spot. Because the real-time, accuracy and reliability is very important to the system, the SCM is adopted as controller on spot to achieve collections, operation and controlling of carbon monoxide density and send density data to PC.

Because the space of program and digital operation is wider, CPU chooses to use the SCM -AT89C51 of the ATMEL Company as the center and use EOROM as storage of the system program.

3.4 Forward Passage

Forward passage is mainly made up of sensor, transducer, low pass and voltage isolator. Voltage isolator is able to isolate digital signal from analog signal in order to eliminate common-mode interference. In the system, the gas sensor adopts Galva no-chemistry carbon monoxide sensor, which has many merits, such as high degree of accuracy and high resolution, voltage/current linear output, wide range ability, no power consumption, free from moisture and life-time dilatation. [4]

3.5 Acousto-Optic Alarm

The acousto-optic alarm system consists of oscillator, gate circuit, driver, musical chip and alarm indicator lamp. The musical chip adopts single record-playback chip produced by the American ISD company. This chip gathers voice processing and memorizing in integral whole. Further more, it will save the information when power-fail and can be controlled either by manual or SCM .It is very convenient to use it. The indicator lamp adopts LED and carries out audible alarm and indicator lamp alarm. The indicator lamp is red on the alarm state and green on the normal operating condition. [5]

3.6 The Monitoring and Controlling of Formality Function

The watchdog circuit of Max813 is used as the monitoring and controlling of formality function, it can receive the reset signal from the controller at a certain time when the system works normally, which can make reset end of The watchdog circuit invalid. Once the formality is out of control or run into endless loop, the circuit would send a reset signal to make the system reset and rework as soon as possible.

3.7 Keyboard/ Display

The display interface of the system keyboard adopts 8279 slug, scanning the keyboard and display by hardware. It is composed of alarm set key, clock set key, shift light key, confirmed key, facility open key, running key etc. Users can accomplish all kinds of operation of input interface though keyboard which is working in interrupt mode. When pressing the key, 8279 will intermit the application and the CPU turns into disposal formality of keyboard monitoring and controlling. The data of the system detected will pass the I/O driving circuit after being disposed by the SCM -AT89C51.

The LED displays the carbon monoxide density and clock on spot. The electrical source adopts safeguard of abnormal electrical source, under the normal condition, the system power supply is from the AC, the pile is on the condition of floating charge at the same time, if the AC supply is intermitted, the pile will do.

4 Discussion

4.1 Stability

Measurement result of density has no relation with the stability and aging degree of impulse infrared light supply, it only affects intensity of the light signal received by detector but the measurement result, which can not achieve by measurement system of two times tight in original way.

For the optics pollution and signal excursion caused by variety of temperatures and the aging of detectors do not affect the measurement results, the request of the cleanness maintained work of the whole machine is lower, it can not attain by density measuring system of the structure of two time tight way or two space tight way.

4.2 Certainty Measurement

Certainty measurement is the most important qualification of this device, which lies on optical-mechanical-electrical and other factors. Except what were discussed above such as monochromatic of the infrared light supply, well amplification of the circuit, constancy of the gas circuit temperature and compensation for the restriction of the absorption law are also the main factors of the instrument.

We can solve the monochromatic problem of the infrared impulse source by putting a narrow band interferometer filter before the detector and choosing the high performance ZnS detector and integrated circuit block chip to solve the linear magnification problem of circuit. For example, voltage follower circuit adopts LM324 four operational amplifier block, which does not need to do zero adjustment and has the merits of temperature compensation, high-gain and internal compensation. Further more, A/D circuit adopts MC14433 integrated package, which has the merits of high precision, well linearity, less influenced by temperature, wide voltage range and power supply protection etc. By adopting these integrated packages, we also settle the linear magnification of the signal [6].

The changes of the temperature in the measuring gas circuit will conduce to the changes of the density of the continuous-flow gas. So it is one of the key points of influencing the measurement accuracy and need to compensate. This instrument carries out temperature compensation by putting a heater and a thermostat to adjust the temperature in the gas circuit. To assure the measurement accuracy under the circumstance of high concentration, we have to compensate the output of the detector.

5 Conclusion

This article discusses a new system which is based on the principal of infrared spectrum imbibitions and adopts density measuring system consisted of double-photo source and double-detector with the SCM as the host. All of these make up an advanced carbon monoxide density measurement. The instrument adopts high-accuracy A/D circuit, which turns the problem of density measurement into frequency measurement to get the merits of high accuracy, well stability and fine anti-interference ability. The carbon monitoring system adopts diode monitored control system consisted of industrial control computer, SCM and actuating mechanism. This system realizes dynamic monitoring and controlling and adds the function of error auto correction. For the system works in high reliability, safety and high sensitivity in multiple experiments, it has much applied value and provides an ideal facility for industries and households in quantitative analyzing and monitoring density of the carbon monoxide. The aviation equipments research fund subsidizes item.

References

1. Yi, G.T.: The latest trend of the gas sensor. J. Infrared (6), 31–36 (2004)
2. Tetsuya, A., Takeyuka, K., Takashi, T.: Measurements of temperature and OH radical concentration in combustion gases bysorption spectroscopy with a diode laser. In: Procedings of the 1999 Pacific Rim Conference on Lasers Electro-Optice, Seoul, Korea, pp. 567–568 (1999)
3. Hu, R.: Intelligent measurement and control system. Xi'an Traffic University, Xi'an (2006)
4. Hou, R., Sun, L., Ren, L., et al.: Fault diagnosis of a tbrbo-unit based on wavelet packet theory. International Journal of Plant Engineering and Management 7(4), 198–203 (2002)
5. Hou, R., Sun, L., Ren, L., et al.: The diagnosis of a generator set based on modern nonlinear theory. In: Proceedings of the 5th Internation Conference on Vibration Engineering, Nanjing China, pp. 651–654 (2002)
6. Bauman, R.P.: Absorpion Spectroscopy, ch. 10. John Wiey and Sons Inc, New York (1992)

source and double-detector with the SCM as the host. All of these make up an advanced carbon monoxide density measurement. The instrument adopts high-accuracy A/D circuit, which turns the problem of density measurement into frequency measurement to get the merits of high accuracy, well stability and fine anti-interference ability. The carbon monitoring system accepts diode monitored control system consisted of industrial control computer, SCM and actuating mechanism. This system realizes dynamic monitoring and controlling and adds the function of error auto correction. For the system works in high reliability, safety and high sensitivity in multiple experiments, it has much applied value and provides an ideal facility for industries and households in quantitative analyzing and monitoring density of the carbon monoxide. The aviation equipment research fund subsidizes item.

References

1. Yu, G.Z.: The latest trend of the gas sensor. J. Infrared (6), 31–36 (2004)
2. Teruya, A., Takeyuka, K., Takashi, T.: Measurements of temperature and OH radical concentration in combustion gases hysorption spectroscopy with a diode laser. In: Proceedings of the 1999 Pacific Rim Conference on Lasers Electr. Optics, Seoul, Korea, pp. 567–568 (1999)
3. Hu, R.: Intelligent measurement and control system. Xi'an Traffic University, Xi'an (2006)
4. Hou, R., Sun, L., Ren, L., et al.: Fault diagnosis of a turbo-unit based on wavelet packet theory. International Journal of Plant Engineering and Management 7(4), 198–203 (2002)
5. Hou, R., Sun, L., Ren, L., et al.: The diagnosis of a generator set based on modern nonlinear theory. In: Proceedings of the 5th International Conference on Vibration Engineering, Nanjing China, pp. 651–654 (2002)
6. Bauman, R.P.: Absorption Spectroscopy, ch. 10. John Wiley and Sons Inc, New York (1992)

The Design of Intelligent Check Instrument for Airplane Voice Warning System

Zeng Xianlin, Xu Hengcheng, and Li Lizhen

First Aeronautical College of Air Force, Xinyang, 464000 China
hnxyzq3@163.com

Abstract. We design a kind of intelligent check instrument for a certain modern battle plane voice warning system, the hardware and software of this instrument and the basic principle is introduced in this paper. This instrument is highly integrative and has powerful function, which can automatically accomplish all the auto-test. For the adoption of the computer numerical value signals processing, the measurement accuracy is greatly improved. Not only can it accomplish the departure test and the home position test of the airborne voice warning system, but also has the comprehensive self-checking function and a friendly man-machine interface. The reliability and compatibility of this instrument is great, it also has many other advantages such as multi-function, simple operation, high degree of measurement accuracy and automation.

Keywords: airplane, voice warning system, self-check function, embedded control, check instrument.

1 Introduction

Voice warning system is one of security monitoring equipment on the plane. It can notify the flying emergency and criticality to the aircrew and ground crew, and replay the standard dictate according to the terminal input signal. The capability has a direct effect on the exertion of the flight safety and the tactical performance. If the monitoring equipment of the original voice warning system are used to accomplished the capability test and malfunction diagnosis of the voice warning system and the matching facilities, it not only needs the cooperation of the oscilloscope, signal generator, universal equipment, but also has the problem with heavy detecting workload and low autoimmunization. Aiming at the problems above, we develop a new single intelligent testing equipment to replace the original voice warning system and auxiliary instruments. [1, 3]

2 Component and Basic Principle of the Check Instrument

This instrument is single intelligent testing equipment which consists of hardware and software and the embedded industrial PC is adopted as the control center.

M. Zhu (Ed.): Electrical Engineering and Control, LNEE 98, pp. 415–420.
springerlink.com © Springer-Verlag Berlin Heidelberg 2011

2.1 Hardware and Basic Principle

The hardware system consists of the embedded industrial PC, wipe record test incentive source, digital I/O, the response signal analysis measurement circuits, port adapter circuits audio signal processing circuits, self-check circuits, etc, the principle structure is as figure1.

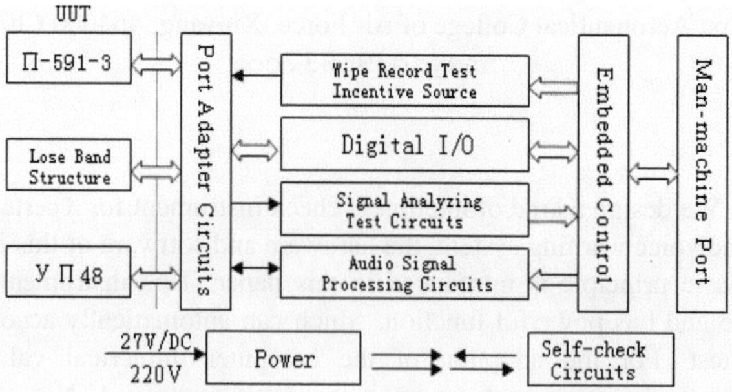

Fig. 1. The structure of voice warning system check instrument

1) Embedded Industrial PC: The embedded industrial PC consists of PC/104, multi-function card, solid state electronics dish, display and keyboard. It is the core of the check instrument. With the cooperation of the management software, it can control part of circuits and acquire instrument panel switch state, instrument working status setting, control measured equipments excitation signal generation and trends, it also can accomplish the output response and related parameters of the acquisition and processing, checking the output of results etc. Furthermore, the embedded industrial PC and fault diagnosis software constitute a fault diagnostic system, which can accomplish the fault diagnosis of the check instrument and the UUT, diagnosis and positioning. Multi-function card, display and keyboard constitute man-machine port.[2]

2) Wipe Record Test Incentive Source: Wipe record test incentive source consists of the preamplifier, audio voltage amplifier, program-controlled source, power amplifier and analog switches. It is used to create audio signal and accessory signals which the lose band structure wipe sound/recording testing is needed. And input warning signals from the microphone or line are used as audio signal which is the source of the inter-belt mechanism. The accessory includes: 100Hz tracking signal, 1000Hz accessory testing signal and 42 KHz high frequency signal which is used as the partial magnetic and wipe sound signals, all of them are created by the program-controlled source.

3) Signal Analysis Test Circuits and Digital I/O: Signal analysis test circuits include that time measurement circuits, sampling-holding circuits and ADC circuits. This part of circuits is to track and analyze the output single of UUT, parameters sampling measurement, and send results of measurement into the computer system for analyzing and verdict. I/O circuits consist of circuit latches, buffer, drive etc. It is the main data exchange access of the check instrument and UUT, which is in charge of reading the controlling signals, importing commend signals and exporting the state signals.

4) Audio Signal Processing Circuits and Port Adapter Circuits: Audio signal processing circuits consists of the preamplifier, voltage amplifier, power amplifier, band-pass filter appliances, analog switches, gain control circuits and automatic-static demising circuits. Audio signal processing circuits is used to process the audio signals from the inter-belt mechanism and replaying device, which can ensure that the output signals of the check instrument is clear, stable and having a high SNR. Port adapter circuits is the bridge between the check instrument and UUT, which is to complete the switch quantity level sectoring, analog quantities level comparing, the guide and controlling of signal trend, power supply of the UUT. The guide and controlling of signal trend is implemented by the input data from the I/O circuits.

5) Self-check Circuits and Power circuits: The process of self-check is that the microcomputer sets the circuits into a certain state by conveying control instructions to the function board, and reading the logic state of the circuits through the data port in turn. Compared with the standard condition it analyzes the integrated results and makes the diagnosis at last. The self-check can be implemented in offline work condition or online work condition. In the online work, the self-check and UUT identifying is processing at the same time, the UUT identifying will be carried on once the self-check is normal. If the check instrument has some fault, the fault diagnostic system will get to work automatically, and the results of self-check will be displayed on the screen in the form of text. The fault diagnosis will be located to a band or a functional circuit. The power will supply all sorts of voltage for the check instrument and the UUT(27V\24V\±12V\±5V). There are two choices for power supply, one is 27V/DC, and the other is 220V/AC. The input voltage of 220V/AC can be transformed into 27V/DC by the AC/DC circuits, and it will be transformed into all sorts of input voltage which are applicable to check instrument by DC/DC circuits. Power control circuits is used to monitor and control the polarity of input voltage, fluctuating range and load condition in real-time, it can protect the check instrument and UUT. The function of the power isolating circuits is to ensure the segregation between measured equipment and power network and to keep the relative independence of the UUT. As long as the measure circumstance is satisfied, the UUT will be charged.

2.2 Software Component and Basic Principle

Software system consists of self-check and UUT recognition module, performance testing module, data management module, fault diagnosis module and system documentation. It is written by the Visual C++ in the DOS environment, the component is as figure 2.

1) Self-check and UUT Recognition Module: In the design of this software, the self-check and UUT recognition is organic unified up; the process is as the figure 3. The fault diagnosis and isolation of check instrument are relied on the self-check program and BITE circuits of the inner system, users can get to know the work condition of the check instrument and exclude the fault in time based on the display results. The automatic identification of UUT can not only be convenient to users to detect the UUT, but also can prevent the adverse consequences from improper connecting, so that the security of the whole system is improved. [4]

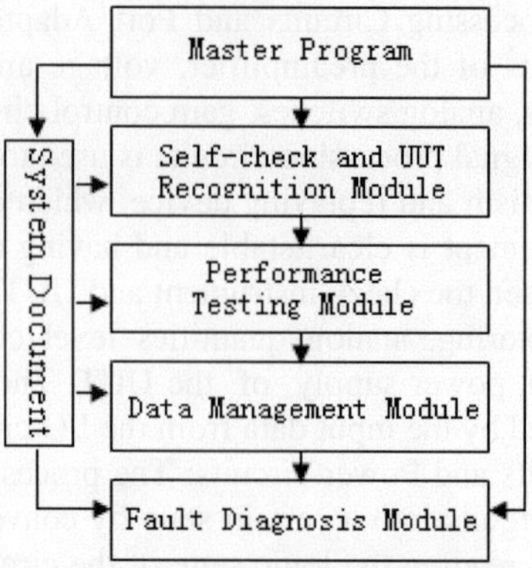

Fig. 2. Software system structure

2) Performance Testing Module: The performance testing module consists of three son modules which are УП48, ЛПМ and П-591, one of them contains several testing items, after choosing the corresponding projects, the program will execute the corresponding function, the structure is as the figure 4. The automatic test can accomplish the tests for the items which are prescribed by the performance testing process; the options test is to test the single item.

3) Data Management Module: The function of this module is to make unified management of testing results which are got form the performance testing modules in form of data base documents, and supply those testing results for users in graphical interfaces to check, browse, modify and print.

4) Fault Diagnostic Module: Based on the BITE and according to the setting testing path, after the reasoning judgment, the check instrument can finish the self-check of its own system. Then, the testing results got form the data management module can be processed by the expert system and data fusion as the foundation to analyze, judge, and figure out the potential failure of the UUT.

5) System document: System document consists of two parts. Using document mainly supply electronic document of the system for users, such as the technical description, circuit principle diagrams, testing process card. Users can achieve help information such as operating methods from the help document by designated buttons.

3 Major Technology

3.1 Virtual Instrument and Intelligent Testing Instrument Technology

To finish the regular inspection of the warning system, original check instrument generally needs five common instruments to work together, but if we adopt the virtual instrument and intelligent testing technology to produce this intelligent testing instrument, the regular inspection and fault diagnosis can be done by it alone, so that the work efficiency and ratio are greatly improved. [5]

3.2 Information Processing and Automatic Testing Technology

In order to realize the automatic test of the voice warning system, information processing and automatic testing technology is adopted in the design of this instrument. It can automatically complete many functions such as the self-check, recognition of UUT, data collection, input incentive creating, output response analyzing and judging, displaying of the testing results.

3.3 Data Fusion and Artificial Intelligence Technology

In order to enhance the ability of fault analysis and judgment, data fusion and artificial intelligence technology are adopted in the design of this instrument. By adopting this technology, the fault of the instrument itself can be located to a certain level circuit and the typical fault of UUT can be located to a certain device. All the information about the fault will be displayed in form of Chinese on the screen.

3.4 Voice Synthesis and Microcomputer Control Technology

For adopting voice synthesizing and microcomputer controlling technology to write voice warning information and aircraft number into the solid voice circuits, the computer will separately extract both of them and merger into intact warning information when we need them. Then the microcomputer system will be started and control the standard voice base to work, and it can rewrite the content momentarily. So that the recording operation of the inter-belt mechanism can be accomplished automatically, the serious recording defects of the original instrument are solved effectively, such as personnel and environment restrictions.

3.5 Dynamic Network Matching Technology

Embedded industrial PC control circuits will change the matching network automatically according to the different working state, so that the contradiction between wipe sounds and partial magnetic source load changing greatly and high requirement of frequency spectrum purity is solved very well.

4 Conclusion

After using by the expert committee and troops, it shows that the hardware design of this intelligent check instrument is reasonable and the software is also reliable. Compared with the original instrument, not only is the function expanded greatly, but also there is a huge technical progress. The main features are as follow: 1.High degree of integration and powerful function. This instrument is an integration of the original instrument and five supporting universal equipments, in addition, it also adds many other functions such as power testing, channel number calibrating and fault diagnosis; 2.High degree of automatic. All the tests can be accomplished automatically; 3.High precision of measurement. The precision of measurement is enhanced an order of magnitude by adopting the computer signal processing technology; 4.Easy operation. This instrument can accomplish Lebanon test of the airborne voice warning system and the in situ test. With a kind man-machine interface, its self-check function is perfect, so it is very easy to operate. Repairing efficiency and precision can be greatly improved by using this instrument, so it improves the combat effectiveness of army. [6]

References

1. Zheng, z., Zeng, x.: Airplane navigation equipment automatic test system. Beijing: Automatic computer measurement and control (4), 46–48 (2000)
2. Ruan, d.: Automatic test instrument technology and computer system design. Xi an University of Electronic Science and Technology Publishers, Xi an (1997)
3. Yang, x.: The voice signals digital processing. Publishers of electronics industry, Beijing (1995)
4. Morgan, D.P., Scofield, C.L.: Neural Networks and Speech Processing. Kluwer Academic Publishers, Dordrecht (2007)
5. Measurement and Automation Catalogue.National Instruments (1999)
6. Wang, t.: The applications of PC in measurement and control. Harbin industry university publishers, Harbin (2005)

Design of a Relaxation Oscillator with Low Power-Sensitivity and High Temperature-Stability

Qi Yu, Zhentao Xu, Ning Ning, Yunchao Guan, and Bijiang Chen

State Key Lab. of Ele. Thin Films and Integrated Devices,
Univ. of Electronics Science & Technology,
Chengdu, Sichuan, 610054, P.R. China
nightkidxzt@yahoo.com.cn

Abstract. A relaxation oscillator for high accuracy application is proposed in this paper. A Dynamic-Threshold (DT) concept is introduced to obtain low power-sensitivity characteristics of frequency, Switched-Resistor (SR) is adopted for frequency stability to temperature and current trimming is also utilized to compensate process drift. Simulation with typical frequency of 5MHz was conducted in 0.18μm CMOS process and shows that the peak error of oscillator's frequency is 0.41% with supply variation of ±10% and temperature drift is only 8.2ppm/°C. The frequency inaccuracy is less than 0.3% via digital trimming for 20% offset of resistances.

Keywords: relaxation oscillator, low Power-Sensitivity, Dynamic-Threshold, Switched-Resistor, high Temperature-Stability.

1 Introduction

Nowadays, as the ICs are developing toward SOC, more and more functional circuits tend to adopt chip integration, while discrete-components circuits (off-chip) are less valuable due to their larger area occupation and higher cost. As a clock generator, the oscillator has to employ full integration for cost saving. The integrated oscillator category consists of Ring Oscillator and Relaxation Oscillator, since the former suffers excessive phase noise, extreme power consumption [1] and high sensitivity to process, power supply and temperature [2], the latter is more preferable.

For high accuracy applications, such as in ADCs, DACs, PLLs etc., the performance of oscillator is crucial to the whole system, therefore it is undesirable that the frequency of oscillator is sensitive to supply drift and temperature variation. In actual implementation, however, the randomness of process is another factor attenuating the performance of oscillator.

The classical topology of relaxation oscillator is composed of two comparators and needs two voltages as high threshold V_{OH} and low threshold V_{OL}, which are derived by resistor divider, are excessively sensitive to supply, while those obtained by reference

M. Zhu (Ed.): Electrical Engineering and Control, LNEE 98, pp. 421–429.
springerlink.com © Springer-Verlag Berlin Heidelberg 2011

circuit [3, 4] reduce the sensitiveness as well as improve the temperature characteristics at the expense of increasing the chip area and design cost. Another way of enhancing the temperature stability of oscillation frequency is to employ poly-resistors with opposite temperature coefficient (TC) [5], however, the special process is necessary for the poly-resistor with positive TC.

In this paper, a Dynamic Threshold (DT) concept is proposed to obtain a low power-dependence frequency and Switched-Resistor (SR) is applied to compensate TC of frequency, which can be widely utilized. Current trimming is also employed for compensation of process variation.

2 The Architecture of Oscillator

A. Power Supply Characteristics and DT Concept

It is favorable that the oscillating frequency is stable with power supply which, in actual application, can deviate by as much as ±10%. Therefore, a concept of threshold generation called DT is proposed to obtain the oscillating period which is independent from supply and the voltage reference circuit is not necessary.

Fig. 1 shows the concept of DT in which all currents duplicate the same current source and operate as current-steering, and the comparator adopts two-stage structure to obtain fast response speed as well as enough amplitude amplification. The mechanism is described as follows:

1) At the initial state, V_B is lower than V_A, so V_{out} stands for low level and the two switches are open. At the moment, capacitor is charged by I until the voltage V_B grows to $V_{AH} = V_{DD} - nI{\cdot}R$, then the output of comparator, V_{out}, turns to be high.

2) The two switches are closed due to the fact that V_{out} is high, thus the current with an amount of mI flows over R and V_A jumps down to $V_{AL}=V_{DD} - (nI{\cdot}R + mI{\cdot}R)$. Meanwhile the net current through B, I_B, equals to $2I - I$, so the result is to discharge the capacitor by I.

3) Therefore, the V_B declines continually till $V_B' = V_{AL}$. Then the output of comparator, V_{out}, flips over and the two switches in Fig.1 open. As a result, the whole system goes back to the initial state and starts cycling.

The period of the relaxation oscillator with above working mechanism can be written as:

$$\Phi = 2\frac{C(V_{OH} - V_{OL})}{I} = 2\frac{C(V_{AH} - V_{AL})}{I} = \frac{2mCIR}{I} = 2mCR \quad . \tag{1}$$

As $V_{OH} - V_{OL} = V_{AH} - V_{AL} = mI{\cdot}R$, in spite of the variation of V_{OH} and V_{OL}, once the current is stable, the difference between them is a constant, which explains the concept of Dynamic Threshold (DT). Even more important, Eq.1 indicates that the period of this oscillator is independent from supply, so the sensitivity of oscillator's frequency to supply drift is reduced.

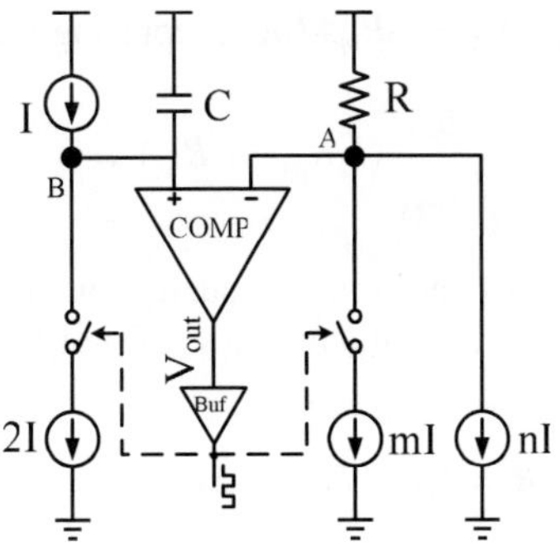

Fig. 1. Proposed DT Concept

B. Temperature Characteristics with Switched-Resistor

Although the DT avoids the influence of supply drift on the oscillating frequency, the temperature characteristics of oscillator is not improved.

For the purpose of making the oscillator achieve favorable temperature characteristics and broad application, the Switched-Resistor (SR) made of two-type poly-resistors is utilized owing to their better accuracy and lower TC, the mechanism of SR is presented in Fig. 2.

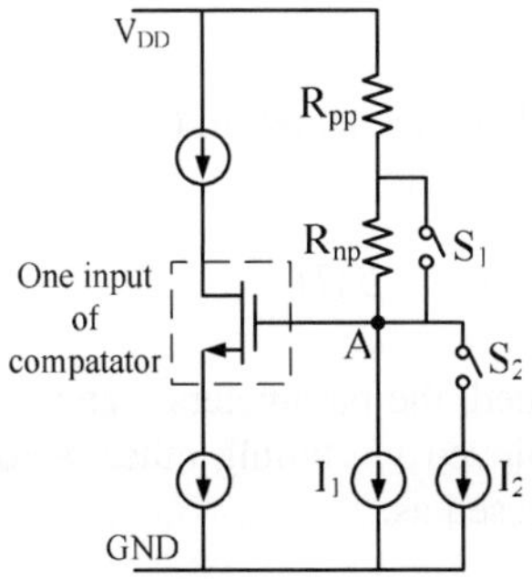

Fig. 2. Circuit Schematic of SR

The Switched-Resistor divides the resistor R in Fig.1 into two resistors with different negative TCs, among which P-poly resistor is represented by R_{pp} and N-poly resistor is represented by R_{np}. The two switches S_1 and S_2 in Fig. 2 are controlled by the same signal. On the basis of the above structures, the maximal and the minimal voltages at A, V_{AH} and V_{AL} are given by:

$$V_{AH} = V_{DD} - I_1(R_{pp} + R_{np}) \ . \tag{2}$$

$$V_{AL} = V_{DD} - R_{pp}(I_1 + I_2) \ . \tag{3}$$

Therefore, $V_{OH} - V_{OL} = V_{AH} - V_{AL} = I_2R_{pp} - I_1R_{np}$, replacing formula (1) with it gives:

$$\Phi = \frac{2C}{I_{charge}}(I_2R_{pp} - I_1R_{np}) \ . \tag{4}$$

Since the second order TC of resistors is usually much smaller than the first order TC of them, the temperature characteristics of poly-resistors are approximately illustrated as:

$$R = R_0(1 + TC_1|_R \cdot \Delta T) \ . \tag{5}$$

Where R_0 is the ideal resistance of R, $TC_1|_R$ is the first order TC of resistor and $\Delta T = T - T_0$ (T_0 is the room temperature). Assuming that $I_{charge} = \alpha I_{ref}$, $I_1 = \beta I_{ref}$ and $I_2 = \gamma I_{ref}$ for they are the replicas of the same current, then the TC of current is reducible regardless of its form, the following formula is attained through replacing Eq. (4) with the current terms and Eq. (5):

$$\Phi = \frac{2C}{\alpha}\left[\gamma R_{pp0}\left(1 + TC_1|_{R_{pp}} \cdot \Delta T\right) - \beta R_{np0}\left(1 + TC_1|_{R_{np}} \cdot \Delta T\right)\right] \ . \tag{6}$$

Suppose $R_{np0} = m \cdot R_{pp0}$ and $TC_1|_{Rnp} = n \cdot TC_1|_{Rpp}$ ($n \neq 1$), Eq. (6) can be rewritten as:

$$\Phi = \frac{2CR_{pp0}}{\alpha}\left[(\gamma - m \cdot \beta) + (\gamma - m \cdot n \cdot \beta)TC_1|_{R_{pp}} \cdot \Delta T\right] \ . \tag{7}$$

The oscillating cycle of oscillator presented in (7) is temperature-independent as long as the following term is tenable:

$$(\gamma - m \cdot n \cdot \beta)TC_1|_{R_{pp}} = 0 \ . \tag{8}$$

When the process is determined, the parameters n and $TC_1|_{Rpp}$ ($TC_1|_{Rpp} \neq 0$) are fixed, then zero TC of cycle is achieved via tuning other factors and the oscillating period stable to temperature is expressed as:

$$\Phi = \frac{2CR_{pp0}}{\alpha}(n-1)m \cdot \beta \ . \tag{9}$$

C. Frequency Tuning and Process Compensation

The temperature stability of oscillating cycle discussed above is based on an ideal condition. In practice, however, the temperature characteristics of the cycle are attenuated by devices' offset caused by process. Assuming the offset of the resistors is represented by ΔR, then the parameter m needs to be modified to m^*:

$$m^* = \frac{(1 + \Delta R_{np})}{(1 + \Delta R_{pp})} m \ . \tag{10}$$

In order to obtain zero TC of oscillation cycle with resistors' offset, substituting m with m^* into Eq. (8) gives:

$$\left(\gamma - \frac{(1+\Delta R_{np})}{(1+\Delta R_{pp})} m \cdot n \cdot \beta \right) TC_1 |_{R_{pp}} = 0 . \tag{11}$$

Therefore, the factor γ must increase to $(1+\Delta R_{np})/(1+\Delta R_{pp})$ times, while the period is $(1+\Delta R_{np})$ times the original value:

$$\Phi = \frac{2CR_{pp0}}{\alpha}(n-1)m \cdot \beta(1+\Delta R_{np}) . \tag{12}$$

According to formula (10), (11) and (12), the resistors' offset not only degrades the period's temperature characteristics but also varies the natural oscillating frequency. Besides, the mismatch of current induces the similar problems. In order to overcome the above difficulties, current trimming is adopted and the current factors α and γ are implemented by binary codes to achieve small occupation.

Based on Eq. (7) ~ (12), the influences of resistors' offset resulted from process on period's temperature characteristics and natural oscillation period are shown in the table below:

Table 1. Cycle Variation with Resistors' Offset

ΔR_{pp} (%)	ΔR_{np} (%)	T C	$(\Phi-\Phi_0)/\Phi_0$
0	0	0	0
20	20	0	0.2
20	-20	0. 4λ	0.2δ
-20	20	-0 .4λ	-0.2δ

Where $\lambda=(2C\cdot m\cdot n\cdot\beta\cdot R_{pp0}\cdot TC_1|_{Rpp})/\alpha$ and $\delta=[1+(1+2TC_1|_{Rpp}\cdot\Delta T)\cdot n]/(n-1)$. The table gives the instruction on tuning of the factors α and γ, in which small λ helps reduce the effect of resistors' offset on period's temperature characteristics. Besides, it implies that the higher the oscillation frequency is, the less sensitive the temperature characteristics are to resistors' offset.

3 Complete Circuit and Detail Design

Given the above aspects, the practical implementation of the relaxation oscillator is presented in Fig. 3. The tuning of both I_{charge} and I_{20} is conducted separately by i and m binary-weighted current sources. All switches in Fig. 3 adopt the current-steering structure composed of M_1 and M_2 for reducing the spikes resulted from current switching, for the spikes attenuate the oscillator's performance [6].

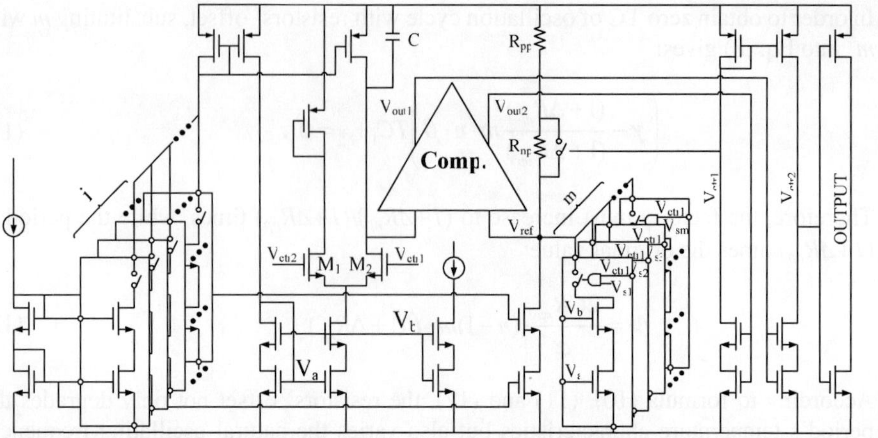

Fig. 3. Complete Schematic of Relaxation Oscillator

4 Simulation Results

The simulation was carried out by HSPICE with 0.18μm CMOS process and the natural frequency is designed as 5MHz. According to the process, MIM capacitor and poly-resistors with maximal offset of 18% are applied. Given the robustness of circuits, 20% of resistors' offset is utilized in the simulation.

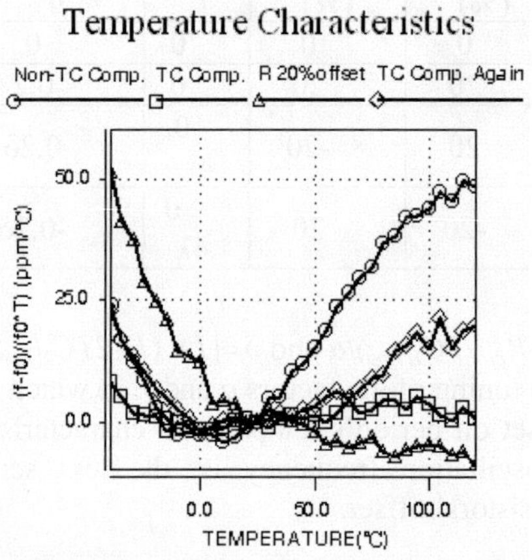

Fig. 4. Oscillator Temperature Drifts under Different Conditions. *Non-TC comp.* indicates TC of oscillator without SR application; *R 20% offset* represents that R_{pp} varies for +20% while R_{np} varies for -20%; *TC comp. Again* stands for TC compensation under *R 20% offset*.

Fig. 4 demonstrates the temperature characteristics of oscillator over temperature range of -40°C to 120°C. The results show that the TC of oscillator is as small as

8.2ppm/°C. Additionally, the curves of *TC comp.* and *TC comp. Again* indicate that the first-order TC of resistor is compensated while its second-order TC which was ignored in the previous discussion remains.

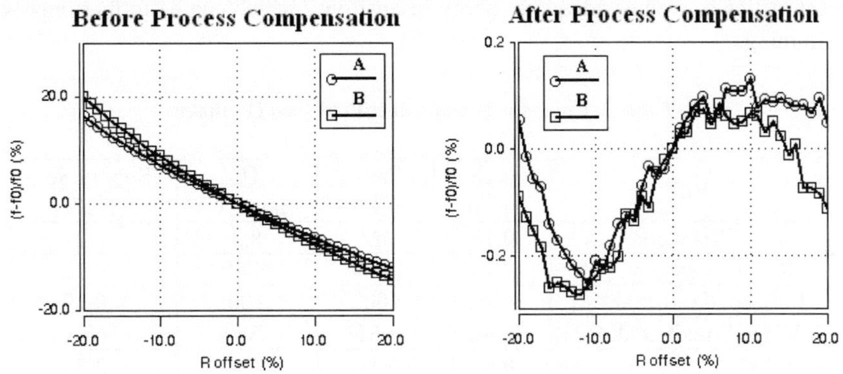

Fig. 5. Influences of Resistor Offset on Frequency without & with Trimming.

Fig.5 illustrates the correlation between the variation of oscillation frequency and that of the resistors' offset, in which *A* represents that both offsets of R_{pp} & R_{np} vary from -20% to +20%; *B* indicates that the offset of R_{pp} varies from +20% to -20% while that of R_{np} varies from -20% to +20%.The simulation results of *Before Process Compensation* in Fig. 5 are well matched with the data in Table 1. Even more important, the frequency deviation from f_0 after trimming decreases significantly compared to that without process compensation.

Table 2. Frequency Variation with Supply Drift

Supply Voltage (V)	Frequency (MHz)	$(\Phi-\Phi_0)/\Phi_0$ (%)
3.0	5.00741	-0.35
3.1	5.01127	-0.27
3.2	5.01598	-0.19
3.3	5.02494 (Φ_0)	0
3.4	5.03059	0.11
3.5	5.03633	0.23
3.6	5.04574	0.41

Table 2 displays the variation of oscillation frequency with changes of supply, in which the frequency varies respectively for +0.41% and -0.35% when supply changes for ±10%. Based on the discussion in Part 2-A, the oscillating frequency's coefficient to supply is zero. In practice, however, the resistance of poly-resistors and the capacitance of capacitor change with the supply voltage respectively due to polysilicon-depletion-effects and capacitor's voltage-coefficient, but the variation of frequency can be minimized by current trimming.

The chief performance of the oscillator designed in this paper and those recently published are listed in Table 3. The proposed topology enjoys favorable temperature characteristics as well as adequate supply-insensitivity which can be further improved by trimming.

Table 3. Comparison with Other Published Oscillator

Ref.	Tech. (μm)	Freq. (Hz)	TC (ppm/°C)	Supply Sens. (%/V)
This work	0.18	5M	8.2	1.27
K. Lasanen et al. [7]	0.35	200k	900	0.78
F. Sebastiano et al. [8]	0.065	100k	100	0.37
V. De Smedt et al. [5]	0.065	6M	86	N/A
U. Denier [9]	0.35	3.3k	<260	2.33

5 Conclusion

In this paper, a high accuracy oscillator which utilizes DT to reduce the dependence of frequency on supply and adopts SR to stabilize temperature characteristics of oscillating frequency as well as employs current trimming to overcome the attenuation caused by resistors' offset has been proposed. Simulation results demonstrate that the frequency variation is 0.76% for supply change from 3V to 3.6V and is 8.2ppm/°C for temperature coverage of 160°C. After fine-tuning, a favorable accuracy of oscillating frequency is guaranteed even if the resistors' offset ranges from -20% to 20%.

References

1. Allam, A., Filanovsky, I.M., Oliveira, L.B., Fernandes, J.R.: Experimental comparison of coupling effects on the performance of quadrature CMOS LC and RC oscillators. In: 50th Midwest. Symp. Circuits and Systems (MWSCAS), pp. 606–609 (2007)
2. Yuan, F., Soltani, N.: Low-voltage low VDD sensitivity relaxation oscillator for passive wireless Microsystems. IET J. Electronics Letters 45(21), 1057–1058 (2009)
3. Qinyu, Z., Liming, L., Guican, C.: An RC Oscillator with Temperature Compensation for Accurate Delay Control in Electronic Detonators. In: IEEE Int. Electron Devices and Solid-State Circuits, 2009 (EDSSC 2009), pp. 274–277 (2009)
4. Barnett, R., Liu, J.: A 0.8V 1.52MHz MSVC Relaxation Oscillator with Inverted Mirror Feedback Reference for UHF RFID. In: IEEE Int. Conf. Custom Integrated Circuits, 2006 (CICC 2006), pp. 769–772 (2006)
5. De Smedt, V., De Wit, P., Vereecken, W., Steyaert, M.: A 66 μW 86 ppm/°C Fully-Integrated 6 MHz Wienbridge Oscillator with a 172 dB Phase Noise FOM. IEEE J. Solid-State Circuits 44(7), 1990–2001 (2009)

6. Raroiu, M., Mialtu, R.: Design Optimizations In High Accuracy Relaxation Oscillators. In: IEEE Int. Semiconductor Conference, CAS 2009, vol. 2, pp. 485–488 (2009)
7. Lasanen, K., Kostamovaara, J.: A 1.2-V CMOS RC Oscillator for Capacitive and Resistive Sensor Applications. IEEE Trans. Instrumentation and Measurement 57(12), 2792–2800 (2008)
8. Sebastiano, F., Breems, L., Makinwa, K., Drago, S., Leenaerts, D., Nauta, B.: A low-voltage mobility-based frequency reference for crystal-less ULP radios. IEEE J. Solid-State Circuits 44(7), 2002–(2009)
9. Denier, U.: Analysis and Design of an Ultralow-Power CMOS Relaxation Oscillator. IEEE Trans. Circuits and Systems I, Reg. Paper 57(8), 1973–1982 (2010)

Automatic Parameters Calculation of Controllers for Photovoltaic dc/dc Converters

E. Arango, C.A. Ramos-Paja, D. Gonzalez, S. Serna, and G. Petrone

Facultad de Minas, Universidad Nacional de Colombia, Colombia
Instituto Tecnológico Metropolitano, Colombia
DIIIE, Università di Salerno, Italy
{eiarangoz,dgonzalm,caramosp}@unal.edu.co

Abstract. In the photovoltaic (PV) systems, the main objective of the control strategy is to extract the maximum power from the source; usually such objective is obtained by combining the action of the MPPT algorithm with a voltage regulator used to attenuate the disturbances at the source terminals. This work proposes an automatic procedure for designing the parameters of the voltage compensation network by solving a set of two nonlinear equations in which, the system phase and gain margins are previously assigned. The method is based on the modeling of the complete PV system (source, converter and load) and takes into account the main parasitic components also. Thanks to this global approach, the design of control strategies to regulate the PV panel output voltage can be optimized. The method has been theoretically analyzed and validated by using PSIM simulations. The designed voltage control loop has been tested also together with a P&O MPPT algorithm in order to regulate the PV voltage in a double stage grid-connected inverter in presence of a wide voltage oscillation at the DC-bulk and with fast irradiance variation.

Keywords: Photovoltaic systems, grid connected, state space model, control systems, automatic controller calculation.

1 Introduction

The strong growth of the worldwide demand of electricity and the need to decrease global warming fosters the interest for using the renewable power supplies. Among these, the Photovoltaic (PV) systems are attractive because they can be dimensioned for a wide range of power rating and are suitable for both stand-alone and grid-connected applications [1].

In the PV system, the main objective of the control strategy is to extract the maximum power from the source (Maximum Power Point Tracking - MPPT) in whichever environmental condition. There are a lot of MPPT algorithms which can be used to force the PV source to work in its maximum power operating point [2], among these, the Perturb and Observe (P&O) MPPT algorithm, based on the PV voltage regulation, is the most common solution due to its efficiency,

M. Zhu (Ed.): Electrical Engineering and Control, LNEE 98, pp. 431–440.
springerlink.com © Springer-Verlag Berlin Heidelberg 2011

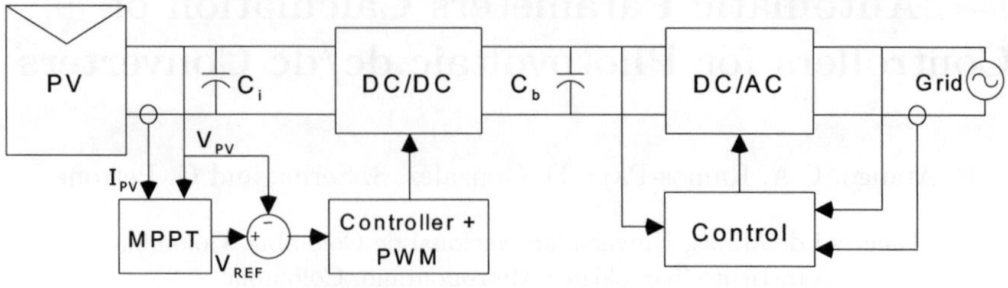

Fig. 1. Grid-connected PV system.

simplicity and robustness; as a drawback this technique requires an additional effort concerning the design of a voltage regulator. The PV voltage, which corresponds to the voltage at the input of a dc/dc converter (or a dc/ac for the single stage PV inverter), can be regulated introducing a feedback compensation network able to drive the switching converter to fix its input voltage.

For improving the MPPT capability, the voltage closed-loop must be designed to ensure a fast dynamic response with respect to the reference signal (defined by the MPPT algorithm) and to reject efficiently the disturbances at input and output of the dc/dc converter. In PV systems the main sources of disturbances are the environmental changes (especially the fast irradiance variation) and load variation that can be modeled as voltage and/or current variation. A typical PV power conversion chain for grid-connected systems is shown in Fig. 1 [3], [4].

The problems related to the noise rejection in the PV grid-connected systems are well described in [4] where the case of the double stage inverter has been analyzed. In that case, the output voltage of the dc/dc power converter is affected by a sinusoidal oscillation at twice the grid frequency around a given DC value. While the DC value is regulated by the inverter control strategy, the amplitude of the sinusoidal oscillation is directly proportional to the average power injected in to the grid and inversely proportional to the C_b value. To increase the system reliability, the new generation of the PV inverters avoid the use of electrolytic capacitor, which assuring high C_b value, thus the sinusoidal oscillation exhibit significant amplitude that may also be transferred to the input of the dc/dc converter causing a mismatch in the maximum power point calculation. The same problems arising in the PV distributed MPPT (DMPPT), where the dc/dc converter is connected to each one of PV panel of the photovoltaic field in order to perform the MPPT locally [5]. In such applications, the disturbance rejection assumes a fundamental role for avoiding that the changes in the operating conditions of a single PV panel (due to, for example, the shading) affect, trough the dc/dc converters, the performances of the other panels.

Unfortunately the strong non-linearity introduced by the PV source [3] and by the switching converter make the optimal design of the compensation network a not trivial task. This is due essentially to the fact that the dynamic performance can change significantly for the presence of the wide variation in both environmental operating condition and load. In [4] and [5], the feedback

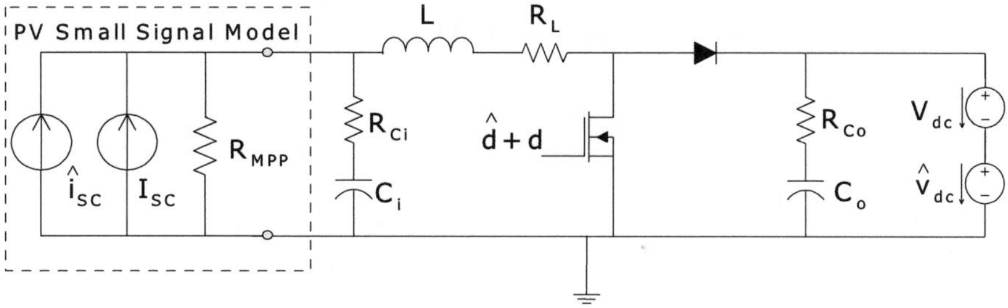

Fig. 2. PV panel and power converter model considering a DC voltage control of bulk capacitor.

compensation network has been designed by using an ideal model of the dc/dc converter so that, the effect of the parasitic parameters have been neglected and the worst-case steady state operating conditions have been considered.

This work proposes, instead, an automatic procedure for designing the parameters of the voltage compensation network based on the modeling of the complete PV chain (source, converter and load) which takes into account the main parasitic components also. The control parameters are obtained by solving a set of two nonlinear equations in which, the system phase and gain margins are previously assigned. Thanks to this global approach, the design of control strategies to regulate the PV panel output voltage can be optimized.

In section 2 the state-space model of the complete PV system has been carried out, while the procedure for designing a PI control network is shown in section 3. The proposed methodology has been theoretically analyzed and validated by using PSIM simulations. In section 4 the voltage control loop, together with a P&O MPPT algorithm, has been tested for regulating the PV voltage in the double stage grid-connected inverter in presence of a wide voltage oscillations at the bulk capacitor (input of the inverter) and with fast irradiance (PV short-circuit current) variation. Conclusions close this work.

2 Modeling of the PV System

This model considers the regulation of the bulk capacitor DC voltage. The inverter is modeled as a voltage source with both DC and 100 Hz components as depicted in Fig. 2. Such a circuit also considers the small signal model of the PV panel at its Maximum Power Point (MPP) for a given irradiance [3], [6]. The inductor current i_L, input v_{Ci} and output v_{Co} capacitor voltages are the state variables. The control variable is the dc/dc converter duty-cycle $\hat{d}$. The PV panel short-circuit current $(I_{sc} + \hat{i}_{sc})$ and the inverter voltage $(V_{dc} + \hat{v}_{dc})$ are the perturbations of the system.

The state-space behavior of the proposed system is described by (1) and (2),

$$\dot{\mathbf{X}} = \mathbf{A}_{\mathbf{ss}}\mathbf{X} + \mathbf{B}_{\mathbf{ss}}\mathbf{U} \quad \wedge \quad \mathbf{Y} = \mathbf{C}_{\mathbf{ss}}\mathbf{X} + \mathbf{D}_{\mathbf{ss}}\mathbf{U} \tag{1}$$

$$\mathbf{X} = \left[\hat{i}_L\ \hat{v}_{Ci}\ \hat{v}_{Co}\right]^T \quad \wedge \quad \mathbf{U} = \left[\hat{d}\ \hat{i}_{sc}\ \hat{v}_{dc}\right]^T \tag{2}$$

where

$$\mathbf{A_{ss}} = \begin{bmatrix} -\frac{\sigma}{L} & \frac{\lambda}{L} & 0 \\ -\frac{\lambda}{C_i} & -\frac{1}{C_i}\cdot\left(\frac{1}{R_{Ci}\cdot R_{MPP}}\right) & 0 \\ 0 & 0 & -\frac{1}{C_o\cdot R_{Co}} \end{bmatrix} \tag{3}$$

$$\mathbf{B_{ss}} = \begin{bmatrix} \frac{V_{dc}}{L} & \frac{\beta}{L} & -\frac{1-d}{L} \\ 0 & \frac{\lambda}{C_i} & 0 \\ 0 & 0 & \frac{1}{C_o\cdot R_{Co}} \end{bmatrix} \tag{4}$$

$$\lambda = \frac{R_{MPP}}{R_{MPP}+R_{Ci}}\ ,\ \beta = \frac{R_{Ci}\cdot R_{MPP}}{R_{MPP}+R_{Ci}} \tag{5}$$

$$\sigma = \frac{R_{Ci}\cdot R_{MPP} + R_L\cdot R_{MPP} + R_{Ci}\cdot R_L}{R_{MPP}+R_{Ci}} \tag{6}$$

Moreover, **C** matrix imposes the PV voltage as the system output Y:

$$Y = v_{PV} = R_{Ci}\cdot i_{Ci} + v_{Ci} \tag{7}$$

obtaining:

$$\mathbf{C_{ss}} = \left[-\frac{R_{MPP}\cdot R_{Ci}}{R_{MPP}+R_{Ci}}\ \frac{R_{MPP}}{R_{MPP}+R_{Ci}}\ 0\right] \wedge \mathbf{D_{ss}} = \left[0\ \frac{R_{MPP}\cdot R_{Ci}}{R_{MPP}+R_{Ci}}\ 0\right] \tag{8}$$

The resulting system is non-linear, therefore, it must be linearized [7]. For that matter, the system equilibrium point is found:

$$\mathbf{P_{ss}}\cdot\mathbf{X_{ss}} = \mathbf{Q_{ss}}\ ,\ \mathbf{X_{ss}} = \left[I_L\ V_{ci}\ D\right]^T \tag{9}$$

$$\mathbf{P_{ss}} = \begin{bmatrix} -\varphi & R_{MPP} & V_{dc}\cdot(R_{MPP}+R_{Ci}) \\ -R_{MPP} & -1 & 0 \\ R_{MPP}\cdot R_{Ci} & R_{MPP} & 0 \end{bmatrix} \tag{10}$$

$$\mathbf{Q_{ss}} = \begin{bmatrix} -I_{SC}\cdot R_{MPP}\cdot R_{Ci} + V_{dc}\cdot(R_{MPP}\cdot R_{Ci}) \\ -I_{SC}\cdot R_{MPP} \\ V_{PV}\cdot(R_{MPP}\cdot R_{Ci}) - I_{SC}\cdot R_{MPP}\cdot R_{Ci} \end{bmatrix} \tag{11}$$

$$\varphi = R_{Ci}\cdot R_{MPP} + R_L\cdot R_{MPP} + R_{Ci}\cdot R_L \tag{12}$$

The solution of the system (9)-(12) is obtained as:

$$\mathbf{X_{ss}} = \mathbf{P_{ss}}^{-1}\cdot\mathbf{Q_{ss}} \tag{13}$$

where values for $\mathbf{X_{ss}}$ at the equilibrium point are calculated.

3 Control Design Using Phase Margin and Gain Margin Constraints

From the state space model described in section 2, the transfer function of the PV system is given by:

$$Pl(s) = \frac{A \cdot s + B}{C \cdot s^2 + D \cdot s + E} \tag{14}$$

$$A = -R_{Ci} \cdot R_{MPP} \cdot V_{dc} \cdot (R_{Ci} \cdot {R_{MPP}}^2 \cdot C_i + {R_{Ci}}^2 \cdot R_{MPP} \cdot C_i) \tag{15}$$

$$B = -R_{Ci} \cdot R_{MPP} \cdot V_{dc} \cdot (R_{Ci} + R_{MPP} + {R_{MPP}}^2) \tag{16}$$

$$C = R_{Ci} \cdot C_i \cdot R_{MPP} \cdot L \cdot ({R_{MPP}}^2 + 2 \cdot R_{Ci} \cdot R_{MPP} + {R_{Ci}}^2) \tag{17}$$

$$D = R_{Ci} \cdot C_i \cdot R_L \cdot R_{MPP} \cdot [R_{MPP} \cdot (R_{Ci} \cdot R_{MPP} + {R_{Ci}}^2 + R_L \cdot R_{MPP} \\ +2 \cdot R_{Ci} \cdot R_L) + {R_{Ci}}^2 \cdot R_L] + L \cdot ({R_{Ci}}^2 + 2 \cdot R_{Ci} \cdot R_{MPP} + {R_{MPP}}^2) \tag{18}$$

$$E = R_{MPP} \cdot (2 \cdot R_{Ci} \cdot R_L + {R_{Ci}}^2 + R_{Ci} \cdot {R_{MPP}}^2 \\ +R_{Ci} \cdot R_{MPP} + R_L \cdot R_{MPP}) + {R_{Ci}}^2 \cdot R_L \tag{19}$$

This work considers a PI controller (20) due to its wide use in industry, but the proposed methodology is also applicable to any controller structure.

$$Co(s) = \frac{x \cdot (1 + y \cdot s)}{s} \tag{20}$$

where the proportional gain (Kp) and integral gain (Ti) parameters are

$$Kp = x \cdot y \ \wedge \ Ti = 1/x \tag{21}$$

Then, the plant-controller loop transfer function $H(s)$ is given by

$$H(s) = Pl(s) \cdot Co(s) = \frac{A \cdot x \cdot y \cdot s^2 + (B \cdot x \cdot y + A \cdot x) \cdot s + B \cdot x}{C \cdot s^3 + D \cdot s^2 + E \cdot s} \tag{22}$$

In the imaginary domain of the frequency plane:

$$H(jw) = \frac{(B \cdot x - A \cdot x \cdot y \cdot w^2) + j \cdot (B \cdot x \cdot y + A \cdot x) \cdot w}{-D \cdot w^2 + j \cdot (E \cdot w - C \cdot w^3)} \tag{23}$$

where the phase and magnitude of (23) are, respectively,

$$\angle H(jw) = \arctan\left(\frac{(B \cdot x \cdot y + A \cdot x) \cdot w}{B \cdot x - A \cdot x \cdot y \cdot w^2}\right) - \arctan\left(\frac{(E \cdot w - C \cdot w^3)}{-D \cdot w^2}\right) \tag{24}$$

$$|H(jw)| = \frac{\sqrt{(B \cdot x \cdot y + A \cdot x)^2 \cdot w^2 + (B \cdot x - A \cdot x \cdot y \cdot w^2)^2}}{\sqrt{(E \cdot w - C \cdot w^3)^2 + (-D \cdot w^2)^2}} \tag{25}$$

The gain margin is measured when equation (24) is equal to $-\pi$ [8], and applying trigonometric operations on both sides of the equation:

$$\frac{(B\cdot y+A)\cdot w_1}{B-A\cdot y\cdot {w_1}^2}=\frac{E-C\cdot {w_1}^2}{-D\cdot w_1} \tag{26}$$

Expanding equation (26):

$$(A\cdot C\cdot y)\cdot {w_1}^4+\delta\cdot {w_1}^2+B\cdot E=0 \tag{27}$$

$$\delta=B\cdot D\cdot y+A\cdot D-B\cdot C-A\cdot E\cdot y \tag{28}$$

Solving (27) as a biquadratic polynom, four roots are obtained:

$$\mathbf{w}_1=\begin{bmatrix}\sqrt{\frac{-\delta+\sqrt{\delta^2-4\cdot(A\cdot C\cdot y)\cdot(B\cdot E)}}{2\cdot(A\cdot C\cdot y)}}\\ -\sqrt{\frac{-\delta+\sqrt{\delta^2-4\cdot(A\cdot C\cdot y)\cdot(B\cdot E)}}{2\cdot(A\cdot C\cdot y)}}\\ \sqrt{\frac{-\delta-\sqrt{\delta^2-4\cdot(A\cdot C\cdot y)\cdot(B\cdot E)}}{2\cdot(A\cdot C\cdot y)}}\\ -\sqrt{\frac{-\delta-\sqrt{\delta^2-4\cdot(A\cdot C\cdot y)\cdot(B\cdot E)}}{2\cdot(A\cdot C\cdot y)}}\end{bmatrix} \tag{29}$$

The higher positive real root on vector $\mathbf{w}_1$ (29) is the frequency where the gain margin is measured. Such a solution used in $|H(jw)|$ gives:

$$|H(jw_1)|=\left.\frac{\sqrt{(B\cdot x\cdot y+A\cdot x)^2\cdot w^2+(B\cdot x-A\cdot x\cdot y\cdot w^2)^2}}{\sqrt{(E\cdot w-C\cdot w^3)^2+(-D\cdot w^2)^2}}\right|_{w=w_1} \tag{30}$$

The inverse of $|H(jw)|$ is the gain margin (Gm) as given in (31).

$$Gm=\frac{1}{|H(jw_1)|} \tag{31}$$

Similarly, the phase margin is measured at the frequency w_2 where the magnitude (25) is equal to 1 [8]. Therefore, w_2 is calculated from

$$\frac{\sqrt{(B\cdot x\cdot y+A\cdot x)^2\cdot {w_2}^2+(B\cdot x-A\cdot x\cdot y\cdot {w_2}^2)^2}}{\sqrt{(E\cdot w_2-C\cdot {w_2}^3)^2+(-D\cdot {w_2}^2)^2}}=1 \tag{32}$$

from which

$$C^2\cdot {w_2}^6+\rho\cdot {w_2}^4+\varphi\cdot {w_2}^2-(B\cdot x)^2=0 \tag{33}$$

$$\rho=D^2-2\cdot E\cdot C-(A\cdot x\cdot y)^2\ ,\ \varphi=E^2-(B\cdot x\cdot y+A\cdot x)^2+2\cdot A\cdot B\cdot x^2\cdot y \tag{34}$$

Solving equation (33), the following vector of possible solutions is found:

$$\mathbf{w}_2 = \begin{bmatrix} \sqrt{-\frac{\rho}{3\cdot C^2} - \frac{1}{3\cdot C^2}\sqrt[3]{\frac{1}{2}\cdot(P1+P2)} - \frac{1}{3\cdot C^2}\sqrt[3]{\frac{1}{2}\cdot(P1-P2)}} \\ -\sqrt{-\frac{\rho}{3\cdot C^2} - \frac{1}{3\cdot C^2}\sqrt[3]{\frac{1}{2}\cdot(P1+P2)} - \frac{1}{3\cdot C^2}\sqrt[3]{\frac{1}{2}\cdot(P1-P2)}} \\ \sqrt{-\frac{\rho}{3\cdot C^2} + \frac{1+j\cdot\sqrt{3}}{6\cdot C^2}\sqrt[3]{\frac{1}{2}\cdot(P1+P2)} + \frac{1-j\cdot\sqrt{3}}{6\cdot C^2}\sqrt[3]{\frac{1}{2}\cdot(P1-P2)}} \\ -\sqrt{-\frac{\rho}{3\cdot C^2} + \frac{1+j\cdot\sqrt{3}}{6\cdot C^2}\sqrt[3]{\frac{1}{2}\cdot(P1+P2)} + \frac{1-j\cdot\sqrt{3}}{6\cdot C^2}\sqrt[3]{\frac{1}{2}\cdot(P1-P2)}} \\ \sqrt{-\frac{\rho}{3\cdot C^2} + \frac{1-j\cdot\sqrt{3}}{6\cdot C^2}\sqrt[3]{\frac{1}{2}\cdot(P1+P2)} + \frac{1+j\cdot\sqrt{3}}{6\cdot C^2}\sqrt[3]{\frac{1}{2}\cdot(P1-P2)}} \\ -\sqrt{-\frac{\rho}{3\cdot C^2} + \frac{1-j\cdot\sqrt{3}}{6\cdot C^2}\sqrt[3]{\frac{1}{2}\cdot(P1+P2)} + \frac{1+j\cdot\sqrt{3}}{6\cdot C^2}\sqrt[3]{\frac{1}{2}\cdot(P1-P2)}} \end{bmatrix} \tag{35}$$

$$P1 = 2\cdot\rho^3 - 9\cdot C^2\rho\cdot\varphi + 27\cdot C^4\cdot(B\cdot x)^2\ ,\ P2 = \sqrt{P1^2 - 4\cdot(\rho^2 - 3\cdot C^2\cdot\varphi)^3} \tag{36}$$

Similar to the gain margin analysis, the higher positive and real frequency is the solution ω_2. Using it in the phase of $H(jw)$:

$$\angle H(jw_2) = \arctan\left(\frac{(B\cdot y + A)\cdot w}{B - A\cdot y\cdot w^2}\right) - \arctan\left(\frac{(E - C\cdot w^2)}{-D\cdot w}\right)\Bigg|_{w=w_2} \tag{37}$$

Finally, the phase margin is given by:

$$Pm = \angle H(jw_2) + \pi \tag{38}$$

From the equations of gain margin (31) and phase margin (38) in terms of the PI controller parameters, a controller that ensures a desired phase Pm_{des} and gain Gm_{des} margins can be calculated. Therefore, equation in gain (39) and equation in phase (40) define a nonlinear set of equations that allow the PI parameters calculation.

$$Gm_{des}\cdot|H(jw_1)| - 1 = 0 \tag{39}$$

$$Pm_{des} - \angle H(jw_2) - \pi = 0 \tag{40}$$

The equations system solution $\{x, y\}$ defined in (20) and (21) can be calculated by means of a numerical solution method, e.g. Matlab fsolve(). The procedure described in this section is summarized in flow chart of Fig. 3.

4 Simulation Results

To illustrate the proposed solution, the following parameters are adopted: $L = 56\ \mu H$, $Ci = 44\ \mu F$, $Co = 44\ \mu F$, $Vdc = 33.15\ V$, $Isc = 4.7\ A$, $Rmp = 81.87\ \Omega$, $RL = 0.3\ \Omega$, $RCi = 0.17\ \Omega$ and $RCo = 0.17\ \Omega$.

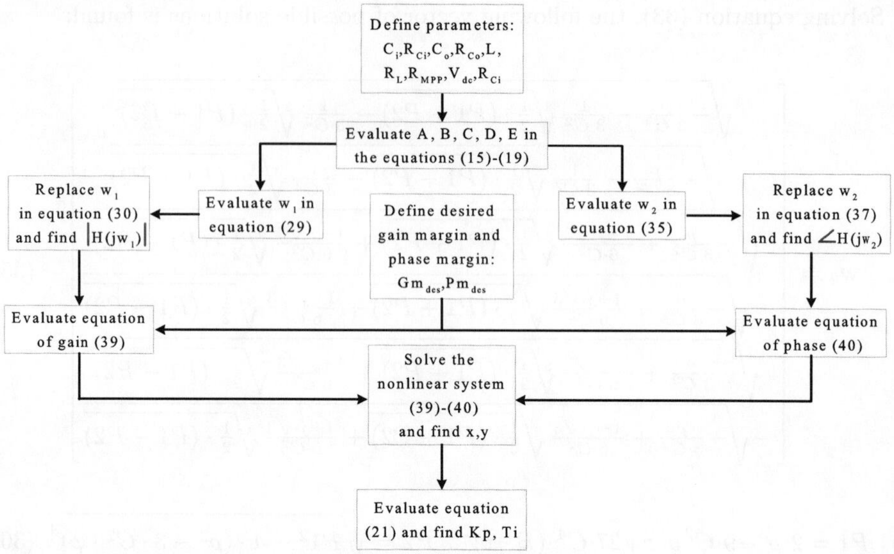

Fig. 3. Flow Chart to calculate PI controller parameters.

Following flow chart of Fig. 3, the controller parameters $Kp = -2.3916$ and $Ti = -7.5552 \times 10^{-6}$ were calculated, obtaining $G_{Co}(s)$ controller given in (41) for a given $Gm_{des} = -27.9\ dB$ and $Pm_{des} = 74.5°$ as observed in Fig. 4(a).

$$G_{Co}(s) = -2.3918\frac{s + 5.5340 \times 10^4}{s} \tag{41}$$

Closed loop transfer functions T_{vci-D}, $T_{vci-Isc}$ and T_{vci-Vb} describe the system dynamics regarding the reference voltage, short circuit current and converter output voltage defined by the inverter. The frequency responses of those transfer functions are shown in Fig. 4(b), where a satisfactory reference tracking is observed on T_{vci-D}, and effective disturbances rejection on $T_{vci-Isc}$ and T_{vci-Vb} are also exhibited.

The controller was evaluated by considering a nonlinear PV model [3], which initial irradiance is $S_1 = 960\ W/m^2$, then a step-type disturbance occurs at $t = 25\ ms$ to $S_2 = 360\ W/m^2$, returning to S_1 at $t = 45\ ms$. The test also considers a bulk capacitor voltage oscillation of 30% of the desired DC voltage at 100 Hz. Therefore, the designed controller was evaluated for both load and irradiance perturbations. In addition, a Perturb and Observer (P&O) based MPPT controller was considered, whose control parameters were calculated as described in [6].

Fig. 5 shows the system transient response in PSIM, where figure 5(a) shows the desired tracking of the voltage reference provided by the MPPT controller. There is also observed a satisfactory rejection of both bulk capacitor and irradiance disturbances. In addition, the maximum power point tracking is achieved.

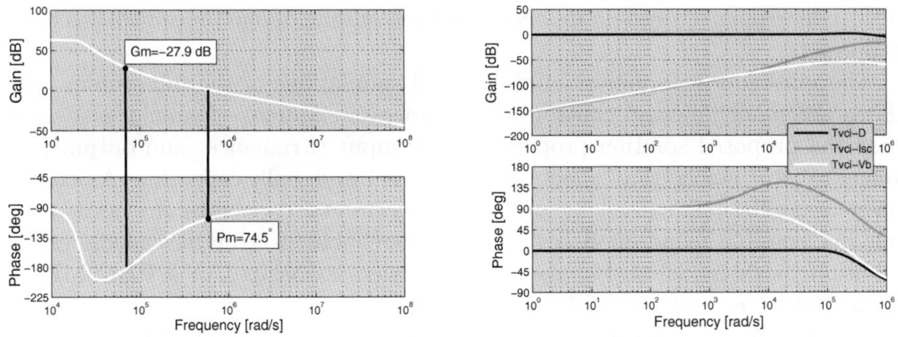

(a) Open loop Bode diagram with margins (b) Closed loop Bode diagram

Fig. 4. Frequency responses of PV system.

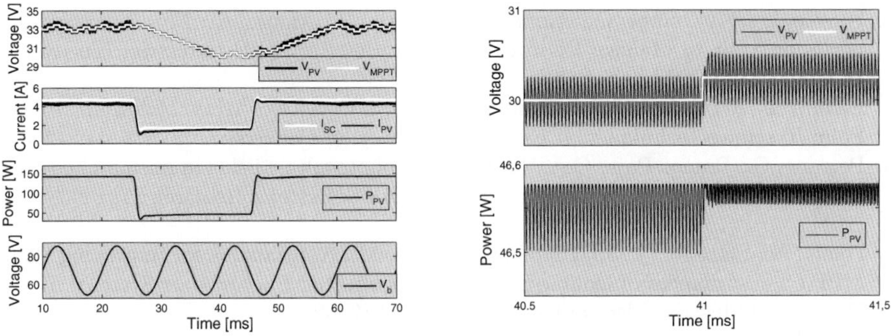

(a) System transient response (b) Zoom from 40.5 ms to 41.5 ms.

Fig. 5. System transient response simulation by using the controller G_{Co}.

Figure 5(b) shows a zoom of the data in Fig. 5(a) from 40.5 ms to 41.5 ms, where a satisfactory transient response is observed.

5 Conclusions

This paper proposes a modeling of the PV system composed by a PV panel associated to a power conversion stage. This full-system modeling allows a simple design of control strategies to regulate input and output power converter capacitor voltages, therefore the maximum power delivered by the PV panel. In addition, an analytical design of a PI controller has been performed, obtaining expressions to calculate the controller parameters depending on desired gain and phase margin requirements. Furthermore, the proposed procedure could be refined for running on a microcontroller in order to adjust in real time, on the basis of the current operating conditions, the parameters of the compensation network implemented in digital form. In any case, in this paper only the aspects concerning the system modeling and the off-line design procedure has been proposed.

The mathematical analysis and circuital simulations of the full-system validates the modelling and control design processes. Moreover, simulations show that the proposed solution properly rejects input (irradiance) and output (inverter) disturbances that affect the PV panel and bulk capacitor. As consequence, oscillations caused by using non-electrolytic capacitors are mitigated, making possible to adopt it instead of non-reliable electrolytic capacitors. In addition, both model and control strategy were associated to a P&O MPPT algorithm to maximize the PV power.

Finally, the proposed technique is a good candidate for mitigating operation noise generated by DMPPT system interaction, which is an active research topic.

References

1. Reynaud, J.F., Gantet, O., Alonsi, P., Estibals, B., Alonso, C.: New Adaptive Supervision Unit to Manage Photovoltaic Batteries. In: IEEE Industrial Electronics Conference, IECON 2009 (2009)
2. Esram, T., Chapman, P.L.: Comparison of Photovoltaic Array Maximum Power Point Tracking Techniques. IEEE Trans. on Energy Conversion 22, 439–449 (2007)
3. Petrone, G., Ramos-Paja, C.A.: Modeling of photovoltaic fields in mismatched conditions for energy yield evaluations. Electric Power Systems Research 81, 1003–1013 (2011)
4. Femia, N., Petrone, G., Spagnuolo, G., Vitelli, M.: A Technique for Improving P&O MPPT Performances of Double-Stage Grid-Connected Photovoltaic Systems. IEEE Trans. on Industrial Electronics 56, 4473–4482 (2009)
5. Adinolfi, G., Femia, N., Petrone, G., Spagnuolo, G., Vitelli, M.: Design of dc/dc Converters for DMPPT PV Applications Based on the Concept of Energetic Efficiency. Journal of Solar Energy Engineering 132 (2010)
6. Femia, N., Petrone, G., Spagnuolo, G., Vitelli, M.: Optimization of Perturb and Observe Maximum Power Point Tracking Method. IEEE Trans. on Power Electronics 20, 963–973 (2005)
7. Leyva, R., Martinez-Salamero, L., Valderrama-Blavi, H., Maixe, J., Giral, R., Guinjoan, F.: Linear State-Feedback Control of a Boost Converter for Large-Signal Stability. IEEE Trans. on Circuits And Systems-I: Fundamental Theory And Applications 48, 418–424 (2001)
8. Ogata, K.: Modern control engineering, 4th edn. Prentince-Hall, Upper Saddle River (2002)

Modeling and Control of Ćuk Converter Operating in DCM

E. Arango, C.A. Ramos-Paja, R. Giral, S. Serna, and G. Petrone

Facultad de Minas, Universidad Nacional de Colombia, Colombia
DEEEA, Universitat Rovira i Virgili, Spain
Instituto Tecnológico Metropolitano, Colombia
DIIIE, Università di Salerno, Italy
{eiarangoz,caramosp}@unal.edu.co, roberto.giral@urv.cat

Abstract. After modelling accurately the Ćuk converter operating in discontinuous inductor conduction mode, a state-feedback control strategy based in the subsequent linearized small-signal full-order model is designed. The resulting closed-loop dc-dc converter exhibits good output voltage regulation even in presence of step and high frequency perturbations in input voltage, load and control signal. The analytical results have been verified by means of circuital simulation in PSIM.

Keywords: Switch-mode dc-dc voltage regulator, discontinuous conduction mode, full-order Cuk converter DCM model, LQR control.

1 Introduction

First published in 1977 [1], the optimum topology dc-dc switching converter also known as Ćuk converter, its variants, and their applications in discontinuous conduction mode (DCM) are still currently under study. In addition to easily achieved unity power factor in power factor correction (PFC) applications, some additional advantages of DCM Ćuk-based PFCs are zero-current turn-on in the power switches, zero-current turn-off in the output diode and small complexity of the control circuitry [2].

Being a fourth order dc-dc converter, the analysis of the basic Ćuk structure in all its DCM modes is complex, as corroborates the number of works proposing reduced and/or full-order models of the basic converters [3]-[5], which have also studied the DCM Ćuk converter in more or less detail. All these works have the common objective of extending the frequency margin of validity of the small-signal models of dc-dc converters, for instance, up to the one third of the switching frequency of analogous models in continuous conduction mode (CCM).

The work here reported will have a similar goal in case of the Ćuk converter with both inductors operating in DCM. To overcome the limitations of previous approaches, this work uses a similar procedure to the one reported in [6]-[7], which was applied to a family of several high order converters exhibiting inherent

M. Zhu (Ed.): Electrical Engineering and Control, LNEE 98, pp. 441–449.
springerlink.com © Springer-Verlag Berlin Heidelberg 2011

DCM. A key point of the procedure is the determination of the duty cycle of one of the subintervals from the difference in average values of the inductor currents.

The application of the proposed method to the DCM Ćuk converter is presented in section 2. The full-order theoretical results obtained are verified by means of simulation in the same section. The linearization of the proposed model permits to obtain a small-signal description of the system in the frequency domain that is used to design the state-feedback control strategy that is tested in section 3. Finally, section 4 gives conclusions and proposals for future works.

2 Modeling of Ćuk Converter in DCM

The modeling procedure considers the Ćuk converter operating in deep DCM, where both inductor currents exhibit discontinuous conduction mode. Figure 1(a) shows the circuital scheme of Ćuk converter, and figure 2 shows the waveforms of the converter operating in DCM, where T denotes the switching period.

The first topology of the converter occurs when the Mosfet is set ON, which implies an inverse polarization of the Diode, generating the topology 1 depicted in figure 1(b). In such a topology, which duration is d_1T and its waveforms are depicted in figure 2, the slope of inductor L_1 current, named i_{L_1}, is positive while the slope of inductor L_2 current, named i_{L_2}, is negative. The resulting C_1 capacitor current is therefore negative and its voltage decrease. But, the current on C_2 capacitor exhibits both positive and negative values during d_1T, then its voltage increases and decreases in the first topology.

The second topology occurs when the Mosfet is set OFF, which implies a positive Diode current equal to $i_{L_1} - i_{L_2}$, generating the topology 2 depicted in figure 1(c). This topology has a duration of d_2T and its waveforms are also depicted in figure 2. In such a topology, the slope of i_{L_1} is negative while the slope of i_{L_2} is positive. C_1 current is positive, therefore its voltage increase, and again C_2 current exhibits both positive and negative values during d_2T, then its voltage decreases and increases.

Since the steady state L_1 current is higher than the steady state L_2 current, as depicted in figure 2, the third topology occurs when the decrement on i_{L_1} and the increment on i_{L_2} cause $i_{L_1} = i_{L_2}$, which implies no current by the Diode, generating the topology 3 depicted in figure 1(d) during d_3T, where $d_3 = 1 - d_1 - d_2$. In this topology, the inductor currents value is named i_X.

Defining the state vector $\mathbf{x} = \left[i_{L_1}\, i_{L_2}\, v_{C_1}\, v_{C_2}\right]^T$, the dynamic of the system is given by:

$$\dot{\mathbf{x}} = \begin{bmatrix} 0 & 0 & 0 & 0 \\ 0 & 0 & -\frac{1}{L_2} & -\frac{1}{L_2} \\ 0 & \frac{1}{C_1} & 0 & 0 \\ 0 & \frac{1}{C_2} & 0 & -\frac{1}{R \cdot C_2} \end{bmatrix} \mathbf{x} + \begin{bmatrix} \frac{1}{L_1} \\ 0 \\ 0 \\ 0 \end{bmatrix} V_g \, , \; \forall\, t \in [0, d_1T] \tag{1}$$

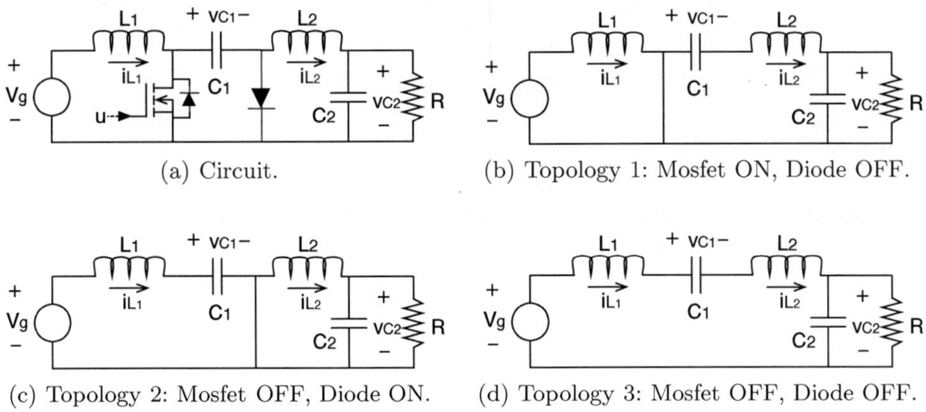

(a) Circuit. (b) Topology 1: Mosfet ON, Diode OFF.

(c) Topology 2: Mosfet OFF, Diode ON. (d) Topology 3: Mosfet OFF, Diode OFF.

Fig. 1. Ćuk converter circuit and topologies on DCM operation.

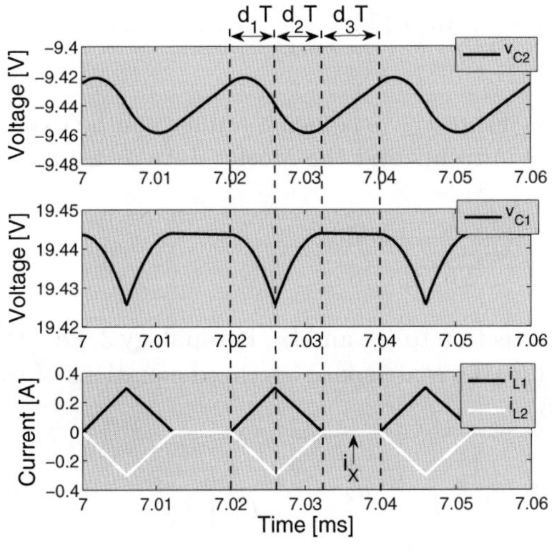

Fig. 2. Ćuk converter waveforms in DCM.

$$\dot{\mathbf{x}} = \begin{bmatrix} 0 & 0 & -\frac{1}{L_1} & 0 \\ 0 & 0 & 0 & -\frac{1}{L_2} \\ \frac{1}{C_1} & 0 & 0 & 0 \\ 0 & \frac{1}{C_2} & 0 & -\frac{1}{R \cdot C_2} \end{bmatrix} \mathbf{x} + \begin{bmatrix} \frac{1}{L_1} \\ 0 \\ 0 \\ 0 \end{bmatrix} V_g \ , \ \forall \, t \in [d_1 T, (d_1 + d_2) \, T] \tag{2}$$

$$\dot{\mathbf{x}} = \begin{bmatrix} 0 & 0 & -\frac{1}{L_1+L_2} & -\frac{1}{L_1+L_2} \\ 0 & 0 & -\frac{1}{L_1+L_2} & -\frac{1}{L_1+L_2} \\ \frac{1}{C_1} & 0 & 0 & 0 \\ 0 & \frac{1}{C_2} & 0 & -\frac{1}{R\cdot C_2} \end{bmatrix} \mathbf{x} + \begin{bmatrix} \frac{1}{L_1+L_2} \\ \frac{1}{L_1+L_2} \\ 0 \\ 0 \end{bmatrix} V_g\,,\ \forall\, t \in [(d_1+d_2)T, T] \quad (3)$$

In topology 3 the inductor currents enter on DCM and the traditional small-ripple approximation is not accurate to describe its dynamics, therefore to apply the averaging method a correction on the inductor currents must be introduced [7]. Taking into account that both inductor currents exhibit triangular waveforms as depicted in figure 2, the average inductor currents are calculated:

$$\bar{i}_{L_1} = \frac{1}{T}\int_0^T i_{L_1}\,dt = i_X + \frac{(d_1+d_2)\,(d_1 T)\,V_g}{2L_1} \quad (4)$$

$$\bar{i}_{L_2} = \frac{1}{T}\int_0^T i_{L_2}\,dt = i_X - \frac{(d_1+d_2)\,(d_1 T)\,(v_{C_1}+v_{C_2})}{2L_2} \quad (5)$$

where the currents component i_X is calculated as

$$i_X = \bar{i}_{L_1} - \frac{(d_1+d_2)\,(d_1 T)\,V_g}{2L_1} = \bar{i}_{L_2} + \frac{(d_1+d_2)\,(d_1 T)\,(v_{C_1}+v_{C_2})}{2L_2} \quad (6)$$

which depends on d_2. From $\bar{i}_{L_1} - \bar{i}_{L_2}$ such a duty can be expressed in terms of the average inductors currents and the Mosfet duty d_1 as:

$$d_2 = \frac{2\,(\bar{i}_{L_1} - \bar{i}_{L_2})}{d_1 T\left[\frac{V_g}{L_1} + \frac{v_{C_1}+v_{C_2}}{L_2}\right]} - d_1 \quad (7)$$

From the above correction to i_{L_1} and i_{L_2} in topology 3, i.e. (4) and (5), and the calculation of d_2 in (7), the state space averaged equations of the Ćuk converter are formulated in (8).

$$\begin{cases} \dfrac{d\bar{i}_{L_1}}{dt} = \dfrac{V_g}{L_1}d_1 + \dfrac{V_g - \bar{v}_{C_1}}{L_1}d_2 + \dfrac{V_g - \bar{v}_{C_1} - \bar{v}_{C_2}}{L_1+L_2}(1-d_1-d_2) \\ \dfrac{d\bar{i}_{L_2}}{dt} = -\dfrac{\bar{v}_{C_1}+\bar{v}_{C_2}}{L_2}d_1 - \dfrac{\bar{v}_{C_2}}{L_2}d_2 + \dfrac{V_g - \bar{v}_{C_1} - \bar{v}_{C_2}}{L_1+L_2}(1-d_1-d_2) \\ \dfrac{d\bar{v}_{C_1}}{dt} = \dfrac{\bar{i}_{L_2}}{C_1}d_1 + \dfrac{\bar{i}_{L_1}}{C_1}d_2 + \dfrac{i_X}{C_1}(1-d_1-d_2) \\ \dfrac{d\bar{v}_{C_2}}{dt} = \dfrac{\bar{i}_{L_2} - \bar{v}_{C_2}/R}{C_2} \end{cases} \quad (8)$$

The dynamics of the system are therefore modeled by equations (6), (7) and (8). The steady state operating point is obtained assuming the derivatives of (8) equal to zero and solving the equations system.

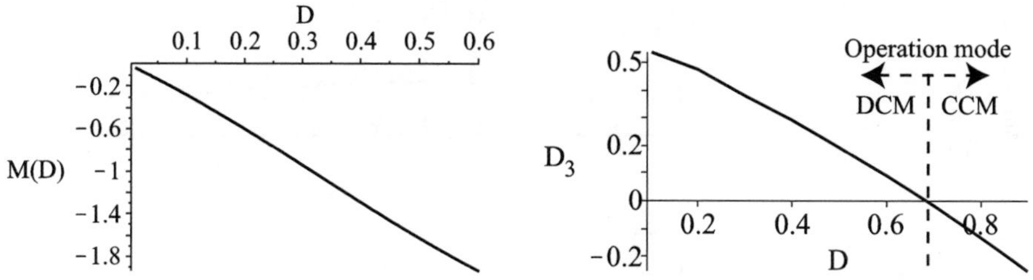

(a) Voltage conversion ratio. (b) DCM operation range.

Fig. 3. Cúk converter steady state characteristics in DCM.

From the operating point analysis, the voltage conversion ratio $M(D)$ of the Ćuk converter operating in DCM can be calculated. But such a relation depends on the system parameters due to its DCM condition, therefore to illustrate $M(D)$ the following parameters have been adopted: $L_1 = L_2 = 200\ \mu H$, $C_1 = 50\ \mu F$, $C_2 = 23.5\ \mu F$, $R = 100\ \Omega$ and $T = 20\ \mu s$. Figure 3(a) shows the steady state $M(D)$ for the given parameters, where a nearly linear behavior is observed. Similarly, the analysis of the steady state duty D_3 of the third topology is depicted in figure 3(b) for the same parameters, where it is observed the steady state Mosfet duty $D_1 = D$ that ensures a DCM operation ($D_3 > 0$) of the Ćuk converter.

The small signal model is obtained by using standard linearization techniques and defining the small signal state vector $\hat{\mathbf{x}} = \left[\hat{i}_{L_1}\ \hat{i}_{L_2}\ \hat{v}_{C_1}\ \hat{v}_{C_2}\right]^T$ and the input vector $\hat{\mathbf{u}} = \left[\hat{d}\ \hat{V}_g\right]^T$. Since such a system exhibit four order matrices with large terms, e.g. forty operands, a numerical example is used to illustrate the small signal model by adopting the previous parameters:

$$\dot{\hat{\mathbf{x}}} = \begin{bmatrix} -1.57\times10^5 & 1.57\times10^5 & -1085 & 515.8 \\ 1.57\times10^5 & -1.57\times10^5 & -3914 & -5514 \\ 16400 & 3596 & -22.23 & -22.23 \\ 0 & 42550 & 0 & -425.5 \end{bmatrix} \hat{\mathbf{x}} + \begin{bmatrix} 1.95\times10^5 & 5511 \\ -1.95\times10^5 & -511.4 \\ -7564 & -92.92 \\ 0 & 0 \end{bmatrix} \hat{\mathbf{u}} \quad (9)$$

Figure 4 compares the frequency response of the small signal model (9) and the PSIM circuital simulation of the Ćuk converter with the adopted parameters. The comparison has been performed for the four states variables of the system: L_1 and L_2 currents, and C_1 and C_2 voltages. It is noted that the both model and circuit frequency responses exhibit a satisfactory agreement for all the states. The frequency range adopted for the Bode diagrams of figure 4 is constrained by the half of the switching frequency.

Such a result shows that the proposed model accurately describes the Ćuk converter dynamics, therefore it is suitable for control purposes.

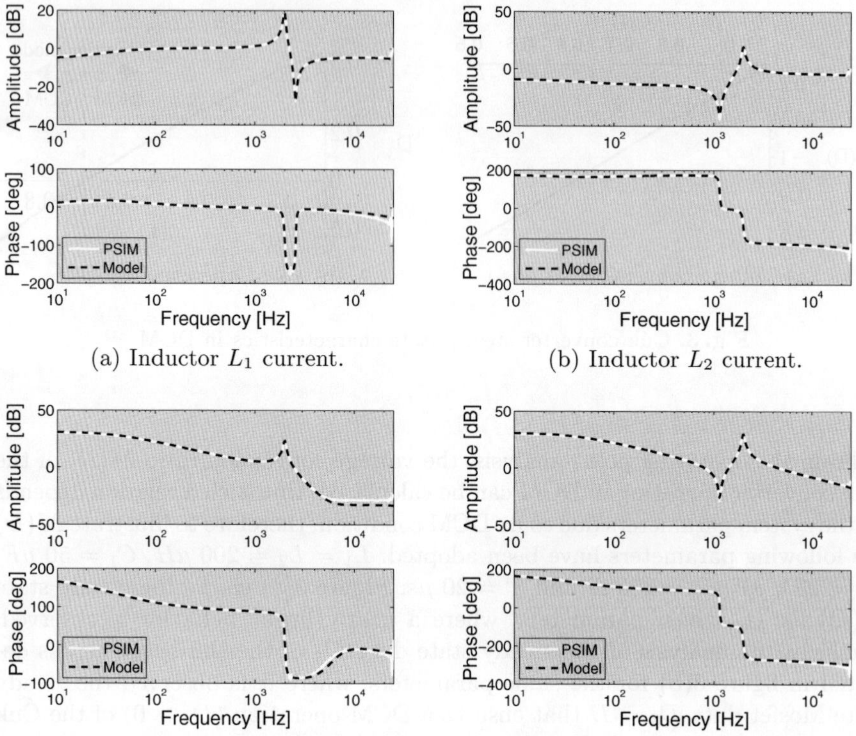

(a) Inductor L_1 current. (b) Inductor L_2 current.

(c) Capacitor C_1 voltage. (d) Capacitor C_2 voltage.

Fig. 4. Comparison of model and PSIM circuit frequency responses.

3 Model Application Example: Controller Design

The proposed controller is intended to illustrate the usefulness of the developed model, but any other control structure can be adopted. This work considers the control of the Ćuk converter by means of the Linear Quadratic Regulator (LQR) technique [8] as depicted in figure 5, where the ΔV_g and Δd represent perturbations on the input voltage and duty cycle, respectively.

In addition, the control structure considers the integral of the output voltage error, $V_{REF} - v_{C_2}$ where V_{REF} is the reference, as an additional state to guarantee null steady-state error. The feedback vector gains $\mathbf{K}$ has been calculated using (9) and the Matlab Control System Toolbox.

Figure 6 shows the frequency responses of the closed loop Ćuk converter for perturbations on the duty cycle and the input voltage, where satisfactory attenuation of such perturbations are observed. In particular, low frequency perturbations are accurately rejected, as well as perturbations near the half of the switching frequency. In the middle of the frequency range the perturbations are

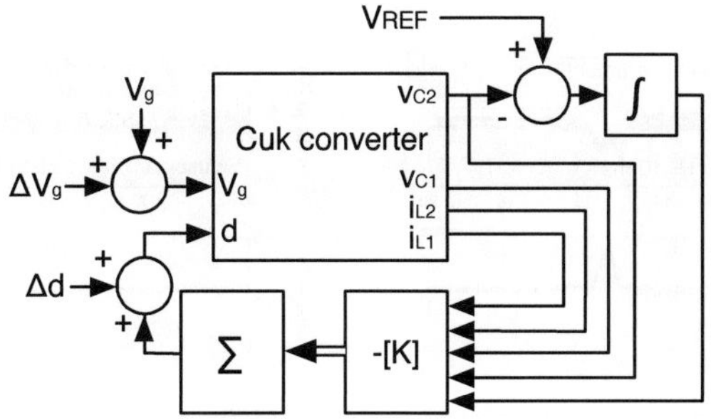

Fig. 5. Control structure.

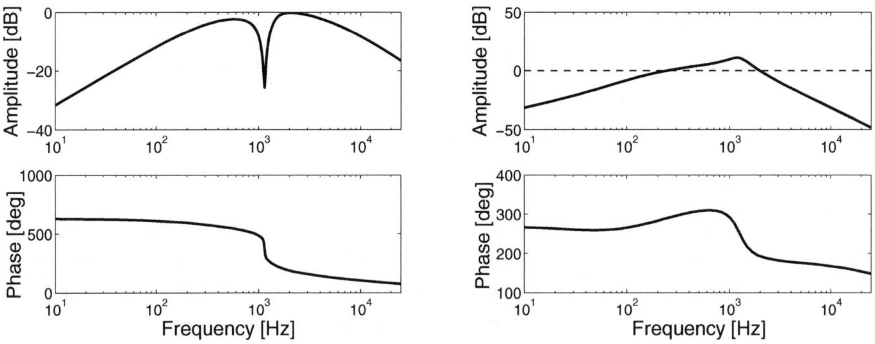

(a) Sensitivity to perturbations on d. (b) Sensitivity to perturbations on V_g.

Fig. 6. Closed loop frequency responses.

not well attenuated, but other more complex control structures can be adopted to address this aspect.

Figure 7(a) shows the closed loop system response to transient load perturbations, which exhibit a step behavior of 100 % of the load current. The simulation shows a satisfactory converter response, where the output voltage does not leaves the 2 % band. Similarly, figure 7(b) shows the system response to permanent sinusoidal perturbations in d and V_g, where the controller exhibits a properly operation. Simulations of figure 7 were performed using a circuital implementation of the Ćuk converter in PSIM, and considering the adopted parameters and vector **K** calculated in Matlab.

4 Conclusions

A full-order modelling procedure has been applied to the discontinuous current conduction mode dc-dc Ćuk converter. After the linearization of the non-linear model of the average state variables, the accuracy of the proposed model was validated by the comparison of the frequency responses of both model and circuital simulation of the Ćuk converter. The satisfactory behavior of the model

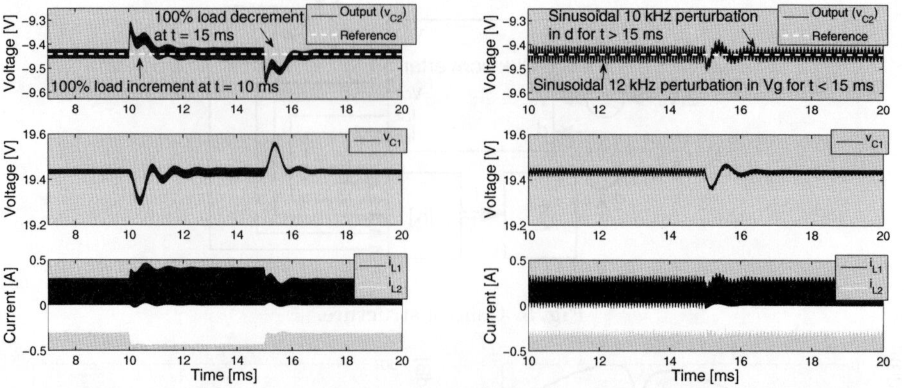

(a) Transient perturbations. (b) Permanent perturbations.

Fig. 7. Closed loop system responses.

has permitted to design an output voltage regulator using LQR control and obtaining a proper reference tracking and disturbances rejection. Future works will consider the application of the modelling procedure and subsequent control design to other high order converters like Sepic and Zeta.

Acknowledgments. This work was supported by the GAUNAL and GITA research groups of the Universidad Nacional de Colombia and the GIT group of the Instituto Tecnológico Metropolitano under the project DACOGEN-PV, and by the Spanish Ministerio de Ciencia e Innovación under grants TEC2009-13172 and CSD2009-00046.

References

1. Ćuk, S., Middlebrook, R.D.: Advances in switched-mode power conversion, vol II. Teslaco, Pasadena (1981)
2. Sabzali, A.J., Ismail, E.H., Al-Saffar, M.A., Fardoun, A.A.: New Bridgeless DCM Sepic and Cuk PFC Rectifiers With Low Conduction and Switching Losses. IEEE Trans. Industry Applications 47, 873–881 (2011)
3. Vorperian, V.: Simplified analysis of PWM converters using model of PWM switch. II. Discontinuous conduction mode. IEEE Trans. Aerospace and Electronic Systems 26, 497–505 (1990)
4. Femia, N., Tucci, V.: On the modeling of PWM converters for large signal analysis in discontinuous conduction mode. IEEE Trans. Power Electronics 9, 487–496 (1994)
5. Sun, J., Mitchell, D.M., Greuel, M.F., Krein, P.T., Bass, R.M.: Averaged modeling of PWM converters operating in discontinuous conduction mode. IEEE Trans. Power Electronics 16, 482–492 (2001)

6. Giral, R., Arango, E., Calvente, J., Martinez-Salamero, L.: Inherent DCM operation of the asymmetrical interleaved dual buck-boost. In: IEEE 2002 28th Annual Conference of the Industrial Electronics Society, vol. 1, pp. 129–134 (2002)
7. Arango, E., Calvente, J., Giral, R.: Asymmetric interleaved DC-DC switching converters: Generation, modelling and control. Lambert Academic Publishing, Saarbrucken (2010)
8. Arango, E., Ramos-Paja, C., Calvente, J., Giral, R., Romero, A., Martinez-Salamero, L.: Fuel cell power output using a LQR controlled AIDB converter. In: IEEE International Conference on Clean Electrical Power, pp. 492–499 (2007)

6. Giral, R., Arango, E., Calvente, J., Martinez-Salamero, L.: Inherent DCM operation of the asymmetrical interleaved dual buck-boost. In: IECON 2002 28th Annual Conference of the Industrial Electronics Society, vol. 1, pp. 129–134 (2002)
7. Arango, E., Calvente, J., Giral, R.: Asymmetric interleaved DC-DC switching converters: Generation, modeling and control. Lambert Academic Publishing, Saarbrücken (2010)
8. Arango, E., Ramos-Paja, C., Calvente, J., Giral, R., Romero, A., Martinez-Salamero, L.: Fuel cell power output using a LQR controlled AIDB converter. In: IEEE International Conference on Clean Electrical Power, pp. 492–499 (2007)

Comparative Analysis of Neural and Non-neural Approach of Syllable Segmentation in Marathi TTS

S.P. Kawachale[1] and J.S. Chitode[2]

[1] Senior Lecturer, E & TC Dept., MIT, Pune, India
[2] Honorary Professor, BVP, COE, Pune, India
smt_prs@yahoo.co.in
js_chitode@rediffmail.com

Abstract. In this paper two approaches neural and non-neural are presented for segmentation of syllables in speech synthesis of Marathi TTS. Speech Synthesis must be capable of automatically producing speech from given text by storing segments of speech. Speech synthesis can be done by making use of different units of speech like phones, di-phones, syllables and words. TTS synthesizers making use of hybrid approach which is combination of mixture of different speech units are also used in various applications. Speech unit's study shows that syllable results in more naturalness. But as compare to phone, di-phone, syllable need more storage space and results in large database for most TTS systems. It has been demonstrated that reliable segmentation of spontaneous speech into syllabic entities is useful for speech synthesizers. It generates more number of words based on very small database. As existing syllables can be used to form new words hence original database is not large. For proper cutting of syllables, detection of vowels is important and it can be done by calculating energy of sound file. As vowels have more energy as compare to consonants, syllables can be cut very easily by making use of this property. But manual segmentation is very time consuming and also it doesn't result in naturalness. To improve the naturalness of resulting segments in speech synthesis, neural network can be helpful. The proposed paper demonstrates the comparative analysis of neural and non-neural approaches for segmentation. Two algorithms are discussed, in non-neural approach slope detection algorithm is demonstrated and in neural approach k-means algorithm is presented.

Keywords: NN- Neural Network, TTS-Text to Speech System, Synthesized Speech- Artificially generated speech.

1 Introduction

A Text to Speech system converts some language text into speech. Artificial production of speech results in speech synthesis. A computer program used for this purpose is known as speech synthesizer. Synthesized speech is generated by concatenating pieces of pre-recorded speech that are stored in a database. TTS systems differ in the size of stored speech units. A system that stores phones or di-phones provides largest output range, but may lack clarity. If large unit size like word or

M. Zhu (Ed.): Electrical Engineering and Control, LNEE 98, pp. 451–460.
springerlink.com © Springer-Verlag Berlin Heidelberg 2011

sentence is used, it allows for high-quality output but memory requirement is very large. A system that stores syllables provides good quality output with less storage space. The main objective is to develop small segments (syllables) from different words. The proposed system uses both non-neural and neural network approach for formation of proper syllables. Slope detection concept is used as a non-neural approach and k-means algorithm is used under neural approach. It is observed that neural approach gives more better results as compared to non-neural approach. The accuracy of neural k-means algorithm is tested for all types of syllables and more than three hundred words.

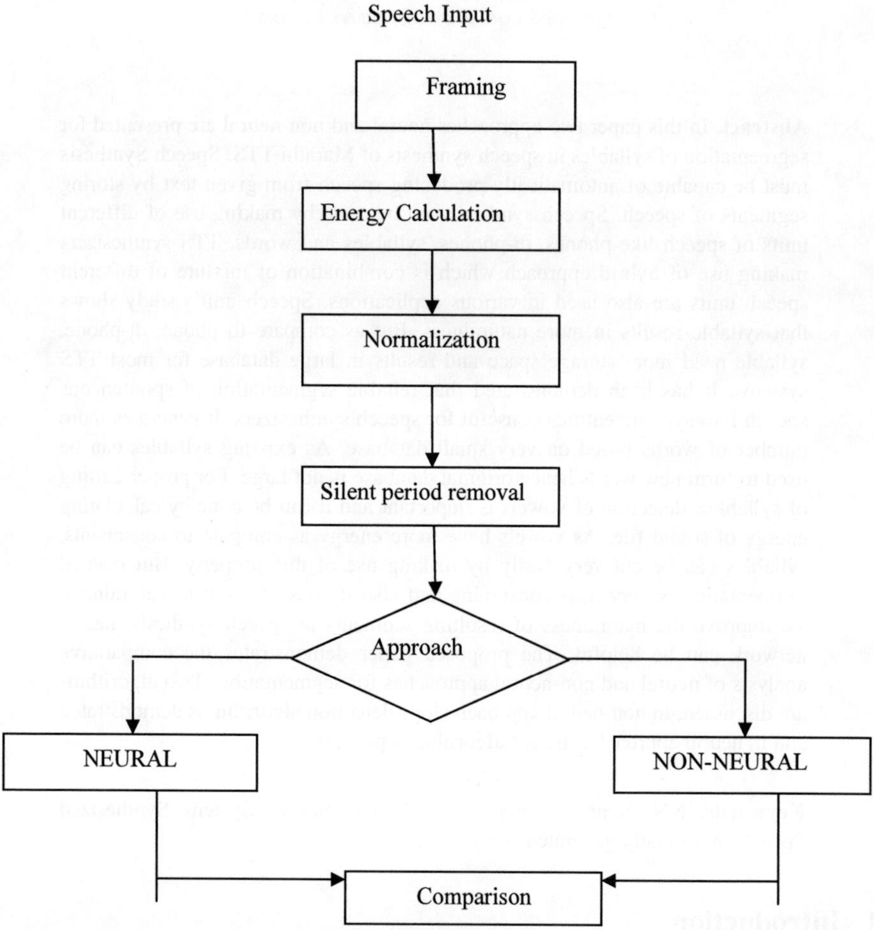

Fig. 1. System Block Diagram

2 Block Diagram

Figure 1 shows the block diagram of system. The speech waveform is converted into energy waveform as vowel has more energy than consonants and this property can be used very easily for syllable detection and hence segmentation. The energy input is given to both neural and non-neural approaches and the results are compared.

The speech signal is dynamic in nature. So it is divided into frames of size 110 samples. Energy of each of these frames is calculated. The energy varies in each frame depending upon the speech signal; it is therefore normalized to give a smooth curve. Silent period is removed after this step. Then starts the process of syllabification .This is done using neural and non-neural approaches.

3 Slope Detection Algorithm

1. The energy plot obtained after normalization and smoothing procedure is used for syllable cutting in non-neural approach.
2. Obtain slope of the curve from energy diagram.
3. Slope is found out by subtracting the next energy sample from previous one with respect to sample numbers.
4. The objective is to locate the point of inflection on the energy plot where the slope changes from negative to positive.

3.1 Results of Slope Detection Algorithm

Following figures shows results of slope detection (non-neural) algorithm for 4 and 3 syllable words.

A) Four-Syllable word: _hm^maV

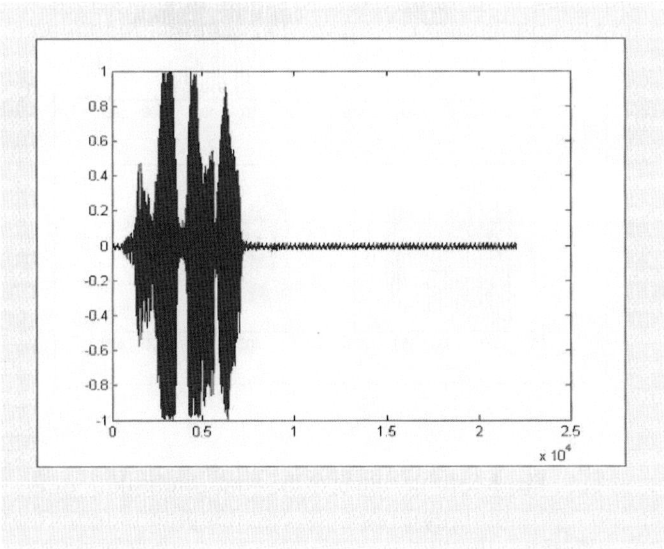

Fig. 2. Recorded Signal

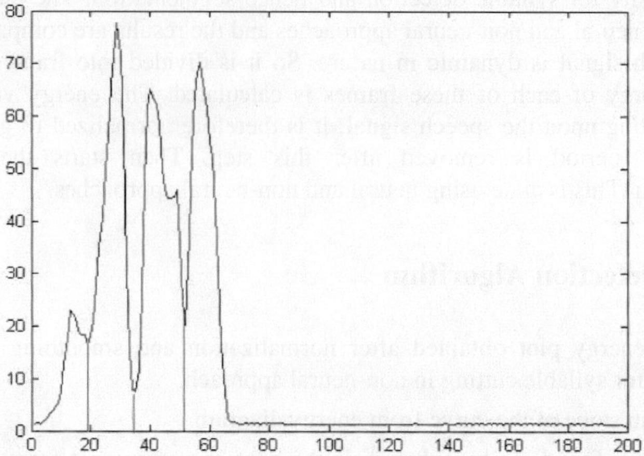

Fig. 3. Energy Plot

Points of cutting are frame number: 19, 36, and 52. These are minima's or segment locations of input word.

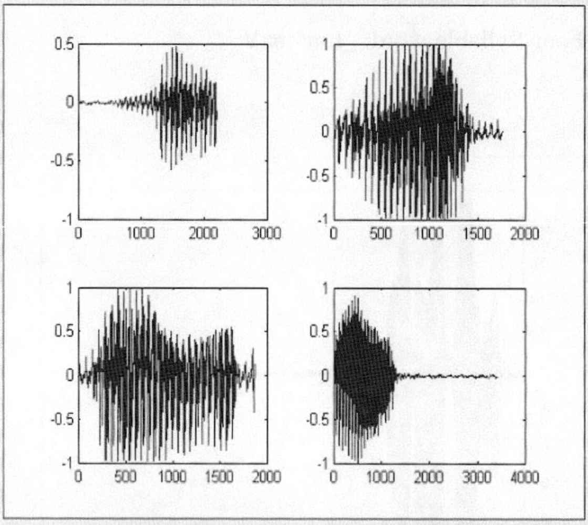

Fig. 4. Syllable cutting of the word _hm^maV

_hm^maV is a 4 syllable word cut into _,hm,^m,aV from left to right in the figure above.

B) Three-Syllable word: A{^Zd

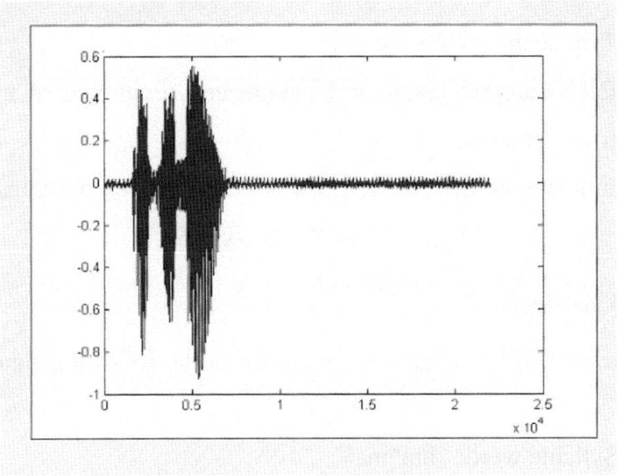

Fig. 5. Recorded signal

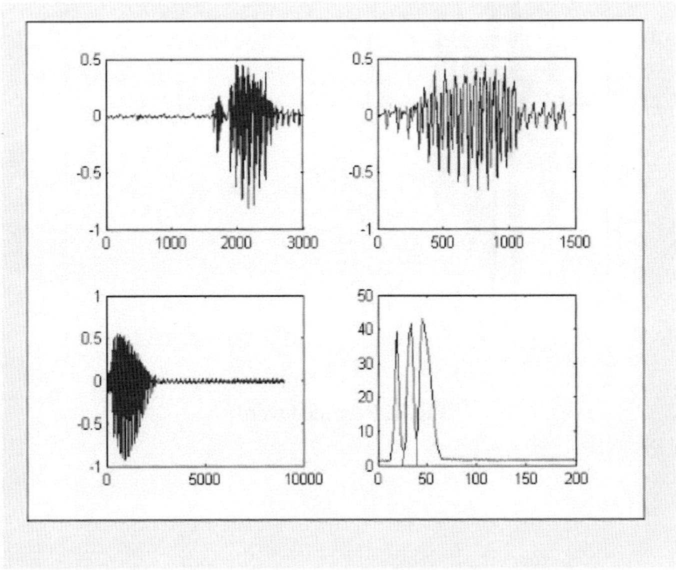

Fig. 6. Syllable Cutting

A{^Zd Word is cut into A,{^,Zd from left to right in the figure above and 4th diagram is energy plot. Points of cutting are frame numbers: 25, 45 in the energy plot.

4 K-Means Algorithm

1. Two centroids are selected.
2. Each centroid contains 2 parameters: amplitude of energy plot and frame number.
3. In K-means, distance is calculated as shown by the equation (1) below

$$D = \sqrt{(x1-x2)^2 + (y1-y2)^2} \,. \tag{1}$$

4.1 Results of K-Means:

Results of k-means (neural) approach are shown in the following figures for 4 and 3 syllable words.

A) Four-Syllable word: _hm^maV

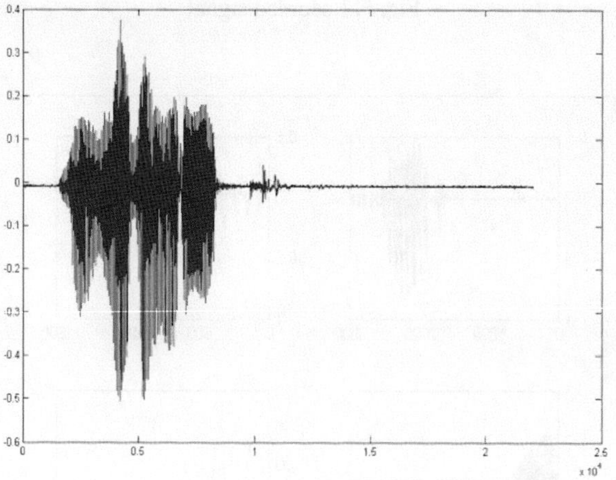

Fig. 7. Recorded word

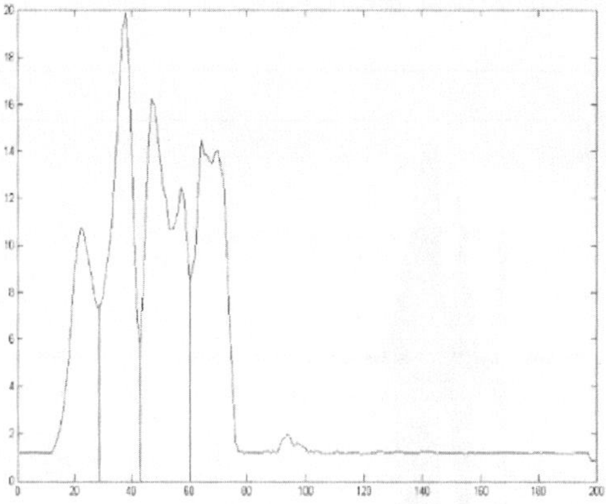

Fig. 8. Energy Plot

Points of cutting are frame numbers 28, 42, and 60. These are segment locations.

A) Syllable cutting of _hm^maV

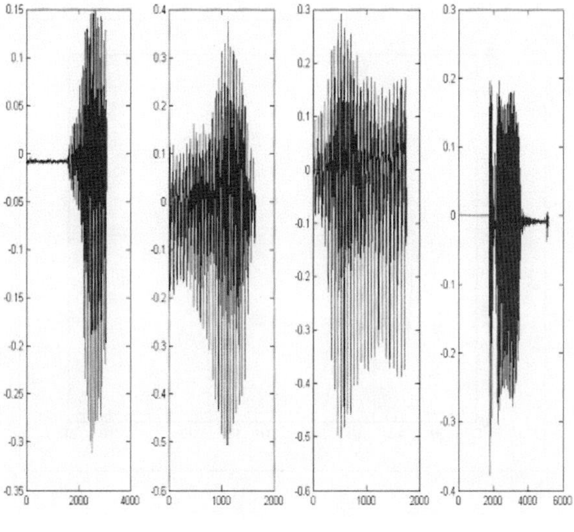

Fig. 9. Cutting of syllable

B) Three-Syllable word: A{^Zd

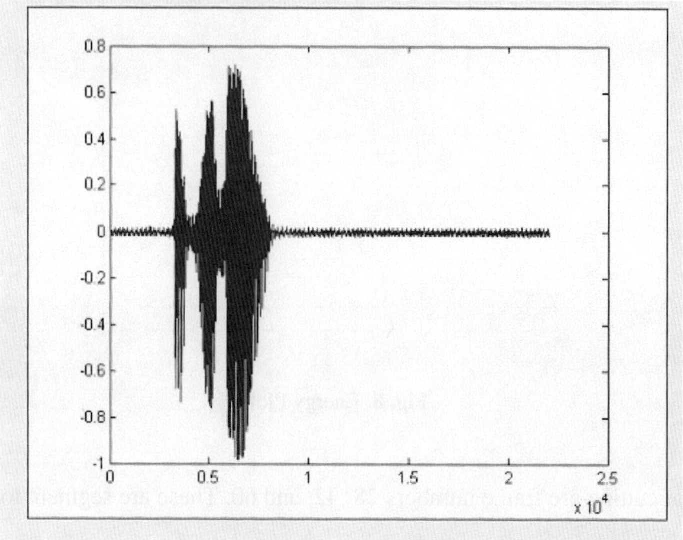

Fig. 10. Recorded Word

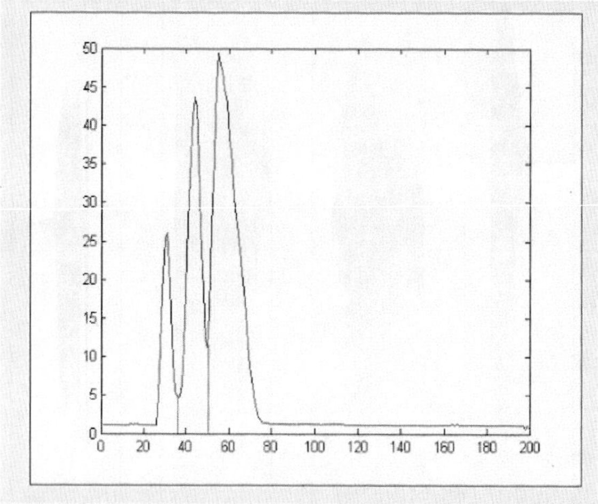

Fig. 11. Energy plot

Points of cutting frame number: 38, 50.
Syllable cutting of A{^Zd

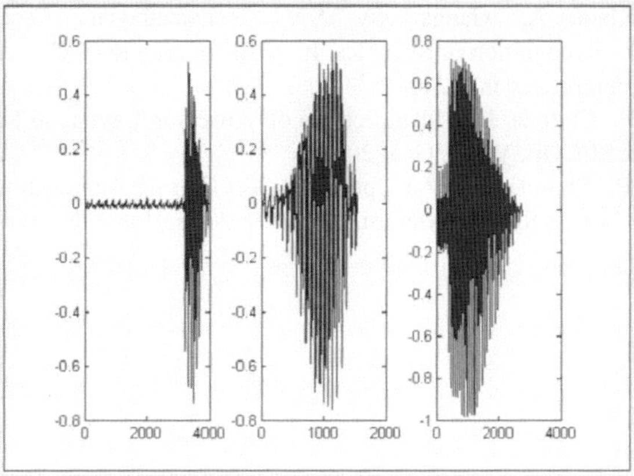

Fig. 12. Syllable cutting of word

From the above examples of 3 and 4 syllables, it is clear that slope detection and k-means can be used effectively for segmentation of syllables.

4 Conclusion

It has been observed that the syllable cutting accuracy is more in case of k-means algorithm as compare to slope detection algorithm. Testing of 200 to 300 words have shown that neural network approach works better for segmentation than non-neural approach.

References

1. Nagarjan, T., Murthy, H.: Sub band-Based Group Delay Segmentation of Spontaneous Speech into Syllable-Like Units. EURASIP Journal on Applied Signal Processing (2004)
2. Noetzel, A.: Robust syllable segmentation of continuous speech using Neural Network. IEEE Electro International (1991)
3. Sreenivasa Rao, K., Yegnanarayana, B.: Modeling syllable duration in Indian languages using Neural network. In: IEEE International Conference on Acoustic, Speech and Signal Processing, vol. 5 (2004)
4. Finster, H.: Automatic speech segmentation using neural network and phonetic transcription. In: IEEE International Joint Conference on Neural Network, vol. 4 (1992)
5. Sharma, M., Mammone, R.: Automatic speech segmentation using Neural tree networks. In: IEEE Workshop on Neural Network for Signal Processing (1995)
6. Toledano, D.T.: Neural Network Boundary refining for Automatic Speech Segmentation. In: IEEE International Conference on Acoustic, Speech and Signal Processing (2000)
7. Rahim, M.G.: A Self-learning Neural Tree Network for Recognition of Speech Features. IEEE, Los Alamitos (1993)

8. Nayeemulla Khan, A., Gangashetty, S.V., Yegnanarayana, B.: Neural Network Preprocessor for Recognition of Syllables. In: IEEE International Conference on Acoustic, Speech and Signal Processing (2004)
9. Kawachale, S.P., Chitode, J.S.: Identification of Vowels in Devnagari Script Using Energy Calculation. In: PCCOE, Verna, Goa (2006)
10. Kawachale, S.P., Chitode, J.S.: An Optimized Soft Cutting Approach to Derive Syllables from Words in Text to Speech Synthesizer. In: SIP 2006, Honolulu, Hawai, U.S.A (2006)

Predictive Control Simulation on Binocular Visual Servo Seam Tracking System

YiBo Deng[1], Hua Zhang[1,2], and GuoHong Ma[1,2]

[1] NanChang University, Provincial key Laboratory of Robot and Weld Automation, JiangXi, 330031,China
isaacisisaac@163.com

[2] NanChang University, School of Mechatronics, JiangXi, 330031, China
Zhanghua_lab@163.com, Ghma2006@gmail.com

Abstract. As a multidisciplinary technique, intelligent seam tracking is the key element for the realization of weld automation and flexibility. In this article a new type of predictive controller is proposed and applied on the binocular passive visual servo weld system. With the abundant information passive visual sensor provides no extra hardware was needed. A co-simulation was run to implement a sine curved seam tracking based on the powerful virtual prototype software ADAMS and Popular control software SIMULINK. In the co-simulation the prime influential factors like control period, weld velocity and predictive distance was discussed. Under the same tracking precision requirement of 0.5 mm, control period could be expanded from 200 ms to 900ms and the weld velocity could be increased from 2 mm/s to 5 mm/s after using predictive controller. The co-simulation result also provides a lot of other valuable data for the mechanical and control design of the experimental system, thus shortening designing cycle and improving the quality.

Keywords: Seam Tracking, Co-Simulation, Predictive Control, Visual Servo.

1 Introduction

Intelligent weld seam tracking is a multidisciplinary technique, it involves electronics, computer, welding, mechanics, material, optics, electromagnetics and so on. A lot of world researchers are working in this field. Weld automation and flexibility is the trend of the future weld technique. A weld system is usually close-looped with three main components: sensors, control unit and manipulators. To realize the mechanization and automation of welding, a variety of sensors were developed, such as electro-magnetic, Photoelectrical, ultra sonic, infrared and CCD. Accompanied with the progress in visual sensor, DIP and AI, computer vision was widely used due to its rich information、high precision and flexibility. In fact weld workers depend mainly on vision while welding[1,2].

According to the principle welding visual sensor systems are divided into three types: structured light, scanned laser and direct photo. Structured light and scanned laser are categorized into active vision servo. Direct photo is called passive vision servo, though image processing in this case is much more complicated and

M. Zhu (Ed.): Electrical Engineering and Control, LNEE 98, pp. 461–468.
springerlink.com © Springer-Verlag Berlin Heidelberg 2011

time-consuming, it contains comprehensive information, it also enables the inspection for the weld pool as the tracking is undergoing, therefore helps to improve the weld quality [3]. Also the development in PC and DIP technique makes people more convinced that it's the most promising vision servo type.

In this article a new type of predictive controller is proposed and applied on the binocular passive visual servo weld system with no extra hardware needed. A co-simulation was run to implement a sine curved seam tracking based on the powerful virtual prototype software ADAMS and Popular control software SIMULINK [4]. In the co-simulation the prime influential factors like control period, weld velocity and predictive distance was discussed. Optimal parameters value was achieved through the simulation.

2 System Modeling

Vision servo system in this article consists of welding cart, cross slider, binocular CCD cameras, weld power supply, analog signal conditioning and acquiring devices, image acquiring card, PC, Servo motor controller. cross slider can move the weld torch in abscissa and ordinate directions. Refer to Fig. 1. magnetic wheels of the cart was driven to realize the forward movement of the system, two DC servo motor was used to actuate the abscissa and ordinate slider. The intended objective is to track a curved planar seam and then a 3D seam.

2.1 Building Mechanical Model

Since we need to build a relatively complex model. UG was chosen as the mechanical modeling tool [5,6]. The final mechanic model was shown in Fig.1. with some details was omitted or simplified.

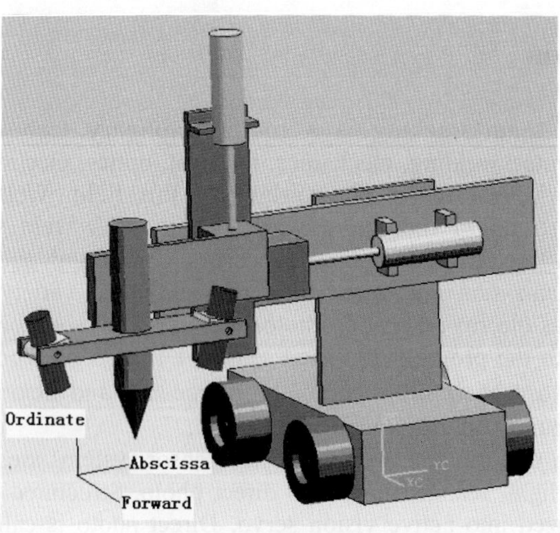

Fig. 1. Mechanical model

2.2 Model Exported to ADAMS

The UG model was exported using Parasolid interface with extension name .x_t, and then imported into ADAMS. After adding constrains, defining input and output variables [7]. The co-simulation can be carried out.

2.3 Adding Joints and Motions Creating System Input and Output Variables

The data exchange between ADAMS/Controls and control software SIMULINK was accomplished by the means of state variables. Firstly we define system input and output variables which include the velocity control value of the abscissa cross slider servo motor, Weld Velocity for cart motor, current tracking error and the error of prescribed position in distance.

2.4 Exporting PLANT Control Model

After Invoking Controls plug-in modal and exporting the PLANT model, ADAMS will automatically create 4 file in the working directory. Among them the one with extension name '.m' is ADAMS/Matlab interface file. Run it in Matlab environment will import data needed for SIMULINK. and then run adams_sys to create mechanical system control model adams_sub.

3 Predictive Controller Principle

During many control situation, delay often existed. in our case it's the data acquisition and processing. It will cause larger over-shoot and longer rectifying time, so the control systems with delay are considered a big problem in many cases [8,9]. In our seam tracking system, a passive vision servo was utilized. Because the time consumed in image acquisition and processing are relatively large, when controlling parameter was calculated a lot of time has passed since the image was acquired. Time delayed could be as large as hundreds of milliseconds or even seconds, which make it impossible for fast welding or large slope seam tracking. And then the performance of system will deteriorate. In this case predictive controller is a good way to improve the system performance. In 1958, smith proposed a new controller which compensate the delay, i.e. Smith predictive controller. It was applied widely in chemical engineering with remarkable effectiveness. Instead of using the PID output as the predictor input, a new predictive controller was proposed which take the input from the sensor directly based on our passive vision servo architecture. Because the image of passive vision sensor contains more information except the current position of weld torch with respect to weld seam, it is possible for us to extract the seam position in advance. With this variable as the input of our predictive controller we achieve a satisfying result [10] . The control diagram are shown in Fig. 2. (b).

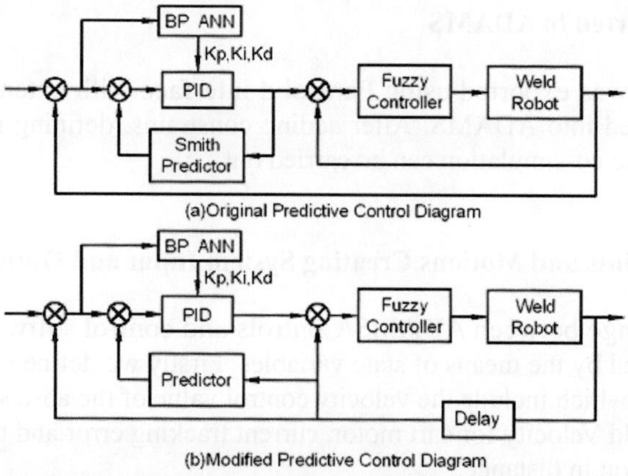

Fig. 2. Predictive control diagram

In the diagram ANN(artificial neural network) adjust the PID parameters dynamically to realize the control optimization. ANN's outputs corresponds to the parameters of the PID controller: kp、ki、kd，by learning and weighted coefficients rectifying. At the same time advancing control theory like fuzzy control could also be added into the system to improve the robustness and interference immunity [3,11,12]. For the current research stage, ANN and Fuzzy control are not included so far.

4 Co-simulation Analysis

Our co-simulation are based on the following conditions: welding workpieces are 5-mm-thick aluminum plates, sine shape weld seam has spatial equation: y = 50sin(x/50), x direction coincident with the cart advancing direction, maximum slope of seam is 1, speed of the cart is indicated with symbol Vw. Before the welding torch will be located to the seam start point by the vision servo system, welding direction will also be extracted at the beginning, initial error are less than 1mm. Welding system belongs to the typical discrete and continuous hybrid control system. Within one of the control period the following things must be done: acquiring image, seam recognition, calculate the feedback value and so on, while the first two will occupy most of time. Control period are indicated as symbol Ct.

4.1 SIMULINK Control Model Diagram

The control diagram was created as Fig.3., it consists of input, controller, mechanical model. Controller employs Predictive controller and classic PID algorithm. System inputs are the current bias and predicted position of seam in advance calculated from the seam image processing. Seam image was processed at every specific periodic time to get the velocity value that is being assigned to the cross slider servo motor.

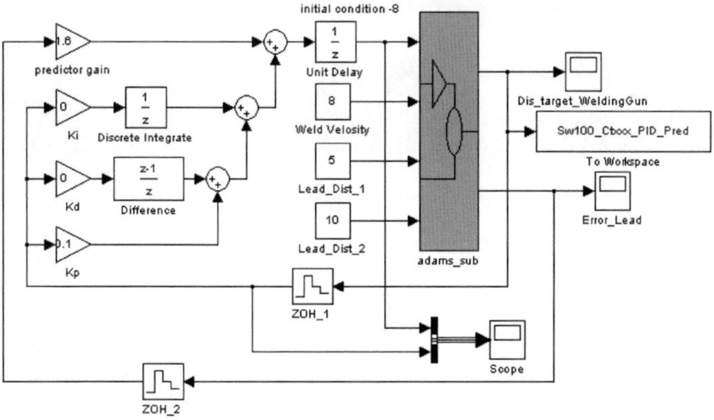

Fig. 3. SIMULINK control diagram

4.2 Comparison between PID Controller and Predictive Controller

Specify the Vw as 2mm/s, Ct as 200ms and leading distance as 5mm. Tracking errors with and without predictive controller are compared in figure4. We can easily tell that the tracking error is decreasing from 0.6mm to 0.15mm after using the presented predictive controller. Overshoot drops from 0.8mm to 0.3mm.

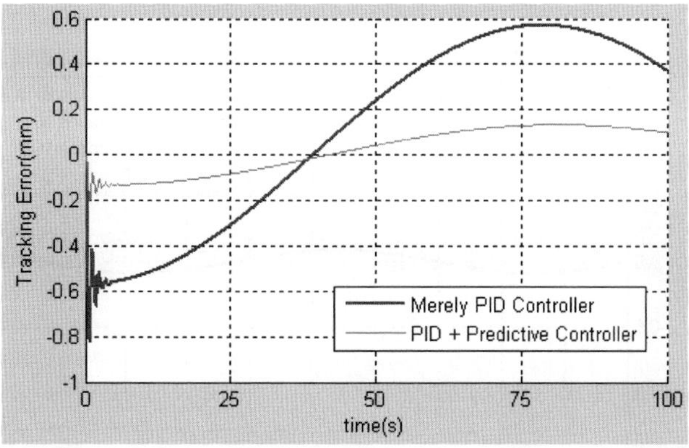

Fig. 4. Comparison between PID controller and Predictive controller

4.3 Tracking Error under Different Welding Cart Speed

Let the velocity of cart be 3, 5, 8 and 10mm/s, co-simulation shows corresponding required Ct would be 1400ms、900ms、550ms and 450ms respectively to assure the systematic stability and maximum tracking error 0.5 mm. Results are illustrated in figure5. We can easily tell that it's easier to realize system dynamics with the slower cart speed.

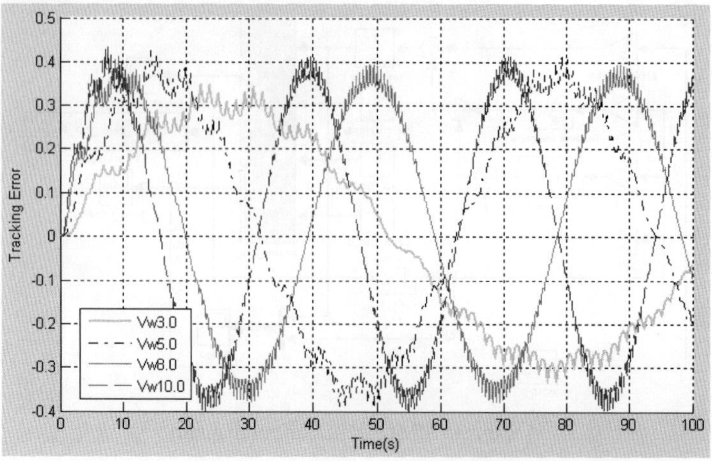

Fig. 5. Graphs under different welding cart speed

4.4 Tracking Results under Different Predictive Distance

When Vw equals 5 mm/s and Ct equals 600ms, change the predictive distance in the mechanical model to 5、10、15、20mm, tracking results are achieved as figure6. it shows when predictive distance is larger then 15mm, tracking error would exceed 0.5 mm.

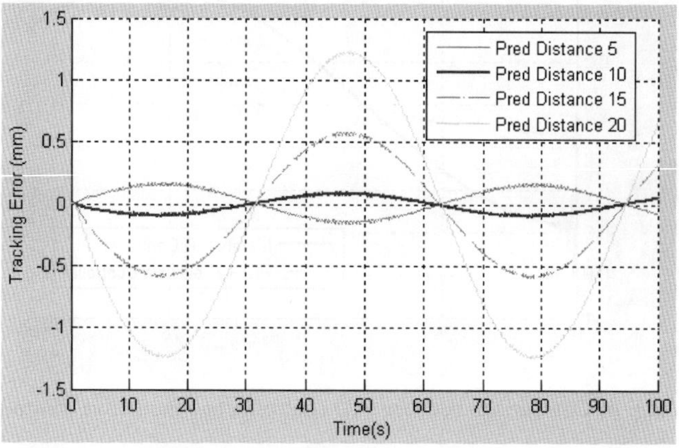

Fig. 6. Tracking results under different Predictive distance

5 Conclusion

In most of the passive vision servo seam tracking systems, control inputs are always calculated far behind the image was taken, which deteriorate the system performance greatly. In this article a predictive controller was proposed. By the means of co-simulation, we compared it with classic PID controller, result shows that Predictive controller could improve tracking precision dramatically; lower the dynamical requirement and allow a faster weld speed. Moreover algorithm is based on the existing information of the image so no extra hardware was needed. During the simulation maximum tracking error was set to 0.5mm. when using pure PID controller, control period must be less than 200 ms while the cart speed lower than 2mm/s; after the predictive controller was adopted, we can make it possible to weld under 900 ms control period and 5 mm/s cart speed, which is appropriate for most cases of aluminum plate welding. Analysis also reveal that tracking precision and control period drops accelerative with the increasing of cart speed or predictive distance. The Co-simulation technique employed in this article take the advantages of ADAMS and SIMULINK all together and united the mechanical and control model of the system. With the seamless communication interface between them, users can do the system debug over and over again till a satisfactory result was get. And the whole procedure can be viewed visually. With the rapid development of the computer tech, co-simulation has become a very powerful method while designing mechanical and control system , it is one of the important trend and getting more and more attentions.

Acknowledgments

This article is Supported by the National Natural Science Fund of China (50705041).

References

1. Yi, J.T., Fang, M.Z., Jiang, C.: Trends and Achievements in welding robot technology. Welding Machine 3, 6–10 (2006)
2. Chen, Q., Sun, Z.: Computer Vision Sensing Technology and Its Application in Welding. Transactions of the China Welding Institution 1, 83–90 (2001)
3. Chen, S.B., et al.: Intelligent methodology for sensing, modeling and control of pulsed GTAW. Welding Journal 79(6) (2000)
4. Shen, J., Song, J.: ADAMS and Simulink Co-Simulation for ABS Control Algorithm Research. Journal of System Simulation 3, 1141–1144 (2007)
5. Wei, D.G.: A tutorial on virtual prototype and ADAMS application. Press of Beijing aviation and aerospace university (2008)
6. Qin, C.: Virtual Prototyping Study Based on Proe /Adams /Matlab for Excavator. Machine Tool & Hydraulics 36(9), 133–135 (2008)
7. Zhen, K., Ren, X.H., Lu, M.C.: Advanced application examples of ADAMS 2005. Press of mechanical engineering (2006)
8. Zhao, D., Zou, T., Wang, Z.: Progress in study of Smith-predictor controller. Chemical Industry And Engineering Prograss (8), 1406–1411 (2010)
9. Martelli, G.: Stability of PID-controlled second-order time-delay feedback systems. Tomatica 45(11), 2718–2722 (2009)
10. Jin, X.R.: Fuzzy Predictive Control For Conic Smashing System Based on ANN Rectified. Mineral Mechanics 35(4), 41–43 (2007)

11. Yamene, S., et al.: Fuzzy control in seam tracking of the welding robots using sensor fusion, pp. 1741–1747. IEEE, Los Alamitos (1994); 0-7803-1993 - 194
12. Gao, Y.F., Zhang, H.: Predictive Fuzzy Control for a Mobile Welding Robot Seam Tracking. In: Proceedings of the 7th World Congress on Intelligent Control and Automation, Chongqing, China, June 25-27, pp. 2271–2276 (2008)

The Application of Savitzky-Golay Filter on Optical Measurement of Atmospheric Compositions

Wenjun Li

Bio-Incubator, Bessemer Building, Imperial College London,
Prince Consort Road, South Kensington, London, UK, SW7 2BP
jutlwj@126.com

Abstract. In optical measurement of atmospheric compositions using differential absorption technologies, traditional moving window average does not match the broadband background well, and will generate large measurement errors. To overcome the said drawbacks of moving window average, Savitzky-Golay filter is used in this paper. The principles of moving window average and Savitzky-Golay filter are introduced in the paper. Their fitting performances to the real intensity spectrum are compared. The results show that with moving window average, even with moving window of 21 points, the fitted background still does not match the real background well and is lower than the real background because of its heavily averaging effect, but for Savitzky-Golay filter, with moving window of 101 points, the fitted background still matches the real background well. The results also show that small window of Savitzky-Golay filter will lead to under fitted background and inability to detect the absorption signatures. Therefore large moving window is recommended for Savitzky-Golay filter. The test results for NO shows that with Savitzky-Golay filter the measurement accuracy can be improved from the best results of 18.3% and 4.3% which moving window average can get with 21 points, to around 3.5% and 1% at low and high concentration levels respectively.

Keywords: optical measurement, atmospheric composition, Savitzky-Golay filter, moving window average.

1 Introduction

Differential absorption spectroscopy is an ideal optical method to measure atmospheric concentrations because of its advantage of being able to measure multiple species fast and accurate. It measures the concentrations by taking differential signals between the intensity of the light source and the measured intensity after absorption. However the intensity of the light source is usually unknown. The solution is to filter out the broadband features of the light source by applying a low pass filter on the measured intensity and calculate the differentials between the received intensity and the filtered out broadband features. The filter or the fitting algorithm selected will affect the measurement accuracy and the performance of the differential absorption instrument significantly. Therefore it is vitally important to find the ideal low pass filter that suits this application. Normally moving window average is used in fitting the broadband features of the light source.

M. Zhu (Ed.): Electrical Engineering and Control, LNEE 98, pp. 469–480.
springerlink.com © Springer-Verlag Berlin Heidelberg 2011

However it lowers down the spectra intensity by averaging the points heavily in a moving window with each point in the window having the same weight. Therefore large fitting errors can be introduced. In this paper, Savitzky-Golay filter is used to improve the fitting result and the measurement accuracy.

2 Principle of Differential Absorption Technology

In Fig.1, the intensity of the light source is $I_0(\lambda)$, which is function of wavelength λ. The transmitted light intensity after absorption is I(λ). According to Lambert-Beer Law, the absorption by the sample is function of the optical path length between the light source and the detector, the cross section and the concentration of the detected sample as follows,

$$\tau = \frac{I(\lambda)}{I_0(\lambda)} = e^{-\sigma c l} \tag{1}$$

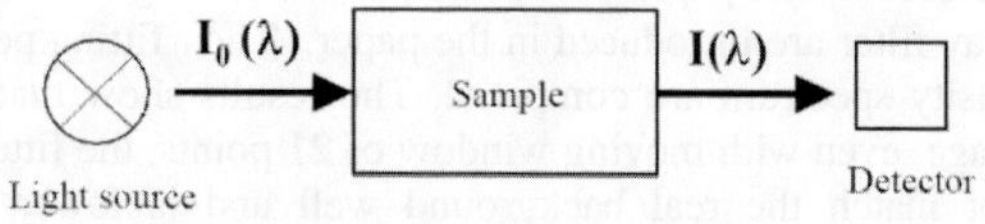

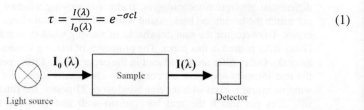

Fig. 1. Sample schematic to aid explanation of Lambert-Beer Law

In which σ is the cross section of the sample, c is the concentration of the sample and l is the optical path length between the light source and the detector. Applying logarithm on both sides of equation (1),

$$A = -ln\tau = \sigma c l \tag{2}$$

Usually *A* in equation (2) is called the absorption or optical density.

3 Retrieval Method of Differential Absorption Technology

Based on the discussion above, as usually the cross section of the sample and the optical path length are already known, as long as the absorption of the sample can be calculated, the concentration of the sample can be worked out according to equation (1) and (2). The problem is the intensity of the light source is usually unknown. It is difficult to use equation (1) and (2) directly. As the light source usually has broad features and the absorption of the sample usually shows narrow features, the intensity of the light source can be fitted by a polynomial P(λ) and the absorption can be worked out by taking the differential between the measured intensity and the polynomial P(λ) as follows,

$$A = ln\frac{P(\lambda)}{I(\lambda)} = \sigma c l \tag{3}$$

Once the absorption A is known, the concentration of the sample can be calculated out based on equation (3). Fig.2 shows an example of the absorption $A(\lambda)$, fitted background $P(\lambda)$ and the measured intensity $I(\lambda)$, in the retrieval method.

Usually moving window average is used to get the fitted $P(\lambda)$. However, in this paper, a more ideal filter which can fit the background better, Savitzky-Golay filter will be used to get the broadband background features of the light source.

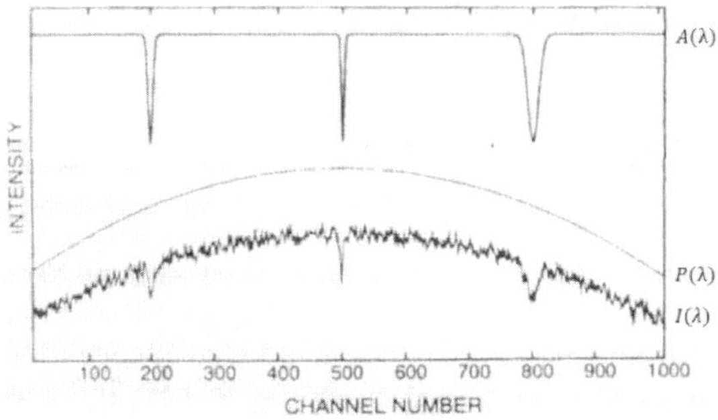

Fig. 2. An example of the absorption, fitted background and the measured intensity

4 Requirements for Fitting Algorithms

According to equation (3), the differentials are taken between the received light intensity and the fitted waveform $P(\lambda)$, the fitting algorithm which is used to get the fitted waveform $P(\lambda)$ decides directly the retrieval accuracy and hence, is the most important part of the whole retrieval procedure. As the light source usually shows broad features, the fitted waveform $P(\lambda)$ in equation (3) should be smooth enough to reflect the real broad features of the light source. Otherwise the fitting errors would be generated and affect the retrieval accuracy dramatically.

Fig.3 shows a typical detailed peak based on the smooth underlying background. If the width of the peak is n points, the width of the fitting window is N, and the depth of the peak is d, if all the points share the same weight in the fitting window, and the depths of all the n points in the peak added together is md , then the fitted background and the peak will be d1 less than it should be, and d1 can be calculated as,

$$d1 = \frac{\sum_{i=0}^{n} d_i}{N} = \frac{md}{N} \tag{4}$$

Usually the value of m is around $n/2$. That is to say, the underlying background is fitted by $\frac{n}{2N}d$ less than it should be. Therefore, the fitting error with respect to the peak depth is $n/2N$. If the width of a peak is 4 points, and 31 points is used in the fitting process, then the fitting error will be around 6%. If the peak is caused by the real gas absorption, then measurement error of at least 6% can be expected. Therefore the fitting algorithm should have the ability to generate the fitted broad band background smooth enough to approach the real lamp features as much as possible.

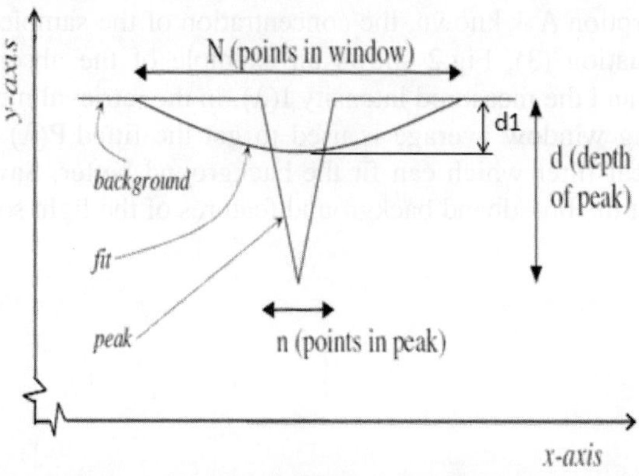

Fig. 3. A typical peak in the received intensity waveform

To make sure the fitted background is smooth is not enough for the high accuracy measurement. If the underlying broad feature is so heavily over averaged that the fitted background does not match the shape of the real background, as Fig.4 shows, although the fitted curve is smooth enough it lowers down the whole background and all the narrow features which appear in the raw data will be taken as the absorption with their full depth in the process of the calculation of differentials between the raw data and the fitted background, while in fact many of the peaks appearing in the raw data come from the narrow features of the lamp or noise. Furthermore, as the fitted background does not follow the shape of the real back ground, extra broad features will also be generated in the differential between the raw data and the fitted background, and hence, larger values will be expected than they should be in the measurement of the concentrations. Therefore, the fitting algorithm should be able to make sure the broadband background is not over fitted and the fitted curve matches the shape of the real background.

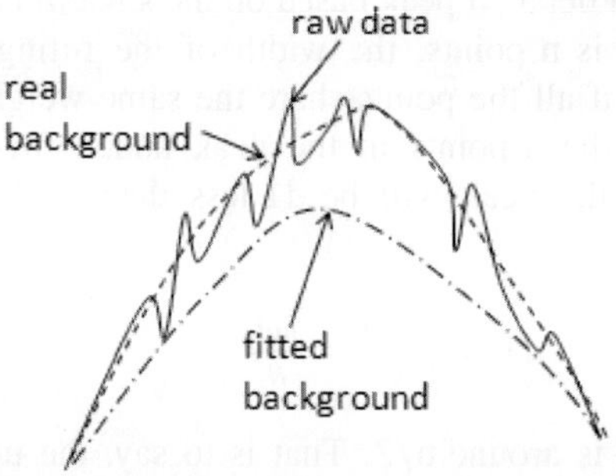

Fig. 4. A typical over fitted curve that does not match the real background

5 Principle of Moving Window Average

Usually moving window average is used as the fitting algorithm in getting the broadband background. For general digital filters applied to a data set with equally spaced values, the simple way is to replace each data d_i with a linear combination g_i of itself and some neighbouring data values, as follows,

$$g_i = \sum_{n=-n_L}^{n_R} c_n d_{i+n} \quad (5)$$

Here n_L is the number of values to the left of the data d_i and n_R is the number of values used in the filter to the right of the data d_i, and the data set between n_L and n_R is called the moving window of the filter. $n_L + n_R$ is called the window width.

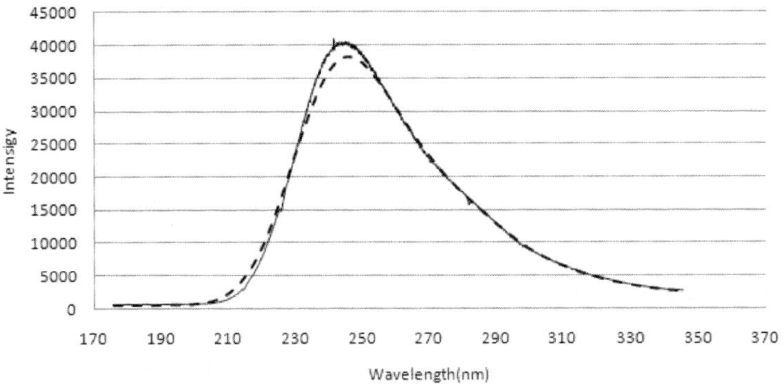

Fig. 5. Background fitting using moving window average with window width of 101 points

For a moving window average, the coefficient c_n is constant, which is $1/(n_L + n_R + 1)$. The big values are lowered down by small values heavily via averaging. The advantage of the moving window average is that the fitted curve is relatively smoother. If the underlying function is constant or linear, there is no bias introduced. If the underlying function has nonzero second order derivatives, bias will be introduced by the moving window average. Usually the background curve of the spectral data is non-constant or linear and contains second or higher order derivatives. Therefore at points where the gradient of the spectral curve changes fast, the moving window average will generate obvious fitting errors, the result of which is the fitted curve does not match the shape of the background as Fig.5 shows. In Fig.5, the solid line is the raw spectral data and the dashed line is the fitted curve by moving window average with window width of 101 points. It is obvious that at points where the raw data contains higher order derivatives, the fitted curve does not match the shape of the real background any more. Using very small moving window may lead to better shape matching, but on the one hand, there is risk that the background is not smooth enough to reflect the real broadband background as Fig.3 shows. On the other hand, even with small moving window it is still not guaranteed that the fitted curve will match what the real background should be. Fig.6 shows the fitting to a real spectrum waveform with window width of 21 points. It can be seen that even with as small fitting window as 21 points, the fitted curve is still lower than the ideal background.

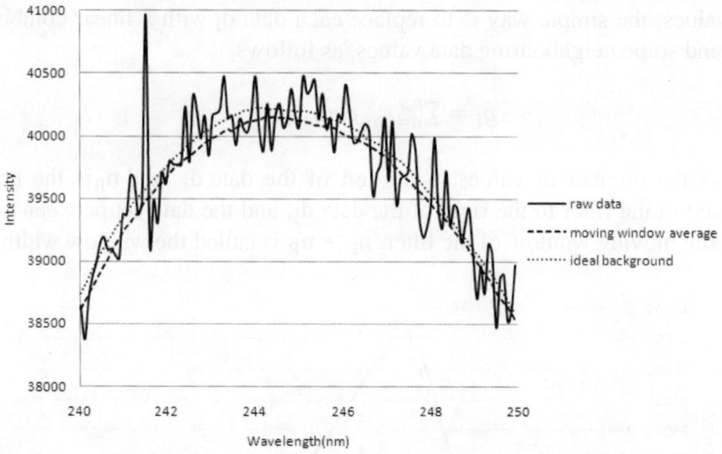

Fig. 6. Fitting to a real spectrum waveform with window width of 21 points

6 Principle of Savitzky-Golay Filter

The bias introduced by the moving window average when the raw data contains higher order derivatives is very undesirable. A more subtle use of averaging is required if the essential background feature in the spectra data is to be retained. Compared with moving window average, Savitzky-Golay filter is an ideal way to apply the averaging process, but to match the exact essential shape the spectra data.

The idea of Savitzky-Golay filter is to replace each value in a moving window with a new value which is obtained from a polynomial fit of degree M to $n_L + n_R$ neighbouring points, including the point to be smoothed. That is to say, we want to fit a polynomial of degree M in i , namely $a_0 + a_1 i + a_2 i^2 + \cdots + a_M i^M$, to data points in the moving window d_{-n_L} to d_{n_R}. Then g_0 in equation (5) will be the value of the polynomial at $i = 0$, namely a_0. The design matrix for the polynomial fitting is

$$A_{ij} = i^j \quad i = -n_L,,,,,,, n_R,\ j = 0,,,,,,, M \tag{6}$$

The polynomial fitting equation can be written as,

$$Aa = d \tag{7}$$

In which $a = [a_0, a_1, a_2,,, a_M]^T$ is the coefficient vector and $d = [d_{-n_L},,,, d_{n_R}]^T$ is the vector of the data set in the moving window. Based on equation (7) the coefficient a can be solved as follows,

$$a = (A^T A)^{-1} A^T d \tag{8}$$

In fact the only coefficient of interest is a_0, and a_0 is only decided by the first row of $(A^T A)^{-1} A^T$. Furthermore, once the window length is decided, the design matrix A is known, therefore the first row of $(A^TA)^{-1}A^T$ can be decided beforehand. As discussed above, a_0 is in fact g_0 in equation (5). Comparing equation (8) and (5), the elements of the first row of $(A^TA)^{-1}A^T$ in fact correspond to the coefficients c_n in equation (5) when g_0 is being solved. Therefore, in Savitzky-Golay filter, the coefficients in equation (5) can be calculated before the filtering is really taking place, which makes the filtering process extremely fast.

Fig.7 shows a typical background fitting using Savitzky-Golay filter with the same fitting window of 101 points as the moving window average in Fig.5. It is obvious that with Savitzky-Golay filter, the fitted curve matches exactly what the real background should be at any points.

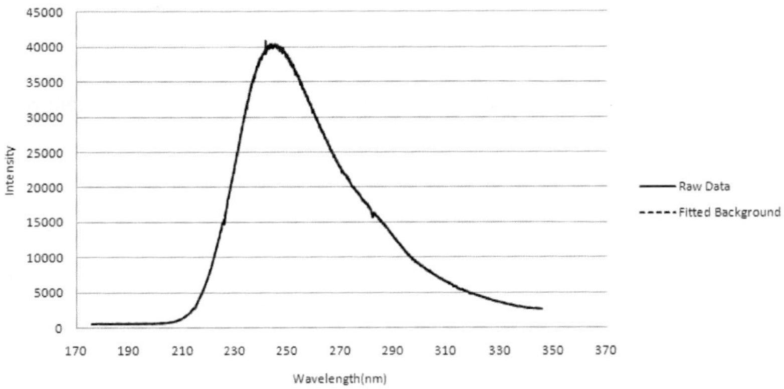

Fig. 7. Background fitting using Savitzky-Golay filter with window width of 101 points

Fig.8 shows the calculated differentials with moving window average and Savitzky-Golay filter. The solid line is the differential with Savitzky-Golay filter and the dashed line is that of moving window average. It can be seen that with Savitzky-Golay filter, there are only narrow features in the differential waveform, while with moving window average there are obvious extra broad signatures as well due to the fact that the fitted background does not match the real shape the background.

Fig.9 compares the two differentials in frequency domain. From the figure it can be seen that the differential of moving window average has higher magnitude at both high and low frequencies. The higher magnitude is mainly caused by the broad features of the differential signal due to the fitting algorithm's inability to match the shape of the background. As the fitted background with moving window average is lower than the actual shape of the background, in the differentiating process, all the peaks in the raw data will be left with their full depth, while with Savitzky-Golay filter, as the fitted curve matches exactly the real background, and runs through the bottom of the peaks, therefore, the bottom part of some of the peaks will be chopped away, which is why lower magnitude is seen in the differential of Savitzky-Golay filter at higher frequencies. Usually, peaks at higher frequencies are either caused by interferences or noises. Lower peaks at high frequencies mean lower noises or interferences. Therefore, lower noises or interferences can be expected with Savitzky-Golay filter.

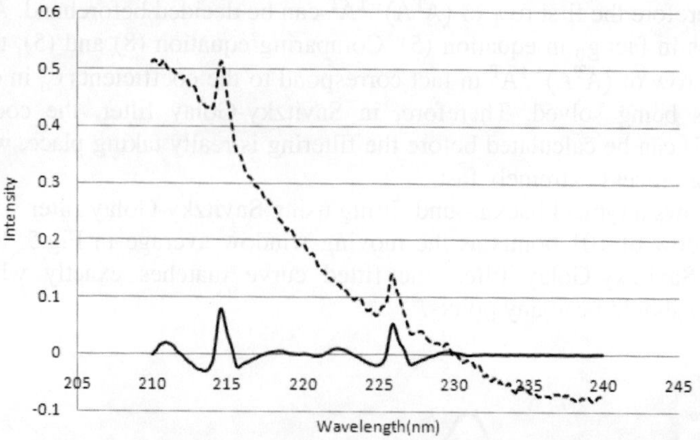

Fig. 8. Comparison of the differentials with Savitzky-Golay filter and moving window average

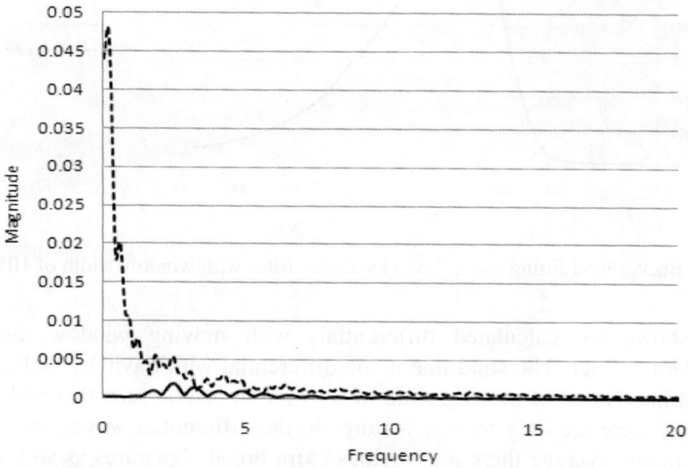

Fig. 9. Comparison of the differentials in frequency domain with Savitzky-Golay filter and moving window average

7 Selection of Window Width

As discussed above, the advantage of Savitzky-Golay filter is that it can match the real broadband feature better than moving window average, but with low window width, there is risk that contrary to moving window average it is so lightly averaged or under fitted that it matches the shape of the raw data, rather than the broadband background. Therefore choosing the right window width is vitally important to the fitting process.

As discussed above, the fitting algorithm should be as smooth as possible and should have the ability to match the exact shape of the background as well. As

Savitzky-Golay filter utilize the combination of all the points in the moving window with different weights on the points to get the averaged background signal, it is obvious that the more points are used in the filtering, the smoother the fitted background will be.

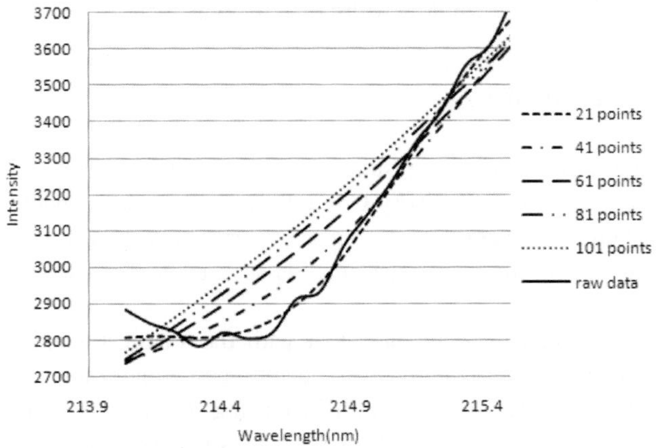

Fig. 10. Fitting of the background with different window width

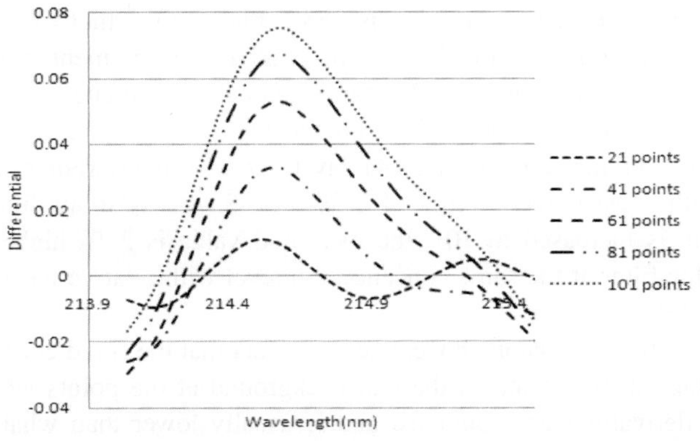

Fig. 11. Fitting of a real peak of differential waveform with different window width

Fig.10 shows the fitting results of Savitzky-Golay filter with different window widths to part of a real intensity waveform. The dent is caused by the absorption of NO. It can be seen that with moving window of 21 points, the fitted curve matches the shape of the raw data rather than the broadband background. Therefore with 21 points, it hardly has the ability to fit the background. With the moving window increasing, the fitted curve approaches the ideal background nearer and nearer. Fig.11

shows the corresponding NO absorption signature fitted out by Savitzky-Golay filter in Fig.10. It can be seen that with 21 points, there is no NO signature at all. With the moving window getting larger and larger, the NO signature in Fig.10 becomes more and more obvious. The fitting with window width of 101 points shows the most significant NO signature. Given that it has been confirmed in Fig.7 above that with 101 points, the fitted curve still matches the real broadband background, it is ideal that the filter should use the window width of 101 points here.

8 Experiment Results

Table1 compares the experiment results of moving window average with different window widths and those of Savitzky-Golay Filter with window width of 101 points in the measurement of NO at different concentration levels. MWA is used in the table to represent moving window average and SG is used to represent Savitzky-Golay filter.

From the table it can be seen that even with the smallest window width of 21 points, the results of moving window average are much higher than those of Savitzky-Golay filter, and with the window width increasing, the results of moving window average also increases dramatically. When 21 points window width is used, at around 100 ppb level, the value of moving window average is 22.8% higher than that of Savitzky-Golay filter with relative measurement error of 18.3% compared with measurement error of only 3.6% for Savitzky-Golay filter. When the window width is increased to 101, the measured value is 258% higher than that of Savitzky-Golay filter at the same concentration level with relative measurement error as high as 254%. At high concentration levels, same dramatic difference is also seen. At concentration level of 1000 ppb, with 21 points window, the result of moving window average is 5.6% higher than Savitzky-Golay filter with measurement error of 4.3% compared with measurement error of 1.2% for Savitzky-Golay filter. When the window width is increased to 101, the measured value is 27% higher than that of Savitzky-Golay filter at the same concentration level with relative measurement error as high as 25.5%.

According to the discussion above, due to the fact that the fitted curve with moving window average does not match the real background at the points where second or higher order derivatives are contained and is usually lower than what it should be, there are extra broad features and higher peaks in the differentials which will lead to the higher values than Savitzky-Golay filter seen in Table1.

Table2 compares the test results of Savitzky-Golay filter with different window widths. It can be seen that the more points are used, the bigger values are measured and the higher accuracies are gotten, which is consistent with the discussion in section 7. At all concentration levels it can hardly detect the gas with 21 points due to the heavily under fitted background, while with 101 points, the measurement accuracy can be as high as 3.6% at 100 ppb level and as high as 1.2% at 1000 ppb level as said above benefitting from its well matching to the real background.

Table 1. Comparison of concentration results between moving window average with different window widths and Savitzky-Golay filter with window width of 101 points

MWA 21 points	MWA 41 points	MWA 61 points	MWA 81 points	MWA 101 points	SG 101 Points	Gas analyzer
125.00	153.89	211.29	286.09	364.89	101.80	105.64
235.58	263.47	318.63	392.98	472.65	208.96	214.37
332.02	359.06	412.08	486.36	566.61	302.03	308.92
432.58	458.92	509.29	583.82	664.73	398.87	406.27
543.61	567.92	616.32	690.48	772.50	506.18	514.98
647.03	670.42	716.40	790.45	873.01	606.55	615.26
742.88	766.41	809.75	883.91	967.08	699.12	711.02
850.44	873.68	914.57	988.73	1072.71	803.43	815.06
951.80	973.48	1012.78	1086.37	1171.21	900.89	913.11
1057.82	1077.16	1114.06	1188.05	1272.62	1001.87	1014.18

Table 2. Comparison of concentration results of Savitzky-Golay Filter with different window widths

SG 21 points	SG 41 points	SG 61 points	SG 81 points	SG 101 points	Gas analyzer
2.22	33.32	71.16	96.49	101.80	105.64
4.38	53.86	123.51	176.86	208.96	214.37
6.14	70.94	168.30	246.62	302.03	308.92
7.93	87.64	213.67	318.59	398.87	406.27
10.00	106.27	263.71	398.67	506.18	514.98
11.73	123.31	310.88	472.31	606.55	615.26
13.43	139.40	354.26	540.23	699.12	711.02
15.61	159.81	403.82	617.80	803.43	815.06
17.37	178.88	453.24	692.81	900.89	913.11
18.96	196.73	502.92	771.11	1001.87	1014.18

9 Conclusion

Traditional method of moving window average cannot match the broadband features of the intensity waveform ideally and will generate higher peaks for both absorption and interference. Therefore, the measurement result of moving window average is usually higher than the real concentration and has larger measurement error. The more points are used in the fitting, the higher measurement result and larger measurement error are expected. Test results show that at low concentration levels, the measurement error can go as high as 18% with small window and 250% with large window. At high concentration levels, the measurement error can go as high as 4% with small window and 25% with large window.

Compared with moving window average, with Savitzky-Golay filter, the fitted curve matches the real background much better. Therefore, there are no extra broad features and lower interferences in the differentials. However, with small moving windows, the fitted curve will match the raw data better rather than the wanted background and the absorptions of the gas cannot be detected with the under fitted background. Therefore, large moving window should be used. Test results show that the measurement accuracy can be improved to around 3.5% and 1% at low and high concentration levels respectively.

References

1. Bian, J., Ma, L.: Reconstruction of NDVI time-series datasets of MODIS based on Savitzky-Golay filter. Journal of Remote Sensing 14(4), 22–27 (2010)
2. Chen, J., Jonsson, P., Tamura, M., Gu, Z.H., Matsushita, B., Eklundh, L.: A simple method for reconstructing a high quality NDVI time-series data set based on the Savitzky-Golay filter. Remote Sensing of Enviroment, pp. 35-42 (2007)
3. Pundt, I., Mettendorf, K.U.: Multibeam long-path differential optical absorption spectroscopy instrument: a device for simultateous measurements along multiple light paths. Applied Optics 44(23), 4985–4994 (2005)
4. Lee, J.S.: Development of a Differential Optical Absorption Spectroscopy (DOAS) System for the Detection of Atmospheric Trace Gas Species: NO2, SO2, and O3. Journal of the Korean Physical Society 41(5), 693–698 (2002)
5. Slezak, V., Santiago, G., Peuriot, A.: Photoacustic detection of NO2 traces with CW pulsed green lasers. Optics and Lasers in Engineering 40, 33–41 (2003)
6. Martin, P.N.: Measurements of Atmospheric Trace Gases Using Open Path Differential UV Absorption Spectroscopy for Urban Pollution Monitoring. PhD dissertation, Imperial College London (2003)
7. Jin, X., Li, J., Schmidt, C.C.: Retrieval of Total Column Ozone From Imagers Onboard Geostationary Satellites. IEEE Transactions on Geoscience and remote sensing 46(2), 479–488 (2008)
8. Levelt, P.F., Hilsenrath, E., Leppelmeier, G.W.: Science Objectives of the Ozone Monitoring Instrument. IEEE Transactions on Geoscience and remote sensing 44(5), 1199–1208 (2006)
9. Dobber, M.R., Dirksen, R.J., Levelt, P.F.: Ozone Monitoring Instrument Calibration. IEEE Transactions on Geoscience and remote sensing 44(5), 1209–1238 (2006)
10. Wolfram, E.A., Salvador, J., D'Elia, R., et al.: New differential absorption lidar for stratospheric ozone monitoring in Patagonia, South Argentina. Journal of Optics 10(10), 14–21 (2008)
11. Drobnik, M., Latour, T.: Application of the differential absorption UV-VIS spectrum to assay some of humic compounds in therapeutic peats. Rocz Panstw Zakl Hig 60(3), 221–228 (2009)
12. Repasky, K.S., Humphries, S.: Differential absorption measurements of carbon dioxide using a temperature tunable distributed feedback diode laser. Review of Scientific Instruments 77, 113107, 0–5 (2006)

Dark Current Suppression for Optical Measurement of Atmospheric Compositions

Wenjun Li

Bio-Incubator, Bessemer Building,
Imperial College London, Prince Consort Road,
South Kensington, London, UK, SW7 2BP
jutlwj@126.com

Abstract. To suppress the dark current of spectrometer which is one of the main factors that limit the detection limit in the optical measurement of atmospheric compositions based on differential absorption technologies, in this paper, the effect of dark current on the gas retrieval procedure is analyzed, the dependency of dark current on temperature and integration time is tested, and two steps to suppress the dark current are introduced. Firstly, dark current is corrected with temperature based two-point correction and both temperature and integration time based four-point correction. Then running average is applied. Test results show that the detection limit can be improved by 0.6 ppb and 0.9 ppb for two-point and four-point correction respectively. When 10-point running average is applied, further improvement of 1.2 ppb can be gotten. With the increasing of the moving window, further improvement can be expected. Generally speaking, the measures introduced in this paper can decrease the dark current and improve the detection limit dramatically.

Keywords: optical measurement, atmospheric composition, dark current suppression, running average.

1 Introduction

In optical atmospheric measurement, differential absorption spectroscopy has long been used as an ideal method to detect atmospheric compositions with high measurement accuracy to ppb level. Dark current of spectrometer is one of the main noises in gas measurement technologies based on differential absorption technologies. It is one of the most significant factors that limit the ability of differential absorption system to improve the detection limit. The existence of dark current will influence the calculated differentials and therefore affect the system's ability to detect low concentration gases. It is assumed that dark current is dependent on temperature and integration time. By modelling the dependency of dark current on temperature and integration time, it is assumed that dark current can be corrected based on temperature and integration time and can be filtered with low pass filters.

M. Zhu (Ed.): Electrical Engineering and Control, LNEE 98, pp. 481–490.
springerlink.com © Springer-Verlag Berlin Heidelberg 2011

2 Principle of Differential Absorption Technology and the Retrieval Method

2.1 Lambert-Beer law

In a typical optical gas detection system as Fig.1 shows, the light from the lamp will be absorbed by the samples when it pass through a space with length L which contains the samples. The relationship between the lamp intensity $I_0(\lambda)$ and the received intensity $I(\lambda)$ by the detector follows Lambert-Beer law,

$$I(\lambda) = I_0(\lambda)\exp\left[-\sigma(\lambda)cL\right] \tag{1}$$

In which $\sigma(\lambda)$ is the cross section of the species and c is concentration of the species.

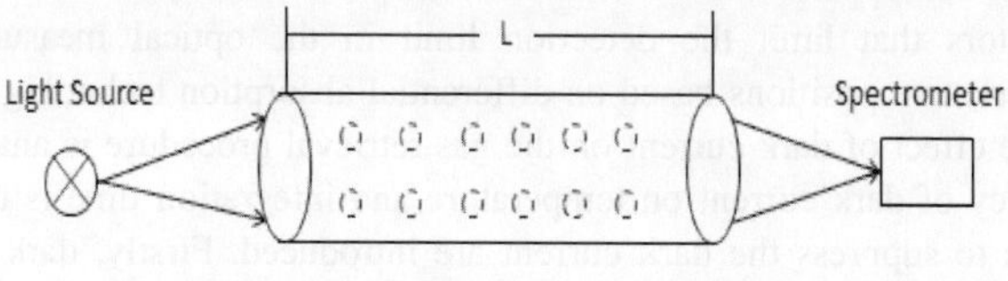

Fig. 1. Schematic of a typical optical gas detection system which makes use of Lmabert-Beer Law

2.2 Retrieval Method

It is very difficult to make use of Lambert-Beer law directly as the intensity of the light source is usually unknown. Most light source shows broad band features compared with the narrow band features of the target samples. Therefore it is assumed that the broad band lamp features can be fitted by a polynomial $I_P(\lambda)$,

$$I_P(\lambda) \approx I_0(\lambda) \tag{2}$$

Then apply logarithm on both sides of equation (1),

$$\log\left(\frac{I_P(\lambda)}{I(\lambda)}\right) = \sigma(\lambda)cL = A(\lambda) \tag{3}$$

In case of multi-species,

$$A(\lambda) = L\sum \sigma_i(\lambda)c_i(\lambda) \tag{4}$$

$A(\lambda)$ is called the absorption or optical depth. Once the absorption is worked out, the concentrations of the target species can be retrieved with equation (3) and (4).

3 Influence of Dark Current on Retrieval Result

Dark current will be added in the intensity signal and taken as the absorption of gases by the algorithm. The mean value of the dark current can be corrected. Therefore it is the residual dark current that finally affects the retrieval results most significantly. A typical dark current waveform is shown in Fig.2, and the residual dark current after correction with standard deviation 10 is shown in Fig.3.

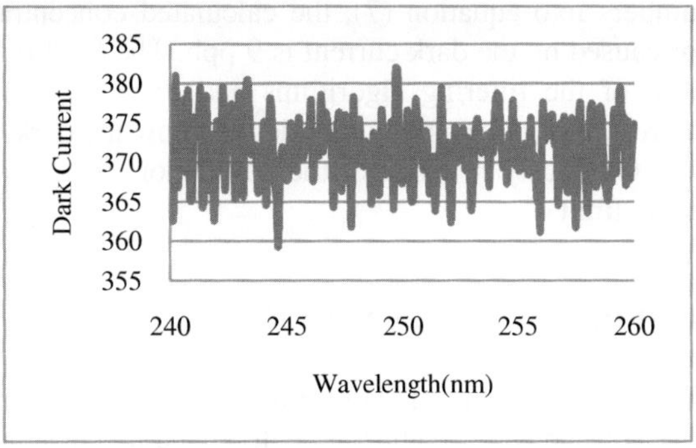

Fig. 2. Typical dark current waveform

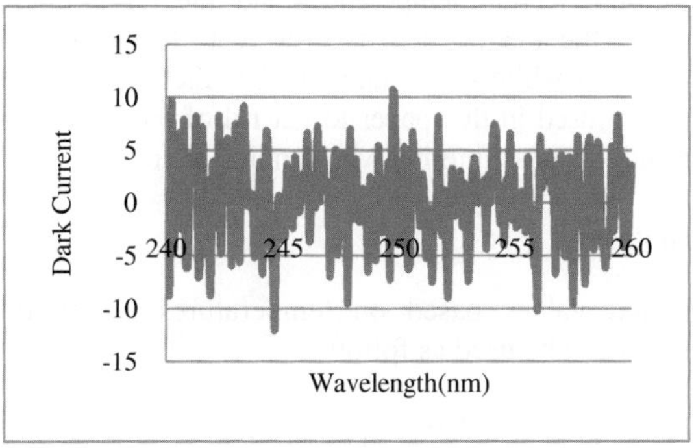

Fig. 3. Residual dark current after correction

If the standard deviation of a dark current waveform is $\delta(\lambda)$,and the smoothed signal is $I_P(\lambda)$, then according to equation (4),

$$\log\left(\frac{I_P(\lambda)}{I_P(\lambda)+\delta(\lambda)}\right) = L\sum \sigma_i(\lambda)c_i(\lambda) \tag{5}$$

If there is only one gas in measurement, the above function becomes,

$$\log\left(\frac{I_P(\lambda)}{I_P(\lambda)+\delta(\lambda)}\right) = L(\lambda)\sigma(\lambda)c \tag{6}$$

The measurement error caused by the dark current can be written as

$$c = \frac{\log\left(\frac{I_P(\lambda)}{I_P(\lambda)+\delta(\lambda)}\right)}{L\sigma(\lambda)} \tag{7}$$

Assuming NO is being measured, δ(λ)=10, and the intensity of the signal is 40000, L=15m, we will have the initial idea of how dark current would affect the retrieval result by looking at the measurement result under one wavelength, for example, λ=211nm. The cross section of NO at wavelength 211nm is 5.66E-19cm2/molecule. Take all these numbers into equation (7), the calculated concentration value or the measurement error caused by the dark current is 9 ppb. The final result may be not as bad as this because of the filtering algorithms and the least square fitting on all wavelengths of interest, but it does give us some idea how the dark current will affect the retrieval results. Obviously it is one of the key factors that hamper the efforts to improve the detection limit.

4 Methods to Suppress Dark Current

It is assumed that the mean value of dark current is dependent on temperature and integration time linearly, and can be calibrated beforehand by changing temperature and integration time when there is no entry light. The simple way to remove dark current is to get the linear interpolated value based on the current temperature, integration time and the calibrated values under certain temperatures and integration times, and then subtract the intensity signal by the interpolated dark current. The residual dark current after correction can be seen as white noise, and can be filtered out with a low pass filter. Therefore two steps are used in this paper to get rid of the dark current, correction via linear interpolation and low pass filtering with running average.

4.1 Linear Interpolation

If only the calibrated values based on temperatures are used, two-point linear interpolated correction can be used as follows,

$$DC = DC_1\left(\frac{T-T_2}{T_1-T_2}\right) + DC_2\left(1-\frac{T-T_2}{T_1-T_2}\right) \tag{8}$$

In which DC_1 is the dark current at temperature T_1 and DC_2 is the dark current at temperature T_2.

If calibrated values based on both temperature and integration time are used, four-point linear interpolated correction can be implemented as follows,

$$DC_{IA} = DC_{1A}\left(\frac{T-T_2}{T_1-T_2}\right) + DC_{2A}\left(1-\frac{T-T_2}{T_1-T_2}\right) \tag{9}$$

$$DC_{IB} = DC_{1B}\left(\frac{T-T_2}{T_1-T_2}\right) + DC_{2B}\left(1-\frac{T-T_2}{T_1-T_2}\right) \tag{10}$$

$$DC = DC_{IA}\left(\frac{I-I_B}{I_A-I_B}\right) + DC_{IB}\left(1-\frac{I-I_B}{I_A-I_B}\right) \tag{11}$$

In above equations all the items with subscript *A* are the values at integration time I_A, and the items with subscript *B* are the dark current at integration time I_B.

4.2 Running Average

Linear interpolation cannot remove dark current completely. The residual dark current after correction will work as noise in the intensity spectrum and will affect the measurement accuracy by being taken into the differential signal. Low pass filters are needed to filter out the dark current. Here running average is used. The principle of overlapping average is as follows.

Suppose there are n data points, $d_0, d_1, d_2 \ldots\ldots d_n$ in a data set, the mean value of the data set is

$$RA_n = \frac{1}{n}\sum_{i=0}^{n-1} d_i \tag{12}$$

The filtered RA_{n+1} of the successive point d_{n+1} can be calculated by adding the new point d_{n+1} to the sum and dropping the old d_n out of the sum as follows,

$$RA_{n+1} = RA_n - \frac{d_n}{n} + \frac{d_{n+1}}{n} \tag{13}$$

Here the number of the data set n is called the running window. Running average in fact is a type of finite impulse response filter and the running window decides the filtering effect.

5 Experiment Results and Analysis

5.1 Linearity of Dark Current against Temperature and Integration Time

Fig.4 shows the dark current values of one pixel under different temperatures from 10°C to 40°C. The solid line is the raw data and the dashed line is the trend line.

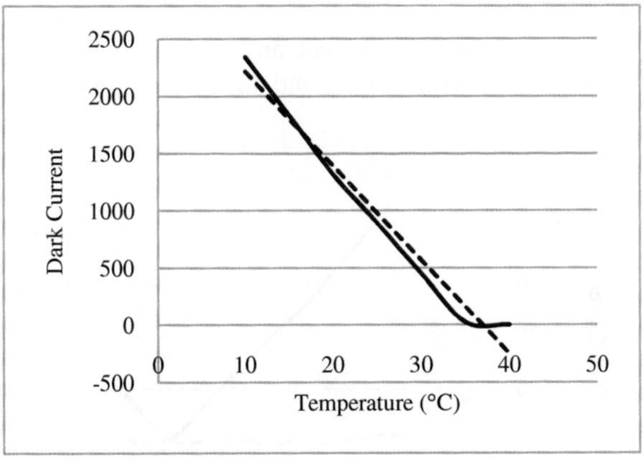

Fig. 4. Dark current values of one pixel under different temperatures

Between 10°C and 35°C, it is clear that generally the dark current decreases linearly against temperature with slight deviation, as the dashed trend line shows. The dark current value falls to zero when the temperature goes higher than 35°C, and does not need to be corrected any more.

Fig.5 shows the dark current changes of three pixels against integration time respectively. The linearity of dark current against integration time is not significantly obvious. Compared with the dark current changes against temperature, there are no dramatic changes observed against integration time.

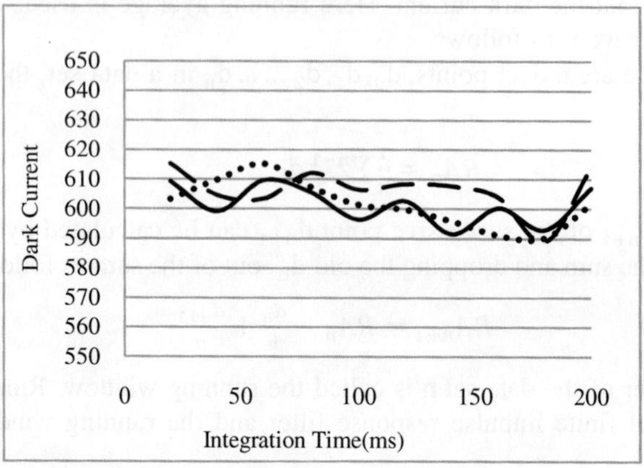

Fig. 5. Dark current changes against integration time

5.2 Residual Dark Current after Correction

Fig.6 shows the residual dark current of one pixel after two-point temperature based correction. The solid line is the raw dark current data and the line with square dot is the residual dark current after correction. When the temperature is below 35°C, it shows good correction result. The residual dark current drops to very low level, with mean value 8. When temperature goes higher than 35°C, because the dark current drops to zero, there is no need to correct any more, which is why we see high correction error in the range between 35°C and 40°C in Figure 6.

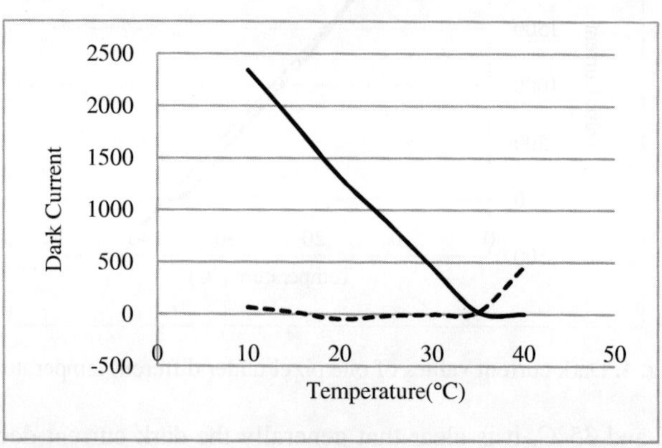

Fig. 6. Residual dark current of one pixel after temperature correction

Fig.7 shows the comparison of residual dark current of one pixel with two-point temperature based correction and four-point temperature and integration time based correction. The solid line is the raw dark current data. The dash line is the residual dark current after four-point correction and the round dot is the residual dark current after two-point correction. Compared with two-point correction, four-point correction lowers down the residual dark current further to negative level with mean value -20 at temperatures between 20°C and 25°C, and keeps the level of around 0 with mean value 3 at temperatures between 25°C and 30°C. The comparison shows that, generally speaking, the four-point correction decreases the dark current further although the linearity of dark current against integration time is not significantly obvious.

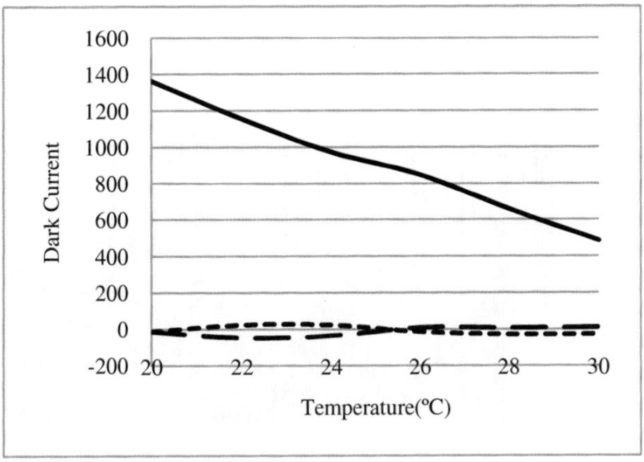

Fig.7. Comparison of residual dark current after two-point and four-point correction

5.3 Filtering Results with Running Average

Fig.8 shows the standard deviation of the residual dark current after filtering against the width of running window. It is obvious that the more points are used in the running average, the smaller the standard deviation is. Fig.9 compares the dark current before and after filtering with running average of 30 points in frequency domain. The dashed line is the original dark current and the solid line is the residual dark current after filtering. It is obvious that the magnitude of the residual dark current is much lower than the original dark current at higher frequencies, which means the dark current noise at high frequency is decreased dramatically. Figure10 compares the dark current after filtering with running average of 30 points and 50 points respectively in frequency domain. The solid line is the residual dark current of running average with 50 points and the dashed line is that of running average with 30 points. It can be seen that with 50 points, the dark current at higher frequency is decreased further compared with 30-point running average.

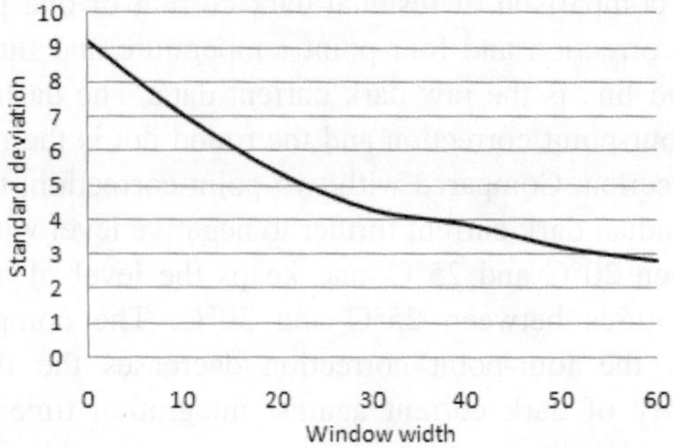

Fig. 8. Changes of standard deviation of the residual dark current against the window width of running average

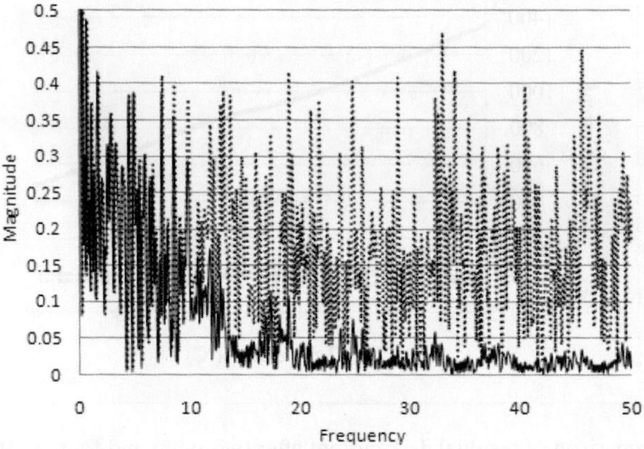

Fig. 9. Comparison of the dark current before and after filtering with running average of 30 points in frequency domain

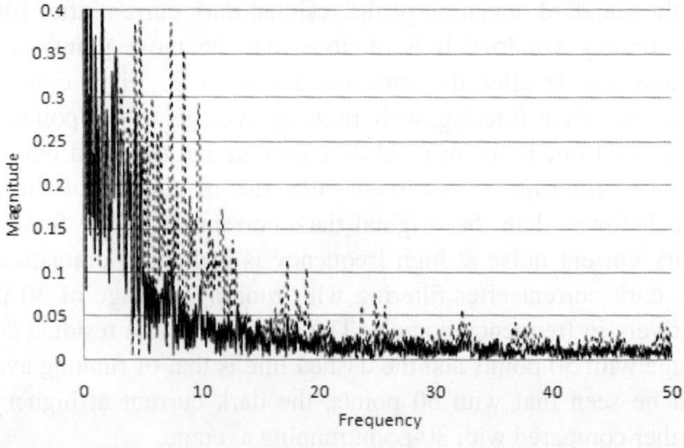

Fig. 10. Comparison of the dark current after filtering with running average of 30 points and 50 points in frequency domain

5.4 Measurement Results of Concentrations

Table 1 compares the measured concentration values of NO at very low concentration levels with two-point correction, four-point correction and running average with different window widths. RA is used to represent running average in the table. A gas analyzer is used to measure the NO concentrations and functions as the reference for other measurement results.

From the table, it can be seen that when the measured results with two-point temperature based correction is about 0.6 ppb higher than the results when no correction is applied, which means the detection limit is improved by around 0.6 ppb with the correction. Comparing the results of four-point temperature and integration time based correction and the two-point temperature based correction, the detection limit is improved further by around 0.3 ppb. The relatively small improvement for detection limit with linear interpolation is because the standard deviation of the dark current is not decreased significantly. When 10-point running average is applied, the detection limit is improved instantly by around 1.2 ppb because of the obvious decreasing of the standard deviation of the dark current. With more points being used, as more dark current is filtered out at higher frequencies and the signal to noise ratio increases accordingly, further detection limit improvements of around 1.7 ppb and 1.3 ppb are obtained with 30-point running average and 50-point running average respectively.

Table 1. Comparison of concentration results (unit: ppb)

No correction	Two-point correction	Four-point correction	10-point RA	30-point RA	50-point RA	Analyzer
-0.8	-0.17	0.18	1.33	3.12	4.43	4.77
-0.3	0.34	0.68	1.77	3.54	4.79	5.06
0.15	0.78	1.12	2.29	4.03	5.32	5.55
0.68	1.31	1.65	2.83	4.59	5.95	6.18
0.86	1.82	2.17	3.55	5.26	6.62	6.92
2.01	2.63	2.98	4.02	5.78	7.10	7.51
3.48	4.10	4.45	5.62	7.44	8.68	9.02
5.17	5.79	6.14	7.37	9.09	10.40	10.88
6.78	7.40	7.75	9.16	9.88	11.22	11.70
7.62	8.24	8.59	9.88	11.49	11.90	12.16
8.83	9.50	9.77	10.91	12.68	13.88	13.21
9.77	10.35	10.63	11.85	13.57	14.79	15.18
10.80	11.46	11.74	13.02	14.72	16.03	16.57
11.91	12.62	12.88	14.21	15.89	17.25	17.69

6 Conclusion

Generally speaking, the methods introduced in the paper to reduce dark current are effective. Temperature and integration time based correction does not decrease the standard deviation of the dark current significantly. Therefore small improvement of detection limit is expected, with improvements of 0.6 ppb and 0.3 ppb for two-point and four-point correction respectively. Running average can decrease the standard deviation of dark current dramatically because of its filtering ability. Detection limit improvement of around 1.2 ppb can be expected with 10-point running average. With the increasing of the moving window, further improvement for detection limit can be expected as more dark current at high frequencies is filtered out and the signal to noise ratio is increased. Test results show that further detection limit improvements of around 1.7 ppb and 1.3 ppb can be obtained with 30-point running average and 50-point running average respectively.

References

1. Onat, B.M., Huang, W., Masaun, N.: Ultra Low Dark Current InGaAs Technology for Focal Plane Arrays for Low-Light Level Visible-Shortwave Infrared Imaging. In: Proc. of SPIE Infrared Technology and Applications, vol. 6542, pp. 65420–1-9 (2007)
2. Widenhorn, R., Dunlap, J.C., Bodegom, E.: Exposure Time Dependence of Dark Current in CCD Imagers. IEEE Trans. on Electron Devices 57(3), 581–587 (2010)
3. Peters, I.M., Bogaart, E.W., Hoekstra, W.: Very-low Dark Current in FF-CCDs. IEEE Trans. on Electron Devices 37, 26–30 (2009)
4. Takayanagi, et al.: Dark current reduction in stacked-type CMOS-APS for charged particle imaging. IEEE Trans.Electron Devices 50, 70–76 (2003)
5. Burke, B.E., Gajar, S.A.: Dynamic Suppression of Interface-State Dark Current in Burried-Channel CCD's. IEEE Trans. Electron Devices 38, 285–290 (1991)
6. Etteh, N.E.I., Harrison, P.: First principles calculations ofthe dark current in quantum well infrared photodetectors. Physica E 13, 381–384 (2002)
7. Manoury, E.-J., et al.: A 36×48mm2 48M-pixel CCD imager for professional DSC applications. Tech. Digest IEDM (2008)
8. Mellqvist, J., Rosen, A.: DOAS for flue gas monitoring-temperature effects in the UV/visible absorption spectra of NO, NO2, SO2, and NH3. J. Spectrose Radiat. Transfer. 56(2), 187–208 (1996)
9. Pundt, I., Mettendorf, K.U.: Multibeam long-path diferential optical absorption spectroscopy instrument: a device for simultaneous measurements along multiple light paths. Applied Optics 44(23), 4985–4994 (2005)
10. Martin, P.N.: Measurements of Atmospheric Trace Gases Using Open Path Differential UV Absorption Spectroscopy for Urban Pollution Monitoring, PhD thesis, Imperial College London (2003)
11. Platt, U., Stutz, J.: Differential Optical Absorption Spectroscopy-Principles and Applications. Springer press, Heidelberg (2008)
12. Jenouvrier, A., Coquart, B.: The NO2 Absorption Spectrum.III:The 200-300nm Region at Ambient Temperature. Journal of Atmospheric Chmistry 25, 21–32 (1996)

Research on Topology Control Based on Partition Node Elimination in Ad Hoc Networks

Jun Liu, Jing Jiang, Ning Ye, and Weiyan Ren

School of Information Science and Engineering,
Northeastern University, 110004, Shenyang, China
liujun@ise.neu.edu.cn, tornadojj@163.com,
yening265@126.com

Abstract. In Ad hoc networks, unreasonable network topology will reduce network capacity, increase packet transmission delay and weaken network robustness. Nodes which connect several separated sub-nets are defined as partition nodes in this paper. Failures of partition nodes will cause network partition. A topology control algorithm focusing on eliminating partition nodes is proposed. Every node judges whether it is a partition node according to reachable relationships collected. The representation nodes chosen from each subset will be connected in the way of chordal ring connection. Some edges are moved away to limit its neighbors when connectivity reaches or exceeds the threshold value. Simulation software NS2 has been used and simulation results demonstrate that the algorithm could enhance the network invulnerability.

Keywords: Ad hoc networks, topology control, partition node.

1 Introduction

With the characteristics such as limited resource, wireless communications and node mobility, the performance of Ad Hoc networks is closely related to topology structure [1]. Unreasonable network topology will lessen network capacity, increase packet transmission delay and reduce network robustness facing to node failures. According to setting special nodes and adjusting node power to affect the links between nodes, the problems about topology control are proposed. However, the method of link control is just one aspect for topology control. In terms of the aims, topology control schemes can be classified as energy efficient topology control, fault tolerant topology control and so on [2-5].

In Ad hoc networks, nodes are equal in function, but in fact that some nodes are special to the whole network [6]. For instance, the nodes act as a row connecting two or more independent subnets are important to the structure of topology. Their failures will cause the network partition. Additionally, the nodes would effect on several properties in dynamic environment, when they become the bottleneck of the network. In this paper, based on eliminating key nodes, a simple, efficient and distributed topology control algorithm has been supposed to optimize the topology in Ad hoc networks.

M. Zhu (Ed.): Electrical Engineering and Control, LNEE 98, pp. 491–498.
springerlink.com © Springer-Verlag Berlin Heidelberg 2011

2 Network Model and Problem Description

Definition 1 (the location relationship). In the network, if node A could find node B by sending a routing message, then the relationship between node A and B is the location relationship, represented by A->B. The location relationship is neither transitive nor symmetrical in common.

Definition 2 (the reachable relationship). In the network, if node A could locate node B and node B could locate node C, then the relationship between node A and node C is the reachable relationship, represented by A->->C. The reachable relationship is transitive, but not symmetrical.

Definition 3 (partition node). In the network, if node C is deleted, its neighbors will be divided into several unreachable subsets, and then node C is a partition node.

Table 1 introduces the variables used.

Table 1. Variables and their meanings

Variables	Meanings
$P_{transmission}(I,J)$	Transmission power of node I to node J
$P_{receiving}(J,I)$	Received power of node J. $P_{receiving}(J,I) = P_{transmission}(I,J) \times \frac{c}{d^{\alpha}}$
$P_{min}(I,J)$	The min-transmission power from node I to J $P_{min}(I,J) = P_{rt}(J) \times \frac{P_{transmission}(I,J)}{P_{receiving}(J,I)}$
$P_{max}(I)$	The max transmission power of node I
$P_{rt}(J)$	The threshold of node J's receiving power

In free space transmission model, constant c is determined by the gain and wavelength of antenna, as well as the wastage of system. However, in two-ray transmission model, it depends on the gain and height of antenna and the wastage of system. Variable d stands for the European distance between the sender and receiver. For α, in free space transmission model it is 2, while in two-ray transmission model it is about 4[7].

The network topology can be denoted as a graph $G(V,E)$ in 2-dimensional or 3-dimensional Euclidean space. V is the set of all nodes. E contains each edge from one node to another if they can directly communicate. The link from I to J can be expressed as $\vec{l}_{ij}$ only when $P_{max}(I) \geq P_{min}(I,J)$. Otherwise, every directional link owns a sole weight, just like $Weight(\vec{l}_{ij})$, which can be ensured by the following rules. Supposing there are two directional links $\vec{l}_{ij}$ and $\vec{l}_{mn}$:

(1) $P_{transmission}(I,J) > P_{transmission}(M,N) \Rightarrow Weight(\vec{l}_{ij}) > Weight(\vec{l}_{mn})$.

(2) $P_{transmission}(I,J) = P_{transmission}(M,N)$ & $i>m \Rightarrow Weight(\vec{l}_{ij}) > Weight(\vec{l}_{mn})$.

(3) $P_{transmission}(I,J) = P_{transmission}(M,N)$ & $i=m$ & $j>n \Rightarrow Weight(\vec{l}_{ij}) > Weight(\vec{l}_{mn})$.

Definition 4: Only when P_{max}(I) and P_{max}(J) ≥ P_{min}(J,I), node J can be called a neighbor of node I.

Definition 5: Node I's neighbor set is NG(i), and i∈ NG(i).

The problems of topology control can be defined formally as follow:

Input: (1) Initial topology $\vec{G} = (V, \vec{E})$

(2) Link $\vec{l_{ij}} \in \vec{E}$, $Weight(\vec{l_{ij}}) = P_{max}(I)$

Output: (1) New topology $\vec{G}' = (V', \vec{E}')$, in which $V' = V$ and $\vec{E}' \subseteq \vec{E}$

(2) New link $\vec{l'_{ij}} \in \vec{E}'$, $Weight(\vec{l'_{ij}}) > Weight(\vec{l_{ij}})$

Goal: (1) Minimize the weight of each link and then $\sum_{\vec{l_{ij}}' \in \vec{E}'} Weight(\vec{l_{ij}}')$

(2) Keep the connectivity level consistenting with initial topology (strongly or weakly).

3 A Novel Topology Control Algorithm

The frame of the algorithm is shown in Figure 1.

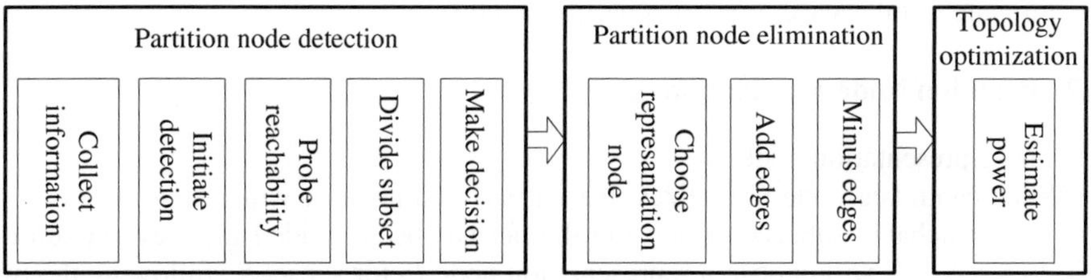

Fig. 1. Three phases of the topology control algorithm based on the key nodes elimination

3.1 Partition Node Detection

a. Collect information.

Node I broadcasts a Hello packet, which includes node I's id, $P_{transmission}$(I,J) and P_{rt}(I), at P_{max}(I) periodically to all the other nodes, like node J. According to these information, node J calculates P_{min}(J,I). If P_{min}(J,I) is smaller than P_{max}(J), node J will put node I into its neighbor set. All the nodes execute this neighbor discovery algorithm singly and asynchronously. Finally, every node will build its neighbor set, NG(i).

b. Initiate detection.

Node I sends a *Msg_Init* to its neighbors (N_1, N_2, ^, N_n) to start the detection. Anyone who receives the message must reply I a *Msg_Response*, which includes its id, bandwidth, delay time, connection degree, transmission power. If I hasn't received the response from N_i, that means node N_i has failed, then node I will delete this neighbor in the routing table.

c. Probe reachability.

Node I sends a *Msg_Probe* to all its neighbors, which includes the address of node I, TTL and its neighbors' id. In Ad hoc networks, every node keeps a "connection list" corresponding to each candidate node. The list includes candidate node's id and neighbors' id. So in the connection list of node N, the details corresponding to candidate node C will tell us the reachable relationships between its neighbors. When a new id N_i appears in the received *Msg_Probe*, node N will send it to other neighbors and add it to the connection list of candidate node C. If there are two or more neighbors' id in the list, the candidate node will receive a *Msg_Arrival* to tell it who are reachable neighbors. The scope of *Msg_Probe* is non-overlapping (just send back a *Msg_Arrival* to inform the reachable relationship at the intersection node), avoiding overlapping reachability probe.

d. Divide subsets.

Candidate node C collects *Msg_Arrival* continually, estimates the reachable relationships between its neighbors and then divides the reachable neighbors into the same subset. In dynamic Ad hoc networks, a few messages may lose, so a timeout value is needed. After timeout, we can regard it as the final division.

e. Make decision.

Candidate node C decides whether it is a partition node by counting how many subsets it has. If there is more than one subset, it is a partition node and then we need to eliminate it; otherwise, it is not a partition node.

3.2 Partition Node Elimination

a. Choose presentation node.

When candidate node C realizes itself a partition node, it will add some edges to merge unreachable subsets. Supposing the net has been divided into several subsets (S_1, S_2,… , S_n), C will choose a representation node P_i for subset S_i, following the two principles below:

(1) Make the node degree higher than the floor to improve fault-tolerant of system (that means do best to improve the minimum connection degree of the nodes).

(2) Converge load factor to develop heterogeneous node according to each node's ability. The calculating formula for load factor will be changed according to services. Set a weight value *w* for every node. Load factor is computed by node degree/*w*. For example, to the business sensitive to bandwidth, *w* is bandwidth; to the business sensitive to time delay, *w* means the distance to goal subset.

Due to the rules upon, partition node C finds P_i whose node degree is the lowest in the subset and also lower than the *Min_Degree*; or else, it will calculate everyone's load factor and choose the lowest one as the representation node.

b. Add edges.

The easiest way to connect these representation nodes is liner chain connection, but it is also the most fragile way. So in this design we propose chordal ring connection. That means representation nodes make a ring firstly, and then add a chord opposite. There're two cases in this phase. If the number of representation nodes is even, node

P_i will be connected to node $P_{i+n/2}$; else node P_i will be connected to nodes $P_{i+(n+1)/2}$ and $P_{i+(n-1)/2}$.

c. Minus edges to limit neighbors.

The number of node connection (node degree) a node can preserve is limited. The more the number is, the more the expense is needed to update self-adaption routing list. This phase must abide by the following principles:

(1) The edges added to eliminating the partition nodes couldn't be cut down
(2) Reckoning the factor of each link, the heaviest one will be cut down. To special business needs, relevant parameters will be considered.

Remark: Principle (1) is prior to Principle (2). Otherwise the network will be divided again because of minus edges.

3.3 Topology Optimization

In each subset, nodes send probe packets to initiate topology detection. The nodes' id, max-transmission power and receiving power threshold are all in the probe packets. When a node receives a probe packet from a neighbor, it will estimate two sides' min-transmission power and broadcast a probe reply packet after a random delay, which includes sponsor id, reply id, max-transmission power and receiving power threshold of reply node, as well as min-transmission power from reply node to sponsor.

A relay node need to meet two requirements: first, its id must be bigger than reply node; second, sponsor node and reply node are both its neighbors. Its duty is to calculate the min-transmission power to reply node. After a random delay, relay node will broadcast a relay packet including sponsor id, reply id, relay id, the min-transmission power from relay node to reply node and the min-transmission power on the other way round. Besides, in order to save expense we always use a bigger node (whose id number is bigger than the others') to relay the probe packets.

During two detection processes, every node keeps a "power list" containing all neighbor links to replace the static min-transmission power. Citing node I, supposing node J is one of its neighbors and its power list includes the id of node J, Prt(J), $P_{\min}$(I,J), $T_{\text{last\_update}}$ and T_{next}. $T_{\text{last\_update}}$ means the recently updated time. T_{next} is the next initiate time. When $T_{\text{next}} < T_{\text{next}}$, transmission power can be computed by equation (1).

$$P_T(I,J) = P_{\min}(I,J) + \frac{P_{\max(I)} - P_{\min(I,J)}}{T_{next} - T_{\text{last\_update}}} \times (T - T_{\text{last\_update}}) \tag{1}$$

Whenever node I receives a packet from node J, it will estimate a new lower minimum transmission power $P_{\min}$(I,J) and update the power list.

3.4 Analysis on Expense

During the process of partition node detecting, every node sends probe packets to different neighbors in non-overlapping areas to avoid repetitive arrival. Supposing there are n nodes in the network, and the mean node degree is m, TTL is equal to t. The detecting expense (the amount of packets needed) of each node can be signified as $\min(O(m^t),O(nm))$, so the total detecting expense is $\min(O(nm^t),O(n^2m))$.

If one node confirms itself a partition node and it has k subsets, then the expense of adding edges will be $O(k)$, no matter in the way of liner chain or chordal ring connection. As k must be less than d *(the node degree of partition node)*, so the expense also can be signified as $O(d)$. In the worst environment, every node is partition node, so the expense will be $O(nm)$. Similarly, the expense of minus edges is $O(nm)$. Therefore, the total expense of partition node elimination is $O(nm)$, much less than that of detection. That is to say, $\min(O(nm^t),O(n^2m))$ is the total expense of the proposed algorithm, mainly generated in the detection process. In addition, the total expense will be $O(n)$ when m and t both are low.

4 Simulation and Performance Analysis

The network simulation software NS2 has been adopted to evaluate the proposed topology control algorithm by setting different node motion scenes. The network contains 1000 nodes and its application layer adopts CBR data flows in which each message owns 512 bytes. Contrasting existing time of partition node in the network, the shorter it is, the stronger the network will be. Packet successfully delivery ratio and control expense in the network are used to analysis the network availability.

4.1 Existing Time of Partition Node

In Fig.2, after partition node elimination the existing time of partition node is obviously shorter than before, proving that the algorithm can efficiently eliminate partition node. But as the dynamic topology is changing all the time, so even after eliminating there will still remain few partition nodes in the network.

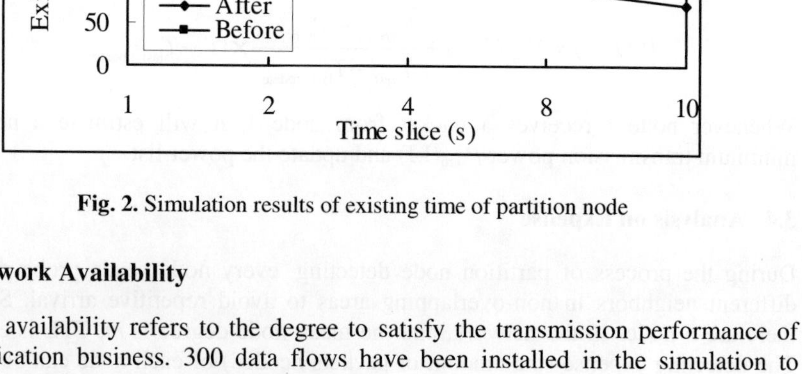

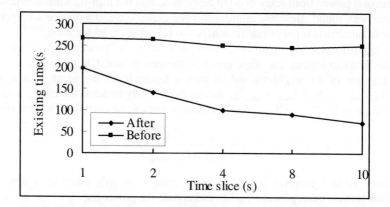

Fig. 2. Simulation results of existing time of partition node

4.2 Network Availability

Network availability refers to the degree to satisfy the transmission performance of communication business. 300 data flows have been installed in the simulation to contrast the performance with changing load.

In Fig.3 and Fig.4, when packet sending rate is low, the packet successfully delivery ratios of the three cases have few differences, but the expense in network after reconfiguration will be much more than before. Because when the network load is light, partition node is difficult to form bottleneck leading to jam, yet, detecting, eliminating partition nodes and optimizing topology will bring much expense.

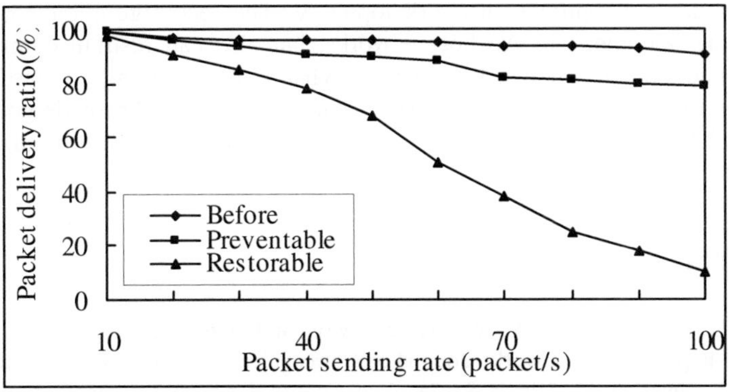

Fig. 3. Packet delivery ratio

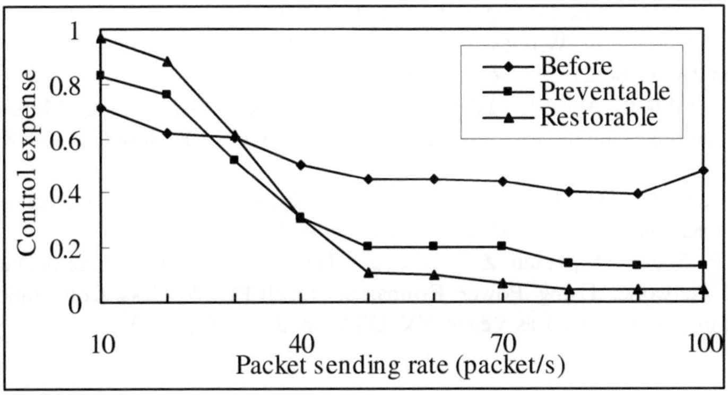

Fig. 4. Control expense

Comparing with before topology reconfiguration and after preventable topology reconfiguration, restorable topology reconfiguration can obtain the highest packet successfully delivery ratio and the fastest speed to decrease expense, along with increasing load in network. When network load is heavy, a mount of partition nodes will be the bottleneck for data transmission business, jamming the network. That will lead to plenty of packets couldn't reach destination nodes and the delivery ratio will reduce fast. At the same time, control expense must sharply increase and network transmission efficiency must be low, because the network is full of path finding requests and response packets caused by rerouting and routing error messages.

5 Conclusions

Partition nodes are the key nodes in Ad hoc networks, which connect several separated sub-nets. Their failures will result in network partition. So a topology control algorithm based on partition node elimination in Ad hoc networks has been proposed. Simulation results demonstrate that it is an efficient algorithm to eliminate partition nodes and optimize network topology. The reconfiguration not only will wipe off the bottleneck in Ad hoc networks, but also can restrain communication jamming and enhance network capacity. What's more, owing to partition node elimination the topology structure becomes balanced. That will provide good basis to routing and load balance, improve communication efficiency and optimize network performances.

References

1. Zhang, L., Zhan, N., Zhao, Y., Wu, W.: Study on Optimum Objective of Topology Control in Multi-hop Wireless Networks. Journal on Communications. 26(3), 117–123 (2005)
2. Bernd, T., Heinfich, M., Ulrich, S.: Topology Control For Fault-Tolerant Communication in Wireless Ad Hoc Networks. J. Wireless Networks 16(2), 387–404 (2010)
3. Seol, J.-Y., Kim, S.-L.: Mobility-Based Topology Control for Wireless Ad Hoc Networks. J. IEICE Transactions on Communications. 17(8), 1443–1450 (2010)
4. Moraes, R.E.N., Ribeiro, C.C., Duhamel, C.: Optimal Solutions for Fault-Tolerant Topology Control in Wireless Ad Hoc Networks. J. IEEE Transactions on Wireless Communications 8(12), 5970–5981 (2009)
5. Rahman, Ashikur, Ahmed, Mahmuda, Zerin, Shobnom.: Analysis of Minimum-Energy Path-Preserving Graphs for Ad-hoc Wireless Networks. J. Simulation 85(8), 511–523 (2009)
6. Li, Z., Chen, G., Qiu, T.: Partition Nodes: Topologically-Critical Nodes of Unstructured Peer-to-Peer Networks. Journal of Software. 19(9), 2376–2388 (2008)
7. Han, X.F., Shen, C.C., Huan, Z.C.: Adaptive Topology Control for Heterogeneous Mobile Ad hoc Networks Using Power Estimation. In: IEEE Wireless Communications and Networking Conference, Las Vegas, NV, USA, pp. 392–399 (2006)

Reaserch on Characteristic of the Flywheel Energy Storage Based on the Rotating Electromagnetic Voltage Converter

Baoquan Kou, Haichuan Cao, Jiwei Cao, and Xuzhen Huang

School of Electrical Engineering and Automation, Harbin Institute of Technology, 150001, Harbin, Heilongjiang Province, China
koubq@hit.edu.cn, haichuan.c@gmail.com, givenmoon@163.com, sangyu317@163.com

Abstract. Rotating electromagnetic voltage converter (REVC) is a new structure voltage conversion device. It has many functions including voltage change, reactive power compensation, harmonics isolation and suppression. Flywheel energy storage device possesses the characteristic of high-power and fast-speed discharge. With loading the flywheel, REVC can output mechanical energy of the flywheel when sudden power failure or voltage sag happens in the power grid, maintain uninterrupted power output to ensure the normal operation of the load. Appling mathematical model, simulation of the working status of the REVC with flywheel is completed, and the characteristics of REVC with Flywheel (REVCF) when a sudden power failure happens in the power grid are summarized. Finally, the prototype experiment proves the correctness of the simulation results.

Keywords: Rotating electromagnetic voltage converter, flywheel energy storage, mathematical model, simulation, characteristics.

1 Introduction

Power transformer is widely used in the application of modern power system. It is power equipment with highly reliable, high-efficiency, which is integrated and modulated by different types of electrical equipment. . New types of voltage conversion device are being developed rapidly to reduce manufacturing costs, improve operating efficiency. [1].

Large inertia flywheel rotating at high speed can store energy in the form of kinetic energy. When in need of emergency to provide energy, flywheel will decelerate as the prime mover to release the stored kinetic energy.

With the characteristics of high energy density, small size, light weight, long life, non-polluting, flywheel energy storage technology is considered the most promising and competitive energy storage technologies, and has very broad application prospects. [2–4].

M. Zhu (Ed.): Electrical Engineering and Control, LNEE 98, pp. 499–508.
springerlink.com © Springer-Verlag Berlin Heidelberg 2011

REVC is a new structure of voltage converter, with dual-winding stator and wound rotor which is equipped with functions of AC voltage transformation and isolation, regulation of reactive power grid side, etc. REVC has a great practical significance and potential in application. [5].

REVC can output mechanical energy of the flywheel when sudden power failure or voltage sag happens in the power grid, maintain uninterrupted power output to ensure the normal operation of the load. It provides enough time for the system to switch to other power supply system, or access to protected status. In addition, REVCF can maintain the load side output voltage when the voltage sag happens in the power grid, and automatically pull into the synchronous operation after the voltage is back to normal. [6].

In this paper, the mathematical model of REVC is derived. The operation state of REVCF is simulated and analyzed by Matlab software. the research on the prototype can prove the validity of the flywheel energy storage in practical applications is proved.

2 Operating Principle of REVC

REVC is an energy converter which is composed by two PWM power inverters and a synchronous motor with dual-winding stator. REVC has two sets of mutually insulated stator windings. One of them connects to the line side and the other connects to the load side. Rotor windings connect to AC of power converter VSC1 through the slip ring and the electric brush. Large inertia flywheel directly connects to the REVC the rotor shaft. The basic configuration of REVC is shown in Figure. 1.

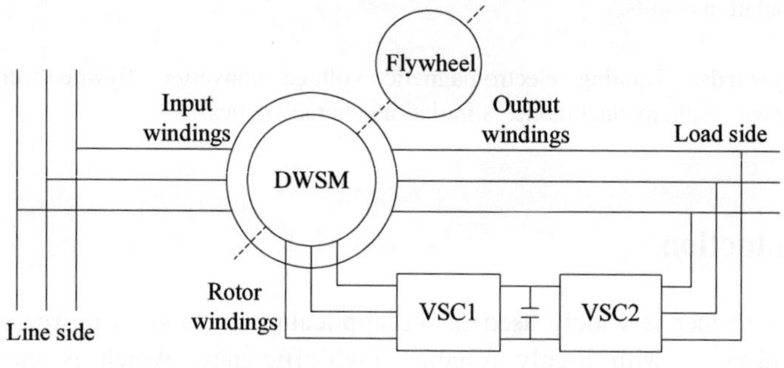

Fig. 1. The basic structure of REVCF.

During normal operation, DC excitation current flowing through the rotorand rotor rotating at synchronous speed. Meanwhile, the flywheel can rotate at synchronous speed, the energy stored in the form of kinetic energy. When sudden power failure or voltage sag happens in the power grid, flywheel releases the stored energy and keep the speed down slowly, and the main field is produced by the DC excitation current of rotor. Continuous outputing power to load, it can maintain the load within a certain time of normal operation.

3 The Mathematical Model of REVC

For analyzing the performance of REVC under condition of loading flywheel or not, research on mathematical model and the parameters of REVC is necessary. Because REVC work in the synchronous state, mathematical model on the dq coordinate is adopted in the analysis [7, 8]. Ignored the motor leakage inductance, inductance parameter matrix can be obtained.

$$\left[L_{dq}\right]=\begin{bmatrix} L_{11d} & 0 & M_{12d} & 0 & M_{1fd} & 0 \\ 0 & L_{11q} & 0 & L_{12q} & 0 & M_{1fq} \\ M_{21d} & 0 & L_{22d} & 0 & M_{2fd} & 0 \\ 0 & M_{21q} & 0 & L_{22q} & 0 & M_{2fq} \\ M_{f1d} & 0 & M_{f2d} & 0 & L_{ffd} & 0 \\ 0 & M_{f1q} & 0 & M_{f2q} & 0 & L_{ffq} \end{bmatrix} \tag{1}$$

Where subscript 1 is the variable of the input winding, and subscript 2 is the variable of the output windings.

Voltage equations can be obtained from the Park equation

$$[u_{dq}]=[L_{dq}]\mathrm{p}[i_{dq}]+[\mathrm{A}][L_{dq}][i_{dq}]+[R][i_{dq}] \tag{2}$$

Where:

$$[\mathrm{A}]=\begin{bmatrix} 0 & -\omega_e & 0 & 0 & 0 & 0 \\ \omega_e & 0 & 0 & 0 & 0 & 0 \\ 0 & 0 & 0 & -\omega_e & 0 & 0 \\ 0 & 0 & \omega_e & 0 & 0 & 0 \\ 0 & 0 & 0 & 0 & 0 & \omega_e-\omega \\ 0 & 0 & 0 & 0 & -(\omega_e-\omega) & 0 \end{bmatrix} \tag{3}$$

$$i_{dq}=[i_{1d} \quad i_{1q} \quad i_{2d} \quad i_{2q} \quad i_{fd} \quad i_{fq}]^{\mathrm{T}} \tag{4}$$

$$u_{dq}=[u_{1d} \quad u_{1q} \quad u_{2d} \quad u_{2q} \quad u_{fd} \quad u_{fq}]^{\mathrm{T}} \tag{5}$$

The equation of state of the phase current can be obtained from equation (2).

$$\mathrm{p}[i_{dq}]=[L_{dq}]^{-1}\cdot\left\{[u_{dq}]-\left([\mathrm{A}]\cdot[L_{dq}]+[R]\right)\cdot[i_{dq}]\right\} \tag{6}$$

And electromagnetic torque is as shown:

$$T_e = p_n\left(\Psi_{1d} i_{1q} - \Psi_{1q} i_{1d} + \Psi_{2d} i_{2q} - \Psi_{2q} i_{2d}\right) \tag{7}$$

Where: p_n is the number of pole pairs of REVC.

Rotor mechanical equations is

$$p\Omega = \frac{1}{J}(T_e - T_L) \tag{8}$$

Where: $\Omega = p\omega$ is the mechanical angular velocity of rotor. $J = J_r + J_{fw}$ is the total moment of inertia of rotor and flywheel.

Equation (6) and (8) is the mathematical model on dq coordinates of REVC.

4 Simulation of the Mathematical Model of REVCF

Mathematical model of REVC is a first-order differential equations with current and speed as state variable. So it can be simulated by S-Function in Matlab Software. Prototype parameters are presented as fellow: Voltage is 340V/220V, 1100VA. Moment of inertia of rotor is $J_r = 0.04\text{kg}\cdot\text{m}^2$. Moment of inertia of flywheel is $J_{fw} = 0.316\text{kg}\cdot\text{m}^2$. System block diagram of S-Function is shown in Figure 2, there U_1 is the input voltage, U_f is the excitation voltage of rotor, J_{fw} is the moment of inertia of the flywheel.

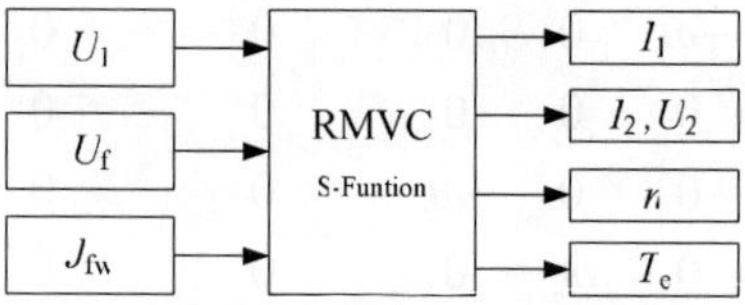

Fig. 2. The system block diagram of REVCF.

First REVCF is started then stabilized at synchronous speed. Rated DC excitation current $I_{fN} = 14\text{A}$ flow thought rotor winding, and the three-phase rated load is connected to output side. Applying $U_1 = 0\text{V}$ to mimic the case of sudden power failure in the power grid at $t = 4$ s, the decline curve of speed is obtained in Figure. 3.

It can be seen, after the sudden power failure in the power grid, REVCF is equivalent to an excitation generator, in which main field is provided by the rotor excitation current, and energy is provided by the kinetic energy which stored in flywheel. With mechanical energy transforming into electrical energy, the rotor speed decreased. The speed and moment of inertia of flywheel determines the amount of energy stored in flywheel. A larger-inertia flywheel should be used because in REVCF the flywheel has low speed.

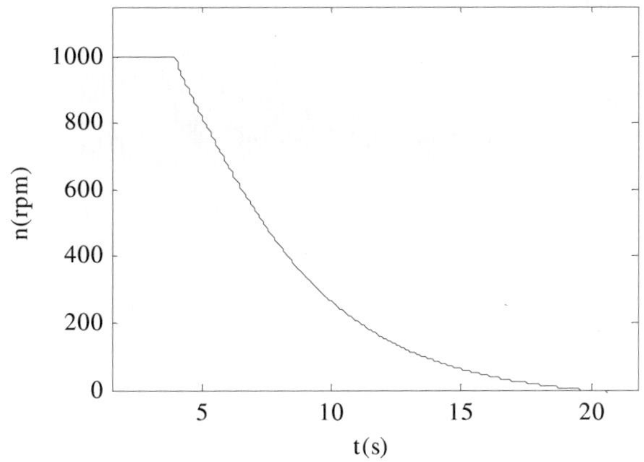

Fig. 3. The curve of rotor speed of REVCF when sudden power failure in the power grid.

Simulation on sudden power failure is shown in Figure. 4. It can be seen that at 10th electrical cycle after power failure the output current is reduced to 80% of rated current, then decrease slowly and can keep a long time. REVCF can provide enough time for the system to switch to backup power supply or other protective measures.

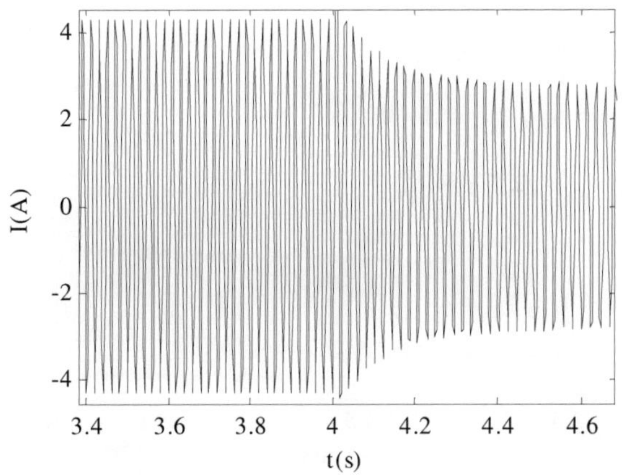

Fig. 4. The waveform of output current of REVCF when sudden power failure in power grid.

When voltage sag happens in the power grid, input voltage waveform of REVCF is shown in Figure 5. It can be seen that the time of voltage sag is 0.25s. When voltage sag, Input induced voltage which is generated by the rotor magnetic field should drop against rotor speed decreased. When the power grid is restored, the input voltage can return to the grid voltage.

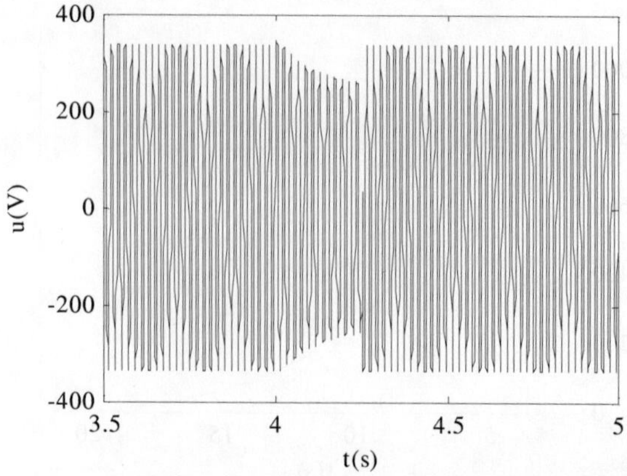

Fig. 5. The curve of input voltage of REVCF when voltage sag happens in the power grid.

When voltage sag happens in the power grid, output current waveform of REVCF is shown in Figure 6. Then the grid voltage is restored, synchronous motor was forcibly led into the power grid because DC rotor excitation current, so the output current waveform has a wave process, then stabilize after about 1 second.

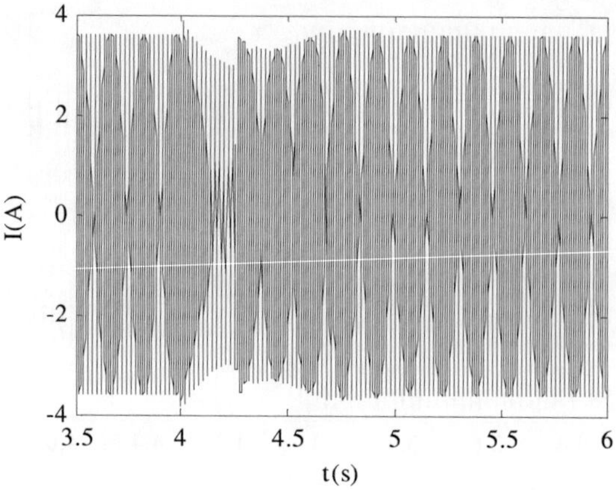

Fig. 6. The waveform of output current of REVCF when voltage sag happens in the power grid.

5 Experiment of REVCF

A prototype of REVC and a large inertia of the flywheel are manufactured. The REVCF system is established and experimented. Experimental platform is shown in Figure. 7. Input winding of prototype is connected to the three-phase AC voltage regulator as the input voltage source, and the load is three-phase symmetrical resistive load. The rotor excitation current is provided by a programmable DC power supply. The rotor winding has two states as short connected and DC excitation, which can be shifted by single-pole double-throw switch.

Fig. 7. The Experimental platform of REVCF.

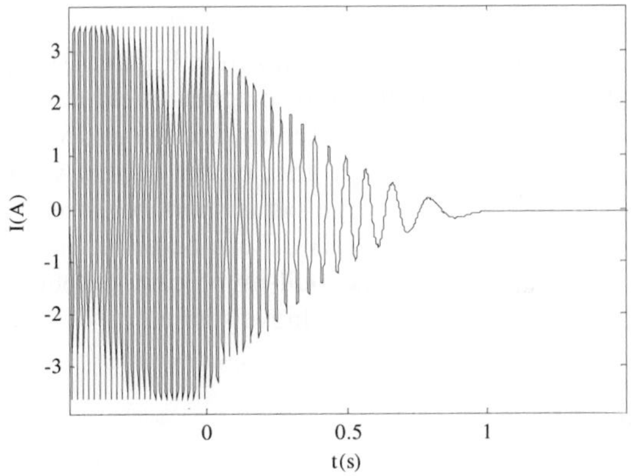

Fig. 8. The waveform of output current of REVC when sudden power failure in power grid.

With no load flywheel, the output current when sudden power failure is tested. The result is shown in Figure 8.

It can be seen that the output current fall rapidly to zero in 1 second because the moment of inertia of the REVC rotor itself is smaller and the stored energy is also smaller. The friction torque of slip ring and brush is another reason that speed quickly decline.

With flywheel, the output current waveform of REVCF when sudden power failure is shown in Figure. 9. It can be seen that until 1s or 50th electrical cycles the output current decreased to only 80% of rated current, so it can provide the time required by the protection of the system intervention.

When REVCF loaded flywheel, the total moment of inertia of the rotor is 8.9 times than that without flywheel. This can greatly extended the time for maintaining the load power.

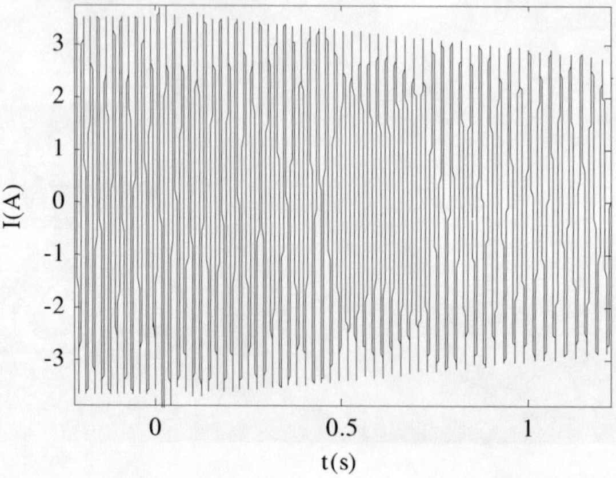

Fig. 9. The waveform of output current of REVCF when sudden power failure in power grid.

The input voltage waveform of REVCF is shown in Figure. 10 when sudden power failure. Input voltage is generated by the rotor magnetic field at this condition. If the power grid is restored, the input voltage has returned to the grid voltage.

When voltage sag happens in the power grid, output current waveform of REVCF is shown in Figure 11. If the time of voltage sag is short, because the drop of REVCF speed is not much and slip is too small, REVCF can be pulled into synchronous after the grid voltage is restored. During the pull-in process output current would be oscillation during about 2 second and then achieve stable.

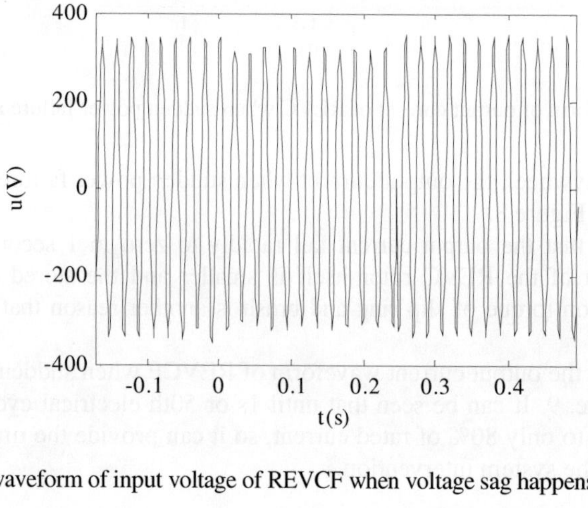

Fig. 10. The waveform of input voltage of REVCF when voltage sag happens in the power grid.

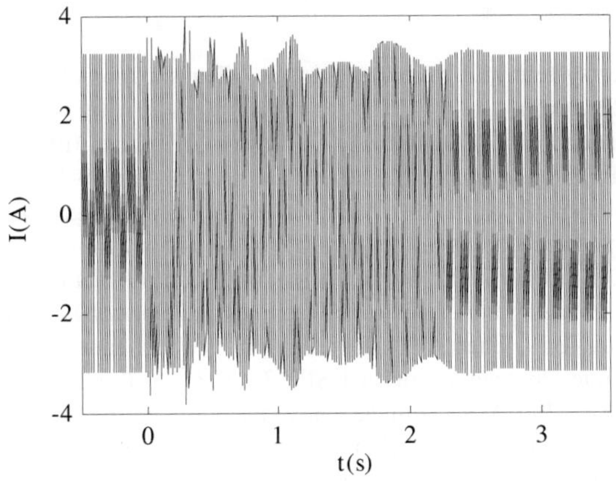

Fig. 11. The waveform of output current of REVCF when voltage sag happens in power grid.

But at the same condition, REVC without flywheel cannot pull into synchronous because speed of REVCF drops rapidly. This result proves advantages of REVCF.

As in the previous simulation, the core non-linear, saturation, friction torque and other factors are ignored, so the results are slightly different from experiment. But the overall trend is the same as the experiment result. This also shows that the mathematical model needs to be further improved.

6 Conclusion

The REVCF is proposed in this paper which has the function to maintain power of the load adopting storage energy in flywheel, achieve another function of REVC besides function for voltage transformation, reactive power compensation, harmonic suppression and isolation.

The mathematical model of REVCF is simulated and a prototype of REVCF is manufactured. And this paper analyzes the performance of REVC under condition of loading flywheel or not.

The results show that until 10th electrical cycles after sudden power failure the output current decreased to only 80% of rated current. When voltage sag happens in the power grid, REVCF can maintain the output current for a short time, and automatically pull into synchronization after the grid voltage is restored.

Acknowledgments. The work described in this paper was supported by Natural Science Foundation of China (No. 50777011) and National High Technology Research and Development Program of China (No.2009AA05Z446). Their powerful supports are grateful.

References

1. Ma, W.M.: Power system integration technique. J. Transactions of China Electrotechnical Society. 20, 16–20 (2005)
2. Singh, B., Al-Haddad, K., Chandra, A.: A review of active filters for power quality improvement. IEEE Transactions on Industrial Electronics 46, 960–971 (1999)
3. Ribeiro, P.F., Johnson, B.K., Crow, M.L., et al.: Energy storage systems for advanced power applications. Proceedings of the IEEE 89, 1744–1756 (2001)
4. Xun, S.F., Li, T.C., Zhou, Z.Y.: Study on passivity control method for flywheel energy storage system discharging unit. J. Electric Machines and Control 14, 7–12 (2010)
5. Zhang, J.C., Huang, L.P., Chen, Z.Y.: Research on flywheel energy storage system and its controlling technique. Proceedings of the CSEE 23, 108–111 (2003)
6. Kou, B.Q., Cao, H.C., Bai, Y.R.: Voltage Converter Based on the Rotating Electromagnetic. P. R. China, ZL, 200910072063.7 (2009)
7. Zhu, J.X., Jiang, X.J., Huang, L.P.: Topologies and charging strategies of the dynam ic voltage restorer with flywheel energy storage. J. Electric Machines and Control 13, 317–321 (2009)
8. Ye, W.F., Hu, Y.W., Huang, W.X.: Simulation and research of capability on dual stator-winding induction generator system. J. Electric Machines and Control 9, 419–424 (2005)

Study of Traffic Flow Short-Time Prediction Based on Wavelet Neural Network

Yan Ge* and Guijia Wang

Dept. of Information Science and Technology,
Qingdao University of Science and Technology, Qingdao,
Shandong 266061, China
gygeyan@163.com

Abstract. In order to improve the accuracy of traffic flow short-time prediction, a traffic flow short-time forecasting model is presented based on wavelet neural network. Firstly, a traffic flow short-time prediction model is established based on improved BP neural network. Secondly, a wavelet neural network prediction model is created to improve slow convergence speed and low forecasting precision of BP neural network. The excitation function of hidden layer use wavelet function instead of sigmoid function. Wavelet neural network combines local characteristics of wavelet transform with self-learning capability of neural network. So it has strong approximation and tolerance. Finally, the two models are used to solve traffic flow short-time prediction separately; simulation results show that wavelet neural network is better than BP neural network. Wavelet neural network has high convergence speed and forecasting precision.

Keywords: Intelligent transportation system, Traffic flow short-time prediction, BP neural network, Wavelet neural network.

1 Introduction

With the development of the economy of society and the improvement of people's living standard, more and more traffic tools are used. So it causes many increasingly serious problems such as traffic congestion, traffic accident, energy consumption and pollution. Intelligent transportation system can effectively relieve these traffic problems [1]. The traffic flow short-time forecast is a very important theory, in several important part of intelligent transportation system. It is one of the difficult problems that the current field of transportation is solving. Traffic flow short-time prediction is the premise of the control and abduction of dynamic traffic. It uses testing equipment to obtain real-time information. It gets forecasting information by forecasting model. It provides real-time effective information for drivers. It can realize dynamic road abduction. It also can save travel time, alleviate road congestion, save energy and reduce pollution for driver.

* This work is supported by the National Natural Science Foundation of China (grant no.60802042), the Natural Science Foundation of Shandong Province, China (grant no. ZR2009GQ013).

M. Zhu (Ed.): Electrical Engineering and Control, LNEE 98, pp. 509–516.
springerlink.com © Springer-Verlag Berlin Heidelberg 2011

At moment t, traffic flow prediction makes real-time prediction for the next decision moment $t+\Delta t$ even after several moments of traffic flow. Normally, the forecasting span is no more than 15 minutes (even less than 5 minutes) between t and $t+\Delta t$. We call this prediction as traffic flow short-time forecast [2].

The methods of traffic flow short-time prediction can be divided into two categories. One type forecasting model is based on traditional mathematic (such as mathematical statistics and calculus) and physical methods. This model demands higher mathematical theory knowledge for modelers. It is difficult to realize for non-professional people. The other type takes modern scientific technology and method (such as neural network and fuzzy theory) as main research means. The model doesn't pursue strictly mathematical derivation and definite physical meaning. It pays more attention to the fitting effect of the real traffic flow phenomena [3].

The neural network model can renew network by real-time traffic information [4-9]. It has instantaneity. It is available for the areas with many influence factors. It can undertake offline training. The applicable scope of this model is traffic system with complicated traffic situation. Traffic system is a complicated system including road, car and people. It has highly non-linear and uncertainty. And neural network can effectively handle the uncertainty. The BP neural network easily comes into local minimum, and it has lower precision. Wavelet neural network can, to some extent, avoid network come into local minimum. It can improve prediction accuracy. This paper presents a traffic flow short-time prediction model based on wavelet neural network.

2 Wavelet Neural Network

The BP neural network is a prior neural network. It is one of the most research neural networks. The BP neural network includes three layers of network structure such as input layer, hidden and output layer. It has one input layer and one output layer and has one or more hidden layers. Adjacent layers carry out absolute connection. Internal neurons of each layer don't connect.

The wavelet neural network is a new form of network by combining wavelet theory with principles of artificial neural networks [10-12]. It combines local characteristics of wavelet transform with self-learning capability of neural network. So it has strong approximation and tolerance. It has higher convergence speed and good prediction effect. The whole network also doesn't easily come into local minimum. The wavelet neural network usually has three layers. The excitation function of hidden layer use wavelet function instead of sigmoid function, compared with BP neural network.

The wavelet neural network has been successfully used in the field of signal processing, mechanical fault detection and diagnosis. A lot of achievements are obtained. This paper uses wavelet neural network in the field of traffic flow prediction.

3 Traffic Flow Short-Time Prediction Model Based on Wavelet Neural Network

This paper takes 5 minutes as an interval. It uses previous 15 minutes to predict later 5 minutes. This paper need to determine the structure of wavelet neural network, wavelet function and training method.

3.1 The Structure of Wavelet Neural Network

The wavelet neural network uses three layers of network structure of BP neural network, such as input layer, hidden layer and output layer. According to the requirements of traffic flow short-time forecast, input layer chooses 3 neurons, hidden layer chooses 9 neurons, and output layer chooses 1 neuron. The structure of wavelet neural network is shown in figure 1.

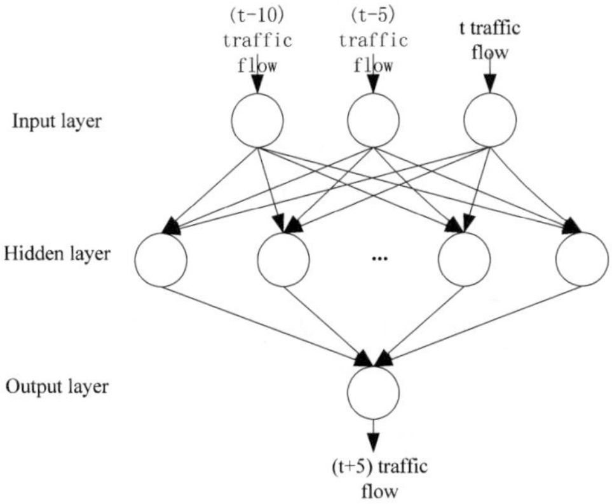

Fig. 1. The model of BP neural network

The input vector X= $(x_1, x_2, \cdots, x_m)$. m is the number of neurons of input layer. y_k is real output. d_k is expected output. K is the number of samples.

The input and output of each neuron of hidden and output layer are defined as follows:

The input of each neuron of hidden layer can be described as formula (1).

$$x'_j = \sum_{i=1}^{m} \omega_{ij} x_i - \theta_j \tag{1}$$

Where, θ_j is the threshold of each neuron of hidden layer. ω_{ij} is the weight from input layer to hidden layer.

The output of each neuron of hidden layer can be described as formula (2).

$$O'_j = f((x'_j - b_j) / a_j)(j = 1, 2, \cdots, n) \tag{2}$$

Where, the function $f(\bullet)$ is morlet wavelet function ($f(x) = \cos(0.75x) + e^{-0.5x^2}$). n is the number of neurons of hidden layer. a_j is the telescopic factor. b_j is the translational factor.

The input of each neuron of output layer can be described as formula (3).

$$x''_k = \sum_{j=1}^{n} \omega'_{jk} O'_j - \theta'_k \tag{3}$$

Where, ω'_{jk} is the weight from hidden layer to output layer. θ'_k is the threshold of each neuron of hidden layer.

The output of each neuron of output layer can be described as formula (4).

$$y_k = g(x''_k) \tag{4}$$

Where, $g(\bullet)$ is sigmoid function ($g(x) = 1/(1+e^{-x})$).

To sum up, the input and output of the whole wavelet neural network can be described as formula (5).

$$y_k = g(\sum_{j=1}^{n} \omega'_{jk} f((\sum_{i=1}^{m} \omega_{ij} x_i - \theta_j - b_j)/a_j) - \theta'_k) \tag{5}$$

The error function (energy function) of wavelet neural network can be described as formula (6).

$$E = \frac{1}{2}\sum(d_k - y_k)^2 \tag{6}$$

The wavelet neural network uses improved gradient descent method to adjust weight, threshold, telescopic factor and translational factor.

3.2 The Improved Training Algorithm of Wavelet Neural Network

Wavelet neural network usually adopts error back propagation algorithm (gradient descending method). It belongs to teacher learning. The basic principle of gradient descent method is that learning process includes signal positive dissemination and error back propagation [13]. The training network's input and desired output is given. Then it calculates step by step to obtain actual output of network. If actual output is not equal to expected output, it will take deviation to spread along the opposite direction of network. That is, it modifies connected weights begin from output layer until meets requirements of error.

Many improved gradient descent methods are presented, because the basic gradient descent method exists prone to local minimum, low learning efficiency, slow convergence speed. They are the change of learning rate, momentum method and the adoption of new excitation function. This paper adopts momentum method to speed up training convergence speed of wavelet neural network.

Momentum method includes momentum factor α (0< α <1). In the process of error back propagation, it adds error generated by previous adjustment of weights to this adjustment of weights to accelerate the speed of convergence.

In improved gradient descent method, the new formula for adjustment of weights can be described as formula (7).

$$\omega(t+1)=\omega(t)+\left(-\eta\nabla E(t)\right)+\alpha\left(\omega(t)-\omega(t-1)\right) \tag{7}$$

The formula for adjustment of each layer's weights by back propagation as follows:

The formula for adjustment of weights from output layer to hidden layer can be described as formula (8).

$$\begin{aligned}\omega'_{jk}(t+1)=\omega'_{jk}(t)+\eta y_k(1-y_k)(d_k-y_k)O'_j\\ +\alpha\left(\omega'_{jk}(t)-\omega'_{jk}(t-1)\right)\end{aligned} \tag{8}$$

The formula for adjustment of weights from hidden layer to input layer can be described as formula (9).

$$\begin{aligned}\omega_{ij}(t+1)=\omega_{ij}(t)+\eta O'_j(1-O'_j)\sum\delta_k\omega'_{jk}x_i\\ +\alpha\left(\omega_{ij}(t)-\omega_{ij}(t-1)\right)\end{aligned} \tag{9}$$

Where, η is learning rate. δ_k is error of the k th sample, $\delta_k = y_k(1-y_k)(d_k-y_k)$.

4 Simulation Experiment

This paper analyzes traffic data (obtained with a unit of 5 minutes) of a main road in Qingdao city. And we can obtain that the time quantum 11:00-13:00 is the rush hour of traffic flow. This paper takes traffic data in this time quantum as sample data, and 21 samples can be got. This paper takes 16 samples as training samples, and the remaining as testing samples. Samples are normalized (divided by a constant) to get samples as table 1 and table 2 shows.

Table 1. Training samples

input sample			expected output
0.096	0.131	0.094	0.125
0.131	0.084	0.125	0.122
0.094	0.125	0.122	0.106
0.122	0.106	0.123	0.117
0.106	0.123	0.117	0.100

Table 2. Forecasting samples

input sample			expected output
0.125	0.122	0.106	0.123
0.117	0.100	0.145	0.120
0.119	0.112	0.105	0.095
0.091	0.082	0.104	0.099
0.106	0.085	0.085	0.084

This paper uses MATLAB to simulate. The error curve obtained by training BP neural network is shown in figure2. The error cure obtained by training wavelet neural network is shown in figure3. We can see from figure 2 and figure3. The error of wavelet neural network is close to 0 before 500 times and the error of BP neural network is close to 0 after 600 times. Consequently, for the time of training, wavelet neural work is obviously less than BP neural network.

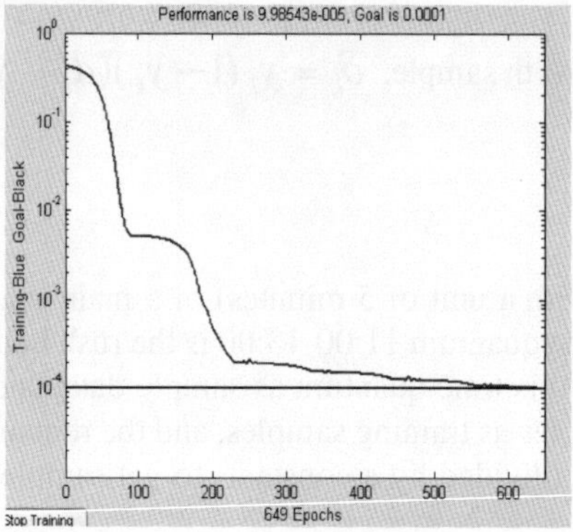

Fig. 2. The error curve of BP neural network

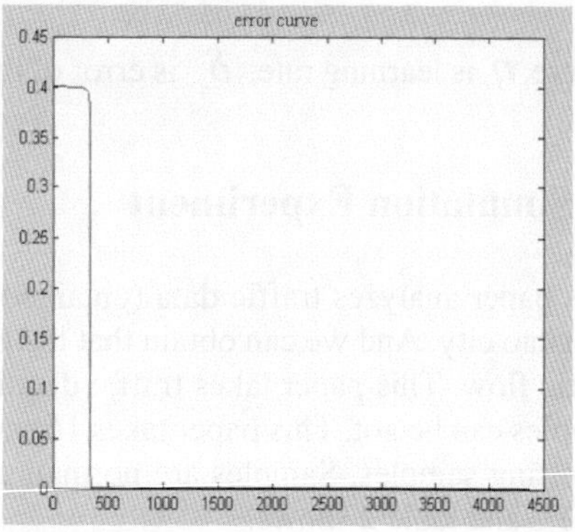

Fig.3. the error curve of wavelet neural network

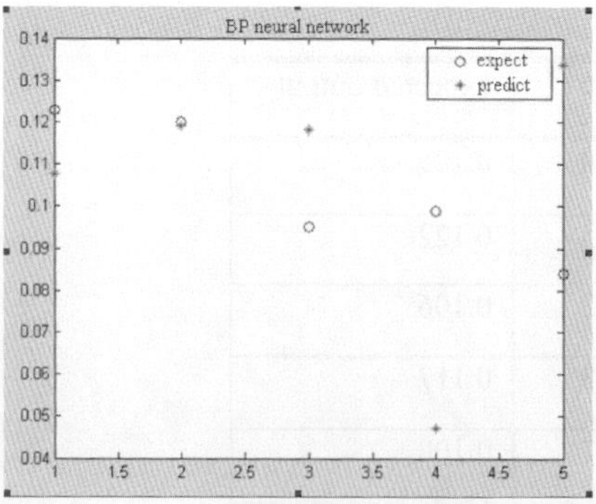

Fig. 4. The results of BP neural network

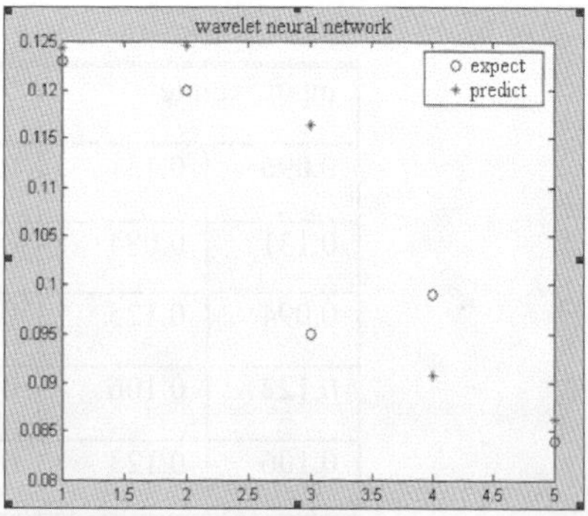

Fig. 5. The results of wavelet neural network

This paper uses the trained network to predict 5 samples. The result of BP neural network is shown in figure 4. The result of wavelet neural network is shown in figure 5. The average error of 5 samples is 0.0278 obtained from figure4. The average error of 5 samples is 0.0092 obtained from figure 5. Therefore, the prediction accuracy of wavelet neural network is obviously better than BP neural network.

5 Conclusions

The road traffic flow short-time prediction is the very important theoretical basis of intelligent transportation system. It is the premise of dynamic traffic control and induction.

The neural network model can predict the future of traffic flow within short-time based on historical traffic data. BP neural network model has slow convergence speed and low forecasting precision. In order to improve the accuracy of traffic flow short-time forecasting, this paper uses wavelet function replace of sigmoid function. So, this paper presents traffic flow short-time prediction model based on wavelet neural network.

Compared with BP neural network, wavelet neural network has higher convergence speed and better prediction accuracy. And wavelet neural network is not easily into the local minimum.

Finally, this paper uses BP neural network and wavelet neural network to predict traffic flow respectively. The results of simulation show that the convergence speed and forecasting precision of wavelet neural network are both better than BP neural network.

References

1. Zhang, Z.: The Overview of the Intelligent Transportation System and developing situation at home and abroad. Friend of Science Amateurs, 97–99 (2010)
2. Smith, B.L., Demetsky, M.J.: Traffic Flow Forecasting: Comparison of Modeling Approaches. Journal of Transportation Engineering 123(4), 261–266 (1997)
3. Ge, Y., Wang, J., Meng, Y., Jiang, F.: Research Progress on Dynamic Route Planning of Vehicle Navigation. Journal of Highway and Transportation Research and Development 27(11), 113–117 (2010)
4. Messai, Nadhir, Thomas, Philippe: A neural network approach for freeway traffic flow prediction. In: IEEE Conference on Control Applications- Proceedings, vol. 2, pp. 984–989 (2002)
5. Yu, J., Shi, Q.: The Applied Review of Neural Network Model in the Short-Term Traffic Flow Prediction. Journal of Henan University of Science and Technology (Natural Science) 26(2), 22–26 (2005)
6. Hobeika, A.G., Kim, C.K.: Traffic Flow Prediction Systems Based on Upstream Traffic. In: Vehicle Navigation & Information System Conference (1994)
7. Nihan, N.L., Holmesland, K.O.: Use of the box and Jenkins time series technique in traffic forecasting. Transportation Research Record 9(2), 25–143 (1980)
8. Ma, J., Li, X., Meng, Y.: Research of Urban Traffic Flow Forecasting Based on Neural Network. Acta Electronica Sinica 37(5), 1092–1094 (2009)

9. Liu, M., Wu, J.: A genetic-algorithm-based neural network approach for short-term traffic flow forecasting. In: Wang, J., Liao, X.-F., Yi, Z. (eds.) ISNN 2005. LNCS, vol. 3498, pp. 965–970. Springer, Heidelberg (2005)
10. Delyon, B., Judilsky, A., Benvenisie, A.: Accuracy Analysis for Wavelet Approximations. IEEE Trans on Neural Network 6(2), 332–348 (1995)
11. Li, J., Li, Q., Hou, H., Yang, L.: Traffic Flow Prediction based on the Wavelet Neural Network with Genetic Algorithm. Journal of Shandong University(Engineering Science) 37(2), 109–112 (2007)
12. Wang, L., Zhang, J., Jin, P.: Prediction of Strength of Corroded Reinforcement Based on Wavelet Neural Network. Journal of Highway and Transportation Research and Development 27(3), 69–74 (2010)
13. Xu, L., Fu, H.: Intelligent Prediction Theory and Methods of Traffic Information. Science Press, Beijing (2009)

A Novel Knowledge Protection Technique Base on Support Vector Machine Model for Anti-classification

Tung-Shou Chen[1], Jeanne Chen[1,*], Yung-Ching Lin[2], Ying-Chih Tsai[2], Yuan-Hung Kao[3], and Keshou Wu[4]

[1] Department of Computer Science and Information Engineering, National Taichung Institute of Technology, Taichung City 404, Taiwan
{tschen,jeanne}@ntit.edu.tw
[2] Graduate School of Computer Science and Technology, National Taichung Institute of Technology, Taichung City 404, Taiwan
{king5240,bigear42}@gmail.com
[3] Department of Information Engineering and Computer Science, Feng Chia University, Taichung City 407, Taiwan
kylemis@gmail.com
[4] Department of Electronic and Electrical Engineering, Xiamen University of Technology, Xiamen, China
kswu@xmut.edu.cn

Abstract. Information security issues are concerned with the disclosure of important privacy information by unethical employees. In this paper, we proposed the concept of noise data in anti-data mining to protect knowledge in data. The scheme is an anti-classification algorithm which is based on support vector machine (SVM). Experimental results showed that the noise data generated from prediction model of SVM can effectively disturb various classification algorithms which make predictions of those classification algorithms less accurate. The scheme effectively protects classified knowledge.

Keywords: Privacy-knowledge; Support vector machine; Anti-data mining; Anti-classification.

1 Introduction

Most enterprises apply data mining technique for data identification [3], access control [12], data authentication [2] and data encryption [18] to ensure information security. These methods mainly aimed at protecting the usability, completeness and confidentiality of the databases. However, these methods leaked information to undesirable persons about valuable information while assuring users that important information is protected. Although enterprises provide employees access authority to databases and have strict rules against mining information, it is difficult to prevent users executing illegal mining.

* Corresponding author.

M. Zhu (Ed.): Electrical Engineering and Control, LNEE 98, pp. 517–524.
springerlink.com © Springer-Verlag Berlin Heidelberg 2011

Information in databases is very important and knowledge hidden in the information is ever more valuable. In this paper, we propose using the anti-data mining (ADM) [5] as the basis to design the novel anti-classification algorithm [6] with support vector machine [16]. This new technique is anti-classification based on SVM model to provide the retrievable characteristics and non-destructive protection method for the sorted and valuable database. The SVM is a well-known classification method in the classification algorithm. We apply the SVM in the database, then extracting the supporting vectors characteristic for classification and analysis models to generate the disturbing noise data which is similar with the original ones so that the disturbing data can efficiently lower the accuracy of classification prediction.

The proposed method has the following characteristics: (a) the protected database can be restored to the original one by keys; (b) data in the databases can be accessed normally which acts in misleading illegal users that the databases are unprotected (c) the new disturbing data are difficult to distinguish from the original ones; (d) users are allowed to customize rate of generating noise data and predict of the lowering expecting accuracy.

2 Related Work

2.1 Privacy-Preserving Data Mining

Discussion on data mining techniques threatening the databases was first proposed by Clifton *et al.* [9] in 1996. The main purpose is that the privacy information was removed from the published databases before starting data mining to prevent from the disclosure of the personal and sensitive information. This kind of researches is called privacy-preserving data mining (PPDM). However, the research of PPDM is mostly involved in protecting the personal privacy information in databases or to prevent the sensitive knowledge from mining [8][10][11][20]. The disclosure of private information by data mining is undesirable.

Currently, research on PPDM include: using sampling method to resolve disclosure of sensitive information [15], using transactions decomposition to remove sensitive items [13] and lowering support of the sensitive items to hide them [17]. However, these methods are suited for published databases for the data mining. Sensitive information and knowledge of enterprises are unprotected from unauthorized parties interested in acquiring both by data mining.

2.2 Anti-Data Mining

The ADM is a technique to protect the knowledge of database from being revealed in data mining. ADM is aimed at protecting against two data mining techniques: one is the anti-clustering [5], which is a method to generate camouflage information by various clustering algorithms to deviate the clustering knowledge from the datasets whereby knowledge is protected; the other is the anti-classification [6], which generates camouflage information of incorrect classes from the datasets to lower the classification expecting accuracy of the protected datasets by various classification tools. The ADM has an advantage in which it is retrievable [5][6], and can be user customized for different protection effects. Due to the nature of the published datasets by PPDM, there is no consideration for retrieval. ADM is proposed to protect information in an enterprise.

The proposed method is based on anti-classification and refers to the classification model of SVM [16]. The records are picked randomly from the classification model. Distances between the supporting vectors of other different classes of picked records are calculated to find the shortest-distance. The predicted records are then used to generate noise data. The noise data generated based on the characteristic selection can effectively disturb the original database compared to noise generated randomly.

3 Anti-Classification Based on SVM Model

The proposed method applies SVM to constructing a classification predicting model from the original data which is the basis for generating noise data to disturb the classifiers. The proposed method focuses on generating noise data so that the similarity between the noise data and the original ones is the critical point preventing noise data from being removed and easily located. The generated noise data will conform to records in the original data.

3.1 Protecting Procedure of Database

In order to have the flexible adjusting protection effect for users, the proposed method designs three parameters which include the noise data ratio (R), expect accuracy (EA) and executive loop T for users setting that effects of protection flexibility. The R is the certain amount of the noise data; EA represents the top limit of the expecting accuracy of the classification model from the protected datasets; T is the top limit of the loops which preventing the algorithm having endless loops. The details of executive steps as following:

Input: *Seed* value of the virtual random, generating data rate R, expecting accuracy EA, executive loop T and the original database D, $D = \{d_i, i = 1, 2, 3, \ldots, n\}$, n is the data amount of D.

Output: The last executive loop T', the protected database D', $D' = \{d'_i, i = 1, 2, 3, \ldots, n'\}$, $n' = n + (T' \times p)$, p is the noise data generated from every loop, the total number of the noise data is $T' \times p$.

Step 1. Import D to the SVM classification algorithm to obtain the classification prediction model X, $X=\{x_1, x_2, x_3, \ldots, x_h\}$, h is the number of the support vectors. Using the k-fold cross validation (k-fold CV) [7] to calculate the expect accuracy of the original dataset OA.

Step 2. Set the protected dataset $D' = D$, calculating time $t = 1$, the generating number of the noise data from each loop $p=n\times R$, and the *Seed* value of the random generator. Make the random generator is the function $Rand()$, the range of the random is [0,1).

Step 3. Pick p record randomly and repeatedly from the X, make d_i' is the picked data, $i \in [1, p]$, if D' has m columns, then $d_i' = (d_{i1}', d_{i2}', d_{i3}', \ldots, d_{im}')$, apply Eq. (1) to the picked data d_i' to find out the shortest-distance x_l which has the different class attribution from d_i' adding or subtracting randomly with d_i' to generate the noise data y_i as Eq. (2).

$$Dist_{\min}(d_{ij}', x_{lj}) = \sqrt{\sum_{j=1}^{m} \left(d_{ij}' - x_{lj}\right)}, \; l \in [1, h]. \tag{1}$$

$$y_{ij} = d_{ij}' \pm x_{lj}, for\ each\ 1 \le j \le m. \tag{2}$$

Step 4. Set the noise data y_i which has the different class from d_i'. Set r is the total class number of the original dataset D, $g(d_i')$ represents the class of d_i', randomly assign the class of y_i is one of the [1,2,..., $g(d_i')$-1, $g(d_i')$+1,...,r]. put the noise data with indicated class y_i into the position of the original dataset, which position is calculated by $Rand() \times (n + t \times p)$, then put the p noise data into the protected dataset D' in the final.

Step 5. Calculate the protected dataset D' by the k-fold CV in the same classification algorithm, to generate the classification predicting accuracy (*PA*) of D'.

Step 6. Decide the stop condition. If $PA \le EA$ or t=T, then record $T' = t$ and algorithm ends; else, set t=t+1 and return to the *Step 3*.

The proposed method has the effect in that (a) the protected database can be restored to the original one by input the correct parameter (illustrated in steps 3.2); (b) data in the database can be accessed normally, because the noise data affects nothing in the database; (c) differentiating between the noise data and the original one is difficult; and (d) users may customized the generated noise data rate and the lower expected accuracy based on requirements.

3.2 Restoring Procedure of Database

Due to the proposed method using adding noise data to disturb the classification expecting accuracy, removing the noise data in protected dataset D' can restore the D' to the original dataset. Details follow.

Input: Virtual random *Seed*, generate data rate *R*, executive loop T' and the protected database $D' = \{d_i', i = 1,2,3,\ldots,n'\}$, n' is the data number of D'.

Output: Original database *D*.

Step 1. Calculate the generated noise data number p from each executive loop. According to the mentioned formula $n' = n + (T' \times p)$ and $p = n \times R$, then

$n = \frac{n'}{1+T'\times R}$, $p = \frac{n'}{1+T'\times R}\times R$. Set the executive time $t = T'$ and the *Seed* value of the random generator, set generating random range of the random generator $Rand()$ is [0,1).

Step 2. Obtain the $Rand()$ of all of the noise data in the original database D and put into the array B. set array B is $B = \{b_1, b_2, b_3, \ldots, b_{T'}\}$, where $b_i = (b_{i1}, b_{i2}, b_{i3}, \ldots, b_{ip})$, $i \in [1, T']$.

Step 3. Remove the noise data from the tth loop. Find the positions of the noise data in tth loop, calculate method is $b_{tj} \times n'$, $j \in [1, p]$, n' is the data number of D' in tth loop. Obtain the positions of the noise data putting into the tth loop $\{b_{t1} \times n', b_{t2} \times n', \ldots, b_{tp} \times n'\}$, then remove them.

Step 4. If t=1, then algorithm ends, else, t=t-1 and $n' = n' - p$, return to the *Step 3*.

The protection method of the proposed method is to add the noise data gradually by executive loops t. The positions are calculated which the total noise data number n' after tth loop have to be included into the protected dataset, by function *Rand()* of the *Seed* value to hide the noise data. Therefore, restore has to start from the last executive loop reversely, to remove the noise data from tth loop gradually. The protected dataset D' is fragile. The dataset cannot be correctly restored if the dataset is modified. Changes include process such as increasing or decreasing data and/or changing data order.

4 Experimental Results and Effect Analyses

Experimental testing includes testing on searches on four famous classification algorithms for the tools: MySVM [19], LibSVM [4], CLC [7] and KNNR [14]. There are 15 two-classes sample databases in the experimental data which is from the UCI Machine Learning Repository (UCI) [1]. Details are listed in Table 1.

The result of the experiment is to compare and discuss between the 15 databases after confirming the original accuracy and the protected databases by increasing 10%, 20%, 30%,…, 100% noise data. Because of the huge experimental data, we integrate the data to the trend chart as Fig.1. The axle x in Fig.1 represents the different rate of the noise data, the axle y is the lowering accuracy. The accuracy value is the average of the accuracy of the 15 databases; the purpose is to understand objectively about the effect of the original databases and the rate of the generating noise data. The more noise data the more obvious effect of the prediction accuracy which seen in the chart.

Table 1. The different types of examples databases in UCI website.

No.	Databases	No. of instances	No. of attributes	No. of classes
1	Australian	690	14	2
2	Breast-cancer	683	10	2
3	Diabetes	759	8	2
4	Fourclass	855	2	2
5	Haberman's Survival Data	306	3	2
6	Heart	270	13	2
7	Hepatitis Database	80	19	2
8	MUSK Databases	476	166	2
9	Pima Indians Diabetes Database	768	8	2
10	Sonar	207	60	2
11	Splice	1000	60	2
12	Svmguide3	1105	22	2
13	Wisconsin Breast Cancer Databases-wdbcm	569	31	2
14	Wisconsin Breast Cancer Databases-wpbc	194	34	2
15	Databases-breast-cancer-wisconsin	683	10	2

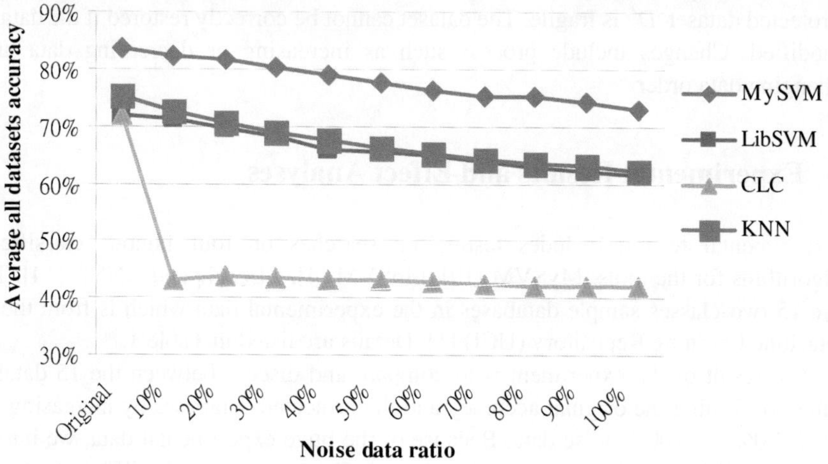

Fig. 1. Comparison on all difference average accuracy of the various classifiers.

The effect degree of the classifiers by the proposed method is in this order; CLC, KNNR, MySVM and LibSVM. The effect is highest for CLC. When the noise data is generated to 10%, then all accuracies are decreased to 28%; the former is generated to 100%, the latter is decreased to 30%. In KNNR, the noise data is generated to 10%, and all accuracies are decreased to 2%; the former is generated to 100%, the latter is decreased to 13%. All accuracies are decreased to 10% in the two classifiers MySVM and LibSVM.

Next, the proposed method is analyzed using the paired T-test to test if there is obvious difference in their expected accuracies between the classification models of the original database and the protected one. The paired T-test is for verifying whether the proposed method affects the classifiers and the suggestion to the generating rate of noise data, the result is as shown in Table 2.

In Table 2, the proposed method increases 10% noise data. There is 0.01 obvious differences test by the classifiers MySVM, CLC and KNNR, and increase 20% noise for LibSVM tests with 0.01 differences. Therefore, the proposed method suggests adding more than 10% noise data to prevent the classifiers MySVM, CLC and KNNR, and more than 20% noise data to the LibSVM to obtain 0.01 obvious differences respectively.

Table 2. T-test results from the before and after protection

	10%	20%	30%	40%	50%	60%	70%	80%	90%	100%
MySVM	0.001**	0**	0**	0**	0**	0.001**	0.001**	0.001**	0.001**	0**
LibSVM	0.068	0.005**	0.003*	0.005**	0.002**	0.001**	0**	0**	0**	0**
CLC	0**	0**	0**	0**	0**	0**	0**	0**	0**	0**
KNN	0**	0**	0**	0**	0**	0**	0**	0**	0**	0**

*Significant at the 0.05 level.
**Significant at the 0.01 level.
H0: Every database before-and-after protection does not have obvious difference.
H1: Every database before-and-after protection has obvious difference.

5 Conclusions

The proposed method provides the classification knowledge protection method based on the SVM model in anti-classification. The database by SVM is trained to extract information from the trained model and to generate the noise data according to the characteristic of the supporting vector. Experimental results showed that the proposed method can efficiently protect information from various classifiers and the generated noise data are similar with the original ones in databases. The proposed method provides a technique to restore the protected database to the original one by inputting the correct parameter keys. Users can also custom-set the rate of generating noise data and lowering the accuracy depending on their expectation to mislead the illegal users.

Acknowledgements. This work was supported partially by the National Science Council of Republic of China under grant NSC 99-2221-E-025-004.

References

1. Asuncion, A., Newman, D.J.: UCI Machine Learning Repository (2007), `http://www.ics.uci.edu/~mlearn/MLRepository.html`
2. Bertino, E., Castano, S., Ferrari, E., Mesiti, M.: Protection and Administration of XML Data Sources. Data & Knowledge Engineering 43, 237–260 (2002)

3. Bertino, E.: Data Security. Data & Knowledge Engineering 25, 199–216 (1998)
4. Chang, C.C., Lin, C.J.: LIBSVM: A Library for Support Vector Machines, Software available at (2001), http://www.csie.ntu.edu.tw/~cjlin/libsvm
5. Chen, T.S., Chen, J., Kao, Y.H.: A Novel Hybrid Protection Technique of Privacy-preserving Data Mining and Anti-data Mining. Information Technology Journal 9, 500–505 (2010)
6. Chen, T.S., Chen, J., Lin, Y.C., Tsai, Y.C.: Research to Protect Database by Shaking Random Sampling Interference (SRSI). In: Proceedings of the 2009 Global Congress on Intelligent Systems, pp. 569–572 (2009)
7. Chen, T.-S., Lin, C.-C., Chiu, Y.-H., Lin, H.-L., Chen, R.-C.: A New Binary Classifier: Clustering-Launched Classification. In: Huang, D.-S., Li, K., Irwin, G.W. (eds.) ICIC 2006. LNCS (LNAI), vol. 4114, pp. 278–283. Springer, Heidelberg (2006)
8. Clifton, C., Kantarcioglu, M., Vaidya, J., Lin, X., Zhu, M.: Tools for Privacy Preserving Distributed Data Mining. ACM SIGKDD Explorations 4, 23–28 (2003)
9. Clifton, C., Marks, D.: Security and Privacy Implications of Data Mining. In: Proceedings of the ACM SIGMOD Workshop on Data Mining and Knowledge Discovery, pp. 15–19 (1996)
10. Fung, B.C.M., Wangb, K., Wanga, L., Hung, P.C.K.: Privacy-preserving Data Publishing for Cluster Analysis. Data & Knowledge Engineering 68, 552–575 (2009)
11. Gkoulalas-Divanis, A., Verykios, V.S.: Exact Knowledge Hiding Through Database Extension. IEEE Transactions on Knowledge and Data Engineering 21, 699–713 (2009)
12. Hwang, M.S., Lee, C.H.: Secure Access Schemes in Mobile Database Systems. European Transactions on Telecommunications 12, 303–310 (2001)
13. Li, X.B., Sarkar, S.: A Tree-based Data Perturbation Approach for Privacy-preserving Data Mining. IEEE Transactions on Knowledge and Data Engineering 18, 1278–1283 (2006)
14. Li, L., Umbach, D.M., Terry, P., Taylor, J.A.: Application of the GA/KNNR Method to SELDI Proteomics Data. Bioinformatics 20, 1638–1640 (2004)
15. Liu, K., Kargupta, H.: Random Projection-based Multiplicative Data Perturbation for Privacy Preserving Distributed Data Mining. IEEE Transactions on Knowledge and Data Engineering 18, 92–106 (2006)
16. Martens, D., Baesens, B., Gestel, T.V.: Decompositional Rule Extraction from Support Vector Machines by Active Learning. IEEE Transactions on Knowledge and Data Engineering 21, 178–191 (2009)
17. Martens, D., Bruynseels, L., Baesens, B., Willekens, M., Vanthienen, J.: Predicting Going Concern Opinion with Data Mining. Decision Support Systems 45(4), 765–777 (2008)
18. Rowan, T.: VPN Technology: IPSEC vs SSL. Network Security 2007, 13–17 (2007)
19. Rüping, S.: mySVM-Manual, vol. 8. University of Dortmund, Lehrstuhl Informatik (2000)
20. Verykios, V.S., Elmagarmid, A.K., Bertino, E., Saygin, Y., Dasseni, E.: Association Rule Hiding. IEEE Transactions on Knowledge and Data Engineering 16, 434–447 (2004)

A DAWP Technique for Audio Authentication

Tung-Shou Chen[1], Jeanne Chen[1,*], Jiun-Lin Tang[1], and Keshou Wu[2]

[1] Department of Computer Science and Information Engineering, National Taichung Institute of Technology, Taichung City 404, Taiwan
{tschen,jeanne}@ntit.edu.tw,
bb6bbjs@gmail.com

[2] Department of Electronic and Electrical Engineering, Xiamen University of Technology, Xiamen, China
kswu@xmut.edu.cn

Abstract. Digital audio content protection is part of information security for audio authentication and audio integrity evaluation. Watermarking is widely used in copyright protection. However, watermark requires a third party authentication. It is desirable for the watermark to be robust. In this paper, we proposed a DAWP technique to extract audio characteristics using MCFF and PSL. The combined biometric methods are used to complement the watermark to protect the integrity of the audio and to fingerprint audio. Experimental results showed that the audio watermark is robust. Results also detected tamper attacks such as cropping and adding noise.

Keywords: Audio authentication, Audio Watermarking, copyright protection.

1 Introduction

Audio authentication can be used to ensure audio source and audio content integrity. Pirated multimedia product in the market and the fraudulent event on the speaker is mostly due to audio source. Audio integrity can be used to ensure whether the content is changed or not. Audio authentication is very important, and past research has been done using watermark for intellectual right protection. It can also be used in the certification and recognition of the speaker's voice [5][6]. Progress in the technology has resulted in different audio format, and the compressed file occupies smaller storage space, which forms strict challenge to watermarking embedding and extraction technology [8][3]. The audio file is easy to be duplicated, which results in serious piracy issues especially in the music industry. To prevent piracy and illegal downloading, the use of audio watermarking technology is an effective tool to protect the intellectual property right [2][10][1]. Figure 1 shows the proposed digital audio watermark protection (DAWP) method for authentication of audio signal publisher and to detect illegal change to the audio signal. The flow chart shows A the audio signal publisher or speaker, B the audio signal receiver and C the third party trustable authentication organization.

* Corresponding author.

M. Zhu (Ed.): Electrical Engineering and Control, LNEE 98, pp. 525–534.
springerlink.com © Springer-Verlag Berlin Heidelberg 2011

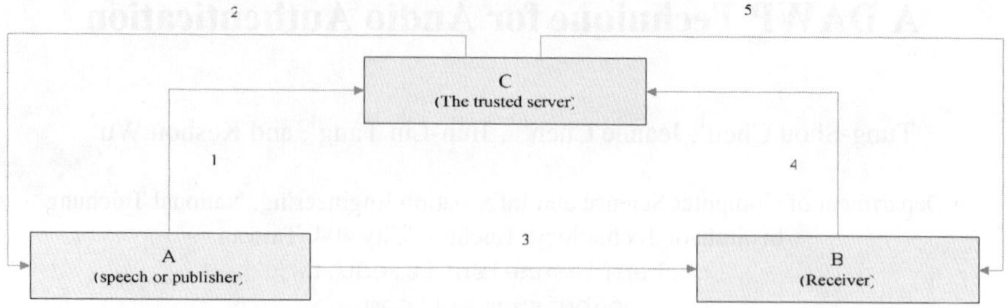

Fig. 1. Three-side architecture flow chart of audio authentication

2 Review on the Related Work Regarding Audio Signal Watermark and Audio Authentication

The protection of the intellectual property of digital multimedia is an important issue. Chang et al. [2] proposed a copyright scheme which made use of db4 Daubechies discrete wavelet transform to transform into frequency domain. A watermark image was embedded into the frequency domain, and counter-propagation neural networks algorithm is used to reinforce the robustness of the watermark. The watermark can be used to prove who the publisher is, but appeared incapable of detecting whether the audio signal is interpolated or not. Park et al. [7] used a fragile circulated watermark to be embedded into the audio signal of the speaker. If the content is changed, the fragile watermark will be distorted [8]. Changes in the audio can be detected and the audio signal integrity can be preserved. However, it is not possible to identify the speaker in a tamper attempt.

In this research, a robust watermark is added into the audio signal to enhance the identification authentication capability of the speaker.

3 The Construction of Audio Watermark and Audio Fingerprint

In DAWP, a watermark fingerprint is added to audio signal (see Fig. 2). The watermark is embedded into the sampled audio signal as recorded by a PCM equipment from a speaker. Each audio signal sample is cut into appropriate size. Masking action in the acoustics is used to find out the way to enhance imperceptibility of the human auditory. The most appropriate candidate area for embedding is calculated. The discrete wavelet transform (DWT) is used to transform the candidate area from spatial domain to frequency domain. The pre-treated binary watermark image is embedded into the low frequency areas. After the frequency domain is transformed into spatial domain, then the audio signal in the sectioned area is combined into a complete audio signal. The audio signal containing watermark is extracted with audio signal feature parameter using Mel-frequency cepstral coefficient. The spectrogram containing watermark audio signal and Mel-frequency cepstral is integrated into audio fingerprint containing audio signal feature.

(1) Watermark preprocessing
We acquired original watermark image WA and calculate the size of WA to become $WA1$.

$$WA1 = \{wa1(x, y),\ \ 0 \le x < m,\ \ 0 \le y < n\} \tag{1}$$

Then disperse $WA1$ two dimensional image into two dimensional image value of one dimensional sequence, that is $WA2$.

$$WA2 = \{wa2(k) = wa1(x, y),\ \ 0 \le x < m, 0 \le y < n,\ \ k = x \times n + y, wa2(k) \in \{1,0\}\} \tag{2}$$

Ai is the length of the entire audio signal, which contains lots of $ai(k)$ set. k must be smaller than the cutting length Li, i must be smaller than the original audio signal length.

$$Let Ai = \{ai(k),\ \ 0 \le k \le Li\}(0 \le i \le S) \tag{3}$$

(2) Watermarked Embedding
In the embedding of watermark, we have referred to the embedding method and regulation of researchers [6]. The following are the steps for the embedding process.

Step 1: Select an audio signal sectioned area *Ai*, which contains *Ai*(*k*) sets (k=0,1,2,..,m×n-1), resulting in $Li/m \times n$ samples.

Step 2: Then the audio signal section Ai is performed with discrete wavelet transform DWT, and here we have used 3 order Daubechies discrete wavelet transform to transform the spatial domain into frequency domain, then we get.

$$Ai(k)^H = \{ai(k)(t)^H,\ \ k = 0,1,\ldots, M \times N - 1,\ \ 0 \le t < Li / M \times N \times 2^H \tag{4}$$

Step 3: Calculate the mean value of $Ai(k)^H$, as in $\overline{Ai(k)^H}$.

Step 4: Embedding the watermark: In the statistical feature of frequency domain, tiny change in some wavelet coefficients does not change the original audio signal carrier. Therefore watermark bit embedded into the corresponding audio section *Ai(k)* by quantizing the mean value $\overline{Ai(k)^H}$, and the rule is given by

$$ai(k)(t)^H = \begin{cases} \overline{ai(k)(t)^H - \overline{Ai(k)(t)^H} + q} & if\ \ wa2(k)=1, \\ ai(k)(t)^H - Ai(k)(t)^H - q & if\ \ wa2(k)=0, \end{cases} \tag{5}$$

$$(k = 0,1,\ldots, M \times N - 1,\ \ 0 \le t < \mathrm{Li} / (M \times N \times 2^H)) \tag{6}$$

(3) Digital Audio watermark-Print
Research on audio signal is to classify and explore it content, such as, environmental sound, machine noise, music, animal sound, voice and other non-language sound. The classification is based on audio signal content and audio frequency search which is basically audio model identification problem [5]. MCFF and LSP is used to find out audio features and to combine (see Figs 2 and 3).

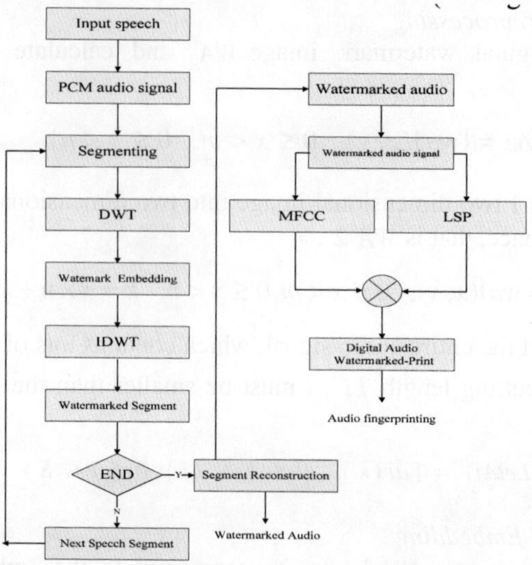

Fig. 2. DAWP technique construction flow chart

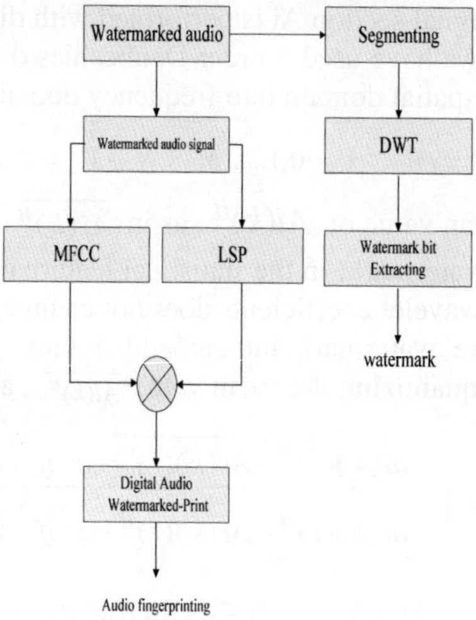

Fig. 3. Watermark extraction and fingerprint comparison

(1) Mel-frequency cepstral coefficient(MFCC)

Using the Mel-frequency cepstral coefficient (MFCC), Dai et al. [4] proposed a feature extraction method through the psychological acoustics masking effect. A two dimensional filter is constructed to reinforce voice recognition system. Fig. 4 shows the Mel cepstral results of MCFF.

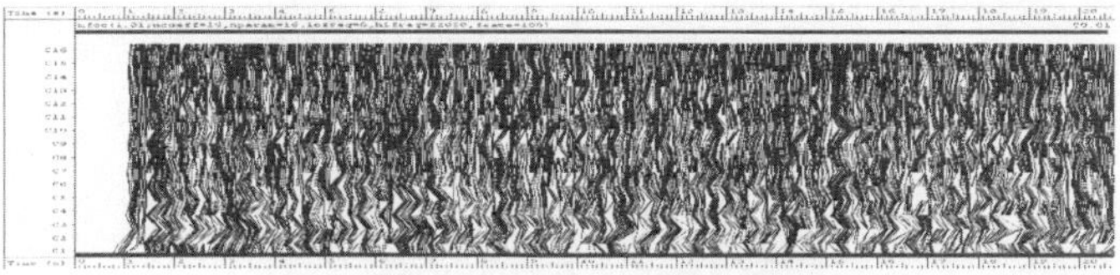

Fig. 4. We use MCFF algorithm on an audio signal to get 16 filters Mel cepstral.

$$Bm(k) = \begin{cases} 0 & k < f_{m-1} \\ \dfrac{k - f_{m-1}}{f_m - f_{m-1}}, & f_{m-1} < k < f_m \\ \dfrac{f_{m+1} - k}{f_{m+1} - f_{m-1}}, & f_{m-1} < k < f_m \\ 0 & f_{m+1} < k \end{cases} \quad 1 \le m \le M \tag{7}$$

$Bm(K)$ means triangular filter of mth band, fm is the central frequency of mth band, fm+1 and fm-1 is the central frequency of two neighboring frequency bands, M is the total number of frequency band.

$$Y[m] = \log\left\{ \sum_{k=f_{m-1}}^{f_{m+1}} |X(k)|^2 B_m(k) \right\} \tag{8}$$

$$c_x^*[n] = \frac{1}{M}\sum_{m=1}^{M} Y[m]\cos\left(\frac{\pi n(m-\frac{1}{2})}{M}\right) \tag{9}$$

The energy at each frequency |X(k)|2 is calculated in Eq.(8) and accumulated to become the energy passing through the filter. The discrete cosine transform (DCT) is performed for the output logarithm energy of all M filters to get the Mel cepstral (Eq.(9). $C^*x[n]$ is the Mel-frequency cepstral coefficient (MCFF) of signal $x[n]$.

(2) Line spectral Pairs (LSP)

The Line spectral Pairs (LSP) is used to represent linear prediction coefficient, filter stability and the representativeness of feature value. Currently, Line Spectral Pairs is widely used in voice and speaker identification [9]. LSP is an algorithm widely used in audio coding, synthesis and recondition. LSP represents the peak position of signal spectrum (see Figs 5 and 6).

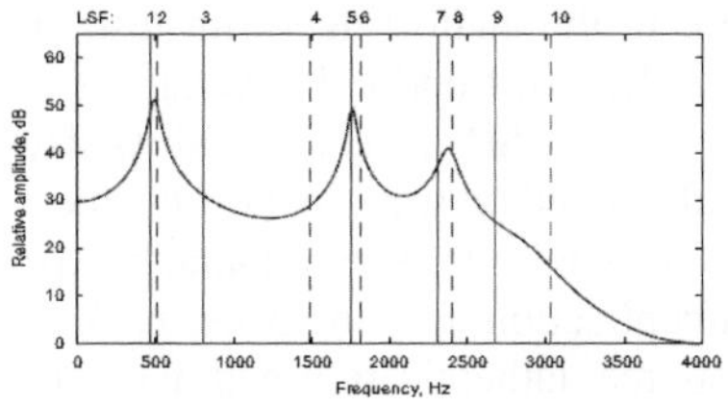

Fig. 5. LPC spectrum peak and frequency [9].

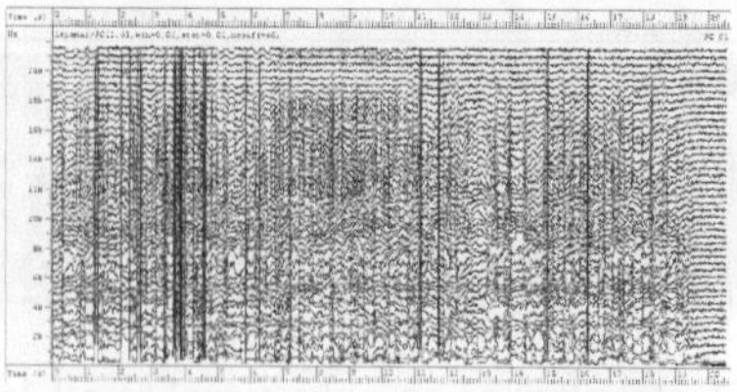

Fig. 6. LSP algorithm for vertical feature.

4 Watermark Extraction and Watermark Fingerprint Comparison

When a trustable third party authentication unit (CA) receives the request for authentication service from the receiver (Figure 3), the following is performed.

(1) MCFF and LSP are used to find out audio feature from the audio signal. Both are then combined into audio fingerprint containing watermark feature as in Figure 9.

(2) The audio signal $Ai*$ containing watermark will be segmented into n sections audio signal block, and each block will contain lots of $Ai*(k)$ sets.

(3) Each block size will be performed using H order discrete wavelet transform.

$$Ai^*(k)^H, D^*(k)^H, D^*(k)^{H-1}, ..., D^*(k)^1. \tag{10}$$

The spatial domain will then be changed into frequency domain. The average value $Ai^*(k)^H$ in the frequency domain will be calculated and use the following rules to extract watermark:

$$WA'2(k) = \begin{cases} 1 & if\ \overline{A^*(k)^H} > 0, \\ -1 & if\ \overline{A^*(k)^H} \le 0, \end{cases} \tag{11}$$
$$(k = 0,1,...., M \times N - 1).$$

(4) Transform the one dimensional sequence value of $WA'2(k) \in \{1, 0\}$ into two dimensional image watermark $WA'1$.

5 Experimental Results

To realize the audio authentication experiment, we used the audio test signal WAV file. The sampling in each second is 44.1 Khz, the sampling accuracy is 16 bits, and 20 seconds of length is used. The watermark image to be embedded is 32×32bit two dimensional value image, and the wavelet transform used is Daubechies third order discrete wavelet transform. In the audio fingerprint test, sfs audio signal processing software provided by the college of London University is used. In MCFF, number of coefficients is set up at 16, and in LSP, analysis window size is set up as 20 ms.

Three experimental results based on watermark robustness, watermark interpolation attack experiment and audio signal interpolation attack experiment. The experimental results show that audio authentication by using watermark and audio fingerprint can be achieved. That is, audio source authentication and audio integrity authentication can then be realized. Figures 7 and 8 show the watermark and the audio signal use in the experimental tests.

Fig. 7. Original watermark image and audio signal

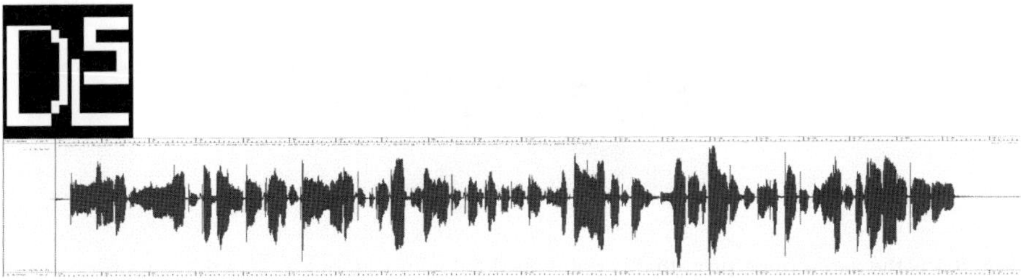

Fig. 8. Extracting the watermark image and the embedded watermark audio signal

To identify the watermark robustness as issued by the authenticated side, we have adopted Normalization Coefficient (NC). After watermark is embedded into the audio signal and after audio signal attack such as cutting, low pass filtering, mp3 compression and amplitude amplification, the extracted watermark is compared with the original watermark (12).

$$NC = \frac{\sum_i \sum_j W(i,j) w'(i,j)}{\sum_i \sum_j [w(i,j)]^2} \tag{12}$$

The peak signal noise ratio (psnr) of the music signal is used to evaluate the comparison when watermark is embedded into the audio signal to the original signal (13).

$$psnr(dB) = 10 \log_{10} \frac{A^2 perk}{\sigma e^2}, \quad \sigma e^2 = (\frac{1}{L}) \sum_{i=1}^{L} (A(i) - A'(i))^2 \tag{13}$$

The one dimensional audio signal is transformed, through feature value transform, into audio fingerprint containing watermark, then through image processing, it is transformed into two dimensional image *I*. We then compare it with audio fingerprint image *J*, which is transformed from the original audio signal that does not contain watermark using the same feature value. Eq.(14) calculates the Peak signal-to-Noise Ratio (PSNR) using in our comparison tests.

$$PSNR = 10\log_{10}\frac{255^2}{MSE}$$
$$MSE = \frac{1}{N^2}\sum_{x=0}^{N-1}\sum_{y=0}^{N-1}(I(x,y)-J(x,y))^2 \tag{14}$$

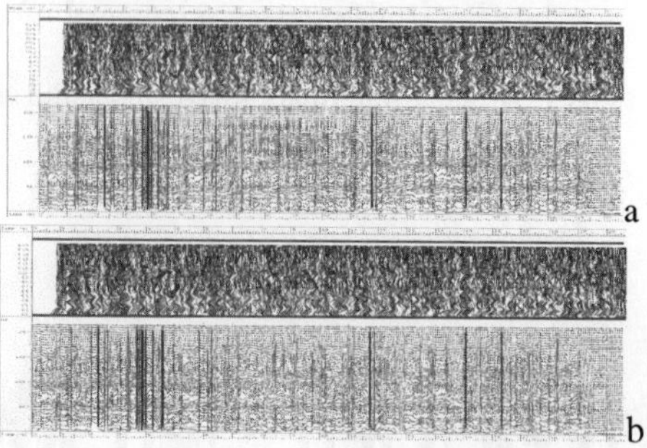

Fig. 9 **a** is the audio fingerprint that is not embedded with watermark. **b** is the audio fingerprint that is embedded with watermark

With the difference values from Figure 9, we will get a feature audio fingerprint. Here there is lots of uniqueness, which can be used in the authentication of audio signal. For example, since the speaking feature of each person is different, the feature found out by Mel cepstral coefficient is also different. In addition, LSP assists the interception of different coefficient feature, when a comparison is made, we will find feature audio fingerprint.

Table 1. Signal processing watermark robustness test

	Attack free	Low- pass filtering	Additive noise	MP3-128kb	Echo addition
watermark					
NC	1	0.9939	0.9970	0.6128	1
psnr	36.1594	23.8008	27.3750	28.1991	27.1862
	Cropping10%front	Cropping10% After	Cropping10%middle	Re-sampling	Amplitude
watermark					
NC	0.7652	1	0.7866	1	0.8537
psnr	29.35	31.6815	31.6815	36.1594	28.3864

(1) Watermark robustness test

In order to describe the signal processing of watermark image during the watermark researches performed by regular researchers, we have added noise and low pass filtering into the experiment. The de-synchronization attack includes amplitude change, pitch change and cutting.

From table 1, it can be seen that our constructed watermark and the robustness test proposed by Chang et al. [6] showed robustness needed by the experiment. Through the embedding of audio signal carrier, when the signal is transferred from the publisher or the speaker to the receiver, the watermark will not be destroyed.

(2) Watermark interpolation attack detection

We also use the music wav format to embed watermark. The proposed DAWP method is used to transform it into DSL audio fingerprint. It is then compared to S original audio fingerprint to get PSNR value of 13.6127, the feature audio fingerprint (see Figure 10). The white part is the feature watermark is arranged in the audio signal, and such feature includes a combination of watermark size and audio size. In order to understand when watermark is interpolated, the audio fingerprint can be obviously compared, we have embedded 32×32 watermark E, its figure is different, but when it is embedded in the same audio, human eye cannot detect it. Through audio fingerprint comparison, PSNR value is 11.8764, which is obviously different than the audio PSNR value of DSL, which means that the audio signal is changed, and from the feature fingerprint, it can be obviously seen that black and white alternate empty spacing arrangement is different. Therefore, if there is no authentication mechanism between audio signal sender A and receiver B, it will be thought that the audio signal owned by E sender, but in fact, it is owned by A DSL sender, which is shown in Figure 11.

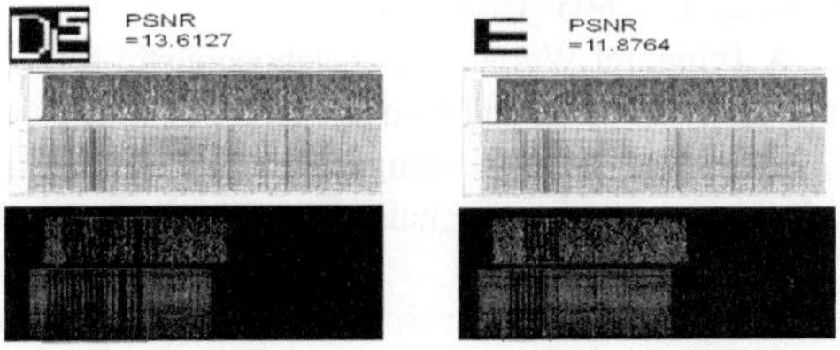

Fig. 10. DSL and E interpolated attack audio fingerprint

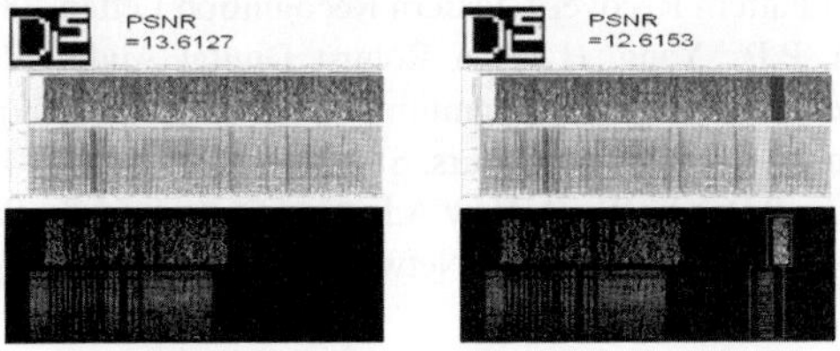

Fig. 11. Audio signal interpolation attack

(3) Audio signal interpolation attack detection

When sender A sends audio to receiver B, the transfer process will include the interpolation of watermark DSL audio, but watermark is still good. If it is not through authentication mechanism, B will think that it is the complete message as given by A, which is as in Figure 11.

6 Conclusion

In this paper, a complete audio authentication plan has been proposed. Through experimental results, we can see that the robust watermark can be used in audio source authentication. Audio fingerprint supports audio integrity authentication. The combination of the audio watermark and fingerprint supplements the purpose of audio authentication.

Acknowledgements. This work was supported by the National Science Council of Republic of China under the grant NSC 99-2221-E-025-004.

References

1. Baluja, S., Covell, M.: Waveprint: Efficient Wavelet-Based Audio Fingerprinting. Pattern Recognition 41, 3467–3480 (2008)
2. Chang, C.Y., Wang, H.J., Shen, W.C.: Copyright-Proving Scheme for Audio with Counter-Propagation Neural Networks. Digital Signal Processing 20, 1087–1101 (2010)
3. Chen, O.T.C., Liu, C.-H.: Content-dependent Watermarking Scheme in Compressed Speech With Identifying Manner and Location of Attacks. IEEE Transactions on Audio, Speech, and Language Processing 15, 1605–1616 (2007)
4. Dai, P., Soon, I.Y.: A Temporal Warped 2D Psychoacoustic Modeling for Robust Speech Recognition System. Speech Communication 53, 229–241 (2011)
5. Dhanalakshmi, P., Palanivel, S., Ramalingam, V.: Pattern Classification Mdels for Classifying and Indexing Audio Signals. Engineering Applications of Artificial Intelligence 24, 350–357 (2011)
6. Hofbauer, K., Kubin, G., Kleijn, W.B.: Speech Watermarking for Analog Flat- fading Bandpass Channels. IEEE Transactions on Audio, Speech, and Language Processing 17, 1624–1637 (2009)
7. Park, C.M., Thapa, D., Wang, G.N.: Speech Authentication System Using Digital Watermarking and Pattern Recovery. Pattern Recognition Letters 28, 931–938 (2007)
8. Wang, X.Y., Niu, P.P., Yang, H.Y.: A Robust Digital Audio Watermarking Based on Statistics Characteristics. Pattern Recognition 42, 3057–3064 (2009)
9. Wang, X.Y., Zhao, H.: Line Spectral Pairs. Signal Processing 88, 449–466 (2008)
10. Wanga, X.Y., Niua, P.P., Qi, W.: A New Adaptive Digital Audio Watermarking Based on Support Vector Machine. A Journal of Network and Computer Applications 31, 735–749 (2008)

PV Array Model with Maximum Power Point Tracking Based on Immunity Optimization Algorithm

Ruidong Xu, Xiaoyan Sun, and Hao Liu

School of Information and Electrical Engineering, China University of Mining and Technology, Jiangsu Xuzhou, 22111
ruidongxu@163.com

Abstract. In order to rapidly tracking the maximum power point of the photovoltaic (PV) system, a model using the immunity optimization algorithm is presented in this paper. The strategy of the integration of immunity algorithm to the maximum power point tracking is given, and the corresponding simulation models are explained. Under the changes of the illumination and temperature, the proposed model can be exactly and timely tracking the maximum output power; therefore, the simulation results show the effective of the algorithm.

Keywords: photovoltaic array; maximum power point tracking; immunity algorithm; model simulation.

1 Introduction

The Environmental pollution is not optimistic for China as a biggest country of coal - consuming in the world which treat the coal as the main energy structure. With the issues of energy shortages and security around the worldwide becoming more and more prominent, the use of renewable energy has attracted wide attention. The photovoltaic (PV) generation system which directly converts solar into electrical energy is one of the most useful and promising technologies in the many of solar applications, and an important content of material and energy revolutions from the twentieth century [1].

The output power of PV array is in connection with the solar radiation, the temperature and the load. The maximum power point tracking (MPPT) circuit is adopted to draw the maximum power of the PV array. The normal algorithms of MPPT [2] include perturb and observation and the incremental conductance algorithm. In the simulation of PV inverters, these algorithms are used to draw the maximum power based on the simulation of the PV array, the output of MPPT circuit is converted into AC power. If the simulation of the PV array and the simulation of the MPPT algorithms are combined, the solar radiation and the temperature of PV simulation can be adjusted, and also the PV array can track the maximum power, the simulation will be simple. The design of the circuit can be guided by such simulations.

In this paper the immunity algorithm is adopted as the MPPT method for the simulation of PV array. The simulation of PV array can track the maximum power point rapidly and right when the radiation and the temperature are changed.

M. Zhu (Ed.): Electrical Engineering and Control, LNEE 98, pp. 535–542.
springerlink.com © Springer-Verlag Berlin Heidelberg 2011

2 Related Work

The equivalent circuit of a PV array is shown as Fig. 1, the output current is as follows:

$$I_o = I_g - I_d \left\{ \exp\left[\frac{q}{AKT} \left(V_o + I_o R_s \right) \right] - 1 \right\} - \frac{V_o + I_o R_s}{R_{sh}} \tag{1}$$

Where I_O is the output current of PV array, I_g is the short current, I_d is the saturation current, q is the charge constant, A is PN ties' coefficient, K is the Boltzmann's constant T is the absolute temperature, U is the output voltage of the PV array.

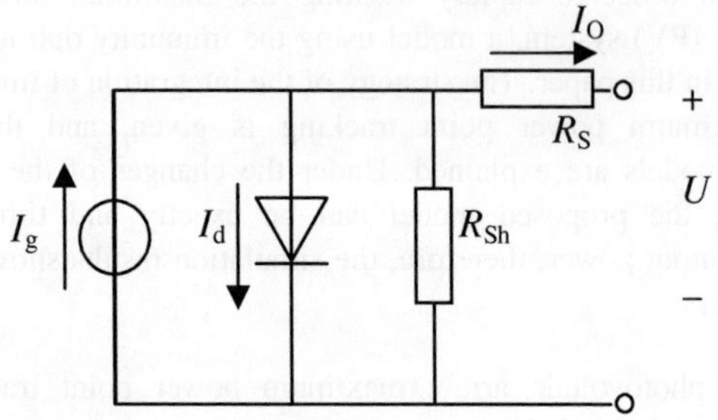

Fig. 1. The equivalent circuit of a PV array.

Artificial immunity systems simulate mechanism of biologic immunity, including clone, hyper-mutation immune, the combination of the antibody and the antigen, the immunity memory course, the fast convergence speed, and best diversity. During the edit, the cell receptor is allowed to be changed in the special condition, which makes the antigenic receptor not easily fall into extreme value in some part.

The immunity algorithms often define the objective functions as antigens and the optimal solutions as antibodies. The affinity stands for the identification degree between an antigen and an antibody or their similar degree. The main process of such algorithms includes: (1) identification of the antigens, the immunity system confirms the antigen's intrusion; (2)generating the initial antibodies, deleting previous antigens and accelerating to produce antibody; (3)calculating the affinity; (4)dividing memory cell, adding the antibody which has big affinity with the antigen to the memory cell ;(5)promoting and inhibiting of the antibody; (6)producing new antibodies. The immunity algorithms have been widely applied to automatic control, fault diagnosis, pattern recognition, machine learning, web security and optimization learning [5].

3 Maximum Power Point Tracking with Immunity Algorithm

In the maximum power point tracking, T and S are viewed as antigens, and the output voltage of the PV array is the antibody. When a new antigen appears (a combination of T and S), the most suitable antibody (V_o) with the fastest speed and the maximum output power of the PV is searched for.

3.1 Trigger Condition of Immunity Response

During the battery's working, the values of *T* and *S* will be changing and a strategy is given here to avoid frequently immunity responses. The new immunity response will be started until the changes of *T* or *S* reaching a certain level. A step function $[x]$ is defined as the largest integer not more than the value of *x*, ε_T as temperature sensitivity, ε_S as luminous sensitivity. In our paper, we set $\varepsilon_T = 1K$ and $\varepsilon_S = 15W/m^2$. *T* and *S* present temperature and light intensity, and T_0 and S_0 are the nearest trigger condition. The immunity response is started when the following equation is satisfied.

$$\left[\frac{|T-T_0|}{\varepsilon_T}\right]+\left[\frac{|S-S_0|}{\varepsilon_S}\right]>0 \tag{2}$$

3.2 Generation of the Initial Antibody Population

The initial antibody population mainly consists of three parts: injecting immunity vaccine, picking up memory cell and randomly generated antibodies. Collecting immunity vaccine means to gather a priori data. The antibodies will be firstly selected from the immunity vaccines when the initial antibody population is generated.

After the immunity response, a certain proportion of vaccine will be chosen as the antibody's initial population according to the affinity degree of the current antigens and vaccines. The greater affinity value is, the larger possibility for antibodies being selected as a member of the initial population. The affinity degree of our algorithm is calculated as the following Eq.3.

$$\theta(T,S)=\frac{\alpha_T}{|T-T_{vac}|+\varepsilon}+\frac{\alpha_S}{|S-S_{vac}|+\varepsilon} \tag{3}$$

Where *T* and *S* are the antibodies of current immunity response; T_{vac} and S_{vac} are antigens corresponding to the antibodies; α_T and α_S are the weighting coefficients, determining the affecting levels of temperature and illuminations to the affinity degree; ε is a small enough integer promising the denominator not being zero.

After the vaccines are injected into the initial antibody population, then picking up a certain ratio of memory cells are picked up and joined to the initial population. The Eq.3 is also used to be a standard to select the memory cells as antibodies. The other antibodies of the initial population are randomly generated in a certain range until the expected size of the population is satisfied.

3.3 Derivation of Antibody Population

The density of an antibody in the population lies on the affinity of this antibody and its antigen. Here, the affinity means the output power *P* of the PV array when the working voltage is represented by a certain antibody. The bigger output power P is, the higher the affinity of the antibody and antigen is. The value of *P* is gained from the multiplication of the practical output voltage and current of the PV array measured through sensors. In our study, we can get the value of *P* from the established PV array stimulation models.

During the derivation course of the antibody population, the density of the *i*-th antibody δ_i can be calculated as follows:

$$\delta_i = P_i / \sum_{j=1}^{n} P_j \tag{4}$$

Where P_i is the corresponding output power of the *i*-th antibody and *n* is the size of the population.

The antibody is sifted from high to low according to the value of the affinity, and the antibody with higher affinity will be cloned automatically into the next generation. Meanwhile the selected antibodies will produce τ mutated ones to go into the next generation. The size of the population is constant and the elimination rate of each antibody is calculated according to the following equation:

$$\sigma = \frac{\tau}{\tau+1} \times 100\% \tag{5}$$

Here, we set $\tau = 3$ and so the elimination rate is $\sigma = 75\%$.

The mutation rule of the antibodies is given as Eq.6.

$$\omega_i(k+1) = Rand(-0.5, 0.5) \times \lambda + \omega_i(k) \tag{6}$$

Where $\omega_i(k)$ is the i-th antibody of k -th generation population; $Rand(a,b)$ will produce a random number between *a* and *b*; λ is the fluctuation range of the mutation, deciding the mutation range and fluctuation level.

3.4 Immunity Corresponding Termination Condition

In the PV generation system, the final aim is, the PV array always keeps working in the maximum power point during the changing of environment. We still can fix the environment variation during the dozen -hundreds of milliseconds of each time immunity corresponding. Therefore, keeping antibody not change, we can treat the output power if reaching into maximum value or not as the condition of judging whether the immunity corresponding is finished or not.

In the simulation, an viewer is adopted to record two values, one is the maximum output power $P_{\max}(k)$ of *k*-th generation and the other is the maximum output power of the previous generation $P_{max}(k-1)$. A comparing unit used to compare the above two values is started after each update of the evolutionary population. If $P_{max}(k) \le P_{max}(k-1)$, the counter consumes 1, otherwise it is cleared to zero. The optimal antibody corresponding to the current antigens is obtained when the value of the counter meets a priori. The current immune response is terminated and the optimal antibody is preserved to the bank of memory cells.

4 Simulation Results and Analysis

4.1 PV Array Simulation Model

In practice, the values of R_s and R_{sh} are not greatly varied and they can be viewed as constants. Besides, the value of R_{sh} is very large, then the last term of Eq.1 can be ignored and it is simplified as Eq.7.

$$I_o = I_g - I_d \left\{ \exp\left[\frac{q}{AKT}(V_o + I_o R_s) \right] - 1 \right\} \tag{7}$$

For

$$I_d = I_{dr} \left(\frac{T}{T_r} \right)^3 \exp\left[\frac{qE_{Go}}{BK} \left(\frac{1}{T_r} - \frac{1}{T} \right) \right] \tag{8}$$

$$I_g = \frac{S}{1000} \left[I_{SCR} + K_I (T - T_r) \right] \tag{9}$$

According to the above three equations, the simulation model of PV array modeled in Matlab/Simulink is shown as Fig.2, the I_g is simulated as Fig.3 and the I_d is simulated as Fig.4.

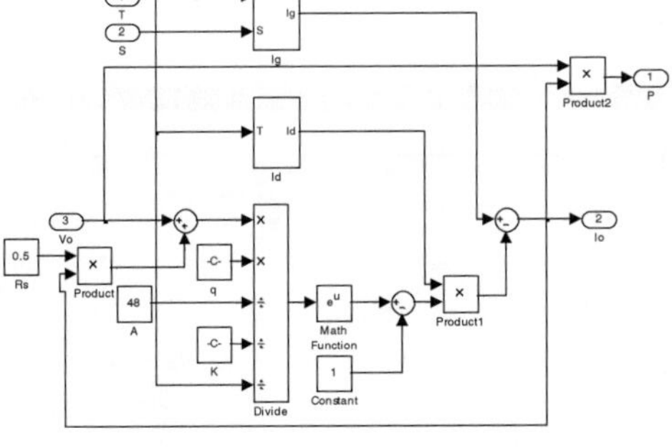

Fig. 2. PV array stimulation model.

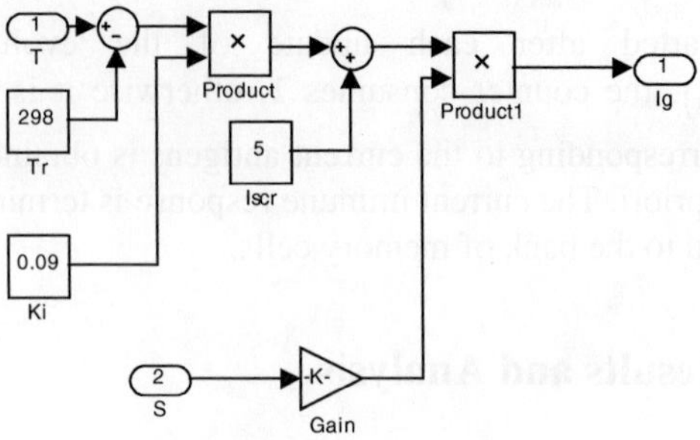

Fig. 3. I_g stimulation model.

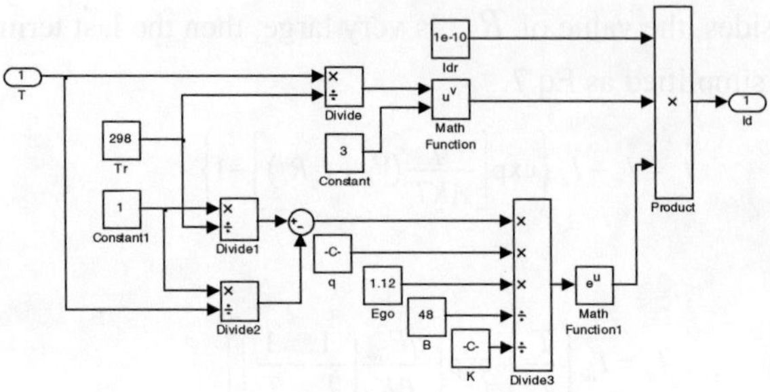

Fig. 4. I_d stimulation model

Making the above immunity optimum algorithm integrate with solar energy battery each stimulation model, this composes PV array stimulation model with maximum power point tracking function, in different illumination, its output is just voltage and current in the maximum power point, in order to change PV array's illumination intensity and working temperature in real time, we designed the PV array alternative interface (refer to Fig.5.)

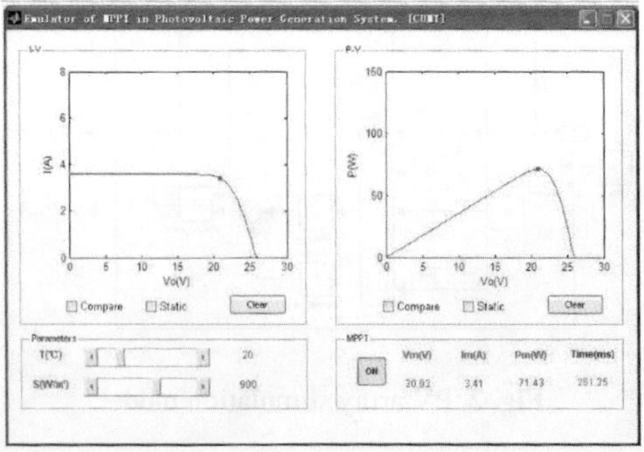

Fig. 5. Alternation interface of PV array stimulation model

4.2 Result and Analysis

In order to completely test the effectiveness of the algorithm, after given a group of regular changing antibody (*T* and *S*), we get maximum power point tracking effect (refer to Fig.6.). The figure includes immunity corresponding result after many-times antibody invasions, every curve represents the P-V relations of the PV array during on certain antigen invasions. The marking points on the curve represents optimal antibody which is produced after this immunity response, as well as corresponding maximum output power point.

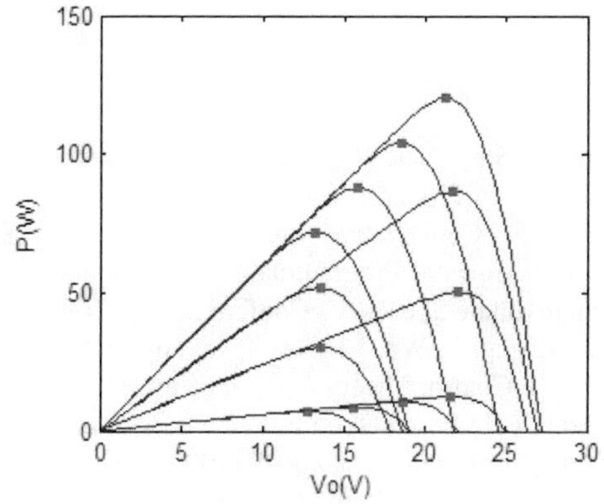

Fig. 6. MPPT stimulation result picture using immunity optimum algorithm

In the figure, the marking points are always appearing exactly in the peak value position of the corresponding P-V curve, which just proves that the proposed algorithm is effective for tracking the maximum power point.

In order to verify the computational complexity of the immunity optimal algorithm, the consumed time is tracked for some random temperatures and illuminations on condition of obtaining the maximum power point. The results are shown as Fig. 7. For 200 times experiments, the average value of the consumed time is about 150ms even if some extreme values. This average value is enough for responding the change of the environment, which means that our algorithm is practical and feasible.

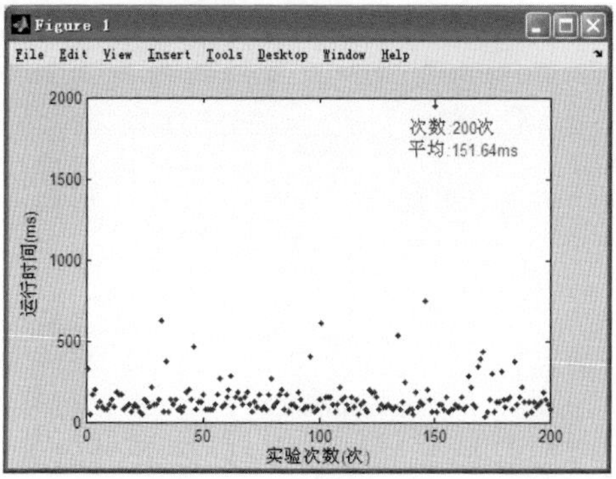

Fig. 7. Consumed time of the algorithm

5 Conclusions

In this paper, we adopted an immunity optimal algorithm to track the maximum power point of PV array on the base of traditional study of solar PV generation system maximum power point tracking algorithm. The simulation models of PV system with maximum power point tracking based on the immunity algorithm is studied. The simulation models are presented in detail and the simulation results are analyzed. From this research, we can conclude that this algorithm can rapidly track the maximum power point of PV array using very short time.

References

1. Zhang, Y.M.: The Current Status and Prospects of solar photovoltaic industry in China. J. Energy Research & Utilization 113(1), 1–6 (2007)
2. Hohm, D.P., Ropp, M.E.: Comparative Study Of Maximum Power Point Tracking Algorithms Using an Experimental Programmable, Maximum Power Point Tracking Test Bed. In: IEEE 28th Photovoltaic Specialists Conference, pp. 1699–1792 (2000)
3. Liu, Z.H., Zhang, Y.J., Zhang, J., Wu, J.H.: A Novel Binary-State Immune Particle Swarm Optimization Algorithm. J. Control Theory & Applications 28(1), 65–75 (2011)
4. Hu, F.X., Guo, H.J., Sun, Y.F.: Research on Immune Algorithm Theory and its Application. J. Computer & Digital Engineering. 37(7), 46–49 (2009)
5. Wang, Q., Lv, W., Ren, W.J.: Immune Genetic Algorithm and Applications in Optimization. J. Application Research of Computers. 26(12), 4428–4431 (2009)

Ground Clutter Analysis and Suppression of Airborne Weather Radar*

Shuai Zhang, Jian-xin He, and Zhao Shi

CMA Key Laboratory of Atmospheric Sounding, Chengdu University of Information Technology
Shuangliu, 610225, Chengdu, China
zs1101@163.com

Abstract. In this paper, Calculates, simulates and analyzes the ground clutter power spectrum of airborne weather radar; describes the simulation process and gives a calculation formula. Analyzes the power spectrum of the main lobe ground clutter; Based on the analysis, designs the complex coefficient filter to filter out the main lobe clutter, and gives the simulation results.

Keywords: Airborne weather radar, Ground clutter Doppler frequency shift, Spectral width, Complex coefficient filter.

1 Introduction

Airborne meteorological radar observation system mainly focuses the target detection on the air-base platform. It is the middle level of the meteorological observation for ground, air and space based levels. By observation on the space platform, the object of interest such as torrential rains, typhoons and other weather phenomena can be observed with fine structure of high selectivity and high mobility, and improve the detection capability of weather systems. However, the application of airborne weather radar faces very complex clutter environment. In the look down detection mode, High performance signal processing technology must be used to suppress ground clutter. To effectively suppress the clutter, it must be full understanding of the power spectrum and spatial characteristics of clutter. Therefore, it is important to calculate the ground clutter spectrum and analysis the clutter power spectral. Relative movement between airborne radar and ground, combined with the impact of the antenna pattern, making ground clutter spectrum of airborne radar have significant changes. Ground clutter of airborne radar is divided into the main lobe clutter, sidelobe clutter and the altitude clutter. The mainlobe clutter is the clutter that generated when the main lobe of the radar antenna illuminate the ground. The sidelobe clutter is the clutter that generated when sidelobe beam irradiate to the ground. The altitude clutter is the clutter that generated when sidelobe beam irradiation to the ground along vertical direction.

* This work was partially supported by the National Science Foundation of China (41075010).

M. Zhu (Ed.): Electrical Engineering and Control, LNEE 98, pp. 543–549.
springerlink.com © Springer-Verlag Berlin Heidelberg 2011

2 Analysis and Calculation of Ground Clutter

2.1 The Factors Affecting the Power Spectrum of the Ground Clutter

(1)The impact of topography
Usually, the ground clutter is non-uniform, the clutter model commonly used statistical models. Backscattering coefficients of ground clutter obey a certain distribution. In this paper, using the modified surface scattering model, the relationship between the backscattering coefficients of ground clutter δ and the angle of radar beam and ground ϕ is defined as: the distribution of δ consists of two components:

$$\sigma(\phi) = \delta_{od} \sin\phi + \delta_{os} \exp(-(90-\phi)^2 / \phi_0^{\ 2}) \tag{1}$$

Where, the first term of formula is the diffuse component; the second is the specular component, which constitutes the altitude clutter of airborne weather radar. δ_{od} and δ_{os} is a known quantity and $\delta_{od} << \delta_{os}$.
(2) The impact of meteorological conditions
Different weather conditions make signal attenuation, scattering, and reflection strength of the radar transmitter very different.
(3)The impact of radar and set machine parameters
Band of radar, pulse repetition rate, beam scanning, signal processing, radar carrier altitude and flight speed of the carrier aircraft, the angle of the antenna beam and the velocity have a significant impact on the clutter spectrum.

2.2 Calculation of Ground Clutter

Make the following assumptions:
Ground reflection is uniform. Unit area equivalent to the cross section area of the clutter is only concerned with the grazing angle.
The ground is approximately flat. Considering the actual radius of the earth, we should give a certain effective radius of the Earth model, but compared with the radius of the earth, carrier aircraft altitude and detection range is very small, thus, the curvature of the Earth is negligible.
Then, the two-dimensional power spectrum of the ground clutter is[5]

$$s(f_d, R) = \frac{P_t \lambda^3 G^2(\theta,\phi)\sigma(\phi)L}{(4\pi)^3 R^3 V_a \cos D \cos\psi \sqrt{1-(\dfrac{\lambda f_d - 2V_a \sin D \sin\psi}{2V_a \cos D \cos\psi})^2}} \tag{2}$$

Where, P_t is the peak power of the transmitter, $G(\theta,\psi)$ is antenna gain, R is the distance between antenna and the ground clutter, λ is wavelength of the radar, L is the propagation loss factor of the system, V_a is the speed of carrier aircraft, f_d is Clutter Doppler shift of the clutter unit, D is the carrier aircraft dive angle, θ is azimuth.

Figure 1 and figure 2 shows the simulation result of the power spectrum of the clutter, Figure 1 shows the of two-dimensional power spectrum of the ground clutter, figure 2 shows the location of the mainlobe clutter and its spectral width.

The simulation results shows two-dimensional spectrum of the main lobe clutter is influenced by beam width. The mainlobe clutter appears between 40km and 65km, the center frequency changes from 850Hz to 950Hz.

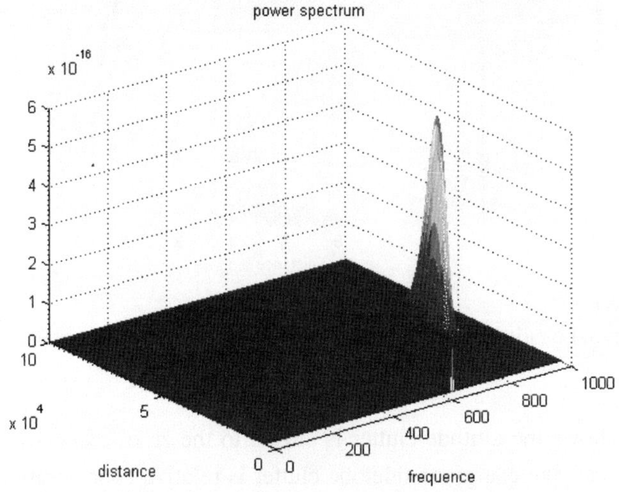

Fig. 1. Two-dimensional power spectral of ground clutter

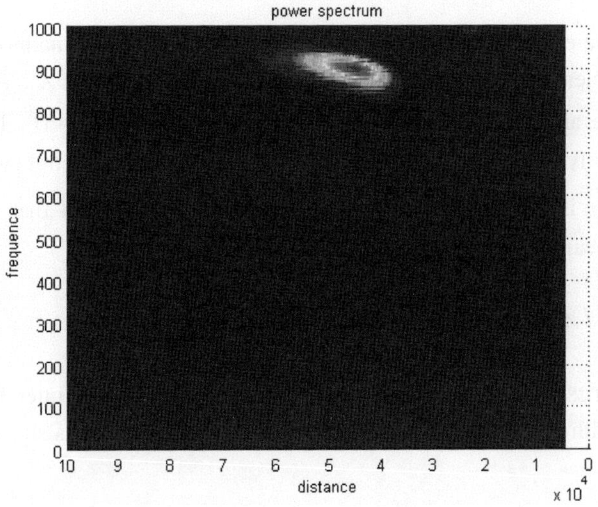

Fig. 2. The location of the main clutter and its spectral width

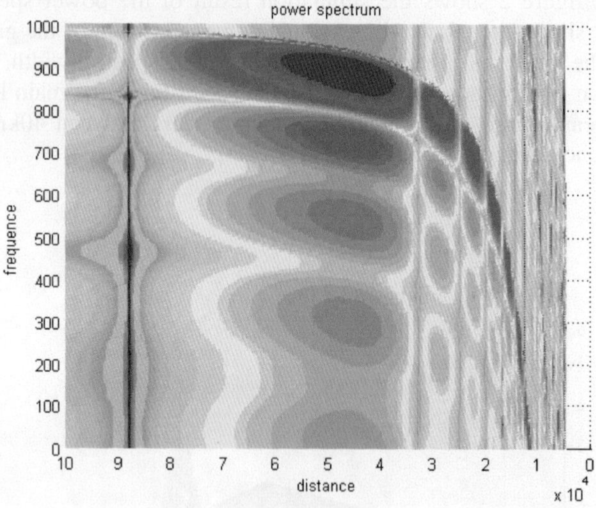

Fig. 3. Logarithmic two-dimensional power spectrum of the ground clutter

Figure 3 shows the altitude clutter is closed to the zero frequency, and appears at the nearby areas; the energy of sidelobe clutter is relative concentration. Strength and Doppler frequency of the ground clutter changes with the distance.

2.3 The Analysis of Mainlobe Clutter Spectrum

According to the effect of doppler, if V is radial velocity of the mainlobe irradiated area. The doppler shift of the mainlobe irradiated area is $2V/\lambda$. As shown in Figure 4, if the aircraft flies along the horizontal direction, $D = 0^0$ .The speed of carrier aircraft is V_a ,then, $V = 2V_a / \lambda * \cos\theta \cos\psi$. Spectral width of the mainlobe clutter is the doppler shift of the beam front substract that of the beam back[1], if the beam width is ψ_b, the spectral width is

$$2V_a / \lambda * \cos\theta(\cos(\psi - \psi_b / 2) - \cos(\psi + \psi_b / 2)) \tag{3}$$

Besides, $\psi = \arcsin(H / R)$, we can come to that the ground clutter Doppler shift and spectral width changes with the distance. The simulation results shows in figure 5 and figure 6.

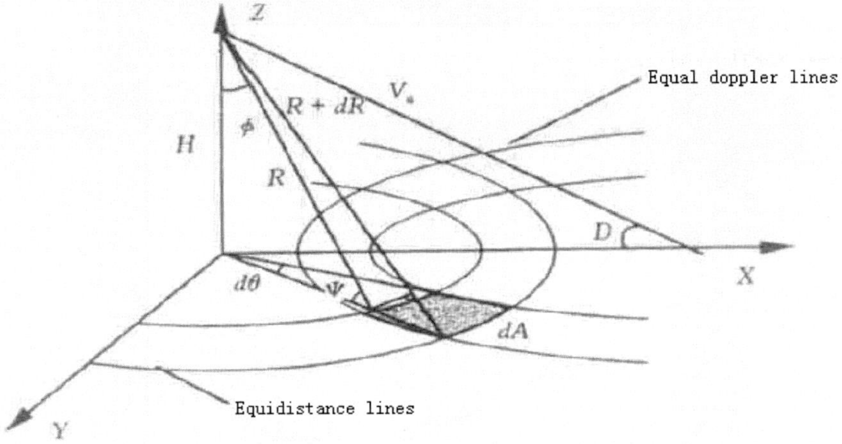

Fig. 4. The relationship between carrier aircraft and the ground clutter uint

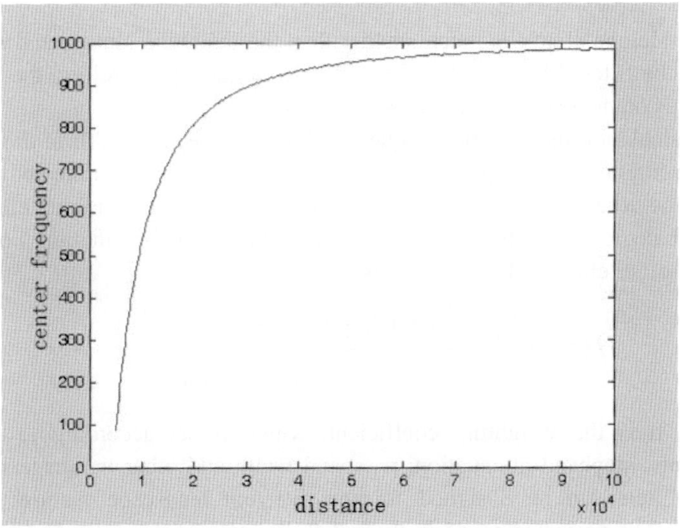

Fig. 5. The center frequency of the mailobe clutter changes with distance

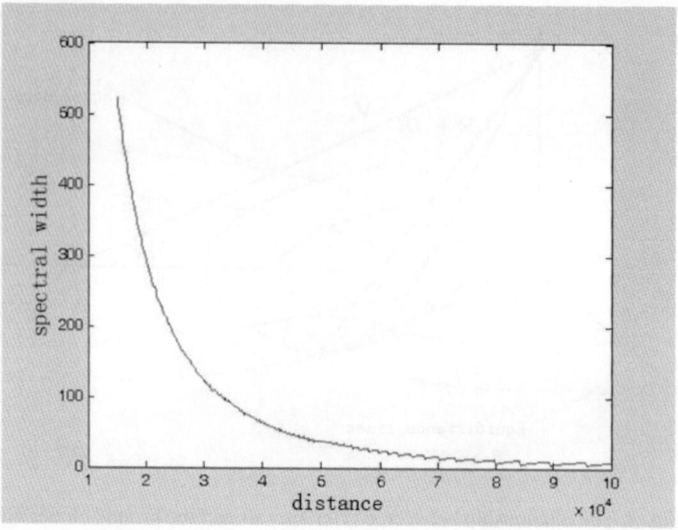

Fig. 6. The spectral width of the mailobe clutter changes with distance

3 Filter Design

Since the sidelobe clutter is much smaller than the mainlobe clutter, and the latitude clutter can be filtered by the filter fixed at zero frequency, therefore, the main work should focus on the main lobe clutter suppression.

After calculation the Doppler frequency shift and spectral width of the main lobe clutter, the filter can be designed to suppress the ground clutter. Finite and the infinite filters can be used to create the weight coefficients library. The Infinite elliptic filters have a high degree of inhibition and the smaller amount of computation. For example, the Fourth-order elliptic filter function is

$$H(z)=\frac{b_0+b_1z^{-1}+b_2z^{-2}+b_3z^{-3}+b_4z^{-4}}{1-a_1z^{-1}-a_2z^{-2}-a_3z^{-3}-a_4z^{-4}} \tag{4}$$

Where, a, b is the weighting coefficient, which is set according to the design requirements, stopband attenuation, passband width and other parameters. Different bandstop filters can be designed for the different landscape features and radar parameters. The projected speed in the radial direction of the radar can be estimated by the signal of the sensor on the plane, then, the complex coefficient can be calculated. The selected weight coefficients library multiplied by the complex coefficients products the coefficients of the ground clutter filter[2]. The filter completed by the convolution of the input sequence and the filter coefficients. Figure7 is the map of the power spectrum after the filter using the complex coefficients filter.

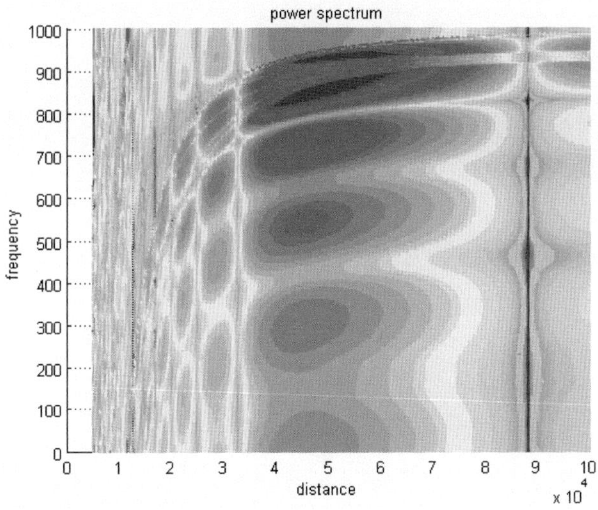

Fig. 7. The power spectrum of the ground clutter filtered by the complex coefficients filter

4 Concluding Remarks

The paper analyzes the ground clutter characteristics of the airborne weather radar, gives the formula for calculating the clutter. The calculation of the power spectrum is very important to the filter design of the ground clutter. Besides, this paper also designs the complex coefficients filter based on the analog of the power spectrum of the ground clutter and gives the simulation of the power spectrum after the filter. The mainlobe ground clutter of the airborne weather radar is effectively suppressed by the designed filter.

References

1. Stimon, G.W.: Introduction to airborne radar, Mendham. SciTech Publishing, Inc., New Jersey (1998)
2. Jamora, D.A.: Designing Clutter Rejection Filters With Complex Coefficients for Airborne Pulsed Doppler Weather Radar, NASA Contractor Report 4550 DOT/FAA/RD-93/24 (1993)
3. Gao, Y.-C.: The Research for Key Techniques of Airborne Meteorological Radar Observation System (2009)
4. Yang, J.-H.: Research on Observation method of Phased Array Weather Radar (2008)
5. Lu, X.-Y., Huang, H.-M.: Airborne Radar Clutter Simulation and Analysis (2002)

Research on Mobile Internet Services Personalization Principles

Anliang Ning[1], Xiaojing Li[1], Chunxian Wang[1], Ping Wang[2], and Pengfei Song[3]

[1] Computing Center, Tianjin Polytechnic University. 300160, Tianjin, China
[2] School of Electronic Engineering & Aotumation, Tianjin Polytechnic University
[3] School of Computer and Information Science, Southwest University, Chongqing, China
nanliang@acm.org, yiran_ning@163.com

Abstract. Service personalization has recently become a very promising research area because it highly sensitive to the immediate environment and requirements of the mobile user. The complexity with many aspects and issues that need to be resolved, such as contents and contexts play critical role with mobility-afforded and its usability and accessibility was presented in the paper. The design principles and creation approach of service personalization also be proposed, which focus on user-centric, simplifying and common ways to be completed for the benefit of mobile user.

Keywords: Mobile Services; Services Personalization; Mobile Internet.

1 Introduction

Service personalization has recently become a very promising research area because it paves the way to a more widespread usage, creation, and customization of services. Its goal is to bring users from a passive role (i.e., content and service consumer) to an active role (i.e., content producer and service creator). Service creation technology is the key criterion which represents how user interacts with the system to create or compose a new service. Since ease-of-use is a key success factor of a service creation approach, different service creation approaches in increasing order of ease-of-use, namely: (i) Script-based, (ii) Rule-based, (iii) Choreography-based, (iv)Template-based, and (v) Natural-language based.

Service personalization performed by means of domain-specific scripting languages [5] is generally hard to use for developers. Rule-based approaches based on ECA rules [4] or policy languages [3] have been used in telecom world for personalization of value-added services. While rules are easy to set, it is quite difficult for end users to foresee possible undesired side effects due to rules conflicts. Choreography-based approaches are gaining their momentum in the end-user service creation world, as proved by the increasing popularity of mashups and the birth of various environments for the intuitive non-expert creation of Web based information services, driven by the big companies, such as Yahoo! Pipes or Microsoft Popfly. These environments present graphical tools based on drag-and-drop interfaces which allow the user to create this little web-based applications even without any computing knowledge.

M. Zhu (Ed.): Electrical Engineering and Control, LNEE 98, pp. 551–558.
springerlink.com © Springer-Verlag Berlin Heidelberg 2011

The natural language approach aims at deriving the formal specification for a service composition starting from an informal request expressed in natural language. Such formal, machine-understandable specification can then be used as input for providing the composite service through automated reasoning, like backward-chaining [8]. Unfortunately, natural-language based techniques such as [14] are still limited to the predefined vocabularies and ontology concepts, scalability, and the intrinsic difficulties in handling the ambiguities of natural languages.

2 Mobile Internet Services

The needs of the wireless and mobile user regarding information access and services are quite different than those of the desktop user. This need is not about browsing the Web but about receiving personalized services that are highly sensitive to the immediate environment and requirements of the user. The mobile internet services are positioned as a unique mobile service and not as 'the Internet on your phone's [10], thereby avoiding unfavorable comparisons with service from the fixed Internet. To this end, and in simple terms, the mobile internet system should devise a flexible middleware and protocols to handle the diversity of content structures and their semantics.

In the Mobile World, delivering targeted, timely information and services is essential. The mobile wireless and Internet are clearly driving the need for more content that is varied, customizable, and available anywhere, anytime, and at low cost. With the penetration of mobile devices reaching 60% or more in many countries, it is only natural that an increasing number of people will access the Internet and invoke services and applications from such mobile devices. Mobile web and data services such as LBS (Location Based Services) and streaming media will drive the development of future mobile networks, in large part, by content and Web-based services, and user experience [1]. The development of these applications and services is driven by consumer behaviors. And the key to commercial success lies in understanding consumers, their lifestyles and attitudes, and in creating the product-service combinations that match their wants and needs.

2.1 Key Issues for Personalized Service

The problem of personalization is a complex one with many aspects and issues that need to be resolved. Some of these issues become even more complicated once viewed from a wireless perspective. Such issues include, but are not limited to, the following:

• What content to present to the user. How to decide what to show, using user profiles, using the user history to predict future needs, etc [4]. When using user profiles we must address the need for (1) storing the interests of the user in a format that is easy to be used, be updated or moved,and (2) relating interests and items based on a semantics level.

• How to show the content to the user. Many users want to see the same things but the form they want the data presented might be different. In the wireless environment this also relates to the mobile device used and its specific characteristics.

• How to ensure the user's privacy. Every personalizing system needs (and records) information about the habits of each user. This leads to privacy concerns as well as legal issues [4]. It also leads to lack of user trust and could result in the failure of the system due to the avoidance of its use.

• How to create a global personalization scheme. The user does not care if a set of sites can be personalized but that at each one of them he has to repeat the personalization process. This is especially annoying and cumbersome for the user on the move carrying a resource poor mobile device.

These are the major issues of personalization. They could be summarized in the following phrase: "What, how and for everything". The variability in computing resources, display terminal, and communication channel require intelligent support on personalized delivery of relevant data and services to mobile users. Personalized service provisioning presents several research challenges on context information management, service creation, and inherent limitations of mobile devices. In order to create and provide/deliver personalized services for the mobile Web, several key research questions should be handled in the first place.

- How to retrieve and manage context information from various context providers is at the very core of these issues.

- Secondly, apart from rendering rich and intuitive graphical service creation environment, non-expert users need considerable help in making a sound composite service. Some inherent complexity of programming should be shielded from these users.

- Need to handle the short-of-resource feature of mobile devices. Mobile devices posses, to a certain extent, limited resources (e.g., battery power and input capabilities). Therefore, mobile devices should better act as passive listeners (e.g., receiving the results) than as active tools for service invocation, so that the computational power and battery life of the devices can be extended.

3 Contents and Contexts

3.1 Personalized Contents

Today, content plays a critical role in the success of the Internet, and will be even more important when the access goes broadband wireless with mobility [4]. The ability to access the Internet anytime, anywhere, anyhow (from a multitude of devices) will give impetus to new services and applications. Emerging systems of 3G and beyond make it possible to determine the context of users, places or objects, by collecting information from sensors, systems and (mobile) devices. The information can be used to adapt the behavior of services automatically, resulting in so-called context-aware services [12]. Context awareness implies the information regarding the user's environment (context) is used to adapt services to his or her current situation and needs. An important type of context-aware service is location-based services that adapt to the current location of the user, including the popular in-car navigation services. Other context-aware mobile services use social context-related information, e.g. presence services like instant messaging. These examples illustrate how context awareness may increase both the usefulness and usability of mobile services, and add value. Context-awareness is being considered in order both to drive user interaction and its services, and to select and tailor services to the context of the usage and system.

3.2 Context Information

Mobile Information System should have capable of adapting to context change, providing flexible context-aware information and service access, and a flexible execution environment [14]. The main goal of the design of adaptive information services is to consider all possible interfaces, keeping in mind the possibility of adapting to different physical devices, the context of use, and user preferences. Some important issues are the usability of the system and its accessibility, which means that a system should be usable by everyone, accommodating variety in technology and diversity of users. In order to personalize the service creation and provision processes some context information is needed. This information can be categorized into three major sets: user context, device context, and service context.

User Context. User context refers to the information related to a user such as preferences, calendar or location. This information is also referred to in the literature as Personal Identifiable Information (PII). PII is any piece of information that—related both digital and real identities—that can potentially be used to uniquely identify, contact, or locate a single person [10, 7]. Telecom operators have gathered loads of information about their customers over the years, thus creating rich users' profiles. They are also able to retrieve dynamic personal information such as customers' location or presence status. This information is exposed by means of Web services interfaces, which are standardized as OMA enablers.

Device Context. The device context information includes hardware and software characteristics of the user's devices, which could be used to drive the service execution or tailor the contents that a service will present to the user. The World Wide Web Consortium (W3C) has created the Composite Capability/Preference Profiles (CC/PP) Recommendation [12], which allows defining capabilities and preferences of users' mobile devices. OMA has specified the User Agent Profile (UAProf) [18], capturing capabilities and preference information for wireless devices. UAProf allows describing device and user agent capabilities as an XML description, and is fully compatible with CC/PP. Moreover, OMA has also specified the Mobile Client Environment enabler, which is specifically chartered to be responsible for base content types, with the intention of enabling the creation and use of data services on mobile hand held devices, including mobile telephones.

Service Context. The service context includes information related to a service. Examples of service context include: (i) service location, (ii) service status (e.g., available, busy), and (iii) Quality of Service (QoS) attributes (e.g., price, availability). The service context information can be obtained via a monitoring application that oversees the execution status of the service (e.g., exceptions).

Context Privacy and Security. The Internet experience has demonstrated that personal data (i.e., user context information) is a valuable asset that can be used incorrectly or fraudulently, and thereafter is prone to be abused and misused. In the technical plane, Identity Management (IM) is the discipline that tries to address all these issues: it refers to the processes involved with management and selective disclosure of PII within an institution or between several of them, while preserving and enforcing privacy and security needs. If different companies have established trust relationships between themselves the process does not require a common root authority and is called Federated Identity Management. Each company maintains its own customer list and some identity attributes or PII, but they can securely exchange them while preserving users' privacy. IM is related to network security, service provisioning, customer management, Single Sign On (SSO), Single Logout (SLO) and the means to share identity information [9]. A federated approach for IM in a Web Services environment is supported by various standards and frameworks, among which the most important ones are Security Assertion Markup Language (SAML),[7] Liberty Alliance,[8] and WS-Federation [2].

4 Design Principles

Service Provision principles and technologies, that leveraging Web services and software agents in combination with publish/subscribe systems, provides the foundation to enable effective access to integrated services in mobile Web environments.

Web services provide the pillars for evolving the Internet into a service-oriented integration platform of unprecedented scale and agility. More specifically, Web service technology is characterized by two aspects that are relevant to accessing heterogeneous resources and applications. From a technology perspective, all interacting entities are represented as services, whether they providing or requesting services. This allows uniformity that is needed for the interaction among heterogeneous applications and resources such as mobile devices. Web services can bring about the convergence of wired and wireless applications and services.

Agents are software entities that exhibit certain autonomy when interacting with other entities [14]. Agents use their internal policies and knowledge to decide when to take actions that are needed to realize a specific goal. Internal policies can use context information (e.g., user preferences, device characteristics, and user location) to enable agents to adapt to different computing and user activities. Agents can also be used to pro-actively perform actions in dynamic environments.

The combination of services and agents provide a self-managing infrastructure. Agents extend services by embedding extensible knowledge and capabilities (e.g., context aware execution and exception handling policies) making them capable of providing personalized and adaptive service provisioning in dynamic environments.

The publish/subscribe paradigm offers a communication infrastructure where senders and receivers of messages interact by producing and consuming messages via designated shared spaces [11]. The communication is asynchronous in the sense that it completely decouples the senders and receivers in both space and time. This enables mobile users to disconnect at any time—either voluntarily to save e.g., communication cost and battery power, or involuntarily due to breakdowns of connections—and re-synchronize with the underlying infrastructure upon reconnection. This form of decoupling is of paramount importance in mobile Web environments, where the entities (e.g., mobile users) involved in communication change frequently due to their movement or connectivity patterns. This communication paradigm caters for loosely coupled interactions among service agents, which has been identified as an ideal platform for a variety of Internet applications, especially in wireless environments.

4.1 Personalized Service Creation Approach

The approach is centered on the concept of process templates. Users specify their needs by reusing and adjusting existing process templates, rather than building their own services from scratch. Process templates are reusable business process skeletons that are devised to reach particular goals such as arrangement of a trip. Each process template has one or more tasks and each task is associated with a service operation. Process templates are modeled using process definition languages such as statecharts [18] and BPEL4WS [3].

In the approach, values that users supply as input parameters are handled differently from the values obtained from user profiles. Indeed, because mobile devices are resource-constrained, values that can be obtained from user profiles should not be requested from users. Users only supply the values for data items that are labeled compulsory. However, in a process template specification, the template provider only indicates which input parameters users have to supply a value. It is the responsibility of the user to specify, during the configuration phase, if the value will be provided manually or derived from her profile.

4.2 Process Templates Configurations

Personalization implies making adjustment according to user preferences. Three kinds of user preferences are associated for each process template's task:

•**Execution Constraints** can be divided into temporal and spatial constraints, which respectively indicate when and where the user would like to see a task executed. Generally, a temporal constraint involves current time, ct, comparison operator, co (e.g., =, ≤, and between), and a user-specified time, ut, which is either an absolute time, a relative time (e.g., termination time of a task), or a time interval. A temporal constraint means that a task can be triggered only if the condition ct co ut is evaluated to true. Similarly, a spatial constraint involves current location, cl and a user-specified location, ul . A spatial constraint means that a task can be fired only when the condition cl = ul is evaluated to true. A location is considered the same as another location if the distance between two locations does not exceed a certain value (e.g., 2 m). It should be noted that the temporal and spatial constraints can be empty, meaning that the corresponding task can be executed at anytime and at anywhere.

•**Data Supply and Delivery** preferences are related to supplying values to the input parameters and delivering values of output parameters of the task. As stated before, the values of some input parameters of a task can be obtained from a user's profile. The user proceeds in two steps: (i) identify which input parameter values can be derived from her profile, and (ii) supply the location of the profile and the corresponding attribute names. Similarly, for the output parameters of a task, a user may specify which parameter values need to be delivered to her.

•**Execution Policies** are related to the preferences on service selection (for communities) and service migration during the execution of a task. The execution policies include the service selection policy and the service migration policy. For a specific task, users can specify how to select a service for this task. The service can be a fixed one (the task always uses this service), or can be selected from a specific service community [6] or a public directory (e.g., UDDI) based on certain criteria (e.g., location of the mobile user). Furthermore, users can specify whether to migrate the services to their mobile devices (e.g., if mobile devices have enough computing resources) or to the sites near the users current location for the execution. Policies can be specified using policy languages like Houdini [19] and Ponder [2].

5 Conclusion

There are many approaches [12] to personalization and each one of them usually focuses on a specific area, such as profile creation, machine learning and pattern matching, data and web mining or personalized navigation. So far, to our knowledge, there has not been an approach towards a total solution or more importantly focusing specifically to the wireless/mobile user. Process templates-based approach aims at simplifying choreography approach, offering different common templates to be completed by user's data, thus providing an easier-to-use interface. Users can select a process template which can be seen as a pre-defined standard workflow (i.e. service composition).

We deemed appropriate to put forward novel solutions and alternatives for the benefit of mobile users. The system builds upon the building blocks of Web services, agents, and publish/subscribe systems and provides a platform through which services can be offered to mobile users. One possible extension to our current system is a mechanism for seamlessly accessing services among multiple computing devices. Indeed, during the invocation of a Web service, especially one having long running business activities or with complex tasks (e.g., composite services), users are more likely to be switching from device to device (e.g., from office PC to PDA).

References

1. Bouwman, et al.: Mobile Service Innovation and Business Models. Springer, Berlin (2008)
2. Johan, Z.: Implementing Value-Added Telecom Services. Artech House Inc., Boston (2006)
3. Hansen, M., Schwartz, A., Cooper, A.: Privacy and Identity Management. IEEE Security and Privacy 6(2), 38–45 (2008)
4. Forbrig, P., et al.: Developing a User-Centered Mobile Service Interface Based on a Cognitive Model of Attention Allocation. In: Human-Computer Interaction, pp. 50–57. Springer, Boston (2010)
5. Hong, S.-J., Thong, J.Y., Moon, J.-Y., Tam, K.-Y.: Understanding the behavior of mobile data services consumers. Information Systems Frontiers 10, 431–445 (2008)
6. Kuo, Y.-F., Yen, S.N.: Towards an understanding of the behavioral intention to use 3G mobile value-added services. Computers in Human Behavior 25, 103–110 (2009)
7. Friedewald, M., Raabe, O.: Ubiquitous computing: An overview of technology impacts. Telemat. Inf. 28(2), 55–65 (2011)
8. Verkasalo, H.: Contextual patterns in mobile service usage. Personal Ubiquitous Comput 13(5), 331–342 (2009)

9. Costaet, D., et al.: Designing a configurable services platform for mobile context-aware applications. International Journal of Pervasive Computing and Communications, 13–25 (2008)
10. Genco, A., et al.: An agent-based service network for personal mobile devices. IEEE Pervasive Computing 5(2), 54–61 (2006)
11. Griffin, D., Pesch, D.: Service provision for next generation mobile communication systems - the Telecommunication Service Exchange. IEEE Transactions on Network and Service Management 3(2), 2–12 (2006)
12. Sun, C.-a., et al.: Modeling and managing the variability of Web service-based systems. Journal of Systems and Software 83(3), 502–516 (2011)
13. Fortier, A., et al.: Dealing with variability in context-aware mobile software. Journal of Systems and Software 83(6), 915–936 (2011)
14. Sama, M., et al.: Multi-layer faults in the architectures of mobile, context-aware adaptive applications. Journal of Systems and Software 83(6), 906–914 (2011)
15. Biel, B., Grill, T., Gruhn, V.: Exploring the benefits of the combination of a software architecture analysis and a usability evaluation of a mobile application. Journal of Systems and Software 83(11), 2031–2044 (2011)
16. Bae, J., Lee, J.-Y., Kim, B.-C., Ryu, S.: Next Generation Mobile Service Environment and Evolution of Context Aware Services. In: Sha, E., Han, S.-K., Xu, C.-Z., Kim, M.-H., Yang, L.T., Xiao, B. (eds.) EUC 2006. LNCS, vol. 4096, pp. 591–600. Springer, Heidelberg (2006)
17. Hassan, M.: Mobile Web service provisioning in peer to peer environments. In: IEEE International Conference on Service-Oriented Computing and Applications, SOCA (2009)
18. Wireless Application Forum. Wireless Application Protocol User Agent Profile Specification, http://www.openmobilealliance.org/tech/affiliates/wap/wap-248-uaprof-20011020-a.pdf (visited on June 27, 2008)
19. IDATE. Mobile Internet services. IDATE News, 455. Retrieved from (January 16 ,2009), http://www.idate.fr/2009/pages/?all=f_actualite&id=561&idl=22

A Specialized Random Multi-parent Crossover Operator Embedded into a Genetic Algorithm for Gene Selection and Classification Problems

Edmundo Bonilla-Huerta, José Crispín Hernández Hernández, and Roberto Morales-Caporal

LITI, Instituto Tecnológico de Apizaco,
Av. Instituto Tecnológico s/n. 93000, Apizaco, Tlaxcala, México
{edbonn,josechh,roberto-morales}itapizaco.edu.mx

Abstract. The microarray data classification problem is a recent complex pattern recognition problem. The most important goal in supervised classification of microarray data, is to select a small number of relevant genes from the initial data in order to obtain high predictive classification accuracy. With the framework of a embedded filter-wrapper, we study the role of the multi-parent recombination operator. For this purpose, we introduce a Random Multi Parent crossover (RMPX) and we analyze their effects in a genetic algorithm (GA) which is combined with Fisher's Linear Discriminant Analysis (LDA). This embedded algorithm has the major characteristic that the GA uses not only a LDA classifier in its fitness function, but also LDA's discriminant coefficients to integrate a multi-parent specialized crossover and mutation operation to improve the performance of gene selection. In the experimental results it is observed that RPMX operator work very well by achieving lower classification error rates.

Keywords: RPMX, multi-parent recombination, hybrid, filter-wrapper, linear discriminant analyze, genetic algorithm, gene selection, microarray.

1 Introduction

Recently in many evolutionary algorithms (EAs), different multi-parent recombinations have been proposed to create offspring. Scanning crossover and diagonal crossover are proposed in [6,7]. These two methods allows to adopt more than two parents in the process of recombination. In [12] is proposed the gene pool recombination operator, where the gene pool consists of several pre-selected parents. In [20] is developed the real-coded center of mass crossover (CMX) and the multi-parent feature-wise crossover (MFX) as two multi-parents recombination operators that can lead to obtain a better performance than the crossover operators used in the genetic algorithms. In [9] is proposed a fitness-weighted crossover (FWX) with an original random threshold mechanism which is used to determine the parent-numbers to reproduce offspring. In [17] are proposed two multi-parent

M. Zhu (Ed.): Electrical Engineering and Control, LNEE 98, pp. 559–566.
springerlink.com © Springer-Verlag Berlin Heidelberg 2011

operators : 1) MPX (multi-parent crossover with polynomial distribution) and 2) MLX (multi-parent cross-over with lognormal distribution) to solve multi-objective optimization problems by using a genetic algorithm NGSC-II. Multi-parent partially mapped crossover operator MPPMX is proposed in [18], which generalizes the partially mapped crossover operator (MPX) to a multi-parent crossover. A Multi-parent uniform crossover operator with short term memory is reported in [10]. Two multi-parent recombination operators (MLX and MLX) are proposed in [3] to solve multimodal problems. In all these methods the parents number plays a key role to keep the population diversity and to avoid the premature convergence, although the optimal parents number is actually a problem still open.

The DNA microarray technique has made possible to monitor and to measure simultaneous thousands of gene expressions in a cell mixture. This technology enables to consider cancer diagnosis based on gene expressions [2,4,1,8]. Given the very high number of genes, it is useful to select a limited number of relevant genes for classifying tissue samples.

This paper presents a multi-parent operators LDA-GA based for gene selection. In this paper, we analyze the effect of the number of parents for our Random Multi Parent crossover (RMPX), this number varies from 2 to 12 parents. We argue that more parents are used more good solutions (gene subset) are obtained, because more parents explore and exploit into space search. Nevertheless, we need to know how many parents are necessary to scape from local minimum and to obtain high classification accuracy solutions with a minimum number of genes. For this study a gene selection filter is used on the hybrid approach and the Fisher's Linear Discriminant Analysis (LDA) is used to provide useful information to a Genetic Algorithm (GA) for an efficient exploration of gene subsets space. LDA is a well-known method of dimension reduction and classification, where the data vectors are transformed into a low-dimensional subspace such that the class centroids are spread out as much as possible. It has been used for several classification problems and recently for microarray data [5].

This paper is organized as follows: Section 2 presents our wrapper LDA-based GA for gene selection. Experimental results are shown in Section 3. Finally, conclusions are drawn in Section 4.

2 Filter-Wrapper Method

In this section we describe our filter-wrapper method LDA-based Genetic Algorithm (LDA-GA) for gene subset selection. First, we apply a filter BSS/WSS (B/W) [5] to retain a group G_p of different p top ranking genes (p=50, 100, 150, 200, 250 and p=300). Then, the LDA-based GA is used to reduce the search space of 2^p. The purpose of this search is to find good solutions (gene subsets) with high classification performance. In what follows, we present the general procedure and then show the components of the LDA-based Genetic Algorithm. In particular, we explain how LDA is combined with the Genetic Algorithm.

2.1 General GA Procedure

2 Our LDA-based Genetic Algorithm is defined as follows:

- Initial population: The initial population is generated randomly in such a way that each chromosome contains a number of genes ranging from $p \times 0.6$ to $p \times 0.75$. The population size is fixed at 100 in this work.
- Evolution: The chromosomes of the current population P are sorted according to the fitness function (see Section 2.3). The "best" 10% chromosomes of P are directly copied to the next population P' and removed from P. The remaing 90% chromosomes of P' are then generated by using crossover and mutation.
- Crossover and mutation: Mating chromosomes are determined from the remaining chromosomes of P by considering each pair of adjacent chromosomes. By applying our multi-parent recombination operator (see Section 2.4), one child is created each time. This child undergoes then a mutation operation (see Section 2.5) before joining the next population P'.
- Stop condition: The evolution process ends when a pre-defined number of generations is reached (fixed at 400 generations in this work).

2.2 Chromosome Encoding

In our model, a chromosome encodes: 1) a gene subset (τ) and 2) Their LDA coefficients(ϕ). Both are defined as:

$$I = (\tau; \phi)$$

where τ and ϕ have the following meaning. The first part (τ) is a *binary vector* and represents effectively a *candidate gene subset.* Each allele τ_i indicates whether the corresponding gene g_i is selected (τ_i=1) or not selected (τ_i=0). The second part of the chromosome (ϕ) is a real-valued vector where each ϕ_i corresponds to the *discriminant coefficient* of the eigen vector for gene g_i. We use the LDA discriminant coefficient to define the contribution of gene g_i to the projection axis w_{opt}. A chromosome can be thus represented as follows:

$$I = (\tau_1, \tau_2, \ldots, \tau_p; \phi_1, \phi_2, \ldots, \phi_p)$$

The length of τ and ϕ is defined by the number of the pre-selected genes (p) obtained with the filter BSS/WSS.

2.3 Fitness Evaluation

The purpose of the genetic search in our embedded approach is to find relevant gene subsets having the minimal size and the highest prediction accuracy. To achieve this double objective, we devise a fitness function taking into account these (somewhat conflicting) criteria.

To evaluate a chromosome I=(τ;ϕ), the fitness function considers the classification accuracy of the chromosome (f_1) and the number of selected genes in the chromosome (f_2). More precisely, f_1 is obtained by evaluating the gene subset τ using the LDA classifier on the training dataset with replacement from the original dataset through the 10-fold cross validation method. The second part of the fitness function f_2 is calculated by the formula:

$$f_2(I) = \left(1 - \frac{m_\tau}{p}\right) \quad (1)$$

where m_τ is the number of bits having the value "1" in the candidate gene subset τ, *i.e.* the number of selected genes; p is the length of the chromosome corresponding to the number of the pre-selected genes from the filter ranking.

Then the fitness function f is defined as the following weighted aggregation:

$$f(I) = \alpha f_1(I) + (1-\alpha) f_2(I)$$
$$\text{subject to } 0 < \alpha < 1$$

where α is a parameter that allows us to allocate a relative importance factor to f_1 or f_2. Assigning to α a value greater than 0.5 will push the genetic search toward solutions of high classification accuracy (probably at the expense of having more selected genes). Inversely, using small values of α helps the search toward small sized gene subsets. So varying α will change the search direction of the genetic algorithm.

2.4 Multi-parent Recombination

We use the discriminant coefficients from the LDA classifier to design our crossover and mutation operators. Here, we explain how our LDA-based specialized genetic operators operates. Our method is based on a modified random threshold [9] to select the parents of crossover operation (denoted by RMPX hereafter). The number of parents np involved in the recombination. The number of parents is determined by the random threshold as shown below:

$$np = \begin{cases} 2 \text{ if } \theta \leq 0.2 \\ 3 \text{ if } 0.2 \geq \theta \geq 0.4 \\ 4 \text{ if } 0.4 \geq \theta \geq 0.6 \\ 5 \text{ if } 0.6 \geq \theta \geq 0.8 \\ 6 \text{ if } \theta \geq 0.8 \end{cases}$$

RMPX combines randomly different parent chromosomes $I^1 \dots I^{np}$, we take the majority of parental genes following the definition of OB-SCAN [19] to create a new chromosome I^c. Given np parents based in the random threshold, OB-SCAN reproduce the child I^c as follows:

$$I_i^{c'} = \begin{cases} 0 & \text{if } \sum_{j=1}^{np} (I_j^c)_i < \frac{n}{2} \\ 1 & \text{if } \sum_{j=1}^{hp} (I_j^c)_i > \frac{n}{2} \\ rand(0,1) \; otherwise & \end{cases}$$

where $rand(0,1)$ denotes a binary random function and $(I_j^c)_i$ the ith gene of the chromosome (I_j^c). In order to obtain a good subset of informative genes, we propose to create a specialized recombination operator RMPX. This genetic operator preserves the genes obtained by the multi-parent crossover (I^c_j) and the genes that have the most frequently appearing genes by the LDA coefficients (J^c_j). That is denoted as $K^c = I^c \otimes J^c$, where K^c is the child that contains the best information of the multi-parent recombination using LDA coefficients. Before inserting the child into the next population, K^c undergoes a mutation operation based in the LDA-coefficients to remove the gene having the lowest discriminant coefficients.

2.5 LDA-Based Mutation

In a conventional GA, the purpose of mutation is to introduce new genetic materials for diversifying the population by making local changes in a given chromosome. For binary coded GAs, this is typically realized by flipping the value of some bits ($1 \rightarrow 0$, or $0 \rightarrow 1$). In our case, mutation is used for dimension reduction; each application of mutation eliminates a single gene ($1 \rightarrow 0$). To determine which gene is discarded, one criterion is used, leading to the next mutation operator.

– *Mutation using discriminant coefficient (M1)*: Given a chromosome $K=(\tau;\phi)$, we identify the smallest LDA discriminant coefficient in ϕ and remove the corresponding gene, that is, the least informative gene among the current candidate gene subset τ).

3 Experiments on Microarray Datasets

3.1 Microarray Gene Expression Datasets

The proposed embedded approach was evaluated on seven DNA microarray datasets that have been used in the literature for the diagnosis of different cancers. A brief description of these data sets is given in Table 1:

Table 1. Summary of datasets used for experimentation

Dataset	Genes	Samples	References
Leukemia	7129	72	Golub et al [8]
Colon	2000	62	Alon et al [2]
Lung	12533	181	Gordon et al [11]
Prostate	12600	109	Singh et al [14]
CNS	7129	60	Pomeroy et al [16]
Ovarian	15154	253	Petricoin et al [15]
DLBCL	4026	47	Alizadeh et al [1]

3.2 Experimental Results

All experiments were made on a DELL precision M4500 laptop with Intel Core i7, 1.87 Ghz processor and 4 GB of RAM. Our model was implemented in MATLAB. The following parameters were used in the experiments: a) population size $|P| = 50$, b) maximal number of generations is fixed at 250, c) individual length (number of pre-selected genes) $p = 50, 100, 150, 200, 250$ and $p = 300$ are evaluated in this experimental protocol. Finally our LDA based crossover operator are explained in subsection 2.4 and 2.5.

We study the effects of multi-parent LDA-based operators (crossover and mutation) within the same embedded GA/LDA framework. Figure 1, list the clear influence of the number of genes used in each dataset. We use several p values (50,100,150,200,250 and 300) of the top-ranking genes obtained by the filter BSS/WSS. More genes are used more high is the rate classification of our model in 6 of 7 datasets. Only ovarian dataset offers a best performance by using a minimal number of genes by using p=50.

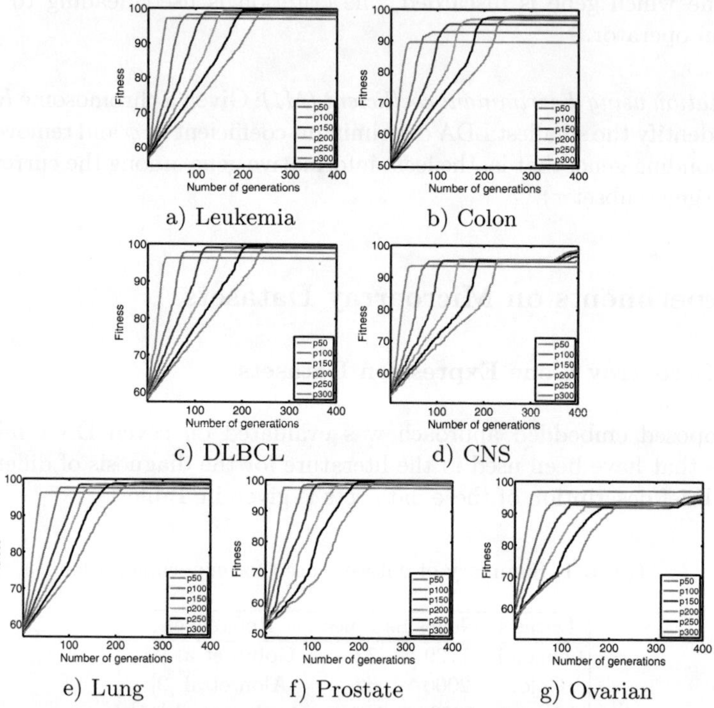

Fig. 1. A comparison between different values of selected genes with $\alpha = 0.50$

Table 2 summarizes the accuracies obtained by our model in each dataset. The first column indicates the dataset used. Second column shows the number of selected genes (p) according to filter method BSS/WSS. Third column list the minimal selected gene subset obtained through 10 executions of our model.

Finally last column show the accuracy obtained from LDA by applying 10-fold cross-validation technique.

Table 2. Results of our model by using $p = 300$ and $\alpha = 0.50$

Dataset	p	Minimal gene subset	Accuracy
Leukemia	300	1941,2288,5598	**99.50%**
Colon	300	451,576,792,907,912,950,1212	**98.83%**
DLBCL	300	2081,3075,3212	**99.50%**
CNS	300	2149,4829,6135,6332	**99.33%**
Lung	300	2549,3789,12196	**99.17%**
Prostate	300	6185,8892,11200	**99.50%**
Ovarian	300	4104,4243,4276,4594,11506	**97.43%**

We observe that our model achieves the highest accuracy with p=300 in all datasets. For leukemia we obtain 99.50% of accuracy with a gene subset of 3 genes. Colon tumor offers a very good performance 98.83% with 7 genes. DLBCL provides a reduced gene subset with only three genes and a classification of 99.50%. The number of genes for the CNS dataset is 4 with a high recognition rate (99.33%). Lung and Prostate cancer with three genes give a classification performance greater to 99%. For the Ovarian cancer we obtain with 5 genes a perfomance of 97.43.

4 Conclusions and Discussion

In this paper we proposed a embedded framework with specialized genetic operator RMXP for the gene selection and classification of microarray gene expression. Our model work very well when the number of parents is randomly selected from the set $\{4, 5, \ldots, 12\}$. We confirm that more than two parents and less than 10 parents are randomly used we obtain better performances of our RPMX-LDA based genetic-operator. We confirm that our model is very competitive using more top-ranking genes that leads to obtain a very small gene subset with high performance.

Acknowledgments. This work is partially supported by the PROMEP project ITAPIEXB-000.

References

1. Alizadeh, A., Eisen, M.B., et al.: Distinct types of diffuse large (b)-cell lymphoma identified by gene expression profiling. J. Nature 403, 503–511 (2000)
2. Alon, U., Barkai, N., et al.: Broad patterns of gene expression revealed by clustering analysis of tumor and normal colon tissues probed by oligonucleotide arrays. Proc. Nat. Acad. Sci. USA 96, 6745–6750 (1999)
3. Bongirwar, V.K., Agarwal, V.H., Raghuwanshi, M.M.: Multimodal optimization using Real Coded Self-Adaptive Genetic Algorithm. International Journal of Engineering Science and Technology 1, 61–66 (2011)

4. Ben-Dor, A., Bruhn, L., et al.: Tissue classification with gene expression profiles. J. Computational Biology 7(3-4), 559–583 (2000)
5. Dudoit, S., Fridlyand, J., Speed, T.P.: Comparison of discrimination methods for the classification of tumors using gene expression data. J. The American Statistical Association 97(457), 77–87 (2002)
6. Eiben, A.E., Raue, P.E., Ruttkay, Z.: Genetic algorithms with multi-parent recombination. In: Davidor, Y., Männer, R., Schwefel, H.-P. (eds.) PPSN 1994. LNCS, vol. 866, pp. 78–87. Springer, Heidelberg (1994)
7. Eiben, A.E.: Multiparent recombination in evolutionary computing. In: Ghosh, A., Tsutsui, S. (eds.) Advances in Evolutionary Computing: theory and applications 2003, pp. 175–192. Springer, Heidelberg (2003)
8. Golub, T., Slonim, D., et al.: Molecular classification of cancer: Class discovery and class prediction by gene expression monitoring. J. Science 286(5439), 531–537 (1999)
9. Gong, D., Ruan, X.: A new multi-parent recombination genetic algorithm. In: Fifth World Congress on Intelligent Control and Automation, pp. 531–537. IEEE Press, New York (2004)
10. Garcia-Martinez, C., Lozano, M.: Evaluating a Local Genetic Algorithm as Context-Independent Local Search Operator for Metaheuristics. J. Soft Computing 14(10), 1117–1139 (2010)
11. Gordon, G.J., Jensen, R.V., et al.: Translation of microarray data into clinically relevant cancer diagnostic tests using gene expression ratios in lung cancer and mesothelioma. J. Cancer Research 17(62), 4963–4967 (2002)
12. Muhlenbein, H., Voigt, H.M.: Gene Pool Recombination for the Breeder Genetic Algorithm. In: The Metaheuristics International Conference, pp. 19–25. Kluwer Academic Publishers, Norwell (1995)
13. Patel, R., Raghuwanshi, M.M.: Multi-objective optimization using multi parent crossover operators. Journal of Emerging Trends in Computing and Information Sciences. 2(2), 33–39 (2010)
14. Singh, D., Febbo, P., Ross, K., Jackson, D., Manola, J., Ladd, C., Tamayo, P., Renshaw, A., D'Amico, A., Richie, J.: Gene expression correlates of clinical prostate cancer behavior. J. Cancer Cell 1, 203–209 (2002)
15. Petricoin, E.F., Ardekani, A.M., Hitt, B.A., Levine, P.J., Fusaro, V.A., Mills, G.B., Simone, C., Fishman, D.A., Kohn, E.C., Liotta, L.A.: Use of proteomic patterns in serum to identify ovarian cancer. J. Lancet 359(9306), 572–577 (2002)
16. Pomeroy, S.L., Tamayo, P., et al.: Prediction of central nervous system embryonal tumour outcome based on gene expression. J. Nature 415, 436–442 (2002)
17. Singh, D., Febbo, P., Ross, K., Jackson, D., Manola, J., Ladd, C., Tamayo, P., Renshaw, A., D'Amico, A., Richie, J.: Gene expression correlates of clinical prostate cancer behavior. J. Cancer Cell 1, 203–209 (2002)
18. Ting, C.K., Su, C.H., Lee, C.N.: Multi-parent extension of partially mapped crossover for combinatorial optimization problems. J. Expert Systems with Applications 37(3), 1879–1886 (2010)
19. Ting, C.K.: On the convergence of multi-parent genetic algorithms. In: The IEEE Congress on Evolutionary Computation, pp. 396–403. IEEE Press, Edinburgh (2005)
20. Tsutsui, S., Ghosh, A.: A study of the effect of multi-parent recombination with simplex crossover in real coded genetic algorithms. In: The IEEE World Congress on Computational Intelligence, pp. 828–833. IEEE Press, Anchorage (1998)

A New Method Based on Genetic-Dynamic Programming Technique for Multiple DNA Sequence Alignment

José Crispín Hernández-Hernández, Edmundo Bonilla-Huerta, and Roberto Morales-Caporal

LITI, Instituto Tecnológico de Apizaco,
Av. Instituto Tecnológico s/n. 93000, Apizaco, Tlaxcala, México
{josechh,edbonn,roberto-morales}itapizaco.edu.mx

Abstract. The sequence alignment are gathered into subgroups on the basis of sequence similarity to allow the representation and comparison of two or more sequences, strings of DNA (deoxyribonucleic acid), RNA (ribonucleic acid) or protein primary structures to highlight their areas of similarity, which could indicate functional or evolutionary relationships between genes or proteins and to identify changes in its structure. The aligned sequences are written with letters (A-actin,C-cysteine, G-guanine, T-thymine) in rows of a matrix in which spaces are inserted for areas with identical or similar structure alignment. This leads to exponentially grow the aligned sequences causing a difficult problem to solve manually. In this paper we tackle this problem by using a Genetic Algorithm combined Dynamic Programming. We obtain competitive results when applied to databases available online.

Keywords: Sequence alignment, genetic algorithm, dynamic programming, amino acids.

1 Introduction

Over the past 10 years, computers have played a role in the areas of biology and medicine. Computing has focused on the analysis of biological sequences, and today many problems such as: pattern discovery, a suitable metric, pairwise and multiple sequence alignments, gene annotation, classification, similarity search and many others remain unsolved. Within the last three years has been proposed many research in the identification of genome sequences, such as fly, and the first attempts in the project of identifying the human genome sequence [4]. These new technologies and high performance as other arrangements of DNA (deoxyribonucleic acid) and mass spectrometry have made considerable progress, these powerful technologies are able to quickly generate data sets in the order of terabytes making it impossible to prosecute them for methods traditional computing [3,8,9]. These data are new challenges for researchers in the areas of computer science and biology.

M. Zhu (Ed.): Electrical Engineering and Control, LNEE 98, pp. 567–574.
springerlink.com © Springer-Verlag Berlin Heidelberg 2011

Therefore it is necessary to use machine learning techniques to create new algorithms toward a more effective and efficient use of information [1,2,5,10].

The DNA array technology (called, arranged by Deoxyribonucleic Acid) helps you review and analyze a number of biochemical mechanisms contained within the cells. Although genes get more attention, the proteins are those that perform many functions of life and form the majority of cellular structures of our body. Proteins are large complex molecules made of smaller units called amino acids. The chemical properties that distinguish the 20 different amino acids cause the protein chains are clustered in three-dimensional structures that define their particular functions in the cell. The proteins in different organisms that are related to each other by their evolution from a common ancestor are called homologous. This relationship can be recognized by comparing multiple sequences. A similar primary structure leads to a three-dimensional (3-D), resulting in similar functionality of proteins. To carry out this sequence searching, we propose a genetic algorithm and a heuristic method with dynamic programming to find the best local optimum alignment.

Recently, Sakakibara [3], has argued the benefits of visualizing biological sequences representing DNA, RNA or protein as a sentence resulting from a formal grammar. When viewing DNA, RNA or protein sequences as a string or formal language alphabets of four nucleotides or 20 amino acids, a grammatical representation and grammatical inference methods can be applied to several problems in biological sequence analysis. So our main problem to solve is the alignment of DNA sequences.

2 Proposed Method

2.1 Genetic Algorithm (GA) for DNA Sequence Alignment

We propose a new method based on genetic-dynamic programming technique for multiple DNA sequence alignment to generate several multiple alignments (AMS) by re-accommodations, simulating the insertion of gaps (holes) and actions of recombination during replication to obtain higher scores for alignment [6]. Given the way this algorithm operates is not guaranteed that the end result is optimal or the highest to be achieved, therefore the need to use another method to optimize the alignment, as in the case of dynamic programming. The success of genetic algorithm seems to lie in the steps taken in the rearrangement of sequences, many of which simulate changes in the evolutionary process of the protein family [7]. Although this algorithm can generate large alignments. The genetic algorithm (GA) was performed according to the following flowchart (see figure 1.a):

The sample sizes are approximately 80,000 sequences of relatively large size, to align automatically.

Initialisation. The first step begins by inputting to the genetic algorithm pairs of sequences, for which only analyze the first five pairs of strings. Our method generate the initial population using this pairs. Each pair is used to obtain as output the best alignment.

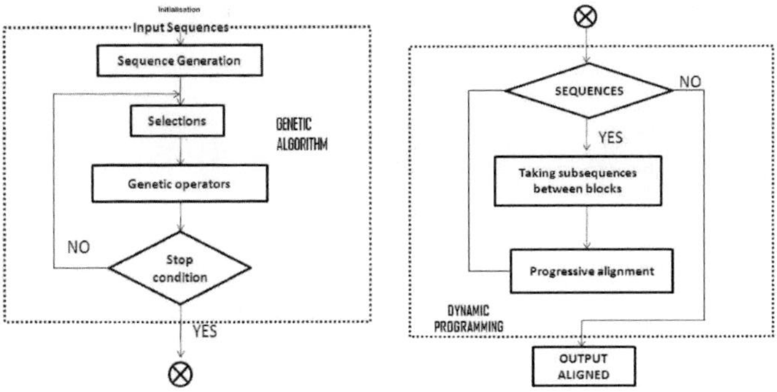

a) Genetic algorithm b) Dynamic programming

Fig. 1. Flowchart of our algorithm

1. Take the first pair of sequences, and define the number of individuals (NI) to generate, and the number of generations to evaluate (NG).

2. Search the longest chain (x) the pair of sequences, calculate the length of this (L = length (x)), and gives a factor ($F = L * 25\%$). This factor gives the spaces (gaps) extras that are being used to move the chains, more specifically, it generates a random number between 0 and F, F is the number of gaps to insert at the beginning and after the final gaps were inserted to make chains have the same length. This results in possible new alignments. This is done (N_i) times turned down due to the number of individuals.

3. Get the score of each alignment, this scores are computed using the qualifications of the blosum62 matrix [4] shown in Figure 2, whose equation is:

$$a_{ij} = \left(\frac{1}{\lambda}\right) log \left(\frac{p_{ij}}{q_i * q_j}\right) \tag{1}$$

Where p_{ij} is the probability that two amino acids i and j replace one another in sequence homology, whereas q_i and q_j are the last chance of finding the amino acids i and j in any protein sequence in random order. The factor λ is merely a scale factor to ensure that, after implementation and necessary rounding to the nearest integer, the array contains integer values dispersed and easily treatable.

Using this score is selected half of alignments to be used later in the process of selection and reproduction.

Genetic Operators. 4. Then the other half of the alignments pass through a mutation process, we obtain the total size of the sequences (TT) taking into account the above input spaces, is obtained as the difference in the maximum number of variables (nmax) and smaller number of variables (Nmin) of the pair of alignments, and generating random numbers between 0 and that rank (A = random (nmax-Nmin)) indicating the number of gaps to insert, then according to the total size sequences, is generated by each gap to add a random number between 1 and (TT) that indicate the position where each space will be inserted (A) carried out this process with the pairs of sequences.

Ala	4																			
Arg	-1	5																		
Asn	-2	0	6																	
Asp	-2	-2	1	6																
Cys	0	-3	-3	-3	9															
Gln	-1	1	0	0	-3	5														
Glu	-1	0	0	2	-4	2	5													
Gly	0	-2	0	-1	-3	-2	-2	6												
His	-2	0	1	-1	-3	0	0	-2	8											
Ile	-1	-3	-3	-3	-1	-3	-3	-4	-3	4										
Leu	-1	-2	-3	-4	-1	-2	-3	-4	-3	2	4									
Lys	-1	2	0	-1	-3	1	1	-2	-1	-3	-2	5								
Met	-1	-1	-2	-3	-1	0	-2	-3	-2	1	2	-1	5							
Phe	-2	-3	-3	-3	-2	-3	-3	-3	-1	0	0	-3	0	6						
Pro	-1	-2	-2	-1	-3	-1	-1	-2	-2	-3	-3	-1	-2	-4	7					
Ser	1	-1	1	0	-1	0	0	0	-1	-2	-2	0	-1	-2	-1	4				
Thr	0	-1	0	-1	-1	-1	-1	-2	-2	-1	-1	-1	-1	-2	-1	1	5			
Trp	-3	-3	-4	-4	-2	-2	-3	-2	-2	-3	-2	-3	-1	1	-4	-3	-2	11		
Tyr	-2	-2	-2	-3	-2	-1	-2	-3	2	-1	-1	-2	-1	3	-3	-2	-2	2	7	
Val	0	-3	-3	-3	-1	-2	-2	-3	-3	3	1	-2	1	-1	-2	-2	0	-3	-1	4
	Ala	Arg	Asn	Asp	Cys	Gln	Glu	Gly	His	Ile	Leu	Lys	Met	Phe	Pro	Ser	Thr	Trp	Tyr	Val

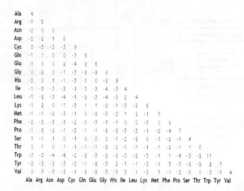

Fig. 2. The BLOSUM62 amino acid substitution matrix

And finally we get the score considering the new alignments obtained by the blosum62 matrix (Figure 2). We compare the score obtained with the previous alignments, if a new alignment score is higher, new alignments are selected, otherwise these steps are repeated iteratively, until obtain sequences with higher score.

The above process is looking more likely to give strong alignments to be elected.

5. After both the first half of the top ranking as the second half of the alignments mutated together, and then are labeled and reordered to be used in the process of reproduction. That means that half alignment will survive in the next generation.

6. Eventually, in the possible alignment gaps can be found, to solve this, we eliminate all columns of the gaps to prevent that a sequence grow exponentially. A tournament selection is carried to selected pairs of alignments; here who has a best score will be chosen as a parent of each pair of sequences. Getting parents to the process of reproduction.

7. From these alignments it generates a random number within the range of one to the length of the shorter chain (Nmin), this will be the cutoff point for each pair of sequences. This random number (AC), one count each character without spaces.

8. Then swap the first part of the upper die with the second bottom and vice versa, and you get the score for each child.

9. It chooses the child having the highest rating and is added to the new generation.

According to the number of the population (NI), we repeat the steps 7, 8 and 9.

The idea is to work with the sequences of chromosomes, with a population of N individuals and N generations.

Figure 3.a shows the input alignments and Figure 3.b shows the output obtained with the genetic algorithm.

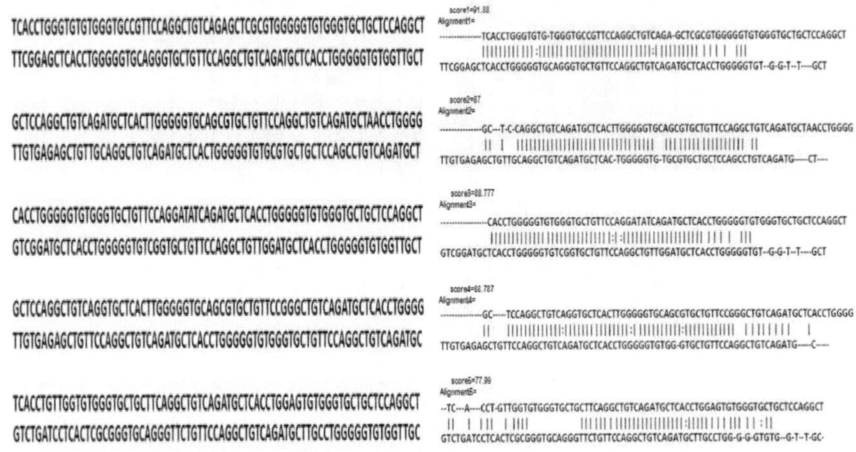

a) Input alignments b) Results of Genetic Algorithm

Fig. 3. Result obtained by genetic algorithm

The alignment obtained is introduced as input to dynamic programming. This process is described in the next section.

2.2 Dynamic Programming (DP) for the Alignment of Sequences

The Dynamic Programming (DP) is an alternative decomposition for smaller subproblems these are solved and then come together at each stage assuming that future will take the right decisions. Its flowchart is shown in Figure 1.b.

The steps defined in the figure 1.b:

1. For this example, the two sequences to be globally aligned are: A T C C G C T T A C (sequence #1) and C T C C T A G (sequence #2). So M = 10 and N = 7 (the length of sequence #1 and sequence #2, respectively). A simple scoring scheme is assumed where $S_{i,j} = 1$ if the residue at position i of sequence #1 is the same as the residue at position j of sequence #2 (match score); otherwise $S_{i,j} = 0$ (mismatch score) and $w = 0$ (gap penalty). Observe figure 4.a.

There are 3 types of ways to choose the sequences

Type 1. As shown in Figure 7, which can be selecting a step to the right and likewise down, forming a diagonal path from the top left, as shown in figure 4.b.

Type 2. The other option is a right way and a number down, as shown in figure 4.c.

Type 3. The last option is a step down and several to the right, as shown in figure 4.d.

To find the best alignment between 2 sequences is to search the way you get the best score.

There are many possible paths and their number increases exponentially with the length of the sequences.

The dynamic programming strategy is to divide the overall problem into subproblems. This first looks for the best path that starts in each of the boxes.

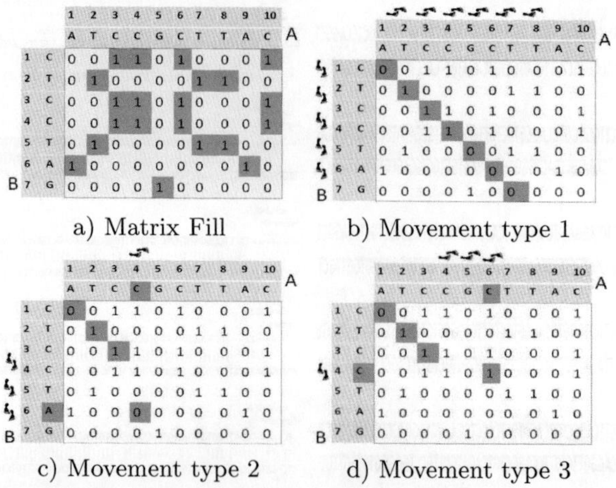

Fig. 4. Creating the Matrix

We started the last row. In this row the score of the best path for each cell line with the score of lace. Check the boxes in the penultimate row of the best way is just one step. To compute the squares of the remaining rows also enough to give a single step because each square that we can move has already gained the best score you can get through it.

Thus the problem of choosing the best path that starts in a cell is reduced to choose the next highest scoring box.

The most points for the fields marked in gray and is added to the score of lace as shown in figure 5.a, and the result of this process is shown in figure 5.b.

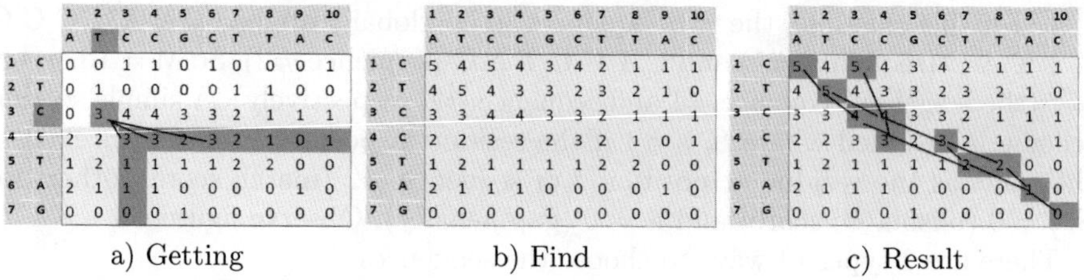

Fig. 5. Filling the Matrix by dynamic programming

To align sequence, the traceback step begins in the 1,1 position in the matrix, i.e. the position that leads to the maximal score. In this case, there is a 5 in that location. Traceback takes the current cell and looks to the neighbor cells that could be direct predacessors. This means it looks to the neighbor to the left (gap in sequence #2), the diagonal neighbor (match/mismatch), and the neighbor above it (gap in sequence #1). The algorithm for traceback chooses as the next cell in the sequence one of the possible predecessors. The highest score is chosen as shown in figure 5.c, and this is done with each pair of sequences. Applying the above example of DNA sequences, the following alignments obtained are illustrated in figure 6.a.

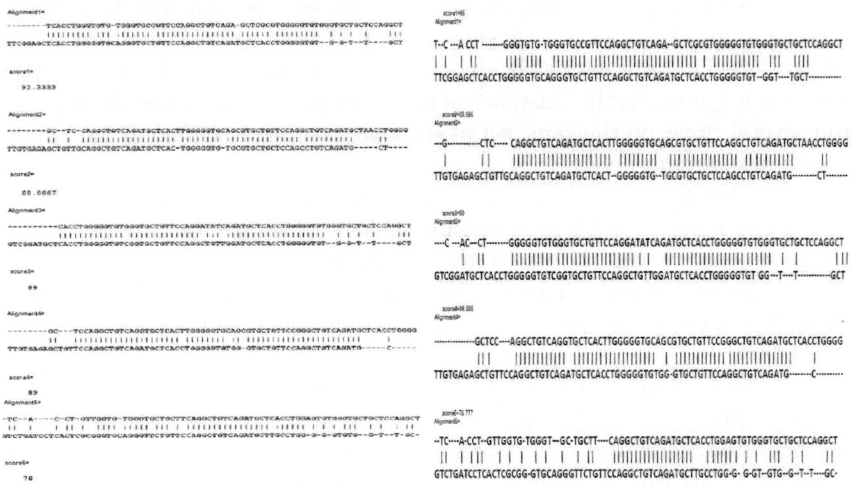

a) Using DP b) Using GA combined with DP

Fig. 6. Our Results

In figure 6.b, we present the results of introducing the sequences in a genetic algorithm combined with dynamic programming.

In order to show that our method was able to obtain a better performance than a GA and DP techniques, we show in the table 1 the results of this comparison. We note that our proposed method using a GA combined with DP allows us to have better alignment of sequences. We note that our proposed method using a GA combined with DP allows us to have better alignment of sequences.

Table 1. Results in the alignment

Genetic Algorithm	Dynamic Programming	GA-DP
91.88%	92.333%	95%
87%	86.667%	89.888%
88.777%	89%	90%
88.787%	89%	89.888%
77.99%	78%	78.777%

3 Conclusion

Due to the large amount of information that is processed to handle strings of DNA, is tuning to various methods and computational techniques in machine learning or the manipulation of that information for analysis so we can get more information about the sequences that are DNA of humans and to develop systems to facilitate information management and access to specialists in the field of molecular biology.

After considering these computer systems as an essential tool to help with different groups of sequences, we show several aspects to be taken into account to obtain good results with this algorithm. As one of the problems of genetic algorithms in general is the existence of local maxima in some cases the algorithm is run several times on the same set of sequences and choose among the different runs the best result.

On the other hand, it is necessary to play around with the algorithm parameters such as number and length of gaps (holes), which can be inserted into the mutated sequences, and the size of the population.

This variation in the parameters is very important because it is not the same to work with sequences that are very similar to each other, to work with sequences highly uneven.

With detector tests could also be that when performing the implementation of the algorithm, it is vitally important to prevent the strings increase in size in an exaggerated manner as if such precautions are not taken, it may happen that all alignments are full of sequences A high percentage are gaps, and if this error is the result of the algorithm spreads can be very poor. As the proposal is to apply another filter to these aligned sequences obtained from genetic algorithm, dynamic programming through to re-align, so you can optimize your alignment.

What you have is a system that can show different solutions to a problem and that is where the ability to interpret these proposals can make a difference.

References

1. Keedwell, E., Narayanan, A.: Discovering gene networks with a neural-genetic hybrid. J. Computational Biology and Bioinformatics 2(3), 231–242 (2005)
2. Mitra, S., Hayashi, Y.: Bioinformatics with soft computing. J. Systems, Man, and Cybernetics, Part C: Applications and Reviews 36(5), 616–635 (2006)
3. Sakakibara, Y.: Grammatical inference in bioinformatics. J. Pattern Analysis and Machine Intelligence 27(7), 1051–1062 (2005)
4. Attwood, T.K., Parry-Smith, D.J.: Introduction to bioinformatics. Pearson Education, South Asia (2007)
5. Mitchell, T.M.: Machine learning. McGraw-Hill, Boston (1997)
6. Notredame, C., Higgins, D.G.: SAGA: Sequence Alignment by Genetic Algorithm. J. Nucleic Acids Research 24(8), 1515–1524 (1996)
7. Cai, L., Juedes, D., Liakhovitch, E.: Evolutionary computation techniques for multiple sequence alignment. In: Evolutionary Computation, vol. (2), pp. 829–835. IEEE Press, La Jolla (2000)
8. Pasanen, T., Saarela, J., Saakiro, I., Toivanen, T.: DNA Microarray Data Analysis. CSC Scientific Computing Ltd., Helsinki (2003)
9. Lu, Y., Tian, Q., Liu, F., Sanchez, M., Wang, Y.: Interactive Semisupervised Learning for Microarray Analysis. J. Computational Biology and Bioinformatics 4(2), 190–203 (2007)
10. Ray, S.S., Bandyopadhyay, S., Mitra, P., Pal, S.K.: Bioinformatics in neurocomputing framework. Circuits, Devices and Systems 152(5), 556–564 (2005)

Digital Image Processing Used for Zero-Order Image Elimination in Digital Holography

Wenwen Liu [1], Yunhai Du[1], and Xiaoyuan He[2]

[1] Department of Engineering Mechanics, Zhengzhou University, Zhengzhou, 450001
[2] Department of Engineering Mechanics, Southeast University, Nanjing 210096
Liuww22@163.com

Abstract. Aiming at the question that the existence of zero-order image has certain effect on the quality of reconstructed images in digital holography, by analysis of spectrum characteristic as well as record and represent theory, two digital image processing algorithm --- differential substitution and the FIR filter are proposed for zero-order image elimination in digital holography. These two methods are free of any other extra equipment, recording only one hologram. The theoretical analysis, digital simulation and experiment verifying results presented that the zero-order image is eliminated and the quality of reconstructed image enhanced. The methods which are proved to be fast and simple can be used in the real time inspection.

Keywords: Digital image processing; Digital holography; Zero-order image elimination; differential substitution; FIR filter.

1 Introduction

In recent years, people take more and more concern about digital holography. It has many applications in areas such as three-dimensional object recognition, flow measuring digital watermarks, security, deformation measuring, vibration measuring, as well as Computer holography [1-4]. After digital reconstruction, the zero-order image, conjugate image and the true image are on the computer screen at the same time. The zero-order image, which includes most of the energy, generates a big bright speckle in the center of the image, and causes a great decrease in the resolution.

To overcome this problem, many methods has been proposed to eliminate or weaken the zero-order image and conjugate image, including gray linear transformation method [5,6], average intensity subtraction method[7], phase-shifting method[8], spectrum filter method [9], and so on. However, these methods have inadequate respectively. Two former methods do not have perfect effect on eliminating the zero-order diffraction. Phase-shifting method needs at least four holograms, increasing the complexity of the device. Spectrum filtering method's reconstruction speed is slow. Two digital image processing algorithm --- differential substitution and the FIR filter are proposed. They only need to record one digital hologram and simple digital image processing.

M. Zhu (Ed.): Electrical Engineering and Control, LNEE 98, pp. 575–582.
springerlink.com © Springer-Verlag Berlin Heidelberg 2011

2 Hologram Acquisition, Reconstruction and Spatial Spectrum

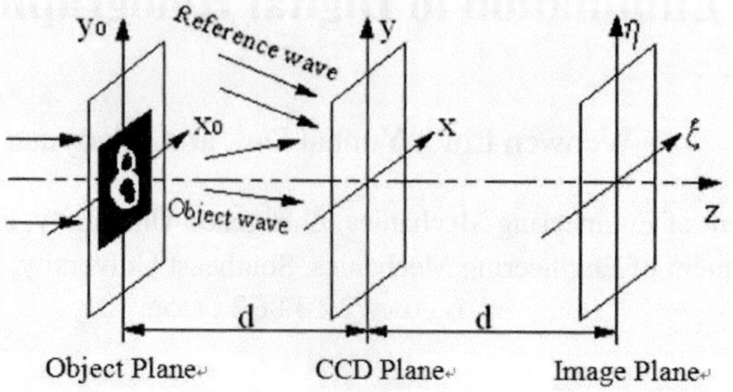

Fig. 1. Principle of digital off-axis holography.

Considering the optical path of digital off-axis holography as shown in Fig. 1, a plane reference wave and the diffusely reflected object wave are made to interfere at the *x-y* plane, where put the CCD used to record the hologram. Under the condition of Fresnel approximation, the object wave can be reconstructed at the ξ-η plane.

In the hologram plane *x-y* the interference between the object wave *O* and the plane reference wave *R* produces a distribution of intensity, which is generally written as a sum of four terms,

$$I_H(x,y) = I_R + I_O(x.y) + R^*O + RO^* \tag{1}$$

where I_R is the intensity of the reference wave and $I_O(x, y)$ is the intensity of the object wave, R^*O and RO^* are the interference terms, and R^* and O^* are the complex conjugates of the two waves. Let us assume that the hologram reconstruction is performed by illumination with a plane wave *U*. Then, as in classical holography, the reconstructed wave front in the hologram plane is given by

$$\Psi(x,y) = UI_R + UI_O(x.y) + UR^*O + URO^* \tag{2}$$

The first two terms of Equation (2) form the zero-order diffraction. The third and the fourth term generate the twin images of the specimen. The third term UR^*O produces a virtual image located at the position initially occupied by the specimen, and the fourth term URO^* produces a real image located on the other side of the hologram. In off-axis holography the object wave *O* and the reference wave R arrive in the hologram plane with separated directions. If we assume that a reference wave with obliquity of θ is in the form of $R(x,y) = \sqrt{I_R}\exp(ik_0x)$, where $k_0 = 2\pi\sin\theta/\lambda$, the hologram intensity becomes

$$I_H(x,y) = I_R + I_O(x.y) + \sqrt{I_R}\exp(-ik_0x)O + \sqrt{I_R}\exp(ik_0x)O^* \tag{3}$$

Performing a Fourier transformation on both sides of Equation (5), we have

$$\tilde{I}_H(f_x,f_y) = \tilde{I}_R + \tilde{I}_O(f_x,f_y) + \sqrt{I_R}\tilde{O}(f_x - k_0, f_y) + \sqrt{I_R}\tilde{O}^*(f_x + k_0, f_y) \tag{4}$$

Digital hologram after FFT transform is shown in Fig. 2(a). We can see clearly that there are three parts in the picture, the spatial frequencies corresponding to the zero-order image are located in the center of it, and the other two parts are the spatial frequencies corresponding to the conjugate image and real image. Its one dimension condition is shown in Fig. 2(b), the first term is a δ function located at the origin of the spatial-frequency plane. The second term, being proportional to the autocorrelation function of $\tilde{O}$ (f_x, f_y), is centered on the origin too, with twice the extent of the object wave spectrum. The third term is the object wave spectrum shifted upwards a distance k_0, while the fourth term is the conjugate of the object wave spectrum, shifted downwards the same distance k_0. Since it is necessary to have at least two sampled points for each fringe in a holographic recording, the maximum spatial frequency of a hologram is f max=1/2Δ when a CCD with sensors spaced by Δ is used for hologram recording. For Δ=4.65 μm (the actual spacing for the CCD sensors used in this work), f max is about 107mm-1. It can be seen from Fig. 2(b) that the maximum spectral width of the object wave spectrum is only 53 mm^{-1}.

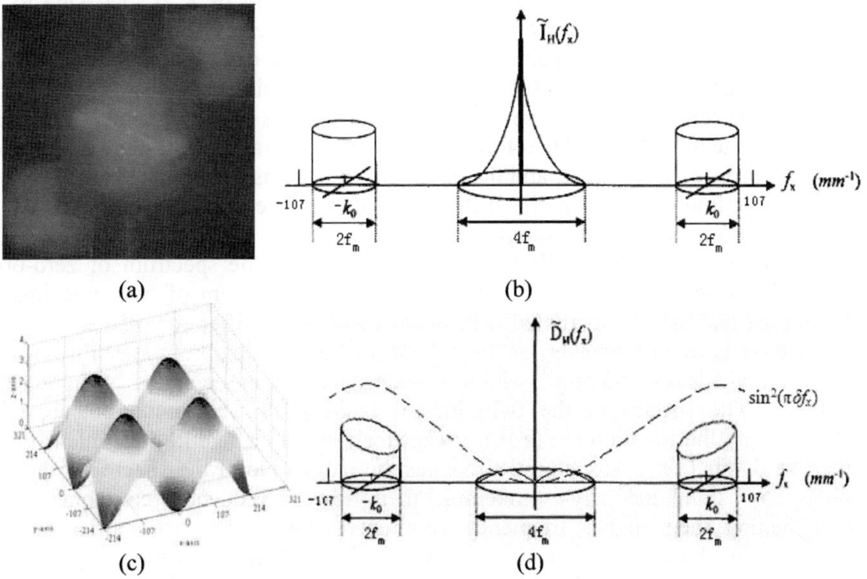

Fig. 2. Schematic diagrams of differential dispose of the hologram intensity. (a) Digital hologram after FFT transform; (b) Schematic of the spatial spectrum of the digital hologram; (c) 3-dimension schematic diagram of $K(f_x, f_y)$; (d) 1-dimension schematic diagram of the spatial spectrum of the disposed digital hologram intensity

3 Method One–Differential Substitution

3.1 Differential of Digital Off-Axis Hologram

The differential of the hologram intensity is defined as

$$D_H(x,y) = \left(\frac{\partial I_H(x,y)}{\partial x}\right) + \left(\frac{\partial I_H(x,y)}{\partial y}\right) = \frac{I_H(x+\delta,y) + I_H(x,y+\delta) - 2I_H(x,y)}{\delta} \tag{5}$$

where δ is the distance between neighboring pixels.

Performing a Fourier transformation on both sides of Equation (5) to consider the spectrum of digital processed hologram, we have

$$DFT[D_H(x,y)] = \frac{\tilde{I}_H(f_x,f_y)[\exp(i2\pi\delta f_x)+\exp(i2\pi\delta f_y)-2]}{\delta} \quad (6)$$

where $\hat{I}_H(f_x, f_y)$ is the Fourier transform of $I_H(x, y)$. Considering real part of Equation (6), we have

$$DFT[D_H(x,y)]_{real} = \tilde{D}_H(f_x,f_y)_{real} = \frac{\tilde{I}_H(f_x,f_y)[\cos(2\pi\delta f_x)+\cos(2\pi\delta f_y)-2]}{\delta}$$

$$= \frac{\tilde{I}_H(f_x,f_y)[2\sin^2(\pi\delta f_x)+2\sin^2(\pi\delta f_y)]}{\delta} \doteq \tilde{I}_H(f_x,f_y)\times K(f_x,f_y) \quad (7)$$

Fig. 2(c) presents the 3-dimension schematic diagram of $K(f_x, f_y)$. Due to the separability of this equation, we need only investigate the one dimensional spectrum $\check{D}_H(f_x)_{real}$, which is shown schematically in Fig. 2(d). We can see that the spectrum of the zero-order diffraction is efficiently suppressed by $\sin^2(\pi\delta f_x)$ function, which is figured by broken line, and the spectra of the twin images are almost intact except for a small change in intensity. Therefore, the zero-order diffraction will almost disappear from the reconstructed image, and the quality of the reconstructed twin images will be significantly improved when the differential of the detected hologram is used for reconstruction instead of the hologram itself.

It can be seen in Fig. 2(d) that the suppression of the spectrum of zero-order diffraction is accompanied by a modification in the spectrum of two twin images. This means that the reconstructed twin images will be modified to some extent when this method is used. However, as mentioned above, the maximum spectral width of the twin image is only 53 mm^{-1}, which is much smaller than the period of the function $\sin^2(\pi\delta f_x)$. The spectra of the twin images undergo no substantial change when multiplied by the function $\sin^2(\pi\delta f_x)$, except for a small change in their intensity. On other hand, $\sin^2(\pi\delta f_x)$ only slightly changes the magnitude of the spectra of the twin images by simultaneously weakening their lower frequency components and strengthening their higher frequency components. According to Fourier imaging principles, this kind of modification only results in an alteration of the image contrast. So, when we reconstruct a hologram with this method, we do slightly alter the image quality of the twin images. But in comparison with the significant improvement in image quality due to the elimination of the zero-order diffraction, this alteration is practically imperceptible.

3.2 Experimental Results

To verify the efficiency of this method, a hologram was reconstructed with and without the elimination of the zero-order diffraction. The experimental setup is shown in Fig. 3, with a dice as specimen. A He-Ne laser is used as the light source, with wavelength of 632.8nm. The laser, linearly polarized plane wave fronts are produced by a beam expander Be including a pinhole for spatial filtering. After get through a beam splitter, the wave fronts become two beams, one illuminates the specimen to be tested, and the other illuminates the reference plane. Then the two beams both come back to the beam splitter respectively after reflecting, and interfere at the target of a CCD camera to form the hologram, which is shown in Fig. 4(a).

The distance between the CCD camera and the specimen is 40 cm. The image in Fig.4 (a) contains 1024 *1024 pixels, and the area of the sensitive chip of the CCD is 4.65*4.65 μm. The interference fringes characterizing the off-axis geometry are clearly observable in this image. Fig. 4(b) shows the amplitude-contrast image obtained by numerical reconstruction of the original hologram shown in Fig. 4(a). Fig. 4(c) shows the differential of the original hologram. Fig. 4(d) presents the amplitude-contrast image obtained by numerical reconstruction of the differential of the hologram. It is obvious that the zero-order diffraction has been efficiently eliminated especially in the first quadrant where the real image is located in, and the resolution of the reconstructed image is significantly improved.

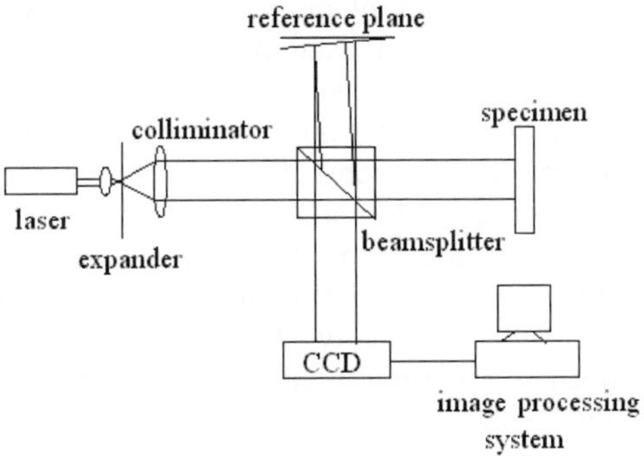

Fig. 3. Experimental setup

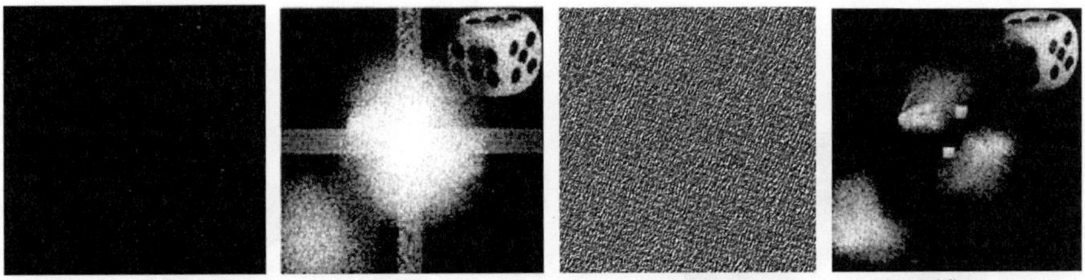

(a) (b) (c) (d)

Fig 4. Elimination of the zero order of diffraction in digital holography. (a) Original off-axis hologram; (b) amplitude-contrast image obtained by numerical reconstruction of the original hologram; (c) the disposed hologram; (d) amplitude-contrast image obtained by numerical reconstruction of the differential of the original hologram

4 Method Two—FIR Filter

4.1 The Principle of Using FIR Filter to Eliminate Zero-Order Image

FIR(Finite Impulse Response) filter is a basic filter used in the digital signal processing (DSP) [10]. It can be arbitrary in the design of any range of filter frequency characteristics. At the same time, it can ensure accurate and strict linear phase characteristics. It has many advantages such as simple, suitable for multi-sample rate conversion, with the ideal number features, can be decimally achieved, and so on.

2-D FIR filter can be achieved in non-recursive way, or can also be achieved with convolution (direct or faster) method. They are always stable, easy to be designed to zero phase shift or linear phase shift. So it has important applications in the image processing [11].

In the digital holographic experiment, the hologram is received by CCD image sensor, and stored in a computer in digital image formats. So, it can be seen as a discrete digital signal. Before the numerical representation, we can design the FIR filter based on the spectrum of the digital hologram collected in the experiment (shown in Figure 5), pre-professing it by application window Fourier transform in air space. According to the analysis in section II, we can see that in numerical representation, the biggest factor that impact on the quality of reconstructed image is the disturb on the ±1-order diffraction from both the self-correlation item of object wave and the light wave directly come through the hologram (the zero-order diffraction). Aiming at this digital holographic characteristic, we apply FIR high-pass filter to enhance the contrast degree of digital hologram, so as to enhance the diffraction efficiency of the hologram, eliminating the impact of zero-order diffraction spot on the reconstructed image. Usually, 2-D FIR filter designed using window function method, frequency sampling method, linear programming method, the frequency transformation method, and so on. In this article, we used the window function method. Matlab provides a function of fwind2 to achieve designing 2-D FIR filter in two-dimensional window method. Its parameters are in the form of the following:

$$H=\text{fwind2}\ (Hd,Win) \tag{8}$$

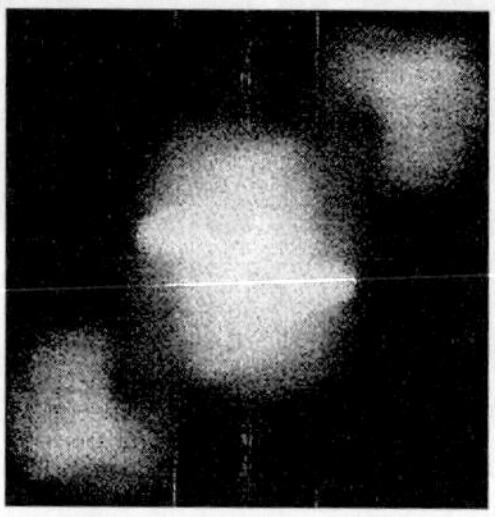

Fig. 5. Schematic diagram of spectrum of experimental recorded hologram

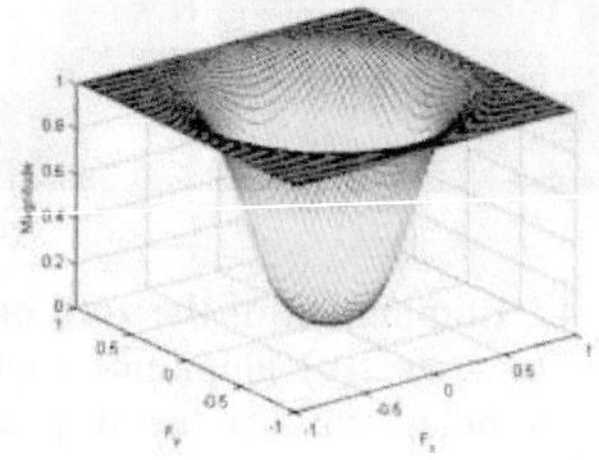

Fig. 6. Schematic diagram of frequency response of 2-D FIR highpass filter

Fwind2 can use the inverse Fourier transform of required frequency response Hd to multiply the window Win to generate two-dimensional FIR filters. Hd is the required frequency response of points with the same interval in Cartesian plane. The size of the Win controls the size of H. According to the spectrum of the digital hologram, Fwind2 is used to design two-dimensional FIR high-pass filter in this article. Its frequency response is shown in Figure 6. Comparing Figures 5 and 6, we can see that after filtering the digital hologram by the designed filter, then exactly the same reconstruction steps, the zero-order image will be eliminated.

4.2 Digital Simulation and Experimental Validation

To verify the correctness of this analysis, we apply the above-mentioned FIR digital filter on simulated and experimental film received two digital holograms. And the Matlab language is used to reconstruction the digital hologram. Figure 7 shows the hologram of letter "F" by digital simulation, the reconstructed image from the original hologram, and the reconstructed image from the filtered hologram. Figure 8 shows the experimental recorded hologram, the reconstructed image from the original hologram, and the reconstructed image from the filtered hologram.

Comparing Figure 7 (b) and 7 (c), Figure 8 (b) and 8 (c) respectively, we can see clearly that the zero-order image is very bright in untreated reconstructed hologram, while the corresponding part in the reconstructed hologram after filtering is most eliminated. Although the strength of the reconstructed image be weakened slightly in theory, the affect is little, and we can use tools such as Matlab to adjust it in experiment. Compared to the number subtraction method and gray-linear transform method proposed in the early literature, FIR filtering method is not limited to the static-field, it is suitable for both the static and dynamic object fields. It not only has the advantages of increase the diffraction efficiency of hologram just like the gray-linear transform method, but also able to weaken the impact of the zero-order diffraction spot, achieving real-time operation.

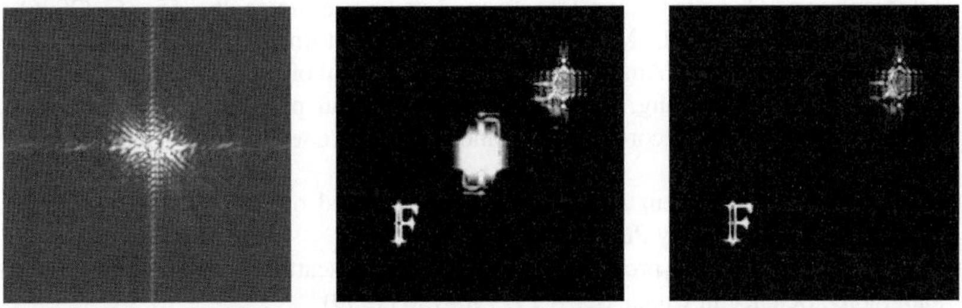

Fig. 7. Digital simulative hologram and reconstructed images:(a)Simulative hologram,(b) Reconstructed image from the original hologram,(c)Reconstructed image from the filtered hologram.

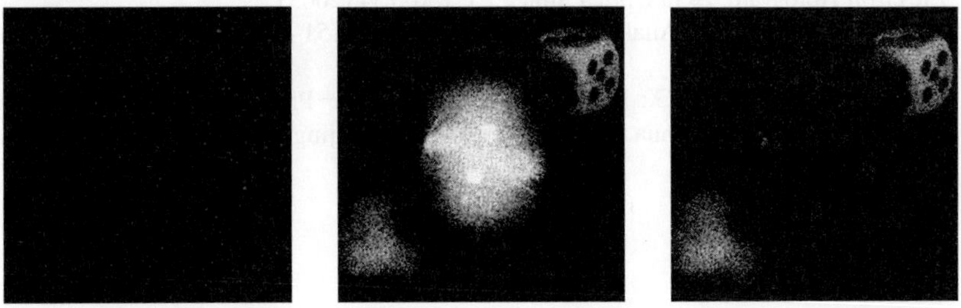

Fig. 8. Experimental recorded hologram and reconstructed images: (a) Experimental recorded hologram,(b) Reconstructed image from the original hologram,(c)Reconstructed image from the filtered hologram.

5 Conclusion

Aiming at the characteristics of numerical reconstructing the hologram, two digital image processing algorithm --- differential substitution and the FIR filter are proposed to eliminate the zero-order image in the reconstructed hologram. The digital simulation and experiment verifying results presented that the zero-order image is eliminated and the quality of reconstructed image enhanced. These methods which are proved to be fast and simple can be used in the real time inspection. They are based on digital image processing free of any extra optical element. They only need to record one digital hologram, can improve the image quality significantly and give better resolution and higher accuracy of the reconstructed image. The main advantages of the two methods are their simplicity in experimental requirements and convenience in data processing.

References

1. Ge, B., Zou, J., Lv, Q.: Surface Morphometry by Digital Holography Based on 4f System. Journal of Tianjing University 39, 712–716 (2006)
2. Yu, C., Gu, J., Liu, W., et al.: An Image Digital Watermark Technique Based on Digital Holography and Discrete Cosine Transform. Acta Optica Sinica 26, 355–361 (2006)
3. LJun, G.H., He, H., et al.: Measuring Weak Phase Information Based on Digital Holographic Phase Difference Amplification. Chinese Journal of Lasers 33, 526–530 (2006)
4. Zheng, C., Han, P., Chang, H.: Four-quadrant sp atial phase-shiftingFourier transform digital holography for recording of cosine transform coefficients. Chinese Optics Letters 4, 145–147 (2006)
5. Wang, X., Wu, C.: System Analysis and Design Based onMATLAB-Image Processing, pp. 7–8. Xidian University Publishers, Xi'an (2000)
6. Hou, B., Chen, G.: Image processing in image through scattering media using fs electronic holography. Science in China, Ser.A 29, 750–756 (1999)
7. Kreis, T.M., Jüptner, W.P.O.: Suppression of the dc term in digital holography. Opt. Eng. 36, 2357–2360 (1997)
8. Takaki, Y., Kawai, H., Ohzu, H.: Hybrid holographic microscopy free of conjug ateand zero- order images. Appl. Opt. 38, 4990–4996 (1999)
9. Liu, C., Liu, Z., Cheng, X., et al.: Spatial-Filtering Method for Digital Reconstruction of Ele ctron Hologram. Acta Optica Sinica 23, 150–154 (2003)
10. Xu, K., Hu, G.: Signal Analysis and Processing, pp. 151–154. Tsinghua University Publishers, Beijing (2006)
11. Liu, M., Xu, H., Ning, G.: Digital Signal Processing—Principle and Arithmetic realization, pp. 197–201. Tsinghua University Publishers, Beijing (2006)

The Preliminary Research of Pressure Control System Danymic Simulation for Ap1000 Pressurizer Based on Parameter Adaptive Fuzzy Pid Control Algorithm

Wei Zhou and Xinli Zhang

Shanghai nuclear engineering research and design institute
zhouw@snerdi.com.cn

Abstract. According to the characteristics of AP1000 pressurizer dynamic process, using AP1000 nuclear plant parameters, a model of pressurizer pressure control system for nuclear power plant is established in this paper. Simulation experiments have been performed using acslX software according to AP1000 pressurizer pressure control system design. Parameter adaptive fuzzy PID controllers have been designed to improve the control function based on the original PID controllers. The comparison between the simulating results indicates that the improved controllers give better performance than the original ones and has a reference value for further control strategies optimization of the pressurizer pressure control system.

Keywords: AP1000, nuclear power plant, pressurizer, fuzzy control, acslX, simulation.

1 Foreword

Recently, China has accelerated the introduction of the AP1000 technology from Westinghouse in the United States. The large advanced pressured water nuclear power plant will be developed based on the innovation of this technology. As an important system in nuclear power plant, AP1000 I&C system share a common hardware platform and execution conception. The system is functional integrated, which will increase the ability to respond to the plant transients[1]. For the new generation of nuclear power plants, it is necessary to verify and make a reasonable improvement of the control logic and control algorithm of AP1000 nuclear power plant through computer simulation, which include the pressurizer control logic verification.

This paper uses acslX software to model the pressurizer behaves. The pressurizer model includes a physical model, and a control model. The physical model uses the traditional vapor-liquid two-phase model, and the control model is established in accordance with Westinghouse AP1000 pressurizer control logic diagrams and control requirements, so the simulation experiments will better reflect the actual control condition of the nuclear power plant. Parameter adaptive fuzzy PID controllers have been designed to improve the control function based on the original PID controllers.

M. Zhu (Ed.): Electrical Engineering and Control, LNEE 98, pp. 583–591.
springerlink.com © Springer-Verlag Berlin Heidelberg 2011

2 AP1000 Pressurizer Mathematical Model

The pressurizer is divided into two parts in the mathematical model. The upper part is the steam region and the lower part is the liquid water region. The AP1000 pressurizer mathematical model is established according with the mass balance equations and the energy balance equations in the two regions.

The mass balance equations:

$$\frac{dM_1}{dt} = W_{FL} - W_{CSP} - W_{CHL} - W_{SV}$$

$$\frac{dM_2}{dt} = W_{SP} + W_{CSP} + W_{CHL} + W_{SU} - W_{FL}$$

Where: M_1, M_2 represent the mass of the steam region and liquid water region; W_{FL} represents the flashing mass flow from the liquid water region into the steam region; W_{SP} represents the spray mass flow; W_{CSP} represents the mass flow of the condensation due to the spray; W_{CHL} represents the mass flow of condensation due to the heat loss; W_{SV} represents the mass flow through the safety valves; W_{SU} represents the surge mass flow.

The energy balance equations:

$$\frac{d(M_1 h_1)}{dt} = W_{FL}h_g + W_{SP}(h_{SP} - h_f) - W_{CSP}h_f - W_{CHL}h_f - W_{SV}h_1 - q_{SL} + M_1 v_1 \frac{dp}{dt}$$

$$\frac{d(M_2 h_2)}{dt} = (W_{SP} + W_{CSP} + W_{CHL})h_f + W_{SUS}h_{SU} - W_{FL}h_g + Q - q_{WL} + M_2 v_2 \frac{dp}{dt}$$

Where: h_g represents the enthalpy of the saturated steam; h_f represents the enthalpy of the saturated water; h_{SP} represents the enthalpy of the spray water; q_{SL} represents the heat loss of the steam region; q_{WL} represents the heat loss of the liquid water region; Q represents the heating power: v_1, v_2 represent the specific volume of the steam region and the liquid water region.

The pressurizer Conservation equation:

$$\frac{dV}{dt} = \sum_{n=1}^{3} \frac{d(M_i v_i)}{dt} = 0$$

The pressure equation can be deduced from the above equations:

$$\frac{dP}{dt} = \sum_{i=1}^{2}\left(k_i \frac{\partial v_i}{\partial h_i} + v_i \frac{dM_i}{dt}\right) \Big/ \sum_{i=1}^{2} M_i \left(\frac{\partial v_i}{\partial P} + v_i \frac{\partial v_i}{\partial h_i}\right) = 0$$

The pressure P cannot be directly calculated from the above formula. k_i is related to the spray flow and the heating power, which is determined by the control function. The other items in the right part of the formula are related to the thermodynamic properties of water, which will be calculated by several empirical formulas. Taking account of these two points, the pressure P can be calculated by the corresponding numerical algorithm.

3 AP1000 Pressurizer Pressure Control Function

The function of pressurizer pressure control system: Automatically maintain the pressure of the reactor coolant system at the setting value during steady state; automatically maintain the pressure within the acceptable operating range in plant transients, and return the pressure to the setting value after transients[2]. The pressure control is completed through the spray valves in the top of the pressurizer and the heater in the bottom of the pressurizer. The AP1000 pressurizer includes two spray valves, four groups of backup heaters and one group of proportional heater. The spray valves and the backup heaters are closed in normal state. The proportional heater operates to compensate the heat loss of the pressurizer. The system pressure is control by the spray valves and the heaters. The deviation between the measured pressure and the set value is the input of the pressure control system, which is used to control the spray valves and the heaters.

4 Parameter Adaptive Fuzzy PID Control Algorithm of the AP1000 Pressurizer Pressure Control

4.1 The Principle of the Parameter Adaptive Fuzzy PID Control Algorithm

Traditional PID controller is widely used in the pressure control system of nuclear power plant due to its simple algorithm, good Stability and high reliability. It has good performance in linear time-invariant systems. But the load of the controlled object changes frequently in actual process, and the interference factors are so complex that we need to adjust the PID parameters online[3]. The method often used is to adjust the parameters of PID controller based on the online identification of the controlled process. This method works if the mathematical model of the controlled object is precise, or else it does not[4]. As fuzzy controller does not have such a high requirement of the mathematical model, it is reasonable and feasible to use fuzzy controller to adjust the parameters of the PID controller.

The deviation of the pressure and the rate of change of the deviation are the inputs to the fuzzy controller, then the fuzzy controller can be used to adjust the parameters online according to the fuzzy control rules, the structure of the parameter adaptive fuzzy PID controller is as shown below:

The fuzzy controller consists of three sub-controllers in the figure, which adjust their respective parameter:K_P, K_I, K_D. The deviation e and the rate of change of the deviation ec are the inputs to the fuzzy controller. The compensation: ΔK_P, ΔK_I

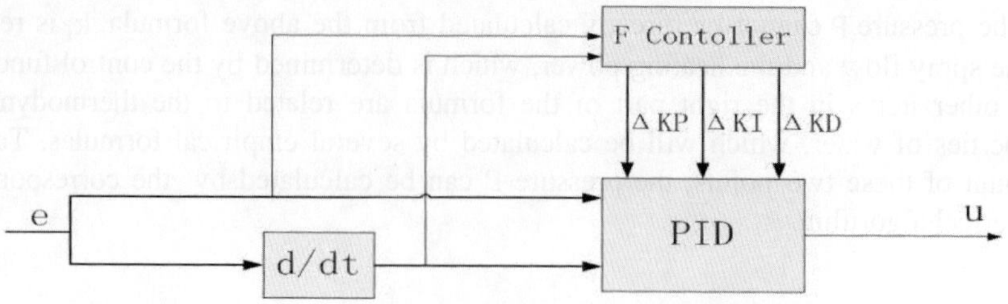

Fig. 1. The structure of the parameter adaptive fuzzy PID controller

, ΔK_D will be obtained after fuzzification, interface mechanism operation and defuzzification. As the additional inputs to the PID controller, ΔK_P, ΔK_I , ΔK_D is used to compensate the original K_P, K_I, K_D.

4.2 The Design of the Parameter Adaptive Fuzzy PID Controller

The deviation e and the rate of change of the deviation ec are chosen as inputs of the fuzzy controller, and the compensation: ΔK_P, ΔK_I , ΔK_D are chosen as the outputs of the fuzzy controller, suppose that there are 7 fuzzy sets for the inputs, and another 7 fuzzy sets for the outputs. They are: {NB, NM, NS, ZO, PS, PM, PB}. We choose membership function for inputs and outputs as shown below:

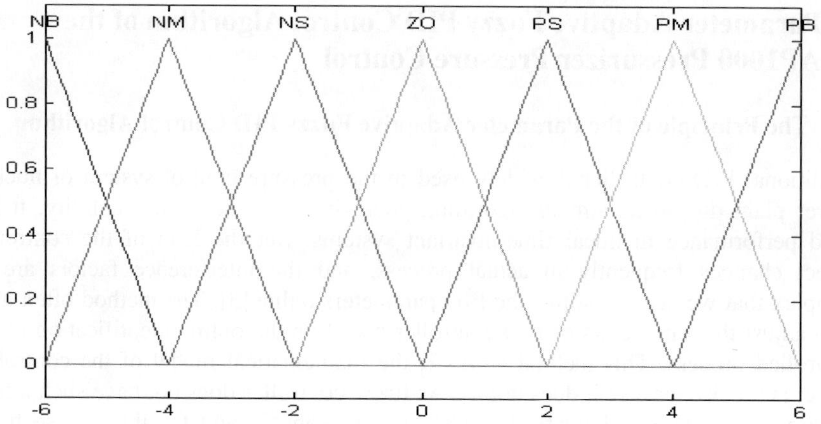

Fig. 2. Membership function for inputs and outputs

The spray valves and the proportional heater in AP1000 pressurizer use the same PI controller. The backup heaters use another PID controller. The rules between e, ec and regulation method of K_P, K_I , K_D can be deduced according to the experience of the operator, the rules are as shown below:

Table 1. The control rules of ΔK_P of the spray valves and proportional heater

EC \ E	NB	NM	NS	ZO	PS	PM	PB
NB	PM	PS	PM	PM	ZO	ZO	ZO
NM	PB	PM	PB	PB	PS	PS	PS
NS	PB	PM	PB	PB	PM	PS	PM
ZO	PB	PM	PB	ZO	PB	PM	PB
PS	PM	PS	PM	PB	PB	PM	PB
PM	PS	PS	PS	PB	PB	PM	PB
PB	ZO	ZO	ZO	PM	PM	PS	PM

Table 2. The control rules of ΔK_I of the spray valves and proportional heater

EC \ E	NB	NM	NS	ZO	PS	PM	PB
NB	ZO	PS	PM	PB	PS	ZO	ZO
NM	ZO	PS	PB	PB	PM	ZO	ZO
NS	ZO	ZO	PB	PB	PM	ZO	ZO
ZO	ZO	ZO	PB	ZO	ZO	ZO	ZO
PS	ZO	ZO	PM	PM	PB	ZO	ZO
PM	ZO	ZO	PM	PM	PB	PS	ZO
PB	ZO	ZO	PS	PM	PB	PS	ZO

Table 3. The control rules of Δ K_P of the backup heaters

EC \ E	NB	NM	NS	ZO	PS	PM	PB
NB	PM	PS	PM	PM	ZO	ZO	ZO
NM	PB	PM	PB	PB	NS	NS	NS
NS	PB	PM	PB	PB	NM	NS	NM
ZO	PB	PM	PB	ZO	NB	NM	NB
PS	PM	PS	PM	NB	NB	NM	NB
PM	PS	PS	PS	NB	NB	NM	NB
PB	ZO	ZO	ZO	NM	NM	NS	NM

Table 4. The control rules of Δ K_I of the backup heaters

EC \ E	NB	NM	NS	ZO	PS	PM	PB
NB	ZO	PS	PM	PB	NS	ZO	ZO
NM	ZO	PS	PB	PB	NM	ZO	ZO
NS	ZO	ZO	PB	PB	NM	ZO	ZO
ZO	ZO	ZO	PB	ZO	ZO	ZO	ZO
PS	ZO	ZO	PM	NM	NB	ZO	ZO
PM	ZO	ZO	PM	NM	NB	NS	ZO
PB	ZO	ZO	PS	NM	NB	NS	ZO

Table 5. The control rules of ΔK_D of the backup heaters

EC \ E	NB	NM	NS	ZO	PS	PM	PB
NB	PS	PM	ZO	ZO	ZO	NS	ZO
NM	PM	PM	PS	ZO	ZO	NS	NS
NS	PB	PB	PS	PS	ZO	NM	NM
ZO	PB	PB	PS	ZO	ZO	NB	NB
PS	PM	PM	ZO	ZO	NS	NB	NB
PM	PS	PS	ZO	ZO	ZO	NM	NM
PB	ZO	PS	ZO	ZO	ZO	NM	NS

5 Simulation Experiment and Result Analysis

5.1 The Features for the Simulation Software and Program

This paper uses acslX software to simulate the pressurizer model. acslX provides a single Integrated Development Environment (IDE) through which simulation model development, execution, and results analysis are controlled. The IDE is shown below:

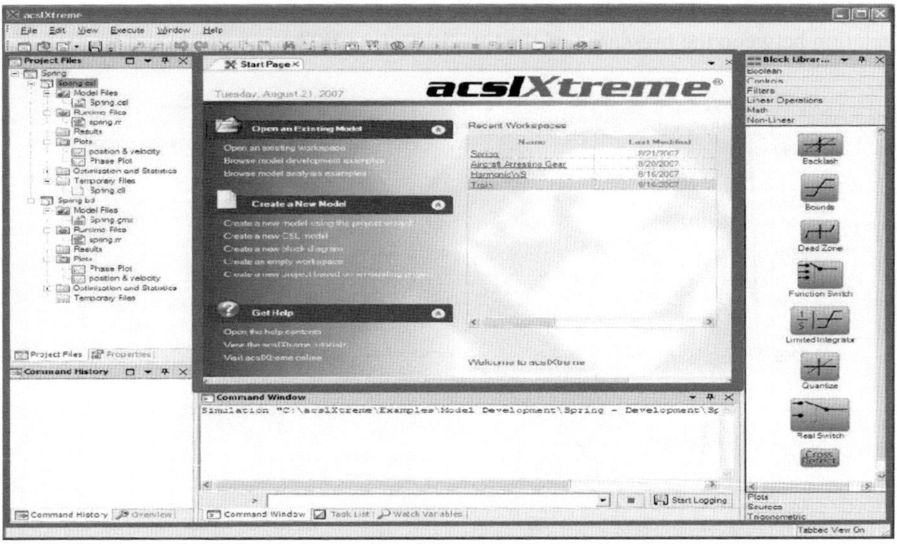

Fig. 3. acslX IDE

The acslX uses Continuous Simulation Language (CSL) to establish models. the structure of CSL is clear, simple and readable, which is easy for user to understand. Users can write the model statements in any order that he likes. acslX can sort the statements automatically[5]. CSL models contain some combination of INITIAL, DYNAMIC (include DERAVITIVE and DISCRETE section), and TERMINAL section[6].

- INITIAL Section

Initial condition values are moved to the state variables (outputs of integrators) at the end of the INITIAL section. Any variables that do not change their values (e.g. constants) during a run can be computed in INITIAL.

- DYNAMIC Section(include DERAVITIVE and DISCRETE section)

The function of DERAVITIVE is to make numerical solution of the differential equations of the pressurizer model. There is a integrated algorithm in the acslX system so that there is no need for the users to write the codes themselves. The function of DERAVITIVE section is to complete the control function of the pressurizer control system.

- TERMINAL sections

Program control transfers to the TERMINAL section when the STOP flag has been set TRUE. On passing out of the last END, control returns to the acslX executive, which reads and processes any further simulation control commands (PLOT, DISPLAY, etc.).

5.2 Results Analysis of the Simulation Experiments

This paper has carried out experiments on the pressurizer model by using the traditional PID control algorithm and the parameter adaptive fuzzy PID control algorithm during the plant transient: power step down from 100% power level to 90%. The surge flow is the input to the pressurizer model. The curves about the relation between the pressure and the time is obtained. The curves are shown below:

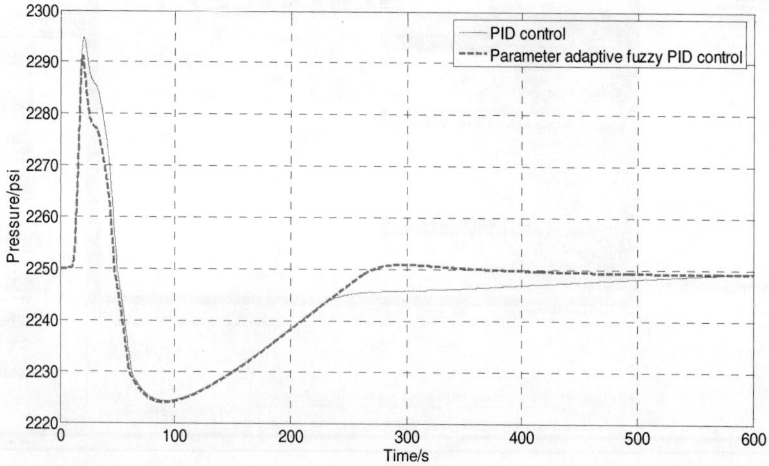

Fig. 4. The curves about the relation between the pressure and the time

As we know that there is a lag between the nuclear power and the electrical power in the nuclear power plant, the nuclear power does not decline immediately when the electrical power declined during the electrical power step down from 100% power level to 90%, which will lead to the mismatch of heat transfer between the primary circuit and the second circuit. The heat from the coolant of the primary circuit cannot transfer to the second circuit immediately, which will lead to the temperature rise and volume expansion of the coolant of the primary circuit. The coolant surges into the pressurizer correspondingly, and the pressure will increase. The spray valves will open to reduce the pressure after the pressure reaches a certain value. The backup heats will operate after the pressure reaches a certain value. The system pressure returns to the set point slowly.

From the figure we can see that both the traditional PID controllers and the parameter adaptive fuzzy PID controllers can all keep the pressure blow 2385psi which is the high pressure trip set point. However, the peak pressure under the control of the parameter adaptive fuzzy PID controllers is lower than the traditional PID controllers, and the time taken to return to the set point is also less than the traditional ones. So we can conclude that the improved controllers make a better performance than the traditional ones. The results show that the

6 Conclusion

In this paper we make an improvement of the traditional PID controllers by adding the parameter adaptive fuzzy PID control algorithm based on the original pressurizer pressure control logic. Experiments have been carried out to simulate the control function during the electrical power step down from 100% power level to 90%. The results show that the parameter adaptive fuzzy PID controllers make a better performance than the traditional ones, which has a reference value for further control strategies optimization of the pressurizer pressure control system in nuclear power plants.

References

1. Lin, C., Yu, Z., Ou, Y.: Chengge Lin, Zusheng Yu, Yangyu Ou: Passive Safety Advanced Nuclear Power Plant AP1000, pp. 423–424. Atomic Energy Press, Beijing (2008)
2. Zhang, G.: Pressurizer Pressure Control System Function Requirement, pp. 1–2. Shanghai Nuclear Engineering Research and Design Institute, Shanghai (2010)
3. Shi, X., Hao, Z.: Fuzzy Control and MATLAB Simulation, pp. 123–124. Tsinghua University Press, Beijing (2008)
4. Lian, X.: Fuzzy Control Technology, pp. 61–62. China Electric Power Press, Beijing (2003)
5. He, Z.: Advanced Continuous System Simulation Language ACSL and its Application, pp. 1–2. Chongqing University Press, Chongqing (1991)
6. AEgis Technologies Group: acslX User's Guide, U.S, pp. 41–46 (2009)

A Diffused and Emerging Clustering Algorithm

Yun-fei Jiang, Chun-yan Yu, and Nan Shen

College of Mathematic and Computer Science, Fuzhou University,
350108 Fuzhou Fujian, China
chunyan.yu@gmail.com

Abstract. Clustering is an important research branch in the field of data mining. Although there are a large number of clustering algorithms, it is difficult to find either a certain clustering algorithm for all the data sets or a best approach for a fixed data set. In this paper, a diffused and emerging clustering algorithm (DECA) is proposed, which uses "similar energy" to make up for the shortcomings of many existing clustering algorithms on similarity, and implements the emergence of classes with high quality. Many experiments demonstrate our algorithm's high accuracy without providing the number of clusters in advance, compared with some typical clustering algorithms such as K-means, Nearest Neighbor and etc.

Keywords: Clustering, Similarity, Diffusion, Emergence, Data Mining.

1 Introduction

Cluster analysis is a very important research direction In the field of data mining. Clustering [1] is the problem of organizing a set of objects into groups, or clusters, in a way as to have similar objects grouped together and dissimilar ones assigned to different groups, according to some similarity measure.

So far, researchers have proposed a large number of clustering algorithms,but each method has its own limitations [2-5]. For example, for partitioning clustering algorithms, the number of clusters is specified in advance; For density-based clustering algorithms, "density" is difficult to determine, etc. Therefore, more clustering algorithms need to be designed so that we can choose the ideal one when faced with a variety of clustering needs. And many clustering techniques, which rely solely on similarity [6], can not produce the desired results. In this paper, a diffused and emerging clustering algorithm (DECA) is proposed, which uses "similar energy" to make up for the shortcomings of many existing clustering algorithms on similarity, and implements the emergence of classes with high quality.

2 A Diffused and Emerging Clustering Algorithm

2.1 Diffusion and Emergence

Data diffusion is a process of the class division, namely that uneven distribution of the data objects are divided into different clusters with the number of data objects

M. Zhu (Ed.): Electrical Engineering and Control, LNEE 98, pp. 593–598.
springerlink.com © Springer-Verlag Berlin Heidelberg 2011

unchanged. Diffusion has two essential preconditions: the total number of data objects remain unchanged and data objects within the class are unevenly distributed. Data diffusion can produce two results: very compact data objects remain the same cluster and relative scattered data objects are divided into different clusters; Number of clusters changes from less to more. The process of diffusion is shown in Figure 1(a).

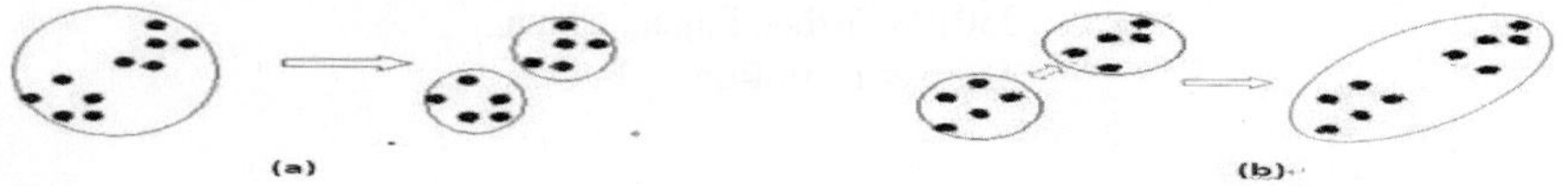

Fig. 1. This shows a figure consisting of diffusion and emergence. The process of diffusion is shown in (a), and (b) depicts the process of emergence.

Emergence is a process that data units interact with each other independently、cooperatively and parallelly, in which data units to meet the condition of combination are combined into the same cluster iteratively and the stable clusters are formed at last. Through emergence, similar data units are combined into a stable class. The behavior of emergence has three important characteristics: independence, cooperation and concurrency. Independence is that each data unit act independently without the influence of other microclusters when data units interact with each other; Cooperation is that data objects within the data unit cooperate with each other to form a whole, which is used to interact with other data units. Concurrency is that data units' actions are carried out simultaneously without the influence of the order of occurrence of behavior. Data unit contains one or more data objects.

To some extent, emergence is the reverse process of diffusion. Thus, the emergence behavior also produces two results: data objects to meet certain conditions are combined into the same class, and number of clusters changes from the more to less. Figure 1(b) depicts the process of emergence.

2.2 Related Definition

Clustering can not avoid the clustering problem of high dimensional data, whose different dimension has different effect on distribution of data points in space. Thus, for high dimensional data, each dimension has different effect on the clustering result. In order that highlight the differences in each dimension of high dimensional data, weighted similarity is introduced to evaluate similarity relation between data objects.

Def. 1. Weighted Similarity. Data set D consists of n m-dimensional data (X_1, $X2$,……, X_n), Xi={x_1, x_2, ……, x_m}.The calculation method of weighted similarity between Data object X_i and X_j is shown in equation (1).

$$s(i,j)=(\sum_{k=1}^{m} W_k(X_{ik}-X_{jk})^2)^{\frac{1}{2}} \tag{1}$$

s(i,j) indicates weighted similarity, W_k is weighting coefficient. Weighting coefficient equals to the ratio of one-dimensional data mean ($mean_k$) and the total data mean ($mean_{total}$),that is the formula (2).

$$W_k = mean_k / mean_{total} \tag{2}$$

$$mean_k = \sum_{i=1}^{n} X_{ik} / n \tag{3}$$

$$mean_{total} = \sum_{i=1}^{n} \sum_{j=1}^{m} X_{ij} / (n * n) \tag{4}$$

Weighted similarity reflects the difference between the dimensional data, expands large influence one-dimensional data has on the whole dataset, and reduce little influence one-dimensional data has on the whole dataset. But the similarity can only provide some numerical value,and not accurately determine scope of numerical value in which data objects are similar by relying solely on the similarity. In the paper,weighted similarity is transformed into"similar energy", and then the problem of similarity measure is solved.

Def. 2. Similar Energy. The calculation method of similar energy is described in equation (5).

$$Energy(i,j) = \frac{\mu}{s(i,j)^A} - \frac{\lambda}{s(i,j)^B} \tag{5}$$

i, j represent data points, λ, μ, A, B are parameters,and B>A, Energy(i,j) is similar energy of data objects.Formula (5)'s image is shown in Figure 2.

According to Figure 2, the relationship between weighted similarity and similar energy is seen clearly. When s> s0,we can not accurately judge which data objects are the same cluster by relying solely on similar energy, but can differentiate to get the maximum of similar energy. When $S = (\lambda B/(\mu A))^{1/(B-A)}$, Energy gets maximum E_max; When s< s0, Energy and s is proportional. When s is very small, energy becomes very small, at this moment, distribution of data points is very close.

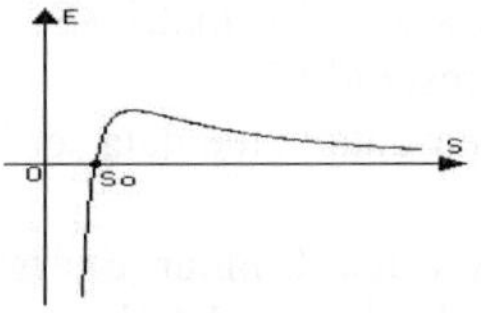

Fig. 2. E is Energy, S is Similarity, So is equilibrium point, that is formula (6). This shows a figure which is described to display a relation between similar energy and similarity.

$$s_0 = \left(\frac{\lambda}{\mu}\right)^{\frac{1}{B-A}} \quad (B>A) \tag{6}$$

Def. 3. Energy threshold. When the data points' distribution is very close, the energy value is very small. In the paper, energy threshold is used to determine scope of numerical value in which data objects are similar. Negative value of similar energy maximum E_max is E'= - E_max. E' is defined as energy threshold.

Def. 4. Emergence Function. Data units to meet merged condition, which is determined by emergence function, are combined into a same cluster. Emergence function is a computing method to calculate the total similar energy between clusters, which formula (7) is used to calculate. x, y are data units, Ci is a cluster label, that is the ith cluster.

$$F(C_i, C_j) = \sum_{x \in C_i, y \in C_j} Energy(x, y), i \neq j \tag{7}$$

Def. 5. Similar data. There are similar energy value between any two data units, if the minimum energy value of the data unit A is the energy value between A and B, and the minimum energy value of the data unit B is also the energy value between B and A, A and B are called "similar data". In the process of the formation of similar data, the behavior of the data units are independent and concurrent, and not all the data units have similar data. If the data objects' number of data unit A and B equals to 1, similar energy between A and B is Energy(A,B);If the data objects' number of data unit A or B is bigger than 1, similar energy between A and B is F(A,B).

2.3 Algorithm Process

In this paper, a diffused and emerging clustering algorithm (DECA) is proposed, that is a process in which data objects diffuse at first and then emerge. Initially, all data objects are seen as one large class, and then the large class make use of diffusion to form a large number of relative small microclusters. At last, similar energy and emergence function are used by microclusters to emerge. The process is as follows:

(I). Diffusion Stage: at this stage, all the data objects in Dataset D diffuse to form a large number of relative small microclusters.

1) Standard data set D' is obtained by normalizing data set D.
2) Calculate weighted similarity matrix S of Standard data set D'.
3) According to weighted similarity matrix S, calculate similar energy matrix "Energy" and energy threshold E'.
4) Get all the pairs of similar data in the data set D'.
5) Repeat
6) Calculate similar energy value (Similar_Energy) of a pair of similar data.
7) If Similar_Energy < E', the similar data is seen as a separate microcluster C.
8) Until all the pairs of similar data are scaned.

(II). Emergence Stage: a large number of microclusters, which is obtained by diffusion, emerge to get final clusters which reach steady state.

1) Repeat
2) According to emergence function, get all pairs of similar data between microclusters.
3) Each pair of similar data is combined to form a new and larger microcluster.

4) Until each cluster has reached steady-state, which is that the number of clusters and data objects in each cluster are no longer changed.
5) Output final clusters reached steady state.

3 Experimental Analysis

3.1 Parameter Settings

In the study, parameters are determined through the analysis of the data set. When mean, variance and other knowledge of dataset are fully taken into account, clustering results are relatively good. In the experiment, the parameters are adjusted so that the value of r_0 floats around s', DECA can get better clustering results.

$$s' = (mean - std)/2 \tag{8}$$

In formula(8), mean is the total mean of dataset, std is the total variance of dataset.

3.2 Experimental Analysis

DECA uses the method proposed by Sun [7] to calculate the correct rate. Correct rate is calculated as formula (9):

$$P = \sum_{i=1,2,...,k} \frac{a_i}{n} \tag{9}$$

In formula(9), P is the correct rate, ai is the number of samples in the ith class, k is the number of classes, n is the total number of samples in data set.

In the experiment, three UCI datasets (Iris、Wine、Image) are used as test data to compare the accuracy of clustering, and then validity of DECA are also verified. According to the result of experiment shown in Table 1, DECA has high accuracy. The algorithm achieved the desired results for Iris and Wine. Although DECA's accuracy rate only reached 60.47% for Image, it is also higher than the other algorithms.Therefore, DECA is effective.

Table 1. Experimzental results table.

Algorithm name	Accuracy			Data sources
	Iris	Wine	Image	
Nearest neighbor	68%	42.7%	30%	References[4]
Furthest neighbor	84.00%	67.40%	39.00%	References[4]
Between groups average	74.70%	61.20%	37.00%	References[4]
Ward Method	89.30%	55.60%	60%	References[4]
K-means	81.6%	87.96%	56%	References[4]
DECA	89.33%	88.20%	60.47%	Experiment

4 Conclusion

A diffused and emerging clustering algorithm (DECA), uses "similar energy" to make up for the shortcomings of many existing clustering algorithms on similarity, and implements the emergence of classes with high quality. Many experiments demonstrate our algorithm's high accuracy without providing the number of clusters in advance.However, clustering of high dimensional data is still worthy of study. Future work will focus on high-dimensional data aggregation, and study how to intelligently adjust the parameters of DECA to facilitate clustering of various data sets.

Acknowledgments. This work is supported by the National Natural Science Foundation of China under Grant No. 60805042, the Natural Science Foundation of Fujian Province under Grant No. 2010J01329,the Industry-University-Research Major Project of Fujian Province under Grant No.2010H6012 and the Program for New Century Talents of Fujian Province under Grant No.XSJRC2007-04.

References

1. Han, I., Kamber, M.: Data Mining: Concepts and Techniques. Morgan Kaufmann Publishers, Berlin (2000)
2. Sun, J.-G., Liu, J., Zhao, L.-Y.: Clustering Algorithms Research. Journal of Software 19(1), 48–61 (2008)
3. Kotsiants, S.B., Pintelas, P.: Recent advances in clustering: A brief survey. WSEAS Trans. on Information Science and Applications 1(1), 73–81 (2004)
4. Jain, A.K., Murty, M.N., Flynn, P.: Data clustering: A review. ACM Computing Surveys 31(3), 264–323 (1999)
5. Peng, J., Tang, C.-J., Cheng, W.-Q.: A Hierarchy Distance Computing Based Clustering Algorithm. Journal of Computers 30(5), 786–795 (2007)
6. Guo, J., Zhao, Y., Bian, W., Li, J.: A Hierarchical Clustering Algorithm Based on Improved Cluster Cohesion and Separation. Journal of Computer Research and Development 45, 202–206 (2008)
7. Sun, Y., Zhu, Q.M., Chen, Z.: An iterative initial- points refinement algorithm for categorical data clustering. Pattern Recognition Letters 23(7), 875–884 (2002)

A Fast Method for Calculating the Node Load Equivalent Impedance Module

Wang Jing-Li

Electronic and Information Engineering College, Liaoning University of Technology, Jinzhou, 121001, P.R. China
wangjingli_lg@163.com

Abstract. This paper proposes a fast method of calculating the node load impedance module. Taking some operation state of power system as the ground state, the true value and the approximate values of the node load impedance module can be worked out in the ground state and other three operation states. The power flow need not to be recalculated under the other operation states, and only need to correct the impedance module by the Lagrange's interpolation in order to rapidly get the result of the nodal load impedance module. The result of the example shows that the method of the calculating the node load impedance module not only guarantees precision, but also increases the speed significantly.

Keywords: Impedance module, Lagrange's interpolation, Power flow calculation.

With the enlargement of the scale of power systems, power grid structure and operation mode of the increasingly complex, the loss caused by voltage collapse getting more and more serious, power system voltage stability problems are becoming more and more important, thus has generally been regarded as one of the important topics. The impedance module method that is a mature method of the judgment of power system voltage stability, but the node load impedance module is closely related to the operation mode, the network topology, the network parameters and the node load, in every operation state the calculation of the nodal load impedance module need to recalculating the power flow, it makes the calculation difficulty and speed slow. How to be precise and fast to get the impedance module on the load side is becoming a very important problem of using impedance module method to analysis voltage stability.

1 The True Value Calculation of the Node Load Impedance Module in the Ground State

In power systems, power flow is calculated if some operation state to be as ground state. The node i is chosen to be researched, that the voltage Vi of the node i is calculated , the node i' s impedance module is noted as $|Z_i|$. The impedance module $|Z_i|$[1] of the node i is calculated by the equation (1)

M. Zhu (Ed.): Electrical Engineering and Control, LNEE 98, pp. 599–603.
springerlink.com © Springer-Verlag Berlin Heidelberg 2011

$$|Z_i| = \frac{|V_i|^2}{S_i} = \frac{V_{ix}^2 + V_{iy}^2}{\sqrt{P_i^2 + Q_i^2}} \tag{1}$$

where, V_{ix} represents the real part of the node voltage *i*, V_{iy} represents the imaginary part of the node voltage i, P_i represents the active power of the node i, Q_i represents the reactive power of the node i, and S_i represents the apparent power of the node *i*.

2 Approximate Calculation of the Node Load Impedance Module

If some node loads are changed, the node load impedance module to be researched will change[2]. $\Delta|Z_i|$ presents its changes. S_j presents the apparent power of the node *j* in the power systems, and ΔS_j presents its changes. The variable quantity of the nodal i load impedance module that will be changed by the changes of the nodal *j* can essentially be described by the equation (2).

$$\Delta|Z_i| = \sum_j \frac{\partial|Z_i|}{\partial S_j}\Delta S_j \tag{2}$$

If the number of the node its load changed in the power system is *n*, the number of the algebraic terms of summation is n in the equation (2)[3]. Then

$$\frac{\partial|Z_i|}{\partial S_j} = \begin{cases} \dfrac{2V_{ix}}{S_i}\dfrac{\partial V_{ix}}{\partial S_j} + \dfrac{2V_{iy}}{S_i}\dfrac{\partial V_{iy}}{\partial S_j} & j \neq i \\ \dfrac{2V_{ix}}{S_i}\dfrac{\partial V_{ix}}{\partial S_j} + \dfrac{2V_{iy}}{S_i}\dfrac{\partial V_{iy}}{\partial S_j} - \dfrac{|V_i|^2}{S_i^2} & j = i \end{cases} \tag{3}$$

where, $\frac{\partial V_{ix}}{\partial S_j}$ and $\frac{\partial V_{iy}}{\partial S_j}$ can be calculated through the equation (4) and (5)

$$\begin{aligned} \frac{\partial V_{ix}}{\partial S_j} &= \frac{1}{\dfrac{\partial S_j}{\partial V_{ix}}} = \frac{1}{\dfrac{P_j}{\sqrt{P_j^2+Q_j^2}}\dfrac{\partial P_j}{\partial V_{ix}} + \dfrac{Q_j}{\sqrt{P_j^2+Q_j^2}}\dfrac{\partial Q_j}{\partial V_{ix}}} \\ &= \frac{1}{\dfrac{P_j}{\sqrt{P_j^2+Q_j^2}}\dfrac{1}{\dfrac{\partial V_{ix}}{\partial P_j}} + \dfrac{Q_j}{\sqrt{P_j^2+Q_j^2}}\dfrac{1}{\dfrac{\partial V_{ix}}{\partial Q_j}}} \end{aligned} \tag{4}$$

$$\frac{\partial V_{iy}}{\partial S_j}=\frac{1}{\dfrac{\partial S_j}{\partial V_{iy}}}=\frac{1}{\dfrac{P_j}{\sqrt{P_j^2+Q_j^2}}\dfrac{\partial P_j}{\partial V_{iy}}+\dfrac{Q_j}{\sqrt{P_j^2+Q_j^2}}\dfrac{\partial Q_j}{\partial V_{iy}}}$$
$$=\frac{1}{\dfrac{P_j}{\sqrt{P_j^2+Q_j^2}}\dfrac{1}{\dfrac{\partial V_{iy}}{\partial P_j}}+\dfrac{Q_j}{\sqrt{P_j^2+Q_j^2}}\dfrac{1}{\dfrac{\partial V_{iy}}{\partial Q_j}}} \tag{5}$$

$\frac{\partial V_{ix}}{\partial P_j}$ 、 $\frac{\partial V_{ix}}{\partial Q_j}$ 、 $\frac{\partial V_{iy}}{\partial P_j}$ 、 $\frac{\partial V_{iy}}{\partial Q_j}$ can be calculated through the small disturbance theorem of the General Tellegen's theorem [4]. The variable quantity of the nodal load impendence module $\Delta|Z_i|$ can be calculated through the equations (2) to (5). The approximate value $|Z_i|'$ of the nodal load impendence module is calculated through equation (6).

$$|Z_i|' = |Z_i|_0 - \Delta|Z_i| \tag{6}$$

3 The Correction of the Node Load Impendence Module

The small disturbance theorem of the General Tellegen's theorem is an approximate calculation in the small disturbance situation. The calculation is linearization. The variable quantity of the load is very small of every times the node load impendence module calculation, otherwise the error is very large, though the speed of the calculation is fast. If the variable quantity of the load through the calculation is big, the correction of the result is needed in order to guarantee to be accurate.

After calculating the node *i* load impendence module in the ground state, and then changing the other nodes load, the difference of the true value and the approximate value of the node load impendence module is calculated in the three operation states of A, B, C, described as e_A, e_B and e_C, the apparent power of its described as S_A, S_B and S_c. The error between the two methods is calculated by the Lagrange's Interpolation. The researching node impendence module can be calculated through the equation (7). The result of the impendence module is very close to the true value that calculated by power flow calculation.

$$|Z_i| = |Z_i|' - e' \tag{7}$$

4 The Steps of the Fast Calculation of Node Load Impendence Module

According to the foregoing thought, the calculation of the basic party in the power system is needed. The steps are as follows:

Step 1: The node load impendence module $|Z_i|_0$ is calculated by means of (1) in the ground state [5].

Step 2: Through changing the node load, the true values of the node load impendence module are calculated by using eq.(1) in the three operation states of the A, B and C.

Step 3: The approximate value of the node load impendence module in the three operation states and the errors e_A, e_B and e_C between the true value and the approximate value will be calculated by using from (2) to (6).

When the operation state has changed, the fast calculation of the node load impendence module in the power system are as follows:

Step1: Use the equations from (2) to (6) to calculate the approximate value of the node load impendence module in the current operation state.

Step2: Calculate the error between the true value and the approximate value of the nodal load impendence module in the current operation state by the Lagrange's Interpolation.

Step3: Use the equation (7) to correct the approximate value of the node load impendence module $|Z_i|$, and $|Z_i|$ is obtained.

5 Numerical Example Analysis

Consider IEEE11 nodes system. The calculated data of the 6th node that load changes constantly are described in the table 1. Let ground state $S_6 = 1.6716$, and the impendence module $|Z_6|_0 = 0.4361$, where ΔS_6 represents the variable quantity of the 6th node load, $|Z_6|$ represents the 6th node load impendence module that calculated by the way this paper proposed, $|Z|$ represents the true value of the nodal load impendence module that calculated by the way of the power flow calculation, and e_s represents the error of the node load impendence module between using the two methods.

Table 1. Results of 11-node power system

ΔS_6	$\|Z_6\|$	$\|Z\|$	e_s
0.1505	0.3692	0.3690	0.054%
0.2136	0.3427	0.3426	0.029%
0.3188	0.2991	0.2993	0.067%
0.3620	0.2811	0.2813	0.071%
0.4319	0.2509	0.2508	0.039%
0.4595	0.2380	0.2378	0.084%
0.5016	0.2150	0.2151	0.046%

It is very clear that the error is very minor between the methods from the table 1. The proposed method is predominates in the time of the calculation. The average calculation speed of this method is 10 times faster than the method of the power flow calculation, and the bigger power system is, the faster speed is.

6 Conclusions

The traditional calculation of the node load impendence module need to power flow calculating in any operation state, and the amount of the calculation is very large. The proposed method need to calculate the base party, then calculate the approximate value of the node load impendence module, at last use the Lagrange's interpolation to correct the approximate value. While the power flow calculation in any operation is not needed. The example shows that the proposed calculation method is concise , fast and practical.

References

1. Xu, B.L., Liu, Z., Wang, Y.G., Ge, W.C.: Property and significance of the power grid nodal load critical impendence module. Harbin Institute of Technology Journal 31(4), 91–95 (1999)
2. Liu, Z.: The impendence analysis of the heavy nodal load in the voltage stable problems. Proceedings of the CSEE, 20(4): 35-39 (2000)
3. He, Y.Z.: Power System Analysis. Huazhong University of Science and Technology Press, Wuhan (1995)
4. Lu, B.C., Guo, Z.Z., Liu, Z.: Base on the general Tellege theorem the calculation of the power system sensitivity. Automatic Control on Electrical Power System 21(10), 17–20 (1997)
5. Niu, H., Guo, Z.Z.: Based on the general Tellege theorem the power flow calculation. Automatic Control on Electrical Power System 22(16), 14–16 (1998)

The Design and Implementation of Anti-interference System in Neural Electrophysiological Experiments

Hong Wan, Xin-Yu Liu, Xiao-Ke Niu, Shu-Li Chen,
Zhi-Zhong Wang, and Li Shi

School of Electrical Engineering, Zhengzhou University, 450001, China
{wanhong,chensl,shili}@zzu.edu.cn,
{lxinyuzzu,niuxiaoke0000}@163.com,
wangzhizhong1982@yahoo.com.cn

Abstract. As the weakness of NRS (Neuronal Response Signals) and complexity of the acquisition instruments, this kind of signal is susceptible to noise that is difficult to be denoised, as a result, it required a very strict condition for such experiment. The paper proposed some efficient improved methods based on those problems existing during the experiment. In addition, based on some practical experience of related electrophysiological experiments, a simple and effective anti-interference system is designed and implemented in the processes of establishing platform for neural electrophysiological experiments and. The anti-interference system is composed of shielding system and grounding system. In the grounding system, we designed a pluggable GCP (Grounding Connection Plate) which was used to connect with the grounding body and the grounding wire of experiment instruments and studied the optimum grounding ways of experiment instruments, which have had a very good result to experiment. After it is applied in the experiment, the signal-to-noise ratio of NRS is enhanced obviously.

Keywords: anti-interference measures, neuronal response signals, grounding connection plate.

1 Introduction

NRS (Neuronal Response Signals) are the most common electrophysiological signals in the visual system research. As it is very weak with the amplitudes just about 5~150 μV, this kind of signal is susceptible to be noised. As a result, Great effort will be spent preventing or removing various possible interference when we recording NRS. Thus, it is important to find some simple and practical anti-interference measures during recording NRS in neural electrophysiological experiments.

The anti-interference measures of the neural electrophysiological experiments are scattered in the related references [1], [2], [3] which were incomprehensive and rarely realized. Based on some practical experience of related electrophysiological experiments, the paper here proposed a simple and effective anti-interference system in the processes of establishing platform for neural electrophysiological experiments.

M. Zhu (Ed.): Electrical Engineering and Control, LNEE 98, pp. 605–611.
springerlink.com © Springer-Verlag Berlin Heidelberg 2011

2 Interference Source

After visual evoked signal is recorded and the preliminarily analyzed, we find that the interference existing in neural electrophysiological experiments mainly includes two kinds with one being the power interference and the other the electromagnetic interference.

The most common power interference is mainly as follows: the voltage fluctuation of power network, the electric sparks and arcing of touch-contact switch or the spike pulse caused by power interruptions. The power interference always appears randomly in anytime and everywhere. The electromagnetic interference sometimes comes from something outside the recording equipment, such as the discharge noise, the noise caused by on-off switching of the power, the industrial frequency interference of high-power transmission line, the electromagnetic radiation of radio equipment, etc. And sometimes from the internal sources such as the AC (Alternating Current) waves, the induction between the different signal, the parasitic oscillation, etc.

3 Anti-interference System

Shielding and grounding are the most common anti-interference measures in electrophysiological experiments with Shielding is mainly used to prevent electromagnetic interference [4] and grounding mostly used to reject the power interference and static interference [5].

3.1 Shielding System

Shielding system is composed of a shielded room and some ancillary facilities. Shielded room is made of copper wire gauze with a mesh size being about 15mm, for its' good electrical conductivity, stable chemical properties and moderate price. The gauze of shielded room (including the four sides and the top) are connected though a metal conductor which is connected with GCP (Grounding Connection Plate), as shown in Figure 2. In the shielded room, the stimulus monitor, the cold light source and the wires with electricity above the laboratory furniture are all wrapped by aluminized paper and then grounded to prevent them from interfering with each other.

3.2 Grounding System

It is very necessary to establish a proper and effective grounding system for laboratorial instruments running normally and reliably. Grounding is first used in high power systems and extended to the weak electricity systems later [6] with its role extending from the safety ground to the signal ground.

Burying of Grounding Body. Grounding body is used for the directly the ground wire with the earth connect. And the burying impacts the effect of the grounding. The grounding body used is buried specially. In order to increase the contact area between the grounding body and the earth and reduce the ground resistance, a flat triangle with 1-meter-long is chosen and laid in relatively moist and shady area near the laboratory about 5m away. Then it is introduced it into the laboratory with a brass wire about

35mm in diameter. In the laboratory, the brass wire is connected with a pluggable GCP, which is made of copper plate of good conductive properties, shown in Figure 1. The grounding terminals of all kinds of equipments are connected to grounding body by the GCP in the laboratory.

Power Grounding. Power interference is one of the largest interference in neural electrophysiological experiments [7], and is also the largest interference source introduced from outside the laboratory. Based on hardware circuit, technologies that

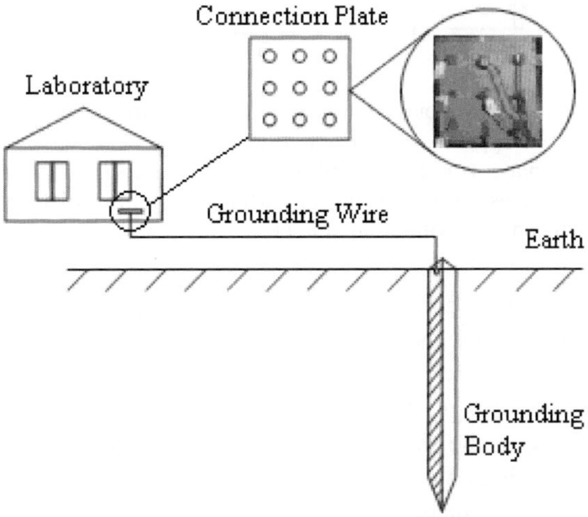

Fig. 1. Diagram of grounding. The picture is the physical map of the GCP in the figure.

removing the 50Hz power frequency interference can be achieved easily [8], but the higher harmonic of peak pulse interference in power network is very diverse, frequent, and of high amplitude which always brings much adverse effect to the AC power-supply electronic system. So it is essential to eliminate the power interference for NRS acquisition.

Generally, the power socket is single-phase and of three line in electrophysiology laboratory. The corresponding regulation of this kind of sockets is as follows: the left is the zero line, the right is the phase (or fire) line, the upper (or middle) is the ground line. Make sure that the earth terminal of the power supply must be grounded, and the earth terminals of strong and weak electricity would be connected with the different grounding bodies separately. During the experiment, it can be found that the instrument plug should be put into the corresponding socket, so as to prevent introducing the power interference from the various instruments to affect the acquisition of NRS.

Instruments Grounding. In neural electrophysiological experiments, the main instruments include anti-vibration experimental table, data acquisition system, micro-manipulation apparatus, stimulator, stereotaxic apparatus, and so on. These instruments, especially those in the shielding mesh, should be grounded which is important for the acquisition of biomedical-signals.

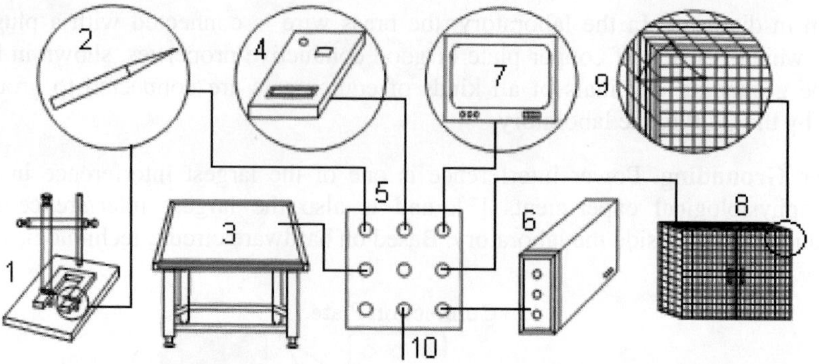

Fig. 2. Diagram of instruments grounding in laboratory. In the figure, 1 is stereotaxic apparatus, 2 is ear bar of stereotaxic apparatus, 3 is laboratory table, 4 is micromanipulator apparatus, 5 is grounding connection plate, 6 is head amplifier of data acquisition system, 7 is stimulation monitor, 8 is shielding mesh, 9 is drawing of partial enlargement of the shielding and 10 is grounding wire which connects with grounding body.

Shielding can reduce the change of magnetic flux of the laboratory, but the change of magnetic flux in internal and external of shielding mesh will also produce some induced voltage on copper wire mesh. So the shielding mesh should be grounded to eliminate the induced voltage. To achieve the grounding of the shielding mesh, the four sides and the top of the shielding mesh are connected respectively with a wire, then a wire from one side connects with the GCP. In addition to the shielding mesh, the electrical equipments, such as head amplifier of data acquisition system, micromanipulation apparatus, stimulating monitor, etc, and some mechanical equipments, such as experimental table, ear bar of stereotaxic apparatus, etc, also should be grounded. Grounding diagram of experimental instruments are shown in Figure 2. Only when these instruments are grounded at the same time, can the interference be removed obviously.

Experimental Animal Grounding. During the experiment, the hair of the experimental animal would happen to rub with other objects easy to produce static electricity. Therefore, in order to eliminate or reduce static and other interference produced by the animal effectively, animal should be grounded. During the experiment, we insert one end of a stainless steel needle into the tail of experimental animal and connect the other end with the GCP, which can reduce the interference caused by animals effectively.

4 Experiment

4.1 Animal Preparation

The data is all recorded in the primary visual cortex(V1) of a healthy female adult SD (Sprague Dawley) rat (11 weeks old, 222g) ,which is raised at the Experiment Animals Center of Zhengzhou University. The rat was firstly anesthetized (10%

chloral hydrate, 0.4ml per 100g) through intraperitoneal injection, and then fixed in the stereotaxic apparatus. The hair above the head is shaved. The skin is removed leaving the skull exposed. The craniotomy is performed above the primary visual cortex with the area of 4mm × 3mm (the centre lies at 3.0mm from the sagittal suture, 7.0mm form Bregma). Then a 2 × 8 array (MicroProbes, platinum iridium electrodes) is implanted in primary visual cortex of depth about 810μm.Body temperature was kept constant, and anesthetic (10% chloral hydrate) was given at a rate of 0.2ml/h. During the entire experiment, the health statement of the rat and the depth of anesthesia were continuously monitored. The data (Cerebus system, Blackrock Microsystems) is recorded 20min after the implantation of electrodes. Two groups of LFPs(Local Field Potentials) are recorded, one of which is with noise while the other is without, as is shown in Figure 3C and 3D. These data were digitized at a sampling rate of 2 kHz, recorded for 2min. The LFPs were extracted from the raw date through Butterworth low-pass-filter with cut-off frequency of 250Hz.

4.2 Data Analysis

The data is LFPs signal of the 27th channel. The data analysis procedures were implemented with the programming language Matlab (Mathworks, Natick, Massachusetts). The row signals in Figure 3A and 3B show that the anti-interference system is effective. It is not obvious that the anti-interference measure is effective only from the comparison between LFPs. Here, we compute the amplitude spectrum of a 60secs LFPs data by taking the fast Fourier transform (FFT):

$$X(k) = \sum_{j=1}^{N} x(j)\omega_N^{(j-1)(k-1)} . \tag{1}$$

where

$$\omega_N = e^{-2\pi i / N} . \tag{2}$$

is an Nth root of unity. N is the sampling numbers. Then the amplitude spectrum is normalized with respect to its norm:

$$Y(k) = \frac{|X(k)|}{(\sum_{k=1}^{N} |X(k)|^2)^{1/2}} . \tag{3}$$

$|X(k)|$ is the complex modulus of $X(k)$, $1 \le k \le N$. Although the LFPs don't indicate the effective of the anti-interference, we are easy to see from the amplitude spectrum of the LFPs, shown in figure 3.

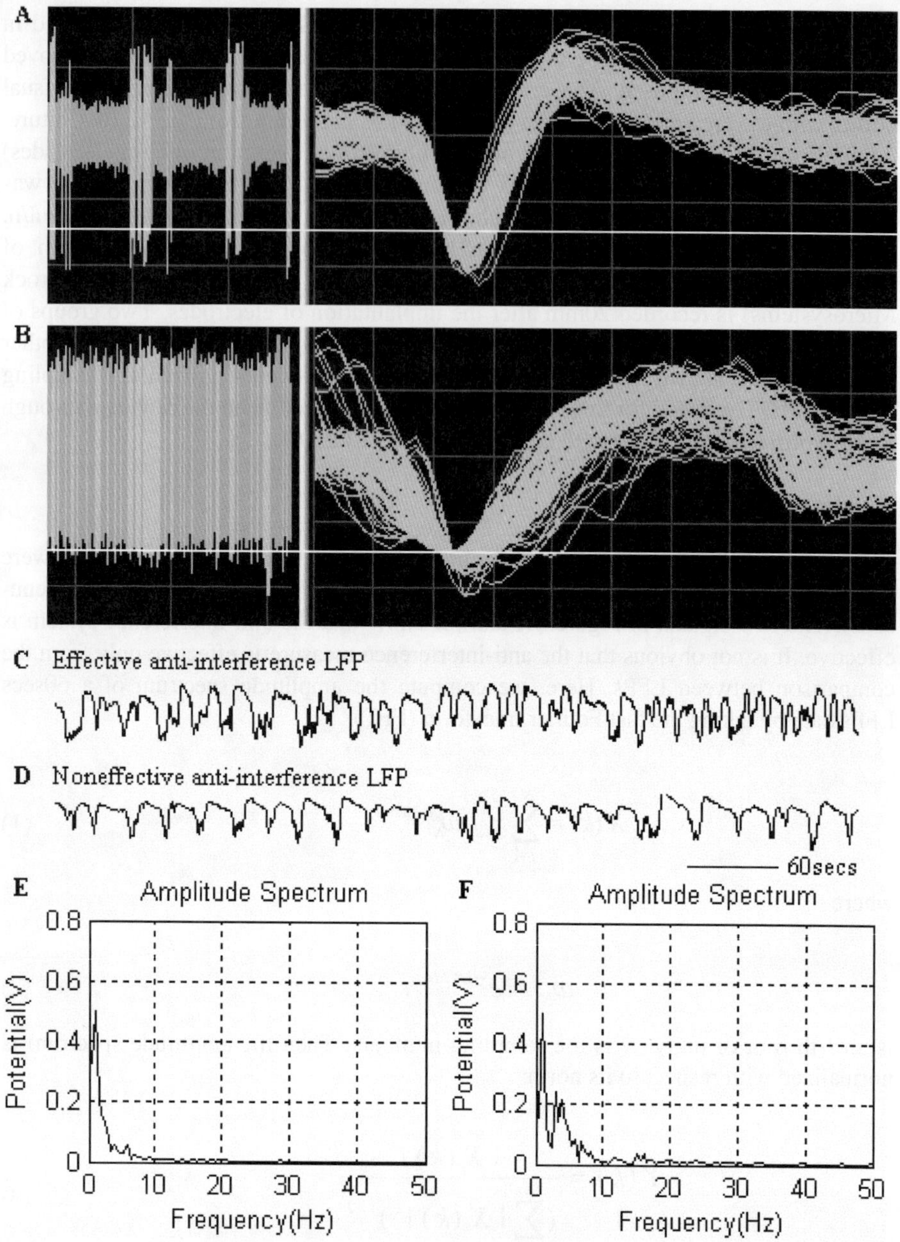

Fig. 3. Diagram of experiment analysis. (A) The row signals and spike of effective anti-interference measures while (B) are non-effective. (C) The LFPs of effective anti-interference measures while (D) are non-effective. (E) The amplitude spectrum of effective anti-interference measures while (F) are non-effective.

5 Conclusion

In this paper, the anti-interference measures in the neural electrophysiological experiments are discussed mainly. We achieved a simple and effective anti-interference system in neural electrophysiological experiments by establishing the neural electrophysiological experiments platform and the experience related to experiments, which have been verified in experiments. According to what we have discussed in this paper, the important points can be summarized as follows:

① It is necessary to bury specifically the grounding body, which can not be shared with other laboratories, in neural electrophysiological experiments, and the grounding bodies of the strong and weak electricity should be buried respectively;

② Make sure the grounding terminal of power connects with the earth;

③ Make sure a variety of laboratory equipments, especially the laboratory equipment in shielding, should be grounded. The machinery equipments and the animals should be grounded too.

Acknowledgements. The present research work was supported by the National Natural Science Foundation of China (No.60841004, No.60971110). The author gratefully acknowledges the support of these institutions.

References

1. Spelman, F.A.: Electrical Interference in Biomedical Systems. IEEE Transactions on Electromagnetic Compatibility 7, 428–436 (1965)
2. Li, C.K., Xu, S.Y., Chen, L.X.: Sources and elimination of interference in patch clamp electrophysiological experiment. J. First Mil. Med. Univ. 22, 656–657 (2002)
3. Zhang, C.F., Mao, H.P.: Analysis and noise reduction measures on background noise and outside interference in the METS. Machinery Design & Manufacture 7, 97–99 (2008)
4. Liu, D.C., Deng, S.Q.: The Study of Metal Meshes on Electromagnetic Interference Shielding Effectiveness. In: Asia-Pacific Conference on Environmental Electromagnetics, pp. 326–332. IEEE Press, New York (2000)
5. Fowler, K.: Grounding and shielding. IEEE Instrumentation & Measurement Magazine 3, 41–48 (2000)
6. Lin, G.: Discussion on power supply disturbance and grounding in laboratory. Telecommunications for Electric Power System 29, 66–69 (2008)
7. Thorp, C.K., Steinmetz, P.N.: Interference and Noise in Human Intracranial Microwire Recordings. IEEE Transactions on Biomedical Engineering 56, 30–36 (2009)
8. Hwang, I.-D., Webster, J.G.: Direct Interference Canceling for Two-Electrode Bio-potential Amplifier. IEEE Transactions on Biomedical Engineering 55, 2620–2627 (2008)

System Design of a Kind of Nodes in Wireless Sensor Network for Environmental Monitoring*

Ying Zhang and Xiaohu Zhao

College of Information Engineering, Shanghai Maritime University,
Shanghai 200135, China
yingzhang@shmtu.edu.cn, zhaoxiaohu320@126.com

Abstract. In order to improve the automation level of data acquisition in various environmental monitoring, a kind of wireless sensor network node had been designed, which can collect some kinds of information in the field depending on various sensors carried. The hardware design of sensor network node based on the MSP430F1611: ultra-low power MCU was introduced, the application system includes: wireless transceiver chip CC2420 and access circuit of external sensors. MSP430F1611 controls the collection procedure for data acquisition by some sensors, which involve: temperature、humidity and light intensity of the environment, etc. After data pretreatment, the data will be sent by the wireless transmitter module to the adjacent nodes, forwarded one by one to the server eventually, the real time environmental monitoring is achieved. The system has the advantage of flexibility to be laid out, and it is suitable to collect information in some harsh environment.

Keywords: Environmental monitoring; wireless sensor network; nodes; MSP430F1611; CC2420.

1 Introduction

With the rapid development of communication technology and microelectronic technology, a kind of intelligent sensor is appeared which is combined with sensor technology, computer technology and communication technology. The application of the sensor in process control and signal monitoring has received tremendous attention and has been a hot research topic in borne and abroad because its advantages such as high accuracy, good reliability and versatility and so on. This paper presents a kind of wireless sensor network node for real-time monitoring the environmental information [1], the environmental parameters can be accurately measured and transmitted reliably with the advantages of sensor systems: digitization, intelligence and wireless.

---

* This project is supported by Science & Technology Program of Shanghai Maritime University (No. 20110036); supported by the key subject construction of Shanghai Education Commission (No. J50602); and supported by the program of international technology transfer of Shanghai Committee of Science and Technology (No. 10510708400).

M. Zhu (Ed.): Electrical Engineering and Control, LNEE 98, pp. 613–620.
springerlink.com © Springer-Verlag Berlin Heidelberg 2011

The core of the node system is MSP430F1611, which is a ultra-low power MCU, with the cooperation of peripheral apparatus of new low-power micro-sensors, it can measure the environmental parameters of temperature, humidity and light intensity in real time. This kind of node uses the AA battery as the supply power. Wireless transmission can avoid laying down wiring harness, and it is suitable for collecting environmental parameters in harsh environment with large areas.

2 Hardware Design

According to management mechanism of wireless sensor network nodes, the nodes are divided into three groups: sensor nodes, cluster head nodes and aggregation nodes. The primary works of cluster head nodes and aggregation nodes are complete the function of data receiving and sending, so the node consist primarily of micro-processing module and wireless transmission module. And the sensor node mainly through the sensors to collect data on the surrounding environment (temperature, humidity, light sensitivity, etc.), then the data will be sent to A/D converter and processed by the processor, at last, processed data is sent by the wireless transmission module to the adjacent nodes, meanwhile, the nodes must also perform data forwarding, which is the neighboring nodes to send the data to the sink node or nodes from the sink node closer. The entire hardware system is mainly divided into 4 modules: data acquisition, data processing, wireless transmission and power supply module. The structure is shown in Fig.1.

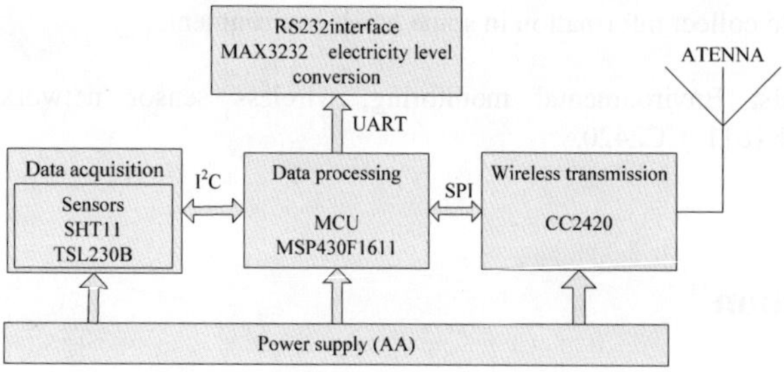

Fig. 1. Node Structure

2.1 Data Acquisition Module

Data acquisition module uses the sensors to collect the information of environmental temperature, humidity, light intensity and other parameters, and then the data will be sent to A/D converter and processed by the processor.

(1) Temperature and Humidity Sensor
SHT11 is produced by Sensirion in Switzerland, which is a digital sensor, and has the I^2C bus interface [2]. SHT11 is a single chip relative humidity and temperature integrated sensor module with a calibrated digital output. It can test the environmental

temperature and humidity. The device includes a capacitive polymer sensing element for relative humidity and a bandgap temperature sensor. It has the characteristics: free testing, free calibration, full exchange and don't need the external circuit.

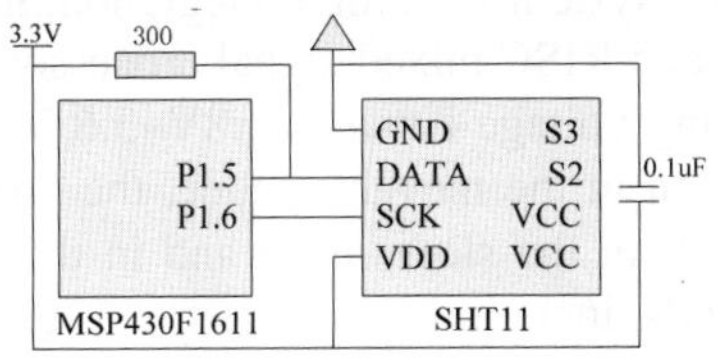

Fig. 2. Connection of SHT11 to MCU

Fig.2 shows the connection between SHT11 and MCU. The MCU has two pins (P1.5 and P1.6), connect with the I2C interface (DATA and SCK) of SHT11, and access SHT11 by simulating the I2C scheduling. The DATA terminal access a pull-up resistor, at the same time VCC and GND terminal access a decoupling capacitor. By the corresponding software design, data collection and transmission can be completed.

(2) Light Intensity Sensor

In the part of light intensity measuring, the chip TSL230B produced by TI (Texas Instruments) is selected to measure the light intensity of the environment. The device uses advanced LinCMOSTM technology. It can complete high-resolution light to frequency conversion doesn't need external components. The spectra of light in surrounding environment can be measured into current, then from the current/frequency converter to convert into the corresponding frequency [3]. This device can output triangle or square waves with certain frequency which are entirely changed by the light amplitude, it has higher resolution and can be directly connected with the MCU. The connection is shown in Fig.3. S0, S1 are sensitivity control terminals of the sensor; S2, S3 are terminals of the frequency coefficient selection; OUT is the output frequency signal port, and sent capture input signal into the MCU, by calculating the numerical difference of count between twice capture time ,we can calculate the value of output frequency value. Light intensity can be got according to relations between the output frequencies and illumination.

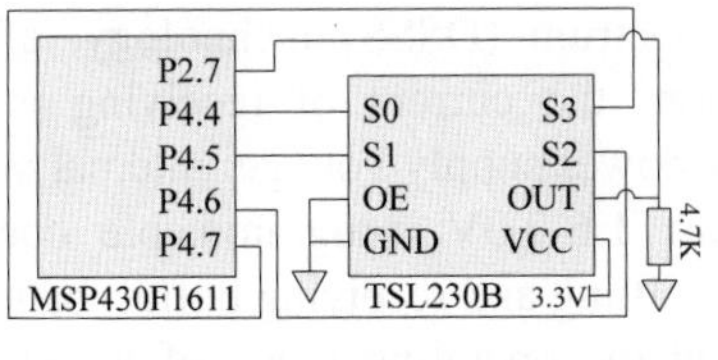

Fig. 3. Connection of TSL230B to MCU

2.2 Data Processing Module

MCU is the heart of the entire node system, so the quality of the MCU's performance decides the whole performance of the entire node. MSP430F1611 is developed by TI company. It is a single chip system with high integration, high-precision, and has 16-bit bus with lowest power and RISC mixed-signal processor [4].

It has the lowest working voltage between 1.8V~3.6 V for regular work, and has minimal power consumption in the movable mode, the operating current only needs 280 μA, only needs 1.6 μA in the sleep mode and in the closed state only requires 0.1A. It has more peripherals, including 8 channels of 12bit A/D and 2 channels of 12 bit D/A, It greatly simplifies the hardware design and save the cost effectively. There are 3 oscillation sources in MSP430 microcontroller, including a high frequency clock, a low-frequency clock and a DCO. The vivid clock selects to make the system can operate under the fairest clock, greatly reduce the power consumption of the system and convenience the design of system; It easy to connect multiple devices because it has a wealth of peripheral interfaces including standard serial port, SPI interface, I^2C interface. Inside of MSP430F1611 has ample store space to access 10kB RAM and 44kB programmable flash memory, it can assure the normal operation of agreement and it is convenient to design and hold the code of protocol. In addition it also has interrupt wake-up function that can interrupt the microcontroller from sleep mode to active mode. So it is very suitable for the design of wireless sensor network node [5].

2.3 Wireless Transmission Module

Wireless transmission module fulfils the communication task between wireless sensor network nodes. With the development of the integration of chip, the circuits system has less energy consumption. Nodes energy mainly consumes in communication. Low-power of communication devices will save node energy and prolong its life time.

The smart RF chip CC2420 produced by TI is used in the design of the scheme. The RF transceiver CC2420 works in slave mode and MSP430F1611 works in master mode through the SPI interface to configure parameters of CC2420 and read/write the data in the buffer. The detailed connection of the pins is shown in Fig.4. The chip works with the protocol of physical layer and data link layer defined by IEEE802.15.4, and works at the frequency point of 2.4 GHz. It has many features: direct sequence spread spectrum (DSSS) technology; effective data transfer rate 250kbps; low consumption: the current of receiving is 19.7mA, the current of launching is 17.4mA; low power supply voltage: internal voltage regulator: 2.1-3.6V, external voltage regulator: 1.6-2.0V; programmable control transmission power; independent 128byte transmitting and receiving buffer; battery power monitoring.

The CC2420 can be set the operating mode of the chip, read/written the status registers and transmit the data by the 4-wire SPI bus (SI, SO, SCLK, CSn).

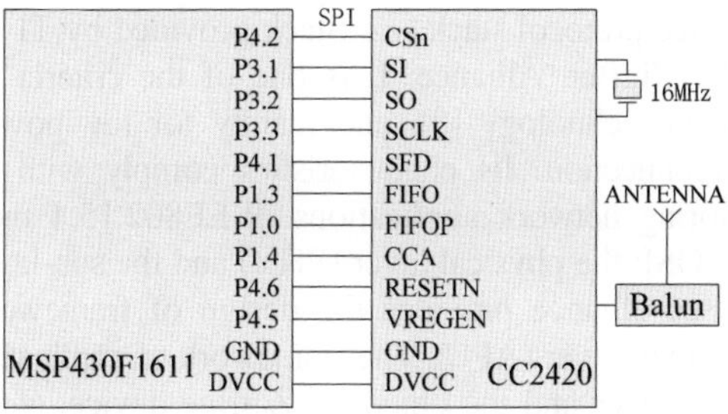

Fig. 4. Connection of CC2420 to MCU

2.4 Power Supply Module

Wireless sensor network nodes are typically arranged in unattended areas or poor environment, sometimes the nodes are mobile. So most of them need to use battery-powered, the selected components of the project need to be considered minimizing power consumption. The power supply of all the parts in the system has been chosen lower voltage supplying. Generally, the range of voltage with 2.4-3.6V can make all modules work normally, therefore, we can use 2 AA batteries for power system which can have the system work at least half a year according to the battery capacity of 1.5 A/h, and can make the system measure the information of temperature, humidity and light illumination 1 time at the sampling interval of 1 min.

2.5 Hardware System Design Essentials

The special application for the wireless sensor network requires that the nodes should be designed as smaller as they can, so the choice of components for the same type need to be selected the smallest type. However due to the node's size is too small, it will be easy to cause interferences between the parts and lines because of the small distances of them in the circuit board. So the design of anti-interference is the key problem of the node for the design. We can design 4 layers PCB board and deposit copper to the region of not laying wires and have them connect to the ground reliably, the bottom of CC2420 board can be used multiple-holes connecting to the ground layer. The filter capacitors should be placed as closer as it can be near to the chip device, meanwhile, in order to achieve anti-electromagnetic interference, the best way is to isolate the digital power to analog power, isolate the digital ground to analog ground, generally the 0 Ω resistance or ferrite beads can be adopted to carry on this kind of separation. In addition, the places of the nodes laying should be avoided trees to reduce the absorption of electromagnetic waves, or it can affect the stability of the transmission [6].

3 Software Design

3.1 Communication Protocol

Communication protocol was chosen as the standard model OSI (Open System Interconnection model) [7]. Considering the generality and to be easy to accomplish development, the ZigBee protocol stack – Z-Stack provided by TI was adopted. ZigBee protocol is defined by ZigBee Alliance. It is one of the criteria used for short-range wireless communication technology [8], and mainly for low-power, low-cost devices with low-speed interconnection. Its characteristics comply with the requirements of environmental monitoring network applications. IEEE802.15.4 meets the bottom two layers that defined by OSI: the physical layer (PHY) and the sub-layer of medium access control (MAC). ZigBee alliance provides the design of framework of network layer (NMK) and the application layer (APL). The framework of application layer includes the application support sub-layer and the object of ZigBee device, etc. The architecture of ZigBee protocol stack is shown in Fig.5. The service access points connect every adjacent layer, including data services and management services, each layer can provide service for the upper or the lower layer by a set of service primitive language.

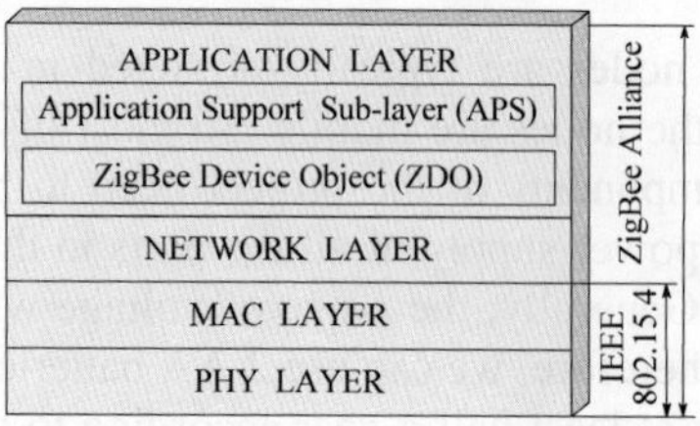

Fig. 5. The architecture of ZigBee protocol stack

Z-Stack is built by the idea of operating system, using the mechanism of round robin in the event. When initialized, the system goes into low power mode. When the event occurs to wake the system, it begins to enter the interruption and deals with the events, and enters the low power mode in the end of the event. If there are several events occurring at the same time, they will be transacted according to their priority. This kind of software architecture can greatly reduce power consumption of the system. The system framework is shown in Fig.6.

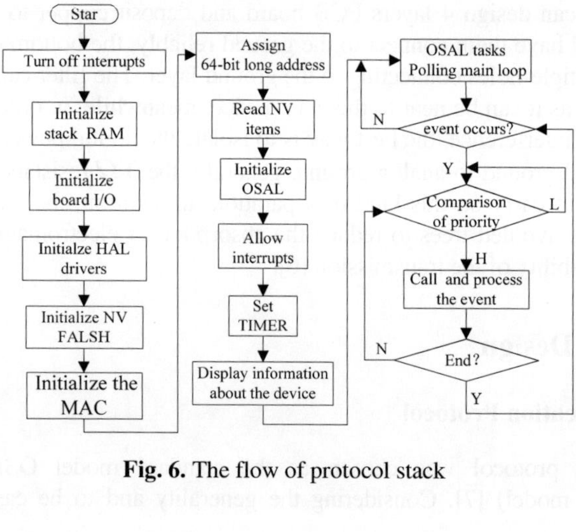

Fig. 6. The flow of protocol stack

3.2 Procedure of Software Design

The software development platform is IAR workbench V4.30. The software is programmed with C language, thus it will increase the efficiency of software design and development, and enhance the reliability, readability and portability of the code. The main flow chart of the nodes is shown in Fig.7.

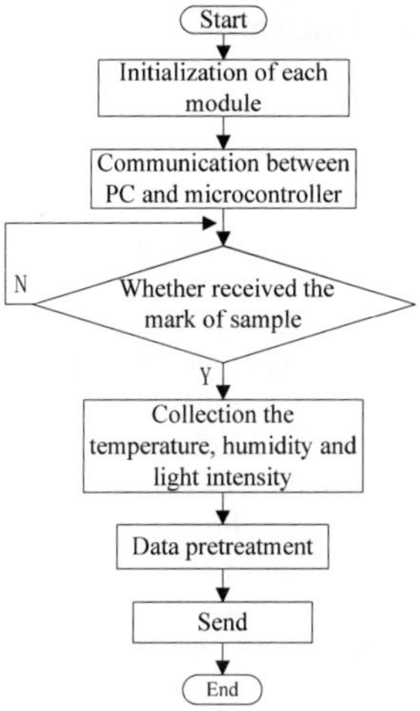

Fig. 7. The software flow of the end devices

4 Nodes Testing

Fig.8 shows the design results of the node of this project, the PC can be connected to the node through the RS232 serial port to monitor the environmental information sent by other nodes, and the program can be downloaded to the memory of the system by the JTAG port. The debugging assistant of serial port received data of the temperature and relative humidity values, the results can be shown in Fig.9, 00 represents temperature collecting, 16 represents the temperature is 22℃; 01 represents humidity collecting, 15 represents the relative humidity is 21% RH.

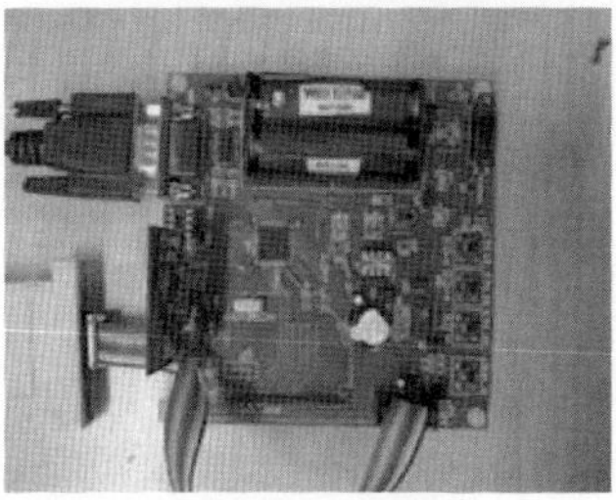

Fig. 8. Wireless sensor network nodes

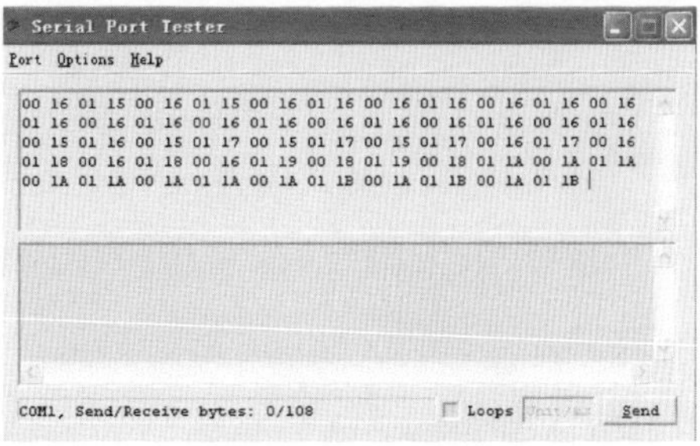

Fig. 9. Test result of communication

5 Conclusion

This paper mainly introduces the system design of wireless sensor network terminal node based on MSP430F1611 and CC2420, each module of the system had been described in detail, the communication protocols and the workflow of the software are also analyzed. The experiment illustrates that the nodes of wireless sensor network in small-scale can work well with good performance. They can fulfill the tasks of signal acquisition, data pretreatment and transmission in some environments.

References

1. Sun, L.-m., Li, J.-z., Chen, Y.: Wireless Sensor Networks. Tsinghua University Press, Beijing (2005)
2. Zhang, Y.-l., Yang, R.-d.: SHT11 Digital Temperature and Humidity Sensor and Its Application. Industry and Mine Automation 3, 113–114 (2007)
3. Zhao, Z.-m.: TSL230 Programmable Light-to-frequency Converters and Its Application. Instrument Technique and Sensor 8, 30–32 (2000)
4. MSP430 Family Software User's Guide. Texas Instruments (1994)
5. Hu, D.-k.: MSP430 Series Ultra-low Power 16-bit FLASH Microcontroller. Beijing University of Aeronautics Press, Beijing (2001)
6. Wang, X.-m., Liu, N.-a.: The Use of 2.4 GHz Radio Frequency Chip CC2420 Achieve Zigbee Wireless Communication Design. International Electronic Components 3, 59–62 (2005)
7. Li, H.: Wireless Network Technology Based on ZigBee and Its Application. Information Technology 1, 12–14 (2008)
8. Man, S.-e., Yu, J.-l., Ma, S.-h.: Energy Model for Evaluating Upper Bound of Energy Consumption in Wireless Sensor Networks. Journal of Jilin University: Information Science Edition 27, 68–72 (2009)

A Non-blind Robust Watermarking Scheme for 3D Models in Spatial Domain

Xinyu Wang, Yongzhao zhan, and Shun Du

School of Computer Science and Communications Engineering, JiangSu University.
212013 Zhenjiang, China
{xywang,yzzhan}@ujs.edu.cn, dushun1000@163.com

Abstract. Improving robustness against various attacks is the main objective for 3D spatial watermarking scheme. Aimed at the goal, a novel spatial robust watermarking scheme is proposed. The Cartesian coordinates of 3D model vertices are converted into spherical coordinates. All vertices are partitioned and sorted by angles, and then vertex sets for watermark embedding are built. The vertex norms in one vertex set are modified to embed a watermark bit and the same watermark bit is embedded repeatedly into various vertex sets. The classification and modification of vertices are separated to avoid de-synchronization problem. In watermark detection, a watermark bit is extracted according to the number of vertices satisfied with specified conditions by comparing the detected model and the original model. Experiments show that the watermarking scheme is robust against attacks such as translation, rotation, uniform scaling, vertex reordering, noise, simplification, cropping, quantization, smoothing, subdivision and their combined attacks.

Keywords: Digital watermarking, 3D model, non-blind watermarking, Spatial watermarking.

1 Introduction

The rapid developments of internet, multimedia techniques and CSCW make people transmit and use various multimedia products such as texts, images, audios, videos and 3D models more and more conveniently. The convenience of Internet has also prompt unauthorized occupancy, duplication, modification and distribution of valuable contents. Digital watermarking has been widely researched as an effective copyright protection for digital products since 1990s. Most previous researches focused on media such as image, video, audio and text watermarking. In recent years watermarking techniques for 3D models have attracted more attentions. Although some watermarking schemes have been proposed, research achievements are immature because of the complexity of representations and the diversity of operations for 3D models.

A watermarking scheme can be classified into spatial and spectral watermarking scheme. Normally, frequency based decomposition of 3D models is considered to be a hard task due to the lack of natural parameterization. Spatial watermarking scheme

M. Zhu (Ed.): Electrical Engineering and Control, LNEE 98, pp. 621–628.
springerlink.com © Springer-Verlag Berlin Heidelberg 2011

is less robust than spectral watermarking scheme, but algorithm is simpler and execution is faster. Consequently, improving the robustness against various attacks is an important goal for spatial watermarking scheme. In spatial watermarking schemes proposed in [1-7] recently, although the robustness of schemes has been enhanced compared with earlier schemes, they are not sufficiently robust against simplification, cropping and also can't resist the combined attacks. Therefore, a new spatial robust watermarking scheme for 3D models is proposed in this paper enlightened by [8]. The scheme builds vertex sets in which vertices are ordered according to angle component of their spherical coordinates and then embeds watermark by modifying the vertex norms within vertex sets. Due to the use of different coordinate components to classify vertices and embed watermark, the alterations of vertex norms will not affect the classification of vertices and the de-synchronization problem will be avoided. By comparing the vertex norms between the detected model and the original model, the number of vertices that satisfy the specified conditions will be counted and then watermark can be extracted using the number of vertices.

2 Main Idea of the Scheme

Embedding watermark into global geometric characters of 3D models will be more robust. Consequently, the vertex norms that are the distances between vertices and model gravity center are selected as [8] to be modified to embed watermark. In order to avoid modifying the components x, y, z of Cartesian coordinate directly, the Cartesian coordinates of vertices will be converted into spherical coordinates (r,θ,φ). The component r that is defined as vertex norm represents approximately the model shape, its modification is supposed to be more robust than a single x, y or z component modification. So the component r of spherical coordinate is selected to be modified to embed watermark.

Attacks on 3D models may lead part vertices to lose or change dramatically that will have an effect on accuracy of watermark detection. Therefore, a moderate watermark detection method is needed. The changes of majority vertices can be regarded as the tendency of changes of all vertices which is less affected by part vertices. To some extent, the number of the vertices that experience some changes can represent the tendency. Based on this idea, in the watermark detection the vertex norms between the detected and the original model are compared to count the vertices that satisfy the specified conditions and then watermark is extracted from the detected model according to the count results.

3 Watermarking Scheme

Let $V=\{V_i\in R^3|0\le i\le N-1\}$ represents the vertex set of a 3D model, where N is the number of vertices, $V_i=(x_i,y_i,z_i)$ is the Cartesian coordinate of a vertex。 The watermark is a binary random sequence $w=(w_0,w_1,\cdots,w_{L-1})$, where L is the length of the watermark sequence, $w_i\in\{0,1\}$, $0\le i\le L-1$.

3.1 Watermark Embedding

Before watermark is embedded the vertices of 3D model should be preprocessed to establish the vertex sets. The steps of pretreatment are as follows:

1. Calculate the model gravity center.
2. Convert the Cartesian coordinate of vertices (x, y, z) to spherical coordinate (r, θ, φ).
3. Construct the vertex sets. First, the interval $[0, 2\pi]$ is divided into M distinct intervals and then all vertices are mapped into these intervals according to their angle component θ of spherical coordinates. In each interval vertices are sorted asendingly by θ. Second, vertex norms of vertices in each interval are normalized into range of [0, 1] using the following equation:

$$\begin{cases} \min r_i = \min\{r_{it} | r_{it} \in Q_i\} \\ \max r_i = \max\{r_{it} | r_{it} \in Q_i\} \quad , \quad 0 \le i \le M-1, \;\; 0 \le j, t \le S_i - 1 \;, \\ \tilde{r}_{ij} = (r_{ij} - \min r_i)/(\max r_i - \min r_i) \end{cases} \tag{1}$$

   where Q_i denotes the i-th interval, r_{it} denotes the vertex norm of the t-th vertex in the interval Q_i, $\min r_i$ is the minimum vertex norm of Q_i, $\max r_i$ is the maximum vertex norm of Q_i, r_{ij} denotes the vertex norm of the j-th vertex in Q_i, $\tilde{r}_{ij}$ denotes the normalized the vertex norm, S_i is the number of the vertices in Q_i. Finally, the vertices in each interval are divided into vertex sets according to the length of the watermark.

By means of the pretreatment mentioned above, all vertices of 3D model are divided into M×L vertex sets and each vertex set is a watermark primitive into which a watermark bit will be embedded. A watermark bit will be embedded M times, thereby the robustness against the simplification and cropping attacks is achieved. When a watermark bit is embedded into a vertex set, each vertex norm in the vertex set is modified using the following equation as [8]:

$$\tilde{r}^{w} = (\tilde{r})^{k}, \; 0<k<\infty, \; k \in R \;, \tag{2}$$

where $\tilde{r}$ is the original vertex norm and $\tilde{r}^{w}$ is the modified vertex norm. In order to control the model distortion, an iteration algorithm[8] is used as follow:

For embedding the watermark bit 0:

1. Initialize the parameters k and Δk.
2. Modify the vertex norm $\tilde{r}$ using the equation (2).
3. If $|\tilde{r}^{w} - \tilde{r}|/\tilde{r} \le \alpha$, α is the strength coefficient , then $k = k + \Delta k$ and go back to 2.
4. End.

For embedding the watermark bit 1, the step 3 is only needed to modify:

3. If $\left|\tilde{r}^w - \tilde{r}\right| / \tilde{r} \le \alpha$, then $k = k - \Delta k$ and go back to 2.

Subsequently, the vertex norms are mapped into the original range by

$$r_{ij}^w = \tilde{r}_{ij}^w \cdot (\max r_i - \min r_i) + \min r_i, \quad 0 \le i \le M-1, \quad 0 \le j \le S_i - 1 , \tag{3}$$

where $\min r_i$ and $\max r_i$ are the minimum and maximum vertex norms in the i-th interval as the same as those used in the pretreatment process.

Finally, the spherical coordinates of vertices are converted to Cartesian coordinates.

3.2 Watermark Detection

If the detected model experiences the geometrical transforms including translation, rotation and uniform scale, a registration step is needed to bring the detected model back to its original location, orientation and scale. The ICP algorithm [9] is used here to register the detected model. Some attacks like simplification, cropping and subdivision will change the number of vertices. Therefore the detected model should be resampled to obtain the same number and order of vertices as the original model.

Then the pretreatments of the original model and the detected model are separately executed in accordance with the method mentioned in 3.1. The vertices of the original model are divided into M×L vertex sets again and the vertices of the detected model are also divided into M×L vertex sets. When the detected model is preprocessed the gravity center of the original model, maximum and minimum vertex norms in original intervals are used to avoid the de-synchronization problem.

Finally, watermark is extracted by analyzing each vertex set of the detected model and its corresponding vertex set of the original model. The steps are as follows:

1. For each vertex set of the detected model and the corresponding vertex set of the original model:
   a) Count the vertices of which the vertex norms in the detected model are less than the corresponding vertex norms in the original model. The result is denoted as C_1. At the same time, count the vertices of which the vertex norms in the detected model are larger than or equal to the corresponding vertex norms in the original model. The result is denoted as C_2.
   b) Record the result of comparison by

$$t_j = t_j + \mathrm{sign}(C_2 - C_1), \quad 0 \le j \le L-1 , \tag{4}$$

   where L is the length of the original watermark, sign() is sign function. Because a watermark bit is embedded M times, the watermark bit can be extracted only by comparing all vertex sets where the watermark bit is embedded. The equation (4) is used to record the results of comparison between the vertex sets of the detected and original model where the j-th watermark bit is embedded. The initial value of t_j is zero.

2. After all vertex sets are processed the watermark $w^d = (w_0^d, w_1^d, \cdots w_{L-1}^d)$ is extracted by

$$w_j^d = \begin{cases} 0, & t_j < 0 \\ 1, & t_j \geq 0 \end{cases}, \quad 0 \leq j \leq L-1 , \tag{5}$$

where L is the length of the original watermark.

4 Experimental Results

The watermarking scheme is implemented using vc++ and the 3D model is the venus model consists of 100,759 vertices downloaded from Stanford. After embed watermark into the model, we analyze the imperceptibility and robustness of the watermarking scheme.

4.1 Imperceptibility

Use SNR(Signal-to-Noise ratio) to measure the visual quality of the watermarked model. After watermark is embedded, the SNR of the watermarked model is 64dB that means the distortion caused by the scheme is less. The distortion of the model can be controlled by the watermark strength factor α in watermark embedding. Reducing the strength coefficient will decrease the distortion but the robustness of the watermark will also be decreased. So the balance between distortion and robustness should be considered in practice.

4.2 Robustness

To test the robustness of the watermarking scheme, the 3-D mesh watermarking benchmark developed by LIRIS laboratory is used to attack the watermarked model including translation, rotation, uniform scale, vertex reordering, noise addition, simplification, cropping, quantization, smoothing, subdivision and some combined attacks. The watermark is then extracted from the attacked model and compared with the original watermark to evaluate the robustness of the scheme. The experimental results are presented with correlation value and bit error ratio (BER). If the correlation value exceeds a chosen correlation threshold, we conclude that the original watermark is presented in the detected model, otherwise is not. An experimental method described in [10] is adopted to select the threshold and the correlation threshold is set to 0.4 in the following experiments.

For translation, rotation and uniform scale, the attacked model can be back to original location, orientation and scale due to the model registration. So the watermark can be extracted accurately from the attacked model.

For vertex reordering, it will not affect the watermark extraction and the watermark can be extracted accurately from the attacked model because all vertices will be partitioned and sorted in watermark detection.

For other attacks, the experimental results will change according to the different attack intensity. For noise attack, a random vector is added to each vertex of the watermarked model. The amplitudes of the noise vector are 0.1%, 0.2% and 0.3% of the average length from vertices to the model center. We perform each noise attack three times and report the median. For simplification attack, the vertices of the model are reduced 50%, 70%, 80% and 90%. For cropping attack, the vertices of the model are removed 10%, 30%, 40% and 50%. For each cropping ratio, three attacked models are generated and the median is reported. For quantization attack, each coordinate of vertices is represented with 11bits, 10bits and 9bits. For smoothing attack, the Laplacian smoothing is performed 5, 10, 15 and 20 times. For subdivision attack, the simple midpoint scheme, the $\sqrt{3}$ scheme and the Loop scheme are performed respectively one time. For combined attacks, the model is cropped 30% and simplified 50% at first, and then noise, quantization, smoothing and subdivision attacks are carried out respectively with a certain intensity. When noise attack is combined, the experiment is performed three times and the median is reported. When subdivision attack is combined, the midpoint scheme, the $\sqrt{3}$ scheme and the Loop scheme are carried out one time respectively and the median is reported. The results of the attacks mentioned above are listed in Table 1.

Table 1. Results of Attacks

Attack	Correlation value	BER(%)
noise 0.1%	1	0
noise 0.2%	0.92	3.96
noise 0.3%	0.42	37.08
simplification 50%	1	0
simplification 70%	1	0
simplification 80%	0.90	5
simplification 90%	0.61	23.75
cropping 10%	1	0
cropping 30%	0.97	1.46
cropping 40%	0.85	8.13
cropping 50%	0.58	26.46
quantization 11bits	1	0
quantization 10bits	1	0
quantization 9bits	0.95	2.5
smoothing 5 iterations	1	0
smoothing 10 iterations	0.91	4.38
smoothing 15 iterations	0.78	11.88
smoothing 20 iterations	0.56	24.38
subdivision(the midpoint scheme)	1	0
subdivision(the $\sqrt{3}$ scheme)	1	0
subdivision(the Loop scheme)	1	0
cropping 30%+simplification 50%	0.94	3.13
cropping 30%+simplification 50%+noise 0.1%	0.81	10.42
cropping 30% simplification 50%+noise 0.2%	0.46	34.58
cropping 30%+simplification 50%+quantization 11bits	0.94	3.13
cropping 30%+simplification 50%+quantization 10bits	0.87	6.88
cropping 30%+simplification 50%+smoothing 5 iterations	0.93	3.75
cropping 30%+simplification 50%+smoothing 10 iterations	0.91	4.38
cropping 30%+simplification 50%+subdivision	0.96	1.88

As can be seen from the Table 1, for noise attack when the amplitude of the noise vector is up to 0.3%, the correlation value of the watermark extracted from the model is still higher than the correlation threshold. For simplification, cropping, quantization and smoothing attack, when 90% of the vertices is simplified, or 50% of the vertices is cropped, or each coordinate of vertices is only represented with 9bits, or smoothing iteration is up to 20 times, the model data and the surface details have been significantly damaged but the watermark extracted from the attacked model still has high correlation value that indicates the watermark scheme is robust enough to against simplification, cropping, quantization and smoothing attacks, especially to simplification attack. For subdivision attack, when the model is subdivided the number of vertices increases 2 to 3 times as against the original. Although lots of new vertices appear, the watermark can be extracted accurately that indicates the watermark scheme is insensitive to subdivision attack. For combined attacks, when the model is cropped 30% and simplified 50%, many vertices have been lost. After other attacks are carried out the model is damaged more seriously but the correlation value of the extracted watermark is still high.

In a word, the watermark scheme proposed in this paper is robust enough to against various common attacks and their combined attacks. Thereby it can prove the existence of the original watermark very well before the model is damaged too serious to recover that means it has lacked its commercial value and is not needed to protect.

5 Conclusion

A novel non-blind watermarking scheme in spatial domain for 3D models is proposed in this paper. The new scheme builds vertex sets by partitioning and sorting vertices according to the angle component of spherical coordinates and then repeatedly embeds watermark via modifying the vertex norms using an iteration algorithm. In watermark detection the vertex norms between the detected and the original model are compared to count the vertices that satisfy the specified conditions, and then watermark is extracted based on the count results. Experiments show that the watermarking scheme is robust enough to against translation, rotation, uniform scaling, vertex re-ordering, noise, simplification, cropping, quantization, smoothing, subdivision and their combined attacks. The future work is to look for a method measuring the influences of attacks on 3D model and then continue to improve the robustness of watermarking scheme against various attacks.

Acknowledgments

This work is supported by National Natural Science Foundation of China (Grant No. 60273040 and No. 61003183), and Graduate Student Research and Innovation Foundation of Jiangsu Province of China (Grant No. 1221170020).

References

1. Liu, Q., Zhang, X.M.: SVD Based Digital Watermarking Algorithm for 3D Models. In: 8th International Conference on Signal Processing, vol. 2, IEEE Press, New York (2006)
2. Zhou, Z.D., Ai, Q.S., Liu, Q.: A SVD-based Digital Watermarking Algorithm for 3D Mesh Models. In: 8th International Conference on Signal Processing, vol. 4. IEEE Press, New York (2006)
3. Nie, X.S., Liu, J., Wang, X.Q., Sun, J.D.: Watermarking for 3D Triangular Meshes based on SVD. In: 2009 Fifth International Conference on Intelligent Information Hiding and Multimedia Signal Processing, pp. 430–433. IEEE Computer Society, NewYork (2009)
4. Salman, M., Ahmad, Z., Worrall, S., Kondoz, A.M.: Robust Watermarking of 3-D Polygonal Models. In: 2008 3rd International Symposium on Communications, Control, and Signal Processing, pp. 340–343. IEEE Press, New York (2008)
5. Han, D.Z., Yang, X.Q., Zhang, C.M.: A Novel Robust 3D Mesh Watermarking Ensuring the Human Visual System. In: 2009 2nd International Workshop on Knowledge Discovery and Data Mining, pp. 705–709. IEEE Press, New York (2009)
6. Kong, X.Z., Yao, Z.Q.: A Novel Double 3D Digital Watermarking Scheme. In: 2009 International Conference on Multimedia Information Networking and Security, pp. 553–556. IEEE Computer Society, New York (2009)
7. Zhang, D.M., Yao, L.: A Non-Blind Watermarking on 3D Model in Spatial Domain. In: 2010 International Conference on Computer Application and System Modeling, vol. 10, pp. 267–269. IEEE Computer Society, New York (2010)
8. Cho, J.W., Prost, R., Jung, H.Y.: An Oblivious Watermarking for 3D Polygonal Meshes using Distribution of Vertex Norms. IEEE Transactions on Signal Processing 55, 142–155 (2007)
9. Ying, S.H., Peng, J.G., Du, S.Y., Qiao, H.: A Scale Stretch Method Based on ICP for 3D Data Registration. IEEE Transactions on Automation Science and Engineering 6, 559–565 (2009)
10. Yin, K.K., Pan, Z.G., Shi, J.Y.: A Robust Mesh Watermarking Algorithm(in Chinese). Journal of Computer Aided Design & Computer Graphics 13, 102–107 (2001)

Infrared Image Segmentation Algorithm Based on Fusion of Multi-Feature

Qiao Kun, Guo Chaoyong, and Shi Jinwei

Department of Basic Courses, Ordnance Engineering College,
NO.97, Heping West Road, 050003, Shijiazhuang, China

Abstract. For the poor contrast and low SNR of infrared images, a new multi-feature fusion image segmentation algorithm is proposed which is based on the features of several traditional image segmentation algorithms. The proposed algorithm segments the target in the region of interest by combining gray, region and boundary information. The proposed algorithm can detect the target from complex background and its performance is superior to traditional algorithms. The experiment results show that the proposed algorithm accurately and effectively segments the target.

Keywords: multi-feature fusion; infrared images; image segmentation; region of interest.

1 Introduction

Target detection technology can provide objective information about the shape and the location in the system of the infrared guidance, so it becomes one of the key technologies necessary. Target detection includes image preprocessing, image segmentation and target identification. Image segmentation is an important step to achieve to target detection, it aims to divide the image into several meaningful regions, to greatly reduce the target identification data to be processed, and to meet accuracy and real-time requirement.

This paper mainly researches how to complete accurate image segmentation in the complex background and the poor contrasts. The traditional image segmentation algorithms include the threshold segmentation algorithm, edge detection algorithm and region segmentation algorithm. The threshold segmentation algorithm divides the infrared image into the target and the background by the selected threshold, neglecting the information of the edge and region of the target. The edge detection is difficult to complete extraction of the target contour edges because of the fuzzy image edges. The region segmentation algorithm is only used for the gray level difference which is not significant. So, it is necessary to improve these algorithms, and achieve more satisfactory segmentation result.

This paper proposed the infrared image segmentation algorithm based on fusion of multi-feature. It uses a bright region as starting points for regional consolidation, until the edge of the merged area combined with the true edge of the target to achieve the best fit, and then we get the target regions.

M. Zhu (Ed.): Electrical Engineering and Control, LNEE 98, pp. 629–634.
springerlink.com © Springer-Verlag Berlin Heidelberg 2011

2 Formulation

2.1 Image Segmentation Based on the Threshold

Otsu's method [1] [2] is a commonly used adaptive threshold algorithm. We suppose the gray levels of the infrared image be [1, 2,…, L]. The number of pixels at level i are dented by n_i and the total number of pixels by $N=n_1+n_2+\ldots+n_L$. In order to simplify the discussion, the gray-level histogram is normalized and regarded as a probability distribution:

$$p_i = \frac{n_i}{N}, p_i \geq 0, \sum_{i=1}^{L} p_i = 1 \tag{1}$$

Now suppose that we dichotomize the pixels into two classes C_0 and C_1 (background and objects, or vice versa) by a threshold at level k; C_0 denotes pixels with levels [1,2,…,k], and C_1 denotes pixels with levels [k+1,…,L].Then the probabilities of class occurrence and the class mean levels, respectively, are given by

$$\omega_0 = \Pr(C_0) = \sum_{i=1}^{k} p_i = \omega(k) \tag{2}$$

$$\omega_1 = \Pr(C_1) = \sum_{i=k+1}^{L} p_i = 1-\omega(k) \tag{3}$$

and

$$\mu_0 = \sum_{i=1}^{k} i \Pr(i \mid C_0) = \sum_{i=1}^{k} \frac{ip_i}{\omega_0} = \frac{\mu(k)}{\omega(k)} \tag{4}$$

Where

$$\omega(k) = \sum_{i=1}^{k} p_i \tag{5}$$

and

$$\mu(k) = \sum_{i=1}^{k} ip_i \tag{6}$$

Are the zeroth- and the first-order cumulative moments of the histogram up to the k-th level, respectively, and

$$\mu_T = \mu(T) = \sum_{i=1}^{L} ip_i \tag{7}$$

The between-class variance is given by

$$\sigma^2 = \omega_0 * \omega_1 * (\mu_1 - \mu_0)^2 \tag{8}$$

The optimal threshold k^* that maximizes σ^2, and the optimal threshold k^* is

$$\sigma^2(k^*) = \max_{1 \le k \le L} \sigma^2(k) \tag{9}$$

When the background is complex and the contrast is poor, the Otsu's method is difficult to dichotomize the pixels into target and background completely. So, we present a general recursive approach for image segmentation by extending Otsu's method [3]. We define the separability factor :

$$SP = \frac{\sigma^2}{\sigma'^2} \tag{10}$$

$$\sigma'^2 = \sum_{i=1}^{L} (i-m) * p_i, 0 < SP < 1 \tag{11}$$

SP is a number ranging from zero to one, indicating the likelihood of separating the class that has lowest intensity in the image from the other class in the same image. We define S=90%.A diagram of the system is presented in Fig. 1.

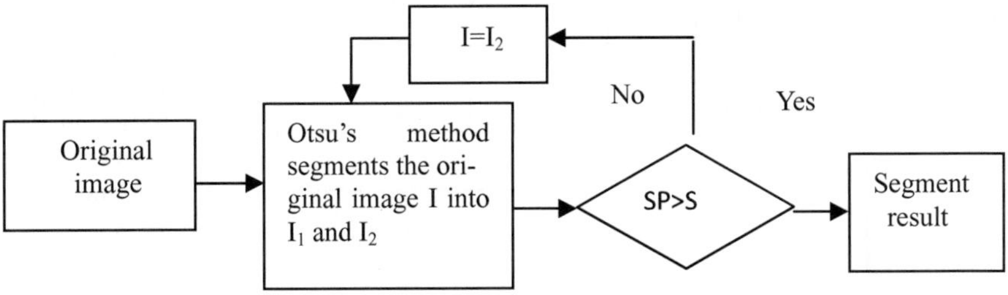

Fig. 1. Diagram of recursive threshold technique

2.2 Edge Detection Based on the Canny and Morphology

Canny [4] edge detection has high positioning accuracy and the ability of resisting noise, but it is difficult to detect the edges of the target without the false edges. Morphology edge detection can remove the false edges, but the edge detected is not single pixel width. So, we propose edge detection based on the Canny and morphology algorithm . The proposed method has the following steps:

1) We detect all the edges E_c of the image by the Canny method, and the edges E_c is single pixel width.
2) We detect the edges E_m of the image by morphology method, and E_m is not single pixels width.
3) We select the edges belonged to E_m in E_c. So, we get E_{cm} which belonged to the target.

2.3 Image Segmentation Based on the Region Growing and Merging [5]

Image segmentation based on the region growing and merging uses adjacent pixel gray consistency as a criterion. It starts from the seeds, develops growing criteria and stopping criteria, and segments targets from image. After we select the seed regions belonged to the target region, we starts from the seed region to merge adjacent small regions. So, we should design the constraint for region merging based on the edge feature [6].We design the constraint

$$E_{\cos t} = E_b + E_t = \frac{N_b}{Area_b} + \frac{N_t}{Area_t} \tag{12}$$

The area of the target and the number of the pixels for target are given by $Area_b$ and N_b, the area of the target and the number of the pixels for background is given by $Area_t$ and

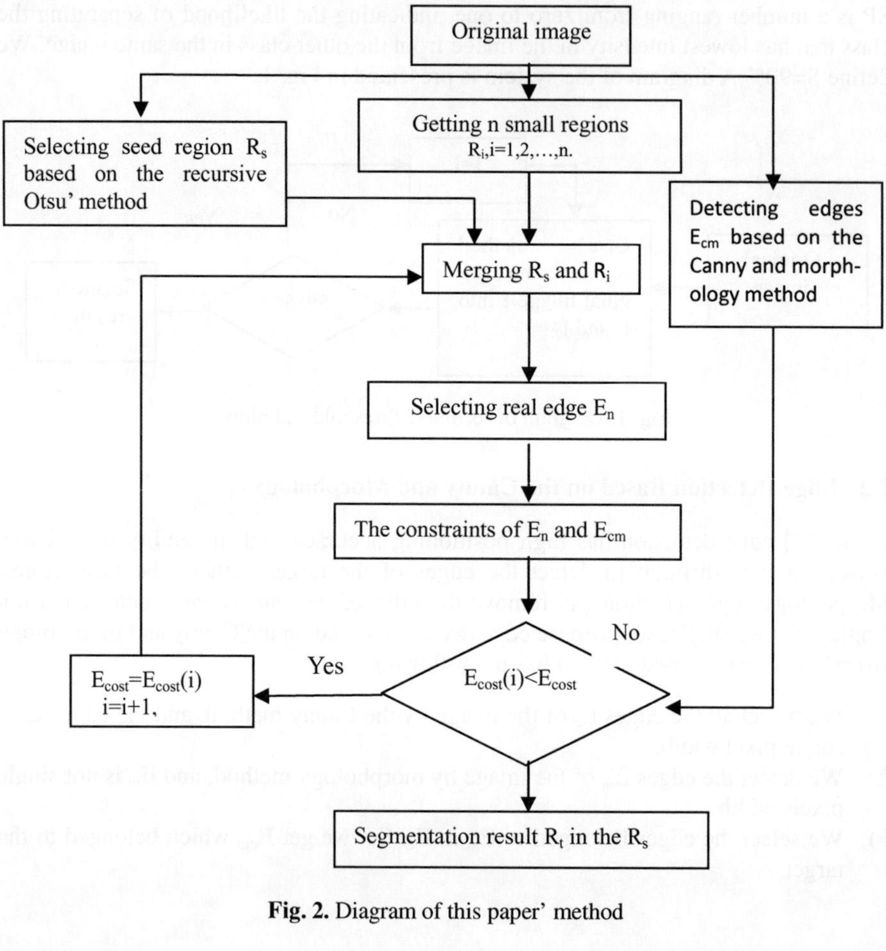

Fig. 2. Diagram of this paper' method

N_t. When the E_{cost} is smaller, the number of the pixels in the region is fewer and the result of segmentation is better. When E_{cost} is zero, we get the ideal segmentation result. So, we give the concrete steps:

1) The selecting of seeds. We select the seed regions R_s based on the recursive Otsu's method, and select the edge of the seed regions E_s.
2) The selecting of the true edge for the target. We detect the true edge of the target from the original image based on the Canny and morphology method.
3) Image segmentation based on the region growing. We divide the image into n small regions R_i , and i=1, 2,…, n.
4) Small regions merging based on Fusion of Multi-feature. We start seed regions to merge adjacent small regions until meet constrains, and we get the true target region R_t. A diagram of the system is presented in Fig. 2.

3 Experimental Results

Several examples of experimental results are shown in Figs.3-5. Fig.3 shows the result of Infrared ship image segmentation.(a) is the original infrared image, (b) is the results based on the Otsu' s method, (c) is the result of this paper's method. From these pictures, Otsu' method only segments the high gray-level region, most of the low gray-level region for ship are not segmented. So, we can't get the shape and size of the real target. It is difficult for us to identify the target. This paper's method can segment complete and accurate target region, and get better segmentation result .

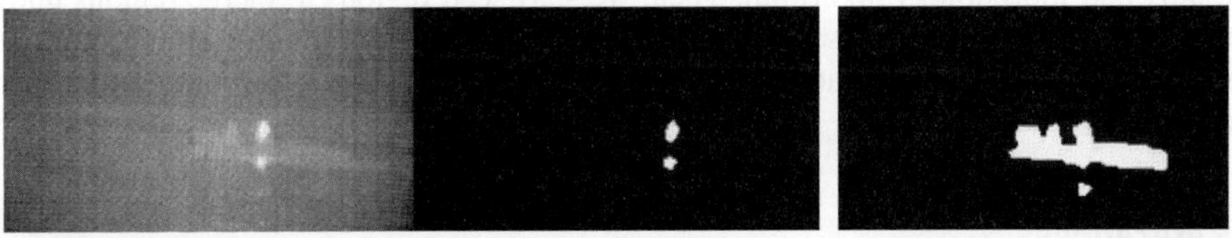

(a)Original infrared ship image (b) Otsu's method (c) This paper's method

Fig. 3. The segmentation results of infrared ship image

Fig.4 shows the segmentation results of Infrared single people. Although the person' s shape changes with people walking, the gray value keeps the same.(a) is the original infrared person image, (b) is the segmentation results based on the Otsu's method.(b) is the segmentation results based on this paper's method. From these pictures, Otsu's method only segment most of the person region, this paper's method segments complete person region.

Fig.5shows the segmentation results of Infrared persons image. (a) is the original infrared persons image, (b)is the segmentation result of Otsu's method,(c)is the segmentation result of this paper's method.

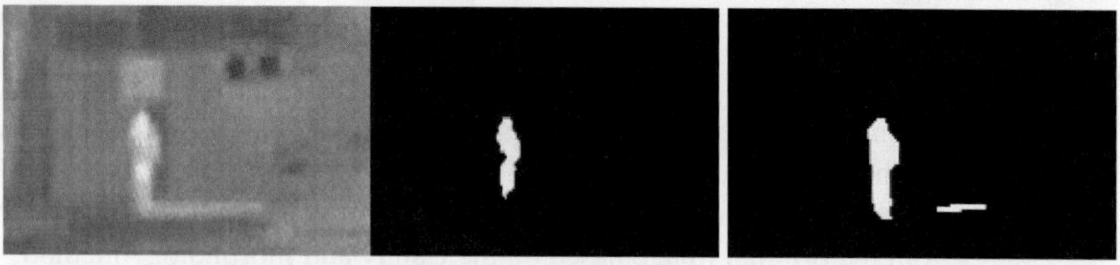

(a)Original infrared person image (b) Otsu's method (c) This paper's method

Fig. 4. The segmentation results of infrared single person image

(a)Original infrared persons image (b) Otsu's method (c) This paper's method

Fig. 5. The segmentation results of infrared persons image

4 Conclusion

This paper proposes a new multi-feature fusion image segmentation algorithm. We selects seed regions by the recursive Otsu's method, and get the real edges of target by Canny and morphology algorithm. Then, we start seed regions to merge regions until the constraint is minimum, so, we get the target regions at last. Experiments show that we can get better segmentation results in the complex background and in the poor contrast.

References

1. Ostu, N.: A threshold selection method from gray-level histograms. IEEE Transactions on Systems, Man and Cybernetics 9(1), 62–66 (1979)
2. Zhang, T.-x., Zhao, G.-z.: Fast recursive algorithm for infrared ship image segmentation 25(4), 691–696 (2006)
3. Cherit, M., Said, C.Y.: A recursive thresholding technique for image segmentation. IEEE Transactions on Image Processing 7(6), 918–921 (1998)
4. Bao, P., Zhang, L., Wu, X.: Canny edge detection enhancement by scale multiplication. IEEE Transactions on Signal processing 5.6(5), 642–655 (1997)
5. Hojjatoleslami, S.A., Kittler, J.: Region growing: A new approach. IEEE Transactions on Signal processing 7(7), 1079–1085 (1998)
6. Hsiao, Y.-T., Chuang, C.-L.: A Contour based Image Segmentation Algorithm using Mophological EdgeDetection. In: 2005 IEEE International Conference on Systems,ManandCybernetics, vol. 3, pp. 2952–2967 (2005)

Adaptive Terminal Sliding Mode Projective Synchronization of Chaotic Systems

Minxiu Yan and Liping Fan

College of Information Engineering, Shengyang University of Chemical Technology, Shengyang 110142, China
{cocoymx,flp}@sohu.com

Abstract. The projective synchronization of a class of second-order chaotic system with unknown uncertainty and external disturbance is discussed. Firstly choosing an improved terminal sliding surface, the singularity phenomenon resulting from improper choice of parameter is solved. Then the adaptive law of unknown upper boundary parameter estimation is given, and a terminal sliding mode adaptive controller is given. The proposed controller has strong robust property since it is beyond the impact of unknown uncertainty and external disturbance. An illustrative simulation result is given to demonstrate the effectiveness of the proposed controller.

Keywords: chaotic system; projective synchronization; terminal sliding mode control; adaptive control; uncertain chaotic systems.

1 Introduction

Chaos as a particular case of nonlinear dynamics has attracted much attention among scientists since the pioneering work of Pecoraet [1]. From then on, many effective methods for chaos synchronization have been provided such as OGY method, differential geometric approach, adaptive control, inverse control, sliding mode control et al[2-5].

Many works mentioned above are related to the complete synchronization. However, in the practical areas, the complete synchronization hardly occurs except under ideal conditions. Recently, an interesting synchronization called the projective synchronization has been proposed in [6]. The complete synchronization and anti-synchronization are the special cases of the projective synchronization. So the projective synchronization has been an important topic because of its practical sense. Sliding mode control (SMC) [7] as one of control methods provides an effective method to deal with uncertain chaotic systems. However, the disadvantage of the SMC method can not ensure a finite convergence time. Gradually terminal sliding mode control (TSMC) [8-10] has been adopted to drive the system to zero in a finite time. From the author's best knowledge, there is little work about the projective synchronization with unknown parameters in presence of external disturbance and model uncertainties by TSMC.

M. Zhu (Ed.): Electrical Engineering and Control, LNEE 98, pp. 635–640.
springerlink.com © Springer-Verlag Berlin Heidelberg 2011

The purpose of this paper is the development of an adaptive terminal sliding mode controller for projective synchronization with unknown upper-bounds of parameter and disturbance. An improved switching surface to get rid of the singularity phenomenon for the synchronization is first proposed. And then, adaptive TSMC is derived to guarantee the occurrence of the sliding motion. Adaptive arithmetic and improved TSMC are to improve tracking precision and robustness.

2 System Description for Uncertain Chaotic System with Proposed Controller

Consider the following drive-response systems described by (1) and (2)

$$\begin{cases} \dot{x}_1 = x_2, \\ \dot{x}_2 = f(x,t), x \in R^2, \end{cases} \tag{1}$$

$$\begin{cases} \dot{y}_1 = y_2, \\ \dot{y}_2 = f(y,t) + \Delta f(y) + d(t) + u(t), \end{cases} \tag{2}$$

where $u(t) \in R$ is control input, $f(\cdot)$ is a continuous nonlinear function of t, x or y, $\Delta f(\cdot)$ is time-varying, not precisely known and uncertain part of chaotic system, $d(t)$ is the external disturbance of system.

Assumption 1. The uncertainty $\Delta f(y)$ and external disturbance $d(t)$ satisfy the following constraints:

$$\|\Delta f(y) + d(t)\| \le l_0 + l_1 \|y\|, \tag{3}$$

where l_0 and l_1 are unknown non-negtive constants.

Define the projective synchronization error as $e_i = y_i + \lambda x_i$, λ is the scale factor. $\lambda = -1$ is the complete synchronization. $\lambda = 1$ is the anti-synchronization. $\lambda < 0$ is the in phase projective synchronization . $\lambda > 0$ is the antiphase projective synchronization. The dynamics of the synchronization error can be expressed as

$$\begin{cases} \dot{e}_1 = e_2, \\ \dot{e}_2 = f(y,t) + \lambda f(x,t) + \Delta f(y) + d(t) + u(t). \end{cases} \tag{4}$$

Our purpose is to design an adaptive TSMC that brings the error to zero in finite time which implies the projective synchronization between (1) and (2) is realized, i.e.

$$\lim_{t \to T} \|y_i + \lambda x_i\| = \lim_{t \to T} \|e_i\| \to 0. \tag{5}$$

In the traditional TSMC, a switching surface representing a desired system dynamics

$$s = e_2 + \beta e_1^{\frac{q}{p}}, \tag{6}$$

where β is a positive design constant, and p and q are positive odd integers which satisfy $p > q$. For system (4), a commonly used controller is

$$u(t) = -(f(y,t) + \lambda f(x,t) + \beta \frac{q}{p} e_1^{\frac{q}{p}-1} e_2 + k_1 s + k_2 \operatorname{sign}(s)). \tag{7}$$

From (7) a singularity will occur because of the second term containing $e_1^{\frac{q}{p}-1} e_2$ if $e_2 \neq 0$ when $e_1 = 0$. In the ideal sliding mode, the problem does not happen because when $s = 0$, $e_2 = -\beta e_1^{\frac{q}{p}-1}$, so the term $e_1^{\frac{q}{p}-1} e_2$ is non-singular. Besides the above fact, the singularity may occur due to others facts such as computation errors and uncertain factors [10].

In order to solve the singularity problem in the traditional TSMC, an improved terminal sliding mode is adopted

$$s = e_1 + \beta \operatorname{sig}(e_2)^{\gamma}, \tag{8}$$

where $\operatorname{sig}(e_2)^{\gamma} = |e_2|^{\gamma} \operatorname{sign}(e_2)$, $\beta > 0$ is a design constant and $1 < \gamma < 2$. The derivative of s along the terms with negative powers does not occur in the derivative of s along the system dynamics.

Choose a fast reachability law as (9) to make the states of error converge from the initial error values which are not zero to the sliding mode.

$$\dot{s}(t) = -k_1 s - k_2 \operatorname{sig}(s)^{\rho}. \tag{9}$$

Having established the suitable sliding mode, the next step is to determine an input guarantee that error system trajectories reach to the sliding mode and stay on it forever. Therefore, an adaptive TSMC is proposed as:

$$u(t) = -(f(y,t) + \lambda f(x,t) + \beta^{-1}\gamma^{-1} \operatorname{sig}(e_2)^{2-\gamma} + k_1 s + k_2 \operatorname{sig}(s^{\rho}) + (\hat{l}_0 + \hat{l}_1 \|y\|) \operatorname{sign}(s(t))). \tag{10}$$

The adaptive laws are

$$\begin{cases} \dot{\hat{l}}_0 = m\|s\|, \\ \dot{\hat{l}}_1 = m\|s\|\|y\|. \end{cases} \tag{11}$$

where $m = \beta\gamma |e_2|^{\gamma-1}$

3 Stability Annlysis

Theorem 1. Consider the error system (4) with unknown parameter and disturbance uncertainties, if the sliding surfaces is designed as (8) and the adaptive terminal sliding mode controller is designed as (10), in which the adaptive law is designed as (11), then the error states will converge to zero in finite time.

Proof. Let

$$\begin{cases} \tilde{l}_0 = \hat{l}_0 - l_0, \\ \tilde{l}_1 = \hat{l}_1 - l_1. \end{cases} \tag{12}$$

Consider the following Lyapunov function candidate

$$V = \frac{1}{2}s^2 + \frac{1}{2}\tilde{l}_0^2 + \frac{1}{2}\tilde{l}_1^2. \tag{13}$$

Define Differentiating (13) with respect to time, it can obtain that

$$\dot{V} = s\dot{s} = s[\dot{e}_1 + m(f(y,t) + \lambda f(x,t) + \Delta f(y) + d(t) + u)] + \tilde{l}_0\dot{\tilde{l}}_0 + \tilde{l}_1\dot{\tilde{l}}_1. \tag{14}$$

According to (10), we have

$$\begin{aligned} \dot{V} &= s\dot{s} + \tilde{l}_0\dot{\tilde{l}}_0 + \tilde{l}_1\dot{\tilde{l}}_1 \\ &= s[\dot{e}_1 + m(f(y,t) + \lambda f(x,t) + \Delta f(y) + d(t) - f(y,t) - \lambda f(x,t) - \\ &\quad \beta^{-1}\lambda^{-1}\operatorname{sig}(e_2)^{2-\gamma} - k_1 s - k_2\operatorname{sig}(s)^{\rho} - (\hat{l}_0 + \hat{l}_1\|y\|)\operatorname{sign}(s(t))] + \tilde{l}_0\dot{\tilde{l}}_0 + \tilde{l}_1\dot{\tilde{l}}_1 \\ &\le s[m(-k_1 s - k_2\operatorname{sig}(s)^{\rho} + m(\Delta f(y) + d(t)) \\ &\quad - m(\hat{l}_0 + \hat{l}_1\|y\|)\operatorname{sign}(s(t))] + \tilde{l}_0\dot{\tilde{l}}_0 + \tilde{l}_1\dot{\tilde{l}}_1 \\ &\le -s(mk_1 s + mk_2\operatorname{sig}(s)^{\rho}) + m\|s\|(l_0 + l_1\|y\|) \\ &\quad - m\|s\|(\hat{l}_0 + \hat{l}_1\|y\|) + \tilde{l}_0\dot{\tilde{l}}_0 + \tilde{l}_1\dot{\tilde{l}}_1 \\ &= -s(mk_1 s + mk_2\operatorname{sig}(s)^{\rho}) - m\|s\|\tilde{l}_0 \\ &\quad - m\|s\|\|y\|\tilde{l}_1 + \tilde{l}_0\dot{\tilde{l}}_0 + \tilde{l}_1\dot{\tilde{l}}_1 \\ &= -s(mk_1 s + mk_2\operatorname{sig}(s)^{\rho}) + \tilde{l}_0(\dot{\tilde{l}}_0 - m\|s\|) \\ &\quad + \tilde{l}(\dot{\tilde{l}}_1 - -m\|s\|\|y\|) \le 0 \end{aligned} \tag{15}$$

4 Numerical Simulation

In this section, the numerical simulations are done using matlab software to check the validity of the proposed method. The Duffing-Holmes system is used as the example.

The drive system is written as follows

$$\begin{cases} \dot{x}_1(t) = x_2(t), \\ \dot{x}_2(t) = -a_1 x_1(t) - a x_2(t) - x_1^3(t) + b\cos(\omega t). \end{cases} \tag{16}$$

The response system with the control input, uncertainty and disturbance in the second state is proposed as follows

$$\begin{cases} \dot{y}_1(t) = y_2(t), \\ \dot{y}_2(t) = -a_1 y_1(t) - a y_2(t) - y_1^3(t) + b\cos(\omega t) + \Delta f(y) + d(t) + u(t). \end{cases} \tag{17}$$

When the parameters of Duffing-Holmes in (16) are selected as $a_1 = -1, a = 0.25,\ b = 0.3, \omega = 1.0$, the drive system has appeared chaotic state.

In this simulation, the initial values of two systems are taken as $x_1(0) = 0.1$, $x_2(0) = 0.2$, $y_1(0) = -0.5$, $y_2(0) = 0.5$. The uncertainty and disturbance are $\Delta f(y) = -0.05 y_1$, $d(t) = 0.2\cos(\pi t)$ .The design parameters of controller are $\gamma = 1.8$, $k_1 = 2.5$, $k_2 = 10$, $\mu = 50$, $\beta = 1$, $\rho = 3/5$.

When $\lambda = -1$ Numerical simulation results are shown in Fig1. When $\lambda = 0.2$ Numerical simulation results are shown in Fig2.

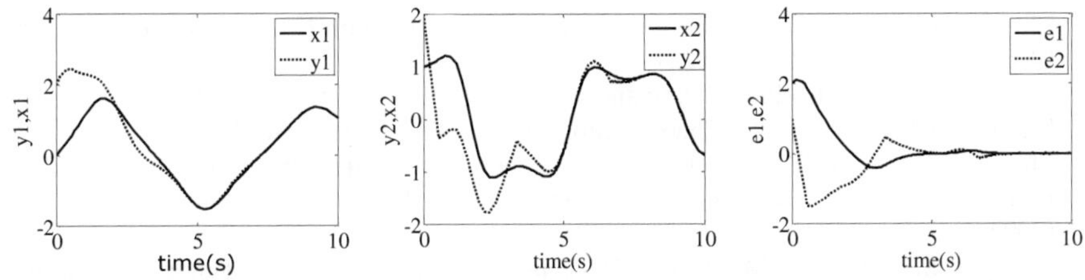

Fig. 1. States of the drive-response system and synchronization errors when $\lambda = -1$

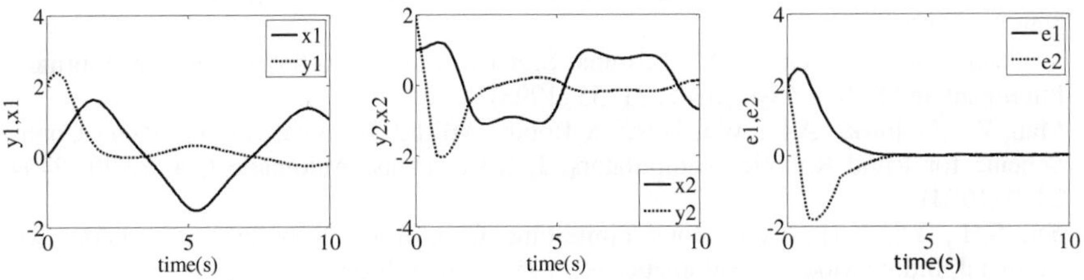

Fig. 2. States of the drive-response system and synchronization errors when $\lambda = 0.2$

From the simulation result, it shows that the obtained theoretic results are feasible and efficient for controlling uncertain chaotic dynamical system.

5 Conclusions

In this paper, an adaptive terminal sliding mode controller has been proposed for projective synchronization of chaotic systems with unknown parameters. By designing only one controller, the method can not only realize chaos synchronization but also gain the identification. The system has strong robustness of external disturbance because of the features of the sliding mode control. Numerical simulations are given to verify the effectiveness of the proposed control scheme.

Acknowledgment

The author gratefully acknowledge the support of the Education Development Foundation of Liaoning Province, Republic of China under Grant (L2010435, L2010436) and the support of the Key Laboratory Project of the Education Development Foundation of Liaoning Province, Republic of China under Grant (LS2010127).

References

1. Pecora, L.M., Carroll, T.L.: Synchronization in Chaotic Systems. J. Physical Review Letters. 64, 821–824 (1990)
2. Efimov, D.V.: Dynamical Adaptive Synchronization. In: 44th IEEE Conference on CDC-ECC 2005, pp. 1861–1866 (2005)
3. Yassen, M.T.: Controlling Chaos and Synchronization for New Chaotic System Using Linear Feedback Control. J. Chaos Solitons & Fractals. 26, 913–920 (2005)
4. Zhang, J., Li, C.G., Zhang, H.B., et al.: Chaos Synchronization Using Single Variable Feedback Based on Back Stepping Method. Chaos Soitons & Fractals 21, 1183–1193 (2004)
5. Bowong, S., Kakmeni, M., Koina, R.: Chaos Synchronization and Duration Time of a Class of Uncertain Chaotic Systems. J. Math. Comput. Simul. 71, 212–228 (2006)
6. Li, G.H.: Projective Synchronization of Chaotic system Using Backstepping Control. J. Chaos Solitons & Fractals 29, 490–494 (2006)
7. Perruquetti, W., Barbot, J.P.: Sliding Mode Control in Engineering. Marcel Dekker Inc., New York (2002)
8. Venkataraman, S.T., Gulati, S.: Terminal Slider Control of Robot Systems. J. Journal of Intelligent and Robotic Systems 7, 31–55 (1993)
9. Man, Z., Paplinski, A.P., Wu, H.R.: A Robust MIMO Terminal Sliding Mode Control Scheme for Rigid Robotic Manipulators. J. IEEE Trans. Automatic Control 39, 2464–2469 (1994)
10. Yu, S.H., Yu, X.H.: Continuous Finite-time Control for Robotic Manipulators with Terminal Sliding Mode. J. Automatica 41, 1957–1964 (2005)

A Detailed Analysis of the Ant Colony Optimization Enhanced Particle Filters

Junpei Zhong[1] and Yu-fai Fung[2]

[1] Department of Computer Science, University of Hamburg,
Vogt Koelln Str. 30
22527 Hamburg, Germany
zhong@informatik.uni-hamburg.de
[2] Department of Electrical Engineering,
The Hong Kong Polytechnic University,
Hung Hom, Kowloon,
Hong Kong
eeyffung@inet.polyu.edu.hk

Abstract. Particle filters, as a kind of non-linear/non-Gaussian estimation method, are suffered from two problems when applied to cases with large states dimensions, namely particle impoverishment and sample size dependency. Previous papers from the authors have proposed a novel particle filtering algorithm that incorporates Ant Colony Optimization (PF_{ACO}), to alleviate effect induced by these problems. In this paper, we will provide a theoretical foundation of this new algorithm. A theorem that validates the PF_{ACO} introduces a smaller Kullback-Leibler Divergence between the proposal distribution and the optimal one when comparing to those produced by the generic PF is discussed.

Keywords: Ant Colony Optimization, Combinatorial Optimization, Metaheuristic Methods, Nonlinear Estimation, Particle Filters.

1 Introduction

Particle Filter (PF) is based on point mass particles that represent the probability densities of the solution space and it is widely used for solving non-linear and non-Gaussian state estimation problems [7]. As an alternative method of Kalman Filter [10,13], it is widely used in applications under non-linear and non-Gaussian environments. The advantage of PF is that it can estimate any probability distribution [4] with an infinite number of samples. Although this optimal estimation is not available in real applications, it can still produce better results in the non-linear/non-Gaussian environment. However, particle impoverishment is inevitably induced due to the random particles prediction and re-sampling applied in generic PF [2], especially for problems that come with a huge number of state dimensions. After a number of iterations, if the generated particles are too far away from the likelihood distribution, their particle weights will approach zero and only a few particles have significant weights, making other particles not efficient to produce accurate estimation results.

M. Zhu (Ed.): Electrical Engineering and Control, LNEE 98, pp. 641–648.
springerlink.com © Springer-Verlag Berlin Heidelberg 2011

Therefore, there are other enhanced PF algorithms that employ different sampling strategies to minimize the impoverishment effect and these strategies include Binary Search [6], Systematic Resampling [9] and Residual Resampling [3]. Those target are copying the important samples and discarding insignificant ones by different calculation and selection methods mainly based on their weights. However, at the meantime, the robustness of the filter is lost, because the diversity of particles is reduced by a certain extent [11]. In [15,14], a metaheuristic method is introduced, in which the Ant Colony Optimization (ACO) is applied to optimize the particle distribution, which will be introduced in the next section. Validating the effectiveness of the PF_{ACO} based on the K-L divergence is included in Section 3 while discussion followed by conclusions, are presented in Section 4 and 5 respectively.

2 Particle Filters

2.1 Generic Particle Filters

Particle filters are algorithms to perform recursive Bayesian estimation using Monte Carlo simulation and importance sampling, in which the posterior density is approximated by the relative density (weights) of particles observed in the state space. The posterior can be approximated by the weighted summation of every particle as follows:

$$p(x_{0:k}|y_{1:k}) \approx \sum_{i=1}^{N} w_k^i \delta(x_{0:k} - x_{0:k}^i) \tag{1}$$

where the weighting value of particle i at time-step k, w_k^i is updated according to Eq. 2.

$$w_k^i \propto w_{k-1}^i \frac{p(y_k|x_k^i)p(x_k^i|x_{k-1}^i)}{q(x_k^i|x_{k-1}^i, y_k)} \tag{2}$$

It can be shown that the approximation (Eq. 1) approaches the true posterior density $p(x_k|y_{1:k})$ [1].

However, in problems that involve a huge number of dimensions, such as the multi-robot SLAM problem, a large number of particles must be included in order to maintain an accurate estimation, the generic resampling method is not sufficient to avoid the impoverishment and size dependence problems. Consequently these problems will become very severe after a number of iterations, rendering a large portion of the particles negligible and reducing the accuracy of the estimation results.

2.2 Ant Colony Optimization Improved PF (PF_{ACO})

In order to optimize the re-sampling step of the generic particle filter, we incorporate ACO into the PF and utilize the ACO before the updating step [15,14]. In this algorithm, the particles will operate as ants and they will move based on the choice of possible routes towards the local peak of the optimal proposal distribution function. The parameter $\tau(t)$, as shown in Eq. 3, is affected by every movement of the particle i by the following equation:

$$\begin{cases} \tau_{i*}(t+1) = (1-\rho)\tau_{i*}(t) + \Delta\tau_{i*}(t) \text{ particles} \in i\text{'s the movement path} \\ \tau_{i*}(t+1) = (1-\rho)\tau_{i*}(t) \text{ otherwise} \end{cases} \tag{3}$$

where $0 < \rho \leq 1$ is the pheromone evaporation rate, $\Delta\tau$ is a constant enhanced value if particle $*$ is located between the starting particle i and the end point. The heuristic function (β) is defined as the reciprocal of the distance between two particles (end points):

$$\eta_{i*}(t) = \frac{1}{d_{i*}} \tag{4}$$

Finally, the optimization step runs iteratively based on a probability function obtained from Eq. 5. It represents the probability of a particle i selecting particle j among $N-1$ particles as the moving direction.

$$p_{ij}(t) = \frac{[\tau_{ij}(t)]_\alpha [\eta_{ij}(t)]_\beta}{\sum\limits_{s \in allparticles} [\tau_{is}(t)]_\alpha [\eta_{is}(t)]_\beta} \tag{5}$$

The initial value of parameter $\alpha(\tau_{i*})$ equals to the particle weight, as stated in Eq. 6.

$$\tau_{i*}(0) = w_* \tag{6}$$

When the ACO algorithm converges and P_{ij} approaches 1 [12], it implies that the particle i re-locates at a closer proximity of particle j, composing a more optimal distribution. A pseudo-program describing the PF_{ACO} algorithm is given below.

Algorithm 1. The PF_{ACO} Algorithm

1: The initialization and prediction steps (these are same as the original PF algorithm) {ACO enhanced PF}
2: **while** the distance between particles' measurement and the true measurement are not within a certain threshold and the iteration number does not exceed the maximum value **do**
3: choose particle i whose distance is exceeding the threshold
4: select the moving target based on the probability (Eq. 5)
5: move towards the target with a constant velocity
6: update the parameters of the ACO (e.g. η, τ), and particle weights
7: **end while**

3 Theoretical Foundation of PF-ACO

In this section, a theorem will be proposed together with its proof in order to elaborate how the PF_{ACO} can produce better solution when compared to the generic PF, which employs a transition function as the proposal distribution.

Theorem. With the convergence nature of ACO, the PF_{ACO} can always achieve the optimal proposal distribution when the ACO converges to an optimal solution.

Proof: In its generic form, a transition model is often employed as the predicted proposal distribution:

$$q(x_k|x_{k-1}, y_k) = p(x_k|x_{k-1})_{tran} \tag{7}$$

while the optimal distribution is defined by Eq. 8.

$$q(x_k|x_{k-1}^i, y_k)_{opt} = p(x_k|x_{k-1}^i, y_k) \tag{8}$$

where the expression $q(x_k|x_{k-1}, y_k)$ in Eq. 8 represents the true distribution of the likelihood of state x with all previous states and observations are given. Since the probability is difficult to be integral, so we usually employ the transition function $p(x_k|x_{k-1})_{tran}$ to approximate the true distribution. The second term $p(x_k|x_{k-1}^i, y_k)$ in an application represents the probability that moving to state x_k in time k, given the samples in previous time step x_{k-1} and the measurement y_k. In ideal cases, the proposal distribution should consider two kinds of noises: noises from the odometer and noise from the sensor. However, the generic transition model only incorporates the probability from motion detector noise. Consequently, the generic transition model can approximately equivalent to the optimal model only if either of following two conditions is satisfied.

1. The odometer has no error in measurement, or
2. the odometer noise has similar noise variance as the observation sensor.

Nevertheless, the above two conditions are difficult to achieve in most of our experiments due to the different variances contributed by various sensors' measurement errors. The observation sensors, such as laser and vision sensors, are getting more accurate, but this is not the case for an odometer. With different magnitude of variance levels, traditional transition model based on the odometer is not as suitable as it used to be, especially in experiments that include observation sensors and motion sensors. In order to prove that ACO is able to solve this problem, Kullback-Leibler divergence (K-L divergence) is introduced. K-L divergence is a non-symmetric measure of the difference between two probability distributions. The approximation of K-L divergence [8] is generated by a set of sample data set: $s_1, s_2, ..., s_N$ based on the model density $p(x)$, so

$$D(p||q) \approx \frac{1}{N}\sum_{i=1}^{N}[\log p(x(n)) - \log q(x(n))] \tag{9}$$

For the generic PF, the above K-L Divergence equals to

$$D(p||q) \approx \frac{1}{N}\sum_{i=1}^{N}[\log p(x_k(n)|x_{k-1}(i), y_k) - \log q(x_k(n)|x_{k-1}(i))] \tag{10}$$

To evaluate the K-L Divergence, we take N Monte Carlo samples in state space for x_k, and calculate their probability density given the condition of particle $x_{k-1}(i)$ and y_k. Based on Eq. 10, it is trivial to derive that the ACO algorithm converges if and only if $p_{ij}(k) = 1$, which indicates the necessary and sufficient conditions of ACO convergence is $d_{ij} = 0$ or $\lim_{k\to\infty} \eta_{ij}(k) = \sum_{s\in \text{allparticles}} \eta_{is}(k)$. Thus, *majority of particles will be located around the peak of the mixture likelihood density function.*

Secondly, assuming that samples $\hat{x}_1, \hat{x}_2, ..., \hat{x}_n$ in the optimal proposal distribution are taken, in order to approach the optimal proposal distribution according to the definition of K-L Divergence, we will derive the relationship between the number of samples and the optimal distribution. If it is necessary to have M samples $(\tilde{x}_k, \tilde{x}_{k+1}, ..., \tilde{x}_{k+M})$ in order to generate N samples $(\tilde{x}_k, \tilde{x}_{k+1}, ..., \tilde{x}_{k+N})$ in the continuous optimal proposal distribution, the number of samples needed to be considered is proportional to the second derivative of the optimal distribution according to the interpolation error [5], which can be illustrated by Fig. 1 and Eq. 11.

$$\begin{aligned} M &= \lambda N[f''(s_k) + f''(s_{k+1}) + \ldots + f''(s_{k+N})]/N \\ &= \lambda[f''(s_k) + f''(s_{k+1}) + \ldots + f''(s_{k+N})] \end{aligned}$$

In Eq. 11, λ is a constant, indicating that *the number of M is proportional to the summation of the second derivatives of all samples in this interval.*

As shown in Fig. 1, k samples in the optimal Gaussian distribution are taken in uniform intervals, and within which, M samples are included in the original discrete distribution, that is $M_{i_k} \in \{M_1, M_2, \ldots, M_k\}$. Similarly, samples in the proposal distribution are also separated into k intervals, that is $N_{i_k} \in \{N_1, N_2, \ldots, N_k\}$. Given the convergence of Ant Colony Optimization algorithm [13], if a certain continuous optimal proposal distribution are divided into M samples, the sample s_t^+ moves closer to these M samples after the ACO improvement. Therefore, we can compare two K-L Divergence before and after the ACO improvement, based on definition of Eq. 9 and 10.

$$D(p||q) \approx \frac{1}{M} \sum_{i_k=1}^{k} \sum_{n\in \text{interval}k} [\log \hat{p}(x_k(n)|x_{k-1}(i), y_k) - \log \tilde{q}(x_k(n)|x_{k-1}(i))] \tag{11}$$

and

$$D(p||q^+) \approx \frac{1}{M} \sum_{i_k=1}^{k} \sum_{n\in \text{interval}k} [\log \hat{p}(x_k(n)|x_{k-1}(i), y_k) - \log \tilde{q}^+(x_k(n)|x_{k-1}(i))] \tag{12}$$

Let the sequence $\hat{M}_{i_k} \in \{\hat{M}_1, \hat{M}_2, \ldots, \hat{M}_k\}$ denote the required particle number in each interval based on Eq. 11. After sufficient iterations to achieve the optimal solution, if in an interval that the required particle number $\hat{M}_{i_k} \le N_{i_k}$, such as $k = 1, 2$ in Fig. 1, it is trivial that

$$D(p||q) = D(p||q^+) \tag{13}$$

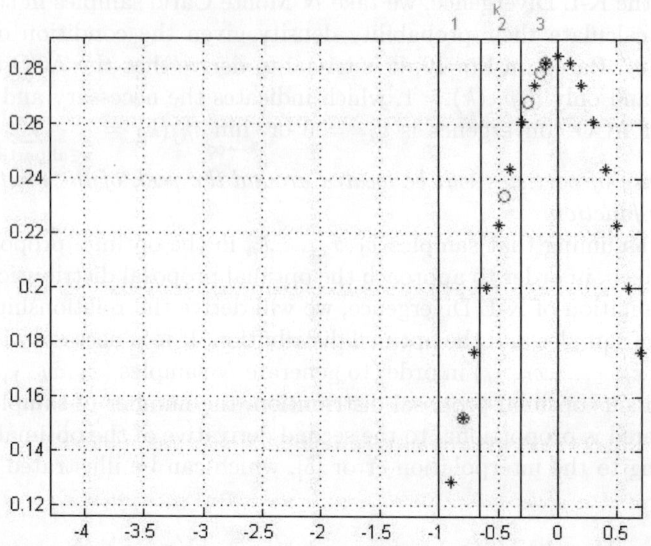

Fig. 1. Within [-1, 0], there are $k = 3$ intervals. In the 1st and 2nd interval, $M_1 = 1$ and $M_2 = 1$ samples may be sufficient to represent the distribution because all the second derivatives in this interval are nearly equal to zero,and $M_3 = 3$ sample are needed to re-construct the distribution.

If within the intervals that the required particle number $\hat{M}_{i_k} > N_{i_k}$, such as $k = 3$ as illustrated in Fig. 1, then

$$\begin{aligned} & D(p||q) - D(p||q^+) \\ \approx & \frac{1}{M}\sum_{n=1}^{M}[\log\frac{p(\hat{x}_k(n)|x_{k-1}(i),y_k)}{q(\tilde{x}_k(n)|x_{k-1}(i))} - \log\frac{p(\hat{x}_k(n)|x_{k-1}(i),y_k)}{q(\tilde{x}_k^+(n)|x_{k-1}(i))})] \\ = & \frac{1}{M}\sum_{n=1}^{M}[\log\frac{p(\hat{x}_k(n)|x_{k-1}(i),y_k)}{q(\tilde{x}_k(n)|x_{k-1}(i))} - \log\frac{p(\hat{x}_k(n)|x_{k-1}(i),y_k)}{p(\hat{x}_k(n)+\varepsilon|x_{k-1}(i),y_k)}] \\ \rightarrow & \frac{1}{M}\sum_{n=1}^{M}[\log\frac{p(\hat{x}_k(n)|x_{k-1}(i),y_k)}{q(\tilde{x}_k(n)|x_{k-1}(i))} - 0] \\ > & 0 \end{aligned} \tag{14}$$

The above convergence comes from the convergence of Ant Colony Optimization [13]. So within the intervals that the required particle number $\hat{M}_{i_k} > N_{i_k}$, we get

$$D(p||q) > D(p||q^+) \tag{15}$$

Given a small number ϵ, with sufficient iterations, we can always achieve arbitrarily small K-L Divergence. Therefore, when we take summation in all intervals for K-L Divergence calculation, we can conclude that $D(p||q) > D(p||q^+)$. □

4 Discussion

The above theorem qualitatively shows that the proposal distribution can ultimately achieve the optimal solution with Ant Colony Optimization.

From the proof presented in Section 3, with reference from the formulation of the combinatorial optimization problem framework, the optimal proposal distribution problem being considered can be classified as a combinatorial optimization problem satisfying Eq. 16.

$$Min(D(q(s_t(i)|s_{t-1}(i), z), p(s_t|s_{t-1}, z)))$$
$$s.t. \begin{cases} \sum_{i=1}^{M} w(i) = 1 \\ w_t(i) = \mu \frac{p(z_t|s_t(i))p(s_t(i)|s_{t-1}(i))}{q(s_t(i)|s_{t-1}(i), z_t)} \end{cases} \tag{16}$$

where $D()$ is the K-L divergence between two distributions. Because the model is not known in advance in the problem, a heuristic method is considered to be one of the possible solutions in this paper. Directly speaking, we know that one important factor of tuning the proposal distribution, so that any similar metaheuristic can also be applied to solve this problem.

5 Conclusions

As a continuous study of the Ant Colony Improved Particle Filter (PF_{ACO}), a theoretical deduction of the improvement process is included in this paper. Our theorem validates that the PF_{ACO} optimizes the proposal distribution to generate a smaller Kullback-Leibler divergence value than that obtained from generic PF. From the theorem, we further discuss a framework to optimize this particle distribution based on combinatorial optimization. Using this framework, metaheurstic methods, e.g. ACO, or other methods can be applied to introduce better estimation results in non-linear/non-Gaussian engineering estimation problems.

Acknowledgments. The work presented in this paper is supported by the Department of Electrical Engineering of the Hong Kong Polytechnic University.

References

1. Arulampalam, M., Maskell, S., Gordon, N., Clapp, T.: A tutorial on particle filters for online nonlinear/non-Gaussian Bayesian tracking. IEEE Transactions on Signal Processing 50(2), 174–188 (2002)

2. Bruno, M., Pavlov, A.: Improved particle filters for ballistic target tracking. In: Proceedings of IEEE International Conference on Acoustics, Speech, and Signal Processing (ICASSP 2004), vol. 2, pp. ii–705. IEEE, Los Alamitos (2004)
3. Cho, J., Jin, S., Dai Pham, X., Jeon, J., Byun, J., Kang, H.: A real-time object tracking system using a particle filter. In: IEEE/RSJ International Conference on Intelligent Robots and Systems, pp. 2822–2827. IEEE, Los Alamitos (2006)
4. Crisan, D., Doucet, A.: A survey of convergence results on particle filtering methods for practitioners. IEEE Transactions on Signal Processing 50(3), 736–746 (2002)
5. Davis, P.J.: Interpolation and approximation (1975)
6. Doucet, A., Godsill, S., Andrieu, C.: On sequential Monte Carlo sampling methods for Bayesian filtering. Statistics and computing 10(3), 197–208 (2000)
7. Gordon, N., Salmond, D., Smith, A.: Novel approach to nonlinear/non-Gaussian Bayesian state estimation. In: IEEE Proceedings Radar and Signal Processing, vol. 140, pp. 107–113. IET (1993)
8. Hershey, J., Olsen, P.: Approximating the Kullback Leibler divergence between Gaussian mixture models. In: IEEE International Conference on Acoustics, Speech and Signal Processing, ICASSP 2007, vol. 4, pp. IV–317. IEEE, Los Alamitos (2007)
9. Isard, M., Blake, A.: Condensation conditional density propagation for visual tracking. International journal of computer vision 29(1), 5–28 (1998)
10. Schmidt, S.F.: Kalman Filter: Its Recognition and Development for Aerospace Applications. J. Guid. and Contr. 4(1), 4–7 (1981)
11. Stachniss, C., Grisetti, G., Burgard, W.: Recovering particle diversity in a Rao-Blackwellized particle filter for SLAM after actively closing loops. In: Proceedings of the 2005 IEEE International Conference on Robotics and Automation, ICRA 2005, pp. 655–660. IEEE, Los Alamitos (2005)
12. Stützle, T., Dorigo, M., et al.: A short convergence proof for a class of ant colony optimization algorithms. IEEE Transactions on Evolutionary Computation 6(4), 358–365 (2002)
13. Wan, E., Van Der Merwe, R.: The unscented Kalman filter for nonlinear estimation. In: The IEEE Adaptive Systems for Signal Processing, Communications, and Control Symposium, AS-SPCC, pp. 153–158. IEEE, Los Alamitos (2000)
14. Zhong, J., Fung, Y., Dai, M.: A biologically inspired improvement strategy for particle filter: Ant colony optimization assisted particle filter. International Journal of Control, Automation and Systems 8(3), 519–526 (2010)
15. Zhong, J.P., Fung, Y.F.: A biological inspired improvement strategy for particle filters. In: Proceedings of the 2009 IEEE International Conference on Industrial Technology, pp. 1–6. IEEE Computer Society, Los Alamitos (2009)

A Compact Dual-Band Microstrip Bandpass Filter Using Meandering Stepped Impedance Resonators

Kai Ye[1,2] and Yu-Liang Dong[1,2]

[1] School of Physical Electronics, University of Electronic Science and Technology of China, Chengdu, 610054, China
[2] EHF Key Laboratory of Fundamental Science, Chengdu, 610054, China
yekai198600@126.com

Abstract. This letter presents a compact dual-band microstrip bandpass filter (BPF) using meandering stepped impedance resonators (SIRs) with a coupling scheme, which exhibited a size reduction of 50% compared with the traditional direct coupling structure at the same frequency. With the new structure, dual-band BPF centered at 10.7/15.8GHz. The isolation between these two passbands is better than 25 dB form 11.5 to 14.5 GHz. It is shown that the measured and simulated performances are in good agreement. The BPFs achieved insertion loss of less than 2 dB and return loss of greater than 20 dB in each band.

Keywords: Bandpass filter(BPF), dual-band, microstrip, stepped-impedance resonator(SIR).

1 Introduction

The blooming of wireless communications for civilian purposed has increased the demands in dual-band function of the circuit components [1]-[9]. In the past, the dual-bandpass filters (BPFs) were realized as two single-band filters. However, such an approach requires large circuit size and external combining networks [10]. Another way for designing dual-band BPFs is developed by using a cascade of open-stub structures to achieve a dual-band performance [11]. However, this approach raised the circuit complexity and increased the circuit size. So, small size and high dual-band performance are still important things for circuit designers.

In this paper, a miniaturized dual-band BPF with input and output ports on line using meandering SIRs is proposed. To reduce the circuit size, make input and output ports on line and enhance the passband selectivity, each hands of SIR is placed in different sides. By determining the impedance ratio and physical length of the SIRs, the dual-band response can be achieved. The measured results of the BPFs are in good agreement with the simulation results.

M. Zhu (Ed.): Electrical Engineering and Control, LNEE 98, pp. 649–653.
springerlink.com © Springer-Verlag Berlin Heidelberg 2011

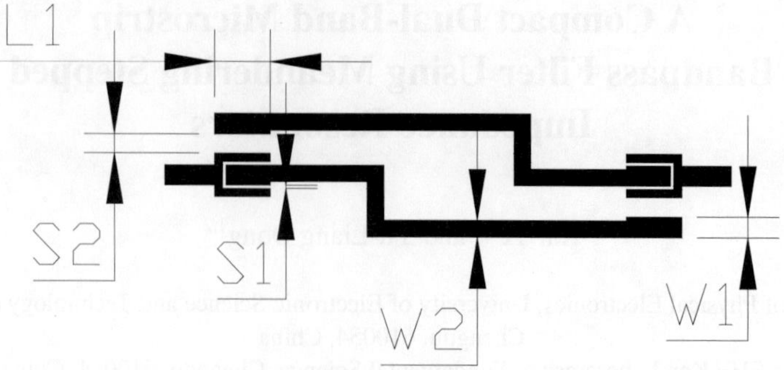

Fig. 1. Layout of the miniaturized dual-band filter

2 Dual-Band Filter Design

Fig.1 shows the geometrical schematic of the proposed dual-band microstrip bandpass filter. First of all, the sinuously shaped SIR resonator is constructed by cascading a long-length high-impedance section in the center with the two short-length low-impedance sections in the two sides. In this design, the widths of those two distinctive sections, i.e.,W_1 and W_2, are pre-selected at 0.8 and 0.66 mm. Following the analysis in [12], the resonant conditions of the normal SIR resonator, as shown in Fig, 2, can be explicitly established. As a result, the two relevant frequencies, f_0 and f_{s1}, can be determined as the two solutions of the following two transcendental equations, respectively:

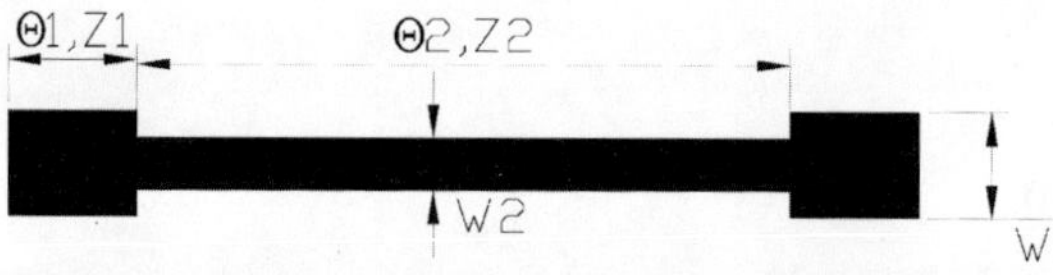

Fig. 2. Basic structure of the stepped-impedance resonator

$$R_Z - \tan\theta_1 \tan\theta_2 = 0 \quad (f = f_0)$$
$$R_Z \tan\theta_1 + \tan\theta_2 = 0 \quad (f = f_{s1})$$

Where R_Z is the ratio of the two characteristic impedances, i.e., $R_Z= R_2 / R_1$. Fig.3 plots the ratio of the first two resonant frequencies, f_{s1}/f_0, as a function of the impedance ratio (R_Z) under three different ratios of the two line lengths in a SIR resonator, i.e., $\theta_2=\theta_1$, θ_2, andθ_3. Additionally, we use the new coupling scheme to input/output of the filter to improve the insertion loss of the dual-band BPF.

The formulas given in [12] remain accurate for the calculation of the W_1 and W_2 values in the new structure, but they obviously become invalid for the calculation of the spacing S_2 between resonators, which is related to the coupling coefficient K.

On the other hand, the coupled spacing (S_2) can be properly tuned to maximize return loss and minimize insertion loss. By adjusting S_2 properly, we could get the required response of the dual-band BPF. Meanwhile, the position of the transmission zero between the two passbands is changed while tuning S_1. Also, we can know that the simulated frequency responses of the dual-band filter under different couplinglength L_1. Therefore, the transmission zero between the two passbands is determined by the coupling degree between the feed lines and SIRs.

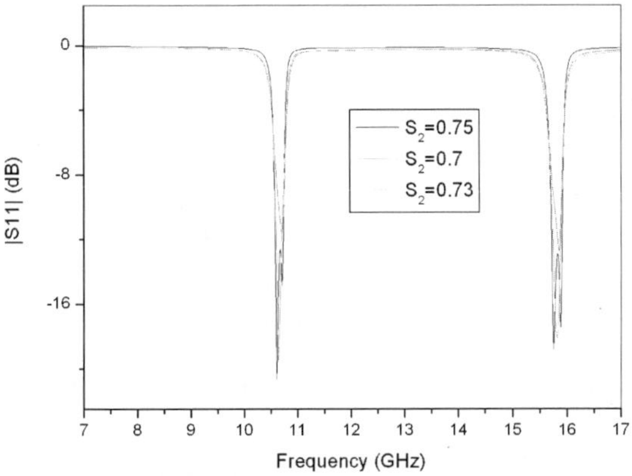

Fig. 3. Frequency responses of |S11| under different coupling spacing S_2

3 Simulation and Measurement Results

This section describes the design and performance of fabricated filters. The proposed filters were designed and fabricated on a common substrate having a thickness of 0.254 mm and a relative dielectric constant of 2.2. The high impedance Z_1 segment is chosen to have 55.57 Ω with a width (W_1) of 0.66mm, and the low impedance Z_2 is chosen to have 50 Ω with a line width (W_2) of 0.8 mm. The filters were simulated using a full-wave EM simulator.

Fig.3 show the frequency responses of the reflection coefficient |S11| under the different selected coupled spacing (S_2). Here, all the other dimensions of this filter are unchanged and they are given in Fig.1. From Fig.3, we can see that |S11| in the two bands is immensely varied in shape. When S_2=7.0mm, |S11| in the two bands is only -12dB. When S_2=7.3mm, |S11| can arrived to -18dB. But when S_2=7.5mm, |S11| in each band is suddenly changed in shape from a single attenuation zero to the double ones because of deficient coupling degree.

Following the above discussion, our optimization design is carried out by taking into account strip thickness of 17 μ m. The design procedure is to at first determine S_1 with satisfactory performance at stop band and than to adjust S_2 toward good transmission in the two pass bands. Fig.4 plots the predicted and measured S-parameters in the frequency range covering the concerned dual bands. From Fig.5 we can see that the S12 of both bands are in very reasonable agreement with each other. And the S11 of measurement has more return loss than simulation in the both bands. The designed filter circuit occupies the overall size of about 20 × 35mm^2. The photograph of the fabricated filter is shown in Fig.5.

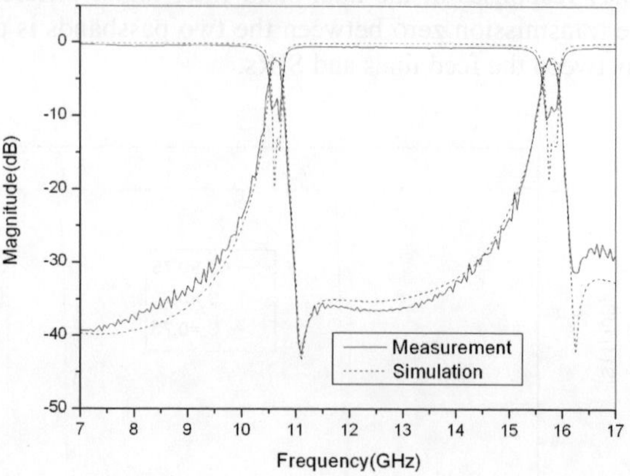

Fig. 4. Simulation and measurement results of proposed dual-band filter

Fig. 5. Photograph of a fabricated dual-band microstrip bandpass filter

4 Conclusion

A novel dual-band bandpass filter using meandering stepped impedance resonators was proposed and analyzed in this paper. The filter can achieve some dual-band applications at 10.7/15.8GHz, and have many characteristics such as compact structure and good stopband. With these features, the proposed filter can be widely used in fields of radar and the communication system.

Acknowledgment

The work described in this paper was supported by "the Fundamental Research Funds for the Central Universities Under grant No.ZYGX2009J041"

References

1. Hashemi, H., Hajimiri, A.: Concurrent multiband low-noise amplifiers-theory, design and applications. IEEE Trans. Microw. Theory Tech. 50(1), 288–301 (2002)
2. Miyke, H., Kitazawa, S., Ishizaki, T., Yamada, T., Nagatomi, Y.: A miniaturized monolithic dual band filter using ceramic lamination technique for dual mode portable telephones. In: IEEE MTT-S Int. Dig., vol. 2, pp. 789–792 (June 1997)
3. Quendo, C., Rius, E., Person, C.: An original topology of dual-band filter with transmission zeros. In: IEEE MTT-S Int. Dig., vol. 2, pp. 1093–1096 (June 2003)
4. Palazzari, V., Pinel, S., Laskar, J., Roselli, L., Tentzeris, M.M.: Design of an asymmetrical dual-band WLAN filter in liquid crystal polymer (LCP) system-on package technology. IEEE Microw. Wireless Compon. Lett. 15(3), 165–167 (2005)
5. Chang, S.F., Chen, J.L., Chang, S.C.: New dual-band bandpass filters with step-impedance resonators comb and hairpin structures. In: Proc. Asia Pacific Microwave Conf., pp. 793–796 (2003)
6. Chang, S.F., Jeng, Y.H., Chen, J.L.: Dual-band step-impedance bandpass filter for multimode wireless LANs. Electron. Lett. 40(l), 38–39 (2004)
7. Lee, H.M., Chen, C.R., Tsai, C.C., Tsai, C.M.: Dual-band coupling and feed structure for microstrip filter design. In: IEEE MTT-S Int. Dig., pp. 1971–1974 (2004)
8. Kuo, J.T., Cheng, H.S.: Design of quasielliptic function filters with a dual-passband response. IEEE Microw. Wireless Compon. Lett. 14(10), 472–474 (2004)
9. Sun, S., Zhu, L.: Coupling dispersion of parallel-coupled microstrip lines for dual-band filters with controllable fractional pass bandwidths. In: IEEE MTT-S Int. Dig. (June 2005)
10. Miyke, H., Kitazawa, S., Ishizaki, T., Yamada, T., Nagatomi, Y.: A miniaturized monolithic dual band filter using ceramic lamination technique for dual mode portable telephones. In: IEEE MTT-S Int. Dig., vol. 2, pp. 789–792 (June 1997)
11. Guan, X., Ma, Z., Cai, P., Kobayashi, Y., Anada, T., Hagiwara, G.: Synthesis of dual-band bandpass filters using successive frequency transformations and circuit conversions. IEEE Microw. Wireless Compon. Lett. 16(3), 110–112 (2006)
12. Makimoto, M., Yamashita, S.: Bandpass filters using parallel coupled stripline stepped impedance resonators. IEEE Trans. Microw. Theory Tech. MTT-28(12), 1413–1417 (1980)

Research on Self-adaptive Sleeping Schedule for WSN

Mingxin Liu, Qian Yu, and Tengfei Xu

College of Information Sicence and Engineering, Yanshan Universiy,
Qinhuangdao City, 066004, P.R. China
liumx@ysu.edu.cn

Abstract. In this paper an improved Sensor-MAC(S-MAC) scheme for Wireless Sensor Networks (WSN) is presented to solve the problem that S-MAC could not adjusting the duty cycle according to the traffic load in WSN. The listen/sleep time in this improved scheme could be self-adaptive adjusting according to the retansmission times in previous cycles. The simulation results demonstrated that this scheme will take more time to listen and improve throughput when network traffic was heavy, and will take more time to sleep and save energy on the opposite condition.

Keywords: Adaptive listening, S-SMAC protocol, Power saving schedule.

1 Introduction

The energy of node is usually supplied by battery that is not easily to be replaced or recharged, so energy efficiency is a critical issue in order to prolong the network lifetime. Besides, because of limited transmit power, data packets are send to sink node normally through relay nodes, which not only consume energy but also result in multi-hop delay.

High energy-efficient medium access protocols (MAC) for WSN were proposed in many literatures to solve the problems as mentoned above.These work are beneficial to optimize network performance and design innovative network mechanisms. IEEE 802.11[2] adopts DCF(Distributed coordination function) and CSMA/CA. Based on IEEE 802.11, SMAC[3] uses periodic listen/sleep cycle for sensor nodes, which reduces idle listening time and energy consumption.Another problem in SMAC is that under lower traffic load, the time of listen channel is changeless and waste more energy, while under heavy traffic load, nodes can not get channel in time.

The main purpose of SMAC[4] for WSN is to improve energy efficiency. In terms of data transmission, SMAC is like IEEE 802.11, but it introduces periodic sleep/listen cycle to make nodes into sleep while the nodes without data transmission. Meanwhile, SMAC has good scalability and collision avoidance.

Every node periodically sleep for certain time, then wake up to decide if need to communicate with other nodes[5]. If not, this node returns to sleep again. This scheme not only reduces idle listening but also improves overhearing.

SMAC introduces periodic listen/sleep schedule and divides time into several fix frames. Regulations following must be observed:

M. Zhu (Ed.): Electrical Engineering and Control, LNEE 98, pp. 655–660.
springerlink.com © Springer-Verlag Berlin Heidelberg 2011

$$\text{listenTime} = \text{syncTime} + \text{dataTime} \tag{1}$$

$$\text{CycleTime} = \text{listenTime} \times \frac{100}{\text{dutyCycle}} + 1 \tag{2}$$

$$\text{sleepTime} = \text{CycleTime} - \text{listenTime} \tag{3}$$

From equation (1), (2) and (3), some conclusions are obtained: listenTime consist of syncTime and dataTime, sleepTime added by listenTime is equal to CycleTime. CycleTime is composed of fix listenTime and sleepTime, although energy wastage is reduced to some extend, duty cycle and active part are fix in SMAC, which cause energy wastage under lower load. While increase packet loss rate and reduce throughput.

The proposed protocol in this paper, S-SMAC, can adaptively adjust listen/sleep time according to the number of retransmissions in previous periods.

2 S-SMAC Protocol Description

S-SMAC can get knowledge of current traffic load by the number of retransmissions in previous five periods, and then adjust duty cycle according to traffic load.

$$\text{sum} = \sum_{i=1}^{5} (\alpha_i \times dutyCyc\_[i]) \tag{4}$$

Where, $dutyCyc\_[i]$ indicate the number of previous five periods; α_i is the retransmission coefficient. Each period of the previous five periods has different correlation with current time. Five coefficients are different and relationship is $\alpha_5 \geq \alpha_4 \geq \alpha_3 \geq \alpha_2 \geq \alpha_1$. Through adaptive adjust, we find that the more periods retransmission need, the nearer the periods is from current adjustion time, the larger relative sum is.

Suppose that transmission follows uniform distribution in WSN, that is to say, from the beginning of the first period to the end of the fifth period, the probability of retry is constant, which is equal to 1/5. When there are two retransmissions taking place in five periods, that is, the probability of retransmission is 2/5, duty cycle should be adopted. Threshold of sum, TH_{sum} is 0.4.

When the sum is larger than TH_{sum}, which indicate channel is busy and duty cycle should be increased to raise listening time. On the contrary, when the sum is smaller than TH_{sum} which indicate traffic load is lower and sleep time should be added up.

S-SMAC requests the nodes adjust the dutycycle under the rules as follows:

$$\text{dutyCycle} = [\frac{a \times \text{sum}}{b - \max}] \times \text{dutyCycle}, \ \text{sum} > TH_{sum} \tag{5}$$

$$\text{dutyCycle} = \frac{2}{3} \times \text{dutyCycle}\,,\ \text{sum} \leq TH_{sum} \tag{6}$$

Where a and b in equation(4) denote adaptive coefficients which can be adjusted according to topology and hops,. Note that $(b - \max) > 0$, $[\frac{a \times \text{sum}}{b - \max}] > 1$ and $b > 5$.

When sum is larger than threshold, duty cycle should be increased by eqution (5), and then add up listening time, improve throughput. When sum is lower than threshold, duty cycle should be reduced to 2/3 of primary duty cycle by eqution (6), which raise sleep time and save energy.The flow chart of S-SMAC protocol is illustrated in Fig.1.

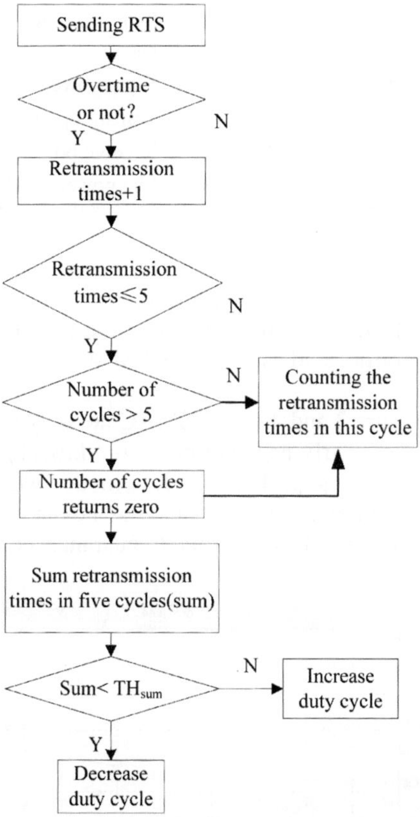

Fig. 1. Flow chart of S-SMAC Protocol

3 Simulation Results and Analysis

We simulate S-SMAC using NS-2. In multi-hop WSN, nodes send data through many paths. Some nodes do not relay packets because they are sleep, which result in more queue delay, and further average delay. So one important point at evaluating sleep strategy in multi-hop WSN is that whether the sleep strategy can seriously affect throughput and average delay at the same time reduce energy wastage. We consider three traditional evaluation metrics: energy consumption, delay, and throughput, and compare S-SMAC with SMAC.

We use a bus topology consisting of eight static nodes. Node 0 is source node, node 7 is sink node and others are intermediate nodes. Distance between two nodes is 200 metres. Simulation lasts 400 s, and node 0 will start to send data at 50 s till 250 s. Simulation parameters are given in Table 1.

Table 1. Parameters seting of NS-2

Parameters		Parameters	
Routing protocol	DSR	Sync window	15 slots
Simulation area	1200×800	Data window	31 slots
Simulation time	400 s	Sleep power	1.0μw
Node number	8	Transmit power	1.0 mw
Node energy	1000 mj	Receive power	1.2 mw
Data length	512 bytes	Idle power	1.0 mw

Figure 2 shows energy consumption under different data transmission intervals. When traffic load is heavy, S-SMAC and SMAC consumes nearly the same energy, because S-SMAC can adaptively adjust duty cycle and increase throughput at the cost of part energy. When traffic load is lower, S-SMAC can raise sleep time to save energy by eqution (5).

We can learn from Figure 3 and 4 that S-SMAC not only efficiently reduce energy wastage, but also have no clearly negative effect on throughput and delay compared with SMAC. When transmiting packets are less, nodes which do not take part in communication will keep in sleep state as much as possible. On the contrary, when channel is busy, nodes should wake up in time to guarantee communication quality.

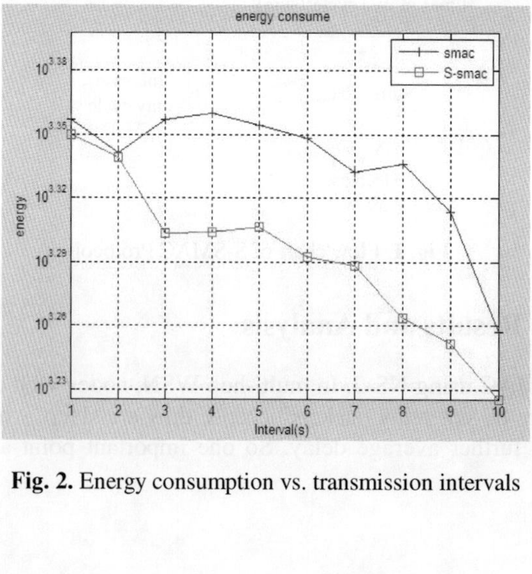

Fig. 2. Energy consumption vs. transmission intervals

From Figure 2,3 and 4 following conclusions are got: compared with traditional SMAC, when traffic load is heavy (transmission interval is 1,2 and 3 seconds respectively), S-SMAC can moderately decrease sleep time and increase listening time to receive more packets and reduce delay; when the traffic load begin to decrease, S-SMAC can reduce listening time to save energy.

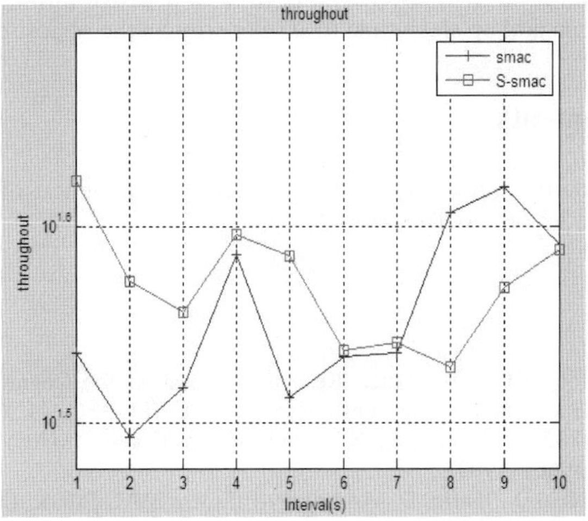

Fig. 3. Average throughput vs. transmission intervals

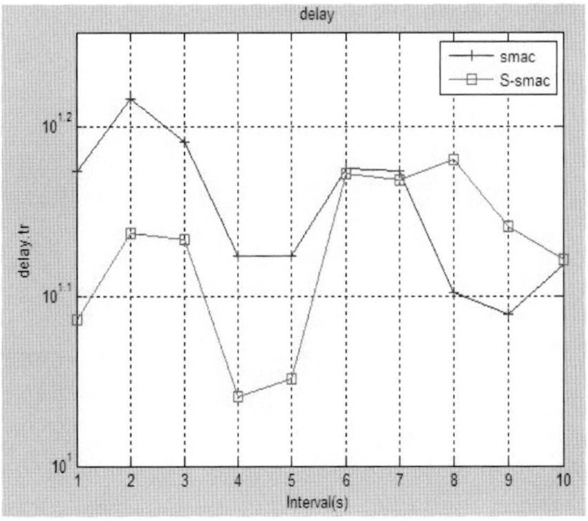

Fig. 4. Average delay vs. transmission intervals

4 Conclusion

How to design an effective power saving mechanism is a hotspot in WSN study. This paper analyzes shortage in SMAC. This protocol is not efficient in the case of various network traffic occur. So we estimate current traffic load according to retransmission times, and then adjust duty cycle according to traffic variation. Simulation result shows that S-SMAC can increase listening time to improve throughput under heavy traffic load, and can add up sleep time to save energy on the opposite condition.

Acknowledgements

This work was partly supported by Natural ScienceFoundation of Hebei Province, China (No. F2011203067, F2011203092).

References

1. Ngo, D.T., Ngoc, T.L.: Distributed Resource Allocation for Cognitive Radio Ad-Hoc Networks with Spectrum-Sharing Constraints. In: IEEE GLOBECOM 2010, pp. 1–6 (2010)
2. IEEE Standard 802.11. Wireless LAN Medium Access Control (MAC) and Physical Layer (PHY) Specifications (1999)
3. Ye, W., Heidemann, J., Estrin, D.: An Energy-Efficient MAC Protocol for Wireless Sensor Networks. In: IEEE INFOCOM, New York, vol. 2, pp. 1567–1576 (2002)
4. Heidemann, J.: Energy-efficient MAC Protocol for Wireless Sensor Networks. In: The 21st Int'l Annual Joint Conf. On Computer and Communications Societies, New York (2002)
5. Demirkol, I., Ersoy, C., Alagöz, F.: MAC Protocols for Wireless Sensor Networks:a Survey. IEEE Communications Magazine (2006)

Design and Application of a New Sun Sensor Based on Optical Fiber

Zhou Wang[1,2], Li Ye[1], and Li Dan[1]

[1] Institute of modern optical technology of Soochow University, Jiangsu Suzhou (215006)
[2] Province Key Laboratory of Modern Optical Technology of Jiangsu, Jiangsu Suzhou (215006)

Abstract. The paper describes the new design and frame of sun sensor using the optical fiber. This scheme can prevent the sun sensor is harming by high energy particles in space or ensure the sun sensor good work in the rain, snow conditions on earth, it is a reason using two twain optical fibers. Sense signals form sunlight is amplified by differential amplifier for obtaining higher SNR. Finally data to measure the location of sun is acquired by MCU. Now this technology has been attempted used to CPV (Concentrating Photovoltaic) for increasing the efficiency of generating electricity by the solar energy.

Keywords: Sun sensor, optical fiber, high precision, differential amplifier.

1 Introduction

The sun sensor widely used in the space or the aviation as a kind of attitude detector, all satellites are equipped with the detector for measuring the sun angle in both azimuth and elevation where sun is its target [1]. Now used technology may divide into two kinds: digital sun sensor and simulate sun sensor [2]. The simulate sun sensor has been washed out because its precision is lower too; the digital sun sensor is become main equipment. According different optical sensor, sun sensor can be separated to linear CCD and area CCD [3] or CMOS-APS [4]. Intending expectancy is its higher accuracy and smaller and lighter and longer life.

These sun sensors working principle nearly is the same. The sunlight beam shoots to photosensitive device through one or more special slits. The different angle of sun shall create the different position of faculae on the surface of device, so the sun moves and the faculae shifts too. This position may be measured by several pixels (X and Y) and their electrical signals or an image data.

Because these photosensitive devices have high sensations and are attainted easily, if running in out space they shall be died by high energy particle or varied radial, finally cause a satellite or a spacecraft miss for controlling. If working in ground they shall be disturbed as rains, snows and sands cover, the bad result is system losing normal task. Above is because sensors are nude in working environments.

Thus a new design want to these sensors be hided inside equipment and well working.

M. Zhu (Ed.): Electrical Engineering and Control, LNEE 98, pp. 661–667.
springerlink.com © Springer-Verlag Berlin Heidelberg 2011

2 Design Principle

This new structure of sun sensor is designed based on several optical fibers which lead sun-light into photosensitive sensors inside, so these devices shall do not be disturbed or mangled. Otherwise its advantages are longer life, simpler structure, smaller size, lower mass, lower cost, wider range of applications. Comparing with the current sun sensor, this design can not only protect the photoelectric sensors, but also increase the accuracy of location especially in a small field of view. A new design has applied for Chinese Invention Patent, application number is 200910264755.1. the title calls "a sun sensor and its measure method".

2.1 Optical Structure

The designing optical structure has multi-fibers as leading sunlight beam into photoelectric sensors, the Fig.1 shows new ideal component where an optical fiber is set in centre of component and its surface is covered with a milky mask for getting the average value of sunlight. Otherwise, there are two pairs of symmetrical distribution optical fibers around, by them sunlight is transmitted the photoelectric sensors are set.

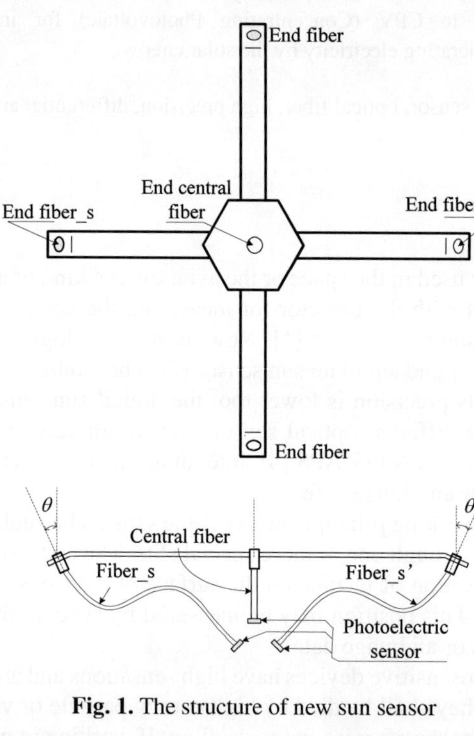

Fig. 1. The structure of new sun sensor

In this component, the angle is θ between the normal of each fiber and central fiber. When sunlight vertically irradiates to the surface of fiber, the photoelectric sensors can receive maximal light intensity and most sensitive electric current. Because there is an angle, so the light energy I through fiber is equal to multiplying sun light energy I_0 and cosine θ together, As $I = I_0 * \cos\theta$. Here the reference value I_0 of average light intensity is obtained by the central fiber.

As each pair fiber is set oppositely, so the final electric signal is a difference volt. Additionally used each pair fiber has the same length and the same material (plastic or glass) and the same waste. The value of difference volt presents the angle between this component and the sun light beam. The measuring result that is the information of the sun vector shall be got through deal with by a micro controller inside the sun sensor.

Comparing with traditional sun sensor, the new structure has higher measuring precision and greater SNR to the method of utilizing the projective image [3]. The data calculating method is similar to a quadrant detector sun sensor.

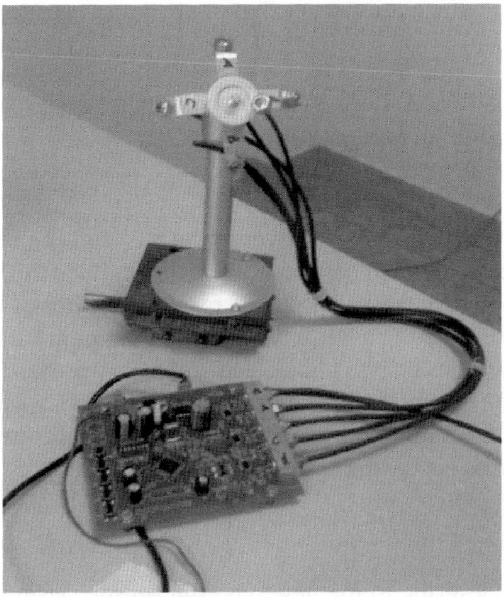

Fig. 2. The photo of the experiment equipment

The result of previously experiments indicates that measuring precision can be able to reach highest resolution ratio ≤ 0.12° and its viewing angle is more than ±60°. Fig.2 shows experiment equipment.

2.2 Calculation Method

First an irradiating light beam L is shown fig.3. Two pair fibers is defined two dimensions (horizontal X and pitching Y), fig.3 only shows X direction for explaining easily. S and S' as shown Fig.3 is partly is the end of an each pair fiber, well then S end or S' end getting the part value L' of light beam L in X direction.

Thus, S end's value should equal:

$$I_S = L' * \cos(\theta - \alpha) \tag{1}$$

Otherwise, S' end's value should equal:

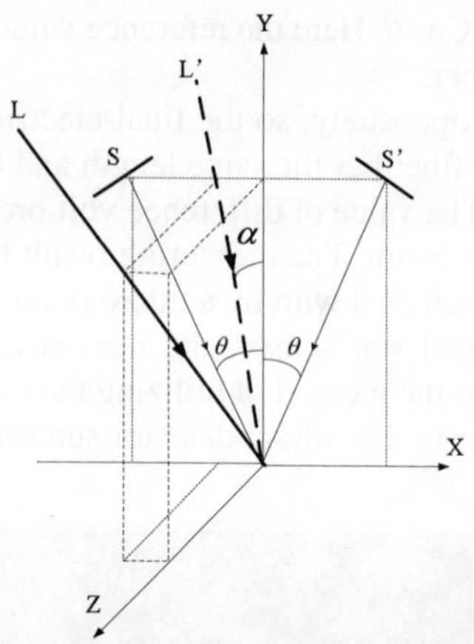

Fig. 3. The calculation method of the angles

$$I_{S'} = L'^*\text{Cos}\,(\theta + \alpha) \tag{2}$$

Even if the angle α may be get by above formula (1) and (2), but calculating method is very complexity to a MCU. There is an easy method that is the looking-up a table which is formed beforehand according to the result of a series of testing.

The other way of calculating angle α still is that looking-up data table, but based on a difference value between each pair photoelectric signals which is form running result of a difference operating amplifier. Final according to angle α, a satellite or a spacecraft control itself attitude of running, example as adjust momentum wheels inside the satellite or jet compress air for the spacecraft navigation.

2.3 Circuit Design

According to above describe to the new type sun sensor, designing electric circuit is shown Fig.4.

In the schematic diagram, IC U1 and U2 are formed a volt follower for increasing input impedance, IC U3 is a difference operating amplifier for obtaining a difference value between a pair photoelectric signals form fibers. And then the difference value (simulate signal) is sent to a MCU U4 and is changed digital signal. The MCU takes on an important task of calculating and controlling in whole system.

For higher dynamic range and lower power dissipation, IC U1 and U2 adopt TLC2254 that is a rail to rail OP and CMOS structure [4] , IC U3 is MAX4208 as an instrument amplifier with CMOS structure.

The MCU U4 is STC-12C5A60S2 made in China, it is a kernel compatible to MCS-51command system and has higher speed compared with tradition one. It is 10 bits high speed ADC result and eight ports and has 2~3 UART function ports. Because it is CMOS structure too, its power dissipation is lower, thus the MCU is enough to bear all task of system need.

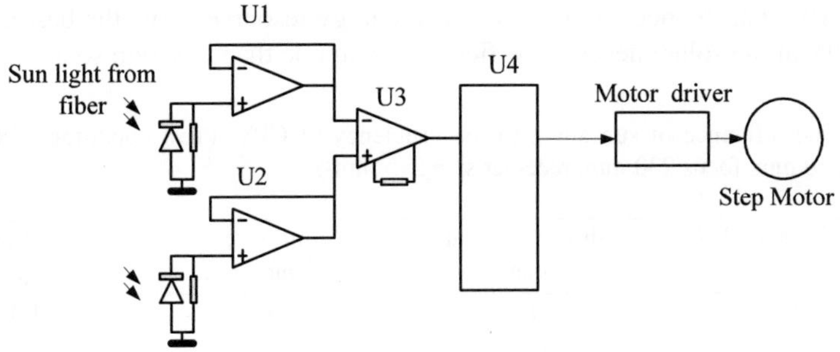

Fig. 4. The schematic diagram of ideal sun sensor

3 Example for Application

The sun sensor mainly used to apply in the space technology, but this type one is provided with some many excellences. The technology is applied to some ground equipments such as CPV (Concentrating Photovoltaic) [6]. CPV technology is different with tradition PV, front one collects the energy of sun light and focuses to a solar cell by a lens in the same time, shown as Fig.5-(a). When light focused dot appears some excursion shown as Fig.5-(b), the efficiency of generating electricity of CPV shall drop, yet a little of current dose not flowing out [7]. The table 1 shows the relation between the efficiency of generating electricity and the solar shoot angle.

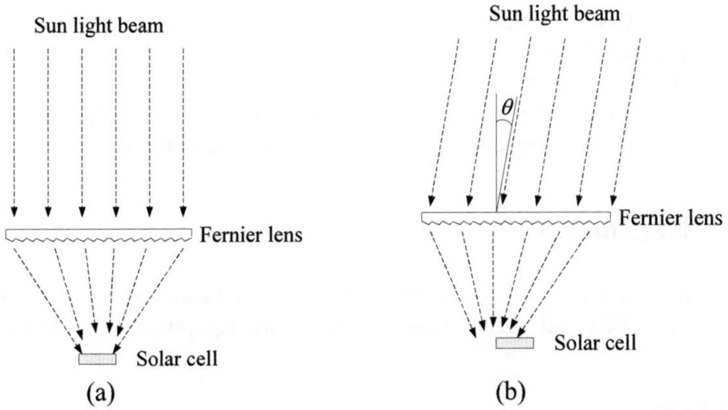

(a) (b)

Fig. 5. The principle of CPV

Tab.1 data is from a calculation by Zemax optical software. Then data shows the efficiency equal zero when the sun incidence angle is great than 3°.

Observably the CPV must rely on a sun tracking system with higher precision [8]. The key part of the system certainly is the sun sensor. Along with application of solar energy, CPV shall be become newest green energy resource. Now the best CPV may obtain 42% of the solar energy, its efficiency is double than tradition solar cell planes.

Table 1. The influence of sun angle to the efficiency of CPV. (Test condition: Fernier lens diameter 150 mm, focus 230 mm, receiver size 5*5 mm^2)

Sun shoot angle θ (°)	Shift of central (mm)	Shift size (mm)	Total efficiency %
0	0	0	100
0.1	0.36	0.6296	96.79
0.15	0.551	1.1783	94.00
0.2	0.727	1.7653	91.01
0.5	1.83	3.254	83.43
1	3.655	17.5074	10.84
3	11.63	19.6349	0.0
5	19.63	19.6349	0.0

4 Conclusion

The paper presents new design and frame of sun sensor based on the optical fiber. This technology has a lots of advantage compared with tradition sun sensor. Those are longer life, smaller size, simpler structure, lower cost and wider range of applications. Now the application project has been looked up such as except the space technology and the CPV (Concentrating Photovoltaic), where the sun is tracking with higher precision and the utilization of solar energy is increased and the reliability of those system used new sun sensor is ensured.

Further objects of application and exploitation are the technology being came into practice and business, especially extend widely application fields.

Acknowledgements

This research work is supported by Institute of Modern Optical Technology of Soochow University and Suzhou Shaoshi Electronic Equipment Limit Duty Company.

References

1. Zabiyakin, A.S., Prasolov, V.O., Baklanov, A.I., El'tsov, A.V., Shalnev, O.V.: Sun sensor for orientation and navigation systems of the spacecraft. In: SPIE, vol. 3901, p. 106 (1999)
2. Li, H., Yihua, H.: Principium and technology development tendency of sun sensors. Electronic Components & Materials 25(9), 5–7 (2006)
3. Chum, J., Vojta, J., Base, J., Hruska, F.: A Simple Low Cost Digital Sun Sensor for Micro-Satellites, `http://www.dlr.de/iaa.symp/Portaldata`
4. Peng, R., Shengli, S., Guilin, C.: New generation of APS-based digital sun sensor. Science Technology and Engineering 8(4), 940–945 (2008)

5. TLC225x, TLC225xA Advanced LinCMOS Rail-to-Rail very Low-Power operational amplifiers. Datasheet of Texas Instruments.USA (1999)
6. Jian, M., Ping, X.: High precision computation for sun position and its application in solar power generation. Water Resources and Power 26(2), 201–204 (2008)
7. Garboushian, V., Roubideaux, D., Yoon, S.: Integrated high-concentration PV near-term alternative for low-cost large-scale solar electric power. Solar Energy Mater Solar Cells 47((1-4), 315–323 (1997)
8. Dasgupta, S., Suwandi, F.W., Sahoo, S.K., Panda, S.K.: Dual Axis Sun Tracking System with PV Cell as the Sensor, Utilizing Hybrid Electrical Characteristics of the Cell to Determine Isolation. In: IEEE ICSET 2010, Kandy, Sri Lanka, December 6-9 (2010)

A 12-b 100-MSps Pipelined ADC*

Xiangzhan Wang, Bijiang Chen, Jun Niu, Wen Luo, Yunchao Guan,
Jun Zhang, and Kejun Wu

State Key Lab. of Ele. Thin Films and Integrated Devices, Univ. of Electronics Science & Technology, Chengdu, Sichuan, 610054, P.R. China
cbj2458@126.com

Abstract. A low power 12-b 100-MSps pipeline analog-to-digital converter (ADC) is implemented in 0.18-μm CMOS process. The op-amp sharing with current selection between the front-end sample-and-hold (S/H) and the first stage multiplying digital-to-analog converter (MDAC) improves the power efficiency dramatically, and the application of range scaling technology decreases the power consumption further. A zero compensation method based on boot-strap technology is introduced in op-amp design to accomplish precise compensation as well as improve the linearity. The simulation result shows the ADC achieves an SFDR of 81 dB with a 50MHz input signal at 100-MSps. The DNL and INL are both within ±0.5LSB. The ADC consumes 46.5 mW from a 1.8V supply.

Keywords: Pipeline, ADC, op-sharing, range-scaling, current switch, charge pump.

1 Introduction

The demand for high performance pipeline ADCs in visual and wireless communication industry is still soaring after many years' consistent increase. To keep power consumption as low as possible is always the key consideration, since the designers could arrange a wider margin for other system specifications when the intrinsic power consumption is lower. Pipeline schemes are becoming a dominant architecture for high-speed high-resolution low-power ADCs [1].

The op-amp sharing [2]-[4] technology is utilized between S/H and first stage MDAC, and a special current selecting circuit is used to allocate different amount of current according to the actual need of SHA and the 1st MDAC. Range scaling [5] [6] technology was brought up recently and it is believed to be efficient to achieve a better power consumption and linearity tradeoff. As the supply voltage drops dramatically in submicron technology, traditional methods of compensating the right-half-plane (RHP) zero in op-amps encounter a problem: the value of MOSFET- made resistor is difficult to adjust when its gate voltage is low. As an improvement, a boot-strap MOSFET is designed to replace the traditional scheme, which could achieve a more accurate compensation. Redundancy digital calibration is used to relieve the effect of some kinds of nonidealities on the linearity of the conversion [1].

* This work is supported by Doctoral Fund of Ministry of Education of China (200806141100).

M. Zhu (Ed.): Electrical Engineering and Control, LNEE 98, pp. 669–676.
springerlink.com © Springer-Verlag Berlin Heidelberg 2011

This paper is structured as follows. Section 2 introduces some low-power technique-es in pipelined ADC. Section 3 presents the general ADC architecture and the detailed circuit implementation. Section 4 describes the prototype ADC simulation results and section 5 ends this paper with conclusion.

2 Low-Power Techniques in Pipelined Adc

2.1 Op-Amp Sharing

Op-amp sharing between the S/H and 1st stage MDAC (MDAC1) is used in proposed ADC with a two-phase non-overlapping clock. As shown in Fig.1, when Φ1 is high, the bootstrapped switch is on, and the S/H samples the input signal and at the same phase, MDAC1 generates the output. Then, as Φ2 becomes high, the S/H produces the output and it is sampled by MDAC1. The principle of op-amp sharing technique is that since the op-amp is not needed during sampling phase, it only needs to be working during the half clock period. So, the op-amp can be used for the S/H when Φ2 is high, otherwise, it can be used for MDAC1. Thus, significant power and chip area can be saved by reducing the number of op-amps. However, it needs additional switches that introduce series resistance and degrades the settling behavior.

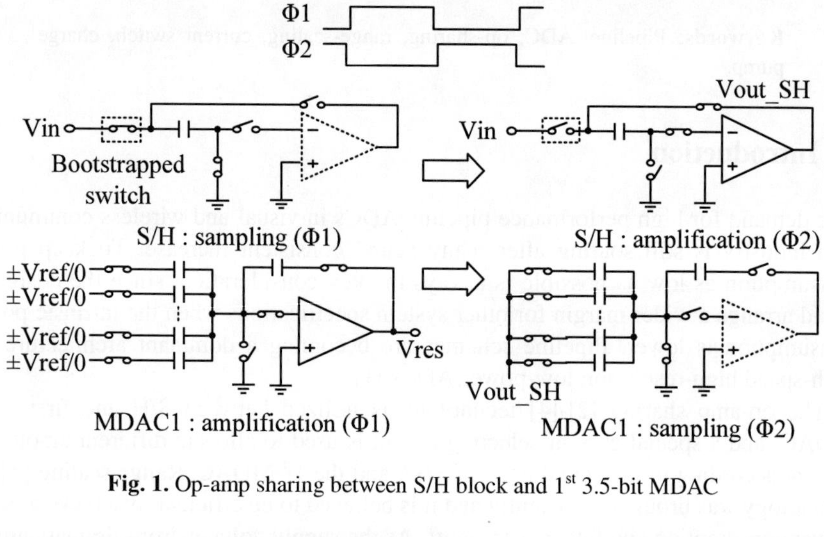

Fig. 1. Op-amp sharing between S/H block and 1st 3.5-bit MDAC

2.2 Current Switch of the Shared Op-Amp

With the introduction of op-amp sharing technique, a single op-amp is worked both in the S/H and the MDAC stage. For placed in different feedback configurations, there are different requirements for the op-amp's performance.

The current consumed by S/H is usually largely greater than that of MDAC1 because the loading of S/H is far greater than that of MDAC1. Therefore it is a waste of power to use the same op-amp in MDAC1 with SH. For this reason we can reduce the current when the op-amp is working as part of MDAC1 to save power.

Current selection is a method to improve the power efficiency of op-sharing technique. Portion of the current is cut off when MDAC1 is active. However, it will take some time for the op-amp to stabilize and it will degrade the settling behavior. The detail analysis of current selection will be done in section 3.

2.3 Range-Scaling of MDAC1

The full scale differential input and output range of the S/H are both 2Vref while the differential output range of MDAC1 is scaled to Vref to conserve power, as shown in Fig.2. All other MDAC except MDAC1 operates in the differential range of Vref. This small voltage range is feasible since the quantization noise is dominant to limit SNR rather than thermal noise.

The current set by GBW is as follows [11].

$$I_{GBW} = \frac{A}{\beta}\left|\ln(D)\right|\left[1+(\frac{\pi}{\ln D})^2\right] \quad ,A=\frac{1}{2\pi}f_{clk}(V_{GS}-V_{TH})C_L \ . \tag{1}$$

Where, D is the settling resolution of MDAC1 and β is its feedback factor. With its weak correlation to D, I_{GBW} is almost half its origin value after range-scaling because of the β being nearly twice its origin one, although D is halved.

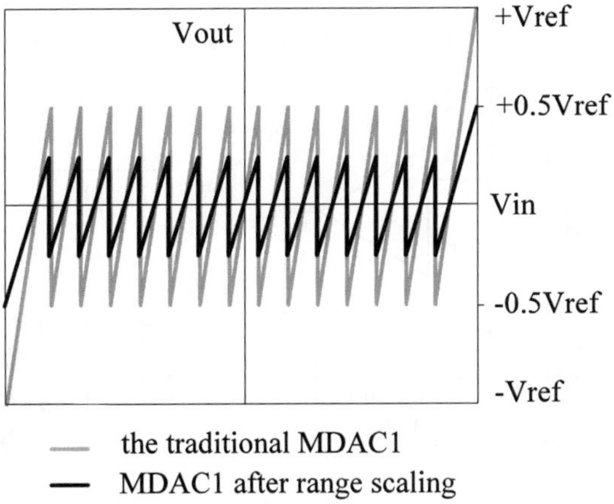

Fig. 2. Transition function of 3.5-bit MDAC1

The current set by the slew rate (I_{SR}) is proportion to the full differential scale output range. I_{SR} is also reduced since the output range is halved because of range-scaling.

After all, a reduction in half of the power consumption in MDAC1 is almost achieved by range-scaling.

The reference voltage and the comparative levels of backend MDAC is reduced to half its origin values. The correction range of the first 3.5-bit MDAC is ±FS/32 [1], while the one of the rest 1.5-bit stages is ±FS/16 after range scaling.

The minimum allowable DAC nonlinearity of MDAC1 remains the same as $FS/2^{N+1}$ [7] after range scaling. The limits on DAC (after the first) nonlinearity reduce to $FS/2^{N-1}$ (half the value without range scaling), and it is still les than its first counterpart.

3 ADC Architecture and Circuit Implementation

The proposed ADC was carefully designed to achieve high sampling rate with required accuracy and low power consumption. The block diagram of the proposed 12-b pipeline ADC is illustrated in Fig.3. The first op-amp is shared between the S/H and MDAC1.The first stage employs a modified 3.5-bit pipeline architecture followed by six 1.5-bit-per-stage pipeline stages and a 3-bit flash ADC.

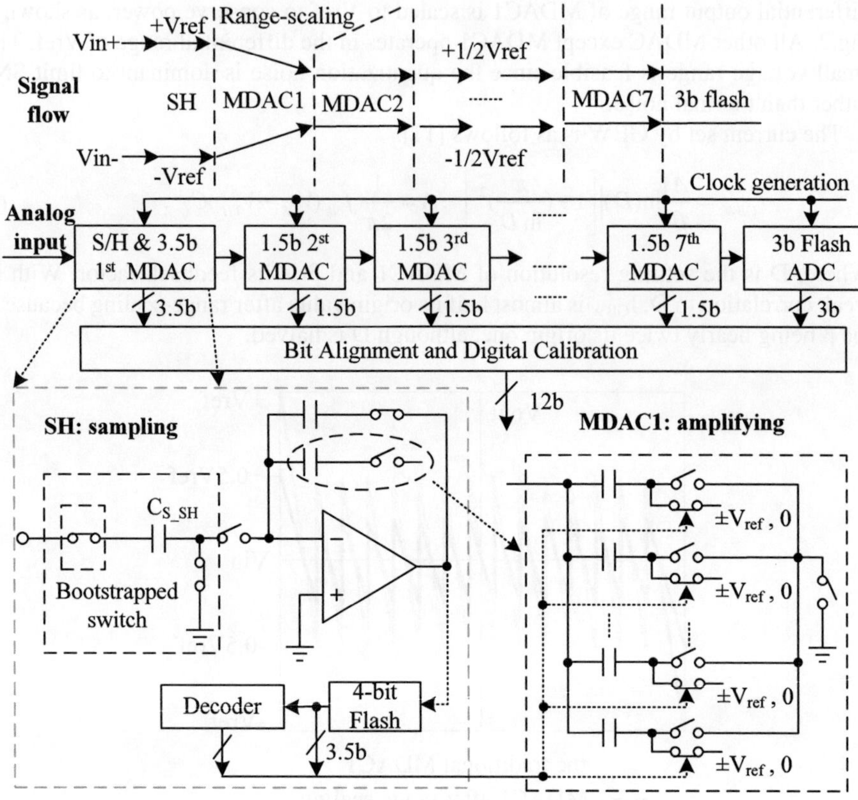

Fig. 3. Block diagram of the pipelined ADC

Multi-bit first stage has been widely employed in the high resolution ADCs for better linearity performance and smaller power consumption. The op-amp sharing technology greatly reduced the power consumption and area.

The full differential signal range is reduced from ±2Vref to ±Vref after range-scaling of MDAC1 to conserve power.

3.1 Range Scaling of MDAC1

The first 3.5-bit MDAC is shown in Fig.4, with a changed feedback capacitance C_F to implement the range scaling. Clk1d is a little delayed than clk1 to implement the bottom plates sampling. As is shown in Fig.4 (a), when clk1 is high and clk2 is low,

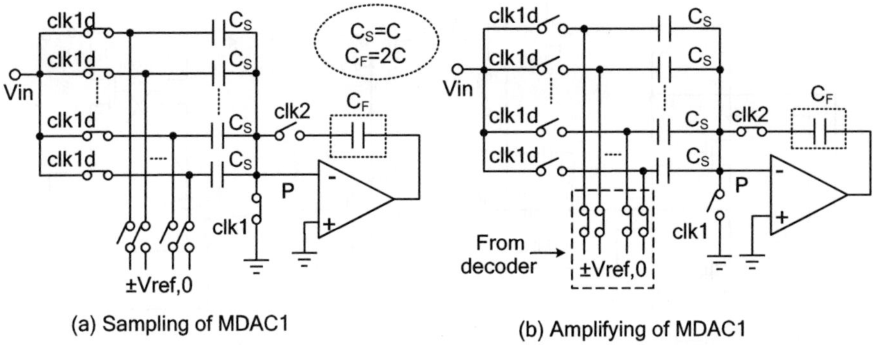

Fig. 4. Range-scaling of MDAC1

the circuit works in the sample phase to reserve the output of the S/H. The eight sample capacitors are charged until clk1 steps down. At this moment the charge reserved in the node P is as follows.

$$Q_S(P) = -8V_{IN}C_S \ . \tag{2}$$

MDAC1 works in the amplifying phase when clk1 is low and clk2 is high shown in Fig.4 (b). The other plate of the sample capacitors are connected to ±Vref or 0 according to the amplitude of V_{IN}. Assume the output of op-amp finally reaches a stabilized voltage V_{OUT}. Then the charge reserved in node P is as follows.

$$Q_A(P) = -V_{OUT}C_F - 8V_{DACi}C_S \ . \tag{3}$$

Where, V_{DACi} is the sub-DAC output to recovery V_{IN} in discrete form. The charge stored in node P remains the same because there's no path to dissipate it. So there is $Q_S(P)=Q_A(P)$. Given $C_S=C$ and $C_F=2C$, V_{OUT} can be determined by solving the above equations simultaneously.

$$V_{OUT} = 4(V_{IN} - V_{DACi}) \ . \tag{4}$$

Equation (4) shows the gain of MDAC1 is 4 rather than 8, and the output range is scaled from 2Vref to Vref by a factor of 2. The "scaling" of the signal range reduces the power consumption a lot.

3.2 Op-Amp

The schematic of the two stage op-amp is shown in Fig.5. Because the trans-conductance ratio of MN2 and MN4 decides phase margin of the op-amp, the preamplifier can be small and low power. The size of MN2 is about one-sixth of MN4's in this design. An advantage of two-stage structure is that high gain can be achieved simply by increasing gain of the preamplifier with a large output swing of the second stage. Lower input parasitic capacitance of the preamplifier helps to increase the feedback factor and reduce memory effect.

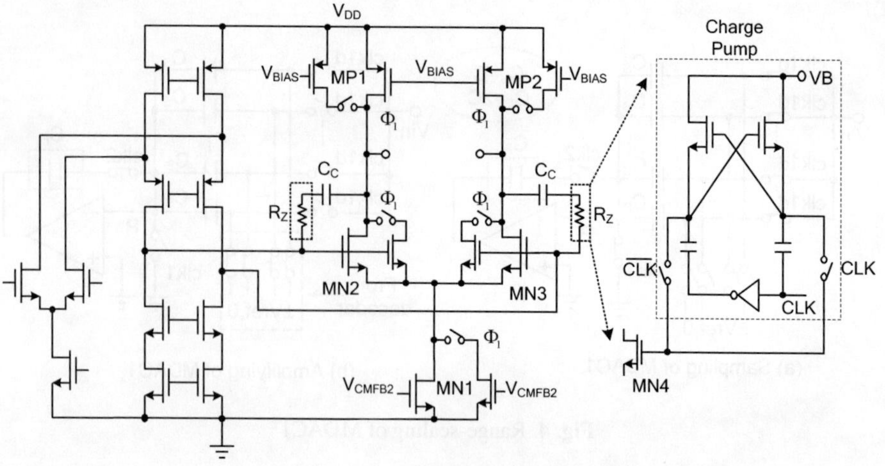

Fig. 5. Schematic of the two-stage op-amp

The main issue of the shared op-amp for the first and second stage is that it is placed in different feedback configurations and has different output loads in phase Φ_1 and Φ_2 as shown in Fig.1. If the op-amp is designed only for the S/H, the poor settling behavior of MDAC1 will degrade the ADC performance. Some fingers of MN1, MN2, MN3, MP1 and MP2 are switched off to conserve power when MDAC1 is in amplifying phase, as shown is Fig.5.

The Miller compensation with zero cancellation is utilized to address the stability issues as shown in Fig 5. When the equation (5) is established, the first non-dominant pole can be eliminated [8].

$$R_Z = \frac{C_L + C_C}{g_{m2} C_C} \tag{5}$$

The R_Z is usually realized by a transistor working in the deep linear region, other than a poly-crystal resistor to relieve the impact of the variation of process condition and temperature.

However in the actual environment, the op-amp common-mode voltage will change with the power supply voltage. Limited by low supply voltage, small changes of the Vgs(MN4) of will result in great changes in its on resistance. Enhancing the Vgs can effectively reduce the change in resistance value of the transistor caused by common mode voltage. To break the supply voltage limit, a charge pump structure [9] [10] to increase the Vgs is introduced as shown in Fig. 5.

$$R_{on}(MN4) = \left[\mu_p C_{OX} \left(\frac{W}{L} \right)_{MN4} \left(V_{GS1} - V_{th1} \right)_{MN4} \right]^{-1},$$
$$\left(\frac{W}{L} \right)_{MN4} = \frac{C_C}{C_C + C_L} \left(\frac{W}{L} \right)_{MN2} \frac{\left(V_{GS} - V_{th} \right)_{MN2}}{\left(V_{GS} - V_{th} \right)_{MN4}} \tag{6}$$

From equation (5) and (6), we can accurate control the resistor of the transistor by increasing the Vgs(MN4).

4 Simulation Result

The pipelined ADC has been simulated in a 0.18-μm 1-poly 6-metal CMOS technology. The power supply voltage of 1.8V is supplied to both analog and digital parts respectively. Simulation results show that the spurious free dynamic range (SFDR) is 81dB when the input signal frequency is 49.7MHz at 100MSPS, as shown in Fig.6 (a). Fig.6 (b) shows the DNL and INL of the ADC are both within ±0.5LSB respectively. The ADC consumes 46.5 mW without the reference buffers.

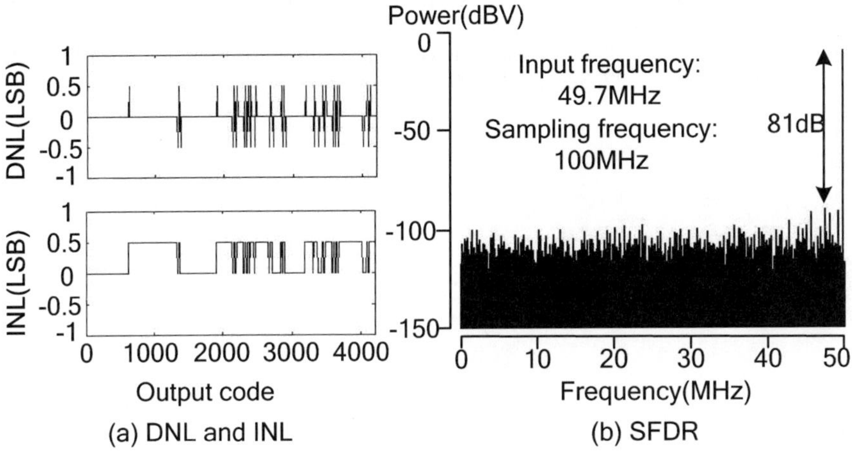

Fig. 6. FFT spectrum and DNL, INL

5 Conclusion

Several power reduction techniques for pipelined ADC are described. The first is the op-amp sharing between the S/H and the first MDAC. The second is current switch of the op-amp of the S/H (the first MDAC). And the third is application of range scaling technology which decreases the power consumption further. Simulation results verify the power efficiency of these techniques.

References

1. Lewis, S., Gray, P.R.: A pipelined 5-Msample/s 9-bit analog-to-digital converter. IEEE J. Solid-State Circuits SC-22, 954–961 (1987)
2. Yu, P.C., Lee, H.S.: 2.5-V, 12-b, 5-Msamples/s pipelined CMOS ADC. IEEE J. Solid-State Circuits 31, 1854–1861 (1996)

3. Kurose, D., Ito, T., Ueno, T., Yamaji, T., Itakura, T.: 55-mW 200-MSPS 10-bit Pipeline ADCs for Wireless Receivers. IEEE J. Solid-State Circuits 41, 1589–1595 (2006)
4. Lee, B.G., Min, B.M., Manganaro, G., Valvano, J.W.: A 14-b 100-MS/s pipelined ADC with a merged SHA and first MDAC. IEEE J. Solid-State Circuits 43, 2613–2619 (2008)
5. Limotyrakis, S., Kulchycki, S.D., Su, D.K., Wooley, B.A.: A 150-MS/s 8-b 71-mW CMOS time-interleaved ADC. IEEE J. Solid-State Circuits 40, 1057–1067 (2005)
6. Van de Vel, H., Buter, B.A.J., van der Ploeg, H., Vertregt, M., Geelen, G.J.G.M., Paulus, E.J.F.: A 1.2-V 250-mW 14-b 100-MS/s Digitally Calibrated Pipeline ADC in 90-nm CMOS. IEEE J. Solid-State Circuits 44, 1047–1056 (2009)
7. Lewis, S.H.: Optimizing the stage resolution in pipelined, multistage, analog-to-digital converter for video-rate applications. IEEE Trans. Circuits Syst. II 39, 516–523 (1992)
8. Razavi, B.: Design of Analog CMOS Integrated Circuits. McGraw-Hill, Boston (2001)
9. Cho, T., Gray, P.: A 10 b, 20 Msample/s, 35 mW pipeline A/D converter. IEEE J. Solid-State Circuits 30, 166–172 (1995)
10. Abo, A.M., Gray, P.R.: A 1.5-V, 10-bit, 14.3-MS/s CMOS pipeline analog-to-digital converter. IEEE J. Solid-State Circuits 34, 599–606 (1999)
11. Yang, H.C., Allstot, D.J.: Considerations for fast settling operational amplifiers. IEEE Trans. Circuits Syst. 37, 326–334 (1990)

Analogue Implementation of Wavelet Transform Using Discrete Time Switched-Current Filters

Mu Li[1,2], Yigang He[1], and Ying Long[1]

[1] College of Electrical & Information Engineering
Hunan University, Changsha 410082, China
[2] College of Information and Electrical Engineering
Hunan University of Science and Technology, Xiangtan 410201, China
limuucn@yaoo.com.cn

Abstract. A novel scheme for the analogue implementation of wavelet transform (WT) using discrete time switched-current (SI) filters is presented. An excellent approximation of the mother wavelet is given by differential evolution (DE) algorithm. The WT circuits consist of the filters whose impulse response is the approximation of the required wavelet. The wavelet filters are designed by a parallel structure with SI integrators as main building blocks. Simulations demonstrate the feasibility of the proposed scheme.

Keywords: wavelet transform, discrete time system, switched-current filters, differential evolution algorithm.

1 Introduction

The wavelet transform is usually implemented by software or digital hardware. Unfortunately, these implementation approaches are not favorable because of the high power consumption and non-real-time processing associated with the required A/D converter. Hence, the analogue implementation of the WT has become an attractive area [1]. Recently, there have been significant advances for implementing the WT in analogue way and its practical applications have been reported [1]-[4]. The low power consumption and small chip area design for analogue WT have been discussed from both the circuit architecture and the wavelet approximation. Among them, reference [1] has applied the switched-capacitor (SC) circuits to implement the WT. However, the SC circuits are not fully compatible with current trends in digital CMOS process, because they require integrated linear floating capacitors (usually double poly) and their performance suffers as supply voltages are scaled down. The work in [2] has proposed a design approach for the WT implementation using current-mode log-domain circuits. In this approach, Padé approximation is used to calculate the transfer function of the wavelet filters. However, the stable transfer function of a wavelet filter does not automatically result from this technique and the wavelet approximation can not be obtained directly in the time domain. Subsequently, L_2 approximation [3] for wavelet function is used. But, because of the existence of local optima in this approximation, the performances greatly depend on the selection of the approximation

M. Zhu (Ed.): Electrical Engineering and Control, LNEE 98, pp. 677–682.
springerlink.com © Springer-Verlag Berlin Heidelberg 2011

starting point. References [3], [4] have presented the analogue WT implementation using Gm-C filters. A major drawback of the designs in [3], [4] is that it is very difficult to implement transconductors with rail-to-rail input capability and thus with maximum dynamic range.

In this paper, we propose a novel scheme for the analogue WT implementation using SI filters. Based on the approximation theory of network function, the model of time-domain wavelet approximation is structure. The global optimum parameters of the model are solved by DE algorithm, and then the wavelet approximation function and the transfer function of wavelet filter are obtained. The circuits of implementing wavelet transform are composed of analog filter bank whose impulse response is the approximation wavelet function and its extensions. Finally, the filters design is realized by a parallel structure with SI integrators as main building blocks. The simulation results show that the proposed method is effectively.

2 Wavelet Function Approximation

The WT is a linear operation that decomposes a signal into components that appear at different scales. The WT of signal $x(t)$ at a scale a and position τ is defined as

$$W(\tau,a)=\frac{1}{\sqrt{a}}\int_{-\infty}^{\infty} x(t)\psi^{*}(\frac{t-\tau}{a})dt \tag{1}$$

where $\psi(t)(\psi(t)\in L^2)$ is the mother wavelet and $*$ denotes the complex conjugation. The main characteristic of the mother wavelet is given by $\int_{-\infty}^{\infty}\psi(t)dt=0$. This means that the mother wavelet is oscillatory and has zero mean value. It is well known that WT usually cannot be implemented exactly in analogue electronic circuits. But we well understand and relatively easy to implement in analog circuits, a possible solution presents itself. The output of a linear filter of finite order with an input signal $x(t)$ is the convolution of that signal with the impulse response $h(t)$ of the linear system, namely $y(t)=\int_{-\infty}^{+\infty} x(t)h(t-\tau)dt$. So when the impulse response $h(t)$ of the system satisfies $h(t)=(1/\sqrt{a})\psi(t/a)$ the analog wavelet transform $W(\tau,a)$ of $x(t)$ is realized.

As an example, a Marr WT system is considered. The Marr mother wavelet is defined as $\psi(t)=(1-t^2)e^{-t^2/2}$. For obvious physical reasons only the hardware implementation of causal stable filters is feasible. However, the system associated with this mother wavelet $\psi(t)$ is non-causal, so the Marr wavelet $\psi(t)$ must be time-shifted to facilitate an accurate approximation of its WT in the time domain. The quality of the analogue implementation of the Marr WT depends mainly on the accuracy of the approximation to $\psi(t-t_0)$, where t_0 denotes the time-shift involved. The choice of the time-shift in the Marr wavelet is $t_0=4$. For the generic situation of stable systems with distinct poles, the impulse response function $h(t)$ of a seventh order filter may typically have the following form:

$$h(t) = k_1 e^{k_2 t} + 2|k_3| e^{k_4 t} \cos(k_5 t + k_6) + 2|k_7| e^{k_8 t} \cos(k_9 t + k_{10}) + 2|k_{11}| e^{k_{12} t} \cos(k_{13} t + k_{14}) \quad (2)$$

where the parameters k_2 , k_4 , k_8 and k_{12} must be strictly negative for reasons of stability. In order to obtain the optimal parameters of the approximation $h(t)$, the optimization model for approximating $h(t)$ in time domain is described as

$$\begin{cases} \min E(k) = \min \sum_{\lambda=0}^{N-1} [h(\lambda \Delta T) - \psi(\lambda \Delta T - 4)]^2 \\ s.t. \;\; k_i < 0, (i = 2,4,8,12) \end{cases} \quad (3)$$

where $E(k)$ is the sum of squares error of the discrete points, N is the sampling total points, ΔT is the sampling time interval. The DE algorithm [5] is applied to optimize the typical nonlinear optimization problem in (3). In the algorithm, we set population size $N_p = 10$, crossover probability constant CR=0.7, difference vector scale factor F=0.85, N=800, maximum evolution generation G_{max}=25000 and $\Delta T = 0.01$. According to the universal steps [5] of the DE algorithm, the problem in (3) can be solved and the seventh order transfer function of Marr wavelet filter (a=1) is obtained

$$H(s) = \frac{0.5583}{s+0.6139} + \frac{-0.7762s+0.7746}{s^2+0.4584s+4.7658} + \frac{0.0741s-1.4489}{s^2+0.3796s+1.5247} + \frac{0.1604s+0.4630}{s^2+0.6404s+9.9515} \quad (4)$$

Fig.1(a) shows the approximation function of Marr wavelet and Fig.1(b) shows the approximation error. In order to describe the approximation performance, we defined an error criterion based on the mean-square error (MSE) which is defined as $\text{MSE} = (1/8)\int_0^8 |h(t) - \psi(t-4)|^2 \, dt$. The MSE of the approximation reaches 1.2088×10^{-5} between the Marr wavelet and approximated function. By the theory of Laplace transforms, different scales transfer functions can be derived from (4). After obtaining the transfer function of wavelet filter, we will design the filter using SI circuits in the next section.

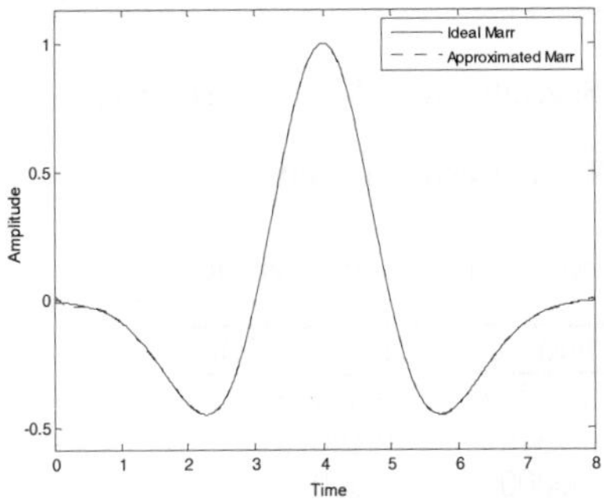

(a) Approximation function of Marr wavelet

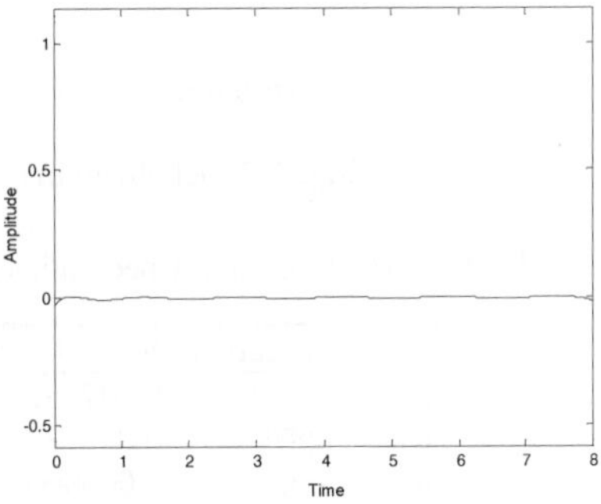

(b) Approximation error

Fig. 1. Marr wavelet approximation using DE algorithm

3 WT Circuit Implementation

The SI is a new analogue sampled-data signal processing technique aiming to solve the low voltage low power problems by operating in the current domain. Moreover, the SI can be implemented using a standard digital CMOS process and is well suited for the WT implementation since the dilation constant across different scales of the transform can be precisely controlled by both the transistor aspect rations and the clock frequency. The SI integrators [6] are used as the basic building blocks for the implementation of the wavelet filters described with the parallel structure. The structure shown as Fig.2(a) has been applied to the system implementation. It is composed of *n* identical seventh order SI filters. The SI filters with one first-order section and three biquads used is shown in Fig.2(b). According to the required response, we can calculate the transistor aspect rations to be used in the SI integrators using the functions in [6]. The transistor aspect ratios (α_1-α_6) of Marr wavelet filters (*a*=1) are given in Table 1. The unlisted transistors have unitary transconductances in the circuits. Adjusting the clock frequency of the SI circuits in Fig.2(b), we can gain the different scales Marr wavelet functions for implementing WT.

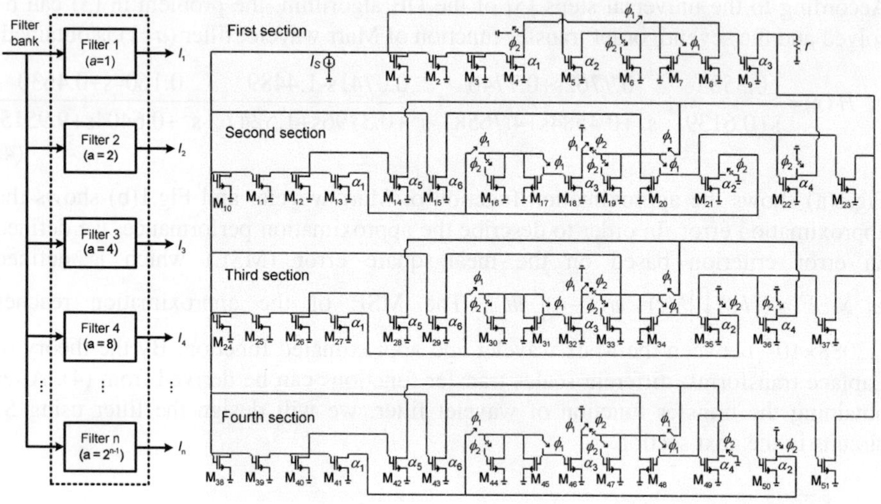

(a) Marr WT filters structure (b) Seventh order Marr wavelet filters

Fig. 2. Block diagram of the WT filters and SI circuits.

Table 1. The transistor aspect ratios of seventh order Marr wavelet filter (*a*=1).

α_i	First section	Second section	Third section	Fourth section
α_1	0.057599	0.007832	0.014712	0.004663
α_2	0.028799	0.048188	0.015482	0.100231
α_3	0.063340	1.000000	1.000000	1.000000
α_4	–	0.046350	0.038545	0.064501
α_5	–	0.078483	0.007524	0.016155
α_6	–	0.041200	0.007440	0.006912

4 Simulation Results

To validate the performance of the wavelet system, the analogue circuit of the wavelet filters is simulated by the ASIZ simulator [7]. The dilation constant of the designed wavelet filters across different scales is controlled by the various clock frequencies. Setting I_s=1A, r=1Ω and clock frequency be 100kHz for the first scale, 50kHz for the second, 25kHz for the third and 12.5kHz for the fourth, respectively, the simulated impulse response of the SI wavelet filters with four scales (a=1,2,4,8) is shown in Fig. 3. Obviously, the impulse response waveforms of the wavelet filters approximate the Marr wavelet well. The response waveforms of the different scale filters achieve the peak value 25.83mA at 0.4ms, 0.8ms, 1.6ms and 3.2ms, respectively. The required gain or attenuation of the wavelet at different scales can be realized by selecting the aspect ratio of the output transistor (e.g., M_{51}).

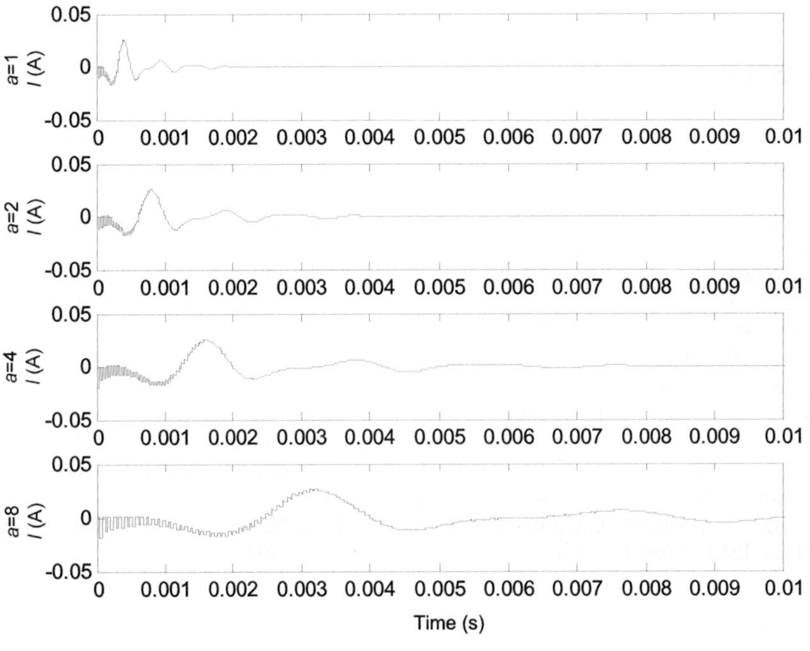

Fig. 3. Impulse responses of the Marr wavelet filters with four scales

5 Conclusion

A new scheme for implementing the WT in an analogue way by means of SI filters has been presented. To achieve an excellent approximation of the mother wavelet, the approximation approach based on DE algorithm is used. The WT filters design is realized by the parallel structure with SI integrators as main building blocks. The SI analogue WT provided is suitable for low-voltage low-power, wide dynamic range application and is compatible with the digital VLSI technology. The results of Marr WT implementation suggest that the shown approach can also be applied to approximate other wavelet bases.

Acknowledgments

This work was funded by the National Natural Science Funds of China for Distinguished Young Scholar under Grant No.50925727, the National Natural Science Foundation of China under Grant No.60876022, the Hunan Provincial Science and Technology Foundation of China under Grant No.2010J4, the Cooperation Project in Industry, Education and Research of Guangdong Province and Ministry of Education of China under Grant No.2009B090300196, and the Project Supported by Scientific Research Fund of Hunan Provincial Education Department No.10C0672.

References

1. Lin, J., Ki, W.H., Edwards, T., Shamma, S.: Analogue VLSI implementations of auditory wavelet transforms using switched-capacitor circuits. IEEE Trans. Circuits and Systems 41(9), 572–583 (1994)
2. Haddad, S.A.P., Bagga, S., Serdijn, W.A.: Log-domain wavelet bases. IEEE Trans. Circuits Syst. 52(10), 2023–2032 (2005)
3. Agostinho, P.R., Haddad, S.A.P., De Lima, J.A., Serdijn, W.A.: An ultra low power CMOS pA/V transconductor and its application to wavelet filters. Analogue integrated Circuits and Signal Processing 57(1-2), 19–27 (2008)
4. Gurrda-Nawarro, M.A., Espinosa-Flores-Verdad, G.: Analogue wavelet transform with single biquad stage per scale. Electronics Letters 46(9), 616–618 (2010)
5. Storn, R., Price, K.: Differential evolution—a simple and efficient heuristic for global optimization over continuous spaces. Journal of Global Optimization 11(4), 341–359 (1997)
6. Hughes, J.B., Bird, N.C., Pattullo, D.M.: Switched-current filters. IEE Proceedings G, Circuits, Devices and Systems 137(2), 156–162 (1990)
7. De Queiroz, A.C.M., Pinheriro, P.R.M., Caloba, L.P.: Nodal analysis of switched-current filters. IEEE Trans. Circuits and Systems 40(1), 10–18 (1993)

A 2D Barcode Recognition System Based on Image Processing

Changnian Zhang, Ling Ma, and Dong Mao

College of Information Engineering, North China University of Technology,
Beijing 100144, China
ml-28wd-23@163.com

Abstract. 2D barcode identification is a hot topic in the field of image processing. This paper shows the analysis of 2D barcode structure and the development at home and abroad. A 2D barcode recognition system based on image processing was proposed. The segmentation and denoising algorithm were improved. The identification and decoding of 2D barcode was realized in VC + + programming environment. The recognition efficiency of the barcode was successfully increased. It pushes the promotion and development of 2D barcode.

Keywords: QR Code; 2D barcode; Image processing.

1 Introduction

2D barcode has received a wide range of attention since the day it appeared for its advantages of large storage capacity, high secrecy, strong resistance to damage and low cost and other features [1]. It has become an important research topic in the field of image processing and pattern recognition.

The overseas research on two-dimensional barcode technology began in the late 1980s and they have developed a variety of code systems such as PDF417, QR Code, DataMatrix, Code 16K and so on. The research in home began in 1993. ANCC (Article Numbering Center of China) translated and studied some commonly used 2D barcode technical specifications. On the basis of digesting some foreign relevant materials, China has made two 2D barcode national standards, GB/T 17172-1997《 FPD417》 and GB/T 18284-2000《 Quick Response Code》.

QR Code (Quick Response Code) released by Japanese Denso Company, had become ISO international standard in 2000 and then recognized as the Chinese national standard GB / T 18284-2000 in 2001 [2]. In addition to the features such as large information capacity, high reliability, the ability of expressing characters and image information and strong security advantages, it also has the following main features, high-speed reading, error correction capability and effectively expressing Chinese and Japan characters.

A QR Code symbol is made up with square modules which are formed into a square array. The symbol consists of coding region and function graphic which includes searching pattern, separator, positioning pattern and correcting pattern. The function graphic can't be used for data encoding. The symbol is surrounded by the white blank area. The sample and the structure are shown in Fig. 1(a) and Fig. 1(b).

M. Zhu (Ed.): Electrical Engineering and Control, LNEE 98, pp. 683–688.
springerlink.com © Springer-Verlag Berlin Heidelberg 2011

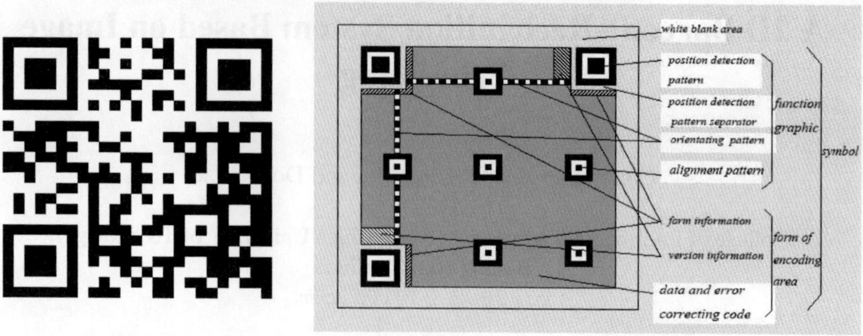

(a) QR Code sample (b) QR Code structure

Fig. 1. QR Code sample and structure

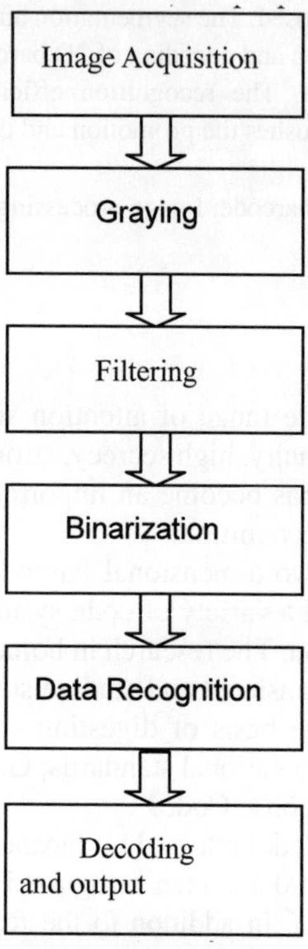

Fig. 2. The system chart of the recognition system

2 Barcode Recognition System Designing

According to the structure of 2D barcode, we propose the system chart. It is shown in Fig. 2.

2.1 Graying

In the experiment, we get a 24-bit color image of the barcode from an ordinary USB camera. We have to convert the color image into gray-scale before we deal with the image because the color image contains a lot of color information. We can improve the speed and efficiency of software in this way. Each pixel of the color image is composed by the RGB components. We have several methods to gray the image, such as maximum method, average method and weighted average method which conversion formula are as follows:

$$Y = \max(R, G, B)$$

$$Y = (R + G + B)/3$$

$$Y = 0.11R + 0.59G + 0.3B$$

There are no significant differences among these three methods. We can choose one of them according the experiment conditions. In our experiment, we used the third one.

2.2 Adaptive Filtering

In the process of image acquisition, the picture we got from the camera contains a lot of noise because of the influence of the acquisition equipment and other factors. Then the image looks fuzzy and spotted. So we need to filter the image before the next step.

In recent years, the anisotropic diffusion equation-based image processing has been attaining more and more attentions. It has been widely applied in the fields of image segmentation, denoising, image matching and feature extraction. As early as 1990s, Perona and Malik [3] had successfully applied anisotropic diffusion equation in image denoising.

The P-M model proposed by Perona and Malik is as follows:

$$\begin{cases} \dfrac{\partial I}{\partial t} = div(c(|\nabla I|))\nabla I \\ I(x, y, 0) = I_0(x, y) \end{cases} \tag{1}$$

Where I is evolution image, and div is divergence operator, and ∇I is gradient, and $c(|\nabla I|)$ is the diffusion coefficient of the anisotropic diffusion equation. $c(\cdot)$ is defined as: $c(s) = e^{-(s/k)^2}$ and $c(s) = \dfrac{1}{1+(s/k)^2}$.

Denoising with anisotropic diffusion equation is an iterative process. The iteration number will affect the denoising result. We combined the structural similarity algorithm and anisotropic diffusion model. We use structure similarity model to assess image quality and then control the number of iteration on the basis of anisotropic diffusion equation. Adaptive filtering can be realized with this method. Literatures [4] and [5] have given an accurate discussion of structural similarity.

2.3 Image Binarization

In many binary algorithms, the Otsu [6] method can extract an ideal threshold in most situations. In some cases, maybe we can't get the best threshold value, but the quality is guaranteed. Some scholars believe that the Otsu is the best choice in adaptive binary algorithms. But in the specific test, we found that we would get some large black areas or even loose the whole image information when there are no obvious differences between the target and the background. To resolve this problem, we could enhance the Otsu through gray value stretching. The principle is to highlight the differences between the foreground and background by increasing the gray level.

In order to simplified calculation, we use the linear stretch to complete the gray stretch. It is shown in Fig. 3. An input image $f(x, y)$ could become output image $g(x, y)$ via mapping function T as follows:

$$g(x, y) = T[f(x, y)]. \tag{2}$$

Suppose the gray range of the input image is $[a, b]$ and output image $[c, d]$. The change can be realized by the following formula.

$$g(x, y) = [(d-c)/(b-a)] + c. \tag{3}$$

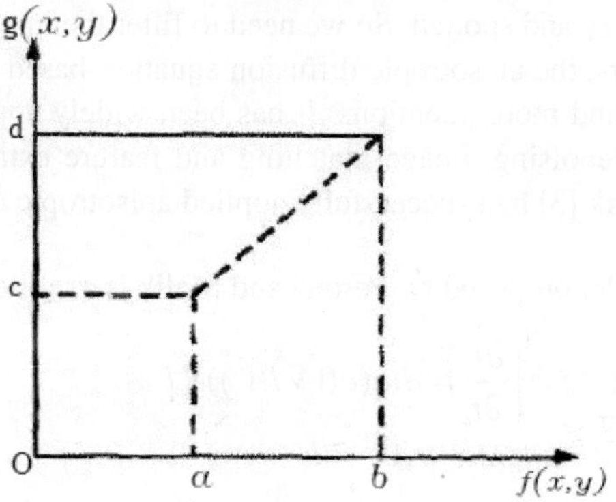

Fig. 3. The schematic diagram of gray stretching

Otsu method is as follows: suppose N is the number of pixels. The image has L gray levels. n_i is the number of the pixels at the gray level i. Thus $N = \sum_{i=1}^{L} n_i$. The probability of each gray level $p_i = \frac{n_i}{N}$. An image is divided into two parts according to the threshold t: C_0 and C_1. They are target and background. C_0 and C_1 correspond to the pixels with gray levels $\{1, 2, \cdots, t\}$ and $\{t+1, t+2, \cdots, L\}$.

The proportion of the target and the background pixels: ω_0, ω_1. Average gray values of the target and the background: μ_0, μ_1. Average gray value of the whole image: μ_T. Otsu can be expressed as:

$$\sigma_B^2 = \omega_0(\mu_0 - \mu_T)^2 + \omega_1(\mu_1 + \mu_T)^2 . \quad (4)$$

Where

$$\omega_0 = \sum_{i=1}^{t} p_i \text{, } \omega_1 = \sum_{i=t+1}^{L} p_i = 1 - \omega_0 \text{, } \mu_0 = \sum_{i=1}^{t} ip_i / \omega_0 \text{, } \mu_1 = \sum_{i=t+1}^{L} ip_i / \omega_1 \text{, } \mu_T = \sum_{i=1}^{L} ip_i .$$

Traversing t from smallest gray value to largest, when σ_B^2 gets its maximum, t is the best threshold.

3 Data Recognition

We need to locate the detection graphic of the barcode after preprocessing to get the version information and decode the barcode. A QR Code symbol contains three detection graphics which have same size and shape. Each of them is composed by three overlapping concentric squares. The module width proportion is 1:1:3:1:1. Since it is unlikely to appear the similar graphic in the symbol image, we can find the approximate regions which meet to the proportion to locate the detection graphic. The barcode is divided into grids on the basis of the detection graphic. Every grid is a module. White module represents 1 while black one represents 0. Now we get the data flow of the QR Code by which we can obtain the code information according the encoding rules. The window of the recognition system is shown in Fig. 4.

Fig. 4. The window of the recognition system

4 Conclusion

There are plenty of image processing methods for 2D barcode. However, the process of designing algorithms is extremely complex in full accordance with the existing algorithms. This article has studied recognition process for the QR code, including graying, binarization, filter, barcode localization and decoding. The denoising and segmentation algorithm are improved for reading QR code accurately and quickly, meanwhile, the system is achieved in VC + +6.0 programming environment to meet real time requirements. It is helpful for the development of embedded system. The application of 2D barcode in mobile phone platform will be a research focus in future with the development of smart phone and the increasing processing capacity.

References

1. Parikh, D., Jancke, G.: Localization and Segmentation of A 2D High Capacity Color Barcode. Applications of Computer Vision, pp. 1–6 (2008)
2. Zhou, J., Liu, Y., Li, P.: Research on Binarization of QR Code Image. In: 2010 International Conference on Multimedia Technology, pp. 1–4 (2010)
3. Perona, P., Malik, J.: Scale Space and Edge Detection Using Anisotropic Diffusion. IEEE Transactions on Pattern Analysis and Machine Intelligence 12, 629–639 (1990)
4. Wang, Z., ConradBovik, A.: A universal Image Quality Index. IEEE Signal Processing Letters 9, 81–84 (2002)
5. Wang, Z., ConradBovik, A., RahimSheikh, H., et al.: Image Quality Assessment: From Error Visibility to Structural Similarity. IEEE Transactions on Image Processing 13, 600–612 (2004)
6. Ostu, N.A.: Threshold Selection Method from Gray-Level Histograms. IEEE Transactions on Systems, Man and Cybernetics 9, 62–66 (1979)
7. GB/T 1828-2000, Chinese National Standard-Quick Respond Code (QR Code), Standards Press of China (2001)

The Research of CNC Communication Based on Industrial Ethernet

Jianqiao Xiong[1], Xiaosong Xiong[2], Xue Li[1], and Bin Yu[1]

[1] School of Mechanical Engineering, Nanjing Institute of Technology, Nanjing Jiangsu 211167, China
[2] City College of Wuhan University of Science and Technology, Wuhan Hubei 430083, China
jqxqj@163.com

Abstract. Firstly, analyzing several interfaces of the current CNC machine tools, then proposing connecting them with Industrial Ethernet and a kind of topologic structure based on Industrial Ethernet.Concretely illuminating the frame structure of digital interface and communication mode of EtherCAT,and analyzing and evaluating the functions and characteristics of the Industrial Ethernet.Finally, a method for designing based on hardware and software of standard CNC platform for Industrial Ethernet.The cnc system proved to be advanced in the aspect of real-time property,reliability and synchronism.

Keywords: Numerical control System, Digital interface, Industrial Ethernet, EtherCAT.

1 Introduction

CNC machining pursues high accuracy and speed.The present cnc equipments have kinds of interfaces,such as,1.Driver interfaces:digital, analog or pulse-type interface and SERCOS field bus interfaces which are used to connect kinds of servo drive system.2.I/O interfaces:including digital I/O interface,analog I/O interface, pulse-type I/O interface and field bus I/O interface which are used to connect the parts of digital input/output,spindle unit and manual pulse generator of machines.3.Network Interfaces:mainly aim at connecting CAD/CAM, FMS/CIMS and Enterprise Intranet.The interface methods of the traditional servo and digital system is that position pulse and analog voltage signal are signaling through the cable,which has shortcomings of various connections,system signal instability and poor immunity,now replaced by the Industrial Ethernet field bus[1].

As to real-time property, current engineering applications have three cases:the first one is the occasion of information integration and low requirements,of which the real-time response time is 100ms or longer.The second one,in most cases,requires the real-time response time of at least 5~10ms.The third one is the occasion of high-performance synchronous control movement,especially the occasion of the servo motion control in the 100 node,of which real-time response time requires less than 1ms and synchronous transmission and vibration time requires less than 1μs[2].For the moment numerical control system at home and abroad are configured

M. Zhu (Ed.): Electrical Engineering and Control, LNEE 98, pp. 689–694.
springerlink.com © Springer-Verlag Berlin Heidelberg 2011

Ethernet.Industrial Ethernet with real-time control of multi-axis CNC machine tools has become an inevitable trend.Industrial Ethernet-based CNC structure has reliability, security, and adaptability, but also has high real time.

2 Communication Topology of CNC System

We design the home-made PC-based CNC system (such as the Huazhong University of Science and Technology's 'Century Star') as a standard platform.The standard platform consists of bus terminal module,communications terminal module,central industrial PC and one or more of the control panel with touch screen.They are connected via Industrial Ethernet.

Real-time Industrial Bus connects the drives and all of the peripheral equipments to the central industrial PC,and Fieldbus and industrial Ethernet technology, and even wireless technology are integrated into the servo drives.Of course, peripherals at first should be standardized as much as possible.The system structure is shown in Figure 1.

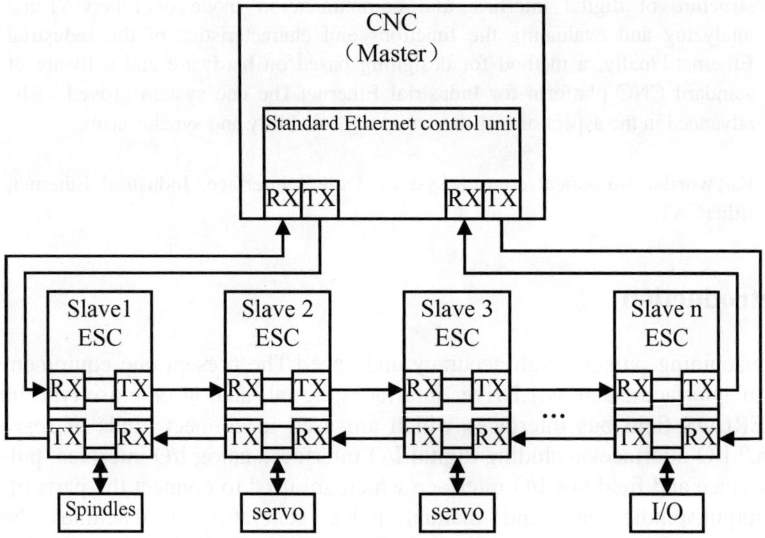

Fig. 1. System structure principle drawing

This design above requires the industrial bus to be able to support any type of topology, including line type,tree type,star type and so on.Also it includes bus structure or linear structure named after the field bus,and And is not limited to the number of cascade switch or hub[3].

3 Real-Time Industrial Ethernet Protocols

As to the requirements of industrial ethernet based open cnc system,the 4th edition of IEC61158 which was drafted in Dec.2007 comprises of 10 industrial ethernet protocol standard: Modbus/TCP, Ethernet/IP, Ethernet Powerlink, EPA, PROFINet RT, EtherCAT, SERCOS-III and PROFINet IRT.Compared with the traditional field bus,it has advantages of fast transmission speed, large amount of data inclusion,long transmission distance.And it uses general ethernet components and has high cost performance.Also it can be plugged into a standard ethernet network end[4].

Concerning these industrial ethernet protocol standard,we can choose EtherCAT,which is completely compatible with the existing broad common SERCOS and CANopen of CNC machine tools and can make the most of the SERCOS and CANopen components able to reuse.

EtherCAT also has good open characteristics and can accommodate other Ethernet-based network services and protocols in the same physical layer.Usually the performance loss can be minimized.It can make different devices such as PLC, hydraulic and drive tasks be implemented on a single platform.

EtherCAT can be delivered directly in the Ethernet frame.The EtherCAT frame structure which meets the IEEE 802.3 [3] is shown in the Figure 2.EtherCAT frame may include a few packets,with each packet corresponding to a specific logical process image memory of which the byte memory is up to 4GB.Data can be freely addressed, from the stations broadcast, multicast and communication can be achieved.The control of any Ethernet protocol stack can be addressed to the EtherCAT system and can also be received through other subnets across routers.

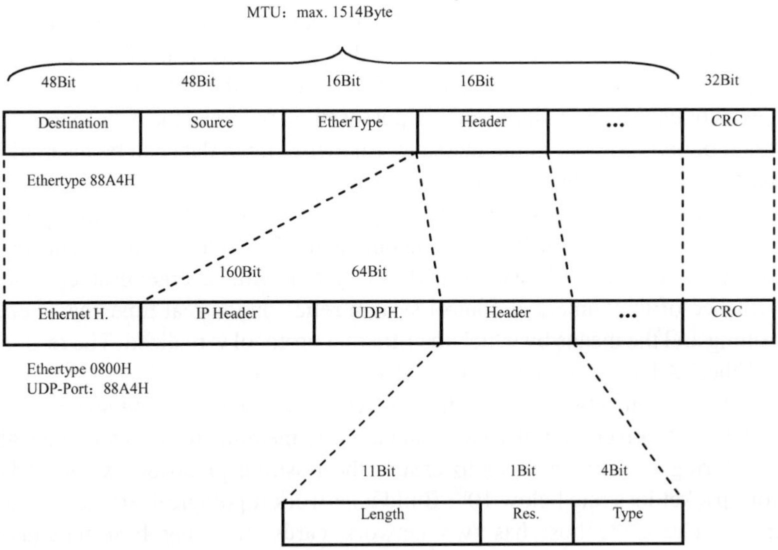

Fig. 2. EtherCAT: Standard Frames according to IEEE 802.3 [3]

4 Performance Evaluation

A numerical control system based on industrial Ethernet puts forward higher demands on network performance compared with normal automatic control applications[5]. They are the several aspects as follows:

(1) Demands on network synchronism. The demands of a numerical control system on synchronism are relatively high. A numerical control system has many CPUs and clock references, while the control instructions and every servo control arithmetic are also discrete. In order to ensure that each shaft in the system executes the instructions in a coordinate and consistent way, we must make sure that the instructions are executed at the same time. EtherCAT uses the distributed clock in order that all devices use the same system time, thereby controls the synchronous execution of tasks of every device. For example, when several servo shafts are executing interpolation tasks, each servo shaft, which is synchronous to one single reference clock, can generate synchronous signals according to synchronous system time. Otherwise, if we use complete synchronism and when there is a telecommunications error, the quality of the synchronous data will be greatly affected. Distributed adjusted clock possesses fault tolerance towards error delays to some extent. Fast Ethernet physical loop structure, where every clock can simply and accurately confirm the real time offset of another, that is to say, it can provide a signal vibration shorter than 1 microsecond in the network extension, is used in data exchange.

(2) Demands on network timeliness. The working beat of a numerical control system depends on the main controller. If the network timeliness is not nice, the period of the working beat will not be smooth and steady, which will lead to fluctuation of the sampling period of servo drivers. Whereas, periodical sampling is the basis of the control arithmetic and that will result in sequential disorder of the servo driving system. EtherCAT is unlike other Ethernets which receive Ethernet data packets at every linking point and then decode and replicate them as process data. When the message is in the process of continual transmission of the devices, the FMMU (Fieldbus memory management unit) in every input and output terminal will read the data appointed to that input and output terminal in the message. Likewise, input data can be inserted into the message as it comes by. The message has a delay of only several nanoseconds. EtherCAT UDP can pack the EtherCAT protocol into the UDP/IP message and the UDP data message will finish unpacking only at the first station. That characteristic can be used in any control addressing EtherCAT system with Ethernet protocol stacks. The performance of the numerical control system relies on the real time characteristics of controlling and the means by which the Ethernet protocol is realized. The response time of the EtherCAT network itself is basically not limited.

(3) Demands on network reliability. Packet loss of the telecommunications data will make the servo drivers out of control and enlarge manufacturing errors, thus affecting manufacturing quality. In order to ensure the position precision, we must keep the network packet loss rate below 10^{-9}. Build a network topological structure as shown in Figure 1. The CNC host has two network cards, realizing host redundancy via Hot-Stand-By. If one of the slave stations breaks down, EtherCAT can still go on working. When the system is tested in a laboratory environment, there are not packet loss phenomena and reliability of data transmission of the system is relatively nice.

We can know from the analysis above that EtherCAT can effectively ensure position and velocity control of a numerical control machine tool and it can even process current (torque) control of distributed drivers.

5 Realization of the Hardware and Software

EtherCAT is quite fit for telecommunications among controllers (host/slave).Freely addressed network variables can be used for process data, parameters, diagnosis, programming and all kinds of remote control service, which will satisfy extensive application demands. Data telecommunications interfaces of host/slave and host/host are the same.

EtherCAT support almost any topological type. It uses only standard Ethernet frame without any compaction, can be sent via any Ethernet MAC and can use standard tools.

Control period is sent from the host, which then sends downstream telegraph, whose maximal valid data length can reach 1498 bytes. Data frame traverses all slave station devices and each device analyses the message that is addressing it as the data frame comes by. It reads or writes data to appointed positions in the message according to the commands in the message heading and slave station hardware adds the work counter of the message by 1, signifying that the data has been processed. The whole process has a delay of only several nanoseconds. After the data frame visits the slave station which is located the last in terms of logical position of the whole system, that slave station sends the processed data frame as upstream telegraph directly to the host. The host will process the returned data after it receives the upstream telegraph and a cycle of telecommunications is over.

The realization of numerical control system based on industrial Ethernet includes the realization of the host and that of the slave stations. The host does not need a specified telecommunications processor while it only requires that there is a passive NIC (Network Interface Card) in the CNC or there is a Ethernet MAC device integrated on the mainboard. The identification and encapsulation of the protocol is realized by means of software in the CNC host. Via the EtherCAT industrial Ethernet,CNC (the host) links several slave station motion controller units, which are composed of a slave station control baseboard, telecommunications cards, input and output modules and motion control cards, each of which controls one servo shaft.

EtherCAT slave station is realized via specified hardware. You can either purchase slave station control chips named ET1100 made by the Beckhoff Company or get authorized binary codes via one-off purchase, realizing the functions of a slave station controller via FPGA. The specified hardware which is used for the realization of a slave station has two MAC addresses, can extend two network interfaces and is very easy to realize cascade connections, which in turn constitute a variety of topological structures.

In terms of the software, users can use the object-oriented language, the C++language, to develop the software that is suitable for his or her own specific system. In order to shorten the developing period, they can also use general software, TwinCAT PLC and TwinCAT NC PTP, provided by the Beckhoff Company. The system operating interface based on TwinCAT is shown as in Figure 3. Versions after TwinCAT V2.10 all support EtherCAT in an all-round way, which provides a range of automatic systems that are comparatively prone to configure and diagnose for standardized application.

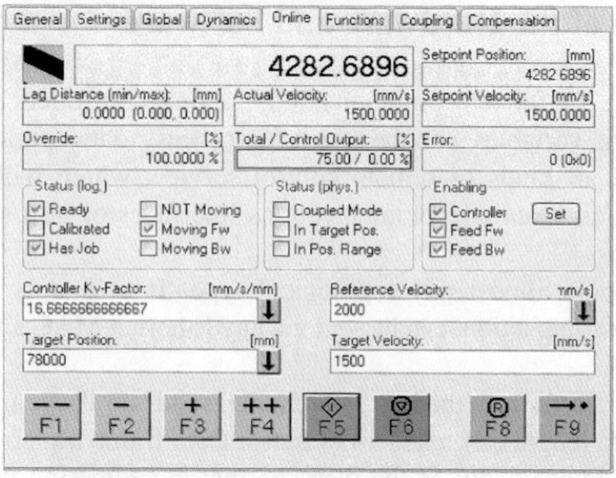

Fig. 3. A system operating interface based on TwinCAT

6 Conclusion

This kind of numerical control system telecommunications method based on industrial Ethernet has been experimented in reality. In the condition that there are ten nodes, telecommunications period of the system is shorter than 0.8 milliseconds and the vibration is shorter than 1 microsecond. That will satisfy the demands of high performance motion control on timeliness, reliability and synchronism of data transmission of numerical control system. Experimental result shows that this method, with its flexible and diversed topological structure, can link devices such as servos, input and output interfaces and so on conveniently, which is safe and reliable, easy to operate, widely open to users, liable to upgrade the software and extend, can configure the hardware at will according to users' demands and has a high cost performance and certain application and generalization value.

Acknowledgements

This work was financially supported by the Nanjing Institute of Technology Foundation (KXJ08133), Jiangsu colleges and universities achievements in scientific research industrialization projects (JHZD09-15).

References

1. Dong, X., Tianmiao, W., Jingmeng, L., Hongxing, W.: Industrial Ethernet-based Integrated Servo System Structure of the NC Machine. Machine Tool & Hydraulics 11, 1–4 (2008)
2. Ping, W., Haofei, X., Qiong, X.: Industrial Ethernet technology. Science Press, BeiJing (2007)
3. Frank, M.: Redundancy with standards in industrial Ethernet LANs. Engineer IT, 72-75 (2006)
4. Yanqiang, L., Ji, H.: Communication protocol for open CNC based on fieldbus. Manufacturing Automation 1, 50–54 (2006)
5. Xie, J., Zhou, Z., Chen, Y.: Time Performance Analysis of Communication in CNC System Based on Profibus. Computer Integrated Manufacturing Systems 9, 285–288 (2003)

An Efficient Speech Enhancement Algorithm Using Conjugate Symmetry of DFT

S.D. Apte[1] and Shridhar[2]

[1] Rajarshi Shahu College of Engineering/E & CE Dept, Pune, India
sdapte@rediffmail.com
[2] Basavashwar Engineering College/E & CE Dept, Bagalkot, India
shridhar.ece@gmail.com

Abstract. In this paper a simple and computationally efficient speech enhancement algorithm is presented. Most of the speech enhancement algorithms use the magnitude of STFT while keeping the phase as it is. In contrast, in this paper the magnitude of STFT of noisy speech is kept as it is while the phase is modified. Modified spectrum of speech is obtained by combining unchanged magnitude spectrum and modified phase spectrum. This modification results into cancellation of low energy components (noise) more than the high-energy (speech) components after speech reconstruction. This results in reduction of noise, which is essential in hearing aids as a front-end signal processing operation. Objective speech quality measures, Subjective listening tests and spectrogram analysis of speech samples show that the proposed method yields improved speech quality. In the proposed method speech enhancement is achieved through modification of phase of DFTs of speech frames through their imaginary parts. This is the driving force for the algorithm to be computationally efficient. This algorithm can be used as a front end signal processing block in hearing aids.

Keywords: Speech enhancement, DFT, Magnitude spectrum, Phase spectrum, Imaginary part.

1 Introduction

Speech enhancement is of paramount importance in hearing aids irrespective of nature of hearing problems[1,2,3]. Assuming the noise to be additive a noisy speech signal is given by

$$x(n)=s(n)+N(n) \tag{1}$$

Where x(n)=Noisy Speech, s(n)=Clean Speech, N(n)=Noise. Taking STFT of equation 1.1 we get

$$X(k)=\sum_{m=-\infty}^{\infty} x(m)w(n-m)e^{-i2\pi km/N} \tag{2}$$

Fourier transform of equation (1.1) gives X(k) = S(k)+N(k). From literature survey it is found that, some of the speech enhancement methods process the magnitude

M. Zhu (Ed.): Electrical Engineering and Control, LNEE 98, pp. 695–701.
springerlink.com © Springer-Verlag Berlin Heidelberg 2011

spectrum of noisy speech, whereas phase spectrum is kept as it is [4,5]. In the proposed work, the magnitude of noisy speech STFT is kept as it is, whereas phase is processed [6,7]. The unprocessed magnitude is combined with processed phase to get modified spectrum [6]. This results into cancellation of low energy (noise) components more than high energy (speech) components in the reconstructed signal [6]. This leads to enhancement of speech signal .Speech quality is tested by objective measures, subjective listening tests and spectrograms.

2 Proposed Method

The principle of the proposed method is "The degree of cancellation or reinforcement of imaginary parts of complex conjugates can be controlled by modifying their phase, in particular through their imaginary parts." The basic assumption of the proposed method is energy of speech components is more than that of noise components. The noisy speech signal x(n) is real; hence its DFT obeys conjugate symmetry[6,7]. The degree of cancellation or reinforcement of imaginary parts is controlled by modifying their phase in particular through their imaginary parts. This leads to computationally efficient algorithm. The block diagram shown in Fig.1 gives the implementation details of the proposed work.

2.1 Phase Modification through Imaginary Parts of DFTs

Assuming N to be even, an imaginary constant C(k) given by

$$C(k) = j\beta\ ;\ 0 \leq k < N/2 \tag{3}$$

$$C(k) = -j\beta\ ;\ N/2 \leq k \leq N-1 \tag{4}$$

is used for phase modification of DFT of speech. It is frequency dependent and antisymmetric about Fs/2. Here it is assumed to be a constant β (dependent on SNR of noisy speech and type of noise), independent of frequency. Using this constant the noisy speech signal of equation (1) is modified as follows.

$$X'(k) = X(k) + C(k) \tag{5}$$

The modified phase of X'(k) is computed and further combined with magnitude of original noisy speech signal to get modified complex spectrum given by

$$S(k) = |X(k)|\, e^{j\angle X'(k)}\ ;\ 0 \leq k \leq (N-1) \tag{6}$$

The degree of cancellation or summation of these complex conjugates can be controlled by modifying their phase. Here phase modification is enforced via imaginary parts of complex conjugates. This approach results in reduced computational complexity. The above explanation can be proved using signal – vector analogy. Considering a pair of complex conjugate numbers $C_1 = X + jY$ and $C_1^* = X - jY$ both having same magnitude

$$M=\sqrt{X^2+Y^2} \tag{7}$$

and Resultant

$$R=\sqrt{2(X^2+Y^2)} \tag{8}$$

and phase angles

$$\phi=\tan^{-1}(Y/X) \text{ and } \phi^*=\tan^{-1}(-Y/X) \tag{9}$$

The modified complex numbers are given by

$$C_{11}=X+jY+j\beta \text{ and } C_{11}^*=X-jY-j\beta. \tag{10}$$

On further simplification, The resultant of above two complex numbers is function of the factor β and is given by

$$R=\sqrt{2(X^2+Y^2)+2(X^2+Y^2)\cos\left(\tan^{-1}\left(\frac{Y+\beta}{X}\right)-\tan^{-1}\left(\frac{Y+\beta}{X}\right)\right)} \tag{11}$$

Case I. Resultant R, when $\beta << \sqrt{X^2+Y^2}$, equation (2.9) reduces to

$$R=\sqrt{2(X^2+Y^2)+2(X^2+Y^2)\cos\left(\tan^{-1}\left(\frac{Y}{X}\right)-\tan^{-1}\left(\frac{Y}{X}\right)\right)} \tag{12}$$

∴ The Resultant

$$R=2\sqrt{X^2+Y^2} \tag{13}$$

This is equal to the resultant of original complex conjugate numbers as shown in the equation (8). The implication of above result is, spectral components having magnitudes more (Speech components) than magnitude of β, the spectral components (speech components) remain unaltered.

Case II. Resultant R, When $\beta >> \sqrt{X^2+Y^2}$
Equation (11) reduces to

$$R=\sqrt{2(X^2+Y^2)+2(X^2+Y^2)\cos\left(\tan^{-1}\left(\frac{\beta}{X}\right)-\tan^{-1}\left(\frac{Y}{X}\right)\right)} \tag{14}$$

On simplification $R=\sqrt{2(X^2+Y^2)+2(X^2+Y^2)\cos\theta}$; and $\cos\theta << 1$

Hence the Resultant R reduces to

$$R << 2\sqrt{X^2+Y^2} \tag{15}$$

The implication of above result is spectral components having magnitudes much smaller (noise components) than magnitude of β, the spectral component gets suppressed more. This leads to enhancement of signal to noise ratio. Empirically determined values of β as a function of input speech SNR for white Gaussian, train and babble noise for which enhancement can be achieved are tabulated in Table. 1.

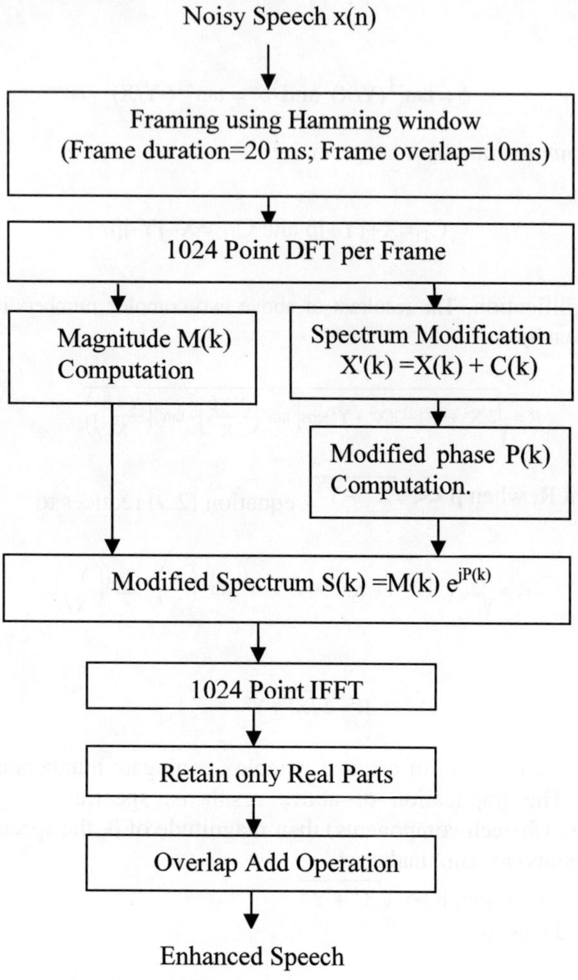

Fig. 1. Block diagram of the proposed speech enhancement scheme

3 Relation between α and β

Enhancement of speech signal can also be achieved by the same principle of phase modification through real parts of DFTs. Let α be the factor used to modify the real parts of complex conjugates similar to the factor β. The relation between α and β is important with respect to computational complexity of the algorithm, which is derived in the following section.

3.1 Lemma1

For any given pair of complex conjugate numbers

$X + jY$ and $X - jY$, If $\tan^{-1}\left|\frac{Y}{X+\alpha}\right| + \tan^{-1}\left|\frac{-Y}{X-\alpha}\right| = \tan^{-1}\left|\frac{Y+\beta}{X}\right| + \tan^{-1}\left|\frac{-Y-\beta}{X}\right|$

Then, $\beta << \alpha$. Where α and β are real.

Proof: Let $\theta = \tan^{-1}\left|\frac{Y}{X+\alpha}\right| + \tan^{-1}\left|\frac{-Y}{X-\alpha}\right| = k$

$$\therefore \alpha = \sqrt{X^2 - Y^2 - \frac{1}{k}2XY} \tag{16}$$

On the same lines

Let $\theta = \tan^{-1}\left|\frac{Y+\beta}{X}\right| + \tan^{-1}\left|\frac{-Y-\beta}{X}\right| = k$

$$\therefore \beta = \frac{-(X+kY) \pm X\sqrt{1+k^2}}{k} \tag{17}$$

From equations (16) and (17) it can be verified that $\beta << \alpha$ for any given values of X, Y and θ.

4 Experimental Details

4.1 Speech Quality Evaluation

In the experimental evaluation the NOIZEUS speech corpus is used..Mean PESQ scores over a subset of NOIZEUS speech data base are calculated and tabulated in Table.2. In addition 10 normal hearing subjects of age group 20-25 years and 10 subjects with moderate hearing loss of age group 50-55 years participated in listening tests. The improvement in Mean Opinion Score (MOS) is plotted in Figure.2 and 3 respectively. Spectrogram analysis is also carried out and is shown in figure.4.

Table 1. Empirically determined values of β as a function of input speech SNR for white Gaussian noise, Train noise and Babble noise.

	NOISE TYPE		
SNR (dB)	AWGN	TRAIN	BABBLE
0.0	0.0070	0.70	0.490
5.0	0.0090	1.00	0.500

Table 2. Mean PESQ scores for white Gaussian noise for spectral subtraction (SS), spectral contrast enhancement (SC), minimum mean square (MM) and proposed (PR) method.

INPUT SPEECH SNR(dB)	NOISY SPEECH	CLEAN SPEECH	METHODS USED			
			SS	SC	MM	PR
0	1.61	4.6	1.74	1.62	1.86	1.92
5	1.86	4.6	2.19	2.02	2.23	2.32

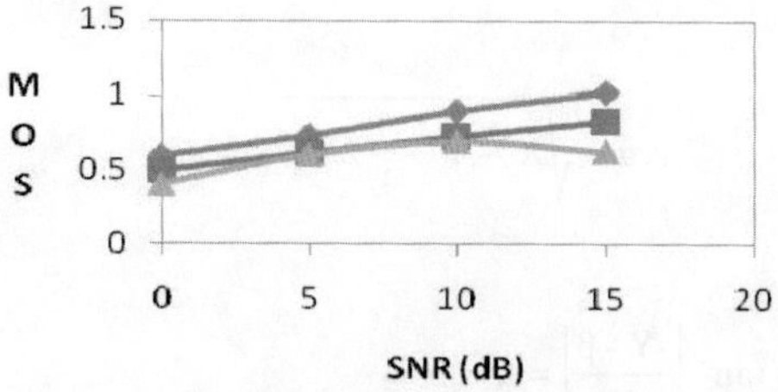

Fig. 2. Mean Opinion Score improvement as a function of input speech SNR for White Gaussian Noise (Top), Train Noise (Middle) and Babble Noise (Bottom) in case of listening tests on Normal Hearing subjects.

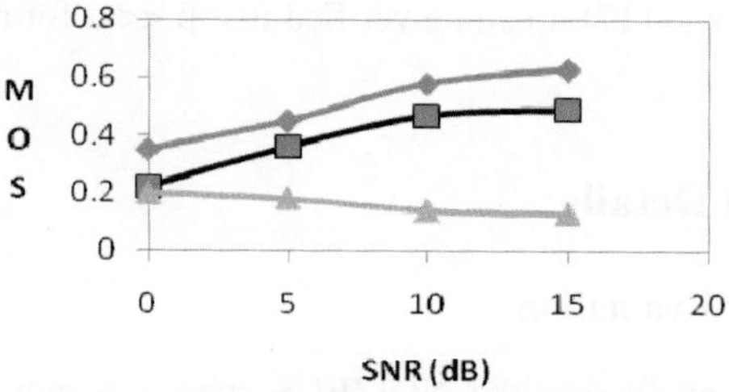

Fig. 3. Mean Opinion Score improvement as a function of input speech SNR for White Gaussian Noise (Top), Train Noise (Middle) and Babble Noise (Bottom) in case of listening tests on subjects with Moderate Hearing Loss.

5 Results

Improvement of Mean Opinion Scores in case of normal hearing subjects and subjects with moderate hearing loss indicate that, the proposed method performs best in case of additive white Gaussian noise. The results of spectrogram analysis shown in Fig. 5 indicate that the enhanced signal in case of babble noise though the noise is suppressed, a small amount of signal distortion is also introduced. The performance of the proposed method is comparable with three well known speech enhancement techniques based on spectral subtraction, Spectral contrast enhancement, and Minimum mean square error methods.

6 Conclusions and Future Scope of Work

In this paper a computationally efficient speech enhancement algorithm is presented. Noisy speech signal magnitude spectrum is combined with modified phase spectrum to produce modified complex spectrum. During signal synthesis low energy (noise) components cancel out more as compared to high energy (speech) components, thus resulting in signal enhancement. The proposed method is validated by objective measures, subjective listening tests and spectrogram analysis. This proposed work can find application in hearing aids as a front end algorithm to suppress the background noise.

(a)

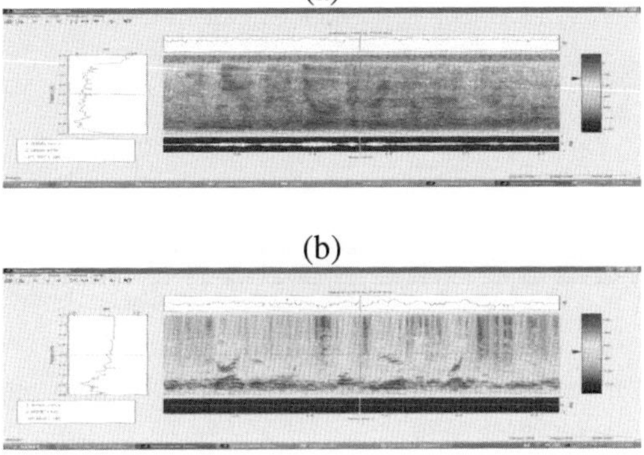

(b)

Fig. 4. Spectrograms of speech sample sp02.wav "He knew the skill of the great young actress." from NOIZEUS speech data base: (a) speech sample corrupted by babble noise (0 dB SNR); (b) corresponding enhanced speech sample

References

1. Baker, R.J., Rosen, S.: Auditory filters nonlinearity in mild/moderate hearing impairment. Journal of Acoustical Society of America 111, 1330–1339 (2002)
2. Rosen, S., Baker, R.J., Darling, A.M.: Auditory filter non linearity at 2kHz in normal listeners. Journal of Acoustical Society of America 103, 2539–2550
3. Moore, B.C.J.: Speech processing for the Hearing impaired: Successes, Failures and Implications for Speech mechanisms. Speech communication 41, 81–89 (2003)
4. Boll, S.: Suppression of acoustic noise in speech using spectral subtraction. IEEE Trans. Acoustics, speech signal processing 27, 113–120 (1979)
5. Ephraim, Y., Malah, D.: Speech enhancement using a minimum mean–square error log-spectral amplitude estimator. IEEE Trans. Acoustics, Speech Signal Processing 33, 443–445 (1985)
6. Kamil, W., Mitar, M., Anthony, S., James, L., Kuldip, P.: Exploiting conjugate symmetry of the short–time Fourier spectrum for speech enhancement. IEEE Signal processing letters 15, 461–464 (2008)
7. Paliwal, K., Basu, A.: A Speech enhancement method based on kalman filtering. In: IEEE Int. Conf. Acoustics, Speech, and Signal Processing (ICASSP 1987), pp. 297–300 (April 1997)

6 Conclusions and Future Scope of Work

In this paper a computationally efficient speech enhancement algorithm is presented. Noisy speech signal magnitude spectrum is combined with modified phase spectrum to produce modified complex spectrum. During signal synthesis low energy (noise) components cancel out more as compared to high energy (speech) components thus resulting in signal enhancement. The proposed method is validated by objective measures, subjective listening tests and spectrogram analysis. This proposed work can find application in hearing aids as a front end algorithm to suppress the background noise.

Fig. 4. Spectrograms of speech sample sp02.wav 'He knew the skill of the great young actress.' from NOIZEUS speech data base (a) speech sample corrupted by babble noise (0 dB SNR) (b) corresponding enhanced speech sample

References

1. Baker, R.J., Rosen, S.: Auditory filter nonlinearity in mild/moderate hearing impairment. Journal of Acoustical Society of America 111, 1330–1339 (2002)
2. Rosen, S., Baker, R.J., Darling, A.M.: Auditory filter non linearity at 2kHz in normal listeners. Journal of Acoustical Society of America 103, 2539–2550
3. Moore, B.C.J.: Speech processing for the Hearing impaired: Successes, Failures and Implications for Speech mechanisms. Speech communication 41, 81–89 (2003)
4. Boll, S.: Suppression of acoustic noise in speech using spectral subtraction. IEEE Trans. Acoustics, speech signal processing 27, 113–120 (1979)
5. Ephraim, Y., Malah, D.: Speech enhancement using a minimum mean square error log-spectral amplitude estimator. IEEE Trans. Acoustics, Speech Signal Processing 33, 443–445 (1985)
6. Kmal, W., Mitar, K., Anthony, S., James, L., Kuldip, P.: Exploiting conjugate symmetry of the short-time Fourier spectrum for speech enhancement. IEEE Signal processing letters 15, 461–464 (2008)
7. Paliwal, K., Basu, A.: A Speech enhancement method based on kalman filtering. In: IEEE Int. Conf. Acoustics, Speech, and Signal Processing (ICASSP 1987), pp. 297–300 (April 1987)

A Look-Ahead Road Grade Determination Method for HEVs

Behnam Ganji and Abbas Z. Kouzani

School of Engineering, Deakin University,
Geelong, Victoria 3217, Australia
{bganj,kouzani}@deakin.edu.au

Abstract. This paper presents a look-ahead road grade determination method for use in energy management of hybrid electric vehicles. Data that is gathered from a digital map and vehicle sensors is used to predict the future road grade and longitudinal forces. The predicted information is employed to specify the near future traction force demand. A simulation is carried out using data associated with a 50 km section of a real highway for a typical vehicle. The results are presented and discussed.

Keywords: HEV, energy management, road-grade prediction.

1 Introduction

One approach to increase the efficiency of power-train is to employ look-ahead control based on the knowledge of future condition of the road. Over the last several years, different systems have been tried in vehicles. For instance, modern vehicles are equipped with global positioning systems (GPS) and mobile phones. The information about the traffic and environment enables the use of predictive energy management controllers. Trajectory forecasting is a group of predictive energy management systems, whereby the decision to optimize the control is made by using a prediction of the road conditions. Such information can be obtained from the combination of road topography maps or GPS, and on-board sensors. This information can be used in a predictive energy management controller, or a rout planner. A predictive energy management controller adapts to current and future road conditions to obtain an optimal performance and reduce energy consumption.

Much of the research in vehicle energy management has focused on the use of standard drive cycles in order to recognize the demand power. However, the effect of the road topology and its influence on fuel efficiency has not been addressed well. Many of the existing methods suffer from the burden of calculations. Although the predicted information can be incorporated in the energy management system, the amount of computational time that would be required for the system to be effectively used may be impractical. A predictive controller requires a significant amount of computational speed in order to make decisions within an acceptable time frame. Even though the on-board computers are enhancing, they are not fast enough to address the stated issue. Therefore, this paper addresses real-time road-slope

M. Zhu (Ed.): Electrical Engineering and Control, LNEE 98, pp. 703–711.
springerlink.com © Springer-Verlag Berlin Heidelberg 2011

prediction for hybrid electric vehicles (HEVs). Then, the longitudinal forces can be calculated using the predicted information. The forecasted information can be implemented in control strategies to reduce fuel consumption.

2 Existing Road Forecasting Methods

Predictive control is an approach to reduce the energy requirement of a HEV. A predictive system requires the actual position of the vehicle, trajectory profile for the mission, relative speed and the distance to the vehicle ahead as well as general information about the traffic status. Such information is provided by telematic systems, including navigation services, traffic information, location-based services, emergency and safety services [1, 2].

Vehicle navigation systems can help alleviate traffic congestion and associated environmental pollution [3]. The traffic information and navigation services are generally provided by GPS and central telematic service providers (TSPs). The combination of this information can provide new possibilities in vehicle control strategies. For instance, this information can predict future road slopes which can be used in future longitudinal load prediction. These vehicles energy management approaches are classified as intelligent transportation [4]. Intelligent transportation and an optimal-trip-based control strategy are in their early stages of investigation.

Ichikawa et al. [5] proposed a system to predict the future driving pattern based on the past driving data. As the recognition of future driving pattern is difficult, they considered a vehicle that commutes to work. Since commuting driving is almost fixed, and run many times, by using the car navigation system, the driver's past driving is recorded in a database. Thereafter, the stored data in the database can be used to extract similar patterns of driving. Then the classified patterns can be used to predict different commuting routes. In order to reduce the number of patterns, the clustering method is used to classify the distance-based velocities.

Knowledge of the upcoming road topography gives a better prediction of the future load, and this can be utilised by look-ahead control to improve the fuel economy and/or safety. A method was developed for automatic deceleration in the entering of the road curves to prevent collision on tense curves [6]. The navigation system was applied to find the safe speed associated to a curve.

A conventional vehicle equipped with telematic facility was considered in Manzie et al investigation [7]. An algorithm used the preview information provided by the telematic to modify the vehicle's speed at each point of the drive-cycle. It was observed that by using in advance information about traffic, the fuel economy of the conventional vehicle could be comparable with the fuel economy of HEV.

An optimal power management method for a plug-in HEV based on dynamic programming (DP) was developed by Bin et al. [8, 9]. Different power cycles were calculated to present the input power demand to the optimization system. The state of charge (SOC) was obtained with linearization of non-linear dynamics of battery for different power split ratios (PSR). To analyse the sensitivity of fuel economy to the different segments and road grades, six sections with different length and three grades were evaluated. The results showed the fuel efficiency was less sensitive to the segment's length. However, these assumptions cannot be valid when there are changes in the road segments. Moreover, the best obtained calculation time was too large.

A challenge in look-ahead control strategies is the estimation of road information such as road slope (grade). A general approach for the measurement of the instantaneous road grade is using a sensor to directly estimate the grade [10]. In recent contributions, GPS receivers are used to achieve road grade estimates. Bae et al. [11] compares two method of grade estimation by using a GPS receiver with 3D velocity output. High precision GPS equipment was used in Han and Rizos work [12]. They used geodesy GPS receivers combined with stationary base stations for enhancement of accuracy. The height and the grade were defined as the states, and then a spatial Kalman filter was used to post-process the data. However, all these schemes rely on the existence of a high-quality GPS signal.

Another approach in the estimation of the geographical characteristics of a road is the use of vehicle sensors information in combination with a longitudinal road model of a vehicle [13]. A Kalman filter was used to obtain an accurate estimation of the vehicle mass and road slope. First, the estimation was done for the slope when the vehicle mass was known. Two different sensors were used to measure the speed and specific forces (retardation). A similar process was used by Vahidi et al. [14]. The estimated grade obtains by using a ''recursive least squares'' method.

Real time road mass estimation without GPS was proposed by Fathy et al. [15]. Mass can be applied for calculation of the road grade. The suggested algorithm was built based on the idea of "perturbation theory". The following equation is the perturbed of the longitudinal equation of the vehicle in the high-frequency state:

$$m\delta\dot{v}_x = \delta F_e - \delta F_b \tag{1}$$

Here m is the vehicle mass, δ denotes a small deviation in the given quantity, F_e is the longitudinal force acting on the vehicle and equals to the effective engine force at the wheels, and F_b is the effective braking force at the wheels. Then, the mass via this equation and measurement sensors in the vehicle can be estimated. However, the proposed method relies on the approximation of the longitudinal equation to the Eq. (1) that cannot be correct in all states. Therefore, this method cannot provide precise estimations that need for accurate control actions.

To improve the accuracy, the idea of automatic creation of road maps from GPS traces was developed [16, 17]. The proposed approaches in those studies tend to induce high-precision maps from traces of vehicles equipped with GPS receivers. The emphasis in these schemes is on the applied data mining methods. Both studies use 2D-maps without road grade information. In addition, they do not investigate the possibility of using a vehicle model and on-board sensors to enhance accuracy.

Attempts to automatically devise road geographical characteristics accurate enough for use in look-ahead energy management applications were made in some recent studies [18, 19]. To acquire the required data for forward-looking driving, a data logger was proposed by Carlsson and Reuss [18]. It showed how the information of the road characteristics of a driven route can be automatically produced and continually updated in a vehicle during each drive. The obtained information can then be utilised as foresight information in the predictive driving strategies. The proposed model is illustrated in Fig. 1.

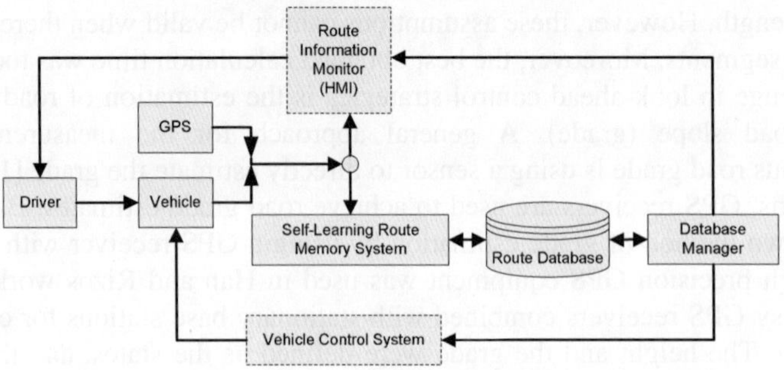

Fig. 1. Schematic system proposed by Clarsson and Reuss.

The database manger coordinates the flow of information between the human machine interface (HMI) and vehicle control system. It was stated that the stored data in the route database can be used to optimize fuel consumption or driving comfort in later trips along the route. However, in this system the method of data acquisition from the sensors is not clear. For example, the curvature of the road was obtained from the yaw rate sensor and the vehicle velocity, but the techniques of calculations were not covered.

To apply look-ahead information in the on-board control, model predictive control (MPC) and receding horizon control (RHC) were used. In MPC, an open-loop optimal problem is solved at each sampling instant and in a finite horizon [20]. The optimization yields an optimal control sequence. MPC was used in Hellstrom et al.'s study [21]. They considered a drive mission for a heavy diesel truck. The combination of an on-board road slope database with the information of a GPS was applied to extract the road geometry ahead. Then, the road look-ahead information was used to obtain the power demand in an optimization formula which was the cost of trip time and fuel consumption.

The key innovation explored in this paper is the method of combining the vehicle sensors such as accelerometer, speedometer and the information of digital map to create a road grade map for use in look-ahead energy management applications.

3 Proposed Road Slope Forecasting Method

Assuming that a moving vehicle at time t is at location $\overrightarrow{x_1}(t)$ and its velocity is $\overrightarrow{v_1}(t)$, the position vector between two consecutive points is:

$$\vec{r}_i(t) = \overrightarrow{x_1}(t + \Delta t) - \overrightarrow{x_1}(t) \tag{2}$$

Here $\overrightarrow{x_1}$ is the position vector in the x-y coordinates. The distance, S between two consecutive points can be written as:

$$s_i(t) = \|\overrightarrow{x_1}(t + \Delta t) - \overrightarrow{x_1}(t)\| \tag{3}$$

In this formula, $\| \ \|$ represents the norm of the specified vector. The direction towards the forward point can be calculated from:

$$\vec{e}_i(t) = \frac{\vec{x}_i(t+\Delta t) - \vec{x}_i(t)}{\|\vec{x}_i(t+\Delta t) - \vec{x}_i(t)\|} \tag{4}$$

The velocity $\vec{v}_i(t)$ can be found as follow:

$$\vec{v}_i(t) = \frac{s_i(t)}{\Delta t}\vec{e}_i(t) \tag{5}$$

Therefore, instantaneous velocity has two orthogonal terms parallel and normal to the road whose magnitude can be represented in the following form:

$$v_x(t) = \left\|\frac{s_i(t)}{\Delta t}\vec{e}_i(t)\right\| \mathrm{Cos}\alpha(t) = \|\vec{v}_i(t)\|\mathrm{Cos}\alpha(t) \tag{6}$$

$$v_y(t) = \left\|\frac{s_i(t)}{\Delta t}\vec{e}_i(t)\right\| \mathrm{Sin}\alpha(t) = \|\vec{v}_i(t)\|\mathrm{Sin}\alpha(t) \tag{7}$$

Here $\alpha(t)$ is the instantaneous slope angle between the direction vector $\vec{e}_i(t)$ and x axis and can be calculated from the following relation:

$$\alpha(t) = \tan^{-1}\left(\frac{y(t+\Delta t)-y(t)}{x(t+\Delta t)-x(t)}\right) \tag{8}$$

The various longitudinal forces in a vehicle are expressed as acceleration, rolling, gravitational and drag. Accordingly, the traction force is derived from the equation of solid body motion in the following form:

$$F_{tr} = M\dot{v}(t) + c_{rr}Mg\mathrm{Cos}\alpha + Mg\mathrm{Sin}\alpha + \frac{1}{2}c_D A\rho(v(t) + v_w(t))^2 \tag{9}$$

The longitudinal forces push the vehicle forward or backward in the x-direction. The vehicle's weight, Mg, acts in the vertical direction through its center of gravity. The weight pulls the vehicle to the ground and its effect in the movement direction, based on the value of $\alpha(t)$, can be either a resistive force or a tractive force. Therefore, by knowing the slope of the road in a known horizon ahead and applying it in the vehicle speed control system a considerable energy saving can be achieved. In the other words, if the variation of gravitational force can be measured in a limited horizon then the speed of vehicle will be changed before uphill and downhill. The speed will decrease before downhill and increase before uphill. Then, the loss of energy due to the brake and operation of the vehicle's engine in high torque regions will be decreased. The saved energy can be significant especially in heavy vehicles.

It is considered that the geographical characteristics of the road is known and can be obtained from digital maps or GIS databases. In regions of low tire slip, the speed of vehicle can obtained from a speed sensor. If the origin considered being the start point of the trip then the slope can be calculated from Eq. (8), and the extracted altitude and longitude from the road database. Afterwards, by employing Eq. (6) and Eq. (7), velocities in x (distance) and y (altitude) directions are calculated. Through numerical integration of V_x, x(t) is determined. Comparing the obtained point with the stored one in the database, the next states x(t+1) and y(t+1) can be achieved. Then, the instantaneous slope can be calculated from Eq. (8) again. This method is illustrated as a conceptual flowchart in Fig. 2.

In order to present the feature of the proposed method, a simulation is carried out in Simulink. Sample road information for a 50km real highway is applied to evaluate the method. The variation of slope for the first 1000m of the real road is presented in Fig. 3. The characteristics of the considered vehicle are shown in Table 1. The results of the simulation are shown in Fig. 4. The figure shows instantaneous slope, gravitational force, vehicle speed and traction force for the considered highway vs. distance. Comparing the instantaneous points with the stored ones in the database, the future slopes can be estimated. A moving average method is a suitable approach to estimate the future slopes. With the movement of vehicle over a meter, a new sample is entered to the look-ahead window and the oldest sample will be discarded from the sampling space so that the slope of the road ahead can be formed by:

$$\alpha(\text{look\_ahead}) = \frac{\sum_{K=1}^{n}(\textbf{sampled date})}{\textbf{n}} - \frac{\textbf{oldest sample}}{\textbf{n}} + \frac{\textbf{new sample}}{\textbf{n}} \tag{10}$$

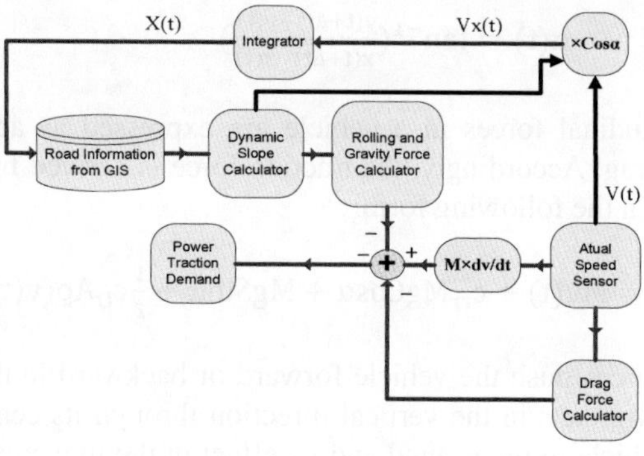

Fig. 2. Flowchart of power demand calculator.

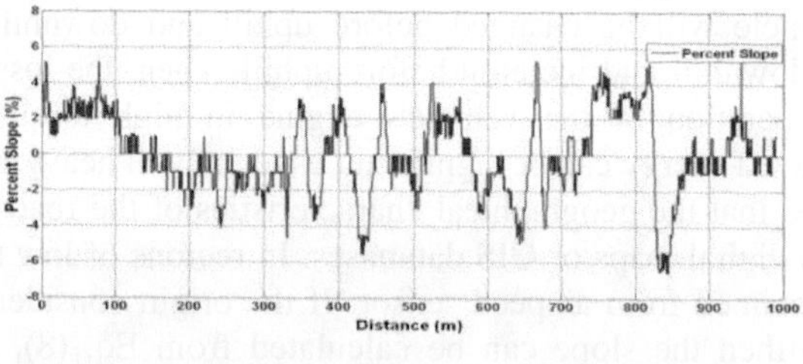

Fig. 3. Variation of the slope for the first 1km of the road.

Table 1. General characteristics of the applied vehicle in simulation.

Description	Symbol	Value	Unit
Vehicle mass	*M*	1678	kg
Rolling resistance	C_{rr}	0.01(1+V/100)*	-
Air drag coefficient	C_D	0.3	-
Frontal area	A	2.25	m^2
Air density	ρ	c	kg/m^3
Gravity	g	9.8	m/s^2

*V is the vehicle speed. The rolling coefficient is considered as a linear function of speed. It is accurate for speeds up to 128 km/h [22].

Then, the in-front slope can be implemented in the calculation of look-ahead gravitational and the estimation of in advance traction force demand. In the proposed method the mass of vehicle is considered to be known. However, in practice mass can be changed particularly in heavy vehicles. The output torque of engine can be measured via engine torque sensor then the traction power can be obtained from:

$$F_{tr} = \frac{T_{engine}}{\eta_{Torque\ Convertor} \times \eta_{Transmission} \times \eta_{differential} \times R_{Tire}} \tag{11}$$

Here, T is torque, η is efficiency and R is radius. With a known traction force, the slope from digital map and speed from speed sensor the vehicle mass can be calculated via longitudinal forces, Eq. (9).

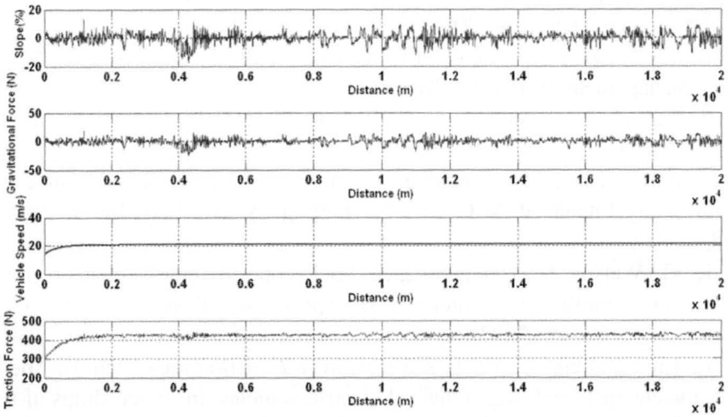

Fig. 4. The results of the simulation for the first 20 km of highway.

4 Conclusions

In this paper, the information from a digital map combined with vehicle sensors was used to estimate road grade and calculate longitudinal forces. The instantaneous obtained road grad was implemented to extract the slopes ahead in a predefined horizon. A moving average method was introduced to achieve the slopes ahead. These slopes can be employed to estimate the near future traction force demand that can be the estimated measurement of a predictive control system. In the proposed approach, it is considered that the mass of vehicle is known. In practice, the mass can be changed with the vehicle load and is significant in heavy vehicles. However, the amount of engine torque is available via engine torque sensor. The traction force can be accurately estimated with the known efficiencies of the different parts of drivetrain. Through, the obtained force, speed of vehicle, the slopes from digital map and the longitudinal dynamic of vehicle mass can be approximated accurately.

Acknowledgments

The authors would like to thank AUTOCRC and Commonwealth of Australia which provide financial support for this research.

References

[1] McQueen, B., McQueen, J.: Intelligent Transportation Systems Architectures. Artech House, Boston (2003)

[2] Peng, Z.-R., Tsou, M.-H.: Internet GIS: Distributed Geographic Information Services for the Internet and Wireless Network. Wiley, Chichester (2003)

[3] García-Ortiz, A., Amin, S.M., Wootton, J.R.: Intelligent transportation systems–Enabling technologies. Mathematical and Computer Modelling 22, 11–81 (1995)

[4] Ghosh, S., Lee, T.: Intelligent Transportation Systems New Principles and Architectures. Taylor & Francis, Abington (2005)

[5] Ichikawa, S., Yokoi, Y., Doki, S., Okuma, S., Naitou, T., Shiimado, T., Miki, N.: Novel energy management system for hybrid electric vehicles utilizing car navigation over a commuting route. In: Proceedings of IEEE Intelligent Vehicles Symposium, Parma, pp. 161–166 (2004)

[6] Sakamoto, H., Imai, M., Tsuchiya, K., Yoshida, T.: Automatic Curve Deceleration System Using Enhanced ACC with Navigation System. SAE Int. J. Passenger Cars 1 (2008)

[7] Manzie, C., Watson, H., Halgamuge, S.: Fuel economy improvements for urban driving: Hybrid vs. intelligent vehicles. Transportation Research Part C: Emerging Technologies 15, 1–16 (2007)

[8] Bin, Y., Li, Y., Gong, Q., Peng, Z.R.: Multi-information integrated trip specific optimal power management for plug-in hybrid electric vehicles. In: Proceedings of the American Control Conference, St. Louis, MO, pp. 4607–4612 (2009)

[9] Bin, Y., Li, C.Y., Peng, Z.: Multiple trip information based spatial domain optimisation for power management of plug-in hybrid electric vehicles. International Journal of Electric and Hybrid Vehicles 2, 259–281 (2010)

[10] Gaeke, E.G.: Road grade sensor,US Patent #3,752,251, G. M. Corporation, Ed. US (1973)

[11] Bae, H.S., Ryu, J., Gerdes, J.C.: Road grade and vehicle parameter estimation for longitudinal control using GPS. In: Proceedings of IEEE Conference on Intelligent Transportation Systems, ITSC, Oakland, CA, pp. 166–171 (2001)

[12] Han, S., Rizos, C.: Road slope information from GPS-derived trajectory data. Journal of Surveying Engineering 125, 59–68 (1999)

[13] Lingman, P., Schmidtbauer, B.: Road slope and vehicle mass estimation using Kalman filtering. In: Proceedings of the17th IAVSD Symposium Copenhagen, Denmark (2001)
[14] Vahidi, A., Stefanopoulou, A.G., Peng, H.: Recursive Least Squares with Forgetting for Online. Estimation of Vehicle Mass and Road Grade: Theory and Experiments. Journal of Vehicle System Dynamics (2005)
[15] Fathy, H.K., Kang, D., Stein, J.L.: Online Vehicle Mass Estimation Using Recursive Least Squares and Supervisory Data Extraction. In: American Control Conference, June 11-13. IEEE, Washington, USA (2008)
[16] Schroedl, S., Wagstaff, K., Rogers, S., Langley, P., Wilson, C.: Mining GPS traces for map refinement, vol. 9. Springer, Heidelberg (2004)
[17] Bruntrup, R., Edelkamp, S., Jabbar, S., Scholz, B.: Incremental Map Generation with GPS Traces. In: IEEE Conference on Intelligent Transportation Systems Vienna, Austria, September 13-16 (2005)
[18] Carlsson, A., Reuss, H.-C.: Implementation of a Self-Learning Route Memory for Forward-Looking Driving. SAE Int. J. Passeng. Cars 1 (2008)
[19] Sahlholm, P., Henrik Johansson, K.: Road grade estimation for look-ahead vehicle control using multiple measurement runs. Control Engineering Practice (2009) (in Press, Corrected Proof)
[20] Mayne, D.Q., Rawlings, J.B., Rao, C.V., Scokaert, P.O.M.: Constrained model predictive control: Stability and optimality. Automatica, 789–814 (2000)
[21] HellstrÃom, E., Ivarsson, M., Ãslund, J., Nielsen, L.: Look-ahead control for heavy trucks to minimize trip time and fuel consumption. Control Engineering Practice 17, 245–254 (2009)
[22] Ehsani, M., Gao, Y., Gays, S.E., Emadi, A.: Modern Electric, Hybrid Electric, and Fuel Cell Vehicles: Fundamentals, Theory, and Design (2005)

A New Method for Analyze Pharmacodynamic Effect of Traditional Chinese Medicine

Bin Nie[1], JianQiang Du[1], RiYue Yu[1,2,3], GuoLiang Xu[1,2,3,*], YueSheng Wang[4], YuHui Liu[2], and LiPing Huang[2]

[1] School of Computer Science, Jiangxi University of Traditional Chinese Medicine, 330006, Nanchang, China
[2] College of Pharmacy, Jiangxi University of Traditional Chinese Medicine, 330006, Nanchang, China
[3] Key Laboratory of Modern Preparation, Ministry of Education, Jiangxi University of Traditional Chinese Medicine, 330006, Nanchang, China
[4] National Pharmaceutical Engineering Centre for Solid Preparation in Chinese Herbal Medicine, 330006, Nanchang, China
xuguoliang6606@126.com,ncunb@163.com

Abstract. The research for pharmacodynamic effect of Traditional Chinese Medicine with Metabolomics has multi-dimensional metabolite data. The paper purpose a new method named arithmetic average value-subtract-ratio to deal with the data and data mining,to analyze the pharmacodynamic effect of Traditional Chinese Medicine. Experiment results show that the method can mine the important information, and can get more satisfactory results. The method was proved to be feasible and effective.

Keywords: metabolomics, pharmacodynamic effect, TCM, arithmetic average, data mining.

1 Introduction

Metabolomics [1-5] is a new discipline of analyzing metabolites qualitatively and quantitatively, metabolomics as a branch of science concerned with the study of systems biology in clinic diagnosis and pharmaceutical.

The common methods[6-9] do with the larger dimensional is difficlut some times,In order to solve the multi-dimensional metabolite data, to analyze pharmacodynamic effect of Traditional Chinese Medicine,In the paper, we put forward a new method. The method consist of four step:the first to analyze the characteristic of metabolite data,the second to structure the m/z mass-to-charge ratio region model of hot natures trade symptoms caused by cold factors,the third to build the math model of analyze pharmacodynamic effect of Traditional Chinese Medicine,the last step to gain the result.

[*] Corresponding author.

M. Zhu (Ed.): Electrical Engineering and Control, LNEE 98, pp. 713–719.
springerlink.com © Springer-Verlag Berlin Heidelberg 2011

2 Analyze the Characteristic of Metabolite Data

The metabolites sample space's data source of key laboratory of modern preparation of Traditional Chinese medicine, ministry of education. The mice as experimental subject, the hot natures treat the cold syndrome group as test purpose.

The data includes twice test mode, the one test mode (model1) consists of the blank controllers which 10 mice, cold syndrome group have 10 mice, hot natures group have the each 10 mice ate pepper high-dose group (PHDG), pepper low-dose group(PLDG), cinnamon high-dose group (CHDG), cinnamon low-dose group(CLDG) respectively. The other test mode (model2) consists of the blank controllers have 10 mice, cold syndrome group have 10 mice, hot natures group which the each 10 mice ate monkshood high-dose(MHD), monkshood low-dose(MLD),zingiberis high-dose(ZHD),zingiberis low-dose(ZLD), Galangal Rhizome high-dose(GRHD), Galangal Rhizome low-dose group(GHLD) ,Evodia rutaecarpa high-dose(ERHD), Evodia rutaecarpa low-dose(ERLD).In the paper, the data are 90*813 dimensions.

3 Structure the M/Z Region Model of Hot Natures Trade Symptoms Caused by Cold Factors

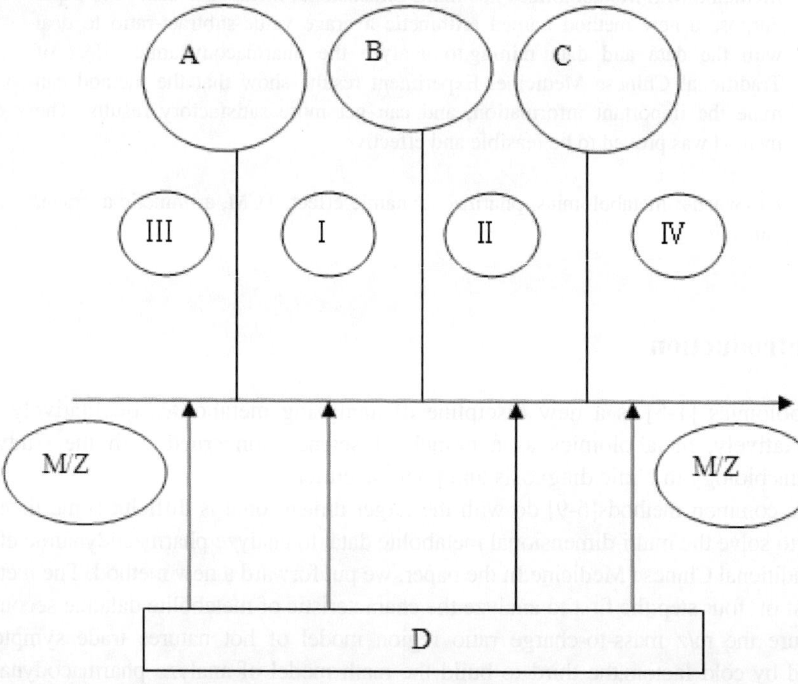

Fig. 1. The M/Z region model of hot natures trade symptoms caused by cold factors

A: the samples M/Z of cold symptoms(smaller)
B:the sampes M/Z of blank
C: the samples M/Z of cold symptoms(larger)
D: the region of hot natures trade symptoms caused by cold factors

Let the samples M/Z is A, if D in the I region indicate pharmacodynamic effect is good, if D in the II or IV region indicate pharmacodynamic effect is better but excessive, if D in the III region indicate pharmacodynamic effect is bad.Let the samples M/Z is C, if D in the II region indicate pharmacodynamic effect is good, if D in the I or III region indicate pharmacodynamic effect is better but excessive, if D in the IV region indicate pharmacodynamic effect is bad.

4 The Math Model and the Algorithm of Analyze Pharmacodynamic Effect of TCM

4.1 Build the Math Model of Analyze Pharmacodynamic Effect of TCM

According to the value of M/Z of the Cold symptoms sample, the blank sample and the hot treatment sample,build the math model of analyze pharmacodynamic effect of TCM show figure 2.

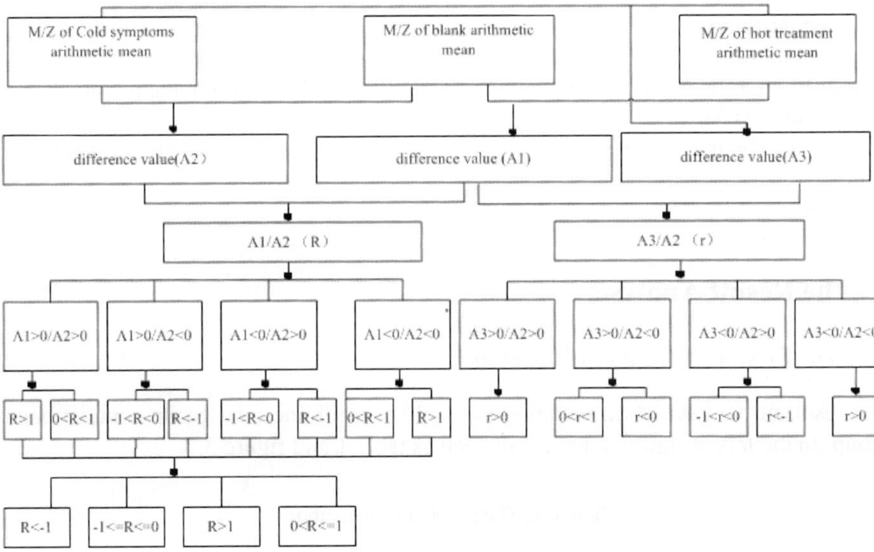

Fig. 2. The math model of analyze pharmacodynamic effect of TCM

difference value A1:the difference value of M/Z of Cold symptoms arithmetic mean and M/Z of blank arithmetic mean. difference value A2: the difference value of M/Z of hot treatment arithmetic mean and M/Z of blank arithmetic mean.difference value

A3: the difference value of M/Z of hot treatment arithmetic mean and M/Z of Cold symptoms arithmetic mean. A1/A2 (R) is the rate of A1 and A2,A3/A2(r) is the rate of A3 and A1.in this paper ,the left of figure 2 call R method,the right of figure 2 call r method.

A2>0 and A1<0:the value of Cold symptoms arithmetic mean less than blank arithmetic mean, the value of hot treatment larger than blank arithmetic mean.in the paper,this situation call model to fall and medicine to raise(MFMR).

A3>0 and A1<0: the value of Cold symptoms arithmetic mean less than blank arithmetic mean, the value of hot treatment larger than Cold symptoms arithmetic mean.in the paper,this situation call model to fall and medicine to raise(MFMR).

A2<0 and A1>0:the value of Cold symptoms arithmetic mean larger than blank arithmetic mean, the value of hot treatment less than blank arithmetic mean.in the paper,this situation call model to raise and medicine to fall (MRMF).

A3<0 and A1>0: the value of Cold symptoms arithmetic mean larger than blank arithmetic mean, the value of hot treatment less than Cold symptoms arithmetic mean.in the paper,this situation call model to raise and medicine to fall (MRMF).

4.2 The Algorithm of Analyze Pharmacodynamic Effect of TCM

Step 1: compute Cold symptoms arithmetic mean,blank arithmetic mean, hot treatment arithmetic mean.
Step 2: compute A1,A2,A3.
Step 3: Extract the M/Z intensity value of MRMF and MFMR,compute the number of MRMF and MFMR.
Step 4:Classfy the different level according to MRMF and MFMR, compute the number.
Step 5:gain the Result.

5 The Result Analysis

5.1 The First Test Result of R Method

The result of cold sympotm model group and hot medicine treat group contrast blank group ,In the left of figure 2, the result such as table 1 and figure 3.

Table 1. The number of R method

The number of M/Z	effective			ineffective
	regular	better	best	
813	499	138	76	100

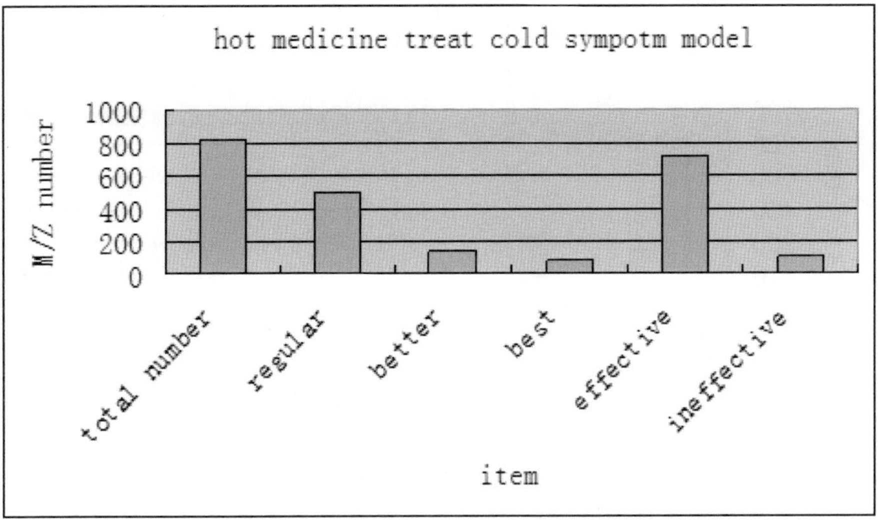

Fig. 3. The number of hot medicine treat cold sympotm model use R method

The result of table 1 and figure 3 indicate hot medicine treat sympotm model is good.

5.2 The First Test Result of R Method

The result of cold model group contrast blank group,hot medicine treat group contrast cold model group, In the right of figure 2, the result such as table 2 and figure 4.

Table 2. The number of r method

total number	effective		ineffective
	MRMF total	MFMR total	
813	238	464	100

5.3 The Second Test Result of R Method

The result of cold model group contrast blank group,hot medicine treat group contrast cold model group, In the right of figure 2, the result such as figure 5.

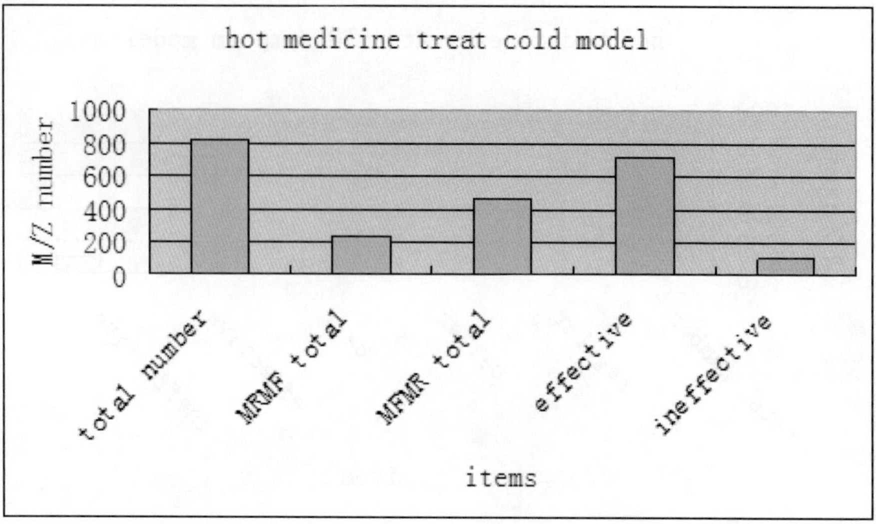

Fig. 4. The number of hot medicine treat cold sympotm model use r method

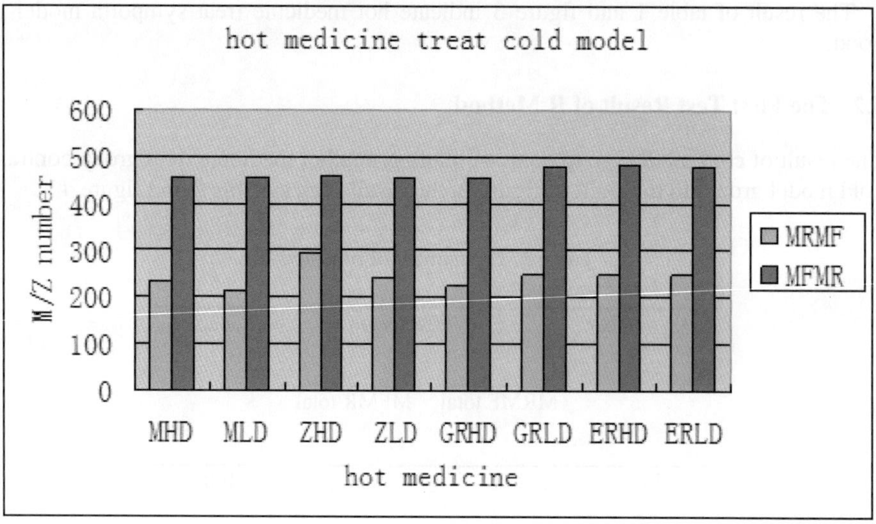

Fig. 5. The result of the second test use r method

monkshood high-dose(MHD), monkshood low-dose(MLD),zingiberis high-dose(ZHD),zingiberis low-dose(ZLD), Galangal Rhizome high-dose(GRHD), Galangal Rhizome low-dose group(GHLD) ,Evodia rutaecarpa high-dose(ERHD), Evodia rutaecarpa low-dose(ERLD).the result inidcate that the hot medicine treat cold sympotm model is effective.

6 Conclusion

In the paper, put forward a new method named arithmetic average value-subtract-ratio to deal with the data and data mining,to analyze the pharmacodynamic effect of Traditional Chinese Medicine. Experiment results show that the method can mine the important information, and can get more satisfactory results. The method was proved to be feasible and effective.

Acknowledgments

The authors wish to express their gratitude to the anonymous reviewers for their valuable comments and suggestions, which have improved the quality of this paper.The team members also include JianJiang Fu,WenHong Li, YuHui Liu,Bo Liu,Fei Qu,YingFang Chen,QiYun Zhang,BingTao Li, Jinmei Peng, Yao Zhang.This work is supported by National Basic Research Program of China(973 Program:2010CB530603,2006CB504702).This work also supported by National Science and Technology Major Project and Grand New Drug Development Program (2009ZX09310), and Supported by Jiangxi province Natural Science Foundation (2009GZS0058), Supported by Technology project of Educational Committee of Jiangxi Province (GJJ11541).

References

1. Holmes, E.: "Metabonomics": understanding the metabolic responses of living systems to path physiological stimul via multivariate statistical analysis of biological NMR spectroscopic data. Xenobiotica 29, 1181–1189 (1999)
2. Nicholson, J.K., Holmes, E., Lindon, J.C., Wilson, I.D.: The challenges of modeling mammalian biocomplexity. Nature Biotechnology 22(10), 1268–1274 (2004)
3. Pognan, F.: Genomics, proteomics and metabonomics in toxicology: Hopefully not 'fashionomics. Pharmacogenomics 5(7), 879–893 (2004)
4. Yin, P.Y., Zhao, X.J., Ki, Q.R., Wang, J.S., Li, J.S., Xu, G.W.: Metabonomics study of intestinal fistulas based on ultraperformance liquid chromatography coupled with Q-TOF mass spectrometry (UPLC/Q-TOF MS). Journal of Proteome Research 5, 2135–2143 (2006)
5. Liu, C.X., Li, C., Lin, D.H., et al.: Significance of metabonomics in drug discovery and development. Asian J. DrugMetab Pharmacokinet 4, 87–96 (2004)
6. Mager, D.E.: Quantitative structure–pharmacokinetic/pharmacodynamic relationships. Advanced Drug Delivery Reviews 58, 1326–1356 (2006)
7. Hamoudeh, M., Salim, H., Barbos, D., et al.: Preparation and characterization of radioactive dirhenium decacarbonyl-loaded PLLA nanoparticles for radionuclide intra-tumoral therapy. European Journal of Pharmaceutics and Biopharmaceutics 67, 597–611 (2007)
8. Lee, J.-J., Jang, C.-S., Wang, S.-W., Liu, C.-W.: Evaluation of potential health risk of arsenic-affected groundwater using indicator kriging and dose response model. Science of the Total Environment 384, 151–162 (2007)
9. Price, S.J., Jena, R., Green, H.A.L., et al.: Early Radiotherapy Dose Response and Lack of Hypersensitivity Effect in Normal Brain Tissue: a Sequential Dynamic Susceptibility Imaging Study of Cerebral Perfusion. Clinical Oncology 19, 577–587 (2007)

Application of the Case-Based Learning Based on KD-Tree in Unmanned Helicopter Control*

Daohui Zhang[1,2], Xingang Zhao[2], and Yang Chen[2,3]

[1] College of Information Science and Engineering,
Shenyang Ligong University, 110159, Shenyang, China
[2] Shenyang Institution of Automation Chinese Academy of Sciences,
110016, Shenyang, China
[3] College of Information Science and Engineering,
Wuhan University of Science and Technology, 430081, Wuhan, China
{zhangdaohui,zhaoxingang,chenyang}@sia.cn

Abstract. The paper proposes a case-based learning based on KD-Tree for the hovering control and path planning of unmanned helicopters. We use nearest neighbor search, a basic case-based learning algorithm, to get the nearest neighbors of the target point in the Euclidean space from the cases to learn. Then we generate the output to control the system using the nearest neighbors. Simulation experiments show that this method performances well in the hovering control, not so well in the path planning. The method brings new ideas to the field of unmanned helicopter's control, and extends the application of the case-based learning based on KD-Tree as well.

Keywords: KD-Tree, nearest neighbor search, case-based learning, hovering control, path planning, unmanned helicopter.

1 Introduction

Nowadays, the control systems of unmanned helicopters are mainly implemented by constructing dynamics models. For example, Zhe Jiang proposed the active-model-based control scheme for flying robot with rotary wing [1]. Jesse Leitner et al. accomplished the analysis of adaptive neural networks for helicopter flight control [2]; Jong A.J. researched helicopter UAV control using classical control theory [3]. Generally, model-based control methods have a higher demand for model's precision. However, all kinds of uncertainty always exist because of unmanned helicopter's inherent nonlinearity and time varying character. As a result, the traditional methods are hard to solve these difficulties.

Recently, path planning methods for unmanned helicopters mainly include methods based on geometrical model searching [4], methods based on virtual potential field and navigation function [5], methods based on mathematic optimization [6] and methods based on biological intelligence [7]. All kinds of

* Liaoning Provincial Natural Science Foundation of China.

 © Springer-Verlag Berlin Heidelberg 2011

methods have been compared in [8] and the analyzed results indicate that many existing methods are difficult to construct models in complex environments with acceptable real-time quality and dynamic constraints.

Case-based learning is a kind of instructor-supervising and memory-based machine learning method [9]. This method is easily understood because the learning process is data-driven and so it shows high real-time quality. Nearest neighbor search is a basic case-based learning method, and KD-Tree is the most common and efficient structure. As a kind of BSP (Binary Space Partitioning), KD-Tree has been improved by many researchers since it was proposed by Jon Louis Bentley in 1975 [10], [11], [12], and it has been applied in many areas.

Therefore, the method of the case-based learning based on KD-Tree in the hovering control and path planning of unmanned helicopters can improve the real-time quality of control systems for it will simplify the control idea without considering complex control processes. This paper will check up the efficiency of the case-based learning based on KD-Tree and generalize it in practice.

This paper is organized as follows. Section 2 of the paper introduces the structure and building process of KD-Tree in detail. Nearest neighbor search and its two algorithms are discussed in Section3. The applications of the method in the hovering control and path planning of unmanned helicopters are provided in Section 4 and Section 5, respectively. The simulations are used to verify the proposed method. At last, Section 6 draws conclusions and discusses further research plans.

2 KD-Tree

KD-Tree is a special binary tree, whose each node can store a k-dimensional vector. Each non-leaf node can be seen as a partitioning hyperplane, which splits the space into two subspaces. The value in some dimension of each level's nodes of the tree is compared with each other, and the dimension is circularly chosen from the k dimensions in turn. The node whose value is less than the pivot value belongs to the left subtree, otherwise the right subtree [13].

In 2-dimesion space (2D-Tree), the nodes compare their X coordinate value in even levels (assuming that the root is in the 0 level), and compare their Y coordinate value in odd levels (see Fig.1).

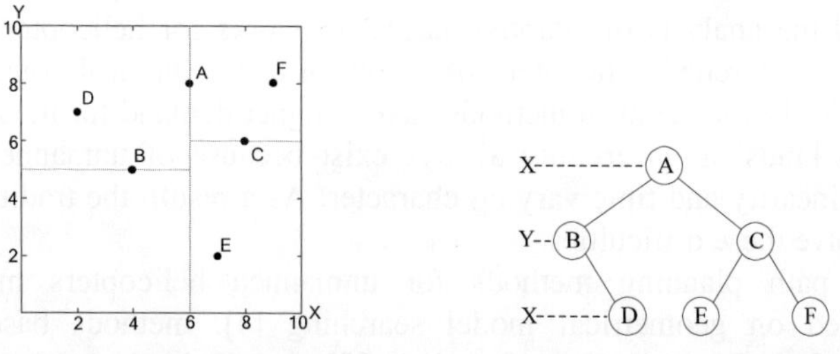

(a) The analyzing figure of 2D-Tree. (b) The structure figure of 2D-Tree.

Fig. 1. 2D-Tree.

The process of building KD-Tree by inserting nodes one by one is given.

1. insert a point, and construct a new node;
2. use the function findparent() to find the node's tentative father node, then insert it as its father node's letf node when its value is less than its father node's value, or as its father node's right node;
3. return to 1, circulate the processes in turn until all nodes are inserted.

The complexity of the KD-Tree which has n nodes is below:
Build: $O(n\log_2 n)$
Insert: $O(n\log_2 n)$
Delete: $O(n\log_2 n)$
Search: $O(n^{1-1/k}+m)$ m---the number of the checked nodes

3 Nearest Neighbor Search

Nearest neighbor search is an optimization problem for finding closest points in the Euclidean space. It can be implemented in the built KD-Tree. This paper gives two algorithms of nearest neighbor search, one is the top-down search, and the other is the bottom-up search [14].

The top-down search firstly finds the tentative father node of the target point using the function findparent(), and the distance from the tentative father node to the target point is as the initial radius of the searching area. The determination of the searching area's initial radius in 2-dimension plane is below (see Fig. 2).

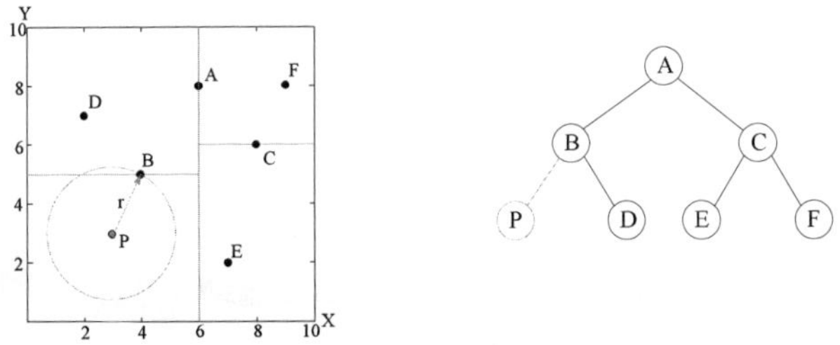

(a) The analyzing figure of 2D-Tree. (b) The structure figure of 2D-Tree.

Fig. 2. The determination of the searching area's initial radius in 2-dimension plane.

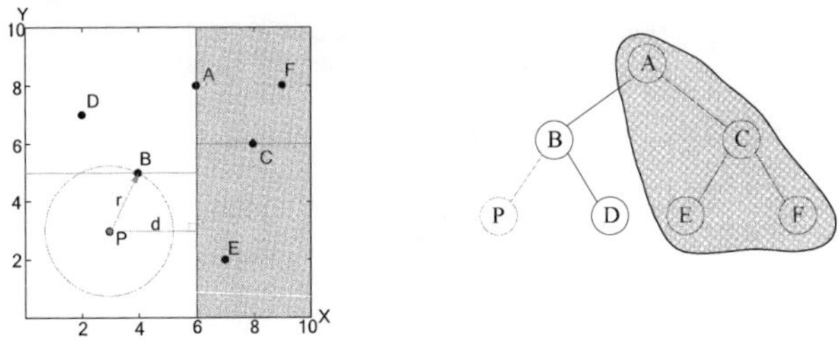

(a) The analyzing figure of 2D-Tree. (b) The structure figure of 2D-Tree.

Fig. 3. The determination of whether the subtrees should be searched in 2-dimension plane.

Then search the nodes down from root node in the KD-Tree, and update the searching area's radius and the current nearest neighbor when the distance from the node to the target point is less than the searching area's radius. The distance (d) from the target point to the partitioning hyperplane is compared with the searching area's radius (r) to determine whether the node's subtrees should be searched. If r<d, there is one subtree outside of the searching area and it should not be searched, or both subtrees are across the searching area and they both should be searched. In 2-dimension plane, if r<d, the node A's right subtree is outside of the searching area, and it should not be searched (see Fig. 3).

When the search ends, the minimum searching area's radius is the minimum distance to the target point in the KD-Tree, and the current nearest neighbor is the nearest neighbor of the target point in the KD-Tree.

The bottom-up search also firstly finds the target point's tentative father node in KD-Tree using the function findparent(), and sets the initial radius. Then, it searches the nodes up from the tentative father node in the KD-Tree, and constructs the minimum searching-hyperrectangle boundaries and the bottom-up search ends when the minimum searching-hyperrectangle boundaries accomplish construction or when all nodes in the KD-Tree have been checked. The bottom-up search also needs to determinate whether the subtrees should be searched by comparing the minimum searching area's radius (r) with the distance (d) from the target point to the partitioning hyperplane. The determination is able to improve the searching efficiency by pruning the searching area.

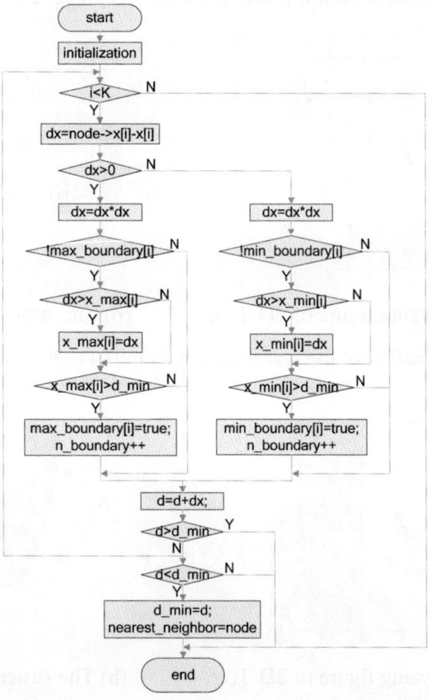

Fig. 4. The flow chart of constructing the minimum searching-hyperrectangle boundaries.

The flow chart of constructing the minimum searching-hyperrectangle boundaries is above (see Fig. 4).

When the number of the KD-Tree's nodes is large enough, the bottom-up search can reduce searching time obviously, because only the fewer nodes need to be searched in the area of the minimum searching-hyperrectangle boundaries rather than all nodes in the entire KD-Tree. Experiments show that the bottom-up search is twice fast than the top-down search on the condition that KD-Tree is large enough. So the bottom-up search is selected in the following experiments.

Sometimes we need k nearest neighbors, the so-called k-nearest neighbor search, and k is the number of the nearest neighbors. In the paper we use nearest neighbor searching to get the nearest neighbor and mark it, then search the new nearest neighbor without considering the marked nearest neighbor, so we can accomplish the k-nearest neighbor search through doing nearest neighbor search k times.

4 Application of the Case-Based Learning in the Hovering Control of Unmanned Helicopters

Nearest neighbor search based on KD-Tree is a kind of case-based learning. It firstly learns the given cases and finds the nearest neighbor of the target point from the given cases. Finally the nearest neighbor is used to generate the output to control the system. This control idea is simply and is easy to operate. Next we will examine the efficiency of this method in simulation experiments.

The input data are 13-dimensional state vectors and the output data are 4-dimensional control vectors in the hovering control of unmanned helicopters. The distance adopts the Euclidean distance. Because the data of different dimensions are diverse, they need to be normalized by the max-min normalization, z-score standardization or decimal scaling.

In the experiment, we adopt k-nearest neighbor search. Three models are provided, the average method, the least square method and kernel regression, for using the 4-dimension control vectors of k nearest neighbors to generate the output. Tests show the average method has lower complexity and operates faster, so we adopt it in the experiment.

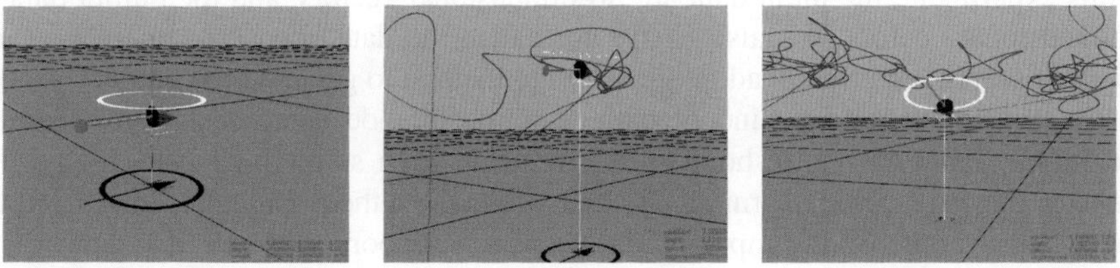

(a) The helicopter just flies off. (b) The helicopter flies 1 min. (c) The helicopter flies 5 min.

Fig. 5. The control effects of the cases.

The control effect of the cases in the VC simulation platform is above (see Fig. 5). The cases come from the active-model-based control method, and the control effect of the cases provides a comparable standard for the k-nearest neighbor search based on KD-Tree.

In the experiment, the value of k has an influence on the control effect. When k=1, the control effect sometimes appears similar to the control effect of the cases, or even is part of the control effect of the cases (see Fig. 6). Although sometimes the control effect is perfect, the control system has low anti-jamming quality, so it is easily out of control (see Fig. 7).

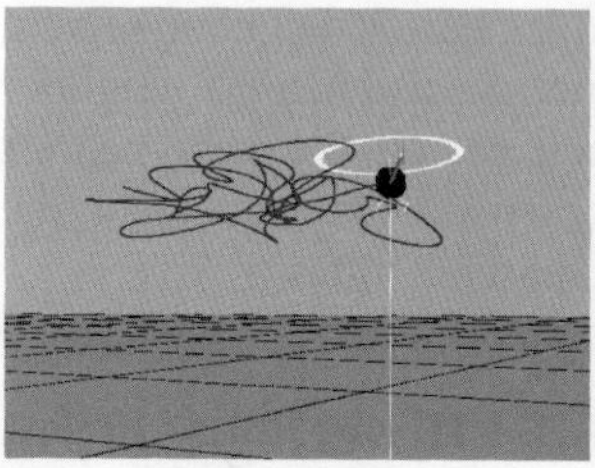
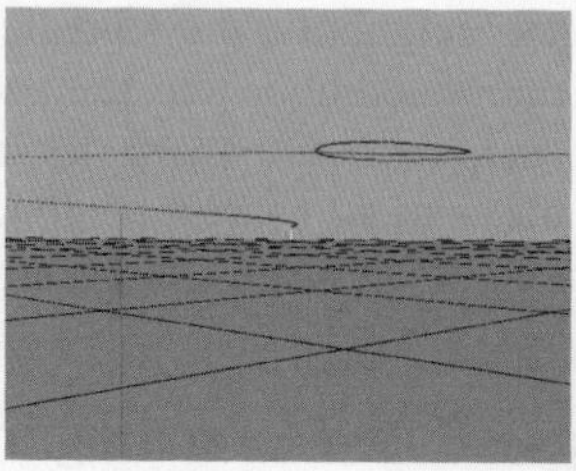

Fig. 6. k=1, program runs 2 min. **Fig. 7.** k=1, out of control. **Fig. 8.** k=10, program runs 2 min.

When k>1, the output is the average of the 4-dimensional control vectors of k nearest neighbors. And the system will have a worse control effect, but higher anti-jamming quality. When k=10, the control effect is shown above (see Fig. 8). Fig. 8 and Fig. 6 choose the same initial state, and we can see that Fig. 8 has a litter worse control effect. The value of k shouldn't be too large, or the searching would cost too much time. Simulation shows that the control effect and the searching time cost both performances well when k=10.

Simulation experiments show that the case-based learning based on KD-Tree performances well on the hovering control of unmanned helicopters, and the further research can be considered to apply it in practice.

5 Application of the Case-Based Learning in the Path Planning of Unmanned Helicopters

In this experiment, the input data are 30-dimensional vectors, and the output data are 3-dimensional vectors. We also firstly normalize the data, then implement k-nearest neighbor search, and finally adopt the average method to generate the output.

We real-timely learn a kind of path planning method using case-based learning, and set the input-error threshold to determine in each step which one will be used between the case-based learning and the original method. Calculate the error that equals the average of the input of the nearest neighbors subtracts the input of the target point, and then compare the error with the threshold, and adopt the learning method when the error is less than the threshold, or adopt the original method.

We adopt a method that combines offline and online learning for the path planning of unmanned helicopters. We insert the points into the KD-Tree which use the original method when the error is above the threshold, and this online learning provides more abundant cases for learning.

Here introduce the simulation experiments of the path planning of unmanned helicopters. The helicopter flies off from a point, the target spirals up from a second point, and three obstacle balls spiral up from a third point. The path planning's aim is to make the helicopter catch up the target in minimum time and distance without bumping the obstacle balls.

In this experiment, k-nearest neighbor search is also adopted. Simulation shows that the value of k has an influence on the control effect but in an uncertain rule. And the value of k can't be too big, or the searching time costs too long. When the threshold's value increases, the number of the nodes which use the learning method increases, but at the same time the failure rate increases owing to the bigger error. The threshold's value is determined based on a comprehensive considering of the failure rate and the participation rate of case-based learning through many experiments.

The effect of the path planning of the original method is given below (see Fig. 9). In Fig. 9, the red-dot track is the helicopter's track, the green-dot track is the target's track, and the blue-dot tracks are the tracks of three obstacle balls. Simulation shows the path planning using the case-based learning based on KD-Tree performances well when k=10, threshold=0.042 (see Fig. 10).

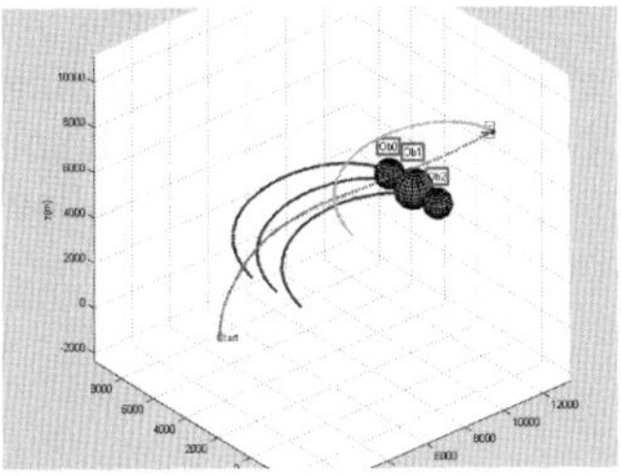

Fig. 9. The original method effect.

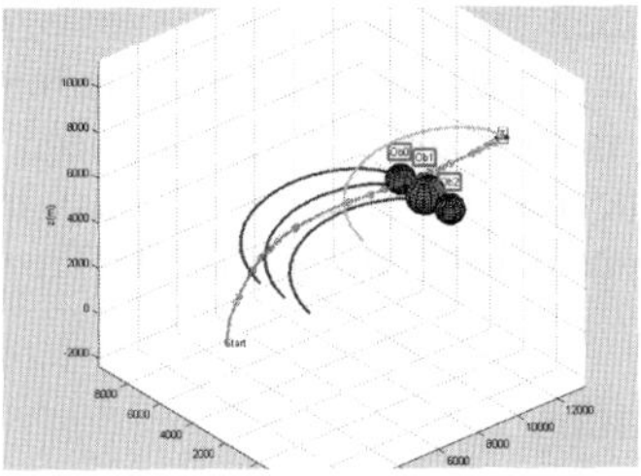

Fig. 10. The effect when k=10, threshold=0.042.

In Fig. 10, the dots with a circle of the helicopter's track are the ones which use the case-based learning based on KD-Tree. The participation of the case-based learning in the path planning is the ratio of the number of the dots with a circle to the total number of the track's dots. We do the simulation experiments 100 times and get the learning method's participation is 23.8/114.

When the threshold's value is given large, the helicopter easily bumps the obstacle balls and the path planning is failure (see Fig. 11); or helicopter cost more time and distance to catch up the target and the path planning is bad (see Fig. 12).

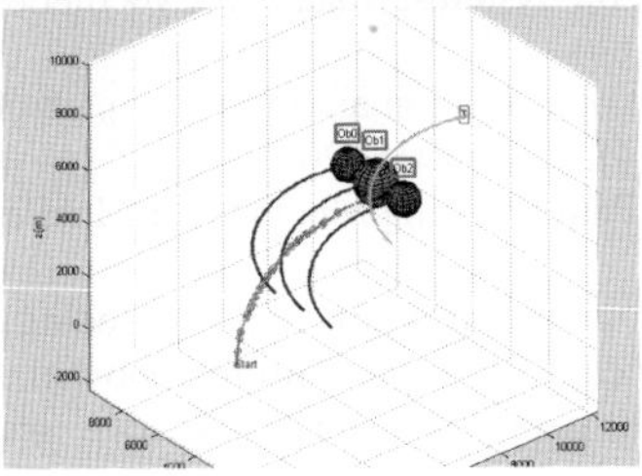

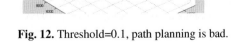

Fig. 11. Threshold=0.2, path planning fails.

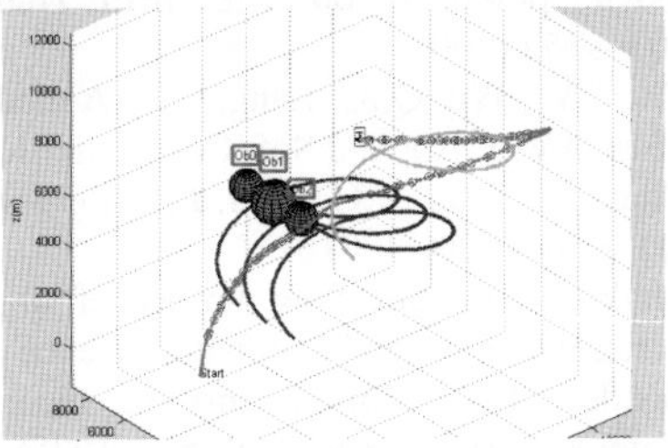

Fig. 12. Threshold=0.1, path planning is bad.

The learning method's participation is not high in this experiment, but it doesn't mean that the case-based learning based on KD-Tree is useless in path planning, because the cases to learn affect the participation and the track of the obstacle balls also play an important role in the participation. Therefore, further work needs to be done in the problem.

6 Conclusions

The case-based learning has been applied in many fields, and its idea is simple, its implement is easy. It can be used to solve many learning problems without considering their complex backgrounds, and only need some good cases to learn. KD-Tree structure and nearest neighbor search are common algorithms of case-based learning, they operate easily and have good expansibility and compatibility.

The hovering control is a difficult basic movement of the flying control of unmanned helicopters, and the case-based learning based on KD-Tree brings a new idea to solve this problem. We can try to add the case-based learning based on KD-Tree to all kinds of path planning methods of the unmanned helicopter in order to improve their efficiency.

There are many waiting for being improved about the case-based learning based on KD-Tree in this paper. More jobs need to be done in the balancing technique of KD-Tree. K-nearest neighbor search may have better algorithms. In the online learning, we can consider rebuilding the KD-Tree in necessary.

References

1. Jiang, Z.: Control Methodology for Flying Robot with Rotary Wing. PhD Thesis, Graduate University Chinese Academy of Sciences (2008)
2. Jesse, L., Anthony, C., Prasad, J.V.: Analysis of Adaptive Neural Networks for Helicopter Flight Control. Journal of Guidance, Control and Dynamics 20(5), 972–979 (1997)
3. Jong, A.J.: Helicopter UAV Control using Classical Control. MSc Thesis, Delft University of Technology (2004)
4. Ladd, A.M., Kavraki, L.E.: Measure theoretic analysis of probabilistic path planning. IEEE Transactions on Robotics Automation 20(2), 229–242 (2004)
5. Kitamura, Y., Tanaka, T., Kishino, F.: 3-D path planning in a dynamic environment using an octree and an artificial potential field. In: International Conference on Intelligent Robots and Systems, pp. 474–481. IEEE, Piscataway (1995)
6. Shih, C.L., Lee, T.T., Gruver, W.A.: A unified approach for robot motion planning with moving polyhedral obstacles. IEEE Transactions on Systems, Man, and Cybernetics 20(4), 903–915 (1990)
7. Chen, M., Wu, Q.X., Jiang, C.S.: A modified ant optimization algorithm for path planning of UCAV. J. Applied Soft Computing Journal 8(4), 1712–1718 (2008)
8. Chen, Y., Zhao, X.G., Han, J.D.: Review of 3D Path Planning Methods for Mobile Robot. J. J. Robot 32(4) (2010)
9. Marko, R.S.: Speeding up Relief algorithms with k-d trees. In: Proceeding of the Electrotechnical and Computer Science Conference, ERK 1998, Portoroz, Slovenia (1998)
10. Bentley, J.L.: Multidimensional binary search trees used for associative searching. Commun. ACM 18(9), 509–517 (1975)

11. Peng, C., Yong, W.: Optimized KD Tree Application in Instance-Based Learning. In: Fifth International Conference on Fuzzy Systems and Knowledge Discovery, pp. 187–191. IEEE Compute Society, Los Alamitos (2008)
12. Panigrahy, R.: Nearest Neighbor Search using Kd-trees. Stanford University, Stanford (2006)
13. kd-tree, http://en.wikipedia.org/wiki/Kd-tree
14. KD Tree, http://www.codeproject.com/KB/architecture/KDTree.aspx

An Average Performance and Scalability Model of Xen System under Computing-Intensive Workload

Jianhua Che[1], Dawei Huang[2], Hongtao Li[3], and Wei Yao[4]

[1] State Grid Electric Power Research Institute, Nanjing 210003, China
[2] College of Computer Science, Zhejiang University, Hangzhou 310027, China
[3] JiNan Bureau of State Land Supervision, Jinan 250014, China
[4] College of Information Science and Technology, Agricultural University of Hebei, Baoding 071001, China
{chejianhua,davidhuang}@zju.edu.cn,
lht@hebnetu.edu.cn,
yaowei@hebau.edu.cn

Abstract. As one of typical paravirtualization hypervisors, Xen has received widespread attentions especially its scaling capability under some kinds of workload. Scalability is an important metric for virtual machine systems all the time. However, the average performance of virtual machines and the scalability number of Xen system are two hands of an antinomy for a specific platform with definite hardware resource. Therefore, how to find a good tradeoff between them becomes very significative. This paper provides an average performance and scalability model of Xen system under computing-intensive workload. The proposed model can figure out not only the average performance of multiple uniform virtual machines concurrently executing computing-intensive workload and the scalability number of Xen system, but also the optimal combination of both opposites for a given platform with definite hardware resource. The experimental result of *LINPACK* showed the validity of this average performance and scalability model of Xen system under computing-intensive workload.

1 Introduction

Virtualization has been a hot research topic as a solution of consolidating computer resources in recent years [1]. There are presently three elementary software virtualization methods: full virtualization, paravirtualization and container-based virtualization [2], and paravirtualization has absorbed numerous eyes for its favorable performance. As a typical representative of paravirtualization hypervisors, Xen holds preferable function and performance especially its scaling capability under some kinds of workload. However, the average performance of multiple uniform virtual machines concurrently executing some kinds of workload and the scalability number of Xen system are two incompatible opposites as the computing resource is given, and their relation is still not explicitly illuminated.

M. Zhu (Ed.): Electrical Engineering and Control, LNEE 98, pp. 731–738.
springerlink.com © Springer-Verlag Berlin Heidelberg 2011

In this paper, we studied three main scheduling algorithms of Xen system(i.e. *BVT*, *SEDF* and *Credit*), and presented an average performance and scalability model of Xen system under computing-intensive workload. With this model, we can figure out the average performance of multiple uniform virtual machines concurrently executing computing-intensive workload and the scalability number of Xen system, and find the optimal tradeoff between them so as to maximize the utilization of a specific platform with definite hardware resource. Finally, the experimental result of *LINPACK* validated the practicability of this model.

The rest of this paper is organized as follows. We first introduce some background knowledge about the Xen hypervisor and the scalability metric in Section 2, and then describe the average performance and scalability model of Xen system under computing-intensive workload in Section 3. The Section 4 analyzes the use of this model, and the Section 5 gives some related work about the performance modeling of Xen system. Finally, we conclude with the future work in Section 6.

2 Background

In this section, we will review some background knowledge about the Xen hypervisor and the scalability metric.

2.1 The Xen Hypervisor

Xen is a paravirtualization hypervisor based on x86 platform originally developed at the University of Cambridge [3] as figure 1, that builds a cooperative mechanism between guest operating system and host operating system by

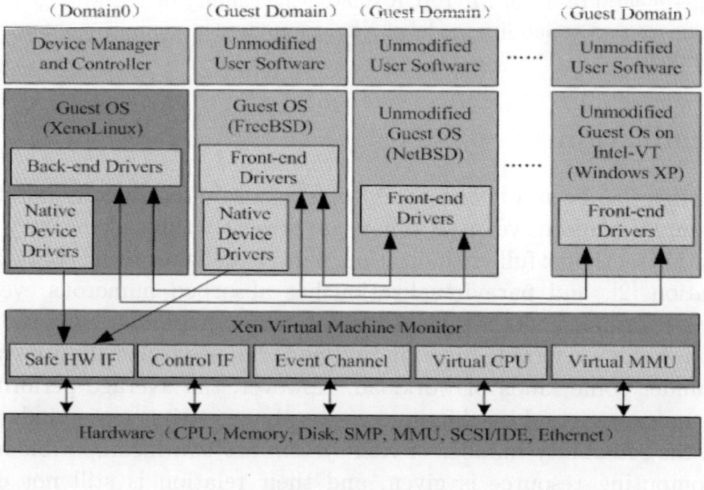

Fig. 1. The architecture of Xen hypervisor

modifying their kernels to promote its virtualization performance. However, existing applications can run without any modification because Xen requires no change to the application binary interface(ABI). Later, Xen has implemented the full virtualization mode along with the advent of virtualizable hardware.

There are three main processor schedulers in Xen system: *BVT*(*Borrowed Virtual Time*) [4], *SEDF*(*Simple Earliest Deadline First*) [5] and *Credit* [6]. *BVT* is a fairness priority scheduling algorithm proposed by KermethJ.Duda in 1999, and divides the time into *real time* recorded by the hardware timer and *virtual time* obtained by computing the *real time* according to some rules. *BVT* monitors the execution time of each process with *virtual time* and schedules the VCPU with the earliest valid *virtual time*, all VCPUs share the processor resource according to their weights representing their respective available processor quotient. *BVT* holds small scheduling overhead and low latency, but does not support non-working-conserving and time-slice-occupying between guest operating systems. *EDF*(*Earliest Deadline First*) is a dynamic optimal scheduling algorithm proposed by Chung Laung Liu in 1973. *SEDF*(*Simple EDF*) creates a 3-tuple(s, p, x) for each guest operating system, and s, p mean that guest operating system executes at least s milliseconds during p milliseconds, x denotes whether guest operating system can occupy the idle time slices remaining in p milliseconds. *SEDF* usually schedules the guest operating systems with the earliest "*latest scheduling time*", and supports working-conserving and non-working-conserving. Therefore, *SEDF* can assign partial processor resource to a guest operating system. *Credit* is the default scheduling algorithm of Xen system since its 3.0 version, and creates a 2-tuple(W, L_U) for each guest operating system. Here, W denotes the weight of each guest operating system that represents its available processor time slice proportion, L_U determines the processor time slice upper limit that a guest operating system can occupy. For example, if L_U=50, the guest operating system with this L_U can only occupy the half of all physical processor time slices; if L_U=100, the guest operating system with this L_U can occupy all physical processor time slices. *Credit* implements non-working-conserving by adjusting the L_U of guest operating system. In addition, *Credit* can manage multiple physical processors in a global view and share all processor resource proportionally and fairly with the SMP manner.

2.2 The Scalability Metric

Scalability is a characteristic of a system, model or function that describes its capability to cope some expanding operational demands [7]. A system, model or function that scales well will be able to maintain or even increase its performance level when tested by larger operational demands. Likewise, *scalability* is also an important metric of a virtual machine system that is defined as the maximum number of virtual machines concurrently running in a virtual machine system without any workload. To our best knowledge, Denali [8] can host more than 10,000 concurrently running virtual machines based on an ordinary commercial platform, and Xen can support hundreds of concurrently running virtual machines based on a common server platform.

To find how many virtual machines Xen system can host based on the current hardware platform, we have made numerous trials to scale up the number of virtual machines in Xen system. Unfortunately, we did not reach the maximum number of virtual machines due to our limited hardware resource. However, we think it's more significant to find the maximum number of virtual machines concurrently executing some kind of workload that Xen system can host. Therefore, many performance testings have been conducted to find the truth.

3 The Average Performance and Scalability Model of Xen System under Computing-Intensive Workload

Based on these numerous performance testings, we found that the average performance of multiple uniform virtual machines concurrently executing computing-intensive workload has a linear relation with that of one same virtual machine executing the same workload in Xen system. By studying the running mechanism of Xen system, we proposed a model that can figure out the average performance of multiple uniform virtual machines concurrently executing computing-intensive workload, and the scalability number of Xen system under computing-intensive workload. The model is as follows:

$$P(m, W) = \frac{\alpha(W)}{m} \cdot P(1, W) + \beta(W) \quad (1)$$

Here, $P(m, W)$ is the average performance of m uniform virtual machines concurrently executing the workload W, $P(1, W)$ is the performance of one same virtual machine executing the workload W, $\alpha(W)$ and $\beta(W)$ are two constant factors related to the workload W. As *Credit* is a fairly sharing processor scheduling algorithm that assigns the same weight to all virtual machines, $\alpha(W)$ and $\beta(W)$ are respectively supposed to 1 and 0 for computing-intensive workload.

To verify this assumption, we measured the average performance of multiple uniform virtual machines concurrently executing a kind of computing-intensive workload-*LINPACK* based on a $DELL^{TM}$ $PowerEdge^{TM}$2900 server. In the experiment, the host operating system is Linux2.6.18-53.1.14.el5xen, the number of virtual machines hosted by Xen system is changed from 1 to 6, the memory allocated for each virtual machine is 256MB, and the problem size of *LINPACK* is changed from 1000 to 10000. The experimental result is shown in the table 1.

For each problem size of *LINPACK*, we use the least square method to regress and analyze the $P(m, W)$ corresponding to the $P(1, W)$. For example, our regression analysis results for the problem size 1000 of *LINPACK* are $\alpha(W) = 1.01375$ and $\beta(W) = -0.071$. Consequently, the equation (1) is turned into the equation (2).

$$P(m, L_{1000}) = \frac{1.01375}{m} \cdot P(1, W) - 0.071 \quad (2)$$

Here, L_{1000} denotes the *LINPACK* workload with the problem size 1000. Our experimental results indicate that $\alpha(W)$ and $\beta(W)$ are almost invariable and

Table 1. The Average Performance of Multiple Uniform Virtual Machines Corresponding to Each Problem Size of *LINPACK*($GFLOPS$)

Size	1VM	2VMs	3VMs	4VMs	5VMs	6VMs
1000	4.4587	2.2159	1.3379	1.1267	0.8374	0.6683
2000	4.5361	2.2694	1.3752	1.1303	0.8493	0.6697
3000	4.6085	2.3109	1.3978	1.1419	0.8504	0.6701
4000	4.6807	2.2965	1.4025	1.1790	0.7757	0.6469
5000	4.7107	2.2883	1.4134	1.1749	0.7535	0.6505
6000	4.7353	2.3379	1.5032	1.1873	0.8174	0.7221
7000	4.6760	2.3563	1.4906	1.1714	0.7318	0.6898
8000	4.6694	2.3372	1.4732	1.1621	0.7299	0.6652
9000	4.6351	2.3198	1.4673	1.1365	0.7354	0.6431
10000	4.6289	2.3075	1.4927	1.1283	0.7263	0.6318

basically regarded as constants for all problem sizes of *LINPACK* as figure 2. In general, $\alpha(W) = 1$ and $\beta(W) = 0$. These also indicate that the average performance of multiple uniform virtual machines concurrently executing *LINPACK* in Xen system presents a linearly degressive rule with the number of virtual machines increasing, i.e. the average performance of m uniform virtual machines concurrently executing *LINPACK* is equivalent to the $1/m$ performance of one same virtual machine executing *LINPACK*.

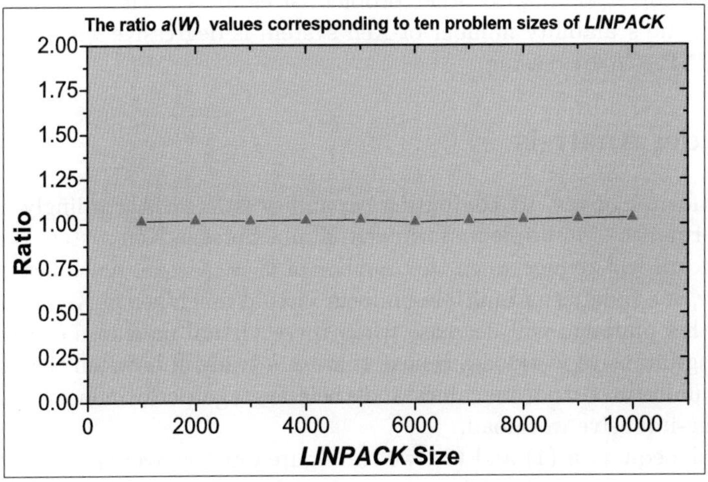

Fig. 2. The value of $\alpha(W)$ corresponding to each problem size of *LINPACK*

From the reverse perspective, the equation (1) can be used to compute the scalability number of Xen system when executing computing-intensive workload. Still taking *LINPACK* as an example, we have already known the values of $\alpha(W)$

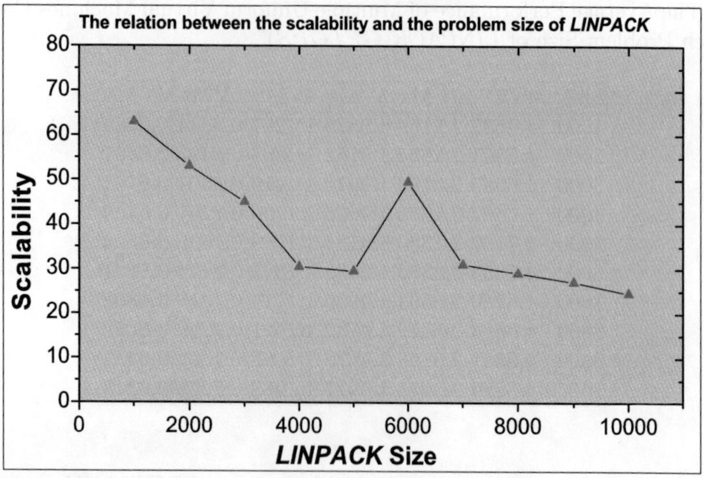

Fig. 3. The scalability number of Xen system corresponding to each problem size of *LINPACK*

and $\beta(W)$ corresponding to *LINPACK*. Assumed that the scalability number of Xen system is m and $P(m, W) > 0$ without losing any generality, we can figure out the scalability number m corresponding to the problem size 1000 of *LINPACK* with the equation (2) to satisfy the inequation $m \leq 63$. The scalability number m corresponding to other problem sizes are shown as figure 3. As we supposed, the scalability number of Xen system is degressive with the problem size of *LINPACK* increasing.

4 Model Analysis

As a platform is given, its computing capability is fixed. Accordingly, the average performance of multiple uniform virtual machines in Xen system running on this platform will reduce when the number of these virtual machines increases. Reversely, the number of multiple uniform virtual machines in Xen system running on this platform will decrease when these virtual machines execute larger computing-intensive workload. Hence, there is a tradeoff between the number of multiple uniform virtual machines and their average performance of executing computing-intensive workload.

With the equation (1) and (2), we can figure out the average performance of multiple uniform virtual machines concurrently executing computing-intensive workload when the number of these virtual machines is given, or the scalability number of Xen system that executes computing-intensive workload and satisfies a certain performance demand. Further, we can find the optimal combination of the scalability number of Xen system and their average performance for a specific hardware platform with two equations. For example, the optimal combination for our $DELL^{TM} PowerEdge^{TM}$ 2900 server is 51 uniform virtual machines and 6000 problem size of *LINPACK*, and the utilization of our $DELL^{TM} PowerEdge^{TM}$2900 server is nearly highest at this time.

5 Related Work

performance modeling of virtualization system has been studied since the first application of virtualization system as the evolution of time sharing and multi-programming in 1960 [9,10,11,12,13,14]. As for the IBM VM/370, Bard et al. [9] gave several models to analyze its performance metrics such as the CPU utilization, and developed a performance predicting tool for this environment. Menasce et al. [10] made a step to predict the performance of virtualization systems with queuing network models, though their models were not verified for all specific virtualization system. Benevenuto et al. [11] proposed specific analytic models for the Xen architecture and validated these models with experimental results.

Recently, Cherkasova et al. [12] presented an approach to measure separately the processor resource overhead of IDD(Isolated Driver Domain) incurred by the I/O operations of a guest domain. The idea is to estimate the processor usage amount of IDD by counting the cost and the number of page exchanges between a certain guest domain and IDD. Further, Menon et al. [13] offered a Xen profiling tool-Xenoprof by modifying the OProfile to diagnose the network I/O performance of Xen system, Gupta et al. [14] described a new performance profiling tool of Xen system-XenMon to monitor the resource usage of Domain0 and guest domains.

In general, all these efforts provided many helpful methodologies and tools to evaluate the overall performance of virtualization system, although involving the fine performance analysis of virtualization system especially the processor resource consumption of Xen system not much.

6 Conclusion and Future Work

In this paper, we first introduced three main scheduling algorithms of Xen system and the scalability metric, and then established a model that can figure out the average performance of virtual machines concurrently executing computing-intensive workload, and the scalability number of Xen system for a specific hardware platform. The optimal combination of the average performance of multiple uniform virtual machines concurrently executing computing-intensive workload and the scalability number of Xen system for a specific hardware platform can be resolved with this model. In the future, we will validate the proposed model with other kinds of workload.

Acknowledgments. This work is supported by the State Key Development Program for Basic Research of China ("973 project", No.2007CB310900) and the 2010 Annual Funding Project of Baoding Association of society and Science(No.20100309). We also would like to thank Dr. Deshi Ye for his helpful advice.

References

1. Menasce, D.: Virtualization: Concepts, Applications, and Performance. In: Proceedings of The Computer Measurement Group 2005 International Conference, Orlando, FL, USA (December 2005)
2. VMware Inc. Understanding Full Virtualization, Paravirtualization and Hardware Assist. Technical report, VMWare Inc. (2007)
3. Barham, P., Dragovic, B., Fraser, K., Hand, S., Harris, T., Ho, A., Neugebauer, R., Pratt, I., Warfield, A.: Xen and the art of virtualization. ACM SIGOPS Operating Systems Review 37(5), 164–177 (2003)
4. Duda, K.J., Cheriton, D.R.: Borrowed-virtual-time(BVT) scheduling: supporting latency-sensitive threads in a general-purpose scheduler. ACM SIGOPS Operating Systems Review 33(5), 261–276 (1999)
5. Leslie, I.M., Mcauley, D., Black, R., Roscoe, T., Barham, P.T., Evers, D., Fairbairns, R., Hyden, E.: The Design and Implementation of an Operating System to Support Distributed Multimedia Applications. IEEE Journal of Selected Areas in Communications (1996)
6. Leslie, I.M., Mcauley, D., Blac, R.: Credit Scheduler, `http://wiki.xensource.com/xenwiki/CreditScheduler`
7. Wikipedia. Scalability, `http://en.wikipedia.org/wiki/Scalability`
8. Whitaker, A., Shaw, M., Gribble, S.D.: Scale and performance in the Denali isolation kernel. ACM SIGOPS Operating Systems Review 36, 195–209 (2002)
9. Bard, Y.: An analytic Model of the VM/370 System. Proc. of IBM Journal of Research and Development 22(5), 498–508 (1978)
10. Menasce, D.A., Dowdy, L.W., Almeida, V.A.F.: Performance by Design: Computer Capacity Planning By Example. Prentice Hall PTR, Upper Saddle River (2004)
11. Benevenuto, F., Fernandes, C., Santos, M., Almeida, V., Almeida, J., Janakiraman, G(J.), Santos, J.R.: Performance models for virtualized applications. In: Min, G., Di Martino, B., Yang, L.T., Guo, M., Rünger, G. (eds.) ISPA Workshops 2006. LNCS, vol. 4331, pp. 427–439. Springer, Heidelberg (2006)
12. Cherkasova, L., Gardner, R.: measuring CPU overhead for I/O processing in the Xen virtual machine monitor. In: Proc. of USENIX Annual Technical Conference (April 2005)
13. Menon, A., Santos, J.R., Turner, Y., Janakiraman, G., Zwaenepoel, W.: Diagnosing Performance Overheads in the Xen Virtual Machine Environment. In: Proc. of First ACM/USENIX Conference on Virtual Execution Environments (VEE 2005), Chicago, IL (June 2005)
14. Gupta, D., Gardner, R., Cherkasova XenMon, L.: QoS Monitoring and Performance Profiling Tool. Technical Report HPL-2005-187, HP Labs (October 2005)

Research on Fuel Flow Control Method Based on Micro Pressure Difference of Orifice Measuring Section

Jianguo Xu and Tianhong Zhang

College of Energy and Power Engineering, NUAA,
Nanjing, China
xujianguo@nuaa.edu.cn

Abstract. The fuel control problem of micro turbine engine (MTE) is studied. The micro pressure difference method is used to measure the fuel flow, which has the advantages of low pressure loss and high accuracy. A new orifice measuring section is designed, and a series of calibration experiments are done with different orifices. The experiment results show that the relationship between fuel flow and the square root of pressure difference is linear in the flow measuring range, which is accordance with theoretical analysis. A PID controller is designed and used in the closed-loop control experiments of fuel flow. In contrast to open-looped fuel control, this new control method is independent of pump motor characteristic, power voltage and outlet pressure. It also has good dynamic performance. The method can be used to improve the quickness and reliability of MTE start-up process.

Keywords: MTE(micro turbine engine); fuel flow control; micro pressure difference; orifice; closed-loop control.

1 Introduction

Fuel flow control is a very important issue in the start-up process of micro turbine engines (MTE). Its accuracy and quickness affect the quality of MTE start-up process directly. At present, the fuel supply for micro turbine engines (MTE) is usually controlled in open loop mode. If pump motor characteristics, battery voltage or outlet pressure are changed, the fuel flow will become uncertain, and the start-up process will become difficult [1-4]. Hence, it is attractive to adopt closed-loop control in fuel supply to improve the quality of MTE start-up process. A new closed-loop control method based on micro pressure difference is presented in this paper, which has the advantages of low pressure loss, high accuracy and quick response. The method can be used to improve the MTE start-up process.

2 Fuel Flow Measurement Based on Micro Pressure Difference

To supply fuel for MTE in closed-loop mode, it is important to measure the real fuel flow accurately. In contrast with civil aero engines, the fuel flow of MTE is small and

M. Zhu (Ed.): Electrical Engineering and Control, LNEE 98, pp. 739–747.
springerlink.com © Springer-Verlag Berlin Heidelberg 2011

is usually driven by an electric pump. This type of pump cannot afford large outlet pressure [5]. At present, the domestic technique for measuring small fuel flow is under developing. Although some foreign commercial flow meters can measure small fuel flow accurately, the measure result is usually given on display screen, and there is not signal interface for control system in most cases [6, 7]. Another disadvantage of foreign commercial flow meters is their poor dynamic response performance. For example, the turbine flow meter made by Eagle Engineering Aerospace Co. Ltd. can measure fluid flow as small as 20ml/min. But its output signal frequency near 20ml/min is only about 2Hz. So the flow measure time may exceed 0.5s and cannot satisfy the requirements of MTE control. Furthermore, the meters also have the disadvantages of large pressure loss, small measure range and high price [8, 9]. Hence it is necessary to study the measure technique of small fuel flow.

According to the principles of fluid dynamics, the flow of liquid in a tube will lead to a pressure loss. Therefore, the flow can be measured according to the pressure loss (micro pressure difference). A flow measure device based on the detection of micro pressure difference is designed as shown in Fig. 1.

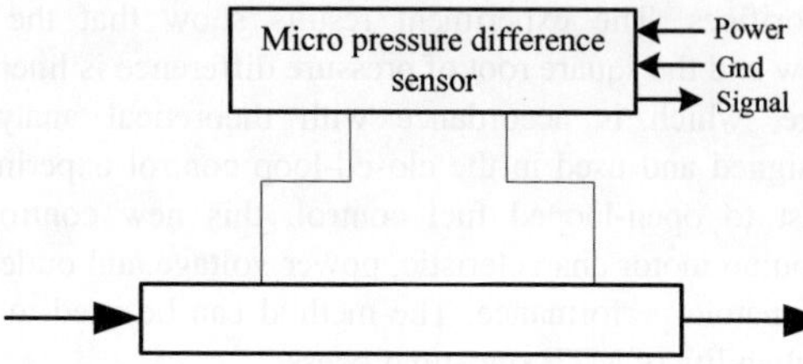

Fig. 1. Measure device based on micro pressure difference

According to Fig. 1, the measure device is comprised of two components: a sensor and a metering section. The sensor transfers pressure difference signals in voltage (0~5V). The precision of the sensor used in the measure device is 0.05%FS, and the scale is 0kPa~10kPa. The micro pressure difference for calculating fuel flow is detected from the two ends of the metering section. Through appropriate design of the structure of metering section, the pressure loss may be very small and the measure precision may be high. Metering section is mainly of two kinds: tube structure and orifice plate structure. Between metering section and sensor, there are two tubes that transfer pressure difference signals [10, 11].

3 Design of Metering Section of Orifice Plate Structure

Metering section is an important component of flow measure device based on micro pressure difference. The type and structure of metering section influence the measurement accuracy of fluid flow significantly. There are mainly two kinds of structure, tube structure and orifice plate structure. Metering section of tube structure is easy to machine, and fluid flow is linear with pressure difference. Its disadvantage is of large volume [10, 11]. Metering section of orifice plate structure has much complex design. But it is small, so it is convenient to use in an aircraft. And orifice plate can be machined and changed easily thus bring convenience to test and research. Therefore, orifice plate structure is chosen as the structure of metering section.

3.1 Relationship between Fluid Flow and Pressure Difference

When the fluid flows through an orifice in a tube, the stream will shrink at the orifice. The fluid velocity will increase, and the static pressure will reduce. Consequently, there will be static pressure difference at the two sides of the orifice, as Fig. 2 shows. The pressure difference is influenced by the flow and physical properties of the fluid. The greater the fluid velocity is, the greater the pressure difference is [12]. Therefore, the fluid flow can be calculated according to the pressure difference.

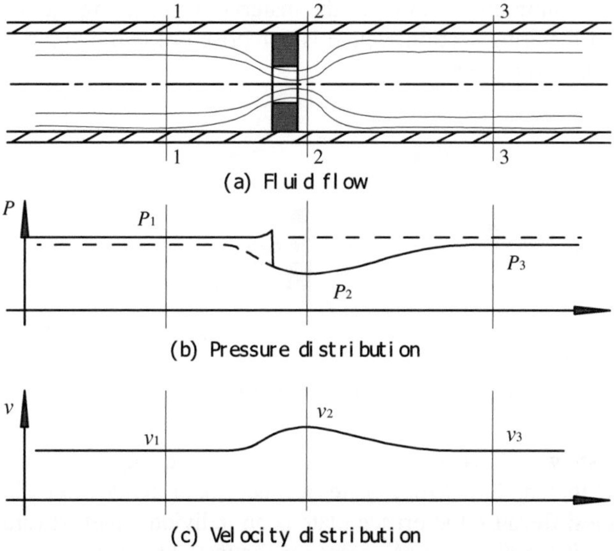

Fig. 2. The forming of pressure difference

Suppose the fluid is incompressible. According to fluid dynamics principles, there are certain mathematical relationship between fluid flow and pressure difference:

$$q_{\mathrm{m}} = \frac{C}{\sqrt{1-\beta^4}} \frac{\pi}{4} d^2 \sqrt{2\Delta P * \rho} \tag{1}$$

$$q_{\mathrm{v}} = q_{\mathrm{m}} / \rho \tag{2}$$

In the above equations, q_{m} is mass flow, q_{v} is volume flow, β is diameter ratio (the ratio of orifice diameter d to tube inner diameter D), C is discharge coefficient,

ΔP is static pressure difference between the two sides of orifice plate, ρ is fluid density.

Discharge coefficient c is an important parameter which represents the primary characteristics of orifice. Usually it is obtained through calibration test. When standard orifice plate structure is adopted, discharge coefficient can also be calculated by Reader-Harris/Gallagher formula (R-G formula) [13]. According to R-G formula, discharge coefficient c is the function of Reynolds number R_e and diameter ratio β, and it will reduce when R_e increases.

3.2 Design of Metering Section Structure

When the structure of metering section is designed, the following elements should be considered carefully: (1) Satisfy the requirement of measuring scale of MTE start-up process. (2) Easy to be machined. (3) Small volume and weight. Based on the above considerations, aluminum is chosen as the material for metering section. Fig. 4 shows the structure of metering section, which is comprised of two cylindrical cavities, one orifice plate, and two nipples.

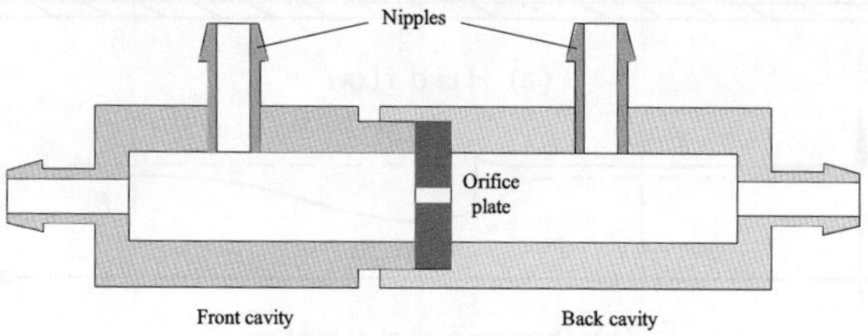

Fig. 3. Metering section of orifice plate structure

As Fig. 4 shows, the front cavity and the back cavity are assembled together through screw thread. The orifice plate between them divides the whole component into two cavities. Because the orifice plate is an individual part, several orifice plates with different orifice diameter are prepared for test. The metering section uses four nipples for connecting to tubing. Two of them are machined as a portion of the two cavities. And the other two are individual parts and are mounted on cavities through screw thread. There is a cone-shaped crossette at the tip of nipple, which has the effect of seal and prevents tubing from slipping off.

3.3 Calibration Tests

Although fluid flow can be calculated from pressure difference according to Equation (1) and Equation (2), discharge coefficient C is obtained through tests. Furthermore, there are phenomena of collision and vortices near the nipples. Thus the results calculated from above equations are not precise. To improve the measuring precision, calibration tests need to be carried, which establish the real relationship between micro pressure difference of metering section and fluid flow through metering section.

Based on theoretic analysis, calibration tests are carried for four orifice plates with high precision flow meter. The diameters of these orifice plates are 1mm、1.2mm、1.5mm and 1.8mm respectively. Fig.4 is the fitting curves of one order on the basis of test data. In Fig. 4, $\sqrt{\Delta P}$ stands for the square root of pressure difference, and Q stands for fuel flow. From the curves, it is obvious that fuel flow is linear with the square root of pressure difference. When orifice diameter increases, the measuring scale also becomes larger. When orifice diameter is 1mm, the measuring range is 30 ml/min – 230 ml/min. So metering section with 1mm orifice diameter can fulfill the flow measuring needs of MTE start-up process.

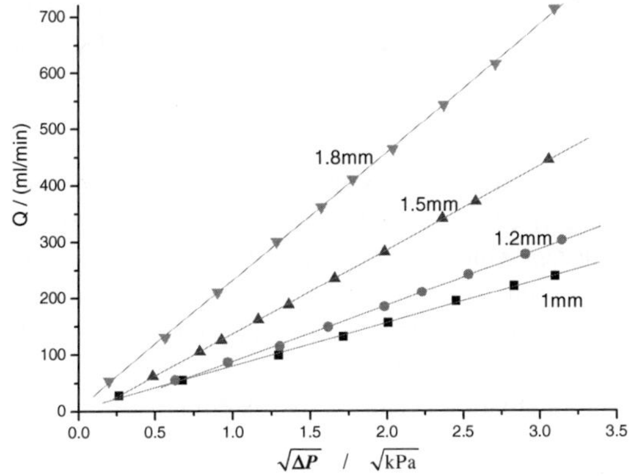

Fig. 4. Square root of pressure difference-fuel flow calibration curves

With the fitting curves in Fig. 4, the fuel flow can be calculated according to pressure difference and can be used as the fuel flow feedback in closed-loop control.

4 Closed-Loop Control of Fuel Flow

4.1 Design of PID Controller

Based on the micro pressure difference measurement device, a closed-loop controller is designed to regulate the power pulse width for pump motor, so that the real fuel flow is consistent with the expected fuel flow. Apparently, the control performance depends on the measurement accuracy of fuel flow. Figure 5 shows the principle of closed-loop control of fuel flow.

PID Control is a popular technique in industry application [14]. The method is independent of the structure and parameters of the controlled object. And it is also simple, robust, reliable and convenient to use. So PID controller is chosen to control fuel flow. The input of PID controller is expected fuel flow, and the output is PWM signal. Amplified by drive circuit, the PWM signal is supplied to pump motor to control the fuel flow.

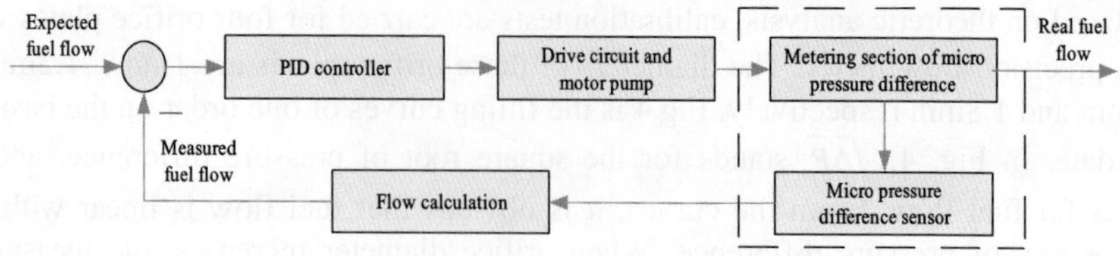

Fig. 5. The schematic diagram of closed-loop fuel flow control

The control step and sample time of the PID controller are 5ms. The PID parameters are firstly obtained through step response curve method. Then these parameters are optimized in the latter experiments as: K_P is 3, T_I is 0.2s and T_D is 0.07s.

4.2 Performance Validation

With the PID controller and a metering section of 1mm orifice diameter, closed-loop fuel flow control tests are carried out to validate the control performance. Fig. 6 is square wave response test curves, which show the real fuel flow *Flow* when the expected fuel flow *Flowset* varies between 100 ml/min and 140 ml/min with a period of 4s. According to Fig. 6, the settling time is less than 0.2s, and the maximum overshoot is less than 5%. After the transient process, the real flow fluctuates slightly in the range of 0.5% around the expected flow. The fluctuation is caused by system noises.

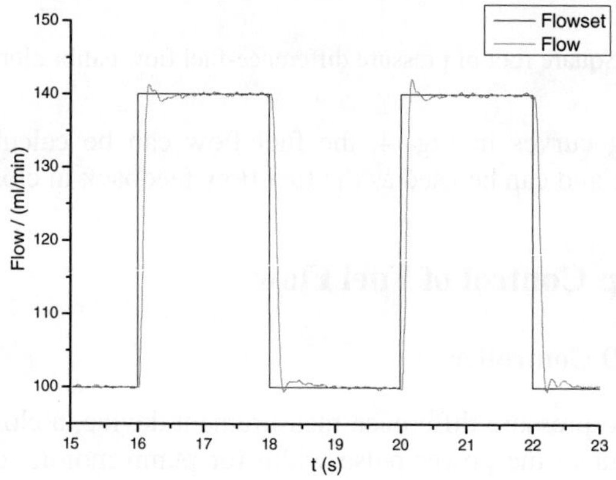

Fig. 6. Square wave response test curves

Fig. 7 is sinusoidal wave response test curves, which show the real fuel flow *Flow* when the expected fuel flow *Flowset* varies with a frequency of 1 Hz in sinusoidal form. According to Fig. 7, the real flow is only 0.628 radian behind the expected flow.

In the grounds of above validation tests, the PID parameters are appropriate. On one hand, the principle steady state errors are eliminated and the steady state accuracy reaches 0.5%. On the other hand, the frequency response bandwidth is more than 4Hz and can satisfy the requirements of MTE start-up control.

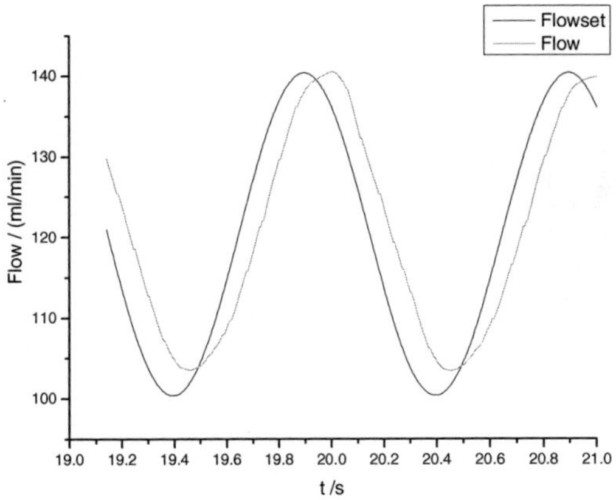

Fig. 7. Sinusoidal wave response test curves

4.3 Performance Compare with Open Loop Method

Performance compare tests are carried out, which further show the superiority of the presented method in this paper.

Step response tests from 0 ml/min to 100 ml/min are carried out by using open loop and closed-loop control methods. Fig. 8(a) is the open loop control test curve. From Fig. 8(a), the settling time is 0.5s. Fig. 8(b) is the closed-loop control test curve with the above PID controller. From Fig. 8(b), the settling time is less than 0.2s, and the maximum overshoot is about 6%. It is obvious that the dynamic performance of closed-loop control is better than open loop control.

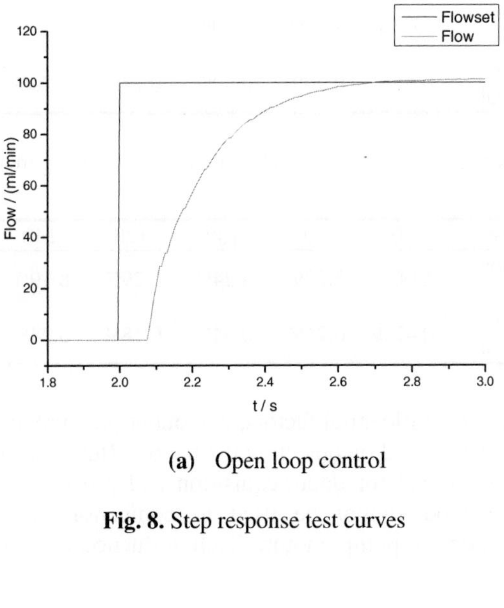

(a) Open loop control

Fig. 8. Step response test curves

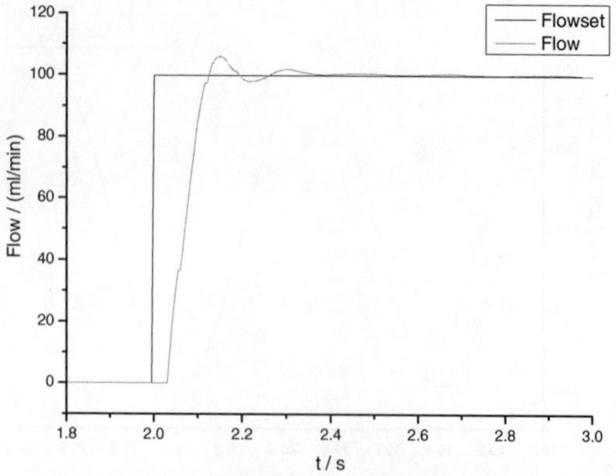

(b) Closed loop control

Fig. 8. (*continued*)

The main disadvantage of open loop control method is that its output can be affected by many factors. According to Table 1, the relative error of fuel flow will be as large as 7.07% when the battery voltage reduces 5%. Base on Table 2, after pump motor runs continuously for 30 minutes, the characteristics of pump motor will change dramatically and the maximum relative error of fuel flow will be 8.95%. In contrast, the two cases have no effect on the closed-loop control, which relative error is less than 0.5% in the whole measuring range.

Table 1. The influence of the variation of power voltage

Expected flow(ml/min)	60	80	100	120	140	160	180
Relative error of open loop control	7.07%	4.66%	4.40%	5.74%	6.82%	6.72%	6.17%
Relative error of closed-loop control	0.35%	0.26%	0.29%	0.24%	0.20%	0.16%	0.24%

Table 2. The influence of the variation of motor characteristics(pump motor runs continuously for 30 minutes)

Expected flow(ml/min)	60	80	100	120	140	160	180
Relative error of open loop control	4.08%	3.23%	4.24%	6.29%	8.19%	8.64%	8.95%
Relative error of closed-loop control	0.42%	0.28%	0.34%	0.18%	0.24%	0.18%	0.23%

The influence of some influential factors, e.g. outlet pressure, can be eliminated by compensating the output based on detecting the factors. But additional sensors need to buy, and the CPU overhead for data acquisition and processing will be increased. Furthermore, the method cannot be used to compensate the influence of the characteristics variation of pump motor. Such influence can only be avoided by closed-loop control.

5 Conclusion

Accurate, rapid fuel flow control is very important for MTE start-up. Open loop control method results in large real fuel flow error because of the influence of battery voltage, outlet pressure and pump motor characteristics. Fuel flow measuring method based on micro pressure difference is studied, metering section of orifice plate structure is designed, and PID controller is designed for closed-loop control of fuel flow. Validation test results show that the above fuel flow control method is independent of the characteristics of pump motor, has high steady state accuracy and good dynamic performance. The method can satisfy the requirements of MTE start-up control.

References

1. Yuan, L., Xianghua, H., Tianhong, Z.: Research on Key Control Technology of Micro Turbine Engine in Centimeter Size. Journal of Aerospace Power 22, 480–484 (2007)
2. Chiang, H.W.D., Hsu, C.N., Lai, A., Lin, R.: An Investigation of Steady and Dynamic Performance of a Small Turbojet Engine. In: Proceedings of ASME Turbo Expo, GT-2002-30577 (2002)
3. Gerendas, M., Pfister, R.: Development of a Very Small Aero Engine. In: 45th ASME International Gas Turbine and Aero Engine Technical Congress and Exposition, -GT-536 (2000)
4. Ying, P.: An Analysis of the Starting Characteristics of Aeroengine. Journal of Aerospace Power 18, 777–782 (2003)
5. Junxia, M., David, R.: Advanced Controller Design for Aircraft Gas Turbine Engines. Control Engineering Practice 13, 1001–1015 (2005)
6. Dongliang, S., Jintang, Z., Wei, K.: Research on the Test Method of Liquid Micro Flowrate. Petroleum Instruments 19, 69–72 (2005)
7. Jianjun, T.: The Development of Standard Micro Flow Measuring Instruments. Industrial Measurement 3, 27–28 (2000)
8. Yanxun, S., Guowei, L., Jian, S.: Flow Measurement and Testing. China Metrology Publishing House, Beijing (2007)
9. Lijun, S., Tao, Z., Gang, L.: A Study on the High Precision Measurement of Fuel Flow. Acta Metrologica Sinica 25, 235–240 (2004)
10. Kunnian, M.: Research on Fuel Metering and Experimental Technology for Micro Turbine Engine. Nanjing University of Aeronautics and Astronautics, Nanjing (2007)
11. Micro Fluid Flow Control Device Independent of Pump Motor Characteristics and of Small Pressure Loss. China, ZL200710024646.3 (2007)
12. Maohuan, D.: Development and Current Situation of Throttle Differential Flowmeter. Industrial Measurement 6, 30–32 (2002)
13. GB/T2624.2-2006 Measuring Fluid Flow with Pressure Difference Device Mounted in Circular Tube, the Second Part: Orifice Plate. Standards Press of China, Beijing (2006)
14. Zaiying, W., Huaixia, L., Yijing, C.: System and Instrument of Process Control. China Machine Press, Beijing (2006)

Dynamic Modeling and Simulation of UPFC

Jieying Song[1], Fei Zhou[2], Hailong Bao[3], and Jun Liu[3]

[1] Department of Electrical Enginneering, NCEPU Baoding Hebei, China
[2] Department of Power Electronics, China Electric Power Research Institute, Beijing, China
[3] Shanghai Municipal Electric Power Company, Shanghai, China
sunnysong006@163.com

Abstract. Unified Power Flow Controller (UPFC) combines function of voltage control, power flow control and oscillation damping. Before installation to a system, analysis for interactions between UPFC and system are needed. Building up steady, transient and dynamic models by present software to realize accurate simulations of UPFC under different system situation is a key step. In this paper, an UPFC model under dq axis including dynamic characteristics on DC side is built by switching function method. After control scheme analysis, effects of different damping strategies are compared. Simulation results show the correctness of the model and the validity of UPFC in controlling power flow and damping system oscillation.

Keywords: UPFC, dynamic modeling, controller design, Oscillation damping.

1 Introduction

Flexible AC Transmission System (FACTS) tremendously improves flexibility, controllability, safty and economy of power system, provides strong technique support for Strong & Smart Grid, thus being research focus of power system in recent years. As typical 3rd generation FACTS device, Unified Power Flow Controller (UPFC) combines voltage control, power flow control and other functions, thus receiving much more attention. Recent research focus on dynamic character analysis and controller design, UPFC are always supposed to be operating well under any system conditions, however, UPFC may be unable to function correctly or even out of service after faults. So a credible UPFC model is important to system simulation, analysis, and controller design.

Power electronic devices modeling usually can be divided into component-level modeling, device level modeling and system-level modeling [2]. When research focus is on interactions between device and system, switching function method can be used to build up system level model, which can describe inner characters of a device without too much complexity. So this paper built a UPFC model which considers whole dynamic characters of the shunt part, series part and DC part by switching function method. Different outer loop controllers together with damping controller are designed based on that model, taking system operation constraints into account. On planning UPFC dynamic control strategies, we hope it can both improve system damping without deterioration of other parts in the system and control the line power under system disturbance or faults.

M. Zhu (Ed.): Electrical Engineering and Control, LNEE 98, pp. 749–756.
springerlink.com © Springer-Verlag Berlin Heidelberg 2011

2 Modeling of UPFC

The equivalent circuits of UPFC and the line where it is installed are shown in fig.1. U_s is system voltage, U_1 is the bus voltage for the shunt part, U_2 is the UPFC access point voltage, U_r is system voltage of the receiving side, X_R is line reactance. Reference directions of all electrical quantities are shown in fig.1.

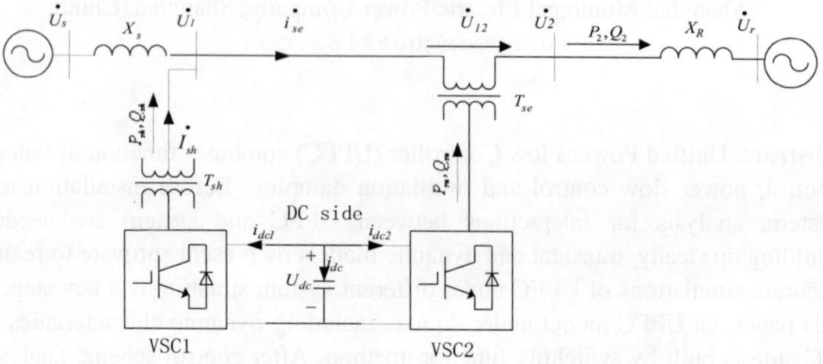

Fig. 1. Equivalent circuit of UPFC and the system

Neglecting main circuit losses, non-saturation characteristic of the transformer and the influence of snubber and protection circuits, the shunt and series converter is equivalent to the series connection of impedance and controllable voltage source. The equivalent impedance of the shunt and series part are $R_{sh}+jX_{sh}$ and $R_{se}+jX_{se}$ respectively, R_{loss} is equivalent losses on the DC side. All parameters used in modeling are reducing calculus to high voltage side. The UPFC model under DQ axis is shown in equation (1).

$$\frac{d}{dt}\begin{bmatrix} i_{shd} \\ i_{shq} \\ i_{sed} \\ i_{seq} \\ u_{dc} \end{bmatrix} = \begin{bmatrix} -\frac{R_{sh}}{L_{sh}} & \omega & 0 & 0 & \frac{S_{dsh}}{L_{sh}} \\ -\omega & -\frac{R_{sh}}{L_{sh}} & 0 & 0 & \frac{S_{qsh}}{L_{sh}} \\ 0 & 0 & -\frac{R_{se}}{L_{se}} & \omega & \frac{S_{dse}}{L_{se}} \\ 0 & 0 & -\omega & -\frac{R_{se}}{L_{se}} & \frac{S_{dse}}{L_{se}} \\ \frac{-\frac{3}{2}S_{dsh}}{C_{dc}} & \frac{-\frac{3}{2}S_{qsh}}{C_{dc}} & \frac{-\frac{3}{2}S_{dse}}{C_{dc}} & \frac{-\frac{3}{2}S_{qse}}{C_{dc}} & 0 \end{bmatrix} \begin{bmatrix} i_{shd} \\ i_{shq} \\ i_{sed} \\ i_{seq} \\ u_{dc} \end{bmatrix} - \begin{bmatrix} \frac{u_{1d}}{L_{sh}} \\ \frac{u_{1q}}{L_{sh}} \\ \frac{u_{12d}}{L_{se}} \\ \frac{u_{12q}}{L_{se}} \\ 0 \end{bmatrix} \tag{1}$$

In equation (1), $S_{dsh}=(m_{sh}/2)\cos\delta_{sh}$, $S_{qsh}=(m_{sh}/2)\sin\delta_{sh}$, $S_{dse}=(m_{se}/2)\cos\delta_{se}$, $S_{qse}=(m_{se}/2)\sin\delta_{sh}$; The instantaneous reactive power injected by the shunt part is:

$$Q_{sh} = \frac{3}{2}(u_{1q}i_{shd} - u_{1d}i_{shq}) = -\frac{3}{2}u_{1d}i_{shq} \tag{2}$$

The sending end power of the transmission line being controlled is :

$$\begin{cases} P_2 = \frac{3}{2}(u_{2d}i_d + u_{2q}i_q) \\ Q_2 = \frac{3}{2}(u_{2q}i_d - u_{2d}i_q) \end{cases} \tag{3}$$

The model of UPFC must fit the following constraints [2]:

Maximum series injected voltage: $U_{12} \leq U_{12max}$

Maximum line current: $I_{se} \leq I_{semax}$

Maximum shunt current: $I_{sh} \leq I_{shmax}$

Access point voltage of UPFC: $U_{1min} \leq U_1 \leq U_{1max}$; $U_{2min} \leq U_2 \leq U_{2max}$;

Active power exchange on the DC side: $|P_{dc}| \leq P_{dcmax}$

The line sending end P-Q operation curve under UPFC control is shown in fig.2, their common part is actual operation region under above mentioned constraints. UPFC is a multifunctional device which contains several different control objects. However, the maximum voltage and current UPFC can withstand are limited. When system operation state changes, not all reference values can be satisfied. So how to make a compromise of different reference control objectives and maximize UPFC system functions under non-ideal situations should be considered before modeling and simulation.

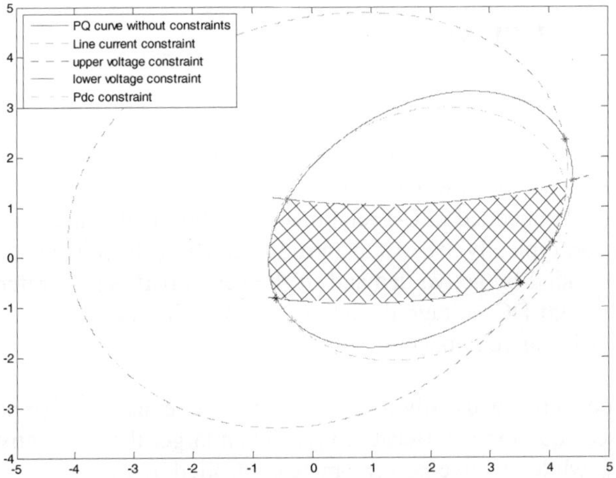

Fig. 2. Line sending end P-Q region curves under different constraints

Through preliminary study of operation constraints for UPFC, limits calculating units and constraints judging element should be added to controllers. The flowchart of constraints solving is shown in fig.3. When UPFC exceeds boundaries of feasible operation region, controller holds injected voltage amplitude constant while changes its phase to find a new feasible operating point.

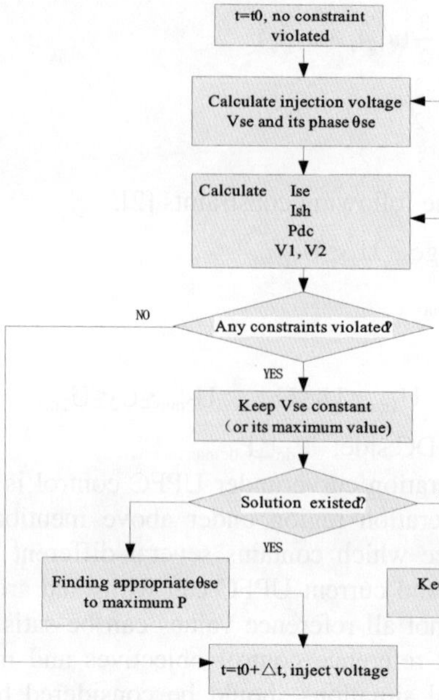

Fig. 3. Flowchart for the solving of constraints

3 Control Strategy Analysis

3.1 Controller Design

The shunt part of UPFC can compensate reactive power, back up bus voltage and provide necessary active power for series converter to hold DC voltage constant. A droop character is imbedded in U-I relationship of the shunt converter, so an deviation adjustment factor, typically 3%~5% is introduced. DC voltage is mainly controlled by i_{shd}, which is decided by active demand of the series part, so i_{shd} reference value can be obtained through DC voltage deviation; on the other hand, reactive power of the converter has a linear relationship with i_{shq}, so reactive power can be controlled by i_{shq}.

The series converter is usually used to control active and reactive power of the line sending end, see equation (3). Because v_{2d} is much larger than v_{2q}, active power can be controlled by i_d while reactive power can be controlled by i_q.

Except power flow control capability, the series part possesses impedance control, phase control and line voltage control ability. Impedance control is equivalent to a series connection of changeable impedance, phase control can change line voltage phase without changing its amplitude while voltage control changes line voltage amplitude without changing its phase. These three controllers directly give reference voltages.

The outer loop controller of both the shunt and series converter can be realized by feed forward and feedback control through PI regulators. From equation (1) we can see d axis current and q axis current is coupled by leakage inductance, and this is not conducive to control, so dq currents decoupling control method is used. By current feedback control and voltage feed forward compensation, the inner loop controller can directly control the output currents, thus realizes fast dynamic response together with the function to limit output current.

Damping control is an auxiliary control function of UPFC, and its design always consults PSS. Signals which can reflect system oscillation go through blocking element, amplify element, lead-lag element and cut ridge element, then feed to controller reference values of the outer loop. Because angular frequency variation □ω and electromagnetic power variation $\triangle P_e$ of a generator are uneasy to measure, line power and the phase difference between line sending and receiving end bus voltage can be used as damping signals. Outputs of damping controller feed back to power or voltage reference values.

3.2 Dynamic Control Strategy

Maintaining line power to a specified value is the control object of UPFC under steady state; when disturbance occurs, suppressing power oscillation should be concerned about. However, either operation mode or control parameter changes can both impact power systems, so command changes should not occur too fast. According to above analysis, the dynamic control strategy is summarized below:

(1) Under steady state, UPFC work on automatic power flow control mode, adjusting the line power to fit system demand.

(2) When small disturbance occurs, auxiliary damping control is added, and different damping schemes can be implemented according to system demand.

(3) When serious outer fault occurs, the series converter of UPFC may be bypassed according to the value of line current. The shunt converter can keep on working to provide reactive power. When system voltage recovers, UPFC goes back to previous operation mode and adds damping control.

4 Simulation Analysis

The dynamic characters of UPFC are testified by a simulation model under PSCAD/EMTDC, neglecting the output harmonic influences. Taking the dynamic processes of shunt converter, series converter and DC link into account, a UPFC model which can be used to testify power flow control ability, stable control ability and voltage control ability, together with fault response characters under unbalance disturbance is built.

Main parameters of UPFC are shown in table 1, besides, the equivalent DC capacitor in the simulation model is 750uF, and DC voltage reference value is 50kV.

Table. 1 Parameters of the UPFC simulation model

UPFC	rated capacity (MVA)	Rated voltage (kV)	Xt (p.u)	Transformer ratio
Series part	100	50	0.1	2 : 1
Shunt part	100	500	0.1	20 : 1

Under steady state, UPFC can realize independent control of line active and reactive power, as shown in fig.4. When serious line faults happen, UPFC should quit during fault, and return to normal operation state after fault is cleared as soon as possible. Fault response of UPFC is shown in fig.5, when single ground fault happens at the end of a line, A phase voltage drops, the series part of UPFC are bypassed when over current is tested. 6 fundamental cycles after fault, phase A breaker operates, and system voltage recovers, then bypass is lifted. 9 fundamental cycles after fault, line breaker reclosed, and system returns to the operating state before fault. The system voltage, line current and series injected voltage waveforms are all shown in fig.5.

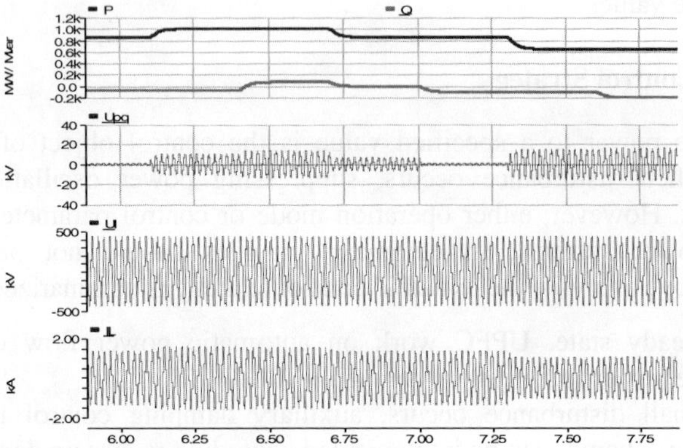

Fig. 5. Power flow control character of UPFC

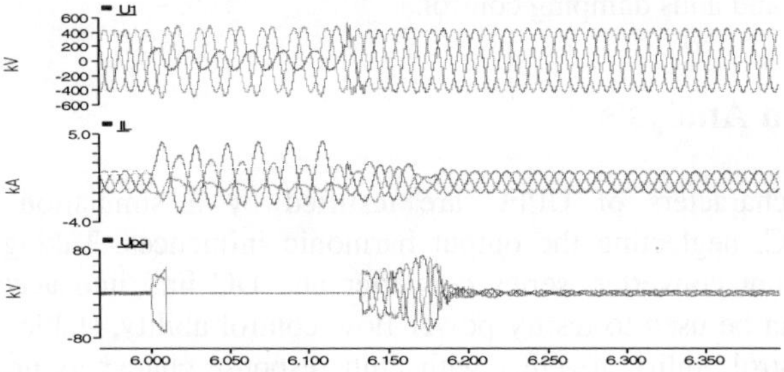

Fig. 6. Single Line resistance ground fault response of UPFC

When dynamic disturbance occurs and brings power oscillations, UPFC can force line power to a specified value thus damping power oscillation. If UPFC is installed on one of double lines, the oscillating power caused by generators flow through the parallel line. Reasonable control of UPFC can inhabit power oscillation of generator. By adding a single phase resistance ground fault on the receiving side bus, electromagnetic power of generator, the angle change rate, the series injected voltage and line current waveforms under different control schemes are shown in Figure 7.

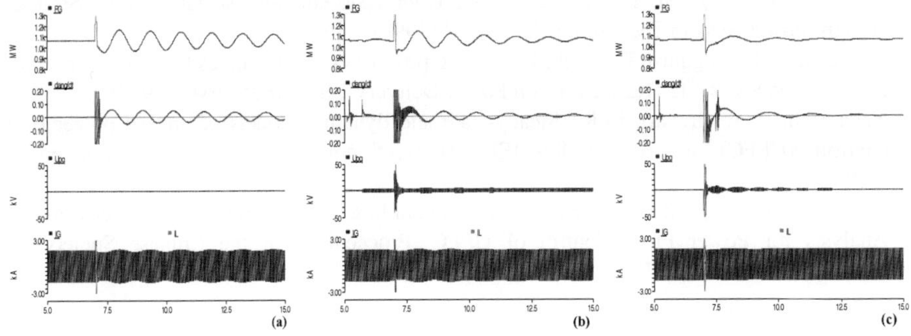

Fig. 7 Waveforms of damping system oscillation by UPFC (a) system oscillation after disturbance without any control; (b) UPFC adopts phase control and damping control to damp oscillation; (c) UPFC adopts power flow control and damping control

The signal used for damping control is the phase difference of two ends of the line being controlled. From fig.7 we can know, when power oscillation happens, different damping methods can be adopted by UPFC with different damping effects.

5 Summary

By switching function method, this paper built a UPFC model under dq axis, which considering operation constraints and including all dynamic process of the whole device. Based on the model, different controllers for the shunt and series converters are designed. Simulation results verified the correctness of the model and the validity of the controllers.

Preliminary compare of different damping strategies are done by this paper, if intelligent methods are used to compute control parameters or combine different operating mode by weighting methods, better damping effects may be get.

References

1. Gyugi, L., Schauder, C.D., Williams, S.L., Rietman, T.R., Torgerson, D.R., Edris, A.: The Unified Power Flow controller: A New Approach To Power Transmission Control. IEEE Transactions on Power Delivery 10(2), 1085–1093 (1995)
2. Bian, J., Rame, D.G., Nelson, R.J., Edris, A.: A Study of Equipment sizes and constraints for A Unified Power Flow controller. IEEE Transactions on Power Delivery 12(3) (1997)
3. Maksimovic, D., Stankovic, A.M., Thottuvelil, V.J., et al.: Modeling and simulation of power electronic converters. Proceedings of the IEEE 89(6), 898–912 (2001)
4. Han, B., Baek, S., Kim, H., Karady, G.: Dynamic Characteristic Analysis of SSSC Based on Multibridge Inverter. IEEE Transactions On Power Delivery 17(2) (April 2002)
5. Limyingcharoen, S., Annakkage, U.D., Pahalawaththa, N.C.: Effects of unified power flow controllers on transient stability. IEEE Proceedings Generation,Transmission and Distribution 145(2), 182–188 (1998)

6. Huang, Z., Diao, Q., Ni, Y., Chen, S.: UPFC Contol System Analysis and Control Strategy Design. Power System Technology 23(7), 3–9 (1999)
7. Noroozian, M., Angquist, L., et al.: Improving power system dynamics by series connected FACTS devices. IEEE Transactions on Power Delivery 12(4), 1635–1641 (1997)
8. Nabavi-Niaki, A., Iravari, M.R.: Steady-state and dynamic models of unified power flow controller(UPFC) for system studies. IEEE Transactionon Power System 11(4) (November 1996)
9. Liu, L., Kang, Y., Chen, J., Zhu, P.: Cross-coupling Control Scheme and Performance Analysis for Power Flow Control of UPFC. Proceedings of the Chinese Society for Electrical Engineering 27(10), 42–48 (2007)

Microturbine Power Generator

Martin Novak, Jaroslav Novak, Ondrej Stanke, and Jan Chysky

Czech Technical University in Prague, Faculty of Mechanical Engineering,
Department of Instrumentation and Control Engineering, Technicka 4,
16607 Prague, Czech Republic
{Martin.Novak2,Jaroslav.Novak,
Ondrej.Stanke,Jan.Chysky}@fs.cvut.cz

Abstract. This paper describes some experience gained with a construction of an experimental testing stand for micro turbine power generation with high-speed permanent magnet synchronous motor and its control. The main goal of this research is to create an affordable small (order of kW) power supply preferably for combined electrical energy and heat production, that could be used in households, portable power generators etc. This approach promises lower fuel consumption, lower emissions and better efficiencies that classical piston based portable power generators. The presented test stand parameters are 40000 RPM, torque 7Nm.

Keywords: Microturbine, High speed permanent magnet motor, Torque control.

1 Introduction

In present days there is a clearly visible trend in new energy sources. One of the principles with good perspectives is the micro energy generation with a micro turbine. This solution uses a miniature gas powered turbine powered e.g. with natural gas, biogas, gasoline etc. The turbine is coupled to a generator providing electrical energy. The advantages over a classical piston based generator are significantly higher efficiency and smaller dimensions for the same output power. They are also less sensitive for fuel impurities. However there are also quite some technological challenges. The turbine is running at very high speeds, usually over 50 000 rpm and with temperatures around 800°C. To couple such a turbine with an electrical generator normally means to use some transmission to reduce the speed and only after the transmission to connect the generator. This however decreases efficiency and produces mechanical problems. One solution is to connect the generator directly to the turbine shaft. In this case the generator has to run at the same speed as the turbine. One promising implementation of the generator is to use a permanent magnet synchronous motor (PMSM) controlled with an inverter. The main problems of PMSM are presently mainly the bearings and also the possible damage of permanent magnets with higher temperatures. Although significant, these issues are however not the scope of this paper. This paper will focus on challenges of inverter control of a high speed PMSM generator and present the experimental setup currently being created.

M. Zhu (Ed.): Electrical Engineering and Control, LNEE 98, pp. 757–764.
springerlink.com © Springer-Verlag Berlin Heidelberg 2011

2 Theoretical Analysis of Torque Control

The lowest feedback control level at high speed drive is the torque control. The standard method for torque control of PMSM goes out from Equation 1.

$$M = 1.5 \cdot p_p \cdot (\psi_d \cdot i_q - \psi_q \cdot i_d) \tag{1}$$

ψ_d is magnetic flux linkage component in axis d, ψ_q is magnetic flux linkage component in axis q, i_d is stator current component in axis d, i_q is stator current component in axis q, p_p is number of pole-pairs on the machine. Using synchronous machine mathematical model equations Equation 1 can be written as

$$\begin{aligned} M &= 1.5 \cdot p_p \cdot [(\psi_f + L_d \cdot i_d) \cdot i_q - L_q \cdot i_q \cdot i_d] = \\ &= 1.5 \cdot p_p \cdot i_q \cdot (\psi_f + L_d \cdot i_d - L_q \cdot i_d) \end{aligned} \tag{2}$$

ψ_f is rotor magnetic flux linkage, L_d is synchronous direct-axis inductance and L_q is synchronous quadrature-axis inductance. If machine works in full magnetic flux mode, the i_d (stator current component in axis d) equals zero and for the torque holds

$$M = 1.5 \cdot p_p \cdot \psi_f \cdot i_q \tag{3}$$

Equation 3 holds for field weakening regime too when L_d=L_q. This equality is usually fulfilled on PMSM. In field weakening mode i_d (stator current component in axis d) acts against permanent magnet flux and enables to operate the machine in high revolutions with constant stator voltage. Equation 3 is expressing an analogy with DC machine.

The armature interference given by i_d component is acting against the induced voltage in this mode and allows using the motor with an increasing speed while maintaining constant RMS stator power supply voltage.

Current and torque control can be performed either in transformed coordinate system or by controlling instantaneous stator current with respect of instantaneous rotor position. In full magnetic flux mode the current of given phase is controlled in such a way that the current amplitude occurs in time when the rotor is perpendicular with this phase.

The basic feedback structure of the implemented controller is shown on Fig. 1. It is a vector control structure with parallel controllers for current components i_d and i_q. In a flux weakening mode the control structure creates a non zero i_d current component. Its magnitude increases with increasing speed, increasing current and decreasing inverter input voltage U_{DC}. Calculation of required i_d current component value is done by a feedback controller Reg |U| and a predictive calculation block. This predictive block is predicting an optimal value i_d^{**} from the instantaneous value of mechanical speed ω and from inverter input voltage U_{DC}. The resulting value of i_d is given by adding the output value from the predictive block and feedback controller. In flux weakening mode this maintains a constant and maximal inverter RMS output voltage as can be seen on Fig. 1.

In this system the key issue is the achievable acceleration of electric motor.

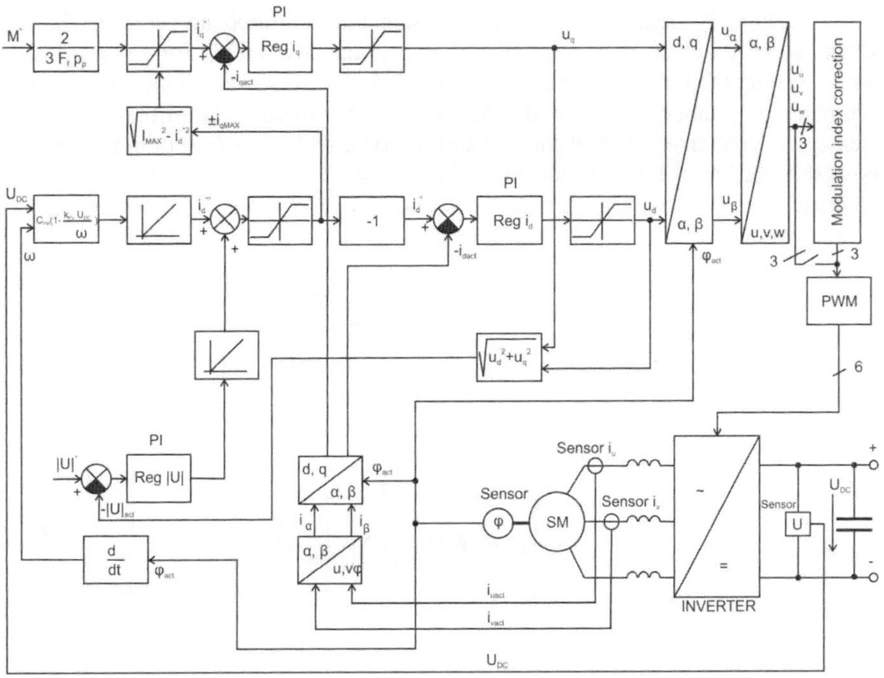

Fig. 1. Torque structure control of high speed drive

3 Experimental System Description

As can be seen form the block diagram on Fig. 2, the experimental system is composed of several blocks. The power network voltage is rectified with an inverter to DC voltage. The inverter has to be able to transfer the energy not only from the power network, but also back. For this reason a standard diode rectifier can not be used, but the structure has to incorporate transistor switches. At the present time this block has not yet been implemented, therefore it is dashed in the diagram.

The DC voltage is supplying an insulated gate bipolar transistor (IGBT) inverter build for this purpose. The inverter is using power module SKM75GD124D and IGBT/MOSFET driver SKHI61 both from Semikron [6]. The inverter is shown on Fig. 3. For the purposes of control and efficiency measurement, the system is equipped with current and voltage sensors in the DC intermediate circuit and on the three phase output of the inverter. The inverter is connected to PMSM with the following parameters: type 2AML406B-S from VUES Brno, nominal voltage 179V, nominal torque 1,2Nm, nominal current 12,2A, nominal speed 25 000 rpm, maximal speed 40 000 rpm, maximal torque 7Nm. As the control algorithm necessitates information about PMSM rotor position a resolver is embedded into the motor and connected to a developed resolver to digital unit. This unit provides 4096 positions per one revolution [1]. A DSP controller based on TMS320F2812 is used for the control of the whole system. A more detailed description can be found in [3].

The turbine is created by means of a standard car turbocharger. This has the advantage of good availability, high reliability and low price. For the purpose of system testing the turbocharger is connected to a compressed air supply, in the future a combustion chamber will be build. Also some other alternatives for the turbine have been considered like a model aircraft turbine or a real aerospace turbine. However those solutions have a big disadvantage in reliability. Model aircraft turbines require complete maintenance only after aprox. 10 hours of service. This is unacceptable for a power generator.

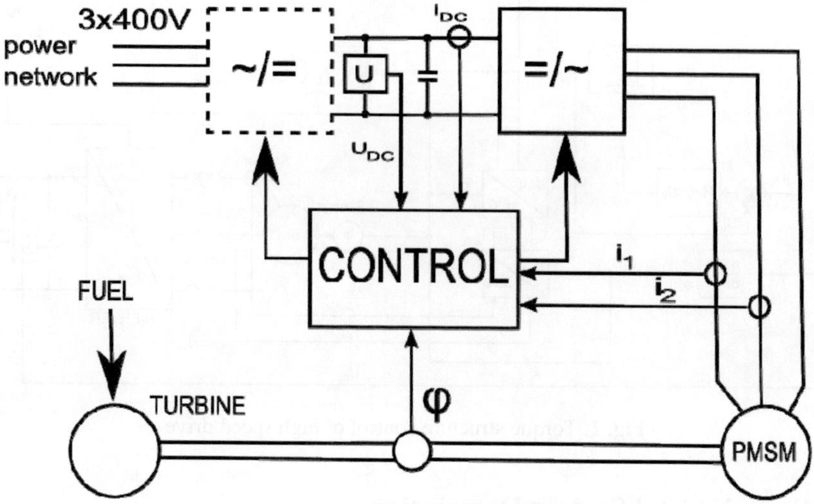

Fig. 2. Block diagram of experimental system

Fig. 3. Experimental IGBT converter

Fig. 4. Coupled high-speed motor with automotive supercharger

An aerospace turbine is better in reliability, however its price starts somewhere above 100 thousand Euros and this is also unacceptable. For this reason the idea is to use an automotive turbocharger and to add a combustion chamber later.

4 Load Angle Limiting in Flux Weakening Mode

The controller structure has to limit the i_q stator current component as not to achieve load angle β over 90°. Also over current protection is required. Therefore also current amplitude has to be limited. The i_q current limiter on Fig. 1 calculates the maximal current based on values of i_q and β. A lower value from those two results is taken.

For high speed motors the i_q limit caused by load angle β maximal value is more common.

The implemented method for i_q stator current component limiting is based on phasor diagram on Fig. 5 where stator resistance R_1 is neglected.

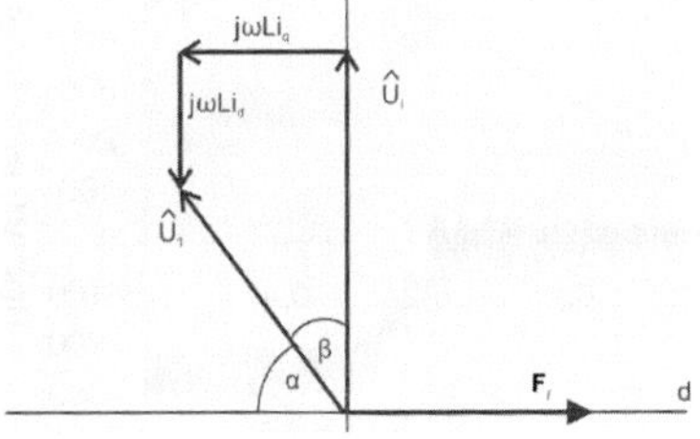

Fig. 5. Principle of load angle limiting

From calculation efficiency point of view and calculation precision and based on the tg function properties the load angle limiting is done by calculation of angle α according to equation:

$$tg\alpha = \frac{U_i - \omega L i_d}{\omega L i_q} = \frac{\omega F_f - \omega L i_d}{\omega L i_q} \quad (4)$$

where U_i is the induced voltage, L stator inductivity, F_f permanent magnet magnetic flux and ω stator voltage angular speed.

By simplification an equation for maximal i_q current value for a given α_{MIN} is obtained. It can be seen that β=90° corresponds to α=0°:

$$i_{qMAX} = \frac{\frac{\Psi_f}{L} - i_d}{tg\alpha_{MIN}} \quad (5)$$

Based on the form of tg function and for reasons of some robustness against imprecision and non linearity caused by changes of inductance L, the value of α_{MIN}=21 was chosen. This corresponds to limiting β to 69° and to lower torque of about 7% from the state where β=90°. There is no feedback control of load angle β when this limiting is activated. The calculation is based only on analytical calculation of mathematical model of motor and motor parameters. For this reason there is some imprecision for limiting angle α calculation in the range from 16° to 26° for a desired value of 21°, tg α_{MIN} = 0,4. This causes torque fluctuations for about ±3%. For this application however this does not pose a problem.

Fig. 6 and Fig. 7 show waveforms for the highest tested acceleration. Value of tg α_{MIN} was set to 0,25, the inverter was powered from a low internal resistance power supply. As can be seen from Fig. 14 there is a relation between individual current component and its absolute value. Also it is visible that when limiting of load angle β and i_q current component occurs this produces |I| limiting too.

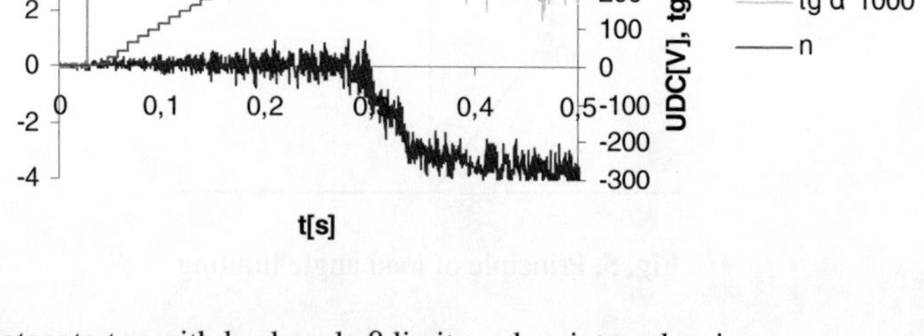

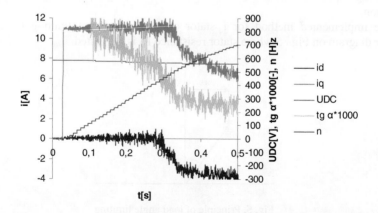

Fig. 6. Motor startup with load angle β limiter – low internal resistance power supply

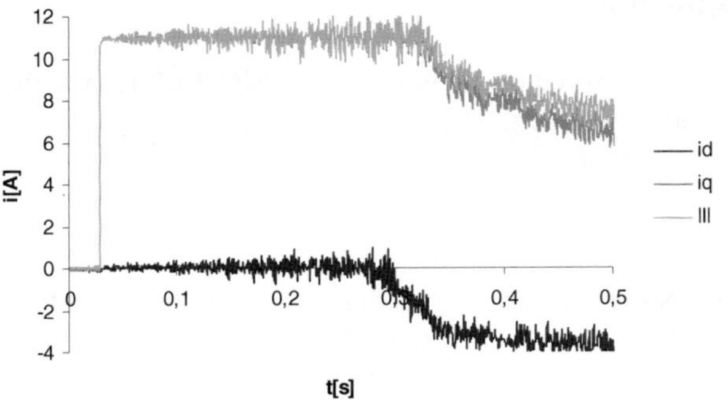

Fig. 7. Motor startup with load angle β limiter – low internal resistance power supply, id, iq current components and current absolute value |I|

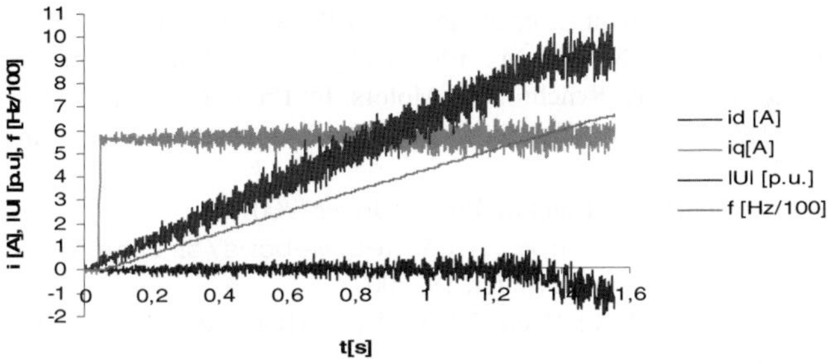

Fig. 8. Current components id , iq , reference voltages and speed(frequency) amplitudes for unloaded startup to 42 000 min

5 Conclusions

This paper presented some results obtained on a high speed micro turbine power generator. System components have been tested individually and the motor was coupled to the turbocharger. Motor control algorithm was tested with an unloaded motor for speeds up to 35 000 RPM. Unfortunately due to mechanical problems with the coupling, the system achieved a resonance and the motor shaft was damaged. Therefore it was not possible to test the motor further. The motor is currently already repaired and it is being mounted back on the test stand. Future work will therefore be the whole system testing at speeds 40 000 rpm with different power generation modes. An advantage of such system is higher efficiency and lower volume as compared to a classical piston generator. The system can also be used for combined electrical energy and heat production. At the time being, a disadvantage of this system is its high price as the components are not produced in larger volumes and a more difficult control requiring a DSP. However it can be expected that if a system like this would be produced industrially, its cost would drop considerably.

Acknowledgments

This research was sponsored by the grant no. MSM6840770035 of the Czech Ministry of Education, Youth and Sport.

References

1. Čambál, M., Novák, M., Novák, J.: Study of Synchronous Motor Rotor Position Measuring Methods, Zagreb, Korema, pp. 62–66 (2005); ISBN 953-6037-42-4
2. Čeřovský, Z., Mindl, P.: Electric Power Divider for Hybrid Car Propulsion Systems. Aalborg. In: 12th European Conference on Power Electronics and Applications EPE (2007); ISBN 9789075815108
3. Čeřovský, Z., Novák, J., Novák, M., Čambál, M.: Digital Controlled High Speed Synchronous Motor. In: EPE PEMC 2008, Poznaň, pp. 997–1002 (2008); ISBN 978-1-4244-1742-1
4. Lettl, J., Ratz, R.: Contribution to Induction Motor Vector Control. Prague. In: XIII. International Symposium on Electric Machinery ISEM 2005 (2005); ISBN 80-01-03328-7
5. Novák, L., Novák, J., Novák, M.: Electrically Driven Compressors on Turbocharged Engines with High Speed Synchronous Motors. In: Proceedings of The 8th Electromotion Conference, EPE Chapter "Electric Drives". HEI Graduate School, Lille; ISBN 978-2-915913-25-5
6. SEMIKRON. Sixpack IGBT and MOSFET Driver (2007), `http://www.semikron.com/internet/webcms/online/pdf/SKHI_61_71.pdf` (accessed March 18, 2007)
7. Novotny, D.W., Lipo, T.A.: Vector Control and Dynamics of AC Drives, (41). Oxford Science Publications (1996)
8. Lettl, J., Fligl, S., Kuzmanovic, D.: Comparison of Different Types of AC/AC Converters. In: Electronics Device and Systems IMAPS CS International Conference EDS 2006 Proceedings, Brno, pp. 427–432 (2006); ISBN 80-214-3246-2
9. Leuchter, J., Bauer, P., Bojda, P., Rerucha, V.: Bi-directional DC-DC Converters for Supercapacitor Based Energy Buffer For Electrical GEN-SETS. In: Proceedings of the 12th European Conference on Power Electronics and Applications (EPE 2007), Aalborg, P.1 – P.10 (2007); ISBN 9789075815108
10. Shepherd, W., Zhang, L.: Power Converter Circuits. Tailor and Francis, USA (2004); ISBN 0-8247-5054-3
11. Čeřovský, Z., Novák, J., Novák, M.: High Speed Synchronous Motor Control for Electrically Driven Compressors on Overcharged Gasoline or Diesel Engines. In: Proceedings of IECON 2009- 35th Annual Conference of the IEEE Industrial Electronics Society, Porto, Portugal (2009); ISBN 978-1-4244-4648-3
12. Novák, J., Čeřovský, Z.: Electrical Drive for Compressor on Turbocharged Engine. In: Proceedings of konference Theoretical and Practical Issues in Transport, Univerzita Pardubice (2010); ISBN 978-80-7395-245-7
13. Wildi, T.: Electrical machines, Drives and Power Systems, 6th edn. Pearson Prentice Hall, London (2006)

A Heuristic Magic Square Algorithm for Optimization of Pixel Pair Modification

Jeanne Chen[1], Tung-Shou Chen[1,*], Yung-Ming Hsu[1], and Keshou Wu[2]

[1] National Taichung Institute of Technology, Department of Computer Science and Information Engineering, Taichung City, Taiwan
{tschen,jeanne,s17973010}@ntit.edu.tw
[2] Department of Electronic and Electrical Engineering, Xiamen University of Technology, Xiamen, China
kswu@xmut.edu.cn

Abstract. Pixel pair modification (PPM) based magic squares is proposed for data hiding. The aim is to improve the drawbacks of payload versus PSNR of LSB. The traditional PPM-related methods did not consider the combination of square contents. In this paper, a heuristic magic square algorithm for PPM is proposed to select the optimal magic square combination for data hiding while the quality of the stego-image is maintained without affecting the amount of payload. Results showed that the heuristic magic algorithm can quickly recommend the suitable combination of magic squares with calculated efficiency on the range of 2% to 20%. The algorithm can enhance PPM based on magic squares and increase the efficiency of data hiding.

Keywords: Data hiding, Pixel pair modification, Magic square, Heuristic.

1 Introduction

The least significant bit (LSB) [4,7,11] is the most simple technique in steganography. Confidential message are embedded into the LSB n-bits of the cover image to get the stego image. The n-bits are adjusted based on choice on visual quality and/or capacity. Capacity is greater when the n is bigger, and visual quality is superior when n is smaller. However, it is difficult to maintain both; one has to be sacrificed for the other. Many improved approaches of LSB have been proposed. Both Chang et al.[1] and Wang et al.[10] proposed effective improvements to the exploiting modification direction (EMD) method as in Eq.(1). The method applied the pixel pair modification (PPM) technique where the confidential message of transformed to base-5 digit system was embedded into the cover image. The capacity of hiding was increased by 1.16bpp. The PPM is the basis for the magic square or array [2,6].

$$\text{bbp} = \log(B)/2\,. \tag{1}$$

B is a base-B digit system.

* Corresponding author.

M. Zhu (Ed.): Electrical Engineering and Control, LNEE 98, pp. 765–772.
springerlink.com © Springer-Verlag Berlin Heidelberg 2011

In PPM for magic square, the square matrix M is $N{\times}N(B)$. The square is formed by unique number of base-B digit system with the magic square $M=\{m_{ij}|i,j=0,1,\ldots,N-1$ and $m_{ij}\in 0,1,\ldots,B-1\}$. The $N{\times}N$ square is expanded by a cycle into $p{\times}p$ reference square of matrix W where p is the scope of the pixel, $W=\{r_{xy}|x,y=0, 1, \ldots, p-1$ and $r_{xy}=m_{(x \bmod N)(y \bmod N)}\}$ and the confidential message is transformed into a $N{\times}N(B)$ *base-B string* $S_B=\{s_0,s_1,s_2,\ldots,s_R\}$. Suppose S_i is the hidden message and the scope of searching is L, $L=[-N/2,N/2]$. In the process of embedding, take out pair-pixel(x,y) from cover image and search $r_{(x+L)(y+L)}$ to make $r_{(x+L)(y+L)}=S_i$. We get a new pair-pixel(x,y), $x'=x+l_x$, $y'=y+l_y$, and replace (x',y') into cover image. The result is a stego image with hidden confidential message S_B. The confidential message was transformed into $N{\times}N(B)$ string. As calculated by Eq.(1) and illustrated in Table 1, when B is bigger, the quantity of hidden data increases which also relatively affected the quality of the stego image. However, in comparison to the LSB method, the capacity can be increased for the same quality.

Table 1. Comparing the LSB, EMD and various N values for hidden capacity.

	LSB	EMD	N=3	N=4	N=5
Embedding efficiency	1	1.161	1.585	2	2.322

Much research has been done on enhancing the visual quality for capacity. Chang et al. proposed changing the cover image pixels by wet paper [3,6] concept. However, more enhancement may be possible for quality and capacity. Improvement may be made by the taking the peak signal of noise ratio (PSNR) to be the performance benchmark of image quality. By taking the PSNR difference between the original image and stego image at the embedding process for the magic square PPM data embedding, the pixel value change can be locate and if the pair-pixel(x,y) value point to the magic square position is the same as in message s_i then data is embedded or only changing a single pixel or pixel total changes is the least, then the resulting stego image has an improved quality with higher PSNR. In this paper, we proposed a combination of the magic square, and apply it by PPM to embed in the cover image at pixel value that is the least affected by the hiding capacity of N.

The $N{\times}N(B)$ magic square has the permutations and combinations of $B!$. For example, a 3x3 magic square has permutations and combinations of 362880. The result is large and all examples cannot be listed especially of the matrix square is extended to 5x5 or larger. Also the process would consume vary large computations and time costs to find the best performance of combination. Therefore, the purpose is to construct an efficient way to apply the magic square to get an optimal or near optimal solution that is fast and less time consuming.

This paper proposed using a heuristic algorithm based on [5,8,9] to evaluate the performance relation between the magic square, cover image and confidential information. The method is based on the $N{\times}N$ magic square to efficiently construct an improved magic square, and to promote quality of the stego image after data is embedded by PPM.

2 The Proposed Method

This study is aimed at the optimal combination of PPM based magic square to develop a heuristic algorithm which is expected to quickly and effectively solve the optimization problems of the magic square. The relationship between confidential message and cover image is referred in promoting the performance of embedded data for the magic square PPM. Confidential message is embedded in nearby embedding position where the change of cover image pixel value is the least. Three influencing factors are considered; first is the position probability of the pair-pixel taken from cover image which points to the magic square, second is transforming to base $N \times N$ of confidential message and the probability of digital characters appearance, and finally is the probability of match by the first and second. This study calculates the probability of confidential message characters appearance and the position probability of the pair-pixel taken from cover image which points to the magic square. It then sorts by the probability to two sets P_B and P_M. An evaluation parameter K is used to decide the computation cost. The value the former K values of P_B are sorted to perform all kinds of permutations and combinations where $K!$ combinations is generated. All combinations and the other values which are not sorted in the setting are filled into the magic square by P_M sequence and to generate $K!$ evaluation combinations. From the calculated evaluating performance of all kinds combination in the magic square, we can then recommend the best performance magic square to improve the performance of magic square to apply for embedding data hidden.

2.1 Confidential Statistical Information

Suppose the confidential message is a base-B digit system, shown as $B=N \times N$, the base-B system is 0,1,2,…,(B-1). Read confidential message one by one, and count the number of occurrences of 0,1,2,…,(B-1) in confidential message, we get $c_0, c_1, c_2, \ldots, c_{(B-1)}$. Let $sum = \sum_{i=0}^{B-1} c_i$ and the digit appearance of the base-B system is:

$$P_{B,i} = c_i / sum. \quad \text{for } 0 \le i \le (B-1). \tag{2}$$

We calculate by probability, calculating the probability of each item appears and sorted them, then got a probability set P_B, the P'_B after sorting is : $P'_B = \{ P'_{B,0}, P'_{B,1}, P'_{B,2}, \ldots, P'_{B,(B-1)} \}$, which $P'_{B,0} > P'_{B,1} > P'_{B,2} > \ldots > P'_{B,(B-1)}$.

2.2 Cover Image Pair-Pixels on the Statistical Probability

First, set a $N \times N$ statistical magic square M_s, $M_s = \{m_{s,t} / t=0,1,\ldots,B-1\}$. Extract each pixel in the $w \times h$ cover image from left to right and top to bottom to become a set Q, $Q = \{q_i | i=0,1,\ldots,(w \times h-1)$ and $q_i \in 0,1,2,\ldots,p-1\}$. Next, extract pair-pixel from Q, and calculate $t=(q_i \bmod N)+(q_{i+1} \bmod N) \times N$, and calculate the statistic $Ms, t = M_{s,t}+1$. Scan Q and to get $m_{s,0}, m_{s,1}, m_{s,2}, \ldots, m_{s,B-1}$, the probability of every item t at magic square is:

$$P_{M,t} = m_{s,t} / w \times h. \quad \text{for } 0 \le t \le (B-1). \tag{3}$$

Finally, we have the probability of cover image pair-pixel P_M:

$P_M = \{ P'_{M,0}, P'_{M,1}, P'_{M,2}, \ldots, P'_{M,(B-1)} \}$, which $P'_{M,0} > P'_{M,1} > P'_{M,2} > \ldots > P'_{M,(B-1)}$

2.3 Generate Evaluation Magic Square

There are some unknown factors if the combination of magic square depends only on P'_B and P_M influence factors. Therefore, the evaluation parameter K is used to disturb the top K values of high probability in P'_B. By increasing the combinations of evaluation magic square, we can increase the effectiveness of the magic square final recommendations. This parameter is also used to calculate cost before evaluating. After setting K value, we sort the combinations which are top K in P'_B set to generate K! combinations. Every combinations of (B-K) are filled into the magic square by P_M sequence. The evaluation magic square set H is generated and the PSNR is calculated resulting in the magic square h_0 as the recommended magic square combinations:

$$H = \{ h_i \mid i = 0,1,...,\ (K!-1)\ \&\ PSNR_{h_0} > PSNR_{h_1} > ... > PSNR_{h_{K!-1}} \}.$$

The K value is set in intervals as the experimental data in for testing in this paper. This parameter directly influences the sorted number P_B. The size of K value will influence the efficiency of the evaluation process. When K value is too small, the magic square is also relatively small, and the algorithm performance is expected to increase. On the other hand, if the K value is too big then the evaluation time efficiency will be reduced. In the 3×3 magic square example, the limit of K value is nine and the evaluation magic square will generate 9! combinations.

3 Experimental Results and Analysis

For making sure that the performance based on heuristic algorithm is accurate, experimental testing is made with Lenna and eleven images taken from the USC-SIPI image database. The twelve images are 512×512 in size as shown in (see Fig.1). The magic square is 3×3 size and the confidential message is transformed into a based-9 digit system. Twenty four confidential messages were generated randomly for testing. Each confidential message has 131072 based-9 digit. This amount has the best embedding capacity for the 512×512 size image using the proposed PPM technique.

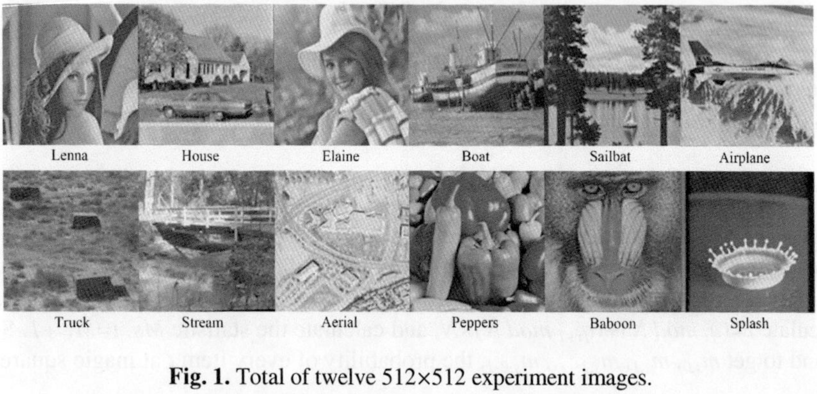

Fig. 1. Total of twelve 512×512 experiment images.

For evaluating the performance of recommendation magic square efficiently, this paper will compare performance for every recommended by the embedding experiment by exhaustive combinations of the 3×3 magic square. The top % is taken to express performance in the percentage of the total, for example, Top 5% express the performance from recommendation magic square is located at top 5 % of all best 9! combinations.

The evaluated factor K is the parameter set to limit the amount of evaluated solution of the algorithm proposed by this paper. This method is increasing progressively by the evaluated factor as seen in Table 2. The number of solving the combination of magic square, for example, if K=3, the amount of permutations and combinations of P_B is 3. Every image using the method of the experiment to embed a confidential message will get the amount of evaluated magic square of 6. In the case that K=5, the number of solution is 120 which is significantly less than the 9! solutions.

Table 2. Derived solutions with K value.

K value	3	4	5	9
The amount of evaluated magic square	6	24	120	362880

Table 3 shows the performance of the evaluated factor K=3, 24 confidential messages with every image in a total of 24 experiments with 288 experimental results. In this paper, we recommended six evaluated magic squares and for the highest PSNR value. 163 recommendation performances fall within the TOP 20% which is approximately 56.7% of the total. Recommendation of top 10% also has 105 times approximately around 36.5% of the total. Furthermore, 12 images have 41.67% recommended top 10% performance.

Table 3. K =3 the performance of recommended magic square.

Image name	TOP 0-10%	TOP 11-20%	TOP 21-30%	Other
Splash	12	4	2	6
Baboon	4	4	8	8
Airplane	4	5	1	14
Sailboat	10	2	2	10
Peppers	8	4	6	6
Aerial	7	6	4	7
Stream	16	5	1	2
Truck	11	4	2	7
Boat	11	6	3	4
Elaine	9	5	3	7
House	7	5	4	8
Lenna	6	8	4	6
Total	105	58	40	85

Table 4 shows the performance of the evaluated factor K = 4. A total of 224 recommendations fall within the TOP 20% which 77.8% of all combinations. Also, there are 166 experiments recommended performance which falls within the TOP 10%, accounting for 57.6% of the total.

Table 4. K =4 the performance of recommended magic square.

Image name	TOP 0-10%	TOP 11-20%	TOP 21-30%	Other
Splash	16	2	4	2
Baboon	9	6	4	5
Airplane	7	9	1	7
Sailboat	12	5	1	6
Peppers	14	2	6	2
Aerial	15	5	2	2
Stream	21	2	1	0
Truck	16	6	0	2
Boat	16	7	1	0
Elaine	16	4	1	3
House	11	7	3	3
Lenna	13	3	6	2
Total	166	58	30	34

Table 5 shows the performance of the evaluated factor K = 5. Each image is embedded a confidential message. The solution magic square requires 120 calculations but gained the best recommendation performance. As the result of experiment, recommended performance achieved top 20% with 266 times, accounting for 92.4% of the total recommended ratio, and the recommendation performance achieving top 10% is 236 time; the total percentage is 81.9%.

Table 6 shows the recommendation performance evaluated for all K values. Results is shown for the number of times and total percentage for top 10% and top 20%. As seen, when K value increases performance also increases at a fixed ratio. This means that the factors proposed in this paper affects performance to be efficient and accurate. The 3×3 size magic square with K value equals 5 has significantly high efficiency and effectiveness.

Table 5. K =5 the performance of recommended magic square.

Image name	TOP 0-10%	TOP 11-20%	TOP 21-30%	Other
Splash	23	1	0	0
Baboon	20	3	0	1
Airplane	15	4	2	3
Sailboat	18	3	2	1
Peppers	21	0	1	2
Aerial	18	3	3	0
Stream	22	2	0	0
Truck	20	2	1	1
Boat	22	2	0	0
Elaine	22	2	0	0
House	18	4	1	1
Lenna	17	4	3	0
Total	236	30	13	9

Table 6. The recommendation performance of all *K* values.

	TOP 10%		TOP 20%	
K=3	105 times	(36.5%)	163 times	(56.6%)
K=4	166 times	(57.6%)	224 times	(77.8%)
K=5	236 times	(81.9%)	266 times	(92.4%)

Figure 2 shows the performance histogram of all experimental analysis. The improve performance condition is clearly shown by increasing *K* values. The optimal result for this paper is for *K*=5. Table 7 shows results for *K*=5 for the different top% recommendations. At top 2% recommendation shows 149 times, which is 51.7% of the total, and the recommendation performance within top 5% is about 67.7%.

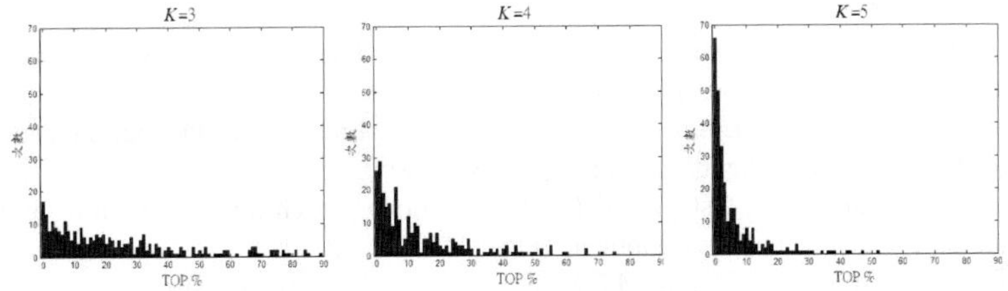

Fig. 2. Histogram for performance analysis.

Table 7. *K*=5 experimental analysis.

0-2%	3-5%	6-10%	11-20%	other	Total
149	46	41	30	22	288

4 Conclusions

The proposed heuristic algorithm is used to optimize the combination of magic square for PPM. Results showed that the right amount of evaluated magic square can be generated without using all permutations and combinations. The best performance magic square combination can be derived effectively and quickly. Different evaluated factor *K*, suggested the high performing magic square combination fall within top 2~20%. When the enhance performance magic square is applied to PPM for embedding data, the quality of image is preserved with high hiding capacity.

Acknowledgements

This work was supported under grant NSC 99-2221-E-025-013 by the National Science Council of Republic of China.

References

1. Chang, C.C., Lee, C.F., Wang, K.H.: An improvement of EMD embedding method for large payloads by pixel segmentation strategy. Image and Vision Calculating 26(12), 1670–1676 (2008)
2. Chang, C.C., Chou, Y.C., Kieu, T.D.: An information hiding scheme using sudoku. In: Proceedings of The Third International Conference on Innovative Calculating, Information and Control, pp. 17–21 (2008)
3. Fridrich, J., Goljan, M., Lisonek, P., Soukal, D.: Writing on wet paper. IEEE Transactions on Signal Processing 53(10), 3923–3935 (2005)
4. Mielikainen, J.: LSB matching revisited. IEEE Signal Processing Letters 13(5), 285–287 (2006)
5. Chen, K., Hwang, K., Chen, G.: Heuristic discovery of role-based trust chains in peer-to-peer networks. IEEE Transactions on Parallel and Distributed Systems 20(1), 83–96 (2009)
6. Li, M.C., Wang, Z.H., Chang, C.C., Kieu, T.D.: A sudoku based wet paper hiding scheme. International Journal of Smart Home 3(2), 1–12 (2009)
7. Wang, R.Z., Lin, C., Lin, J.: Image hiding by optimal LSB substitution and genetic algorithm. Pattern Recognition 34(3), 671–683 (2001)
8. Hong, S.J., Cain, R.G., Ostapko, D.L.: A heuristic approach for logic minimization. IBM Journal of Research and Development 18(5), 443–458 (2010)
9. Kumar, S., Kumar, A., Bhatia, M.P.S., Singh, G.: A heuristic approach to image database search. International Journal of Recent Trends in Engineering 3(2), 173–176 (2010)
10. Wang, S., Zhang, X.: Efficient steganographic embedding by exploiting modification direction. IEEE Communications Letters 10(11), 781–783 (2006)
11. Ni, Z., Shi, Y., Ansari, N., Su, W.: Reversible data hiding. IEEE Transaction on Circuits and Systems for Video Technology 16(3), 354–362 (2006)

A New Data Hiding Method Combining LSB Substitution and Chessboard Prediction

Keshou Wu[1], Zhiqiang Zhu[2], Tung-Shou Chen[3,*], and Jeanne Chen[3]

[1] Department of Computer Science and Technology,
Xiamen University of Technology, Xiamen, China
kswu@xmut.edu.cn

[2] Department of Electronic and Electrical Engineering,
Xiamen University of Technology, Xiamen, China
zzq89boy@126.com

[3] Department of Computer Science and Information Engineering,
National Taichung Institute of Technology, Taichung City 404, Taiwan
{tschen,jeanne}@ntit.edu.tw

Abstract. The embedding capacity and the image quality of a stego image are two important factors to evaluate a data hiding method. The aim is to get the largest embedding capacity with the lowest image distortion. In this paper, a novel mixed data hiding (MDH) method is proposed. MDH combines the non-reversible technique LSB substitution and the reversible technique chessboard prediction to get the benefits out of these two techniques. Experimental results showed that the proposed method achieved significantly higher payload (about 4 times higher) while maintaining low image distortion. Furthermore, comparisons results with other existing methods showed that MDH has significant advantage.

Keywords: non-reversible, reversible, LSB, chessboard prediction, mixed data hiding.

1 Introduction

Data hiding is a concept that embeds secret data by modifying pixels of an image. This concept has been widely used in many applications, such as watermarking, copyright protection, and tamper detection [3][5][11][12][14]. When data are embedded into a cover image, the contents of the pixels of the cover image shall be inevitably modified and thus the image is distorted. The original image which has not been "modified" is called the cover image while the modified image is called the stego image.

Distortion of an image always exists. When we take a photo in natural environment, the camera converts the scene into digital format. The digitalization is lossy. Fig. 1 shows the most original image as the virtual image. Therefore, the photo we take of is

* Corresponding author.

M. Zhu (Ed.): Electrical Engineering and Control, LNEE 98, pp. 773–780.
springerlink.com © Springer-Verlag Berlin Heidelberg 2011

not the original, and is termed an image with high quality and small distortion. In other words the cover image is not the original. Suppose we modify this image a little and keep the PSNR of the stego image at 45 dB. Visually the variations between the cover and the stego images are not easily detectable; it is difficult to differentiate between the two images.

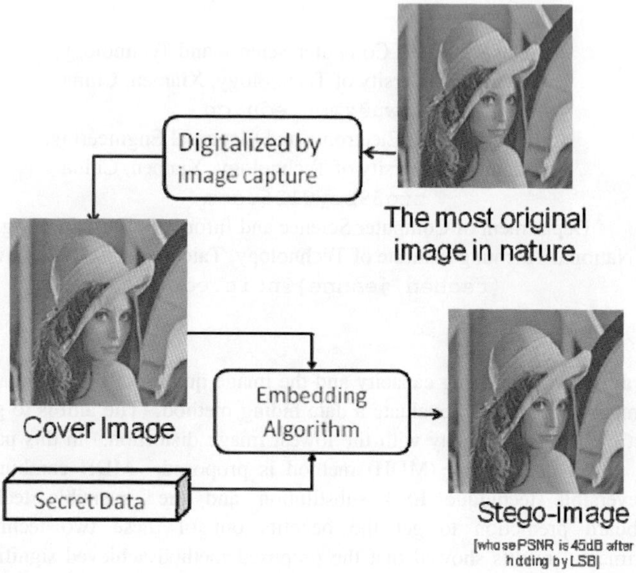

Fig. 1. Flow between the most original image, cover image and stego image

The content of this paper is based on this concept of the images. In the proposed method, the least significant bit (LSB) substitution [9] and chessboard prediction [6] are combined to generate a more efficient mixed data hiding (MDH) method. LSB is done by replacing LSBs of pixels in a cover image with message bits. This method is easy to implement and had low CPU cost. Non-reversible techniques such as LSB matching [1][9], exploiting modification direction [15] and pixel difference [2][13] are characterized as having high data hiding capacity and low image distortion. Reversible techniques are characterized by low capacity for low image distortion [6][7][8]. The PSNR is calculated from the differences between the original, the cover, and/or the stego images. Furthermore, the distortion cannot be visually detected.

In MDH, we first embedded data by LSB substitution to increase embedding capacity while maintaining the PSNR of the stego image to higher than 45dB. Next, we used chessboard prediction [6] to hide data in the stego image. The stego image had already been embedded with data in the first round by LSB substitution. Chessboard prediction predicts a pixel value by its surrounding four pixels, and then makes a histogram of the prediction errors. Data is then embedded in the peak value of the histogram by shifting the prediction errors. For reversible techniques, such as Ni's method [10] and MPE [7], the most significant aspect is that the cover image is recoverable. However, the embedding capacity of reversible techniques is generally smaller in comparison to other non-reversible data hiding. The embedding capacity of reversible data hiding technique is primarily determined by the peak height of the histogram. Fortunately, since the cover image could be recovered, we can apply the reversible data hiding method again and again to embed more data into the image. However, the quality of the stego image may be decreased significantly. MDH adopts the chessboard prediction to embed more data in the stego image, which could be restored to be very similar to the original image.

2 The Proposed Method

The proposed method mixed data hiding (MDH) first employs the non-reversible data hiding technique LSB substitution and makes certain that the PSNR value of the stego image is higher than an acceptable threshold T. After that, MDH adopts the chessboard prediction to increase the payload of the embedded data.

2.1 Data Embedding of MDH

In the first phase, we employ LSB to replace the LSBs of each pixel value. Every transformation of the single pixel should make sure that the PSNR of the stego image is higher than threshold T. After this phase, we apply the reversible data hiding method again and again to increase the payload of the image.

[Phase One: non-reversible data hiding]
Input: cover image I, secret message S, and threshold T.
Output: stego image, payload of the image, the PSNR of stego image, r, $I_{j,k}$,n

Step 1. Transform each pixel of the original image $I_{i,j}$ into 8 bits $b_7b_6b_5b_4b_3b_2b_1b_0$ where $b_r \in [0,1]$ and the pixel value of $I_{i,j}$ equals to $\sum b_r \times 2^r$.

Step 2. When r is 0, replace each pixel's last bit b_0 with secret data s_i. At each modification the MSE is calculated as

$$MSE = \frac{1}{mn}\sum_{i=0}^{m-1}\sum_{j=0}^{n-1} \| I(i,j) - K(i,j) \|^2 \tag{1}$$

If the MSE is smaller than $\frac{MAX^2}{10^{T/10}}$ ($\frac{MAX^2}{10^{T/10}}$ is deduced from the following equation. When MSE is smaller than $\frac{MAX^2}{10^{T/10}}$, the PSNR of the stego image is higher than threshold T),choose the next pixel $I_{i,j+1}$ and go back to step 2; else goto step 3.

$$PSNR = 10 \times \log_{10}\left(\frac{MAX^2}{MSE}\right) \tag{2}$$

Step 3. After modifying the last bit of each pixel in the image; replace each pixel if the MSE is still smaller than $\frac{MAX^2}{10^{T/10}}$; replace the second last bit b_1 (r=1); and even replace for the third last bit b_2 (r=2)) with secret data s_i and then repeat step 2. Also record the location of the last pixel $I_{j,k}$ that was modified.

Step 4. Calculate the PSNR. After performing step 2, we can ensure that the payload of the stego image is the largest with PSNR higher than T.

[Phase Two: reversible data hiding]

Step 5. In the second phase, the chessboard prediction algorithm is applied. If i and j is both odd or even, $I_{i,j}$ represents a black pixel, leave unused. The prediction error is calculated in $E_{i,j} = I_{i,j} - \left\lfloor \frac{I_{i-1,j} + I_{i+1,j} + I_{i,j-1} + I_{i,j+1}}{4} \right\rfloor$. The boundary cases have been considered in [6], and are therefore not discussed.

Step 6. Output the prediction errors in form of histogram, recorded as H_{err}.

Step 7. Find the two peaks p_1 and p_2 of histogram H_{err}. Assume p_1 is bigger than p_2.

Step 8. If prediction error $E_{i,j}$ is bigger than p_1, right shift $E_{i,j}$, that is, $E'_{i,j} = E_{i,j}+1$. If prediction error $E_{i,j}$ is smaller than p_2, left shift $E_{i,j}$, i.e., $E'_{i,j} = E_{i,j}-1$.

Step 9. Embed the secret message by shifting the two peaks p_1 and p_2 of prediction error histogram H_{err}. If secret message bit is 1 and the corresponding prediction error $E_{i,j}$ equals to p_1, right shift $E_{i,j}$, that is, $E'_{i,j} = E_{i,j}+1$. If the secret message bit is 1 and the corresponding prediction error $E_{i,j}$ equals to p_2, left shift $E_{i,j}$, that is, $E'_{i,j} = E_{i,j}-1$.

Step 10. Reverse prediction errors. $I_{i,j}$ is the pixels of the cover image. $E'_{i,j}$ is the modified prediction errors that have been already embedded with secret message. $I'_{i,j}$ is the pixels of the stego image.

$$I'_{i,j} = E'_{i,j} + \left\lfloor \frac{I_{i-1,j} + I_{i+1,j} + I_{i,j-1} + I_{i,j+1}}{4} \right\rfloor$$

Step 11. To increase hiding embedding capacity, repeat steps 4 to 10 repetitively until all secret message are embedded. The number of times of chessboard prediction n must be recorded.

2.2 Extraction and Image Restoration of MDH

In the image restore process, Chessboard prediction is used n times to extract the secret. Next extract the secret message embedded in LSBs before the last pixel $I_{j,k}$.

[Phase One: reversible data extraction]

Input: stego image, r, n , $J_{an,k}$.
Output: the recovered image whose PSNR is T, secret message S

The value of peak p_1 and p_2 must be known before extracting and recovering. Assume p_1 is greater than p_2. Then, we get the prediction errors of black pixels as in 3.1 at Step 5 of Phase Two for white and black pixels had been modified. If i and j are both not even or odd, $I_{i,j}$ is white pixels. Nothing is done for the white pixels. However, bit extractions and recovery is performed for the black pixels as follows.

1. If error is p_1, the embedding secret message is 0.
2. If error is p1+1, the embedding secret message is 1. The prediction error is then left shifted.
3. If error is p2, the secret message is 0.
4. If error is p_2-1, the embedding secret message is 1. The prediction error is then right shifted.

After all the secret message were extracted, we then compute $E_{i,j}$. We left shift $E'_{i,j}$ which is bigger than p_1, that is, $E_{i,j}= E'_{i,j}$ -1 and right shift $E'_{i,j}$ which is smaller than p_2, that is, $E_{i,j}= E'_{i,j}$ +1. Then reverse prediction errors as in 3.1 at Step 10 of Phase Two to recover the black pixels. The white pixels are recovered similar to the black pixels. The reversible data extraction is performed n times.

[Phase Two: non-reversible data extraction]
We use r and $I_{j,k}$ to locate the last bit of the pixels. $I_{j,k}$ represents the last modified pixel and r represents the last modified bit of $I_{j,k}$. The secret message is embedded in the last LSBs of each pixel.

3 Experimental Results and Analyses

In all experiment testing, six 8-bit grayscale images of 512×512 pixels were used. The images were Airplane, Boat, Baboon, Lena, Peppers and Tiffany (see Figure 2). The performance of MDH is compared with other data hiding schemes in terms of payload and distortion. T was set to be 45 dB in our experiments. This threshold does not show visually perceptible differences between the cover and stego images.

Airplane Boat Baboon Lena Peppers Tiffany

Fig. 2. Gray cover images

(1) Payload and image quality
After using LSB substitution, the embedding capacity is significantly high and we also make sure that the PSNRs of stego images were all higher and close to T (i.e., 45 dB). The payload of five stego images embedded by LSB substitution are listed in Table 1. The results conclusively showed high payload from 442,025 bits for Airplane to 466,496 bits for Tiffany.

Table 1. Payload after LSB substitution when T=45dB

Image	Airplane	Boat	Baboon	Lena	Peppers	Tiffany
Payload	442,025	459,510	465,287	465,213	465,222	466,496

Next, we increased payload by using the chessboard prediction. Chessboard is a reversible method and the stego image embedded by chessboard prediction can be reverted into the pre-stego image stage (the stego image after LSB substitution). The increased payload are demonstrated in Table 2. As illustrated the payload is further increase for each of the test images.

Table 2. Increasing payload by chessboard prediction

Images	Airplane	Boat	Baboon	Lena	Peppers	Tiffany
payload (bits)	79,691	36,700	20,971	55,312	39,059	58,982
PSNR(dB)	42.279	42.611	42.944	42.817	42.571	42.295
payload (bits)	135,790	67,895	40,108	92,274	71,041	103,809
PSNR(dB)	39.887	40.26	40.352	40.378	40.235	39.971
payload (bits)	174,850	93,061	56,623	120,323	122,945	137,363
PSNR(dB)	38.06	38.316	38.168	38.326	38.362	38.08
payload (bits)	208,142	114,819	71,303	144,440	145,227	164,364
PSNR(dB)	36.452	36.647	36.508	36.484	36.871	36.124
payload (bits)	235,667	133,693	84,672	163,838	164,102	188,219
PSNR(dB)	34.964	35.053	34.939	34.931	35.353	34.969

(2) Comparison using only LSB

If we used LSB substitution to embed secret data with PSNR limited to T (45dB), the results of the LSB substitution and the proposed method MDH is identical. However, if the payload is to be further increased, then the PSNR had to be downgraded to 40dB or 35dB. This resulted in significant differences in two methods. The payload for the proposed MDH method is smaller than the LSB substitution (see Table 3).

Table 3. Comparing payload between LSB and MDH for T=40dB and 35dB

	Image	Airplane	Boat	Baboon	Lena	Peppers	Tiffany
LSB	Payload(bit) T=40dB	635,329	640,131	640,888	642,061	641,160	641,100
	Payload(bit) T=35dB	830,638	833,572	832,691	833,312	832,491	831,506
MDH	Payload(bit) T=40dB	521,716	496,210	486,258	520,525	504,281	525,478
	Payload(bit) T=35dB	757,383	629,903	570,930	684,363	668,383	713,697

Although the payload of the proposed method is smaller, the LSB stego images showed visually detectable distortions when compared to the MDH stego images for the same PSNRs. Furthermore, the MDH is reversible in the chessboard prediction error phase. The circled region showed distortion in LSB substitution which is not detected in MDH in Fig. 3. The MDH result when compared to the cover image Airplane in Fig.2 showed no detectable difference.

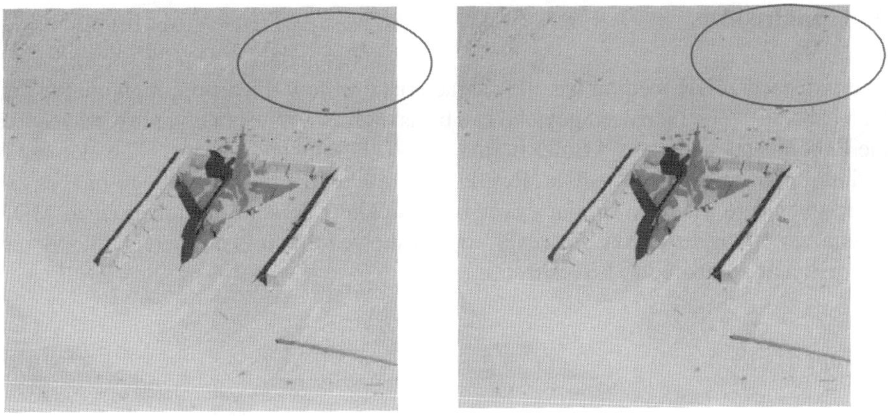

(a) LSB substitution (PSNR=35dB) (b) MDH (PSNR=35dB)

Fig. 3. Comparing distortions between LSB and MDH results

(3) Comparison using only chessboard prediction

Chessboard prediction is reversible; that is, the cover image can be wholly recovered. In the proposed MDH method, LSB substitution is used first to embed some secret message while making sure that the PSNR is maintained to be above T. T must be high enough so that there is no visually detectable distortion on the stego image. Then, Involves the chessboard prediction. Since the chessboard is reversible, the resulting MDH stego image is low distortion image with high payload. Also, the MDH stego images can be recovered with visual quality similar to the original cover image.

The MDH and chessboard prediction resulted in significant difference in payload size. Table 4 shows the payload results of the MDH and chessboard prediction for the stego images at PSNRs of 40dB and 35dB. The payload size in the proposed MDH method can be 8 times higher than the chessboard prediction. For example, Baboon at 40db showed a payload size of 486,258 bits in MDH as opposed to 59,244 bits in chessboard. MDH showed an average of four times more payload than chessboard.

Table 4. The payload of chessboard prediction and MDH when T= 40dB or 35dB

	Image	Airplane	Boat	Baboon	Lena	Peppers	Tiffany
chess-board	Payload(bit) T=40dB	198,705	100,139	59,244	151,989	101,449	158,335
	Payload(bit) T=35dB	289,931	154,927	92,798	225,798	155,189	237,764
MDH	Payload(bit) T=40dB	521,716	496,210	486,258	520,525	504,281	525,478
	Payload(bit) T=35dB	757,383	629,903	570,930	684,363	668,383	713,697

4 Conclusions

The proposed MDH method has the characteristic of low distortion, high quality and high capacity. The increased payload can be as high as 8 times in complex images like the Baboon and as high as 4 times in images like Tiffany and Peppers. The increase is higher for lower PSNR threshold. Results also reflect that the MDH stego images can be recovered to visually similar cover images. The LSB substitution used in MDH resulted in minor distortion. However, results showed that it is not perceptible by raw eye vision as compared to LSB substitution.

References

1. Chan, C.K., Cheng, L.M.: Hiding data in images by simple LSB substitution. Pattern Recognition 37(3), 469–474 (2004)
2. Chao, R.M., Wu, H.C., Lee, C.C., Chu, Y.P.: A Novel Image Data Hiding Scheme with Diamond Encoding. EURASIP Journal on Information Security 2009, Article ID 658047, 1–9 (2009)
3. Feng, J.B., Lin, I.C., Tsai, C.S., Chu, Y.P.: Reversible Watermarking: Current Status and Key Issues. International Journal of Network Security 2(3), 161–171 (2007)
4. Fridrich, J., Goljan, M., Hogea, D., Soukal, D.: Quantitative Steganalysis of Digital Images: estimating the secret message length. Multimedia Systems 9(3), 288–302 (2003)
5. Fridrich, J., Goljan, M., Du, R.: Lossless Data Embedding — New Paradigm in Digital Watermarking. EURASIP Journal on Applied Signal Processing 2002(2), 185–196 (2002)
6. Hong, W.: An Efficient Prediction-and-Shifting Embedding Technique for High Quality Reversible Data Hiding. Eurasip Journal on Advances in Signal Processing 2010, Article ID 104835, 1–12 (2010)
7. Hong, W., Chen, T.S., Shiu, C.W.: Reversible data hiding for high quality images using modification of prediction errors. Journal of Systems and Software 82(11), 1833–1842 (2009)
8. Lin, C.C., Tai, W.L., Chang, C.C.: Multilevel Reversible Data Hiding Based on Histogram Modification of Difference Images. Pattern Recognition 41(12), 3582–3591 (2008)
9. Mielikainen, J.: LSB Matching Revisited. IEEE Signal Processing Letters 13(5), 285–287 (2006)
10. Ni, Z., Shi, Y.Q., Ansari, N., Su, W.: Reversible data hiding. IEEE Transactions on Circuits and Systems for Video Technology 16(3), 354–362 (2006)
11. Perez-Freire, L., Perez-Gonzalez, F., Furon, T., Comesana, P.: Security of Lattice-Based Data Hiding Against the Known Message Attack. IEEE Transactions on Information Forensics and Security 1(4), 421–439 (2006)
12. Rykaczewski, R.: Comments on An SVD-Based Watermarking Scheme for Protecting Rightful Ownership. IEEE Transactions on Multimedia 9(2), 421–423 (2007)
13. Wu, D.C., Tsai, W.H.: A Steganographic Method for Images by Pixel-Value Differencing. Pattern Recognition Letters 24(9-10), 1613–1626 (2003)
14. Wu, N.I., Hwang, M.S.: Data Hiding: Current Status and Key Issues. International Journal of Network Security 4(1), 1–9 (2007)
15. Zhang, X., Wang, S.: Efficient Steganographic Embedding by Exploiting Modification Direction. IEEE Communications Letters 10(11), 781–783 (2006)

Data Mining the Significance of the TCM Prescription for Pharmacodynamic Effect Indexs Based on ANN

Bin Nie[1], JianQiang Du[1], RiYue Yu[1,2,3], YuHui Liu[2], GuoLiang Xu[1,2,3,*], YueSheng Wang[4], and LiPing Huang[2]

[1] School of Computer Science, Jiangxi University of Traditional Chinese Medicine, 330006, Nanchang, China
[2] College of Pharmacy, Jiangxi University of Traditional Chinese Medicine, 330006, Nanchang, China
[3] Key Laboratory of Modern Preparation, Ministry of Education, Jiangxi University of Traditional Chinese Medicine, 330006, Nanchang, China
[4] National Pharmaceutical Engineering Centre for Solid Preparation in Chinese Herbal Medicine, 330006, Nanchang, China
xuguoliang6606@126.com, ncunb@163.com

Abstract. The Traditional Chinese Medicine (TCM) Prescription include king drug,ministerial drug,assistant drug,sacrifice drug.The information may be reflected in the significance of the prescription.In the paper put forward data mining the significance of the TCM Prescription for Pharmacodynamic effect indexs based on artificial neural networks method.The aim to data mining the significance of the Gegen Qinlian Decoction(Radix Puerariae,Radix Scutellariae,Rhizoma Coptidis, Glycyrrhiza uralensis) for the pharmacodynamic effect indexs(HbA1c, CHOL, TG, XHDL, LD, INS),the method is to use artificial neural networks,the result indicate the method can mining the different significance according different Pharmacodynamic effect indexs.

Keywords: Traditional Chinese Medicine, Pharmacodynamic effect indexs, significance, ANN.

1 Introduction

The compatibility of chinese traditional medicine is soul of chinese formulae,the rule of prescription compatibility or the theory of prescription compatibility is the key of the formulas of Chinese medicine.

The artificial neural network (ANN) include back propagation artificial neural network (BP-ANN),radial basis function artificial neural network (RBF-ANN),the group meth od of data handling artificial neural network(GMDH-ANN), self-organizing neural networks(SONN),and etc.It is has the nonlinear approximation and

* Corresponding author.

M. Zhu (Ed.): Electrical Engineering and Control, LNEE 98, pp. 781–786.
springerlink.com © Springer-Verlag Berlin Heidelberg 2011

self-adaption, has made great speedy development,perfection and application in recent years[1-7].

2 Data Sources

In the paper,the data sources of National Basic Research Program of China(973 Program),the experiment using uniform design method, its includes 4 factors Gegen(Radix Puerariae),Huangqin(Radix Scutellariae),Huanglian(Rhizoma Coptidis),Gancao(Glycyrrhiza uralensis),and 16 levels,the total dosage is 120g. experimental subject is mouse, experiment purpose is to analyze the relationship between TCM prescription' dosage and pharmacodynamic effect of treat diabetes mellitus.the pharmacodynamic effect indexs consists of HbA1c, CHOL, TGL, XHDL, LDL, yd(insulinum).

3 The Basic Theory Artificial Neural Network

The artificial neural network (ANN) consists of a number of information processing elements named nodes that are grouped in layers. The input layer nodes receive input information and transmit the information to the next layer.The Basic concepts of artificial neuron show as figure 1:

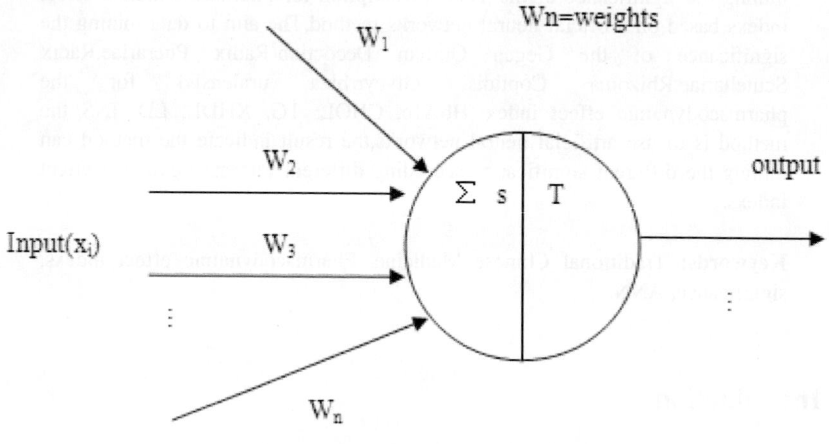

Fig. 1. The Basic concepts of artificial neuron

In the figure 1:input are input information vectors.w1,w2,wn are weight vectors.∑ is the weighted sum of the input and weight. s is the threshold.T is the transmit function.

$$output = f(\sum_{i=1}^{n} w_i * x_i - s) \tag{1}$$

4 Data Mining the Significance of the TCM Prescription for Pharmacodynamic Effect Indexs Based on ANN

This work, data mining the significance of the TCM Prescription for Pharmacodynamic effect indexs based on artificial neural networks method using software PASW 13,the Prescription' Dosage indexs:Radix Puerariae(X1), Radix Scutellariae(X2), Rhizoma Coptidis(X3), Glycyrrhiza uralensis(X4), the pharmacodynamic effect indexs consists of HbA1c, CHOL, TGL, XHDL, LDL, yd(insulinum).let Gegen(Radix Puerariae,X1),Huangqin(Radix Scutellariae,X2),Huanglian(Rhizoma Coptidis,X3),Gancao(Glycyrrhiza uralensis,X4),The different significance according different Pharmacodynamic effect indexs indicate such as figure 2-7.

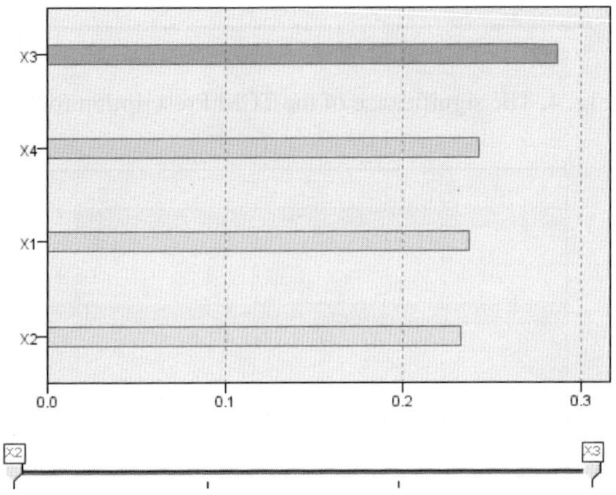

Fig. 2. The significance of the TCM Prescription for HBA1C

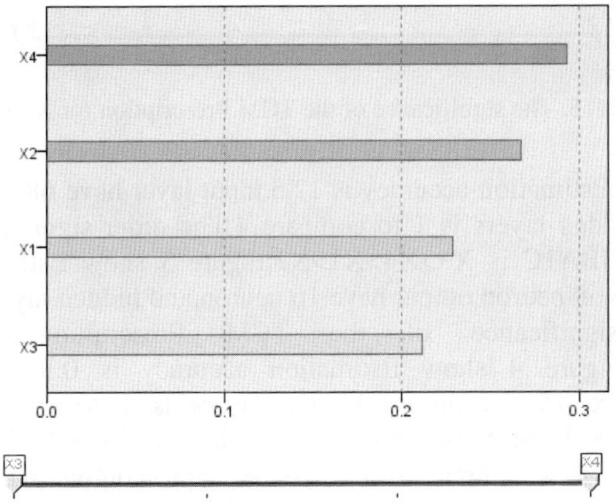

Fig. 3. The significance of the TCM Prescription for CHOL

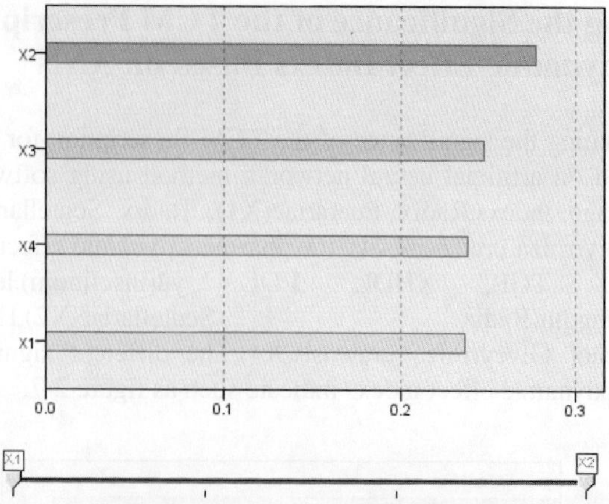

Fig. 4. The significance of the TCM Prescription for TG

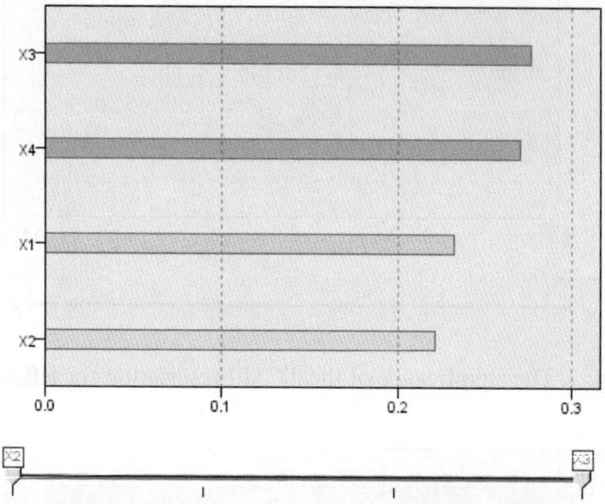

Fig. 5. The significance of the TCM Prescription for XHDL

Figure 2 show Estimation accuracy is 12.5,input layer have 68 neuron,output have 14 neuron,and hidden layers is 1 to compare 4.The order significance of the TCM Prescription for HBA1C is X3>X4>X1>X2;Figure 3 show Estimation accuracy is 0,input layer have 68 neuron,output have 16 neuron,and hidden layers is 1 to compare 4.The order significance of the TCM Prescription for CHOL is X4>X2>X1>X3;Figure 4 show Estimation accuracy is 0,input layer have 68 neuron,output have 15 neuron,and hidden layers is 1 to compare 4.The order significance of the TCM Prescription for TG is X2>X3>X4> X1;Figure 5 show Estimation accuracy is 0,input layer have 68 neuron,output have 13 neuron,and hidden layers is 1 to compare 4.The TCM Prescription for XHDL is X3>X4>X1>X2;Figure 6 show Estimation accuracy is 2.5,input layer have 68 neuron,output have 74 neuron,and hidden layers is 1 to compare 7.The order significance of the TCM Prescription for HBA1C, CHOL, TG, XHDL is X2>X1>X3>X4;Figure 7 show Estimation accuracy is 5.556,input layer have 68 neuron,output have 91 neuron,and hidden layers is 1 to compare 7.The order significance of the TCM Prescription for HBA1C,CHOL,TG,XHDL and yd is X1>X4>X2>X3.

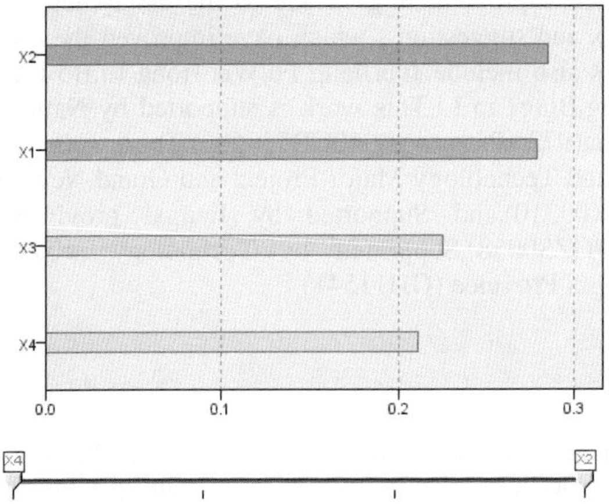

Fig. 6. The significance of the TCM Prescription for HBA1C, CHOL, TG, XHDL

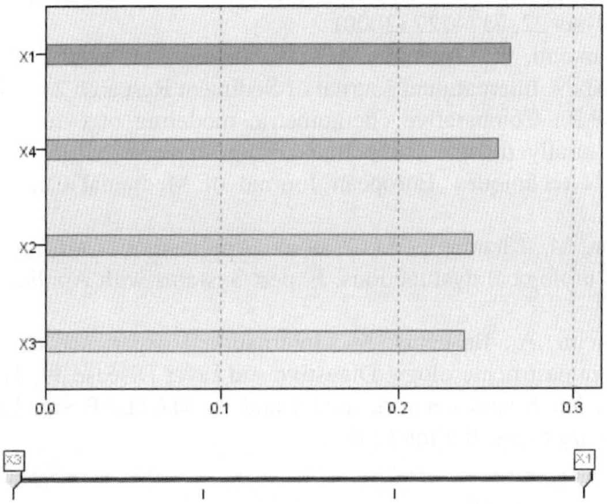

Fig. 7. The significance of the TCM Prescription for HBA1C,CHOL,TG,XHDL and yd

5 Conclusion

The different significance of the TCM prescription according different Pharmacodynamic effect indexs.The paper put forward data mining the significance of the TCM Prescription for Pharmacodynamic effect indexs based on artificial neural networks method. It was proved to be feasible and effective after tested with a database of Gegen Qinlian Decoction experiments.

Acknowledgments

The authors wish to express their gratitude to the anonymous reviewers for their valuable comments and suggestions, which have improved the quality of this paper. The team members also include JianJiang Fu,WenHong Li,Bo Liu,Fei Qu,YingFang Chen,QiYun Zhang,BingTao Li.This work is supported by National Basic Research Program of China(973 Program:2010CB530603).This work also supported by National Science and Technology Major Project and Grand New Drug Development Program (2009ZX09310),and Supported by Jiangxi province Natural Science Foundation (2009GZS0058),Supported by Technology project of Educational Committee of Jiangxi Province (GJJ11541).

References

1. Bourquin, J., Schmidli, H., van Hoogevest, P., Leuenberger, H.: Pitfalls of artificial neural networks (ANN) modelling technique for data sets containing outlier measurements using a study on mixture properties of a direct compressed dosage form. European Journal of Pharmaceutical Sciences 7, 17–28 (1998)
2. Agatonovic-Kustrin, S., Beresford, R.: Basic concepts of artificial neural network (ANN) modeling and its application in pharmaceutical research. Journal of Pharmaceutical and Biomedical Analysis 22, 717–727 (2000)
3. Yang, C.T., Marsooli, R., Aalami, M.T.: Evaluation of total load sediment transport formulas using ANN. International Journal of Sediment Research 24, 274–286 (2009)
4. Roy, K., Roy, P.P.: Comparative chemometric modeling of cytochrome 3A4 inhibitory activity of structurally diverse compounds using stepwise MLR, FA-MLR, PLS, GFA, G/PLS and ANN techniques. European Journal of Medicinal Chemistry 44, 2913–2922 (2009)
5. Gil, D., Johnsson, M., Chamizo, J.M.G., et al.: Application of artificial neural networks in the diagnosis of urological dysfunctions. Expert Systems with Applications 36, 5754–5760 (2009)
6. Grossi, E., Mancini, A., Buscema, M.: International experience on the use of artificial neural networks in gastroenterology. Digestive and Liver Disease 39, 278–285 (2007)
7. Zhou, K., Kang, Y.: Neural network model and its MATLAB simulation program desig. Tsinghua University Press, Beijing (2005)

Design of Compact Dual-Mode Dual-Band Bandpass Filter for Wlan Communication System

Yang Deng, Mengxia Yu, and Zhenzhen Shi

School of Physical Electronics, University of Electronic Science and Technology of China, Chengdu, China
feiren923@163.com

Abstract. A novel approach for designing dual-mode dual-band bandpass filter is proposed and experimentally studied. By embedding a pair of crossed slots in a circular patch resonator, the dual-mode dual-band filter is realized and the structure could be reduced significantly. And coupling between two degenerate modes will be adjusted, as well the band-to-band isolation increased, by loading with two quarter-circular holes, which can be placed at the center of the circular patch. Further improving the performance of the filter, the first and second transmission zero can be controlled by the line-to-ring feeds. Finally, a dual-mode dual-band filter with center frequency at 5.2 GHz and 5.8GHz, which included in WLAN bands, exhibits a band-to-band isolation better than 30dB.

Keywords: dual-mode dual-band, band pass filte, WLAN.

1 Introduction

Recent development in wireless communication systems has created a need for RF circuits with a dual-band operation. Various configurations have been proposed for realizing a dual-band filter [1-5]. In [1], a dual-band filter was implemented as a combination of two individual filters with two specific single passbands. By cascading a wideband passband filter and a stopband filter, a dual-band bandpass filter (BPF) was achieved in [2], but these solutions suffer from high insertion loss and large overall size. The stepped impedance resonator (SIR) had been used in dual-band design too [3,4]. In this case, the second resonance property is dominated by impedance ratio of the SIR.

At the same time, dual-mode filter have been widely used in wireless communications systems because of their advantages in applications, while requiring high quality such as smaller size and lower loss. Dual-mode filters by realizing two resonances in a single resonator and thus cutting the number of required resonators by half have been one of the widely employed ways of miniaturization, which also leading to lower radiation loss. The most two popular geometries for realizing the dual-mode filters have been rings [6-8] or patches [9-11] either in the square shape or circular shape. The patch-based filters are miniaturized by creating modifications on patch, which are usually in the shape of rectangular slots or circular holes, leading to

M. Zhu (Ed.): Electrical Engineering and Control, LNEE 98, pp. 787–792.
springerlink.com © Springer-Verlag Berlin Heidelberg 2011

miniaturization of the patch filters by increasing the current path length. However, modifications, at many cases not only increase the current path length of the fundamental mode, but also of the second higher order mode. So, miniaturizations very frequently come out with the poor isolation between the first two resonant frequencies, while dual-band filters needed. Therefore, there should be a need to propose an approach to design dual-mode dual-band filters with miniaturizations and good band-to –band isolations.

In this letter, a 5.2/5.8 GHz dual-mode microstrip dual-band passband filter is proposed. The dual-mode and dual-band properties derive from a pair of crossed slots, which formed on the circular patch resonator. And the band-to-band isolation improved by two quarter-circular holes located at the center of the resonator. Then, in order to improve the out-of-band performance, a novel line-to-ring feeds have been used [12]. And finally, this proposed filter is verified by simulations and measurements.

2 Design of Dual-Mode Dual-Band Bandpass Filter

The configuration of the proposed dual mode dual-band filter is shown in Figure1. In general, there is an obstacle between miniaturizations and good band-to -band isolations, because the modifications for miniaturization may influence both of the current path length of the two modes and, thereby, the isolation between the two bands, which decided by the two modes respectively.

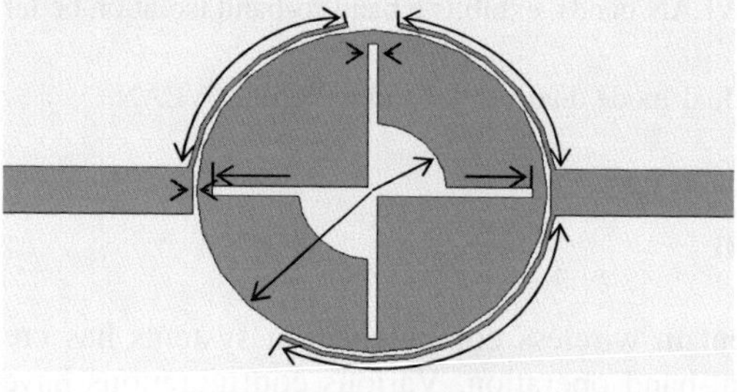

Fig. 1. Configuration of the proposed dual-mode dual-band filter.

A pair of crossed slots are created on the circular patch and oriented as shown in Figure 1. The fundamental mode frequency will drop because of the increase of its current path length. And the second higher order mode frequency, however, will remain almost uninfluenced, as the slots are parallel to the current path, and will not alter its path length. And the two quarter-circular holes at the center have been designed to adjust coupling between the two orthogonal degenerate modes of the modified resonator.

The filter proposed here is fabricated on a FR4 substrate with h=1mm and ξr=4.6. The width of the feed line is chosen to be 1.5mm, which corresponds to the characteristic impedance of 50 Ω. The filter is fed by a pair of line-to-ring feeding lines as shown in Figure 1. The length of the broadside-coupled sections is λ/4 long. Figure 2 presents the simulated results that compare the values of S-parameters of filters with and without line-to-ring feeding structure. The results inform us that the line-to-ring feeding structure can improve the performance of out-of-band obviously.

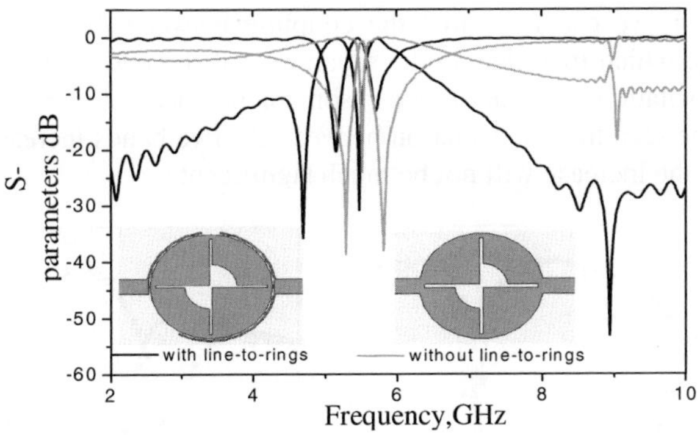

Fig. 2. Simulated results comparing the properties of ring filters with and without line-to-ring.

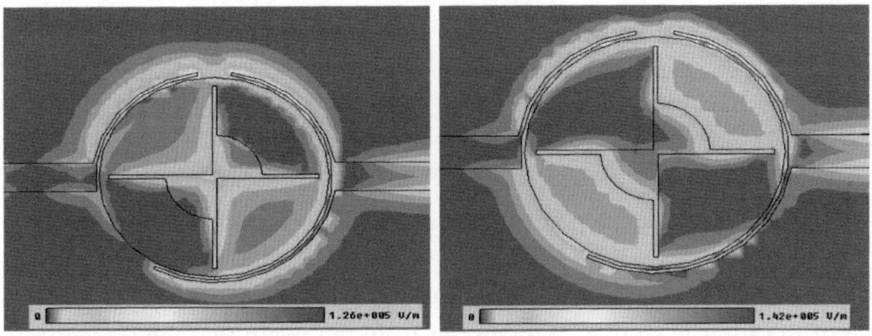

(a) (b)

Fig. 3. Simulated electric field patterns of the proposed dual-mode filter. (a) mode-1(5.72GHz), (b) mode-2(5.85GHz).

Figure 3 shows the simulated electric field patterns by using CST-MWS. It can be seen from Figure 3 (a) that two zeros are located at the right-upper and the left-lower corners. And the electric field pattern shown Figure 3 (a) rotated to 90° is just the pattern shown in Figure 3(b).

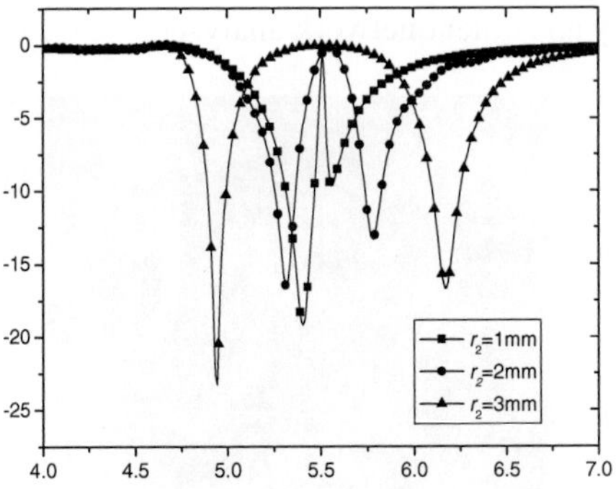

Fig. 4. The frequency response with different size of r_2.

From Figure 4, we can infer that the coupling between the two modes will be influenced by r_2, which then decide the separation of the two bands. To observe the band –to-band isolation, the dual-mode resonator has been simulated with different value r_2. It can be seen that the isolation between the two bands increases as the value r_2 increases, but the increase will not be much significant.

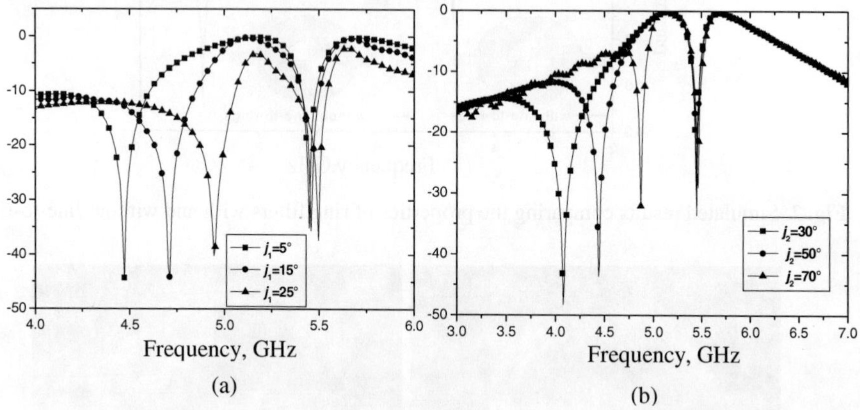

Fig. 5. Resonant properties with different lengths of the line-to-rings.

In Figure 5(a), the influence of j_1 of the line-to-ring feeds on the frequency response of the filter is shown. We can see from the result that the transmission zeros moves toward the higher frequency as the increase of its value. The counterpart dependence properties of the resonator upon the j_2 are verified by the simulated curves shown in Figure 5(b), where other dimensions are fixed. It can be found that j_2 has the same influence of j_1. So it is easy to adjust the transmission zeros of the filter.

3 Results

The dimensions shown in Figure 1 are: l=9.7mm, w=0.3mm, r_1=5.2mm, r_2=2.2mm, j_1=80°, j_2=120°, g=0.15mm. The picture of fabricated sample was shown in Figure 6 and it was measured by an Agilent network analyzer.

Fig. 6. Picture of fabricated dual-mode dual-band filter.

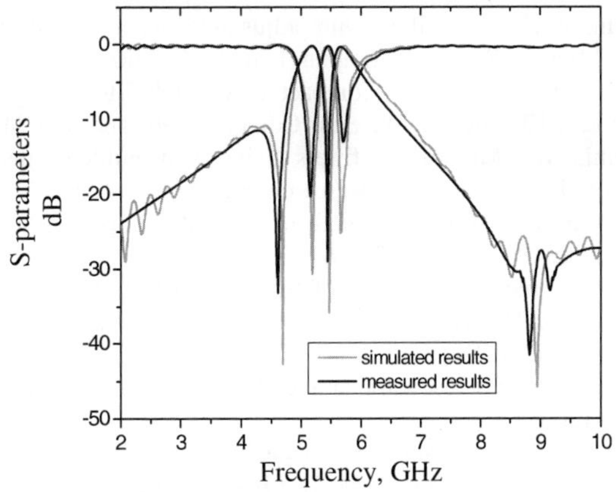

Fig. 7. Simulated and measured results of the dual-mode bandpass filter.

Simulated and measured results are compared in Figure 7 and show good agreement. The lower passband is occurs at 5.2 GHz with a -10dB bandwidth of 5% and the upper passband occurs at 5.8GHz with a -10dB bandwidth of 10%. And the result shows a band-to-band isolation better than 30dB. And the filter will be useful in WLAN communication systems. At the mid-band frequency of each passband, low insertion loss (including the losses from the SMA connectors) of less than 0.8 dB was obtained. Measurements differ from simulations since the SMA board connectors are not calibrated out and thus adversely influence the S-parameters. Those losses are contributed partly by the fabrication inaccuracies, mismatch at the connectors and material losses.

4 Conclusions

A novel dual-mode dual-band filter with center frequency at 5.2/5.8GHz is proposed. The dual-band property derives from a pair of crossed slots, which formed on the circular patch resonator. And the band-to-band isolation improved by two quarter-circular holes located at the center of the resonator. Then, in order to improve the out-of-band performance, a novel line-to-ring feeds have been used. And finally, a very good agreement between the measurements and simulations has been demonstrated.

References

1. Kuo, J.T., Yeh, T.H., Yeh, C.C.: Design of microstrip bandpass filters with a dual-passband response. IEEE Trans. Microw. Theory Tech. 53(4), 1331–1337 (2005)
2. Tsai, L.-C., Huse, C.-W.: Dual-band bandpass filters using equal length coupled-serial-shunted lines and Z-transform techniques. IEEE Trans. Microw. Theoty Tech. 52(4), 1111–1117 (2004)

3. Huang, T.-H., Chen, H.-J., Chang, C.-S., Chen, L.-S., Wang, Y.-H., Houng, M.-P.: A novel compact ring dual-mode filter with adjustable second-passband for dual-band applications. IEEE Microw. Wireless Compon. Lett. 16(6), 360–362 (2006)
4. Wu, B., Liang, C., Qin, P., Li, Q.: Compact dual-band filter using defected stepped impedance resonator. IEEE Microw. Wireless Compon. Lett. 18(10), 674–676 (2008)
5. Khalaj-Amirhosseini, M.: Microwave filters using waveguides filled by multi-layer dielectric. Progress In Electromagnetics Research 66, 105–110 (2006)
6. Djoumessi, E.E., Wu, K.: Multilayer dual-mode dual-bandpass filter. IEEE Microw. Wireless Compon. Lett. 19(1), 21–23 (2009)
7. Mo, S.-G., Yu, Z.-Y., Zhang, L.: Design of triple-mode bandpass filter using improved hexagonal loop resonator. Progress in Electromagnetics Research 96, 117–125 (2009)
8. Wu, G.-L., Mu, W., Dai, X.-W., Jiao, Y.-C.: Design of novel dual-band bandpass filter with microstrip meander-loop resonator and CSRR DGS. Progress In Electromagnetics Research 78, 17–24 (2008)
9. Tu, W.-H., Chang, K.: Miniaturized dual-mode bandpass filter with harmonic control. IEEE Microw. Wireless Compon. Lett. 15(12), 838–840 (2005)
10. Zhu, L., Tan, B.C., Quek, S.J.: Miniaturized dual-mode bandpass filter using inductively loaded cross-slotted patch resonator. IEEE Microw. Wireless Compon. Lett. 15(1), 22–24 (2005)
11. Xiao, J.-K., Li, S.-P., Li, Y.: Novel planar bandpass filters using one single patch resonators with corner cuts. J. of Electrom. Waves and Appl. 20(11), 1481–1493 (2006)
12. Zhu, L., Wu, K.: A joint field/circuit model of line-to-ring coupling structures and its application to the design of microstrip dual-mode filters and ring resonator circuits. IEEE Trans. Microw. Theory Tech. 47(10), 1938–1948 (1999)

AMOS: A New Tool for Management Innovation in IT Industry

Shujun Tang[1] and Liuzhan Jia[2]

[1] Department of Education Science, Yangtze University, Jingzhou, Hubei Province, China
[2] School of Psychology, Central China Normal University, Wuhan, Hubei Province, China

Abstract. Amos is a multi-variable statistical software widely used in organizational behavior research and management innovation. As an effective and efficient tool, Amos is usually used to construct a theoretical model based on data acquired, confirm a proposed theoretical model, compare several alternative models to get one better model, and conduct path analysis to modify the generated model. The vigorous function of Amos and comparative advantage over similar software both predict that and potential application promising prospect in IT management innovation.

Keywords: AMOS, Structural equation modeling, software, Management Innovation.

1 Introduction

The increasing social division of labor makes all elements in a close tie, from employee to executive, from production to marketing, and so the performance of an organization not wholly controlled in just one working unit, but in different and related segments. The executives in IT industry usually complain that they have to deal with so many different affecting factors, outside or inside, besides, what is related to what, and what will cause what, they don't know but eager to get the answer. Analysis of Moment Structure (abbreviated as AMOS) is such a tool, widely used in organizational behavior, human resource management, applied psychology, social research and behavioral science. AMOS can confirm and test variables relationships among observed (measured) variables and latent (unobserved) variables and the relationships among the latent variables themselves. Structural equation modeling contains measurement model and structure model. Each equation represents a causal relationship that generates a hypothesis which can be tested by estimating the structural parameter of the relationship [1]. For many managers in the IT industry, AMOS can help better deal with the growing challenges from outside and the increasing stress from inside, by statistical techniques, all possible affecting variables can put in the structural equation model ,and predict how much they are related , and to what extent they will affect other variable.

M. Zhu (Ed.): Electrical Engineering and Control, LNEE 98, pp. 793–800.
springerlink.com © Springer-Verlag Berlin Heidelberg 2011

2 Overview of AMOS

AMOS is a software developed within the Microsoft Windows interface; it allows the researcher to use AMOS Graphic to work directly from path diagram. Amos Graphic provides the user with all the tools that will be needed in later creating and working. On the menu bar, every tool is represented by a button and performs a particular function. AMOS allows the preliminary users try the evaluation edition in a certain period. When the evaluation period expired, users may have to purchase it to continue the use. This software provides users with friendly but powerful tutor and guider, the users can learn how to use the program by the tutor and examples offered by the software. The software has developed higher version recently and has gained better quality [2].

AMOS is a computer program that used in multivariate data analysis which known as structural equation modeling, causal modeling, and analysis of moment structures. This program incorporates procedures such as multi regression, analysis of variance(ANOVA) and multivariate analysis of variance(MANOVA), path analysis and mediate variable analysis. It was developed by the AMOS Development Corporation that is owned by the Statistical Product and Service Solutions corporation. There are two components included in AMOS that also in all other moment structure analysis programs. The first component is measurement model and the other is structural model. The measurement model involves connecting observed variables to a set of unobserved or latent variables through confirmatory factor analysis, while the structural equation model is the causal relationships among the latent variables themselves.

AMOS is an easy-to-use program for users. Through the software, the researchers can quickly specify, view, and modify their theory model graphically using simple drawing tools. Then the users can evaluate the supposed model's fitness, make any modifications by the generated recommendations, and print out a publication-quality graph of final model. AMOS can perform the computations command quickly and displays the results in seconds.

The availability of AMOS should contribute to popularizing the techniques of structural equation modeling. There are several advantages to recommend it. To begin with firstly, it is inexpensive that implies companies can purchase their own copy without so much money, what is more, the AMOS has been inserted into SPSS software as a menu option. Secondly, it has clear and easy user's guide easy for operation. Any instructor who has tried to learn through the documentation that accompanies Linear Structural Relations packaged will find a pleasure to avoid the tedious symbols which dominated by the use of Greek letters of the Linear Structural Relations.

AMOS also produces tabular output similar to that of SPSS that displays the unstandardized and standardized regression coefficients, the standard error estimates of the unstandardized regression coefficients, and tests of statistical significance of the null hypothesis that each unstandardized regression coefficient equals zero. And users will notice the appearing of a floating toolbar. The toolbar may partially obscure the AMOS drawing area, but users can move it out of the way by dragging it to the side of the computer screen.

Analysis a set of data by AMOS required three basic steps: First, making preparation of an input file that include the data set and model specified. Second, running the AMOS program processing. Third, inspecting the output file of the analysis, and accepting the model or modifying it. A nice feature of AMOS is its high-quality graphical output. The users can take this output and copy it to the windows clipboard, from there you can insert it into a word processor such as Microsoft Word processor or a presentation package like Microsoft Power Point.

3 The Features of AMOS

3.1 The Evaluation of Theory Model

The main application of the AOMS is to evaluate the fitness of a hypothesis model. So the experts have developed many parameters to demonstrate the superiority and inferiority for the model. Much research has contributed to the development of goodness of fitting statistics that address the sensitivity of the chi-square to sample size; all programs could report a smorgasbord of fit indexes. The based indicator is chi-square. The lower value of the chi-square means the better of the theory model. Because the chi-square test of absolute model fit is sensitive to sample size and non-normality in the underlying distribution of the input variables, investigators often turn to various descriptive fit statistics to assess the overall fit a model of the data. In this framework, a model may be rejected on an absolute basis, yet a researcher may still claim that a given model out performs some other baseline model by a substantial amount. Take another way, the argument researchers make in this context is that their chosen model is substantially less false than a baseline model, typically the independence model. A model that is parsimonious, and yet performs well in comparison to other models may be of substantive interest. For example, the Tucker-Lewis Index (TLI) and the Comparative Fit Index (CFI) compare the absolute fit of the specified model to the absolute fit of the independence model. The greater the discrepancy between the overall fit of the two models, the larger the values of these descriptive statistics.

A separate block of the output displays parsimony adjusted fit statistics. These fit statistics are similar to the adjusted R square in multiple regression analysis: the parsimony fit statistics penalize large models with many estimated parameters and few leftover degrees of freedom. The output file contains a large array of model fit statistics. All are designed to test or describe overall model fit. Each researcher has his or her favorite collection of fit statistics to report. Commonly reported fit statistics are the chi-square, its degrees of freedom, its probability value, the Root Mean Square Error of Approximation (RMSEA) accompany its lower and upper confidence interval boundaries. There is also a Standardized Root Mean Residual (Standardized RMR) available through the menu option, but it is important to note that this fit index is only available for complete datasets (it will not be printed for databases containing incomplete data). The analysis process of AMOS was based on a covariance matrix of the variables which concerned. In order to test and compare the supposed model, AMOS uses many fitness indexes to manifest the character of model. Models will be compared statistically by using the chi-square difference test [3].

Amos permits exploratory specification theory searches for the best theoretical model, given an initial model using the following fit function and modification index: chi-square, chi-square divided by the degrees of freedom (C/df), Akaike Information Criteria (AIC), Browne-Cudeck Criterion (BCC), Bayes Information Criterion (BIC), and significance level (p). Maximum likelihood estimation and unbiased covariance (as input to be analyzed) analysis properties were selected default in AMOS to obtain these fitness index above. Before extracting estimates, standardized estimates and modification indices in Amos's output option were selected. To interpret standardized estimates in group comparisons, researchers would need to standardize the entire sample data before calculating the measurement calculation.

3.2 The Construct of Theory Model

Five basic steps characterize most statistical modeling applications: model specification, model identification, model estimation, model testing, and model modification [4].In some substantive situations this may be excessive demands for researchers because theories are often poorly developed or even non-existent. However, theory-implied models are commonly stated in multiple regression, path analysis, confirmatory factor analysis, and structural equation models.

Concerning these modeling demands, there are three approaches to distinguish statistical modeling. The three approaches include: (a) a strictly confirmatory criterion in which a single formulated model is either accepted or rejected; (b) alternative models or competing models situation in which several models are suggested and one of them is selected, and (c) the model generating when an initial model is specified and does not fit the data well, it is modified (re-specified) and repeatedly tested until some fit and is accepted. Overall, these mentioned approaches have been called exploratory versus confirmatory approaches to statistical modeling.

The strict confirmatory approach is rarely in practice because most researchers are simply not willing to reject a proposed model without at least suggesting some alternative models. The researchers have spent much time and energy in formulating the supposed theory model. Modification indices are currently used by researchers mainly in structural equation modeling to instruct model modification. The AMOS software now provides researchers with an easy implementation of model modification in the output file using several well-known fit function criteria in an automated exploratory specification search procedure of earlier version of AMOS. The automated exploratory specification search procedure yields a ranking of the top ten best models by default given several fit function criteria to imply how to make modification of model by add or delete or restrict a path or specify a parameter. However, researcher must ultimately choose one as the best model to retain.

To be beared in mind, no automated specification search can make decision without sound theory. Therefore, as long as researchers keep in mind that the best use of automatic search procedures is to limit their attention to plausible theoretical models, the specification search procedure will never and should not be abused in empirical applications. It will still be the responsibility of the researcher to decide which one to accept as the best theoretical model.

The initial model is generally hypothesized after a review of the literature, supported by theory, and is being analyzed to confirm the initial theory-implied model using sample data. Researchers often discover that the initial model doesn't yield reasonable fit criteria and will therefore modify the model by adding or dropping paths.

Once obtained a model that fits well and is theoretically consistent and provides statistically significant parameter estimates, researcher must interpret it in the light of research questions and then distill the results in written form for publication. AMOS provides two ways for researcher to examine parameter estimates. One method is to use the path diagram output to visually display the parameter estimates while the other approach is to use tables similar to those containing the overall model fit statistics.

Theoretical decision making and model choose is therefore of vital concern because statistical analysis can not be supplanted ground for sound judgment in statistical modeling. The researcher must decide which path model is theoretically more meaningful and has better fitness index. However, we see once again that multiple fit function criteria do not always unanimously, so depending upon the researchers choice, the initial structural equation model could be better or similar to the one discovered by the specification search procedure.

As an evidence of misfit model, the AMOS provides the modification indexes that can be conceptualized as a chi-square statistic with one degree of freedom. Especially, concerning each fixed parameter specified, AMOS provides a modification indexes value of which represents the expected drop in overall chi-square value if the parameter were to be freely estimated in a subsequent run, all freely estimated parameters are supposed have modification indexes values equal to zero. Although this decrease in chi square is expected to approximate the modification indexes value, the actual difference can be larger than expected. However, their absolute magnitude is not as important as their relative size, which can serve as a means to measuring the importance of one against the others in pinpointing possible nonfit parameters.

3.3 The Consideration of Modifying Model

Any time the researchers re-specify or modify a model, he or her implicitly changes its meaning in some fundamental way. In many instances, a change in model specification results in a trivial or unimportant corresponding alteration of the model's substantive meaning, but in other cases model modification can foreshadow a strong shift in the model's meaning from a theoretical standpoint. Therefore, it is crucially important to think through each proposed model modification and ask myself if making the modification is theoretically consistent with the research goals. A second consideration to take into account when modify a model is that researcher is relying on the empirical data rather than theory to help specify the model. The more empirically-based modifications incorporate into the final model, the less likely the model is to replicate in new samples of data. For these reasons, researcher should modify models based upon theory as well as the empirical results provided by the modification indices. As a practical consideration, it is also worth noting that AMOS provides modification index output only when complete data are input into the program. In other words, the one cannot obtain modification index information when use missing data with AMOS.

When choosing the best theoretical model based on the specification search results, all or only a few of the fit function criteria may suggest the best model. Substantive theory and model validation must therefore guide any model modification process. But as noted by many experts, researchers should begin with a substantive theoretical model to avoid misuse in empirical applications.

The modification of an initial model to improve fit index has been termed as specification search. The specification search process is typically undertaken to detect and correct specification errors between an initial theory and implied model to reveal the nature relationship among the variables under study. The research experience has suggested that specification errors would invoke serious consequences and should be corrected. The most common approach for conducting specification searches was to alternate parameter restrictions in the initial model, once at a time to be observed exactly, to examine model fit index improvement.

Traditional modification criterions include F tests, R-squared change, lagrange multiplier tests, modification indices, or expected change statistics to evaluate hypotheses concerning whether a restriction is statistically inconsistent with the data. The fit function criteria for the initial theory-implied model and the selected specification search model are listed for the regression, path, factor, and structural model examples.

4 The Comparison of AMOS with Other SEM Softwares

In addition to AMOS, There are several other famous software also dealing with linear structural relations and moment structure, such as LISREL, EQS and Multiple plus. Although LISREL is the first in the commercial market, and Multiple plus is often considered as powerful, AMOS is most popular for its competitive advantages over the counterparts.

AMOS was some different with others programs in the following three aspects: the preliminary analyses, model specification, parameter estimation, goodness of fit, misspecification, and the treatment of missing, non-normal and categorical data.

4.1 Preliminary Analysis of Data

Concerning the preliminary analysis of data, different software do varying degrees and in varying ways. AMOS: descriptive statistics related to non-normality as well as to detection of outliers, can be requested via an "Analysis Properties" dialogue box that is easy to access. EQS: EQS always reports the univariate and multivariate sample statistics. As in the LISREL, information about the sample statistics must be obtained by using particular program.

4.2 Preliminary Analysis of Data

AMOS can use drawing tool button to specify model in path diagram. As a convention, in the schematic presentation of structural equation models, measured or observed variables are represented in rectangles and unmeasured or latent variables in ellipses or circles. As each of the AMOS, EQS and LISREL program, the maximum likelihood estimation is default. However, other estimate methods are available for users. By the AMOS, this requirement is conveyed to the program by selecting the "Estimation" tab in the "Analysis Properties" dialogue box and then chooses the estimation procedure which is desired.

Thinking about the model assessment, the primary interest in SEM is the extent to which a hypothesized model fit or adequately describes the sample data, evaluation of model fit should derive from various perspectives and be based on several criteria that enable users to assess model fit from a diversity perspectives. In particular, these focus on the adequately of the model as an entirety and the parameter estimates.

In order to assess the fitness of individual parameters in the model, there are three criteria, first is the feasibility of the parameter estimates, second is the appropriateness of the standard errors, and third is the statistical significance of the parameter estimates.

Assessing the model adequacy focuses on two kinds of information, one is the residuals and the other is modification index. The residuals values in SEM represent the deviation between elements in the sample data and in the restricted variance and covariance matrices, one residual represents each pair of observed variables. Supposed a well fitting model, these values should be close to zero and evenly distributed among all observed variables, but the large residuals accompany by particular parameters indicate there exist misspecification in the model, then to affect the whole model fit.

4.3 Treatment of Missing, Non-normal and Categorical Data

The researchers have increasingly recognized a critical issue in SEM which is the presence of missing data, and the process ability of software packages to deal with such incomplete data. As incomplete data can cast seriously bias conclusions drawn from an empirical study, they should be resolved regardless whatever the reason for their absent. The degree to that such conclusions can be bias depends on both the amount and the style of missing values. Unfortunately, so far, there are currently no clear and widely accept guidelines regarding what is exactly a large amount missing data. The single method used in dealing with incomplete data in AMOS represents a direct approach based on the full information maximum likelihood estimation. It is direct in the approach that analysis are employ with no attempt to restore the data matrix to rectangular form, as is the case with methods involved imputation and weighting. In contrast with the commonly used indirect methods, the maximum likelihood approach is regarded to be theoretically grounded robust and has been proved as provided several advantages over other methods. The less responsible method used was to both listwise and pairwise deletion of missing data case. Or with pattern matching imputation, a missing value is replaced by an observed score from another case in the data set for which the response pattern is same of similar through all variables, the primary limitation of this method lies in that in some situation no matching case is determined, then none imputation is undertake.

One of the critically important theory assumption associated the structural equation modeling is that the supposed data should have a multivariate normal distribution. To violate this normal distribution can seriously corrupt statistical hypothesis testing such that norm theory test statistic may not reveal an adequate evaluation of the model under research. The approach of AMOS to resolve the problem of multivariate nonnormal data is to undertake a procedure in terms of "bootstrap". The Bootstrap serves as a re-sampling method by which the original sample in considered representing the population. Multiple subsamples of the same size as the parent sample are then drawn randomly with replacement, from this population. These subsamples then provide the data for empirical investigation of the diversity of parameter estimates and index of fit. The considered advantage of this approach is that it allows the researchers to assess the stability of parameter estimate and then produce their values with a greater extend of accuracy. When data are not normally distributed or are otherwise flawed in some way (almost always the case), larger samples will be compensation. It is difficult to make absolute recommendations what sample sizes is suitable. The general recommendation is thus to obtain more data whenever possible.

Another important assumption supposed with SEM is the requirement that all variables should to be on continuous scale. However, concern about the data in organizational and psychology generally, and assessment data in particular are typically of ordinal scale, this issue has long been focused. Although use of the distribution free method is widely accepted as an appropriate estimation procedure in this regard, so it's very stringent and impractical on the sample size [15]. When concerns to the AMOS, it is unable to identical the categorical scaling of variables. As contract, the EQS program could allows for estimation with normal theory maximum likelihood, and then offers the robust chi-square and standard errors for correct statistical inference. However, this approach is not recommended for all case but the largest sample size.

5 The Review of AMOS

One major important problem in the assessment instrument is the extent to which the measurement scale does measure that which they are supposed to measure, in psychological measurement words, the degree to which their factorial structures are valid. One of the most strictly methodological approaches to ensure for the validity of dimensions structures is the use of confirmatory factor analysis (CFA) within the framework of structural equation modeling. The AMOS has reached this standard and promotes the research of complex organizational variables. The AMOS is better in some aspects than other similar software and has gained advancement of application in many research fields. Many managers intended to promote the management innovation and business process reengineering may choose this software, so it has a promising prospect in the IT industry in the near future.

References

1. Schumacker, L.R.E.: A Beginner's Guide to Structural Equation Modeling. Lawrence Erlbaum, Mahwah (1989)
2. Gilles, E., Gignac, B.P., Tim, B., Con, S.: Differences in Confirmatory Factor Analysis Model Close-Fit Index Estimates Obtained from AMOS 4.0 and AMOS 5.0 via Full Information Maximum Likelihood - No Imputation: Corrections and Extension to Palmer et al. Australian Journal of Psychology 58, 144–150 (2006)
3. Steiger, J.H., Shapiro, A., Browne, M.W.: On the Multivariate Asymptotic Distribution of Sequential Chi-Square Statistics. Psychometrika 50, 253–264 (1985)
4. Schumacker, R.E.: A Comparison of the Mallows Cp and Principal Component Regression Criteria for Best Model Selection in Multiple Regression. Multiple Linear Regression Viewpoints 21, 12–22 (1994)

A Research on the Application of Physiological Status Information to Productivity Enhancement

Qingguo Ma, Qian Shang, Jun Bian, and Huijian Fu

School of management, Zhejiang University, Hangzhou, 310027

Abstract. This paper is grounded in the field of Neuro-Industrial Engineering, focuses on humans, takes human's physiological status data (e.g. EEG, EMG, GSR and Temp) into account and applies biofeedback technology to investigate the application of physiological status in enhancing productivity, and thus improves production efficiency and increases the profitability of certain enterprises, which will be beneficial to the long-term development of the enterprises and harmonious development of the whole society.

Keywords: Neuro-Industrial Engineering (Neuro-IE), physiological status indexes, productivity, Neuromanagement.

1 Introduction

1.1 The Development of Industrial Engineering

Industrial Engineering is a discipline that designs, improves and implements the integrated system formed by people, material, equipment, energy resources, information, etc, which emphasizes on improving productivity, reducing the costs and insuring quality in order to obtain maximum overall effectiveness with the production system operating in the best condition. Thus it has been attached great importance by many countries in recent decades.

The development of Industrial Engineering has experienced three notable stages as "Scientific Management Era", "the Comprehensive Improvement after Wars" and "Human Factors Period", each of which has played an important role in facilitating enterprise production and social development in their own times. In the course of the development of IE, Taylor's Theory of Scientific Management was an important thought about motion study and time research[1], which generated 17 elemental motions and standardized operations procedure; in the stage of "the Comprehensive Improvement after Wars", Enterprise Resource Planning (ERP), Toyota Production System, Total Quality Management and other classic production and management methods have improved productivity by using digital information; And in the stage of "Human Factors Period", with the realization of the effectiveness of human itself, engineers have combined psychology, anatomy and anthropometry with engineering design and operations management, trying their best to improve work environment[2].

M. Zhu (Ed.): Electrical Engineering and Control, LNEE 98, pp. 801–808.
springerlink.com © Springer-Verlag Berlin Heidelberg 2011

1.2 Existing Problems

Traditional researches overemphasized on productivity and standardization, taking humans as machines, even work environment and equipments are improved in the stage of "Human Factors Period", attention has not yet been drawn to human itself, which makes human a "living machine" and causes accidents such as "13 leaps" of Foxconn employees in Shenzhen, 2010. Though there are factors related to failed management and low payment, the advent of expressions such as "sweetshop" and "industrialized apathy" has reflected workers' true conditions in factories and largely discredited Foxconn. Meanwhile, a high degree of fatigue is induced by excess workload, which causes tremendous potential safety hazards. It's reported in the US that the loss is about 18 billion every year caused by work fatigue in industrial production, and at least 12 billion by transportation accidents on expressway.

Traditional production mode could help enterprises gain advantages of high quality, high efficiency, low cost and quick response, but it doesn't consider human emotion and cognition, neglecting people's physiological and psychological factors by taking people merely as a part of the production line. However, with the progress of science and the improvement of living standard, workers' job satisfaction is no more determined only by acquiring economic benefits. People's body structure, physical function, physical and emotional states and emotional changes influence their psychological and physiological status, which further influence people's external performance. Thus it's significant to consider "people's psychological and physiological status" as an important parameter in production management.

2 The Introduction of Neuro-industrial Engineering

Given the deficiencies of traditional research, Neuro-Industrial Engineering engendered by the combination of neuroscience and industrial engineering has provided resolutions. In 2006, Prof. Ma (the supervisor of the Neuromanagement lab at Zhejiang University of China) et al. firstly advanced the concept of "NeuroIE" [3]. Based on people's physiological status, NeuroIE obtains objective and actual data by measuring human brain and physiological indexes with advanced neuroscience tools and biofeedback technology, analyzes the data, adds neural activities, physiological status in production process as new factors into operations management, and finally realizes man-machine integration by adjusting work environment and production system according to people's response to the system, avoiding accidents and improving efficiency and quality.

Biofeedback technology is used in this research. This technology amplifies physiological process and bioelectrical activities inside human body with the help of electronic instrument. Amplified information of electrical activities is presented in visual (e.g. gauge reading) or auditory (e.g. humming) forms, which enable the subject to recognize its own physiological status and learn how to control and rectify abnormal physiological changes to a certain extent. The data collected by Biofeedback instrument includes myoelectricity, skin temperature, brain wave activities, skin conductivity, blood pressure, heart rate, etc, all of which could reflect individual's physiological and psychological status.

This research is grounded on Holley Metering Limited. We are enabled to infer workers' physiological and psychological status after collecting and analyzing the indexes listed above in the course of production, and then make appropriate adjustment and arrangement according to the characters and requirements of certain stations in order to radically remove hidden dangers caused by workers' physiological status (e.g. fatigue) and bad mood (e.g. boredom).

3 Result and Conclusion

This research is based on Holley Metering Limited, and we choose two key stations (i.e. input program station and maintenance station) in the workshop after surveys and interviews, because there are significant differences between these two stations in terms of task characteristics, which mean they could be analyzed as typical stations in the production line. Meanwhile, both new and old employees, between whom there is obvious difference in job performance, are selected in these two stations respectively.

Portable biofeedback instrument is used to measure and record the selected employees' physiological indexes (e.g. EEG, EMG, GSR and Temp). Then the data are analyzed with correspondent biofeedback technology, and the following conclusions are drawn from the analysis.

3.1 The Rule of Physiological and Psychological Changes That Influence Job Performance

EEG data. EEG is mainly used to measure and record brain electrical activities (amplitude and frequency) in different brain areas, and the rhythm and amplitude of brain electrical activities are closely related to emotion and attention[4].

Regarding resource utilization of left and right brain, it occupies more right brain resource in the station of input program because it needs standardized and formalized operations and the procedures are fixed and simple, meaning it needs more spatial thinking, which is exactly the function of right brain. While in maintenance station, it occupies more resource in left brain, because the specific attribute of the station requires richer work experience, more thinking and judgment, which is mainly the function of left brain.

Table 1. The frequency and description of EEG[4]

Brain waves	Frequency	Description of indexes
Θ waves	4~7HZ	Level of fatigue
SMR waves	12~15HZ	Level of attention
B waves	13~40HZ	Level of excitation

Seeing from the brain resource utilization of new and old workers, in input program station, both the variation values of θ and B waves of new worker's are higher than those of old worker's, meaning the new worker has a higher level of fatigue and is emotionally nervous because of nonproficiency in operations. But it's opposite in maintenance station as the variation values of θ and SMR waves of the old worker's are higher, implicating a higher level of fatigue and attention, which is mainly the result of more occupancy of brain resource and attention caused by deeper and more comprehensive thinking.

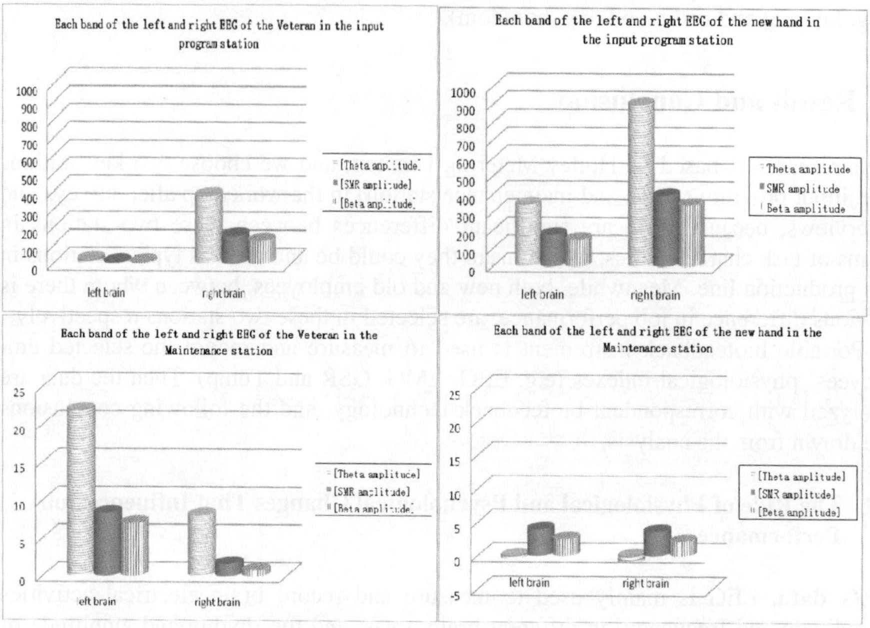

Fig. 1. EEG data of input program station and maintenance station

EMG data. EMG measures myoelectricity voltage on body surface. There is close relation between myoelectricity and the level of muscular tension as EMG rises rapidly when muscles are tense, and vice versa. Thus EMG is a significant indicator of performance intensity and emotion[5].

In view of EMG data, in input program station, the old worker's right myoelectricity voltage is higher than left, while it's opposite for the new worker, because for an old worker, he has to fetch objects in a greater spatial extension with right hand in the process, which raises up right myoelectricity voltage by greater movement range, but for the new worker, a specialized worker is assigned to help him fetch needed objects and the new worker has to put the parts after processing on the conveyor belt, which raises up left myoelectricity voltage. In maintenance station, the old worker's right myoelectricity voltage is higher than left, and the new worker's is opposite, because normally left hands are used to hold and invert the device, while as a result of inexperience, the new worker has to check and test the device repeatedly so as to find out the faults, which raises up left myoelectricity voltage.

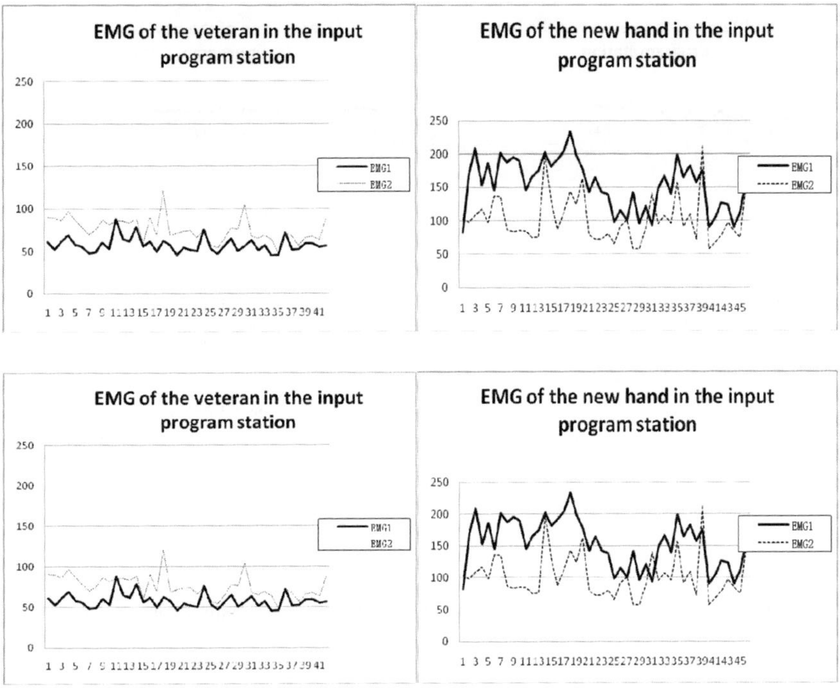

Fig. 2. EMG data of input program station and maintenance station

Comparing EMG data of new and old workers, new workers' myoelectricity voltage are higher than old workers' in both stations, because new workers are inexperienced and need repeated inspections and tests in order to find the correct operation points.

GSR and Temp data. GSR feedback reflects the changes of sympathetic nerves' systematic activities by measuring the changes of sweat glands' activities. In the situation of emotional strain, horror and anxiety, the sweat glands excretes more sweat, which causes the skin conductance to rise and increases GSR. Thus GSR is the most remarkable indicator of emotional changes and could be used to relieve emotional strain. Temp is also an indicator of emotional change that is in accordance with GSR, as when a person is relieved, the excitation of sympathetic nerves declines and blood flow volume at finger tip increases, but when a person is nervous, the excitation of sympathetic nerves escalates and blood flow volume at finger tip decreases, which causes the skin temperature to fall.

In input program station, old worker's GSR value firstly declines and then climbs up, suggesting anxiety induced by fatigue after a period of work, and is notably higher than the new worker on the whole. And seen from the following figure, new and old workers' Temp value changes are small and there is no significant difference between them.

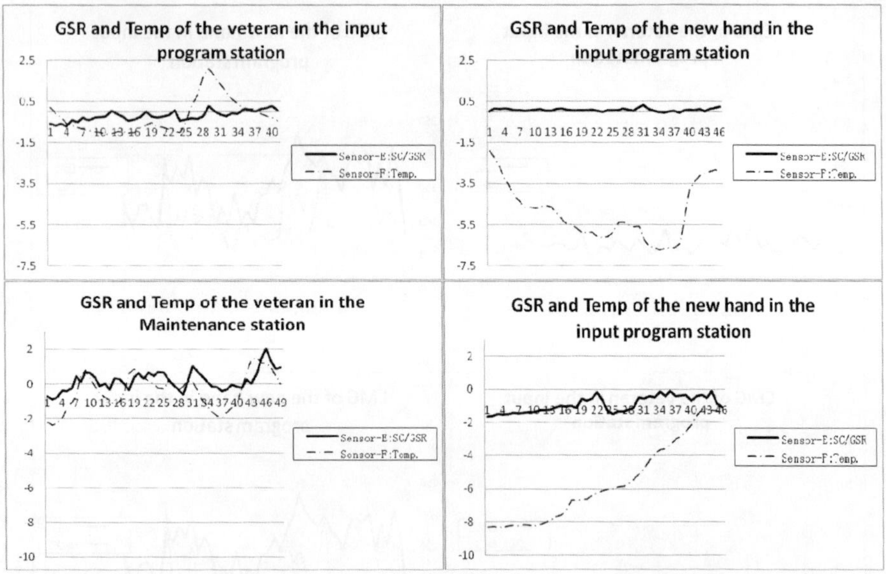

Fig. 3. GSR and Temp data of input program station and maintenance station

In maintenance station, new and old workers' GSR value changes are small and there is no significant difference between them, too. Regarding Temp, old workers' Temp value change is relatively small, while the temperature of the new worker firstly falls dramatically and then recovers gradually, suggesting emotional strain and anxiety at the beginning of maintenance work, and emotional status recovers after entering a stable working state.

3.2 The Best Rest Point to Minimize the Probability of Errors and Realize the Balance between Quality and Efficiency

The design of rest system is generally based on the level of physical and mental fatigue. Physiological and psychological indicators enable us to determine the best rest time point and adjust rest system from a new perspective. Among those brain waves, Theta waves rise with the escalation of fatigue level, and SMR waves with the escalation of attention and vigilance. When a person starts to feel fatigue, the probability of errors increases and it requires more attention and vigilance to keep on working. Based on that rule, a best rest point could be obtained to minimize the probability of errors.

As is shown in the figure, compared with input program station, maintenance station requires more elaborate operations and brain resource. Seen from the variation tendency, in maintenance station, EEG wavebands firstly rise, then fall, and after that climb up again before reaching the peak; but in the station of input program station, EEG wavebands rise placidly, meaning fatigue level rises slowly, and then reach the peak. In traditional management patterns, time point for rest is usually set at the latter fatigue point, neglecting the increase of the probability of errors caused by no rest at the first fatigue point for workers at stations that require more brain resource (e.g. maintenance station). Thus this research facilitates the design of rest system for the management.

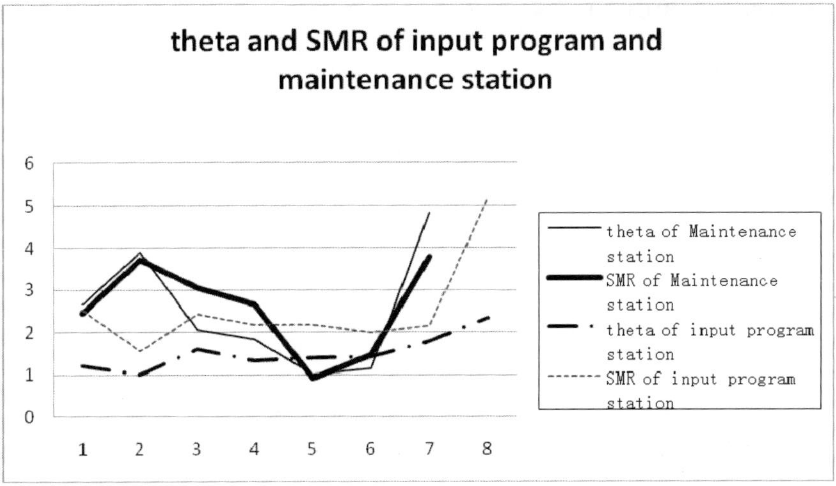

Fig. 4. The variations of Theta vaves and SMR waves of input program station and maintenance station

4 Prospect

The advent of Neuro-IE has pushed forward the development of industrial engineering to a new level, and the application of biofeedback technology in enterprises, which for the first time brings physiological and psychological status of "human itself" as significant factors into the research of production efficiency and provides a new perspective to investigate the way to enhance employees' work efficiency. It plays a major role in the deepening of industrial engineering theory and is of great realistic significance and application value for modern production and social development.

References

1. Shue, W., Jiang, Z.: Conspectus of Industrial Engineering. Mechanical Industry Press, Beijing (2009)
2. Roussel, P.A., Saad, K.N., Erickson, T.J., Third Generation, R.: hird Generation R&D. Harvard Business School Press, Boston (1991)
3. Ma, Q., Wang, X.: Cognitive Neuroscience, Neuroeconomics and Neuromanagement. Management World 10 (2006)
4. Kadir, R.S.S.A., Ismail, N., Rahman, H.A., Taib, M.N., Murat, Z.H., Lias, S.: Analysis of Brainwave Dominant After Horizontal Rotation (HR) Intervention Using EEG for Theta and Delta Frequency Bands. In: 5th International Colloquium on Signal Processing & It's Applications (CSPA), pp. 284–287 (2009)

5. Laparra-Hernández, J., Belda-Lois, J.M., Medina, E., Campos, N., Poveda, R.: EMG and GSR signals for evaluating user's perception of different types. International Journal of Industrial Ergonomics 39, 326–332 (2009)
6. Bundele, M.M., Banerjee, R.: Detection of Fatigue of Vehicular Driver using Skin Conductance and Oximetry Pulse: A Neural Network Approach. In: Proceedings of the iiWAS 2009, pp. 725–730 (2009)

The Impact of Procedural Justice on Collective Action and the Mediation Effect of Anger*

Jia Liuzhan and Ma Hongyu**

Psychology College, Central China Norm University,
Wuhan, the People's Republic of China
mahy@mail.ccnu.edu.cn

Abstract. Through the experiment conditions include 292 university students to examine the relationship among procedure justice, anger and collective action participation intention. The manipulation of procedural justice was to offer the voice or not in the process of price increment decision of university dinner management. The result showed that the subject had higher anger and collective action participation intention in the injustice condition than in the justice conditions. The procedural justice had significance positive effect on collective action participation intention; the mediate effect test confirmed that the anger had served as partly mediate effect between procedure justice and collective action participation intention.

Keywords: procedural justice; collective action; anger.

1 Introduction

The popular protest cry "NO JUSTICE, NO PEACE" implies that perceived injustice is motivator for collective action. Lots research [1] [2] [3] [4] [5] [6] [7] has found the injustice is necessary starting point for social protest. Smith et al. [8] found that subjects who suffered injustice would like to experience negative feeling such as anger, sadness and fear etc., then to engage in attempts to restore justice. Different feeling has various effects on behavior, as the anger would like to motivate protest behavior; fear is a powerful inhibitor of collective action. Since justice had been categorized into distributive justice and procedural justice, Wright and Moghaddam [9] found that procedural justice can influence all the collective members that suffered injustice to cause the collective protest behavior whether distributive justice can influence particular individuals, so the procedural justice would have positive effect on collective action. So this research takes procedural justice as independent variable to explore the relationship between procedural justice and collective action and the influence of anger.

---

* Supported by the Fundamental Research Funds for the Central Universities (Program No: 2010-47); Financial support also was provided by the Hubei Province Social Science Foundation Grant (Program No: [2010]022).

** Corresponding author.

M. Zhu (Ed.): Electrical Engineering and Control, LNEE 98, pp. 809–815.
springerlink.com © Springer-Verlag Berlin Heidelberg 2011

1.1 Procedural Justice

Procedural justice refers to the opinion of fairness in the processes that resolve disputes and allocate resources. The study about procedural justice often use the manipulation of voice which to offer or deprive the voice of subject who was influenced in the process of decision. This manipulation of procedural justice has popular accepted in the research of procedural justice. Smith [8] found that subjects would perceive higher fairness when were provided voice in the decision than deprive voice, this phenomenon was called "Voice Effect". The manipulation of procedural justice by voice effect has proved valid in experience. So this research adopts the voice as the manipulation of procedural justice.

1.2 Collective Action

Collective action refers a group member engages in collective action any time she or he acts as a representative of the group and where the action is directed at improving the conditions of the group as a whole [9].The participants in collective action mostly perceived unfair discrimination to accumulate dissatisfied such as anger, fear and resentment, they took collective action as an opportunity to express their dissatisfied feeling and restore justice. In the actual research mostly takes collective action participation intention as the measurement of collective action because the actual collective action is difficult to predict and stimulate in laboratory conditions. This method has been widely used in the current research, such as the Wright et al.[9][10], Klandermans et al.[1][2][3], Kelloway et al.[4],Van Zomeren et al.[5][6][7].

1.3 Anger

The study of Smith et al. [8] had found it's likely to perceive anger when disadvantaged people suffered injustice; the anger is an important factor to stimulate collective action. Mallet et al. [11] found that anger had positive effect on collective action. When perceived injustice, people would like to feel anger and participant in social protest to change the disadvantaged situation. Then we raise the question for what function of anger between procedural justice and collective action. As injustice and anger are also positive influence factors on collective action, the anger can be caused by injustice, so the study put forward the hypothesis that anger has a mediation factor between procedural justice and collective action.

2 Method

2.1 Participants

This study carried out a pilot study firstly by choosing 70 students in a particular university for required course credit. Participants in the pilot study include 23 male and 44 female, the average age is 19.89 years. Through the pilot study, research found that subjects experienced higher anger ($t=2.420$, $p<.05$), reported higher collective action participation intention ($t=2.133$, $p<.05$) under injustice condition than justice condition. So the manipulation of procedural justice is success and valid. 288 students include 88 male (29.33%) and 200 female (66.67%), mean age 19.95 years participated in the formal study that recruit from class for course credit. The subjects were randomly assigned to justice and injustice condition. At last, the procedural justice condition had 146 participants which contained 54 male and 90 female, the procedural injustice contained 33 male and 106 female.

The independent variable procedural justice had two conditions which were procedural justice and procedural injustice. The study was processed through experiment situation and then took questionnaire measurement of participants. The experiment condition was the increase of food price of university dinner which could draw great attention and familiar with students. The condition which offered voice of student in the increase decision of food price was justice condition whether deprived the voice of student was injustice condition. In the study experimenter told participants that for the price of commodities and CPI has climbing that result in cost rising of dinner management, so the dinner management plan to increase food price to cope the difficult of the cost rising. The condition of procedural justice was to offer voice for students in the food price increase decision, provided opportunity for students to express their opinion in the decision and cooperate with the dinner management to determine whether or not to increase and the range of food price. The condition of procedural injustice was to deprive the voice of student and refuse to provide opportunity for students to express their attitude and make the decision by dinner management arbitrary. When the experiment was finished, researcher took questionnaire measurement of procedural justice, anger and collective action participant intention.

2.2 Questionnaire

After experiment, participants were asked to respond to questionnaires on seven point Likert scale (such as, 1, not at all, 7, all), the higher score indicated higher degree of procedural injustice. First, we measure procedural justice with three items derived from Van Zomeren et al.[5], such as the manner of dinner management is unfairness and neglect the practical difficult of students, α= .81.

Second, the questionnaire of anger was derived from Mackie [12], the scale use seven point Liker type and had four items such as I feel angry/irritated/furious/displeased because the plan of dinner management, α= .85, the higher score meant higher level of anger.

Third, the measurement of collective action participant intention was took form Van Zomeren et al.[7]with five items, such as I would participate in a future demonstration /participate in raising our collective voice/do something together with fellow students/participate in some form of collective action/sing a petition to stop the dinner management, the Cronbach's alpha is .93. The data was input and processed by SPSS15.0, statistics method used were independent samples T test, correlate analysis and regression analysis.

3 Results

3.1 Descriptive Statistics

The key variables in study are show in table 1. Then to make test of the difference of anger, collective action on procedural justice, the independent samples T test has shown that participants experienced higher anger (t=2.023,p<.05), collective action(t=2.056,p<.05) and procedural justice(t=2.409,p<.05) in procedural injustice condition than in procedural injustice condition.

Table 1. α coefficient, mean, standard deviation and correlations between variables (N=292)

VARIABLE	a	M	SD	2	3
1Procedural justice	.882	14.47	4.157		
2 Anger	.906	16.92	5.791	.584(**)	
3 Collection action	.832	19.73	4.930	.418(**)	.450(**)

Note. **p<.01.

Then the study continued to test the manipulation validity of procedural justice. We took two methods to verify the valid manipulation of procedural justice. Firstly, we set an item in questionnaire and ask participants to evaluate the reality of experiment condition, the result revealed that the point of this item(5.07) was higher than the middle value; it can be served a proof of valid of procedural justice. Secondly, this research took procedural justice scale to test the experiment condition, the data shown that participants who assigned to injustice condition report higher score (t=2.409, p<.05) than justice condition. So the manipulation of procedural justice was success and valid.

3.2 The Impact of Procedural Justice on Collective Action Participant Intention

This study took procedural justice as independent variable, collective action as dependent variable to run the regression analysis. Firstly, take age and gender as control variable to run hierarchical regression found that gender had significance effect on collective action participant intention, the male had stronger intention than male participants(t=2.034,p<.05), but the age had no significance impact on collective action participant intention. So we took the gender as control variable to run hierarchical regression, the result indicated that procedural justice had positive impact on collective action, the standard coefficient is .416(p<.000). So we draw the conclusion that procedural justice had positive effect on collective action.

3.3 The Mediation Effect of Anger between Procedural Justice and Collective Action Participant Intention

The research continued to test the mediation effect of anger between procedural justice and collective action. As the former regression shown that collective action had significance difference in gender, so we first took gender as control variable. Following the instruction of Baron and Kenny [13], there were four steps in establishing mediation effect. Step one should confirm the significant direct effect of the predictor variable (procedural injustice) on the dependent variable (collective action). The regression revealed there was a significant total effect of procedural justice on collective action (β=.414, p<.000). So the first step which to establish mediation was satisfied. The second step should to determine whether the independent variable procedural justices significantly affect the mediation variable anger. The regression analysis shown that procedural injustice had positive influence of anger (β=.585, p<.000). Then the second step needed to establish mediation was satisfied. Step three was to determine whether the impact of mediators on the dependent variable was significant. The regression analysis shown that anger had positive effect on collective action (β=.444, p<.000). So the third step for establishing mediation was satisfied for anger. Step four was to establish anger mediated the relationship between procedural injustice and collective action. When the anger and procedural justice were simultaneously entered as predictors of collective action, the standard coefficient was decreased from .414 to .234(p<.000), but the standard coefficient was still significance. The regression suggest the partly mediation of anger between procedural justice and collective action.

4 Discussions

4.1 The Impact of Procedural Justice on Collective Action Participant Intention

The experiment revealed that procedural justice had positive significant effect on collective action. Participants who were assigned to injustice condition were more likely to participant in collective action compared with participants who were in justice condition. This result was agree with the conclusion from the research of Zomeren[5][6] and Miller et al.[14].The research of Pettigrew et al.[15] had found that the ranks of the center among lower social class standing were more easily suffered injustice. The lower social ranks are most numerous among that have limited incomes, have no own homes or apartments or live in smaller quarters. Subject would likely to protest and resume injustice when they suffered injustice, and the collective action is a typical and easier available protest behavior compare with other reaction. It will more easily for them to join collective action when suffered more serious injustice.

4.2 The Mediation Effect of Anger

The study has confirmed the partly mediation effect of anger between procedural justice and collective action. The cognitive approval theory of emotion [16] suggests people would experience anger and protest when they suffered injustice. Relative deprivation theory [17] also emphasis the motivate effect of negative emotions such as anger and resentment on collective action. Since injustice is the cause of collective action, even the effect of anger has include as predictors in regression equation, the direct effect of procedural justice still has significance effect. So the impact of procedural justice could not be replaced by anger.

4.3 The Collective Action Participant Intention in Experiment Conditions

This study has found that participants under justice and injustice conditions all have highly collective action participant intention, the score are 19.17 and 20.27 respectively, and the score are both higher than middle point. This result may be correlative with the particular experiment condition. Because the food price of dinner is closely to the student and the lightly increase of food price would lead to strongly dissatisfied and protest. The students highly demand to express opinion and have voice in the issue closely related themselves. On the contemporary has happened the blooding conflict between Britain college students and government for the tuition increase; at the same time the CPI has climbed to the highest level in history, that has brought heavy pressure on the living cost. All those factors have caused higher collective action intention of students. This study has found that participants under justice and injustice conditions were all reported very high level of collective action may be implicit that decision which harmful to subjects in nature whatever made in justice or injustice style would confront refused and suffered protest.

4.4 Implications for Future Research

Since the justice was categorized into procedural justice and distribution justice, this study has only research the procedural justice, and then the effect of distribution justice on collective action is under research. In the regression equation of procedural justice and anger on collective action, the total explanation percentage of variance is a relative middle level (R^2=.241), it implicates that there were several factors could affect collective action missed in this research. Smith et al. [8] found that cognitive factors had important influence in collective action, such as the evaluation of justice, group efficacy and collective identify. So the cognitive factors are the future variable in collective action research.

This research has found that procedural justice has positive effect on collective action, participants in injustice condition reported higher collective action participant intention than justice condition; anger has partly mediation effect between procedural justice and collective action participant intention.

References

1. Klandermans, B., Jose, M.S., Mauro, R., Marga, D.W.: Identity Processes in Collective Action Participation: Farmers' Identity and Farmers' Protest in the Netherlands and Spain. Pol. Psy. 23, 235–251 (2002)
2. Klandermans, B.: How Group Identification Helps to Overcome the Dilemma of Collective Action. Ame. Beha. Sci. 45, 887–900 (2002)
3. Klandermans, B., Jojanneke, V.T., Jacquelien, V.S.: Embeddings and Identity: How Immigrants Turn Grievances into Action. Ame. Soci. Rev. 73, 992–1012 (2008)
4. Kelloway, E.K., Francis, L., Catano, V.M., Teed, M.: Predicting Protest. Bas. App. Soc. Psy. 29, 13–22 (2007)
5. Van Zomeren, M., Spears, R., Colin, W.L.: Exploring Psychological Mechanisms of Collective Action: Does Relevance of Group Identity Influence How People Cope With Collective Disadvantage? Bri. Jour. Soci. Psy. 47, 353–372 (2008)
6. Van Zomeren, M., Postmes, T., Spears, R.: Toward an Integrative Social Identity Model of Collective Action: A Quantitative Synthesis of Three Socio-Psychological Perspectives. Psy. Bull. 134, 504–535 (2008)
7. Van Zomeren, M., Colin, W.L., Russell, S.: Does Group Efficacy Increase Group Identification? Resolving Their Paradoxical Relationship. J. Exp. Soc. Psy. 46, 1055–1060 (2010)
8. Smith, H.J., Tracey, C., Thomas, K.: Anger, Fear, or Sadness: Faculty Members' Emotional Reactions to Collective Pay Disadvantage. Poli. Psy. 29, 221–246 (2009)

9. Wright, S.C., Taylor, D.M., Moghaddam, F.M.: Responding to Membership in A Disadvantaged Group: From Acceptance to Collective Action. Jour. Per. Soci. Psy. 58, 994–1003 (1990)
10. Wright, S.C.: Ambiguity, Social Influence, and Collective Action: Generating Collective Protest in Response to Tokenism. Per. Soci. Psy. Bull. 23, 1277–1290 (1997)
11. Mallett, R.K., Jeffrey, R.H., Stacey, S., Janet, K.S.: Seeing Through Their Eyes: When Majority Group Members Take Collective Action on Behalf of An Outgroup. Gro. Pro. & Int. Rela 11, 451–470 (2008)
12. Mackie, D.M., Devos, T., Smith, E.R.: Intergroup Emotions: Explaining Offensive Action Tendencies in an Intergroup Context. Jour. Per.Soc. Psy. 79, 602–616 (2000)
13. Baron, R.M., Kenny, D.A.: The Moderator-Mediator Variable Distinction in Social Psychological Research: Conceptual, Strategic, and Statistical Considerations. J. Per. Soc. Psy. 51, 1173–1182 (1986)
14. Miller, D.A., Tracey, C., Amber, L.G., Nyla, R.B.: The Relative Impact of Anger and Efficacy on Collective Action is Affected by Feelings of Fear. Gro. Pro. & Inte. Rel. 12, 445–462 (2009)
15. Lazarus, R.S.: Progress on A Cognitive-Motivational-Relational Theory of Emotion. Ame. Psy. 46, 819–834 (1991)
16. Pettigrew, T., Christ, O., Wagner, U.: Relative Deprivation and Intergroup Prejudice. J. Soc. Iss. 64(2), 385–401 (2008)
17. Crosby, E.J.: A Model of Egoistical Relative Deprivation. Psy. Rev. 83, 85–113 (1976)

Autonomous Pointing Avoidance of Spacecraft Attitude Maneuver Using Backstepping Control Method

Rui Xu[1], Xiaojun Cheng[2], and Hutao Cui[2]

[1] School of Aerospace Eningeering, Beijing Inttitute of Technology,
Beijing, China
[2] School of Astronautics, Harbin Inttitute of Technology,
Harbin, China
xj_cheng@yahoo.cn

Abstract. For the attitude maneuver of spacecraft under pointing constraints, an algorithm based on navigation function and backstepping method is proposed in this paper. By proper construction of navigation function for the constraint attitude control problem, the end of the direction vector of sensor can move in free work space, which is intuitively on the unit spherical surface, and almost globally converges to the desired position. Backstepping to dynamics subsystem, the control input is derived which can autonomously steer the system to the desired state. Simulation results validate the proposed algorithm.

Keywords: attitude control; backstepping; pointing constraint.

1 Introduction

With the development of space exploration, different categories of sensors are used to obtain required information for special mission. However, some sensors can't point to some special regions, which disable arbitrary slewing of the spacecraft. For example, the infrared sensor can not point to the sun[1]. Therefore, the traditional attitude control methods are no longer suitable for this situation because of these constraints.

Artificial Potential Function is often used to navigate robot under complex environment[2-4]. Koditschex and Rimon gave some excellent investigation results such as the possibility of "almost" global navigation function and invariance of navigation properties under diffeomorphism[5]. Additionally, they constructed an "almost" global navigation function which has only one global minimum except for some saddles. Park and Bang designed a controller for orbital target tracking of geostationary satellite under avoidance constraint with this navigation function[6], but there are some ambiguities in construction of navigation function when radius of avoidance area is combined with attitude Modified Rodriguez Parameters(MRP).

Backstepping is a popular nonlinear control design technique[7], and excessive investigation achievements of attitude control using backstepping method are reported due to the cascaded structure of the spacecraft kinematics and dynamics[8-9]. Radice designed a controller based on backstepping method for constraint attitude problem

M. Zhu (Ed.): Electrical Engineering and Control, LNEE 98, pp. 817–825.
springerlink.com © Springer-Verlag Berlin Heidelberg 2011

but only one inadmissible pointing is considered[10]. In this paper, we combine "almost" global navigation function with backstepping method, and propose a new controller for the attitude maneuver of spacecraft to avoid pointing to some regions.

The paper is organized as follows. In the second section, the model for the spacecraft rotation and pointing problem are presented. The detail about the construction of navigation function and derivation of controller is described in the third section. The rest part of the paper gives the simulation results and conclusion.

2 Problem Formulation

This section presents the models of attitude kinematics and dynamics and the detail of pointing constraint for spacecraft slewing.

2.1 Rotation Motion with Quaternion

The spacecraft is assumed to be a rigid body, and the torque imposed on spacecraft is along with the primary axis of the body. For simplicity, the body-fixed frame is chosen to coincide with primary axis. The equations of kinematics and dynamics are given by

$$\dot{\boldsymbol{q}} = 0.5\boldsymbol{Q}\boldsymbol{\omega}\,. \tag{1}$$

$$\dot{\boldsymbol{\omega}} = \boldsymbol{J}^{-1}\left(\boldsymbol{u} - \left[\boldsymbol{\omega}^{\times}\right]\boldsymbol{J}\boldsymbol{\omega}\right). \tag{2}$$

where $\boldsymbol{q} = \begin{bmatrix} q_0 & \underline{\boldsymbol{q}} \end{bmatrix}^{\mathrm{T}} = \begin{bmatrix} q_0 & q_1 & q_2 & q_3 \end{bmatrix}^{\mathrm{T}}$ denotes unit quaternion that represents attitude of the spacecraft with respect to inertial frame, $\boldsymbol{\omega} = \begin{bmatrix} \omega_1 & \omega_2 & \omega_3 \end{bmatrix}^{\mathrm{T}}$ denotes angular velocity of the spacecraft specified in body frame. $\boldsymbol{J} = \mathrm{diag}\left(J_1, J_2, J_3\right)$ denotes body frame referenced inertial matrix of the spacecraft. $\left[\boldsymbol{\omega}^{\times}\right]$ is an antisymmetric matrix with the form,

$$\left[\boldsymbol{\omega}^{\times}\right] = \begin{bmatrix} 0 & -\omega_3 & \omega_2 \\ \omega_3 & 0 & -\omega_1 \\ -\omega_2 & \omega_2 & 0 \end{bmatrix}. \tag{3}$$

and

$$\boldsymbol{Q} = \begin{bmatrix} -q_1 & -q_2 & -q_3 \\ q_0 & -q_3 & q_2 \\ q_3 & q_0 & -q_1 \\ -q_2 & q_1 & q_0 \end{bmatrix}. \tag{4}$$

2.2 Pointing Avoidance Imposed on Attitude Rotation

In general, the light of bright body should not go into the field of view of faint light sensor, such as infrared telescope, to keep the sensor working in a good situation and avoid its damage. Fig.1 depicts the sketch of pointing avoidance imposed by a sensor. The least angle between the direction vector of sensor and vector of bright body mainly depends on the field of view of the sensor. The constraint is denoted as an inequation of the following form,

$$\left\| \boldsymbol{r} - \boldsymbol{r}_{oj} \right\| > l_{r_{oj}} . \tag{5}$$

where $\boldsymbol{r}$ denotes the direction vector of sensor, $\boldsymbol{r}_{oj}$ denotes the direction vector of the jth obstacle celestial body, both of the two vectors are specified in inertial frame. $\|\bullet\|$ represents 2-norm of vector, $l_{r_{oj}}$ is the least 2-norm of difference between above two vectors. Only (5) being satisfied, the sensor is safe, otherwise it will be damaged. $l_{r_{oj}}$ is largely determined by the field of view of the sensor.

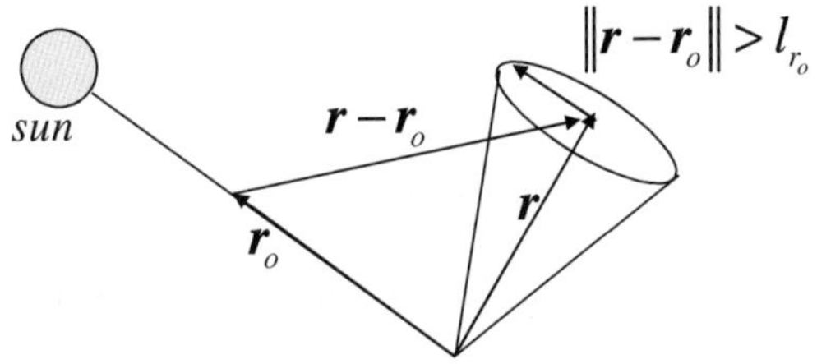

Fig. 1. Sketch of constraint during attitude maneuver of spacecraft

3 Pointing Avoidance of Attitude Control Using Backstepping Method

The objective of the attitude control is to design a controller which can steer the spacecraft from a safe attitude to a desired safe one while guaranteeing the satisfaction of pointing constraints. To solve the problem, we introduce a classical navigation function, and derive a controller based on backstepping method.

3.1 Construction of Navigation Function

Assume that there is only one light sensitive sensor mounted on the spacecraft. We define the bounded workspace,

$$W = \left\{ \boldsymbol{r} \in E^n : \left\| \boldsymbol{r} - \boldsymbol{r}_d \right\|^2 < \rho_0^2 = 4 \right\} \tag{6}$$

where E^n denotes Euclid space, $\boldsymbol{r}_d$ denotes the desired orientation of the sensor, the 2-norm of which is 1 as well as other direction of vectors in this paper. $\boldsymbol{r}$ denotes the current orientation of the sensor. The largest 2-norm of the difference between $\boldsymbol{r}$ and $\boldsymbol{r}_d$ is equal to 2 when $\boldsymbol{r}$ and $\boldsymbol{r}_d$ are reversed.

There are n smaller spheres which bound the obstacles,

$$O_j = \left\{ \boldsymbol{r} \in E^n : \left\| \boldsymbol{r} - \boldsymbol{r}_{oj} \right\|^2 < \rho_j^2 \right\}, j = 1, ..., n \tag{7}$$

The free space can be obtained by removing the obstacles from work space,

$$F = W - \bigcup_{j=1}^{n} O_j \tag{8}$$

Different from general work space which needs to impose some constraints to make F being valid, F is always a valid free work space for the problem in this paper, since the unit spherical surface where the end of direction vector moves is a closed space.

From the point of view of Koditschex and Rimon[5], strict global navigation function is not possible because of saddles introduced by obstacles, but "almost" global navigation function is possible. We use the same navigation function in ref.[5], which has the form,

$$V = \left(\frac{\gamma_d^k}{\gamma_d^k + \beta} \right)^{1/k} \tag{9}$$

where k is a scalar constant, γ_d and β are intentionally designed as follows,

$$\gamma_d = \left\| \boldsymbol{r} - \boldsymbol{r}_d \right\|^2 \tag{10}$$

$$\beta_0 = 4 - \left\| \boldsymbol{r} - \boldsymbol{r}_d \right\|^2 \tag{11}$$

$$\beta_j = \left\| \boldsymbol{r} - \boldsymbol{r}_{oj} \right\|^2 - \rho_j^2 \tag{12}$$

$$\beta = (\beta_0 \beta_1 \cdots \beta_n) \tag{13}$$

3.2 Controller Design Using Backstepping Method

The attitude matrix can be easily derived in (14),

$$\boldsymbol{C}_{Ib}(\boldsymbol{q}) = \left[\left(q_0^2 - \underline{\boldsymbol{q}}^{\mathrm{T}} \underline{\boldsymbol{q}} \right) \boldsymbol{I}_3 + 2 \underline{\boldsymbol{q}} \underline{\boldsymbol{q}}^{\mathrm{T}} - 2 q_0 \left[\underline{\boldsymbol{q}}^{\times} \right] \right]^{\mathrm{T}} \tag{14}$$

then, direction vector of the sensor may be specified in inertial frame by transformation as (15),

$$\boldsymbol{r} = \boldsymbol{C}_{Ib}(\boldsymbol{q})\boldsymbol{r}_b \tag{15}$$

where $\boldsymbol{r}_b$ is the direction vector of sensor specified in body frame.

Differential of (9) is given in (16),

$$\dot{V} = \frac{1}{(\gamma_d^k+\beta)^{2/k}}\left((\gamma_d^k+\beta)^{1/k}\dot{\gamma}_d - \gamma_d\frac{(\gamma_d^k+\beta)^{1/k-1}}{k}(k\gamma_d^{k-1}\dot{\gamma}_d+\dot{\beta})\right) \tag{16}$$

where $\dot{\gamma}_d$ and $\dot{\beta}$ are further given as (17)-(21). For clarity, we use several variables instead of some long formulations.

$$\dot{\boldsymbol{r}} = \dot{\boldsymbol{C}}_{Ib}(\boldsymbol{q})\boldsymbol{r}_b = -\boldsymbol{C}_{Ib}(\boldsymbol{q})\left[\boldsymbol{r}_b^{\times}\right]\boldsymbol{\omega} \tag{17}$$

$$\dot{\gamma}_d = 2(\boldsymbol{r}-\boldsymbol{r}_d)^{\mathrm{T}}\dot{\boldsymbol{r}} = -2(\boldsymbol{r}-\boldsymbol{r}_d)^{\mathrm{T}}\boldsymbol{C}_{Ib}(\boldsymbol{q})\left[\boldsymbol{r}_b^{\times}\right]\boldsymbol{\omega} = \boldsymbol{\Psi}\boldsymbol{\omega} \tag{18}$$

$$\dot{\beta}_0 = -2(\boldsymbol{r}-\boldsymbol{r}_d)^{\mathrm{T}}\dot{\boldsymbol{r}} = 2(\boldsymbol{r}-\boldsymbol{r}_d)^{\mathrm{T}}\boldsymbol{C}_{Ib}(\boldsymbol{q})\left[\boldsymbol{r}_b^{\times}\right]\boldsymbol{\omega} = \boldsymbol{B}_0\boldsymbol{\omega} \tag{19}$$

$$\dot{\beta}_j = 2(\boldsymbol{r}-\boldsymbol{r}_j)^{\mathrm{T}}\dot{\boldsymbol{r}} = -2(\boldsymbol{r}-\boldsymbol{r}_j)^{\mathrm{T}}\boldsymbol{C}_{Ib}(\boldsymbol{q})\left[\boldsymbol{r}_b^{\times}\right]\boldsymbol{\omega} = \boldsymbol{B}_j\boldsymbol{\omega} \tag{20}$$

$$\dot{\beta} = \frac{\mathrm{d}}{\mathrm{d}t}(\beta_0\beta_1\cdots\beta_n) = \sum_{j=0}^{n}\left(\dot{\beta}_j\prod_{\substack{i\neq j\\ i=0}}^{n}\beta_i\right) = \left(\left(\prod_{i=1}^{n}\beta_i\right)\boldsymbol{B}_0 + \sum_{j=1}^{n}\left(\left(\prod_{\substack{i\neq j\\ i=0}}^{n}\beta_i\right)\boldsymbol{B}_j\right)\right)\boldsymbol{\omega} = \boldsymbol{\Xi}\boldsymbol{\omega} \tag{21}$$

Therefore, the derivation of navigation can be reshaped as (22),

$$\dot{V} = \frac{1}{(\gamma_d^k+\beta)^{2/k}}\left(\left((\gamma_d^k+\beta)^{1/k} - (\gamma_d^k+\beta)^{1/k-1}\gamma_d^k\right)\boldsymbol{\Psi} - \gamma_d\frac{(\gamma_d^k+\beta)^{1/k-1}}{k}\boldsymbol{\Xi}\right)\boldsymbol{\omega} = \boldsymbol{Z}^{\mathrm{T}}\boldsymbol{\omega} \tag{22}$$

and the pseudo control $\boldsymbol{\omega}_s$ for the stabilization of kinematics subsystem is written as,

$$\boldsymbol{\omega}_s = -\frac{\boldsymbol{Z}}{\mu}\boldsymbol{\omega}_s = -\frac{1}{\mu}(F\boldsymbol{\Psi} - G\boldsymbol{\Xi}) \tag{23}$$

where μ is a positive constant and

$$F = \left(\gamma_d^k + \beta\right)^{-1/k} - \left(\gamma_d^k + \beta\right)^{-1-1/k} \gamma_d^k \tag{24}$$

$$G = \gamma_d \frac{\left(\gamma_d^k + \beta\right)^{-1/k-1}}{k} \tag{25}$$

The differential of $\boldsymbol{\omega}_s$ is

$$\dot{\boldsymbol{\omega}}_s = -\frac{1}{\mu}\left(\dot{F}\boldsymbol{\Psi} + F\dot{\boldsymbol{\Psi}} - \dot{G}\boldsymbol{\Xi} - G\dot{\boldsymbol{\Xi}}\right) \tag{26}$$

where

$$\begin{aligned}\dot{F} = &\Big((-1/k)\left(\gamma_d^k + \beta\right)^{-1/k-1}\left(k\gamma_d^{k-1}\boldsymbol{\Psi} + \boldsymbol{\Xi}\right) + \\ &(1+1/k)\left(\gamma_d^k + \beta\right)^{-2-1/k}\gamma_d^k\left(k\gamma_d^{k-1}\boldsymbol{\Psi} + \boldsymbol{\Xi}\right) - \left(\gamma_d^k + \beta\right)^{-1-1/k} k\gamma_d^{k-1}\boldsymbol{\Psi}\Big)\boldsymbol{\omega}\end{aligned} \tag{27}$$

$$\dot{\boldsymbol{\Psi}} = -2\left(-\boldsymbol{C}_{Ib}(\boldsymbol{q})\left[\boldsymbol{r}_b^{\times}\right]\boldsymbol{\omega}\right)^{\mathrm{T}}\left(\boldsymbol{C}_{Ib}(\boldsymbol{q})\left[\boldsymbol{r}_b^{\times}\right]\right) - 2\left(\boldsymbol{r} - \boldsymbol{r}_d\right)^{\mathrm{T}}\left(\boldsymbol{C}_{Ib}(\boldsymbol{q})\left[\boldsymbol{\omega}^{\times}\right]\left[\boldsymbol{r}_b^{\times}\right]\right) \tag{28}$$

$$\dot{G} = \left(\frac{\left(\gamma_d^k + \beta\right)^{-1/k-1}}{k}\boldsymbol{\Psi} + \gamma_d \frac{(-1/k-1)\left(\gamma_d^k + \beta\right)^{-1/k-2}\left(k\gamma_d^{k-1}\boldsymbol{\Psi} + \boldsymbol{\Xi}\right)}{k}\right)\boldsymbol{\omega} \tag{29}$$

and

$$\dot{\boldsymbol{\Xi}} = \sum_{j=1}^{n}\left(\boldsymbol{B}_j\boldsymbol{\omega}\prod_{\substack{i=1\\ i\neq j}}^{n}\beta_i\right)\boldsymbol{B}_0 + \left(\prod_{i=1}^{n}\beta_i\right)\dot{\boldsymbol{B}}_0 + \sum_{j=1}^{n}\left(\sum_{\substack{l=0\\ l\neq j}}^{n}\left(\boldsymbol{B}_j\boldsymbol{\omega}\prod_{\substack{i\neq j\\ i\neq l\\ i=0}}^{n}\beta_i\right)\boldsymbol{B}_j + \left(\prod_{\substack{i\neq j\\ i=0}}^{n}\beta_i\right)\dot{\boldsymbol{B}}_j\right) \tag{30}$$

in (30),

$$\dot{\boldsymbol{B}}_0 = 2\left(-\boldsymbol{C}_{Ib}(\boldsymbol{q})\left[\boldsymbol{r}_b^{\times}\right]\boldsymbol{\omega}\right)^{\mathrm{T}}\left(\boldsymbol{C}_{Ib}(\boldsymbol{q})\left[\boldsymbol{r}_b^{\times}\right]\right) + 2\left(\boldsymbol{r} - \boldsymbol{r}_d\right)^{\mathrm{T}}\left(\boldsymbol{C}_{Ib}(\boldsymbol{q})\left[\boldsymbol{\omega}^{\times}\right]\left[\boldsymbol{r}_b^{\times}\right]\right) \tag{31}$$

and

$$\dot{\boldsymbol{B}}_j = -2\dot{\boldsymbol{r}}^{\mathrm{T}}\left(\boldsymbol{C}_{Ib}(\boldsymbol{q})\left[\boldsymbol{r}_b^{\times}\right]\right) - 2\left(\boldsymbol{r} - \boldsymbol{r}_j\right)^{\mathrm{T}}\left(\boldsymbol{C}_{Ib}(\boldsymbol{q})\left[\boldsymbol{\omega}^{\times}\right]\left[\boldsymbol{r}_b^{\times}\right]\right) \tag{32}$$

Backstepping to the dynamics subsystem, we define the potential function,

$$U = V + \frac{1}{2}\eta(\boldsymbol{\omega}-\boldsymbol{\omega}_s)^{\mathrm{T}}(\boldsymbol{\omega}-\boldsymbol{\omega}_s) \tag{33}$$

Differential of U is

$$\dot{U} = -\mu\boldsymbol{\omega}_s^{\mathrm{T}}\boldsymbol{\omega}_s + (\boldsymbol{\omega}-\boldsymbol{\omega}_s)^{\mathrm{T}}\left(-\mu\boldsymbol{\omega}_s + \eta\left(\boldsymbol{J}^{-1}\left(\boldsymbol{u}-\left[\boldsymbol{\omega}^{\times}\right]\boldsymbol{J}\boldsymbol{\omega}\right)-\dot{\boldsymbol{\omega}}_s\right)^{\mathrm{T}}\right) \tag{34}$$

Considering the stability of the system, differential of U should be negative definite. Finally, the controller can be obtained as the form of (35),

$$\boldsymbol{u} = \left[\boldsymbol{\omega}^{\times}\right]\boldsymbol{J}\boldsymbol{\omega} + \boldsymbol{J}\left(\dot{\boldsymbol{\omega}}_s + \frac{\mu}{\eta}\boldsymbol{\omega}_s + s(\boldsymbol{\omega}_s-\boldsymbol{\omega})\right) \tag{35}$$

and

$$\dot{U} = -\mu\boldsymbol{\omega}_s^{\mathrm{T}}\boldsymbol{\omega}_s - s(\boldsymbol{\omega}-\boldsymbol{\omega}_s)^{\mathrm{T}}(\boldsymbol{\omega}-\boldsymbol{\omega}_s) \le 0 \tag{36}$$

4 Numerical Simulation

Supposing that there are four obstacle regions, the direction of sensor mustn't point to these regions in the process of attitude maneuver. The detail information of obstacles is summarized in table 1, controller parameters are presented in table 2, and the information of spacecraft is given in table 3. We mention that all vectors are specified in inertial frame except for some special noted ones.

Table 1. Simulation parameters about obstacles

Number of obstacle	Direction of obstacles	Value of l_{r_o}
No. 1	$[0,1,0]^{\mathrm{T}}$	$\cos(\pi/6)$
No.2	$[0,-1,0]^{\mathrm{T}}$	$\cos(\pi/6)$
No.3	$[1,0,0]^{\mathrm{T}}$	$\cos(\pi/6)$
No.4	$[-1,0,0]^{\mathrm{T}}$	$\cos(\pi/6)$

Table 2. Simulation parameters of controller

Variables	parameters
η	10
μ	10
s	1
k	2

Table 3. Simulation parameters about the spacecraft

	Variables	parameters
Inertial matrix	$\boldsymbol{J}$	$\mathrm{diag}(100,110,120)$
Initial quaternion	$\boldsymbol{q}_{init}$	$[1,0,0,0]^{\mathrm{T}}$
Initial angular velocity	$\boldsymbol{\omega}_{init}$	$[0,0,0]^{\mathrm{T}}$
Direction of sensor in body frame	$\boldsymbol{r}_b$	$[0,0,1]^{\mathrm{T}}$
Terminal direction of sensor	$\boldsymbol{r}_d$	$[\sqrt{2}/2,0,-\sqrt{2}/2]^{\mathrm{T}}$
Terminal angular velocity	$\boldsymbol{\omega}_f$	$[0,0,0]^{\mathrm{T}}$

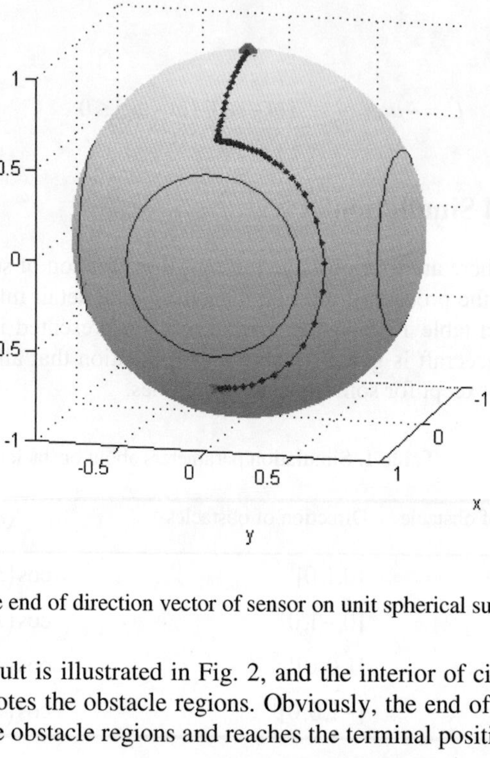

Fig. 2. The end of direction vector of sensor on unit spherical surface

The simulation result is illustrated in Fig. 2, and the interior of circles on the unit spherical surface denotes the obstacle regions. Obviously, the end of direction vector of sensor bypasses the obstacle regions and reaches the terminal position.

5 Conclusions

In this paper, we constructed an navigation function for pointing avoidance of spacecraft attitude maneuver based on the work of Koditschex and Rimon[5]. By transforming the attitude constraint control problem to a path finding problem of point robot on the unit spherical surface, the navigation function was constructed according to the classical one which is an "almost" global navigation function. Finally, a controller using the idea of backstepping method was derived. The simulation result shows that the algorithm of this paper is valid.

Acknowledgment

This work was supported by National Nature Science Foundation of China (60803051).

References

1. McInnes, C.R.: Large-Angle Slew Maneuvers with Autonomous Sun Vector Avoidance. Journal of Guidance Control and Dynamics 17(4), 875–877 (1994)
2. Jurgen, G., Vadim, U.: Sliding Mode Control for Gradient Tracking and Robot Navigation Using Artificial Potential Fields. IEEE Transactions on Robotics and Automation 2(2), 247–254 (1995)
3. Rimon, E., Koditschek, D.E.: Exact Robot Navigation Using Artificial Potential Functions. IEEE Transactions on Robotics and Automation 8(5), 501–518 (1992)
4. Ge, S.S., Cui, Y.J.: Dynamic Motion Planning for Mobile Robots Using Potential Field Method. Autonomous Robots 13(3), 207–222 (2002)
5. Koditschek, D.E., Rimon, E.: Robot Navigation Functions on Manifolds with Boundary. Advances in Applied Mathematics 11(4), 412–442 (1990)
6. Park, Y.-W., Bang, H.: Attitude controller design for orbital target tracking of geostationary satellite under avoidance constraint. Acta Astronautica 68, 830–842 (2011)
7. Khalil, H.K.: Nonlinear Systems, 3rd edn. Prentice Hall, Upper Saddle River (2002)
8. Ali, I., Radice, G., Kim, J.: Backstepping Control Design with Actuator Torque Bound for Spacecraft Attitude Maneuver. Journal of Guidance, Control, and Dynamics 33(1), 254–259 (2010)
9. Kim, K.S., Kim, Y.: Robust backstepping control for slew Maneuver using nonlinear tracking function. IEEE Transactions on Control Systems Technology 11(6), 822–829 (2003)
10. Radice, G., Ali, I.: Autonomous attitude using potential function method under control input saturation. In: 59th International Astronautical Congress Federation, France, pp. 5072–5077 (2008)

Based on TSP Problem the Research of Improved Ant Colony Algorithms

Zhigang Zhang and Xiaojing Li

School of Mathematics and Physics, USTB, Beijing 100083, China
lixiji19@126.com

Abstract. This paper compared and analyzed the time complexity and space complexity of three improved ACO algorithms, ASelite, ACS and MMAS. This paper made simulation experiments with these algorithms on the classic TSP problem with 29 cities. Experimental results show that the improved algorithms were highly efficient than basic ACO. And found that the running time of ACO, ASelite and MMAS are similar; the running time of ACS is relatively longer. The analyses show that the space complexity of the several ant colony algorithms that mentioned in this article is the same. Finally, analyzed the convergence conditions with probability 1 of these mentioned ant colony algorithms, and proved it. Summarized these ant colony algorithm can be used for various TSP problems.

Keywords: Ant colony algorithm; TSP; Complexity; Convergence.

1 Introduction

ACO as known has a strong advantage in solving discrete optimization problems. The time and space complexity of basic ACO have been analyzed, and give its convergence condition with probability 1. But not for its improved algorithms ASelite, ACS and MMAS these most commonly used improved algorithms. Based on the simulation of improved ant colony algorithms, this paper analyzed the solving efficiency of these algorithms on the TSP problem, compared and analyzed the time and space complexity of the improved ant colony algorithms, furthermore gave the conditions and proofs of their convergence with probability 1.

2 Background Knowledge

Biological studies have shown that a group of ants cooperate with each other can find the shortest path between the food sources and the nest, while the single ant can not. ACO is a simulation of this optimized mechanism [1]. The ACO model:

$$p_{ij}^{k}(t) = \begin{cases} \dfrac{[\tau_{ij}(t)]^{\alpha}[\eta_{ij}(t)]^{\beta}}{\sum_{s\in J_k(i)}[\tau_{is}(t)]^{\alpha}[\eta_{is}(t)]^{\beta}}, \forall j\in N_i, if\ \ j\in J_k(i) \\ 0, \qquad otherwise \end{cases} \tag{1}$$

M. Zhu (Ed.): Electrical Engineering and Control, LNEE 98, pp. 827–833.
springerlink.com © Springer-Verlag Berlin Heidelberg 2011

$$\tau_{ij}(t+n) = (1-\rho) * \tau_{ij}(t) + \Delta\tau_{ij}\,,\ \Delta\tau_{ij} = \sum_{k=1}^{m} \Delta\tau_{ij}^{k} \tag{2}$$

$$\Delta\tau_{ij}^{k} = \begin{cases} Q/L_k \quad , & \text{if } ant \text{ k } pass \text{ ij } in\ this\ tour \\ 0 \quad , & otherwise \end{cases} \tag{3}$$

ASelite [2], ACS [3] and MMAS [4] are improved algorithms of ACO from modify the probability selection formula and pheromone updating formula of ACO.

The convergence theorems of basic ant colony algorithm are as follows:

Assume there are m paths which are marked as $AC_1B, \cdots, AC_mB$, and the lengths of which without loss generality $d_1 \le \cdots \le d_m$. Set $q_{i,k}$ as pheromone of AC_iB and $p_{i,k}$ as the probability of ants choosing AC_iB, after pass AC_iB k times.

Theorem 1. When $\alpha \ge 0, \beta \ge 0$, then

$$q_{1,k} \ge q_{2,k} \ge \cdots \ge q_{m,k}\,,\ p_{1,k} \ge p_{2,k} \ge \cdots \ge p_{m,k}\,.$$

Theorem 2. When $\alpha \ge 1$ and $\beta \ge 0$, than $p_{1,k} \ge p_{1,k-1}$.

Theorem 3. When $\alpha \ge 1$ and $\beta \ge 0$, than $\lim_{k\to\infty} p_{1,k} = 1$.

The above show that the average probability to choose AC_1B as the shortest path of ACO approach 1, which means the overall trend of basic ACO is convergent [5].

3 Simulation and Efficiency

The data used in this paper is from the classic TSP test data set, usually used to test the ACO, GA and SA. This paper selected the TSP problem with 29 cities. Used the complex combination algorithm we obtained the shortest path is in the interval $[9.25\times10^3, 9.30\times10^3]$. The number of feasible paths of which is $\frac{(29-1)!}{2}$. The distribution of 29 cities' coordinates is as shown in Fig. 1:

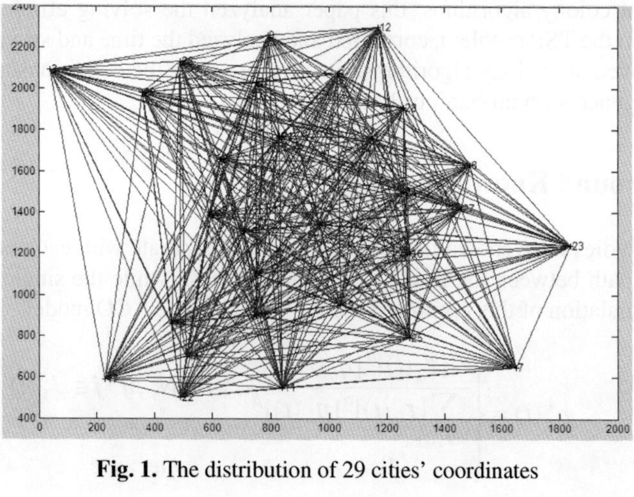

Fig. 1. The distribution of 29 cities' coordinates

Select the number of ants 10. The following were used basic ACO, ASelite, ACS and MMAS to solve the above-mentioned TSP problem with iterate 1000 times and iterate 100 times, the results are shown in Fig. 2 and Fig. 3.

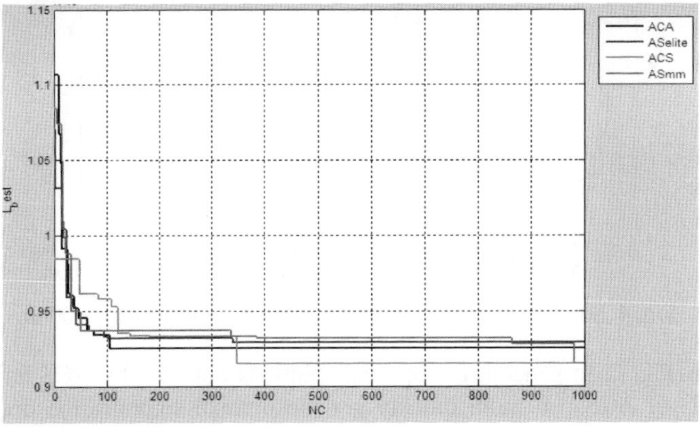

Fig. 2. Solve TSP problem with 29 city nodes with iterate 1000 times

Can be seen from Fig. 2 basic ACO and its improved algorithms are convergent for this TSP with iterate 1000 times. But their convergence speeds are different, so dose the optimal solutions. MMAS and ACS obtained better solutions than basic ACO and ASelite; generally ASelite obtained better solutions than ACO.

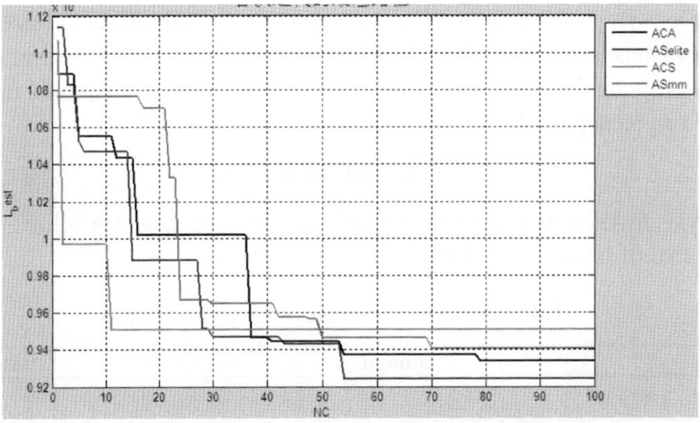

Fig. 3. Solve TSP problem with 29 city nodes with iterate 100 times

Fig. 3 shows the superiority of MMAS and ACS can't be found with fewer iteration times, instead basic ACO and ASelite obtain better solutions. This shows the convergence speed of ASelite and basic ACO is faster than that of MMAS and ACS. In general the convergence speed of ASelite is faster than that of ACO; and the convergence speed of ACS is faster than that of MMAS. Besides, ACO and ASelite would appear stagnation, but MMAS and ACS can avoid this. As Fig. 4:

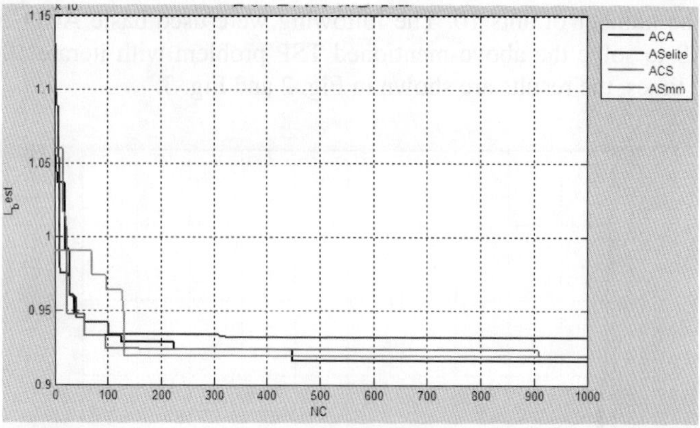

Fig. 4. Solve TSP problem with 29 city nodes with iterate 1000 times

Show as Fig. 2, in general, the optimal solutions obtained by ACS and MMAS are better, and yet ACS can more quickly get the optimal solution; the optimal solution obtained by ASelite is better than that by basic ACO. Therefore, the efficiency of ACS is the best, followed by MMAS, once again the ASelite algorithm, and basic ACO is the least efficient algorithm.

4 Complexity Analysis

4.1 Time Complexity Analysis

There use the asymptotic notation of time complexity to analyze, set the size of TSP as n, the number of ants is m, the number of loop variables is NC, then the time complexity of basic ACO program can be analyzed step by step. When n large enough, the influence of low-power can be negligible, its time complexity is:

$$\mathrm{T}(n) = O(NC \cdot n^2 \cdot m) \tag{4}$$

After analysis to know that the time complexity of MMAS and ASelite is basically the same with that of basic ACO, which also is the formula 4.

Table 1. The time complexity analysis of ACS

step	content	time complexity
1	Initialization	$O(n^2 + m)$
2	Set ants tabu list	$O(m)$
3	Individual ant construct solutions	$O(n^2 m)$
4	Calculation of solution evaluation and the amount of track updates	$O(n^2 m + nm)$
5	The update of pheromone trace concentrations	$O(n^2)$
6	Determine whether reached the algorithm's termination conditions, if not, go to Step 2	$O(nm)$
7	Output results	$O(1)$

When n is large enough, ignore the influence of low-power, we can see from Table 1, the time complexity of ACS is

$$\mathrm{T}(n) = O(NC \cdot (n^2 \cdot m + n \cdot m)) \tag{5}$$

Table 2. The running time of four kinds of ant colony algorithm

Algorithms	Running time(sec)			The average running time(sec)
basic ACO	53.13	52.51	52.66	52.77
ASelite	53.09	53.06	53.13	53.09
ACS	75.67	75.11	74.97	75.25
MMAS	52.66	52.46	52.42	52.51

Through measuring the time of basic ACO and its improved algorithms iterate for 1000 times, the results shown as Table 2. It shows, the running time of basic ACO, ASelite and MMAS are similar; the running time of ACS is relatively longer, which is consistent with the theoretical analysis, such as the formula (4), (2).

4.2 Space Complexity Analysis

If the scale of the TSP problem is n, and the number of ants is m, then the space complexity we can obtained from literature [6] is:

$$S(n) = O(4n^2) + O(2nm) \tag{6}$$

For the improved ant colony algorithms, when the scale of the problem is the same, its space complexity is similar to that of basic ACO.

For ASelite, only more than basic ACO there is a n-order two-dimensional array which represent the pheromone updates' amount of the edge where the optimal ants on, so the space complexity is $S(n) = O(5n^2) + O(2nm)$; For ACS, more than basic ACO there is a n-order one-dimensional array for the state transition, so the space complexity is $S(n) = O(5n^2) + O(3nm)$; MMAS and basic ACO have the same space complexity. The analysis shows the space complexities of the ACOs are very simple for the data storage, so they are easy to implement.

5 Convergence Analysis

For the convergence theorem of basic ACO, the proof of Theorem 1, 2 and 3 has been given in literature [5]. The proof of convergence Theorems that of the improved algorithms ASelite, ACS and MMAS are given below:

(1) For ASelite algorithm compared with the basic ACO the pheromone expressions $q_{i,k}$ are different, there are the following conclusions and the proofs:

① If $\alpha \geq 0$ and $\beta \geq 0$, then $q_{1,k} \geq q_{2,k} \geq \cdots \geq q_{m,k}$, $p_{1,k} \geq p_{2,k} \geq \cdots \geq p_{m,k}$.

Compared with Theorem 1 the different is $q_{i,1} = \rho C + np_{i,0}\frac{Q}{d_1} + \sigma p_{i,0}\frac{Q}{d_1}$ instead of $q_{i,1} = \rho C + np_{i,0}\frac{Q}{d_1}$. Its proof is similar with Theorem 1.

② If $\alpha \geq 1$, $\beta \geq 0$ then $p_{1,k} \geq p_{1,k-1}$.

The different of its proof with that of Theorem 2 is the formula

$$\frac{q_{i,k}}{q_{1,k}} - \frac{q_{i,k-1}}{q_{1,k-1}} = \frac{(nQ+\sigma Q)(\frac{p_{i,k-1}q_{1,k-1}}{d_i} - \frac{p_{1,k-1}q_{i,k-1}}{d_1})}{(pq_{1,k-1} + np_{1,k-1}\frac{Q}{d_1} + \sigma p_{1,k-1}\frac{Q}{d_1})q_{1,k-1}}$$

By reasoning can obtain the conclusion: $p_{1,k} > p_{1,k-1}$.

③ If $\alpha \geq 1$, $\beta \geq 0$ then $\lim_{k\to\infty} p_{1,k} = 1$.

Its proof is the same with Theorem 3.

(2) For ACS compared with basic ACO the selection probability $p_{i,k}$ and pheromone expressions $q_{i,k}$ are all different, there are the conclusions and proofs:

① If $\beta \geq 0$, then $q_{1,k} \geq q_{2,k} \geq \cdots \geq q_{m,k}$, $p_{1,k} \geq p_{2,k} \geq \cdots \geq p_{m,k}$.

There use $p_{i,k} = \frac{(C/d_i^\beta)}{\sum_{j=1}^{m}(C/d_j^\beta)}$ instead of $p_{i,k} = \frac{(C^\alpha/d_i^\beta)}{\sum_{j=1}^{m}(C^\alpha/d_j^\beta)}$ and $q_{i,1} = \rho C + (1-\rho)p_{i,0}\frac{1}{d_1}$ instead of $q_{i,1} = \rho C + np_{i,0}\frac{Q}{d_1}$ in Theorem 1. By reasoning can obtain: $q_{1,k} \geq q_{2,k} \geq \cdots \geq q_{m,k}$, $p_{1,k} \geq p_{2,k} \geq \cdots \geq p_{m,k}$.

② If $\beta \geq 0$ then $p_{1,k} \geq p_{1,k-1}$.

Proof: by

$$\frac{1}{p_{1,k}} - \frac{1}{p_{1,k-1}} = (\frac{d_1}{d_2})^\beta\left[(\frac{q_{2,k}}{q_{1,k}}) - (\frac{q_{2,k-1}}{q_{1,k-1}})\right] + \cdots + (\frac{d_1}{d_m})^\beta\left[(\frac{q_{m,k}}{q_{1,k}}) - (\frac{q_{m,k-1}}{q_{1,k-1}})\right]$$

and $$\frac{q_{i,k}}{q_{1,k}} - \frac{q_{i,k-1}}{q_{1,k-1}} = \frac{(1-\rho)(\frac{p_{i,k-1}q_{1,k-1}}{d_i} - \frac{p_{1,k-1}q_{i,k-1}}{d_1})}{(\rho q_{1,k-1} + (1-\rho)p_{1,k-1}\frac{1}{d_1})q_{1,k-1}}$$

can obtain $p_{1,k} > p_{1,k-1}$.

③ If $\beta \geq 0$ then $\lim_{k\to\infty} p_{1,k} = 1$.

Its proof is the same with Theorem 3.

(3) For MMAS compared with basic ACO the different is using the pheromone expressions $q_{i,1} = \rho C + p_{i,0}\frac{1}{d_1}$ instead of $q_{i,1} = \rho C + np_{i,0}\frac{Q}{d_1}$. So the three Theorems

and their proofs are similar with that of basic ACO, and is simple than that of basic ACO.

After reasoning and proving we obtained, the Theorem 1 to Theorem 3 of improved ant colony algorithms are still holds. It shows that the overall trend of improved ACOs is convergent. So it can be used varieties TSP problem's solution.

6 Conclusion

This paper analyzed the efficiency of basic ACO and three improved algorithms by instance, and found the most efficient algorithm is ACS, followed by MMAS, once again is the ASelite, the least efficient algorithm is basic ACO. One of the common problems of basic ACO is that the excessive concentration of ants search lead to local minimum value. But ACS and MMAS can avoid this problem.

This paper shows the theoretical results matched the actual results that about three improved algorithms' time complexity. This paper analyzed the convergence and conditions of converging with probability 1of these improved algorithms. Find that these three algorithms can be used for various TSP problems.

References

1. Zhang, J., Ao, L.: Improved Ant Colony Algorithm to Solve TSP problems. Chinese Journal of Xidian University 32(5), 681–685 (2005)
2. Huang, X., Zhang, H., Hong, X.: Theorizes and Applications of Modern Intelligent Algorithms. Chinese Science Press 1, 283–383 (2005)
3. Dorigo, M., Gambardella, L.M.: Ant colony system: A cooperative learning approach to the traveling salesman problem. IEEE Transaction on Evolution Computation 1(1), 53–66 (1997)
4. Stutzle, T., Hoos, H.: MAX-MIN Ant system and local search for combinational optimization problems. In: Meta-Heuristics: Advances and Trends in Local Search Paradigms for Optimization, pp. 137–154. Kluwer, Boston (1998)
5. Fang, W., Xiao, Q.: Convergence Analysis of Ant Colony Algorithm and Its Solution on TSP. Computer & Digital Engineering 35(9), 46–48 (2007)
6. Gambardella, L.M., Dorigo, M.: Solving symmetric and asymmetric TSPs by ant colonies. In: Proc of the Int Conf on Evolutionary Computation, pp. 622–627. IEEE, Los Alamitos (1996)

An Achievement of a Shortest Path Arithmetic' Improvement

Zhigang Zhang, Xiaojing Li, and Zhongbing Liu

School of Mathematics and Physics, USTB, Beijing 100083, China
lixiji19@126.com

Abstract. Secondly, this thesis researches the shortest path arithmetic on meshwork, and it optimizes arithmetic Z8-1. It implements the Z8-1 arithmetic, and proves its correction and finally analysis the conclusion and time complexity.

Keywords: Z8-1; the shortcut path; complexity.

1 Introduction

The shortest path algorithm, most of them are based on the improved algorithms of Dijkstra algorithm, Floyd algorithm and Matrix Algorithm, The time complexity of the three basic algorithm were $O(n^2)$, $O(n^3)$ and $O(n^4)$, Zhou Peide [1] proposed the Z8-1 algorithm in his treatise reduced time complexity to $O(n)$, Greatly reduces the time complexity, but the choice of the initial path in this algorithm can also be optimized and improved, the improved algorithm model have been given in the literature [3], this paper will focus on the achievement of the algorithm.

2 Steps of the Algorithm

Implementation of finding the shortest path algorithm as following:

Algorithmimplementation of improved Z8-1 algorithm

Input Network map $G=\{V,E,W\}$, the set of vertices and coordinates is $V=\{v_i(x_i,y_i)\mid i=1,2,...,n\}$, the edge set is $E=\{e_{ij}\mid i,j=1,2,...,n,i\neq j\}$, weight set is $W=\{w_{ij}\mid i,j=1,2,...,n,i\neq j\}$. The texture depth of the adjacent node tree is h , any two specified network nodes v_s and v_t .

Output A shortest path and its length between v_s and v_t .

Step0 Remove the node v_i of degree 1, form the new "input" data in addition to the points v_s and v_t ;

M. Zhu (Ed.): Electrical Engineering and Control, LNEE 98, pp. 835–841.
springerlink.com © Springer-Verlag Berlin Heidelberg 2011

Step1 Connect v_s and v_t into a straight line $\overline{v_s v_t}$, calculate $k = \left|\overline{v_s v_t}\right|$;

Step2 if there's an edge between v_s and v_t, then the edge is the shortest path; break;
Else if there's a side chain on $\overline{v_s v_t}$ then the shortest path is the side chain; break; else continue

Step3 Starting from v_s and v_t respectively, structure the adjacent node tree $T(v_s)$ and $T(v_t)$ of v_s and v_t:

Step3.1 Calculation

$$Choose(v_s) = \{v_i \mid e_{si} = v_s v_i \;\&\; 0 < w_{si} < \infty\} - Parent0(v_s) \text{ and}$$

$$Choose(v_t) = \{v_i \mid e_{ti} = v_t v_i \;\&\; 0 < w_{ti} < \infty\} - Parent0(v_t);$$

Step3.2 add the $Choose(v_s)$ and $Choose(v_t)$ respectively, to the point $T(v_s)$ and $T(v_t)$;

Step3.3 if the depth of $T(v_s)$ or $T(v_s) < h$ then use the points of $T(v_s)$ and $T(v_t)$ respectively, instead of v_s and v_t;
goto step3.1;
Else goto step4;

Step4 calculated the points $\min\limits_{v_i \in Choose(v_s)} \{\sigma_2(v_i)\}$ and $\min\limits_{v_i \in Choose(v_t)} \{\sigma_2(v_i)\}$ of $T(v_s)$ or $T(v_s)$ to the initial path, mark as A_1 and B_1, the projection points on the $\overline{v_s v_t}$ marked as A_1' and B_1', respectively.

Step5 Calculate $k_{s1} = \left|\overline{v_s A_1'}\right|$ and $k_{t1} = \left|\overline{v_t B_1'}\right|$.
If $k_{s1} + k_{t1} > k$ then will not calculate the side starting from v_t;

Step6 $\overline{v_s v_t}$, v_s and v_t were replaced by $\overline{A_1 B_1}$, A_1 and B_1, repeat step2-step5, the tree in step3 does not need to re-structure, Only use the adjacent node tree generated by last cycle, then added a layer of leaves to deepen, The points of step4 are always project on $\overline{v_s v_t}$. Until A_u, B_u (called rendezvous point) are the same, or A_u, B_u are two end points of the edge e (called join edge). Point range $v_s, A_1, ..., A_u, B_u, ..., B_1, v_t$ constitutes the initial path; calculate the length of the initial path:

$$L = \sum_{i=0}^{u} s_{si} + \sum_{i=0}^{u} s_{ti} \text{ and } \sum_{i=1}^{u} k_{si} + \sum_{i=1}^{u} k_{ti} = k$$

or

$$L = \sum_{i=0}^{u} s_{si} + \sum_{i=0}^{u} s_{ti} + s_{A_u B_u} \text{ and}$$

$$\sum_{i=1}^{u} k_{si} + \sum_{i=1}^{u} k_{ti} + | k_{A_u B_u} | = k$$

Step7 With v_s and v_t as root points, starting from beginning and ending points respectively, put the edge which has the minimum angle on both sides of the line segment between start and end points to the binary tree repeated, until the two trees meet in the points A_u, B_u. During the process of generate binary tree, using the following three steps to reduce the size of binary tree.

Step7.1 if $\sum_{j=0}^{i-1}|A_jA_{j+1}|+|A_iA_{i-1}|+|A_iv_t|>L$ then deletes point A_i;

Step7.2 if $\exists A_x \in Parent(A_i)$ and $\sum_{j=x}^{i}|A_jA_{j+1}|>|A_xA_i|$ then add A_i to A_x as the child node;

Step7.3 if $\exists A_x = A_i$

if $\sum_{j=0}^{x} s_{sj} > \sum_{j=0}^{i} s_{sj}$ then delete A_x;

else delete A_i;

Step8 take $l=\sum_{i=0}^{u} s_{si} + \sum_{i=0}^{u} s_{ti} + s_{A_uB_u}$ as the shortest path;

Repeat the above algorithm $|s'|/2$ times, where $|s'|$ indicates the degree of node v_s. When complete the implementation of the algorithm $i+1$ times, delete the edges of two branches which with v_s as the starting point in the binary tree, which is also true at the $i(=1,2,...,|s'|/2)$ times' execution. When implement the algorithm $|s'|/2$ times, would calculate the shortest path with all nodes which adjacent to v_s as the starting side. Then select the shortest path of them as the required shortest path.

The following proves the correctness of the algorithm.

Theorem 1. The initial path's length of improved algorithm is less than or equal to that of the original algorithm;

Proof: The measuring function $\sigma_2(A_i)$ means the change scale of the sum of $|\beta_i|d_i$ which is the child node A_i of A_{i-1} on the stroke unit k_i, in the adjacent node tree with A_{i-1} as the root node. According to the definition of the selection standard about $\sigma_2(A_i)$, there are two cases:

① If $\exists A_i$, makes $\sigma_2(A_i) \geq 0$, select the point A_i that $\sigma_2(A_i) \to 0+$ add in the initial path. Therefore can obtain $|\beta_i|d_i/k_i \to 0+ \Rightarrow |\beta_i| \to 0+$, $d_i \to 0+$, $k_i \to +\infty$. As figure 1, $|\beta_i| \to 0+$ means the direction of $A_{i-1}A_i$ close to $A_{i-1}v_t$, therefore A_i is closer to the target point v_t; $d_i \to 0+$ means the direction of $A_{i-1}A_i$ close to $A_{i-1}v_t$, therefore A_i is closer to the target point v_t; $k_i \to +\infty$ means the father of stroke the closer A_i to v_t. Therefore, the path length at this time is smaller.

② If $\forall A_i$, makes $\sigma_2(A_i) < 0$, select the point A_i that $\sigma_2(A_i) \to -\infty$ add in the initial path. Therefore can obtain $|\beta_i| d_i / k_i \to -\infty \Rightarrow |\beta_i| \to +\infty$, $d_i \to +\infty$, $k_i \to 0-$. As figure 2, this means A_i deviate from v_t, these three means A_i is lease deviate from v_t, and therefore, the length of total path is smaller.

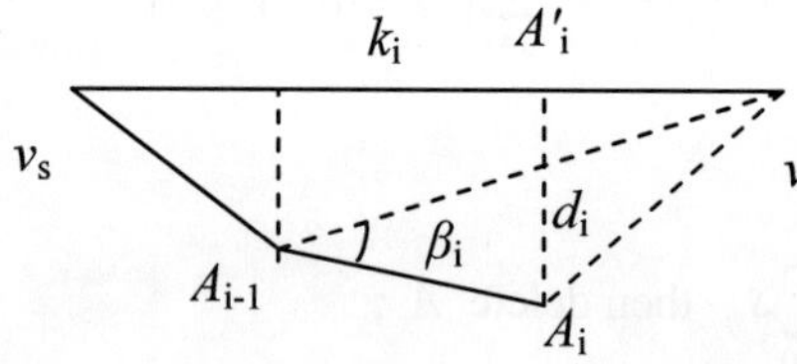

Fig. 1.

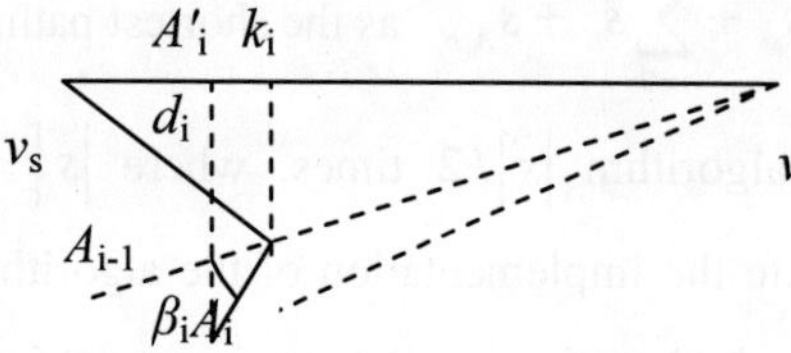

Fig. 2.

3 Conclusion Analysis

Compared the improved algorithm and Z8-1 algorithm from the landscape orientation, that improved algorithm has been optimized.

Conclusion. With the increases of h, the time of finding the initial path is increasing; the time of finding the shortest path is decreasing. And the total time must be less than or equal to the certain amount of time before improvements.

Proof: Set the time function of finding the initial path about h is $t_1 = f(h)$, the domain is $h \in Z^+$, the range is $t_1 \in (0, +\infty)$, as with the h increases, the adjacent node tree increased in size, so the time of finding the initial path increased, therefore t_1 is increasing. Set the time function of Z8-1 algorithm to find the initial path as $\overleftarrow{t_1} = \overleftarrow{f}(h)$, due to its standard of finding initial path is the minimum angle between the current point, thus corresponds to construct the adjacent node tree of $h = 1$, Therefore, the time to find the initial path should be $\overleftarrow{t_1} = \overleftarrow{f}(1)$, and $f(1) = \overleftarrow{f}(1)$. As with h increases, the size of the tree adjacent nodes increases, therefore $f(h) \geq \overleftarrow{f}(1)$.

Set the function of finding the shortest path about h as $t_2 = g(h)$, the domain is $h \in Z^+$, the range is $t_2 \in (0,+\infty)$, as with h increases, the extent of prediction is deep, the length of the initial path will decrease, so the size of binary tree will be reduced, therefore t_2 is decreasing. The same as above set the time function of Z8-1 algorithm of finding the shortest path as $\overleftarrow{t_2} = \overleftarrow{g}(1)$. As with h increases, the size of the binary tree reduced, therefore $g(h) \leq \overleftarrow{g}(1)$.

The total time for finding the shortest path is $t = t_1 + t_2 = f(h) + g(h)$. For $f(h)$ is increasing, $g(h)$ is decreasing, hence, in according to

$$\begin{cases} t_1 = f(h) \\ t_2 = g(h) \\ t_1 = t_2 \end{cases} \qquad (1)$$

we can obtain $h = h_0$. As Figure 3

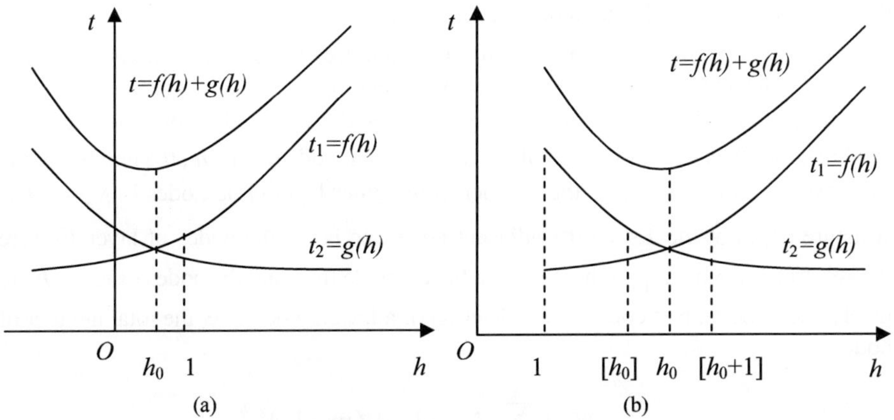

Fig. 3. Time function curve

The time when $h_0 \in (-\infty,1)$, as Figure 3(a), $t = f(h) + g(h)$ has the minimum value at $h_0 = 1$. At this point the execution time of the improved algorithm is less than or equal to that of the algorithm before improved. That is:

$$t = f(1) + g(1) \leq \overleftarrow{t} = \overleftarrow{f}(1) + \overleftarrow{g}(1) \tag{2}$$

The time when $h_0 \in [1, +\infty)$, as Figure 3(b), $t = f(h) + g(h)$ has the minimum value at $[h_0]$ or $[h_0+1]$. At this point the minimum execution time of the improved algorithm is:

$$t_{\min} = \min\{f([h_0]) + g[h_0]), f([h_0+1]) + g([h_0+1])\} \tag{3}$$

therefore

$$t_{\min} < f(1) + g(1) \leq \overleftarrow{t} = \overleftarrow{f}(1) + \overleftarrow{g}(1) \tag{4}$$

In conclusion, we obtain $t \leq \overleftarrow{t}$. Finish.

4 The Time Complexity Analysis

First, let's analyze the execution time formula $f(h)$ of the adjacent node tree,

Execution time of the initial path is in direct proportion to the scale of adjacent node tree, execution time of the shortest path is in direct proportion to the scale of binary tree. Set t_0 as the average execution time of a node in adjacent tree, set the average number of adjacent nodes in current network node graph as $N_0 \in Z^+$, the number of nodes for the initial path is $m+2$, the initial path is $v_s A_1 A_2 ... A_m v_t$.

Based on construction method of adjacent node tree we can see, totally need construct m adjacent node trees to select m nodes adding to the initial path, The adjacent node tree of the i th node is that add a layer child nodes to the adjacent node tree of the $i-1$ th node. The structure depth of the first adjacent node tree is h, the number of the first layer child nodes is N_0, the number of the other layer child nodes is $N_0 - 1$. The structure depth of the $1 < i \leq m$ adjacent node tree is h, the nodes of layer $h-1$ are the sub-tree branch A_{i-1} of the $i-1$ th adjacent node tree, and the nodes of layer h are newly added node. Not consider the deletion of adjacent node tree, the total number of nodes is:

$$N_0 + \sum_{i=2}^{h} (N_0 - 1)^i + (m-1)N_0{}^h \tag{5}$$

Therefore, the time of finding the initial path is:

$$f(h) = t_0 \left[N_0 + \sum_{i=2}^{h} (N_0 - 1)^i + (m-1)N_0{}^h \right] \tag{6}$$

The execution time of Z8-1 algorithm is $\overleftarrow{f}(1) = f(1) = N_0 m$.

The following analyzed the time complexity of improved algorithm.

Suppose the network nodes constitute a square mesh, if v_s, v_t are located at the Non-adjacent sides of the square, the number of the binary tree leaf nodes that are generated at this time at most is $\frac{\sqrt{n}}{2}$, therefore the number of binary tree nodes is:

$$2(1+2+\cdots+\frac{\sqrt{n}}{2})=\frac{\sqrt{n}}{2}(\frac{\sqrt{n}}{2}+1)=\frac{n}{4}+\frac{\sqrt{n}}{2}=O(n)$$

The step0-step2 of algorithm requires constant time. According to conclusion 4, let $N_0=4$, $m=\sqrt{n}$ then step3-step6 costs $t_0\left[4+\sum_{i=2}^{h}3^i+(\sqrt{n}-1)4^h\right]=O(\sqrt{n})$, step7 costs $O(n)$, step8 costs $O(\sqrt{n})$ addition and the finding path time $O(n)$.

Therefore, the complexity of algorithm is:

$$O(\sqrt{n})+O(n)+O(\sqrt{n})+O(n)=O(n)$$

Repeat the above algorithm $\frac{|s'|}{2}$ times, its complexity is

$$O(n\times\frac{|s'|}{2})=O(n)$$

Also, because the deletion rules and nodes duplication of the adjacent node tree and binary tree, so the time complexity close to linear. The time complexities of algorithm before and after improved are the same, but by shortening the initial path reduced the size of the binary tree, the total execution time decreases.

References

1. Zhou, P.: Computational Geometry-Design and Analysis of Algorithms, vol. (2), pp. 294–300. Chinese Tsinghua University Press, Beijing (2005)
2. An, H., Hu, G., He, Y.: The Dynamic Algorithm of Shortest Path. Chinese Computer Engineering and Applications 180, 173–174 (2003)
3. Zhang, Z., Liu, Z., Ai, D.: An Optimization of a Shortest-Path Algorithm. In: Proceedings of 2008 International Conference on Humanized Systems (ICHS 2008), Beijing, China, October18-22 (2008)
4. Li, F., Zhang, J.: The Improvement of Network Shortest Path Algorithm and its Implementation. Chinese Journal of Xiamen University (Natural Science) 44(1), 236–238 (2005)
5. Song, T., Fan, D.: Graph Theory Solution of the Shortest Path Problem in GIS Analysis. Surveying and Mapping of Sichuan 25(4), 179–182 (2002)
6. Luo, Y., Li, L., Zhu, D., Zheng, L.: The Data Model of Shortest Path Calculate in Vehicle Navigation System. Journal of Kunming University of Science and Technology (Science and Technology) 29(3), 106–109 (2004)

Energy Management Strategy for Hybrid Electric Tracked Vehicle Based on Dynamic Programming

Rui Chen, Yuan Zou, and Shi-jie Hou

School of Mechanical Engineering, Beijing Institute of Technology, Beijing 100081, China
chenruihappy@sohu.com

Abstract. The dynamic programming (DP) theory is applied in designing energy management control strategy of Hybrid Tracked Vehicle. As hybrid tracked vehicles use dual-energy sources, the engine-generator sets and power batteries. We applied the dynamic programming theory to optimize the distribution between the dual-energy sources, so we can make the fuel consumption minimize. Through the analysis of the optimization results, near-optimal rules are extracted. Then we do some simulations to verify the control rules based on MATLAB software. It was found that the fuel economic performance improved significant, compared to former rule-based control strategy.

Keywords: hybrid tracked vehicle、energy management、dynamic programming.

1 Introduction

The hybrid electric tracked vehicles [1] use dual-energy sources, which are engine - generator sets and power batteries. Therefore, it is a key technology to distribute the power output between the two energy-sources, to achieve the best fuel economy. We named this technology as energy management. The former energy management strategies of hybrid electric tracked vehicles [2] are fuzzy-control based on engine speed, the engine-generator constant-speed control, hybrid power system load power-tracking control, constant-voltage control, constant-SOC fuzzy control and so on. They are mainly rule-based control strategy with the advantage of simple design, real time, and good robustness. While the disadvantages of the rule-based methods are that we need to rely on engineering experience to make rules and often could not obtain the optimal result. Recently, people keep their eyes on applying the global optimization theories to design energy management control strategy [3]. A popular method used is dynamic programming [4], which calculates the optimal control signals over a given driving cycle. In this paper, we will apply dynamic programming theory to design the optimal energy management strategy for hybrid electric tracked vehicles.

2 Hybrid Electric Tracked Vehicle Model

The configuration of hybrid electric tracked vehicle is presented in Figure1 [5], [6]. The feature of this configuration is that two electrical powers are coupled together in the

M. Zhu (Ed.): Electrical Engineering and Control, LNEE 98, pp. 843–851.
springerlink.com © Springer-Verlag Berlin Heidelberg 2011

power converter, which works as an electric power coupler to control the power flows from the batteries and generator to the two electric motors, or in the reverse direction, from the electric motors to the batteries. The IC engine, and the generator constitute the primary energy supply and the batteries work as the energy bumper.

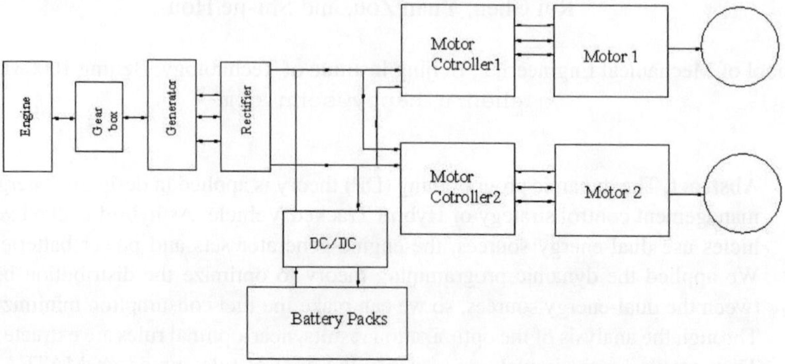

Fig. 1. Structural diagram of hybrid electric tracked vehicles

The hybrid electric tracked vehicle includes the engine - generator set, power battery, two drive motors, DC-DC converters, hybrid power system control unit.

2.1 The Engine - Generator Set Model

The engine - generator sets are composed of high-power diesel engine and permanent magnet synchronous generator and between the engine and generator is a gearbox which used to match the speeds.

Figure 2 is the equivalent circuit of permanent magnet synchronous generator plus the rectifier. Parameters are defined as follows. Where T_m is the electromagnetic torque, ω_m is the generator speed; K_e is the coefficient of induced electromotive force; $K_x\omega_m$ is the equivalent impedance, $K_x = 3PL/\pi$ is resistance coefficient, and P is the number of generator pole pairs. I_{dc} is the generator current, U_{dc} is the output voltage.

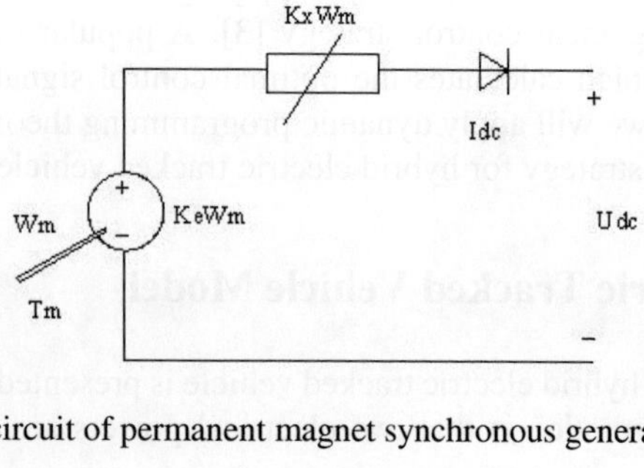

Fig. 2. Equivalent circuit of permanent magnet synchronous generator plus rectifier

The electromagnetic torque equation of the permanent magnet synchronous generators-Rectifier is as follows:

$$\begin{cases} U_{dc} = K_e \omega_m - K_x \omega_m I_{dc} \\ T_m = K_e I_{dc} - K_x I_{dc}^2 \end{cases} . \quad (1)$$

Generator's input torque is come from the diesel engine through a transmission with a transmission ratio of i_{e-g} . According to the following relationship of torque balance, we can get the equation (2).

$$\begin{cases} \frac{T_{eng}}{i_{e-g}} - T_m = \frac{2\pi}{60} \left(\frac{J_e}{i_{e-g}^2} + j_g \right) \frac{dn_g}{dt} \\ n_g = i_{e-g} n_{eng} \end{cases} \quad (2)$$

where the output torque of the engine is T_{eng} , the speed of the engine is n_{eng} , the speed of the generator is n_g , the inertia of the engine is J_e and the inertia of the generator is J_g .

2.2 The Model of Battery Packs

We use the equivalent circuit model to simulate the dynamic characteristics of the battery. The equivalent circuit is composed of resistors, capacitors and other circuit elements based on the battery principle.

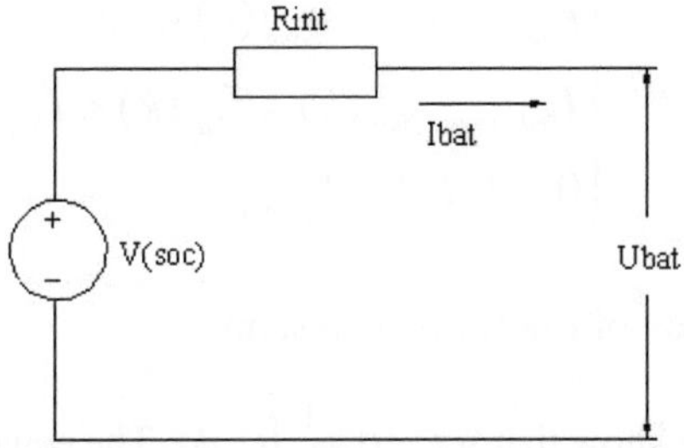

Fig. 3. Equivalent circuit model of battery

The Battery's voltage equation as follows:

$$\begin{cases} U_{bat} = V(soc) - I_{bat} \cdot R_{\text{int}} \\ SOC = (1 - \frac{1}{C} \int I_{bat} dt) \times 100\% \end{cases} \quad (3)$$

3 The Optimal Design of Energy Management Strategy Based on Dynamic Programming

3.1 Discrete Equation of Dynamic System

First, the dynamic equation of hybrid electric tracked vehicle is described as a discrete format.

$$x(k+1) = f(x(k), u(k), P_{dem}(k)) \tag{4}$$

where $x(k)$ is the state vector of the system such as $SOC(k)$ of battery and generator speed $n_g(k)$. $u(k)$ is the control input such as diesel engine electronic throttle $acc(k)$. $P_{dem}(k)$ is the power demand of the system. Our goal is to find the optimal control input to make the fuel consumption minimized. The cost function to be minimized is described as follows:

$$J = \sum_{k=0}^{N-1} fuel(k) \tag{5}$$

where N is the duration of the driving cycle and $fuel(k)$ is the instantaneous fuel cost. Besides, to avoid the engine and battery working in a poor station, we should add some constrains to this system as follows.

$$\begin{cases} 0 < soc(k) < 1 \\ |SOC(N) - SOC(0)| < \Delta SOC \\ n_{eng\_idle} < n_{eng}(k) < n_{eng\_\max} \\ n_{eng}(k+1) - n_{eng}(k) < \Delta n \\ I_{bat\_\max\_char}(k) < I_{bat}(k) < I_{bat\_\max\_disch}(k) \\ 0 < I_g(k) < I_{g\_\max} \end{cases} \tag{6}$$

3.2 Discrete Equation of the Battery System

From equation (3), we know that $SOC = (1 - \frac{1}{C}\int I_{bat} dt)$. The equation can be expressed as a discrete-time format.

$$SOC(k+1) = SOC(k) - \frac{1}{C} I_{bat} \tag{7}$$

where I_{bat} is:

$$I_{bat} = \begin{cases} \dfrac{V(soc) - \sqrt{V(soc)^2 - 4 \cdot R_{\text{int}} P_{dem}}}{2 \cdot R_{\text{int}}}, \mathrm{U}_{dc} > K_e \omega_m \\ \dfrac{V(soc) - U_{dc}}{R_{\text{int}}}, \mathrm{U}_{dc} \le K_e \omega_m \end{cases} \tag{8}$$

3.3 Discrete Equation of the Generator

Equation (2) describe the generator system, we can derivation the discrete-time format of generator speed from it.

$$n_g(k+1) = n_g(k) + \frac{\frac{T_g}{i_{e-g}} - T_m}{\frac{2\pi}{60}\left(\frac{J_g}{i_{e-g}^2} + J_g\right)} \tag{9}$$

3.4 Dynamic Programming

The core of dynamic programming theory [7] is the principle of optimality. It divides a multi-step decision problem into a series of single-step decision problems, and then solves the single-step problems from the last step until the initial step. These processes should follow this principle, no matter what the initial state is, the process from the current step till the last step must be optimal.

Step N-1:

$$J_{N-1}^*(SOC(N-1), n_g(N-1)) = \min_{acc(N-1)} fuel(N-1) \tag{10}$$

Step k, (0<k<N-1):

$$J_k^*(SOC(k), n_g(k)) = \min_{acc(k)} \left[fuel(k) + J_{k+1}^*(SOC(k+1), n_g(k+1) \right] \tag{11}$$

where $J_k^*(SOC(k), n_g(k))$ is the optimal cost function at state $x(k)$ starting from the stage k. It represents the optimal cost from k stage to final stage.

4 The Analyzes Result Based on Dynamic Programming

4.1 Rule-Based Energy Management Strategy [8]

The rule-based strategy is that we set the engine speed as four levels; each level covers a range of power. Then we can switch the engine speed among the four levels by tracking the power demand. This principle is described as figure 4.

The speed switch points are
N0=800r/min,N1=1450r/min,N2=1750r/min,N3=2045r/min.
And corresponding power points are P0=0Kw,P1=35Kw,P2=80Kw,P3=150Kw.

4.2 Simulation Result of Rule-Based Energy Management Strategy

Choose the CYC_HWFET driving cycle as the simulation cycle. This cycle has an average speed of 48.2 Km/h, time 756 seconds and a distance of 10.26 Km. Then we can calculate the power demand by solving the vehicle's dynamic equation (12). The structural parameters are as follow: weight is 15000kg, frontal area (A) is 5.4 m^2, resistance coefficient (Cd) is 1, friction coefficient f is 0.0494, System efficiency η is 0.78. And the simulation results are displayed in figure6.

$$P_{dem} = (0.5C_d AV^2 + mgf + ma) \cdot V \tag{12}$$

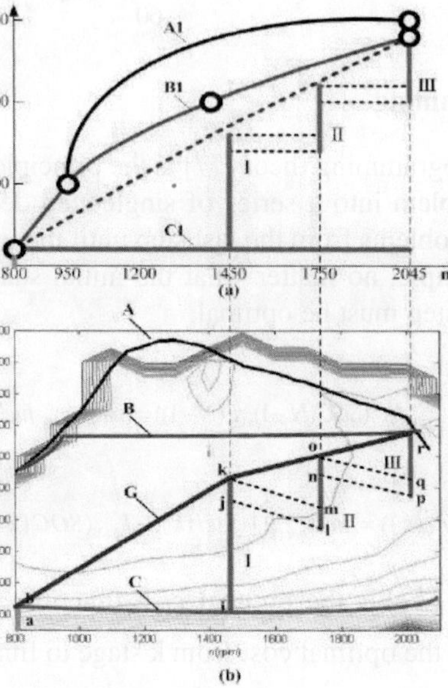

Fig. 4. Rule-based energy management strategy

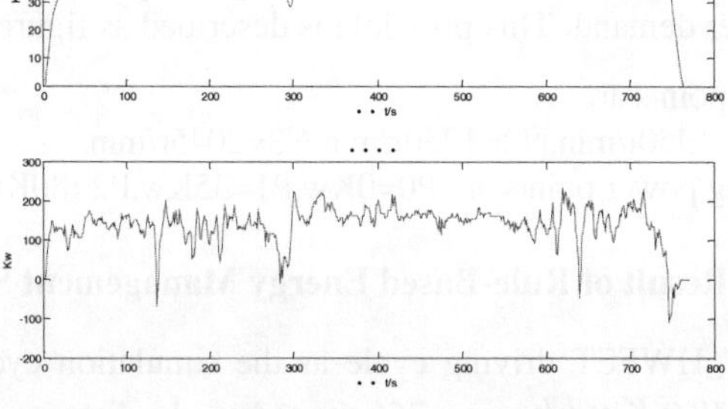

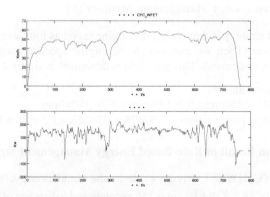

Fig. 5. CYC_HWEFT driving cycle

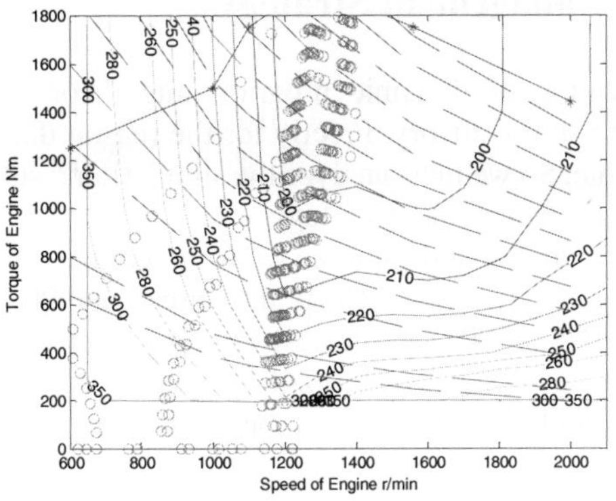

Fig. 6. Engine working speed of rule-based strategy

4.3 Simulation Based on Dynamic Programming Theory

To apply dynamic programming theory, we need to write the vehicle model and the dynamic programming algorithm then use the algorithm solves the vehicle model based on MATLAB software. The optimal results show in figure 7 [9].

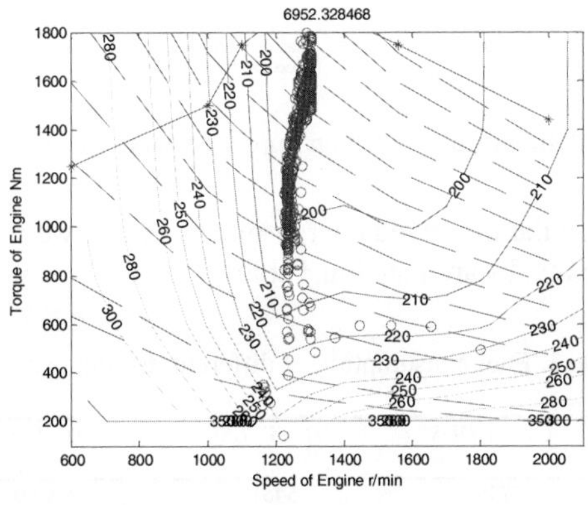

Fig. 7. Optimal engine working speed points and fuel use

By analyzing the simulation results, we can see that the fuel consumption is 7696.2g applying the rules-based strategy and is 6952g after the application of dynamic programming theory. The fuel economy was improved by 9.7%. And from Figure 7 we can see that the engine's working points were optimized.

5 New Energy Management Strategy

Although the application of dynamic programming theory can improves the fuel economy, we cannot use this theory directly for the reason that accurate future power demand is not available. So we must analyze the simulation results carefully and extract the sub-optimal energy management strategy.

We can see from figure 8 that the engine works at a range of speed from 900r/min to 1500r/min. The engine speed switch points are N0=900, N1=1100, N2=1250, N3=1360, N4=1400, N5=1500. And corresponding power are P0=40Kw, P1=80Kw, P2=120Kw, P3=160Kw, P4=200Kw, P5=240Kw. Again we choose the CYC_HWFET driving cycle as the simulation cycle to test the new energy management strategy.

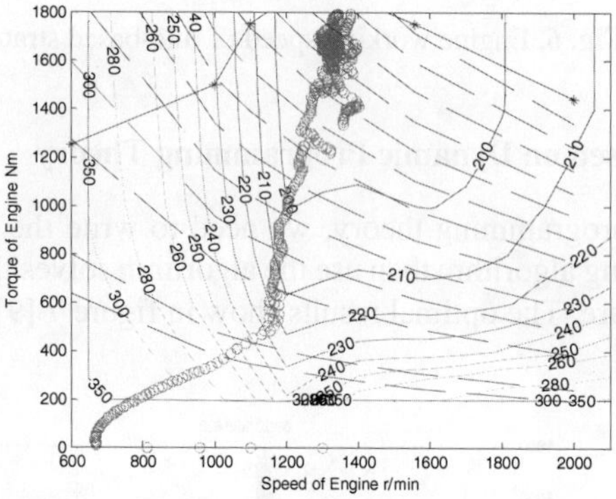

Fig. 8. Simulation results based on improved energy management strategy

In order to verify the improved energy management, we use different driving cycles to simulate. The results are listed in Table1.

Table 1. Fuel consumption of different driving cycles (g)

	CYC_UDDS	*CYC_WVUSUB*	*CYC_SC03*	*CYC_WVUINTER*
Initial strategy	5735.5	5461.6	2790.9	1061.8
DP method	4895.0	4949.7	2375.0	9705.7
Improvement (%)	14.6%	9.4%	14.9%	8.6%
Improved strategy	4948.5	5095.3	2442.6	9764.6
Improvement (%)	12.5%	6.7%	12.5%	8.0%

6 Conclusion

Energy management is one of the key technologies of hybrid electric vehicles. It has a great impact on fuel economy. In this paper, we apply dynamic programming theory to optimize a standard driving cycle and to analyze the optimal results. Then we extract control rules and form the new energy management strategy. As we can see from Table 1, the fuel economy is improved a lot after using the new energy management strategy, compared to the initial energy management strategy. It proved that the dynamic programming theory is useful in optimizing the hybrid system. In addition, we also can see that there are some differences in the amount of improvements for different driving cycles. Because the new energy management strategy was extracted based on optimal results of a given driving cycle, which may not be the optimal for other driving cycles. To avoid this defect, more and more scholars began to apply stochastic dynamic programming theory [10], [11] to the hybrid power system. This will be my future direction.

References

1. Liao, Z.-l., Zang, K.-m., Ma, X.-j., Zhang, Y.-n.: Primary Study on the Technical Status Quo and Key Technology Development of the Electrical Drive of the Armored Vehicles. Journal of Academy of Armored Force Engineering 19(4) (2005) (in Chinese)
2. Li, J-q.: Energy Distribution and Management of Hybrid Power System of Electric Drive Vehicle. Beijing Institute of Technology: PHD Thesis (2005) (in Chinese)
3. Sciarretta, A., Guzzella, L.: Control of hybrid electric vehicles. IEEE Control Systems Magazine 27(2), 60–70 (2007)
4. Lin, C.-C., Peng, H., Grizzle, J., Kang, J.-M.: Power management strategy for a parallel hybrid electric truck. IEEE Transactions on Control Systems Technology 11(6), 839–849 (2003)
5. Dong, Y.-g., Zhang, C.-n., Sun, F.-c., Wu, J.-b.: Engine-Generatorps Control Strategy for Hybrid Electric Transmission Tracked Vehicle. Transactions of Beijing Institute of Technology 30(4) (2010) (in Chinese)
6. Zhang, Y.-n., Ge, Y.-s., Yan, N.-m., Xie, Y.-h.: ModelingControl and Simulation of the Mover for a Tracked. ACTA ARMAM ENTARII. 26(1), 10–14 (2005) (in Chinese)
7. KirK, D.E.: Optimal control theory: an introduction (1970)
8. Wu, J-b.: Research on Control Strategy for Hybrid Electric Transmission Tracked Vehicle Based on Fuel Economy. Beijing Institute of Technology: PHD Thesis (2008) (in Chinese)
9. Sundstrom, O., Guzzella, L.: A Generic Dynamic Programming Matlab Function. In: International Conference on Control Applications (2009)
10. Tate, E., Grizzle, J., Peng, H.: Shortest path stochastic control for hybrid electric vehicles. International Journal of Robust and Nonlinear Control 18, 1409–1429 (2008)
11. Lin, C.-C., Peng, H., Grizzle, J.: A stochastic control strategy for hybrid electric vehicles. In: Proceedings of the American Control Conference (2004)

SISO/MIMO-OFDM Based Power Line Communication Using MRC

Jeonghwa Yoo, Sangho Choe, and Nazcar Pine

The Catholic University of Korea
Bucheon, Korea
mundade@hanmail.net, schoe@catholic.ac.kr, naz@nazcarpine.com

Abstract. In this paper, we present a single-input, single-output (SISO) and multiple-input multiple-output (MIMO) orthogonal frequency division multiplexing (OFDM) (SISO/MIMO-OFDM) based power line communication (PLC) scheme using maximum ratio combining (MRC), which is applicable to various future high-speed data services via power line like Smart Grid. We employ the Zimmerman frequency model as multipath power line fading channel and the Middleton's Class A model as impulse noise channel. In order to improve the 3-phase (MIMO) or single-phase (SISO) PLC performance, we introduce a new MRC effectively summing up both multiple antenna diversity gain and multipath fading diversity gain. Via computer simulation, we proved the performance advantage of the proposed SISO/MIMO over the existing SISO/MIMO.

Keywords: Maximum ratio combining (MRC), MIMO, OFDM, Power line communication (PLC), SISO.

1 Introduction

Smart Grid is the future electric wire network where the information transmission technology is used to exchange information among electric power provider, electricity industry and electric power receivers. With the advent of Smart Grid, the interest in power line communication (PLC) has been increased since it allows data transmission via power line without additional infrastructure unlike cellular, Wi-Fi, very-high-bitrate digital subscriber line (VDSL) and so on. PLC signal experiences channel distortions due to multipath fading and impulse noise in power line channel. In this paper, we employ the Zimmermann frequency model [1] for power line multipath fading channel and the Middleton class A model [2] for impulse noise channel.

In this paper, we present a multiple-input multiple-output orthogonal frequency division multiplexing (MIMO-OFDM) PLC system with antenna and fading MRC (AFMRC). AFMRC, where antenna MRC (AMRC) and fading MRC (FMRC) are merged, enjoys both diversity gains: antenna diversity and multipath fading diversity. This suggested scheme improves performance not only in the 2x2 MIMO PLC via 3-phase 4-wires power line but also in the single-input single-output (SISO) PLC via single-phase 2-wires power line. For SISO PLC, it uses FMRC rather than AMRC.

M. Zhu (Ed.): Electrical Engineering and Control, LNEE 98, pp. 853–860.
springerlink.com © Springer-Verlag Berlin Heidelberg 2011

In this paper, via computer simulation, we convey performance analysis of SISO/MIMO-OFDM under various PLC channel conditions. Through simulation, we prove that MIMO PLC with AFMRC is superior to MIMO PLC with AMRC, in terms of bit error rate (BER). Simulation results verify that the proposed SISO with FMRC is more effective at the indoor single-phase PLC applications than the conventional SISO without MRC. We also evaluate system design parameters through the comparison of BER performance when the impulse noise index A varies.

The contents of this paper are organized as follows. In Section 2, we explain MIMO-OFDM based PLC system model and PLC channel characteristics including multipath fading and impulse noise. We then demonstrate the proposed MRC technologies, FMRC for SISO and AFMRC for MIMO. In Section 3, via computer simulation, we show the suggested system performance results under various PLC channel conditions. Finally, we conclude this paper in Section 4.

2 System Model

In this section, MIMO OFDM PLC system, fading channel, impulse noise and the diversity schemes used in this paper.

2.1 MIMO-OFDM PLC System

In this paper, we design a MIMO-OFDM system which contains *I* transmit antennas and *J* receive antennas. In the OFDM transmitter the modulation signal *S(k)* experiences the following inverse fast Fourier transform (IFFT)

$$s(n) = \frac{1}{N}\sum_{k=0}^{N-1} S(k)e^{j2\pi nk/N}, \ (n = 0,1,...,N-1), \quad (1)$$

where k is the frequency index and n is the time index (= 0, 1, …, N-1). *S(k)* is the kth subcarrier modulation signal sample in the frequency axis, and *s(n)* is the nth sample in the time axis after the IFFT operation.

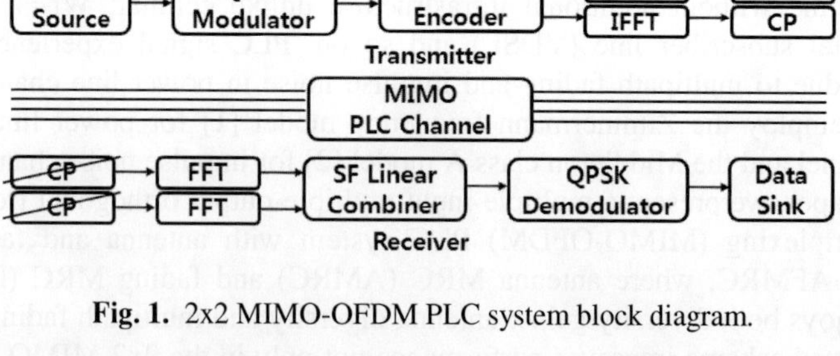

Fig. 1. 2x2 MIMO-OFDM PLC system block diagram.

Unlike wireless communications, in MIMO PLC, a pair of electrical wires is converted into a single antenna channel so that the number of transmitting and receiving antennas is typically limited to two for 3-phase 4-wires and one for single-phase 2-wires. Therefore, MIMO-OFDM is mainly used in either indoor or outdoor with 3-phase 4-wires power line whereas SISO-OFDM is mostly used in indoor with single-phase 2-wires power line. Fig. 1 shows the 2x2 MIMO-OFDM PLC system block diagram, using 3-phase 4-wire power line.

Fig. 2(a) illustrates a typical 3-phase 4-wire power line cross-cut interior structure. Fig. 2(b) demonstrates a 2x2 MIMO system with two antenna paths which consists of single antenna path formed with C_1 and C_2 and another single antenna path made of C = C_3 and C_4; C_0 takes a role as a ground connection. Space frequency (SF) encoder is used to reduce the error probability caused by the interference in MIMO channel. The following two SF encoder vectors, $\mathbf{S_1}$ and $\mathbf{S_2}$, are formed by arranging sub-carrier signal samples in a proper order [4].

$$\mathbf{S}_1 = [S_1(0),...,S_1(\frac{N}{2}-1),S_1(\frac{N}{2}),...,S_1(N-1)]^T; \quad \mathbf{S}_2 = [S_2(\frac{N}{2}),...,S_2(N-1),S_2(0),...,S_2(\frac{N}{2}-1)]^T, \qquad (2)$$

where $S_1[k]= S_2[k]$, (k=0,1, …, N-1), $\mathbf{X}^T$ refers to the transpose of $\mathbf{X}$. $\mathbf{S_1}$ and $\mathbf{S_2}$ are respectively converted to the corresponding time sample vectors $\mathbf{s_1}$ = IFFT {$\mathbf{S_1}$} and $\mathbf{s_2}$ = IFFT {$\mathbf{S_2}$} through the IFFT process (see (1)) and then transmitted to receiver via each antenna path. In Fig. 2(b), this transmitting process is occurred at the signal processor and corresponding receiving process is processed at the linear synthesis and detector.

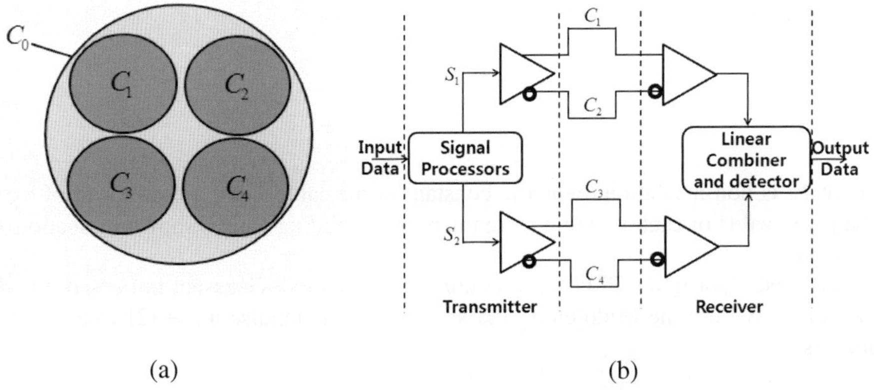

Fig. 2. (a) 3-phase 4-wires power line interior structures, (b) 2x2 MIMO transmission process using 3-phase 4-wires power line.

As shown in Fig. 1, cyclic prefix (CP) is added to the OFDM modulated sample vectors ($\mathbf{s_1}$, $\mathbf{s_2}$) before their transmission to prevent the inter-symbol interference (ISI) due to multipath fading delay. The received signal via fading channel goes through a demodulation process, i.e., fast Fourier transform (FFT) process, to recover its data stream after removing the added CP. It is assumed that there is no coupling between the two antenna paths for simulation simplicity (this assumption is also practically reasonable [3]). In the proposed system model, at the MRC based linear synthesizer, we combine the two received signals via each antenna path in a maximum likelihood manner to improve the system BER performance. In this paper, we propose a linear synthesis scheme, which is herein simply called AFMRC, combining signals via multiple fading paths as well as multiple antenna paths (For details, see Section 2.3).

2.2 Fading Channel and Impulse Noise in PLC

For PLC system design, herein, we define the power line channel characteristics considering channel noise statistics and PLC network topology. PLC channel can be characterized with both noise channel model including Gaussian noise and impulse noise and multipath fading channel model due to multiple signal reflections caused by power line impedance mismatch. In the case of multipath fading channel, if we assume that there is L number of different signal paths, the fading channel transfer function can be expressed as

$$H_i(k) = \sum_{l=1}^{L} H_{i,l}(k); \quad H_{i,l}(k) = g_{i,l} \cdot e^{-(\alpha_0 + \alpha_1 \cdot k^u) \cdot d_{i,l}} \cdot e^{-j2\pi k \cdot (d_{i,l} / v_p)} \tag{3}$$

where $H_i(k)$ is the multipath fading channel transfer function of the kth subcarrier through the ith antenna path. That is, $H_i(k)$ can be represented as the sum of all the lth path fading transfer functions $H_{i,l}(k)$. α_0, α_1 and u are the power line cable parameters, and $|g_{i,l}|$ is the weighting factor of the ith antenna and lth fading path. $d_{i,l}/v_p$ is equivalent to the corresponding path delay $\tau_{i,l}$ (where $d_{i,l}$ represents the length of the ith antenna lth fading path) as follows:

$$\tau_{i,l} = \frac{d_{i,l} \cdot \sqrt{\varepsilon_0}}{c_0} = \frac{d_{i,l}}{v_p}, \tag{4}$$

where ε_0 is non-insulation dielectric constant of the cable and c_0 is the speed of light (the given value of each variable assignment on the fading path is shown in Section 3, Table 1).

The noise floor in the PLC receiver consists of additive Gaussian noise and impulsive noise. We use the Middleton's class A model for impulse noise [2] which is defined as

$$p(x) = \sum_{m=0}^{\infty} \frac{\alpha_m}{2\pi\sigma_m^2} e^{-\frac{|x|^2}{2\sigma_m^2}}; \qquad \alpha_m = e^{-A} \frac{A^m}{m!}. \tag{5}$$

The variance of the Gaussian distribution is denoted as σ_m^2 is defined as

$$\sigma_m^2 = \sigma^2 \frac{m / A + \tau}{1 + \tau} \tag{6}$$

where m is a channel state parameter of Poisson distributed channel characteristics with P(m)= α_m and has an independent value for each noise sample. $\tau = \sigma_g^2 / \sigma_i^2$, where σ_g^2 represents the variance of the white Gaussian noise and σ_i^2 the variance of the impulse noise. σ^2 indicates the variance of the Class A noise which is equal to the sum of white Gaussian noise variance and impulse noise variance (i.e., $\sigma^2 = \sigma_g^2 + \sigma_i^2$), and it also specifies the N_0 value in E_b / N_0 (energy bit per noise spectral density). The impulse noise characteristic is dependent on A, called the impulse index. The frequency axis noise signal $N_i(k)$ is the result of the FFT operation of the time axis noise signal $n_i(n)$ generated using the distribution function in (5) and can be expressed as

$$N_i(k) = \sum_{n=0}^{N-1} n_i(n) e^{-j2\pi nk/N}, \ (k = 0,1,...,N-1). \tag{7}$$

For the convenience of analysis, we assume that the channel is quasi-stationary (i.e., constant within the OFDM symbol time T_s). Thus, the kth subcarrier channel model of $H_i(k)$ which is the fading channel of the ith antenna path in (3) can be multiplied with the kth subcarrier signal $S_i(k)$ and then added with the noise channel model $N_i(k)$ in (7). As a result, the received signal $Y_i(k)$ at the ith antenna path and kth subcarrier can be expressed as follows [3]

$$Y_i(k) = \sqrt{\frac{E_s}{2}} H_i(k) S_i(k) + N_i(k) = \sum_{l=1}^{L} Y_{i,l}(k) = \sum_{l=1}^{L} \sqrt{\frac{E_s}{2}} H_{i,l}(k) S_i(k) + N_i(k), \tag{8}$$

where E_s represents the average energy of the transmit signal. In (8), the lth multipath fading and kth subcarrier of received signal is expressed as $Y_{i,l}(k) = H_{i,l}(k) S_l(k) + N_l(k)$. Hereinafter; the frequency index k is omitted for simple expression.

2.3 Combining Scheme

The signal synthesis techniques at the receiving stage are typically MRC, equal gain combining (EGC) and selection combining (SC). MRC is the best technique among them because it derives the optimal weight considering time delay, phase shift, and fading amplitude, and using this value it compounds and detects the signal. On the other hand, it has higher complexity than EGC and SC because it requires exact information of channel. In this paper, the ideal channel estimation for all paths is assumed for a simple computer simulation.

AMRC is a conventional MRC for multi-antenna path signals that multiplies the received signals through I antenna paths with the complex conjugate of their corresponding estimated channel functions to get antenna diversity. Thus, the combined signal $\tilde{Y}$ using AMRC for the multi-antenna path signals becomes

$$\tilde{Y} = \sum_{i=1}^{I} Y_i H_i^* , \tag{9}$$

where H_i^* is the complex conjugate of the channel function in (8). As a result, the BER performance is influenced by the sum of squared magnitude of channel signals along I antenna paths.

AFMRC is a combining technique of multiple antennas MRC (AMRC) and multiple fading MRC (FMRC). When applying the proposed scheme, AFMRC, to the 3-phase 4-wires 2x2 MIMO PLC channel (i.e., I=2), the received signal can be expressed as

$$\tilde{Y} = \sum_{i=1}^{I} \sum_{l=1}^{L} Y_{i,l} H_{i,l}^* , \tag{10}$$

where $Y_{i,l}$ denotes the received signal at the lth multipath fading path of the ith antenna path. L is the order of multipath fading channel. Compared to existing MRC, the proposed MRC improves the performance of MIMO PLC. However, as shown in (10), its receiver complexity increases L-fold by adding FMRC.

Even in the case of the indoor single-phase 2-wires SISO channel, whereas conventional scheme does not have any diversity gain, the proposed scheme does have FMRC diversity gain.

3 Simulation Result

We carry out the simulation for the proposed system model under the PLC channel conditions mentioned in section 2, and the BER simulation results of the proposed system and the conventional system are compared. We assume that the power line channel is a multipath fading channel with L = 6, and the simulation parameters of each fading channel path are shown in Table 1 [4]. For simplicity, we also assume that the fading channel parameters at the two antenna paths are the same[1].

Fig. 3 compares BER of 2x2 MIMO PLC when the impulse noise index A varies. For this experiment, we set τ = 0.1. For larger A values, noise channel characteristics is similar as Gaussian, for smaller A values, it is similar as impulse noise. As a result, BER decreases as A value increases with over certain E_b / N_0 values (about 27dB), as shown in Fig. 3. For example, it is observable that for A=10, we can get approximately 1dB gain at BER = 10^{-5} compared to A=1, approximately 4dB compared to A=0.3, and about 5.5dB compared to A=0.1. However, when E_b / N_0 is less than 27dB, BER increases as A gets larger due to the reduction of effective signal to noise ratio.

[1] In case of PLC channels using 3-phase 4-wires power line, the changing of channel parameters between the antenna paths is almost negligible in fact [5].

Practically, in PLC channel environment, A has a value within the range of 0.0001 to 0.35 so that we choose $A = 0.3$ for the remaining experiments in this chapter [4].

Table 1. Parameters of power line multipath fading channel.

Path No. l	g_l	d_l(m)	Path No. l	g_l	d_l(m)
1	0.760	200	4	0.098	272
2	0.369	224	5	-0.050	296
3	-0.189	248	6	0.026	320

For a conventional technique, AMRC, the diversity can be obtained by multiplying its optimum weight to different space antenna path. In this paper, we suggest a new hybrid approach (called AFMRC) which combines AMRC, using multi-antenna path diversity, and FMRC, using multipath fading path diversity. Especially, for FMRC, we estimate the most optimum weight value considering the time delay, phase, and fading amplitude, and use it for the frequency-selective multipath faded signal synthesis to achieve diversity gain. In this experiment, while assuming the ideal channel estimation on PLC fading channel, it is likely in the sense of being able to estimate the channel parameters by just having simple measurement (using overhead), unlike the wireless fading channel [1]. In Fig. 4, the performance comparison between conventional AMRC and the suggested AFMRC on SISO and MIMO system is demonstrated. First, in case of MIMO, AFMRC obtains the performance gain of about 2dB at 10^{-5} BER compared to AMRC. Fig. 4 also compares conventional method (without MRC) and the proposed method (using FMRC) on SISO system, and it shows the 2dB improvement based on 10^{-5} BER at the proposed scheme. Simulation results verify that the proposed SISO with FMRC is more effective at the indoor single-phase PLC applications than conventional SISO without MRC.

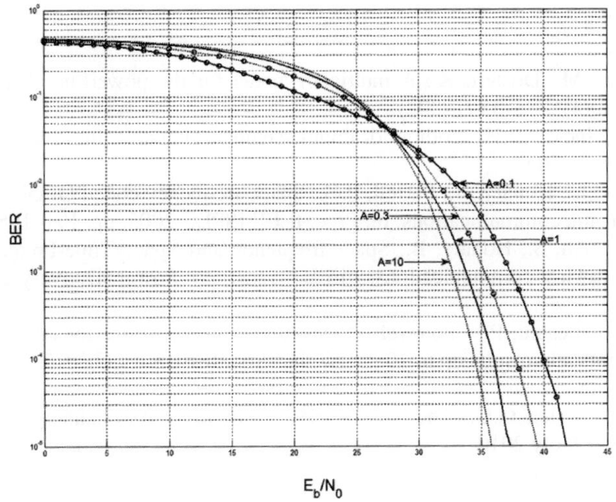

Fig. 3. BER Comparison of 2x2 MIMO PLC when A varies.

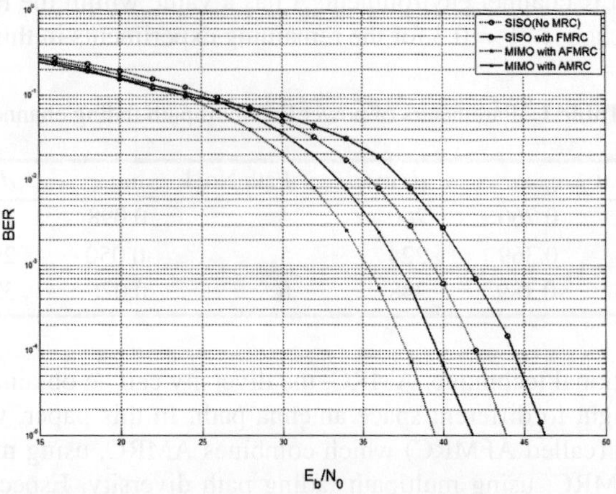

Fig. 4. Performance comparison of SISO/MIMO-OFDM with different MRC schemes (*A*=0.3).

4 Conclusions

In this paper, we have proposed a SISO/MIMO-OFDM based PLC system using MRC. The BER performance was improved by using AFMRC scheme that combines the conventional AMRC and FMRC. Simulation results have verified that the suggested method is effective for both 3-phase 4-wires 2x2 MIMO PLC and single-phase 2-wires SISO PLC. Under the PLC channel with impulse noise plus fading, compared to conventional MIMO, the proposed MIMO has a performance gain of about 2dB at 10^{-5} BER. For future work, we will continue to have further research on MIMO-OFDM-based PLC which considers the coupling effects between antenna links.

References

1. Zimmerman, M., Dostert, K.: A multipath model for the powerline channel. IEEE Trans. Commun. 50(4), 553–559 (2002)
2. Haring, J., Vinck, A.J.H.: Coding for impulsive noise channels. In: Proc. IEEE ISPLC 2001, pp. 103–108 (2001)
3. Giovaneli, C.L., Honary, B., Farrell, P.G.: Space-frequency coded OFDM system for multi-wire power line communications. In: Proc. IEEE ISPLC 2005, pp. 191–195 (April 2005)
4. Adebisi, B., Ali, S., Honary, B.: Space-frequency and Space-Time-Frequency M3FSK for indoor multiwire communications. IEEE Trans. on Power Delivery 24(4), 2361–2367 (2009)
5. Hashmat, R., Pagani, P., Chonavel, T.: MIMO communications for inhome PLC networks: measurements and results up to 100 MHz. In: Proc. IEEE ISPLC 2010, pp. 120–124 (March 2010)
6. Lai, S.W., Messier, G.G.: The wireless/power-line diversity channel. In: Proc. IEEE ICC 2010, pp. 1–5 (July 2010)

Fault Diagnosis of Mine Hoist Based on Multi-dimension Phase Space Reconstruction

Qiang Niu and Xiaoming Liu

School of Computer Science & Technology, China University of Mining and Technology, Xuzhou 221116

Abstract. According to the complex nature of the nonlinear existing in mine hoist fault time series data, the phase space reconstruction method is used to reconstruct the multi-dimension phase space for mine hoist fault time series, and then the generalized correlation dimension and the maximum Lyapunov exponents are calculated. Lyapunov index and correlation dimension are hung together as the characteristic vector for classification, while the wavelet transform method is compared on classification results. Experimental results show that mine hoist fault time series possess chaotic characteristics, and the method of combining Chaos feature extraction and support vector machine (SVM) proposed in this paper has very good diagnosis effect.

Keywords: multi-dimension phase space reconstruction; maximum Lyapunov exponents; generalized correlation dimensions; classification.

1 Introduction

Mine hoist is a large-sized electromechanical equipment of machine, electricity, liquid in coal production process. It takes the responsibility of hoisting Coal and gangue, lowering materials, lifting personnel and equipment. More, it is the only channel of connecting surface and underground, called the "throat" in coal production. The safe and reliable operation conditions of mine hoist directly affect coal miners' life safety and mine production capacity [1].

Mine hoist fault diagnosis should not only emphasize theory but also pay attention to practical application. It is one of the important topics in the development process of mining equipment automation, and scientific and technical workers of domestic and international have done a lot of research work. At home and abroad, studies for mine hoist fault diagnosis technology are three categories mainly based on analytical model, signal processing and knowledge, and different methods in the corresponding areas have obtained good results [2-4]. Reference [2] establishes fault analysis model for the fault of brake failure, overwinding and sliding failure in friction ascension, using tribology, sensing science and mechanical vibration theory. Then it puts forward researching hoist fault by wavelet modulus method. Reference [3] studies wavelet packet analysis band frequency characteristics and the basic law of frequency bands energy leakage. And the methods for hoist rope friction band energy leakage identification and signal feature extraction are proposed. Taking hoist sliding events for

M. Zhu (Ed.): Electrical Engineering and Control, LNEE 98, pp. 861–868.
springerlink.com © Springer-Verlag Berlin Heidelberg 2011

background, reference [4] has established the hoist sliding fault expert knowledge base using the method based on expert system fault diagnosis.

Mine Hoist Fault Time Series contain a large number of nonlinear information and obvious chaos phenomena, so that chaos theory becomes efficient approach on hoist fault feature analysis. The research of Lyapunov index is contributed to reveal the internal regularity of system movement, and correlation dimension reflects system complexity and the irregular movement. Therefore, the methods of Lyapunov index analysis and the correlation dimension analysis are applied to hoist fault diagnosis in this article. The phase space reconstruction method was used to reconstruct the multi-dimension phase space for mine hoist fault time series, and then the generalized correlation dimension and the maximum Lyapunov index are calculated and analyzed in order to diagnose the faults. Then, Lyapunov index and correlation dimension are hung together as the characteristic vector for fault classification. At the same time, experimental results are compared to the results obtained from classification using wavelet transform.

The rest of this paper is organized as follows. Section 2 presents the chaos features of hoist fault. Section 3 describes the experiment of hoist fault diagnosis and evaluates the results by comparing the wavelet method. Section 4 provides some brief concluding remarks.

2 Chaos Theory Analysis on Hoist Fault

2.1 Multidimensional Phase Space Reconstruction on Mine Hoist Time Series

Takens delay embedded theorem laid a reliable basis for reconstructing power system of single variable time series, thus has been widely used in the nonlinear dynamics system research. Theoretically as long as embedded dimension is enough big, single variable signal enough to reconstruct impulsion system. But situation in the actual problem is not exactly like this, the particular observed quantity can not be correctly sure to reconstruct the phase space; especially when observation data contained measurement noise, the difference of nonlinear characteristics are bigger between the simple phase space reconstruction and the original dynamic systems. Due to the multivariate data contain more information about the motive power system, so it is necessary to use multi-dimension phase space reconstruction which can rise to a certain degree of noise reduction and increase the prediction quality role [5-6].

The mine hoist is used as an example to illustrate the process of the multi-dimension phase space reconstruction. Assume in the dynamics system of mine hoist, a time series included M variables were observed:
$\{x_n\}_{n=1}^{N} = \{(x_{1,n}, x_{2,n}, \cdots, x_{M,n})\}_{n=1}^{N}$. It is the measured values of M continuous variables $x(t) = \{x_1(t), x_2(t), \cdots, x_M(t)\}$, namely $x_n = x(t_0 + n\Delta t), n=1,2,\cdots,N$. Among them t_0 is the initial time, Δt is the sampling time. Make time delay reconstruction as the following:

$$V_n = \begin{Bmatrix} x_{1,n} & , & x_{1,n-\tau_1} & , & \cdots & , & x_{1,n-(m_1-1)\tau_1} \\ x_{2,n} & , & x_{2,n-\tau_2} & , & \cdots & , & x_{2,n-(m_2-1)\tau_2} \\ & & & & \cdots & & \\ x_{M,n} & , & x_{M,n-\tau_M} & , & \cdots & , & x_{M,n-(m_M-1)\tau_M} \end{Bmatrix}^T \quad (1)$$

$$n = J_0, J_0+1, \cdots, N \quad J_0 = \max_{1\le i\le M}(m_i - 1)\tau_i + 1$$

In the formula, τ_i and m_i are separately for delay time intervals and embedding dimension. Similar to Takens delay embedded theorem, as long as m_i or $m = \sum_{i=1}^{M} m_i$ be sufficiently large, the mapping existed $F : R^m \to R^m$ made $V_{n+1} = F(V_n)$. Then the evolution of the state space reflects the evolution of mine hoist dynamics system, which means that the geometric features of mine hoist system attractor are equivalent to the geometric features of *M* dimensional state space reconstruction, therefore any of the differential or topology variables in mine hoist system can be calculated in the reconstruction of the state space.

2.2 Maximum Lyapunov Exponents for Mine Hoist Fault Time Series

The maximum Lyapunov exponent λ_i quantitative described system phase space adjacent trajectory's exponentially properties of convergence or divergence, and one of the important parameters to describe the dynamic characteristics of chaos. When the Lyapunov exponent $\lambda < 0$, phase volume contracts, moves stably, and is not sensitive to the initial conditions; When the Lyapunov exponent $\lambda > 0$, trajectory quickly separates, long time behavior is sensitive on the initial conditions, and the Motion state is chaotic; $\lambda = 0$ correspond to stability borders, and belongs to a kind of critical condition. If the Lyapunov exponent $\lambda > 0$, then the system must be chaos [7] . Due to limited sample of hoist fault time series, this paper adopts promotion small data quantity method [8] to compute the maximum Lyapunov exponent.

2.3 Generalized Correlation Dimensions for Mine Hoist Fault Time Series

When hoist equipment goes wrong, its dynamic characteristics often presents complexity and nonlinear, and hoist time series appear non-stationary and exhibit certain degree of statistical between autocorrelation. Fractal science take complicated things which has self-similarity locally and globally as research object, explore the complexity, and naturally can be used to describe hoist fault signal's irregularities and complexity. Fractal dimension is an important index to Measure fractal complexity, and correlation dimension is a relatively effective method to calculate fractal dimension. Here the univariate correlation dimension is extended to multivariate circumstances [9]. In the reconstruction of *M* dimension space,

$$\sum_{j=J_0, j\neq i}^{N} H(r-\|V_i - V_j\|) \tag{2}$$

The above formula shows the number of V_j that the distance between V_i and V_j is less than r beside V_i itself, among which $H(\cdot)$ is *Heavside* function. Define q order generalized correlation integral of time series $\{x_n\}_{n=1}^N = \{(x_{1,n}, x_{2,n}, \cdots, x_{M,n})\}_{n=1}^N$ as

$$C_q^M(r) = C_q^M(r; \tau_1, \cdots, \tau_M; m_1, \cdots, m_M) =$$

$$\{\frac{1}{N-J_0+1}\sum_{i=J_0}^{N}[\frac{1}{N-J_0}\sum_{j=J_0, j\neq i}^{N} H\cdot(r-\|V_i - V_j\|)]^{q-1}\}^{\frac{1}{q-1}} \tag{3}$$

In the formula, q is positive integer, $q \neq 1$.

$C_q^M(r)$ describes the distribution that the distance is less than r, and r can neither be too large, nor too small. If $C_q^M(r) \propto r^{(q-1)D_q}$ existed in a section of r, then D_q is called q order generalized correlation integral, and when $q = 2$, it is the normal G-P correlation dimension D_2.

3 Hoist Fault Diagnosis and Evaluation of Results

3.1 Experimental Methods and Conditions

The data set used in experiment comes from Kailuan Group Jinggezhuang Mine in Tangshan, large motor is the main power of mine hoist, and the main parameters of the motor are motor current and excitation current. The phase space is mainly reconstructed by moter current and excitation current, with extracting characteristic. Each of excitation current and moter current has a data about 10000.

The time series of collected motor current and excitation current are showed in Figure 1 and Figure 2. In order to prevent noise to influence the data during acquisition process, the data used in this paper is dealt by wavelet transform to remove noise before analysis.

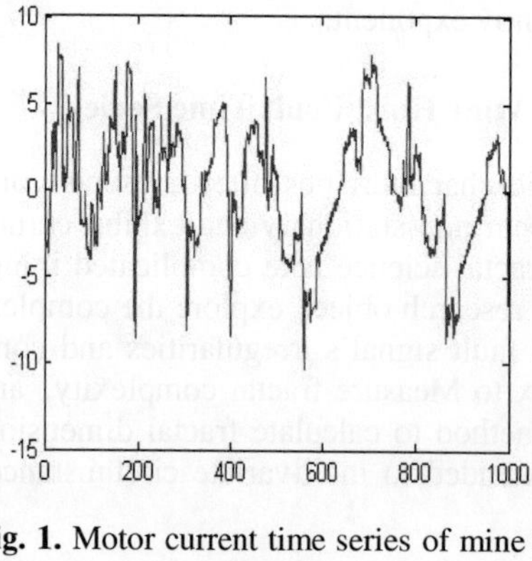

Fig. 1. Motor current time series of mine hoist

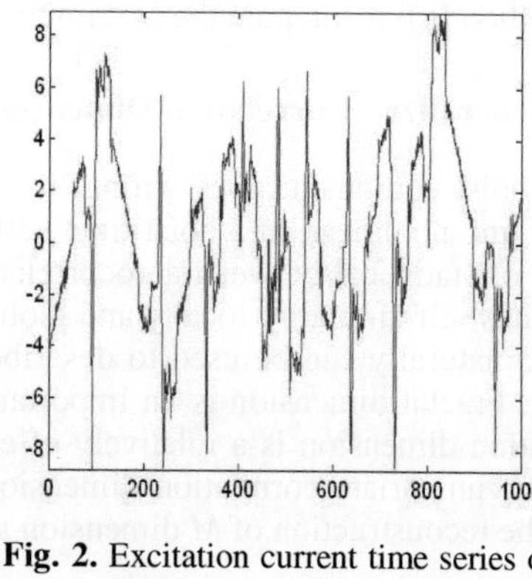

Fig. 2. Excitation current time series of mine hoist

3.2 Hoist Fault Chaos Feature Extraction

Reconstruct the multi-dimension phase space for the two time series from excitation current and magnetic fields current using formula (1), compute Lyapunov index through the generalized small data sets arithmetic, compute the generalized correlation dimension by using generalized correlation dimension method, index spectrum and correlation dimension spectrum of different faults sequence are got as the follows.

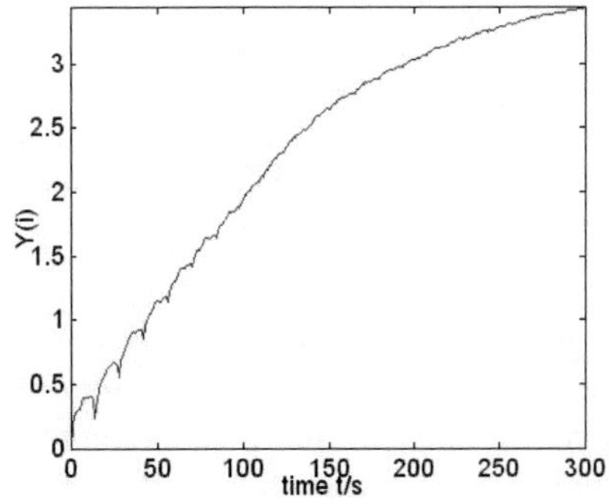

Fig. 3. (a) Lyapunov exponent spectrum of no fault

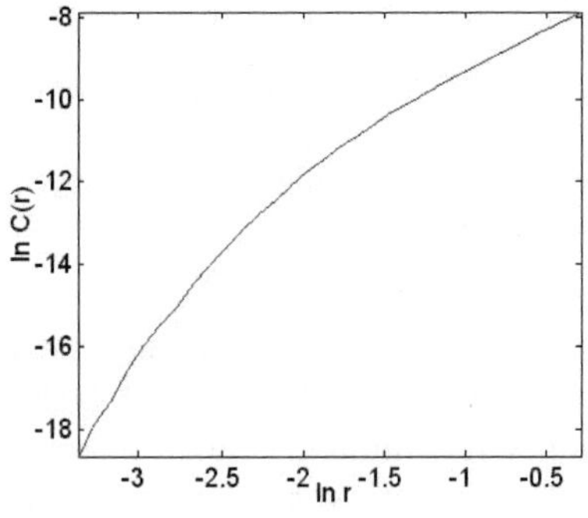

Fig. 3. (b) Generalized correlation dimension spectrum of PLC in soft overwind fault

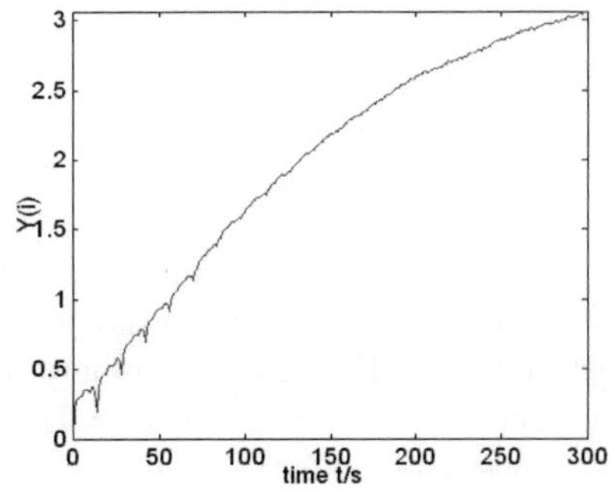

Fig. 4. (a) Lyapunov exponent spectrum of PLC in soft overwinding fault

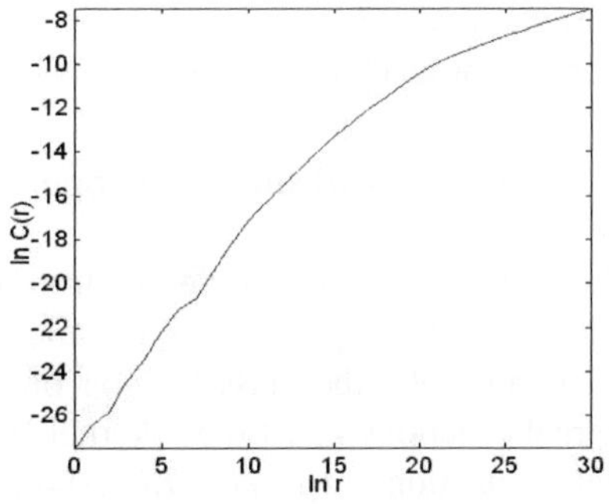

Fig. 4. (b) Generalized correlation dimension spectrum of PLC in soft overwinding fault

Corresponding to different time series of Lyapunov exponent spectrum, figure 3-a, 4-a, 5-a and 6-a show the hoist's feature when have no fault or have fault. There are a number of peak jumping on each curve, but the jumping phase of the peak was located in the same line, with a linear law. The wave of curve fitting, part a straight line slope significantly greater than zero. When fitting a straight line of the peak part of the curve, the slope was significantly greater than 0. The slope of the fitted straight line is the maximum value of Lyapunov index, with a result of L(3-a)=0.7321>0、L(4-a)=0.6099>0、L(5-a)=0.6325>0、L(6-a)=0.6657>0, indicating the exist of chaos mine time series.

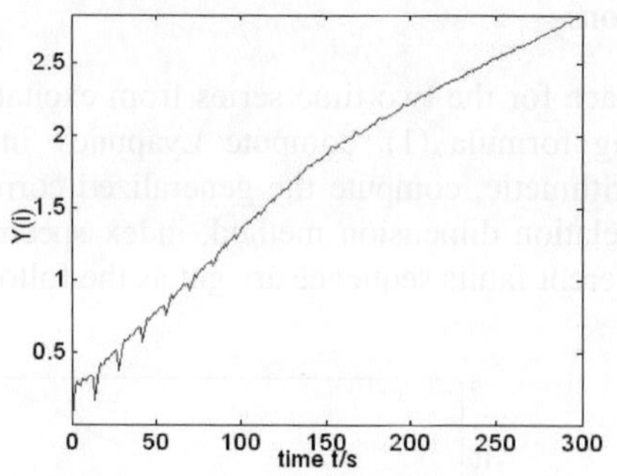

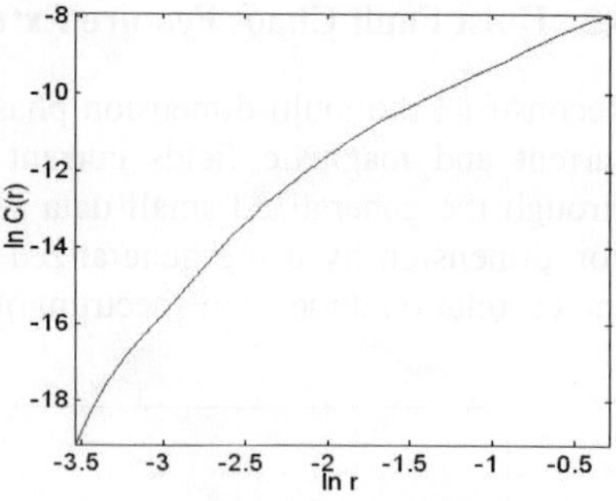

Fig. 5. (a) Lyapunov exponent spectrum of reverse fault

Fig. 5. (b) Generalized correlation dimension spectrum of reverse fault

Corresponding to different time series of generalized correlation dimension spectrum, figure 3-b, 4-b, 5-b and 6-b show the hoist's feature when have no fault or have fault. When fitting a straight line of each curve, the slope of the fitted straight line is the value of generalized correlation dimension , with a result of G(3-b)=0.7616>0、G(4-b)=0.8217>0、G(5-b)=0.8197>0、G(6-b)=0.9415>0.

Conclusion also can be achieved from figure 3 to figure 6: When a fault occurs, Lyapunov exponent less than normal, and correlation dimension greater than normal, then we can know a fault occurs. By joining Lyapunov exponent and generalized correlation dimension together to compose feature vector <L, G> as the input of support vector machine, the fault classification can be done.

3.3 Fault Classification of Support Vector Machine

Support vector machine comes from the optimal classification of linearly separable. The optimal classification of linearly separable not only can separate two samples, but also can make the classification of the largest interval [10]. Now commonly used kernel functions are: Linear kernel function (linear)$K(x_i,x)=(x_i \bullet x)$; Polynomial kernel function (ploy) $K(x_i,x)=(x_i \bullet x)^d$; Gaussian radial basis function (RBF) $K(x_i,x)=e^{-\gamma\|x_i-x\|^2}$; Sigmoid kernel function $K(x_i,x)=\tanh[a(x_i \bullet x)+b]$.

While 60 sets of data is collected of each hoist fault, then the trained support vector machine is used to classify the fault with 30 sets of data served for training sample and other 30 sets of data served for testing sample. In addition to the normal, there are three different faults, four different sample data and two different data must to build a classifier. Table 1 show the diagnosis result when using Chaos-SVM and Wavelet-SVM with kernel function linear. Meanwhile in order to as contrast, table 2 show the diagnosis result with the kernel function RBF.

Table 1. Recognition rate of Chaos-SVM and Wavelet-SVM (kernel function linear)

	Chaos-SVM Recognition rate(%)	Wavelet-SVM Recognition rate(%)
no fault	98.75	93.51
soft overwinding fault of PLC	98.49	93.54
safety circuit fault	99.18	94.05
reverse fault	93.86	88.56

Table 2. Recognition rate of Chaos-SVM and Wavelet-SVM (kerbel function RBF)

	Chaos-SVM Recognition rate(%)	Wavelet-SVM Recognition rate(%)
no fault	98.88	93.87
soft overwinding fault of PLC	98.25	93.52
safety circuit fault	99.32	94.12
reverse fault	93.03	88.87

From table 1 to table 2, the recognition rate of joining chaos features and support vector machine is greater than the recognition rate of joining wavelet transform to extract features and support vector machine obviously. What's more, the recognition rate of joining chaos features and support vector machine is always greater than 94%. So, the method used in this paper has a good diagnosis result.

4 Conclusion

Chaos-SVM is proposed for hoist fault diagnosis in this paper. The method has a good performance on hoist fault recognizing and hoist fault distinguish. It has a recognition rate over 98% on no fault situation, PLC soft overwinding fault, and safety circuit fault situation, particularly a high accuracy distinction on complex fault. Experimental results show that this method is more efficient than Wavelet-SVM. This method is not only useful for hoist system, but also can be extended to other fault diagnosis.

References

1. Editorial committee of Mine hoist breakdown processing and technical reformation. Mine hoist breakdown processing and technical reformation. Mechanical industry press, Beijing (2005)
2. Siemens industrial products [EB/OL], `http://w1.siemens.com/answers/cn/zh/index.htm?stc=cnccc020001`
3. Write, G.S., Zhongzhi, S., Yongquan, L.: Translate. Knowledge engineering and knowledge management. Mechanical industry press, Beijing (2003)
4. Noy, N.F., Mcguinness, D.L.: Ontology development: guide to creating your first ontologyER. Stanford University, Stanford (2001)

5. Prichard, D., Theiler, J.: Generating surrogate data for time series with several simultaneously measured variables. Physical Review Letters 73(7), 951–954 (1994)
6. Rombowts, S.A.R.B., Keunen, R.W.M., Stom, C.: Investigation of nonlinear structure in multichannel EEG. Physics Letter A 202, 352–358 (1995)
7. Jinhu, L., Anju, L., Shihua, C.: Chaos time series analysis and application. Wuhan university press, Wuhan (2002)
8. Haiyan, W., Shan, L.: Chaos time series analysis and application. Science press, Beijing (2006)
9. Zwelee, G.: Wavelet-Based Neural Network for Power Disturbance Recognition an d Classification. IEEE Tram on Power Delivery 19(4), 1560–1568 (2004)
10. Yanwei, C., Yaocai, W., Tao, Zhijie, W.: Fault Diagnosis of a Mine Hoist Using PCA and SVM Techniques. Journal of China Univ Mining&Technol. 18(3), 0327–0331 (2008)

Bistatic SAR through Wall Imaging Using Back-Projection Algorithm

Xin Li[1], Xiao-tao Huang[1], Shi-rui Peng[2], and Yi-chun Pan[2]

[1] UWB Laboratory, School of Electronic Science and Engineering, National University of Defense Technology, DEYA Road 46, KAIFU District, 410073 Changsha, China
[2] Department of Information Countermeasures, Air Force Radar Academy, HUANGPU Street 288, JIANGAN District, 430019 Wuhan, China
lx_airforce@yahoo.com.cn, hxtdh@yahoo.com.cn, psr99@21cn.com, dhhxt@yahoo.com.cn

Abstract. In the BiSAR (Bistatic Synthetic Aperture Radar) through wall imaging system, effective imaging algorithm is of great importance. Back Projection (BP) algorithm can serve almost all cases with good precision. A bistatic configuration with a stationary receiver and a side-looking vehicle-borne transmitter has been designed for through wall imaging in this paper. BP algorithm in this case is developed. The range, Doppler and azimuth resolution formulations are deuced. Imagings of a room with targets in it are simulated, and resolution distributions of the scene under different bistatic configuration are simulated too. it is found out that the bistatic angle affects the imaging of a scene. Simulation results and the theoretical analysis can be reference for bistatic through wall SAR imaging.

Keywords: Bistatic SAR, Through wall imaging, Back projection algorithm, Bistatic resolution, Fixed-receiver.

1 Introduction

Increasing attention has been paid to imaging of a building [1-9], which can provide building layouts and insurgent activity profiles. Bistatic SAR through-wall imaging is a promising new aspect in through-wall sensing, Bistatic radar provides greater diversity in aspect angle than monostatic radar and acquiring more scattering information, which is competent for imaging of a building [1-3].

Under the US DARPA VisiBuilding program [1, 4], one core technical area is imaging the building, in which synthetic apertures are along the two different sides of the building with a fixed-receiver bistatic system or a vehicle-borne system separately. The SAR images for each pass were independently formed and then incoherently combined to generate the overall building map. For the present results, Back Projection (BP) algorithm is widely used in through wall applications [4-12]. In [13], 2-D resolution of a bistatic SAR configuration with a stationary transmitter and a forward-looking receiver is analyzed. Range Doppler algorithm has been modified for imaging in this scenario. Whereas, no theoretical deduction of algorithm implement is presented in detail for bistatic through wall SAR imaging. In this paper, we

M. Zhu (Ed.): Electrical Engineering and Control, LNEE 98, pp. 869–878.
springerlink.com © Springer-Verlag Berlin Heidelberg 2011

investigate BP imaging of a bistatic through wall SAR configuration which has a stationary receiver and a side-looking vehicle-borne transmitter. Sensors are placed on different sides of the building and wall effects are taken into consideration. Resolutions of different configuration are studied by simulations.

2 Analysis of 2-D Resolution of through Wall BiSAR

2.1 BiSAR through Wall Imaging Geometry

Figure 1 shows the spatial geometry of BiSAR through wall imaging of an enclosed room. SAR beam width is θ_B.

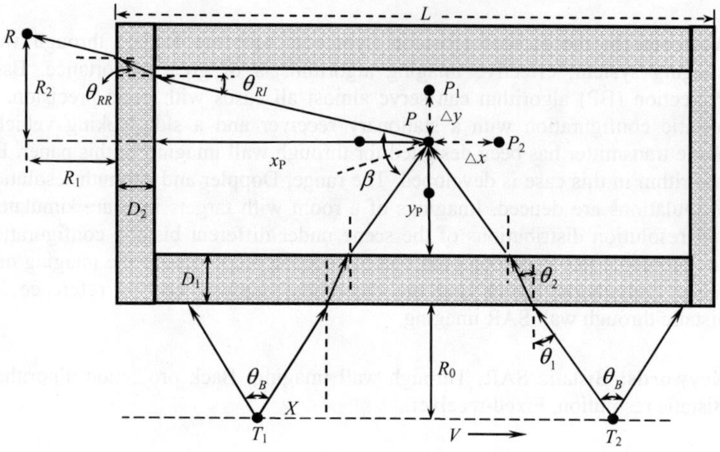

Fig. 1. BiSAR through wall imaging model.

The transmitter is moving along the front wall of the building at velocity V, stand-off distance is R_0, and wall thickness is D_1 and D_2 respectively. The relative permittivity of wall is ε_r. The stationary receiver is located at R, stand-off distance is R_1, for a point scatterer P locates at (x_p, y_p) behind wall, the distance between P and receiver in vertical direction is R_2. The receive beam always points at the scene center.

In engineering, the coordinates of the refraction points can be obtained approximately [14], we have

$$X = V \cdot t \cdot R_0 \left[\sqrt{\varepsilon_r}\left(y_P + D_1 \right) - D_1 \right] \left[\sqrt{\varepsilon_r}\left(R_0 + y_P \right)\left(y_P + D_1 \right) - D_1 R_0 \right]^{-1} \quad (1)$$

where $t \in \left[-\frac{L}{2V} - \frac{R_0}{V} tg \frac{\theta_B}{2}, \frac{L}{2V} + \frac{R_0}{V} tg \frac{\theta_B}{2} \right]$, L is the size of the building in the observation direction.

$$\cos\theta_1 = \left[1 + \left(X / R_0 \right)^2 \right]^{-\frac{1}{2}}, \quad \cos\theta_2 = \left[1 - X^2 \Big/ \left[\varepsilon_r \left(X^2 + R_0^2 \right) \right] \right]^{\frac{1}{2}} \quad (2)$$

θ_1 and θ_2 are the incidence and refraction angle respectively. So, the range from transmitter to scatterer P is

$$R_T(t) = \frac{y_P + R_0}{\cos\theta_1} + \frac{D_1\sqrt{\varepsilon_r}}{\cos\theta_2} = \sqrt{1 + \left(\frac{V\cdot\left[\sqrt{\varepsilon_r}(y_P + D_1) - D_1\right]}{\sqrt{\varepsilon_r}(R_0 + y_P)(y_P + D_1) - D_1R_0}\cdot t\right)^2} \cdot \left(y_P + R_0 + D_1\varepsilon_r\left[(\varepsilon_r - 1)\left(\frac{V\cdot\left[\sqrt{\varepsilon_r}(y_P + D_1) - D_1\right]}{\sqrt{\varepsilon_r}(R_0 + y_P)(y_P + D_1) - D_1R_0}\cdot t\right)^2 + \varepsilon_r\right]^{-\frac{1}{2}}\right) \tag{3}$$

Similarly, the range from scatterer P to receiver is

$$R_R = \sqrt{1 + \left(\frac{R_2\left[\sqrt{\varepsilon_r}(x_P + D_2) - D_2\right]}{\sqrt{\varepsilon_r}(R_1 + x_P)(x_P + D_2) - D_2R_1}\right)^2} \cdot \left(x_P + R_1 + D_2\varepsilon_r\left[(\varepsilon_r - 1)\left(\frac{R_2\left[\sqrt{\varepsilon_r}(x_P + D_2) - D_2\right]}{\sqrt{\varepsilon_r}(R_1 + x_P)(x_P + D_2) - D_2R_1}\right)^2 + \varepsilon_r\right]^{-\frac{1}{2}}\right) \tag{4}$$

For simplicity let, $K = \dfrac{V\left[\sqrt{\varepsilon_r}(y_P + D_1) - D_1\right]}{\sqrt{\varepsilon_r}(R_0 + y_P)(y_P + D_1) - D_1R_0}$, thus, the instantaneous slant range for scatterer P is

$$R(t) = R_T(t) + R_R = \sqrt{1 + (K\cdot t)^2}\left(y_P + R_0 + D_1\varepsilon_r\left[(\varepsilon_r - 1)(K\cdot t)^2 + \varepsilon_r\right]^{-\frac{1}{2}}\right) + R_R \tag{5}$$

2.2 Range Resolution

We use vector gradient method [16] to derive the value and the direction of the bistatic through wall SAR resolutions. The total slant range from the transmitter to point P and from P to the receiver and a surface of constant bistatic range consists of the points satisfying

$$R(t) = |\mathbf{R}_T(t)| + |\mathbf{R}_R(t)| = \text{constant} \tag{6}$$

where $\mathbf{R}_T(t)$ and $\mathbf{R}_R(t)$ are the vectors from P to transmitter and receiver respectively. The gradient of $R(t)$ is

$$|\nabla R(t)| = |\mathbf{u}_T + \mathbf{u}_R| = 2\cos(\alpha_P - \beta_P)/2 = 2\cos\beta/2 \tag{7}$$

where $\mathbf{u}_T$ and $\mathbf{u}_R$ are unit vectors from P to transmitter and receiver respectively. α_P, β_P are the angle between $\mathbf{u}_T$, $\mathbf{u}_R$ and the x-axis respectively, β is the bistatic angle. So, the slant range resolution can be written as

$$\rho_r\left(x_P, y_P; t\right) = \frac{\delta R}{\mathbf{u}_r \cdot \nabla R} = \frac{c}{B \cdot |\nabla R|} = \frac{c}{2 \cdot B \cdot \cos(\beta/2)} \tag{8}$$

The instantaneous bistatic angle is

$$\beta\left(x_P, y_P; t\right) = \arctan\frac{\sqrt{\varepsilon_r}\left(R_0 + y_P\right)\left(y_P + D_1\right) - D_1 R_0}{Vt \cdot \left[\sqrt{\varepsilon_r}\left(y_P + D_1\right) - D_1\right]} + \arctan\frac{R_2\left[\sqrt{\varepsilon_r}\left(x_P + D_2\right) - D_2\right]}{R_1\sqrt{\varepsilon_r}\left(R_1 + x_P\right)\left(x_P + D_2\right) - D_2 R_1^2} \tag{9}$$

2.3 Doppler and Cross Range Resolution

The Doppler shift of the signal received from a point **r** is

$$f_d(\mathbf{r}) = \left[\mathbf{V}_T \cdot \mathbf{u}_T + \mathbf{V}_R \cdot \mathbf{u}_R\right]/\lambda \tag{10}$$

where λ is the wavelength of the transmit wave, and $\mathbf{V}_R$=0. The surface of constant Doppler shift satisfies

$$f_d(\mathbf{r}) = \text{constant} \tag{11}$$

The gradient of $f_d(\mathbf{r})$ at the point **r** is

$$\nabla f_d = \left[\frac{\partial f}{\partial x}\mathbf{i} + \frac{\partial f}{\partial y}\mathbf{j} + \frac{\partial f}{\partial z}\mathbf{k}\right] = \frac{1}{\lambda}\frac{1}{|\mathbf{R}_T|}\left[\mathbf{V}_T - \left(\mathbf{V}_T \cdot \mathbf{u}_T\right)\mathbf{u}_T\right] \tag{12}$$

It can be derived that

$$|\nabla f_d| = V\left(R_0 + \sqrt{\varepsilon_r} D_1 + y_P\right) / \left(\lambda R_T^2(t)\right) \tag{13}$$

Thus, the Doppler resolution at any point in the scene is

$$\rho_a = \frac{1}{T_S \cdot |\nabla f_d(t)|} = \frac{R_T^2(t)}{R_0 + \sqrt{\varepsilon_r} D_1 + y_P}\lambda\left[2\sin\frac{\theta_B}{2}\left(\frac{R_0 + y_P}{\cos\frac{\theta_B}{2}} + \frac{D_1}{\sqrt{\varepsilon_r - \sin^2\frac{\theta_B}{2}}}\right)\right]^{-1} \tag{14}$$

where T_s is the synthetic time of target. Note from Eq. (14) the Doppler resolution only depends on the angular rate of transmitter about the point and wall parameters. The angle between the direction of the Doppler resolution and the *x*-axis is

$$\eta = \text{angle}\left(\nabla f_d\right) = \text{asin}\left[\left(R_0 + \sqrt{\varepsilon_r} D_1 + y_P\right) / R_T(t)\right] \tag{15}$$

When $\varepsilon_r = 1$, t=0, that is there is no wall, the Doppler resolution becomes $\rho_a \approx d_a$, d_a is the dimension of the antenna. That is twice the resolution of monostatic side-looking SAR.

Defining θ_G to be the angle between the gradients, i.e., $\theta_G = \eta - \beta/2$, a representation of the 2-D resolution cell size is its area given by

$$A = \rho_r \cdot \rho_a / \sin\theta_G \tag{16}$$

The cross range resolution can be[15]

$$\rho_{crr} = \rho_a / \sin\theta_G \tag{17}$$

So, small θ_G gives large cell, and the 2-D imaging ability will be poor consequently.

3 BiSAR through Wall BP Imaging

Supposing the transmitter radiates chirp signals, whose carrier frequency is f_0,

$$s(\tau) = rect(\tau/T_P) \cdot \exp\{j\pi K_r \tau^2\} \tag{18}$$

where τ is fast time in range, T_P is pulse duration, rect( ) is rectangle function. K_r is the chirp FM rate in range. The echoes at base band is

$$\begin{aligned} s_r(\tau,t) &= w_r[\tau - R(t)/c] w_a(t) \\ &\cdot \exp\{-j2\pi f_0 R(t)/c\} \cdot \exp\{j\pi K_r (\tau - R(t)/c)^2\} \end{aligned} \tag{19}$$

where t is slow time in azimuth, $w_r(\tau)$ and $w_a(t)$ are the range and azimuth envelop. $R(t)$ is instantaneous bistatic slant range in (5). c is the speed of light in free space.

Range compression can be performed with reference function

$$s_r(\tau) = s^*(-\tau) = rect(\tau/T_P) \cdot \exp(-j\pi K_r \tau^2) \tag{20}$$

The compressed signal is

$$s_{rc}(\tau,t) = s_r(\tau,t) \otimes s_r(\tau) = A \cdot p_r[\tau - R(t)/c] w_a(t) \cdot \exp\{-j2\pi f_0 R(t)/c\} \tag{21}$$

where $p_r(\tau)$ is the self-correlation of $w_r(\tau)$. In order to better understand the algorithm, slow time is written is discrete form $t = t_m = mPRT (m = 0,1,2...)$, PRT is the pulse repetition time, $|m \cdot PRT| \le T_S$.

Then, interpolation is performed in fast time domain. Imaging scene is divided into small pixels, for the mth pulse transmitted, the time delay of pixel (i,j) in the scene is

$$t_d(m;i,j) = R(t_m)/c \tag{22}$$

The instantaneous bistatic slant range is

$$R(t_m) = \sqrt{1 + (K \cdot m \cdot PRT)^2} \left(y_P + R_0 + D_1 \varepsilon_r \left[(\varepsilon_r - 1)(K \cdot m \cdot PRT)^2 + \varepsilon_r \right]^{-\frac{1}{2}} \right) + R_R \tag{23}$$

The corresponding out put of range compression is

$$s_{rc}(\tau,t_m)=A\cdot p_r\left[\tau-t_d(m;i,j)\right]w_a(t_m-t_c)\cdot\exp\{-j2\pi f_0 t_d(m;i,j)\} \quad (24)$$

The phase compensation factor is

$$H_c=\exp\{j2\pi f_0 t_d(m;i,j)\} \quad (25)$$

If there is M echoes in the effective aperture, the value of pixel (i,j) can be constructed by accumulating the compensated $s_{rc}(\tau,t)$ coherently.

$$I(i,j)=\sum_{m=1}^{M}s_{rc}(\tau,t_m)\cdot H_c \quad (26)$$

For every pixel in the scene, this process is performed along the target locus.

4 Simulation Study

Simulation scene is shown Fig.1, and the imaging area is 10m×10m. The scattering intensity of the inner surface of the wall and that of the back wall is 3 dB less than that of the outer surface. There exist five scatterers, the scattering intensity is as much as that of the outer surface of the wall. Components caused by multipathing and reverberation are not taken into consideration due to the great attenuation. Other parameters in the simulation are listed in Table 1.

Table 1. BiSAR through wall imaging simulation parameters.

Parameter	Value	Parameter	Value	Parameter	Value	Parameter	Value
ε_r	6.25	R_0	10m	x_p	4.8m	B	1GHz
D_1	0.3m	R_1	2m	y_p	4.7m	d_a	0.5m
D_2	0.2m	R_2	5m	Carrier frequency	1GHz	V	5.5m/s

We define a bistatic angle β_0 to depict different bistatic scenario, which is the angle between transmitter and receiver line of sight at time t=0, when they both pointing at the center of the scene as shown in Fig. 1.

4.1 Bistatic Range and Doppler Frequency Contours

The equal bistatic slant range and equal Doppler frequency contours of the scene under different bistatic scenarios at various time instants are shown in Fig.2 and Fig.3.

It can be seen that the distribution of bistatic slant range and Doppler frequency contours varies with time and so does the angle between them. At the same instant, distribution of Doppler frequency contours of different bistatic configurations are the same, while the bistatic range contours are not. Smaller β_0 gives smaller resolution cells in the scene.

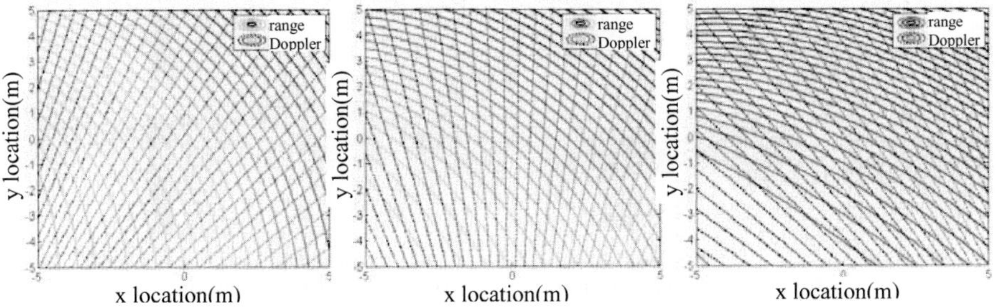

(a) t=-T_s/2 (b) t=0 (c) t=T_s/2

Fig. 2. Bistatic range and Doppler frequency contours. ($\beta_0 = 90°$).

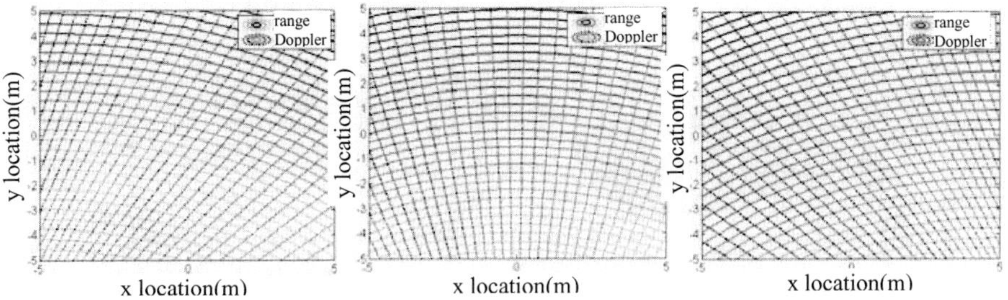

(a) t=-T_s/2 (b) t=0 (c) t=T_s/2

Fig. 3. Bistatic range and Doppler frequency contours. ($\beta_0 = 0°$).

4.2 Resolutions Distribution

Fig. 4 and Fig. 5 shows the bistatic range resolution, Doppler and cross range resolution, and region for which the range and cross range resolution are each less than 1m^2 when $\beta_0 = 90°$ and $\beta_0 = 0°$.

It can be seen that Doppler resolution distributions are same at a certain instant despite the bistatic configuration, and 2-D resolution of the lower left area is relatively poor when $\beta_0 = 90°$. Smaller β_0 gives better resolutions distributions.

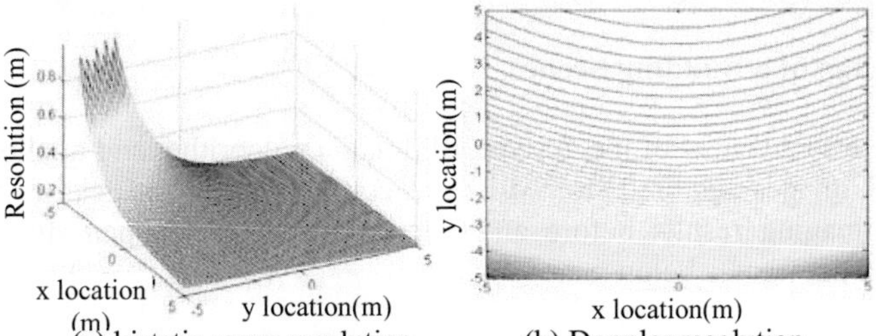

(a) bistatic range resolution (b) Doppler resolution

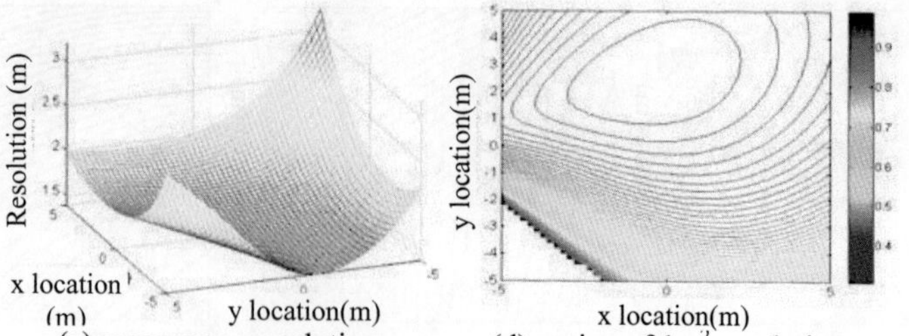

(c) cross range resolution (d) region of $1m^2$ resolution

Fig. 4. Resolution distribution of imaging scene ($\beta_0 = 90°$).

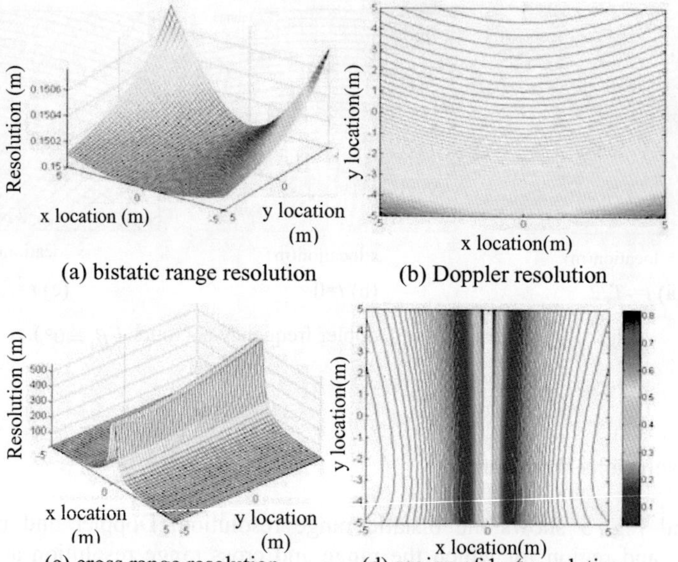

(a) bistatic range resolution (b) Doppler resolution

(c) cross range resolution (d) region of $1m^2$ resolution

Fig. 5. Resolution distribution of imaging scene ($\beta_0 = 0°$).

4.3 Imaging of a Building Scene

To examine effectiveness of the derived BP imaging algorithm and quality of through wall BiSAR images acquired from different configurations, images of the scene are simulated. Imaging results before and after wall effects compensation are shown separately in the figures below.

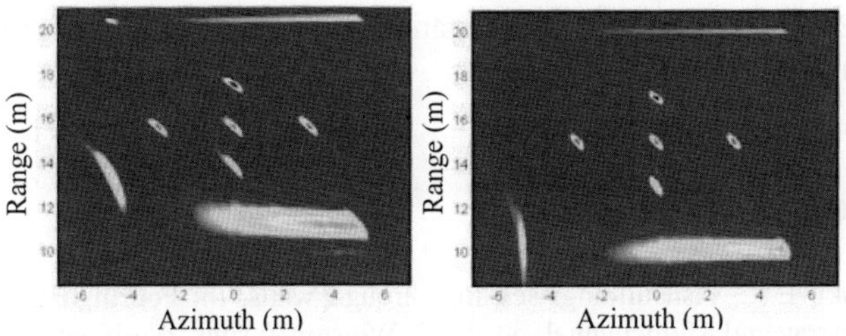

Fig. 6. Images of the scene $\beta_0 = 90°$.

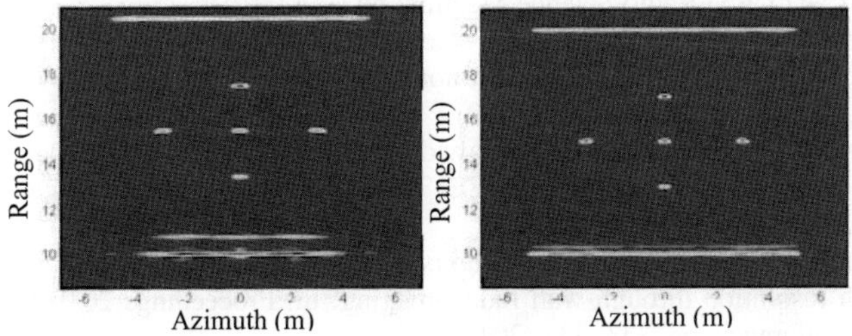

Fig. 7. Images of the scene $\beta_0 = 0°$.

As can be seen that wall effects must be compensated otherwise targets will be misplaced in the image and even the image will be defocused. Images acquired when $\beta_0 = 0°$ is much better than that when $\beta_0 = 90°$. Besides, from the many simulations we conducted it can be found that the smaller the bistatic angle is, the better the images will be focused. So in order to get acceptable images β_0 should be smaller than $90°$.

5 Conclusions

As the imaging ability and good precision of BP imaging algorithm is appreciated, this paper investigates the BP algorithm in bistatic through wall SAR application. A bistatic through wall SAR system with a stationary receiver and a side-looking vehicle-borne transmitter has been studied, and formulas of resolutions and BP algorithm are derived.Resolutions distribution and images of the scene are simulated. Wall effects are taken into consideration and compensated during the imaging. Besides, theoretical analysis and imaging results indicate that to acquire acceptable images, the bistatic angle should be no larger than $90°$.

Acknowledgments. This work is supported by the Program for New Century Excellent Talents in University NCET-07-0223, the National Natural Science Foundation of China under grant 60972121, and a Foundation for the Author of

National Excellent Doctoral Dissertation of the People's Republic of China (FANEDD) grant 201046.

References

1. Baranoski, E.J.: VisiBuilding: sensing through walls. In: Fourth IEEE Workshop IEEE Sensor Array and Multichannel, pp. 1–22. Winston & sons, Washington DC (2006)
2. Zhang, Y., Lavely, E.M., Weichman, P.B.: The VisiBuilding inverse problem (closed session presentation). In: ASAP 2007, June 5-6 (2007)
3. Le, C., Nguyen, L., Dogaru, T.: Radar imaging of a large building based on near-field Xpatch model. In: IEEE International Symposium Antennas and Propagation and CNCUSNC/ URSI Radio Science Meeting, pp. 1–4 (2010)
4. Nguyen, L., Ressler, M., Sichina, J.: Sensing through the wall imaging using the Army Research Lab ultra-wideband synchronous impulse reconstruction (UWB SIRE) radar. In: Proceedings of SPIE, vol. 6947 (2008)
5. Le, C., Dogaru, T., Nguyen, L.: Ultrawideband (UWB) radar imaging of building interior: measurements and predictions. IEEE Trans. Geosci. Remote Sens. 47(5), 1409–1420 (2009)
6. Browne, K.E., Burkholder, R.J., Volakis, J.L.: A novel low-profile portable radar system for high resolution through-wall radar imaging. In: Proceedings 2010 IEEE International Radar Conference, pp. 333–338 (2010)
7. Ertin, E., Moses, R.L.: Through-the-wall SAR attributed scattering center feature estimation. IEEE Transactions on Geoscience and Remote Sensing 47(5), 1338–1348 (2009)
8. Dogaru, T., Le, C.: SAR Images of Rooms and Buildings Based on FDTD Computer Models. IEEE Trans. Geosci. Remote Sens. 47(5), 1388–1401 (2009)
9. Dogaru, T., Sullivan, A., Kenyon, C.: Radar Signature Prediction for Sensing- Through-the-Wall by Xpatch and AFDTD. In: DoD High Performance Computing Modernization Program Users Group Conference, pp. 339–343 (2009)
10. Buonanno, A., D'Urso, M., Ascione, M.: A new method based on a multisensor system for through-the-wall detection of moving targets. In: IEEE International Instrumentation & Measurement Technology Conference - I2MTC 2010, pp. 1521–1525 (2010)
11. D'Urso, M., Buonanno, A., Prisco, G.: Moving Targets Tracking for Homeland Protection Applications: a Multi-sensor Approach. In: Proceedings 2010 IEEE International Radar Conference, pp. 1220–1223 (2010)
12. Masbernat, X.P., Amin, M.G., Ahmad, F.: An MIMO-MTI approach for through-the-wall radar imaging applications. In: 2010 5th International Waveform Diversity and Design Conference (WDD 2010), pp. 188–192 (2010)
13. Qiu, X.L., Hu, D.H., Ding, C.B.: Some reflections on bistatic SAR of forward-looking configuration. IEEE Geoscience and Remote Sensing Letters 5(4), 735–739 (2008)
14. Johansson, E.M., Mast, J.E.: Three-dimensional ground penetrating radar imaging using synthetic aperture time-domain focusing. In: Proc. of SPIE Conference on Advanced Microwave and Millimeter Wave Detectors, Orlando, vol. 2275, pp. 205–214 (1994)
15. Cardillo, G.P.: On the use of the gradient to determine bistatic SAR resolution. Proc. Antennas Propag. Soc. Int. Symp. 2, 1032–1035 (1990)

A Fully-Integrated High Stable Broadband Amplifier MMICs Employing Simple Pre-matching/Stabilizing Circuits

Jang-Hyeon Jeong*, Young-Bae Park, Bo-Ra Jung, Jeong-Gab Ju, Eui-Hoon Jang, and Young-Yun

Department of Radio Communication and Engineering,
Korea Maritime University,
Dong Sam-dong, Young Do-gu, Busan, 606-791, Korea
yunyoung@hhu.ac.kr

Abstract. In this work, using simple low-loss pre-matching/stabilizing circuits, high stable broadband driver amplifier (DA) and medium power amplifier (MPA) MMICs including all the matching and biasing components were developed for K/Ka band applications. The DA and MPA MMICs exhibited good RF performances and stability in a wide frequency range. The layout size of the DA and MPA MMICs were 1.7X0.8 and 2.0X1.0 μm^2, respectively.

Keywords: DA(driver amplifier), MPA(medium power amplifier), MMIC(monolithic microwave integrated circuit), Pre-matching, K/Ka band.

1 Introduction

Recently, demands for fully-integrated RF devices with broadband operation range are increasing in the LMDS market of K/Ka band MMIC. Promising applications for GaAs technology include MMIC chip sets for K/Ka band LMDS system[1]-[4]. Until now, a number of articles concerning amplifier ICs for K/Ka band applications have been reported, but most of them are hybrid ICs which require off-chip biasing/matching components on PCB due to the large size of the components. It has resulted in a high manufacturing cost due to a large module size and a high assembly cost.

In this work, using simple pre-matching/stabilizing circuit, miniaturized high stable broadband driver amplifier (DA) and medium power amplifier (MPA) MMICs including all the matching and biasing components were developed for K/Ka band applications. To reduce a layout size and perform the layout/circuit design efficiently, a CAD-oriented MIM shunt capacitor and its lumped equivalent circuit were also developed.

* Jang-Hyeon Jeong received BS degrees in department of radio communication and engineering of Korea Maritime University in 2010. And Jang-Hyeon Jeong is working toward MS degree in Korea Maritime University, Busan, Korea.

M. Zhu (Ed.): Electrical Engineering and Control, LNEE 98, pp. 879–885.
springerlink.com © Springer-Verlag Berlin Heidelberg 2011

2 Schematic Circuit Configuration of the Amplifier MMICs

The amplifier of this work was designed for medium power application of driver stage. A schematic circuit of the driver amplifier (DA) MMIC(mololithic microwave integrated circuit is shown in Fig. 1. For broadband design of the amplifier MMIC, pre-matching circuits were employed to the gate input and drain output of the FETs. To improve the stability of the amplifier MMIC, parallel RC stabilizing circuit was employed at the input of the amplifier MMIC. To reduce a layout size, MIM shunt capacitors were employed for shunt capacitive matching in place of open stubs.

For the FET, a GaAs HEMT was employed. The gate width (Wg) of the first and second stage FET are 50X4 and 50X6μm, respectively, and the Wg of the final stage amplifier is 50X8μm.

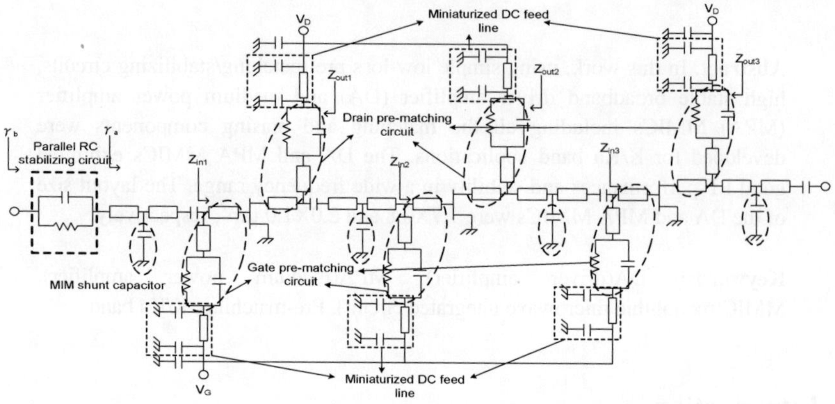

Fig. 1. A schematic circuit of the driver amplifier (DA) MMIC

3 Design of Miniaturized Broadband Amplifier with Simple Pre-matching/Stabilizing Circuit

3.1 Simple Pre-matching Circuits for Miniaturized Broadband Amplifier

In this work, to achieve low-loss broadband design (in K/Ka band) of the amplifier MMIC with a small chip size, simple pre-matching circuits employing parallel RC circuits were used for the gate input and drain output of the FET as shown in Fig.1. The resistance, capacitance, and inductance values of the pre-matching circuits shown in Fig. 1 were selected by the concept of the pre-matching using parasitic elements of FET [5,6]. In this work, the input and output of the FETs were designed to show only real part of impedance in a wide frequency range by removing the reactance part of the impedance of the FETs via the pre-matching circuits. The gate input and drain output pre-matching circuits connected with the parasitic elements of FET are shown in Fig. 2 (a) and (b), respectively. The gate input pre-matching circuits of Fig. 1 are

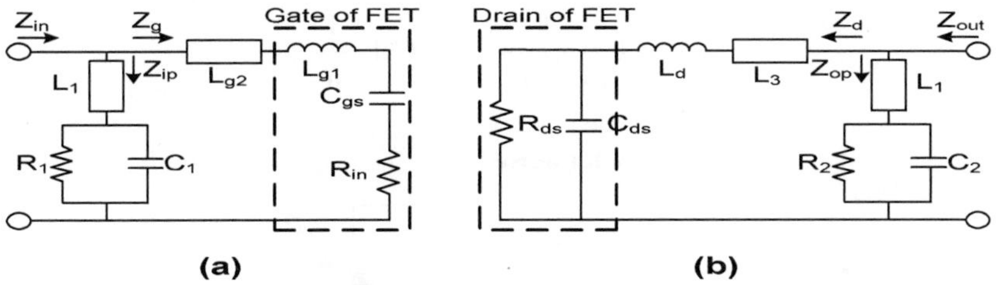

Fig. 2. (a) The gate input pre-matching circuit connected with the parasitic elements of FET (b) The drain output pre-matching circuit connected with the parasitic elements of FET

connected with the parasitic elements of the gate of the FET in parallel as shown in Fig. 2 (a). The L_{g1}, C_{gs}, and R_{in} are parasitic inductance and capacitance of the gate of the FET, and the gate input resistance of the FET, respectively, and L_{g2} is the inductance of stripline inductor, which is externally connected to increase the parasitic inductance of the FET. The input pre-matching circuit consists of the stripline inductors and RC parallel circuit. In this case, the input impedance (Z_{in}) can be expressed as follows.

$$Z_{in} = \frac{1}{1/Z_g + 1/Z_{ip}} = R_{in}(\omega) + j\omega X_{in}(\omega), \tag{1}$$

where Z_g and Z_{ip} are the input impedance of the gate of the FET and the impedance of the prematching circuit, respectively as shown in Fig. 2 (a). To reduce the strong dependency of the Z_{in} on frequency, the reactance part, $X_{in}(\omega)$, should be removed. Therefore, R_1, C_1 and L_1 were selected to remove the reactance part, $X_{in}(\omega)$.

3.2 Highly Improved Stability of the Amplifier MMIC Employing Parallel RC Stabilizing Circuit

For K/Ka band applications, the circuit stability should be guaranteed in a wide frequency range from DC to operation band to prevent an unwanted oscillation. Especially, In this work, in order to improve the stability of the amplifier MMIC, parallel RC stabilizing circuit was connected to the input of the amplifier MMIC as shown in Fig. 1. Without the parallel RC stabilizing circuit at the input port, the amplifier MMIC exhibited conditionally stability ($K < 1$) in a frequency range from 14 to 15 GHz due to the instability of the FET itself. Therefore, the value of resistance R of the RC stabilizing circuit was selected to increase the stability of the amplifier MMIC by improving the input return loss in the above frequency range. The value of the capacitance C was selected to bypass the resistor in the operation band or higher frequency range, and therefore the RF performances in the operation band are not affected by the parallel RC stabilizing circuit. In this work, the values of 0.5 pF and 20 Ω were selected for the C and R, respectively. The input return loss in out-of-band for the amplifier without the RC stabilizing circuit moved toward the high resistive and capacitive region by R′ and C′. Equations of those can be expressed as follows.

$$R' = \frac{R}{1 + (\omega RC)^2}, \tag{2}$$

$$C' = \{ \frac{1}{\omega^2 RC} + RC \} \cdot \frac{1}{R}, \quad (3)$$

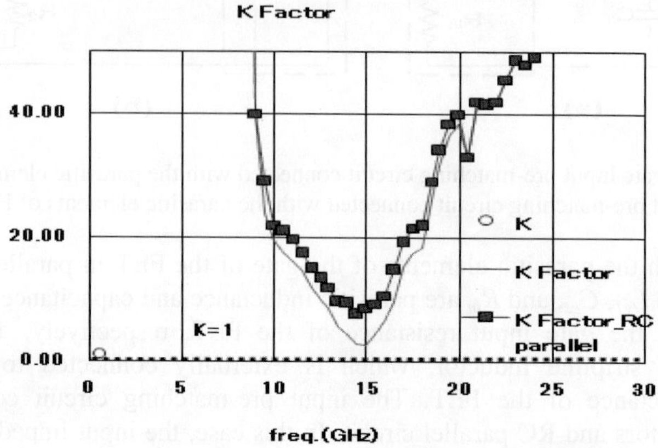

Fig. 3. Measured K factors of the amplifier MMIC

The measured K factors of the amplifier MMICs are shown in Fig. 3, where the open circle line correspond to the results of the amplifier MMIC with the parallel RC stabilizing circuit, and the thick solid line correspond to those without the parallel RC stabilizing circuit. As shown in this figure, the stability in the frequency range from 14 to 15 GHz was highly improved by the parallel RC circuit, and therefore, the K factor of the amplifier MMIC exhibits higher values than 8 in all frequency.

3.3 RF Performances

The photograph of the DA is shown in Fig. 4. The DA exhibited very small chip size of 1.7X0.8 μm^2. The measured and simulated gain, and measured return loss of the DA MMIC are shown in Fig. 5(a). It shows 20±2 dB between 17 GHz and 28 GHz.

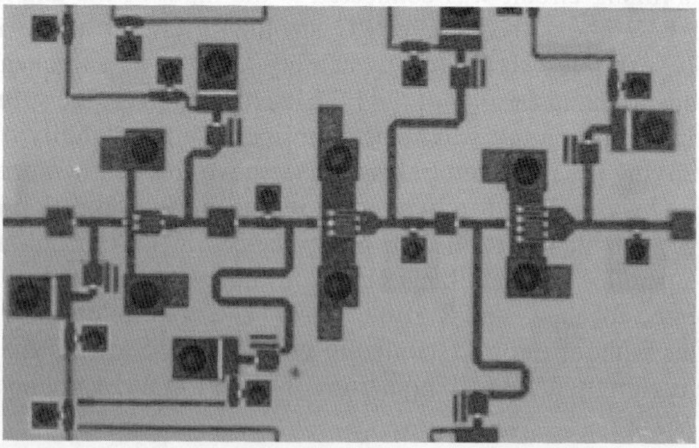

Fig. 4. A photograph of the DA MMIC

The input and output return losses show lower values than –6 dB in the frequency range. The measured P_{1dB} of the DA MMIC exhibited the value of 15±2.5 dBm between 17 GHz and 28 GHz. Measured P_{out}-P_{in} characteristics of the DA MMIC at 25 GHz is shown in Fig.5(b). The measured P_{1dB} at 25 GHz is 17 dBm.

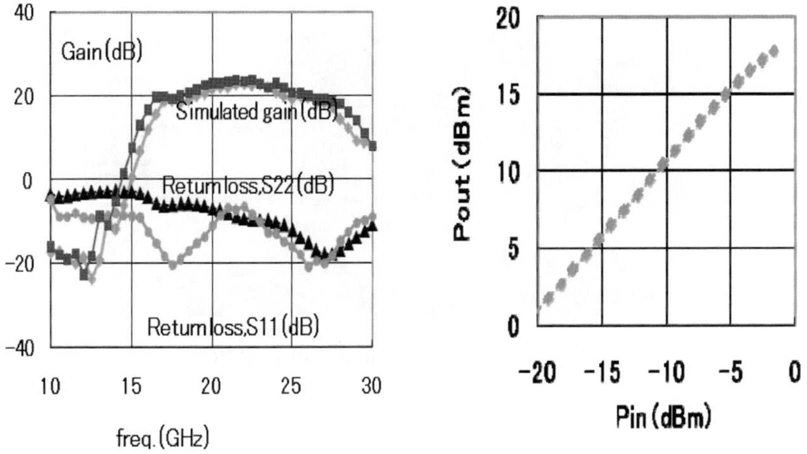

(a) (b)

Fig. 5. (a) Measured and simulated gain of DA MMIC (b)Measured Pout-Pin characteristic of DA MMIC

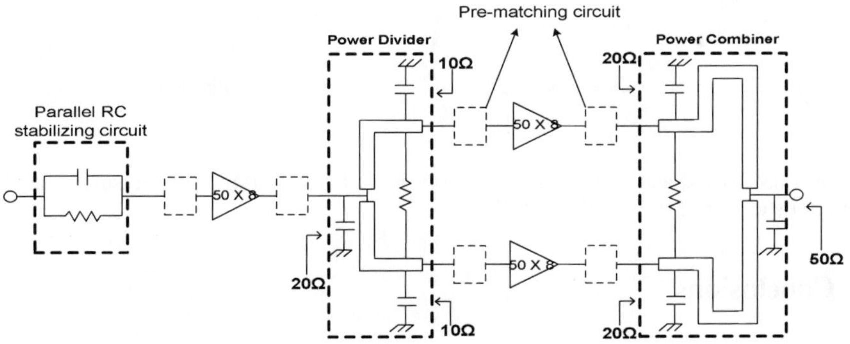

Fig. 6. A schematic circuit of the MPA MMIC

Figure 6 shows a schematic circuit of the medium power amplifier (MPA) MMIC, A Photograph of the MPA MMIC is shown in Fig. 7. The MPA MMIC exhibited 2.0X1.0μm^2. The measured and simulated gain, and measured return loss of the MPA MMIC are shown in Fig. 8(a). The solid line corresponds to the measured, and, open circles and solid triangles correspond to the measured input and output return loss of the MPA MMIC, respectively. The measured gain of the MPA MMIC also exhibits a good flatness in a wide frequency range. It shows 13±2 dB between 19.5 GHz and 32 GHz. The measured P_{out} of the MPA MMIC exhibited the value of 20±2 dBm in the frequency range. Measured P_{out}-P_{in} characteristic of the MPA MMIC at 25 GHz is shown in Fig.8(b). The measured P_{out} at 25 GHz is 21.5 dBm. The MMICs in this work show small chip size and broadband characteristics.

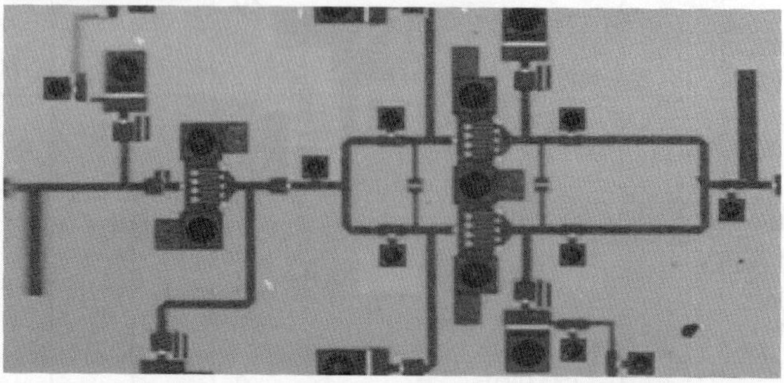

Fig. 7. A photograph of the MPA MMIC

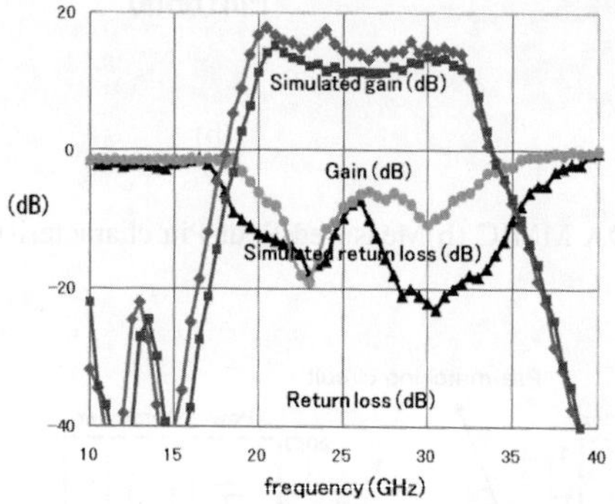

(a)

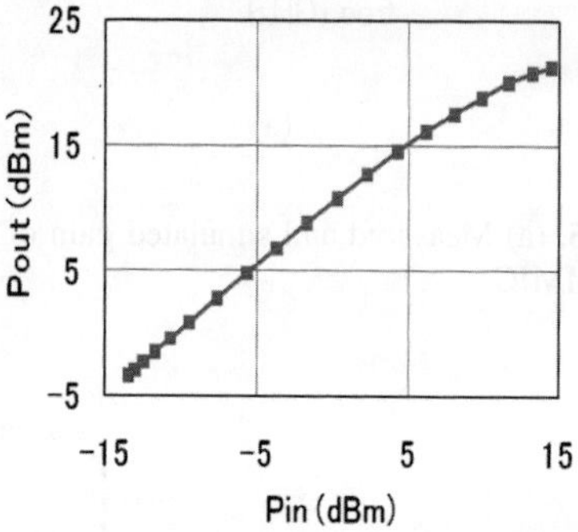

(b)

Fig. 8. (a)Measured and simulated gain, return loss of MPA MMIC (b) Measured Pout-Pin characteristic of the MPA MMIC at 25GHz

4 Conclusions

In this work, using miniaturized simple pre-matching and stabilizing circuits, miniaturized high stable broadband driver amplifier (DA) and medium power amplifier (MPA) MMICs including all the matching and biasing components were developed for K/Ka band applications. The fabricated DA and MPA MMICs showed good RF performances in a wide frequency range, and a good stability from DC to operation band. The DA and MPA MMICs showed very small layout size of 1.7X0.8 and 2X1 μm^2, respectively.

Acknowledgments

This research was supported by the MKE(the Ministry of Knowledge Economy), Korea, under the ITRC(Information Technology Research Center) support program supervised by the NIPA(National IT Industry Promotion Agency)(NIPA-2011-C1090-1121-0015. This work was financially supported by the Ministry of Knowledge Economy (MKE) and the Korea Industrial Technology Foundation (KOTEF) through the Human Resource Training Project for Strategic Technology.

References

1. Siddiqui, M.K., Sharma, A.K., Gallejo, L.G., Lai, R.: A High-Power and High-Efficiency Monolithic Power Amplifier at 28 GHz for LMDS application. IEEE Trans. Microwave Theory Tech. 46, 2226–2232 (1981); Smith, T.F., Waterman, M.S.: Identification of Common Molecular Subsequences. J. Mol. Biol. 147, 195–197 (1981)
2. Satoh, T., Berutto, A.B., Poledrelli, C., Khandavalli, C., Nikaido, J.,, S.: A 68% P.A.E. Power pHEMT for K-Band Satellite Communication System. In: IEEE MTT-S Int. Microwave Symp. Dig., pp. 963–963 (1999); Foster, I., Kesselman, C.: The Grid: Blueprint for a New Computing Infrastructure. Morgan Kaufmann, San Francisco (1999)
3. Ingram, D.L., Stones, D.I., Huang, T.W., Nishimoto, M., Wang, H., Siddiqui, M., Tamura, D., Elliott, J., Lai, R., Biedenbender, M., Yen, H.C., Allen, B.: A 6 Watt Ka-Band MMIC Power Module Using MMIC Power Amplifier. In: IEEE MTT-S Int. Microwave Symp. Dig., pp. 1183–1186 (1997); Foster, I., Kesselman, C., Nick, J., Tuecke, S.: The Physiology of the Grid: an Open Grid Services Architecture for Distributed Systems Integration. Technical report, Global Grid Forum (2002)
4. Matinpour, B., Lal, N., Laskar, J., Leoni, R.E., Whelan, C.S.: K-Band Receiver Front-Ends in a GaAs Metamorphic HEMT Process. IEEE Trans. Microwave Theory Tech. 49, 2459–2463 (2001)
5. Itoh, Y., Takagi, T., Masuno, H., Kohno, M., Hashimoto, T.: Wideband High Power Amplifier Design Using Novel Band-Pass Filters with FET's Parasitic Reactances. IEICE Trans. Electron. E76-C(6), 938–943 (1993)
6. Mochizuki, M., Itoh, Y., Takagi, M.N.T., Mitsui, Y.: Wideband Monolithic Lossy Match Power Amplifier Having an LPF/HPF-Combined Interstage Network. IEICE Trans. Electron. E78-C(9), 1252–1254 (1995)
7. Bahl, I., Bhartia, P.: Microwave Solid State Circuit Design, pp. 509–514. John Wiley & Sons, Inc., Chichester (1988)

High Attenuation Tunable Microwave Band-Stop Filters Using Ferromagnetic Resonance

Baolin Zhao, Xiaojun Liu, Zetao He, and Yu Shi

State Key Laboratory of Electronic Thin Films and integrated Devices,
University of Electronic Science and Technology of China,
Chengdu 610054, China
zblflash@sina.com, liuxiaojun860305@163.com,
marco19801207@126.com, shiyu_aaa@163.com

Abstract. We present some results on Fe-Co-B based microwave band-stop filters. These structures, prepared on GaAs substrates, are compatible in size and growth process with on-chip high-frequency electronics. The absorption notch in transmission can be tuned to various frequencies by varying an external applied magnetic field. For our devices, which incorporate Fe-Co-B as the ferromagnetic material, the resultant FMR frequencies range from 8-16 GHz for applied fields up to only 900 Oe. Comparatively, the frequency range of those devices using permalloy and Fe is substantially lower than Fe-Co-B-based devices for applied the same fields. We constructed devices using monocrystalline Fe-Co-B films grown in a sputtering system. Our devices are of different construction than other dielectric microstrips and show much improvement in terms of notch width and depth. The maximum attenuations of 5 dB/cm and 50dB/cm are observed respectively in two different structures.

Keywords: band-stop filters, Fe-Co-B, ferromagnetic resonance frequency.

1 Introduction

Microwave filters are widely used in both military and civilian communications occasion. It is popular to integrate different electronic components into circuits, but an obvious obstacle to the increased use of microwave technology is the lack of advances in magnetic structures at high frequencies. Tunable filters based on the ferromagnetic dielectric YIG are a well-established technology with many practical applications [1]. The growth of epitaxial Fe film on GaAs was first demonstrated by Prinz and Krebs using molecular beam epitaxy [2]. Deposition of such films by ion-beam sputtering was first reported by Tustison, et al. [3], and their application to microwave filters has previously been discussed by Schloemann, et al [4]. These materials, however, have one significant drawback; their saturation magnetization Ms is low and therefore their operating frequency suits for moderate applied fields. The Fe-Co-B thin film has a much higher Ms than YIG and Fe films. This paper reports the use of eptaxial Fe-Co-B films on (100) GaAs in tunable band-stop filters. An electronically tunable microwave band-stop filter has been realized in Fig.1. The microstrip runs along either a [100] or [110] direction, the “easy” and “hard” directions of magnetization of

M. Zhu (Ed.): Electrical Engineering and Control, LNEE 98, pp. 887–892.
springerlink.com © Springer-Verlag Berlin Heidelberg 2011

the Fe-Co-B film. Theoretical analysis based on the parallel-plate approximation indicates that peak attenuation should occur at the ferromagnetic resonance (FMR) frequency and be proportional to the length of the microstrip line.

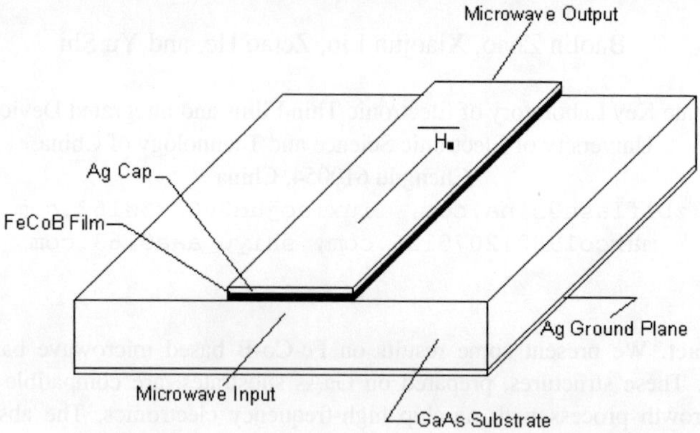

Fig. 1. Architecture of Fe-Co-B based microwave notch filter

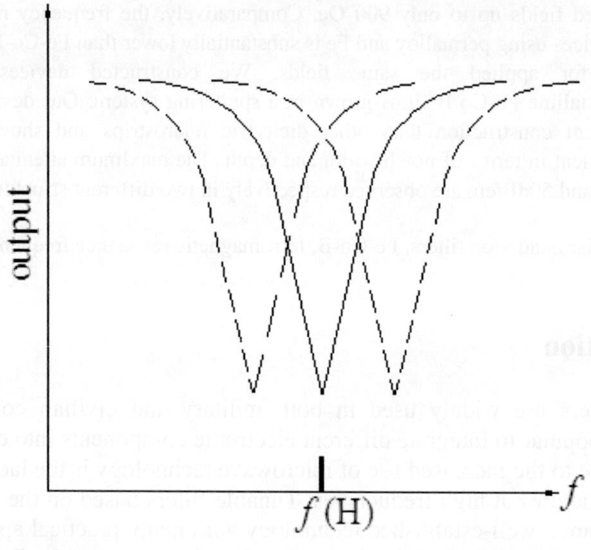

Fig. 2. Electronic tuning of the peak absorption frequency of microwave notch filter

The operational frequency f can be obtained from the ferromagnetic resonance condition and is set by material properties, such as saturation magnetization M_S, anisotropy fields Ha, the gyromagnetic ratio γ, and the magnitude of an applied field H. The FMR frequency f for microstrip lying along the easy and hard direction of magnetization can be shown to occur respectively at

$$f=\gamma[(H+H_a)(H+H_a+4\pi M_s)]^{1/2} \quad (1)$$

$$f=\gamma[(H-H_a)(H+H_a/2+4\pi M_s)]^{1/2} \quad (2)$$

Therefore the resonance can be varied with an external magnetic field. Numerically, γ-2.8 MHz/Oe, Ha~37 Oe, and $4\pi M_s$~24000 Oe. Note that larger M_S Values, such as those in Fe-Co-B and Fe substantially increase the resonance frequency. Thus, at an external magnetic field (H) of 900 Oe, *f*=16 GHz. Fig. 2 depicts the corresponding tuning of the frequency for peak absorption of the microwave power as the magnetic field is varied. Such tuning of the peak absorption has also been analyzed using theoretical models similar to those reported in Refs [6].

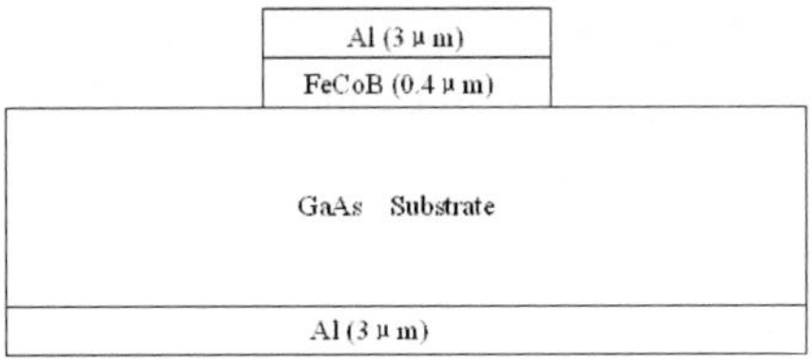

(a)

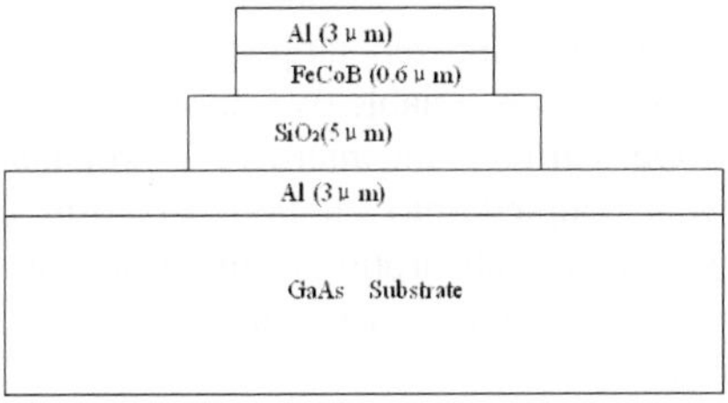

(b)

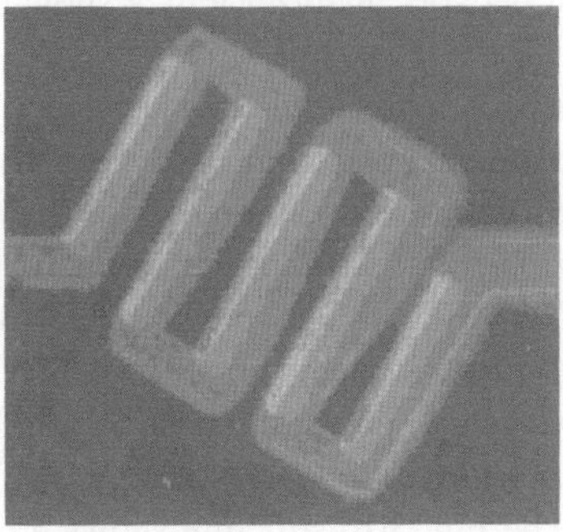

(c)

Fig. 3. (a) The structures of low attenuation, (b) The structure of high attenuation, (c) Meandering line

2 Theory and Experimental Work

We studied two types of transmission line structure. The lines were designed for a nominal characteristic impedance of 50Ω. For the structure of low attenuation, shown in Fig. 3(a), the microstrip is aligned with direction of the Fe-Co-B film and the thickness is 0.6 μm. In order to achieve large peak attenuations, a meandering line is used in which the FeCoB film is present only in the long sections of the meander pattern as seen in the shaded regions of Fig. 3(c). In addition, the Al layer is deposited on the Fe-Co-B layer during the same pump down as protection against oxidation. As shown in Fig. 3(b), we show a cross section of our device. Using sputtering, we deposit the following sequence of materials on GaAs substrates. We begin by growing the bottom electrode: a 2 nm seed layer of Fe, a 3 - μm - thick Al layer, and a Ti layer for adhesion. The rest of the structure is grown through a shadow mask, starting with a 5 μm thick SiO2 film, a thin Ti film for adhesion, Fe-Co-B film (0.3 μm, 0.6 μm, 1 μm thick), followed by a thick Al film (3 μm). The device is patterned by photolithography and dry etched.

3 Results and Discussion

In a previous report [7], it demonstrated a band-stop filter implemented with Fe in a microstrip structure. This device showed large power attenuation at the band-stop frequency. The filter frequency was tunable over a range from 8-16 GHz by means of applied fields up to 900 Oe. An incident microwave propagating along a 50Ω GaAs based microstrip line was coupled into an Fe-Co-B film to excite ferromagnetic resonance, which in turn results in absorption of the propagating microwave. Here we shall present the experimental results of two structure of transmission line and diffrent thickness of Fe-Co-B film.

The performances of our devices were determined by a vector network analyzer. We characterized the microstrip transmission lines at frequencies from 1 to 40 GHz using an automated vector network analyzer. Fig. 4 show the performance observed in experiment filters in which the Fig. 3(a) structure was used. The magnetic field is applied paralled to the long sections of the meander line. The ferromagnetic resonance (FMR) frequency can be tuned over from 8-16 GHz by a magnetic field between 10Oe to 900Oe. For the optimum thickness of the Fe-Co-B film at 0.6 μm, the observed peak attenuation is about 5dB/cm with a background attenuation at 2.3GHz away from resonance of 0.6 dB/cm. The bandwidth at 3dB/cm is 1.3GHz. Fig. 5 show the performance observed in experiment filters in which the Fig. 3(b) structure was used. This structure showed large power attenuation at the band-stop frequency-over 50 dB/cm. The filter frequency was tunable over a range from 8-16 GHz by means of the same applied fields. The Fig. 3(b) structure shows a better performance than The Fig. 3(a) structure except the power transmitted outside of the band-stop region was low.

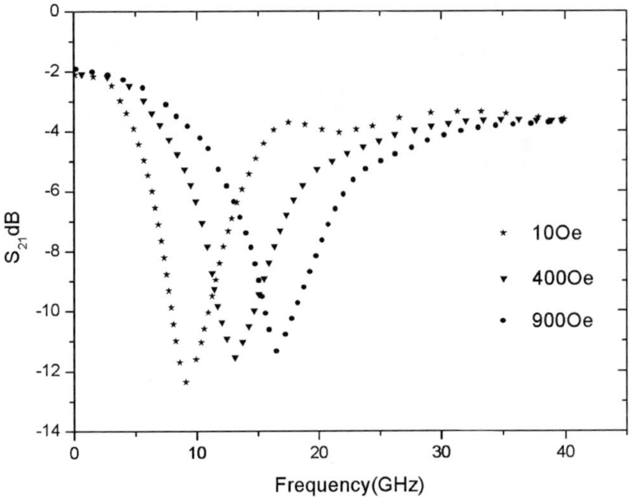

Fig. 4. Result of low attenuation structure versus magnetic field

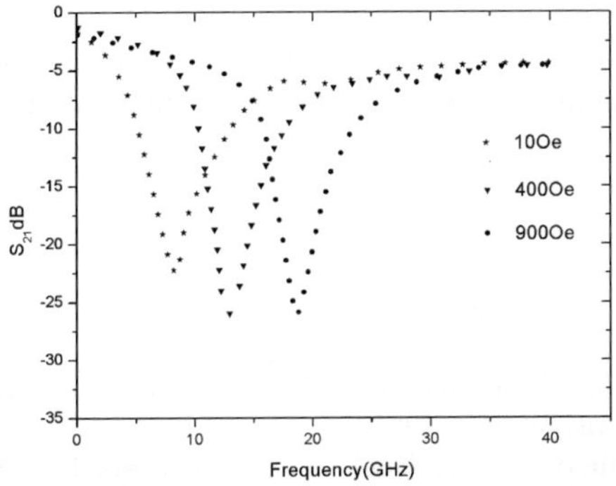

Fig. 5. Result of high attenuation structure versus magnetic field

The ability to alter the shape of the band stop region is crucial in designing a filter. We studied the effect of dielectric thickness, ferromagnet thickness, and external field. We found that the shape of the band stop region can be adjusted to a great extent by varying the thickness of the SiO_2 layer. As shown in Fig. 6, a dielectric thickness of 1 μm provides five times attenuation in band stop region than a thickness of 80 μm, holding all other parameters constant. A band stop filter would have a lower attenuation with a thicker dielectric layer. The skin depth of iron is on the order of 1 μm; therefore, increasing the ferromagnet layer thickness beyond 1 μm has little effect on attenuation. The applied magnetic field has only a very slight effect on minimum attenuation; as it was shown in Fig.c5, it's main influnce is to shift the center of the band stop region in frequency.

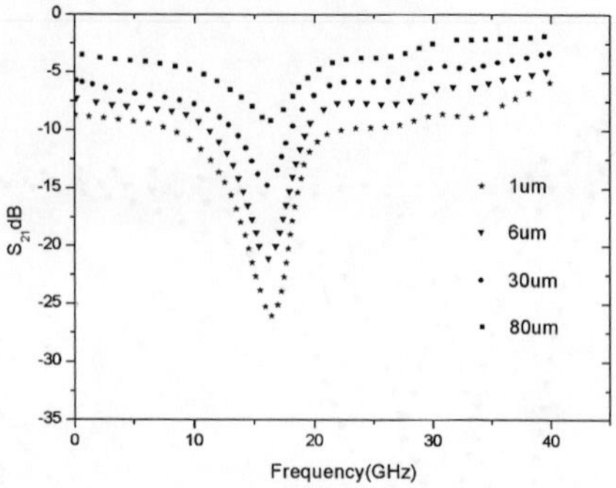

Fig. 6. Effect of dielectric thickness on attenuation.

4 Conclusions

A wideband tunable bandstop filter has been constructed using Fe-Co-B film grown on GaAs substrates. In comparison to the traditional ferrite film, such bandstop filter based on FeCoB film provides a more widely tunable range by the same field. The shape of the band stop region can be modified by adjustment of the physical parameters. These results demonstrate that use of ferromagnetic metallic films in transmission-line structures may have practical application in future microwave devices.

References

1. Ishak, W.S.: Proc. IEEE 76, 171 (1988)
2. Prinz, G.A., Krebs, J.J.: Molecular Beam Epitaxial Growth of Sing-Crystal Fe Films on GaAs. Appl. Phys. Letts. 39, 397 (1981)
3. Tustison, R.W., Varitimos, T., Van Hook, H.J., Schloemann, E.: Epitaxial Fe films on (100) GaAs substrates by ion-beam sputtering. Appl. Phys. Letts. 51, 285 (1987)
4. Schloemann, E., Tustison, R., Weissman, J., Van Hook, H.J.: Epitaxial Fe films on GaAs for hybrid semiconductor-magnetic devices. J. Appl. Phys. 63, 3140 (1988)
5. Prinz, G.A., Krebs, J.J.: Molecular beam epitaxial growth of single-crystal Fe films on GaAs. Appl. Phys. Lett. 39, 397–399 (1981)
6. Liau, V.S., Wong, T., Stacey, W., Aji, S., Schloemann, E.: Tunable Band-stop Filter Based on Epitaxial Fe Film on GaAs. In: IEEE MTT Symp. Digest., pp. 957–960 (1991)
7. Cramer, N., Luic, D., Camley, R.E., Celinski, Z.: High attenuation tunable microwave notch filters utilizing ferromagnetic resonance. J. Appl. Phys. 87, 6911–6913 (2000)

Constraint Satisfaction Solution for Target Segmentation under Complex Background

Liqin Fu[1], Changjiang Wang[1], and Yongmei Zhang[2]

[1] National Key Laboratory for Electronic Measurement Technology, North University of China, Taiyuan, China
Liqin_fu@126.com, Changj.w@126.com
[2] School of Information Engineering, North China University of Technology, Beijing, China

Abstract. Many tasks in artificial intelligence can be seen as constraint satisfaction problems (CSP). Aiming at the target segmentation in low contrast image under complex background, Constraint Satisfaction Neural Network (CSNN) is adopteded. This paper discusses the uncertainty in intelligent perception question firstly. Then, segmentation based CSNN is studied. In CSNN, every layer corresponds to a segment. Neurons in the layers hold the probability that the pixel belongs to the segment represented by corresponding layer. Each neuron synapses to eight neighborhood neurons in the same layer as well as all the other layers. Synaptic weights represent evidences or counter-evidences for segment decisions. Finally, an example was provided to verify the effectiveness of the algorithm .

Keywords: Image segmentation, Intelligent perception, Constraint Satisfaction Neural Network (CSNN), uncertainty.

1 Introduction

Image segmentation, which can be considered as a pixel labeling process in the sense that all pixels that belong to the same homogeneous region are assigned the same label, is an essential process for most subsequent image analysis tasks. The target segmentation in low contrast image under complex background has been the subject of intense investigation for no more than two decades. The imaging guidance technology has been widely applied in modern warfare, but the maximum detection range is critical. The basic problem inherent to extent the detection range is the detection of small, low observable targets in images and in some sense is to find an efficient method for increasing operating range. The difficulties in low contrast image segmentation lie in the follows: the distance is far, the area is small and the object often submerges in the background. Therefore, comparing with other target segmentations, how the targets can be robustly detected under complex background has become a more realistic and challenging research.

Many techniques have been proposed to deal with the image segmentation problem [1]. Histogram-Based Techniques work well only under very strict conditions, such as small noise variance or few and nearly equal size regions. Another problem is the

M. Zhu (Ed.): Electrical Engineering and Control, LNEE 98, pp. 893–899.
springerlink.com © Springer-Verlag Berlin Heidelberg 2011

determination of the number of classes, which is usually assumed to be known. Edge-Based Techniques presents serious difficulties in producing connected, one-pixel wide contours/surfaces. The heart of Region-Based Techniques is the region homogeneity test, usually formulated as a hypothesis testing problem. Markov Random Field-Based Techniques require fairly accurate knowledge of the prior true image distribution and most of them are quite computationally expensive.

In this paper, we present a technique based on Constraint Satisfaction Neural Network (CSNN) for target segmentation under Complex background.

2 Constraint Satisfaction Neural Network Segmentation

2.1 Constraint Satisfaction in Intelligent Perception

Many tasks in artificial intelligence can be seen as constraint satisfaction problems. The task specification can be formulated to consist of a set of variables, each of which must be instantiated in a particular domain, and a set of constraints (predicates) that the values of the variables must simultaneously satisfy. The solution of CSP is a set of value assignments such that all the constraints are satisfied.

Previous methods to resolve uncertainties have included model-based systems that predict the existence of objects, evidential combination schemes, such as Dempster-Shafer theory, and domain dependence techniques that simplify the problem by reducing the range of possible inputs. The model-based descriptions can also take the form of constraints. The system can then manipulate these to try and match objects in the image with the various constraints. The SIGMA image understanding system is a model-based system that also uses contextual clues to focus its attention. Once it develops an understanding, a pass is made to resolve any conflicts. PSEIKI is a vision system developed for robot self-location that uses the Dempster-Shafer formalism to resolve uncertainties. SPAM is another vision system that uses domain dependence as well as confidence values to aid in object labeling.

Constraint Satisfaction Problem is a problem composed of a set of variables and a set of constraints. Values are assigned to the variables such that all the constraints are satisfied. The notion of constraint is basic in operations research. A constraint describes what are the potentially acceptable decisions (the solutions to a problem) and what are the absolutely unacceptable ones: it is an all-or-nothing matter. Moreover, no constraint can be violated, i.e., a constraint is classically considered as imperative. Especially the violation of a constraint cannot be compensated by the satisfaction of another one. If a solution violates a single constraint, it is regarded as unfeasible.

2.2 Constraint Satisfaction Neural Network (CSNN)

Generally a neural network consists of many interconnected parallel processing elements called neural units. These units compute from local information stored and transmitted via connections. In general, a unit consists of two parts: a linear summator and a nonlinear activation function which are serialized (see Fig. 1). The summator of unit receives all activations from connected units, and sums the received activations, weighted with corresponding connection weights together with a bias. The output of summator is the net input of unit. This net input is passed through an activation function resulting in the activation of unit. The summator and the activation function are respectively defined as follows:

$$N_i = \sum_{j=1}^{n} \left(W_{ij} \times A_j\right) + B_i \tag{1}$$

$$A_i = f\left(N_i\right) \tag{2}$$

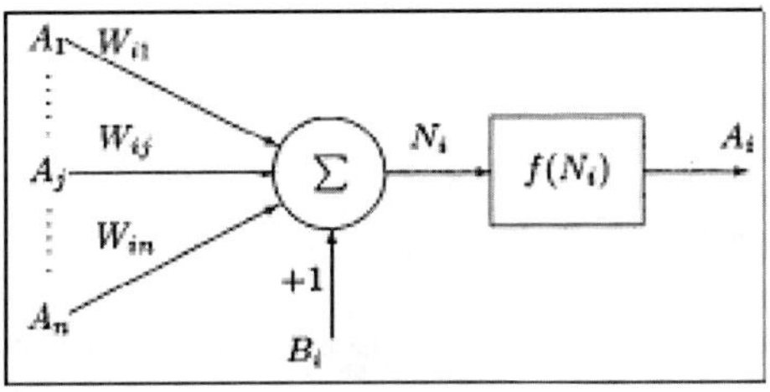

Fig. 1. General neural unit mode

where $w_{i,j}$ is the connection weight from unit j to unit i. Different unit functions are realized by the use of several types of activation function, such as linear threshold function, linear-segmented function and S-shaped function.

CSNN is a circuit network aiming to maximize the activation of its nodes in relation to the constraints existing among them. The constraints are built into the weight connections and biases of the network. An auto-associative backpropagation (auto-BP) learning scheme is initially used to determine the weights and biases of the CSNN. The auto-BP network has an input (I) and output (0) layer with equal number of nodes (N+l). During the training phase, the auto-BP is trained to map any given pattern to itself using the backpropagation learning algorithm. Then, the auto-associative weights and biases are modified slightly to serve as the constraints of the CSNN. The weight modifications satisfy three important structural characteristics of the CSNN:

(i) there is **no** hierarchy among the nodes

(ii) the bidirectional weights are symmetric (i.e. $w_{ij} = w_{ji}$)

(iii) there are **no** reflexive weights (w_{ii} =0).

2.3 Segmentation Based CSNN

The implemented via CSNN can be interpreted as a fuzzy region growing instrumented by means of a neural network that is forced to satisfy certain constraints [2,3]. The neural network structure used in all the resolution levels is depicted in Fig.2

[4]. The fuzzy measure attached to the pixels is their probability to belong to a certain segment. The spatial constraints imposed can be in the form of edge information, neighborhood constraints or multi resolution inheritance constraints. The label probabilities grow or diminish in a winner-take-all style as a result of contention between segments. The global satisfaction of the label assignment in the image domain give rise to the final segmentation.

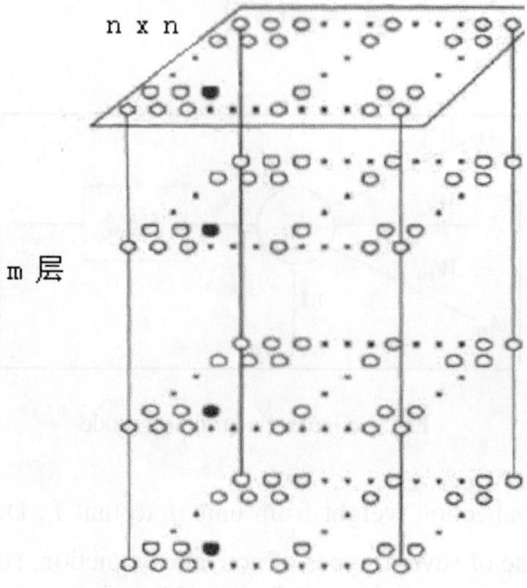

Fig. 2. Constraint satisfaction neural network topology

More specifically every layer in this topology corresponds to a segment. Neurons in the layers hold the probability that the pixel belongs to the segment represented by corresponding layer. Each neuron synapses to eight neighborhood neurons in the same layer as well as all the other layers. Synaptic weights represent evidences or counter-evidences for segment decisions. They are updated in such a manner that a neuron excites those neurons that represent the same label, and inhibits the ones that represent the significantly different labels. The synaptic weights must be determined to guarantee the convergence of the network.

3 Example

3.1 Problem Description

The proposed algorithm is applied to a remote-control toy tank image in grass land. The toy tank is photographed using camera with focus 35mm, as shown in Fig.3. Because the pattern of the toy tank looks like that of grass land, the contrast is very low.

Fig. 3. The raw image of toy tank

3.2 Results

Finding moving objects in image sequences is one of the most important tasks in computer vision. Background subtraction is an old technique for finding moving objects in a video sequence, for example, cars driving on a freeway. The idea is that subtracting the current image from a time-averaged background image will leave only nonstationary objects. It fails with slow-moving objects and doses not distinguish shadows from moving objects.

The results of segmentation are shown in Fig.4. Fig.4(a) is the background image obtained by estimation. Fig.4(b) is the result image after background subtraction. Fig.4(c) is the result by Otsu's method. Fig.4(d) is the result of our method.

(a) The background image

Fig. 4. Segmentation results

(b) The background subtraction result

(c) Otsu segmentation result

(d) CSNN segmentation result

Fig. 4. (*continued*)

4 Conclusion

Intelligent information we obtained is always incomplete and polluted by noise. It brings uncertainties to object detection or recognition problem. In this paper, we have presented a view of intelligent perception using CSNN. In particular, we have discussed the method of CSNN based image segmentation. A remote-control toy tank image in grass land is used to verify the effectiveness of the algorithm. The study reaffirmed the potential of using the CSNN as an effective method in image segmentation.

References

1. Kim, J., Fisher, J.W., et al.: A nonparametric statistical method for image segmentation using information theory and curve evolution. IEEE Transactions on Image Processing 14(10), 1486–1502 (2005)
2. Lin, W.C., Tsao, E.C., Chen, C.T.: Constraint Satisfaction Neural Networks for Image Segmentation. Pattern Recognition 25, 679–693 (1992)
3. Kurugollil, F., Birecik, S., Sezgin, M., Sankur, B.: Image Segmentation Based on Boundary Constraint Neural Network. In: Proceedings IWISP 1996, pp. 353–256 (1996)
4. Adorf, H.M., Johnston, M.D.: A discrete stochastic neural network algorithm for constraint satisfaction problems. In: Proceedings International Joint Conference on Neural Networks, vol. 1(3), pp. 917–924 (1990)

The Integrated Test System of OLED Optical Performance

Yu-jie Zhang, Wenlong Zhang, and Yuanyuan Zhang

College of Electrical and Information Engineering
Shaanxi University of Science and Technology
Xi'an, China
zyy19870215@163.com

Abstract. At present, manual measurement of discrete devices is used to measure optical properties of Organic light-emitting diode (OLED),For this situation, the paper presents an OLED optical performance Integrated Test System which can simultaneously measure optical and electrical property of the light-emitting device and display real-time measurement curve, control measurement environment on a single platform, system has realized automatic optical properties measurement of OLED, has the characteristics of strong comprehensive, easy measurement, high precision and so on.

Keywords: OLED, optical performance, test system.

1 Introduction

Organic light-emitting diode (Organic Light Emitting Diode, referred to as OLED) has many advantages, it is a new type of flat panel display and two-dimensional source. the current - voltage characteristic, brightness - voltage characteristic, temperature - current characteristic, luminous efficiency - voltage characteristic, color coordinate and the EL spectrum of OLED device are all important indicators of light and electrical properties of the device, the external quantum efficiency (External quantum efficiency) and energy conversion efficiency (Power efficiency) [1]. At present, a DC voltmeter and ammeter is used to measure the current and voltage characteristics of the OLED device more generally. Faint light Photometer is used to measure brightness and temperature measure machine to measure the temperature of device. The measurement data will be inputted into computer for subsequent analysis and processing to obtain the optical properties of the sample [2]. Voltage, current, brightness, temperature and other physical quantities can not be achieved simultaneously, this will result more errors and lower efficiency, amount of data collected and collecting density are greatly restricted, as a result, the optical properties of sample are unable to be accurately reflected [3] On these questions, an integrated Test System of OLED optical performance is designed, microcontroller STM32F103RB acts as the core of system. System combines a precise constant (constant current) programmable power supply, voltage and current measurement, temperature measurement, light measurement and device protection together organically. Touch screen can be set to achieve test conditions,

M. Zhu (Ed.): Electrical Engineering and Control, LNEE 98, pp. 901–907.
springerlink.com © Springer-Verlag Berlin Heidelberg 2011

graphical display and other functions, so system realizes the purpose of high accuracy and fast automatic measurement of optical properties and attenuation characteristics of OLED devices. The system will provide more convenient and reliable test conditions for OLED device's Optical properties.

2 Systems Architecture

Overall system block diagram is shown in Figure 1, The STM32F103RB microcontroller acts as system core, system includes programmable DC power supply by IPD-3303SLU, photoelectric conversion circuit, range switching circuit, temperature measurement circuit, touch screen circuit, SD card circuit; internal data processing software Completes system data processing mechanism to process and display the measurement results.

System works as follows: programmable DC power supply's initial voltage value, the voltage termination value, step value and the sampling period are set by Touch screen. The voltage drives OLED device emit light, the sample light emitted after the incident to the photodiode is converted to Current signal from the I / V converter circuit , then current signal is converted to voltage signal, micro-controller controls range switching circuit to switch feedback resistor of I / V converter to output different voltage signal[5]. output voltage is inputted into the microcontroller's A / D channel After the second Amplifiers; the infrared temperature sensor will sample device's temperature and deliver the signal to the microcontroller's A / D converter channel, microcontroller will convert light and temperature analogy signals into digital data. The data will be analyzed and processed, microcontroller then draw the voltage - brightness, current - temperature, voltage - current curve and display on Touch screen. Measurement data is saved to the SD card, and user can easily query historical measurement data.

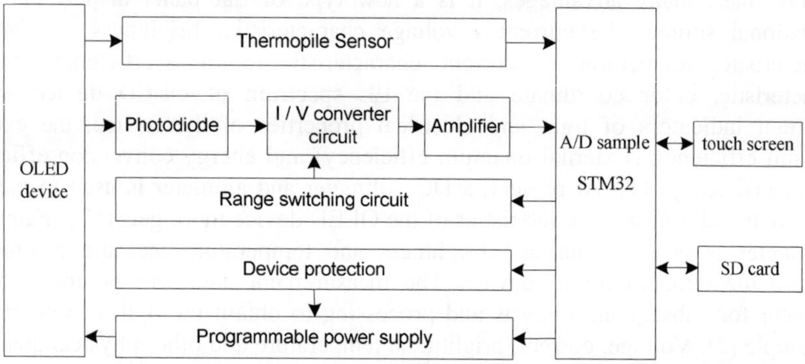

Fig. 1. Block diagram of system

3 System Hardware Design

According to requirements of system, system hardware should include MCU chip selection, brightness measurement, temperature measurement, display part, they are described as follows.

3.1 MCU Chip Selection

Measurement system should be able to quickly and accurately complete the device performance test, also has a high level of integration, stability and reliability. Provide better visual interface, large memory, can support real-time operating system with low power consumption and cost-effective to provide multiple high-precision A / D channel, has a rich communication interface and GPIO. Considering the variety of processor chip including performance, cost, power consumption, development environment support, after sale service, marketing degree, and many other factors, we ultimately choose ST's STM32 family of 32-bit Enhanced processor STM32F103RB as the system microcontroller. It is 30% faster than the same level product ARM7TDMI, the internal integration of I2C, SPI, CAN, UART, USB, providing multi-channel high-precision A / D, and rich GPIO Satisfies the requirement of system measurement, display, storage, rich on chip resources help to reduce system-chip peripherals, improve reliability and cost savings. The MCU has powerful IDE suppliers including Hites , IAR , Keil , Raisonance , Rowley and so on. ST provides a wealth of software support, including the STM32 firmware library, STM32 USB developer's kit, STM32 motor control firmware library, helping users to complete system design in the shortest time, reducing development costs.

3.2 Touch-Screen Circuits

System requires high precision touch screen, touch screen should be able to display real-time measured curve and the Chinese characters. A High-precision 4.3-inch touch screen is used. It can display characters and images and can meet system requirement very well. STM32F103RB's on chip FSMC can drive TFT without external driving chip. STM32F103RB's on chip SPI device can finish TFT'S initialization. FSMC transmit data between TFT and STM32F103RB.MCU's GPIO control the backlight of TFT, simple interface is easy to control .The connection between TFT and STM32F103RB is as Figure2:

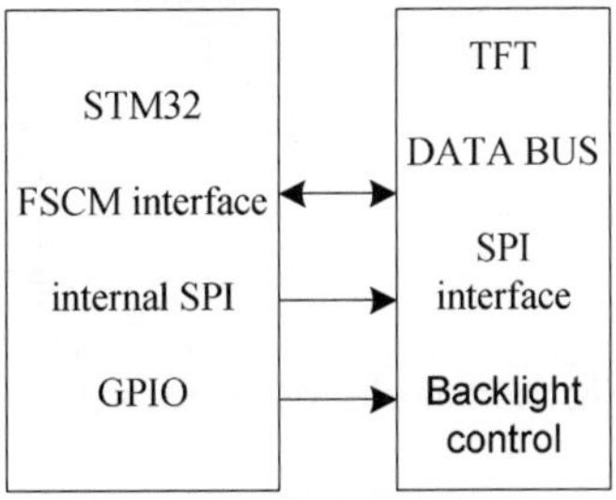

Fig. 2. The connect between TFT and STM32F103RB

3.3 Brightness Measurement

Brightness measurement circuit contains the I / V convert circuit and the secondary Amplifiers circuit , OLED device emitted light incident photodiode GT101, output only 0.002uA ~ 0.2mA current signal, A very low input bias current monolithic electrometer operational amplifiers OP07 is used as / V converter circuit. OP07 has the advantage of low input current and low input offset voltage, therefore it is suitable for very sensitive photodiode amplifiers. As is shown in Figure 3, the R12 can be used to adjust the offset null of amplifier and R2 to compensate bias current of operational amplifier [7].

The Brightness range is 0.1×10-2~200×103cd/m2 in system, analog circuit design can not meet the dynamic range of brightness variations of magnitude 5 transitions, so sub-photometric measurements is needed, by using Multi-scale amplifier can realize the purpose [8]. Brightness range consists of 6 ranges, by switching I / V convert's feedback resistor to switch measure range. 6-way normally closed analog switch MAX312 can be used to switch the feed back resistor. MAX312 has a pretty little on-resistance of 10Ω and 0.35-resistance flatness, does not affect the precision of system, the common port is connected to the output of the amplifier, when input high level ,common port will conduct with normally closed port, gating the feedback resistor connected to it, Otherwise disconnected. Coordinated through the software, only one feedback resistor can be connected to common port.

I/V convert output only 0~200mV voltage signal, if system requires the precision of 0.1, the resolution of each range should be at least 0.1mV, STM32F103RB has integrated a internal 12 bits A/D convert, if refers to internal reference voltage of 3.3V, the resolution is 0.8mV, therefore, the second amplifier needs to amplify signal for at least 8 times. To improve the precision of the system, LF441CN is used as the second amplifier, it amplifies signal for 10 times, signal of 0~200mV will be amplified to 0~2.0V.

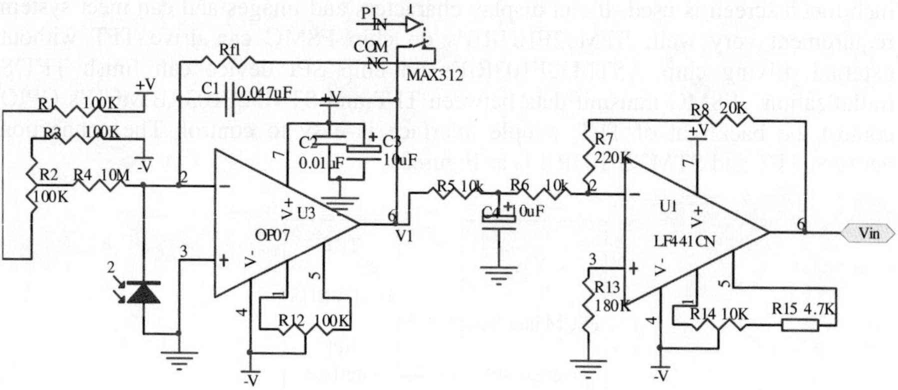

Fig. 3. Photoelectric detection circuit

According to A/D convert result, STM32F103RB will judge the state of amplifier, if the measure range is not appropriate, then STM32F103RB will switch the measure range until appropriate, then sample data.

3.4 Temperature Measurement

OLED device temperature changes with the current changes during measurement. System uses Germany PerkinElmer's non-contact thermopile infrared sensor

A2TPMI334-L5.5OAA300 as temperature sensor, PerkinElmer A2TPMI is a special signal processing within the integrated circuit and temperature compensation circuitry of the multi-purpose infrared thermopile sensors, the target of the thermal radiation into an analog voltage with a sensitivity of 42mV/mw, STM32F103RB's internal 12 bits A/D convert meet the requirement of resolution, so temperature can be measured directly. A2TPMI334-L5.5OAA300 measures the target temperature range of -20 ~ 300 ℃within a response time of 35ms, half-power point response frequency is less than 4Hz, meet the temperature measurement requirements.

4 System Software Design

Fast, real-time, reliable measurement of device characteristics is the basic requirement of the system. Embedded real-time operating system is a perfect selection for system, which can response system events Timely and coordinate all the tasks of system with high stability and reliability, combining with system hardware design, so UC / OS-II RTOS is used in system. UC / OS-II is a preemptive priority-based real-time kernel with readable, reliable and flexible program, which can handle 56 user tasks, task priority can be dynamically adjusted, it also provides semaphores, mailboxes and message queues to communicate between tasks. It can minimize the size of ROM and RAM through the cut of software.

4.1 Software Design Idea

System uses the UC / OS-II embedded operating system, designer just need to design application software on the basis of UC / OS-II, system functions will be conducted by kernel scheduling. The idea of multi-task mechanism is used to design application software, system events are handled by the task execution, and each task is designed as super cycle and waiting for a specific event to execute their own function. Semaphores of the UC / OS-II communication mechanism are used in communication between tasks to coordinate system overall functions, therefore, the quality of task division will directly decide system performance.

4.2 Task Division and Priority Setting

System events include interrupt events and polling events, including touch screen and UART two interrupt events; command analysis, data acquisition, programmable power control, SD card operation, historical data query, drawing six polling events. so screen and UART task should be interrupt level tasks, according to event completion of the events and system data flow direction, polling events are divided into program-controlled power control task, real-time data measurement task and historical data query task, data query task is a collaboration of real-time measurement data acquisition, SD card operations and drawing three events, So the system is divided into command analysis, program-controlled power control, data measurement, historical data query, and device protection 5 tasks; the Relationship between tasks and system is as Figure 4.

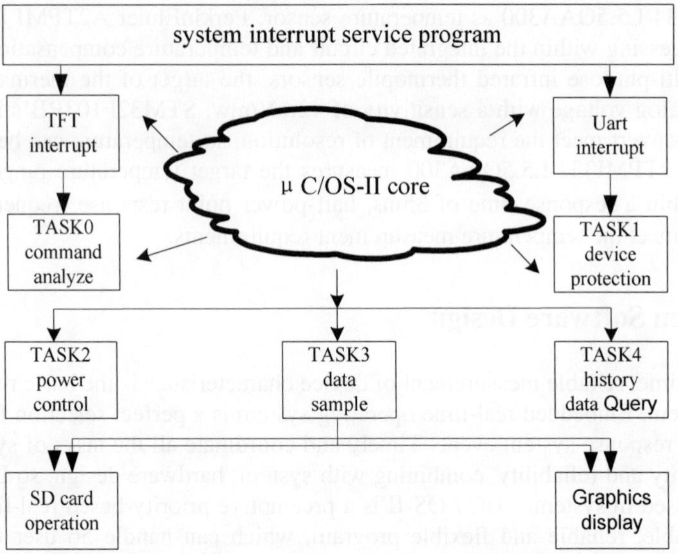

Fig. 4. Relationship between tasks and system

Interrupt-level tasks require the system to respond in the shortest time, they have the most demanding real-time, should be set higher priority, the command analysis task is used more frequently, should be set the highest priority 0, the device protection task is set priority 1. Polling task can set priority according to the sequence of the task execution, in system, the sequence of task execution is programmable power supply control task, real-time measurement task, historical data query task, programmable power supply control task is set priority 2, real-time measurement task priority 3 and Historical data query task priority 4.

4.3 System Functions and Communication between Tasks

Communications should be built between tasks in order to accomplish the overall function of the system. UC / OS-II has semaphores, mailboxes, message queues 3 communication mechanisms and semaphores are used in system design, the producer tasks produce semaphores and the consumer tasks use them. Command analysis task resolute command input through the touch screen, if the command is right, then produce programmable power control or historical data query semaphore , programmable power supply task is activated by semaphore and send power target voltage command to programmable power device , then delay for 50mS, when the voltage is stable enough, data acquisition Semaphore will be produced; data acquisition task is activated by Semaphore and accomplish functions of switching range, data acquisition, data processing, SD card operations and real-time graphics; historical data query task is activated by semaphore ,when the Semaphore arrives, it will read data from SD card and display graphics in touch screen, if the Semaphores does not arrive, tasks waiting for them are blocked, otherwise ,they will be added to the ready task queue and waiting for the kernel scheduler. Tasks of higher priority send Semaphore to lower priority tasks to prevent priority inversion and deadlock the system, improve system stability and reliability.

5 Conclusion

The Integrated Test System of OLED Optical Performance achieves a rapid, accurate, reliable, automated measurement of the optical and electric properties of OLED device and makes great contribution to the improving of work efficiency, plays an important role in study of carrier transport properties, luminescence, luminescence efficiency, Long-term production practice shows that it is pretty value for production process, material evaluation and performance testing of packaged devices.

References

1. Hui, Z.: Study of Integrated Optical performance Test System. Zhejiang University, Hangzhou (2005)
2. Pan, J.G.: LED light color comprehensive performance analysis test theory and equipment. LCD and Display 18(2), 138–140 (2003)
3. CIE publication 127-1997 measurement of LEDs
4. Xu, W.C., Xu, K.X.: Performance analysis of optical test system. Sensors and Actuators 12(4), 679–681 (2007)
5. Dai, Y.s.: Weak signal detection method and equipment, pp. 45–63. National Defence Industry Press, Beijing (1994)
6. Ding, T.: A new OLED brightness uniformity test method. Optical Technology 33(1), 41–43 (2007)
7. Yiu, J.: The Definitive Guide of ARM Cortex-M3. Song Yan Ze Translation. Beijing Aerospace University Press, Beijing (2009)
8. Labrosse, J.J.: Embedded real time operating system μC / OS-II: 2. Shao Bei Bei translation. Beijing Aerospace University Press, Beijing (2003)
9. Ning, Z., Peng, C., Ju, H.P.: The transplant and application of embedded real time operating system μC / OS-II in ARM. Computer Technology and Application (4), 29–31 (2004)

Application of Strong Tracking Kalman Filter in Dead Reckoning of Underwater Vehicle

Ye Li[1,2], Yongjie Pang[1,2], Yanqing Jiang[1,2], and Pengyun Chen[1,2]

[1] State Key Laboratory of Underwater Vehicle,
Harbin Engineering University, Harbin 150001, China
[2] College of Shipbuilding Engineering, Harbin Engineering University,
Harbin 150001, China

Abstract. Based on the dead reckoning of underwater vehicle, an algorithm involving improved strong tracking adaptive Kalman filter is proposed in this paper. That strong tracking adaptive Kalman filter is added with an adaptive factor and estimator of measurement noise covariance to affect the magnitude of fading factor, and it enhances tracking ability of the filter to suit for the lower-speed navigation characteristic of underwater vehicle. The application of this algorithm in dead reckoning of underwater vehicle shows its effectiveness and accuracy.

Keywords: Dead reckoning, Underwater vehicle, Strong tracking Kalman filter, Measurement noise.

1 Introduction

Precision and updating are two significant problems of underwater vehicle navigation. Recent navigation sensors involve inertial navigation system, GPS receiver, Doppler velocity log, gyroscope, acoustic altimeter etc. For the Faraday cage effect of water to radio-frequency signal, GPS signal could not be received in deep seawater. Inertial navigation system can offer excellent strap-down navigation capability, but high-precision INS is expensive and its size is also limited. For long-range navigation of underwater vehicle, recent method is combining GPS signals and results of dead reckoning on the surface and positioning vehicle in deep water with results of dead reckoning based on the sensors of DVL and gyroscope. Consequently, the results of dead reckoning are crucial when underwater vehicle submerges into deep water [1]-[3].

To improve the optimality and convergence of filter, strong tracking Kalman filter adds a fading factor to the equation of calculating time propagation error covariance. This approach effectively enhances tracking ability of filter and restrains divergence of filter. Fading factor adjusts the magnitude of time propagation error covariance to affect filter gains, so fading factor estimation algorithm is the key of strong tracking Kalman filter. But sometimes it can not satisfy precision of filter in practical engineering, especially when fluctuation of system state is small. While the motion

M. Zhu (Ed.): Electrical Engineering and Control, LNEE 98, pp. 909–915.
springerlink.com © Springer-Verlag Berlin Heidelberg 2011

feature of underwater vehicle is low-velocity and low-acceleration. In this condition, the precision of normal strong tracking Kalman filter can not satisfy engineering demand[4]-[5].

After analysis of fading-factor adaptive estimation algorithm, this paper proposes a modified strong tracking Kalman filter which is suitable for the lower-speed navigation characteristic of underwater vehicle. This algorithm adds an adaptive factor and an estimator of measurement noise covariance to improve dynamics performance of filter and enhances its precision. In addition, application of this algorithm in dead reckoning of underwater vehicle shows its effectiveness.

2 Dead Reckoning Model of Underwater Vehicle Based on Strong Tracking Kalman Filter

The status of dead reckoning is crucial in underwater navigation. Its principle is to utilize heading, speed and time at a known position to estimate the next position of the vehicle in the uniform coordinates.

Strategy of filtering is to make the optimal estimation about the results of dead reckoning, and use the strong tracking Kalman filter to estimate real state from noise. Model of dead reckoning is based on one-order Singer Model. For the acceleration of underwater vehicle is small, underwater motion is usually considered as uniform accelerated straight line motion in the factual calculation. So the filter algorithm has to bear this inaccuracy of model.

Underwater vehicle is equipped with acoustic altimeter, so motion model is considered as two dimensions. Exemplified as eastern motion, the sample period h=0.5s, basic system equation is given as:

$$\begin{bmatrix} X_e(k) \\ \dot{X}_e(k) \\ \ddot{X}_e(k) \end{bmatrix} = \begin{bmatrix} 1 & h & \frac{h^2}{2} \\ 0 & 1 & (1-\frac{\alpha h}{2})h \\ 0 & 0 & (1-\alpha h+\frac{\alpha^2 h^2}{2}) \end{bmatrix} \begin{bmatrix} X_e(k-1) \\ \dot{X}_e(k-1) \\ \ddot{X}_e(k-1) \end{bmatrix} + \begin{bmatrix} 0 \\ \frac{\alpha h^2}{2} \\ \alpha(1-\frac{\alpha h}{2})h \end{bmatrix} W'(k) \quad (1)$$

Where $\{W'(k)\}$ is uncorrelated Gaussian white noise sequence, and its mean is zero and covariance is $2\alpha\sigma_A^2/h$. Usually we think the motion of vehicle is uniform accelerated straight line motion and thus α equals 0. Northern motion equation is the same. Thus the system equation is constructed, and system state is $\vec{X}=[x_e \quad \dot{x}_e \quad \ddot{x}_e \quad x_n \quad \dot{x}_n \quad \ddot{x}_n]$.

Measurement equation is:

$$\begin{bmatrix} Y_{xe}(k) \\ Y_{ve}(k) \\ Y_{xn}(k) \\ Y_{ve}(k) \end{bmatrix} = \begin{bmatrix} 1 & 0 & 0 & 0 & 0 & 0 \\ 0 & 1 & 0 & 0 & 0 & 0 \\ 0 & 0 & 0 & 1 & 0 & 0 \\ 0 & 0 & 0 & 0 & 1 & 0 \end{bmatrix} \begin{bmatrix} x_e(k) \\ \dot{x}_e(k) \\ \ddot{x}_e(k) \\ x_n(k) \\ \dot{x}_n(k) \\ \ddot{x}_n(k) \end{bmatrix} + \begin{bmatrix} V_{xe}(k) \\ V_{ve}(k) \\ V_{xn}(k) \\ V_{vn}(k) \end{bmatrix} \quad (2)$$

Where $Y_{xe}(k), Y_{ve}(k), Y_{xn}(k), Y_{vn}(k)$ are eastern position, eastern velocity, northern position and northern velocity.

3 Fading-Factor Adaptive Estimation Algorithm

Consider a linear, discrete time, stochastic multivariable system

$$\begin{aligned} x(k+1) &= \phi(k+1,k)x(k) + \Gamma_k W(k) \\ y(k) &= H(k)x(k) + V(k) \end{aligned} \quad (3)$$

Where $x(k)$ is the $n\times 1$ state vector, $y(k)$ is the $m\times 1$ measurement vector, $\phi(k+1,k)$ and $H(k)$ are state transition matrix and observation matrix respectively.

The sequence of $W(k)$ and $V(k)$ are uncorrelated Gaussian white noise sequence with mean and covariance as follow:

$$\begin{aligned} &E[V(k)] = 0, Cov[V(k), V(j)] = [V(k), V^T(j)] = Q(k)\delta_{kj} \\ &E[W(k)] = 0, Cov[W(k), W(j)] = [W(k), W^T(j)] = R(k)\delta_{kj} \\ &Cov[V(k), W(j)] = E[V(k)W^T(j)] = 0 \end{aligned} \quad (4)$$

Where δ_{kj} denotes $Kronec\ker$ delta function.

The discrete linear strong tracking Kalman filter equation is shown as follow:

$$X(k+1/k) = \Phi(k+1/k)\hat{X}(k) \quad (5)$$

$$\hat{X}(k+1) = \hat{X}(k+1/k) + K(k+1)[Z(k+1) - H(k+1)\hat{X}(k+1/k)] \quad (6)$$

$$K(k+1) = P(k+1/k)H^T(k+1)[H(k+1)P(k+1)H^T(k+1) + R(k+1)]^{-1} \quad (7)$$

$$P(k+1/k) = \lambda(k+1)\Phi(k+1/k)P(k)\Phi^T(k+1/k) + Q(k) \quad (8)$$

$$P(k+1) = [I - K(k+1)H(k+1)]P(k+1/k) \quad (9)$$

Meanwhile, there are three kinds of method to calculate $\lambda(k+1)$, but for the improvement of real-time character of algorithm, the usual choice is a sub-optimal fading factor estimator which is shown as system of equation(5) ~(9) .

$$\lambda(k+1) = \max\{1, tr[N(k+1)]/tr[M(k+1)]\} \quad (10)$$

Where tr is symbol of trace of matrix.

$$\begin{cases} N(k+1) = V_0(k+1) - R(k+1) - H(k+1)Q(k)H^T(k+1) \\ M(k+1) = H(k+1)\phi(k+1,k)P(k/k)\phi^T(k+1,k)H^T(k+1) \end{cases} \quad (11)$$

In equation （11）, $V_0(k+1)=E[\gamma(k+1)\gamma^T(k+1)]$ is covariance of the residuals, but its magnitude is unknown in this adaptive algorithm. It can be estimated by the following method:

$$V_0(k+1)=\begin{cases}\gamma(1)\gamma^T(1), & k=0\\ \dfrac{[\rho V_0(k)+\gamma(k+1)\gamma^T(k+1)]}{1+\rho}, & k\geq 1\end{cases} \tag{12}$$

where $0<\rho\leq 1$ is forgetting factor and usually is 0.95; $\gamma(k+1)$ is new notation which is:

$$\gamma(k+1)=y(k+1)-H(k+1)\hat{X}(k+1/k) \tag{13}$$

From analysis of equations(5) ~(13) and substantial simulation experiments, we know the uncertainty of fading factor depends on two factors: uncertainty of modeling and magnitude of noise estimated value.

(1) Strong tracking Kalman filter has strong robustness about uncertainty of modeling, but for some strong-coupling system moving in uncertain environment, its modeling is so difficult that strong tracking Kalman filter can not suit its uncertainty.

(2) When estimation of system noise and measurement noise covariance is inaccurate and the magnitude of residual $\gamma(k+1)$ cannot counteract the error of noise estimator, the estimation algorithm of fading factor possibly has no effect on the tracking ability. Especially for the measurement noise covariance, the common estimation is based on the error of sensor itself and neglects the influence of environment problem. So algorithm cannot utilize any new information from the measurement and lack the adaptability to the uncertain factor. Its physical meaning performed in the strong tracking filter is that when fluctuation of system state is small, the extent of "activation" about fading factor is not big enough to satisfy the demand of filter, and even fails to stifle the divergence of filter. And this problem also performs in the calculation: the magnitude of $N(k+1)$ is excessively small or even out of the positive definite condition.

Consequently, after analysis in two factors above, the dynamic performance of strong tracking filter can be improved by the estimator of noise and the calculation of $N(k+1)$.

4 Simulation Experiment

The specific approach is given as:

(1) System initialization

(2) Adding related data: velocity data from Doppler Velocity Log, attitude data from fiber-optic gyroscope and the sample period.

(3) Deleting wild point

(4) Dead reckoning in period and filtering its results with the modified strong tracking Kalman filter

The simulation is based on the circle test of minimized underwater vehicle "SY-1" built by the team directed by Harbin Engineering University. That experiment is made in Song Hua Lake. The vehicle is near the free surface and its GPS receiver exposes to the surface to receive the position signal. So for the complicated environment, the tracking filter needs to add the adaptive factor μ .Basic parameter is: system noise covariance is $Q = diag\{0.015,1\times10^{-6},0,0.015,1\times10^{-6},0\}$, initial value of measurement noise is $R(0) = 0.1\times diag\{1,1,1,1\}, \mu = 0.001, U = (0.11)^2, b = 0.96$. Input data came from practical data in circle experiment. All the data are measured in each 0.5 second interval from the sensor.

Simulation 1, Activation extent of fading factor

This simulation is illustrated as eastern velocity.

Figures 1, 2 demonstrate the activation extent of eastern velocity with two methods.

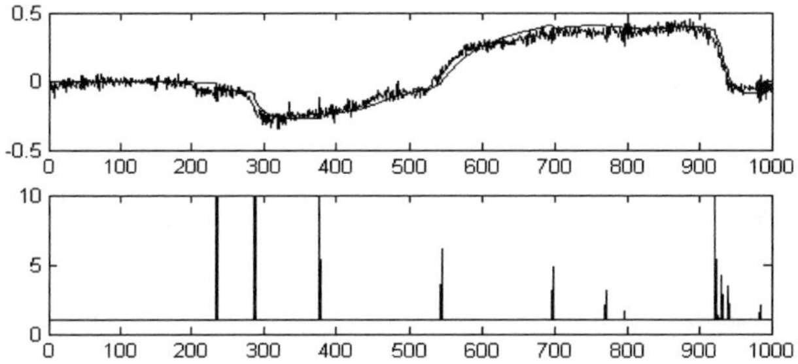

Fig. 1. Result of normal strong tracking filter

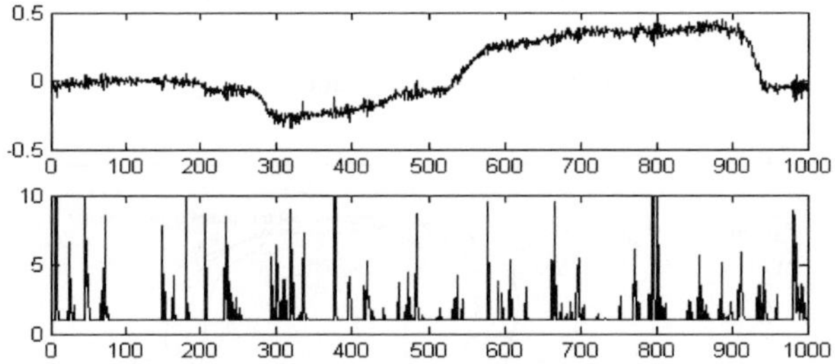

Fig. 2. Result of this paper algorithm

In figures 1, 2, abscissa is time of filtering, ordinate is value of eastern velocity (scale is knot), value of fading factor sequentially.

From above, compared with normal strong tracking Kalman filter in the condition of small fluctuation of system state, algorithm of this paper obviously "activates" the fading factor, and improves the dynamics performance of filter.

Simulation 2, compare normal strong tracking filter and this paper algorithm

This simulation is illustrated as longitude and latitude. Figures 3-6 demonstrate the compare of the results from GPS, normal strong tracking filter and modified strong tracking filter. And Figure 8 shows the absolute error of dead reckoning with two methods.

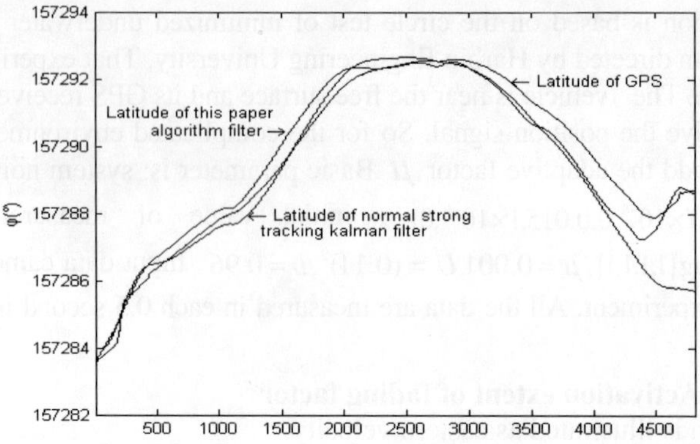

Fig. 3. Compare latitude filtering

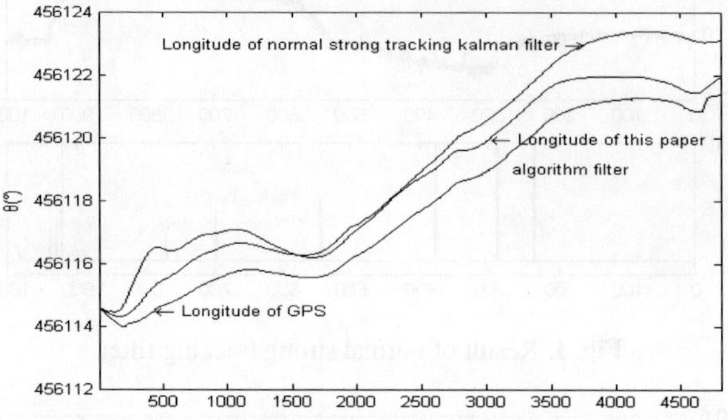

Fig. 4. Compare longitude filtering

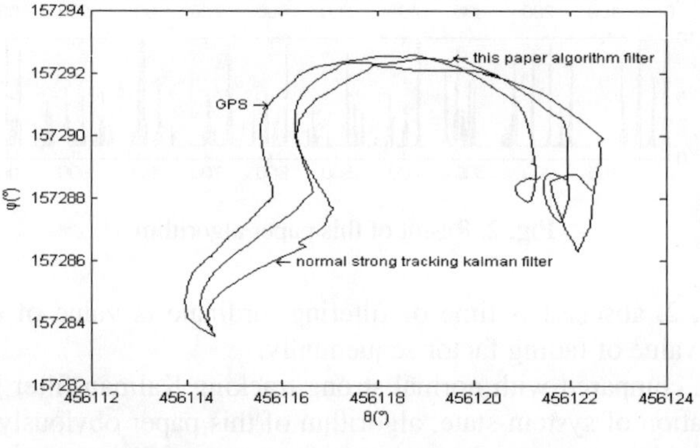

Fig. 5. Compare track of dead reckoning

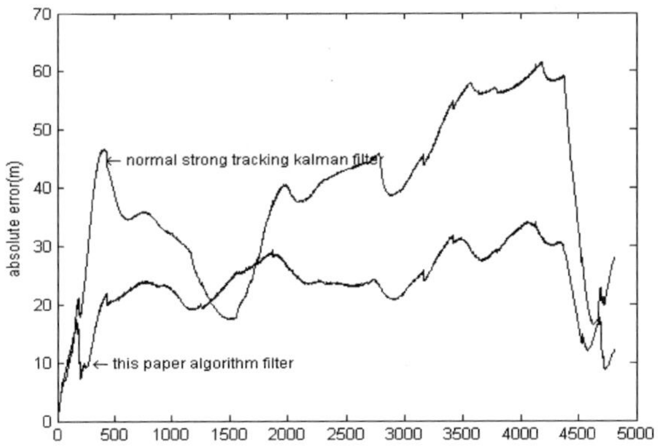

Fig. 6. Compare error of filtering

5 Conclusion

When the fluctuation of system state is small, normal strong tracking Kalman filter performs poor tracking ability. Based the normal strong tracking Kalman filter algorithm, this paper improves its fading factor adaptive estimation algorithm, promotes tracking ability and precision of filter and applies the modified algorithm in dead reckoning of underwater vehicle. The simulation results show that compared with normal strong tracking Kalman filter, the modified algorithm achieves better precision which satisfies the engineering demand of dead reckoning of underwater vehicle.

Acknowledgement

The project was financially supported by the National Natural Science Foundation of China (50909025) and the Basic Research Foundation of Central University (HEUCFZ1003).

References

1. Zilong, F., Jian, L., Liu, K.-z.: Dead Reckoning Method for Autonomous Navigation of Autonomous Underwater Vehicle. Robot 27(2), 168–172 (2005)
2. Lu, L., Lei, W.: Applied navigation system of “SY-1”Small AUV. In: Proceedings of 13th International Symposium on Unmanned Untethered Submersible Technology, New Hampshire USA, August 24-27 (2003)
3. Jun, M., Hai, Z.: Data Fusion Between GPS and Calculation of position Based Ocean Database. Computer Simulation 19(5), 25–29 (2002)
4. Xiao, Q., Rao, M., Ying, Y., Sun, Y.: A new state estimation algorithm-adaptive fading Kalman filter. In: Proceedings of the 31st Conference on Decision and Control, vol. 16, pp. 1216–1221. IEEE, Tucson (1992)
5. Fang, J.-c., Wan, D.-j., Wu, Q.-p., Shen, g.-x.: A new kinematic filtering method of GPS positioning for moving vehicle. Journal of Chinese Inertial Technology 5(2), 4–10 (1997)

The High-Performance and Low-Power CMOS Output Driver Design

Ching-Tsan Cheng[1], Chi-Hsiung Wang[1], Pei-Hsuan Liao[1], Wei-Bin Yang[1], and Yu-Lung Lo[2]

[1] Department of Electrical Engineering, Tamkang University, Taipei, Taiwan, R.O.C
496440149@s96.tku.edu.tw, chwang0327@gmail.com, robin@ee.tku.edu.tw, owenliao0728@yahoo.com.tw
[2] Department of Electrical Engineering, National Kaohsiung Normal University, Kaohsiung, Taiwan, R.O.C
yllo@nknu.edu.tw

Abstract. There are many important points for high-speed CMOS integrated circuit, such as switching speed, power dissipation and full-swing of voltage. In this paper, a group of high performance and low-power bootstrapped CMOS drivers are designed in order to reduce power consumption and enhance the speed of switch for driving a large load, where the drivers will reduce the power consumption by using bootstrap manipulate conditional to input statistics. Moreover, the low swing bootstrapped feedback controlled split path (LBFS) is conducted to reduce the dynamic power dissipation and limiting the voltage swing of gate of the output stage. Finally, the charge transfer feedback controlled split path (CRFS) CMOS buffer is used to restitute and pull down the gate voltage for reducing power consumption and line noise. According to the HSPICE simulation results, the proposed drivers of the CMOS driver are reduced by 20%~40 compared to the conventional design.

Keywords: CMOS buffer, Bootstrapped driver, low-power driver, feedback-controlled.

1 Introduction

Nowadays, thousands of transistors are designed onto a single chip by using the advanced process of semiconductor technology. The majority of design methods can be used to achieve the goal, such as higher chip density, lower power depletion, and more fast operation speed. It has great influence in the high-speed synchronous CMOS integrated circuit, the capacitive load about clock and output buffers will increase from the higher circuit density. When a CMOS integrated circuit drives, the large capacitive load will increase power depletion and circuit delay in the chip [1]-[4]. The popular way to reduce power consumption in the CMOS digital circuit is to cut down the supply voltage, but this way will decrease overdrive voltage (V_{GS}-V_{TH}), current and speed. And it is a difficult skill to design buffers to solve the problem of capacitive load in the efficient pattern. The styles to achieve the goal we can use are very much. For example, the simplest buffer, whose circuit is shown in Fig.1, is one

M. Zhu (Ed.): Electrical Engineering and Control, LNEE 98, pp. 917–925.
springerlink.com © Springer-Verlag Berlin Heidelberg 2011

of the common CMOS tapered buffer, but operation speed will degenerate [5]-[8]. Another is bootstrapped CMOS driver [10], shown in Fig.2. In the bootstrapped CMOS driver, the nodes, p1 and Vn1, were increased to V_{DD} or scaled down to V_{SS} to make switching speed better, but the advance was not display.

The following example is to scale down the turn-on-delay and supply a high speed performance [9] achieved in [7], it's modulated the input voltage of V_{DD} through quick dropper and self-aligned timing generator. However, the circuit also needs a couple of bootstrap capacitors and can deplete a big number of power. Because no matter what the input data is, the driver always be bootstrapped.

In this paper, many bootstrapped CMOS drivers and common drivers are introduced to overcome the disadvantages, such as power consumption. In section 2 describes the circuit structure, operation principle and proposed new bootstrapping feedback-controlled split-path CMOS driver. In section 3, comparison and simulation results are described for proving the advantages in the proposed driver. At last, the conclusion is drawn in section 4.

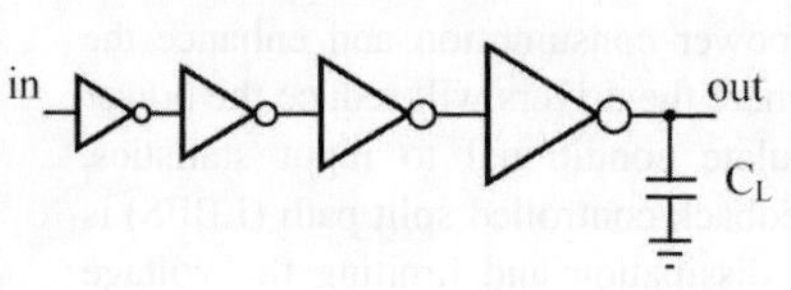

Fig. 1. Tapered Buffer.

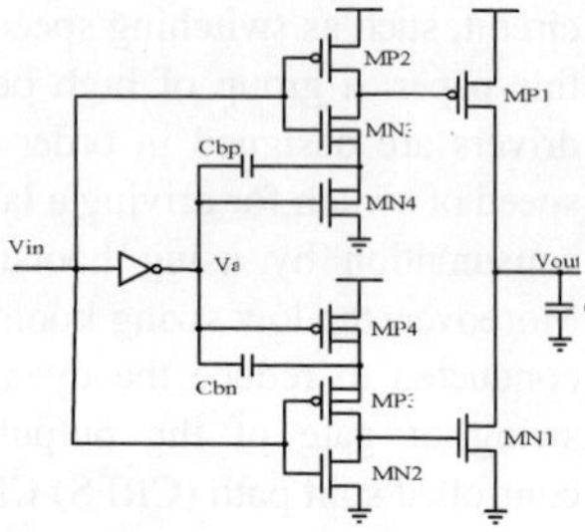

Fig. 2. Conventional bootstrapped CMOS driver.

2 The Circuit Structure and Operation Principle

In the part, many circuit structures and simulation waveform will be introduced, such as low-power inverted-delay-unit (LPID) CMOS buffer, charge-transfer feedback-controlled 4-split-path (CRFS) CMOS buffer and improved conditional-bootstrapping latched CMOS driver imCBLD CMOS buffer [15].

2.1 Power Dissipation in a CMOS Buffer

The major of power dissipation in a CMOS buffer consists of dynamic switching and DC current. The dynamic witching power dissipation is generated when the output and parasitic capacitive loads either is charged or discharged. Owing to the input signal transition, both PMOS and NMOS will be turned on to produce the DC current and then get the extra consumption [13] [14].

The dynamic dissipation P_d can be expressed by the following equation:

$$P_d = \alpha_{0\to1} \times C_{load} \times V_{dd} \times \Delta V_0 \times f \tag{1}$$

Where the $\alpha_{0\to1}$ is the node transition coefficient, C_{load} is the output capacitive load and the ΔV_0 is the voltage swing of the output node.

Another major power dissipation, the short-circuit power dissipation, equation can be showed by the following equation:

$$P_{sc} = \frac{k}{12}(V_{dd} - V_T)^3 \times \frac{\tau_{in}}{T} \tag{2}$$

Where k is the gain of the inverter, V_T is the threshold voltage, τin is the input transition time, and T is the cycle period time. Power consumption will be reduced as design the proper parameter from equation (1) and equation (2).

2.2 Charge-Transfer Feedback-Controlled Split-Path (CFS) CMOS Buffer

The circuit, CFS CMOS buffer [16], was designed to control the split-path driver and charge-transfer diodes. The operation of CFS CMOS buffer is divided into three types, namely output pull-high, output pull-low and charge-transfer operation. The schematic and the waveform are shown in the Fig. 3(a) and (b).

The buffer is manipulated in the output pull-high operations during t_1 - t_9, and pull-low operation during t_{10}-t_{17}, shown in the Fig. 3(b). Both input and output of signal are low in the beginning, and then once input levels up, output comparatively is still low at the time t_1. As can be seen that the node V_{p1} falls at the time t_2, the node V_{p2} rises at the time t_3, and V_{p5} falls between the time t_4-t_6. Therefore, the PMOS (MP) is switched on and its signal output is charged from low to high at t_6. For the reason that the feedback on signal output_b are linked to the NAND gate logic, there is a rise in the node V_{p1} at the time t_7, a fall in the node V_{p2} at the time t_8, and a rise in the node V_{p5} at the time t_9. PMOS (MP) usually stands by and only works within the pull-high operation, however, it shuts down over the pull-high operation. The pull-low operation is almost the same operation mode compared with the output pull-high operation. NMOS (MN) is only turned on in the pull-low operation and keeps off during the other time. The usage of this way will scale the short-circuit power consumption down when the circuit operates.

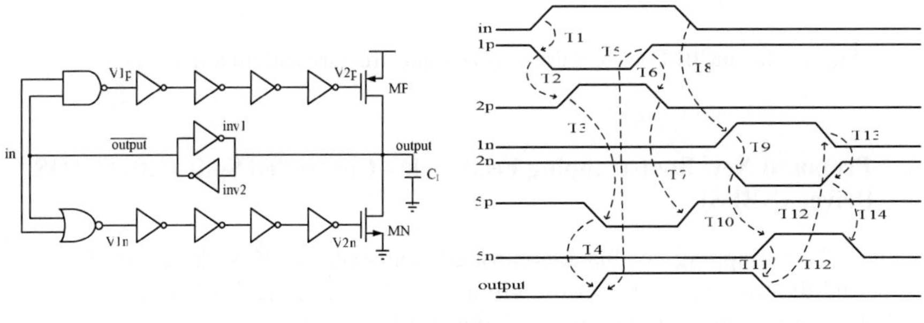

(a) (b)

Fig. 3. The CFS CMOS buffer (a) circuit structure and (b) timing diagram.

2.3 Improved Conditional-Bootstrapping Latched CMOS Driver (imCBLD)

The schematic structure of improved version of CBLD is shown in Fig. 4(a) [12]. This circuit structure improves the nodes, N_{BP} and N_{BN}, by M_{N9} and M_{P9}. The body of M_{N9} is linked to the node N_{CP} substituted for supply voltage, and the body of M_{P9} is linked to the node N_{CN} substituted for ground. No matter what N_{CP} or N_{CN} will not make forward bias.

As shown in Fig. 4(b), the former pulse is continuous motions of UP and the latter is continuous motions of DN. In the period T1, assuming the value of output is low at the beginning, the transistors MN7 and MN8 would be turned on and the transistors MP7 and MP8 would be off. N_{BP} is connected to N_{CP} instead of the supply voltage, after MP2 is turned off and MN3 is turned on. At the same time, MN4 is turned on and MP6 becomes off which makes N_{CN} pulled down. After that, both N_{CP} and N_{BP} fall to be lower than ground by the capacitive C_B. The voltage N_{CN} rises to the top in VDD and falls to the bottom in ground when MN7 becomes off and MP8 on. That is, the operation of bootstrapping is used to improve the speed as MP1 drives. N_{BP} and N_{CP} are back to the previous state at T2. When works at T3, the voltages remains high in N_{BP} and low in N_{CP} within imCBLD, in which N_{BP} and N_{CP} keep dissociated because MN9 is turned off with MP1 being off. In fact, the state of N_{BP} is in a high-impedance if N_{BP} is not influenced by extra noise. However, the condition will not interfere with the state of output to get the high logic state. Nevertheless, if the bootstrapping operation is used, all of the nodes in the circuit will not transit. Hence, we can reduce the maximum power.

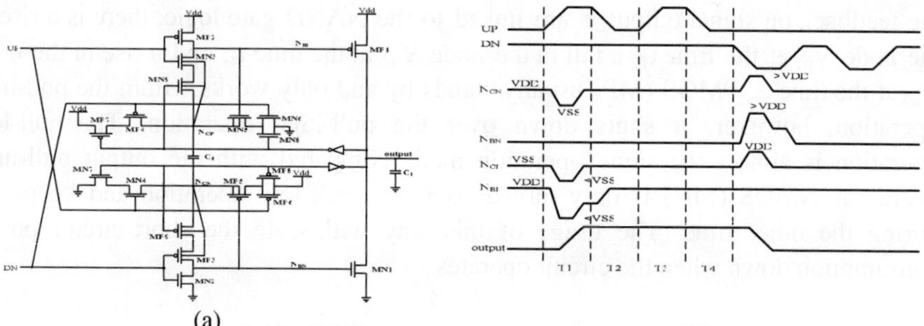

(a) (b)

Fig. 4. The imCBLD CMOS driver (a) circuit structure and (b) timing diagram.

2.4 Proposed New Bootstrapping Feedback – Controlled Split-Path CMOS Driver (NBFS)

The new bootstrapping feedback-controlled split-path CMOS driver is showed in Fig. 5, and the transistors size of our circuit are shown in the Table I. The NBFCB is divided into two types, namely wave shape and bootstrapping circuit and designed to conform to the spectrum of single-ended clock at Intel ICH10. There are several advances of the NBFS driver, such as reducing the power-delay product and improving the speed as switch operation.

Table 1. Transistors size of our circuit

	W(μm)/L(μm)	Parallel Number
MN1	1/0.35	114
MP1	3/0.35	114
MN2~8	2/0.35	5
MP2~8	3/0.35	5

Circuit Structure

The digital logics, such as Inverter, NAND and NOR, are used in the circuit to generate a tri-state period, shown in Fig. 5(a). The short circuit current will be decreased by this circuit, because the design is settled to be not turned on at the same time for the voltage output level. This design is not only reduce the short circuit current, but also the propagation-delay time. Using these ways, it is important to make enough tri-state time for the bootstrapping capacitor to complete charging and discharging. Eliminating the external power consumptions is enforcing at the same time.

After input signal is distributed by the previous digital logic, it is a method to enhance the signal amplitude of the nodes N_{BP} and N_{BN}. To increase the speed of the output level switching, and to decrease power-delay product both of them reduce the power depletion.

Operation Principle

As shown in Fig. 5(b), the input signal and output signal are ground at the beginning. During T1, the nodes of UP at VSS cause the transistors, MP2 and MP6, turned on when the output of NAND at VDD. The transistor, MN5, is turned on owing to the nodes N_{BP} rising to VDD. The nodes of DN at VDD cause the transistors, MN2 and MN6, turned on when the output of NOR is VSS. The transistor, MP5, is turned on owing to the nodes N_{CP} falling to VSS. At the time, the nodes N_{CN} is charged to VDD and N_{CP} is discharged to VSS. During T2, the node UP rise to high and the node DN keep the previous state, because the input signal rises to VDD. As node UP rise, MN3 and MN4 will be turned on. At this time, output remains VSS. The transistors, MN7 and MN9, are turned on so that NBP is equal to NCP. At the same time thecharge stored in the CB discharge to VSS. MP1 open sharply to make output rise to VDD because NCN at VSS and NCP below VSS. In the period of T3, the output rise to high and the input signal keeps the previous state. As the node UP falls to VSS, MP2, MP6 and MP8 will be turned on. After that, NBP at VDD and MN5 is turned on. As the node DN rises to VDD, MN2 and MN6 will be turned on. After that, NBN at VSS and MP5 is turned on. The node NCN and NCP are charged to VDD and VSS by the capacitor, CB. When input signal falls to VSS and output keeps the previous state, the node DN will falls to VSS at T4. And the transistors, MP3, MP4, MP7 and MP9, are also turned on. Following from that, NBN and NCN are the same status and the node NCP is charged above VDD so that output will falls to VSS.

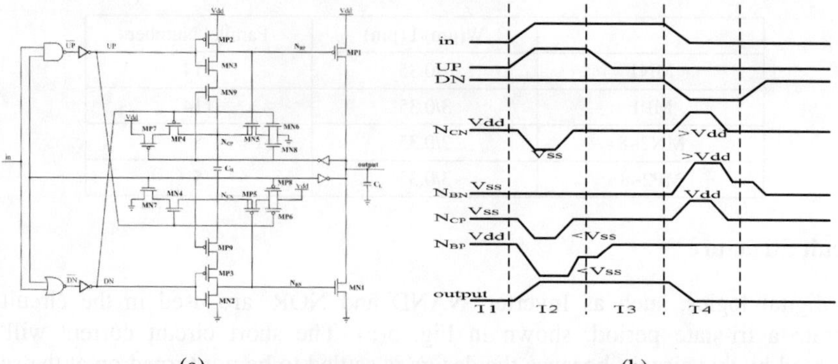

(a) (b)

Fig. 5. NBFS CMOS driver (a) circuit structure and (b) timing diagram.

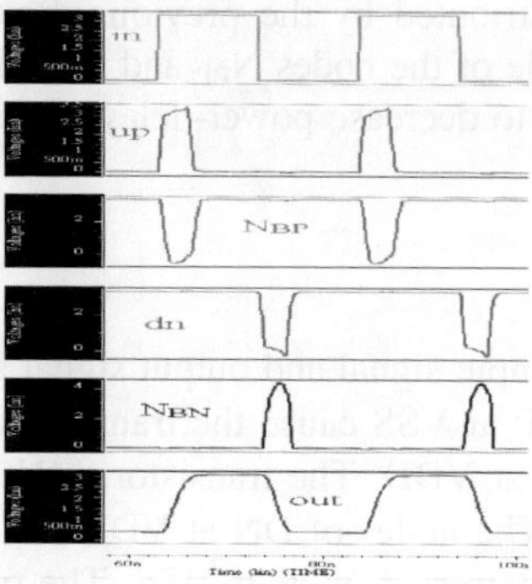

Fig. 6. The post-layout simulation of NBFS CMOS driver.

3 Simulation Results

There are several improved points of the proposed bootstrapped CMOS driver, such as operation speed and power consumption. Both operation speed and power consumption are improved by the proposed bootstrapping CMOS drivers. There are several advantages of the design. In contrast to common tapered drivers, the proposed driver is suit to work at low voltage, and it will reduce power consumption by this way. In contrast to common bootstrapping driver, the speed in switching is scaled down by the structure of imCBLD. It will not transit when the node UP rises, which we can see the simulation result, shown in the Fig. 6. The reasons of power consumption reduction by imCBLD are bootstrapped when the change of output needed and transient response is eliminated which compared with CBLD.

Additionally, the advantage of the proposed driver circuit is designed using the single capacitor for the operations of pull-up and pull-down. The overall integration bootstrap voltages designs follow this viewpoint using the single capacitors for reducing the proportion of capacitor. The proposed driver will get larger bootstrap voltage to obtain the higher switching speed compared with conventional driver [12].

The normalize power-delay product of various voltage supplies of 1.8V, 2.3V, 2.8V and 3.3V at 48MHz will see in the Table 2 to Table 4. The HSPICE simulation results are based on the device parameter of 0.35um 3.3V CMOS process. According to the simulation results, the proposed driver will get the higher performance at low voltage supply, and it is suitable to use at the design of low voltage supply and consumes the lower power dissipation at the same time. Fig. 7 shows the layout of the proposed driver, the chip area without input and output digital driver is 0.68×0.57 mm^2.

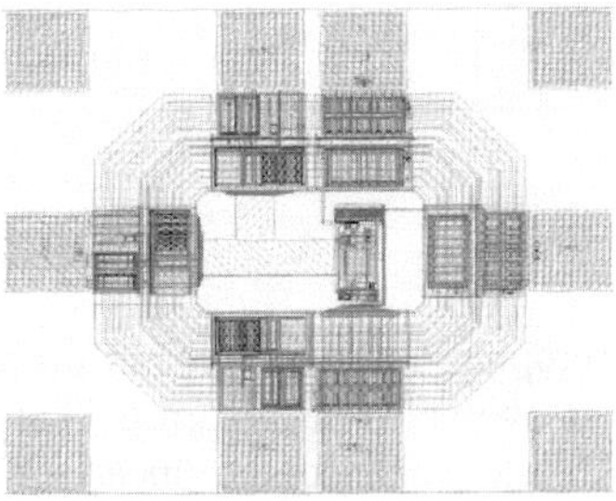

Fig. 7. Layout picture of the proposed NBFS CMOS driver.

Table 2. Simulation results of drivers at 48MHz when Vin is 3.3V

	[5]	[11]	[12]	**This work**
Avg_power(mW)	34.03	34.12	32.84	**28.85**
Propagation-delay(ns)	1.829	3.18	4.224	**1.776**
Power-delay product(pJ)	62.24	108.5	138.7	**51.24**
Normalize Power-delay product by NBFSB	1.215	2.117	2.707	**1**

Table 3. Simulation results of drivers at 48MHz when Vin is 2.8V

	[5]	[11]	[12]	**This work**
Avg_power(mW)	24.36	24.28	22.97	**21.12**
Propagation-delay(ns)	2.055	3.549	4.831	**2.014**
Power-delay product(pJ)	50.07	86.16	111.0	**42.52**
Normalize Power-delay product by NBFSB	1.178	2.026	2.611	**1**

Table 4. Simulation results of drivers at 48MHz when Vin is 2.3V

	[5]	[11]	[12]	**This work**
Avg_power(mW)	16.37	16.00	15.29	**14.52**
Propagation-delay(ns)	2.474	4.320	5.936	**2.402**
Power-delay product(pJ)	40.49	69.13	90.76	**34.88**
\Normalize Power-delay product by NBFSB	1.161	1.982	2.602	**1**

Table 5. Simulation results of drivers at 48MHz when Vin is 1.8V

	[5]	[11]	[12]	**This work**
Avg_power(mW)	9.804	9.222	8.982	**9.192**
Propagation-delay(ns)	3.432	5.902	8.268	**3.269**
Power-delay product(pJ)	33.65	54.43	74.26	**30.05**
Normalize Power-delay product by NBFSB	1.119	1.811	2.417	**1**

4 Conclusions

In this paper, a new bootstrapping CMOS driver is proposed. The design structure provides a higher speed on switching by storing charge more than before, like imCBLD [12] and CFS [16], to enhance the speed of charging and discharging. Another benefit is lower power consumption. By using the high efficiency conditional bootstrapping circuit can cut down time of pull-high and pull-down when operation. Finally, employing the digital logics, such as NAND, INVERTER and NOR, are used to generate the tri-state signal. There is generated short circuit current in the common digital logic. Not only the time of tri-state time is enough, but also the short circuit current will be depleted and propagation-delay will be reduced.

Acknowledgments

The authors would like to thank the National Chip Implementation Center and National Science Council, Taiwan, for fabricating this chip and supporting this work, respectively.

References

1. Bailey, D.W., Benschneider, B.J.: Clock design and analysis for a 600-MHz alpha microprocessor. IEEE J. Solid-State Circuits 33(11), 1633–1637 (1998)
2. Gronowski, P.E., Bowhill, W.I., Preston, R.P., Gowan, M.K., Allmon, R.L.: High-performance microprocessor design. IEEE J. Solid-State Circuits 33(5), 676–686 (1998)
3. Friedman, E.G. (ed.): High Performance Clock Distribution Networks. Kluwer Academic, Dordrecht (1997)
4. Lazorenko, D.I., Chemeris, A.A.: Low-power issues for SoC. In: Proc. IEEE Int. Symp. Consum. Electron, pp. 1–3 (June 2006)

5. Li, N.C., Haviland, G.L., Tuszynski, A.A.: CMOS tapered buffer. IEEE J. Solid-State Circuits 25(4), 1005–1008 (1990)
6. Leung, K.: Control slew rate output buffer. In: IEEE Custom Integrated Circuits Conference, pp.5.5.1–5.5.4 (1988)
7. Wada, T., Eino, M., Anami, K.: Simple noise model and low-noise data-output buffer for ultrahigh-speed memories. IEEE J. Solid-State Circuits 25(6), 1586–1588 (1990)
8. Vemuru, S., Thorbjornsen, A.: Variable-taper CMOS buffer. IEEE J. Solid-State Circuits 26(9), 1265–1269 (1991)
9. Lou, J.H., Kuo, J.B.: A 1.5 V full-swing bootstrapped CMOS large capacitive-load driver circuit suitable for low-voltage CMOS VLSI. IEEE J. Solid- State Circuits 32(1), 119–121 (1997)
10. Law, C.F., Yeo, K.S., Samir, R.S.: Sub-1V bootstrapped CMOS driver for giga-scale-integration era. Electron. Lett. 35(5), 392–393 (1999)
11. Khoo, K.-Y., Willson, A.N.: Low Power CMOS Clock Buffer. In: Proc. Int. Symp. on Circuits and System (ISCAS 1994), May 30-June 2 (1994)
12. Kim, J.-W., Kong, B.-S.: Low-Voltage Bootstrapped CMOS Drivers with Efficient Conditional Bootsrtapping. IEEE Tran. On Circuits and Systems-II 55(6) (June 2008)
13. Veendrick, H.J.: Short-circuit dissipation of static CMOS circuitry and its impact on the design of buffer circuits. IEEE J. Solid-State Circuits SC-19(4), 468–473 (1984)
14. Chandrakasan, A.P., Brodersen, R.W.: Sources of power consumption. In: Low Power Digital CMOS Design, pp. 92–97. Kluwer Academic, London (1995)
15. Cheng, K.H., Yang, W.B.: Circuit Analysis and Design of Low-Power CMOS Tapered Buffer. IEICE Transactions on Electronics E86-C (5), 850–858 (2003)
16. Cheng, K.H., Yang, W.B., Huang, H.Y.: The Charge-Transfer Feedback-Controlled Split-Path CMOS Buffer. IEEE Transactions on Circuits and Systems II 46, 346–348 (2002)

Circularly Polarized Stacked Circular Microstrip Antenna with an Arc Feeding Network

Sitao Chen, Xiaolin Yang, and Haiping Sun

School of Physical Electronics, University of Electronic Science and Technology of China, 610000 Chengdu, China
School of Microelectronics and Solid-State Electronics, University of Electronic Science and Technology of China, 610000 Chengdu, China
chensitao0212@163.com, yxlin@uestc.edu.cn, haipsun@yahoo.com.cn.

Abstract. A novel circularly polarized circular microstrip antenna is proposed for small satellite applications. The circular patch is fed by an arc feeding network. A circular parasitic patch with a circular slot in the centre and two convex rectangular slots on edge is placed above the former patch to improve the impedance matching and bandwidth, as well as the axial ratio (AR). The simulated impedance bandwidth for S11 < -10 dB is 1.485-1.675 GHz (11.9% at 1.59 GHz). The 3-dB AR bandwidth is 1.525-1.660 GHz (8.5% at 1.59 GHz). The gain is 8.5 dBi at 1.59 GHz. The measured impedance bandwidth for S11 < -10 dB is 1.49-1.67 GHz (11.3% at 1.59 GHz) which agrees well with the simulated results.

Keywords: Circularly polarized, arc feeding network, stacked microstrip antenna, circular slot, two convex rectangular slots.

1 Introduction

Circularly polarized (CP) microstrip antenna has been widely used in satellite communications, such as GPS, digital aeronautical mobile satellite and INMARSAT mobile ground terminal. Rectangular and circular patch are traditionally used for circular polarization microstrip antennas [1], [2]. Recently, ring patch is an available candidate. In [3], [4] and [5], annular-ring, elliptical-ring and rectangular-ring are applied respectively.

There are many different ways to realize circular polarization by special feeding network, such as single-fed, multi-fed in [6] and [7]. Double-fed is mostly applied among these multi-fed methods. However, how to design a proper feeding network to realize circular polarization and make the antenna profile as lower as possible, is still an obstacle in double-fed or other multi-fed occasions. Some techniques have been reported to solve this problem. In [8], the feeding network is placed in the centre of a ring, not increasing any extra space.

Stacked microstrip antenna has been studied in recent years. Because it has a dominant advantage that broadens the bandwidth of a microstrip antenna. The reason is that when the two resonating frequencies of the feeding patch and parasitic patch

M. Zhu (Ed.): Electrical Engineering and Control, LNEE 98, pp. 927–932.
springerlink.com © Springer-Verlag Berlin Heidelberg 2011

are close to each other, the bandwidth becomes wider [9]. What's more, stacked patch can improve the impedance matching and enhance the gain of the antenna [10].

In this paper, we propose a novel circular polarization microstrip antenna. A circular patch is fed by an arc power divider which is proved that it can produce good circular polarization. A ring parasitic patch with two convex rectangular slots is placed above the feed patch to improve impedance matching and axial ratio performance.

2 Antenna Design

The configuration of the proposed antenna is shown in Fig.1 (a). It mainly contains three components: a circular patch with an arc feeding network, an annular ring parasitic patch, and a rectangular metallic ground plane. The circular patch and the arc feeding network are printed on a substrate, thickness sub1z = 0.1 mm, and relative permittivity ε r =2.2. There are two foam layers. One is between the circular patch and the ground plane; another is between the circular patch and the parasitic patch, shown in Fig.1 (b). The relative permittivity of the foam is ε r1 = 1.06.

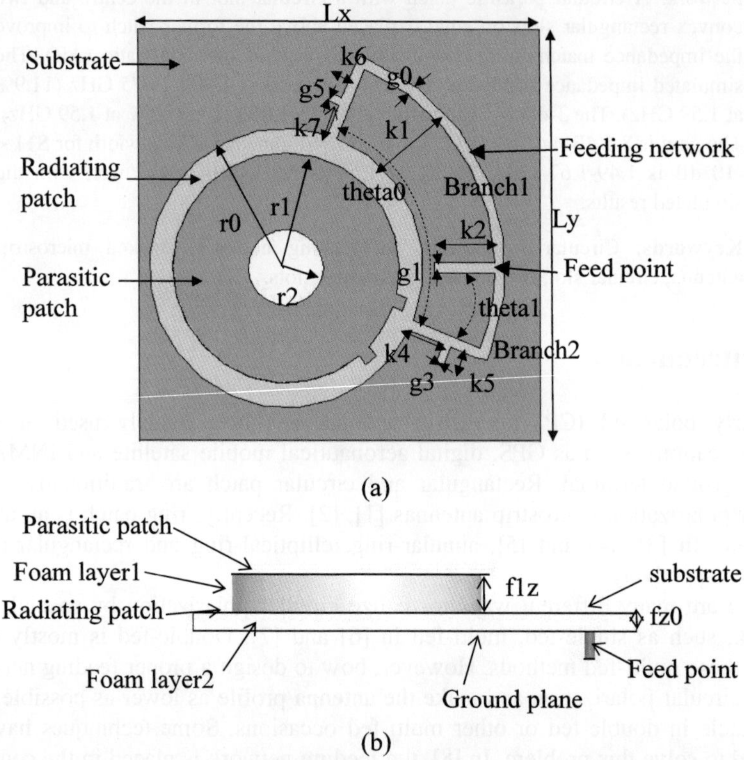

Fig. 1. The configuration of the proposed antenna. (a) Top view. (b) Side view. r0 = 50.36 mm, r1 = 42 mm, r2 = 13.8 mm, fz = 3.94 mm, f1z = 6.9 mm, k1 = 27 mm, k2 =23 mm, k5 = 7.6 mm, theta0 = 88.1 deg, theta1 =22.6 deg, g0 = 4 mm, g1 = 4 mm. Lx = 150 mm, Ly = 150 mm, g3 = 2.5 mm, g5 = 3 mm, k4 = 15.5 mm, k5 = 7.6 mm, k6 = 4 mm, k7 = 11.5 mm.

In order to achieve a performance of circular polarization, we utilize an arc feeding network, whose principle is identical with that of the traditional divider except the shapes. From Fig.1 (a), we can see that the length difference between branch 1 and branch 2 is caused by the angle difference between theta0 and theta1. As we known, the angle difference is 42.9°, and the radius is 77.36 mm which is the sum of r0 and k1. So we can calculate the length difference that is 57.89 mm. And the guided wavelength is λg = 183 mm at 1.59 GHz. So the length difference between branch1 and branch2 is 0.3 λg which is approximately $\lambda g/4$. So a phase shift of 90° between the two branches is produced. And then the circular polarization is realized. A circular slot in the centre and two convex rectangular slots on edge of the parasitic patch, enhance the circular polarization. They mainly have effect on the current of the parasitic patch at the higher band of 1.60-1.66 GHz.

The two patches in different size have different resonating frequencies. It has been reported that when the parasitic patch is stacked close to the fed patch, at the patch distance fz1 < $0.1\lambda g$, the amplitude of the current has two peaks to both sides of the resonant frequency. As the patch distance fz1 increases, the peak value of the current amplitude increases and the two peaks become one [9]. So if the patch distance is fz1< $0.1\lambda g$, the wide bandwidth of the antenna is achieved; if fz1=$0.5\lambda g$, the high gain is achieved. To improve the wide bandwidth and high gain and decrease the profile synthetically, a parasitic patch is placed above the feed patch with the distance of 0.04 λg. And the matching branches in the feeding network make the impedance matching much better. So the input impedance from the port of coaxial cable is around 50Ω at L band. The size of the matching branches is below Fig.1.

3 Results

To confirm the theoretical results, we have fabricated a prototype shown in Fig.2.

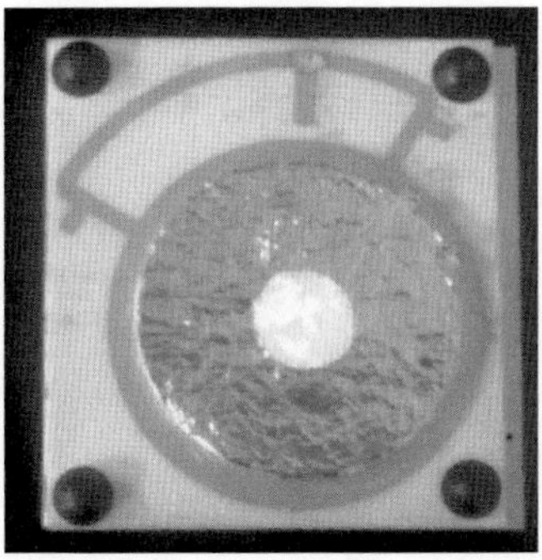

Fig. 2. Prototype of the proposed antenna.

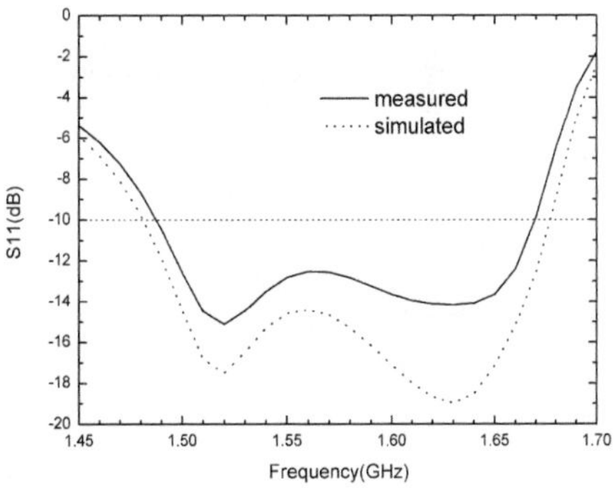

Fig. 3. Simulated and measured S11.

Fig.3 shows the simulated S11 compared to the measured values. The measured S11 < -10 dB is 1.49-1.67 GHz (11.3% at 1.59 GHz), and it agrees well with the simulated one. The difference between the two is mainly caused by the manufacture tolerances.

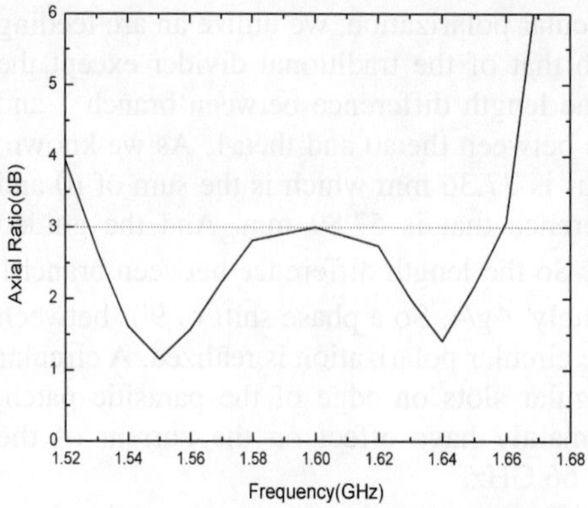

Fig. 4. Axial ratio of the antenna.

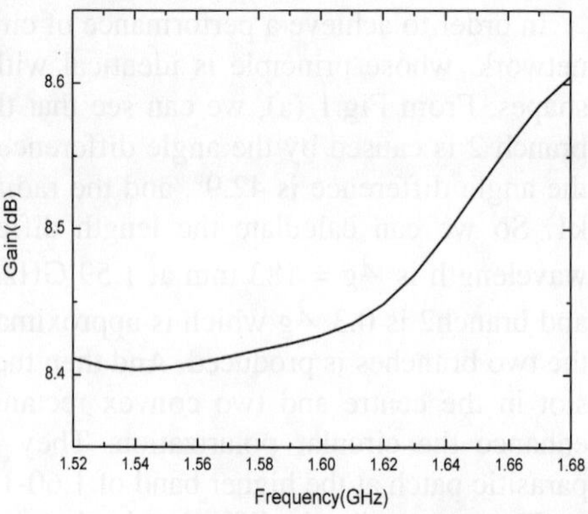

Fig. 5. Gain of the antenna.

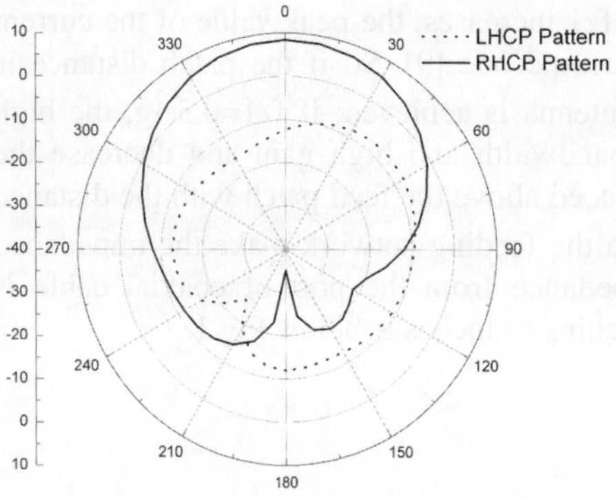

(a) Gain pattern at 1.54 GHz (phi = 0°).

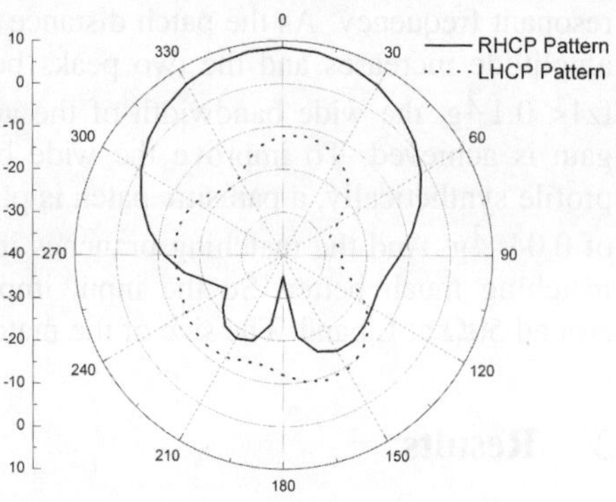

(b) Gain pattern at 1.54 GHz (phi = 90°).

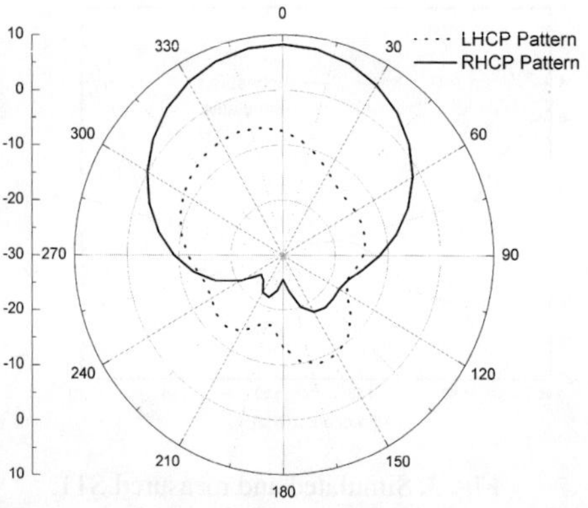

(c) Gain pattern at 1.59 GHz (phi = 0°).

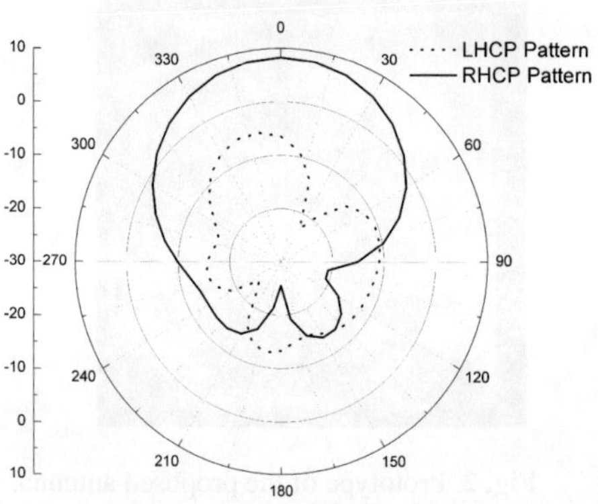

(d) Gain pattern at 1.59 GHz (phi = 90°).

Fig. 6. The gain pattern of the antenna at phi=0° and phi=90°at 1.54 GHz, 1.59 GHz and 1.64 GHz.

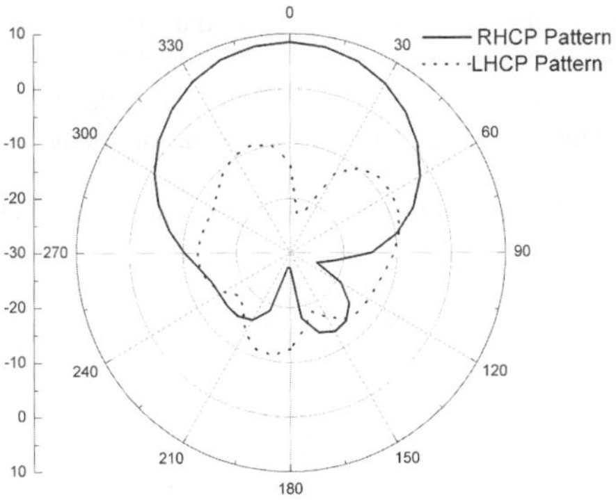

(e) Gain pattern at 1.64 GHz (phi = 0°).

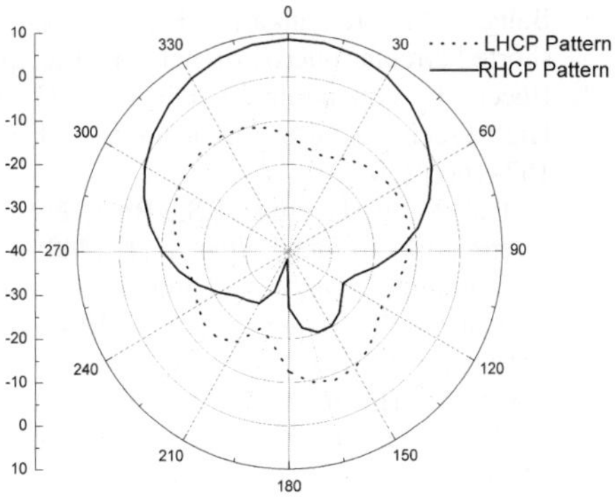

(f) Gain pattern at 1.64 GHz (phi = 90°).

Fig. 6. (*continued*)

Fig.4 shows the simulated axial ratio of the antenna. The simulated AR<3dB is 1.525-1.660GHz (8.5% at 1.59 GHz), which agrees well with the requirement of the digital aeronautical mobile satellite. It proves that this configuration can reach good circular polarization. The simulated gain of the antenna is given in Fig.5. It is seen that the gain is above 8.4dBi from 1.52-1.68 GHz, and it becomes higher along with the frequency. Fig.6 presents the radiation pattern at 1.54 GHz, 1.59 GHz and 1.64 GHz. The half power bandwidth is 50°, 48° and 50° at phi = 90° at 1.54 GHz, 1.59 GHz and 1.64 GHz respectively.

4 Conclusions

A novel circular polarization stacked microstrip antenna is proposed. The circular radiating patch is fed by an arc network, which provides two signals with 90° phase shift. A circular parasitic patch with a circular slot in the centre and two convex rectangular slots is placed above the feed patch to improve the impedance matching and bandwidth, as well as the axial ratio (AR). The measured S11 < -10 dB is 1.49-1.67 GHz (11.3% at 1.59 GHz).The simulated AR < 3 dB is 1.525-1.660GHz (8.5% at 1.59 GHz). The gain is above 8.4 dBi from 1.52-1.68 GHz. The performance is good enough, so this antenna can be used in digital aeronautical mobile satellite applications, and it also can be utilized as a unit of an antenna array of INMARSAT mobile ground terminal.

References

1. Zhou, Y., Chen, C.-C., Volakis, J.L.: Dual Band Proximity-Fed Stacked Patch Antenna for Tri-Band GPS Applications. IEEE Transactions on Antennas and Propagation 55(1), 220–223 (2007)
2. Wong, K.-L., Chang, K. (eds.): Compact CP microstrip antenna. John Wiley & Sons, Inc., Chichester (2002)

3. Baligar, J.S., Revankar, U.K., Acharya, K.V.: Broadband stacked annular ring coupled shorted circular microstrip antenna. Electronic Letters 36(21), 1756–1757 (2000)
4. Boccia, L., Amendola, G., Di Massa, G.: A Dual Frequency Microstrip Patch Antenna for High-Precision GPS Applications. IEEE Antennas and Wireless Propagation Letters 3, 157–160 (2004)
5. Sudha, T., Vedavathy, T.S., Bhat, N.: Wideband single-fed circularly polarized patch antenna. Electronics Letters 40(1) (2004)
6. Nasimuddin, Chen, Z.N., Qing, X.: Dual-Band Circularly Polarized Slotted Patch Antenna With a Small Frequency-Ratio. IEEE Transactions on Antennas and Propagation 58(6), 2112–2115 (2010)
7. Lin, S.-Y., Huang, K.-C.: A Compact Microstrip Antenna for GPS and DCS Application. IEEE Transactions on Antennas and Propagation 53(3), 1227–1229 (2005)
8. Chen, X., Fu, G., Gong, S.-X., Yan, Y.-L., Zhao, W.: Circularly Polarized Stacked Annular-Ring Microstrip Antenna With Integrated Feeding Network for UHF RFID Readers. IEEE Antennas and Wireless Propagation Letters 9, 542–545 (2010)
9. Nishiyama, E., Aikawa, M., Egashira, S.: Stacked microstrip antenna for wideband and high gain. In: IEE Proc.-Microw. Antennas Propag, vol. 151(2), pp. 143–148 (2004)
10. Egashira, S., Nishiyama, E.: Stacked Microstrip Antenna with Wide Bandwidth and High Gain. Transactions on Antennas and Propagation 44(11), 1533–1534 (1996)

Modeling the Employment Using Distributed Agenciesand Data Mining

Bogart Yail Márquez[1], Manuel Castanon-Puga[1], E. Dante Suarez[2], and José Sergio Magdaleno-Palencia[1]

[1] Baja California AutonomousUniversity, Chemistry and EngineeringFaculty, Calzada Universidad 14418, Tijuana, Baja California, México. 22390
bogart@uabc.mx, puga@uabc.edu.mx, jrcastror@uabc.edu.mx
http://fcqi.tij.uabc.mx

[2] Trinity University, Department of Business Administration, One Trinity Place, San Antonio, TX, USA. 78212
esuarez@trinity.edu
http://www.trinity.edu

Abstract. There are several ways to model a complex social system, as is the employment. This work is motivated by need to establish a model for the study of social and economic systems in situations where conventional analysis is insufficient; this paper is to present a methodology applied to complex social problems with soft computing.

Keywords: Complex Social Systems, Data Mining, Neuro-Fuzzy, Distributed Agencies, Employment.

1 Introduction

The primary cause of the existing inequality in Mexico is the employment, which is reflected in the results of economic development.

The purpose of this paper is not to discuss the various determinants factors for the conceptualization of employment, but instead propose a methodology with different computational techniques that aid complex social simulations and provide an option for the analysis of social problems.

The objective of our study is to use the methodology and corresponding computational platform that incorporates available mathematical and computational theories that have not been appropriately considered in models of complex social phenomena[1].

2 Economic Systems

Market economics, with its supply and demand enforcers, represent the law of the urban jungle. Wage rates, housing prices, and the prices of most goods and services, are regulated by markets. Notwithstanding Adam Smith's invisible hand, the market

M. Zhu (Ed.): Electrical Engineering and Control, LNEE 98, pp. 933–940.
springerlink.com © Springer-Verlag Berlin Heidelberg 2011

economy does not necessarily bring about efficient allocations. This is exacerbated when there exists no level playing field, such as when we are in the process of assigning public services in a city with lightning-fast growth. In order to determine how best to allocate public investment, it is necessary to develop a cost-benefit analysis, but such analysis must take into account elements of social justice, human rights and the fulfillment of basic human needs. The analysis must also take into account all relevant externalities, as only then will an appropriate balance be struck between the benefits of the economies of scale that come with agglomeration with the corresponding and unavoidable diseconomies of scale.

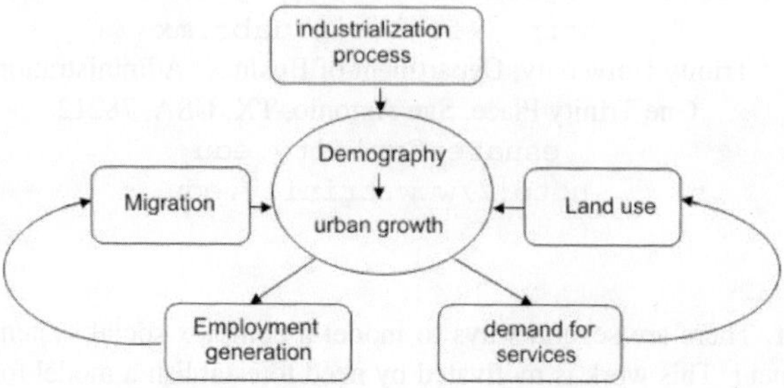

Fig. 1. Model of Urban Growth[2].

Social systems contain many components which depend on many relationships; this makes it difficult to construct models closer to reality[3]. To analyze these systems with a dynamic and multidimensional perspective, we will consider data mining theory, fuzzy logic[4] and distributed agencies.

3 Methodology

The modeling of a realistic social system cannot be achieved by resorting to only one particular type of architecture or methodology. The growing methodology of Distributed Agency (DA) represents a promising research avenue with promising generalized attributes, with potentially groundbreaking applications in engineering and in the social sciences—areas in which it minimizes the natural distances between physical and sociological nonlinear systems.

We consider a disentangled agent that is formed by multiple and relatively independent components. Part of the resulting agent's task is to present alternatives, or 'fields of action' to its components. Correspondingly, the composed agent is itself constrained by a field of action that the superstructure to which it belongs presents. We therefore drop customary assumptions made in traditional social disciplines and MAS about what is considered a decision-making unit. We redefine what a unit of decision is by unscrambling behavioral influences to the point of not being able to clearly delineate what the individual is, who is part of a group and who is not, or where a realm of influence ends; the boundary between an individual self and its social coordinates is dissolved.

Reductionist linear science has concentrated on the study of entities that are clearly delineated, where one could separate what belongs to an agent's nature against the backdrop of what does not. The relevant agent is taken to be exogenous, and therefore disconnected from the system to which it belongs. At their core, these traditional disciplines are based on a selfish and unitary agent, or atom of description. Implicitly or explicitly, these paradigms claim that all aggregate complexity can be traced back to the lower level of the system: the strategies and actions of the selfish agent. In other words, these represent research agendas that purposely de-emphasize the existence of any level other than that of the individual.

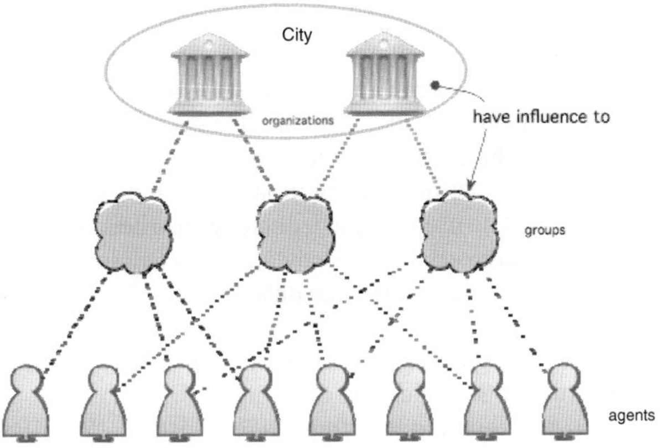

Fig. 2. Multiple levels represented.

On the other hand, the idea of emergence reflects the factthat different and irreducible levels of interaction will naturally arise in complex systems such as the ones studied by social disciplines, and thus the agent, as we define it in this work, is a combination of levels of interaction. It is through this lens that we would like to consider humans, who will partly be independent creatures, possessors of free will, but who are also partly created by an array of upper levels that 'suggest' agreeable utility functions. This conception stems directly from the concept of complexity, in which wholes are more than the sum of their parts. If we believe in that proposition, then we should expect to find a world full of emergent phenomena, with distinctive levels of interaction that have agency of their own. The proposed language redefines agents in two ways. First, there are no obvious atomic agents, for all actors represent the emerging force resulting from the organization of relatively independent subsets. Second, agents are not created in a vacuum, but are rather the result of what an upper level spawns[5].

This model allows us to evaluate a preference for a candidate considering deferent kinds of identity variables, and different uncertainty degrees as to how these identities interact and determine intent. Agents' interactions could create a dynamic behavior if each agent perceives and change her uncertainty values depending on her perception of the behavior of the other agents. The agents adapt their preferences by varying degrees of uncertainty in any of the membership functions of input and output variables.In this work the model is composed by the combination from two theories: for the uncertain and emergente modeling we use fuzzy logic originally introduced by Zaded [4, 6] and ditributed agencies [7].

4 Implementations

Real world simulations such as employment must include some means of validation [8]. In econometrics, studies performed on populations and economy have abundant data, the main problem is finding data sets that are adapted to the desired architecture [9].

With the steady increase in availability of information, from existing projects such as databases needed for social simulation, it becomes essential to use data mining. In our case, for the vast amount of data, we obtain the necessary quantitative information on the most important subject of the social and economic system from government databases. Data mining extract implicit information such as social patterns in order to discover knowledge[10]. The use of these techniques has been wide spread in this field in recent years, most research efforts are dedicated to developing effective and efficient algorithms that can extract knowledge from data[11].

For the particular case of the City of Juarez, in the state of Chihuahua Mexico we use the geographical, demographic and economic indicators provided by the databases of the National Institute of Statistics and Geography (INEGI).

The information for this city is fragmented in 549 “AGEBs”. “The AGEB delimitates urban areas, a whole locality of 2,500 inhabitants or more, or a municipal seat, regardless of the number of people, in groups that typically range from 25 to 50 blocks. Rural AGEB’s frame an area whose land use is predominantly agricultural and these are distributed communities of less than 2,500 inhabitants that for operational purposes, are denominated rural locations[12].

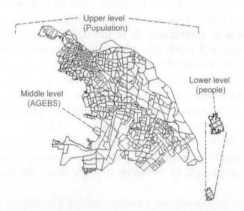

Fig. 3. Levels of agents represented on the social system.

The databases for the model were compiled in a geographic information system that helped in the creation, classification and format of the data layers required. This saves us the issue of having different thematic maps of information; each map, quantifies the spatial structure to display and interpret the different areas and spatial patterns of Juarez.We use Fuzzy Logic for modeling the employment and a NetLogo simulation to model population interaction.

Using NetLogo platform[13]; we are able to simulate social phenomena, model complex systems and give instructions to hundreds or thousands of independent agents all operating holistically. It also allows us to filter information from geographical information system with spatial and statistical data. This makes it possible to explore the relationship and behavior of agents and the patterns that emerge from the interactions within a geographical space.

Each AGEB contains quantitative information about employment, education, income, articles and infrastructure that a home has.Since simulating each person under such circumstances is unattainable, as it is the case of modeling the social development of a city like Juarez, the population total and all those relationships that accompany it, such as migration and birth-rates, must be analyzed with some sort of macro or "top-down" model. On the other hand, all low-level interactions at a micro level, such as partner selection or the decision to form a family with a given number of children, must be captured with a "bottom-up" model.

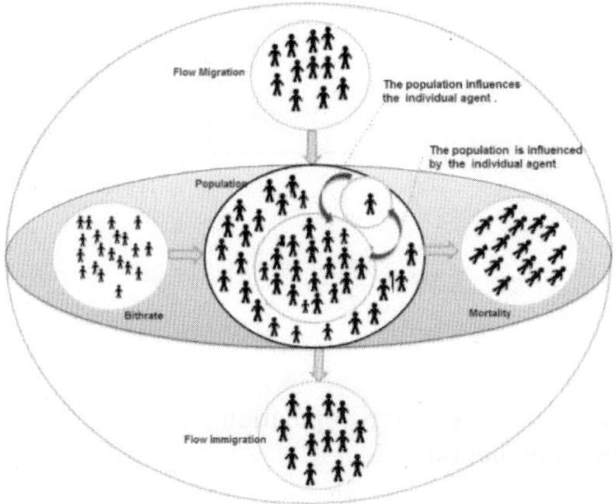

Fig. 4. Bottom-up model and top-down model at population.

Using neuro-fuzzy system to automatically generating the necessary rules, this phase of data mining with a fuzzy system becomes complicated as there is no clear way to determine which variables should be taken into account [14]. This optimization algorithm is widely used and is a numerical method that minimizes an objective function in a multidimensional space, finding the approximate local optimal solution to a problem with N variables [15].

Using this grouping algorithm we obtain the rules, which are assigned to each agent which represents an AGEB. The agent receives inputs from its geographical environment and also must choose an action in an autonomous and flexible way to fulfill its function [16].

The rules obtained from the clustering algorithm can tell us which agents have more income; which are at a higher or lower educational level, and what resources are available.

Taking the compound employment index which measures employment[17] as reference on the three basic dimensions (health, education and an acceptable quality of life) we calculate this factor with.

Studies within the economics are normally performed on a large data platforms, in situations in which there is too much rather tan to few data points. The most problematic aspect of this stage of the modeling process is to define data sets that match the desired architecture[9]. The continuous expansion of available information for social simulation makes the use of data mining unavoidable. In our case study, for example, we have access to many different sources of quantitative and qualitative data describing both economic as well as sociological aspects of reality. Many of these data sets are readily available from governmental sources.Data mining provides us with the process for extracting implicit information, such as social patterns, that reveal ingrained knowledge [10]. The use of these techniques has had significant progress in the last decade[11].

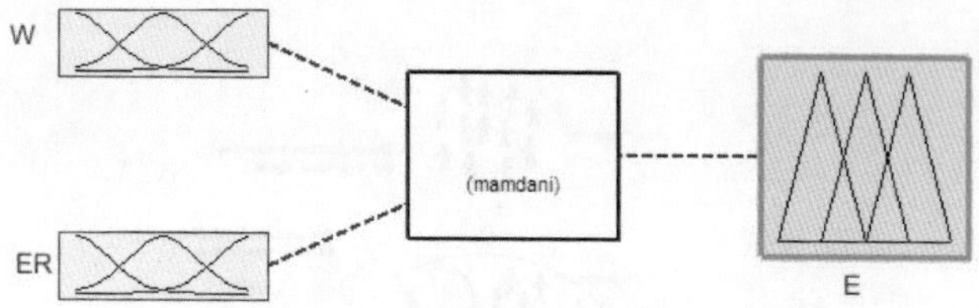

$$W = \frac{EAP}{TP} \quad (1)$$

- W: Workforce
- EAP: Economically active population.
- TP: Total Population

$$ER = \frac{EP}{EAP} \quad (2)$$

- ER: Employment rate.
- EP: Employed population.
- EAP: Economically active population.

Therefore, the next step is for the agencies to try and solve problems distributed among a group of agents, finding the solution as the result of cooperative interaction. Communication facilitates the processes of cooperation; the degree of cooperation between agents can range from complete cooperation to a hostile [18].

5 Conclusions

The employment of a population can be measured as the various multi-levels of population since lower to higher level. Using different techniques can be a powerful tool for any planning process. The use of distributed agencies allows us to unearth new theoretical models and furthers our understanding the relationships found within distinct levels of reality; and helped us to see very concrete cases.

The methodology we are using is developed in a holistic manner, originally focusing on the description and interconnection of different levels of reality, whether these refer to either different dimensions or different time granularities. The applications of the approach are ultimately very general, but they are particularly useful for interdisciplinary analysis, where different disciplines overlap or interact in their description of natural or social phenomena. This general language links together the developments in computational science with those in the social sciences, as they pertain to the nascent paradigm of complexity.

We created a system that is composed of agents where each agent represents an AGEB whose adaptation is the result of complex interactions in nonlinear dynamics, emergent phenomena which arise in the system, and to compare reality with the artificial system and observe the properties, processes and relationships by using different computational methods.

References

1. Márquez, B.Y., et al.: Methodology for the Modeling of Complex Social System Using Neuro-Fuzzy and Distributed Agencies. Journal of Selected Areas in Software Engineering, JSSE (2011)
2. Marquez, B.Y., Castañon-Puga, M., Suarez, E.D.: On The Simulation of a Sustainable System Using Modeling Dynamic Systems and Distributed Agencies. In: INC 2010: 6th International Conference on Networked Computing. IEEE, Gyenongju (2010)
3. Ding, Y.-C., Zhang, Y.-K.: The Simulation of Urban Growth Applying Sleuth Ca Model to the Yilan Delta in Taiwan. Journal Alam Bina, Jilid 09(01) (2007)
4. Zaded, L.A.: Fuzzy sets. Information and Control 8 (1965)
5. Suarez, E.D., Castañon-Puga, M., Marquez, B.Y.: Analyzing the Mexican microfinance industry using multi-level multi-agent systems. In: Proceedings of the 2009 Spring Simulation Multiconference. Society for Computer Simulation International, San Diego (2009)
6. Zadeh, L.A.: From Computing with Numbers to Computing with Words - From Manipulation of Measurements to Manipulation of Perceptions. In: Intelligent Systems and Soft Computing: Prospects, Tools and Applications. Springer, Heidelberg (2000)
7. Suarez, E.D., Rodríguez-Díaz, A., Castañón-Puga, M.: Fuzzy Agents. In: Castillo, O., et al. (eds.) Soft Computing for Hybrid Intelligent Systems, pp. 269–293. Springer, Berlin (2007)
8. Drennan, M.: The Human Science of Simulation: a Robust Hermeneutics for Artificial Societies. Journal of Artificial Societies and Social Simulation 8(1) (2005)
9. Marquez, B.Y., et al.: On the Modeling of a Sustainable System for Urban Development Simulation Using Data Mining and Distributed Agencies. In: 2nd International Conference on Software Engineering and Data Mining 2010. IEEE, Chengdu (2010)
10. Dubey, P., Chen, Z., Shi, Y.: Using Branch-Grafted R-trees for Spatial Data Mining. In: Bubak, M., van Albada, G.D., Sloot, P.M.A., Dongarra, J. (eds.) ICCS 2004. LNCS, vol. 3036, pp. 657–660. Springer, Heidelberg (2004)

11. Peng, Y., et al.: A Descriptive Framework for the Field Of Data Mining and Knowledge Discovery. International Journal of Information Technology & Decision Making 7, 639–682 (2008)
12. INEGI, I Conteo de Población y Vivienda 2005. Instituto Nacional de Estadística Geografía e Informática (2006)
13. Wilensky, U.: NetLogo Software (1999), http://ccl.northwestern.edu/netlogo
14. Castro, J.R., Castillo, O., Martínez, L.G.: Interval Type-2 Fuzzy Logic Toolbox. Engineering Letters (2007)
15. Stefanescu, S.: Applying Nelder Mead's Optimization Algorithm for Multiple Global Minima. Romanian Journal of Economic Forecasting, 97–103 (2007)
16. Gilbert, N.: Agent-Based Models. Sage Publications Inc., Los Angeles (2007)
17. Egghe, L., Rousseau, R., Rousseau, S.: TOP-curves: Research Articles. J. Am. Soc. Inf. Sci. Technol. 58(6), 777–785 (2007)
18. Gilbert, N.: Computational social science: Agent-based social simulation. In: Phan, D., Amblard, F. (eds.) Agent-based Modelling and Simulation, pp. 115–134. Oxford, Bardwell (2007)

Robust Player Tracking and Motion Trajectory Refinement for Broadcast Tennis Videos

Min-Yuan Fang, Chi-Kao Chang, Nai-Chung Yang, I-Chang Jou, and Chung-Ming Kuo*

Department of Information Engineering
I-Shou University
No. 1, Sec. 1, Syuecheng Rd., Dashu Township, Kaohsiung 840, Taiwan
kuocm@isu.edu.tw

Abstract. Player tracking and trajectory detection play an important role in the content analysis of broadcast tennis videos. It is still a challenge because the player size is small and many noises and interference exist in a tennis court, and therefore often results in a detection failure. In this paper, we propose a robust technique for player tracking and trajectory modification using an adaptive Kalman filtering. Experimental results indicate that the proposed method improves the successful rate of player tracking significantly, especially for the upper court players.

Keywords: player tracking, trajectory detection, Kalman filtering.

1 Introduction

In the past few years, significant content analysis has been performed to various kinds of sports such as soccer, tennis, baseball, American football, etc. [1-5]. The motion trajectory of player can provide very useful information for sport content analysis. For example, in tennis sports, events such as net play, baseline rally and ace ball can be detected by referring players' position in the court. Players' tactic in the matches can be also discovered from the players' trajectory. Therefore, player tracking becomes one of most important issue for content analysis of sport videos. For tennis videos, a more challenge task is the detection and tracking of the players on the upper-half court. As shown in Fig. 1(a), we denote the player on upper-half court as upper player, and the player on lower-half court as lower player. Due to the camera's viewpoint, the objects on the upper-half court are much smaller than that of lower-half court.

To explain the difficulty of upper player detection and tracking, we use an example to demonstrate the difference in resolution. If the lengths of the two baselines in image space are detected, we can use simple proportional relationship, as shown in

* Corresponding author.

M. Zhu (Ed.): Electrical Engineering and Control, LNEE 98, pp. 941–949.
springerlink.com © Springer-Verlag Berlin Heidelberg 2011

Fig.1 (b), to estimate the player size in both upper-half and lower-half court. In general, the size of upper player is very small (52×14 pixels to 76×21 pixels), and its size is about 32.4%~24.1% of the lower player. Therefore, the difficulty of the detection and tracking of the upper player increases significantly. In practice, because the interference of the commercial board or the color of a player would be very similar to that of the playfield, thus the practical detectable player size is often much less than the estimate. In this paper, we propose a robust player tracking and trajectory modification method to address the problems mentioned above. It aims at upper player tracking; of course, it is also applied to the lower player. The method is mainly based on an adaptive Kalman filter, in which the parameters are adjusted dynamically according to the detection performance of players.

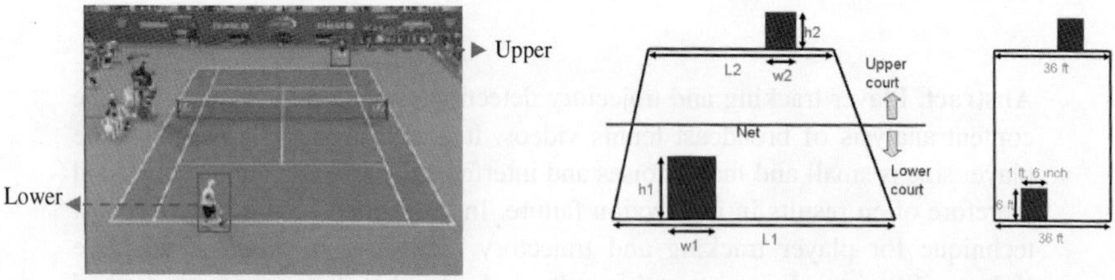

Fig. 1. Lower player and upper players in a court view and the relationship of court baseline length and player size: (a) in image space, (b) in real-world space.

2 Proposed Method

To extract player objects, we first filter out playfield and court line using color feature. Then, we detect players from the remaining image. Finally, the detection result is fed into an adaptive Kalman filter (KF) to estimate the player's position of each frame and modify the detected trajectory. The procedure can be partitioned into two phases conceptually. In the detection phase, object extraction is performed from the first frame of the input video. For the subsequent frames, the tracking phase is conducted with the adaptive Kalman filter.

2.1 Playfield Court Line Detection and Filtering

Court line information provides import reference for the analysis of the court view. Court lines are usually white. Thus, to detect white pixels is quite straightforward and efficient for the detection of court lines. We first transform the RGB color space into HSV space. The detection of court line is then performed in V channel. Through binarization and noise removing, Radon transformation (RT) projects these candidates into peaks in Radon space. By searching these peaks, we can obtain the parameters of court lines, and equations of court lines can be calculated accordingly. we detect the candidate pixels which belong to the court lines. Fig. 2 shows the results of court line detection and noise removing.

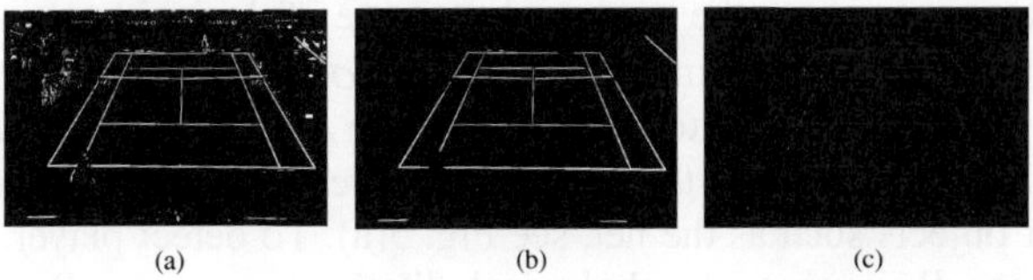

Fig. 2. Illustration of noise removal and thinning

The obtained image is then projected into Radon space using Radon transformation. It is obvious that the estimated lines (marked in green) match the original court lines very well. With the line equations, we can redraw court lines on the image of the court view, as shown in Fig. 4. Fig. 4(a) shows a schematic diagram of a tennis court. Five horizontal lines (h_1-h_5) and five vertical lines (v_1-v_5) construct a tennis court.

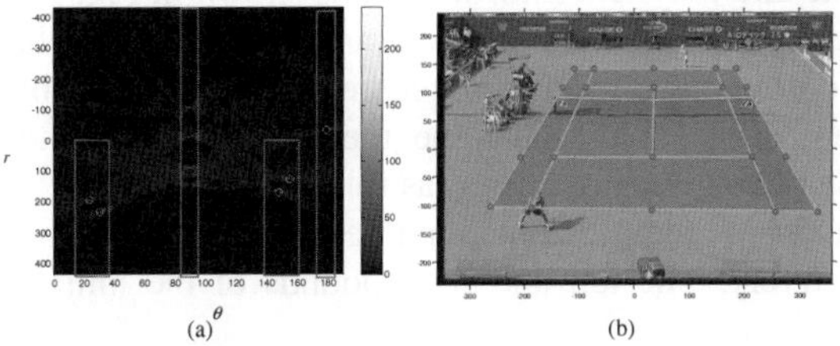

Fig. 3. Court line detection: (a) Radon transformation of the thinning lines (b) estimated court lines (redrawing line in green)

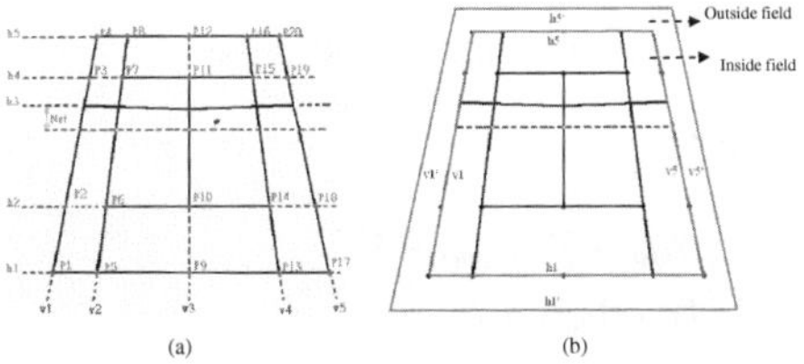

Fig. 4. Schematic diagram of court lines

To filter out the pixels that belong to the playfield, we first convert a RGB court image to the HSV color space. The hue value and intensity value of a pixel at (x,y) are denoted by $hue(x,y)$ and $v(x,y)$, respectively. Then, we filter out the playfield's pixels using dominant colors within the outside and inside fields. The non-dominant color image (B_{NDC}) is defined as

$$B_{\text{NDC}}(x,y)=\begin{cases}0, \text{ if } \left|hue(x,y)-\mu_{\text{Hue}}\right|<\alpha\sigma^2_{\text{Hue}} \\ \quad \text{and } \left|v(x,y)-\mu_{\text{Value}}\right|<\alpha\sigma^2_{\text{Value}} \\ 1, \text{ otherwise.}\end{cases} \tag{1}$$

where μ_{Hue} and σ_{Hue} represent the mean and variance of the inside or outside fields for hue component, respectively. Similarly, μ_{Value} and σ_{Value} are for intensity component. Note that Eq. (1) is only applied to the pixels within F_{inside} and F_{outside} as shown in Fig. 4(b). As a result, B_{NDC} filter out the playfield and reserves only player objects, court lines, and other objects such as the net, see Fig. 5(a). To detect player, we first widen court line images B_{CL} using morphological dilation operation. By subtracting the widened court lines from B_{NDC} as in Eq. (1), we obtain a binary image containing candidates of player objects, denoted by B_{can}, as shown in Fig. 5(b).

$$B_{\text{can}} = B_{\text{NDC}} - (B_{\text{CL}} \oplus SE), \tag{2}$$

where SE is the square structure element of n×n matrix and ⊕ is the dilation operator.

2.2 Player Object Detection

Using the court line and playfield information, as shown in Fig. 3, and the knowledge stated above, we can restrict the initial search area. For the upper-half court, the initial searching area is defined as a rectangle area below the baseline line, h_5. Because the broadcasting style is different in each game, the court might extend to the outside of a frame. Thus, the searching area is defined as follows.

We denote X and Y as the horizontal and vertical coordinate of the intersection of court lines, respectively. The left and right bounds of the initial searching area are defined as,

$$f_{left}^{upper} = \begin{cases} X_{P3}, \text{ if } X_{P3} > 0 \\ 0, \text{ otherwise} \end{cases},$$
$$f_{right}^{upper} = \begin{cases} X_{P19}, \text{ if } X_{P19} < framewidth \\ framewidth, \text{ otherwise} \end{cases} \tag{3}$$

The upper and lower bounds are

$$f_{up}^{upper} = \begin{cases} y_{upper}, \text{ if } y_{upper} > 0 \\ 0, \text{ otherwise} \end{cases},$$
$$f_{down}^{upper} = \begin{cases} \left[\max\left(Y_{P4}, Y_{P20}\right) + 10\right], & \text{if} > 0 \\ 0, \text{ otherwise} \end{cases} \tag{4}$$

Because the noises in lower-half court are much less than those in upper-half court, the initial searching area of the lower-half court can be defined as follows.

$$f_{left}^{lower} = 0, \quad f_{right}^{lower} = framewidth,$$
$$f_{up}^{lower} = Y_{Pc}, \; f_{down}^{lower} = frameheight \tag{5}$$

where Pc is the center edge of net. The initial searching areas for upper-half and lower-half courts are shown in Fig. 6.

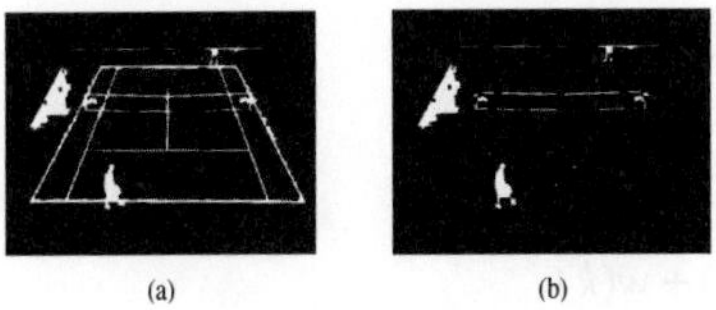

Fig. 5. (a) B_{NDC} , (b) Afte court line filtering.

Fig. 6. Initial search area (green box) and the result of player detection (red box)

2.3 Player's Position Estimation Using Kalman Filter

After the player detection, we use a $\Delta W \times \Delta H$ bounding box to enclose the detected player. To estimate the player's position for the subsequent frames, we define a search window located on the center of the bounding box, as shown in Fig. 7. According to the maximal movement of a player, we define the size of the search window to be $2 \times \Delta W$ in width and $1.4 \times \Delta H$ in height. Within the search window, a full search scheme sliders the bounding box to find a new position which generates maximal object area for each frame.

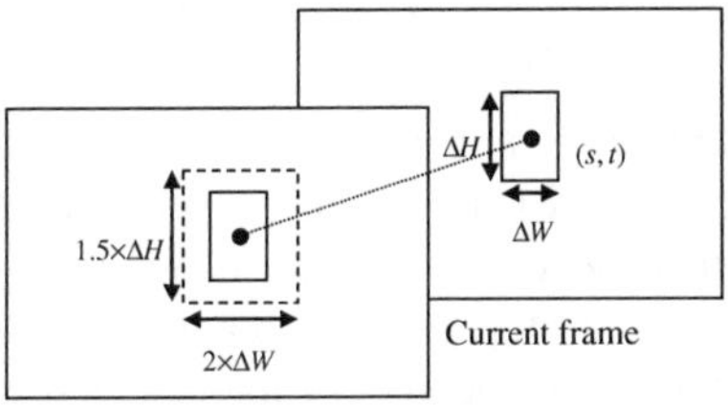

Fig. 7. Search window centered in a bounding box

The upper player is quite small in size, and the background behind the upper player is rather complicated. The player detection process is difficult to obtain a complete player's body. This results in failure of the tracking using the above object search algorithm. In this work, we design an adaptive Kalman filter to solve the problem.

Because the interval between the two consecutive frames is very short, let us assume that the moving speed of a player (moving object) keeps constant. In addition, the *x*-direction and *y*-direction position of the tracking object are assumed mutually independent. Based on the assumptions, we can formulate the x-direction or y-direction position of the player using three subsequent frames k^{th}, $(k-1)^{th}$ and $(k-2)^{th}$ frame as

$$d(k) = d(k-1) + s(k) * 1 + w(k) \tag{6}$$

where $d(k)$ denotes the position of the player and $w(k)$ is the process noise. The speed of a player $s(k)$ can be estimated from the position difference of $(k-1)^{th}$ and $(k-2)^{th}$ frames by

$$s(k) = d(k-1) - d(k-2) \tag{7}$$

Thus, Eq.(6) becomes

$$d(k) = 2d(k-1) - d(k-2) + w(k) \tag{8}$$

Consequently, our proposed state model of Kalman filter can be represented as

$$\mathbf{v}(k) = \mathbf{\Phi}\mathbf{v}(k-1) + \mathbf{\Gamma}\mathbf{w}(k) = \begin{bmatrix} 2 & -1 \\ 1 & 0 \end{bmatrix} \begin{bmatrix} d(k-1) \\ d(k-2) \end{bmatrix} + \begin{bmatrix} 1 \\ 0 \end{bmatrix} w(k),$$

$$\text{where } \mathbf{v}(k) = \begin{bmatrix} d(k) \\ d(k-1) \end{bmatrix}, \mathbf{v}(k-1) = \begin{bmatrix} d(k-1) \\ d(k-2) \end{bmatrix}, \tag{9}$$

$$\mathbf{\Phi} = \begin{bmatrix} 2 & -1 \\ 1 & 0 \end{bmatrix} \text{ and } \mathbf{\Gamma} = \begin{bmatrix} 1 \\ 0 \end{bmatrix}.$$

The measurement model can be represented as

$$\mathbf{z}(k) = \mathbf{H}(k)\mathbf{v}(k) + \mathbf{e}(k) = \begin{bmatrix} 1 & 0 \end{bmatrix} \begin{bmatrix} d(k) \\ d(k-1) \end{bmatrix} + e(k), \tag{10}$$

where $\mathbf{z}(k) = d(k)$, $\mathbf{v}(k)=\begin{bmatrix} d(k) \\ d(k-1) \end{bmatrix}$,and $\mathbf{H} = \begin{bmatrix} 1 & 0 \end{bmatrix}$.

Assume the $\mathbf{w}(k)$ and $\mathbf{e}(k)$ are Gaussian noise with zero mean; that is, $\mathbf{w}(k)=N(0,\mathbf{Q}(k))$ and $\mathbf{e}(k)=N(0,\mathbf{R}(k))$, where Q(k) and R(k) are process error covariance and measurement error covariance matrices, respectively. We apply the occupation ratio of the detected player object to adjust $Q(k-1)$ and $R(k)$ dynamically. The occupation ratio is defined as the area of the detected object divided by the area of the bounding window as

$$\alpha(k) = \frac{area(\text{detected player})}{area(\text{bounding window})} \tag{11}$$

Finally, $Q(k-1)$ and $R(k)$ are simply defined as

$$Q(k-1) = \alpha(k) \text{ and } R(k) = 1 - \alpha(k) \tag{12}$$

The adaptive Kalman filter can reduces the effect of the unreliable measurement, and improves tracking accuracy significantly.

For motion trajectory of upper-court player, because the player size is too small and with too many noises interference, thus the trajectory is very unreasonable. As in Fig. (1), according to the relationship of image space and real-world space, we transfer the motion in image space to real-world space and use the following formula to modify the motion trajectory.

$$PT_{yupper}(t) = \begin{cases} 5, \; when \;\; abs\left[\left(PT_{yupper}(t-1)\right) - \left(PT_{yupper}(t)\right)\right] \; < 5 \\ \frac{1}{10} PT_{yupper}(t), \; when \;\; 100 > abs\left[\left(PT_{yupper}(t-1)\right) - \left(PT_{yupper}(t)\right)\right] \; \geq 5 \\ 100, \; when \;\; abs\left[\left(PT_{yupper}(t-1)\right) - \left(PT_{yupper}(t)\right)\right] \; \geq 100 \end{cases} \quad (13)$$

Because the player's position in vertical direction is more sensitive than that of horizontal, therefore the refinement is on vertical direction only. We use 5 cm and 100 cm as two threshold, and the trajectory is modified according to Eq.(13).

3 Experimental Results

In our experiments, we record several videos of tennis from broadcast channels, including US Opens, French Opens and Wimbledon Opens. All of clips are videos of rallying between two sides, i.e. player's running and stroking. The experimental materials contain 54 clips of 20 seconds in average, from 10 matches, including 48 clips for singles and 6 clips for doubles. The video format is MPEG-2, that is, image resolution is 720×480 and frame rate is 30 fps.

Our proposed Kalman filtering cooperates with the player object detection to improve the tracking performance. The success rate of the tracking is raised from 77% to 94% in average, as shown in Table 1; In other words, the Kalman filtering obtains 17% improvement.

Fig. 8 shows the sequential frames of the successful tracking with Kalman filtering and two tracking misses without Kalman filtering. The test clips are selected from US Opens. Fig. 9 shows the motion trajectory with and without refinement. Obviously, because the trajectory refinement effectively reduces randomness of the player's motion, it performs much reasonable in tracking than the original trajectory.

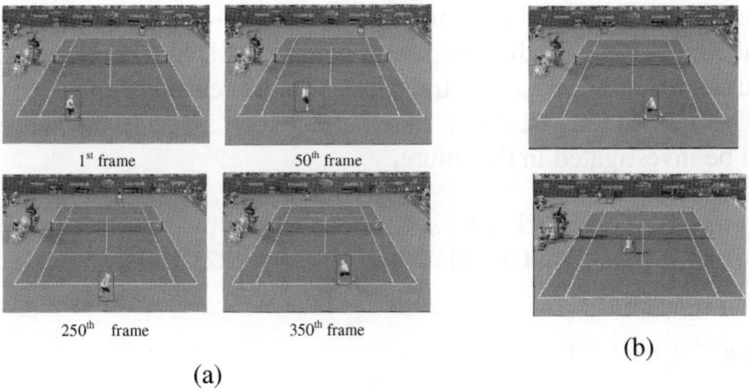

Fig. 8. Tracking result in US Opens (a) with KF (b) two tracking misses (without KF)

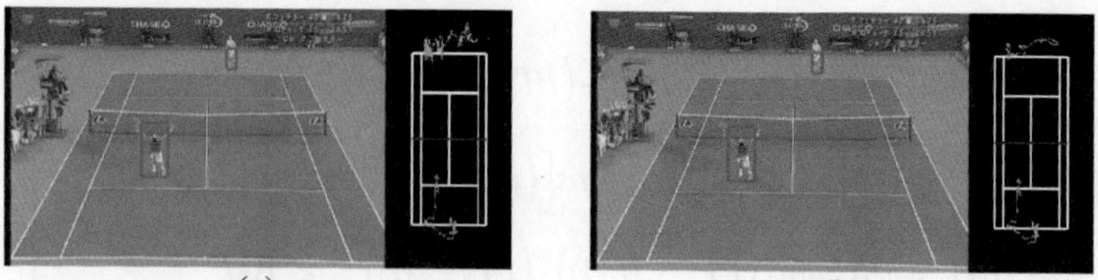

(a) (b)

Fig. 9. The motion trajectory (a) without trajectory refinement, (b) with trajectory refinement

Table 1. Tracking result of singles with(without) Kalman filtering

	A	B	C	success rate(%)
US Opens	35	33(28)	2(7)	94.3(80)
French Opens	7	7(6)	0(1)	100(85.7)
Wimbledon Opens	6	5(3)	1(3)	83.3(50)
Total	48	45(37)	3(11)	94(77)

A: # of video clip; B: # of success; C: # of miss.

4 Conclusion

In this paper, we have presented a robust technique, which is used for broadcast tennis videos, for player tracking and motion trajectory refinement. In the detection phase, the playfield and court line are first filtered out from a court view. Then, the remaining is applied to detect player objects in a delimited searching area. In the tracking phase, a bounding box containing the detected object (player) is used to search the position of the player in the next frame. The utilization of an adaptive Kalman filter greatly corrects the detection (measurement) errors, and then improves the tracking accuracy. Effective mechanisms for automatically adjusting parameters $R(k)$ and $Q(k)$ are developed based on the occupation ratio in the detection phase. Finally, a refinement formula that smooths the motion trajectory is developed. The experimental results indicate that the proposed method achieves average success rate of 94% for tracking, and the motion trajectory is more reasonable than that of original trajectory. The applications based on this method such as event detection and tactics analysis will be investigated in the future.

Acknowledgments. This work was supported in part by National Science Counsel Granted NSC 98-2221-E-214-044-MY2 and NSC 99-2221-E-214-055.

References

1. Ekin, A., Tekalp, A.M., Mehrotra, R.: Automatic Soccer Video Analysis and Summarization. IEEE Trans. on Image Processing 12(7), 796–806 (2003)
2. Zhang, D., Chang, S.F.: Real-time view recognition and event detection for sports video. Journal of Visual Communication and Image Representation 15(3), 330–347 (2004)

3. Huang, C.L., Shih, H.C., Chao, C.Y.: Semantic analysis of soccer video using dynamic Bayesian network. IEEE Trans. on Multimedia 8(4), 749–760 (2006)
4. Xie, L., Xu, P., Chang, S.F., Divakaran, A., Sun, H.: Structure analysis of soccer video with domain knowledge and hidden Markov models. Pattern Recognition Letters 25(7), 767–775 (2004)
5. Leonardi, R., Migliorati, P., Prandini, M.: Semantic indexing of soccer audio-visual sequences: a multimodal approach based on controlled Markov chains. IEEE Trans. on Circuits Syst. Video Techn. 14(5), 634–643 (2004)
6. Pallavi, V., Mukherjee, J., Majumdar, A.K., Sural, S.: Graph-Based Multiplayer Detection and Tracking in Broadcast Soccer Videos. IEEE Transactions on Multimedia 10(5), 794–805 (2008)
7. Hung, M.H., Hsieh, C.H.: Event Detection of Broadcast Baseball Videos. IEEE Transactions on Circuits and Systems for Video Technology 18(12), 1713–1726 (2008)
8. Jiang, Y.C., Hsieh, C.H., Kuo, C.M., Hung, M.H.: Court line detection and reconstruction for broadcast tennis videos. In: Kalra, P.K., Peleg, S. (eds.) ICVGIP 2006. LNCS, vol. 4338, Springer, Heidelberg (2006)
9. Weng, S.K., Kuo, C.M., Tu, S.K.: Video object tracking using adaptive Kalman filter. Journal of Visual Communication and Image Representation 17(6), 1190–1208 (2006)

A DCT-Manipulation Based Algorithm for Scalable Video Coding

Chi-Kao Chang, Min-Yuan Fang, I-Chang Jou, Nai-Chung Yang, and Chung-Ming Kuo*

Department of Information Engineering
I-Shou University
No. 1, Sec. 1, Syuecheng Rd., Dashu Township, Kaohsiung 840, Taiwan
kuocm@isu.edu.tw

Abstract. Scalable video coding (SVC) techniques, which reuse the existing motion vectors, provide a way to convert the compressed information between the delivery systems and terminal devices. In this paper, we aim on the increasing of the visual quality for video downscaling in DCT domain. For arbitrary video downsizing, we propose an efficient algorithm, which manipulates compressed video in the compressed domain, for arbitrary video downsizing. First, the original 8×8 block size is properly enlarged according to the downscaling factor. After the downscaling procedure, the downscaled image blocks are reconstructed to 8×8 matrix to comply with the JPEG format. Finally, we introduce a small-range searching method to increase the accuracy of motion estimation, so that we can approximate the best motion vector. The experimental results indicate that the proposed method provides sufficient improvement for arbitrary video downscaling in the compressed domain.

Keywords: scalable video coding (SVC); arbitrary video downscaling.

1 Introduction

Multimedia communications has become very popular nowadays, many service providers investigate various multimedia applications, and their functionality includes news, sports, entertainments and other media contents to serve the users in heterogeneous networks. Transcoding technique provides a way to convert a certain file format to another one between service providers and client devices; however, it usually suffers from lower coding efficiency and visual quality. With more video applications being provided, the challenge of high visual quality in transcoded video stream is becoming an important issue.

To downscale the motion-compensated video, conventional approach need three steps: 1) decompress the video, 2) re-estimate the downscaled motion vectors (MVs), and 3) re-encode the video. However, this process needs expensive computational efforts due to the motion re-estimation process. A possible way to overcome the computational complexity is to consider scalable video coding (SVC) [1-5], which

* Corresponding author.

M. Zhu (Ed.): Electrical Engineering and Control, LNEE 98, pp. 951–958.
springerlink.com © Springer-Verlag Berlin Heidelberg 2011

reuses the existing motion vectors, for reducing the computational cost without losing significant visual quality.

When original motion vectors are available in the compressed domain, it becomes possible to save computational cost by utilizing the known information. Many fast motion re-estimation algorithms have been proposed without losing quality significantly. Several successful methods such as 1) Average approach (AA) [6,7]; 2) Area weighted average (AWA) [8]; 3) Median approach (MA) [6]; and distance-trimmed filter (DTF) algorithm [9]. These methods change motion vectors from primitive frame (e.g. CIF) to downsizing one (e.g. QCIF) directly. In general, scalable coding in current video coding standards achieves lower computational cost. However, a major problem with this technique is the poor video quality due to the inaccurate MV estimation [4]. Moreover, as the arbitrary downsized video is implemented, a more efficient technique for MV estimation by using the existing MVs is expected.

In this paper, we aim at how to increase the visual quality for video transcoding with arbitrary video downsizing. First, we introduce some existing problems. Then, we propose an object-based MV re-estimation scheme for the downscaled videos, which is further embedded with Kalman filter [10, 11] to improve the accuracy of downsized motion vectors. It will be shown that the proposed method achieves better video performance than that of traditional algorithms.

2 Problem Statement and Proposed Method

2.1 Video Downscaling

In the discrete cosine transform (DCT), which is the basis of the current JPEG standard, source image samples $I(i, j)$ is divided into many blocks, each block is expressed as a 8×8 matrix. An efficient video downscaling algorithm should provide arbitrary-size downscaling flexibly. In the spatial domain, this could be simply solved by interpolation techniques. In the DCT domain, however, this technique leads to mismatch due to the DCT block structure. To solve this problem, we adopt the DCT manipulation technique, which refers to pixel averaging and down sampling (PAD) proposed by Chang and Messerschmitt [12] in the compressed domain. The algorithm allows computing the DCT of the new block directly from the DCT of the original blocks. For example, if we want to downscale to 1/3 size, we first select 3×3 image blocks as shown in Fig. 1.

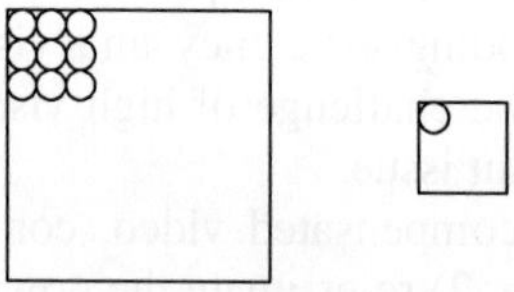

Fig. 1. Relation between the original blocks and its mapped block in down-sized frame

However, each block is composed of 8×8 matrix in DCT algorithm, and 8 is not an exact multiple of the divisor 3. The direct approach is to decode the compressed block, and then manipulates them in the spatial domain. In order to reduce the computational complexity, we first change each image block to a 9×9 matrix, and the all the coefficients of the increased parts is set to "0" as shown in Fig.2. Since the increased coefficients have been put into higher-frequency part, this approximation can be applied to downscale to 1/3 size with ignored loss of information.

In the DCT domain, the distributive property can be written as

$$\mathrm{DCT}(\mathbf{AB})=\mathrm{DCT}(\mathbf{A})\mathrm{DCT}(\mathbf{B}) \tag{1}$$

A new image block contains contributions from nine original neighboring blocks; each of it can be assembled by

$$\mathbf{B}'_i=\mathrm{DCT}(\mathbf{H}_i\mathbf{B}_i\mathbf{W}_i) \tag{2}$$

where $\mathbf{B}_i$ is the input image block, $\mathbf{H}_i$ and $\mathbf{W}_i$ is the pre-matrix and the post-matrix respectively. Multiplying $\mathbf{B}_i$ with a pre-matrix $\mathbf{H}_i$ and then multiplying $\mathbf{W}_i$ with a post-matrix achieves to translate the elements in image block $\mathbf{B}_i$ to assemble the new image block $\mathbf{B}'_i$ as shown in Fig.2(c).

$$\mathbf{H}_i=\begin{bmatrix} a & a & a & 0 & 0 & 0 & 0 & 0 & 0 \\ 0 & 0 & 0 & a & a & a & 0 & 0 & 0 \\ 0 & 0 & 0 & 0 & 0 & 0 & a & a & a \\ b & b & b & 0 & 0 & 0 & 0 & 0 & 0 \\ 0 & 0 & 0 & b & b & b & 0 & 0 & 0 \\ 0 & 0 & 0 & 0 & 0 & 0 & b & b & b \\ c & c & c & 0 & 0 & 0 & 0 & 0 & 0 \\ 0 & 0 & 0 & c & c & c & 0 & 0 & 0 \\ 0 & 0 & 0 & 0 & 0 & 0 & c & c & c \end{bmatrix}, \quad \begin{cases} a=1,\ b=c=0, & \text{for } i=1 \text{ to } 3 \\ b=1,\ a=c=0, & \text{for } i=4 \text{ to } 6 \\ c=1,\ a=b=0, & \text{for } i=7 \text{ to } 9 \end{cases} \tag{3}$$

$$\mathbf{W}_i=\begin{bmatrix} a & 0 & 0 & b & 0 & 0 & c & 0 & 0 \\ a & 0 & 0 & b & 0 & 0 & c & 0 & 0 \\ a & 0 & 0 & b & 0 & 0 & c & 0 & 0 \\ 0 & a & 0 & 0 & b & 0 & 0 & c & 0 \\ 0 & a & 0 & 0 & b & 0 & 0 & c & 0 \\ 0 & a & 0 & 0 & b & 0 & 0 & c & 0 \\ 0 & 0 & a & 0 & 0 & b & 0 & 0 & c \\ 0 & 0 & a & 0 & 0 & b & 0 & 0 & c \\ 0 & 0 & a & 0 & 0 & b & 0 & 0 & c \end{bmatrix}, \quad \begin{cases} a=1,\ b=c=0, & \text{for } i=1,4,7 \\ b=1,\ a=c=0, & \text{for } i=2,5,8 \\ c=1,\ a=b=0, & \text{for } i=3,6,9 \end{cases} \tag{4}$$

We can calculate coefficients of the downscaled block DCT($\mathbf{B}'$) by

$$\mathbf{B}'=\frac{1}{9}\sum_{i=1}^{9}\mathbf{H}_i\mathbf{B}_i\mathbf{W}_i \tag{5}$$

For the purpose of coding the downscaled block to DCT format, the 9×9 matrix is cut back into 8×8 matrix as shown in Fig.2 (e). In this case, errors can be ignored because there are no significant contributions in the high-frequency bands.

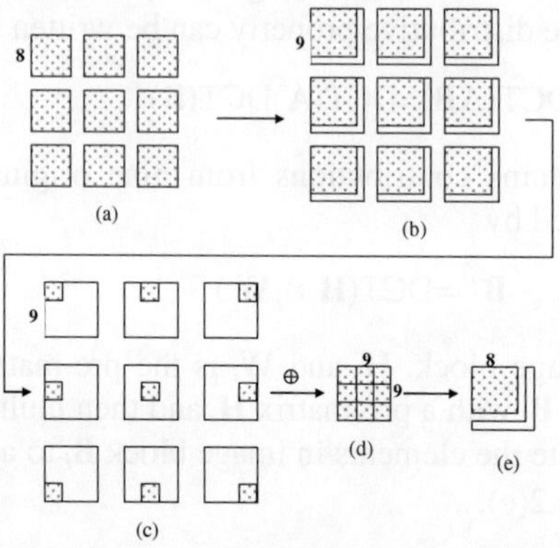

Fig. 2. Reconstructing the image blocks based on ignoring the high-frequency bands

2.2 Motion Estimation

Consider a new image block **b** which consists of four original neighboring blocks ($\mathbf{b_1}$ to $\mathbf{b_4}$) as shown in Fig. 3. The new image block can be divided to four sub-blocks, b_{14}, b_{23}, b_{32} and b_{41}. We can reassemble the DCT of the new block in the DCT domain by

$$\mathbf{B}'=\frac{1}{9}\sum_{i=1}^{9}\mathbf{H}_i\mathbf{B}_i\mathbf{W}_i \tag{6}$$

where $\mathbf{H}_{hi}, \mathbf{H}_{wi}, \mathbf{I}_h$ and $\mathbf{I}_w$ are identity matrices with size *h*×*h* and *w*×*w*, respectively.

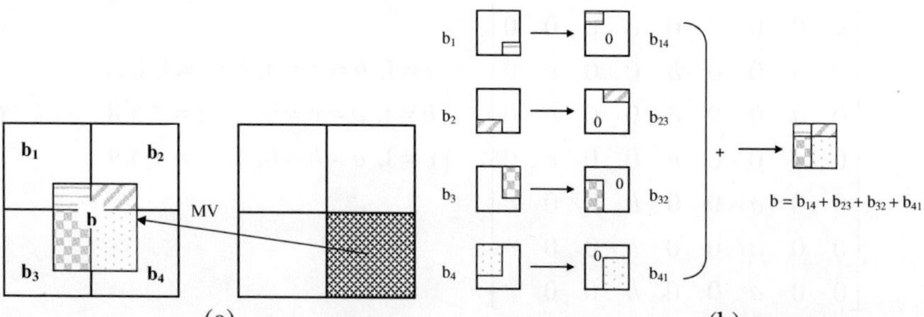

Fig. 3. (a)Relationship between the new image block **b** and the four original neighboring blocks ($\mathbf{b_1}$ to $\mathbf{b_4}$) (b) Reconstruction of the new image block from the four sub-blocks

The bi-linear interpolation can be used in the DCT domain for half-pixel accuracy. In Fig. 4, for example, the black dots represent the original DCT coefficients. To determine the interpolated coefficients in the horizontal direction, we can calculate DCT($\mathbf{B}_h$) as

$$\mathrm{DCT}(\mathbf{B}_h)=\mathrm{DCT}(\mathbf{b})\mathrm{DCT}(\mathbf{h}) \tag{7}$$

Similarly, the interpolated coefficients in the vertical direction ($\mathbf{B}_v$) and both the two directions ($\mathbf{B}_{hv}$) are given by:

$$\mathrm{DCT}(\mathbf{B}_v)=\mathrm{DCT}(\mathbf{v})\mathrm{DCT}(\mathbf{b}) \tag{8}$$

and

$$\mathrm{DCT}(\mathbf{B}_{hv})=\mathrm{DCT}(\mathbf{v})\mathrm{DCT}(\mathbf{b})\mathrm{DCT}(\mathbf{h}) \tag{9}$$

In Eq.(7) to (9), **h** and **v** are the matrices for computing the interpolated coefficients:

$$\mathbf{h}=\begin{bmatrix} 0.5 & 0 & 0 & 0 & 0 & 0 & 0 & 0 \\ 0.5 & 0.5 & 0 & 0 & 0 & 0 & 0 & 0 \\ 0 & 0.5 & 0.5 & 0 & 0 & 0 & 0 & 0 \\ 0 & 0 & 0.5 & 0.5 & 0 & 0 & 0 & 0 \\ 0 & 0 & 0 & 0.5 & 0.5 & 0 & 0 & 0 \\ 0 & 0 & 0 & 0 & 0.5 & 0.5 & 0 & 0 \\ 0 & 0 & 0 & 0 & 0 & 0.5 & 0.5 & 0 \\ 0 & 0 & 0 & 0 & 0 & 0 & 0.5 & 1 \end{bmatrix} \quad \mathbf{v}=\begin{bmatrix} 0.5 & 0.5 & 0 & 0 & 0 & 0 & 0 & 0 \\ 0 & 0.5 & 0.5 & 0 & 0 & 0 & 0 & 0 \\ 0 & 0 & 0.5 & 0.5 & 0 & 0 & 0 & 0 \\ 0 & 0 & 0 & 0.5 & 0.5 & 0 & 0 & 0 \\ 0 & 0 & 0 & 0 & 0.5 & 0.5 & 0 & 0 \\ 0 & 0 & 0 & 0 & 0 & 0.5 & 0.5 & 0 \\ 0 & 0 & 0 & 0 & 0 & 0 & 0.5 & 0.5 \\ 0 & 0 & 0 & 0 & 0 & 0 & 0 & 1 \end{bmatrix} \tag{10}$$

Note that the process of calculating the DCT of the matrices **h** and **v** can be accomplished in the preprocessing step, and hence the algorithm keeps low computational complexity.

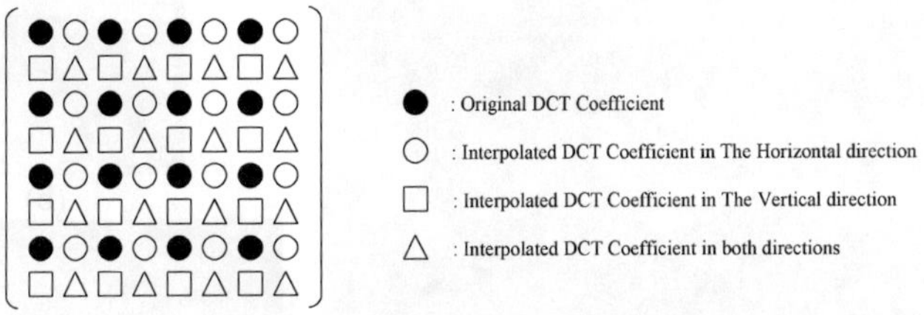

Fig. 4. Interpolations of the DCT coefficients

Theoretically, the fast ME algorithms are able to find MVs very close to their exact value. However, in our experiments we found that if we move the final position one pixel along the direction of FS result, then a better performance could be achieved. It suggests that most of the MVs from the fast algorithms are under-predicted. Therefore, we propose a small-range searching method to increase the accuracy of motion estimation. In Fig. 5, assume that the MV points to black dot. Since the best MV could be located on the region near the point, we can simply search its neighborhood to refine the best MV.

In this MC recalculating step, the new motion vectors can be interpolated from the old motion vectors in the DCT domain [13]. In this work, we use a 3×3 search window and compare their sum of absolute differences (SADs) to obtain the best-matching MV.

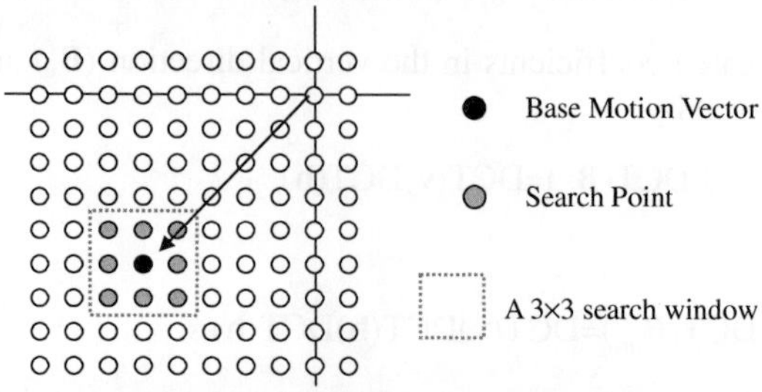

Fig. 5. Small-range searching

3 Experimental Results

In this section we demonstrate the usefulness of our algorithm for arbitrary video downscaling in the compressed domain. In this experiment, various fast downscaling algorithms including AA, MDN, AWA and DTMF are tested to evaluate their performance. Adopted video sequences (Foreman, Football, Paris, Mother & Daughter and Bream) are used for the simulations; frame format is CIF 352×288, each 30 frames/sec with 101 frames long. A fixed block size of 16×16 is chosen.

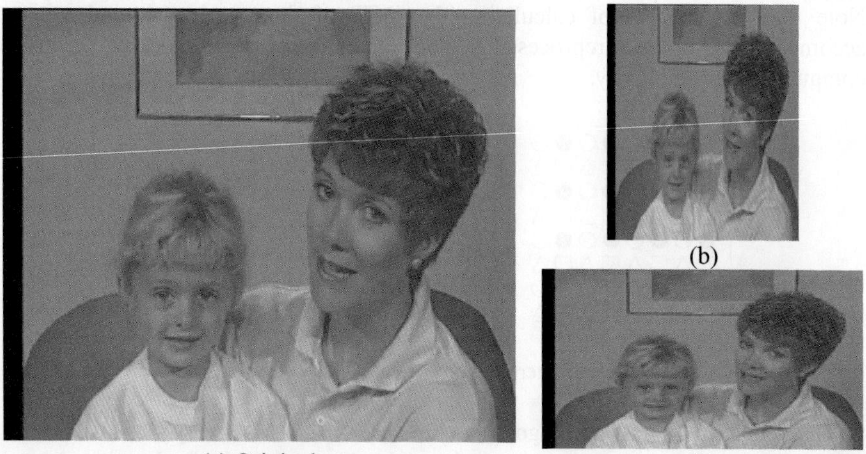

(a) Original (b) (c)

Fig. 6. M&D. (a) Original. (b) Downsized by downscaling factor 2/3 (c) Downsized by downscaling factor 3/2

To demonstrate the flexibility of the proposed method for arbitrary-size downscaling, two downsized frames are shown in Fig.6. for different scale factors.

Table 1 summarizes the simulation results on the six test sequences, where AA, MDN, AWA, DTMF and their combinations with the proposed method are listed. It can be seen that proposed method improves performance significantly. Note that the downscaled results of the AA and AWA are equivalent for CIF format with a downscaling factor of 2. For the Foreman sequence, for example, the proposed method outperforms AA, MDN, AWA and DTMF by 4.24, 3.58, 4.24 and 3.94 dB, respectively. On average, there is a 3.02 dB gain. The demonstrations reveal that the proposed method performs better computational efficiency than the conventional algorithm, which has to decompress the video, re-estimate the downscaled MVs, and re-encode the video for downsizing.

Table 1. PSNR comparison for different test sequences

		Foreman	Football	Mobile	Paris	M&D	Bream
AA	Standard	29.856	22.567	22.690	28.546	36.295	28.499
	Proposed	34.098	24.654	23.660	30.739	39.518	30.622
MDN	Standard	30.318	23.082	23.257	28.973	37.865	29.285
	Proposed	33.898	24.457	23.713	30.669	39.393	30.576
AWA	Standard	29.856	22.567	22.690	28.546	36.295	28.499
	Proposed	34.098	24.654	23.660	30.739	39.518	30.622
DTMF	Standard	30.311	23.113	23.265	29.914	38.312	28.713
	Proposed	34.247	24.999	23.770	30.794	39.975	30.675

In Section 2, it has been mentioned that we use a small search window and compare their sum of absolute differences (SADs) to obtain the best-matching MV. Table 2 lists the ratio of the refined MVs to the total MVs. For example, we now examine the performance of Algorithm AA in the case of Foreman. It can be observed that there are 48.3% MVs which are not the best-matching MVs. By using the proposed algorithm, all these MVs can be properly refined so that we can achieve a 0.442 dB gain for the sequence (Foreman).

Table 2. The ratio of the refined MVs to the total MVs.

	Foreman	Football	Mobile	Paris	M&D	Bream
AA	48.3%	83.4%	17.3%	14.6%	17.4%	24.8%
MDN	42.6%	73.6%	13.7%	12.6%	9.2%	21.2%
AWA	48.3%	83.4%	17.3%	14.6%	17.4%	24.8%
DTMF	45.2%	80.4%	13.9%	13.4%	9.3%	22.37%

4 Conclusion

In this paper, we proposed a method that provides arbitrary-size downscaling in the DCT domain. The method overcomes the mismatch problem due to the DCT block structure for downsizing. To improve video quality, a small-range searching method is used to obtain an optimal estimation of motion vector. The simulation results indicate that the proposed algorithm achieves better performance. Furthermore, our algorithm does not significantly increase the computational cost.

Acknowledgments

This work was supported in part by National Science Counsel Granted NSC 98-2221-E-214-044-MY2 and NSC 99-2221-E-214-055.

References

1. Assuncao, P.A.A., Ghanbari, M.: Transcoding of single-layer MPEG video into lower rates. Proc. Inst. Elect. Eng.—Vision, Image, Signal Processing 144(6), 377–383 (1997)
2. Bjork, N., Christopoulos, C.: Transcoder architecture for video coding. IEEE Trans. Consumer Electron 44, 88–98 (1998)
3. Shen, B., Sethi, I.K., Vasudev, B.: Adaptive motion-vector resampling for compressed video downsizing. IEEE Trans. Circuits Syst. Video Technol. 9, 929–936 (1999)
4. Youn, J., Sun, M.T., Lin, C.W.: Motion vector refinement for high performance transcoding. IEEE Trans. Multimedia 1, 30–40 (1999)
5. Schwarz, H., Marpe, D., Wiegand, T.: Overview of the scalable video coding extension of the H.264/AVC standard. IEEE Trans. on Circuits and Systems for Video Tech. 17, 1103–1120 (2007)
6. Shanableh, T., Ghanbari, M.: Heterogeneous video transcoding to lower spatio-temporal resolutions and different encoding formats. IEEE Trans. on Multimedia 2, 101–110 (2000)
7. Bjork, N., Christopoulos, C.: Transcoder architectures for video coding. IEEE Trans. on Consumer Electronics 44, 88–98 (1998)
8. Tan, Y.P., Sun, H.W.: Fast motion re-estimation for arbitrary downsizing video transcoding using H.264/AVC standard. IEEE Trans. on Consumer Electronics 50, 887–894 (2004)
9. Liu, C.S., Yang, N.C., Kuo, C.M., Wei, W.L.: Motion Vector Re-estimation for Video Transcoding with Arbitrary Video Downsizing. International Journal of Innovative Computing, Information and Control 6(6), 2657–2670 (2010)
10. Kuo, C.M., Hsieh, C.H., Jou, Y.D., Lin, H.C., Lu, P.C.: Motion estimation for video compression using Kalman filtering. IEEE Trans. on Broadcasting 42, 110–116 (1996)
11. Kuo, C.M., Chao, C.P., Hsieh, C.H.: An Efficient Motion Estimation Algorithm for Video Coding Using Kalman Filter. Real-Time Imaging 8, 253–264 (2002)
12. Chang, S.F., Messerschmitt, D.G.: A new approach to decoding and compositing motion-compensated DCT based images. In: Proceedings of the ICASSP, Minneapolos, MN, pp. V.421–V.424 (1993)
13. Chang, S.F., Messerschmitt, D.G.: Manipulation and Compositing of MC-DCT Compressed Video. IEEE Journal Selected Areas in Commun. 13, 1–11 (1995)

Co-channels Interference Rejection in an OFDM System Using Beamforming

Md. Rajibur Rahaman Khan, Modar Safir Shbat, and Vyacheslav Tuzlukov

College of IT Engineering, Electronics Engineering Department
Kyungpook National University,
1370 Sankyuk-dong, Buk-gu, Daegu 702-701, South Korea
rajibur_ckt@yahoo.com, modboss80@knu.ac.kr,
tuzlukov@ee.knu.ac.kr

Abstract. Orthogonal frequency-division multiplexing (OFDM) is one of the promising techniques for modern wireless communication systems. This paper focuses on a key issue for adaptive beamformer algorithm. In this paper, we propose an adaptive beamformer algorithm that is based on least square-constant modulus (LS-CM) property of the OFDM signals which allowing us to effectively suppress co-channel interference (CCI). Simulation results confirm theoretical study of the proposed beamforming algorithm.

Keywords: OFDM, adaptive beamformer, co-channel interference.

1 Introduction

Orthogonal frequency-division multiplexing (OFDM) is a bandwidth-efficient signaling scheme for wideband digital communications system [1] and [2]. Therefore, it has been applied to worldwide standards of wireless local area networks (WLANs) and considered to be a promising technique for high data-rate transmission in wireless communication systems [3] and [4].

The principal advantage of OFDM is that it converts effectively a frequency selective fading channel into a set of parallel flat-fading channels. Both the intersymbol interference (ISI) and intercarrier interference (ICI) can be completely eliminated by inserting between symbols of small time interval known as a guard interval. The guard interval length is made equal to or greater than the delay spread of the channel. If the symbol signal waveform is extended periodically in the guard interval (cyclic prefix (CP)), then orthogonality of the carrier is maintained over the symbol period, and thus eliminating ICI. Also, ISI is eliminated owing to that the successive symbols do not overlap due to the guard interval. Hence, there is no need to perform channel equalization at the receiver and the complexity of the receiver is quite low.

However, one main problem affecting the OFDM wireless communication systems performance is the co-channel interference (CCI) impact. If CCI in OFDM wireless communication system are high and burst, the conventional single channel OFDM

M. Zhu (Ed.): Electrical Engineering and Control, LNEE 98, pp. 959–966.
springerlink.com © Springer-Verlag Berlin Heidelberg 2011

receivers are not able to cancel efficiently those interferences. Recently, the adaptive beamforming technology has been combined with OFDM technique to suppress CCI [1] and [5].

By adding adaptive array at the receiver structure, the beamforming system is able to separate the desired signal from interfering signals which originate based on different spatial locations. Therefore, the wireless communication system capacity is increased. For OFDM-adaptive array wireless communication system the beamforming can be applied to either in the time-domain or frequency-domain [6]. The time-domain beamforming is called the Pre-FFT because the array processing is done before the FFT step in the time domain and the frequency-domain beamforming can be called the Post-FFT because the array processing is done after the FFT step in the frequency domain. The Post-FFT requires a set of weights for each array branch. On the other hand, the Pre-FFT array antenna requires only one weight set and it can reduce the computational complexity [7] and [8].

In this paper, we investigate the proposed adaptive beamforming algorithm with the purpose to cancel CCI existing in the OFDM wireless communication system. The beamforming algorithm is based on the least square-constant modulus (LS-CM) property of the OFDM signals. The proposed beamformer can effectively suppress CCI and improve the OFDM wireless communication system performance at high-level interference. The simulation results confirm our theoretical investigation.

The remainder of this paper is organized as follows. Section 2 describes the OFDM wireless communication system model. In Section 3, we have discussed the proposed beamforming algorithm. To evaluate the proposed approach, a number of computer simulation results are presented in Section 4. Finally, some conclusions are discussed in Section 5.

2 System Model

The block diagram of a modern OFDM transmitter system with adaptive beamformer is shown in Fig. 1, which consists of pilot symbols, zero carriers, and phase-shift keying (PSK) waveforms modulated by the data symbols. Firstly, the high-speed data to be transmitted are converted in N parallel subchannels sending the signal through a serial-to-parallel (S/P) converter. Next, the inverse fast Fourier transform (IFFT) converts the S/P converter output data. The IFFT output signal is periodically extended to form a guard interval and then passes through a parallel-to-serial (P/S) converter. The P/S converter output is the digital-to-analog (D/A) converted signal at the output of the transmit antenna. The transmitted signal comes in at each receiver antenna element of the beamformer with the corresponding direction of arrival (DOA).

Let the S/P converter output in the transmitter can be presented as

$$\mathbf{Y}_k = \left[y(k,0), y(k,1), \dots, y(k,N-1) \right]^T , \tag{1}$$

where $y(k,n)$ is the nth symbol of the kth user, $n = 1,2,\dots,N$; $k = 1,2,\dots,K$, and T denotes the transpose. The $N \times N$ unitary matrix performing the FFT operation is denoted by $\mathbf{U}_k$ with elements

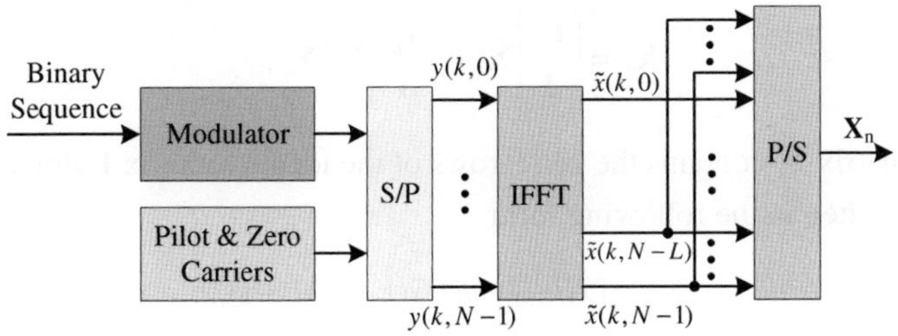

Fig. 1. OFDM transmitter system.

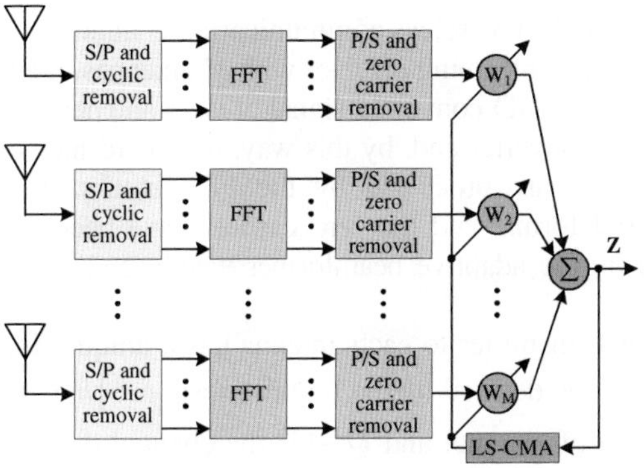

Fig. 2. OFDM receiver system.

$$\mathbf{U}_{kl} = e^{-j2\pi(k-1)(l-1)/N} \tag{2}$$

The corresponding IFFT matrix is the matrix $\mathbf{U}_k^H$, where the superscript denotes the Hermitian transpose. Therefore, the IFFT output can be expressed in the following form

$$\tilde{\mathbf{X}}_k = \mathbf{U}_k^H \mathbf{Y}_k . \tag{3}$$

Additionally, we can present the IFFT output as

$$\tilde{\mathbf{X}}_k = \left[\tilde{x}(k,0), \tilde{x}(k,1), ..., \tilde{x}(k,N-1)\right]^T . \tag{4}$$

The FFT operation matrix for the *k*th user is defined in the following form

$$\mathbf{U}_k = \begin{bmatrix} 1 & 1 & \cdots & 1 \\ 1 & e^{-j2\pi(1)(1)/N} & \cdots & e^{-j2\pi(1)(N-1)/N} \\ \vdots & \vdots & \ddots & \\ 1 & e^{-j2\pi(N-1)(1)/N} & \cdots & e^{-j2\pi(N-1)(N-1)/N} \end{bmatrix} . \tag{5}$$

The IFFT output is cyclically extended to generate the IFFT output signal $\mathbf{X}_k$ by copying the last L elements of sequence $\tilde{\mathbf{X}}_k$ into the beginning (as illustrated in the Fig.1). The IFFT output signal for the *k*th user can be written in the following matrix form:

$$\mathbf{X}_k = \begin{bmatrix} \mathbf{I}_L \\ \mathbf{I} \end{bmatrix} \tilde{\mathbf{X}}_k = \begin{bmatrix} \mathbf{I}_L \\ \mathbf{I} \end{bmatrix} \mathbf{U}_k^H \mathbf{Y}_k , \tag{6}$$

where the matrix $\mathbf{I}_L$ contains the last L rows of the identity matrix $\mathbf{I}$ with a size N and $\mathbf{X}_k$ can be written in the following form

$$\mathbf{X}_k = [\tilde{x}(k, N-L), \tilde{x}(k, N-L+1), \ldots \\ \ldots, \tilde{x}(k, N-1), \tilde{x}(k,0), \tilde{x}(k,1), \ldots, \tilde{x}(k, N-1)]^T . \tag{7}$$

Now, the parallel data $\mathbf{X}_k$ are converted to the serial format signal $\mathbf{X}_n$ through the P/S converter and transmitted by wireless communication channel.

A block diagram of beamforming receiver with M antennas indexed by m is shown in Fig.2. In the receiver, the A/D converted signal at each antenna element passes through the corresponding S/P converter and, by this way, the guard interval data are cyclically removed. These signals are processed by the FFT blocks. Thus, the symbols are converted into a serial format and the zero carriers have been removed. Each branch signal is multiplied by the adaptive beamformer weight and added up to generate the desired signal.

The path from the transmitter to each antenna has a unique finite impulse response (FIR) channel that can be defined by the impulse response $\{h_m(0), h_m(1), \ldots, h_m(Q-1)\}$, where Q is the length of channel and $Q-1$ is the channel delay. We assume that the channel length Q does not exceed the guard interval, i.e. $Q \le L$ and the extended length L of the channel impulse response is appending zeros.

After the discarded CP the received signal vector at the mth antenna element of the beamforming receiver can be presented in the following form

$$\mathbf{V}_{mk} = a_m(\theta)\mathbf{H}_m \mathbf{U}_k^H \mathbf{Y}_k + \mathbf{w}_m , \tag{8}$$

where $a_m(\theta)$ is the mth element of the array steering vector and given by

$$a_m(\theta) = e^{-j2\pi(m-1)d\sin(\theta)/\lambda} \tag{9}$$

In (8) and (9) we use the following notations: $d = \lambda/2$ is the inter-element space for the liner array; λ is wavelength; θ is the angle of arrival (AOA); $\mathbf{H}_m$ is the $N \times N$ channel impulse response circulant matrix with the first column given by

$$\mathbf{h}_m = \left[h_m(0), h_m(1), \ldots, h_m(L-1), 0, \ldots, 0\right]^T . \tag{10}$$

The elements of the noise vector, $\mathbf{w}_m$, are independent and identically distributed (i.i.d) Gaussian variables, each with zero mean and variance σ^2, and can be presented in following form:

$$\mathbf{w}_m = \left[b_m(k,0), b_m(k,1), \ldots, b_m(k, N-1)\right]^T . \tag{11}$$

The received signal vector in (8) at the mth antenna element is processed by the N-point FFT to generate the signal vector in frequency domain and is given by

$$\mathbf{R}_m = \mathbf{U}_k \mathbf{V}_{mk}$$
$$= \mathbf{U}_k a_m(\theta)\mathbf{H}_m \mathbf{U}_k^H \mathbf{Y}_k + \mathbf{N}_m$$
$$= a_m(\theta)\mathbf{\Lambda}_m \mathbf{Y}_k + \mathbf{N}_m, \tag{12}$$

where $\mathbf{N}_m$ is the filtered noise and it can be written in the following form

$$\mathbf{N}_m = \mathbf{U}_k \mathbf{w}_m . \tag{13}$$

The diagonal matrix contains the channel frequency response for the mth antenna element. Therefore,

$$\Lambda_m = \mathrm{diag}\mathbf{U}_k \mathbf{h}_m . \tag{14}$$

Finally, the co-channel signal model with K users can be represented by

$$\mathbf{R}_m = \sum_{i=1}^{K} a_{mi}(\theta)\mathbf{\Lambda}_{mi}\mathbf{Y}_{ki} + \mathbf{N}_m . \tag{15}$$

3 Adaptive Algorithm

The overall output

$$\mathbf{y}_k = [\mathbf{y}(k,0), \mathbf{y}(k,1), ..., \mathbf{y}(k,S-1)]^T \tag{16}$$

of the beamformer for the kth user is obtained by summing the weighted signals across the branches as given by

$$\mathbf{y}_k = \sum_{m=1}^{M} \mathbf{W}_k^* \mathbf{R}_m = \mathbf{W}_k^H \mathbf{R}_m , \tag{17}$$

where the superscript * denotes complex conjugation and $\mathbf{W}_k$ are the tall block matrices with the size $MS \times S$ that allow to couple the same subcarriers of different antenna elements while avoiding the interference of other subcarriers and can be written in the following form:

$$\mathbf{W}_k = [\mathbf{W}_{k1}, \mathbf{W}_{k2}, ..., \mathbf{W}_{kM}]^T . \tag{18}$$

The goal of the adaptive algorithm is to adjust the weights in order to restore the transmitted signal property, such as the constant modulus (CM) property of binary phase-shift keying (BPSK) and quadrature phase-shift keying (QPSK) signals. So, we can define the least square (LS) cost function in the following form [10]

$$\xi = \sum_{n=0}^{R-1} E\left[\left| |y(k,n)|^2 - 1 \right|^2 \right], \tag{19}$$

which stimulates the elements of the vector $\mathbf{y}_k$ to have a constant modulus restoring the CM property of the transmitted data by this way.

The weight update for a set of K users is obtained by computing the gradient of an instantaneous value of the cost function in (19) with respect to the weight matrix $\mathbf{W}_k$ and can be determined in the following form [11].

$$\mathbf{W}_{k+1} = \mathbf{W}_k + \mu \mathbf{J}_k^* \mathbf{R}_k , \quad (20)$$

where μ is a positive step size, which controls the convergence rate and steady-state properties of the algorithm. The resulting beamformer output $\mathbf{Z}$ is processed by a conventional OFDM receiver to decode the data. The block diagonal matrix $\mathbf{J}_k$ of the size $MS \times MS$ has identical blocks with diagonal elements taken from the following vector

$$\mathbf{j}_k = [\mathbf{I} - \text{diag}[y(k) \otimes y^*(k)]] y(k) , \quad (21)$$

where $\otimes$ performs the Schur product. So, we can write the block diagonal matrix $\mathbf{J}_k$ in the following form

$$\mathbf{J}_k = \text{diag}[\mathbf{j}_k^T, ..., \mathbf{j}_k^T]^T . \quad (22)$$

4 Computer Simulation

In this section, we present computer simulation examples to evaluate the performance of the proposed beamforming algorithm that is based on eight uniform linear array (ULA) antennas with $\lambda/2$ distance, and we employed 128 subcarriers OFDM system and 4-QAM modulation. Fig.3 shows the null pattern response of the adaptive array OFDM system versus the azimuth angle in the cases of AWGN fading, and interfering signal. We see that, the main beam have been detected at -40° and four interferences has been located at $-80^\circ, 0^\circ, 40^\circ$ and 55°.

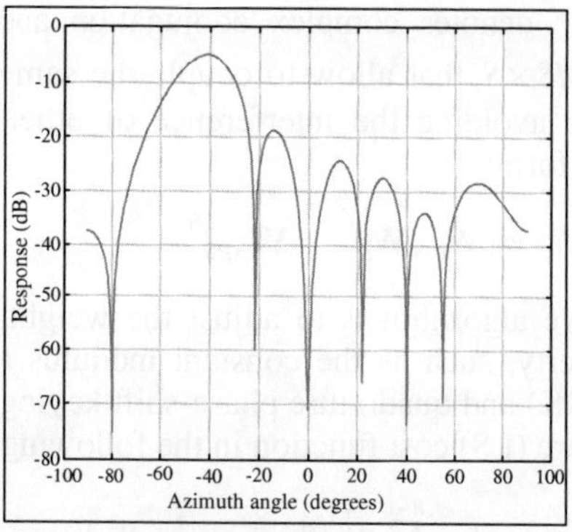

Fig. 3. The beam pattern at the beamformer output. Deep nulls are observed in the directions at $-80^\circ, 0^\circ, 40^\circ$ and 55°.

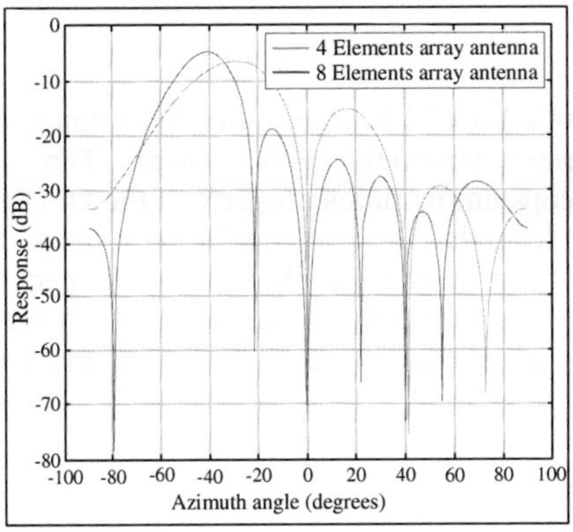

Fig. 4. The beam pattern at the beamformer output in the adaptive array antenna with 4 elements and 8 elements.

Figure 4 shows the beamformer performance versus the azimuth angle under changing the number of antenna elements from 8 to 4. As we can see from Fig. 4, the DOA of the desired signal is at -40° and DOA of four interferers are given at $-80^{\circ}, 0^{\circ}, 40^{\circ}$ and 55° in the case of 8 elements antenna. However, in the case of 4 elements, the beamformer was not able to cancel all the interfering users, i.e. there is no null at the spatial location at 55°, and the DOA of the interferers are given at 0° and 40°.

Figure 5 represents the bit error rate (BER) performance of the investigated OFDM system with the proposed beamforming algorithm versus SNR and signal-to-interference ratio (SIR) as a parameter in the case of AWGN. The proposed beamformer presents the better performance in comparison with the conventional receiver discussed in [10].

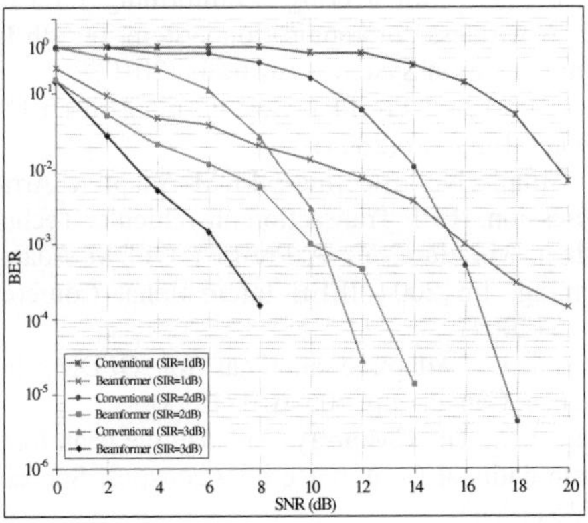

Fig. 5. BER performance versus SNR for AWGN channel.

5 Conclusion

In this paper, we present a LS-CM beamforming algorithm with the purpose to cancel CCI in OFDM wireless communication system. The proposed beamforming algorithm exploits the constant modulus property of the OFDM signals and makes the beamformer to be able to update the weights in the blind way that can decrease the system dependence on the pilot symbols. We analyze the performance of the proposed beamforming algorithm via computer simulations which confirm an ability of the proposed beamforming algorithm to suppress effectively CCI in OFDM wireless communication systems.

Acknowledgments

This research was supported by the Kyungpook National University Research Grand, 2009, and Industry-Academic Cooperation Foundation, Kyungpook National University and SL Light Corporation Joint Research Grant (the Grant No. 201014590000).

References

1. Khan, R.R., Tuzlukov, V.: Beamforming for rejection of co-channels interference in an OFDM system. In: 3rd International Congress on Image and Signal Processing, Yantai, China, pp. 3318–3322 (2010)
2. Xiaodong Wang, H., Poor, V.: Wireless Communication Systems:Advanced Techniques for Signal Reception. Communications Engineering and Emerging Technologies Series. Prentice-Hall, Englewood Cliffs (2002)
3. Van Nee, R., Awater, G., Morikura, M.: New High-rate Wireless LAN Standards. IEEE Communications Magazine 37, 82–88 (1999)
4. Chuang, J., Sollenberger, N.: Beyond 3G: Wideband Wireless Data Access Based on OFDM and Dynamic Packet Assignment. IEEE Communications Magazine 38, 78–87 (2000)
5. Khan, R.R., Tuzlukov, V.: Null-steering beamforming for cancellation of co-channel interference in CDMA wireless communication system. In: 4th Int. Conference on Signal Processing and Communication Systems, Australia (2010)
6. Lei, Z., Chin, F.P.S.: Post and Pre-FFT Beamforming in an OFDM System. Vehicular Technology Conference 1, 39–43 (2004)
7. Hara, S., Hara, Y.: Simple Null-Steering OFDM Adaptive Array Antenna for Doppler-Shifted Signal Suppression. IEEE Transaction on vehicular Technology 54, 91–99 (2005)
8. Hara, S., Budsabathon, M., Hara, Y.: A Pre-FFT OFDM Adaptive Antenna Array with Eigenvector Combining. In: 2004 IEEE International Conference on communication, vol. 4, pp. 2412–2416 (2004)
9. Vook, F.W., Baum, K.L.: Adaptive antennas for OFDM. In: IEEE 48th Vehicular Technology Conference, Canada, pp. 606–610 (1998)
10. Gooch, R.P., Lundell, J.D.: The CM array: An adaptive beamformer for constant modulus signals. In: IEEE International Conference on Acoustics, Speech, and Signal Processing, Japan, pp. 2523–2526 (1984)
11. Venkataraman, V., Shynk, J.J.: Adaptive algorithms for OFDMA wireless Ad Hoc networks with multiple antennas. In: 38th Asilomar Conference on Signals, Systems, and Computers, Pacific Grove, CA, pp. 110–114 (2004)

Research of Super Capacitors SOC Algorithms

Hao Guoliang, Liu Jun, Li Yansong, Zhang Qiong, and Guo Shifan

School of electrical & electronic Engineering
North China Electric Power University (Beijing)
102206, Beijing, China
haoguoliangde@163.com

Abstract. SOC is one of important parameters in energy control of super capacitors. Analyzed and compared the advantages and disadvantages in existing SOC algorithms between super capacitors with traditional batteries. This paper established first order nonlinear model of super capacitors which were used as power system storage devices. And considered the impact of equivalent capacitance value produced by terminal voltage and self-discharge effects. Based on first order nonlinear model and according to the optimal Kalman filtering, this paper designed a new algorithm which combined Kalman filter theory, A-H method and terminal voltage.

Keywords: super capacitors, first order nonlinear model, SOC, Kalman filtering.

1 Introduction

Super capacitor has great power density, can be charged and discharged fast with high current. It also has long cycle life, high safety factor, wide suitable temperature range, green material [1] and many other advantages. Because of the benefits, it becomes the first choice of energy storage in microgrid.

SOC of the super capacitor which is the ratio remaining power to the total, could be defined with charge [2] or energy [3]. The former is used widely, and adopted by this paper. Computational formula as followed.

$$SOC = 1 - Q / Q_{\max} \tag{1}$$

Where, $Q_{\max}$ presents the maximum charge can be released by super capacitor from fully charged until fully discharged. $Q = \int i(t)dt$ presents the charge of super capacitor which is charged in a period (the current is positive when discharged and negative when charged).

Currently, there are many accuracy and easily used SOC algorithms which are used in traditional batteries, such as Open circuit voltage approach, A-H approach , internal impedance approach, fuzzy logic-based approach, discharge approach, terminal voltage approach, neural network approach Kalman filtering approach and so on.

The approaches which used in calculating SOC of traditional batteries could be used for reference of super capacitor, while, there are many differences in characters

M. Zhu (Ed.): Electrical Engineering and Control, LNEE 98, pp. 967–973.
springerlink.com © Springer-Verlag Berlin Heidelberg 2011

of charge and discharge between traditional batteries and super capacitors. So the approaches applied to super capacitors in some papers are A-H approach [4], terminal voltage approach [5] and Kalman filtering approach [6].

Considering the application that super capacitors were used as energy storage devices in microgrid and the character of charge and discharge, this paper designed a new SOC algorithm which combined A-H approach, terminal voltage approach and Kalaman filtering approach. It was a closed-loop SOC algorithm can improve the accuracy significantly.

2 State Space Model of Super Capacitor

Establishing model of super capacitor is the base of SOC calculating with Kalman filtering algorithm. Considering equivalent capacitance of super capacitors is not constant but varies with its voltage, and the character of self-discharge, this paper improved first-order nonlinear model of a super capacitor.

2.1 Equivalent Circuit Model of Super Capacitor

The equivalent circuit model of super capacitor as shown:

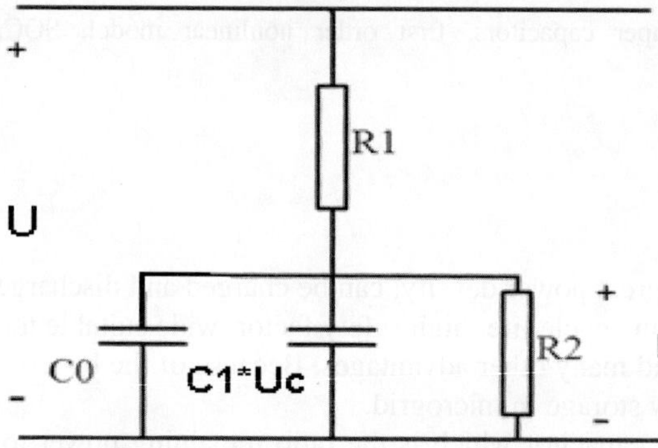

Fig. 1. The equivalent circuit model of super capacitor

$$U(t) = R_1 I(t) + U_C(t) \quad (2)$$

$$I(t) = C\frac{dU_C(t)}{dt} + \frac{U_C(t)}{R_2} \quad (3)$$

$$SOC(t) = SOC(t_0) - \frac{1}{Q_0}\int_{t_0} I(t)dt \quad (4)$$

$$U_C(t) = F[SOC(t)] \tag{5}$$

Where, $C = C_0 + C_1 * U_C$,represented the value of ideal capacitor varies with terminal voltage of super capacitor, is variable.

Formulas (2) and (3) reflected the electrical relationship of first order nonlinear model. $SOC(t)$ represented the state of super capacitor charge at the time t. Q_0 was the full capacity. $F[SOC(t)]$ was the nonlinear function about SOC and terminal voltage of super capacitor. We got a mathematical model of the super capacitor which considered the current integration.

Then, discretized the model above and got the state space model convenient for computing in the computer:

$$\begin{bmatrix} SOC(k+1) \\ U_C(k+1) \end{bmatrix} = \begin{bmatrix} 1 & 0 \\ 0 & \exp(-\frac{\Delta t}{R_2 C}) \end{bmatrix} \begin{bmatrix} SOC(k) \\ U_C(k) \end{bmatrix} + \begin{bmatrix} -\frac{\Delta t}{Q_0} \\ R_2(1-\exp(-\frac{\Delta t}{R_2 C})) \end{bmatrix} I(k) + \begin{bmatrix} w_1(k) \\ w_2(k) \end{bmatrix} \tag{6}$$

$$U(k) = R_1 I(k) + F[SOC(k)] + v(k) \tag{7}$$

The input of the state space model was the current $I(k)$, the output was terminal voltage $U(k)$, and the state variables were $X(k) = \begin{bmatrix} SOC(k) \\ U_C(k) \end{bmatrix}$. $w(k) = \begin{bmatrix} w_1[k] \\ w_2[k] \end{bmatrix}$ were disturbances of state variables brought by some unmeasurable random variables and $v(k)$ was measurement noise of super capacitor.

First order nonlinear model can describe the character that the equivalent capacitance varies with terminal voltage, and the self-discharge effect. This model considered the current integration on the basis of first order nonlinear model, while put the SOC as part of state variables. By using this model, the value of SOC can be obtained directly through state estimation which made use of recursive Kalman filtering.

2.2 Model Parameters Identification

Currently, linear least squares and nonlinear least squares methods are used in model parameter identification, while, these methods requires a large number of experiments. This paper designed a circuit analysis approach to identify the parameters of first order nonlinear model. This method is simpler and easier to implement. This circuit analysis method required that the current and voltage measurements must be accurate, otherwise bigger error would emerge. This method adopted constant current [7] to charge and discharge the super capacitor.

1) R_1

We charged the super capacitor at the time $t=t_0$, the initial voltage $v_0=0$ and the initial charge $Q_0=0$.

At the time $t=t_1$, the charging current was increasing and the terminal of super capacitor could be seen as the voltage drop on the R_1.Measured the voltage V_1 and the current I_1 on the super capacitor, then R_1 was calculated as follows:

$$R_1 = \frac{V_1}{I_1} . \tag{8}$$

2) Constant capacitance C_0

When the time $t=t_2$, the current stabilized, and measure the voltage of super capacitor V_2. This point was the initial stage of charging, and the voltage of equivalent capacitor was so small that could be neglected, so the voltage of super capacitor V_2 was equal to the voltage drop on the resistance R_1 and the voltage drop on constant capacitance C_0.According to these, the C_0 was calculated as:

$$C_0 = \frac{\int_0^{t_2} i(t)dt}{V_2 - I_c \times R_1} \tag{9}$$

3) Variable capacitor C_1

When the time $t=t_3$, the voltage drop on the super capacitor reached the rated voltage V_3, and stop charging. At this moment, $V_3 = V_{C1} + I_C * R_1$,the charge Q_T which the super capacitor stored was equal to the charge in rising current period and constant current period. Shown in the following equation:

$$Q_T = I_C \times (t_3 - t_2) + \int_0^{t_2} i(t)dt . \tag{10}$$

And according to the equation:

$$Q_T = \int C_1 dv = C_0 U_C + \frac{C_1 U_C^2}{2} \tag{11}$$

Then the value of C_1 was got by solving the simultaneous two equations.

$$C_1 = \frac{2}{V_{C1}} \left(\frac{I_C (t_3 - t_2) + \int_0^{t_2} i(t)dt}{V_{C1}} - C_0 \right) \cdot \tag{12}$$

4) R_2

The resistance R_2 reflected the self-discharge effect. Because the resistance was very large, could be neglected in the stage of charging. Super capacitor began to self-discharge after stopping charging. A period later, When the time $t=t_4$, the terminal voltage of the super capacitor was V_4, and after a period of time, when the time $t=t_5$, the terminal voltage was V_5.R_2 is calculated as followed:

$$R_2 = \frac{\int_{t_4}^{t_5} U_C(t)dt}{C_0 U_C + \frac{C_1 U_C^2}{2}} \tag{13}$$

$$Q_T = \int_{t_4}^{t_5} \frac{U_c(t)}{R_2} dt = \int [C_0 + C_1 U_C(t)] dU_C(t) \tag{14}$$

3 Kalman Filtering Algorithm Based on First Order Nonlinear Model

Making minimum variance optimal estimate to the state of dynamic system was the core idea of Kalman filter theory. In the super capacitor SOC calculating, the super capacitor could be seen as the dynamic system and SOC was the internal state of the system. General mathematical form of the super capacitor model was followed as:
Equation of state:

$$x_{k+1} = A_k x_k + B_k u_k + \omega_k = f(x_k, u_k) + \omega_k \, . \tag{15}$$

Observation equation:

$$y_k = C_k x_k + v_k = g(x_k, u_k) + v_k \tag{16}$$

The input vector of system was u_k, which included the current, the temperature, the remaining capacity, the variables such as resistance and so on. The output vector of system was the operating voltage of super capacitor. SOC was included in the state variable of the system which was x_k. $f(x_k, u_k)$ and $g(x_k, u_k)$ were nonlinear equations which determined by the super capacitor model, and must be linearized. The center of calculating SOC was a set of recursive equation which included SOC estimates and the covariance matrix. The covariance matrix reflected the estimated error and gave the range of estimated error.

The state equation (6) of the state space model of super capacitor was linear, while only the function $F[SOC(t)]$ in the output equation (7) was nonlinear. So that, we just linearized the output equation [8] then got the following matrix A, B and C corresponding to the Kalman filter algorithm equations (15) and (16).

$$A(k) = \begin{bmatrix} 1 & 0 \\ 0 & \exp(-\frac{\Delta t}{R_2 C}) \end{bmatrix}, \quad B(k) = \begin{bmatrix} -\frac{\Delta t}{Q_0} \\ R_2[1 - \exp(-\frac{\Delta t}{R_2 C})] \end{bmatrix},$$

$$C(k) = \left[\frac{\partial F(SOC_K)}{\partial SOC} \quad -1 \right] \Bigg|_{X(k) = \hat{X}(k|k-1)}$$

Kalman filter algorithm was composed by filter calculating and the gain of filter calculating. First, the filter got the predicted value of state variable ($\hat{X}(k \mid k-1)$) according to the filtering results ($\hat{X}(k-1)$) last step, then got the $\hat{U}(k)$ according to the output equation (7), and then calculated the predicted error on the basis of actual measurement, finally modified the new filtering results of state variables predicted value. That, we got the filtering result through the cycle "prediction-correction-prediction ".

The gain of filter $K(k)$ was the weight of measured value of terminal voltage in revising the SOC. If the error of initial state variable value was large, the gain of filter was large too, and also the weight of measured value of terminal voltage. Now, the gain of filter did more for the correction of SOC. With the decrease of error, the output values were closer to the measured values, and the $K(k)$ became smaller, so the gain of filter did less for the correction of SOC.

Assumed that the initial expectation of initial state variable value was $\hat{X}(0)$, the variance was $P(0)$, and the variance matrix of state variable predicted error and filtering error are $P(k \mid k-1)$ and $P(k)$ .The flow chart of Kalman filter calculation was shown as below:

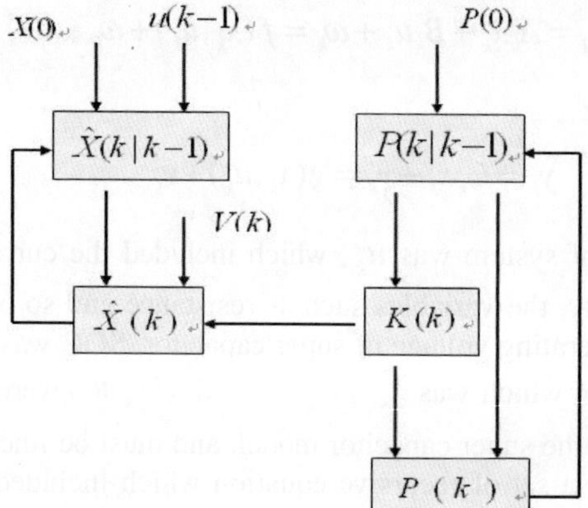

Fig. 2. The flow chart of Kalman filter calculations

4 Conculsion

This paper presented the first order nonlinear model of super capacitor which considered the self-discharge effect, and identified the parameters of this model. Based on this model, a new method was designed,which combined A-H approach , terminal voltage approach and Kalman filtering approach. SOC was calculated with A-H approach, at the same time, the terminal voltage of super capacitor was estimated, as the basis for correcting the SOC. Taking into account the size of noise and the error to determine the filter gain of every step (K(k)) and get the weight of terminal voltage approach in SOC calculating. Having done these, we got the optimal estimation of SOC. Through combining the methods of A-H and terminal voltage, the terminal voltage method overcame the shortcoming of A-H method that it had cumulative error and made the calculating SOC with closed-loop come true. Meanwhile, because of considering the effect of noise, this algorithm had a strong inhibiting effect.

References

1. Wang, X., Guo, J.-h., Xie, Q.-h., Huang, W.: Applications of Super Capacitor in Microgrid. Power System and Clean Energy 25(6) (June 2009)
2. Ma, Y.-l., Chen, Q.-s., Zhu, Y.: Research on the definition of SOC of batteries in the condition of changing currents. Battery Bimonthly 31(1) (2001)
3. Minghui, H., Datong, Q., Hong, S., Bin, P.: The evaluation of SOC in hybrid cars battery management system. Journal of Chongqing University 26(4), 8 (2003)
4. Wenbing, X.: Design and research on energy monitoring system for super capacitor. Shanghai Jiao Tong University (2008)
5. Qiang, G.-b., Li, Z.-x., Chen, J.: Modeling of Ultracapacitor Energy Storage Used in Hybrid Electric Vehicle. Energy Technology 26(2) (2005)
6. Plett, G.L.: Extended Kalman filtering for battery management systems of LiPB-based HEV battery packs. Journal of Power Sources 134, 252–261 (2004)
7. Haidong, L.: Study on Supercapacitor modular Technology. Institute of electrical engineering, Chinese academy of sciences (2006)
8. Xia, C.-y., Zhang, S., Sun, H.-t.: A strategy of estimating state of charge based on extended Kalman filter. Chinese Journal of Power Sources 31(5) (2007)

The Survey of Information System Security Classified Protection

Zhihong Tian, Bailing Wang, Zhiwei Ye, and Hongli Zhang

Research Center of Computer Network and Information Security Technology
Harbin Institute of Technology
Haerbin, Heilongjiang Province, China
{tianzhihong,wbl,yezhiwei,zhl}@hit.edu.cn

Abstract. Information security has become very important in most organizations. The state has carried out classified protection system to safeguard the information system. This article discussed the classified protection of information security criteria and technological essentials in the implementation for classified security protection.

Keywords: classified protection, TCB, access control, identification and authentication.

1 Introduction

Security comes into play once computer systems process sensitive data or run sensitive services. The classified protection of information system security classify systems into five broad hierarchical classes of enhanced security protection. They provide a basis for the evaluation of effectiveness of security controls built into automatic data processing system products. It is a protection system carried by the state to safeguard the information system [1].

In general, secure systems will control, through use of specific security features, access to information such that only properly authorized individuals, or processes operating on their behalf, will have access to read, write, create, or delete information.

2 Criteria

The criteria are divided into five classes ordered in a hierarchical manner. Each class represents a major improvement in the overall confidence one can place in the system for the protection of sensitive information. Assurance of correct and complete design and implementation for these systems is gained mostly through testing of the security-relevant portions of the system. The security-relevant portions of a system are referred to as the Trusted Computing Base (TCB) [2].

2.1 Discretionary Security Protection

The Trusted Computing Base (TCB) of a class (C1) system nominally satisfies the discretionary security requirements by providing separation of users and data. It

M. Zhu (Ed.): Electrical Engineering and Control, LNEE 98, pp. 975–980.
springerlink.com © Springer-Verlag Berlin Heidelberg 2011

incorporates some form of credible controls capable of enforcing access limitation on an individual basis, i.e. ostensibly suitable for allowing users to be able to protect project or private information and to keep other users from accidentally reading or destroying their data. The class (C1) environment is expected to be one of cooperating users processing data at the same level of sensitivity. The following are minimal requirements for systems assigned a class (C1) rating [3]:

1) Security Policy: Discretionary Access Control.
2) Accountability: Identification and Authentication.

2.2 Controlled Access Protection

Systems in this class enforce a more finely grained discretionary access control than (C1) systems, making users individually accountable for their actions through login procedures, auditing of security-relevant events, and resource isolation. The following are minimal requirements for systems assigned a class (C2) rating [4]:

1) Security Policy: Discretionary Access Control, Object Reuse.
2) Accountability: Identification and Authentication, Audit.

2.3 Labeled Security Protection

Class (B1) systems require all the features required for class (C2). In addition, an informal statement of the security policy model, data labeling, and mandatory access control over named subjects and objects must be present. The capability must exist for accurately labeling exported information. Any flaws identified by testing must be removed. The following are minimal requirements for systems assigned a class (B1) rating:

1) Security Policy [5]: Discretionary Access Control, Object Reuse, Labels, Mandatory Access Control.
2) Accountability: Identification and Authentication, Audit.

2.4 Structured Protection

In class (B2) systems, the TCB is based on a clearly defined and documented formal security policy model that requires the discretionary and mandatory access control enforcement found in class (B1) systems be extended to all subjects and objects in the ADP system. In addition, covert channels are addressed. The TCB must be carefully structured into protection-critical and non- protection-critical elements. The TCB interface is well-defined and the TCB design and implementation enable it to be subjected to more thorough testing and more complete review. Authentication mechanisms are strengthened, trusted facility management is provided in the form of support for system administrator and operator functions, and stringent configuration management controls are imposed. The system is relatively resistant to penetration. The following are minimal requirements for systems assigned a class (B2) rating [6]:

1) Security Policy: Discretionary Access Control, Object Reuse, Labels, Mandatory Access Control.
2) Accountability: Identification and Authentication, Audit.

2.5 Verified Protection

The class (B3) TCB must satisfy the reference monitor requirements that it mediate all accesses of subjects to objects, be tamperproof, and be small enough to be subjected to analysis and tests. To this end, the TCB is structured to exclude code not essential to security policy enforcement, with significant system engineering during TCB design and implementation directed toward minimizing its complexity. A security administrator is supported, audit mechanisms are expanded to signal security-relevant events, and system recovery procedures are required. The system is highly resistant to penetration [7]. The following are minimal requirements for systems assigned a class (B2) rating:

1) Security Policy: Discretionary Access Control, Object Reuse, Labels, Mandatory Access Control.

2) Accountability: Identification and Authentication, Audit.

3 Technological Essentials

3.1 The Trusted Computing Base

In order to encourage the widespread commercial availability of trusted computer systems, these evaluation criteria have been designed to address those systems in which a security kernel is specifically implemented as well as those in which a security kernel has been implemented. The latter case includes those systems in which objective is not fully supported because of the size or complexity of the reference validation mechanism. For convenience, these evaluation criteria use the term Trusted Computing Base to refer to the reference validation mechanism, be it a security kernel, front-end security filter, or the entire computer system [8].

The heart of a trusted computer system is the Trusted Computing Bas (TCB) which contains all of the elements of the system responsible for supporting the security policy and supporting the isolation of objects (code and data) on which the protection is based. The bounds of the TCB equate to the "security perimeter" referenced in some computer security literature. In the interest of understandable and maintainable protection, a TCB should be as simple as possible consistent with the functions it has to perform. Thus the TCB includes hardware, firmware, and software critical to protection and must be designed and implemented such that system elements excluded from it need not be trusted to maintain protection. Identification of the interface and elements of the TCB along with their correct functionality therefore forms the basis for evaluation [9].

For general-purpose systems, the TCB will include key elements of the operating system and may include all of the operating system. For embedded systems, the security policy may deal with objects in a way that is meaningful at the application level rather than at the operating system level. Thus, the protection policy may be enforced in the application software rather than in the underlying operating system. The TCB will necessarily include all those portions of the operating system and application software essential to the support of the policy. Note that, as the amount of code in the TCB increases, it becomes harder to be confident that the TCB enforces the reference monitor requirements under all circumstances [10].

3.2 Access Control

The purpose of access control is to limit the actions or operations that a legitimate user of a computer system can perform. Access control constrains what a user can do directly, as well as what programs executing on behalf of the users are allowed to do. In this way access control seeks to prevent activity that could lead to breach of security [11].

Discretionary access control policies assign the right to access protected resources to individual users. In the traditional formulation, a policy is given as a matrix recording permitted operations directly for each (subject, object) pair. (This presupposes that there has already been a mapping from current subjects to principals.) In actual implementations, the policy is typically stored in the form of access control lists for each object, stating the principals allowed to access the object, together with the permitted mode of access. Capabilities are a design alternative where each subject is associated with a data structure recording the objects the subject is permitted to access. Groups are an elementary layer of indirection in user-centric access control [12].

In mandatory access control, subjects and objects are labeled with security levels. Security levels form a (partially) ordered set. Traditional security levels are unclassified, confidential, secret, and top secret; more sophisticated policies refer to lattices of security levels. Security levels constitute a layer of indirection between subjects and objects. The policies for protecting the confidentiality of classified data demand that a subject can only read objects at its own or at a lower level (no read up). To prevent unauthorized declassification of information, a subject may only write to objects at its own or higher levels.

3.3 Identification and Authentication

For the system to be secure, the system must assure that only authorized users can log in and that they log in only as they are authorized to log in. Identification is the mechanism by which, via the login name, the system recognizes a user as legitimate. Authentication is the mechanism by which, via the password, the system verifies the identity of a user. Together, these mechanisms are known as Identification and Authentication (I & A) [13].

The algorithm for I & A checks the expiration date associated with the user ID. In addition, you are strongly encouraged to assign unique user logins, rather than permit the use of "group" logins or the sharing of a login by more than one user.

When a user is identified and authenticated by the system, certain information about the user's access to information is also revealed, namely the access attributes of the user. Only if the user's access attributes meet the requirements for accessing an object can the user access the object.

The I & A mechanisms prevent unauthorized users from logging in to your system, and they ensure that users log in only to areas for which they are authorized. These mechanisms supply the ``who" information to the system so that the system can make decisions about and enforce the ``who can access what" parts of the security policy [14].

3.4 Security Audit

Information security audit is an audit of how the confidentiality, availability and integrity of an organization's information is assured. It is one of the best ways to determine the security of an organization's information without incurring the cost and other associated damages of a security incident.

IS security auditing involves providing independent evaluations of an organization's policies, procedures, standards, measures, and practices for safeguarding electronic information from loss, damage, unintended disclosure, or denial of availability. The broadest scope of work includes the assessment of general and application controls. The current state of technology requires audit steps that relate to testing controls of access paths resulting from the connectivity of local-area networks, wide-area networks, intranet, Internet, etc., in the IT environment.

3.5 Object Reuse

Object reuse refers to the allocation or reallocation of system resources (storage objects) to a subject. Security requires that no system resource can be used to pass data from one process to another in violation of the security policy. This includes internal system resources not normally visible to users such as buffers and caches[15]. The administrator needs to be aware of some methods for reusing removable media such as floppy disks and tapes. As physical objects, these are not under the control of the system software, and it is the responsibility of the administrator to manage their reuse in a secure fashion.

All removable media should be physically labeled with a description of the sensitivity of the data they contain.

If a removable storage medium needs to be reused at a level other than that described on the physical label, such as might occur with a tape used for backups of sensitive system files being recycled for other uses when no longer needed, the medium should be bulk erased and relabeled to prevent subsequent users from retrieving data from the medium.

3.6 Security Evaluation

Customers purchasing and deploying an IT system may desire to get some assurance about its security. Security evaluation is meant to address this need, analyzing the level of security provided by a product given a set of generic security requirements. The results are of potential interest for a larger number of customers. Certification analyses a product with respect to the requirements of a specific customer. Accreditation is the customer's decision to deploy specific products.

The Trusted Computer Systems Evaluation Criteria (Orange Book) was a first influential contri-bution to security evaluation. It was driven by the requirements of the defense applications of the 1970s and in essence written for evaluation of operating sys-tems. The demands of commercial security triggered developments in security evaluation that led to the Common Criteria. The Common Criteria have been applied to security controls in the core of IT systems and to add-on security components.

Smart cards are a success story of security evaluation. High assurance evaluation has revealed flaws in products. Factors contributing to this success are market demand, functionality that is by and large fixed so evaluation results do not immediately get out of date, and the fact that smart cards are primarily security devices. The security evaluation of operating systems has not been a similar success. There is a wide variety of user requirements. Functionality is complex and evolving so evaluations are almost by default out of date. High assurance evaluation is hardly feasible and lower assurance evaluation hardly finds flaws. Security mechanisms are increasingly placed outside the operating system, e.g., in browsers. Hence, the use of an evaluated operating system may not deliver the desired security guarantees.

Evaluation of application software is even more challenging. Applications are heavily customized so rapid and adaptable evaluation methodologies are needed. Security at the application layer is moving closer to the end user; hence, unsophisticated users have to be considered when assessing the usability of security features.

Acknowledgment

This paper is supported by the National Natural Science Foundation of China under Grant No.60903166; the National High-Tech Research and Development Plan of China under Grant Nos. 2009AA01Z437. the PLA General Armament Department Pre-Research Foundation of China under Grant No.9140A150601090Z08. the Fundamental Research Funds For the Central Universities under Grant No. HIT.NSRIF.2010041.

References

1. GB 17859-1999. Classified criteria for security protection of Computer information system
2. GBT 22239-2008. Information security technology–Baseline for classified protection of information system security
3. GBT 22240-2008. Information security technology–Classification guide for classified protection of information system security
4. GBT XXXX-XXXX. Testing and Evaluation Criteria for Security Classified Protection of Information System
5. GB/T XXXX-XXXX. Information Security Technology–Guide of Implementation for Classified Security Protection of Information System
6. U.S. Department of Defense (DoD). Trusted Computer System Evaluation Criteria
7. Common Criteria Editorial Board. Common Criteria for Information Technology Security Evaluation
8. Dieter Gollmann. Computer security
9. Sandhu, R.S.: Pierangela Samarati. Access Control: Principles and Practice
10. Lampson, B., Abadi, M., Burrows, M.: Authentication in Distributed Systems: Theory and Practice
11. Hayes, B.: Conducting a Security Audit
12. Sandhu, R.S.: Role-Based Access Control Models
13. Za, H., Fan, H.: The application and program of Classified Protection system, http://www.phei.com.cn/bookshop/bookinfo.asp?booktype=main&bookcode=TP102620
14. Fan, H, Li, J.: http://www.venustech.com.cn/NewsInfo/346/5863.Html
15. Liang, Z.: The application of system security based on the Classified Protection, http://www.cnki.com.cn/Article/CJFDTOTAL-XXDL200908082.htm

Optimization of a Fuzzy PID Controller

Dongxu Liu, Kai Zhang, and Jinping Dong

Dept. Engineering Physics, Tsinghua University
Beijing, China

Abstract. There are strong nonlinear characteristics in an active magnetic bearing (AMB) system. For a simple magnetic levitation ball model with one degree of freedom (DOF), there are problems that are difficult to be solved by using linear control methods (such as PID control method). Generally a fuzzy PID controller can perform better than a PID controller during the initial levitation process. But it's difficult to design and optimize a fuzzy PID controller, because it mostly depends on experts' knowledge and experiences. For a one DOF levitation ball model, the membership functions (MFs) were modified, the fuzzy rules were optimized, and some other ways were also tried to improve the performances of the fuzzy PID controller. And it turned out that the most effective one is to optimize the fuzzy rules. The modified fuzzy rules enhanced the anti-shock abilities of the fuzzy PID controller greatly, and at the same time the controller still performs well during the initial levitation process.

Keywords: Nonlinear; fuzzy control; anti-shock abilities.

1 Introduction

Compared with traditional bearing technologies, AMB technology is a novel one because it has the following advantages—no contact, no lubrication, no wear etc... So it is increasingly being used in various high-precision machineries or extreme environments [1]. However, since there are strong nonlinear characteristics in an AMB system, there always are some problems difficult to be solved in its controller design, and it restricts this technology from wider application.

Among all the control theories, linear control methods are relatively easy to be realized and are relatively mature. So for an AMB controller design, a simple method is to linearize the nonlinear magnetic model near an operating point, and using PID or other linear control methods to design a controller [2]. However, for an AMB system, using a linear method must meet two small offset assumption conditions: firstly, the rotor's displacement offset near the operating point should be small enough; secondly, compared with a bias current, the control current should also be small enough; thus it will limit the applications for a wider range of industrial plants. Among the nonlinear control theories, fuzzy control is a relatively easy one. So it seems that fuzzy logic can be used to get better performances in a controller. It has a variety of application forms. The fuzzy logic can be used for PID parameter tuning [3], or directly build a T-S fuzzy model [4]; the input variables of a fuzzy inference system can be the displacement and velocity signals [5], it can also be the displacement and current

M. Zhu (Ed.): Electrical Engineering and Control, LNEE 98, pp. 981–988.
springerlink.com © Springer-Verlag Berlin Heidelberg 2011

signals [6]. No matter in which ways Fuzzy Control method is used to design a controller, the nonlinear effects of the AMB system can be compensated more or less, and better performances can be approached than using a PID controller. For there are strong nonlinear characteristics during the initial levitation process, it requires comprehensive performances of the controller. So the controller performances can be evaluated during the initial levitation process. However, a fuzzy controller is designed mostly based on experiences, and there are few general ways to follow. It's difficult to evaluate the performances of a controller, and there are few general ways to follow to optimize the fuzzy PID controller parameters. Some ways to optimize the fuzzy controller were discussed: the simulation of the fuzzy PID controller with the self-tuning scaling factors shows a better performance in the transient and steady state response, and the MFs are also tuned to optimize the controller [7]; a concept of upper and lower MFs are introduced, and the fuzzy controller with upper and lower MFs can handle the rule uncertainties [8]; genetic algorithms (GA's) are introduced to design the MFs and fuzzy rules, and robustness examination has been done in [9]. Generally speaking, optimizing a fuzzy controller is more difficult than designing one.

2 Simulation Model

As shown in Figure 1, the simulation model was applied. The ball mass m=0.284kg, the gravity coefficient g=10N/kg, the gap between the ball and the magnet is 0.125mm, the coil turns n=50, the magnet pole area A_l=32mm^2. The equilibrium position of the ball is set as an operating point, that is, s_0= 0.125mm, select a bias current i_0=2A. In order to prevent the ball from damage, a backup bearing was used (not shown in the figure). And the gap between the ball and the backup bearing is 0.075mm. So the maximum displacement for the ball is 0.075mm, not 0.125mm. Neglecting the magnetic leakage flux, saturation and other factors, a magnetic force formula can be obtained for the dual-pole AMB system in differential mode.

$$f = k[\frac{(i_0+i_x)^2}{(s_0-x)^2} - \frac{(i_0-i_x)^2}{(s_0+x)^2}],\ k = \frac{1}{4}\mu_0 n^2 A_l\ . \tag{1}$$

Assume that the stiffness K=k_s, and a theoretical P-value of natural stiffness can be obtained by the following formula.

$$P = (K + k_s)/k_i = 2\frac{k_s}{k_i}\ . \tag{2}$$

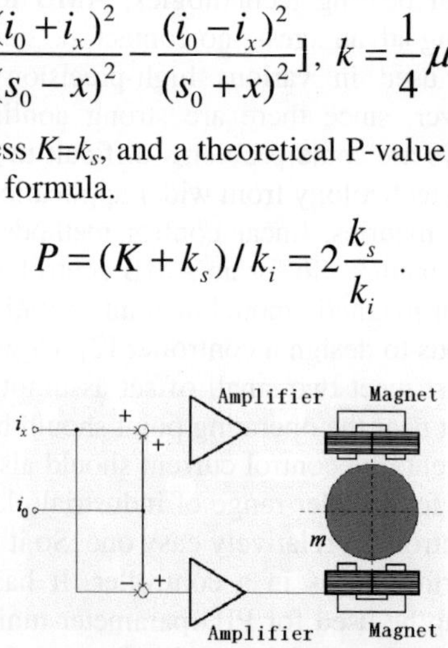

Fig. 1. Dual-pole AMB system

3 Fuzzy PID Controller

A fuzzy PID controller is obtained combined the fuzzy logic block with the PID controller. The fuzzy inference system can be used to tune PID parameters according to the system's states, and better performances can be approached by the fuzzy PID controller. The simulation structure is shown in Figure 2: the structure of the whole AMB system is shown in Fig. a, the structure of the fuzzy PID controller system is shown in Fig. b, and the structure of the magnetic bearing system is shown in Fig. c. As shown in Fig. b, the displacement error signal and its differentiation signal are the input of the fuzzy logic block, and the PID parameters are obtained through the fuzzy processing. The control signal is the output of the fuzzy PID block. As shown in Fig. c, the nonlinear magnetic force model is used in the simulation. A state-space model is used to build a second-order integrator. So it's possible to set the ball's initial displacements and initial velocities. It should be noted that the displacement in Fig. c cannot be used for magnetic force formula directly, so the "Fcn" block is used for converting.

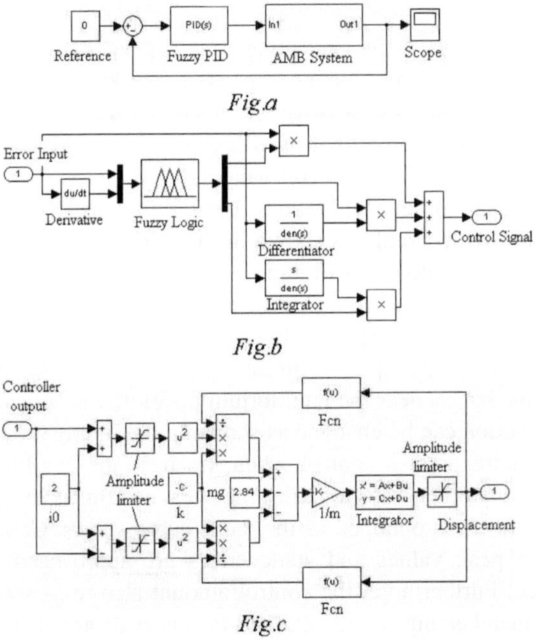

Fig. 2. System structure (Fig. a), the fuzzy PID controller model (Fig. b) and the magnetic force model (Fig. c)

3.1 Membership Functions

In the fuzzy inference system, membership functions are used to convert a crisp input value to a fuzzy value. Theoretically, the more MFs are set, the better performances

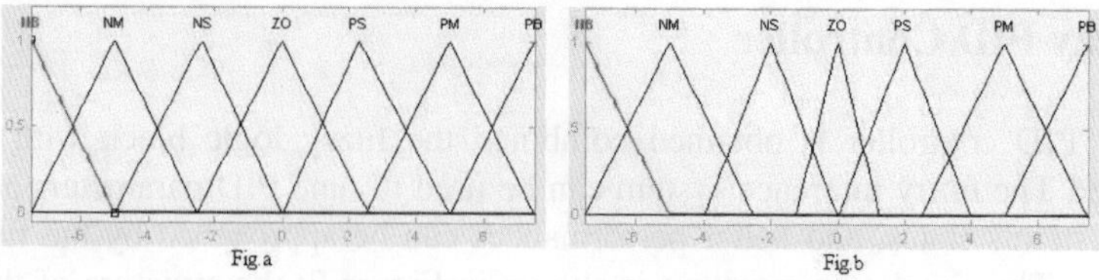

Fig. 3. Default MFs (Fig. a) and modified MFs (Fig. b)

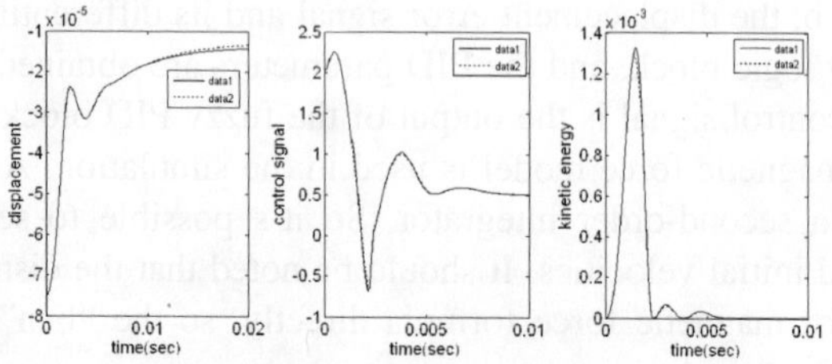

Fig. 4. Simulation results of the fuzzy PID controller with different MF settings

will be obtained. But more MFs need long time for processing. More MFs lead to better control characteristics and longer response time. As the range and the shape of the MFs can both affect the result of fuzzy processing and the performances of the fuzzy PID controller, so if the number of MFs cannot be added, then the shape can be adjusted appropriately to get better performances.

Generally, triangular and trapezoidal membership functions (NB, NM, NS, ZO, PS, PM, PB) are set to divide evenly as shown in Fig. a of Figure 3. And the simulation result is the data1 curve shown in Figure 4. As the most common shape of MFs is triangular, the general shape of MFs won't be changed. But the MF turning point of e can be modified as shown in Fig. b of Figure 3. Usually under the steady states the ball is not far away from the equilibrium position. If better control characteristics are concerned when the ball is near the equilibrium position, the MFs (NS, ZO, PS) near the equilibrium position can be changed as shown in the figure 3. This will ensure that the ball can get more precise control when it's near the equilibrium position. Of course, there are many other ways to adjust the MFs, and this is just one of them.

The simulation result is obtained as the data2 curve shows. Compared with data1, smaller oscillation peak values and static errors are approached when the turning points are changed. Furthermore, the control amount also turns smaller. As the MFs are changed, the membership of NS, ZO and PS intervals are also changed when the ball is near the equilibrium position. Thus, the result of fuzzy processing is also changed. Although it indeed affects the control characteristics, actually it is generally less effective than the number of MFs.

3.2 Fuzzy Rules

The next task is to set the fuzzy rules of the fuzzy inference system. It's the key point to design a controller, and it's also the most difficult part. Whether the fuzzy rules are set appropriately determines the characteristics of the fuzzy PID controller directly. In order to solve the large overshoot problem during the initial levitation process, the following principles can be applied: When the ball is quite far from the equilibrium position, the small P and D are set. Then the ball gets lower energy, and the levitation speed and the oscillation peak are small. As the distance between the ball and the equilibrium position gets smaller, larger P and appropriate D can be set. So appropriate stiffness with large damping effects can reduce the static error of the system, and would not lead to a large overshoot at the same time. When the ball is close enough from the equilibrium position to meet two small offset assumption conditions, theoretical PID parameters can be used to achieve a stable performance near the equilibrium position.

If the ball's dynamic characteristics are mainly concerned, the parameter "I" can be set to 0 during the simulation. Because compared with the initial levitation process, the time to eliminate the static error with the integral feedback effects is so long that the integral feedback effects can be ignored. So actually the fuzzy PID controller means the fuzzy PD controller here.

For simplicity, the velocity of the ball (ec) is not introduced as the input of the fuzzy logic block. So the error signal (e) is the only input of the fuzzy inference system. 7 MFs are set for both input and output variables, and just 7 fuzzy rules are enough for reasoning.

The rules (fuzzy1) are shown as follows:

Table 1. Fuzzy rules with just one input signal (e)

e	NB	NM	NS	ZO	PS	PM	PB
P/D	NB/NB	NB/NS	PS/ZO	ZO/ZO	PS/ZO	NB/NS	NB/NB

A fuzzy PID controller was designed with the fuzzy rules above (the fuzzy domain was divided evenly, same for the fuzzy PID controller mentioned below). As shown in Figure 5, data1 is the simulation result of a PID controller, and data2 is the simulation result of the fuzzy PID controller with 7 MFs. Better performances can be obtained compared with the PID controller. Furthermore, the control signal and the oscillation peak value are also much smaller. However, there are still some oscillation peaks during the initial levitation process. It's not good enough, and the fuzzy rules need more optimizations.

Generally, the error signal (e) and the velocity signal (ec) of the ball are both used as the input of the fuzzy inference system. If ec is introduced as another input variable, the fuzzy rules can be adjusted according to the movement direction of the ball.

The fuzzy rules (fuzzy2) are developed as follows:

A fuzzy PID controller was designed with the fuzzy rules above, and data3 in Figure 5 is the simulation result. Compared with the other two curves, the oscillation peaks become less, and the overshoot and the control signal become smaller. Obviously, the two-input fuzzy inference system performs better than the one-input fuzzy inference system and the PID controller.

Table 2. Fuzzy rules with two input signals (e and ec)

ec / e	NB	NM	NS	ZO	PS	PM	PB
NB	NB/NB	NS/NB	PS/ZO	ZO/ZO	ZO/PM	NB/NS	NB/NB
NM	NB/NB	NS/NB	PS/ZO	ZO/ZO	ZO/PM	NB/NS	NB/NB
NS	NB/NB	NS/NB	PS/ZO	ZO/ZO	ZO/PM	NB/NS	NB/NB
ZO	NB/NB	NB/NS	PS/ZO	ZO/ZO	PS/ZO	NB/NS	NB/NB
PS	NB/NB	NB/NS	ZO/PM	ZO/ZO	PS/ZO	NS/NB	NB/NB
PM	NB/NB	NB/NS	ZO/PM	ZO/ZO	PS/ZO	NS/NB	NB/NB
PB	NB/NB	NB/NS	ZO/PM	ZO/ZO	PS/ZO	NS/NB	NB/NB

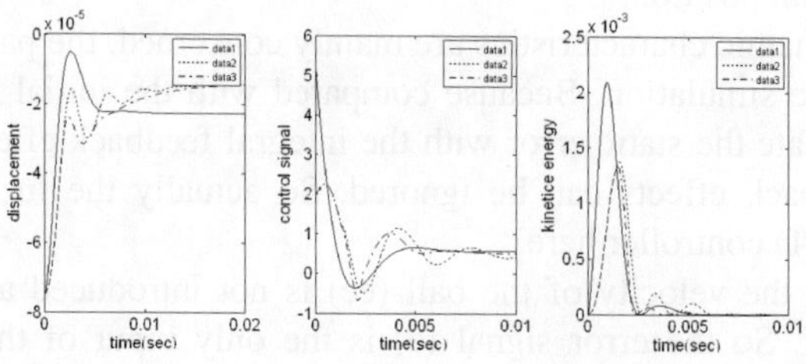

Fig. 5. Simulation results of the PID controller and two different fuzzy PID controllers

4 Response Analysis with the Rectangular Excitation Force

It's very important to perform well during the initial levitation process for an AMB system. However, the initial levitation is a quite short process and for most of the time the AMB system is in steady states. It's even more important for the AMB system to have better anti-shock abilities in steady states. So the response analysis was carried out. The anti-shock abilities of different controllers were compared through the simulation results.

The comparison between the fuzzy PID controller with fuzzy2 rules and the PID controller is shown in Fig. a of Figure 6. The equilibrium position is set as the initial position. In the figure, data1 is the simulation result of the PID controller, and data2 is the simulation result of the fuzzy PID controller. As the result shows, the response amplitude of the fuzzy PID controller is much greater than the PID controller. And it even collides with the backup bearings in some cases. In the table of fuzzy2 rules, the P value is smaller than the theoretical value for most cases. And smaller P value leads to smaller dynamic stiffness of the AMB system. So it's easier to be affected by the rectangular excitation force.

In order to solve the problem, fuzzy2 rules can be modified further. If the ball moves toward the equilibrium position, the P value isn't changed; if it moves away from the equilibrium position, the P values are all changed much larger. And the ZO interval of e and ec is divided into two parts: one is the plus part; the other is the minus part. Then for all intervals the parameters can be set according to the movement direction of the ball, including the ZO interval.

On the one hand, when the ball moves away from the equilibrium position because of the rectangular excitation force, a very large P will be applied. Large P leads to large dynamic stiffness, and large dynamic stiffness leads to large restoring force. This will strengthen the anti-shock abilities of the AMB system to get the ball back to the equilibrium position in a short time. On the other hand, when the ball is moving toward the equilibrium position, the fuzzy rules are not changed. So theoretically the performances during the initial levitation process won't be affected.

The modified fuzzy rules (fuzzy3) are shown as follows:

Table 3. Modified fuzzy rules to enhance the anti-shock abilities, with ZO- and ZO+ fuzzy intervals added.

e \ ec	NB	NM	NS	ZO-	ZO+	PS	PM	PB
NB	NB/NB	PB/NS	PB/PB	PB/ZO	ZO/ZO	ZO/ZO	NB/NB	NB/NB
NM	NB/NB	PB/NS	PB/PB	PB/ZO	ZO/ZO	ZO/ZO	NB/NB	NB/NB
NS	NB/NB	PB/NS	PB/PB	PB/ZO	ZO/ZO	ZO/ZO	NB/NB	NB/NB
ZO-	NB/NB	PB/NS	PB/PB	PB/ZO	ZO/ZO	ZO/ZO	NB/NB	NB/NB
ZO+	NB/NB	NB/NB	ZO/ZO	ZO/ZO	PB/ZO	PB/PB	PB/NS	NB/NB
PS	NB/NB	NB/NB	ZO/ZO	ZO/ZO	PB/ZO	PB/PB	PB/NS	NB/NB
PM	NB/NB	NB/NB	ZO/ZO	ZO/ZO	PB/ZO	PB/PB	PB/NS	NB/NB
PB	NB/NB	NB/NB	ZO/ZO	ZO/ZO	PB/ZO	PB/PB	PB/NS	NB/NB

A fuzzy PID controller was designed using the fuzzy3 rules. The simulation result comparison between this fuzzy PID controller and the PID controller is shown in Fig. b of Figure 6.

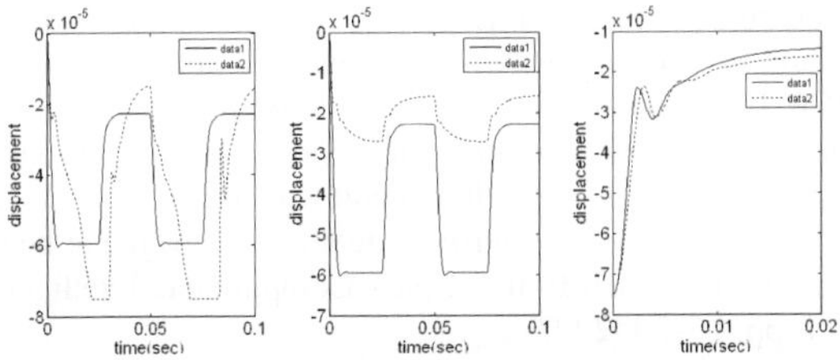

Fig. 6 Simulation results to examine the anti-shock abilities of the PID and fuzzy PID controllers, and the performance during the initial levitation process of the two fuzzy PID controllers.

In the figure, data1 is the simulation result of the PID controller, and data2 is the simulation result of the fuzzy PID controller. Obviously, the fuzzy PID controller performs much better under the same rectangular excitation force. The simulation results during the initial levitation process are shown in Fig. c of Figure 6. Data1 is the simulation result of the fuzzy PID controller with fuzzy2 rules, and data2 is the simulation result of the fuzzy PID with fuzzy3 rules. As the figure shows, there are no big differences between them. So the modified fuzzy PID controller performs better than the PID controller during the initial levitation process, and there are much stronger anti-shock abilities for a fuzzy PID controller.

5 Conclusion

Based on the analysis above, appropriate modifications of MFs and fuzzy rules can optimize the performances of the fuzzy PID controller. There are two aspects of the optimization: one is to decrease the overshoots during the initial levitation process, the other is to enhance the anti-shock abilities of the AMB system. Theoretically, the control effect will become more precisely if more MFs are set. But generally the MFs cannot be set too many. However, there are still ways to improve the performances without increasing the MFs, one of which is to modify the turning points of the MFs. But the improvement is not very significant compared with adding more MFs. The fuzzy rules determine the performances of the fuzzy PID controller directly. As the fuzzy rules can be set according to the displacement and the velocity of the ball, large P is set when the ball moves away from the equilibrium position while the other rules are not changed. Then the modified fuzzy rules increase the dynamic stiffness when the ball moves away from the equilibrium position, which will enhance the anti-shock abilities. And at the same time the controller still performs well during the initial levitation process. Considered as a whole, the comprehensive performances are improved by modifying the MFs and fuzzy rules appropriately.

References

1. Schweitzer, G., Bleuler, H., Traxle, A.: Active magnetic bearings basics properties and applications, Vdf Hochschulverlag AG an der ETH, Zurich (1994)
2. Humphris, R.R., Kelm, R.D., Lewis, D.W., Allaire, P.E.: Effect of control algorithms on magnetic journal bearing properties. J. Eng. Gas Turbines Power 108, 624–632 (1986)
3. Hartavi, A.E., Ustun, O., Tuncay, R.N.: A comparative approach on PD and fuzzy control of AMB using RCP. IEEE Electric Machines and Drives Conf., pp. 1507-1510 (2003)
4. Huang, S.-J., Lin, L.-C.: Fuzzy modeling and control for conical magnetic bearings using linear matrix inequality. Journal of Intelligent and Robotic Systems 37, 209–232 (2002)
5. Habib, M.K., Inayat-Hussain, J.I.: Control of dual acting magnetic bearing actuator system using fuzzy logic. In: Proc. IEEE Int. Conf. Computional Intelligence in Robotics and Automation, vol. 1, pp. 16–20 (2003)
6. Hung, J.Y.: Magnetic bearing control using fuzzy logic. IEEE Trans. Industry Applications 31(6), 1492–1497 (1995)
7. Woo, Z.-W., Chung, H.-Y., Lin, J.-J.: A PID type fuzzy controller with self-tuning scaling factors. Fuzzy Sets and Systems, 321–326 (2000)
8. Liang, Q., Mendel, J.M.: Interval type-2 fuzzy logic systems: theory and design. IEEE Trans. Fuzzy Systems 8(5), 535–550 (2000)
9. Homaifar, A., McCormick.: Simultaneous design of membership functions and rule sets for fuzzy controllers using genetic algorithms. IEEE Trans. Fuzzy Systems 3(2), 129–139 (1995)

Study on the Application of Data Mining Technique in Intrusion Detection

Hongxia Wang, Tao Guan, and Yan Wang

Department of Computer Science and Application
Zhengzhou Institute of Aeronautical Industry Management
450015, Zhengzhou, China
whxpc@163.com

Abstract. Intrusion detection is one of the key techniques in Internet security. However, there are still many problems at present, especially the intrusion detection system with abilities of self-adaptation and self-learning need to be improved. It will help solve these problems to apply data mining algorithms, such as association analysis and sequential pattern analysis into the intrusion detection system.

Keywords: Internet security, intrusion detection, data mining.

1 Introduction

With the rapid development of Internet, more and more network attacks appear, which have brought about greater and greater damage. It has become more serious for Internet and information security. Traditional means for Internet security, e.g. fire wall technique, are unable to defend various and continuous attacks. Therefore, as an active defense technique, intrusion detection system (IDS) is attracting a great deal of attention, which may fill up the deficiency of fire wall technique, provide real-time intrusion detection, prevent internal attacks and solve the problem of back door safety for application layer.

2 Survey of Intrusion Detection Techniques

Intrusion detection refers to find out behaviors that violate safety strategy and evidence that is being attacked on network or in system by means of collecting information from several key points in computer network or system and analyzing them, so as to take measures and response properly in accordance with the analytical results. The intrusion detection system is the combination of software and hardware that are working during the intrusion detection. From the methods of data analysis, intrusion detection is classified into two categories: misuse detection and anomaly detection.

(1) Misuse Detection
Based on the analysis of the features of various attacking means, misuse detection analyses and processes current data and then matches their features; if a match that

M. Zhu (Ed.): Electrical Engineering and Control, LNEE 98, pp. 989–994.
springerlink.com © Springer-Verlag Berlin Heidelberg 2011

satisfies the condition is found out, then it indicates one attack. The detection targets the types of attacking means, so it is widely used in many commercial systems. Moreover, misuse detection has high detecting probability and low false alarm probability; and it is more convenient and easier to develop rule base and feature set.

(2) Anomaly Detection
Anomaly detection first establishes a state model according to the user's normal activity, and then observes the difference between current activity and normal activity; if the difference exceeded the set threshold value, it indicates an unauthorized attack. This method may help detect unknown intrusion behaviors with low false negative, but high false positive.

An excellent IDS should be effective, self-adaptive and extendable. A great deal of information is to be processed during the intrusion detection, and the establishment of the system depends on whether the rules are complete. It is not always possible to establish a complete knowledge base for a large network system, which requires a systematic upgrading by professionals, otherwise it will be very difficult to update the rules; this has limited the requirements of effectiveness, self-adaptation and extendibility. In practice, intrusion detection model can only process a special audit data source, whose updating cost is high and speed is relatively slow.

3 Survey of Data Mining Techniques

Data mining refers to extract knowledge that people are interested in from large data bases or data warehouses, which is usually known as concept, rule, law, pattern, etc. Data mining aims at helping the decision-maker find out potential association between data and the factors that are neglected; and the information may be very helpful to trend forecast and decision behavior. Applying data mining into the intrusion detection system may overcome the limits of traditional system. Data mining techniques that are often used include association analysis, sequential analysis, classification, clustering and so on.

(1) Association analysis is used to determine the association between fields in data records. The algorithm makes an association analysis of system features in audit data, e.g. the association between command and parameter fields in the user's audit data may be helpful to the establishment of a record for normal user's behaviors. The popular algorithms of association analysis include Apriori, AprioriTid, etc.
(2) Sequential analysis is used to determine the affinity between different data records. The algorithm can help to find out the sequential patterns for some audit events that often appear in data set in accordance with the time sequence, e.g. through the analysis of sequential patterns of audit data that are exposed to denial of service on Internet, it will improve detection performance to add some statistic features in the detection model based on each host computer or service type. The popular algorithms of sequential analysis include AprioriAll, DynamicSome, etc.
(3) Classification is used to classify specific data items into some category that is defined beforehand. The algorithm usually generates some kind of "classifier" in the end. To the intrusion detection, the final object of classification algorithm is to correctly identify those new audit records, whether they are "normal" or "abnormal". The classification algorithms that are often used include RIPPER, C4.5, Nearest Neighbor, etc.

4 Intrusion Detection System Based on Data Mining

4.1 The Process of Intrusion Detection

According to the characteristics of IDS and data mining technique, the IDS structure based on the application of data mining technique is described in Figure 1.

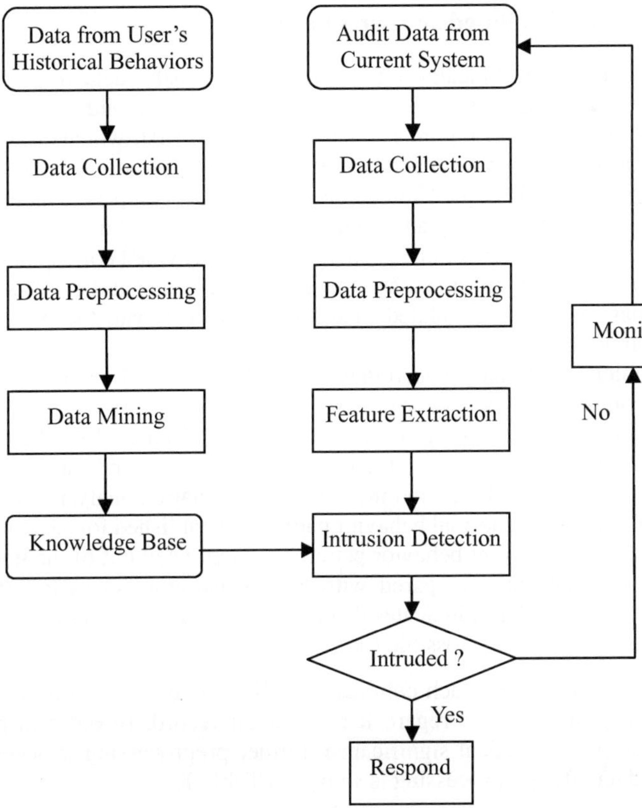

Fig. 1. IDS Structure based on Data Mining Technique

(1) Data Collection. It includes the collection of training data and those from current behaviors; and available tools include TCPDUMP, sniffer, etc.

(2) Data Preprocessing. It refers to format the data acquired and convert them into the form that is able to be processed by the algorithms of data mining.

(3) Data Mining. It refers to extract behavior features and rules from the system's data by means of data mining technique, so as to establish the abnormal or normal patterns for Internet security.

(4) Knowledge Base. The abnormal or normal patterns that are necessary for the system are stored in knowledge base; and the IDS compares them with the features of user's behaviors to assess weather the latter are intrusion behaviors.

(5) Feature Extraction. It refers to extract the features of user's behaviors from the user's behavior data by means of a technique similar to data mining.

(6) Intrusion Detection. According to some algorithms, the system extracts relative rules and data from knowledge base to detect the features of current user's behaviors. Corresponding behaviors will be done on the basis of detecting results: if it is an intrusion behavior, the system will alarm, take some measures to stop the intrusion and keep the evidence; if it is normal, then the system continues monitoring the user's behaviors.

4.2 Anomaly Detection Based on User's Behaviors

Intrusion detection is technically classified into anomaly detection and misuse detection. Here the anomaly detection is used for running command shell during user's Telnet session. According to Figure 1, data are collected first, which are from secure audit sources on Internet that were announced in 1998 by Defense Advanced Research Projects Agency (DARPA) of the United States. These data comprise a large number of Telnet session records, which are analyzed to generate a model of anomaly detection for user's behaviors based on command shell. DARPA provided corresponding data in 1999 and 2000 respectively for the assessment of IDS. Aiming at the user's behaviors, Telnet abnormal login was simulated in the model to estimate the performance of anomaly detection engine.

The key problem to anomaly detection is to establish a normal model and how to use this model to compare and assess current user's behaviors.

At the training stage, according to the command shell entered by the user during Telnet session, affinities and rules in the user's running commands are extracted with the help of the algorithms of data mining, such as association analysis and sequential pattern analysis, and then a normal behavior pattern is established for each user. During the practical detection, current behavior patterns are extracted out of the user's Telnet sessions, and then they are compared with history patterns, so as to calculate the similarity. The more similar, it indicates that the current pattern is more identical to the history behavior, and it is less possible for anomaly.

(1) Preprocess data. For each command Shell entered by the user as well as its relative properties, we regard it as an audit record. In order to mine more contents with practical significance, further preprocessing is necessary. The audit data after preprocessing is shown as Table 1.

Table 1. Examples of Audit Records by Command Shell

User Name	Time	Host IP	User IP	Command	Command Parameter
Jane	nt	210.45.147.1	192.168.0.1	Cp	File
Jane	nt	210.45.147.1	192.168.0.1	Ls	Dir
John	pm	210.45.147.1	202.108.122.38	Fg	
John	Pm	210.45.147.1	202.108.122.38	Cat	Dir
John	Pm	210.45.147.1	202.108.122.38	Kill	
John	Pm	210.45.147.1	202.108.122.38	Jobs	
John	Pm	210.45.147.1	202.108.122.38	Bg	

(2) Establish a normal behavior pattern for each user by means of the algorithms in data mining. The user's behaviors are classified into two patterns: association pattern and sequential pattern. According to the algorithm of Apriori, the association pattern is shown as Table 2.

Table 2. Association Pattern of User's Behaviors

Support(min.)	Quantity of Rules	Example	Note
65%	2	192.168.10.1 ~202.138.1.4 ~nt	The user usually logs in the host computer addressed 202.138.1.4 from 192.168.10.1 at night.
50%	4	192.168.10.1 ~202.138.1.4 ~nt~cat	The user usually logs in the host computer addressed 202.138.1.4 from 192.168.10.1 at night and often runs cat, the shell command.
70%	11	192.168.10.1 ~202.138.1.4 ~nt~chown	The user usually logs in the host computer addressed 202.138.1.4 from 192.168.10.1 at night and often runs chown, the shell command.

Similarly, according to the algorithm of sequential pattern, we find out the patterns shown as Table 3.

Table 3. Sequential Pattern of John's Behaviors

Support(min.)	Quantity of Rules	Example	Note
4%	3	Chown->etc/ aliases ->mail	The user often runs the command sequence: Chown->etc/aliases->mail.
8%	7	Chown->mesg->talk	The user often runs the command sequence: Chown->mesg->talk.
5%	21	Chown->mesg->talk ->wall->nice->ps	The user often runs the command sequence: Chown->mesg->talk->wall->nice->ps.
3%	33	Ls->mail->su-> Tcsh->ls->df	The user often runs the command sequence: Ls->mail->su->tcsh->ls->df.

(3) Match patterns. If the normal testing data are used to mine the user's current pattern and the association rules and sequential patterns are calculated by means of the algorithms of pattern comparison, i.e. complete sequential comparison and correlation function respectively, the similarity is usually very high; if the abnormal session is adopted and the algorithm of Apriori is used, the result will be opposite.

5 Conclusion

To the similarity of history behaviors, there is obvious difference between normal and abnormal user's behaviors. The experiment has proved that it may effectively disclose the rules of user's behavior patterns to mine his commands by means of association rules and sequential patterns in data mining technique; and it may correctly assess user's behaviors to calculate the similarity between the user's historical and current patterns by means of sequence matching, such as correlation function. The efficiency and accuracy of intrusion detection are improved effectively, so the network system becomes more secure.

Acknowledgments

The research work is supported by the Foundation of He'nan Educational Committee (No.2011B520038) and the Key Scientific and Technological Project of He'nan Province of China(Grant NO.112102210024).

References

1. Tang, Z., Li, J.: Intrusion Detection Technique. Tsinghua University Press, Beijing (2010)
2. Bi, Z., Xu, S.: Actuality and Development of Intrusion Detective Technique. Software Guide (2010)
3. Xu, X., Fu, H., Xiong, Z.: Intrusion Detection Technology Research Based on Data Mining. Microcomputer Information (2007)
4. Shi, Z.: Study on the Application of Data Mining Based on Intrusion Detection System. Sci-Tech Information Development & Economy (2005)

Fault Recovery Based on Parallel Recomputing in Transactional Memory System*

Wei Song and Jia Jia

National Laboratory for Parallel and Distributed Processing, School of Computer, National University of Defense Technology, Changsha, 410073, China
frank1997@gmail.com,
morpheux@163.com

Abstract. This paper addresses the issue of fault recovery in transactional memory, and proposes a method of fault recovery based on parallel recomputing in transactional memory system. This method utilizes the data-versioning mechanism of transactional memory system to avoid the extra cost of state saving, rolls back a single transaction to avoid wasting the computing time of the fault-free transactions, and adopts the parallel recomputing method to reduce the cost of fault recovery. This paper applies this method to OpenTM programs, and proposes the implementation method of parallel recomputing in OpenTM. At last, this paper tests the performance of this method through a test program. The experimental results show that, compared with the fault recovery method of rolling back a single transaction, the parallel recomputing method in transactional memory system can execute the fault recovery quickly and accurately and the method has a well scalability.

Keywords: Fault tolerance; Parallel recomputing; Fault Recovery; Transactional memory.

1 Introduction

With the development of the microprocessor, which shifts from the traditional single-core high-frequency to the shared memory multi-core architecture, developing thread-level parallelism has attracted more and more attention. And the traditional multi-threaded programing models, such as locks, show more and more obvious disadvantages of complexity and fallibility in avoiding deadlock and in achieving high scalability. To the contrast, transactional memory, with its easy programing and high scalability, and as a promising mechanism to address concurrent access to shared data in multi-core processors, has attracted more and more attention by researchers [1,2]. On the other hand, in recent years a large number of multi-core processors are used in high-performance computer system. Among the Top 500 high-performance computer systems issued in Nov. 2009, all the top 5 systems except for the second one

* Supported by the National Natural Science Foundation of China under Grants No. 60921062 and 61003087.

M. Zhu (Ed.): Electrical Engineering and Control, LNEE 98, pp. 995–1002.
springerlink.com © Springer-Verlag Berlin Heidelberg 2011

use homogeneous multi-core processors [3], this makes the using of transactional memory in high-performance computer systems become possible. As a result, the reliability problems in high-performance systems make the fault-tolerance in transactional memory system become a concerning issue.

Checkpointing, as a fault recovery technique, is widely used in the domain of large-scale scientific computing. And it has problems of wasting computing time of fault-free process and saving a large amount of program state data. To solve these problems we put forward a fault-recovery method based on parallel recomputing in [5]. In this paper, based on the characteristics of transactional memory system, we propose a method of parallel recomputing for transactional memory system. This method utilizes the data-versioning mechanism of transactional memory to avoid the extra cost of state saving, rolls back a single transaction to avoid wasting the computing time of the fault-free transactions, and adopts the parallel recomputing method to reduce the cost of fault recovery.

The contributions of this paper are as follows:

- Proposes a fault-recovery method based on parallel recomputing in transactional memory system.
- Proposes the implementation of parallel recomputing in OpenTM programs.
- Validates the performance of parallel recomputing in transactional memory system through a test program.

The organization of this paper is as follows. Section 2 introduces the basic idea of the parallel recomputing in transactional memory system. Section 3 proposes the implementation of parallel recomputing in OpenTM programs. Section 4 validates the performance of parallel recomputing in transactional memory system through a test program. Section 5 reviews the related work. The last section concludes our work.

2 Basic Idea

2.1 Basic Idea of Parallel Recomputing

In traditional checkpointing technique, it needs to save a checkpoint before entering a fault-tolerant section. If a fault occurs, all processes will have to roll back to the last checkpoint and restart the computation. One hand, it wastes the computing time of fault-free processes. On the other hand, it makes that the fault-recovery time cannot be less than the time interval between the last checkpoint and the fault point.

In order to improve the fault-tolerant performance, we put forward a fault-recovery thought of parallel recomputing[5]. Different from checkpointing, this method utilizes multiple processors to recompute the workload of the fault process. Before entering the fault-tolerant section, the parallel recomputing may need to save a small amount of data. But when a fault occurred, the parallel recomputing only need to roll back the fault process and this avoids wasting the computing time of fault-free processes. Because of using multiple processors to recover the fault, the fault-recovery time could be than the time interval between the last checkpoint and the fault point. Fig. 1 shows the basic idea of parallel recomputing.

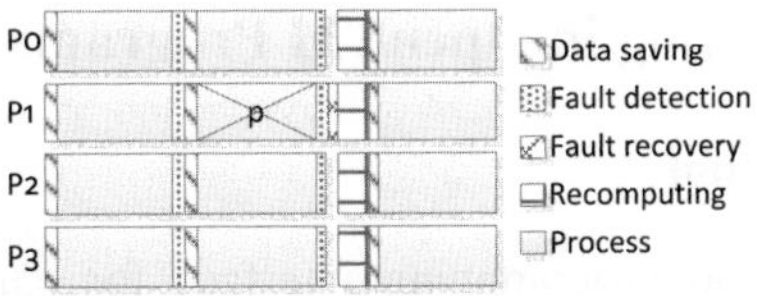

Fig. 1. Basic idea of parallel recomputing

We need to indicate that, because the parallel recomputing only rolls back the fault process, we need to consider the fault isolation in the selection of recomputing section to ensure that the fault could not spread outside the recomputing section.

2.2 Parallel Recomputing in Transactional Memory System

In transactional memory system, the execution of transactions has atomicity and isolation, so rolling back a single transaction will not affect the correct execution of other transactions. Therefore, a transaction can be regard as a natural recomputing section and the selection of recomputing section, which is in the unit of transactions, doesn't need to consider the fault isolation. In addition， the data-versioning mechanism of transactional memory system provides a mechanism similar to the checkpoint mechanism. Therefore, in transactional memory system, the parallel recomputing of transactions doesn't need an extra data saving and avoids the fault-tolerant cost under the fault-free execution. Fig. 2 shows the principle of the parallel recomputing in transactional memory system.

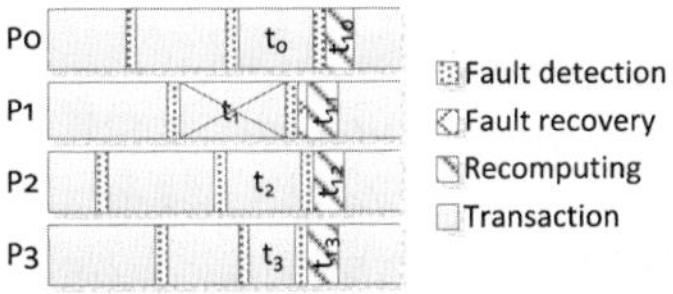

Fig. 2. Supposing there are 4 processor cores in the system, and the fault detection is performed before the committing of every transactions. If transaction t1 executing on P1 is detected a fault before committing, as showed in Fig. 2, then the system invokes the fault handler to roll back t1, and divides it into 4 transactions in smaller granularity, denoted by t10, t11, t12and t13. t11 is executed on P1, and the other three are placed into the transaction pool waiting for being executed. After P3 completing the current transaction t3, it checks the transaction pool and gets t13 out to execute. Similarly, after P2 and P0 completing the current transactions t2 and t0, they get t12 and t10 out of the transaction pool to execute. And the recomputation of t1 is completed.

We can see that if we don't consider the cost of fault detection, the parallel recomputing in transactional memory system won't introduce any extra cost under fault-free execution. And when there occurs a fault, the rollback of fault transaction won't affect the execution of the fault-free transactions.

3 Parallel Recomputing in OpenTM Programs

3.1 OpenTM Introduction

OpenTM is an application programming interface for parallel programming with transactions, proposed by Stanford University[6]. It is designed as an extension of OpenMP to support non-blocking synchronization and speculative parallelization using transactional techniques. OpenTM inherits the advantages of OpenMP and provides a simple, high-level interface to express transactional parallelism. And OpenTM code can be efficiently mapped to a variety of HTM, STM and hybrid TM implementations.

The following are the main compiler directives that OpenTM introduces:

- #pragma omp transaction *[clause[[,] clause]…]*
- #pragma omp transfor *[clause[[,] clause] …]*
- #pragma omp transsections *[clause[[,] clause]…]*

Besides, OpenTM also provides a few advanced constructs, such as the alternative execution paths: orelse, the transaction handlers: handler, and so on.

The syntax of orelse is as follow:

#pragma omp transaction
structured-block 1
#pragma omp orelse
structured-block 2

If the transaction for block 1 aborts for any reason, the code for block 2 will be executed as a transaction. Otherwise, the code for block 2 will never be executed.

The syntax of handler is as follow:

#pragma omp transaction *[clause[[,] clause]…]*
structured-block 1
#pragma omp handler *clause*
structured-block 2

Where clause can be one of the following: commit, abort and violation. The difference between violation and abort is that, violation refers to a rollback triggered by a dependency, while abort is invoked by the transaction itself.

3.2 Parallel Recomputing in OpenTM Programs

In the previous section, we introduce the main programming interfaces of OpenTM. We can see that the programming interfaces of OpenTM and OpenMP have a lot of similarities. It should have a good acceptance for the programmers who are familiar to shared memory programming. And OpenTM may be a promising application programming interface for transactional memory.

As shown in Fig. 3, in this section we will introduce the implementation method of parallel recomputing in OpenTM programs through a kernel code.

As an example, we use a core code of two dimension n integer matrix A and B multiplication generating dimension n matrix C. Supposing we regard the solution of every 16 elements of matrix C as a transaction, as shown in Fig. 3 (a).

Fig. 3 (b) provides the code which is the program in Fig. 3 (a) with the parallel recomputing mechanism. As we mentioned above, transactions are the fault-tolerant recomputing section of parallel recomputing in transactional memory system, so the error detection should be performed before the committing of the transaction. We suppose there is a fault detection section before the code of line 12 in Fig. 3 (b), and the flag variable error_detected will be set to true if any fault has been detected.

The data-versioning mechanism that transactional memory system provides can be regarded as a mechanism similar to the checkpoint mechanism, so we can recover the program state through rolling back the fault transaction. And the interface omp_abort() provided by OpenTM can roll back a transaction explicitly. As shown in line 13, 14, if the error_detected is true, then roll the fault transaction back.

```
1  #pragma omp transfor
2  for(i=0;i<n;i++){
3          for(j=0;j<n;j+=16){
4                  #pragma omp transaction
5                  {
6                          for(m=j;m<j+16&&m<n;m++){
7                                  C(i,j+m)=0;
8                                  for(k=0;k<n;k++){
9                                          C(i,j+m)+=A(i,k)*B(k,j+m);
10                                 }
11                         }
12                 }
13         }
14 }
```

(a)

```
1  #pragma omp transfor
2  for(i=0;i<n;i++){
3          for(j=0;j<n;j+=16){
4                  #pragma omp transaction
5                  {
6                          for(m=j;m<j+16&&m<n;m++){
7                                  C(i,j+m)=0;
8                                  for(k=0;k<n;k++){
9                                          C(i,j+m)+=A(i,k)*B(k,j+m);
10                                 }
11                         }
12                 /*error detecting*/
13                         if(error_detected)
14                                 omp_abort();
15                 }
16                 /*parallel computing*/
17                 #pragma omp handler abort
18                 {
19                         #pragma omp transfor
20                         for(m=j;m<j+16&&m<n;m++){
21                                 C(i,j+m)=0;
22                                 for(k=0;k<n;k++){
23                                         C(i,j+m)+=A(i,k)*B(k,j+m);
24                                 }
25                         }
26                 }
27         }
28 }
```

(b)

Fig. 3. Parallel recomputing in OpenTM

Line 16-26 is the procedure of parallel recomputing. The key problems of implementing are catching the fault-rollback operation of transactions and dividing the transaction to execute in parallel.

In section 3.1, we have mentioned that OpenTM provides two advanced constructs, orelse and handler, which can catch the rollback operation of transactions. As we know that there're two kinds of rollback operations in OpenTM. One is the violation operation which is triggered by a dependency, and the other is abort operation which is invoked by the programmer explicitly. Obviously, the operation we want to catch is the abort operation. For the orelse constructor doesn't distinguish violation operation and abort operation, we chose the handler constructor to catch the fault-rollback operation, as shown in line 17.

For the problem of dividing the transaction to execute in parallel, because the main body of the transaction in Fig. 3 is a for loop, we can use the compiler directive #pragma omp transfor to divide the transaction and execute it in parallel.

From Fig. 3 we can see that it is simple to transform an OpenTM program into a parallel recomputing program by using the constructor handler to catch the fault-rollback operation. And for the transaction whose main body is a loop, it's easy to accomplish the transformation by automatic tools. Here we do not make much description.

4 Experiments

In this section, we will validate the performance of our parallel recomputing mechanism in transactional memory system through a test case. Our test platform is an 8 cores transactional memory system simulated by simics3.0.29+gems2.1, and the frequency of each processor core is 75MHZ, memory size is 2GB, disk size is 8GB. Our test case is a 60 dimension matrix multiplication program, and we regard the solution of every column of the result matrix as a transaction.

In the experiment, we first execute the program in the condition of fault-free. Then we assume one of the transactions occurs a fault, and we recover the fault by rolling back a single transaction. Finally, we recover the fault transaction by using the parallel recomputing method. Therefore, there are three sets of data in our experiment: the execution time of original transactional program, the execution time of the program base on rolling back a single transaction and the execution time of the program with the parallel recomputing mechanism. Fig. 4 shows the experimental results.

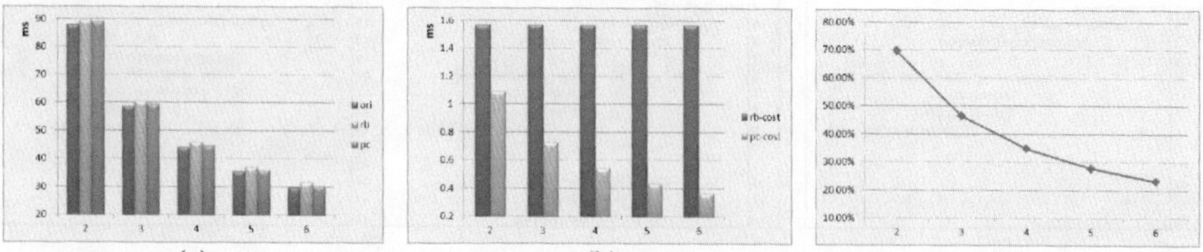

(a) (b) (c)

Fig. 4. The experimental results

We perform the test on 2 processor cores to 6 processor cores, calculate the transaction execution time (in nanosecond) by using the clock cycles achieved by gems. Fig. 4 (a) shows the trends of the test data with the increase of the number of processor cores. And the illustration ori represents the execution time of original transactional program, illustration rb represents the execution time of the program base on rolling back a single transaction, illustration pc represents the execution time of the program with the parallel recomputing mechanism. We can see that the execution time of the program with the parallel recomputing mechanism is obviously less than the execution time of the program base on rolling back a single transaction.

Fig. 4 (b) further shows the comparison between the fault recovery cost of the program base on rolling back a single transaction and the fault recovery cost of the program with the parallel recomputing mechanism. We can see that the fault recovery cost of the program base on rolling back a single transaction is almost a constant value, it's each fault recovery cost is the rollback and recomputation time of one transaction. And the fault recovery cost of the program with the parallel recomputing mechanism obviously decreases with the increase of the number of processor cores. It illustrates that, for the programs with good scalability, increasing the number of processor cores taking part in the computation could effectively reduce the fault recovery cost, in programs with parallel recomputing mechanism.

In Fig. 4 (c), we presents the trend of the proportion of the fault recovery cost of the program with the parallel recomputing mechanism and the fault recovery cost of the program base on rolling back a single transaction with the increase of the number of processor cores. We can see that the fault recovery method based on parallel recomputing in transactional memory system has good scalability.

5 Related Work

The rollback-recovery technique is widely used in the domain of fault tolerant[4]. A detailed survey of these techniques can be found in [4]. Checkpointing, as a primary rollback-recovery technique, can be generally divided into system-level checkpointing and application-level checkpointing. System-level checkpointing requires all processes periodically checkpoint themselves by saving the content of their address space (including all values in the stack, heap and global variables), registers and the state of communication library to stable storage. Upon failure, all processors load the last checkpoints and restart from there. In a system including a large number of processors, the cost of saving checkpoints is very great and the cost cannot be avoided even in a fault-free execution. To reduce the overhead of writing checkpoint files to stable storage, Plank presented diskless checkpointing[7]. This technique uses high-speed memory instead of low-speed disk to save checkpoint files. But the memory capacity limits its application in large-scale scientific computation.

Application-level checkpointing is an effective method to reduce the checkpoint size and solve the I/O bottleneck[8]. It requires the programmers to save the minimum amount of data necessary to recover program state. A further optimization approach is to combine the above two techniques: the programmer writes checkpointing code to save the necessary information to memory[9].

The existing research of rollback-recovery technique mainly aimed at reducing checkpoint size and fast saving the checkpoint. Compared with these methods, the method of parallel recomputing in transactional memory system mentioned in this paper utilizes the characteristic of transactional memory system to avoid the extra cost of checkpoint saving in one hand. In the other hand, it reduces the fault recovery time by using parallel recomputing, and realizes a more efficient fault recovery in transactional memory system.

6 Conclusion

This paper proposes a fault recovery method based on parallel recomputing in transactional memory system. This method has the following advantages:

1) It utilizes the data-versioning mechanism of transactional memory system to avoid the extra cost of state saving.
2) It rolls back a single transaction to avoid wasting the computing time of the fault-free transactions.
3) It adopts the parallel recomputing method to reduce the cost of fault recovery.

Furthermore, this paper applies this method to OpenTM programs, and proposes the implementation method of parallel recomputing in OpenTM.

Finally, we validate the performance of this method through a 60 dimension matrix multiplication program. The experimental results show that, compared with the fault recovery method of rolling back a single transaction, the parallel recomputing method in transactional memory system can execute the fault recovery quickly and accurately and the method has a well scalability.

References

1. Harris, T., Cristal, A., Unsal, O.S., Ayguadé, E., Gagliardi, F., Smith, B., Valero, M.: Transactional Memory: an Overview. IEEE Micro Special Issue: Hot Tutorials (May/June 2007)
2. Larus, J., Kozyrakis, C.: Transactional Memory. Commun. ACM 51(7), 80–88 (2008)
3. TOP500 Supercomputing Site, `http://www.top500.org`
4. Elnozahy, E.N., Alvisi, L., Wang, Y., Johnson, D.: A survey of rollback-recovery protocols in message-passing systems. ACM Computing Surveys 34(3), 375–408 (2002)
5. Yang, X., Du, Y., Wang, P., Fu, H., Jia, J., Wang, Z., Suo, G.: The fault tolerant parallel algorithm: the parallel recomputing based failure recovery. In: Proceedings of the 16th International Conference on Parallel Architectures and Compilation Techniques (2007)
6. Baek, W., Minh, C.C., Trautmann, M., Kozyrakis, C., Olukotun, K.: The OpenTM transactional application programming interface. In: Proceedings of the 16th International Conference on Parallel Architecture and Compilation Techniques (PACT 2007) pp. 376–387. IEEE Computer Society, Washington, DC, USA (2007)
7. Plank, J.S., Li, K., Puening, M.A.: Diskless checkpointing. IEEE Trans. Parallel Distrib. Syst. 9(10), 972–986 (1998)
8. Beguelin, A., Seligman, E., Stephan, P.: Application level fault tolerance in heterogeneous networks of workstations. Journal of Parallel and Distributed Computing 43(2), 147–155 (1997)
9. Chen, G.E., Fagg, E., Gabriel, J., Langou, T., Angskun, G.: Fault tolerant high performance computing by a coding approach. In: ACM SIGPLAN Symposium on Principles and Practice of Parallel Programming (PPoPP 2005), Chicago, Illinois, pp. 213–223 (June 2005)

Forecast of Hourly Average Wind Speed Using ARMA Model with Discrete Probability Transformation

Juan M. Lujano-Rojas, José L. Bernal-Agustín, Rodolfo Dufo-López, and José A. Domínguez-Navarro

Department of Electrical Engineering, University of Zaragoza, Calle María de Luna 3, 50018, Spain
lujano.juan@gmail.com,
{jlbernal,rdufo,jadona}@unizar.es

Abstract. In this paper the methodology for wind speed forecasting with ARMA model is revised. The transformation, standardization, estimation and diagnostic checking processes are analyzed and a discrete probability transformation is introduced. Using time series historical data of three weather stations of the Royal Netherlands Meteorological Institute, the forecasting accuracy is evaluated for prediction intervals between 1 and 10 hours ahead and compared with artificial neural network training by back-propagation algorithm (BP-ANN). The results show that for the wind speed time series under study, in certain cases the ARMA model with discrete probability transformation can improve the BP-ANN at least 17.71%.

Keywords: Wind Speed Forecasting, Autoregressive Moving Average Model (ARMA), Artificial Neural Network.

1 Introduction

Wind speed and power predictions are important aspects for the wholesale energy market, wind farm owners and power system operators [1]. The autoregressive moving average model (ARMA) is a conventional statistical method for wind speed prediction that is based on analyzing historical data. This model frequently achieves accurate results in short-term prediction [2]. Brown [3] proposed a general model that considers the non-Gaussian shape of the probability distribution function (PDF), the diurnal non-stationarity and the autocorrelation of wind speed. This model is used by Daniel and Chen [4] for evaluating the forecasting accuracy of wind speed in Jamaica. Torres et al. [5] compared the forecasting accuracy between the ARMA and persistence models using time series data of five weather stations in Spain for prediction intervals between 1 and 10 hours ahead. The results show that for an interval prediction of 1 hour in advance the ARMA model improved the persistence model from 2–5% and, for an interval of 10 hours, 12–20%. In [6-8], comparative studies between ARMA and artificial neural network (ANN) models are shown. However, the ARMA models used in these studies do not consider the non-Gaussian shape of the PDF. This must be considered because the PDF of wind speed time series follows

M. Zhu (Ed.): Electrical Engineering and Control, LNEE 98, pp. 1003–1010.
springerlink.com © Springer-Verlag Berlin Heidelberg 2011

a Weibull distribution, making the direct application of stochastic models difficult for its analysis [9]. Reference [10] presents a comparative analysis of three types of ANN for forecasts 1 hour ahead, concluding that the most suitable ANN depends on the time series under study and the index error used to evaluate the forecasting accuracy.

In this paper, the methodology for wind speed forecasting with the ARMA model is revised. The transformation, standardization, estimation and diagnostic checking processes are analyzed and a Discrete Probability Transformation (DPT) is introduced. The forecasting error is evaluated for prediction intervals between 1 and 10 hours ahead using hourly average wind speed data of three weather stations in Netherlands. The ARMA model with discrete probability transformation (ARMA+DPT) is compared with an ANN training with back-propagation algorithm (BP-ANN).

2 Materials and Methods

The hourly average wind speed time series of weather stations Volkel (January 2007, 2008 and 2009), K13 (February 2007, 2008 and 2009) and Valkenburg (March 2007, 2008 and 2009) are used to fit the ARMA model (2007 and 2008 for each station) and evaluate the forecasting error (2009 for each station). These time series are provided by the Royal Netherlands Meteorological Institute [11].

2.1 ARMA Model

The mathematical formulation of the autoregressive moving average (ARMA) model is:

$$\hat{W}_{(t)} = \varphi_1 \hat{W}_{(t-1)} + \ldots + \varphi_p \hat{W}_{(t-p)} + a_{(t)} - \theta_1 \varepsilon_{(t-1)} - \ldots - \theta_q \varepsilon_{(t-q)} \tag{1}$$

Where $\hat{W}_{(t)}$ is the transformed and standardized wind speed of hour t, $\varphi_1,\ldots, \varphi_p$ are the autoregressive parameters, $\theta_1,\ldots, \theta_q$ are the moving average parameters, $\varepsilon_{(t)}$, $\varepsilon_{(t-1)},\ldots, \varepsilon_{(-qt)}$ are random variables with an average value of zero and a standard deviation of σ. The fit of the ARMA model to the wind speed time series of interest requires the transformation, standardization, estimation and diagnostic checking processes.

2.1.1 Transformation and Standardization

The Weibull PDF is frequently used in the time series analysis. In order to analyze a determined wind speed time series it is necessary to transform the series to another one that has a Gaussian PDF. This transformation is [3]:

$$U_{T(t)} = U_{(t)}^{\ m} \text{ with } t = 1,\ldots,n \tag{2}$$

Where $U_{(t)}$ is the wind speed time series of interest, $U_{T(t)}$ is the transformed wind speed time series with Gaussian PDF, m is the power transformation and n is the number of hourly wind speed observations. The power transformation m is calculated from Weibull shape k. Dubey (1967) [12] showed that for shape factors between 3.26 and 3.60, the Weibull PDF is similar to the Gaussian PDF, thus the power transformation of equation (2) with *m* between *k*/3.60 and *k*/3.26 adequately describes the Gaussian PDF. The selected value of this range is *m* value, for which the coefficient of skewness (*SK*) is closest to zero. The coefficient of skewness is calculated by equation (3) [13].

$$SK = \frac{Q_3 + Q_1 - 2Q_2}{Q_3 - Q_1} \tag{3}$$

Where $Q_1 = C^{-1}(0.25)$, $Q_2 = C^{-1}(0.5)$ and $Q_3 = C^{-1}(0.75)$ are the first, second and third quartiles, respectively. $C^{-1}(.)$ is the inverse of the Cumulative Distribution Function (CDF) calculated by the algorithm of the Fig. 3 (Section 2.1.4).

The transformed wind speed time series is not stationary. In order to be stationary it is necessary to subtract the hourly average and divide per the hourly standard deviations. If $\mu_h(h)$ and $\sigma_h(h)$ with h=1,2,..., 24 are the hourly average and standard deviation of transformed wind speed series, respectively, it is assumed that these functions are periodic: $\mu_h(t=25)=\mu_h(h=1)$ and $\sigma_h(t=25)=\sigma_h(h=1)$ [3]. The transformed and standardized series ($W_{(t)}$) is:

$$W_{(t)} = \frac{U_{T(t)} - \mu_{h(t)}}{\sigma_{h(t)}} \tag{4}$$

2.1.2 Estimation

The plots of autocorrelation function (ACF) and partial autocorrelation function (PACF) of the series of Equation (4) have useful information to determine the value of order *p* and *q* of the ARMA model [14].

Others approach for selecting the *p* and *q* values are Bayesian Information Criterion (BIC) [15] and Akaike Information Criterion (AIC) [16]. The optimal value of *p* and *q* are selected minimizing the respective criterion BIC or AIC of the equations (5) and (6), respectively.

$$BIC(p,q) = n\log(\sigma_{(p,q)}{}^2) + (p+q)\log(n) \tag{5}$$

$$AIC(p,q) = n\log(\sigma_{(p,q)}{}^2) + 2(p+q) \tag{6}$$

$$\sigma_{(p,q)}{}^2 = \frac{\sum_{t=1}^{n}(W_{(t)} - \hat{W}_{(t)})^2}{n-(p+q)} \tag{7}$$

Where $\sigma_{(p,q)}{}^2$ is the variance of the residual, $\hat{W}_{(t)}$ is the value calculated by the ARMA model. The autoregressive and moving average parameters are calculated by minimization of the quadratic prediction error criterion [17].

2.1.3 Diagnostic Checking

The statistical checking is made by the Ljung-Box test [18]. In this test the statistical S is compared with the chi-square distribution χ_α^2 with L-p-q degrees of freedom. If $\chi_\alpha^2 > S$ then the model is adequate for the significance level α. The statistical S is calculated by Equation (8).

$$S = n(n+2)\sum_{\Omega=1}^{L} \frac{r_\Omega^{\ 2}}{(n-\Omega)} \tag{8}$$

Where L is a number of lags considered and r_Ω is the correlation coefficient of the residuals corresponding to lag Ω.

2.1.4 Forecasting with ARMA Model

The forecasting process starts by evaluating expressions (9) and (10).

$$\hat{W}_{(t+l)} = \varphi_1 W_{(t-1+l)} + \ldots + \varphi_p W_{(t-p+l)} + \varepsilon_{(t+l)} - \theta_1 \varepsilon_{(t-1+l)} - \ldots - \theta_q \varepsilon_{(t-q+l)} \tag{9}$$

$$\hat{U}_{T(t+l)} = \mu_{h(t+l)} + \sigma_{h(t+l)} \hat{W}_{(t+l)} \tag{10}$$

Where l=1,…,hrp, hrp are the hours ahead of forecasting. An important fact is that the values of equation (10) can be negative, so depending of the m value, undoing the transformation process using the power $1/m$ is difficult. For this reason, in this paper the undoing of the transformation process is made with the probability transformation in the discrete form. Using this transformation process, the PDF of the equation (10) will be the PDF of the original wind speed time series. This probability trans-formation is based in Equation (11) [19] and the forecasting values $\hat{U}_{(t+l)}$ are calcula-ted by Equation (12).

$$C(\hat{U}_{(t+l)}) = u_{(t+l)} = \hat{C}_T(\hat{U}_{T(t+l)}) \tag{11}$$

$$\hat{U}_{(t+l)} = C^{-1}(\hat{C}_T(\hat{U}_{T(t+l)})) \tag{12}$$

Where C is the CDF in discrete form of the original wind speed time series $U_{(t)}$, $\hat{C}_T$ is the CDF of Equation (10), assuming a Gaussian CDF with average and standard deviation equal to the transformed time series and $u_{(t+l)}$ is a variable with uniform density function in [0,1]. Once $u_{(t+l)}$ is calculated with Equation (11), it is necessary to calculate the value of $\hat{U}_{(t+l)}=C^{-1}(u_{(t+l)})$. The first step is to build the vector $\boldsymbol{x}$, whose elements are zero until *ceil*(max($U_{(t)}$)) with t=1,2,…,n in step Δx. The next step is to build the vector $\boldsymbol{C}$, whose elements are the cumulative probabilities of each element of vector $\boldsymbol{x}$. Finally, the forecasted value of wind speed $\hat{U}_{(t+l)}$ is calculated by the algorithm of Figure 1. The function *size*($\boldsymbol{C}$) calculates the number of rows of vector $\boldsymbol{C}$.

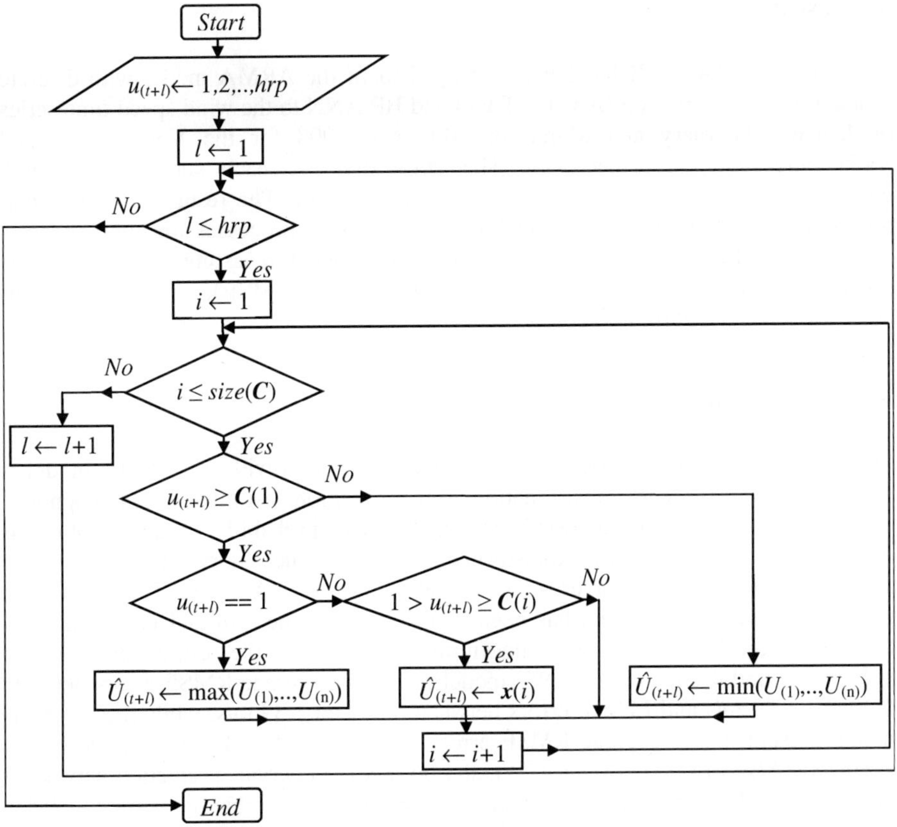

Fig. 1. Algorithm to calculate the inverse of the CDF in discrete form.

2.1.5 BP-ANN Model

The artificial neural network training by back-propagation algorithm (BP-ANN) used for hourly average wind speed forecasting is shown in Fig. 2.

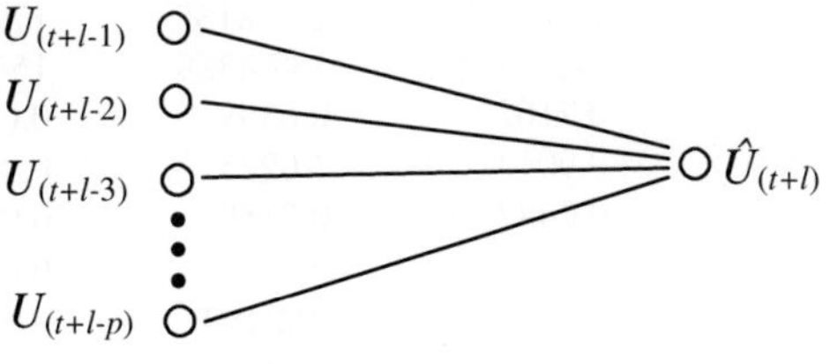

Fig. 2. Artificial neural network for wind speed forecasting.

The data used for training the artificial neural network is scaled between 0.1 and 0.9 to avoid the saturation regions of the sigmoid function.

3 Results

A program in MATLAB has been developed to fit the ARMA model with discrete probability transformation (ARMA+DPT) and BP-ANN to the wind speed time series of January, February and March of 2007 and 2008 for the Volkel, K13 and Valkenburg stations, respectively. The 2009 data for each station and month combination were used to evaluate the forecasting error. The results of transformation, standardization, estimation and diagnostic checking processes are summarized in Table 1. In tables 2, 3 and 4, the comparison among forecasting errors indices is shown: mean bias error (MBE), root mean square error (RMSE), mean absolute bias error (MABE) and correlation coefficient (R).

4 Conclusion

In this paper, the methodology for wind speed prediction with the ARMA model is carefully reviewed. The transformation, standardization, estimation and diagnostic checking processes are analyzed and a discrete probability transformation is introduced for undoing the transformation process necessary to calculate the autoregressive and moving average parameters of the ARMA model. This model was used as a prediction tool and has been compared with the artificial neural network training by the back-propagation algorithm. According to the results shown in the Tables 2, 3 and 4, the ARMA+DPT model present in average RMSE errors between 0.90 and 1.89 m/s and MABE errors between 0.63 and 1.39 m/s, while the BP-ANN model present higher average RMSE errors, between 0.95 and 2.05 m/s and higher average MABE errors, between 0.72 and 1.56 m/s for intervals prediction between 1 and 10 hours ahead, respectively. Figures 17, 18 and 19 show that the ARMA+DPT in certain cases can improve the BP-ANN by as much as 17.71%; however, in other cases the BP-ANN error can be 0.9% smaller than the ARMA+DPT model.

Table 1. Parameters of ARMA+DPT model for each station.

Parameter/Station	Volkel	K13	Valkenburg
(p,q)	(2,0)	(5,0)	(7,0)
L-p-q	147	132	142
$\chi_{0.05}^2$	176.2938	159.8135	170.8092
S	143.1374	137.8389	154.9721
m	0.7162	0.6899	0.6269
φ_1	0.9014	1.0225	0.9210
φ_2	0.0487	0.0150	0.0173
φ_3	-----	0.0024	0.0359
φ_4	-----	-0.0480	-0.0210
φ_5	-----	-0.0211	-0.0295
φ_6	-----	-----	0.0438
φ_7	-----	-----	-0.0174
k	2.3353	2.25	2.0436
Criterion	AIC	AIC	AIC
Δx		0.0001	

Table 2. Forecasting errors (Volkel).

$l(h)$	Volkel					
	BP-ANN			ARMA+DPT		
	RMSE	MABE	R	RMSE	MABE	R
1	0.9750	0.7123	0.9142	0.8691	0.5714	0.9241
2	1.1982	0.8692	0.8656	1.0547	0.7097	0.8879
3	1.3531	0.9805	0.8243	1.1596	0.7828	0.8632
4	1.4959	1.0858	0.7825	1.2793	0.8751	0.8326
5	1.6316	1.1950	0.7364	1.3899	0.9382	0.8024
6	1.7451	1.2642	0.6930	1.4960	1.0168	0.7698
7	1.8598	1.3561	0.6431	1.5865	1.0750	0.7419
8	1.9457	1.4107	0.6137	1.6018	1.1096	0.7431
9	2.0668	1.5201	0.5407	1.7507	1.2042	0.6842
10	2.1068	1.5652	0.4943	1.8121	1.2774	0.6492

Table 3. Forecasting errors (K13).

$l(h)$	K13					
	BP-ANN			ARMA+DPT		
	RMSE	MABE	R	RMSE	MABE	R
1	0.8802	0.6747	0.9671	0.8789	0.5989	0.9674
2	1.0489	0.7972	0.9528	1.0705	0.7533	0.9515
3	1.2238	0.9270	0.9353	1.2795	0.9069	0.9313
4	1.3589	1.0109	0.9195	1.4179	1.0146	0.9146
5	1.4839	1.1252	0.9044	1.5478	1.1445	0.8999
6	1.6193	1.2432	0.8838	1.6471	1.2304	0.8823
7	1.8601	1.3519	0.8483	1.9063	1.3775	0.8457
8	1.8986	1.4056	0.8421	1.9441	1.4217	0.8420
9	2.0141	1.5466	0.8232	1.9005	1.4359	0.8458
10	1.9873	1.4650	0.8234	2.0244	1.5064	0.8214

Table 4. Forecasting errors (Valkenburg).

$l(h)$	Valkenburg					
	BP-ANN			ARMA+DPT		
	RMSE	MABE	R	RMSE	MABE	R
1	1.0081	0.7779	0.9380	0.9521	0.7048	0.9449
2	1.1910	0.9330	0.9128	1.0935	0.8157	0.9256
3	1.3945	1.0750	0.8765	1.2158	0.9056	0.9086
4	1.5402	1.2119	0.8467	1.3302	1.0016	0.8877
5	1.6996	1.3384	0.8080	1.4590	1.1203	0.8639
6	1.8082	1.4060	0.7793	1.5904	1.1965	0.8389
7	1.9382	1.5316	0.7404	1.6971	1.2914	0.8112
8	2.0466	1.6378	0.7042	1.7202	1.3218	0.8067
9	2.0129	1.6190	0.7154	1.7490	1.3378	0.7979
10	2.0649	1.6530	0.6967	1.8301	1.3936	0.7748

Acknowledgments

This work was supported by the "Ministerio de Ciencia e Innovación" of the Spanish Government under Project ENE2009-14582-C02-01.

References

1. Barthelmie, R.J., Murray, F., Pryor, S.C.: The economic benefit of short-term forecasting for wind energy in the UK electricity market. Energy Policy 36, 1687–1696 (2008)
2. Lei, M., Shiyan, L., Chuanwen, J., Hongling, L., Yan, Z.: A review on the forecasting of wind speed and generated power. Renewable and Sustainable Energy Reviews 13, 915–920 (2009)
3. Brown, B.G.: Time series models to simulate and forecast wind speed and wind power. Journal of Climate and Applied Meteorology 23, 1184–1195 (1984)
4. Daniel, A.R., Chen, A.A.: Stochastic simulation and forecasting of hourly average wind speed sequences in Jamaica. Solar Energy 33, 571–579 (1991)
5. Torres, J.L., García, A., De Blas, M., De Francisco, A.: Forecast of hourly average wind speed with ARMA models in Navarre (Spain). Solar Energy 79, 65–77 (2005)
6. Sfetsos, A.A.: A comparison of various forecasting techniques applied to mean hourly wind speed time series. Renewable Energy 21, 23–35 (2000)
7. Cadenas, E., Rivera, W.: Wind speed forecasting in the south coast of Oaxaca, México. Renewable Energy 32, 2116–2128 (2007)
8. Palomares-Salas, J.C., De la Rosa, J.J.G., Ramiro, J.G., Melgar, J., Agüera, A., Moreno, A.: ARIMA vs. Neural Networks for wind speed forecasting. In: IEEE International Conference on Computational Intelligence for Measurement Systems and Applications, pp. 129–133. IEEE Computer Society Press, Los Alamitos (2009)
9. Nfaoui, H., Buret, J., Sayigh, A.A.M.: Stochastic simulation of hourly average wind speed sequences in Tangiers (Morocco). Solar Energy 56, 301–314 (1996)
10. Gong, L., Jing, S.: On comparing three artificial neural networks for wind forecasting. Applied Energy 87, 2313–2320 (2010)
11. Royal Netherlands Meteorological Institute, `http://www.knmi.nl/`
12. Dubey, S.D.: Normal and Weibull distributions. Naval Research Logistics Quarterly 14, 69–79 (1967)
13. Kim, T.H., White, H.: On more robust estimation of skewness and kurtosis. Finance Research Letters 1, 56–73 (2004)
14. Box, J.E.P., Jenkins, G.M.: Time series analysis: forecasting and control. Prentice-Hall, Inc., Englewood Cliffs (1976)
15. Schwarz, G.: Estimating the dimension of a model. The Annals of Statistics 6, 461–464 (1978)
16. Akaike, H.: A new look at the statistical model identification. IEEE Transactions on Automatic Control 19, 716–723 (1974)
17. Ljung, L.: System Identification. Theory for the user. PTRInformation and System Sciences Series. Prentice-Hall, Englewood Cliffs (1999)
18. Ljung, G.M., Box, G.E.P.: On a measure of lack of fit in time series models. Biometrika 65, 297–303 (1978)
19. Rosenblatt, M.: Remarks on a multivariate transformation. Annals of Mathematical Statistics 23, 470–472 (1952)
20. Rojas, R.: Neural Networks. Springer, Berlin (1996)

Multi-Layer Methodology Applied to Multi-period and Multi-Objective Design of Power Distribution Systems

José L. Bernal-Agustín[1], Rodolfo Dufo-López[1], Franklin Mendoza[2], and José A. Domínguez-Navarro[1]

[1] Department of Electrical Engineering, University of Zaragoza, Calle María de Luna 3, 50018, Spain
jlbernal@unizar.es, rdufo@unizar.es, jadona@unizar.es
[2] Department of Electrical Engineering, UNEXPO, Puerto Ordaz, Venezuela
fmendoza@unexpo.edu.ve

Abstract. This paper presents, for the first time, an application of a Multi-Layer methodology to solve the Multi-Period and Multi-Objective design of power distribution systems. The methodology consists of two steps. In the first, an improved version of the Strength Pareto Evolutionary Algorithm (SPEA) is applied to the Multi-Objective design of Power Distribution Systems, obtaining the Pareto Front for each period of the Multi-Period problem. The layer concept is used to obtain an initial and valid solution for the last period. In the second step, a procedure is applied to find a possible sequence of solutions for all periods. Results are presented in terms of the topology required for each period.

Keywords: Multi-Objective optimization, Power Distribution, Multi-Layer.

1 Introduction

Planning distribution systems can be described as selecting a system configuration, or a sequence of systems configurations for successive years, which allows meeting the requirements of the load while taking into account its future growth [1].

There are two overall approaches to the distribution planning problem: single stage and Multi-Stage [2]. The single stage model, usually called the static model, considers that energy demand will be static during the period considered for the design. On the other hand, the Multi-Stage model is solved in several stages, years, or periods; this is why it is known as Multi-Stage planning or Multi-Period planning.

This paper presents the development of Multi-Period planning for power distribution systems networks, through a modification of the dynamic methodology suggested in [3] to Multi-Period optimization and [4] to Multi-Layer of optical-telecommunication networks.

In the proposed methodology, each problem is solved using the mathematical model found in [5], taking into account the demand variation of each node as well as the different costs involved for each. The search of optimal solutions is carried out minimizing two objectives: Cost and Energy Not Served (*ENS*) simultaneously using Multi-Objective Evolutionary Algorithms (MOEAs) [6] In this paper, the MOEA is the SPEA [7] to determine possible solutions for each particular period. The results and conclusions demonstrate the practical applicability of this design methodology.

M. Zhu (Ed.): Electrical Engineering and Control, LNEE 98, pp. 1011–1018.
springerlink.com © Springer-Verlag Berlin Heidelberg 2011

2 Pareto Optimal Set Concept

A solution is Non-Dominated [6] if the corresponding objective vector cannot be improved in any dimension without degrading another. The solution set of a Multi-Objective optimization problem consists of all Non-Dominated solutions and is known as Pareto Optimal Set or Pareto Optimal Front. The concept of Pareto Optimal Set for a minimization problem (objectives $F_1(x)$ and $F_2(x)$) is illustrated in Figure 1.

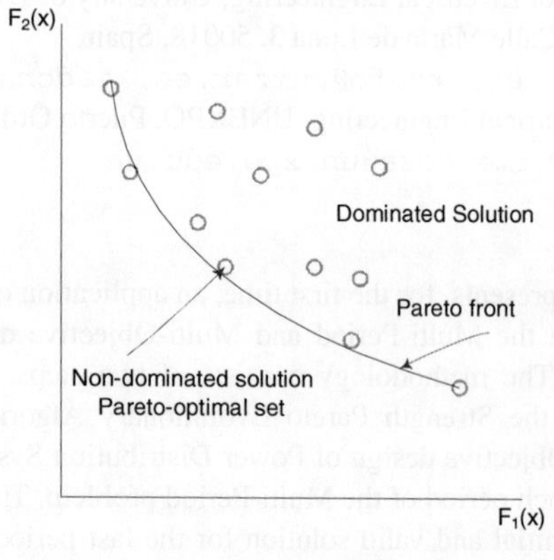

Fig. 1. Pareto Front. Example for a minimization Multi-Objective optimization problem

3 Strength Pareto Evolutionary Algorithm

MOEAs are based on using processes observable in nature (crossing, mutation, evolution, adaptation to the environment) and then applying these concepts to the solutions of a problem, assigning fitness to each one of the possible solutions, and optimizing several objectives simultaneously [8].

The MOEA that has been applied in this paper is the SPEA. SPEA was proposed in [7] as an approach that incorporates several of the desirable features of other well-know Multi-Objective evolutionary algorithms. The basic characteristics of the MOEA (coding, evaluation function, crossing method, mutation, etc.) are equal to those used in another previous work by the authors.[5]

SPEA has obtained very good results recently in applications to complex problems in electrical engineering [5,9-12].

4 Multi-Period and Multi-Objective Design of Power Distribution Systems

The problem of Multi-Period and Multi-Objective design of power distribution systems is considered an optimization problem in which it is necessary to minimize two functions: *Costs* and *ENS* [5].

The cost function includes fixed cost, $C_{FIXED}(x)$, and variable cost, $C_{VARIABLE}(x)$. Next, it is presented with the objective function to minimize the total cost:

$$C_{NETWORKS}(x) = C_{FIXED}(x) + C_{VARIABLE}(x) \tag{1}$$

The reliability function can be expressed by the *ENS* [13]:

$$ENS(x) = \sum_{i=1}^{n} \sum_{j \in m(j)} \lambda_j r_j p_i \tag{2}$$

where, n is the number of load nodes, λ_j is the failure rate of element j, r_j is its average outage time, and p_i is the load at node i affected for the failure.

The Multi-Objective problem can be formulated as:

$$\min \ f(x) = \left[C_{NETWORKS}(x),\ ENS(x)\right] \tag{3}$$

subject to technical constraints,[13] which are Kirchhoff's current law constraints for all the nodes, the capacity constraints for the feeders and substations, and the voltage drop constraints.

Distribution systems are usually operated radially. However, the mathematical model used in this paper is valid for solving a Multi-Objective optimization problem, the solutions for which can be topologically meshed because it includes not only the feeders in operation, but also the feeders in reserve.

5 Proposed Multi-Layer and Multi-Period Methodology

The method presented in this paper is based on [3] and [4] in which the research works were focused on the design of telecommunications networks.

The planning and expansion of distribution systems have a Multi-Period dynamic nature. This means that the decision to build a feeder or a substation in a planning period can affect decisions in the next planning periods [14].

It has been considered that the Multi-Period problem has n periods or stages (T_1. T_2,..., T_n), simultaneously minimizing two objectives in the optimization process: *Costs* and *ENS*.

The proposed Multi-Layer and Multi-Period methodology consists of two steps.

Step 1: The optimizations of the n-1 first periods (from T_1 to T_{n-1}) are carried out using SPEA, obtaining the Pareto Fronts for each of the periods. Next, five solutions from each Pareto Front are selected. In this work in particular, the solution obtained applying the Max-Min approach [15] and the four solutions nearest to the Max-Min one have been chosen. All these solutions are compared, using the layer concept [4], determining the common routes (routes that appear in the solutions of all the periods). The optimization of the n period is carried out, forcing the common routes to appear in the obtained solutions.

To illustrate how the layer concept is used to determine the common routes, we offer an example for an 8-nodes expansion distribution network showing common routes obtained in figure 2. The common routes for each period are:

T_1: 1-2 and 1-4 (comparing layers L_1, L_2 and L_3);
T_2: 3-5 (comparing layers L_2 and L_3); and
T_3: 4-7 and 7-8 (in this case, only layer L_3 is considered).
Therefore, the common routes are 1-2, 1-4, 3-5, 4-7, and 7-8.

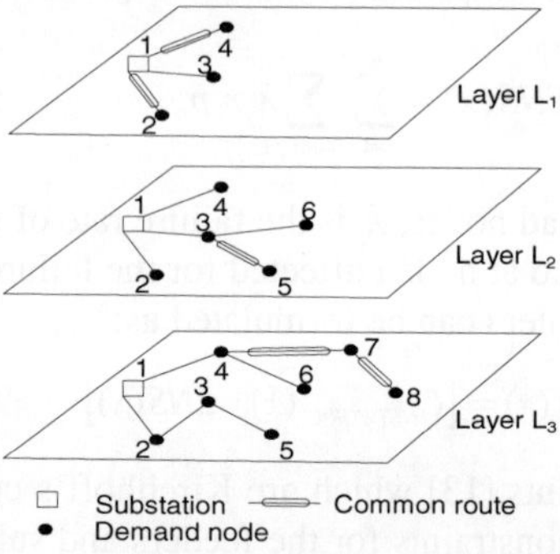

Fig. 2. Determination of common routes

Step 2: From the Pareto front of the period *n*, a solution is selected, such as that obtained by applying the Max-Min approach. This solution of the period *n* is used to obtain an initial solution for the period *n*-1, applying these procedures.

a) The nodes and substations that did not exist in the period *n*-1, but that existed in the period *n*, are eliminated, as are the lines that are connected to these nodes and substations.
b) If some nodes remain isolated from the network, it is necessary to add feeders that will connect the isolated nodes to the network.
c) If the solution does not fulfill the technical constraints (for example, the thermal limit of the feeders), it is necessary to increase the section of the feeders until the technical constraints are fulfilled.
d) If a node is connected to the network only using reserve feeders, these feeders will be turned on.

The described procedure is applied successively to obtain solutions for all the other periods (from T_{n-2} to T_1), so we get one solution for each period, as well as coordinated results for all the periods.

If on b), c) or d) a new feeder appears in period T_i, then it will have to be included in all the later periods (T_{i+1} to T_n) like reserve feeders. If a feeder changes size, then its size will have to be modified in all the later periods. In any case, the technical constraints such as the thermal limit of the feeders must be fulfilled.

6 Computational Results

The network used is described in [16]. Initially, this network is made up of 10 nodes and 2 substations (nodes A and B), at a 10 kV voltage level. It expands in a time period of 9 years until it achieves a maximum of 20 nodes and 4 substations. In Figure 3, the test system is shown. The sizes of the existing distribution substations are 20 MVA. Table 1 gives the annual peak power demands of the distribution network nodes, and we have proposed two feeder sizes, LA110 and LA56.

It can be seen that one of the differences from the results presented in [16] is that in the present work, the achieved solutions incorporate reserve feeders. This permits the optimization of the *Costs* and *ENS* simultaneously (Multi-Objective optimization).

Results obtained through the proposed Multi-Layer technique are shown in Table 2. The "feeder" column indicates feeders and their sizes that are fixed in the expansion process. It is possible for some routes to show different sizes during the expansion process due to the circuit reconfiguration by the added substation in some periods. "X" indicates than the corresponding feeder is an "operation feeder," and "R" indicates than the corresponding feeder is a "reserve feeder."

From the Pareto front of period 9, the solution that is obtained applying the Max-Min approach is selected. Then, by applying the procedure explained in step 2 of section 3, the 9 solutions of the 9 periods are obtained (see Figure 4). The proposed and applied methodology gives coordinated results for all the periods. As mentioned in section 2, from a Multi-Objective point of view, we can obtain a set of solutions (Pareto Front). No solution from the Pareto Front is better than any other, showing that the results are a "technical and valid" design, but it is not an optimal design. When solving a Multi-Period and Multi-Objective design problem, we cannot speak about optimal solutions, only about technically valid and coordinated solutions for all the periods.

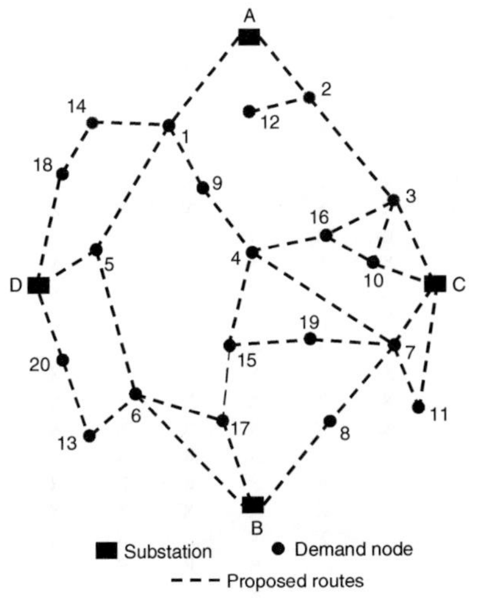

Fig. 3. Proposed routes and nodes

Table 1. Annual Power Demand Requirements, in MVA.

Node	Year 1	Year 2	Year 3	Year 4	Year 5	Year 6	Year 7	Year 8	Year 9
1	4.05	4.43	4.85	3.45	3.78	4.13	4.52	4.95	5.42
2	0.78	0.86	0.70	0.77	0.84	0.92	1.01	1.11	1.21
3	2.58	2.83	3.09	3.38	2.78	3.04	3.33	3.64	3.98
4	0.32	0.35	0.38	0.41	0.34	0.37	0.41	0.45	0.49
5	0.28	0.31	0.34	0.37	0.33	0.36	0.39	0.43	0.47
6	1.17	1.28	1.05	0.92	1.00	1.10	1.20	1.32	1.44
7	4.04	3.09	3.38	3.70	4.05	3.33	3.64	3.98	4.36
8	0.72	0.67	0.73	0.60	0.65	0.72	0.78	0.86	0.94
9	1.14	1.25	1.03	1.12	1.23	1.35	1.47	1.61	1.77
10	1.56	1.70	1.86	2.04	2.23	1.83	2.00	2.19	2.40
11		1.67	1.78	1.91	2.05	2.21	2.38	2.58	2.80
12			0.88	0.93	0.99	1.05	1.13	1.20	1.29
13			1.12	1.15	1.18	1.22	1.26	1.31	1.35
14				3.05	2.52	2.66	2.81	2.97	3.16
15				1.62	1.62	1.62	1.62	1.62	1.62
16				--	0.94	0.99	1.05	1.13	1.22
17				2.16	2.20	2.24	2.29	2.35	2.40
18					1.89	1.94	1.99	2.04	2.10
19						1.45	1.55	1.67	1.81
20							3.79	3.79	3.79

Table 2. Determination of common routes.

Route	T_1	T_2	T_3	T_4	T_5	T_6	T_7	T_8	T_9	Feeder
A-1	X	X	X	X	X	X	X	X	X	LA110
B-6	X	X	X	X	X	X	X	X	X	LA110
1-9	X	X	X		X	X	X	X	X	
1-5	R	R	R		R	R	R	R	R	
2-3	X	X	R	R	R	R				
A-2	X	X	X	X	X	X	X	X	X	LA56
3-10	X	X	R	R		R		R		
4-7	R	R	X	X						
4-9	X	X	R	R	X	R	R	R	R	
5-6	X	X	X	X	X					
7-8	X	X	R	R	R	R				
B-8	X	X	X	X	X	X	X	X	X	LA56
7-11			R	R	R	X			R	
2-12			X	X	X	X	X	X	X	LA56
C-3			X	X	X	X	X	X	X	LA56
6-13			X	X	X	X	X	X	X	LA56
C-7			X	X	X	X	X	X	X	LA110
C-10			X	X	X	X	X	X	X	LA110
C-11			X	X	X	X	X	X	X	LA56
1-14			X	X	X	X	X	X	X	
4-15				X	X	R	R		R	
15-17				R	R	R	R	R	R	LA56
B-17				X	X	X	X	X	X	LA110
6-17				X	X	X	X	X	X	
3-16					R	R	R			
4-16					R				R	LA56
10-16					X	X	X	X	X	LA56
14-18					X	X	X	X	X	LA56
D-5					X	X	X	X	X	LA110
7-19						X	X	X	X	
D-18						R	X	X	X	LA110
15-19						X	X	X	X	
D-20										LA110

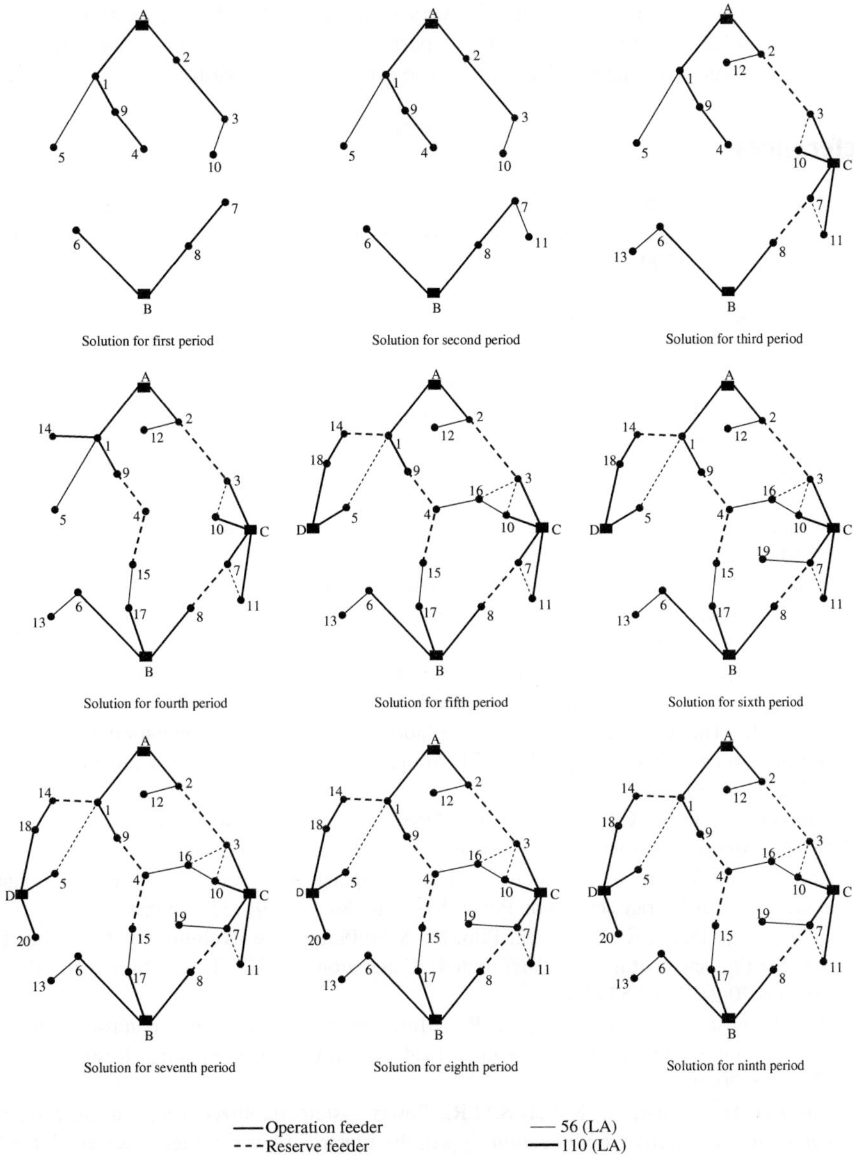

Fig. 4. Solutions for the nine periods

7 Conclusions

A novel methodology for the Multi-Objective and Multi-Period optimization of a distribution power system has been presented. The Pareto Front of each period is achieved using the SPEA. In the optimization process, for each solution and period, the cost and the *ENS* are calculated, obtaining valid solutions that fulfilled all the technical constraints. The concept of layers is used to identify the common routes from some selected solutions of all the periods. Finally, a novel procedure gives coordinated solutions. The results validate the proposed methodology.

References

1. Skok, M.: The evolutionary algorithm for the dynamic planning of distribution networks. Ph.D. dissertation, Department of Power Systems Faculty of Elec. Eng. and Comp., Univ. Zagreb, Croatia (2005)
2. Vaziri, M., Tomsovic, K., Bose, A.: A Directed Graph Formulation of the Multistage Distribution Expansion Problem. IEEE Transactions on Power Delivery 19(3), 1335–1341 (2004)
3. Geary, N., Antonopoulos, A., O'Really, J., Drakopoulos, E.: Planning techniques for Multi-Period Optical Network Designs. In: London Communications Symposium, London, pp. 14–15 (2000)
4. Decocq, B., et al.: Planning of Optical Networks. EURESCOM Project P709 Deliverable (March 3, 2000), `http://www.eurescom.de/public/projectresults/P700- series/709d3.htm`
5. Mendoza, F., Bernal-Agustin, J.L., Dominguez-Navarro, J.A.: NSGA and SPEA Applied to Multi-Objective Design of Power Distribution Systems. IEEE Transactions on Power Systems 21(4), 1938–1945 (2006)
6. Coello, C.A., Veldhuizen, D.A.V., Lamont, G.B.: Evolutionary Algorithms for Solving Multi-Objective Problems. Kluwer Academic/Plenum Publishers (2002)
7. Zitzler, E., Thiele, L.: Multiobjective evolutionary algorithms: a comparative case study and the strength Pareto approach. IEEE Transactions on Evolutionary Computation 3(4), 257–271 (1999)
8. Srinivas, N., Deb, K.: Multi-objective Optimization Using Non-dominated Sorting in Genetic Algorithms. Evolutionary Computation 2(3), 221–248 (1994)
9. Abido, M.A.: Environmental/economic power dispatch using multiobjective evolutionary algorithms. IEEE Transactions on Power Systems 18(4), 1529–1537 (2003)
10. Celli, C.G., Ghiani, E., Mocci, S., Pilo, F.: A Multiobjective Evolutionary Algorithm for the Sizing and Siting of Distributed Generation. IEEE Transactions on Power Systems 20(2), 750–757 (2005)
11. Bernal-Agustín, J.L., Dufo-López, R.: Multi-objective design and control of hybrid systems minimizing costs and unmet load. Electric Power Systems Research 79(1), 170–180 (2009)
12. Yassami, H., Darabi, A., Rafiei, S.M.R.: Power system stabilizer design using Strength Pareto multi-objective optimization approach. Electric Power Systems Research 80(7), 838–846 (2010)
13. Ramirez-Rosado, I.J., Bernal-Agustin, J.L.: Reliability and costs optimization for distribution networks expansion using an evolutionary algorithm. IEEE Transactions on Power Systems 16(1), 111–118 (2001)
14. Silvestre, A., Braga, D., Saraiva, J.T.: Multiyear Dynamic Approach for Transmission Expansion Planning and Long-Term Marginal Cost Computation. IEEE Transactions on Power Systems 20(3), 1631–1639 (2005)
15. Lai, Y.L., Hwang, C.L.: Fuzzy Mathematical Programming. Springer, Berlin (1992)
16. Gonen, T., Ramirez-Rosado, I.J.: Review of distribution systems planning models: a model for optimal multistage planning. IEEE Proceedings 133(7) (1986)

Dynamics of Land Cover and Land Use Change in Quanzhou City of SE China from Landsat Observations*

Weihua Pan[1,2], Hanqiu Xu[2], Hui Chen[1], Chungui Zhang[1], and Jiajin Chen[1]

[1] Institute of Meteorological Science of Fujian Province, Fuzhou, 350001, China
[2] College of Environmental and Resources Engineering,
Fuzhou Unversity, Fuzhou, 350002, China
Weihuapan80@163.com

Abstract. Urbanization has significantly modified the landscape during the reform periods in China, which has significant climatic implications across all scales due to the simultaneous removal of natural land cover and introduction of urban materials. Change detection from remote sensing data can provide important information to study this change in land cover and land use (LCLU). A study on the LCLU in Quanzhou of China was carried out using remote sensing technology. By using the Landsat data archive, we have created a two epoch time-series for LULC for the period 1989-2000. The map is based on a image supervised-classification approach for monitoring LULC change filtered with a detailed image post-classification. To quantitatively compare LULC of different year, two indexes called Land-Use-Degree Ratio Index (L) and Land-Use Change Ratio Index (R) are created. Results show that the arable land has decreased stupendously and build-up land has increased greatly during the 11 years in Quanzhou city, which stimulated the competition between build-up land and arable land and caused the decrease of arable land in quantity and quality. Comparisons with census data indicate that the great change of LULC is forced by human-induced factors response to specific economics, demographic, or environmental conditions.

Keywords: Urbanization, dynamics change, LULC, remote sensing, Quanzhou city.

1 Introduction

Urbanization is a major event in China during the reform periods, which has altered the natural landscape and resulted in a series of climatic and social problems. Urban growth and sprawl have drastically modified the biophysical environment (Auer, 1978 [1], Roth, 1989 [2], Kaiser et al, 1995 [3]). The noteworthy is the replacement of soil and vegetation with impervious urban materials, such as concrete, asphalt, and buildings, thus significantly impact the land-atmosphere energy exchange processes

* Sponsored by the National Natural Science Foundation of China (40371107) and key project program of Fujian provincial (2009N0030) and projects of FMB (2006K04 and 2009q02).

M. Zhu (Ed.): Electrical Engineering and Control, LNEE 98, pp. 1019–1027.
springerlink.com © Springer-Verlag Berlin Heidelberg 2011

and several noticeable phenomenon that have arisen as a result of urban expansion are that urban climates are warmer and more polluted than the rural counterparts and more and more land is deserted rapidly short of precipitation (Owen, 1998 [4], Carlson, 2000 [5]). As we all know, China is a country with the maximum population in the world and with very limited arable land resource to support people, and the resource shortage pressure forced us to save and use the land with reason than before.

From a broader perspective urbanization is one of many ways in which human are altering the land-cover of the globe, and the changes are estimated to have significantly altered more than 80% of the Earth's land area over the last several centuries (Thomas 1956 [6], Vitosek *et al.* 1997 [7]).

Land cover and land use (LCLU) change detection provides a fundament input for planning, management and environmental studies, such as landscape dynamics or natural risks and impacts (European Commission 1998 [8]). One technology which offers considerable promise for monitoring landcover change is satellite remote sensing, because of its temporal resolution, provides an excellent historical framework for estimating the spatial extent of LCLU change and a repetitive measurement of earth surface conditions relevant to climatology, hydrology, oceanography and land cover monitoring (Kostmayer, 1989 [9], Jensen and Cowen, 1999 [10]). One mission in particular, the Landsat series begun in 1972, was designed and continues to operate with the objective of tracking changes in land-cover conditions. The high spatial resolutions and regular revisit times of the Landsat mission are well suited to monitoring and studies of LCLU change.

Of critical importance is linking these observed changes in land-cover to the driving socio-economic or environmental origins. In particular, the LCLU change offers a graphic depiction of the interplay between economics, political systems, and the environment. In this paper, we take the Quanzhou city of SE China as a example to study the regional LCLU change during the reformed periods in China and analyse its main driving forces, which provides lots of referenced data for local government agencies to maintain a long and continuable development.

2 Study Site and Data

The study area is the Quanzhou city, which lies in southeast of Fujian province, PRC. The Quanzhou city is one of three cities comprising the famous Golden Triangle Area of South Fujian. Moreover, the urbanization process of Quanzhou city has stupendously developed during the reform period from 1989 to 2000, and the GDP of Quanzhou city is top in rank of those of all cities in Fujian Province.

The Quanzhou city lies in a coastal plain named Jinjiang Plain replenished with silt, clay and water, where grows plenty of crops, forest and vegetation in the study area. To generate and monitor our time-series of LCLU change in Quanzhou city, we obtained Landsat scenes from 1989 and 2000. Both scenes were imaged during the April-June regional growing season to minimize seasonal variability. Spring or earlier summer, when vegetation is in the stage of vigorous growth, is the preferred season for the Landsat scenes. The images are registered in a NAD27 projection, and use manual selection of ground control points (GCPs) followed by a linear (affine) transformation using nearest neighbor resampling. RMS errors for the GCP sets were less than 0.50 pixels in all cases.

3 Land Cover and Land Use Mapping

For this study, a common legend is established based on the official census and actual land-cover characters of Quanzhou city. In the case of 2000, nine LCLU categories are added to the image classification scheme: paddy field, dry land, forest, grass, orchard land, urban surface (village, etc), cultivated/exposed land, beach, river and ocean. For 1989 the same classes as in 2000 are used plus two new categories: cloud and shadow because of air conditions.

Traditionally, classification strategies have been divided into two broad categories: supervised classification and unsupervised classification. Considering the characteristics of the study area, the classification was performed by means of a hybrid classifier. In the case the procedure works by starting with the unsupervised process to obtain many classes, while the supervised process is used to objectively assign these classes to the final categories. An unsupervised approach known as ISODATA (Iterative Self Organizing Data Analysis) is adopted for image classification, and it makes use of the minimum-distance method to identify homogeneous spectral clusters iteratively according to the number of clusters specified (Jensen, 1996 [11]). Based on the unsupervised classification map, the training areas for supervised classification are defined with assist of some ground data knowledge. In the present, The most commonly applied supervised classification method is the maximum likelihood procedure because of its robustness (Markam and Townshend, 1981 [12], Irons et al, 1985 [13], Haack et al 1987 [14]). In order to achieve an accurate classification, another important step of the process is to select optimal band clusters. We sample lots of the DN values of image pixels in the training areas of multifarious land-cover, statistic the mean values of them (table 1 and 2) and draw out their spectrum plots based on their mean values. it shows that the mean values spectrum plot present different coherent range of reflectance values in the red band, near-infrared band, middle-infrared band, and the distribution in spectrum values can be used to classify the LCLU categories in image classification, so four bands of Landsat image (band 3, band 4, band 5 and band 7) are used in supervised approach.

Once the two LCLU classification maps (1989s and 2000s) have been obtained, a major problem is detected, apparently caused not by the classification method but by the inherent problem of spectral confusion, such the phenomenon as the homogeneous objects have different spectrum values but the heterogeneous objects have the same spectrum values in images. The problem is the appearance of areas erroneously classified as arable land instead of orchard land, urban surfaces instead of dry land, and so on. This is a consequence of the confusion between different land-cover spectrums, perhaps because they have a very similar radiometric response due to seasonal difference. The solution is to visual examination in classification maps with the help of digital topography map, referenced ground data and unsupervised classification maps. Finally, the boundary errors caused by spectral mixing as well as spectral confusion due to similarity of spectral signatures of several land-use and land-cover classes have all been corrected and the resultant classification maps are shown in figure 1 (left and middle).

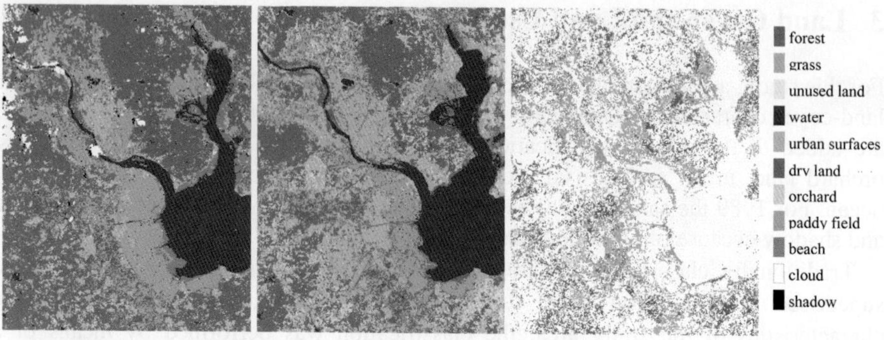

Fig. 1. Classified images of 1989 and 2000 and changed image from 1989 to 2000 (left: 1989; middle: 2000; right: changed image)

Table 1. Error matrix from the 1989s image. Both classification results (in rows) and ground truth (in columns), in pixels.

Classified Data	Referred Data										
	1	2	3	4	5	6	7	8	9	Total	UA(%)
1.dry land	54	2	3	2	3	0	0	2	2	68	79.41
2.paddy land	2	29	0	0	0	2	1	0	0	34	85.29
3.forest	2	0	45	3	0	0	0	0	3	53	84.91
4.grassland	3	0	2	35	0	0	0	0	2	42	83.33
5.urban surfaces	3	0	1	0	54	0	1	2	1	62	87.09
6.water	0	1	1	0	0	38	2	0	0	42	90.47
7.beach	0	1	0	0	0	2	25	0	0	28	89.28
8.unused land	2	0	0	0	2	0	0	18	0	22	81.81
9.orchard	2	0	2	2	0	0	0	0	37	43	86.04
Total	68	33	54	42	59	42	29	22	45	394	
PrA(%)	79.41	87.87	83.33	83.33	91.52	90.47	86.21	81.81	82.22		

Table 2. Error matrix from the 2000s image. Both classification results (in rows) and ground truth (in columns), in pixels.

Classified Data	Referred Data										
	1	2	3	4	5	6	7	8	9	Total	UA(%)
1.dry land	42	2	2	3	3	0	0	2	4	58	72.41
2.paddy land	2	45	0	1	0	2	2	0	0	52	86.53
3.forest	2	1	52	3	0	0	0	0	3	61	85.24
4.grassland	2	1	2	45	1	0	0	0	2	53	84.90
5.urban surfaces	2	0	1	0	54	0	1	2	2	62	87.09
6.water	0	1	0	1	0	47	2	0	0	51	92.15
7.beach	0	1	0	0	2	2	28	1	0	34	82.35
8.unused land	2	0	0	0	3	0	0	22	0	27	81.48
9.orchard	3	0	3	1	0	0	0	0	45	52	86.53
Total	55	51	60	54	63	51	33	27	56	450	
PrA(%)	78.36	88.23	86.66	83.33	85.71	92.15	84.84	81.48	80.35		

Accuracy assessment is performed for the 1989 and the 2000 land-cover/land-use classification maps. Based on a stratified random sample of 394 pixels selected from the 1989 map, an overall accuracy of 85 percent is obtained. In terms of producer's and user's accuracies, a minimum of 80 percent is reached for eight classes (except that the accuracy of dry land is 79.41 percent). As a result, Kappa indices of the supervised classification are 82.93% (PrA = Producer's Accuracy in %; UA = User's

Accuracy in %). As for the 2000 land-use/land-cover map produced, a stratified random sample of 450 pixels revealed an overall accuracy of 84.44 percent and an Kappa indices of 82.38 percent. Both producer's and user's accuracies are over 80 percent (except that the accuracy of dry land is 72.41 percent). Therefore, the two land-use and land-cover maps are comparable in accuracy despite differences in the type of Landsat images used (the 1989s image is acquired by Landsat 5, the 2000s image is acquired by landsat 7).

4 Strategies for Change Detection

Change detection and monitoring involve the use of multi-date images to evaluate differences in land cover due to environmental conditions and human actions between the acquisition dates of images (Singh 1989 [15]). A variety of techniques for land-cover change detection exist for satellite imagery. Broadly, these may be separated into two approaches: (1) detection of changes in independently-produced classifications and (2) determining change directly from radiometry (Kam, 1995 [16], Ridd and Liu, 1998 [17]). Because there lies spectral difference due to different season acquired and confusion between different kinds of land-cover change, we select the first approach to detect the LCLU change and determine the semantic meaning of the land-cover change obviously (table 3). In this case the change of LCLU between the 1989s and the 2000s are detected based on the following formula as:

```
If I1!=I2 THEN I2 ELSE NULL
```

Where I1 is the 1989s classification map, I2 is the 2000s classification map. Obviously, by using the formula we can eliminate the unchangeable regions between the two maps and acquire the changed land-use and land-cover information needed. Considering the cloud and shadow only lies in 1989s map and unable to compare them with reason, we ameliorate the formula to solve the problem:

```
IF INREGION(I3) THEN NULL ELSE IF I1!=I2 THEN I2 ELSE
NULL
```

Where inregion (I3) denotes the regions where clouds and shadows exist. After this, we can acquire the changed map shown as figure 1(right part) during the 11 years urbanization periods

Table 3. Change detection matrix from 1989 to 2000 according to LCLU categories for both years, in km^2.

1989 \ 2000	1	2	3	4	5	6	7	8	9
1.Urban surface	36.68	0.21	0.64	1.39	2.64	8.01	0.33	1.08	1.85
2 Dry land	12.42	5.57	0.30	2.47	6.34	25.48	0.78	6.97	3.16
3.Water body	0.69	0.26	75.12	1.47	0.09	1.26	0.03	0.90	0.10
4.Beach land	1.21	0.29	4.71	3.96	0.17	2.55	0.03	1.69	0.42
5.Orchard	4.91	0	0	0.03	3.89	6.76	0.19	0	0.97
6.Forest land	33.92	0.52	0.29	1.12	40.24	100.73	3.82	2.36	12.62
7.Unused land	3.43	0.12	0.11	0.11	1.07	2.24	0.11	0.23	0.53
8.Paddy land	15.85	4.88	0.46	1.45	1.69	12.60	0.52	25.93	2.30
9.Grass land	7.09	0.83	0.29	0.93	2.21	6.95	0.39	4.02	1.58
10.Cloud / shadow	1.85	0.32	0.58	0.33	0.91	7.63	0.02	1.15	1.23

5 Results and Discussion

5.1 LCLU Change Degree Estimation

To measure the rates of LCLU change, a change analysis technique has to be applied to the two classified maps. Based on the exploitation degree of the land-cover, the land-use classes have been grouped into 4 levels in this study. Level 1 is mainly distributed in the unused land because of its origin, and level 2 is mainly composed by forest, water body and grassland. Accordingly, level 3 is made up of arable land, orchard land and man-made lawn, and the highest 4 level includes mainly urban surfaces, villages, mine and transportation roads, and so on (table 4). Moreover, the further computations of the land-use-degree ratio index and land-use–change formula are based on the following:

$$L = 100 \times \sum_{i=1}^{n} A_i \times C_i \qquad (1)$$

$$R = \frac{\sum_{i=1}^{n}(A_i \times C_{i,b}) - \sum_{i=1}^{n}(A_i \times C_{i,a})}{\sum_{i=1}^{n}(A_i \times C_{i,a})} \qquad (2)$$

Where L is land-use-degree ratio index, A_i is the level of land-use type, C_i is the percentage of the corresponding levels in the total study areas, $C_{i,a}$ and $C_{i,b}$ is the 1989s and 2000s percentage of the corresponding levels in the total study areas respectively. The calculation of the L and R is on the basis of the ratio of LCLU area to the total study area with the consideration of weighted values of each land-use type level. For estimation, the value of R is bigger than zero value means that the LCLU of the study area is in the developing stage. On the contrary, it means in a adjusting or declining stage. As a result, the L values of the 1989s and the 2000s and the value of R in Quanzhou city are calculated as followed in table 5. From the table 5, We can make a judgment that the LCLU of Quanzhou city is in a developing stage.

Table 4. Classified grade according to LCLU categories of Quanzhou city

Classified grade	Unused cultivation	Extensive cultivation	Intensive cultivation	Town assemble exploitation
Land use type	Unused land	Forest, water, grass	Arable land, orchard, man-made lawn	Urban surfaces, villages, mine, traffic roads
Exploitation degree	1	2	3	4

Table 5. The extent and ratio index of changed matrix from 1989 to 2000

Classified grade	Unused cultivation	Extensive cultivation	Intensive cultivation	Town assemble exploitation
Land use type	Unused land	Forest, water, grass	Arable land, orchard, man-made lawn	Urban surfaces, villages, mine, traffic roads
Exploitation degree	1	2	3	4

5.2 LCLU Change Results Analysis

Our study indicates the LCLU of Quanzhou city has been developing during the past 11 years, and the urban area has grown by 62.77 km^2. Overlaying classified images from 1989 and 2000 gives an indication of the land-cover conversion associated with the rapid urbanization development. With the rapid development in economics and further advance in reform policy, the urban area has enlarged from 55.42 km^2, to 118.19 km^2 in 2000, or 6.77 rate $year^{-1}$ on average. While residential development has strongly favored the occupying of arable land, the clearing of forest and the filling of water body. Commercial development has converted both agricultural and forested land. This may reflect consumer preference for watered homesites, or it may reflect a lack of arable land adjacent to existing communities where new residential growth is likely to occur, which caused the competition between the arable land and the construction land. It is worth noting that the arable land has been in a decline progress both in quantity and quality. The total amount of arable land keeps 129.35 km^2 in 1989, but in 2000, it has decreased stupendously to 58.01 km^2, mainly occupied by new residential growth. Another factor leads to arable land reduction lies in the people's concept transformation. Because of the high market value of breeding aquatics and planting fruit trees instead of planting traditional crops, such as rice, wheat and earthnut. Nevertheless, to our great pleasure, the nation had called for protecting the arable land and actualized strict policies to forbid misusing land resource. Moreover, the substantial parts of the urban area have reforested, most likely as cleared agricultural land fall out of use near developed areas.

The smart change of LCLU in Quanzhou city highlights regional economic and political variations. It is evident that the economics of Quanzhou city increases in a high speed, and GDP of the city ranks No.1 in Fujian province. At present, the levels of urbanization and industrialization are rather high because of economics extraversion. Based on the census data of local jurisdictions, larger amount of straight foreign investments swarm into the city and capital construction such as factory, transportation road building appear much more pronounced than before. The correlation between population increase and LCLU change appears particularly strong for the Quanzhou area. On one hand, these quick growth rates of LCLU correlate with regional population suggesting more people's inhabitation to build, which drives the land-cover to change. On the other hand, with the increase of personal income data, it certainly requests that the development of the region should keep pace with the major economic and developing trend. Indeed, it is fairly reliable to consider that these underlying population variations may be reflected in land-cover changes.

6 Conclusions

For the past 11 years from 1989 to 2000, the Quanzhou city had undergone dramatic change in LCLU that had resulted in loss of cropland and forest, thus drastically altering the land surface characteristics. The two Landsat images have been used to map and extract land use and land cover information of Quanzhou for 1989 and 2000. By means of supervised image classification technology supported by unsupervised image classification approach, grounded referenced data and census of local government, the overall, producer's and user's accuracies of the classified maps meet mostly the standard of at least 80 percent. The land-use and land-cover maps reveal a great increase in urban use at the expense of forest, water and cropland. To further study, it has shown that satellite observation of LCLU change is related to underlying socio-economic trends and the outcome of local policies. With the rapid development of economics and increase of population, more and more land are used to satisfied the needs of people, resulted in lots of urban surfaces instead of croplands and forest. To seek high income benefits drive people to plant high-value-added crops replace of traditional rice and wheat, which further impulses the LCLU change of Quanzhou. However, the significant change of LCLU has greatly affect our existence and lives, to maintain the persistent development of economics and society, strict land-protected policies and logical land-used schemes should be put into practice all over the counties and cities, even to the whole nation. Moreover, the dynamics monitoring of LCLU change through remote sensing technology provides important information for us to discover and analyze the estate of LCLU during the urbanization process.

References

1. Auer, A.H.: Correlation of land use and cover with meteorological anomalies. Journal of Applied Meteorology 17, 636–643 (1978)
2. Roth, M., Oke, T.R., Emery, W.J.: Satellite derived urban heat islands from three coastal cities and the utilization of such data in urban climatology. International Journal of Remote Sensing 10(11), 1699–1720 (1989)
3. Kaiser, E., Godschalk, D., Chapin Jr., S.F.: Urban Land Use Planning. University of Illinois, Urbana (1995)
4. Owen, T.W., Carlson, T.N., Gillies, R.R.: Assessment of satellite remotely-sensed land cover parameters in quantitatively describing the climate effect of urbanization. International Journal of Remote Sensing 19(9), 1663–1681 (1998)
5. Carlson, T.N., Arthur, S.T.: The impact of land use-land cover changes due to urbanization on surface microclimate and hydrology: a satellite perspective. Global and Planetary Changes 25, 49–56 (2000)
6. Thomas Jr., W.L.: Man's role in changing the face of the earth. Univerisity of Chicago Press, Chicago (1956)
7. Vitosek, P.M., Mooney, H.A., Lubchenco, J., Melillo, J.M.: Human domination of Earth's ecosystems. Science 277, 494–499 (1997)
8. European Commission: Remote sensing of Mediterranean desertification and environmental changes (resmedes), Luxembourg: Office for official publications of the European communities (1998)
9. Kostmayer, P.H.: The American landscape in the 21st century. Congressional Record 135, 9963 (1989)
10. Jensen, J.R., Cowen, D.C.: Remote sensing of urban/suburban infrastructure and socio-economic attributes. Photogrammetric Engineering and Remote Sensing 65(5), 611–622 (1999)
11. Jensen, J.R.: Introductory digital image processing: a remote sensing perspective, p. 316. Prentice-Hall, Englewood Cliffs (1996)
12. Markam, B.L., Townshend, R.G.: Land cover classification accuracy as a f unction of sensor spatial resolution. In: Proceedings of the Fifteenth International Symposium on Remote Sensing of Environment held in Ann Arbor, Michigan, pp. 1075–1085 (May 1981)
13. Irons, J.R.: Introductory Digital Image Processing: a remote sensing perspective. Prentice-Hall, New Jersey (1995)
14. Haack, B., Bryant, N., Adams, S.: An assessment of Landsat MSS and TM data for urban and near-urban land-cover digital classification. Remote Sensing of Environment 21, 201–213 (1987)

15. Singh, A.: Review article-digital change detection techniques using remotely-sensed data. International Journal of Remote Sensing 10, 989–1003 (1989)
16. Kam, T.S.: Intergrating GIS and remote sensing techniques for urban land-cover and land-use analysis. Geocarto International 10, 39–49 (1995)
17. Ridd, M.K., Liu, J.J.: A comparison of four algorithms for change detection in an urban environment. Remote Sensing of Environment 63, 95–100 (1998)

A SOA-Based Model for Unified Retrieval System

Hui Li and Li Wang

College of Information Science and Technology,
Beijing University of Chemical Technology, Beijing, China
ray@mail.buct.edu.cn, wangli1008@gmail.com

Abstract. Regarding present problems such as lacking of a uniform access strategy, poor extension, tight coupling and low resource utilization, which on the processing of heterogeneous data in unified retrieval system, this paper presents a SOA-based model for unified retrieval system. Based on SOA, the model adopts SDO specifications in data programming to access and manipulate data uniformly in multiple heterogeneous data sources. Moreover, by using SCA specification, the interfaces of data retrieving can be encapsulated as a series of services. Each service is independent and can be uniformly accessed, which guarantees the loose coupling of the system. By applying SOA design pattern and web service technology into heterogeneous retrieval system, the flexibility and expansibility of data integration can be improved greatly. Therefore, data sources in unified retrieval system are "plug and play".

Keywords: Data retrieval, Unified retrieval system, Data integration, SOA, SCA, SDO.

1 Introduction

In recent years, digital library construction has achieved a tremendously rapid development. To meet the needs of customers, library builds a digital information source by means of purchasing network database and self-build database. These databases are often provided by different providers, thus the usage and retrieval interfaces are different. So readers need to log in and enter keywords repeatedly in order to retrieve data in several distributed databases. It's inconvenient for readers, thus, a bottleneck will be inevitably caused in the effective use of database. Therefore, it is necessary to integrate these isolated digital resources. The unified retrieval system provides a unified data retrieval interface, which enable users to retrieve data in different heterogeneous data sources. After finishing data combination, filtration and sorting, the retrieval result is given.

In order to implement a loose coupling and high expansibility data retrieval system, this paper presents a SOA-based model for unified retrieval system. By using Service Data Object (SDO) as the unified data programming architecture, the model can provide a uniform manner to access and manipulate data from heterogeneous data sources. Among systems, business data is converted to data graph, which realize the

M. Zhu (Ed.): Electrical Engineering and Control, LNEE 98, pp. 1029–1037.
springerlink.com © Springer-Verlag Berlin Heidelberg 2011

unified data format. Furthermore, on the basis of Service Component Architecture (SCA), each interface of data retrieving is encapsulated as service to achieve the separation between business logic and implementation technique. Services are reusable and can be accessed in a uniform manner.

2 SOA and Related Technology

Acting as a component model, SOA connects different functional units of application through well defined interfaces and contracts. The interface is defined by a neutral manner and it should be independent on service implement hardware platform, operation system and program language. Thus, service constructed in a variety of systems could be interacted in a uniform and common way [1]. SOA is based on the idea that business functions are separated, encapsulated and published as services. Services can be reused or joined together to implement a specific business functions, which greatly saves the enterprise overhead and improves the resources utilization.

2.1 Web Service

Web service is an online application that is published by the service provider to realize specific business functions, so it can be accessed and used through Internet. Since web service is independent, loose coupled and reusable, it has become the major way to achieve SOA.

SOAP (Simple Object Access Protocol) is a simple, lightweight, XML-based protocol that facilitates message exchanging in a loose and distributed environment, and the specific transport protocol may be HTTP (Hyper-Text Transfer Protocol), SMTP (Simple Mail Transfer Protocol) or IIOP (Internet Inter-ORB Protocol) [2-3]. WSDL (Web Services Description Language) is an XML- based language that can be used to describe web service. It specifies the service location, operations, port type, and so on. After the service is created, it needs to be registered and published on the Internet through UDDI (Universal Description Discovery & Integration), so that, service requesters can access to the web service.

As shown in Fig. 1, the architecture of web service is consisted of three roles: service provider, service register and service requester. Service provider can publish their services which are described with WSDL and respond to their own service request. Service registry is used to classify and register services. Service requester could then search the UDDI directory to find the service interface and then invoke it.

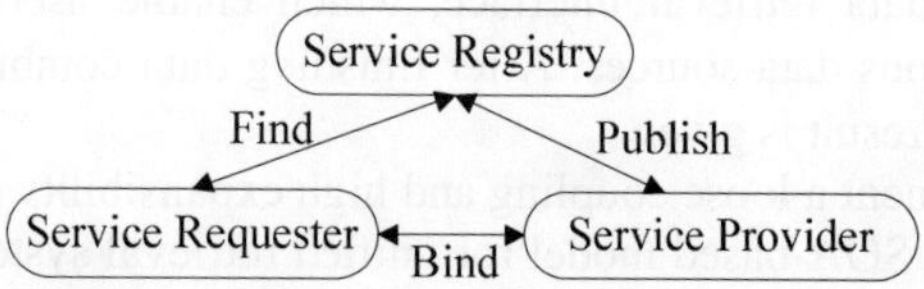

Fig. 1. Web services architecture

2.2 SCA and SDO

Service Component Architecture (SCA) is a set of specifications which provides a programming model for building applications and systems using a Service-Oriented Architecture [4].

SCA provides the service encapsulation capability through a service-oriented interface called service, supporting both java interface and WSDL interface. The service interface can be implemented by using any one of these programming languages, such as C++, Java, COBOL, PHP, BPEL, XSLT, SQL and XQuery [5]. SCA component also offers a service-oriented interface called reference to invoke the external services, including both SCA component services and other services. Binding is used by the service and reference of SCA component. Service uses binding to describe the access mechanisms about how to invoke this service. However, reference uses binding to describe the access mechanisms about how to invoke other external services.

SCA emphasizes the decoupling of service implementation and service assembly to guarantees the loose coupling among the service components. Each business function is wrapped in SCA service component, and then it can be promoted as standard service through the service-oriented interface. Thus, applications can uniformly access to the services.

Service-Oriented Architecture needs a data model which is based on open standards, with back-end data sources irrelevant and be able to represent business data clearly. Acting as a data programming model, Service Data Object (SDO) has played an important role to solve this problem.

SDO aims at providing a unified data programming model for heterogeneous data sources, including relational databases, web services, XML data sources and enterprise information systems, which enable applications to access and manipulate heterogeneous data in a uniform manner. Data graph is a core concept of SDO. A data graph is a collection of tree-structured or graph-structured data objects [5]. As a unified data programming model, SDO not only provides the data mediator services to connect to difference data sources, but also provides a metadata API to convert data model to a data graph. Therefore, client applications can connect to the data sources through data mediator services and get a data graph in response. Among the systems, all the applications share the unified data format, which have realized the seamless integration.

As show in Fig. 2, client applications use data mediator services to connect to the data source and get a data graph in response. After manipulating the data graph, client applications will send an updated data graph to the data mediator service, and then the data mediator service will check the updates and apply the updates to the original data source.

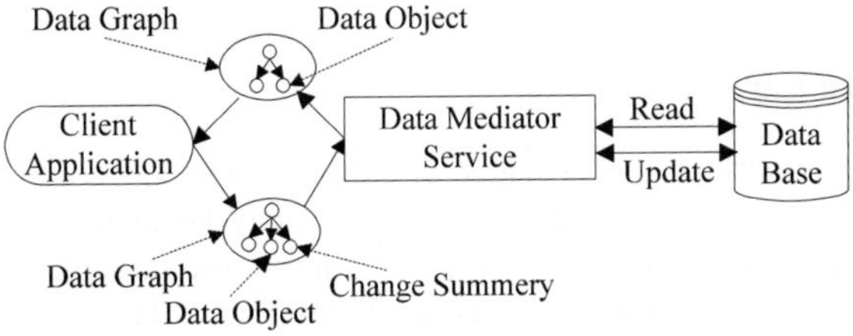

Fig. 2. Data access framework of SDO

SCA provides a uniform manner to access services, while SDO provides a unified model for data programming. Thus, in order to improve its functionality, SCA promotes the use of SDO to represent the business data that forms the parameters and return values of services [6].

3 Design Philosophy for SOA-Based Unified Retrieval System

Based on SOA and web service technology, the interfaces of data retrieving are provided as web services, thus, the SOA-based unified retrieval system can complete data retrieving by calling the retrieval services. The design philosophy for the SOA-based unified retrieval system is as follows:

Step 1: The digital resources consist of several heterogeneous data sources, and the interfaces of data retrieving are provided by each data source.

Step 2: Data sources are heterogeneous, thus, for the purpose of realizing the unified data programming, it's essential to design the specific adapter for different data sources.

Step 3: Each interface of data retrieving is encapsulated as web service, so as to ensure the loose coupling of the system. After the retrieval service is published, it can be discovered, reused and combined with other services.

Step 4: In order to achieve specific data retrieval functions, the data retrieval services can be assembled into a data retrieval process by using BPEL.

Step 5: The retrieval system should provide a unified retrieval interface. Through the unified retrieval interface, users can retrieve data in multiple distributed databases, which make the data retrieval become more convenient.

Step 6: Each retrieval service will return the result data. After finishing data combination, filtration and sorting, the retrieval results will be given.

4 A Model for SOA-Based Unified Retrieval System

This paper presents a SOA-based model for unified retrieval system. By using SDO and SCA specification, the model can not only simplify and unify the way of service encapsulation but also enable application to handle data from heterogeneous data sources in a uniform way. This system can effectively retrieve data in multiple distributed databases. Meanwhile, it brings more convenience and efficiency to customers.

As shown in Fig. 3, the proposed model is composed of 5 layers: data source layer, adaptation layer, retrieval service layer, retrieval process layer and unified retrieval application layer.

4.1 Data Source Layer

Data Source Layer acts as the provider of business data. According to the existing international standards, resources in digital library may be in line with OAI standard, Z39.50 standard or other non-standard [7]. In addition, the relational databases such as Oracle and SQL Server are frequently used in digital library. Therefore, data source layer usually includes OAI standard data source, Z39.50 standard data source, Oracle database, SQL Server database and other data sources. Each data source is a separate module.

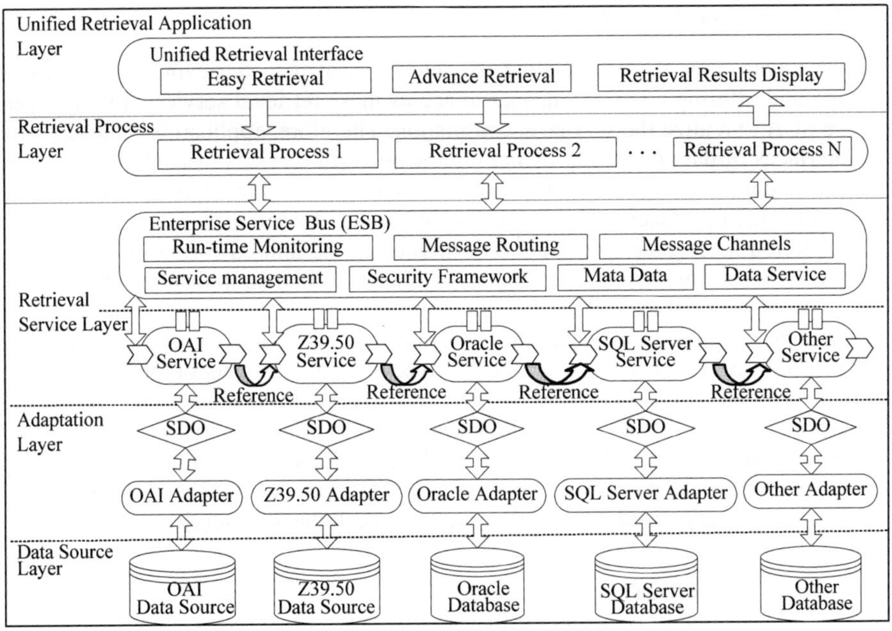

Fig. 3. SOA-based model for unified retrieval system

The major challenges of heterogeneous data integration which enterprises face nowadays are how to adapt to the complex information demand and how to describe the heterogeneous data in a unified format so as to realize the effective data sharing and exchanging.

4.2 Adaptation Layer

Adaptation Layer aims at offering a unified way to access and manipulate data from heterogeneous data sources. SDO allows applications to connect to each data source through data mediator services, and the retrieval results that returned by each data source will be converted to a unified format of data graph. Therefore, in each layer, heterogeneous data will be transferred in a unified format of data graph. Acting as a data programming architecture and API, SDO realizes unified data programming in heterogeneous data sources, thus, applications can handle data in a uniform manner. By using SDO specification, the following data sources can be adapted [8]:

(1) Database adapter, supports DB2, Oracle, SQL Server, Sybase and other popular databases.
(2) File adapter, supports Word, Excel, Text, XML and other common file formats.
(3) JMS adapter, supports a variety of message data provided by the message server.

XML has the feature of universality and cross-platform-ability, thus XSD is often used to model the data object. SDO framework can perfectly map the XML data to the SDO data object, and provide a flexible and powerful API for SDO data processing [9].

4.3 Retrieval Service Layer

Service Layer aims at encapsulating the interfaces of data retrieving as a series of services and offering a unified manner to access these retrieval services. In this layer, we use SCA to realize the service encapsulation. Therefore, applications can access to the interfaces of data retrieving by calling the retrieval services instead of calling the traditional data access API. Each data provider can use the SCA specification to encapsulate the interfaces of data retrieving as the stander services.

Adaptation layer guarantees data transmission in the unified format of data graph. Therefore, according to business logic, each layer will manipulate the data graph instead of the original data.

Since the business data that form the parameters and return values of services are represented by Service Data Objects, the data type of input parameter or return value need to be converted into DataObject, as shown in Fig. 4. Moreover, in the service implementation, it needs to obtain relevant information from the input parameter whose data type is DataObject, as shown in Fig. 5.If the data type of input parameter in the SCA service interface is DataObject, services users need to create a DataObject instance firstly, and then set the customized data to this instance and call the SCA service.

```
package LiteratureModule;
import commonj.sdo.DataObject;
import org.osoa.sca.annotations.Remotable;

@Remotable
public interface LiteratureInfoInterface {
public String getLiteratureTitle (DataObject literature);
}
```

Fig. 4. Service interface definition

```
package LiteratureModule;
import commonj.sdo.DataObject;
public class LiteratureInfoImpl implements LiteratureInfoInterface {
        @Override
        public String getLiteratureTitle(DataObject literature)
        {
                return literature.getString("title");
        }
}
```

Fig. 5. Service interface implement

The configuration information of SCA is mainly in a file whose file extension is .composite. As shown in Fig. 6, the composite file defines a service named Literature firstly. This service is used to promote the SCA component named LiteratureServiceComponent. The SCA component implements this web service by adding a binding (binding.ws), and the web service URL is given by uri attribute. The service interface is implemented by a class named LiteratureInfoImpl.

```
<?xml version="1.0" encoding="UTF-8"?>
<composite xmlns="http://www.osoa.org/xmlns/sca/1.0" xmlns:t="http://tuscany.apache.org/xmlns/sca/1.0"
xlnms:c="http://Literature" targetNamespace="http://Literature" name="Literature">
<service name="Literature" promote="LiteratureServiceComponent">
<binding.ws uri="http://localhost:8080/Literature"></binding.ws>
</service>
<component name="LiteratureServiceComponent">
<implementation.java class="LiteratureModule.LiteratureInfoImpl"</implementation.java>
</component>
</composite>
```

Fig. 6. SCA composite file

After the interfaces of data retrieving are encapsulated as web services, it is time to realize the unified service management, including service registration, verification, debugging, deployment and storage. For the purpose of making up for the shortage of traditional point to point connection between client and service provider, enterprise service bus is used for service integration and management.

ESB is a standards-based middleware that provides secure interoperability among enterprise applications [10]. Based on the web service and XML technology, ESB supports synchronous and asynchronous messaging [11]. ESB provides a series of capabilities including protocol translation, message transformation, message routing, service management and security framework. More significantly, ESB supports service hot plug, thus service components can be easily integrated to ESB platform. Each service component communicates with each other by the standardized communication mechanism. When adding a new data source or original data source has changed, the data source can be integrated to the original retrieval system as much as possible without affecting the existing system.

4.4 Retrieval Process Layer

Business Process Layer aims at creating data retrieval processes by orchestrating web services to achieve the specific data retrieval functions. Business Process Execution Language for Web Services (BPEL4WS), also known as BPEL, is the emerging standard for assembling several discrete services into an end-to-end process flow [12]. The cost and complexity of process integration can radically reduce. BPEL provides a XML-based grammar for describing the business logic, therefore, web services that participating in a process flow can be controlled and coordinated efficiently [13].

As shown in Fig. 7, it is a BPEL business processes. According to the request of data retrieving submitted by the user, this process determines which retrieval service should be invoked. When users submit the request, BPEL provides a basic activity named Receive to get the request message and create a BPEL process instance, then Assign activity will be used to assign values to the input variable. According to the request message, BPEL process can determine which data source is requested and which retrieval service should be invoked, after that, Assign activity will be used to assign values to the input parameter of the retrieval service interface. These retrieval services including OAI retrieval service, Z39.50 retrieval service, Oracle retrieval service, SQL

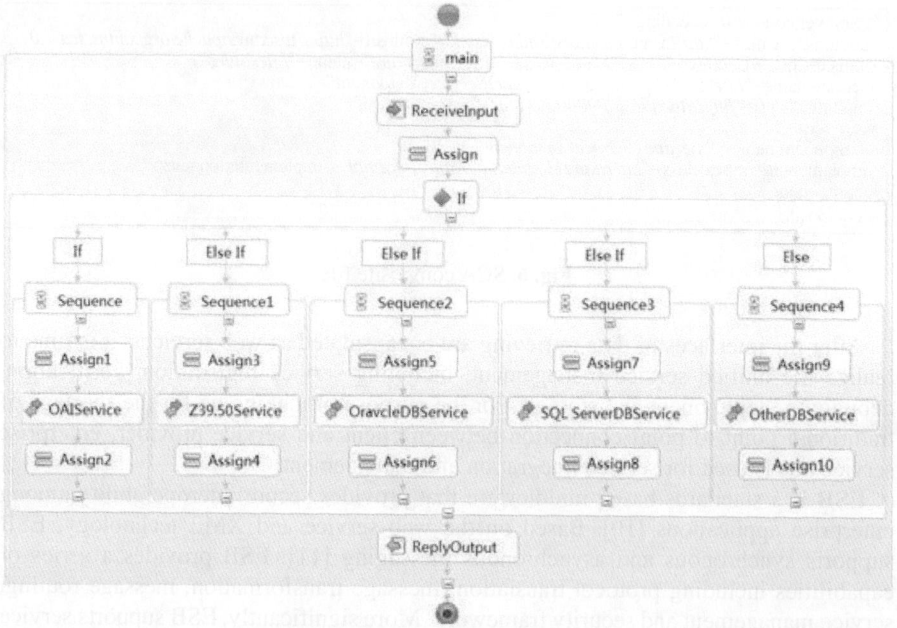

Fig. 7. BPEL process for data retrieval

Server retrieval service, and so on. At last, after the result data have being copied to the output variable through Assign activity, Reply activity will return the retrieval results to client applications.

4.5 Unified Retrieval Application Layer

Unified retrieval application layer is the top level of the retrieval system. It provides a unified data retrieval interface for external users or other systems. Thus, users can retrieve data in multiple heterogeneous data sources through the unified retrieval interface, and the retrieval results will be displayed after finishing data combination, filtration and sorting.

5 Conclusions

In this paper we have presented an SOA-based model for unified retrieval system. By using SCA and SDO specification, the model provides a unified data programming model and realizes the unified service encapsulation. In order to guarantee the loose coupling of the retrieval system, interfaces of data retrieving are provided as a series of services. Furthermore, applications could deal with data from heterogeneous data sources in a uniform manner. The SOA-based unified retrieval system could realize the heterogeneous data sharing and integration efficiently. Meanwhile, the scalability of the system is improved greatly.

References

1. David Campbell.: Service Oriented Database Architecture: APP Server-Lite. J.SIGMOD.857-862(2005).
2. Jun Wang, AiRong Yu, XiaoYi Zhang, Lei Qu.: A Dynamic Data Integration Model Based on SOA.J. Computing, Communication, Control, and Management.196-199(2009).
3. W. T. Tsai, X. Wei, R. Paul, J. Xu, Q. Huang, and B. Xiao.: Single Model, Multiple Analyses (SMMA) for Service-Oriented System Engineering (SOSE).In: Proc. of IEEE FTDCS. (2007).
4. SOA specification, http://gocom.primeton.com/modules/osoa/?see=SCA-specs&PHPSESSID =%3C.
5. What is SDO, http://www.osoa.org/pages/viewpage.action?pageId=48.
6. What is SCA, http://www.osoa.org/pages/viewpage.action?pageId=46.
7. YanPing Zhang.: System Analysis of Information Resources Integration Platform in Libraries of Colleges and Universities. J. China Information Review. 8, 44-48(2007).
8. LiDing Chen, WenHao Zhang.: Information exchange platform design Based on SOA.J. Microcomputer Information.25, 6-8(2009).
9. ZiYao Wang.: SOA core technology and application. Publishing house of electronics industry, Beijing (2008).
10. ESB, http://www.webopedia.com/TERM/E/ESB.html.
11. WenYing Zeng,YueLong Zhao,DeYu Qi. ESB principle, architecture, implementation and application. J. Computer Engineering and Applications.25, 225-228(2008).
12. Vivek Ranadivé.: The power to predict: how real-time businesses anticipate customer needs, create opportunities, and beat the competition. J. McGraw-Hill Professional. 147-150(2006).
13. What is BPEL and why is it so important to my business, http://www.softcare.com/whitepapers/ wp_whatis_bpel.php.

Clinical Sign-Based Fish Disease Diagnosis Aid System

Chang-Min Han[1], Soo-Yung Yang[2], Heeteak Ceong[1], and Jeong-Seon Park[1]

[1] Chonnam National University, Chonnam, Korea
[2] Elsys Co.Ltd., Chonnam, Korea
hans81@nate.com, hi@elsys.info, {htceong, jpark}@chonnam.ac.kr

Abstract. This paper presents a fish disease diagnosis aid system capable of faster treatment of infected fish in order to prevent the spread of disease. The proposed system is a clinical sign-based diagnosis from the initial sketchy observations of the symptoms. The system suggests candidate diseases with parsed selection based on water temperature, growth phase of the diseased fish, external clinical signs, internal clinical signs and microscopic observations and provides the information treatment and prevention methods on the diagnosed disease by a connected web server through internet. The designed fish disease diagnosis system was implemented to diagnose 14 diseases of flatfish. The system can support fish farmers and fish doctors by providing easy diagnosis of diseased flatfish.

Keywords: Fish disease diagnosis, computer-aided expert system, flatfish, clinical sign-based diagnosis.

1 Introduction

With the advances in information technology (IT), various possible applications to aquaculture have been developed, including monitoring systems for water environments, recording systems for productions history and tele-diagnosis system of fish diseases [1], [2], [3]. Fish disease is a serious problem due to its ability to spread rapidly through water to neighboring aqua-farms. Therefore, rapid and accurate diagnosis is required to control such diseases, prevent their spread and limit excessive use of antibiotics [4], [5], [6].

Traditionally, fish diseases have been diagnosed by using the accumulated experience of fish farmers or fish doctors. However, the accuracy of such diagnosis ultimately depends on individual skill and experience and the time spent studying each disease. In order to overcome this limitation, IT has been applied to the diagnosis of fish diseases. In earlier applications, some drug companies or research institutes provided a web-based guide for diagnosing fish diseases by answering a series of questions about the symptoms of diseased Koi carp or freshwater fish [7], [8], [9], [10].

With the development of further applications, various expert systems for fish disease diagnosis have been reported, such as web-based, expert systems which help fish doctors or fish farmers in the tele-diagnosis of fish diseases [1], [2], [3], [11].

M. Zhu (Ed.): Electrical Engineering and Control, LNEE 98, pp. 1039–1046.
springerlink.com © Springer-Verlag Berlin Heidelberg 2011

However, these methods also depend on the subjective decisions of fish doctors or fish farmers.

The goal of this research is to design and implement an easy diagnosis aid system which suggests candidate diseases from the selected clinical signs. The system suggests candidate diseases with parsed selection based on water temperature, growth phase of the diseased fish, external clinical signs, internal clinical signs and microscopic observations. The designed fish disease diagnosis aid system was implemented to diagnose 14 kinds of flatfish disease. The information on the diagnosed disease, treatment and prevention methods are provided by a connected web server through internet. The system will support fish farmers and fish doctors by providing easy diagnosis of diseased flatfish.

The next section briefly overviews the proposed fish disease diagnosis aid system based on clinical signs. Section 3 describes the detail procedure of the clinical sign-based diagnosis aid system for providing candidate diseases from the series of clinical signs. Finally, the conclusions and directions for future research are discussed in Section 4.

2 System Overview

The developed fish disease diagnosis aid system supports the earlier treatment of diseased fish and thereby prevents the spread of disease. The clinical sign-based diagnosis aid system is used to suggest candidate diseases with parsed selection based on water temperature, growth phase of the diseased fish, external clinical signs, internal clinical signs and microscopic observations. It also indicates the cause of the disease, and suggests prevention and treatment methods for each candidate disease.

Fig. 1 shows an example flow of the designed fish disease diagnosis aid system. When unusual symptoms of fish or dead fish appear in the aqua-farm of flatfish, fish farmers or doctors can investigate key factors from external clinical signs to microscopic observations. In the figure, there are three external clinical signs (ulcer on the skin, ulcer on the proboscis and hemorrhage on the skin), no internal clinical signs and something small on the muscle. Of course only external clinical signs can be investigated by fish farmers. From this series of observations, the system decides that the diseased flatfish may be infected with parasites such as scuticociliatosis, trichodina and white spot, and then provides the detailed information of the each candidate diseases.

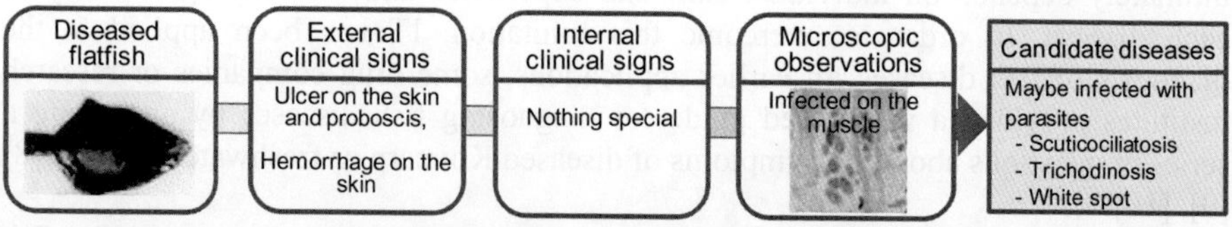

Fig. 1. Diagnosis example of the designed fish disease diagnosis system based on clinical signs

In order to implement the designed diagnosis process, we configured an aid system containing 2 kinds of information database, as shown in Fig. 2. As described above, the clinical sign-based disease diagnosis system provides candidate diseases by comparing the observed and selected clinical signs to the information on the flatfish disease database. From the candidate diseases, various information such as disease cause, diagnosis, treatment and prevention method are provided by retrieving the remote fish disease database in the fish information system.

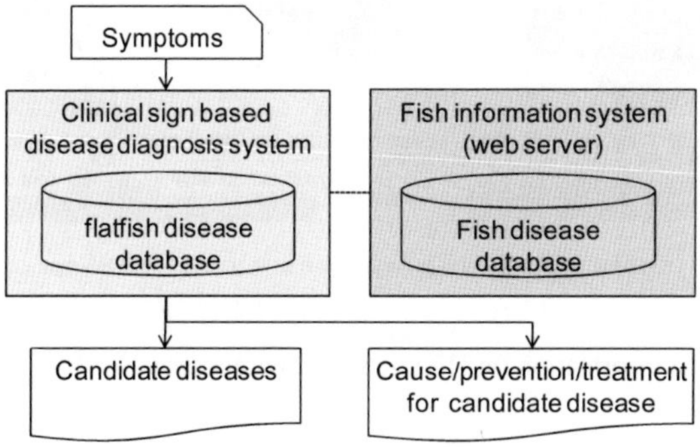

Fig. 2. Configuration of the proposed fish disease diagnosis aid system based on clinical signs

The fish information system on the remote server provides various search schemes to gather information on 115 diseases of fish. The developed system can automatically connect and search the database to provide helpful information for the diagnosed fish disease. The fish disease database and flatfish disease database are briefly described in the following subsections. The procedures for the clinical sign-based disease diagnosis are described in detail in the next section.

2.1 Fish Disease Database

To provide helpful information on the diagnosed disease, the proposed system can automatically connect to the fish disease database on the web server of Eco Aquafarm Research Center [12]. This database contains the information on 115 diseases of fish and shellfish such as the name, pathogen, main infection species, cause, symptoms (external symptoms and infection part), diagnosis and treatment method and environmental condition of disease, and sample images of diseased fish. The knowledge of the fish disease database was summarized from the literature [4], [5], and [6] by several experts of aqua-life medicine. Fig. 3 shows an example of the retrieved information on Hirame rhabdoviral disease of flatfish from the fish disease database.

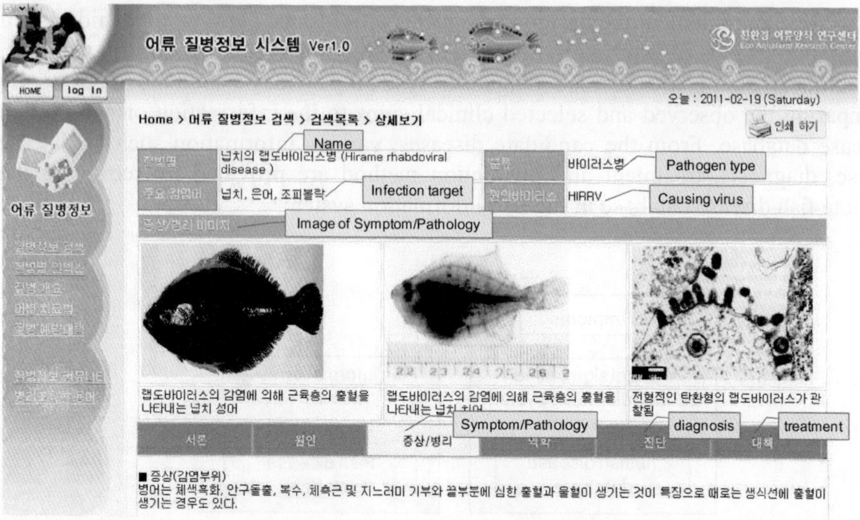

Fig. 3. Example of the information on Hirame rhabdoviral disease of flatfish retrieved from the fish disease database.

2.2 Flatfish Disease Database

In order to implement a clinical-based fish disease diagnosis aid system, we first setup the standard process for diagnosing 14 kinds of flatfish disease, as summarized by several experts of aqua-life medicine based on disease information in the literature [4], [5] and [6]. Only flatfish disease is considered because flatfish is one of the most frequently cultured species in Korea. The standard diagnosis process is explained in detail in the next section. Then we construct a flatfish disease database for the clinical sign-based diagnosis aid system. The database contains the diagnosis information based on the water temperature, growth phase of the diseased fish, external clinical signs, internal clinical signs and microscopic observations for each disease and typical image for each symptom. Table 1 shows 14 kinds of flatfish disease which are classified into 3 categories according to the causative pathogens: 5 bacterial diseases, 3 parasitic diseases and 6 viral diseases.

Table 1. 14 kinds of flatfish disease classified into 3 categories according to the kind of causative pathogen

Causative pathogen	Diseases
Bacterium	Bacterial white enteritis, edwardsiellosis, gliding bacterial disease, streptococcosis, vibriosis
Parasite	Scuticociliatosis, trichodinosis, white spot(or ich)
Virus	Aquabirnavirus disease, hirame rhabdoviral disease, lymphocysti disease, viral epithelial hyperplasia disease, viral hemorrhagic septicemia, viral nervous necrosis

3 Clinical Sign-Based Disease Diagnosis Procedure

To design the clinical sign-based fish disease diagnosis aid system, the standard diagnosis process of the 14 flatfish diseases was setup, and the four-step diagnosis system implemented. The factors for discriminating diseases are water temperature, growth phase (baby, young, adult), skin color, abnormal swimming, ulcer on the proboscis, ascites, muscle infections, kidney hypertrophy, and so on. These factors are summarized as the four-step clinical sign-based diagnosis process (Table 2).

Table 2. Basic condition or clinical signs for discriminating 14 kinds of flatfish disease in the proposed, four-step clinical sign-based diagnosis process

Step	Key factors
Basic condition	- water temperature(℃): 10~15, 10~18, 10~25, 12~, 15~20, 18~20, 20~, 20~22, 23~ - growth phase: baby(less than 30 days), young, adult
External clinical signs	Fish's activities, growth state, external injury on the head, exophthalmos, changes in skin color, external injury on the proboscis, abnormal fins, abnormal muscles, abnormal anus, over secretion of mucus
Internal clinical signs	Ascites, abnormal liver/digestive apparatus/muscle/genital gland/gall/kidney/spleen/air bladder/brain
Microscopic observations	Abnormal epithelial cell, parasite exist or not on the region of tumer or ulcer, mesentery bacteria exist or not, able to isolate bacteria, species of the observed bacteria: gram(+, -) bacillus/mesentery/linkage group bacteria

Based on the aforementioned standard process for clinical sign-based disease diagnosis, the clinical signs of 14 kinds of flatfish disease are collected in the flatfish disease database as described in the previous section. Some examples of clinical signs of flatfish diseases are shown in Table 3.

Table 3. Example of clinical signs of the disease stored in the flatfish disease database

Step \ Disease		Vibriosis	White spot	Lymphocysti disease
Basic condition	growth phase	baby ~ adult	adult	Baby, adult
	water temperature	15 ~ 20℃	12℃~	15 ~ 25℃
External clinical signs		Hemorrhage on the gills/fins/anus, melanized skin, skin ulcer	White points on the skin, discolored fins, over secretion of mucus	Tumor on the surface/fins/ proboscis
Internal clinical signs		.	.	.
Microscopic observations	observation	.	parasite on the gills/skin	giant cell on the region of tumor
	bacteria isolation	+	.	.
	isolated bacteria	Gram(-) bacillus	.	.
Guide for final diagnosis		TCBS germiculture	Discrimination of parasite from microscopic images	Observation of tumor cells with microscope

From the information collected for diagnosing flatfish diseases, we implemented the clinical sign-based, disease diagnosis aid system shown in Fig. 4. To diagnose the disease, the basic condition of water temperature is inputted and the growth phase of the diseased fish selected. Then sequentially, the following three categories of key factors for diagnosing disease are selected: external clinical signs, internal clinical signs and microscopic observations. After confirming the selected clinical signs from all the observed clinical signs, the system provides a maximum of five presumed disease candidates in order of preference. Once a user selects one disease from the list, a summarized guide for final diagnosis and treatment method is provided. If a user requests more detail information for the disease, the system automatically connects to the web server of the fish information system, retrieves the information on the disease, and presents the treatment and prevention methods on the browser.

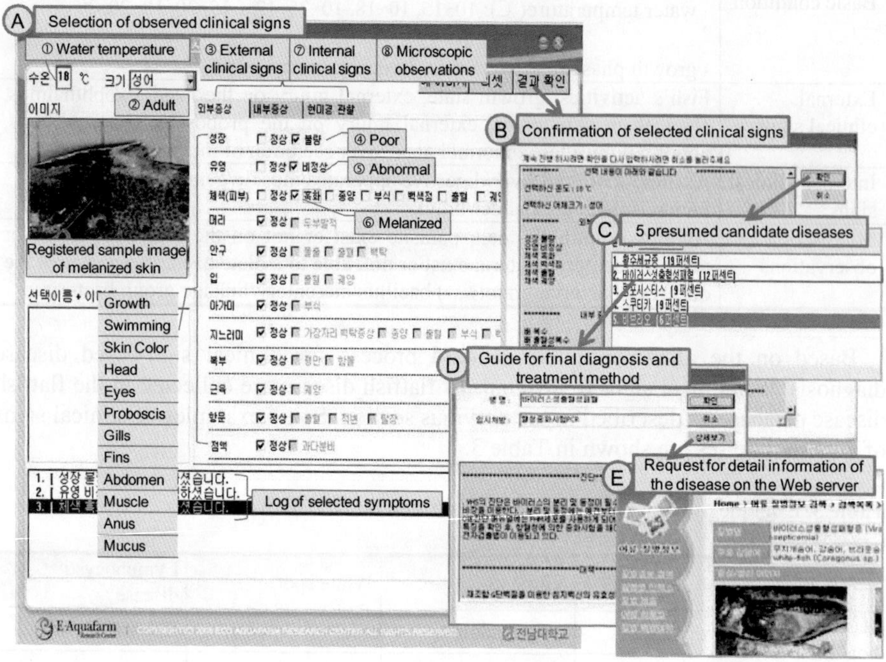

Fig. 4. Example of using the developed clinical sign-based diagnosis aid system.

3.1 Basic Condition Input

Water temperature is one of the strongest environmental factors that cause disease and its effect varies according to each disease as shown in Table 3. Water temperature is directly inputted by a user and compared to the range of temperature in the flatfish disease database for each disease. Another factor is growth phase because different diseases occur at different growth phases of flatfish. In the system a user can select one of the provided three levels: baby(less than 30 days), young and adult.

3.2 External Clinical Sign Selection

For easy selection of observed clinical signs, external clinical signs are listed according to the following regions and conditions: growth(poor), swimming(abnormal), skin(melanized, tumor, corroded, white point, hemorrhage, ulcer), head(rubefaction), eyes(protrusion, hemorrhage, nebula), proboscis(hemorrhage, ulcer), gills(corroded), fins(nebula on the edge, tumor, hemorrhage, corroded, white point, ulcer), abdomen(inflated, dent), muscle(ulcer), anus(hemorrhage, red excreta, rupture) and mucus(over secretion).

To select a clinical sign, a user first activates each clinical sign by unchecking the default "normal" check box. When a user selects the observed signs of flatfish from the listed options, the system shows the registered sample image for the selected clinical signs. Once a user has the image for the sign, the two images can be compared by loading the image. An example of selecting external clinical signs is shown in Fig. 4, which presents a registered sample image of the melanized skin.

3.3 Internal Clinical Sign Selection

Internal clinical signs are symptoms observable by dissecting a diseased fish. The regions of internal clinical signs are stomach (ascites, hemorrhage, bleeding ascites, clean ascites), liver (discoloration, brown, abscess, hemorrhage), digestive apparatus (nebula, inflated, atrophy), muscle (hemorrhage, abscess), genital gland (hemorrhage), gall (yellow), kidney (hypertrophy, white focus), spleen (hypertrophy, white focus), air bladder (hemorrhage) and brain (hemorrhage).

The selection method of internal clinical signs is the same as that of the external clinical signs. If dissection of a diseased fish is impossible, this selection step can be omitted.

3.4 Microscopic Observation Selection

Microscopic observation selection is composed of two parts: selecting the observation symptoms and selecting the isolated bacteria. The factors for the first part are surface (hyperplasia of epithelial cells, existence of parasite), fins (hyperplasia of epithelial cells), regions of tumor (giant cells), gills (hyperplasia of epithelial cells, existence of parasite), regions of ulcer (existence of parasite) and mesentery bacteria (observed with 400x magnification). The factors for the second part are gram (isolation failure, +, -) and species of bacteria (bacillus, mesentery, linkage group).

The selection method of microscopic observations is the same as that of the external clinical signs. If microscopic observation is impossible, this selection step can be omitted.

4 Conclusions

In this paper, we have proposed a fish disease diagnosis aid system capable of faster treatment of infected fish and hence greater prevention of the spread of disease. The proposed system is a clinical sign-based diagnosis from the initial sketchy observations of the symptoms. The system suggests candidate diseases with parsed selection based on water temperature, growth phase of the diseased fish, external clinical signs, internal clinical signs and microscopic observations and provides the information of the candidate diseases. The designed fish disease diagnosis system was implemented to diagnose 14 diseases of flatfish. The detail information, treatment and prevention methods on diagnosed disease were provided by a connected web server through internet. The system will support fish farmers and fish doctors by providing easy diagnosis of diseased flatfish.

Further studies in which the clinical sign-based diagnosis aid system is used by many fish farmers and fish doctors are required before the promise of the system can be confirmed for real applications. Furthermore, the knowledge of many real diagnosis cases should be reflected in the system.

Acknowledgments. This research was financially supported by the Ministry of Education, Science Technology (MEST) and National Research Foundation of Korea (NRF) through the Human Resource Training Project for Regional Innovation. We would like to thank several experts of aqua-life medicine in Eco Aquafarm Research center for collecting huge amount of information about fish diseases and for providing fish information service.

References

1. Duan, Y., Fu, Z., Li, D.: Toward Developing and Using Web-based Tele-Diagnosis in Aquaculture. Expert System ith Applications 25(2), 247–254 (2003)
2. Zhang, X., Fu, Z., Wang, R.: Development of the ES-FDD: Expert System for Fish Disease Diagnosis. In: MTS/IEEE Techno-Ocean 2004, Kobe, Japan, pp. 482–487 (2004)
3. Li, D., et al.: Toward Developing a Tele-Diagnosis System on Fish Disease. In: Bramer, M. (ed.) Artificial Intelligence in Theory and Practice. Book Series IFIP International Federation for Information Processing, vol. 217, pp. 445–454. Springer, Boston (2006)
4. Kim, J.W., et al.: A Guideline for Aqua-life Disease, HangulGraphics, Busan, Korea (2005)
5. Kim, J.-H., et al.: Infections and Parasitic Diseases of Fish and Shellfish. Life science publishing Co., Seoul (2006)
6. Park, S.W., Oh, M.J.: Medical Science for Aqua-life, Bioscience, Seoul, Korea (2008)
7. Fishdoc Co. Ltd, `http://www.fishdoc.co.uk`
8. National Fish Pharmaceuticals, `http://www.fishyfarmacy.com/symptoms.html`
9. Active Window Productions, Inc., `http://fins.actwin.com/disease/chart1.php`
10. FishVet, Inc., `http://www.fishvet.com/sw_parasite_chart.htm`
11. Xing, B., Li, D., Wang, J., Duan, Q., Wen, J.: An Early warning System for Flounder Disease. In: IFIP International Federation for Information Processing Computer and Computing Technologies in Agriculture II, vol. 294, pp. 1011–1018 (2009)
12. Fish Disease Information System of Eco Aquafarm Research Center, `http://earc.chonnam.ac.kr/disease/index.php` (in Korean)

A Fully-Integrated Amplifier MMIC Employing a CAD-Oriented MIM Shunt Capacitor

Young Yun*, Jang-Hyeon Jeong, Young-Bae Park, Bo-Ra Jung, Jeong-Gab Ju, and Eui-Hoon Jang

Department of Radio Communication and Engineering, Korea Maritime University, Dong Sam-dong, Young Do-gu, Busan, 606-791, Korea
yunyoung@hhu.ac.kr

Abstract. In this work, using a CAD-oriented MIM shunt capacitor appropriate for circuit/layout design in millimeter-wave applications, a highly integrated broadband amplifier MMIC including all the matching components was developed for K band applications. Concretely, bulky power divider and combiner were highly miniaturized using the CAD-oriented MIM shunt capacitor, and they were all integrated on MMIC. In addition, the CAD-oriented MIM shunt capacitors were employed for capacitive matching in place of bulky open stubs. The amplifier MMIC exhibited good RF performances and good stability in a wide frequency range from 17 to 28 GHz. The layout size of the amplifier including all passive components was 2.0X1.0 mm^2.

Keywords: broadband amplifier, MMIC, pre-matching circuit, MIM shunt capacitor, K band.

1 Introduction

The K band MMIC amplifiers have been reported in various literatures [1]-[5]. For a realization of low cost k band communication system, demands for fully integrated MMIC including all passive matching components are increasing. Generally, in the millimeter-wave frequency range, open stubs mainly have been used for shunt capacitive matching. However, the open stub occupies a large layout area in MMICs.

---

* Young Yun received the B.S. degree in electronic engineering from Yonsei University, 1993, and the M.S. degree in electrical and electronic engineering from Pohang University of Science and Technology, 1995, and Ph.D. in electrical engineering from Osaka University, 1999, respectively. From 1999 to 2003, he worked as an engineer in Matsushita Electric Industrial Company Ltd., Osaka, Japan, where he has been engaged in the research and development of monolithic microwave ICs (MMICs) for wireless communications. In 2003, he joined Dept. of Radio Communication and Engineering, Korea Maritime University. Now he is currently an assistant professor, and his research interests include design and measurement for RF/microwave and millimeter-wave IC, and design and fabrication for HEMT and HBT. He is corresponding author of this paper.

M. Zhu (Ed.): Electrical Engineering and Control, LNEE 98, pp. 1047–1054.
springerlink.com © Springer-Verlag Berlin Heidelberg 2011

For small layout size, MIM shunt capacitors are more suitable than the open stubs. In the millimeter-wave frequency, however, the MIM shunt capacitors and their accurate lumped equivalent models reflecting layout information for an efficient circuit/layout design have hardly been reported yet [6].

In this work, to reduce a layout size and perform the layout/circuit design efficiently, a CAD-oriented MIM shunt capacitor and its lumped equivalent circuit were developed for millimeter wave applications. Using the MIM capacitor, we perform the layout/circuit design of miniaturized passive components, and fabricated a fully-integrated amplifier MMIC.

2 An Efficient CAD-Oriented MIM Shunt Capacitor and Its Lumped Equivalent Circuit Suitable for Millimeter-Wave Circuit/Layout Design

Due to the CAD-oriented MIM shunt capacitor, we don't need an EM simulation during a circuit design. All capacitance and inductance values of the equivalent circuit of the MIM shunt capacitor are expressed by simple closed-form equations depending only the length of electrode. Therefore, only the length of electrode is optimized during circuit design without EM simulation, which makes the circuit/layout design very simple and greatly reduces the simulation time. A top view and a cross-sectional view (according to the dotted line X-X) of the MIM shunt capacitor are shown in Fig. 1 (a) and (b), respectively, and its lumped equivalent circuit is shown in Fig. 2. In the lumped equivalent circuit of Fig. 2, the C_a correspond to the capacitance of the shunt capacitive part, which is shown in Fig. 1 (b), the capacitor consisting of upper and lower electrode, and insulator). The C_a is required for the shunt capacitive impedance matching. The C_b corresponds to the parasitic capacitance between the lower electrode and backside metal, which is shown in Fig. 1 (b). The L_a is the inductance of the parasitic inductor originating from the upper and lower electrode. The EM simulation data were used for the parasitic elements from the via hole and the interconnection part between the via hole and upper electrode.

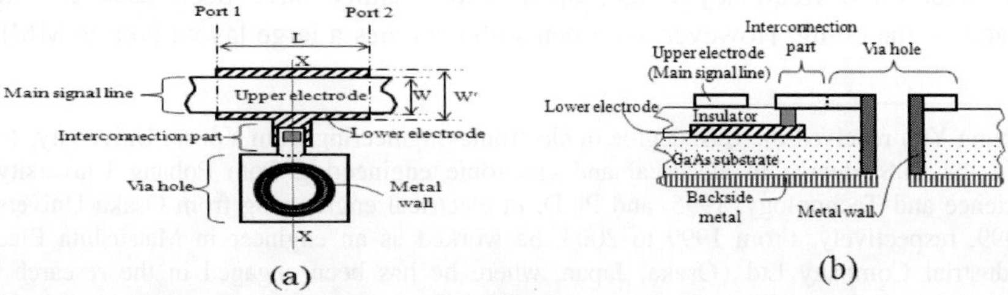

Fig. 1. (a) A top view of the CAD-oriented MIM shunt capacitor (b) A cross-sectional view of the MIM shunt capacitor

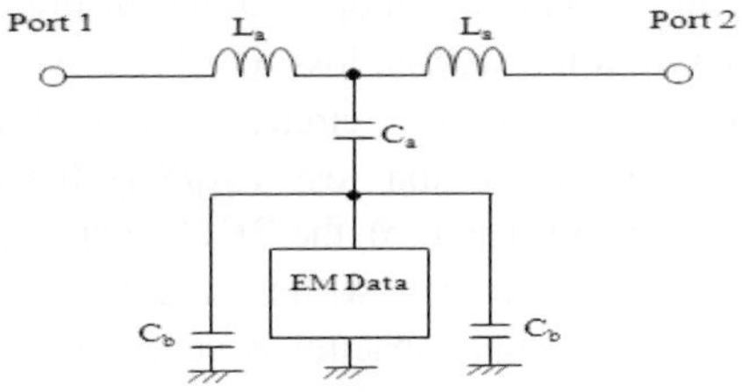

Fig. 2. The lumped equivalent circuit of the CAD-oriented MIM shunt capacitor

The characteristics of the MIM shunt capacitor and its lumped equivalent circuit are summarized as follows.

(1) The lower electrode and insulator exist under the main signal line as shown in Fig. 1 (a) and (b), and the main signal line was employed as the upper electrode. Other passive elements or transistors are connected to the port 1 or 2.

(2) The width W of the upper electrode (main signal line) was fixed to 20 μm, and the structures and size of the via hole and the interconnection part were also fixed. Therefore, all the capacitance and inductance values (C_a, C_b and L_a) of the equivalent circuit depend on only the length L of the lower electrode (See Fig. 1 (a)). Therefore, all the capacitance and inductance values are expressed by simple closed-form equations depending only the length L, which simplifies the layout/circuit design process.

(3) For the equivalent circuit shown in Fig. 2, the EM simulation data were used for the parasitic elements from the via hole and the interconnection part, and the structures and size of the via hole and the interconnection part were fixed. Therefore, the additional EM simulation is not required even though the capacitance value C_a is changed by adjusting the length L of the lower electrode. It is very important for the usefulness of the model.

If the width of the upper and lower electrode are fixed as 20 and 25μm, respectively, and silicon nitride (SiN) having a thickness of 200 nm is employed as the insulator on the GaAs substrate, all the parameters (C_a, C_b and L_a) of the equivalent circuit are dependent on only the length L of the lower electrode, which is expressed as follows.

$$C_a = 6.198X10^{-3}L \quad pF, \tag{1}$$

$$C_b = 14.2X10^{-3}L \quad fF, \tag{2}$$

$$L_a = 0.54375(L-52) + 6.45 \quad pH, \tag{3}$$

where the unit of the length L is μm. The RF characteristics calculated from the equivalent circuit employing the Eqs. (1) ~ (3) were compared with the measured data. Figure 3 (a) and (b) show the measured and calculated RF characteristics of the

MIM shunt capacitor. Thick solid line and open circles of Fig. 3 (a) correspond to the simulated and measured return/insertion loss (S_{11}/S_{21}) for a length L of 50μm, respectively. Thick solid line and open circles of Fig. 3 (b) correspond to the simulated and measured return/insertion loss (S_{11}/S_{21}) for a length L of 80μm, respectively. As shown in Fig. 3 (a) and (b), the RF characteristics calculated from the equivalent circuit shows a good agreement with the measured data.

The layout size of the MIM shunt capacitor is very small in comparison with for conventional open stub. For example, for a shunt capacitance value of 0.31 pF, the total layout size of the MIM shunt capacitor is 3,000μm^2, and for a shunt capacitance value of 0.5 pF, it is 4,800 μm^2. On the other hand, for a shunt capacitance value of 0.31 pF, the layout size of the open stub is 16,000 μm^2, and for a shunt capacitance value of 0.5 pF, it is 20,000 μm^2. Above results indicate that the MIM shunt capacitor and its equivalent circuit is very efficient for the RF circuit/layout design. In this work, without EM simulation, the above equations and equivalent circuit were used for circuit/layout design, and only the length L of the lower electrode was optimized for impedance matching, which makes the circuit/layout design process very simple.

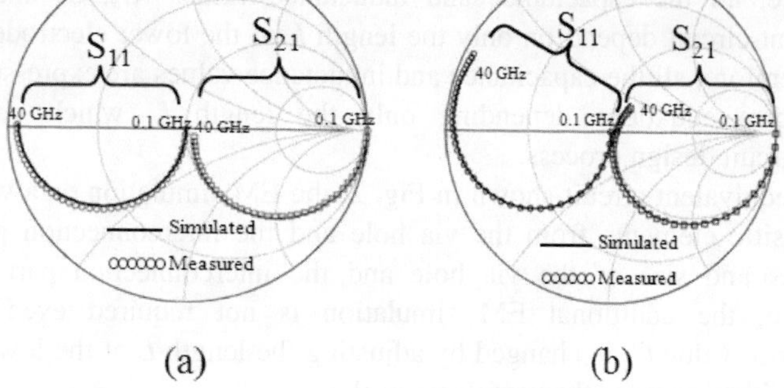

Fig. 3. (a) Measured and calculated RF characteristics of the MIM shunt capacitor for a length L of 50μm and (b) for a length L of 80 μm.

3 Schematic Circuit Configuration of the Amplifier MMIC

Figure 4 shows a schematic circuit of the medium power amplifier (MPA) MMIC. As shown in this figure, the amplifier includes all passive components. Concretely, MIM shunt capacitors were employed for shunt capacitive matching in place of open stubs for a reduction of layout size. Especially, miniaturized power divider and combiner employing the CAD oriented capacitors. Using miniaturized π-type transmission line with parallel-connected shunt capacitors, the size of conventional λ/4 Wilkinson power divider was reduced to a great extent, and the miniaturized power divider/combiner was integrated on the MMIC. Fig. 5 (a) and (b) show the λ/4 transmission line and the π-type miniaturized transmission line with parallel-connected capacitance. The 2 port Y parameters of Fig. 5 (a) and (b) can be easily calculated from the 2 port network theory, and from the results, we can see that the 2 port Y-parameters of the Fig. 5 (b) agree with those of 1/4λ transmission line (Fig. 5 (a)) if following conditions are satisfied.

$$Z = \frac{Z_C}{\sin(\frac{2\pi L}{\lambda_g})} \tag{4}$$

$$C = \frac{\cos(\frac{2\pi L}{\lambda_g})}{2\pi f Z_C} \tag{5}$$

where Z and Zc are the characteristic impedance of miniaturized transmission line and λ/4 transmission line, respectively, and L and λ_g are the length of the miniaturized stripline and wavelength, respectively. In this work, the CAD-oriented MIM capacitor was employed for the parallel-connected shunt capacitors. As shown in Fig. 4, the length of transmission line of divider and combiner was reduced to λ/12 and λ/24, respectively.

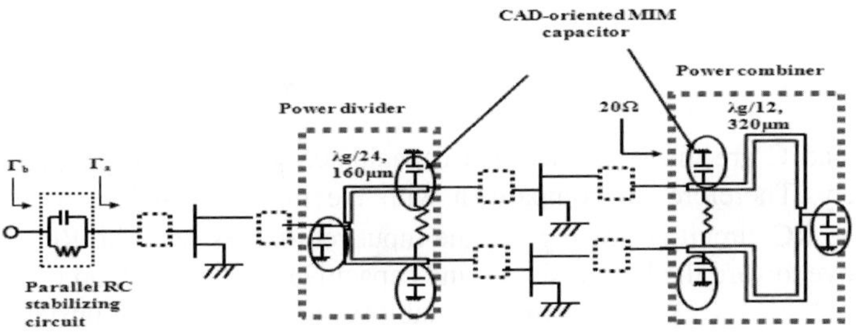

Fig. 4. A schematic circuit of the medium power amplifier.

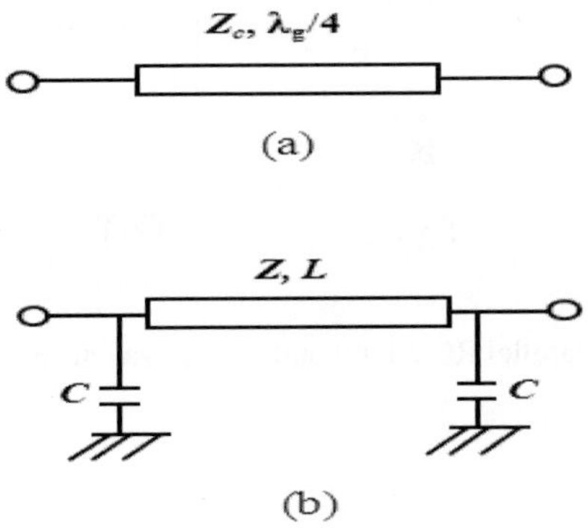

Fig. 5. (a) λ/4 transmission line and (b) π-type miniaturized transmission line with parallel-connected capacitance.

4 A Highly Improved Stability of the Amplifier MMIC Employing RC Circuit

For K band MMICs, the stability in a wide frequency range should be guaranteed from DC to operation band in order to prevent an unwanted. In this work, in order to improve the stability of the amplifier MMIC, parallel RC circuit was connected to the input of the amplifier MMIC as shown in Fig. 4. The value of resistance R of the parallel RC circuit was selected to stabilize the amplifier MMIC in the unstable frequency range by improving the input return loss in the frequency range. The value of the capacitance C was selected to bypass the resistor in the operation band or higher frequency range, and therefore the RF performances in the operation band are not affected by the parallel RC circuit. As shown in Fig. 6 (a) and (b), parallel RC circuit is equivalent to the series R'C' circuit, and in the case, the resistance R'and capacitance C'can be expressed as follows.

$$R' = \frac{R}{1 + (\omega RC)^2}, \tag{6}$$

$$C' = \{\frac{1}{\omega^2 RC} + RC\} \cdot \frac{1}{R}, \tag{7}$$

where R and C are the resistance and capacitance value of the parallel RC circuit, respectively. Therefore, in comparison with the input return loss of the amplifier without the RC circuit (Γ_a of Fig. 4), the input return loss with the RC circuit (Γ_b of Fig. 4) move toward the high resistive and capacitive region by R'and C'of the parallel RC circuit. In this work, to improve the stability and return loss in unstable frequency range, the values of 0.5 pF and 20 Ω were selected for the C and R, respectively.

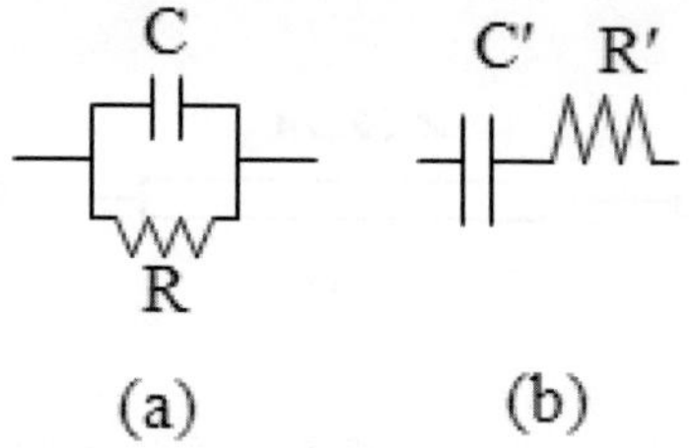

Fig. 6. (a) Parallel RC circuit and (b) equivalent series R'C'circuit

5 RF Performances

A Photograph of the MPA MMIC is shown in Fig. 7. The MPA MMIC exhibited 2.0X1.0 mm^2. The measured and simulated gain, and measured return loss of the

MPA MMIC are shown in Fig. 8 (a). The solid line corresponds to the measured, and, open circles and solid triangles correspond to the measured input and output return loss of the MPA MMIC, respectively. The measured gain of the MPA MMIC also exhibits a good flatness in a wide frequency range. It shows 13±2 dB between 19.5 GHz and 32 GHz. The measured P_{out} of the MPA MMIC exhibited the value of 20±2 dBm in the frequency range. Measured P_{out}-P_{in} characteristic of the MPA MMIC at 25 GHz is shown in Fig.8(b). The measured P_{out} at 25 GHz is 21.5 dBm. The MMICs in this work show small chip size and broadband characteristics.

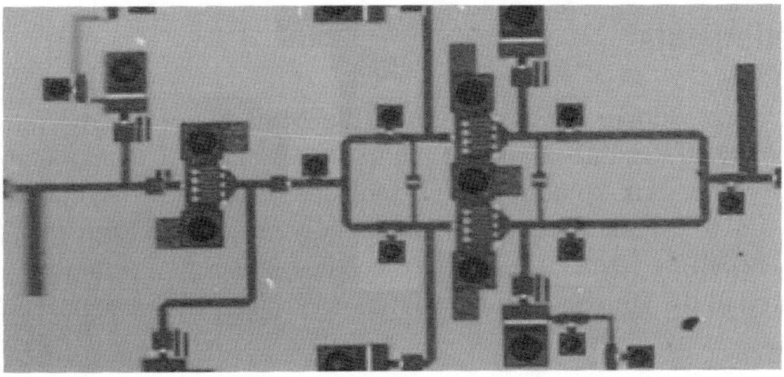

Fig. 7. A photograph of the MPA MMIC

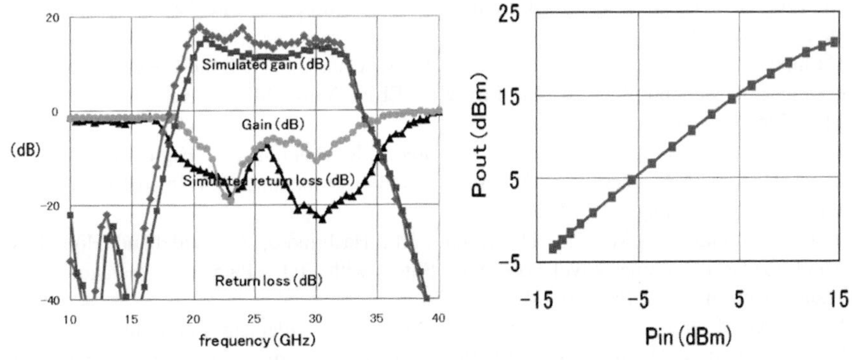

(a) (b)

Fig. 8. (a)Measured and simulated gain, return loss of MPA MMIC (b) Measured Pout-Pin characteristic of the MPA MMIC at 25GHz

6 Conclusions

In this work, we developed a CAD-oriented MIM shunt capacitor for application to circuit/layout design in millimeter-wave. All equations of equivalent circuit were only dependent on the length of electrode, and only the length of the electrode was optimized for impedance matching, which makes the circuit/layout design process very simple. Using the CAD-oriented MIM shunt capacitor, we developed a highly integrated broadband amplifier MMIC including all the matching components for K band applications. Concretely, bulky power divider and combiner were highly miniaturized using the CAD-oriented MIM shunt capacitor, and they were all integrated on MMIC. In addition, the CAD-oriented MIM shunt capacitors were employed for capacitive matching in place of bulky open stubs. The amplifier MMIC exhibited good RF performances and good stability in a wide frequency range from 17 to 28 GHz. The layout size of the amplifier including all passive components was 2.0X1.0 mm^2.

Acknowledgments

This research was supported by the MKE(the Ministry of Knowledge Economy), Korea, under the ITRC(Information Technology Research Center) support program supervised by the NIPA(National IT Industry Promotion Agency)(NIPA-2011-C1090-1121-0015). This work was financially supported by the Ministry of Knowledge Economy (MKE) and the Korea Industrial Technology Foundation (KOTEF) through the Human Resource Training Project for Strategic Technology.

References

1. Udomoto, J., Ishida, T., Akaishi, A., Araki, T., Kadowaki, N., Komaru, M., Mitsui, Y.: A 50 PAE K-band power MMIC amplifier. In: European Microwave Conf., vol. 1, pp. 263–266 (1999)
2. Matinpour, B., Lal, N., Laskar, J., Leoni, R.E., Whelan, C.S.: K-Band Receiver Front-Ends in a GaAs Metamorphic HEMT Process. IEEE Trans. Microwave Theory Tech. 49, 2459–2463 (2001)
3. Mimino, Y., Hirata, M., Nakamura, K., Sakamoto, K., Aoki, Y.,, S.: High Gain-Density K-Band pHEMT LNA MMIC for LMDS and Satellite Communication. In: IEEE MTT-S Int. Microwave Symp. Dig., pp. 17–20 (2000)
4. Itoh, Y., Takagi, T., Masuno, H., Kohno, M., Hashimoto, T.: Wideband High Power Amplifier Design Using Novel Band-Pass Filters with FET's Parasitic Reactances. IEICE Trans. Electron. E76-C(6), 938–943 (1993)
5. Mochizuki, M., Itoh, Y., Nii, M., Takagi, T., Mitsui, Y.: Wideband Monolithic Lossy Match Power Amplifier Having an LPF/HPF-Combined Interstage Network. IEICE Trans. Electron. E78-C(9), 1252–1254 (1995)
6. Bahl, I., Bhartia, P.: Microwave Solid State Circuit Design, pp. 509–514. John Wiley & Sons, Inc., Chichester (1988)

A Design of Adaptive Optical Current Transducer Digital Interface Based on IEC61850

Zhang Qiong, Li Yansong, Liu Jun, and Hao Guoliang

School of Electrical & Electronic Engineering
North China Electric Power University (Beijing), 102206, Beijing, China
z111qiong@126.com

Abstract. This paper introduces the related theory of Adaptive Optical Current Transformer(AOCT), and gives one design method of its digital output interface using hardware platform DPS2812Mv2 on this basis. The interface circuit of Ethernet controller CS8900A and the method to compose driver software are given in details, and also the implementation of IEC61850-9 protocol is discussed. This solution has practical engineering value.

Keywords: AOCT; CS8900A; IEC61850-9.

1 Introduction

Recently, with the expansion of power grid and the rising of voltage level, the establishing and extending of digital substation which is based on IEC61850 standards will be an inevitalbe etrend. Researching electronic transformer and its digital interface in digital substation process layer will be of great significance and engineering value. In order to meet the needs of the digital secondary equipment based on IEC61850 standards, the interface between the electronic transformers and protective devices must be first insolved. Therefore the International Electrotechnical Commission established IEC60044-7/8 and IEC61850-9-1/2 standards [1-4], so as to standardize the digital output interface of electronic transformers.

The present paper is based on a new kind of Adaptive Optical Current Transducer(abbreviation AOCT) proposed by predecessors. The AOCT applly adaptive kalman filter and wavelet theory to form the model reference adaptive control system. Since the disturbance of temperature and stress was eliminated in the AOCT, the accuracy in the steady state and transient state of the AOCT is improved. Since the AOCT take on preceding feedback characteristic, the structure of OCT is able to be simplified to improve the stability of AOCT. And AOCT have been installed in the substation BaoDing city HeBei Province [5].

2 Adaptive Optical Current Transducer

2.1 The Basic Principle of Optical Current Transformer

The Optical Current Transducer following Faraday magneto-optic effect [6-7] has been the mainstream of the optical transducing technology persistently. The current

M. Zhu (Ed.): Electrical Engineering and Control, LNEE 98, pp. 1055–1061.
springerlink.com © Springer-Verlag Berlin Heidelberg 2011

which OCT measuring is by means of the line integral of the magnetic intensity that is created by the measured current. According to the Faraday magneto-optic effect, when linearly polarized light pass the mediator such as crystals and optical glass across the external magnetic field which is paralleled its direction of propagation, its polarization plane will generate deflection, the angle of deflection θ shows as:

$$\theta = \mu \cdot V \cdot \int_L H \cdot dl \tag{1}$$

u is the permeability of the Faraday magneto-optic material, V is Verdet constant of the Faraday magneto-optic material, it is related to the media characteristics, optical source wavelength, outside temperature and so on. H is the intensity density on the Faraday magneto-optic material and L is the optical distance length of polarized. light across the Faraday magneto-optic material, as Fig. 1 shows. To realize the current measuring, linearly polarized light is looping-in surrounded the current, according to Ampere loop principle:

$$\theta = \mu V \oint_L \vec{H} \cdot d\vec{l} = \mu \cdot V \cdot N \cdot i \tag{2}$$

N is the loop number that linearly polarized light surrounding the current, i is the measured current.

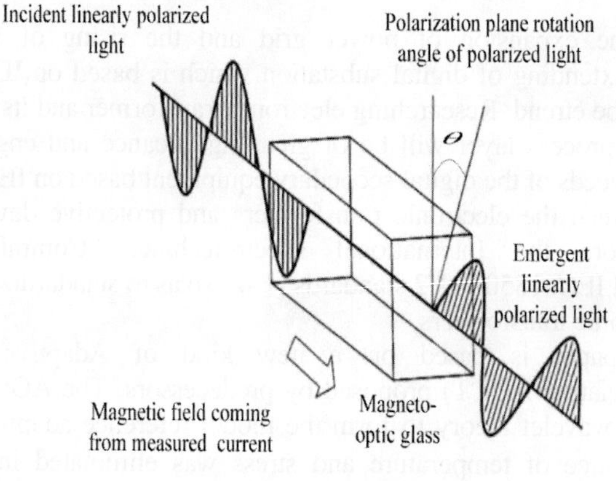

Fig. 1. Schematic diagram of Faraday magneto-optic effect

2.2 The Current Measurement Principle of AOCT

The Adaptive Optical Current Transformer (AOCT) combines the CT based on the principle of electromagnetic induction with the OCT based on Faraday magneto-optic effect principle [8]. The AOCT uses the high-accuracy current value measured by CT to correct the output of AOCT in the steady state, and output transient current of the OCT by the good dynamic performance in the transient state. With the combination of the good dynamic performance of OCT and the high steady-state measurement accuracy of CT, the AOCT has good characteristies both in the steady state and transient state. This paper adopted the Adaptive Optical Transducing Principle and the Solenoid Collecting Magnetic Field Optical Path proposed in [9]. Fig. 2 gives the system principle diagram of AOCT.

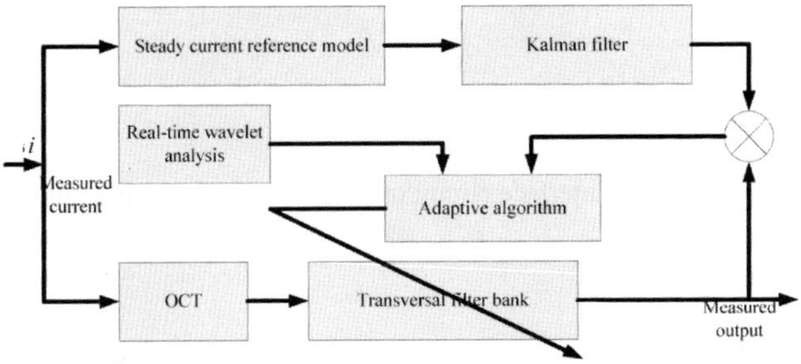

Fig. 2. The structure of the adaptive optical current transducer

3 Digital Output Interface Module

The output of AOCT signal processing system has two forms: analog signals and digital signals. This paper mainly introduces how to transmit the data calculated by DSP2812 to Ethernet by CS8900A based on IEC61850-9, so as to realize the Ethernet communication function.

3.1 General Description of CS8900A

The CS8900A [10-11] is a low-cost Ethernet LAN Controller optimized for Industry Standard Architecture(ISA) Personal Computers. Its highly-integrated design eliminates the need for costly external components required by other Ethernet controllers. The CS8900A includes on-chip RAM, 10BASE-T transmit and receive filters,and a direct ISA-Bus interface with 24 mA Drivers.

The CS8900A's Ethernet Media Access Control(MAC) engine is fully compliant with the IEEE802.3 Ethernet standard (ISO/IEC 8802-3, 1993),and supports full-duplex operation. It handles all aspects of Ethernet frame transmission and reception, including: collision detection, preamble generation and detection, and CRC generation and test Programmable MAC features include automatic retransmission on collision, and automatic padding of transmitted frames.

3.2 Hardware Circuit Design

The Ethernet interface circuit of CS8900A is composed of network isolation transformer H1102 and RJ45, as Fig. 3 shows.

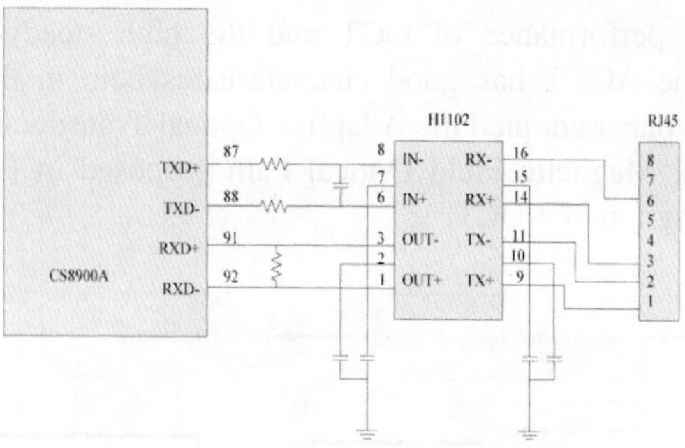

Fig. 3. Hardware interface circuit of Ethernet communication module

3.3 Software Design

The sampling value are mapped to underlying the actual object and communication protocols in IEC61850-9-1, but mapped to ISO/IEC8802-3 in IEC61850-9-2 which is also called special communication service mapping based on process bus. The standard IEC61850-9-1 don't have to think about the collision and detection problems in Ethernet data transmission, and there is also no important and minor points for data transmission information. However, it introduces the concept of VLAN (virtual local area network) and level of data transmission priority which are necessary for the realization of the process bus transmission in IEC61850-9-2. So it will be very easy to make the transition for the devices following IEC61850-9-1 to work in the process bus network defined by IEC61850-9-2. In this paper,the frame is packeted with the format prescribed by IEC61850-9-1.

1) Initialization of LAN Controller CS8900A
The initialization of the chip CS8900A, setting parameters of related registers, is very important.

- LINECTL(LineCTL, Read/Write, Address:PacketPage base + 0112h) LineCTL determines the configuration of the MAC engine and physical interface. In this design, set the initial value to 00d3H ,choose 10BASE-T physical interface and enable transmit and receive control bits.
- RXCTL(RxCTL, Read/Write, Address: PacketPage base +0104h) RxCTL has two functions: define what types of frames to accept and configure the Destination Address filter. In this design ,setting the initial value to 0d05H means that accept frames with Destination Address(DA) equal to physical addresses and broadcast frames.
- RxCFG(RxCFG, Read/Write, Address: PacketPage base + 0102h) RxCFG determines how frames will be transferred to the host and what frame types will cause interrupts. In this design ,setting the initial value to 0103H means that receive a correct data will cause an interrupt.
- BUSCT (BusCTL, Read/Write, Address: PacketPage base + 0116h) BusCTL controls the operation of the ISA-bus interface.

- ISQ (ISQ, Read-only, Address: PacketPage base + 0120h) The Interrupt Status Queue Register is used in both Memory Mode and I/O Mode to provide the host with interrupt information.
- TxCMD(Write-only, Address: PacketPage base + 0144h) The word written to PacketPage base + 0144h tells the CS8900A how the next packet should be transmitted.
- TXLENG (Write-only, Address: PacketPage base + 0146h) This register is used in conjunction with register TxCMD. When a transmission is initiated via a command in TxCMD, then the length of the transmitted frame is written into this register.

The communication between CS8900A and TMS320F2812 can adopt the memory mode or the I/O mode. This article adopts the I/O mode.The I/O base address defaults to 0x0300H. When configured for I/O Mode operation, the CS8900A is accessed through eight, 16-bit I/O ports that are mapped into sixteen contiguous I/O locations in the host system's I/O space. I/O Mode is the default configuration for the CS8900A and is lways enabled.

2) Transmission of Frame

Packet frame

The frame is packeted with the format prescribed by IEC61850-9-1, as show in Fig. 4. However, a structural variable type must be defined.

PR	SD	DA	SA	TPID	TCI	TYPE	PDU	APDU	FCS
56bit	8bit	48bit	48bit	16bit	16bit	16bit	64bit	584bit	32bit

Fig. 4. The structure of Ethernet frame based on IEC61850-9-1

- PR: Preamble, using for the synchronization between transmitter and receiver, it is 1010101010… of 56 bits.
- SD: Frame start, is 10101011 of 8 bits. The last two bits means that the following data is what the host transmits.
- DA: Destination Address, is set to the broadcast address(FFFFFFFFFFFF).
- SA: Source Address, should be as the only signs in the local area network and can be defined by own.
- TPID: Tag Protocol Identifier, Marker of priority, based on priority and VLAN Ethernet IEEE802 1Q , is set to 0x8100h.
- TCI: the signs of control information, priority and VLAN markers
- TYPE: set to Ethernet type 0x88ba, information of IEC61850 registered by the IEEE Copyright Registration Institution.
- PDU: Protocol Data Unit
- APDU: Application Protocol Data Unit consists of Application Protocol Control Information (32 bits) and Application Service Data Unit. Application Service Data Unit is made up of basic data set(ASDU,46 words) and status data set(ASDU,23 words). However, LNName(Logical Node Name) and DataSetName defined in the

basic data set are required in the IEC61850-9-2 which is necessary for the transition from IEC61850-9-1 to IEC61850-9-2. In IEC61850-9-2, the data set can be defined in application at any time, but they are set to fixed value in IEC61850-9-1 because of without realizing interoperability.

- FCS: Frame Check Sequence, also called CRC(32bits).

Transmit operations of frame

I/O mode transmit operations occur in the following order (using interrupts):

- The host bids for storage of the frame by writing the Transmit Command to the TxCMD Port (I/O base + 0004h) and the transmit frame length to the TxLength Port (I/O base + 0006h).
- The host reads the BusST register (Register 18)to see if the Rdy4TxNOW bit (Bit 8) is set. To read the BusST register, the host must first set the PacketPage Pointer at the correct location by writing 0138h to the PacketPage Pointer Port (I/O base + 000Ah). It can then read the BusST register from the PacketPage Data Port (I/O base + 000Ch). If Rdy4TxNOW is set, the frame can be written. If clear, the host must wait for CS8900A buffer memory to become available. If Rdy4TxiE (Register B, BufCFG,Bit 8) is set, the host will be interrupted when Rdy4Tx (Register C, BufEvent, Bit 8) becomes set. If the TxBidErr bit (Register 18, BusST, Bit7) is set, the transmit length is not valid.
- Once the CS8900A is ready to accept the frame, the host executes repetitive write instructions (REP OUT) to the Receive/Transmit Data Port (I/O base + 0000h) to transfer the entire frame from host memory to CS8900A memory.

4 Conclusion

The digital interface of electronic transformers is an important part of information exchange between process layer and bay layer in digital substation. Adaptive Optical Current Transducer has high measuring accuracy and good performance of long-term operation. On this basis, the paper has developed one design method of its digital output interface using hardware platform DPS2812Mv2 and development environment of CCS software based on IEC61850. The design of interface is with high reliability and quick transmission speed ratio, and the program structure is simple and with high efficiency. It can satisfy the requirements of the electronic transformer digital interface.

References

1. IEC60044-7 Instrument transformers-Part7: electrical voltage transducers (2002)
2. IEC60044-8 Instrument transformers-Part8: electrical current transducers (2002)
3. IEC61850-9-1 Communication Networks and Systems in Substations Part9-1: Specific Communication Service MapPing(SCSM)-Serial Unidirectional Multidrop Point to Point Link (2003)
4. IEC61850-9-2 Communication Networks and Systems in Substations Part9-2: SPecific Communication Service MapPing(SCSM)-For Process Bus (2003)

5. Li, Y-s., Zhang, G-q., Yu, W-b., Yang, Y-h.: Adaptive optical current transducer. Proceedings of the CSEE 23(11), 100–105 (2003)
6. Zhang, S.-y., Wang, Y.-h., Huang, J.: Anoptical current transducer based on the electronic speckle shearing pattern interferometer. Automation of Electric Power Systems 31(17), 83–86 (2007)
7. Wang, X.-x., Zhang, C.-x., Zhang, C.-y.: A new all digital closed-loop fiber optic currentt transformer. Automation of Electric Power Systems 30(16), 77–80 (2006)
8. Li, Y.-s., Zhang, G.-q., Yu, W.-b.: Combined method to improve the accuracy of opt ical current transducer. Automation of Electric Power Systems 27(19), 43–47 (2003)
9. Li, Y.-s., Guo, Z.-z., Yang, Y.-h.: Research on the basic theory of adaptive optical current transducer. Proceedings of the CSEE 25(22), 21–26 (2005)
10. CS8900A Product Data Sheet Cirrus Logic, Inc. (1999)
11. CS8900A technical reference manual. Cirrus Logic, Inc. (2001)

Performance Analysis of DSR for Manets in Discrete Time Markov Chain Model with N-Spatial Reuse*

Xi Hu[1,**], Guilin Lu[2], and Hanxing Wang[2]

[1] School of Fundamental Studies, Shanghai University of Engineering Science, Shanghai, 201620

[2] Dept. of Mathematics and Statistics, Shanghai Lixin University of Commerce, Shanghai, 201620

Abstract. In this paper, we analyze the performance of the typical reactive routing protocol DSR for mobile ad hoc networks in a discrete time Markov chain model with N-spatial reuse. Some performance parameters are studied such as the probability generation function of the flooding distance, the average flooding distance, the probability that a request packet finds a τ-time-valid route. The results obtained in this paper are theoretically useful to optimize and evaluate the routing protocols.

Keywords: Mobile ad hoc networks, Markov chain, Routing protocols, Performance analysis.

1 Introduction

A mobile ad hoc network (Manet) is a self-organized mobile wireless network without the aid of any base stations or fixed network infrastructures. It should be applied widely in various fields and now has become a hot issue in the field of the 4th generation wireless mobile communication networks. In such a network, each mobile node may communicate with any other node in the network by initiating a route discovery process, whether directly reachable within the limited wireless transmission range or not. Thus the routing protocol is one of the crucial issues for a Manet [1-3].

Performance analysis is very important for a network, either to design or to optimize a protocol will involve how to define and estimate some basic performance parameters of the network. Because it's very difficult to research these parameters in a Manet, people would like to study them by means of simulation experiment rather than theoretical research. It is very important to establish corresponding suitable model for studying these parameters. Jacquet and Laouiti did the pioneering work on the theoretical research of the flooding distance (FD) under the random graph model without spatial reuse, and obtained the analytical expressions of the FD [4]. But they only consider the case without spatial reuse which is similar to indoor wireless

---

* Supported by the China Natural Sciences Foundation under Grant No. 60872060 and the Foundation of Shanghai University of Engineering Science.

** Corresponding author. Xi Hu, female, doctor, e-mail: xih_xih@163.com

M. Zhu (Ed.): Electrical Engineering and Control, LNEE 98, pp. 1063–1070.
springerlink.com © Springer-Verlag Berlin Heidelberg 2011

networks. We proposed a discrete time Markov chain models with spatial reuse in the reference [5]. In this paper, we further analyze the performance of the typical reactive routing protocol DSR for a Manet in the discrete time Markov chain model with N-spatial reuse. Some performance parameters are studied such as the probability generation function of the FD, the average flooding distance, the probability that a request packet finds a τ-time-valid route. The results are theoretically useful to optimize and evaluate the routing protocols.

2 Performance Analysis

In the following we study the performance of the reactive routing protocol DSR under the discrete time Markov chain model with N-spatial reuse (referred to as N-SRRWM model) [5].

Lemma 1. In the N-SRRWM model, the Markov chain $\{d_t\}_{t\in Z^+}$ has the unique stationary probability distribution

$$\pi(0)=(1+\sum_{i=0}^{a-1}\frac{p_i\cdots p_0}{q_{i+1}\cdots q_1})^{-1},\pi(k)=\frac{p_{k-1}\cdots p_0}{q_k\cdots q_1}\pi(0),1\le k\le a \quad . \tag{1}$$

Especially when $p_0=p_1=\cdots=p_{a-1}=\tilde{p}$, $q_1=q_2=\cdots=q_a=\tilde{q}$, we have

$$\pi(0)=(1+\sum_{i=0}^{a-1}(\frac{\tilde{p}}{\tilde{q}})^{i+1})^{-1},\pi(k)=(\frac{\tilde{p}}{\tilde{q}})^k\pi(0),1\le k\le a\,. \tag{2}$$

By the definition of stationary distribution, we know that the probability distribution of each state of $\{d_t\}$ will keep in the stationary distribution after a long time of dynamical changes in the Manet. i.e. $P(d_t=k)=\pi(k)$, $0\le k\le a$. Especially, if the initial distribution is just the stationary distribution: $P(d_0=k)=\pi(k)$, $0\le k\le a$, then $\{d_t\}$ is a stationary Markov chain. That is to say, for $\forall t\in Z^+$, $P(d_t=k)=\pi(k)$, $0\le k\le a$. We assume that $\{d_t\}$ is a stationary Markov chain with the initial distribution as $\pi(0),\pi(1),\cdots,\pi(a)$.

Lemma 2. In the N-SRRWM model with n nodes in the network, if the transmission radius is ρ, and the Markov chain $\{d_t\}$ has the stationary distribution with the initial distribution as (2) , then the probability of any two nodes v and v' to communicate with each other in one hop is

$$p=\sum_{k=0}^{\rho}\pi(k)\,; \tag{3}$$

and the average neighbor number E_n of any node is

$$E_n = (n-1)p. \tag{4}$$

2.1 Performance Analysis of Routing Discovery

The flooding distance (FD) between two random nodes is the number of hops or links contained in the route between the two nodes. Since the nodes in the network are randomly distributed and the intermediate nodes are randomly selected to retransmit, the FD between any two nodes is a random variable. The FD is naturally an important performance parameter of the network, and other performance parameters are often relevant to it.

For the N -SRRWM model, let ξ_N denote the FD between two given nodes in the network, and $\left\{p_i^{(N)}\right\}_{i=0}^{\infty}$ be the distribution law of the FD such that $p_i^{(N)} = P(\xi_N = i)$. Denote the probability generating function (p.g.f.) of the FD between any two nodes in the network as $G_N(s)$, by the definition of p.g.f , we have

$$G_N(s) = E(s^{\xi_N}) = \sum_{i=0}^{\infty} p_i^{(N)} \cdot s^i.$$

Theorem 1. For the N -SRRWM model, the p.g.f. of the FD between any two nodes in the network is

$$G_N(s) = ps + \frac{s(1-q^N)}{q^{N-1}} \sum_{i=0}^{\infty} \frac{spq^{(i+1)N}}{1-q^{(i+1)N}} \prod_{j=0}^{i-1} \frac{1-[1-s(1-q^N)]q^{(j+1)N}}{q^{N-1}(1-q^{(j+1)N})}, \tag{5}$$

where $q = 1-p$.

Proof: Because $\{d_t\}_{t\in Z^+}$ has the initial distribution with $\pi(k)$ $(0 \le k \le a)$, $\{d_t\}_{t\in Z^+}$ is a stationary Markov chain. By the lemma 2, the probability of any two nodes in the network to be neighbors is $p = \sum_{k=0}^{\rho} \pi(k)$, In other words, when a node send a message, the other node in the network will receive it correctly with probability of p . And by the hypothesis 7 of the N -SRRWM model in [5], each random process $\{d_t(v,v')\big|_{t\in Z^+} : v,v' \in V\}$ is independent of each other.

For the N -SRRWM model, we consider two nodes, one of which is considered as the source and another as the destination. In the flooding process, the query retransmissions can be classified such that the retransmissions made at the same time belong to the same class. We can give the order to the query retransmission classes according to their retransmission time order. Because the network is dense and the space can be reused N times, the k -th class has and only has N nodes which make the k -th retransmission. Especially the 0 class is the first transmission (or 0 retransmission) made by the source node. Let B be the random event that the destination node receives the request packet from the first k classes. Obviously, the probability of occurrence in B is $P(B)=1-qq^{(k-1)N}$ where $q=1-p$. So the conditional probability generating function (c.p.g.f.) of the FD conditioned on B is

$$G_{N}^{k}(s)=E(s^{\xi_N}/B)=\frac{1}{1-qq^{(k-1)N}}\sum_{i=0}^{\infty}p_i^{(N)}s^i.$$

It can be easily proved that each of the N nodes which belong to the k -th class has the same conditional distribution law of the FD by induction. In fact, the conditional distribution law of the FD of a node conditioned on the event that the node makes the first retransmission is centered at a single point. For any integer k , we suppose that each of the N nodes which belong to the k -th class has the same conditional distribution law of the FD as that of the other $N-1$ nodes. For the N nodes which belong to the $k+1$ -th class, We enumerate these N nodes from 1 to N , and denote the FD of the i -th node by $\xi_{N}^{(i)}(1\le i\le N)$. Then $\xi_{N}^{(i)}-1(1\le i\le N)$ can be regarded as the FD of the i -th node which makes the k -th retransmission for its source node. By the assumption, the conditional distribution law of the FD of each node of the N nodes belong to the $k+1$ -th class is the same as that of other $N-1$ nodes.

Since there is a common conditional distribution law of the FDs of the N nodes which belong to the k -th class, we can denote the c.p.g.f. of the FD of a node which belongs to the k -th class by $P_k(s)$. Obviously, $P_0(s)=1$ and $P_1(s)=s$. The node which belongs to the k -th class must have received the query at least once from the k first classes. Thus the conditional probability $P(i/k)$ $(i<k)$ that a node receives for the first time the query from the i -th class conditioned on the event that the node belongs to the k -th class is

$$\begin{cases} P(0/k)=\dfrac{p}{1-qq^{(k-1)N}}, i=0 \\ P(i/k)=\dfrac{qq^{(i-1)N}(1-q^N)}{1-qq^{(k-1)N}}, 1\le i<k. \end{cases} \tag{6}$$

Furthermore, the c.p.g.f. of the FD of any two nodes in the network is $sP_i(s)$ when it receives for the first time the query from the i -th class. Noting that $P_0(s)=1$, we obtain the recursion by the formula of total expectation

$$P_k(s)=s\sum_{i=1}^{k-1}\frac{qq^{(i-1)N}(1-q^N)}{1-qq^{(k-1)N}}P_i(s)+\frac{ps}{1-qq^{(k-1)N}}.$$

That is $(q^N - qq^{kN})P_k(s) - psq^N = s\sum_{i=1}^{k-1} qq^{iN}(1-q^N)P_i(s).$

Let $P(z,s) = \sum_{i=1}^{\infty} P_i(s)z^i$, $A(z) = \dfrac{1-[1-s(1-q^N)]z}{q^{N-1}(1-z)}$ and $B(z) = \dfrac{psz}{1-z}$,

thus the above equation is

$$P(z,s) = B(z) + A(z)P(q^N z, s).$$

Then we have

$$P(z,s) = \sum_{i=0}^{k-1} B(q^{iN} z)\prod_{j=0}^{i-1} A(q^{jN} z) + P(q^{(k+1)N} z, s)\prod_{j=0}^{k} A(q^{jN} z).$$

Noting that for any $0 \le z < 1$,

$$\lim_{k\to\infty}\sum_{i=0}^{k-1} B(q^{iN} z)\prod_{j=0}^{i-1} A(q^{jN} z) < \infty \text{ and } \lim_{k\to\infty} P(q^{kN} z, s)\prod_{j=0}^{k-1} A(q^{jN} z) = 0,$$

we obtain $P(z,s) = \sum_{i=0}^{\infty} B(q^{iN} z)\prod_{j=0}^{i-1} A(q^{jN} z).$ Substituting $A(q^{jN} z)$ and $B(q^{iN} z)$, we have

$$P(q^N, s) = \frac{psq^N}{1-q^N} + \sum_{i=1}^{\infty} \frac{spq^{(i+1)N}}{1-q^{(i+1)N}} \prod_{j=0}^{i-1} \frac{1-[1-(1-q^N)s)]q^{(j+1)N}}{q^{N-1}(1-q^{(j+1)N})}.$$

The destination would receive the query for the first time from the i-th class with probability $qq^{(i-1)N}(1-q^N)$ when $i \ge 1$ and p when $i = 0$, in this case the FD of the destination equals to the FD of this node plus one. Thus the p.g.f. of the FD conditioned on the fact that the destination receives the query from the i-th class is $sP_i(s)$. The unconditional p.g.f of the FD is

$$G_N(s) = ps + \sum_{i=1}^{\infty} qq^{(i-1)N}(1-q^N)sP_i(s) = ps + \frac{s(1-q^N)}{q^{N-1}} P(q^N, s)$$

$$= ps + \frac{s(1-q^N)}{q^{N-1}} \sum_{i=0}^{\infty} \frac{spq^{(i+1)N}}{1-q^{(i+1)N}} \prod_{j=0}^{i-1} \frac{1-[1-(1-q^N)s)]q^{(j+1)N}}{q^{N-1}(1-q^{(j+1)N})}.$$

The theorem is proved.

Let D_N be the average flooding distance between any two nodes in the network, then $D_N = E\xi_N$. By the relationship between the p.g.f. and the expectation of the

FD, we have: $D_N = E\xi_N = G'_N(1)$ (where $G'_N(s)$ is the first-order derivative of $G_N(s)$) . It's easy to have the following conclusion.

Theorem 2. For the N -SRRWM model, the AFD D_N between any two nodes is

$$D_N = 1 + q(1-q^N)\sum_{k=0}^{\infty}\frac{q^{(N+1)k}}{1-q^{(k+1)N}},$$

and

$$\lim_{N\to\infty} D_N = 1+q.$$

The DSR protocol takes the following measure to avoid redundant copies of the routing request (RREQ) during a route discovery process. Each RREQ packet contains a field of TTL (Time-to-live) which is used to limit the number of intermediate nodes allowed to forward the copy of the RREQ packet. This number of TTL is decremented and the request packet will be discarded if the number reaches zero before finding the target. Different number of TTL can be used to implement different algorithms to improve the performance of Manets. One example is that a node may set the TTL number as one to implement a "non-propagating" route request as an initial phase of a route discovery, another example is that a node may use the TTL number to implement an "expanding ring" searching for the target. Although a query packet with TTL may control the spread of its copies in the network, the probability that it can be received by its destination would decrease. The following theorem gives the probability that a query packet with TTL as H is received by its destination.

Theorem 3. For the N -SRRWM model, the probability $P_{hl}(H)$ that a route is discovered by a RREQ with TTL assumed as H is $P_{hl}(H) = \sum_{k=0}^{H} p_k^{(N)}$ where $p_k^{(N)} (k \ge 0)$ are evaluated by the Laplace reverse transform of $G_N(s)$ shown in the theorem 1.

Proof: A route is discovered successfully by a RREQ packet with TTL assumed as H if and only if the FD ξ_N of the packet satisfies that $\xi_N \le H$. Thus

$$P_{hl}(H) = P(\xi_N \le H) = \sum_{k=0}^{H} P(\xi_N = k) = \sum_{k=0}^{H} p_k^{(N)}$$

where $p_k^{(N)} (k \ge 0)$ are evaluated by the Laplace reverse transform of $G_N(s)$ shown in the theorem 1.

2.2 Performance Analysis of Routing Maintenance

The topology structure of a Manet may change frequently for the properties of node's mobility and the transmission power fluctuations. The link's instability should reduce the probability that a packet is successfully transmitted over a link. It is clear that the effect of link instability is magnified as route length increases. In order to describe the route stability, which naturally reflects the stability of the whole network, we here introduce the concept of a route being τ -time-valid. The probability of a flooding route being τ -time-valid is one of the performance metrics of the network to describe the stability of the route constructed via flooding.

Definition. We call a link τ -time-valid if the time needed to maintain the link valid is no less than τ units of time, and a route τ -time-valid if the time needed to maintain the route valid is no less than τ units of time.

For any state $j \in S$, let $\gamma_j = \inf\{t \geq 0 : d_t = j\}$ be the first time for the process $\{d_t, t \in Z^+\}$ to hit the state j, and its distribution be

$$F_{kj}(x) = P_k(\gamma_j \leq x) = P(\gamma_j \leq x\big|_{d_0=k}).$$

Theorem 4. For the N -SRRWM model, the probability p_τ that a link in the network is τ -time-valid is $p_\tau = \sum_{k=0}^{\rho} \pi(k)[1 - F_{k\rho+1}(\tau)]$ where $\pi(k)$ is given by the lemma 1.

Proof: A given link is τ -time-valid if and only if the length of the link $d_t \leq \rho, (0 \leq t \leq \tau)$, then

$$p_\tau = P(d_t \leq \rho, 0 \leq t \leq \tau) = P(\gamma_{\rho+1} > \tau, d_0 \leq \rho)$$

$$= \sum_{k=0}^{\rho} P(d_0 = k) P_k(\gamma_{\rho+1} > \tau) = \sum_{k=0}^{\rho} \pi(k)[1 - F_{k\rho+1}(\tau)].$$

The theorem is proved.

Note: By Lemmon 2, for any given node in the network there are k τ -time-valid links connected to the node with probability $C_{n-1}^{k} p_\tau^k (1-p_\tau)^{n-1-k}$, and in the average sense there are $(n-1)p_\tau$ links connected to the node, where $q = 1 - p$.

Theorem 5. For the model of N -SRRWM, the probability $\psi_N(p, p_\tau)$ that a flooding route is τ-time-valid is given by

$$\psi_N(p, p_\tau) = pp_\tau + \frac{p_\tau(1-q^N)}{q^{N-1}} \sum_{i=0}^{\infty} \frac{pp_\tau q^{(i+1)N}}{1-q^{(i+1)N}} \prod_{j=0}^{i-1} \frac{1-[1-p_\tau(1-q^N)]q^{(j+1)N}}{q^{N-1}(1-q^{(j+1)N})}$$

Proof: A flooding route with the length of k between a given source and its destination is τ-time-valid with probability p_τ^k. The probability $\psi_N(p, p_\tau)$ is given by $E(p_\tau^\xi)$ with ξ being the FD between the source and the destination, and the proof is completed by the theorem 1.

3 Conclusion

Manets have become a hot issue in the field of the 4th generation wireless mobile communication network. Especially on the study of routing protocol in Manets, various routing protocols based on different performance scales have been proposed. Therefore, the comparison, evaluation, improvement and standardization of routing protocols are starting to show an increasing significance. With the improvement and perfection of simulation tools and platforms, the simulation research becomes simple and convenient as a result of which we can obtain abundant research documents in this field. Despite the difficulties we may encounter, theoretical modelling and analysis are still worthy of our efforts due to their incomparable superb abstraction and comprehension capabilities. This paper introduced and studied some network performance parameters of the routing discovery and routing maintenance process under the typical reactive routing protocol DSR, the results achieved under the N-SRRWM model in this paper may be theoretically useful to optimize and evaluate the routing protocols.

References

1. Das, S.R., Perkins, C.E., Royer, E.M.: Performance comparison of two on-demand routing protocols for ad hoc networks. In: Proceeding of INFOCOM, Tel Aviv, Israel, pp. 3–12 (2000)
2. Ramanathan, S., Steenstrup, M.: A survey of routing techniques for mobile communication networks. Mobile Networks and Appl. 1(2), 89–104 (1996)
3. Boppana, R.V., Marina, M.K., Konduru, S.: Analysis of routing techniques for mobile ad hoc networks. In: Banerjee, P., Prasanna, V.K., Sinha, B.P. (eds.) HiPC 1999. LNCS, vol. 1745, pp. 239–245. Springer, Heidelberg (1999)
4. Jacquet, P., Laouiti, A.: Analysis of mobile ad hoc network routing protocols in random graph mobiles. Rapport de Recherche no. 3835. Institut national de Recherche en Informatique et en Automatique (1999)
5. Xi, H., Hanxing, W., Shengguo, G.: A discrete-time Markov chain model with N-spatial reuse for mobile ad hoc networks. OR Transactions 13(3) (2009) (in chinese)

Equivalent Circuit Model of Comb-type Capacitive Transmission Line on MMIC for Application to the RF Component Design in Millimeter-Wave Wireless Communication System

Eui-Hoon Jang*, Young-Bae Park, Bo-Ra Jung, Jang-Hyeon Jeong, Jeong-Gab Ju, and Young Yun

Dept. of Radio Communication and Engineering
Korea Maritime University
Busan, Republic of Korea
yunyoung@hhu.ac.kr

Abstract. In this work, the equivalent circuit of comb-type capacitive transmission line (CCTL) structure was investigated using theoretical analysis. The equivalent circuits for CCTL cell were extracted, and all lumped circuit parameters were expressed by closed-form equation. A fairly good agreement between calculated and measured results were observed up to 50 GHz. Above results indicate that the proposed equivalent circuit model can be efficiently used for application to millimeter-wave wireless communication system.

Keywords: Comb-type capacitive transmission line (CCTL), equivalent circuit, MMIC(monolithic microwave integrated circuit), Wireless communication system, LED.

1 Introduction

RF transmission line is a key element in the hardware of wireless communication system, and most of RF passive components of wireless communication system is fabricated using the RF transmission line [1]-[7]. However, bulky RF passive components such as conventional impedance transformers and dividers have been fabricated outside of MMIC (monolithic microwave integrated circuit) due to their large sizes[2-6]. To solve this problem, several papers dealing with short wavelength transmission line employing periodic structure have been published on GaAs substrate [8]-[11]. According to the results, the microstrip line employing periodic structure showed slow-wave characteristics such as short wavelength. Of all slow-wave structures [8-11], the transmission line employing (CCTL) structure showed the shortest wavelength. Concretely, the wavelength of the CCTL structure was only 10 %

* The B.S. degree in Radio Communication and Engineering from Korea Maritime University, 2011, respectively, and is currently MMIC is a major, are working to the M.S. degree at Korea Maritime University.

M. Zhu (Ed.): Electrical Engineering and Control, LNEE 98, pp. 1071–1077.
springerlink.com © Springer-Verlag Berlin Heidelberg 2011

of conventional microstrip line on MMIC [2]. For application to the design of RF passive components in wireless communication system, the equivalent circuit of the CCTL structure should be extracted. However, an extensive investigation on equivalent circuit of the CCTL structure has not been performed yet. In this work, the CCTL structure was fabricated on GaAs substrate, and an extensive investigation of equivalent circuit for the CCTL structure was performed for application to wireless communication systems.

2 The Short Wavelength Microstrip Line Employing CCTL

Figure 1 and 2 show the microstrip line employing CCTL. At here, CCTL means comb-type capacitive transmission line. As shown in Fig. 1 and 2, comb-type line exists on the semiconducting substrate. The CCTL structure consists of line and PMS (periodic metal strips), and the periodic metal strips are connected to the both sides of the line. Upper layer metal (ULM) exists on the top side of the CCTL structure, and it was electrically connected to backside ground plane through the via-holes. Therefore, the ULM serves as ground plane with backside ground plane. The SiN film exists between ULM and PMS for tight coupling. As is well known, conventional microstrip line has only a periodical capacitance C_a (C_a is shown in Fig. 2) between line and backside ground plane, while the CCTL structure has additional capacitance C_c as well as C_a due to a strong electromagnetic coupling between PMS and ULM.

Therefore, as shown in Figure 3, the CCTL structure exhibits much shorter wavelength (λ_g) than conventional ones, because λ_g are inversely proportional to the periodical capacitance, in other words,

$$\lambda_g = \frac{1}{f\sqrt{LC}} \qquad (1)$$

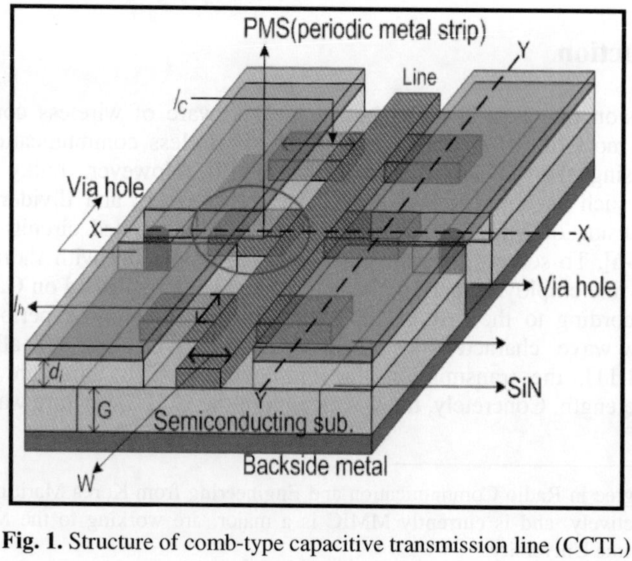

Fig. 1. Structure of comb-type capacitive transmission line (CCTL)

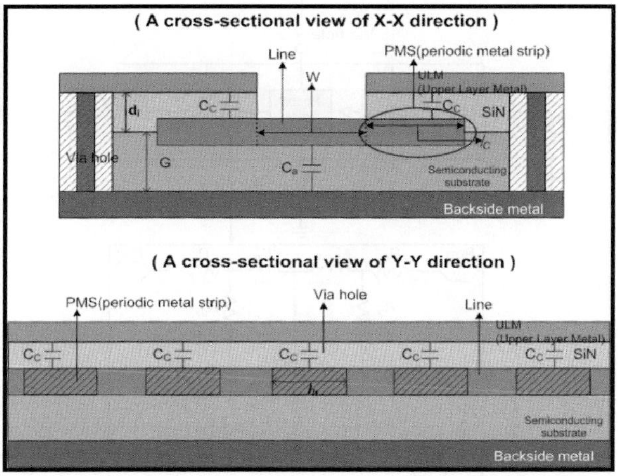

Fig. 2. Cross sectional view of CCTL structure

In Figure 3, a spacing L between strips and line width W, which are shown in Fig. 1, are all 20 μm. The length of PMS l_c (l_c is shown in Figure 1 and 2) is 10 - 30 μm. The CCTL structure was fabricated on GaAs substrate with a height of 100 μm. Compared with the conventional microstrip line, the CCTL structure shows very strong slow-wave characteristic due to its periodic structure. Therefore, as shown in Figure 3, the CCTL structure exhibits wavelength shorter than conventional one. Concretely, at 10 GHz, the wavelength of the CCTL structure and conventional microstrip line are 1.06 and 10.56 mm, respectively. Therefore, the wavelength of the CCTL structure is only 10 % of conventional microstrip line on MMIC. From the Fig.1, we can see that an increase of l_c results in an enhancement of periodical capacitance C_c due to an increase of capacitive area.

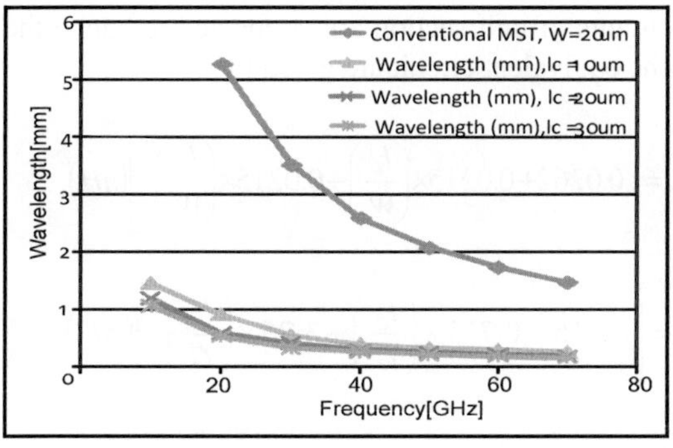

Fig. 3. The wavelength variation of CCTL structure and conventional one.

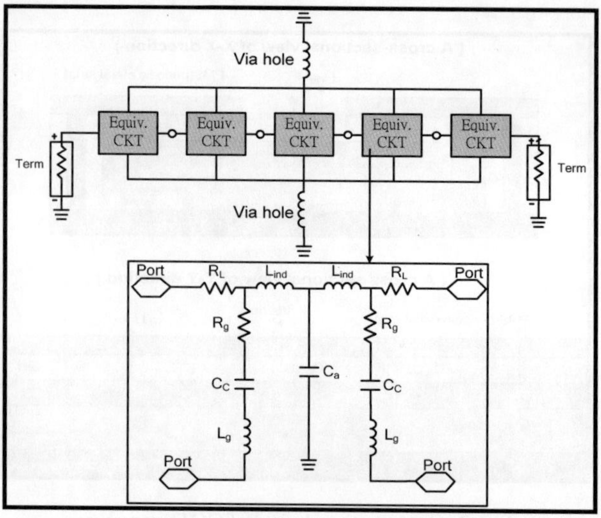

Fig. 4. The equivalent circuit of CCTL structure

3 The Equivalent Circuit of the CCTL Structure

Figure 4 shows the equivalent circuit of the CCTL structure, which corresponds to the equivalent circuit of the Nth unit section of the periodic structure surrounded by rectangular box. C_C corresponds to the capacitance between ULM (Upper Layer Metal) and PMS (Periodic Metal Strip), which is shown in Figure 2, and it is proportional to the cross area $l_h \cdot l_c$ (As shown in Figure 1 and 2, l_h and l_c are the width and length and the periodic strips of PMS, respectively, and l_h was fixed to 20 μm). R_g and L_g are resistance and inductance originating from the loss and current flow of the PMS with a length of l_c, respectively. C_a corresponds to the capacitance between line and ground metal. L_{ind} and R_L are parasitic inductance and resistance originating from the top line. It shows an equivalent circuit of the CCTL structure. As shown in this figure, a number of the equivalent circuits of unit section are connected to each other. The capacitance and inductance of the equivalent circuit are given by,

$$L_{ind} = \left[0.0262 + 0.0215 \times \left(\frac{l_C}{W} \right) - 0.0215 \times \left(\frac{l_C}{W} \right)^2 \right] (nH) \quad (2)$$

$$C_a = \left[0.215 - 0.775 \times \left(\frac{l_C}{G} \right) + 3.25 \times \left(\frac{l_C}{G} \right)^2 \right] (pF) \quad (3)$$

$$C_C = \left[0.45 - 0.194 \times 10^{-2} \times \left(\frac{l_C}{d_i} \right) + 0.48 \times 10^{-5} \times \left(\frac{l_C}{d_i} \right)^2 \right] (pF) \quad (4)$$

$$R_L = \left[-0.05 \times \left(\frac{W}{l_C} \right) + 0.6 \right] (\Omega) \tag{5}$$

$$R_g = \left[0.885 - 0.38 \times \left(\frac{l_C}{l_h} \right) + 0.26 \times \left(\frac{l_C}{l_h} \right)^2 \right] (\Omega) \tag{6}$$

$$L_g = \left[-0.0646 + 0.1364 \times \left(\frac{l_h}{l_C} \right) + 0.0398 \times \left(\frac{l_h}{l_C} \right)^2 \right] (nH), \tag{7}$$

where W, l_c and d_i are top line width, length of periodic metal strip and the thickness of SiN (See Fig. 1). G is length of between line and ground, l_h is width of PMS. Especially, equation C_C includes $(l_c/d_i)^2$ because there are non-linearity originating from parasitic due to fringing capacitance. The whole equivalent circuit is shown in Figure 4. As shown in this figure, a number of the equivalent circuits of unit section are connected to each other. Using the equivalent circuit and closed-form equations, we calculated RF characteristics of the CCTL structure, which was compared with measured results. Figure 5, 6 show the measured and calculated data for insertion loss S_{21} of the CCTL structure with various l_c values. For the calculation result, equivalent circuit and closed form equations were used. As shown in these figures, we can observe a fairly good agreement between calculated and measured results up to 50 GHz. Above the results indicate that the proposed equivalent circuit of CCTL structure can be efficient used in millimeter wireless communication systems.

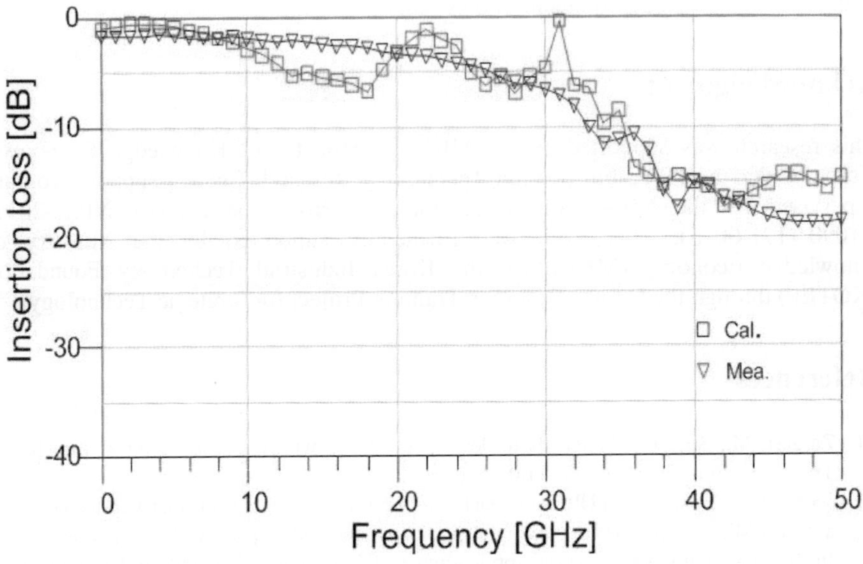

Fig. 5. Insertion loss of the CCTL structure (l_c =10um)

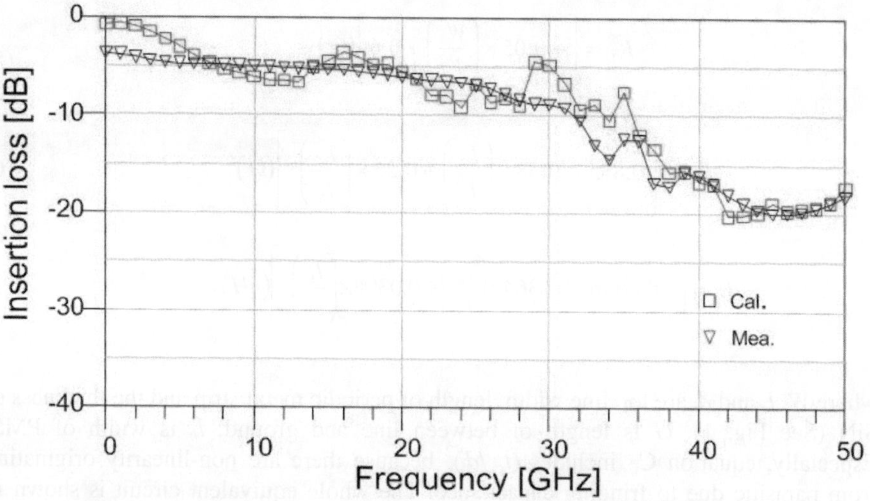

Fig. 6. Insertion loss of the CCTL structure (l_c =20um)

4 Conclusions

In this work, we investigated the equivalent circuit of the CCTL structure using theoretical analysis. The equivalent circuits for CCTL cell were extracted, and all lumped circuit parameters were expressed by closed-form equation. We observed a fairly good agreement between calculated and measured results up to 50 GHz. Above results indicate that the proposed equivalent circuit model can be efficiently used for application to millimeter-wave wireless communication system.

Acknowledgment

This research was supported by the MKE(the Ministry of Knowledge Economy), Korea, under the ITRC(Information Technology Research Center) support program supervised by the NIPA(National IT Industry Promotion Agency)(NIPA-2011-C1090-1121-0015). This work was financially supported by the Ministry of Knowledge Economy (MKE) and the Korea Industrial Technology Foundation (KOTEF) through the Human Resource Training Project for Strategic Technology.

References

1. Zargari, M., Su, D.: Challenges in designing CMOS wireless systems on a chip. IEICE Trans. Electron. E90-C, 1142–1148 (2007)
2. Jeong, J.H., Yun, Y.: An Ultra-compact On-chip Impedance Transformer Fabricated Using a Novel Microstrip Line Employing Periodically Arrayed Capacitive Elements on MMIC. In: International Conference on Applications of Electromagnetism and Student Innovation Competition Awards, Taipei, Taiwan, August 11-13, pp. 102–106 (2011)

3. Matsunaga, K., Miura, I., Iwata, N.: A CW 4-W Ka-Band Power Amplifier Utilizing MMIC Multichip Technology. IEEE J. Solid State Circuits 35, 1293–1297 (2000)
4. Webster, D.R., Ataei, G., Haigh, D.G.: Low-Distortion MMIC Power Amplifier Using a New Form of Derivative Superposition. IEEE Trans. Microwave Theory and Tech. 49, 328–332 (2001)
5. Itoh, Y., Nii, M., Takeuchi, N., Tsukahara, Y., Kurebayashi, H.: MMIC/Super-MIC/MIC-Combined C- to Ku-Band 2W Balanced Amplifier Multi-Chip Module. IEICE Trans. Electron E80-C(6), 757–762 (1997)
6. Yun, Y., Fukuda, T., Kunihisa, T., Ishikawa, O.: A High Performance Downconverter MMIC for DBS Applications. IEICE Trans. Electron. E84-C(11), 1679–1688 (2001)
7. Yun, Y., Nishijima, M., Katsuno, M., Ishida, H., Minagawa, K., Nobusada, T., Tanaka, T.: A Fully-Integrated Broadband Amplifier MMIC Employing a Novel Chip Size Package. IEEE Trans. Microwave Theory Tech. 50, 2930–2937 (2002)
8. Yun, Y.: Miniaturised, low impedance ratrace fabricated by microstrip line employing PPGM on MMIC. IEE Electronics Letters 40(9), 540–541 (2004)
9. Yun, Y.: A Novel Microstrip Line Structure Employing a Periodically Perforated Ground Metal and Its Application to Highly Miniaturized and Low Impedance Passive Components Fabricated on GaAs MMIC. IEEE Transactions On Microwave Theory and Technique 53(6), 1951–1959 (2005)
10. Yun, Y., Lee, K.S., Kim, C.R., Kim, K.M., Jung, J.W.: Basic RF Characteristics of the Microstrip Line Employing Periodically Perforated Ground Metal and Its Application to Highly Miniaturized On-Chip Passive Components on GaAs MMIC. IEEE Transactions On Microwave Theory and Technique 54(10), 3805–3817 (2006)
11. Yun, Y., Jung, J.W., Kim, K.M., Kim, H.C., Chang, W.J., Ji, H.G., Ahn, H.K.: Experimental Study on Isolation Characteristics between Adjacent Microstrip Lines Employing Periodically Perforated Ground Metal for Application to Highly Integrated GaAs MMICs. IEEE Microwave and Wireless Components Letters 17(10), 703–705 (2007)
12. Pozar, D.M.: Microwave engineering, 3rd edn. Addison-Wesley, Reading (2005)

Author Index